C000154469

ISBN 978-0-331-52166-5
PIBN 11081943

FB &c Ltd, Dalton House, 60 Windsor Avenue, London, SW19 2RR.
Company number 08720141. Registered in England and Wales.

For support please visit www.forgottenbooks.com

DEPARTMENT OF THE INTERIOR
FRANKLIN K. LANE, Secretary

UNITED STATES GEOLOGICAL SURVEY
GEORGE OTIS SMITH, Director

Bulletin 696

A CATALOGUE

OF THE

MESOZOIC AND CENOZOIC PLANTS OF NORTH AMERICA

BY

F. H. KNOWLTON

WASHINGTON
GOVERNMENT PRINTING OFFICE
1919

	Page.
Introduction	7
Bibliography	13
Catalogue	45
Biologic classification of genera	663
Index of genera and families	679
Floral lists of North American Mesozoic and Cenozoic plant-bearing formations	697
Triassic floras	697
Triassic (?) flora (Moqui villages, Ariz.)	699
Jurassic floras	700
California and Oregon, formations not settled	700
Foreman formation, California	701
Mariposa slate, California	702
Slate Springs, Calif	702
Boulder, Colo	702
Bridger Creek, near Bozeman, Mont	702
Yakutat group (Lias ?), Alaska	702
Lower Jurassic (Lias), Matanuska Valley, Alaska	702
Cape Lisburne, Alaska	702
Copper River region, Alaska	703
Herendeen Bay, Alaska	703
Tuxedni sandstone (Middle Jurassic), Tuxedni Bay, Alaska	703
Cretaceous floras	703
North and Middle Atlantic regions	703
Potomac group (Lower Cretaceous)	703
Patuxent formation	703
Arundel formation	706
Patapsco formation	707
Potomac group, formation not identified	709
Raritan formation	709
Magothy formation	714
Matawan formation	720
Monmouth formation	720
South Atlantic and Gulf regions	720
Tuscaloosa formation	720
Eutaw formation	723
Black Creek formation	724
Middendorf arkose member of Black Creek formation, South Carolina	725
Ripley formation	726
Cusseta sand member of Ripley formation, Georgia	727
Trinity group	727
Upson clay (Upper Cretaceous)	727
Bingen sand, Arkansas	728

Floral lists of North American Mesozoic and Cenozoic plant-bearing formations—Continued.
Cretaceous floras—Continued. Page.
Rocky Mountain region 728
Lower Cretaceous floras 728
Kootenai formation 728
Morrison formation, Freezeout Hills, Wyo 730
Cloverly formation, Wyoming 730
Lakota sandstone 730
Fuson formation 731
Cheyenne sandstone 732
Lower Cretaceous (Lakota age?), Hovenweep Canyon, southwestern Colorado 732
Comanche series, formation not identified 732
Lower Cretaceous rocks, formations not identified 732
Upper Cretaceous floras 732
Dakota sandstone 732
Woodbine sand, Arthurs Bluff, Lamar County, Tex 742
Bear River formation, western Wyoming 742
Colorado group 742
Frontier formation, Wyoming 742
Graneros formation, Nebraska 743
Mill Creek formation, Mill Creek, British Columbia 743
Montana group 743
Eagle sandstone near Virgelle, Mont 743
Pierre shale, Canada 743
Mesaverde formation 743
Belly River formation, Canada 745
Judith River formation, Montana 745
Trinidad sandstone 746
Vermejo formation 746
Fruitland formation, San Juan Basin, N. Mex 748
Kirtland formation, San Juan Basin, N. Mex 749
Adaville formation, Hodges Pass, Wyo 749
Fox Hills formation, northeastern Colorado 749
Montana group, formations not identified 749
Laramie formation, Denver Basin, Colo 750
Laramie (?) formation, Hodges Pass, Wyo 752
"Laramie" formation, Yellowstone National Park 752
Medicine Bow ("Lower Laramie") formation, Carbon County, Wyo 752
Pacific coast region 753
Knoxville formation 753
Horsetown formation 754
Chico formation 755
Chignik formation (Upper Cretaceous), Alaska 755
Upper Cretaceous rocks, Vancouver Island, British Columbia 755
Cretaceous rocks (formations not identified) 756

l lists of North American Mesozoic and Cenozoic plant-bearing formations—Continued.

Page.
Tertiary floras.......... 757
Eocene floras.......... 757
South Atlantic and Gulf regions.......... 757
Midway (?) formation, Earle, Bexar County, Tex.......... 757
Wilcox group.......... 757
Barnwell formation (Twiggs clay member), Georgia.......... 764
Claiborne group.......... 764
Yegua formation.......... 764
Lisbon formation.......... 764
Formation not identified.......... 764
Rocky Mountain region.......... 764
Lance formation.......... 764
Post-Laramie formation, Black Buttes, Wyo.......... 767
Post-Laramie (?) formation, Medicine Bow, Wyo.......... 768
Fort Union formation.......... 768
Paskapoo formation, Canada.......... 773
Raton formation.......... 774
Denver formation.......... 777
Dawson arkose, Colorado.......... 780
Evanston formation.......... 780
Livingston formation.......... 781
Hanna formation, Carbon County, Wyo.......... 781
Green River formation.......... 782
Wasatch (?) formation, Clark Fork basin, Wyo.......... 783
Pacific coast region.......... 783
Puget group.......... 783
Swauk formation.......... 784
Payette formation.......... 784
Clarno formation (lower part).......... 785
Clarno formation (upper part).......... 785
Ione formation, California.......... 786
Kenai formation, Alaska.......... 786
Edmonton formation, Canada.......... 789
Eocene rocks, formations not identified.......... 789
Later Tertiary floras.......... 791
"Loup Fork," Long Island, Kans.......... 791
"Monument Creek formation," Colorado.......... 791
Catahoula sandstone (Oligocene).......... 791
Oligocene rocks, formations not identified.......... 791
Miocene floras.......... 793
Brandon, Vt.......... 793
Calvert formation, Atlantic coast.......... 796
Elko, Nev.......... 796
Florissant lake beds, Colorado.......... 797
Intermediate flora, Yellowstone National Park.......... 801
Flora of Lamar River, Yellowstone National Park.......... 802
Esmeralda formation, Nevada.......... 803
Mascall formation, John Day Basin, Oreg.......... 803
Ellensburg formation, Washington.......... 805
Auriferous gravels, California.......... 805
San Pablo formation, California.......... 807
Calistoga, Calif.......... 807
Miocene rocks, formations not identified.......... 807

Floral lists of North American Mesozoic and Cenozoic plant-bearing formations—Continued.
Tertiary floras—Continued.
Later Tertiary floras—Continued. Page.
Pliocene floras 808
Citronelle formation, Alabama 808
Merced formation, California 809
Pliocene rocks, formations not identified 809
Tertiary rocks, formations not identified 809
Pleistocene floras 810
Talbot formation, middle Atlantic coast 810
Sunderland formation, middle Atlantic coast 810
Chowan formation, North Carolina 810
Pleistocene deposits, formations not identified 810

A CATALOGUE OF THE MESOZOIC AND CENOZOIC PLANTS OF NORTH AMERICA.

By F. H. Knowlton.

INTRODUCTION.

In 1898 was published what may be considered as in part a first edition of the present catalogue under the title "A catalogue of the Cretaceous and Tertiary plants of North America." [1] That catalogue was compiled as a result of personal needs and was presented, even in its somewhat imperfect form, in the hope that it might be of assistance to other workers in this and allied fields. Its favorable reception appears to have justified this hope. It was intended to publish a new edition, or at least a supplement bringing it up to date, at 10-year intervals, but pressure of other duties prevented, and the present catalogue represents an interval of 20 years. During this time there has been great activity in North American paleobotany, with the result that the number of described species has been nearly doubled.

As may be noted from its title, the scope of the present catalogue has been extended to its logical limits—that is, to include the whole of the Mesozoic as well as the Cenozoic plants of North America. It is, however, exclusive of Greenland and Mexico.

The first catalogue of approximately this scope was published by Lesquereux in 1878 [2] under the title "Catalogue of the Cretaceous and Tertiary plants of North America, with references to the descriptions." In this catalogue the species are arranged in botanic sequence. The Cretaceous flora known at that date embraced only 157 species and the Tertiary flora 549 species, or a total of 706 for these two great geologic systems.

The only other similar catalogue is that of S. A. Miller, published in 1881 under the title "North American Mesozoic and Cænozoic geology and palæontology; or an abridged history of our knowledge of the Triassic, Jurassic, Cretaceous, and Tertiary formations of this continent." This is hardly more than a crude compilation from that of Lesquereux and from certain publications that had appeared between the date of Lesquereux's work and the year 1881.

[1] Knowlton, F. H., U. S. Geol. Survey Bull. 152, 1898.

[2] Lesquereux, Leo, U. S. Geol. and Geog. Survey Terr. Ann. Rept. for 1876, pp. 487-520, 1878.

A number of lists without references or synonymy which embrace simply the names and in part the distribution of the species have been published—for example, Lesquereux's "Table of distribution of the species of fossil plants in the Tertiary formations of North America";[1] "Table showing the geographical distribution of the fossils of the Laramie group, collected during the season of 1877";[2] "Table of distribution of the plants of the Cretaceous Cenomanian formation";[3] "Table of distribution of the species of the Laramie group";[4] "Table of distribution of the plants of the Green River and White River groups";[5] "Table of distribution of North American Miocene fossil plants";[6] "Table showing the distribution of the plants of the Dakota group";[7] Ward's "Table of distribution of Laramie, Senonian, and Eocene plants";[8] Berry's "Correlation of the Potomac formations,"[9] with lists and distribution of species; and Berry's "The Upper Cretaceous floras of the world,"[10] including lists of all species known from American as well as foreign formations.

A brief explanation may be given of certain nomenclatoral features attempted in the present catalogue. I have given for each genus the original place and date of publication. In respect to exclusively fossil genera much care has been expended in verifying these points, and though it is perhaps too much to claim that all the citations are absolutely correct, it is hoped and believed that the errors will prove to be few. Especial care has been taken in regard to genera established on American material, and for these the type species also has been given. For each American genus which is not monotypic or for which the type was not designated by the author, the type has been fixed by the American rules of botanic nomenclature. No attempt has been made, however, to fix the type of Old World genera. For the living genera that have also been found fossil the date and place of publication have been taken from various sources, and many of them have not been verified. When my first catalogue was published botanic nomenclature, at least in this country, was in a state of great unrest, with two or three more or less opposing points of view. One school wished to make the introduction of binomial nomenclature, or the publication of Linnaeus's "Species

---

1 Lesquereux, Leo, U. S. Geol. and Geog. Survey Terr. Ann. Rept. for 1871, pp. 307–309, 1872; Idem for 1872, pp. 410–417, 1873.

2 Lesquereux, Leo, Idem for 1877, p. 255, 1879.

3 Lesquereux, Leo, The Cretaceous and Tertiary floras: U. S. Geol. Survey Terr. Rept., vol. 8, pp. 93–103, 1883.

4 Idem, pp. 115–120.

5 Idem, pp. 206–212.

6 Idem, pp. 266–272.

7 Lesquereux, Leo, U. S. Geol. Survey Mon. 17, pp. 222–225, 1891.

8 Ward, L. F., U. S. Geol. Survey Sixth Ann. Rept., for 1884–85, pp. 443–515, 1886.

9 Berry, E. W., Maryland Geol. Survey, Lower Cretaceous, pp. 153–171, 1911.

10 Berry, E. W., Maryland Geol. Survey, Upper Cretaceous, pp. 183–212, 1916.

Plantarum" (1753), the initial point for both genera and species, and another school argued for the publication of Linnaeus's "Systema," in 1735, as the starting point for genera. Other botanists advocated going still further back, to Tournefort—who has been called the father of genera—or even to the remote place in literature where a genus was first printed, no matter in what connection. Pending the settlement of these questions, and purely as a matter of convenience in procuring the data, the date selected in the important Kew Index—1735—was taken for living genera in my first catalogue. Since that time, however, botanists have very generally agreed to adopt 1753 as the starting point for genera as well as for species.

For each American form that is known only in a fossil state, the original date and place of publication are given. This is followed by all or in any event by the most important references, especially such as refer to descriptions and figures. For each Old World form that is recognized in North American strata the original date and place of publication are followed by a reference to the publication in which the form is first recorded as American. Then follow in chronologic order the American references.

For each living species that has been found fossil only the authority is given, and this is followed by the first reference to the species in a fossil state and then by such other references as are important for one reason or another—for example, "*Quercus lyrata* Walter. Berry, Jour. Geology, vol. 15, 1907, p. 343."

The synonymy is placed in its proper order under the species to which it belongs. Each synonym, however, is entered in its proper alphabetic place, followed by an indication of the form to which it is now referred, thus, "*Ficus? fimbriata* Lesquereux = *Eremophyllum fimbriatum*."

In the reports on fossil plants by Lesquereux published in the annual reports of the Hayden Survey he described many species supposed to be new to science and also identified numerous foreign species that he did not allude to afterward. There are probably nearly 100 species coming under this heading that are hardly entitled to be recognized as belonging to our floras. When Lesquereux came to study these new or foreign species more carefully, in preparing his final monographs, he discarded them or merged them with other species and made no record of such disposition. As a result, they stand in the books as apparently valid species, but in reality they are hardly more than nomina nuda. The specimens on which they were based are now lost or included under other names in such a way that there is absolutely no traceable connection between them. In the present catalogue, as in its predecessor, these species have been indicated by a bracketed statement at the end, thus: "*Cercis eocenica* Lesquereux, U. S. Geol. and Geog. Survey Terr. Ann. Re[p.] 1872 [1873], p. 384 [not afterward recognized by its author]."

In the first edition of the catalogue, published in 1898, I t[ook] occasion to say that owing to disagreement among geologists [and] paleontologists as to the age of certain beds, the position of [the] species was stated, whenever it was possible to do so without circ[um]locution, in such a way as to fix the relative position without go[ing] too far. Thus, such terms as "Belly River series," "Denver bed[s]," "Fort Union beds," "Amboy clays," and "Potomac formatio[n]" indicated sufficiently the geologic position without attempting [a] discrimination that was perhaps unwarranted by the existing st[ate] of our knowledge. It is of course too much to say that compl[ete] agreement has been reached on all these points, but during [the] last 20 years there has been a great advance in our understandi[ng] of the stratigraphic position of many of the plant-bearing beds, a[nd] above all there has come a clarity and precision in the definition [of] the various stratigraphic units. For example, the "Laramie grou[p]" is now the Laramie formation; the "Denver beds" and "Fort Un[ion] beds" have become the Denver and Fort Union formations; t[he] "Amboy clays" are now mainly distributed between the Rarit[an] and Magothy formations; the "Potomac formation" is now t[he] Potomac group, which includes the Patuxent, Arundel, and Pataps[co] formations. In the catalogue the word "formation" has be[en] omitted in order to save space, and where such terms as Larami[e,] Denver, Fort Union, Lance, Raritan, Magothy, Patuxent, Arund[el,] and Patapsco are used, "formation" is to be understood. For t[he] species of which our knowledge is not yet sufficient to warra[nt] reference to a definite formation, the more general terms Triassi[c,] Jurassic, Upper Cretaceous, Eocene, Oligocene, Miocene, Pleistocen[e,] etc., are sufficient indications of position.

The approximate stratigraphic position of the North America[n] Mesozoic and Cenozoic plant-bearing beds, as understood by th[e] writer, is graphically presented in the accompanying correlatio[n] table. This table includes, in addition to all the plant-bearing units, a few well-known non plant-bearing marine units in orde[r] that the relative positions may be held well in mind. In certain columns—for example, that devoted to the Pacific coast region—the arrangement of formations is crowded and not strictly synchronous. It would have required numerous additional columns to make the table absolutely correlative, but the approximate position is indicated.

In the original catalogue it was frankly stated that no attempt was made to give an absolutely complete account of the geographic distribution of the species.

plant-bearing.

Dakot

al

k. (Arikaree.)

(White Riv

Fort Union

Lance.

Fox Hills.

(Pierre.)

(Niobrara.)
(Carlile.)
(Greenhorn.
Graneros.

Dakota.

Fuson.
(Minnewast
Lakota.

Morrison.

(Unkpapa.)
(Sundance.)

It would have required a monographic revision of each species to determine accurately its areal distribution. The distribution given is taken mainly from the books, with little or no revision, and while it is undoubtedly correct in a majority of cases, it is not to be taken as absolutely so.

During the last 20 years a great deal of this necessary revision of our knowledge of the areal distribution of the floras has been accomplished, and the present catalogue is much more complete and authoritative in this respect. But even so, it is still far short of what it should and ultimately will be. As it stands, it embodies practically all the published data as well as a very considerable amount of information from unpublished sources.

As already stated, the original catalogue was the outgrowth of my own personal needs. It was begun without the thought that it would be published, and hence many of the references were more in the nature of convenient catch titles than in proper bibliographic form. In the present catalogue the references have been entirely rewritten and brought into uniformity. Each reference is so far complete in itself that no key or further explanation seems necessary.

The bibliography consists simply of a list of North American works and papers that have been consulted in preparing this catalogue, together with such others as contain original descriptions or figures of our plants. It has not been thought necessary to give a bibliography of the references to such foreign species as have been identified in our flora, with the single exception of those found in the several parts of Heer's "Flora fossilis arctica."

It has been recognized for a long time that there are a considerable number of homonyms among the specific names of our Mesozoic and Cenozoic plants. A few of these have been detected and renamed in recent years by Berry, Hollick, and others. A few years ago Prof. T. D. A. Cockerell, of the University of Colorado, compiled a full list of these homonyms with a suggested name to replace each one. This list he has generously placed at my disposal, and they stand in this catalogue with the authority "Knowlton and Cockerell," though most of the credit for them should be given to Prof. Cockerell.

In observing the various uses that were made of the original catalogue it has been noted that one of the things often sought is a complete list of the genera and species of the several formational units. The data for such a list are of course present in the catalogue, but to obtain this information would necessitate going through the catalogue again and again and require much time. To supply this need I have brought the genera and species of each unit together in separate alphabetic lists. In these lists the systems are arranged in their geologic sequence—for example, Triassic, Jurassic, Cretaceous, Tertiary.

An asterisk (*) before a locality name in the main catalogue indicates the type locality. This plan has not been carried out uniformly throughout the catalogue for the reason that it was not suggested until most of the work of cataloguing had been completed, and to have applied it uniformly would have required going over the literature a second time and delayed the preparation of the catalogue beyond the time that could be allotted to it.

The enumeration of the fossil plants of the Canadian Provinces is still in a somewhat unsatisfactory state. Since the death of Sir William Dawson and Prof. D. P. Penhallow work has practically ceased on these floras, with the result that our knowledge is fragmentary and in part misleading. Moreover, many of the early illustrations were crude and unsatisfactory, and a considerable proportion of the identifications were obviously erroneous. The matter can be settled only by a thorough revision of the several floras involved.

BIBLIOGRAPHY.

Ami, H. M. Determinations of fossil plants from various localities in British Columbia and the Northwest Territories, by Prof. D. P. Penhallow, of McGill University, Montreal, with notes on the geological horizons indicated, by H. M. Ami.

Canada Geol. Survey Ann. Rept., new ser., vol. 16, 1904 [1906], pp. 389A–392A.

—— Notes on the geological horizons indicated by the fossil plants recently determined by Prof. Penhallow from various localities in British Columbia and the Northwest Territories, Canada.

Reprint from Canada Geol. Survey Ann. Rept., new ser., vol. 16, 1904 [1906], dated September, 1905, pp. 6–10.

Barbour, Erwin Hinckley. Chalcedony-lime nuts from the Badlands.

Proc. and Collections Nebraska State Historical Soc., 2d ser., vol. 2, 1898, pp. 272, 273, pl. 5, text fig. 3.

Berry Edward W. The flora of the Matawan formation ["Crosswicks clays"].

New York Bot. Gard. Bull., vol. 3, 1903, pp. 45–103, pls. 43–57.

—— The American species referred to *Thinnfeldia*.

Bull. Torrey Bot. Club, vol. 30, 1903, pp. 438–445.

—— New species of plants from the Matawan formation.

Am. Naturalist, vol. 37, 1903, pp. 677–684, figs. 1–8 (on p. 678).

—— Additions to the flora of the Matawan formation.

Bull. Torrey Bot. Club, vol. 31, 1904, pp. 67–82, pls. 1–5.

—— A palm from the Mid-Cretaceous.

Torreya, vol. 5, 1905, pp. 30–33, figs. 1, 2.

—— An old swamp bottom.

Torreya, vol. 5, 1905, pp. 179–182, fig. 1.

—— Additions to the fossil flora from Cliffwood, N. J.

Bull. Torrey Bot. Club, vol. 32, 1905, pp. 43–48, pls. 1–2.

—— A *Ficus* confused with *Proteoides*.

Bull. Torrey Bot. Club, vol. 32, 1905, pp. 327–330, pl. 21.

—— The flora of the Cliffwood clays.

New Jersey Geol. Survey Ann. Rept., 1905 [1906], pp 135–172, pls. 19–26.

Berry, Edward W. Living and fossil species of *Comptonia*.
Am. Naturalist, vol. 40, 1906, pp. 485–520, pls. 1–4.

—— Contributions to the Mesozoic flora of the Atlantic Coastal Plain—I.
Bull. Torrey Bot. Club, vol. 33, 1906, pp. 163–182, pls. 7–9.

—— Contributions to the Pleistocene flora of North Carolina.
Jour. Geology, vol. 15, 1907, pp. 338–349.

—— A *Tilia* from the New Jersey Pleistocene.
Torreya, vol. 7, No. 4, April, 1907, pp. 81, 82.

—— Pleistocene plants from Virginia.
Torreya, vol. 6, 1906, pp. 88–90.

—— Pleistocene plants from Alabama.
Am. Naturalist, vol. 41, 1907, pp. 689–697, pls. 1, 2.

—— Contributions to the Mesozoic flora of the Atlantic Coastal Plain—II, North Carolina.
Bull. Torrey Bot. Club, vol. 34, 1907, pp. 185–206, pls. 11–16.

—— Paleobotanical notes: Cretaceous floras in North and South Carolina; New species of plants from the Magothy formation; The stomata in ***Protophyllocladus subintegrifolius*** (Lesquereux) Berry.
Johns Hopkins Univ. Circular, new ser., 1907, No. 7, pp. 79–91, figs. 1–6 (on p. 85).

—— Some araucarian remains from the Atlantic Coastal Plain.
Bull. Torrey Bot. Club, vol. 35, 1908, pp. 249–260, pls. 11–16.

—— A new Cretaceous *Bauhinia*.
Torreya, vol. 8, 1908, pp. 218, 219, text fig. 3.

—— A mid-Cretaceous species of *Torreya*.
Am. Jour. Sci., 4th ser., vol. 25, 1908, pp. 382–386, text figs. 1–3.

—— Additions to the Pleistocene flora of North Carolina.
Torreya, vol. 9, April, 1909, pp. 71–73, text figs. 1, 2.

—— Juglandaceae from the Pleistocene of Maryland.
Torreya, vol. 9, May, 1909, pp. 96–99, text figs. 1–6.

—— A Miocene flora from the Virginia Coastal Plain.
Jour. Geology, vol. 17, 1909, pp. 19–30, text figs. 1–11.

—— Pleistocene swamp deposits in Virginia.
Am. Naturalist, vol. 43, 1909, pp. 432–436.

—— Contributions to the Mesozoic flora of the Atlantic Coastal Plain—III, New Jersey.
Bull. Torrey Bot. Club, vol. 36, 1909, pp. 245–264, pls. 18, 18a.

—— A new species of *Dewalquea* from the American Cretaceous.
Torreya, vol. 10, Feb., 1910, pp. 34–38, text fig. 1.

—— Additions to the Pleistocene flora of New Jersey.
Torreya, vol. 10, 1910, pp. 261–267, text figs. 1, 2.

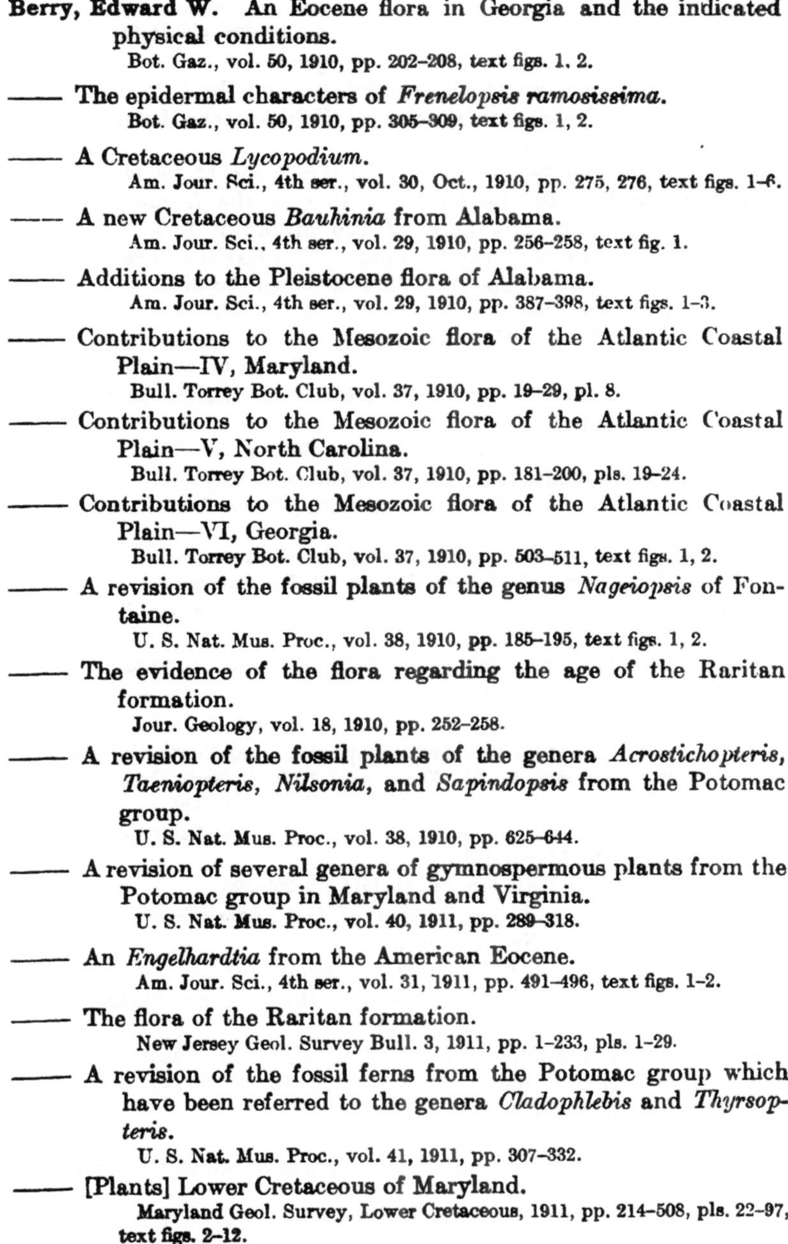

Berry, Edward W. An Eocene flora in Georgia and the indicated physical conditions.
Bot. Gaz., vol. 50, 1910, pp. 202–208, text figs. 1, 2.

—— The epidermal characters of *Frenelopsis ramosissima.*
Bot. Gaz., vol. 50, 1910, pp. 305–309, text figs. 1, 2.

—— A Cretaceous *Lycopodium.*
Am. Jour. Sci., 4th ser., vol. 30, Oct., 1910, pp. 275, 276, text figs. 1–6.

—— A new Cretaceous *Bauhinia* from Alabama.
Am. Jour. Sci., 4th ser., vol. 29, 1910, pp. 256–258, text fig. 1.

—— Additions to the Pleistocene flora of Alabama.
Am. Jour. Sci., 4th ser., vol. 29, 1910, pp. 387–398, text figs. 1–3.

—— Contributions to the Mesozoic flora of the Atlantic Coastal Plain—IV, Maryland.
Bull. Torrey Bot. Club, vol. 37, 1910, pp. 19–29, pl. 8.

—— Contributions to the Mesozoic flora of the Atlantic Coastal Plain—V, North Carolina.
Bull. Torrey Bot. Club, vol. 37, 1910, pp. 181–200, pls. 19–24.

—— Contributions to the Mesozoic flora of the Atlantic Coastal Plain—VI, Georgia.
Bull. Torrey Bot. Club, vol. 37, 1910, pp. 503–511, text figs. 1, 2.

—— A revision of the fossil plants of the genus *Nageiopsis* of Fontaine.
U. S. Nat. Mus. Proc., vol. 38, 1910, pp. 185–195, text figs. 1, 2.

—— The evidence of the flora regarding the age of the Raritan formation.
Jour. Geology, vol. 18, 1910, pp. 252–258.

—— A revision of the fossil plants of the genera *Acrostichopteris, Taeniopteris, Nilsonia,* and *Sapindopsis* from the Potomac group.
U. S. Nat. Mus. Proc., vol. 38, 1910, pp. 625–644.

—— A revision of several genera of gymnospermous plants from the Potomac group in Maryland and Virginia.
U. S. Nat. Mus. Proc., vol. 40, 1911, pp. 289–318.

—— An *Engelhardtia* from the American Eocene.
Am. Jour. Sci., 4th ser., vol. 31, 1911, pp. 491–496, text figs. 1–2.

—— The flora of the Raritan formation.
New Jersey Geol. Survey Bull. 3, 1911, pp. 1–233, pls. 1–29.

—— A revision of the fossil ferns from the Potomac group which have been referred to the genera *Cladophlebis* and *Thyrsopteris.*
U. S. Nat. Mus. Proc., vol. 41, 1911, pp. 307–332.

—— [Plants] Lower Cretaceous of Maryland.
Maryland Geol. Survey, Lower Cretaceous, 1911, pp. 214–508, pls. 22–97, text figs. 2–12.

Berry, Edward W. Notes on the genus *Widdringtonites*.
Bull. Torrey Bot. Club, vol. 39, 1911, pp. 341–348, pls. 24, 25.

—— Notes on the ancestry of the bald cypress.
Plant World, vol. 14, 1911, pp. 39–45 [1–7], text figs. 1, 2.

—— A Lower Cretaceous species of Schizaeaceae from eastern North America.
Annals of Botany, vol. 25, 1911, pp. 193–198, text fig. and pl. 12.

—— Contributions to the Mesozoic flora of the Atlantic Coastal Plain—VII, The flora of the Matawan formation; The flora of the Magothy formation; The Raritan flora of Maryland; Upper Cretaceous flora of South Carolina.
Bull. Torrey Bot. Club, vol. 38, 1911, pp. 399–424, pls. 18, 19.

—— American Triassic Neocalamites.
Bot. Gaz., vol. 53, Feb., 1912, pp. 174–180, text fig. 1, pl. 17.

—— The age of the plant-bearing shales of the Richmond coal field [Virginia].
Am. Jour. Sci., 4th ser., vol. 34, Aug., 1912, pp. 224, 225.

—— Pleistocene plants from the Blue Ridge in Virginia.
Am. Jour. Sci., 4th ser., vol. 34, Aug., 1912, pp. 218–223, figs. 1–5 (on p. 220).

—— Notes on the geological history of the walnuts and hickories.
Plant World, vol. 15, Oct., 1912, pp. 225–240, text figs. 1–4.

—— Contributions to the Mesozoic flora of the Atlantic Coastal Plain—VIII, Texas.
Bull. Torrey Bot. Club, vol. 39, 1912, pp. 387–406, pls. 30–32.

—— A fossil flower from the Eocene.
U. S. Nat. Mus. Proc., vol. 45, 1913, pp. 261–263, pl. 21.

—— [Lists of Cretaceous and Tertiary plants from North Carolina].
In Stephenson, L. W., The Coastal Plain of North Carolina: North Carolina Geol. and Econ. Survey, vol. 3, 1912, pp. 73–145.

—— Contributions to the Mesozoic flora of the Atlantic Coastal Plain—X, Maryland.
Bull. Torrey Bot. Club, vol. 41, 1914, pp. 295–300.

—— The affinities and distribution of the lower Eocene flora of southeastern North America.
Am. Philos. Soc. Proc., vol. 53, 1914, pp. 129–250.

—— Additions to the Pleistocene flora of the Southern States.
Torreya, vol. 14, 1914, pp. 159–162.

—— Fruits of a date palm in the Tertiary deposits of eastern Texas.
Am. Jour. Sci., 4th ser., vol. 37, 1914, pp. 403–406, text figs. 1, 2.

—— Two new Tertiary species of *Trapa*.
Torreya, vol. 14, 1914, pp. 105-108, text figs. 1–5.

—— The Upper Cretaceous and Eocene floras of South Carolina and Georgia.
U. S. Geol. Survey Prof. Paper 84, 1914, pp. 1–200, pls. 1–29.

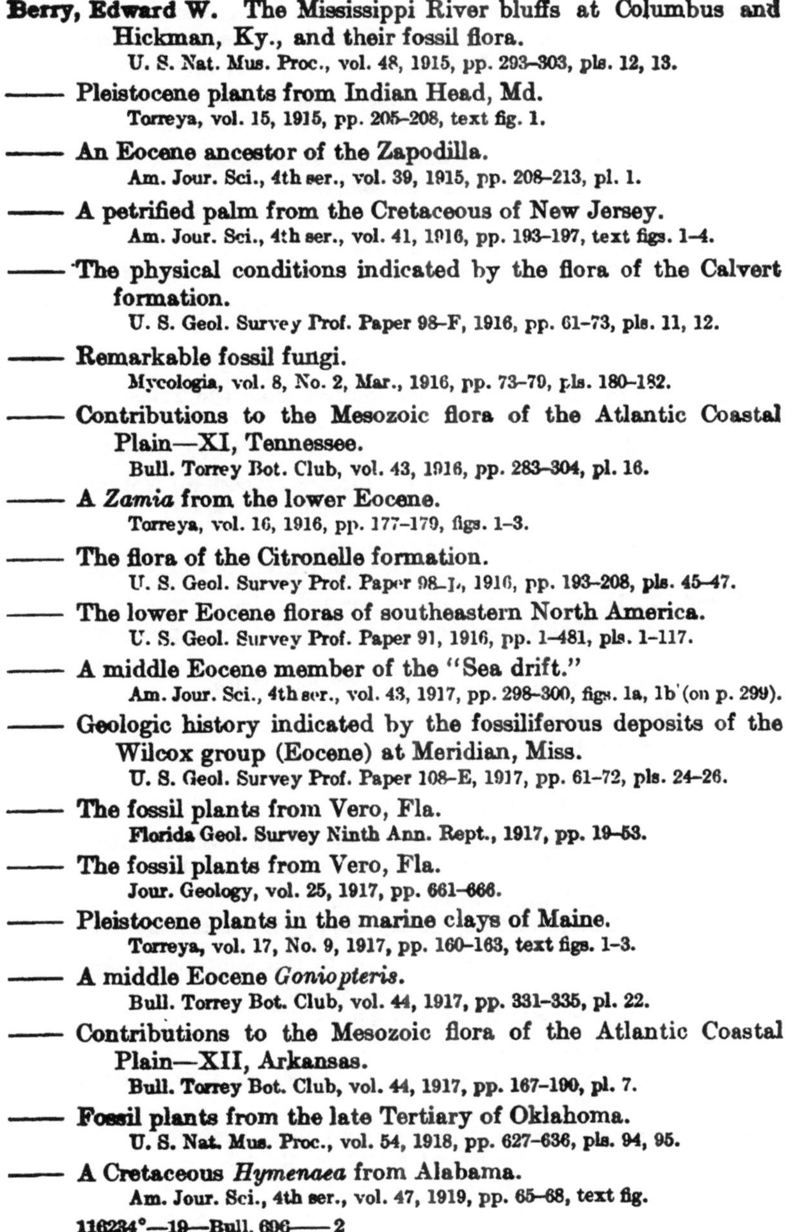

Berry, Edward W. The Mississippi River bluffs at Columbus and Hickman, Ky., and their fossil flora.
U. S. Nat. Mus. Proc., vol. 48, 1915, pp. 293–303, pls. 12, 13.

—— Pleistocene plants from Indian Head, Md.
Torreya, vol. 15, 1915, pp. 205–208, text fig. 1.

—— An Eocene ancestor of the Zapodilla.
Am. Jour. Sci., 4th ser., vol. 39, 1915, pp. 208–213, pl. 1.

—— A petrified palm from the Cretaceous of New Jersey.
Am. Jour. Sci., 4th ser., vol. 41, 1916, pp. 193–197, text figs. 1–4.

—— The physical conditions indicated by the flora of the Calvert formation.
U. S. Geol. Survey Prof. Paper 98–F, 1916, pp. 61–73, pls. 11, 12.

—— Remarkable fossil fungi.
Mycologia, vol. 8, No. 2, Mar., 1916, pp. 73–79, pls. 180–182.

—— Contributions to the Mesozoic flora of the Atlantic Coastal Plain—XI, Tennessee.
Bull. Torrey Bot. Club, vol. 43, 1916, pp. 283–304, pl. 16.

—— A *Zamia* from the lower Eocene.
Torreya, vol. 16, 1916, pp. 177–179, figs. 1–3.

—— The flora of the Citronelle formation.
U. S. Geol. Survey Prof. Paper 98–L, 1916, pp. 193–208, pls. 45–47.

—— The lower Eocene floras of southeastern North America.
U. S. Geol. Survey Prof. Paper 91, 1916, pp. 1–481, pls. 1–117.

—— A middle Eocene member of the "Sea drift."
Am. Jour. Sci., 4th ser., vol. 43, 1917, pp. 298–300, figs. 1a, 1b (on p. 299).

—— Geologic history indicated by the fossiliferous deposits of the Wilcox group (Eocene) at Meridian, Miss.
U. S. Geol. Survey Prof. Paper 108–E, 1917, pp. 61–72, pls. 24–26.

—— The fossil plants from Vero, Fla.
Florida Geol. Survey Ninth Ann. Rept., 1917, pp. 19–53.

—— The fossil plants from Vero, Fla.
Jour. Geology, vol. 25, 1917, pp. 661–666.

—— Pleistocene plants in the marine clays of Maine.
Torreya, vol. 17, No. 9, 1917, pp. 160–163, text figs. 1–3.

—— A middle Eocene *Goniopteris*.
Bull. Torrey Bot. Club, vol. 44, 1917, pp. 331–335, pl. 22.

—— Contributions to the Mesozoic flora of the Atlantic Coastal Plain—XII, Arkansas.
Bull. Torrey Bot. Club, vol. 44, 1917, pp. 167–190, pl. 7.

—— Fossil plants from the late Tertiary of Oklahoma.
U. S. Nat. Mus. Proc., vol. 54, 1918, pp. 627–636, pls. 94, 95.

—— A Cretaceous *Hymenaea* from Alabama.
Am. Jour. Sci., 4th ser., vol. 47, 1919, pp. 65–68, text fig.

Berry, Edward W. Upper Cretaceous floras of the eastern Gulf region in Tennessee, Mississippi, Alabama, and Georgia.
U. S. Geol. Survey Prof. Paper 112, 1919, 177 pp., pls. 5–33.

Britton, Elizabeth G. A new Tertiary fossil moss.
Bull. Torrey Bot. Club, vol. 26, 1899, pp. 79–81, text fig.

—— and **Hollick, Arthur.** American fossil mosses, with description of a new species from Florissant, Colo.
Bull. Torrey Bot. Club, vol. 34, 1907, pp. 139–142, pl. 9.

—— —— A new American fossil moss.
Bull. Torrey Bot. Club, vol. 42, 1915, pp. 9–10, text figs. 1, 2.

Britton, N. L. Geological notes in western Virginia, North Carolina, and eastern Tennessee.
New York Acad. Sci. Trans., vol. 5, 1886, pp. 215–223.

Brown, Amos P. New cycads and conifers from the Trias of Pennsylvania.
Acad. Nat. Sci. Philadelphia Proc., vol. 63, 1911, pp. 17–21, pls. 1–5.

Capellini, J., and Heer, O. Les phyllites crétacées du Nebraska.
Soc. helvétique sci. nat. Mém., 1866, pp. 1–22, pls. 1–4.

C[hapin], J. H. *Cycadinocarpus chapinii.*
Sci. Assoc. Meriden, Conn., Proc. and Trans., vol. 4, 1889–90, p. 62.

Cockerell, T. D. A. Contributions toward a list of the fauna and flora of Wet Mountain Valley, Colo. [Fossil plant renamed on p. 154.]
West American Scientist, vol. 6, 1889, pp. 153–155.

—— Publication of rejected names.
Science, new ser., vol. 17, 1903, p. 189.

—— *Rhus* and its allies.
Torreya, vol. 6, 1906, pp. 11, 12.

—— The fossil fauna and flora of the Florissant (Colo.) shales.
Univ. Colorado Studies, vol. 3, 1906, pp. 169–176, 1 pl.

—— Fossil plants from Florissant, Colo.
Bull. Torrey Bot. Club, vol. 33, 1906, pp. 307–312, text figs. 1–6.

—— Some Old World types of insects in the Miocene of Colorado. [Describes plants near end of article.]
Science, new ser., vol. 26, 1907, pp. 446, 447.

—— The genus *Crataegus* in Colorado.
Univ. Colorado Studies, vol. 5, 1907 [1908], pp. 41–45. [Fossil species on pp. 42, 43.]

—— A new plant (*Ficus*) from the Fox Hills Cretaceous.
Univ. Colorado Studies, vol. 4, 1907, pp. 152, 153.

—— The fossil flora of Florissant, Colo.
Am. Mus. Nat. Hist. Bull., vol. 24, 1908, pp. 71–110, pls. 6–10.

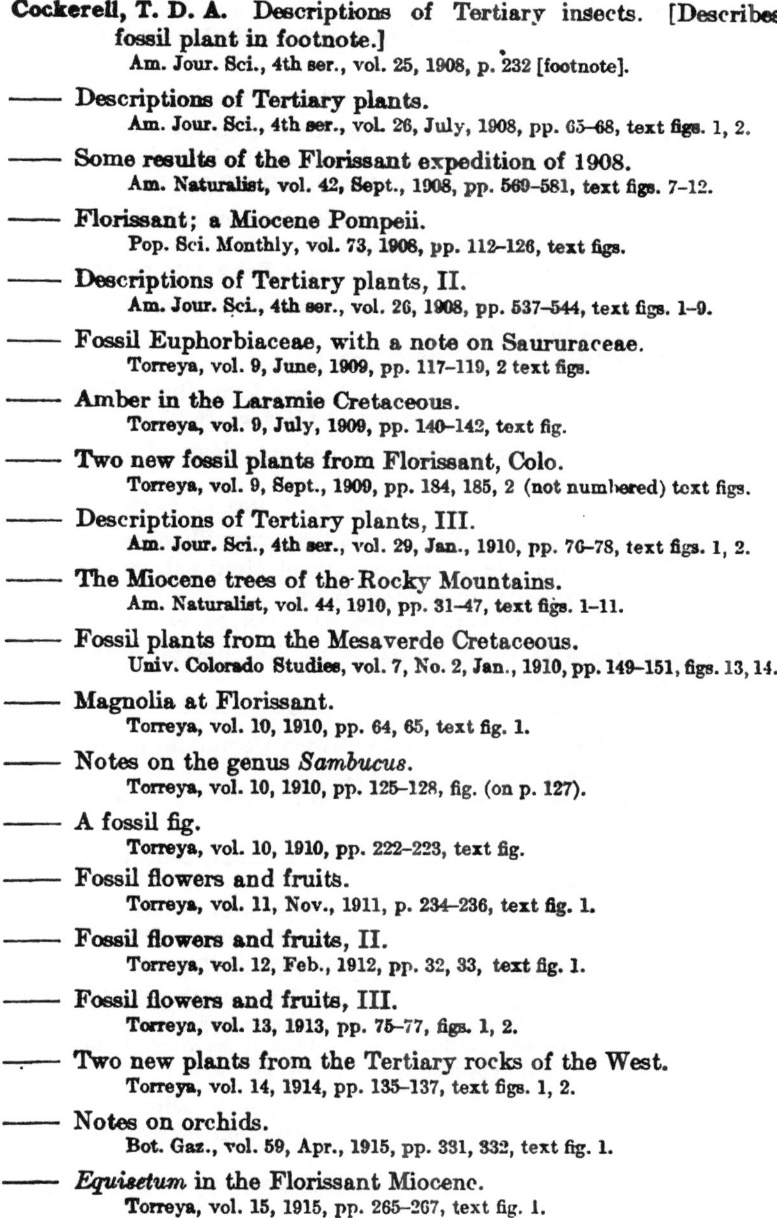

Cockerell, T. D. A. Descriptions of Tertiary insects. [Describes fossil plant in footnote.]
Am. Jour. Sci., 4th ser., vol. 25, 1908, p. 232 [footnote].

—— Descriptions of Tertiary plants.
Am. Jour. Sci., 4th ser., vol. 26, July, 1908, pp. 65–68, text figs. 1, 2.

—— Some results of the Florissant expedition of 1908.
Am. Naturalist, vol. 42, Sept., 1908, pp. 569–581, text figs. 7–12.

—— Florissant; a Miocene Pompeii.
Pop. Sci. Monthly, vol. 73, 1908, pp. 112–126, text figs.

—— Descriptions of Tertiary plants, II.
Am. Jour. Sci., 4th ser., vol. 26, 1908, pp. 537–544, text figs. 1–9.

—— Fossil Euphorbiaceae, with a note on Saururaceae.
Torreya, vol. 9, June, 1909, pp. 117–119, 2 text figs.

—— Amber in the Laramie Cretaceous.
Torreya, vol. 9, July, 1909, pp. 140–142, text fig.

—— Two new fossil plants from Florissant, Colo.
Torreya, vol. 9, Sept., 1909, pp. 184, 185, 2 (not numbered) text figs.

—— Descriptions of Tertiary plants, III.
Am. Jour. Sci., 4th ser., vol. 29, Jan., 1910, pp. 76–78, text figs. 1, 2.

—— The Miocene trees of the Rocky Mountains.
Am. Naturalist, vol. 44, 1910, pp. 31–47, text figs. 1–11.

—— Fossil plants from the Mesaverde Cretaceous.
Univ. Colorado Studies, vol. 7, No. 2, Jan., 1910, pp. 149–151, figs. 13, 14.

—— Magnolia at Florissant.
Torreya, vol. 10, 1910, pp. 64, 65, text fig. 1.

—— Notes on the genus *Sambucus*.
Torreya, vol. 10, 1910, pp. 125–128, fig. (on p. 127).

—— A fossil fig.
Torreya, vol. 10, 1910, pp. 222–223, text fig.

—— Fossil flowers and fruits.
Torreya, vol. 11, Nov., 1911, p. 234–236, text fig. 1.

—— Fossil flowers and fruits, II.
Torreya, vol. 12, Feb., 1912, pp. 32, 33, text fig. 1.

—— Fossil flowers and fruits, III.
Torreya, vol. 13, 1913, pp. 75–77, figs. 1, 2.

—— Two new plants from the Tertiary rocks of the West.
Torreya, vol. 14, 1914, pp. 135–137, text figs. 1, 2.

—— Notes on orchids.
Bot. Gaz., vol. 59, Apr., 1915, pp. 331, 332, text fig. 1.

—— *Equisetum* in the Florissant Miocene.
Torreya, vol. 15, 1915, pp. 265–267, text fig. 1.

Cockerell, T. D. A. A Lower Cretaceous flora in Colorado.
Washington Acad. Sci. Jour., vol. 6, 1916, pp. 109–112, figs. 1, 2.

—— Colorado a million years ago.
Am. Mus. Jour., vol. 16, 1916, pp. 443–450, figs. in text.

Conrad, T. A. Notes on American fossiliferous strata. Description of a new cycadaceous plant from the Raritan clay.
Am. Jour. Sci., 2d ser., vol. 47, 1869, pp. 361–363, text fig.

Conwentz, H. *Cupressinoxylon taxodioides*, ein vorveltliches cypressenähnliches Holz aus Californien.
Schriften Naturf. Gesell. in Danzig, vol. 4, Heft 3, 1878, pp. 122–124.

—— Ueber ein tertiäres Vorkommen cypressenartiger Hölzer bei Calistoga in Californien.
Neues Jahrb., 1878, pp. 800–813, pls. 13, 14.

Cook, George H. Red sandstone or Triassic formation. [Fossils.]
Geology of New Jersey, Ann. Rept. for 1879, pp. 26, 27.

Cragin, F. W. Contributions to the palaeontology of the plains, No. 1.
Washburn College Laboratory Bull., vol. 2, 1889, pp. 65–68.

Dawson, J. W. (later Sir William). On the newer Pliocene and post-Pliocene deposits of the vicinity of Montreal, with notices of fossils recently discovered there.
Canadian Naturalist and Geologist, vol. 2, 1857, pp. 401–426.

—— The evidence of fossil plants as to the climate of the post-Pliocene period in Canada.
Canadian Naturalist, new ser., vol. 3, 1868, pp. 69–76.

—— Note on the fossil plants from British Columbia, collected by Mr. James Richardson in 1872.
Canada Geol. Survey, Rept. Progress for 1872–73, Appendix I, 1873, pp. 66–71, pl. 1.

—— Note on fossil woods from British Columbia, collected by Mr. Richardson.
Am. Jour. Sci., 3d ser., vol. 7, Jan., 1874, pp. 47–51. [Note.—This article is an exact reprint of that last mentioned, and is apparently later.]

—— Note on the plants, collected by Mr. G. M. Dawson, from the lignite Tertiary deposits near the forty-ninth parallel.
British N. A. Bound. Com. (Report on the geology and resources in the vicinity of the 49th parallel, from the Lake of the Woods to the Rocky Mountains), 1875, Appendix A, pp. 327–331, pls. 15, 16.

—— Note by Principal Dawson, on fossil plants collected by Dr. Selwyn in the lignite Tertiary formation of Roche Percée, Souris River.
Canada Geol. Survey, Rept. Prog. 1879–80, Appendix ii, 1881, pp. 51A–55A.

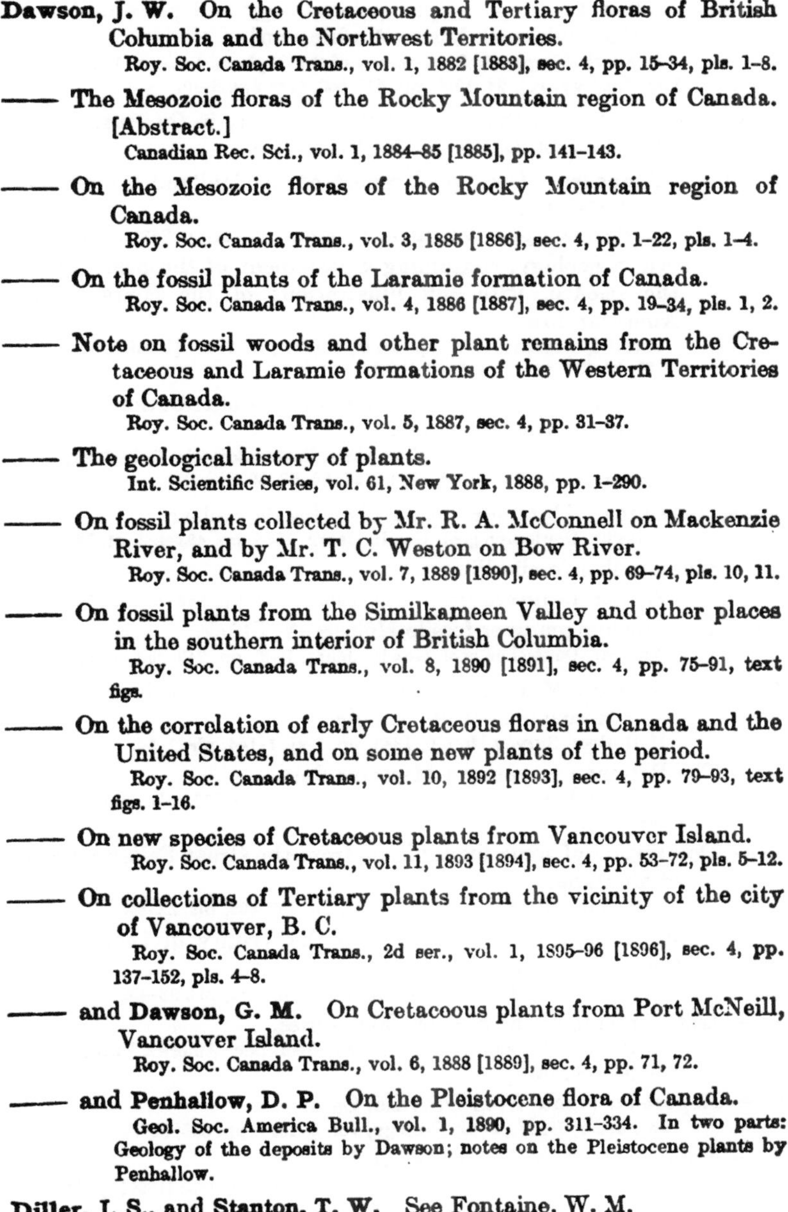

Dawson, J. W. On the Cretaceous and Tertiary floras of British Columbia and the Northwest Territories.
Roy. Soc. Canada Trans., vol. 1, 1882 [1883], sec. 4, pp. 15–34, pls. 1–8.

—— The Mesozoic floras of the Rocky Mountain region of Canada. [Abstract.]
Canadian Rec. Sci., vol. 1, 1884–85 [1885], pp. 141–143.

—— On the Mesozoic floras of the Rocky Mountain region of Canada.
Roy. Soc. Canada Trans., vol. 3, 1885 [1886], sec. 4, pp. 1–22, pls. 1–4.

—— On the fossil plants of the Laramie formation of Canada.
Roy. Soc. Canada Trans., vol. 4, 1886 [1887], sec. 4, pp. 19–34, pls. 1, 2.

—— Note on fossil woods and other plant remains from the Cretaceous and Laramie formations of the Western Territories of Canada.
Roy. Soc. Canada Trans., vol. 5, 1887, sec. 4, pp. 31–37.

—— The geological history of plants.
Int. Scientific Series, vol. 61, New York, 1888, pp. 1–290.

—— On fossil plants collected by Mr. R. A. McConnell on Mackenzie River, and by Mr. T. C. Weston on Bow River.
Roy. Soc. Canada Trans., vol. 7, 1889 [1890], sec. 4, pp. 69–74, pls. 10, 11.

—— On fossil plants from the Similkameen Valley and other places in the southern interior of British Columbia.
Roy. Soc. Canada Trans., vol. 8, 1890 [1891], sec. 4, pp. 75–91, text figs.

—— On the correlation of early Cretaceous floras in Canada and the United States, and on some new plants of the period.
Roy. Soc. Canada Trans., vol. 10, 1892 [1893], sec. 4, pp. 79–93, text figs. 1–16.

—— On new species of Cretaceous plants from Vancouver Island.
Roy. Soc. Canada Trans., vol. 11, 1893 [1894], sec. 4, pp. 53–72, pls. 5–12.

—— On collections of Tertiary plants from the vicinity of the city of Vancouver, B. C.
Roy. Soc. Canada Trans., 2d ser., vol. 1, 1895–96 [1896], sec. 4, pp. 137–152, pls. 4–8.

—— and **Dawson, G. M.** On Cretaceous plants from Port McNeill, Vancouver Island.
Roy. Soc. Canada Trans., vol. 6, 1888 [1889], sec. 4, pp. 71, 72.

—— and **Penhallow, D. P.** On the Pleistocene flora of Canada.
Geol. Soc. America Bull., vol. 1, 1890, pp. 311–334. In two parts: Geology of the deposits by Dawson; notes on the Pleistocene plants by Penhallow.

Diller, J. S., and Stanton, T. W. See Fontaine, W. M.

Duror, Caroline A. Report upon paleontology and paleophytology [of the Skykomish basin, Washington].
In Smith, Warren S., Stratigraphy of the Skykomish basin, Washington: Jour. Geology, vol. 24, 1916, pp. 570–580, text figs. 3–10.

Eames, Arthur J. On the origin of the broad ray in *Quercus*.
Bot. Gaz., vol. 49, 1910, pp. 161–166, pls. 8, 9.

Emmons, Ebenezer. Geological report of the Midland Counties of North Carolina.
Raleigh, 1856, pp. 1–351, pls. 1–13. [Fossil plants, pp. 287–293, 324–335.]

—— American geology, containing a statement of the principles of the science, with full illustrations of the characteristic American fossils.
Part VI, Albany, 1857, pp. i–vii; 1–152, pls. 3–8. [Fossil plants, pp. 34–39, 99–134, text figs. 8, 68–102.]

Ettingshausen, Dr. **Constantin,** Baron von. Contributions to the Tertiary flora of Australia.
Geol. Survey New South Wales Mem. Palaeontol., No. 2, Sydney, 1888, footnotes, pp. 25, 26, 45, 48, 49, 51, 57, 58, 68, 69.

Felix, Johannes. Untersuchungen über fossile Hölzer.
Zeitschr. Deutschen geol. Gesell., Jahrg. 1886, pp. 483–492, pl. 12.

—— Studien über fossile Pilze.
Zeitschr. Deutschen geol. Gesell., Jahrg. 1894, pp. 269–280, pl. 19.

—— Untersuchungen über fossile Hölzer—V, Stück.
Zeitschr. Deutschen geol. Gesell., Jahrg. 1896, pp. 249–260, pl. 6.

Fontaine, William Morris. Contributions to the knowledge of the older Mesozoic flora of Virginia.
U. S. Geol. Survey Mon. 6, 1883, pp. i–xi, 1–144, pls. 1–54.

—— The Potomac or younger Mesozoic flora.
U. S. Geol. Survey Mon. 15, 1889, pp. i–xiv, 1–377, pls. 1–180 (bound separately).

—— Description of some fossil plants from the Great Falls coal field of Montana.
U. S. Nat. Mus. Proc., vol. 15, 1892, pp. 487–495, pls. 82–84.

—— Notes on some fossil plants from the Trinity division of the Comanche series of Texas.
U. S. Nat. Mus. Proc., vol. 16, 1893, pp. 261–282, pls. 36–43.

—— List of plants of the Shasta-Chico series.
In Diller, J. S., and Stanton, T. W., Geol. Soc. America Bull., vol. 5, 1894, pp. 450, 451.

—— Report on plants from Oroville, Calif.
In Turner, H. W., Jour. Geology, vol. 3, 1895, pp. 395, 396.

—— Note on plants of Horsetown beds.
In Stanton, T. W., U. S. Geol. Survey Bull. 133, 1895 [1896], p. 22.

—— Notes on some Mesozoic plants from near Oroville, Calif.
Am. Jour. Sci., 4th ser., vol. 2, 1896, pp. 273–275.

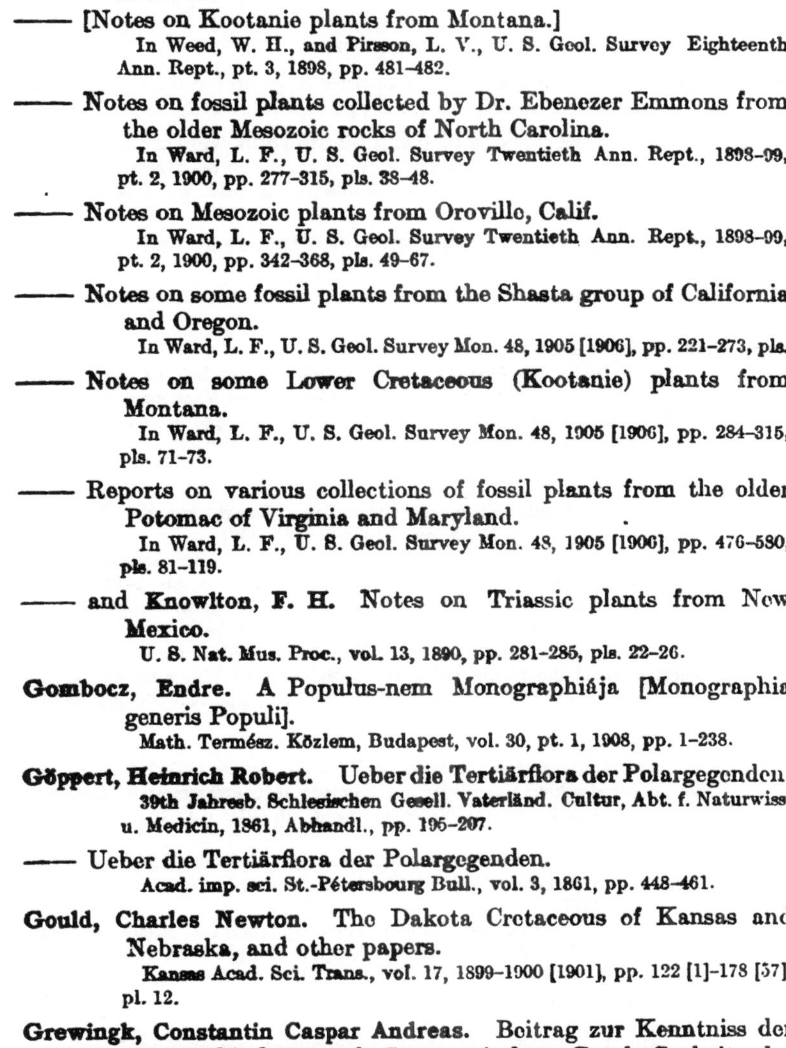

Fontaine, William Morris. Notes on Lower Cretaceous plants from the Hay Creek coal field, Crook County, Wyo.
U. S. Geol. Survey Nineteenth Ann. Rept., 1897-98 [1899], pt. 2, pp. 645-702.

—— [Notes on Kootanie plants from Montana.]
In Weed, W. H., and Pirsson, L. V., U. S. Geol. Survey Eighteenth Ann. Rept., pt. 3, 1898, pp. 481-482.

—— Notes on fossil plants collected by Dr. Ebenezer Emmons from the older Mesozoic rocks of North Carolina.
In Ward, L. F., U. S. Geol. Survey Twentieth Ann. Rept., 1898-99, pt. 2, 1900, pp. 277-315, pls. 38-48.

—— Notes on Mesozoic plants from Oroville, Calif.
In Ward, L. F., U. S. Geol. Survey Twentieth Ann. Rept., 1898-99, pt. 2, 1900, pp. 342-368, pls. 49-67.

—— Notes on some fossil plants from the Shasta group of California and Oregon.
In Ward, L. F., U. S. Geol. Survey Mon. 48, 1905 [1906], pp. 221-273, pls.

—— Notes on some Lower Cretaceous (Kootanie) plants from Montana.
In Ward, L. F., U. S. Geol. Survey Mon. 48, 1905 [1906], pp. 284-315, pls. 71-73.

—— Reports on various collections of fossil plants from the older Potomac of Virginia and Maryland.
In Ward, L. F., U. S. Geol. Survey Mon. 48, 1905 [1906], pp. 476-580, pls. 81-119.

—— and **Knowlton, F. H.** Notes on Triassic plants from New Mexico.
U. S. Nat. Mus. Proc., vol. 13, 1890, pp. 281-285, pls. 22-26.

Gombocz, Endre. A Populus-nem Monographiája [Monographia generis Populi].
Math. Termész. Közlem, Budapest, vol. 30, pt. 1, 1908, pp. 1-238.

Göppert, Heinrich Robert. Ueber die Tertiärflora der Polargegenden.
39th Jahresb. Schlesischen Gesell. Vaterländ. Cultur, Abt. f. Naturwiss. u. Medicin, 1861, Abhandl., pp. 195-207.

—— Ueber die Tertiärflora der Polargegenden.
Acad. imp. sci. St.-Pétersbourg Bull., vol. 3, 1861, pp. 448-461.

Gould, Charles Newton. The Dakota Cretaceous of Kansas and Nebraska, and other papers.
Kansas Acad. Sci. Trans., vol. 17, 1899-1900 [1901], pp. 122 [1]-178 [57], pl. 12.

Grewingk, Constantin Caspar Andreas. Beitrag zur Kenntniss der Orographischen und Geognostischen Beschaffenheit der Nord-West-Küste Amerikas mit Anliegenden Inseln.
Verhandl. Russ-kais. mineralog. Gesell. St. Petersburg, 1848-49, St. Petersburg, 1850, pp. 41, 93, 97, 124.

Hannibal, Harold. A Pliocene flora from the Coast Ranges of California.
Bull. Torrey Bot. Club, vol. 38, 1911, pp. 329–342, pl. 15.

—— Jura-Cretaceous stonewort and Limneas, supposedly from Arkansas.
Science, new ser., vol. 48, 1918, p. 578.

Hatcher, J. B. *Sabal rigida;* a new species of palm from the Laramie.
Carnegie Mus. Annals, vol. 1, 1901, pp. 263–264, text fig. 1.

Heer, Oswald. Descriptions of fossil plants from Nebraska.
In Meek, F. B., and Hayden, F. V., Remarks on the Lower Cretaceous beds of Kansas and Nebraska, etc.: Philadelphia Acad. Sci. Proc., vol. 10, 1858, pp. 265, 266.

—— Ueber einige fossile Pflanzen von Vancouver und Britisch-Columbien.
Neue Denkschr. d. Allgem. schweiz. Gesell., vol. 21, 1865, pp. 1–10, pls. 1, 2.

—— Flora fossilis arctica.
Vol. I, 1868. Die fossile Flora der Polarländer, enthaltend die in Nord Grönland, auf der Melville-Insel, im Banksland, am Mackenzie, in Island und in Spitzbergen entdeckten fossilen Pflanzen, pp. 1–192, pls. 1–50. Mit zinem Anhang über versteinerte Hölzer der artischen Zone, von Dr. Carl Cramer, pp. 167–180, pls. 34–42.

Vol. II, 1871:

Abt. 1. Fossile Flora der Bären-Insel: Kongl. svenska Vetensk. Akad. Handl., vol. 9, No. 5, pp. 1–51, pls. 1–15, 1871.

Abt. 2. Flora fossilis alaskana; Fossile Flora von Alaska: Kongl. svenska Vetensk. Akad. Handl., vol. 8, No. 4, pp. 1–41, pls. 1–10, 1869.

Abt. 3. Die Miocene Flora und Fauna Spitzbergens: Kongl. svenska Vetensk. Akad. Handl., vol. 8, No. 7, pp. 1–98, pls. 1–16, 1870.

Abt. 4. Contributions to the fossil flora of North Greenland, being a description of the plants collected by Mr. Edward Whymper during the summer of 1867: Phil. Trans. London, vol. 159, pp. 445–488, pls. 39–56, 1869.

Vol. III, 1874.

Abt. 1. Nachträge zur Miocene Flora Grönlands, enthaltend die von der Schwedischen Expedition in Sommer 1870 gesammelten Miocenen Pflanzen: Kongl. svenska Vetensk. Akad. Handl., vol. 13, No. 2, pp. 1–29, pls. 1–5, 1874.

Abt. 2. Die Kreide-Flora der arctischen Zone: Kongl. svenska Vetensk. Akad. Handl., vol. 12, No. 6, pp. 1–138, pls. 1–38, 1874.

Vol. IV, 1877:

Abt. 1. Beiträge zur fossilen Flora Spitzbergens: Kongl. svenska Vetensk. Akad. Handl., vol. 14, No. 5, pp. 1–141, pls. 1–32, 1876.

Abt. 2. Beiträge zur Jura-Flora Ostsibiriens und des Amurlandes: Mém. Acad. imp. sci. St.-Pétersbourg, 7th ser., vol. 22, pp. 1–222, pls. 1–31, 1876.

Abt. 3. Ueber die Pflanzen-Versteinerungen von Andö in Norwegen, pp. 1–15, pls. 1, 2, 1877.

Vol. V, 1878:

Abt. 1. Die Miocene Flora des Grinnell-Landes, pp. 1–38, pls. 1–9, 1878.

Heer, Oswald. Flora fossilis arctica—Continued.

Abt. 2. Beiträge zur fossilen Flora Sibiriens und des Amurlandes: Mém. Acad. imp. sci. St.-Pétersbourg, 7th ser., vol. 25, No. 6, pp. 1–58, pls. 1–15, 1878.

Abt. 3. Primitiae florae fossilis sachalinensis. Miocene Flora des Insel Sachalin: Mém. Acad. imp. sci. St.-Pétersbourg, 7th ser., vol. 25, No. 7, pp. 1–61, pls. 1–15, 1878.

Abt. 4. Beiträge zur Miocene Flora von Sachalin: Kongl. svenska Vetensk. Akad. Handl., vol. 15, No. 4, pp. 1–11, pls. 1–4, 1878.

Abt. 5. Über Fossile Pflanzen von Novaja Semlja: Kongl. svenska Vetensk. Akad. Handl., vol. 15, No. 3, pp. 1–6, pl. 1, 1878.

Vol. VI, 1882:

Abt. 1:

1. Nachträge zur Jura-Flora Sibiriens: Mém. Acad. imp. sci. St.-Pétersbourg, 7th ser., vol. 32, No. 10, pp. 1–34, pls. 1–9, 1880.

2. Nachträge zur fossilen Flora Grönlands: Kongl. svenska Vetensk. Akad. Handl., vol. 18, No. 2, pp. 1–17, pls. 1–6, 1880.

3. Beiträge zur Miocene Flora von Nord-Canada, pp. 1–17, pls. 1–3, 1880.

4. Untersuchung über fossile Hölzer aus der arctischen Zone, von C. Schroeter, pp. 1–38, pls. 1–3, 1880.

Abt. 2. Den ersten Theil der fossilen Flora Grönlands, pp. 1–12, pls. 1–47, 1882.

Vol. VII, 1883. Flora fossilis grönlandica. Die Fossile Flora Grönlands, pp. 1–275, pls. 48–109, 1883.

Henderson, Junius. The sandstone of Fossil Ridge in northern Colorado and its fauna. [Fossil plant on p. 186.]

Univ. Colorado Studies, vol. 5, 1908, p. 186.

Herrick, Francis H. Microscopical examination of wood from the buried forest, Muir Inlet, Alaska.

Nat. Geog. Mag., vol. 4, 1892, pp. 75–78, text figs. 4, 5.

Hill, Robert T. Paleontology of the Cretaceous formations of Texas. The invertebrate paleontology of the Trinity division.

Biol. Soc. Washington Proc., vol. 8, 1893, pp. 39, 40, pl. 1, fig. 1–1d.

Hinde George Jennings. The glacial and interglacial strata of Scarboro Heights and other localities near Toronto, Ontario.

Canadian Jour. (Canadian Institute), new ser., vol. 15, 1878, pp. 388–413. (Plants, p. 399.)

Hitchcock, Charles H. A new species of *Carpolithes.*

Portland Soc. Nat. Hist. Proc., vol. 1, pt. 1, 1862, p. 95, pl. 1, fig. 5.

Hitchcock, Edward. Description of a brown coal deposit in Brandon, Vt., with an attempt to determine the geological ages of the principal hematite ore beds in the United States.

Am. Jour. Sci., 2d ser., vol. 15, 1853, pp. 95–104.

—— Description of a new species of *Clathropteris*, discovered in the Connecticut Valley sandstone.

Am. Jour. Sci., 2d ser., vol. 20, 1855, pp. 22–25, text figs. 1, 2.

Hitchcock, Edward. Fossil fruits [from Brandon, Vt.].
Geology of Vermont, vol. 1, 1861, pp. 226–234, figs. 111–160.

Holden, Ruth. Cretaceous Pityoxyla from Cliffwood, N. J.
Am. Acad. Arts and Sci. Proc., vol. 48, 1913, pp. 609–623, pls. 1–4.

—— Contributions to the anatomy of Mesozoic conifers—No. 2, Cretaceous lignites from Cliffwood, N. J.
Bot. Gaz., vol. 58, 1914, pp. 168–177, pls. 12–15.

—— Some fossil plants from eastern Canada.
Annals of Botany, vol. 27, April, 1913, pp. 243–255, pls. 22, 23.

Hollick, Arthur. The paleontology of the Cretaceous formations on Staten Island.
New York Acad. Sci. Trans., vol. 11, 1892, pp. 96–103, pls. 1–4.

—— Additions to the paleobotany of the Cretaceous formation on Staten Island.
New York Acad. Sci. Trans., vol. 12, 1892, pp. 28–39, pls. 1–3.

—— Paleobotany of the yellow gravel at Bridgeton, N. J.
Bull. Torrey Bot. Club, vol. 19, 1892, pp. 330–333.

—— Preliminary contribution to our knowledge of the Cretaceous formation in Long Island and eastward.
New York Acad. Sci. Trans., vol. 12, 1892, pp. 222–237, pls. 5–7.

—— A new fossil palm from the Cretaceous formation at Glen Cove, Long Island.
Bull. Torrey Bot. Club, vol. 20, 1893, pp. 168, 169, pl. 149.

—— Some further notes upon *Serenopsis kempii*.
Bull. Torrey Bot. Club, vol. 20, 1893, pp. 334, 335, pl. 144.

—— Additions to the paleobotany of the Cretaceous formation on Long Island.
Bull. Torrey Bot. Club, vol. 21, 1894, pp. 49–65, pls. 174–180.

—— Plants from the Cretaceous clays at Northport, Long Island.
In Ries, Heinrich, School of Mines Quart., vol. 15, 1894, p. 355.

—— Some further notes on the geology of the north shore of Long Island.
New York Acad. Sci. Trans., vol. 13, 1894, pp. 122–132.

—— Fossil Salvinias, including description of a new species.
Bull. Torrey Bot. Club, vol. 21, 1894, pp. 253–257, pl. 205.

—— Winglike appendage on the petioles of *Liriophyllum populoides* Lesq. and *Liriodendron alatum* Newb., with description of the latter.
Bull. Torrey Bot. Club, vol. 21, 1894, pp. 467–475, pls. 220, 221.

—— A new fossil *Nelumbo* from the Laramie group of Florence, Colo.
Bull. Torrey Bot. Club, vol. 21, 1894, pp. 307–310, text figs.

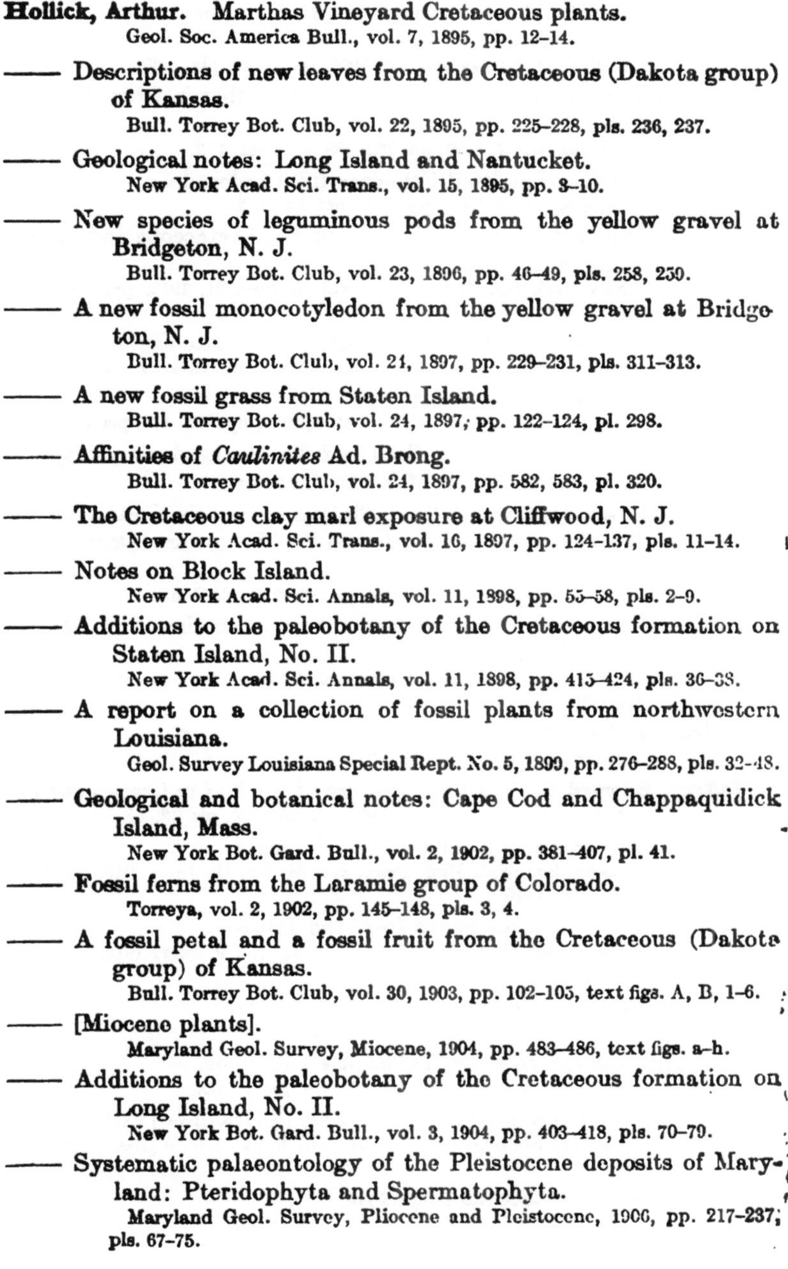

Hollick, Arthur. Marthas Vineyard Cretaceous plants.
Geol. Soc. America Bull., vol. 7, 1895, pp. 12–14.

—— Descriptions of new leaves from the Cretaceous (Dakota group) of Kansas.
Bull. Torrey Bot. Club, vol. 22, 1895, pp. 225–228, pls. 236, 237.

—— Geological notes: Long Island and Nantucket.
New York Acad. Sci. Trans., vol. 15, 1895, pp. 8–10.

—— New species of leguminous pods from the yellow gravel at Bridgeton, N. J.
Bull. Torrey Bot. Club, vol. 23, 1896, pp. 46–49, pls. 258, 259.

—— A new fossil monocotyledon from the yellow gravel at Bridgeton, N. J.
Bull. Torrey Bot. Club, vol. 24, 1897, pp. 229–231, pls. 311–313.

—— A new fossil grass from Staten Island.
Bull. Torrey Bot. Club, vol. 24, 1897, pp. 122–124, pl. 298.

—— Affinities of *Caulinites* Ad. Brong.
Bull. Torrey Bot. Club, vol. 24, 1897, pp. 582, 583, pl. 320.

—— The Cretaceous clay marl exposure at Cliffwood, N. J.
New York Acad. Sci. Trans., vol. 16, 1897, pp. 124-137, pls. 11–14.

—— Notes on Block Island.
New York Acad. Sci. Annals, vol. 11, 1898, pp. 55–58, pls. 2–9.

—— Additions to the paleobotany of the Cretaceous formation on Staten Island, No. II.
New York Acad. Sci. Annals, vol. 11, 1898, pp. 415–424, pls. 36–38.

—— A report on a collection of fossil plants from northwestern Louisiana.
Geol. Survey Louisiana Special Rept. No. 5, 1899, pp. 276–288, pls. 32–48.

—— Geological and botanical notes: Cape Cod and Chappaquidick Island, Mass.
New York Bot. Gard. Bull., vol. 2, 1902, pp. 381–407, pl. 41.

—— Fossil ferns from the Laramie group of Colorado.
Torreya, vol. 2, 1902, pp. 145–148, pls. 3, 4.

—— A fossil petal and a fossil fruit from the Cretaceous (Dakota group) of Kansas.
Bull. Torrey Bot. Club, vol. 30, 1903, pp. 102–105, text figs. A, B, 1–6.

—— [Miocene plants].
Maryland Geol. Survey, Miocene, 1904, pp. 483–486, text figs. a–h.

—— Additions to the paleobotany of the Cretaceous formation on Long Island, No. II.
New York Bot. Gard. Bull., vol. 3, 1904, pp. 403–418, pls. 70–79.

—— Systematic palaeontology of the Pleistocene deposits of Maryland: Pteridophyta and Spermatophyta.
Maryland Geol. Survey, Pliocene and Pleistocene, 1906, pp. 217–237, pls. 67–75.

Hollick, Arthur. The Cretaceous flora of southern New York and New England.
U. S. Geol. Survey Mon. 50, 1906, pp. 1–219, pls. 1–40.

—— Description of a new Tertiary fossil flower from Florissant, Colo.
Torreya, vol. 7, 1907, pp. 182, 183, text figs. 1, 2.

—— A new genus of fossil Fagaceae from Colorado.
Torreya, vol. 9, 1909, pp. 1–3, text figs. 1, 2.

—— Additions to the paleobotany of the Cretaceous formation on Long Island, No. 3.
New York Bot. Gard. Bull., vol. 8, Nov. 21, 1912, pp. 154–170, pls. 162–170.

—— A new fossil species of *Ficus* and its climatic significance.
New York Bot. Gard. Jour., vol. 16, 1915, pp. 43–47, 2 pls.

—— A fossil fern monstrosity.
New York Bot. Gard. Mem., vol. 6, 1916, pp. 473, 474, pls. 31, 32.

—— See also Newberry, J. S., The flora of the Amboy clays; The later extinct floras of North America.

—— and **Jeffrey, Edward C.** Affinities of certain Cretaceous plant remains commonly referred to the genera *Dammara* and *Brachyphyllum*.
Am. Naturalist, vol. 40, 1906, pp. 189–216, pls. 1–5.

—— —— Studies of Cretaceous coniferous remains from Kreischerville, N. Y.
New York Bot. Gard. Mem., vol. 3, May 20, 1909, pp. i–viii; 1–138, pls. 1–29.

Holzinger, John M. On some fossil mosses.
Biologist, vol. 6, 1903, pp. 93, 94.

Jeffrey, Edward C. On the structure of the leaf in Cretaceous pines.
Annals of Botany, vol. 22, 1908, pp. 207–220, pls. 13, 14.

—— A new *Prepinus* from Marthas Vineyard.
Boston Soc. Nat. Hist. Proc., vol. 34, No. 10, July, 1910, pp. 333–338, pl. 33, figs. 1–10.

—— A new araucarian genus from the Triassic.
Boston Soc. Nat. Hist. Proc., vol. 34, No. 9, July, 1910, pp. 325–332, pls. 31, 32.

—— A fossil *Sequoia* from the Sierra Nevada.
Bot. Gaz., vol. 38, 1904, pp. 321–332.

—— *Araucariopitys*, a new genus of araucarians.
Bot. Gaz., vol. 44, 1907, pp. 435–444, pls. 28–30.

—— and **Chrysler, M. A.** The lignites of Brandon [Vt.].
Vermont State Geologist Fifth Ann. Rept., 1905–6 [1906], pp. 195–201, pls. 49–51.

—— —— On Cretaceous Pityoxyla.
Bot. Gaz., vol. 42, 1906, pp. 1–14, pls. 1, 2.

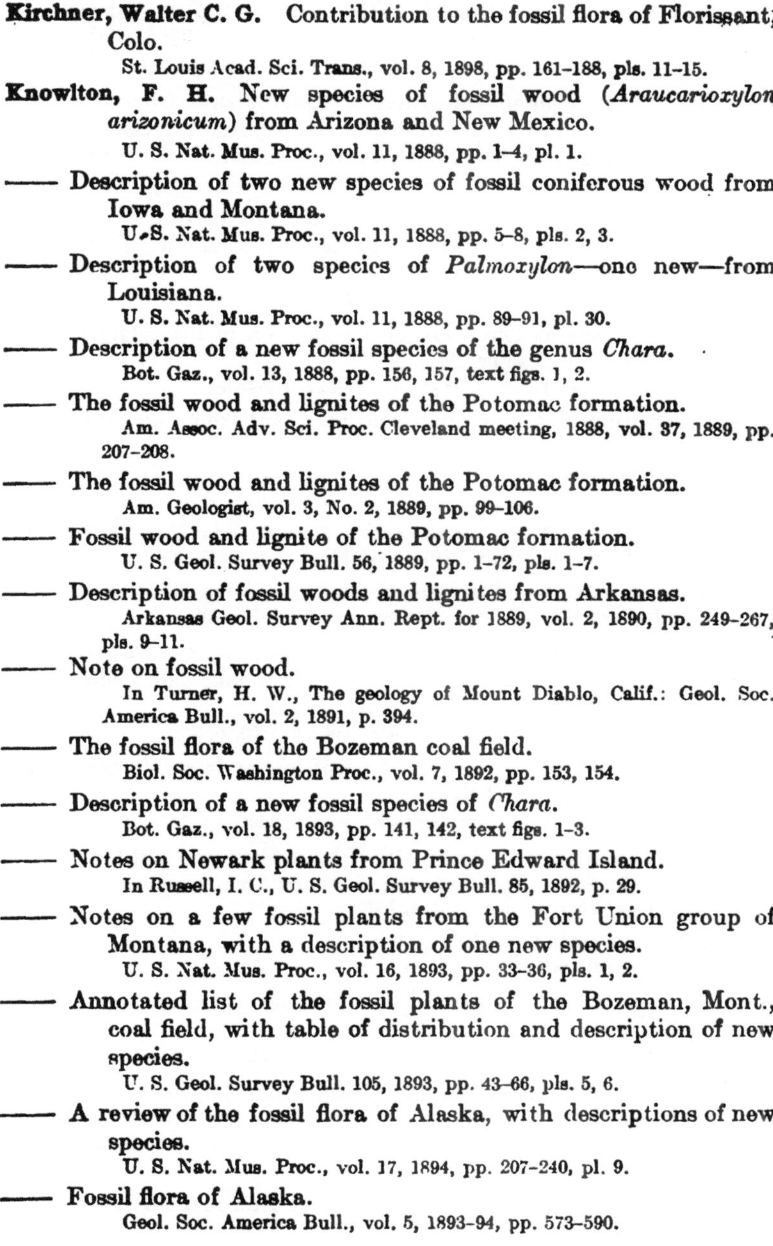

Kirchner, Walter C. G. Contribution to the fossil flora of Florissant, Colo.

St. Louis Acad. Sci. Trans., vol. 8, 1898, pp. 161–188, pls. 11–15.

Knowlton, F. H. New species of fossil wood (*Araucarioxylon arizonicum*) from Arizona and New Mexico.

U. S. Nat. Mus. Proc., vol. 11, 1888, pp. 1–4, pl. 1.

—— Description of two new species of fossil coniferous wood from Iowa and Montana.

U. S. Nat. Mus. Proc., vol. 11, 1888, pp. 5–8, pls. 2, 3.

—— Description of two species of *Palmoxylon*—one new—from Louisiana.

U. S. Nat. Mus. Proc., vol. 11, 1888, pp. 89–91, pl. 30.

—— Description of a new fossil species of the genus *Chara*.

Bot. Gaz., vol. 13, 1888, pp. 156, 157, text figs. 1, 2.

—— The fossil wood and lignites of the Potomac formation.

Am. Assoc. Adv. Sci. Proc. Cleveland meeting, 1888, vol. 37, 1889, pp. 207–208.

—— The fossil wood and lignites of the Potomac formation.

Am. Geologist, vol. 3, No. 2, 1889, pp. 99–106.

—— Fossil wood and lignite of the Potomac formation.

U. S. Geol. Survey Bull. 56, 1889, pp. 1–72, pls. 1–7.

—— Description of fossil woods and lignites from Arkansas.

Arkansas Geol. Survey Ann. Rept. for 1889, vol. 2, 1890, pp. 249–267, pls. 9–11.

—— Note on fossil wood.

In Turner, H. W., The geology of Mount Diablo, Calif.: Geol. Soc. America Bull., vol. 2, 1891, p. 394.

—— The fossil flora of the Bozeman coal field.

Biol. Soc. Washington Proc., vol. 7, 1892, pp. 153, 154.

—— Description of a new fossil species of *Chara*.

Bot. Gaz., vol. 18, 1893, pp. 141, 142, text figs. 1–3.

—— Notes on Newark plants from Prince Edward Island.

In Russell, I. C., U. S. Geol. Survey Bull. 85, 1892, p. 29.

—— Notes on a few fossil plants from the Fort Union group of Montana, with a description of one new species.

U. S. Nat. Mus. Proc., vol. 16, 1893, pp. 33–36, pls. 1, 2.

—— Annotated list of the fossil plants of the Bozeman, Mont., coal field, with table of distribution and description of new species.

U. S. Geol. Survey Bull. 105, 1893, pp. 43–66, pls. 5, 6.

—— A review of the fossil flora of Alaska, with descriptions of new species.

U. S. Nat. Mus. Proc., vol. 17, 1894, pp. 207–240, pl. 9.

—— Fossil flora of Alaska.

Geol. Soc. America Bull., vol. 5, 1893–94, pp. 573–590.

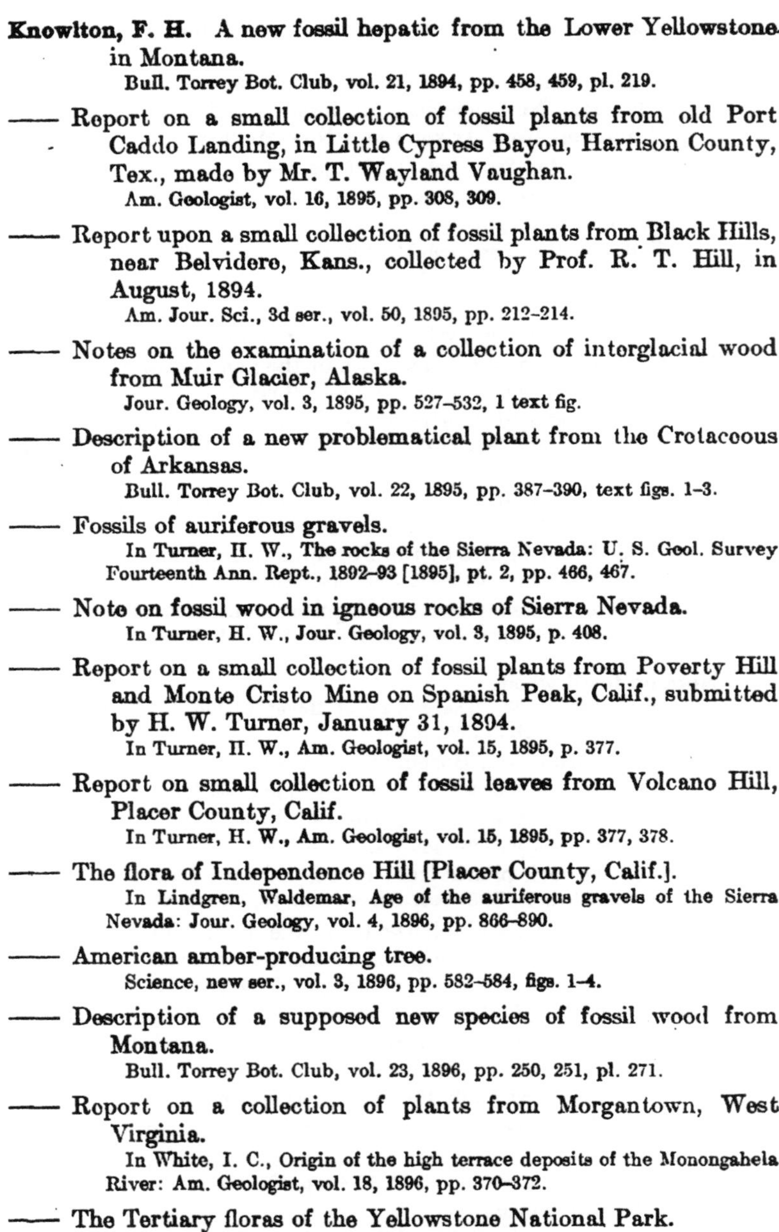

Knowlton, F. H. A new fossil hepatic from the Lower Yellowstone in Montana.
Bull. Torrey Bot. Club, vol. 21, 1894, pp. 458, 459, pl. 219.

—— Report on a small collection of fossil plants from old Port Caddo Landing, in Little Cypress Bayou, Harrison County, Tex., made by Mr. T. Wayland Vaughan.
Am. Geologist, vol. 16, 1895, pp. 308, 309.

—— Report upon a small collection of fossil plants from Black Hills, near Belvidere, Kans., collected by Prof. R. T. Hill, in August, 1894.
Am. Jour. Sci., 3d ser., vol. 50, 1895, pp. 212–214.

—— Notes on the examination of a collection of interglacial wood from Muir Glacier, Alaska.
Jour. Geology, vol. 3, 1895, pp. 527–532, 1 text fig.

—— Description of a new problematical plant from the Cretaceous of Arkansas.
Bull. Torrey Bot. Club, vol. 22, 1895, pp. 387–390, text figs. 1–3.

—— Fossils of auriferous gravels.
In Turner, H. W., The rocks of the Sierra Nevada: U. S. Geol. Survey Fourteenth Ann. Rept., 1892–93 [1895], pt. 2, pp. 466, 467.

—— Note on fossil wood in igneous rocks of Sierra Nevada.
In Turner, H. W., Jour. Geology, vol. 3, 1895, p. 408.

—— Report on a small collection of fossil plants from Poverty Hill and Monte Cristo Mine on Spanish Peak, Calif., submitted by H. W. Turner, January 31, 1894.
In Turner, H. W., Am. Geologist, vol. 15, 1895, p. 377.

—— Report on small collection of fossil leaves from Volcano Hill, Placer County, Calif.
In Turner, H. W., Am. Geologist, vol. 15, 1895, pp. 377, 378.

—— The flora of Independence Hill [Placer County, Calif.].
In Lindgren, Waldemar, Age of the auriferous gravels of the Sierra Nevada: Jour. Geology, vol. 4, 1896, pp. 866–890.

—— American amber-producing tree.
Science, new ser., vol. 3, 1896, pp. 582–584, figs. 1–4.

—— Description of a supposed new species of fossil wood from Montana.
Bull. Torrey Bot. Club, vol. 23, 1896, pp. 250, 251, pl. 271.

—— Report on a collection of plants from Morgantown, West Virginia.
In White, I. C., Origin of the high terrace deposits of the Monongahela River: Am. Geologist, vol. 18, 1896, pp. 370–372.

—— The Tertiary floras of the Yellowstone National Park.
Am. Jour. Sci., 4th ser., vol. 2, 1896, pp. 51–58.

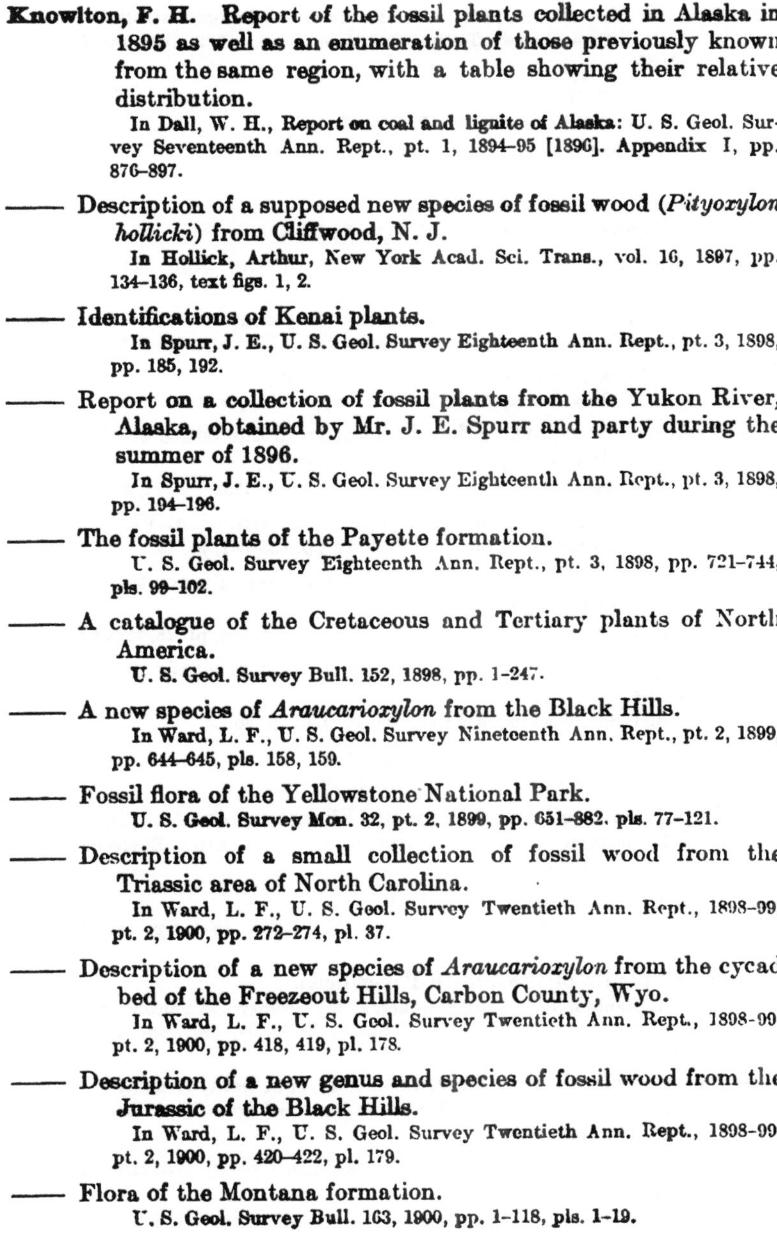

Knowlton, F. H. Report of the fossil plants collected in Alaska in 1895 as well as an enumeration of those previously known from the same region, with a table showing their relative distribution.

In Dall, W. H., Report on coal and lignite of Alaska: U. S. Geol. Survey Seventeenth Ann. Rept., pt. 1, 1894–95 [1896]. Appendix I, pp. 876–897.

—— Description of a supposed new species of fossil wood (*Pityoxylon hollicki*) from Cliffwood, N. J.

In Hollick, Arthur, New York Acad. Sci. Trans., vol. 16, 1897, pp. 134–136, text figs. 1, 2.

—— Identifications of Kenai plants.

In Spurr, J. E., U. S. Geol. Survey Eighteenth Ann. Rept., pt. 3, 1898, pp. 185, 192.

—— Report on a collection of fossil plants from the Yukon River, Alaska, obtained by Mr. J. E. Spurr and party during the summer of 1896.

In Spurr, J. E., U. S. Geol. Survey Eighteenth Ann. Rept., pt. 3, 1898, pp. 194–196.

—— The fossil plants of the Payette formation.

U. S. Geol. Survey Eighteenth Ann. Rept., pt. 3, 1898, pp. 721–744, pls. 99–102.

—— A catalogue of the Cretaceous and Tertiary plants of North America.

U. S. Geol. Survey Bull. 152, 1898, pp. 1–247.

—— A new species of *Araucarioxylon* from the Black Hills.

In Ward, L. F., U. S. Geol. Survey Nineteenth Ann. Rept., pt. 2, 1899, pp. 644–645, pls. 158, 159.

—— Fossil flora of the Yellowstone National Park.

U. S. Geol. Survey Mon. 32, pt. 2, 1899, pp. 651–882, pls. 77–121.

—— Description of a small collection of fossil wood from the Triassic area of North Carolina.

In Ward, L. F., U. S. Geol. Survey Twentieth Ann. Rept., 1898–99, pt. 2, 1900, pp. 272–274, pl. 37.

—— Description of a new species of *Araucarioxylon* from the cycad bed of the Freezeout Hills, Carbon County, Wyo.

In Ward, L. F., U. S. Geol. Survey Twentieth Ann. Rept., 1898–99, pt. 2, 1900, pp. 418, 419, pl. 178.

—— Description of a new genus and species of fossil wood from the Jurassic of the Black Hills.

In Ward, L. F., U. S. Geol. Survey Twentieth Ann. Rept., 1898–99, pt. 2, 1900, pp. 420–422, pl. 179.

—— Flora of the Montana formation.

U. S. Geol. Survey Bull. 163, 1900, pp. 1–118, pls. 1–19.

Knowlton, F. H. Fossil plants of the Esmeralda formation.
U. S. Geol. Survey Twenty-first Ann. Rept., pt. 2, 1901, pp. 209–222, pl. 30.

—— Fossil hickory nuts.
Plant World, vol. 4, 1901, pp. 51, 52.

—— A fossil flower.
Plant World, vol. 4, 1901, pp. 73, 74.

—— A fossil nut pine from Idaho.
Torreya, vol. 1, 1901, pp. 113–115, figs. 1–3.

—— Preliminary report on fossil flora of the John Day basin, Oreg.
In Merriam, J. C., A contribution to the geology of the John Day basin: California Univ. Dept. Geology Bull., vol. 2, 1901, pp. 269–314.

—— Fossil flora of the John Day Basin, Oreg.
U. S. Geol. Survey Bull. 204, 1902, pp. 1–153, pls. 1–17.

—— Preliminary report on fossil plants from the State of Washington, collected by Henry Landes, 1901.
Washington Geol. Survey Ann. Rept., vol. 1, 1902, pp. 32, 33.

—— Notes on the fossil fruits and lignites of Brandon, Vt.
Bull. Torrey Bot. Club, vol. 29, 1902, pp. 635–641, pl. 25.

—— Report on a small collection of fossil plants from the vicinity of Porcupine Butte, Mont.
Bull. Torrey Bot. Club, vol. 29, 1902, pp. 705–709, pl. 26, text fig.

—— Change of name of *Ficus? hesperia* from vicinity of Ashland, Oreg.
Biol. Soc. Washington Proc., vol. 15, 1902, p. 86.

—— The publication of rejected names.
Science, new ser., vol. 17, 1903, pp. 506–508.

—— Report on fossil plants from the Ellensburg quadrangle.
In Smith, Geo. Otis, U. S. Geol. Survey Geol. Atlas, Ellensburg folio (No. 86), 1903, p. 3.

—— Fossil plants from Kukak Bay.
Alaska, vol. 4, Harriman Alaska Expedition, 1904, pp. 149–162, pls. 22–32.

—— Fossil plants of the Judith River beds.
In Stanton, T. W., and Hatcher, J. B., Geology and paleontology of the Judith River beds: U. S. Geol. Survey Bull. 257, 1905, pp. 129–168, pls. 14–19.

—— [Lists of plants from Herendeen Bay coal field, Alaska.]
In Paige, Sidney, The Herendeen Bay coal field: U. S. Geol. Survey Bull. 284, 1906, p. 103.

—— [Lists of Kenai plants from Chickaloon Creek, Matanuska coal field, Alaska.]
In Martin, G. C., The Matanuska coal field, Alaska: U. S. Geol. Survey Bull. 289, 1906, pp. 14, 15.

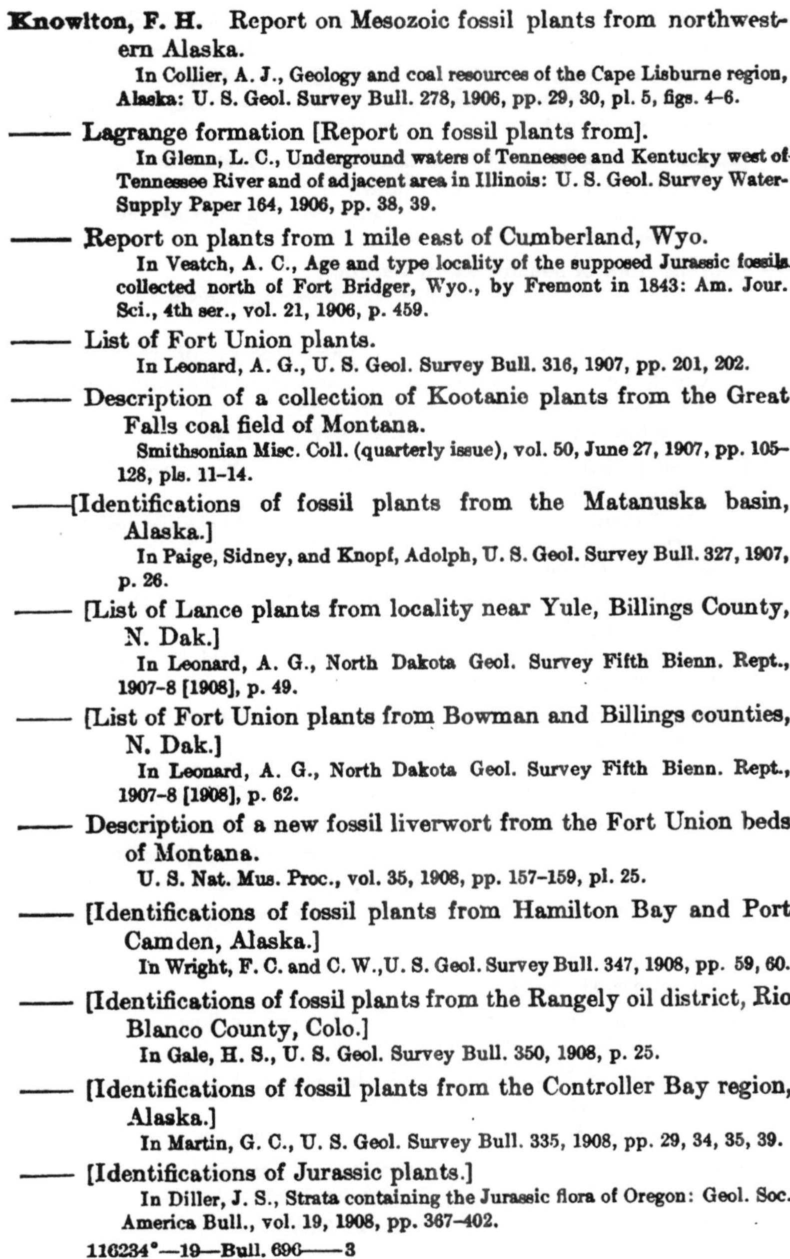

Knowlton, F. H. Report on Mesozoic fossil plants from northwestern Alaska.

In Collier, A. J., Geology and coal resources of the Cape Lisburne region, Alaska: U. S. Geol. Survey Bull. 278, 1906, pp. 29, 30, pl. 5, figs. 4–6.

—— Lagrange formation [Report on fossil plants from].

In Glenn, L. C., Underground waters of Tennessee and Kentucky west of Tennessee River and of adjacent area in Illinois: U. S. Geol. Survey Water-Supply Paper 164, 1906, pp. 38, 39.

—— Report on plants from 1 mile east of Cumberland, Wyo.

In Veatch, A. C., Age and type locality of the supposed Jurassic fossils collected north of Fort Bridger, Wyo., by Fremont in 1843: Am. Jour. Sci., 4th ser., vol. 21, 1906, p. 459.

—— List of Fort Union plants.

In Leonard, A. G., U. S. Geol. Survey Bull. 316, 1907, pp. 201, 202.

—— Description of a collection of Kootanie plants from the Great Falls coal field of Montana.

Smithsonian Misc. Coll. (quarterly issue), vol. 50, June 27, 1907, pp. 105–128, pls. 11–14.

——[Identifications of fossil plants from the Matanuska basin, Alaska.]

In Paige, Sidney, and Knopf, Adolph, U. S. Geol. Survey Bull. 327, 1907, p. 26.

—— [List of Lance plants from locality near Yule, Billings County, N. Dak.]

In Leonard, A. G., North Dakota Geol. Survey Fifth Bienn. Rept., 1907–8 [1908], p. 49.

—— [List of Fort Union plants from Bowman and Billings counties, N. Dak.]

In Leonard, A. G., North Dakota Geol. Survey Fifth Bienn. Rept., 1907–8 [1908], p. 62.

—— Description of a new fossil liverwort from the Fort Union beds of Montana.

U. S. Nat. Mus. Proc., vol. 35, 1908, pp. 157–159, pl. 25.

—— [Identifications of fossil plants from Hamilton Bay and Port Camden, Alaska.]

In Wright, F. C. and C. W., U. S. Geol. Survey Bull. 347, 1908, pp. 59, 60.

—— [Identifications of fossil plants from the Rangely oil district, Rio Blanco County, Colo.]

In Gale, H. S., U. S. Geol. Survey Bull. 350, 1908, p. 25.

—— [Identifications of fossil plants from the Controller Bay region, Alaska.]

In Martin, G. C., U. S. Geol. Survey Bull. 335, 1908, pp. 29, 34, 35, 39.

—— [Identifications of Jurassic plants.]

In Diller, J. S., Strata containing the Jurassic flora of Oregon: Geol. Soc. America Bull., vol. 19, 1908, pp. 367–402.

116234°—19—Bull. 696——3

Knowlton, F. H. Report on fossil plants collected for J. S. Diller by James Storrs, February, 1908, on the western side of the Sacramento Valley, Tehama County, Calif.
In Diller, J. S., Strata containing the Jurassic flora of Oregon: Geol. Soc. America Bull., vol. 19, 1908, pp. 384–386.

—— [Report on fossil plants from the Foreman formation (Jurassic) of the Taylorsville region, Calif.]
In Diller, J. S., Geology of the Taylorsville region, California: U. S. Geol. Survey Bull. 353, 1908, p. 55.

—— [Eocene flora of southwestern Oregon.]
In Diller, J. S., Geology of the Taylorsville region, California: U. S. Geol. Survey Bull. 353, 1908, p. 77.

—— The stratigraphic relations and paleontology of the "Hell Creek beds," "*Ceratops* beds," and equivalents, and their reference to the Fort Union formation.
Washington Acad. Sci. Proc., vol. 11, 1909, pp. 179–238.

—— [Lists of plants from Dakota sandstone and Mesaverde formation in Utah.]
In Richardson, G. B., Reconnaissance of the Book Cliffs coal field, between Grand River, Colo., and Sunnyside, Utah: U. S. Geol. Survey Bull. 371, 1909, pp. 14, 17, 18.

—— [Lists of fossil plants from the Kootenai formation of the Great Falls coal field, Mont.]
In Fisher, C. A., Geology of the Great Falls coal field, Montana: U. S. Geol. Survey Bull. 356, 1909, pp. 33–35.

—— [Lists of plants from the Tuxedni sandstone and Chinitna shale, Iliamna region, Alaska.]
In Martin, G. C., and Katz, F. J., A geological reconnaissance of the Iliamna region, Alaska: U. S. Geol. Survey Bull. 485, 1912, pp. 63, 64.

—— Descriptions of fossil plants from the Mesozoic and Cenozoic of North America—I: 1, Two new chain ferns (*Woodwardia*) from Oregon and Wyoming; 2, A new name for *Davallia tenuifolia* Swartz, as identified by Dawson, and *Asplenium tenerum* Lesquereux.
Smithsonian Misc. Coll., quarterly issue, vol. 52, pt. 4, Jan. 11, 1910, pp. 489–496, pls. 63, 64.

—— Succession and range of Mesozoic and Tertiary floras.
Jour. Geology, vol. 18, 1910, pp. 105–116.

—— The climate of North America in later glacial and subsequent postglacial time.
Postglaziale Klimaveränderungen, Stockholm, 1910, pp. 367–369.

—— The Jurassic age of the "Jurassic flora of Oregon."
Am. Jour. Sci., 4th ser., vol. 30, 1910, pp. 33–64.

—— [Lance plants from Moreau River, S. Dak.]
In Stanton, T. W., Am. Jour. Sci., 4th ser., vol. 30, 1910, p. 176.

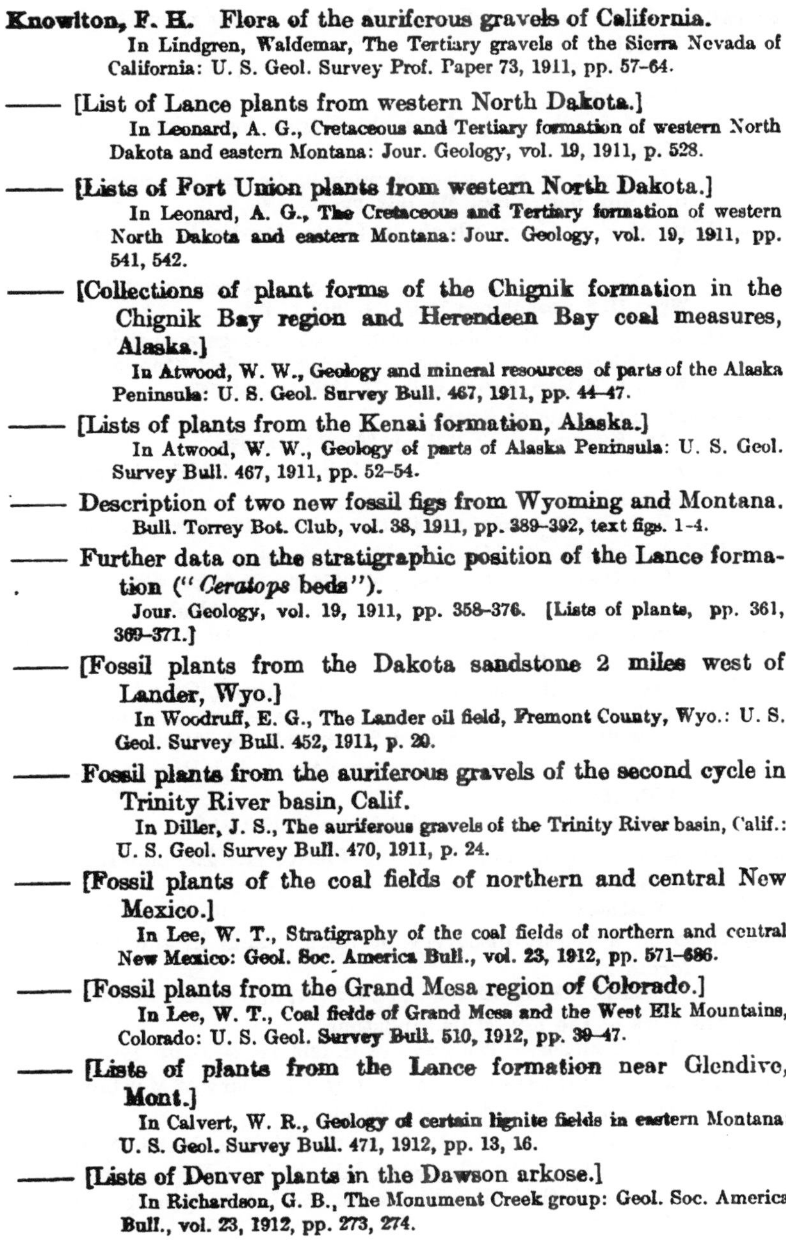

Knowlton, F. H. Flora of the auriferous gravels of California.
In Lindgren, Waldemar, The Tertiary gravels of the Sierra Nevada of California: U. S. Geol. Survey Prof. Paper 73, 1911, pp. 57–64.

—— [List of Lance plants from western North Dakota.]
In Leonard, A. G., Cretaceous and Tertiary formation of western North Dakota and eastern Montana: Jour. Geology, vol. 19, 1911, p. 528.

—— [Lists of Fort Union plants from western North Dakota.]
In Leonard, A. G., The Cretaceous and Tertiary formation of western North Dakota and eastern Montana: Jour. Geology, vol. 19, 1911, pp. 541, 542.

—— [Collections of plant forms of the Chignik formation in the Chignik Bay region and Herendeen Bay coal measures, Alaska.]
In Atwood, W. W., Geology and mineral resources of parts of the Alaska Peninsula: U. S. Geol. Survey Bull. 467, 1911, pp. 44–47.

—— [Lists of plants from the Kenai formation, Alaska.]
In Atwood, W. W., Geology of parts of Alaska Peninsula: U. S. Geol. Survey Bull. 467, 1911, pp. 52–54.

—— Description of two new fossil figs from Wyoming and Montana.
Bull. Torrey Bot. Club, vol. 38, 1911, pp. 389–392, text figs. 1–4.

—— Further data on the stratigraphic position of the Lance formation ("*Ceratops* beds").
Jour. Geology, vol. 19, 1911, pp. 358–376. [Lists of plants, pp. 361, 369–371.]

—— [Fossil plants from the Dakota sandstone 2 miles west of Lander, Wyo.]
In Woodruff, E. G., The Lander oil field, Fremont County, Wyo.: U. S. Geol. Survey Bull. 452, 1911, p. 20.

—— Fossil plants from the auriferous gravels of the second cycle in Trinity River basin, Calif.
In Diller, J. S., The auriferous gravels of the Trinity River basin, Calif.: U. S. Geol. Survey Bull. 470, 1911, p. 24.

—— [Fossil plants of the coal fields of northern and central New Mexico.]
In Lee, W. T., Stratigraphy of the coal fields of northern and central New Mexico: Geol. Soc. America Bull., vol. 23, 1912, pp. 571–686.

—— [Fossil plants from the Grand Mesa region of Colorado.]
In Lee, W. T., Coal fields of Grand Mesa and the West Elk Mountains, Colorado: U. S. Geol. Survey Bull. 510, 1912, pp. 39–47.

—— [Lists of plants from the Lance formation near Glendive, Mont.]
In Calvert, W. R., Geology of certain lignite fields in eastern Montana: U. S. Geol. Survey Bull. 471, 1912, pp. 13, 16.

—— [Lists of Denver plants in the Dawson arkose.]
In Richardson, G. B., The Monument Creek group: Geol. Soc. America Bull., vol. 23, 1912, pp. 273, 274.

Knowlton, F. H. [Lists of Triassic (Newark) plants from Dan River, N. C.]
In Stone, R. W., Coal on Dan River, N. C.: U. S. Geol. Survey Bull. 471, 1912, p. 143.

—— [Lists of fossil plants from the Lance formation of eastern Montana.]
In Calvert, W. R., Geology of certain lignite fields in eastern Montana: U. S. Geol. Survey Bull. 471, 1912, pp. 195, 198.

—— [Lists of Fort Union plants in Little Powder River coal field, Wyo.]
In Davis, J. A., The Little Powder River coal field, Campbell County, Wyo.: U. S. Geol. Survey Bull. 471, 1912, p. 426.

—— [Lists of Fort Union plants from Lost Springs coal field, Wyo.]
In Winchester, D. E., The Lost Springs coal field, Converse County, Wyo.: U. S. Geol. Survey Bull. 471, 1912, pp. 481, 482.

—— The fossil forests of Arizona.
Am. Forestry, vol. 19, 1913, pp. 208–218, text figs.

—— Description of a new fossil fern of the genus *Gleichenia* from the Upper Cretaceous of Wyoming.
U. S. Nat. Mus. Proc., vol. 45, 1913, pp. 555–558, pl. 44, figs. 1, 2.

—— The Jurassic flora of Cape Lisburne, Alaska.
U. S. Geol. Survey Prof. Paper 85, 1914, pp. 39–64, pls. 5–8.

—— [Collection of fossil plants from Port Wells, Alaska.]
In Johnson, B. L., The Port Wells gold-lode district: U. S. Geol. Survey Bull. 592, 1914, p. 209.

—— [Report on fossil plants collected near Eugene, Oreg.]
In Washburne, C. W., Reconnaissance of the geology and oil prospects of northwestern Oregon: U. S. Geol. Survey Bull. 590, 1914, p. 36.

—— Description of a new fossil fern from the Judith River formation of Montana.
Torreya, vol. 15, No. 4, April, 1915, pp. 67–70, text figs. 1–5.

—— Notes on two conifers from the Pleistocene Rancho La Brea asphalt deposits, near Los Angeles, Calif.
Washington Acad. Sci. Jour., vol. 6, No. 4, Feb. 10, 1916, pp. 85, 86.

—— Note on a recent discovery of fossil plants in the Morrison formation.
Washington Acad. Sci. Jour., vol. 6, No. 7, Apr. 4, 1916, pp. 180, 181.

—— The flora of the Fox Hills sandstone.
U. S. Geol. Survey Prof. Paper 98–H, 1916, pp. 85–93, pls. 15–18.

—— A new fossil *Selaginella* from the lower Tertiary of Montana.
Torreya, vol. 16, 1916, pp. 201–203, pl. 1.

—— A review of the fossil plants in the United States National Museum from the Florissant lake beds at Florissant, Colo., with descriptions of new species and list of type specimens.
U. S. Nat. Mus. Proc., vol. 51, 1916, pp. 241–297, pls. 12–27.

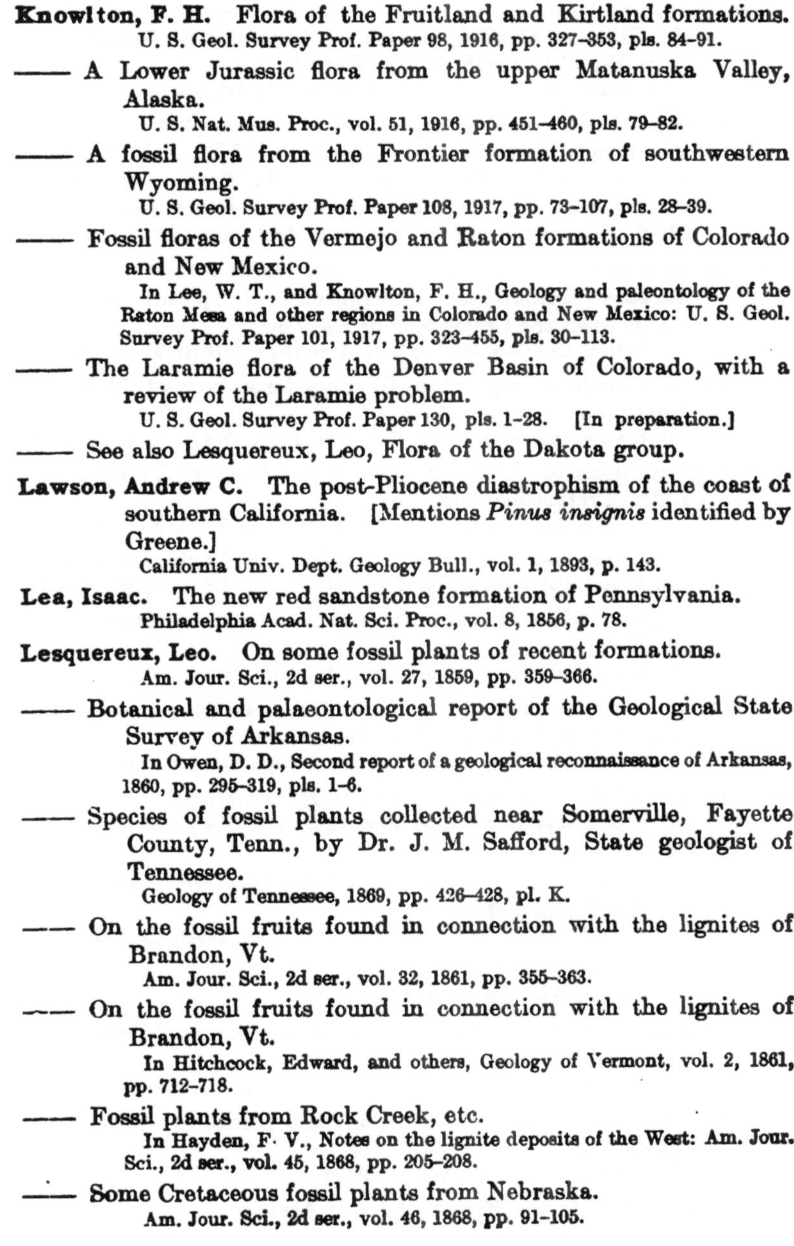

Knowlton, F. H. Flora of the Fruitland and Kirtland formations.
U. S. Geol. Survey Prof. Paper 98, 1916, pp. 327–353, pls. 84–91.

—— A Lower Jurassic flora from the upper Matanuska Valley, Alaska.
U. S. Nat. Mus. Proc., vol. 51, 1916, pp. 451–460, pls. 79–82.

—— A fossil flora from the Frontier formation of southwestern Wyoming.
U. S. Geol. Survey Prof. Paper 108, 1917, pp. 73–107, pls. 28–39.

—— Fossil floras of the Vermejo and Raton formations of Colorado and New Mexico.
In Lee, W. T., and Knowlton, F. H., Geology and paleontology of the Raton Mesa and other regions in Colorado and New Mexico: U. S. Geol. Survey Prof. Paper 101, 1917, pp. 323–455, pls. 30–113.

—— The Laramie flora of the Denver Basin of Colorado, with a review of the Laramie problem.
U. S. Geol. Survey Prof. Paper 130, pls. 1–28. [In preparation.]

—— See also Lesquereux, Leo, Flora of the Dakota group.

Lawson, Andrew C. The post-Pliocene diastrophism of the coast of southern California. [Mentions *Pinus insignis* identified by Greene.]
California Univ. Dept. Geology Bull., vol. 1, 1893, p. 143.

Lea, Isaac. The new red sandstone formation of Pennsylvania.
Philadelphia Acad. Nat. Sci. Proc., vol. 8, 1856, p. 78.

Lesquereux, Leo. On some fossil plants of recent formations.
Am. Jour. Sci., 2d ser., vol. 27, 1859, pp. 359–366.

—— Botanical and palaeontological report of the Geological State Survey of Arkansas.
In Owen, D. D., Second report of a geological reconnaissance of Arkansas, 1860, pp. 295–319, pls. 1–6.

—— Species of fossil plants collected near Somerville, Fayette County, Tenn., by Dr. J. M. Safford, State geologist of Tennessee.
Geology of Tennessee, 1869, pp. 426–428, pl. K.

—— On the fossil fruits found in connection with the lignites of Brandon, Vt.
Am. Jour. Sci., 2d ser., vol. 32, 1861, pp. 355–363.

—— On the fossil fruits found in connection with the lignites of Brandon, Vt.
In Hitchcock, Edward, and others, Geology of Vermont, vol. 2, 1861, pp. 712–718.

—— Fossil plants from Rock Creek, etc.
In Hayden, F. V., Notes on the lignite deposits of the West: Am. Jour. Sci., 2d ser., vol. 45, 1868, pp. 205–208.

—— Some Cretaceous fossil plants from Nebraska.
Am. Jour. Sci., 2d ser., vol. 46, 1868, pp. 91–105.

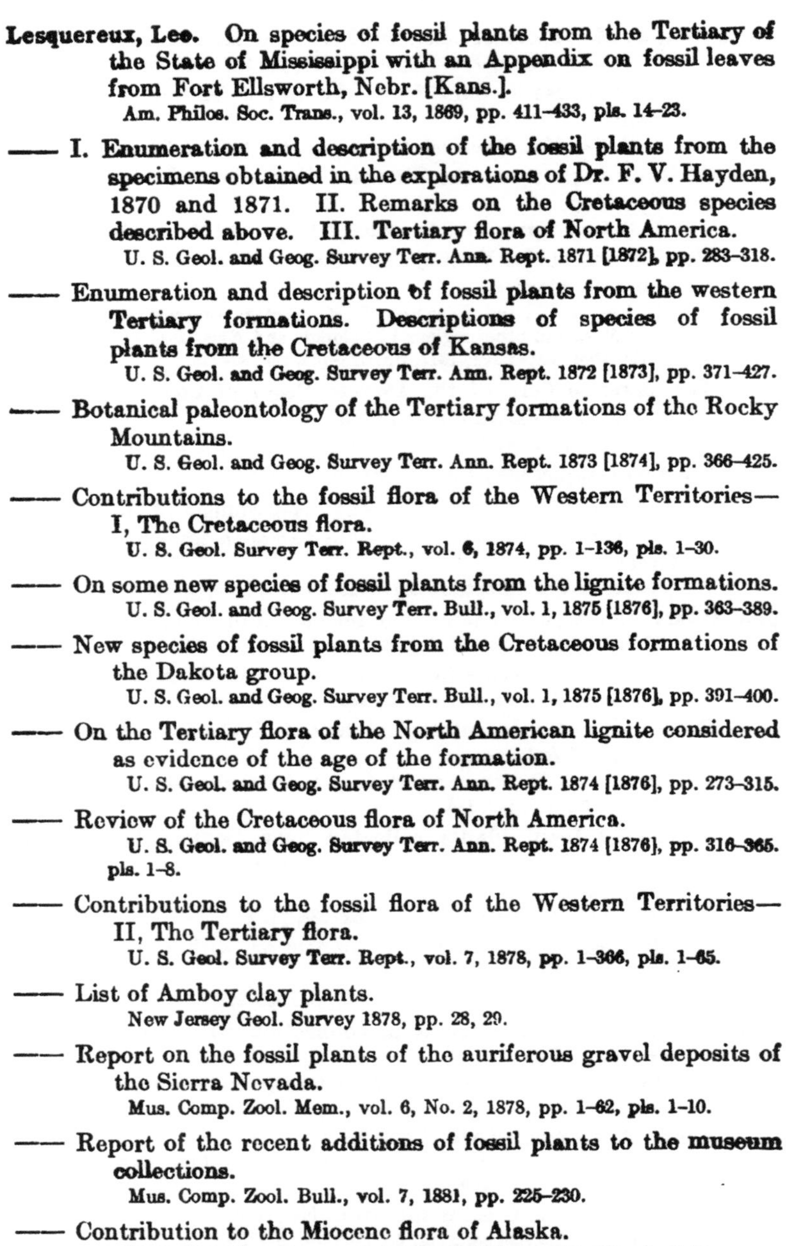

Lesquereux, Leo. On species of fossil plants from the Tertiary of the State of Mississippi with an Appendix on fossil leaves from Fort Ellsworth, Nebr. [Kans.].
Am. Philos. Soc. Trans., vol. 13, 1869, pp. 411–433, pls. 14–23.

—— I. Enumeration and description of the fossil plants from the specimens obtained in the explorations of Dr. F. V. Hayden, 1870 and 1871. II. Remarks on the Cretaceous species described above. III. Tertiary flora of North America.
U. S. Geol. and Geog. Survey Terr. Ann. Rept. 1871 [1872], pp. 283–318.

—— Enumeration and description of fossil plants from the western Tertiary formations. Descriptions of species of fossil plants from the Cretaceous of Kansas.
U. S. Geol. and Geog. Survey Terr. Ann. Rept. 1872 [1873], pp. 371–427.

—— Botanical paleontology of the Tertiary formations of the Rocky Mountains.
U. S. Geol. and Geog. Survey Terr. Ann. Rept. 1873 [1874], pp. 366–425.

—— Contributions to the fossil flora of the Western Territories—I, The Cretaceous flora.
U. S. Geol. Survey Terr. Rept., vol. 6, 1874, pp. 1–136, pls. 1–30.

—— On some new species of fossil plants from the lignite formations.
U. S. Geol. and Geog. Survey Terr. Bull., vol. 1, 1875 [1876], pp. 363–389.

—— New species of fossil plants from the Cretaceous formations of the Dakota group.
U. S. Geol. and Geog. Survey Terr. Bull., vol. 1, 1875 [1876], pp. 391–400.

—— On the Tertiary flora of the North American lignite considered as evidence of the age of the formation.
U. S. Geol. and Geog. Survey Terr. Ann. Rept. 1874 [1876], pp. 273–315.

—— Review of the Cretaceous flora of North America.
U. S. Geol. and Geog. Survey Terr. Ann. Rept. 1874 [1876], pp. 316–365. pls. 1–8.

—— Contributions to the fossil flora of the Western Territories—II, The Tertiary flora.
U. S. Geol. Survey Terr. Rept., vol. 7, 1878, pp. 1–366, pls. 1–65.

—— List of Amboy clay plants.
New Jersey Geol. Survey 1878, pp. 28, 29.

—— Report on the fossil plants of the auriferous gravel deposits of the Sierra Nevada.
Mus. Comp. Zool. Mem., vol. 6, No. 2, 1878, pp. 1–62, pls. 1–10.

—— Report of the recent additions of fossil plants to the museum collections.
Mus. Comp. Zool. Bull., vol. 7, 1881, pp. 225–230.

—— Contribution to the Miocene flora of Alaska.
U. S. Nat. Mus. Proc., vol. 5, 1882 [1883], pp. 443–449, pls. 6–10.

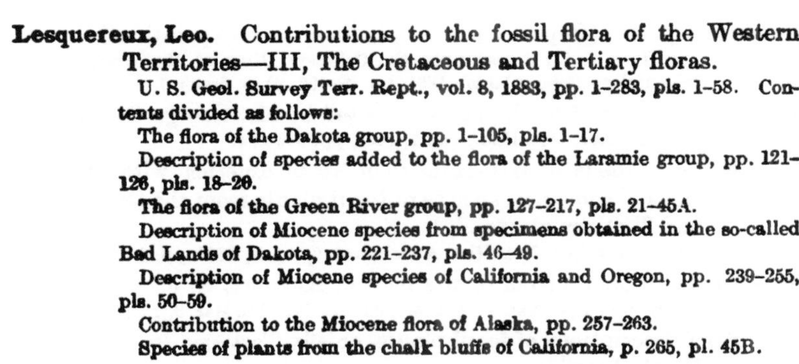

Lesquereux, Leo. Contributions to the fossil flora of the Western Territories—III, The Cretaceous and Tertiary floras.
U. S. Geol. Survey Terr. Rept., vol. 8, 1883, pp. 1–283, pls. 1–58. Contents divided as follows:
The flora of the Dakota group, pp. 1–105, pls. 1–17.
Description of species added to the flora of the Laramie group, pp. 121–126, pls. 18–20.
The flora of the Green River group, pp. 127–217, pls. 21–45A.
Description of Miocene species from specimens obtained in the so-called Bad Lands of Dakota, pp. 221–237, pls. 46–49.
Description of Miocene species of California and Oregon, pp. 239–255, pls. 50–59.
Contribution to the Miocene flora of Alaska, pp. 257–263.
Species of plants from the chalk bluffs of California, p. 265, pl. 45B.

—— List of recently identified fossil plants belonging to the United States National Museum, with descriptions of several new species.
U. S. Nat. Mus. Proc., vol. 10, 1887, pp. 21–46, pls. 1–4.

—— Recent determinations of fossil plants from Kentucky, Louisiana, Oregon, California, Alaska, Greenland, etc., with descriptions of new species.
U. S. Nat. Mus. Proc., vol. 11, 1888, pp. 11–38, pls. 4–16.

—— The flora of the Dakota group; a posthumous work, edited by F. H. Knowlton.
U. S. Geol. Survey Mon. 17, 1891 [1892], pp. 1–400, pls. 1–66.

—— Cretaceous fossil plants from Minnesota.
Minnesota Geol. and Nat. Hist. Survey Final Rept., vol. 3, pt. 1, author's edition issued Feb. 15, 1893 [date of complete volume 1895], pp. 1–22, pls. A, B.

—— The genus *Winchellia*.
Am. Geologist, vol. 12, 1893, pp. 209–213, pls. 8, 9.

Lewis, Henry Carvill. On a new fucoidal plant from the Trias.
Philadelphia Acad. Nat. Sci. Proc., vol. 32, 1880, pp. 293, 294, text fig.

Marcou, Jules. The Triassic flora of Richmond, Va.
Am. Geologist, vol. 10, 1890, pp. 160–174.

McBride, Thomas H. A new cycad.
Am. Geologist, vol. 12, 1893, pp. 248–250, pl. 11; Bull. Lab. Nat. Hist. Iowa State Univ., vol. 2, 1893, pp. 391–393, pl. 12.

—— On certain fossil plant remains in the Iowa herbarium.
Davenport Acad. Sci. Proc., vol. 10, 1907, pp. 153–162, pls. 1–12.

Meehan, Thomas, in Mercer, Henry C. The bone cave at Port Kennedy, Pa., and its partial excavation in 1894, 1895, and 1896.
Philadelphia Acad. Nat. Sci. Jour., 2d ser., vol. 11, 1899, pp. 277, 278, 279, text fig. 8 (1)–(16).

Mitchell, Guy E. The petrified forests of Yellowstone Park.
Travel, New York, vol. 23, 1914, pp. 19–21, 7 unnumbered text figs.

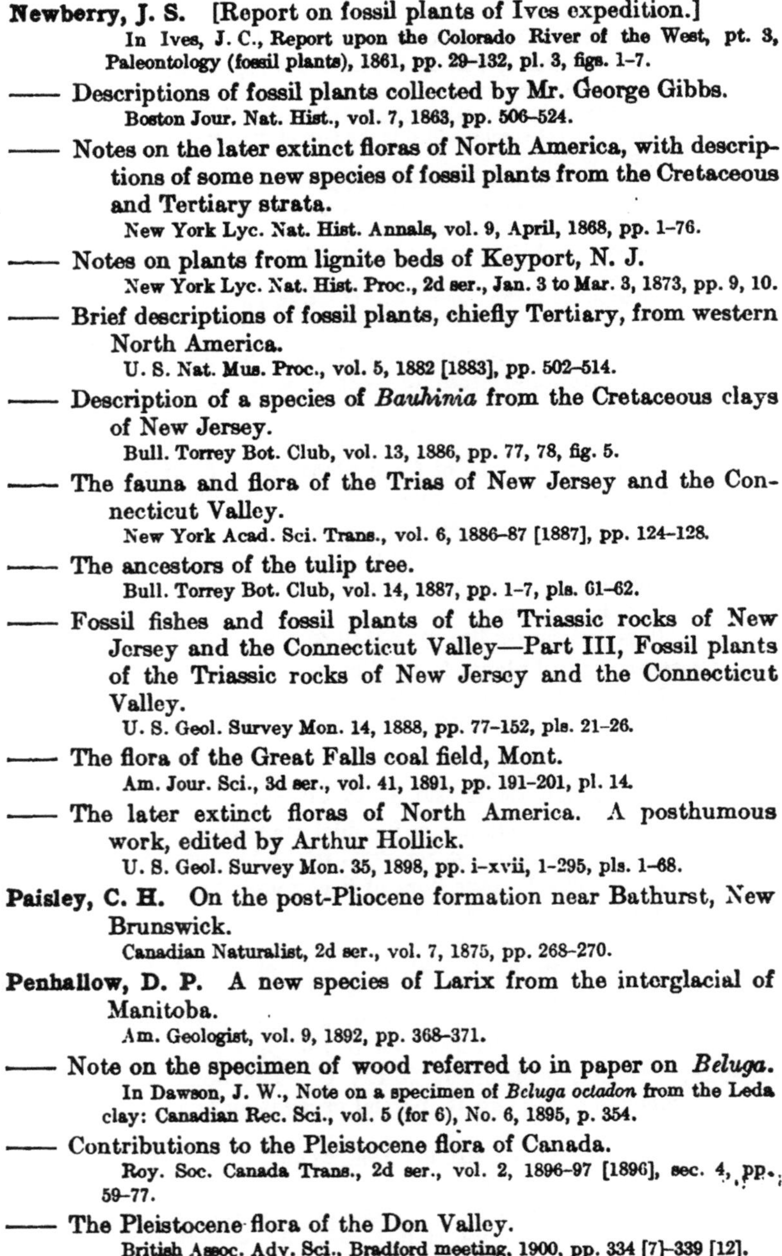

Newberry, J. S. [Report on fossil plants of Ives expedition.]
In Ives, J. C., Report upon the Colorado River of the West, pt. 3, Paleontology (fossil plants), 1861, pp. 29–132, pl. 3, figs. 1–7.

—— Descriptions of fossil plants collected by Mr. George Gibbs.
Boston Jour. Nat. Hist., vol. 7, 1863, pp. 506–524.

—— Notes on the later extinct floras of North America, with descriptions of some new species of fossil plants from the Cretaceous and Tertiary strata.
New York Lyc. Nat. Hist. Annals, vol. 9, April, 1868, pp. 1–76.

—— Notes on plants from lignite beds of Keyport, N. J.
New York Lyc. Nat. Hist. Proc., 2d ser., Jan. 3 to Mar. 3, 1873, pp. 9, 10.

—— Brief descriptions of fossil plants, chiefly Tertiary, from western North America.
U. S. Nat. Mus. Proc., vol. 5, 1882 [1883], pp. 502–514.

—— Description of a species of *Bauhinia* from the Cretaceous clays of New Jersey.
Bull. Torrey Bot. Club, vol. 13, 1886, pp. 77, 78, fig. 5.

—— The fauna and flora of the Trias of New Jersey and the Connecticut Valley.
New York Acad. Sci. Trans., vol. 6, 1886–87 [1887], pp. 124–128.

—— The ancestors of the tulip tree.
Bull. Torrey Bot. Club, vol. 14, 1887, pp. 1–7, pls. 61–62.

—— Fossil fishes and fossil plants of the Triassic rocks of New Jersey and the Connecticut Valley—Part III, Fossil plants of the Triassic rocks of New Jersey and the Connecticut Valley.
U. S. Geol. Survey Mon. 14, 1888, pp. 77–152, pls. 21–26.

—— The flora of the Great Falls coal field, Mont.
Am. Jour. Sci., 3d ser., vol. 41, 1891, pp. 191–201, pl. 14.

—— The later extinct floras of North America. A posthumous work, edited by Arthur Hollick.
U. S. Geol. Survey Mon. 35, 1898, pp. i–xvii, 1–295, pls. 1–68.

Paisley, C. H. On the post-Pliocene formation near Bathurst, New Brunswick.
Canadian Naturalist, 2d ser., vol. 7, 1875, pp. 268–270.

Penhallow, D. P. A new species of Larix from the interglacial of Manitoba.
Am. Geologist, vol. 9, 1892, pp. 368–371.

—— Note on the specimen of wood referred to in paper on *Beluga*.
In Dawson, J. W., Note on a specimen of *Beluga octadon* from the Leda clay: Canadian Rec. Sci., vol. 5 (for 6), No. 6, 1895, p. 354.

—— Contributions to the Pleistocene flora of Canada.
Roy. Soc. Canada Trans., 2d ser., vol. 2, 1896–97 [1896], sec. 4, pp. 59–77.

—— The Pleistocene flora of the Don Valley.
British Assoc. Adv. Sci., Bradford meeting, 1900, pp. 334 [7]–339 [12].

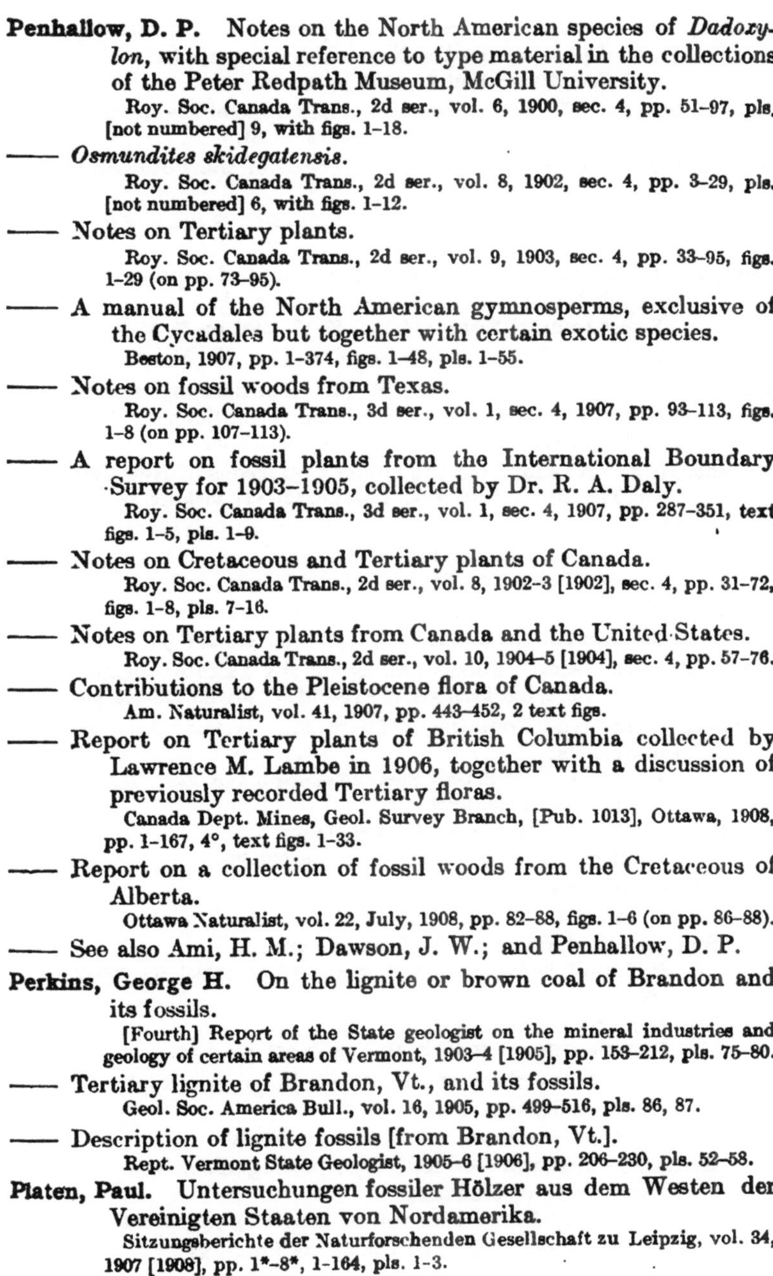

Penhallow, D. P. Notes on the North American species of *Dadoxylon*, with special reference to type material in the collections of the Peter Redpath Museum, McGill University.
Roy. Soc. Canada Trans., 2d ser., vol. 6, 1900, sec. 4, pp. 51–97, pls. [not numbered] 9, with figs. 1–18.

—— *Osmundites skidegatensis.*
Roy. Soc. Canada Trans., 2d ser., vol. 8, 1902, sec. 4, pp. 3–29, pls. [not numbered] 6, with figs. 1–12.

—— Notes on Tertiary plants.
Roy. Soc. Canada Trans., 2d ser., vol. 9, 1903, sec. 4, pp. 33–95, figs. 1–29 (on pp. 73–95).

—— A manual of the North American gymnosperms, exclusive of the Cycadales but together with certain exotic species.
Boston, 1907, pp. 1–374, figs. 1–48, pls. 1–55.

—— Notes on fossil woods from Texas.
Roy. Soc. Canada Trans., 3d ser., vol. 1, sec. 4, 1907, pp. 93–113, figs. 1–8 (on pp. 107–113).

—— A report on fossil plants from the International Boundary Survey for 1903–1905, collected by Dr. R. A. Daly.
Roy. Soc. Canada Trans., 3d ser., vol. 1, sec. 4, 1907, pp. 287–351, text figs. 1–5, pls. 1–9.

—— Notes on Cretaceous and Tertiary plants of Canada.
Roy. Soc. Canada Trans., 2d ser., vol. 8, 1902–3 [1902], sec. 4, pp. 31–72, figs. 1–8, pls. 7–16.

—— Notes on Tertiary plants from Canada and the United States.
Roy. Soc. Canada Trans., 2d ser., vol. 10, 1904–5 [1904], sec. 4, pp. 57–76.

—— Contributions to the Pleistocene flora of Canada.
Am. Naturalist, vol. 41, 1907, pp. 443–452, 2 text figs.

—— Report on Tertiary plants of British Columbia collected by Lawrence M. Lambe in 1906, together with a discussion of previously recorded Tertiary floras.
Canada Dept. Mines, Geol. Survey Branch, [Pub. 1013], Ottawa, 1908, pp. 1–167, 4°, text figs. 1–33.

—— Report on a collection of fossil woods from the Cretaceous of Alberta.
Ottawa Naturalist, vol. 22, July, 1908, pp. 82–88, figs. 1–6 (on pp. 86–88).

—— See also Ami, H. M.; Dawson, J. W.; and Penhallow, D. P.

Perkins, George H. On the lignite or brown coal of Brandon and its fossils.
[Fourth] Report of the State geologist on the mineral industries and geology of certain areas of Vermont, 1903–4 [1905], pp. 153–212, pls. 75–80.

—— Tertiary lignite of Brandon, Vt., and its fossils.
Geol. Soc. America Bull., vol. 16, 1905, pp. 499–516, pls. 86, 87.

—— Description of lignite fossils [from Brandon, Vt.].
Rept. Vermont State Geologist, 1905–6 [1906], pp. 206–230, pls. 52–58.

Platen, Paul. Untersuchungen fossiler Hölzer aus dem Westen der Vereinigten Staaten von Nordamerika.
Sitzungsberichte der Naturforschenden Gesellschaft zu Leipzig, vol. 34, 1907 [1908], pp. 1*–8*, 1–164, pls. 1–3.

Russell, Israel Cook. Correlation papers: The Newark system. [List of plants from the Newark system, pp. 64, 65.]
U. S. Geol. Survey Bull. 85, 1892, pp. 62–65.

Sinnott, Edmund W. *Paracedroxylon*, a new type of araucarian wood.
Rhodora, vol. 11, 1909, pp. 165–173, pls. 80, 81.

—— and **Bartlett, H. H.** Coniferous woods of the Potomac formation.
Am. Jour. Sci., 4th ser., vol. 41, 1916, pp. 276–293, figs. 1–18 (on pp. 281, 283, 289).

Snow, F. H. On the discovery and significance of stipules in certain dicotyledonous leaves of the Dakota rocks.
Kansas Acad. Sci. Trans., vol. 11, 1887–88 [1889], pp. 33–35, text figs. 1–8.

Stanton, T. W., and **Diller, J. S.** See Fontaine, W. M.

Stanton, T. W., and **Knowlton, F. H.** Stratigraphy and paleontology of the Laramie and related formations in Wyoming.
Geol. Soc. America Bull., vol. 8, 1897, pp. 127–156.

Stevens, N. E. A palm from the Upper Cretaceous of New Jersey.
Am. Jour. Sci., vol. 34, Nov., 1912, pp. 421–436, text figs. 1–24.

Stopes, Marie C. On the true nature of the Cretaceous plant *Ophioglossum granulatum* Heer.
Annals of Botany, vol. 25, 1911, pp. 903–907, figs. 1, 2.

Trelease, William. The ancient oaks of America.
Brooklyn Bot. Gard. Mem., vol. 1, 1918, pp. 492–501, pls. 13–22.

Ulrich, Edward Oscar. Fossils and age of the Yakutat formation.
Alaska, vol. 4, Harriman Alaska Expedition, 1904, pp. 125–146, pls. 11–21.

Wanner, Atreus, and **Fontaine, William M.** Triassic flora of York County, Pa.
In Ward, L. F., U. S. Geol. Survey Twentieth Ann. Rept., 1898–99, pt. 2, 1900, pp. 233–255, pls. 21–34.

Ward, Lester F. On Mesozoic dicotyledons.
Am. Jour. Sci., 3d ser., vol. 27, 1884, pp. 292–303.

—— The ginkgo tree.
Science, vol. 5, No. 124, 1885, pp. 495–497.

—— Synopsis of the flora of the Laramie group.
U. S. Geol. Survey Sixth Ann. Rept., 1884–85 [1886], pp. 399–557, pls. 31–65.

—— Types of the Laramie flora.
U. S. Geol. Survey Bull. 37, 1887, pp. 1–353, pls. 1–57.

—— Synopsis of the flora of the Laramie group: Reply to Lesquereux's discussion of same.
Am. Jour. Sci., 3d ser., vol. 34, 1887, pp. 488, 489.

—— Remarks on an undescribed vegetable organism from the Fort Union group of Montana.
Am. Assoc. Adv. Sci. Proc., vol. 37, 1888 [1889], pp. 199–201.

—— The paleontologic history of the genus *Platanus*.
Am. Assoc. Adv. Sci. Proc., vol. 37, 1888 [1889], pp. 201, 202.

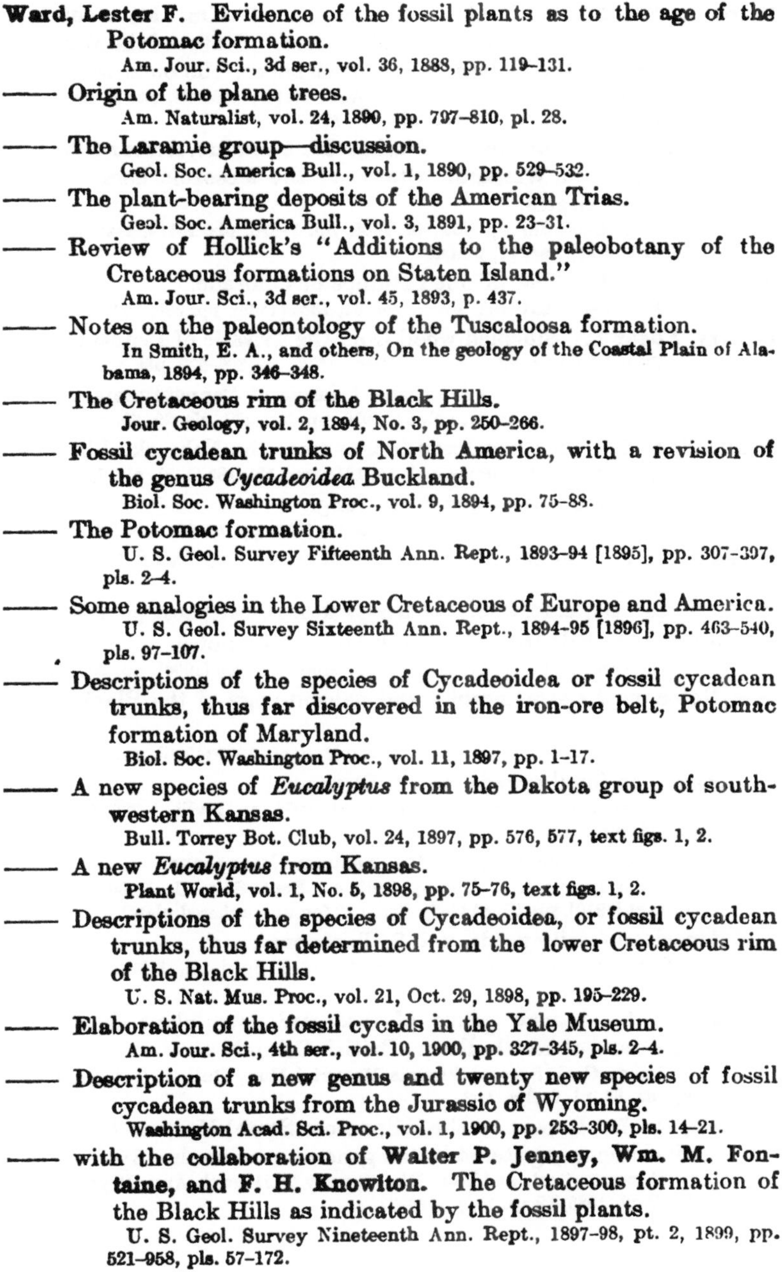

Ward, Lester F. Evidence of the fossil plants as to the age of the Potomac formation.
Am. Jour. Sci., 3d ser., vol. 36, 1888, pp. 119–131.

—— Origin of the plane trees.
Am. Naturalist, vol. 24, 1890, pp. 797–810, pl. 28.

—— The Laramie group—discussion.
Geol. Soc. America Bull., vol. 1, 1890, pp. 529–532.

—— The plant-bearing deposits of the American Trias.
Geol. Soc. America Bull., vol. 3, 1891, pp. 23-31.

—— Review of Hollick's "Additions to the paleobotany of the Cretaceous formations on Staten Island."
Am. Jour. Sci., 3d ser., vol. 45, 1893, p. 437.

—— Notes on the paleontology of the Tuscaloosa formation.
In Smith, E. A., and others, On the geology of the Coastal Plain of Alabama, 1894, pp. 346–348.

—— The Cretaceous rim of the Black Hills.
Jour. Geology, vol. 2, 1894, No. 3, pp. 250–266.

—— Fossil cycadean trunks of North America, with a revision of the genus *Cycadeoidea* Buckland.
Biol. Soc. Washington Proc., vol. 9, 1894, pp. 75–88.

—— The Potomac formation.
U. S. Geol. Survey Fifteenth Ann. Rept., 1893–94 [1895], pp. 307–397, pls. 2–4.

—— Some analogies in the Lower Cretaceous of Europe and America.
U. S. Geol. Survey Sixteenth Ann. Rept., 1894-95 [1896], pp. 463–540, pls. 97–107.

—— Descriptions of the species of Cycadeoidea or fossil cycadean trunks, thus far discovered in the iron-ore belt, Potomac formation of Maryland.
Biol. Soc. Washington Proc., vol. 11, 1897, pp. 1–17.

—— A new species of *Eucalyptus* from the Dakota group of southwestern Kansas.
Bull. Torrey Bot. Club, vol. 24, 1897, pp. 576, 577, text figs. 1, 2.

—— A new *Eucalyptus* from Kansas.
Plant World, vol. 1, No. 5, 1898, pp. 75–76, text figs. 1, 2.

—— Descriptions of the species of Cycadeoidea, or fossil cycadean trunks, thus far determined from the lower Cretaceous rim of the Black Hills.
U. S. Nat. Mus. Proc., vol. 21, Oct. 29, 1898, pp. 195–229.

—— Elaboration of the fossil cycads in the Yale Museum.
Am. Jour. Sci., 4th ser., vol. 10, 1900, pp. 327–345, pls. 2–4.

—— Description of a new genus and twenty new species of fossil cycadean trunks from the Jurassic of Wyoming.
Washington Acad. Sci. Proc., vol. 1, 1900, pp. 253–300, pls. 14–21.

—— with the collaboration of **Walter P. Jenney, Wm. M. Fontaine, and F. H. Knowlton.** The Cretaceous formation of the Black Hills as indicated by the fossil plants.
U. S. Geol. Survey Nineteenth Ann. Rept., 1897–98, pt. 2, 1899, pp. 521–958, pls. 57–172.

Ward, Lester F. With the collaboration of **Wm. M. Fontaine, Atreus Wanner,** and **F. H. Knowlton.** Status of the Mesozoic floras of the United States, first paper, The older Mesozoic.
U. S. Geol. Survey Twentieth Ann. Rept., 1898-99, pt. 2, 1900, pp. 211-748, index, pp. 931-953, pls. 21-179.

—— with the collaboration of **William M. Fontaine, Arthur Bibbins,** and **G. R. Wieland.** Status of the Mesozoic floras of the United States.
U. S. Geol. Survey Mon. 48, 1905, pp. 1-616, pls. 1-119, text figs. 1-9. Part I, text; Part II, plates.

Warder, Robt. B. The silicified stumps of Colorado.
Am. Assoc. Adv. Sci. Proc., vol. 31, 1882 [1883], pp. 398, 399.

Wherry, Edgar T. Silicified wood from the Triassic of Pennsylvania.
Philadelphia Acad. Nat. Sci. Proc., 1912, pp. 366-372, pls. 3, 4.

—— Two new fossil plants from the Triassic of Pennsylvania.
U. S. Nat. Mus. Proc., vol. 51, 1916, pp. 327-329, pls. 29, 30.

Whitford, A. C. On a new fossil fungus from the Nebraska Pliocene.
Nebraska Univ. Studies, vol. 14, 1914, pp. 181-183 (1-3), pls. 1, 2.

—— A description of two new fossil fungi.
Nebraska Geol. Survey, vol. 7, pt. 13, 1916, pp. 85-92, text figs. 1-13.

Wieland, G. R. On the foliage of the Jurassic cycads of the genus *Cycadella*.
In Ward, L. F., U. S. Geol. Survey Mon. 48, 1905 [1906], pp. 198-203.

—— Jurassic cycads from the Black Hills; field notes.
In Ward, L. F., U. S. Geol. Survey Mon. 48, 1905 [1906], pp. 205-207.

—— Notes on the stratigraphy and paleontology of the Black Hills rim.
In Ward, L. F., U. S. Geol. Survey Mon. 48, 1905 [1906], pp. 317-326.

—— American fossil cycads, vol. 1, Structure.
Carnegie Inst. Washington Pub. 34, 1906, pp. i-viii, 1-296, text figs. 1-138, pls. 1-50.

—— Two new Araucarias from the western Cretaceous.
South Dakota Geol. Survey Bull. 4. Rept. State Geologist for 1908, reprint [1910], pp. 77-81 [1-7], figs. 1, 2 [on unnumbered plate].

—— American fossil cycads, vol. 2, Taxonomy.
Carnegie Inst. Washington Pub. 34, 1916, pp. i-vii, 1-277, text figs. 1-97, pls. 1-58.

Wilson, W. J. A new genus of dicotyledonous plant from the Tertiary of Kettle River, British Columbia.
Canada Geol. Survey, Victoria Memorial Mus., Bull. 1, Oct. 23, 1913, pp. 87, 88, pl. 9, figs. 1, 2.

Winchell, N. H. Report on the [Minnesota State] Museum for 1884: Specimens registered in the general museum in 1884.
Minnesota Geol. and Nat. Hist. Survey Thirteenth Ann. Rept., 1884. St. Paul, 1885, Chap. 11, pp. 76-77.

CATALOGUE.

ABIES (Tournefort) Adanson, Fam. Pl., vol. 2, 1763, p. 480.

Abies balsamea (Linné) Miller. Penhallow, Roy. Soc. Canada, Trans., 2d ser., vol. 2, 1896, p. 70; Jour. Geol., vol. 3, 1895, p. 625; Brit. Assoc. Adv. Sci., Bradford meeting, 1900, sec. C, p. 335[8]; Man. N. A. Gym., 1907, p. 228.

Pleistocene: Scarborough Heights, Ontario, Canada.

Abies linkii Roemer = *Abietites linkii.*

Abies nevadensis Lesquereux = *Taxodium distichum miocenum.*

Abies setigera Lesquereux = *Abietites setiger.*

Abies sp., Grewingk, Verhandl. d. Russ. K. mineralog. Gesell., St. Petersb., 1848–49 [1850], p. 166.

Kenai: Kodiak Island, Alaska.

ABIETITES Hisinger, Lethaea suecica, 1837, p. 110.

Abietites angusticarpus Fontaine = *Abietites macrocarpus.*

Abietites angusticarpus Fontaine. Fontaine, in Diller and Stanton, Geol. Soc. Am., Bull., vol. 4, 1894, p. 450 = *Abietites macrocarpus.*

Abietites californicus Fontaine = *Abietites macrocarpus.*

Abietites carolinensis (Fontaine) Fontaine, in Ward, U. S. Geol. Surv., Twentieth Ann. Rept., pt. 2, 1900, p. 309, pl. xlvii, f. 1.

Pachypteris sp.? Emmons, American geology, pt. 6, 1857, p. 112, f. 80.

Palissya carolinensis Fontaine, U. S. Geol. Surv. Mon. 6, 1883, p. 109, pl. li, f. 5.

Triassic (Newark): North Carolina.

Abietites cretacea Newberry, U. S. Geol. Surv., Mon. 35, 1898, p. 18, pl. xiv, f. 5.

Dakota: Whetstone Creek, N. Mex.

Abietites dubius Lesquereux, Am. Jour. Sci., 2d ser., vol. 45, 1868, p. 207; U. S. Geol. and Geog. Surv. Terr., Ann. Rept., 1869 [reprint 1873], p. 196; Rept. U. S. Geol. Surv. Terr., vol. 7 (Tert. Fl.), 1878, p. 81, pl. vii, f. 24 [not f. 20, 21, 21a].—Knowlton, U. S. Geol. Surv., Prof. Paper 101, 1917, p. 249.

Livingston: Bozeman coal field, Mont.

Vermejo: Raton Mountains, N. Mex.; near Trinidad, Colo.

Montana?: Upper Kanab Valley, Utah.

——: Two miles east of Hayden, Colo.

Abietites ellipticus Fontaine = *Abietites macrocarpus.*

Abietites ernestinae Lesquereux, Rept. U. S. Geol. Surv. Terr., vol. 6 (Cret. Fl.), 1874, p. 49, pl. i, f. 7.

Pterophyllum haydenii Lesquereux, ex. p. Am. Jour. Sci., 2d ser., vol. 46, 1868, p. 91.

Dakota: Decatur, Nebr.

Abietites foliosus (Fontaine) Berry, U. S. Nat. Mus., Proc., vol. 40, 1911, p. 314; Md. Geol. Surv., Lower Cret., 1911, p. 408; U. S. Geol. Surv., Prof. Paper 112, 1919, p. 64.

Leptostrobus foliosus Fontaine, U. S. Geol. Surv. Mon. 15, 1890, p. 230, pl. ci, f. 4; pl. ciii, f. 5; pl. civ, f. 1.—Fontaine, in Ward, U. S. Geol. Surv. Mon. 48, 1906, p. 482.

Patuxent: Dutch Gap, Va.
Patapsco: Brooke, Va.

Abietites linkii (Roemer) Dunker, Mongr. d. Norddeutschen Wealdenb., 1846, p. 18, pl. ix, f. 11a–c.—Fontaine, U. S. Nat. Mus., Proc., vol. 16, 1893, p. 268, pl. xxxvii, f. 2.

Abies linkii Roemer, N. deutsch. Oolith.-Form. Nachtr., 1830, p. 10, pl. xvii, f. 2.

Trinity: Glen Rose, Tex.
Tuscaloosa: Snow place, Tuscaloosa County, Ala.
Patuxent: Dutch Gap, Va.
Patapsco: Brooke, Va.
Kootenai: Great Falls, Mont.

Abietites longifolius (Fontaine) Berry, U. S. Nat. Mus., Proc., vol. 40, 1911, p. 315; Md. Geol. Surv., Lower Cret., 1911, p. 407, pl. lxvii, f. 7.

Leptostrobus longifolius Fontaine, U. S. Geol. Surv., Mon. 15, 1890, p. 228, pl. ci, f. 2, 3; pl. cii, f. 1–4; pl. ciii, f. 6–12; pl. civ, f. 6; in Ward, U. S. Geol. Surv., 19th Ann. Rept., pt. 2, 1899, p. 671, pl. clxiii, f. 15; pl. clxv, f. 3; U. S. Geol. Surv., Mon. 48, 1906, pp. 281, 481, 482, 491, 506, 528, 557, pl. cx, f. 11; pl. cxvi, f. 1.

Patuxent: Fredericksburg, Potomac River, and Dutch Gap, Va.
Patapsco: Brooke, 72d milepost, Mount Vernon and Hill Hole, Va.; Fort Foote, Vinegar Hill and Federal Hill (Baltimore), Md.
Fuson: Pine Creek, Oak Creek and Dull Creek, Crook County, Wyo.

Abietites macrocarpus Fontaine, U. S. Geol. Surv., Mon. 15, 1890, p. 262, pl. cxxxii, f. 7; in Ward, U. S. Geol. Surv., Mon. 48, 1906, pp. 261, 547, pl. lxviii, f. 15, 16; pl. cxv, f. 2, 3.—Berry, U. S. Nat. Mus., Proc., vol. 40, 1911, p. 313; Md. Geol. Surv., Lower Cret., 1911, p. 405, pl. lxvii, f. 1–4.

Abietites ellipticus Fontaine, U. S. Geol. Surv., Mon. 15, 1890, p. 263, pl. cxxxii, f. 8, 9; pl. cxxxiii, f. 2–4; pl. clxviii, f. 8; in Ward, idem, Mon. 48, 1905 [1906], p. 260, pl. lxviii, f. 14.

Abietites macrocarpus—Continued.

Abietites angusticarpus Fontaine, U. S. Geol. Surv., Mon. 15, 1890, p. 263, pl. cxxxiii, f. 1; in Ward, U. S. Geol. Surv., 19th Ann. Rept., pt. 2, 1899, p. 671, pl. clxiii, f. 14; U. S. Geol. Surv., Mon. 48, 1906, pp. 528, 538, 556, 572, pl. cxiv, f. 10.

Williamsonia? bibbinsi Ward. Fontaine, in Ward, U. S. Geol. Surv., Mon. 48, 1906, p. 554, pl. cxv, f. 11.

Abietites californicus Fontaine, in Diller and Stanton, Geol. Soc. Am., Bull., vol. 5, 1894, p. 450.

Patuxent: *Fredericksburg and Dutch Gap, Va.; Broad Creek, Md.

Arundel: Arlington and near Lansdown, Md.

Patapsco: Vinegar Hill and Fort Foote, Md.

Knoxville: Calif.

Horsetown: Shasta County, Calif.

Fuson: Oak Creek, Crook County, Wyo.

Abietites marylandicus Fontaine, in Ward, U. S. Geol. Surv., Mon. 48, 1905 [1906], p. 549, pl. cxv, f. 4, 5.—Berry, U. S. Nat. Mus., Proc., vol. 40, 1911, p. 314; Md. Geol. Surv., Lower Cret., 1911, p. 406, pl. lxvii, f. 5, 6.

Patapsco: *Vinegar Hill, Relay, Md.

Abietites setiger (Lesquereux) Lesquereux, Rept. U. S. Geol. Surv. Terr., vol. 7 (Tert. Fl.), 1878, p. 82, pl. vii, f. 17, 18.

Abies setigera Lesquereux, U. S. Geol. and Geog. Surv. Terr., Ann. Rept., 1872 [1874], p. 404.

Livingston: Meadow Creek, 12 miles east of Bozeman, Mont.

Abietites tyrrellii Dawson, Roy. Soc. Canada, Trans., vol. 3, 1885 [1886], sec. 4, p. 17.

Pierre: Berry Creek, Alberta.

Abietites sp., Fontaine, U. S. Nat. Mus., Proc., vol. 16, 1893, p. 277, pl. xliii, f. 4.

Trinity: Glen Rose, Tex.

Abietites? sp., Fontaine (Cone), in Diller and Stanton, Geol. Soc. Am., Bull., vol. 5, 1894, p. 450 [name]; in Ward, U. S. Geol. Surv., Mon. 48, 1905 [1906], p. 262, pl. lxviii, f. 17.

Knoxville: Near Lowrey, Tehama County, Calif.

ACACIA Adanson, Fam. Pl., vol. 2, 1763, p. 319.

Acacia lamarensis Knowlton, U. S. Geol. Surv., Mon. 32, pt. 2, 1899, p. 730, pl. xcviii, f. 6.

Miocene: Lamar River, Yellowstone National Park.

Acacia macrosperma Knowlton, U. S. Geol. Surv., Mon. 32, pt. 2, 1899, p. 729, pl. xcviii, f. 8.

Miocene: Fossil Forest Ridge, Yellowstone National Park.

Acacia oregoniana Lesquereux, U. S. Nat. Mus., Proc., vol. 11, 1888, p. 14, pl. v, f. 4.—Knowlton, U. S. Geol. Surv., Bull. 204, 1902, p. 69.
Mascall: Van Horn's ranch, South Fork John Day River, 12 miles west of Mount Vernon, Grant County, Oreg.

Acacia septentrionalis Lesquereux, U. S. Geol. and Geog. Surv. Terr., Ann. Rept., 1873 [1874], p. 418; Rept. U. S. Geol. Surv. Terr., vol. 7 (Tert. Fl.), 1878, p. 299, pl. lix, f. 9, 9a.
Miocene: Costello's ranch, Florissant, Colo.

Acacia wardii Knowlton, U. S. Geol. Surv., Mon. 32, pt. 2, 1899, p. 730, pl. xcviii, f. 7.
Miocene: Fossil Forest Ridge, Yellowstone National Park.

Acacia wilcoxensis Berry, U. S. Geol. Surv., Prof. Paper 91, 1916, p. 222, pl. lv, f. 1, 2.
Wilcox (Holly Springs): Oxford, Miss.

Acacia sp., Knowlton, in Glenn, U. S. Geol. Surv., Water-Supply Paper 164, 1906, p. 38.
Lagrange: Near Grand Junction, Tenn.

Acacia sp., Knowlton, Washington Acad. Sci., Proc., vol. 11, 1909, p. 211.
Lance: Near Ranchester, Wyo.

ACACIAEPHYLLUM Fontaine, U. S. Geol. Surv., Mon. 15, 1889, p. 279.

Acaciaephyllum ellipticum Fontaine, in Ward, U. S. Geol. Surv., Mon. 48, 1905 [1906], p. 269, pl. lxix, f. 18.
Horsetown: Byron Gulch, near Ono, Shasta County, Calif.

Acaciaephyllum longifolium Fontaine = *Thinnfeldia granulata.*

Acaciaephyllum microphyllum Fontaine = *Thinnfeldia granulata.*

Acaciaephyllum pachyphyllum Fontaine, in Ward, U. S. Geol. Surv., Mon. 48, 1905 [1906], p. 270, pl. lxix, f. 20.
Knoxville: 1 mile west of Lowrey, Tehama County, Calif.

Acaciaephyllum spatulatum Fontaine = *Thinnfeldia granulata.*

?Acaciaephyllum variabile Fontaine = *Thinnfeldia granulata.*

ACACIAPHYLLITES Berry, U. S. Geol. Surv., Prof. Paper 84, 1914, p. 45. [Type, *A. grevilleoides.*]

Acaciaphyllites grevilleoides Berry, U. S. Geol. Surv., Prof. Paper 84, 1914, p. 45, pl. ix, f. 9, 10.
Black Creek (Middendorf): Middendorf, S. C.

ACALYPHA Linné, Sp. pl., 1753, p. 1003.

Acalypha myricina Cockerell, Torreya, vol. 9, 1909, p. 117, text f. (not numbered).
Miocene: *Florissant, Colo.

ACER (Tournefort) Linné, Sp. pl., 1753, p. 1054.

Acer aequidentatum Lesquereux, Mus. Comp. Zool., Mem., vol. 6, No. 2, 1878, p. 26, pl. vii, f. 4, 5.—Knowlton, in Lindgren, Jour. Geol., vol. 4, 1896, p. 890.

Miocene (auriferous gravels): Chalk Bluffs, Nevada County and Independence Hill, Placer County, Calif.

Acer aequidentatum Lesquereux, U. S. Geol. Surv. Terr., Rept. vol. 7 (Tert. Fl.), 1878, p. 262, pl. xlviii, f. 1, 3 = *Platanus guillelmae*. [Said to have been found "Near the confluence of White and Green Rivers, Utah," but to judge from the matrix and associated forms it came most probably from Carbon, Wyo.]

Acer amboyense Newberry, U. S. Geol. Surv., Mon. 26, 1895 [1896], p. 106, pl. xlvi, f. 5–8.—Berry, Geol. Surv. N. J., Bull. 3, 1911, p. 181.

Raritan: Woodbridge and South Amboy, N. J.

Acer arcticum Heer, Fl. foss. arct., vol. 4, abt. 1, 1877, p. 86, pl. xxii; pl. xxiii; pl. xxiv, f. 1, 2; pl. xxv, f. 1–3.—Lesquereux, Mus. Comp. Zool., Mem., vol. 6, No. 2, 1878, p. 60; ?U. S. Geol. Surv. Terr., Rept., vol. 8 (Cret. and Tert. Fl.), 1883, p. 233, pl. xlix, f. 8, 9.

Miocene (auriferous gravels): Oregon Creek near Forest City, Nevada County, Calif.

Fort Union: Bad Lands, N. Dak.

Acer bendirei Lesquereux, U. S. Nat. Mus., Proc., vol. 11, 1888, p. 14, pl. v, f. 5; pl. vi, f. 1; pl. vii, f. 1; pl. viii, f. 1.—Knowlton, U. S. Geol. Surv., Bull. 204, 1902, p. 73.

Acer trilobatum productum (Al. Braun) Heer. Lesquereux, Rept. U. S. Geol. Surv. Terr., vol. 8 (Cret. and Tert. Fl.), 1883, p. 253, pl. lix, f. 1, 2, 4 [not f. 3, which is *Platanus dissecta*].—Penhallow, Rept. Tert. Pl. Brit. Col., 1908, p. 35.

Mascall: Van Horn's ranch, John Day Basin, Oreg.

Miocene: Spanish Ranch, Calif.; Horsefly River, British Columbia.

Acer bolanderi Lesquereux, Mus. Comp. Zool., Mem., vol. 6, No. 2, 1878, p. 27, pl. vii, f. 7–11.

Miocene (auriferous gravels): Table Mountain, Tuolumne County, Calif.

Acer dimorphum Lesquereux, U. S. Nat. Mus., Proc., vol. 11, 1888, p. 15, pl. ix, f. 1.—Knowlton, U. S. Geol. Surv., Bull. 204, 1902, p. 74.

Mascall: Van Horn's ranch, John Day Valley, Oreg.

Acer dubium Penhallow, Roy. Soc. Canada, Trans., 2d ser., vol. 8, 1902, p. 70 [name]; Rept. Tert. Pl. Brit. Col., 1908, p. 35.

Oligocene: Horsefly River, British Columbia.

Acer florigerum Cockerell, Am. Jour. Sci., 4th ser., vol. 26, 1908, p. 66, f. 1c.
Miocene: *Florissant, Colo.

Acer florissanti Kirchner, St. Louis Acad. Sci., Trans., vol. 8, 1898, p. 181, pl. xi, f. 1.—Knowlton, U. S. Nat. Mus., Proc., vol. 51, 1916, p. 282.
Miocene: *Florissant, Colo.

Acer fragilis Knowlton, U. S. Geol. Surv., Prof. Paper 101, 1917, p. 330, pl. ci, f. 1, 2.
Raton: Near Trinidad, Colo., and Raton, N. Mex.

Acer gigas Knowlton, U. S. Geol. Surv., Bull. 204, 1902, p. 76, pl. xiv, f. 1.
Mascall: *Near Van Horn's ranch, John Day Basin, Oreg.

Acer gracilescens Lesquereux, Rept. U. S. Geol. Surv. Terr., vol. 8 (Cret. and Tert. Fl.), 1883, p. 234, pl. xlix, f. 7 [6 ?].
Fort Union: *Badlands, N. Dak.

Acer grahamensis Knowlton and Cockerell, n. name.
Acer macropterum Heer [homonym, Visiani, 1860], Kongl. Vetenskaps-Akad. Förhandl., 1868, p. 68; K. Sven. Vetenskaps-Akad. Handl., vol. 8, 1869, p. 37, pl. ix, f. 7–9 [in Fl. foss. arct., vol. 2, Abt. 2, 1871, p. 37, pl. ix, f. 7–9].—Knowlton, U. S. Nat. Mus., Proc., vol. 17, 1894, p. 227; Geol. Soc. Am., Bull., vol. 5, 1893, p. 585.
Kenai: Port Graham, Alaska.

Acer grossedentatum Heer, Fl. Tert. Helv., vol. 3, 1859, p. 54, pl. cxii, f. 17–23.—Dawson, Geol. Surv. Canada, 1875–76, p. 259.—Penhallow, Rept. Tert. Pl. Brit. Col., 1908, p. 35.
Eocene: Quesnel, British Columbia.

Acer indivisum Lesquereux = *Acer lesquereuxii*.

Acer indivisum Weber. Ward = *Platanus guillelmae*.

Acer kirchnerianum Knowlton, U. S. Nat. Mus., Proc., vol. 51, 1916, p. 282.
Miocene: *Florissant, Colo.

Acer lesquereuxii Knowlton, U. S. Geol. Surv., Bull. 152, 1898, p. 26.
Acer indivisum Lesquereux, Rept. U. S. Geol. Surv. Terr., vol. 8 (Cret. and Tert. Fl.), 1883, p. 180, pl. xxxvi, f. 6, 9. [Homonym, Weber, 1852.]
Green River: Uinta County (formerly wrongly called Randolph County), Wyo.

Acer medianum Knowlton, U. S. Geol. Surv., Bull. 204, 1902, p. 76, pl. xiv, f. 4, 5.
Mascall: Near Van Horn's ranch, John Day Basin, Oreg.

Acer merriami Knowlton, U. S. Geol. Surv., Bull. 204, 1902, p. 74, pl. xiv, f. 7.
Acer n. sp., Knowlton, in Merriam, Univ. Calif., Bull. Dept. Geol., vol. 2, 1901, p. 309.
Mascall: Van Horn's ranch, John Day Basin, Oreg.

Acer minor Knowlton, U. S. Geol. Surv., Bull. 204, 1902, p. 76, pl. xiv, f. 2, 3.

Acer, fruits of, Lesquereux, U. S. Nat. Mus., Proc., vol. 11, 1888, p. 15, pl. vii, f. 2.

Mascall: Near Van Horn's ranch, John Day Basin, Oreg.

Acer minutum Hollick, N. Y. Acad. Sci., Trans., vol. 12, 1892, p. 35, pl. iii, f. 6; U. S. Geol. Surv., Mon. 50, 1906, p. 89, pl. xxxiii, f. 14.

Raritan: Tottenville, Staten Island, N. Y.

Acer mysticum Kirchner, St. Louis Acad. Sci., Trans., vol. 8, 1898, p. 181, pl. xi, f. 2.—Knowlton, U. S. Nat. Mus., Proc., vol. 51, 1916, p. 283.

Miocene: *Florissant, Colo.

Acer obtusilobum? Unger. Lesquereux [1868] = *Menispermites salinae.*

Acer oregonianum Knowlton, U. S. Geol. Surv., Bull. 204, 1902, p. 75, pl. xiii, f. 5–8.

Acer, fruits of, Lesquereux, U. S. Nat. Mus., Proc., vol. 11, 1888, p. 15, pl. vi, f. 2, 3.

Mascall: *Van Horn's ranch, John Day Basin, Oreg.

Acer osmonti Knowlton, U. S. Geol. Surv., Bull. 204, 1902, p. 72, pl. xiii, f. 3.

Clarno (upper part): *Bridge Creek, Crook County, and Trout Creek, Harney County, Oreg.

Acer paucidentatum Hollick, N. Y. Acad. Sci., Trans., vol. 16, 1897, p. 132, pl. xiv, f. 2, 3.—Berry, N. Y. Bot. Gard., Bull., vol. 3, 1903, p. 82; Geol. Surv. N. J., Ann. Rept. 1905 [1906], p. 138.

Magothy: *Cliffwood and Morgan, N. J.

Acer perditum Cockerell, Am. Jour. Sci., 4th ser., vol. 26, 1908, p. 65, f. 13.

Miocene: *Florissant, Colo.

Acer pleistocenicum Penhallow, Geol. Soc. Am., Bull., vol. 1, 1890, p. 327, text f. 1; Roy. Soc. Canada, Trans., 2d ser., vol. 10, 1904, sec. 4, p. 72; Am. Nat., vol. 41, 1907, p. 443 text f. 1.

Pleistocene: *Don River valley, Toronto, Canada.

Acer pseudo-chrysophylla Lesquereux, Mus. Comp. Zool., Mem., vol. 6, No. 2, 1878, p. 60.

Miocene (auriferous gravels): North Fork Oregon Creek, near Forest City, Calif.

Acer rubrum Linné. Berry, Am. Jour. Sci., 4th ser., vol. 29, 1910, p. 397; Jour. Geol., vol. 25, 1917, p. 662.

Pleistocene: Black Warrior River, 328½ miles above Mobile, Ala.; Vero, Fla.

Acer saccharinum Linné. Berry, Am. Jour. Sci., 4th ser., vol. 29, 1910, p. 397.

Pleistocene: Black Warrior River, 356 miles above Mobile, and Abercrombie Landing, Chattahoochee River, Ala.

Acer saccharinum Wangenheim = *Acer saccharum* Marshall.

Acer saccharum Marshall.

Acer saccharinum Wangenheim. Penhallow, Geol. Soc. Am., Bull., vol. 1, 1890, p. 329; Roy. Soc. Can., Trans., 2d ser., vol. 2, 1896, p. 70; Brit. Assoc. Adv. Sci., Bradford meeting, 1900, p. 335[8].

Acer spicatum Lamarck. Dawson, Can. Nat., n. s. vol. 3, 1866, p. 71; idem, vol. 6, 1869, p. 403; Geol. Hist. Pl., 1888, p. 228.

Pleistocene: Greens Creek, Ottawa River, Canada.

Acer saskatchewense Dawson, Roy. Soc. Canada, Trans., vol. 3, 1885 [1886], sec. 4, p. 16.

Belly River?: *Medicine Hat, Alberta.

Acer secreta Lesquereux, U. S. Geol. and Geog. Surv. Terr., Ann. Rept. 1871 [1872]. Suppl., p. 12. [Not afterward mentioned by author.]

Evanston: Evanston, Wyo.

Acer spicatum Lamarck. Penhallow, Brit. Assoc. Adv. Sci., Bradford meeting, 1900, p. 335.

Pleistocene: Don Valley, Toronto, Canada.

Acer spicatum Lamarck. Dawson, Can. Nat., new ser., vol. 3, 1866, p. 71 = *Acer saccharum.*

Acer torontoniensis Penhallow, Am. Nat., vol. 41, 1907, p. 444, text f. 2.

Pleistocene: *Don Valley, Toronto, Canada.

Acer tricuspidatum Al. Braun = *Acer trilobatum tricuspidatum.*

Acer trilobatum? (Sternberg) Al. Braun. Lesquereux, Am. Jour. Sci., vol. 27, 1859, p. 361; U. S. Geol. and Geog. Surv. Terr., Ann. Rept. 1871 [1872], p. 293. [This is a homonym, but as it is a European species not certainly identified in North America the name has not been changed.]

Phyllites trilobatus Sternberg, Flora d. Vorwelt, Index Iconum, 1825, pl. i, f. 2.

Puget: Bellingham Bay, Wash.
Evanston: ?Evanston, Wyo.

Acer trilobatum productum? (Al. Braun) Heer. Lesquereux, U. S. Geol. and Geog. Surv. Terr., Ann. Rept. 1872 [1873], p. 388; Rept. U. S. Geol. Surv. Terr. (Tert. Fl.), 1874, p. 261, pl. xlviii, f. 2, 3a.—Knowlton, U. S. Nat. Mus., Proc., vol. 17, 1894, p. 227, pl. ix, f. 3; Geol. Soc. Am., Bull., vol. 5, 1893, p. 586.

Acer trilobatum Al. Braun. Lesquereux, U. S. Geol. and Geog. Surv. Terr., Ann. Rept. 1873 [1874], p. 408.

Dawson: Sedalia, Colo.
——? Carbon, Wyo.
Kenai: Herendeen Bay, Alaska.

Acer trilobatum productum (Al. Braun) Heer. Lesquereux, Rept. U. S. Geol. Surv. Terr., vol. 8 (Cret. and Tert. Fl.), 1883, p. 253, pl. lix, f. 1, 2, 4 = *Acer bendirei.*

Acer trilobatum productum (Al. Braun) Heer. Lesquereux, Rept. U. S. Geol. Surv. Terr., vol. 8 (Cret. and Tert. Fl.), 1883, p. 253, pl. lix, f. 3 = *Platanus dissecta.*

Acer trilobatum tricuspidatum (Al. Braun) Heer, Fl. Tert. Helv., vol. 3, 1859, p. 49, pl. cxiii, f. 3–10.—Ward, U. S. Geol. Surv., 6th Ann. Rept., 1884–85 [1886], p. 554, pl. xlix, f. 8, 9; U. S. Geol. Surv., Bull. 37, 1887, p. 66, pl. xxix, f. 3, 4.

Acer tricuspidatum Al. Braun, Neues Jahrb. f. Min., 1845, p. 172.

Fort Union: Clear Creek, near Glendive, Mont.; Little Missouri River, N. Dak.

Acer trilobatum var., Knowlton, Harriman Alaska Exped., vol. 4, 1904, p. 155, pl. xxix, f. 2.

Kenai: Kukak Bay, Alaska.

Acer vitifolium Al. Braun = *Acer vitiphyllum.*

Acer vitiphyllum Knowlton and Cockerell, n. name.

Acer vitifolium Al. Braun, Neues Jahrb., 1845, p. 172.—Lesquereux, U. S. Nat. Mus., Proc., vol. 10, 1887, p. 44. [Homonym, Opiz, 1829.]

Pleistocene ?: Wytheville, Va.

Acer vivarium Knowlton, U. S. Geol. Surv., Mon. 32, pt. 2, 1899, p. 735, pl. xcviii, f. 4.

Miocene: Fossil Forest Ridge, Yellowstone National Park.

Acer sp., Berry, Am. Jour. Sci., 4th ser., vol. 34, 1912, p. 222.

Pleistocene: (Buena Vista, Va.)

Acer, fruit of, Dawson, Roy. Soc. Canada, Trans., vol. 8, 1890, sec. 4, p. 87, text f. 20.

Eocene ?: Stump Lake, British Columbia.

Acer sp. ? Hollick, Maryland Geol. Surv., Pliocene and Pleistocene, 1906, p. 234, pl. lxxi, f. 7, 8.

Pleistocene (Sunderland): Head of Island Creek, Calvert County, Md.

Acer sp., fruit of, Hollick, U. S. Geol. Surv., Mon. 50, 1906, p. 89, pl. xxxiii, f. 12, 13.

Magothy: Gay Head, Marthas Vineyard, Mass.

Acer sp., Knowlton, in Lindgren, Jour. Geol., vol. 4, 1896, p. 890.

Miocene (auriferous gravels): Independence Hill, Placer County, Calif.

Acer, fruit of, Knowlton, U. S. Geol. Surv., Mon. 32, pt. 2, 1899, p. 736, pl. xcviii, f. 5.

Fort Union: Crescent Hill, Yellowstone National Park.

Acer sp., Knowlton, in Merriam, Univ. Calif., Bull. Dept. Geol., vol. 2, 1901, p. 289.—Knowlton, U. S. Geol. Surv., Bull. 204, p. 73, pl. xiii, f. 1, 2.
Clarno (upper part): 1½ miles east of Clarnos Ferry, Oreg.

Acer n. sp., Knowlton, in Merriam, Univ. Calif., Bull. Dept. Geol., vol. 2, 1901, p. 309 = *Acer merriami.*

Acer sp., Knowlton, Wash. Geol. Surv., Ann. Rept., vol. 1, 1902, p. 33.
Miocene: Black River Junction, King County, Wash.

Acer sp., Lesquereux, U. S. Geol. and Geog. Surv. Terr., Ann. Rept., 1871 [1872], p. 286. [Not afterward recognized.]
Green River ?: Barrel Springs, Wyo.

Acer sp., Lesquereux, Mus. Comp. Zool., Mem., vol. 6, No. 2, 1878, p. 60.
Miocene (auriferous gravels): Oregon Creek near Forest City, Nevada County, Calif.

Acer sp., Lesquereux, Rept. U. S. Geol. Surv. Terr., vol. 8 (Cret. and Tert. Fl.), 1883, p. 181, pl. xxxvi, f. 7, 8.
Miocene: Florissant, Colo.

Acer, branches of, Lesquereux, U. S. Nat. Mus., Proc., vol. 11, 1888, p. 15.
Mascall: John Day Basin, Oreg.

Acer, fruits of, Lesquereux, U. S. Nat. Mus., Proc., vol. 11, 1888, p. 15, pl. vi, f. 2, 3 = *Acer oregonianum.*

Acer, fruits of, Lesquereux, U. S. Nat. Mus., Proc., vol. 11, 1888, p. 15, pl. vii, f. 2 = *Acer minor.*

Acer sp. ?, fruit of, Newberry, U. S. Geol. Surv., Mon. 35, 1898, p. 115, pl. xlvi, f. 8.
Clarno (upper part): Bridge Creek, Crook County, Oreg.

Acer sp., Penhallow, Rept. Tert. Pl. Brit. Col., 1908, p. 33, f. 1–4.
Oligocene: Stump Lake, Tulameen Lake, Horsefly River, Quesnel, and Burrard Inlet, British Columbia.

ACERATES Elliott, Sketch Bot. South Carolina, vol. 1, 1817, p. 316.

Acerates amboyensis Berry, Bull. Torr. Bot. Club, vol. 36, 1909, p. 263; Geol. Surv. N. J., Bull. 3, 1911, p. 214; U. S. Geol. Surv., Prof. Paper 112, 1919, p. 136.
Acerates sp., Hollick, in Newberry, U. S. Geol. Surv., Mon. 26, 1896, p. 121, pl. xxxii, f. 17; pl. xli, f. 4, 5.—Berry, Bull. Torr. Bot. Club, vol. 34, 1907, p. 205.
Tuscaloosa: Sherleys Mill and Sanders Ferry Bluff, Ala.
Raritan: South Amboy, N. J.
Black Creek: Blackmans Bluff, Neuse River, N. C.

Acerates fructifer Cockerell, Am. Nat., vol. 42, 1908, p. 580, text f. 10.
Miocene: *Florissant, Colo.

Acerates sp. Hollick, in Newberry = *Acerates amboyensis.*

Aceriphyllum aralioides Fontaine = *Araliaephyllum magnifolium.*

ACERITES Viviani, Mém. Soc. géol. France, vol. 1, 1833, p. 131.

Acerites menispermifolia Lesquereux = *Menispermites menispermifolia.*

Acerites multiformis Lesquereux, U. S. Geol. Surv., Mon. 17, 1891 [1892], p. 156, pl. xxxiv, f. 1–9.
Dakota: Near Delphos, Kans.

Acerites negundifolium Dawson, Roy. Soc. Canada, Trans., vol. 8, sec. 4, 1890, p. 86, text f. 19.—Penhallow, Rept. Tert. Pl. Brit. Col., 1908, p. 35.
Oligocene?: Stump Lake, British Columbia.

Acerites pristinus Newberry = *Liquidambar obtusilobatus.*

ACORUS Linné, Sp. pl., 1753, p. 324.

Acorus affinis Lesquereux, U. S. Geol. and Geog. Surv. Terr., Ann. Rept., 1873 [1874], p. 410. [Not afterward recognized.]
Miocene: *Florissant, Colo.

Acorus brachystachys Heer, K. Sven. Vetenskaps-Akad. Handl., vol. 8, No. 7, 1870 [Fl. foss. arct., vol. 2, abt. 3, 1871], p. 51, pl. viii, f. 7, 8.—Lesquereux, U. S. Geol. and Geog. Surv. Terr., Ann. Rept., 1871 [1872], p. 288; U. S. Geol. Surv. Terr., Rept., vol. 7 (Tert. Fl.), 1878, p. 105, pl. xiv, f. 12–15.
Denver: Sedalia, Colo.
Hanna: *Creston and Carbon, Wyo.

Acorus calamus Linné. Lesquereux, Am. Jour. Sci., 2d ser., vol. 27, 1859, p. 365.
Pleistocene: Banks of Mississippi River, near Columbus, Ky.

Acrostichides Fontaine, U. S. Geol. Surv., Mon. 6, 1883, p. 24 = *Acrostichites* Göppert.

Acrostichides egyptiacus (Emmons) Fontaine = *Sphenopteris egyptiaca.*

Acrostichides rhombifolius Fontaine = *Acrostichites tenuifolius.*

Acrostichides rhombifolius rarinervis Fontaine = *Acrostichites tenuifolius rarinervis.*

ACROSTICHITES Göppert, Nov. Act. Acad. Caes. Leop.- Car., vol. 17, Suppl. (Sys. Fil. Foss.), 1836, p. 284.

Acrostichites? brevipennis Ward, in Fontaine, in Ward, U. S. Geol. Surv., Twentieth Ann. Rept., pt. 2, 1900, p. 334.—Knowlton, in Diller, U. S. Geol. Surv., Bull. 353, 1908, p. 55.
Foreman: Taylorsville region, Calif.

Acrostichites? coniopteroides Ward, in Fontaine, in Ward, U. S. Geol. Surv., Twentieth Ann. Rept., pt. 2, 1900, p. 333.
Foreman: Taylorsville region, Calif.

Acrostichites densifolius Fontaine, U. S. Geol. Surv., Mon. 6, 1883, p. 34, pl. x, f. 1–1c.
Triassic (Newark): Clover Hill, Va.

Acrostichites? fructifer Ward, in Fontaine, in Ward, U. S. Geol. Surv., Twentieth Ann. Rept., pt. 2, 1900, p. 333.
Foreman: Taylorsville region, Calif.

Acrostichites linnaeaefolius (Bunbury) Fontaine, U. S. Geol. Surv., Mon. 6, 1883, p. 25, pl. vi, f. 3; pl. vii, f. 1–4; pl. viii, f. 1, 1a; pl. ix, f. 1, 1a; in Ward, U. S. Geol. Surv., Twentieth Ann. Rept., pt. 2, 1900, pp. 240, 287, pl. xxv, f. 7, 8.
Neuropteris linnaeaefolia Bunbury, Quart. Jour. Geol. Soc., vol. 3, pt. 1, 1847, p. 281, 288, pl. x.
Triassic (Newark): Gowry and Blackheath shafts, near Midlothian Chesterfield County, Va.; York County, Pa.; North Carolina.

Acrostichites microphyllus Fontaine, U. S. Geol. Surv., Mon. 6, 1883, p. 33, pl. vii, f. 5; pl. x, f. 2; pl. xi, f. 4; pl. xii, f. 3; in Ward, U. S. Geol. Surv., Twentieth Ann. Rept., pt. 2, 1900, p. 240, pl. xxv, f. 9, 10.
Triassic (Newark): Clover Hill, Va.; York County, Pa.

Acrostichites oblongus Emmons = *Lonchopteris oblonga.*

Acrostichites? princeps (Presl) Schenk. Fontaine, in Ward, U. S. Geol. Surv., Twentieth Ann. Rept., pt. 2, 1900, p. 334.—Diller, Geol. Soc. Am., Bull., vol. 3, 1893, p. 374.—Knowlton, in Diller, U. S. Geol. Surv., Bull. 353, 1908, p. 55.
Foreman: Taylorsville region, Calif.

Acrostichites tenuifolius (Emmons) Fontaine, in Ward, U. S. Geol. Surv., Twentieth Ann. Rept., pt. 2, 1900, p. 287, pl. xxxix, f 4.
Odontopteris tenuifolius Emmons, American geology, pt. 6, 1857, p. 105, pl. iii, f. 5.
Acrostichides rhombifolius Fontaine, U. S. Geol. Surv., Mon. 6, 1883, pp. 29, 105, pl. viii, f. 2, 3; pl. xi, f. 1–3; pl. xii, f. 1, 2; pl. xiii, f. 1, 2; pl. xiv, f. 1, 2; pl. xlix, f. 7.
Triassic (Newark): Carbon Hill and Clover Hill, Va.; Ellingtons, N. C.

Acrostichites tenuifolius rarinervis (Fontaine) Ward, U. S. Geol. Surv., Twentieth Ann. Rept., pt. 2, 1900, p. 423.
Acrostichides rhombifolius rarinervis Fontaine, U. S. Geol. Surv., Mon. 6, 1883, p. 32, pl. xiii, f. 3.
Triassic (Newark): Clover Hill, Va.

ACROSTICHOPTERIS Fontaine, U. S. Geol. Surv., Mon. 15, 1889, p. 106. [Type *A. longipennis.*]

Acrostichopteris adiantifolia (Fontaine) Berry, U. S. Nat. Mus. Proc., vol. 38, 1910, p. 629; Md. Geol. Surv., Lower Cret., 1911, p. 224, pl. xxiv, f. 2, 3.

Baieropsis adiantifolia Fontaine, U. S. Geol. Surv., Mon. 15, 1889, p. 211, pl. xcii, f. 8, 9; pl. xciii, f. 1–3; pl. xciv, f. 2, 3.—Fontaine, in Ward, U. S. Geol. Surv., Nineteenth Ann. Rept., pt. 2, 1899, p. 684, pl. clxviii, f. 8; U. S. Geol. Surv., Mon. 48, 1905 [1906], pp. 510, 528, 538.

Patuxent: Fredericksburg, Potomac Run, Telegraph, Va.
Patapsco: Fort Foote?, Md.; Chinkapin Hollow?, Va.
Arundel: Arlington?, Md.
Lakota: Barrett and Hay Creek, Crook County, Wyo.

Acrostichopteris cyclopteroides Fontaine emend., Berry, U. S. Nat. Mus., Proc., vol. 38, 1910, p. 630; Md. Geol. Surv., Lower Cret., 1911, p. 226, pl. xxiv, f. 1.

Acrostichopteris cyclopteroides Fontaine, U. S. Geol. Surv., Mon. 15, 1889, p. 109, pl. xciv, f. 8.

Baieropsis denticulata Fontaine, U. S. Geol. Surv., Mon. 15, 1889, p. 210, pl. xciii, f. 7.

Patuxent: Dutch Gap, Va.

Acrostichopteris densifolia Fontaine [in part] = *Acrostichopteris longipennis.*

Acrostichopteris densifolia Fontaine [in part] = *Acrostichopteris parvifolia.*

Acrostichopteris expansa (Fontaine) Berry, Md. Geol. Surv., Lower Cret., 1911, p. 229.

Baieropsis expansa Fontaine, U. S. Geol. Surv., Mon. 15, 1890, p. 207, pl. lxxxix, f. 3; pl. xc, f. 1; pl. xci, f. 2; pl. xcii, f. 5 [not pl. lxxxix, f. 1, which=*Schizaeopsis americana*].

Patuxent: Fredericksburg, Dutch Gap, and Trents Reach, Va.

Acrostichopteris fimbriata Knowlton, Smithsonian Misc. Coll., vol. 50, 1907, p. 110, pl. xi, f. 3, 3a.

Kootenai: 6 miles southwest of Geyser, Cascade County, Mont.

Acrostichopteris longipennis Fontaine, emend. Berry, U. S. Nat. Mus., Proc., vol. 38, 1910, p. 627; Md. Geol. Surv., Lower Cret., 1911, p. 223, pl. xxiii, f. 1, 2; pl. xxiv, f. 7.

Acrostichopteris longipennis Fontaine, U. S. Geol. Surv., Mon. 15, 1889 [1890], p. 107, pl. clxx, f. 10; pl. clxxi, f. 1, 5, 7.—Fontaine, in Ward, U. S. Geol. Surv., Mon. 48, 1905 [1906], p. 557.

Acrostichopteris densifolia Fontaine, U. S. Geol. Surv., Mon. 15, 1889 [1890], p. 107, pl. clxx, f. 2; pl. clxxi, f. 2, 6; pl. clxxii, f. 13; pl. cxiv, f. 4.

Acrostichopteris parvifolia Fontaine. Fontaine, in Ward, U. S. Geol. Surv., Mon. 48, 1905 [1906], p. 558, pl. cxvi, f. 5.

Baieropsis foliosa Fontaine, U. S. Geol. Surv., Mon. 15, 1889, p. 209, pl. xciii, f. 4–6; in Ward, U. S. Geol. Surv., Mon. 48, 1905 [1906], pp. 481, 482, 489, 504, 508, pl. cx, f. 9.

Acrostichopteris longipennis—Continued.

Baieropsis denticulata angustifolia Fontaine, U. S. Geol. Surv., Mon. 15, 1889, p. 210, pl. cxii, f. 7; in Ward, U. S. Geol. Surv., Mon. 48, 1905 [1906], p. 491.

Patapsco: Near Wellhams, Federal Hill, Md.; near Brooke, 72d milepost, Hell Hole, mouth of Hell Hole ?, White House Bluff, Dumfries Landing, Aquia Creek cut, and Mount Vernon, Va.

Acrostichopteris parceloba Fontaine = *Acrostichopteris parvifolia.*

Acrostichopteris parvifolia Fontaine, emend. Berry, U. S. Nat. Mus., Proc., vol. 38, 1910, p. 630; Md. Geol. Surv., Lower Cret., 1911, p. 226, pl. xxiv, f. 4, 5.

Acrostichopteris parvifolia Fontaine, U. S. Geol. Surv., Mon. 15, 1889 [1890], p. 108, pl. xciv, f. 5, 9, 10, 12; pl. clxxi, f. 4 [not Fontaine, in Ward, 1906].

Acrostichopteris densifolia Fontaine, U. S. Geol. Surv., Mon. 15, 1889 [1890], p. 107, pl. xciv, f. 4 [not remaining figures].

Baieropsis adiantifolia minor Fontaine, U. S. Geol. Surv., Mon. 15, 1889 [1890], p. 212, pl. xciv, f. 1.

Acrostichopteris parcelobata Fontaine, U. S. Geol. Surv., Mon. 15, 1889 [1890], p. 108, pl. xciv, f. 6, 7, 11, 14.

Patuxent: Dutch Gap, Fredericksburg, Trents Reach, Potomac Run, Va. [Not Federal Hill, Md.]

Acrostichopteris parvifolia Fontaine. Fontaine, in Ward = *Acrostichopteris longipennis.*

Acrostichopteris pluripartita (Fontaine) Berry, U. S. Nat. Mus., Proc., vol. 38, 1910, p. 631; Md. Geol. Surv., Lower Cret., 1911, p. 227, pl. xxiv, f. 6.

Baieropsis pluripartita Fontaine, U. S. Geol. Surv., Mon. 15, 1889 [1890], p. 208, pl. lxxxix, f. 4; pl. xc, f. 2–5; pl. xci, f. 1, 3, 4, 7; pl. xcii, f. 1, 2, 6?; in Ward, U. S. Geol. Surv., Nineteenth Ann. Rept., pt. 2, 1899, p. 685, pl. clxviii, f. 9–12; U. S. Geol. Surv., Mon. 48, 1905 [1906], pp. 479, 481, 482, 505, pl. cvii, f. 1.

Baieropsis pluripartita minor Fontaine, U. S. Geol. Surv., Mon. 15, 1889 [1890], p. 208, pl. xci, f. 5; pl. xcii, f. 3, 4.

Baieropsis longifolia Fontaine, U. S. Geol. Surv., Mon. 15, 1889 [1890], p. 210, pl. xci, f. 6; in Ward, U. S. Geol. Surv., Mon. 48, 1905 [1906], pp. 505, 517, pl. cxi, f. 3.

Patuxent: Fredericksburg, Trents Reach, and Dutch Gap, Va.; New Reservoir, D. C.

Patapsco: Hell Hole, 72d milepost, near Brooke, Va.; Overlook Inn ?, Md.

Lakota: Barrell, Crook County, Wyo.

Fuson: Pine Creek, Crook County, Wyo.

ACROSTICHUM Linné, Sp. pl., 1753, p. 1067.

Acrostichum crassifolium Fontaine, U. S. Geol. Surv., Mon. 15, 1889 [1890], p. 105, pl. xvi, f. 7.

Patuxent: Fredericksburg, Va.

Acrostichum georgianum Berry, U. S. Geol. Surv., Prof. Paper 84, 1914, p. 133, pl. xxvii, f. 1.
Barnwell (Twiggs): Phinizy Gully, Columbia County, Ga.

Acrostichum haddeni Hollick, Torreya, vol. 2, 1902, p. 146, pl. 4, f. 3–6.—Knowlton, U.S. Geol. Surv., Prof. Paper 101, 1917, p. 244.
Vermejo: *Florence and Walsenburg, Colo.

Acrostichum hesperium Newberry, U. S. Nat. Mus., Proc., vol. 5, 1882 [1883], p. 503; U.S. Geol. Surv., Mon. 35, 1898, p. 6, pl. lxi, f. 2–5.
Green River: Green River, Wyo.

ACTINOPTERIS Schenk, Foss. Fl. Grenzsch. Keup. u. Lias Frankens, Wiesbaden, 1865, p. 23.

Actinopteris quadrifolia (Emmons) Fontaine, in Ward, U. S. Geol. Surv., Twentieth Ann. Rept., pt. 2, 1900, p. 310, pl. xlvii, f. 2.
Sphenoglossum quadrifolium Emmons, Geol. Rept. Midland Counties, N. C., 1856, p. 335, pl. i, f. 2; American geology, pt. 6, 1857, p. 134, pl. v, f. 2.
Actinopteris quadrifoliata Fontaine, U. S. Geol. Surv., Mon. 6, 1883, p. 121, pl. lii, f. 3.
Triassic (Newark): North Carolina.

Actinopteris quadrifoliata Fontaine = *Actinopteris quadrifolia.*

ADIANTITES, Göppert, Syst. Filic. Foss., 1836, p. 216.

Adiantites amurensis Heer = *Coniopteris hymenophylloides*

Adiantites gracillimus Lesquereux = *Anemia? gracillima.*

Adiantites nympharum Heer, Fl. foss. arct., vol. 4, 1876, pt. 2, p. 93, pl. xvii, f. 5, 5b.—?Fontaine, in Ward, U. S. Geol. Surv., Mon. 48, 1905 [1906], p. 76, pl. xii, f. 9–11.
Jurassic: Douglas County, Oreg.

Adiantites orovillensis Fontaine, Am. Jour. Sci., 4th ser., vol. 2, 1896, p. 274; in Ward, U. S. Geol. Surv., Twentieth Ann. Rept., pt. 2, 1900, p. 344, pl. xlix, f. 2, 3.
Jurassic: Oroville, Calif.

Adiantites parvifolius Fontaine = *Onychiopsis nervosa.*

Adiantites praelongus Dawson, Roy. Soc. Canada, Trans., vol. 1, sec. iv, 1882 [1883], p. 25, pl. v, f. 19.
Upper Cretaceous: Baynes Sound, Vancouver Island.

ADIANTUM (Tournefort) Linné, Sp. pl., 1753, p. 1094.

Adiantum montanensis Knowlton, Smithsonian Misc. Coll., vol. 50, 1907, p. 112, pl. xii, f. 1, 2.
Kootenai: 6 miles southwest of Geyser, Cascade County, Mont.

AESCULOPHYLLUM Dawson, Roy. Soc. Canada, Trans., 2d ser., vol. 1, sec. 4, 1895 [1896], p. 149. [Type *A. hastingsense.*]

Aesculophyllum hastingsense Dawson, Roy. Soc. Canada, Trans., 2d ser., vol. 1, sec. 4, 1895 [1896], p. 149, pl. viii, f. 16.—Penhallow, Rept. Tert. Pl. Brit. Col., 1908, p. 35.
Eocene: Burrard Inlet, British Columbia.

AESCULUS Linné, Sp. pl., 1753, p. 344.

Aesculus antiquus Dawson, Brit. N. A. Bound. Com. (Rept. Geol. and Rec. Vic. 49th Parallel), 1875, App. A, p. 330, pl. xvi, f. 8, 9.—Penhallow, Rept. Tert. Pl. Brit. Col., 1908, p. 36.
Paskapoo (Fort Union): Badlands west of Woody Mountain; Porcupine Creek, Saskatchewan.

Aesculus arctica Knowlton, Harriman Alaska Exped., vol. 4, 1904, p. 155, pl. xxx.
Kenai: Kukak Bay, Alaska.

Aesculus? simulata Knowlton, U. S. Geol. Surv., Bull. 204, 1902, p. 78, pl. xv, f. 1, 2.
Mascall: Van Horn's ranch, John Day Valley, Oreg.

Aesculus sp., Dawson, Roy. Soc. Canada, Trans., vol. 4, 1886, sec. 4, p. 29, pl. ii, f. 16.—Penhallow, Rept. Tert. Pl. Brit. Col., 1908, p. 35.
Oligocene: Tulameen River, British Columbia.
Paskapoo (Fort Union): Porcupine Creek, Saskatchewan.

Aesculus sp., Knowlton, in Lindgren, Jour. Geol., vol. 4, 1896, p. 890.
Miocene (auriferous gravels): Independence Hill, Placer County, Calif.

AGARICITES Meschinelli, in Saccardo, Sylloge Fung., vol. 10, 1892, p. 745[9]. [Type, *Agaricus wardianus*.]

Agaricites conwentzi Platen, Sitzungsb. Naturf. Ges., Liepzig, vol. 34, 1907 [1908], p. 15, pl. i, f. 2, 3 in part.
Tertiary: Calistoga, Calif.

AILANTHOPHYLLUM Dawson, Roy. Soc. Canada, Trans., vol. 8, sec. 4, 1890 [1891], p. 88. [Type, *A. incertum*.]

Ailanthophyllum incertum Dawson, Roy. Soc. Canada, Trans., vol. 8, sec. 4, 1890 [1891], p. 88, text f. 25.—Penhallow, Rept. Tert. Pl. Brit. Col., 1908, p. 36.
Paskapoo (Fort Union): Tranquille River, British Columbia.

AILANTHUS Desfontaines, Act. Acad., Paris, 1786, p. 265.

Ailanthus americana Cockerell, Am. Jour. Sci., 4th ser., vol. 26, 1908, p. 539, f. 3.
Miocene: *Florissant, Colo.

Ailanthus longe-petiolata Lesquereux, Rept. U. S. Geol. Surv. Terr., vol. 8 (Cret. and Tert. Fl.), 1883, p. 197, pl. xl, f. 6, 7.
Green River: *Uinta County (formerly wrongly called Randolph County), Wyo.

Ailanthus ovata Lesquereux, Rept. U. S. Geol. Surv. Terr., vol. 8 (Cret. and Tert. Fl.), 1883, p. 254, pl. li, f. 7, 8.—Knowlton, U. S. Geol. Surv., Bull. 204, 1902, p. 69.
Clarno (upper part): *Bridge Creek, Oreg.

Albertia latifolia? Emmons = ***Otozamites carolinensis***.

Aleurites eocenica Lesquereux = *Grewiopsis eocenica.*

Algae gen. et sp.? Penhallow, Canadian Nat., n. s., vol. 3, p. 75; Brit. Assoc. Adv. Sci., Bradford meeting, 1900, p. 335.
Pleistocene: Greens Wharf, Ottawa River, and Montreal, Canada.

ALGITES Seward, Wealden Flora, pt. 1, 1894, p. 4.

Algites americana Berry, Bull. Torr. Bot. Club, vol. 38, 1911, p. 401; U. S. Geol. Surv., Prof. Paper 84, 1914, p. 14; Md. Geol. Surv., Upper Cret. 1916, p. 758, pl. 1, f. 1.
Magothy: *Severn River and Round Bay, Md.
Black Creek: Florence, S. C.; Black River, N. C.

ALISMACITES Saporta, Ann. sci. nat., 4th ser., Bot., vol. 17 (Études), No. 1, 1862, p. 228 [75].

Alismacites dakotensis Lesquereux, U. S. Geol. Surv., Mon. 17, 1891 [1892], p. 37, pl. ii, f. 10.
Dakota: *Ellsworth County, Kans.

ALISMAPHYLLITES Knowlton, U. S. Geol. Surv., Prof. Paper 101, 1917, p. 286. [Type, *A. crassifolium.*]

Alismaphyllites crassifolium Knowlton, U. S. Geol. Surv., Prof. Paper 101, 1917, p. 286, pl. lv, f. 1.
Raton: Fishers Peak mine, 3 miles southeast of Trinidad, Colo.

ALISMAPHYLLUM Berry, Md. Geol. Surv., Lower Cret., 1912, p. 452. [Type, *Sagittaria victor-masoni.*]

Alismaphyllum victor-masoni (Ward) Berry, Md. Geol. Surv., Lower Cret., 1912, p. 453, pl. lxxix, f. 5.
Sagittaria victor-masoni Ward, U. S. Geol. Surv., Fifteenth Ann. Rept., 1895, p. 354, pl. iii, f. 5.—Fontaine, in Ward, U. S. Geol. Surv., Mon. 48, 1905 [1906], p. 491.
Patapsco: White House Bluff, Va.

ALLANTODIOPSIS Knowlton and Maxon, n. gen. [Type, *Pteris erosa* Lesquereux.]

Allantodiopsis erosa (Lesquereux) Knowlton and Maxon, n. comb.
Pteris erosa Lesquereux, U. S. Geol. and Geog. Surv. Terr., Ann. Rept. 1871, Suppl. 1872, p. 12; idem, 1873 [1874], p. 392; Rept. U. S. Geol. Surv. Terr., vol. 7 (Tert. Fl.), 1878, p. 53, pl. iv, f. 8; idem, vol. 8 (Cret. and Tert. Fl.), 1883, p. 121, pl. xix, f. 5.—Knowlton, U. S. Geol. Surv., Mon. 32, pt. 2, 1899, p. 668, f. 6; U. S. Geol. Surv., Prof. Paper 101, 1917, p. 244.
Pteris subsimplex Lesquereux, U. S. Geol. and Geog. Surv. Terr., Ann. Rept. 1873 [1874], p. 392; Rept. U. S. Geol. Surv. Terr., vol. 7 (Tert. Fl.), 1878, p. 52, pl. iv, f. 5–7.
Osmunda major Lesquereux, Rept. U. S. Geol. Surv. Terr., vol. 8 (Cret. and Tert. Fl.), 1883, p. 121, pl. xviii, f. 5.
Pteris undulata Lesquereux, Mus. Comp. Zool., Bull., vol. 16, 1888, p. 43.

Allantodiopsis erosa—Continued.

Asplenium erosum (Lesquereux) Knowlton, U. S. Geol. Surv., Bull. 152, 1898, p. 45.

Asplenium subsimplex (Lesquereux) Knowlton, U. S. Geol. Surv., Bull. 152, 1898, p. 45; Bull. 204, 1902, p. 21.

Vermejo: Rockvale, Colo.
Denver: Golden, Colo.
Clarno (lower part): Cherry Creek, Crook County, Oreg.
Raton: *Fishers Peak, Colo.
Fort Union: Yellowstone National Park.

ALNITES Hisinger, Lethaea svecica, 1837, p. 111.

Alnites crassus Lesquereux, Geol. and Nat. Hist. Surv. Minnesota, vol. 3, 1893, p. 13, pl. B, f. 4.
Dakota: New Ulm, Minn.

Alnites curta Dawson, Roy. Soc. Canada, Trans., vol. 8, 1890 [1891], sec. 4, p. 86, text f. 18–18c.—Penhallow, idem, 2d ser., vol. 8, 1902, p. 70; Rept. Tert. Fl. Brit. Col., 1908, p. 36.
Eocene: North Fork Similkameen River and Horsefly River, British Columbia.
Eocene: Tulameen River, British Columbia.

Alnites grandifolia (Newberry). Newberry, N. Y. Lyc. Nat. Hist., Ann., vol. 9, 1868, p. 9 [name].—[Lesquereux], U. S. Geol. and Geog. Surv. Terr. (Ill. Cret. and Tert. Fl.), 1878, pl. iv, f. 2.—Newberry, U. S. Geol. Surv., Mon. 35, 1898, p. 67, pl. iv, f. 2; ?Roy. Soc. Canada, Trans., 2d ser., vol. 8, sec. 4, 1902, p. 60; Rept. Tert. Pl. Brit. Col., 1908, p. 36.

Alnus grandifolia Newberry, in Reynolds, Rept., 1869, p. 164 [nomen]; U. S. Nat. Mus., Proc., vol. 5, 1882 [1883], p. 509; Knowlton, idem, vol. 17, 1894, p. 220; Geol. Soc. Am., Bull., vol. 5, 1893, p. 582.

Dakota: Blackbird Hill, Nebr.
Paskapoo (Fort Union): Red Deer River, Alberta (Penhallow).
Kenai(?): Cook Inlet, Alaska.

Alnites insignis Dawson, Roy. Soc. Canada, Trans., vol. 1, 1882 [1883], p. 28, pl. viii, f. 36.
Upper Cretaceous: Nanaimo, Vancouver Island.
Mill Creek: Mill Creek, British Columbia.

Alnites kefersteinii Göppert = *Alnus kefersteinii*.

Alnites macquarrii Forbes = *Corylus macquarrii*.

Alnites quadrangularis Lesquereux = *Hamamelites quadrangularis*.

Alnites unequilateralis Lesquereux = *Alnus inaequilateralis*.

ALNUS Gaertner, Fr. and Sem., vol. 2, 1791, p. 54, pl. xc.

Alnus alaskana Newberry, U. S. Nat. Mus., Proc., vol. 5, 1882 [1883], p. 509; U. S. Geol. Surv., Mon. 35, 1898, p. 65, pl. xlviii, f. 8.—Knowlton, U. S. Nat. Mus., Proc., vol. 17, 1894, p. 220; Geol. Soc. Am., Bull., vol. 5, 1893, p. 582.—Penhallow, Rept. Tert. Pl. Brit. Col., 1908, p. 37.
Kenai?: Kootznahoo Archipelago, Alaska.
Eocene: Tulameen River, British Columbia.

Alnus americana Ettingshausen, Denks. d. K. Akad. d. W., Math.-naturw. Classe, vol. 47, 1883, p. 115 [15]; Geol. Surv. N. S. W., Mem. Palaeont. No. 2 (Tert. Fl. Australia), 1888, p. 25.
Betula göpperti Lesquereux, Rept. Geol. and Geog. Surv. Terr., vol. 7 (Tert. Fl.), 1878, p. 138, pl. xvii, f. 23, 23a [note f. 21, 22].
Evanston: Evanston, Wyo.

Alnus auraria Knowlton and Cockerell, n. name.
Alnus rugosa Lesquereux, Mus. Comp. Zool., Bull., vol. 16, 1888, p. 45. [Homonym, *A. rugosa* Koch, 1872.]
Denver: Golden, Colo.

Alnus carpinifolia Lesquereux, Mus. Comp. Zool., Bull., vol. 16, 1888, p. 45.
Denver: Golden, Colo.

Alnus carpinoides Lesquereux, Rept. U. S. Geol. Surv. Terr., vol. 8 (Cret. and Tert. Fl.), 1883, p. 243, pl. l, f. 11; pl. li, f. 4, 4a, 5.—Knowlton, U. S. Geol. Surv., Bull. 204, 1902, p. 42.—Penhallow, Rept. Tert. Pl. Brit. Col., 1908, p. 37.
Clarno (upper part): Bridge Creek, Oreg.
Eocene: Horsefly and Tranquille rivers, British Columbia.

Alnus cordata Lesquereux = *Alnus praecordata.*

Alnus corrallina Lesquereux, Rept. U. S. Geol. Surv. Terr., vol. 8 (Cret. and Tert. Fl.), 1883, p. 243, pl. li, f. 1–3.
Miocene: Corrall Hollow, south of Mount Diablo, Calif. [incorrectly given as John Day Valley, Oreg.].

Alnus corylifolia Lesquereux = *Alnus corylina.*

Alnus corylina Knowlton and Cockerell, n. name.
Alnus corylifolia Lesquereux, U. S. Nat. Mus., Proc., vol. 5, 1882 [1883], p. 446, pl. vii, f. 1–4; Rept. U. S. Geol. and Geog. Surv. Terr., vol. 8 (Cret. and Tert. Fl.), 1883, p. 258.—Knowlton, U. S. Nat. Mus., Proc., vol. 17, 1894, p. 220; Geol. Soc. Am., Bull., vol. 5, 1893, p. 583; Harriman Alaska Exp., vol. 4, 1904, p. 155.
Kenai: Kachemak Bay, Cook Inlet, Kukak Bay, Alaska Peninsula, Alaska.

Alnus grandifolia Newberry = *Alnites grandifolia.*

Alnus grandiflora Newberry, U. S. Nat. Mus., Proc., vol. 5, 1882 [1883], p. 509.
Kenai: Cook Inlet, Alaska.

Alnus grewiopsis Ward, U. S. Geol. Surv., Sixth Ann. Rept. 1884–85 [1886], p. 551, pl. xxxix, f. 8; U. S. Geol. Surv., Bull. 37, 1887, p. 30, pl. xiv, f. 1.
Laramie?: Hodges Pass, Wyo.

Alnus inaequilateralis Lesquereux, Rept. U. S. Geol. Surv. Terr., vol. 8 (Cret. and Tert. Fl.), 1883, p. 151.
Alnites unequilateralis Lesquereux, U. S. Geol. and Geog. Surv. Terr., Bull.; vol. 1, 1875 [1876], p. 381; idem, Ann. Rept., 1874 [1876], p. 307; Rept. U. S. Geol. Surv. Terr., vol. 7 (Tert. Fl.), 1878, p. 141, pl. lxii, f. 1–4.
Green River: Alkali station, Wyo.

Alnus kanseana Lesquereux = *Hamamelites kansaseanus* = *Quercus kanseana*.

Alnus kefersteinii (Göppert) Unger. Heer, Fl. foss. arct., vol. 2, Abt. 2, 1869, p. 29, pl. iii, f. 7, 8.—Lesquereux, U. S. Geol. and Geog. Surv. Terr., Ann. Rept. 1871 [1872], p. 292; Rept. U. S. Geol. Surv. Terr., vol. 7 (Tert. Fl.), 1878, p. 140, pl. xviii, f. 6–8; pl. lxiv, f. 11.—Knowlton, U. S. Nat. Mus., Proc., vol. 17, 1894, p. 220; Geol. Soc. Am., Bull., vol. 5, 1893, p. 582.—Lesquereux, U. S. Nat. Mus., Proc., vol. 5, 1882 [1883], p. 446.—Knowlton, U. S. Geol. Surv., Bull. 204, 1902, p. 43.
Alnites kefersteinii Göppert, De Floribus in Statu. Fossili, 1837, p. 564, pl. xli, f. 1–7.
Kenai: Alaska.
Mascall: Van Horn's ranch, John Day Basin, Oreg.

Alnus kefersteinii (Göppert) Heer, var., Heer, Fl. foss. Alask.; K. Sv. Vetenskaps-Akad. Handl., vol. 8, No. 4, 1869, p. 28, pl. v, f. 9 [in Fl. foss. arct., vol. 2, Abt. ii, 1871].—Knowlton, U. S. Nat. Mus., Proc., vol. 17, 1894, p. 220; Geol. Soc. Am., Bull., vol. 5, 1893, p. 582.
Kenai: Port Graham, Alaska.

Alnus macrodonta Knowlton, U. S. Geol. Surv., Bull. 204, 1902, p. 42, pl. iv, f. 1.
Clarno (upper part): Bridge Creek, Oreg.

Alnus praecordata Cockerell, Am. Mus. Nat. Hist., Bull., vol. 24, 1908, p. 84.—Knowlton, U. S. Nat. Mus., Proc., vol. 51, 1916, p. 264, pl. xix, f. 1.
Alnus cordata Lesquereux, Rept. U. S. Geol. Surv. Terr., vol. 8 (Cret. and Tert. Fl.), 1883, p. 151. [Homonym, Desfontaines, 1827.]
Miocene: *Florissant, Colo.

Alnus pseudo-glutinosa Göppert, Alaska, 1861 = *Corylus macquarrii*.

Alnus rhamnifolia Nuttall. Hannibal, Bull. Torr. Bot. Club, vol. 38, 1911, p. 336, pl. xv, f. 6.
Pliocene: Near Portiola and Calabasas Canyon, Santa Cruz Mountains, and near Hollister, Gavilan Range, Calif.

Alnus rubra Bongard. Knowlton, Jour. Geol., vol. 3, 1895, p. 532; Geol. Soc. Am., Bull., vol. 5, 1893, p. 583; U. S. Nat. Mus., Proc., vol. 17, 1894, p. 220.
Pleistocene?: Muir Glacier, Alaska.

Alnus rugosa (Du Roi) K. Koch. Hollick, Md. Geol. Surv., Pliocene and Pleistocene, 1906, p. 225, pl. lxix, f. 1-3.
Pleistocene (Talbot): Drum Point, Calvert County, Md.

Alnus rugosa Lesquereux = *Alnus auraria.*

Alnus serrata Newberry, N. Y. Lyc. Nat. Hist., Ann., vol. 9, 1868, p. 55.—[Lesquereux], U. S. Geol. and Geog. Surv. Terr., Ill. Cret. and Tert. Fl., 1878, pl. xvi, f. 11.—Newberry, U. S. Geol. Surv., Mon. 35, 1898, p. 66, pl. xxxiii, f. 11.
Fort Union: Banks of Yellowstone River, Mont.

Alnus serrulata fossilis Newberry, U. S. Geol. Surv., Mon. 35, 1898, p. 66, pl. xlvi, f. 6.—Knowlton, U. S. Geol. Surv., Bull. 204, 1902, p. 42.—Penhallow, Rept. Tert. Pl. Brit. Col., 1908, p. 37.
Clarno (upper part): Bridge Creek and Trout Creek, Harney County, Oreg.
Oligocene: Quilchena, British Columbia.

Alnus truncata Lesquereux, Rept. U. S. Geol. Surv., vol. 8 (Cret. and Tert. Fl.), 1883, explanation of pl. xxviii, f. 7, 8; error for *Betula truncata.*

Alnus sp., Grewingk, Verhandl. d. Russ.-k. mineralog. Gesell. St. Petersb., 1848–49 [1850], pp. 114, 364.
Kenai: Tschugatsk (Kenai) Peninsula, Alaska.

Alnus sp., Knowlton, in Turner, Jour. Geol., vol. 6, 1898, p. 498.
San Pablo: Kirker Pass, north of Mount Diablo, Contra Costa County, Calif.

Alnus sp., Knowlton, in Smith, G. O., U. S. Geol. Surv., Folio 86, Ellensburg, Wash., 1903, p. 3.
Ellensburg: Ellensburg, Wash.

Alnus sp., Knowlton, Harriman Alaska Exped., vol. 4, 1904, p. 155, pl. xxviii, f. 2; pl. xxxiii, f. 4.
Kenai: Kukak Bay, Alaska.

Alnus, sp., Knowlton, U. S. Nat. Mus., Proc., vol. 51, 1916, p. 264, pl. xvi, f. 2.
Miocene: *Florissant, Colo.

Alnus sp.? fruit of, Newberry, U. S. Geol. Surv., Mon. 35, 1898, p. 67, pl. xlvi, f. 7.—Knowlton, U. S. Geol. Surv., Bull. 204, 1902, p. 42.
Clarno (upper part): Bridge Creek, Oreg.

Alnus sp., Penhallow, Roy. Soc. Canada, Trans., 2d ser., vol. 2, 1896, sec. 4, p. 70.
Pleistocene: Scarboro Heights, Ontario.

Alnus sp., Penhallow, Brit. Assoc. Adv. Sci., Bradford meeting, 1900, p. 335.
Pleistocene: Greens Wharf, Ottawa River, Canada.

Alnus sp., Penhallow, Rept. Tert. Pl. Brit. Col., 1908, p. 37.
Oligocene: Quilchena, British Columbia.

AMELANCHIER Medicus, Phil. Bot., vol. 1, 1789, p. 155.

Amelanchier alnifolia Nuttall. Hannibal, Bull. Torr. Bot. Club, vol. 38, 1911, p. 337.
Pliocene: Calabasas Canyon, Santa Cruz Mountains, Calif.

Amelanchier canadensis rotundifolia (Michaux) Torrey and Gray = *Amelanchier rotundifolia.*

Amelanchier obovata Knowlton, U. S. Geol. Surv., Prof. Paper 101, 1917, p. 269, pl. xlii, f. 4.
Vermejo: Coal Creek, near Rockvale, Colo.

Amelanchier peritula Cockerell, Am. Mus. Nat. Hist., Bull., vol. 24, 1908, p. 95, pl. vi, f. 6.—Knowlton, U. S. Nat. Mus., Proc., vol. 51, 1916, p. 273.
Miocene: *Florissant, Colo.

Amelanchier rotundifolia (Michaux) Roemer.
Amelanchier canadensis (Linné) Medicus, var. *rotundifolia* (Michaux) Torrey and Gray. Hollick, Bull. Torr. Bot. Club, vol. 19, 1892, p. 331.
Pleistocene: Bridgeton, N. J.

Amelanchier scudderi Cockerell, Bull. Torr. Bot. Club., vol. 33, 1906, p. 310, text f. 4.
Miocene: Florissant, Colo.

Amelanchier similis Newberry, N. Y. Lyc. Nat. Hist., Ann., vol. 9, 1868, p. 48.—[Lesquereux], U. S. Geol. and Geog. Surv. Terr. (Ill. Cret. and Tert. Fl.), 1878, pl. xxv, f. 6.—Newberry, U. S. Geol. Surv., Mon. 35, 1898, p. 111, pl. xl, f. 6.
Fort Union: Yellowstone River, Mont.

Amelanchier typica Lesquereux, Rept. U. S. Geol. Surv. Terr., vol. 8 (Cret. and Tert. Fl.), 1883, p. 198, pl. xl, f. 11.—Cockerell, Am. Nat., vol. 42, 1908, p. 580, text f. 11.
Miocene: Florissant, Colo.

Amelanchier typica var. Lesquereux, Mus. Comp. Zool., Bull., vol. 16, 1888, p. 57.
Denver: Golden, Colo.

Amelanchier whitei Hollick, U. S. Geol. Surv., Mon. 50, 1906, p. 83, pl. xxxii, f. 1.
Magothy: Gay Head, Marthas Vineyard, Mass.

AMPELOPHYLLITES Knowlton, n. name.

Ampelophyllum Lesquereux, U. S. Geol. and Geog. Surv. Terr., Bull., vol. 1, 1875 [1876], p. 395. [Not *Massalongo*, Syllabus Pl. Foss., 1859, p. 81.]

Ampelophyllites attenuatus (Lesquereux) Knowlton, n. comb.

Ampelophyllum attenuatum Lesquereux, U. S. Geol. and Geog. Surv. Terr., Bull., vol. 1, 1875 [1876], p. 396; U. S. Geol. and Geog. Surv. Terr., Ann. Rept. 1874 [1876], p. 354, pl. ii, f. 3.

Dakota: Fort Harker, Kans.

Ampelophyllites firmus (Lesquereux) Knowlton, n. comb.

Ampelophyllum firmum Lesquereux, U. S. Geol. and Geog. Surv. Terr., Bull., vol. 1, 1875 [1876], p. 396.

Dakota: Kansas.

Ampelophyllites ovatus (Lesquereux) Knowlton, n. comb.

Ampelophyllum ovatum (Lesquereux) Lesquereux, U. S. Geol. and Geog. Surv. Terr., Ann. Rept. 1874 [1876], p. 355.

Populites ovata Lesquereux, Am. Jour. Sci., 2d ser., vol. 46, 1868, p. 94.

Celtis? ovata Lesquereux, Rept. U. S. Geol. Surv. Terr., vol. 6 (Cret. Fl.), 1874, p. 66, pl. iv, f. 2, 3.

Dakota: Decatur, Nebr.

Ampelophyllum attenuatum Lesquereux = *Ampelophyllites attenuatus.*

Ampelophyllum firmum Lesquereux = *Ampelophyllites firmus.*

Ampelophyllum ovatum (Lesquereux) Lesquereux = *Ampelophyllites ovatus.*

AMPELOPSIS Michaux, Fl. Bor. Am., vol. 1, 1803, p. 159.

Ampelopsis bruneri (Ward) Cockerell, Am. Mus. Nat. Hist., Bull., vol. 24, 1908, p. 103.

Vitis bruneri Ward, U. S. Geol. Surv., Sixth Ann. Rept. 1884–85 [1886], p. 554, pl. li, f. 4, 5; U. S. Geol. Surv., Bull. 37, p. 69, pl. xxxii, f. 1, 2.

Hanna: Carbon, Wyo.

Ampelopsis bruneri carbonensis (Ward) Cockerell, Am. Mus. Nat. Hist., Bull., vol. 24, 1908, p. 103.

Vitis carbonensis Ward, U. S. Geol. Surv., Sixth Ann. Rept., 1884–85 [1886], p. 554, pl. li, f. 6; U. S. Geol. Surv., Bull. 37, 1887, p. 70, pl. xxxii, f. 3.

Hanna: Carbon, Wyo.

Ampelopsis montanensis Cockerell, Am. Mus. Nat. Hist., Bull., vol. 24, 1908, p. 103.

Vitis cuspidata Ward, U. S. Geol. Surv., Sixth Ann. Rept., 1884–85 [1886], p. 554, pl. li, f. 9–11; U. S. Geol. Surv., Bull. 37, 1887, p. 71, pl. xxxii, f. 6–8.

Fort Union: Burns Ranch and Sevenmile Creek, near Glendive, Mont.

Ampelopsis quinquefolia Michaux. Meehan, in Mercer, Acad. Nat. Sci. Philadelphia, Jour., 2d ser., vol. 11, 1899, pp. 277, 281.

Pleistocene: Port Kennedy, Pa.

Ampelopsis tertiaria Lesquereux, U. S. Geol. and Geog. Surv. Terr., Ann. Rept. 1871 [1872], Suppl., p. 7; Rept. U. S. Geol. Surv. Terr., vol. 7 (Tert. Fl.), 1878, p. 242, pl. xliii, f. 1.

Green River: Green River station, Wyo.

Ampelopsis xantholithensis (Ward) Cockerell, Am. Mus. Nat. Hist., Bull., vol. 24, 1908, p. 103.

Vitis xantholithensis Ward, U. S. Geol. Surv., Sixth Ann. Rept., 1884-85 [1886], p. 554, pl. li, f. 7, 8; U. S. Geol. Surv., Bull. 37, 1887, p. 71, pl. xxxii, f. 4, 5.—Knowlton, Washington Acad. Sci., Proc., vol. 11, 1909, p. 194.

Fort Union: Burns ranch, near Glendive, Mont.

Lance: Near Melville, Mont.

AMYGDALUS Linné, Sp. pl., 1753, p. 472.

Amygdalus gracilis Lesquereux, Rept. U. S. Geol. Surv. Terr., vol. 8 (Cret. and Tert. Fl.), 1883, p. 199, pl. xl, f. 12-15; pl. xliv, f. 6.—Penhallow, Rept. Tert. Pl. Brit. Col., 1908, p. 37.

Miocene: Florissant, Colo.

Green River: Uinta County (formerly incorrectly called Randolph County), Wyo.

Eocene: Quilchena, British Columbia.

ANABACAULUS Emmons, American geology, pt. 2, 1857, p. 25.

Anabacaulus duplicatus Emmons, American geology, pt. 6, 1857, p. 26, f. 6.—Ward, U. S. Geol. Surv., Twentieth Ann. Rept., pt. 2, 1900, p. 423.

Triassic (Newark): Chatham County, N. C.

Anabacaulus sulcatus Emmons, American geology, pt. 6, 1857, p. 26, f. 7.—Ward, U. S. Geol. Surv., Twentieth Ann. Rept., pt. 2, 1900, p. 423.

Triassic (Newark): Egypt, N. C.

ANACARDIOXYLON Felix, Stud. Foss. Hölz., Leipzig, 1882, p. 70.

Anacardioxylon magniporosum Platen, Sitzungsb. Naturf. Ges., Leipzig, vol. 34, 1907 [1908], p. 57.

Miocene (auriferous gravels): Nevada County, Calif.

ANACARDITES Saporta, in Heer, Recherches clim. vég. pays tert., 1861, p. 149.

Anacardites antiquus Lesquereux, U. S. Geol. Surv., Mon. 17, 1891 [1892], p. 156, pl. lvii, f. 1.

Dakota: Near Fort Harker, Kans.

Anacardites falcatus Berry, U. S. Geol. Surv., Prof. Paper 91, 1916, p. 261, pl. lix, f. 6.

Lagrange: Puryear, Henry County, Tenn.

Anacardites grevillaefolia Berry, U. S. Geol. Surv., Prof. Paper 91, 1916, p. 262, pl. lvii, f. 5; pl. lviii, f. 5.

Wilcox (Grenada): Grenada, Grenada County, Miss.

Lagrange: Puryear, Henry County, Tenn.

Anacardites marshallensis Berry, U. S. Geol. Surv., Prof. Paper 91, 1916, p. 261, pl. lviii, f. 6.
Wilcox (Holly Springs): Holly Springs, Marshall County, Miss.

Anacardites metopifolia Berry, U. S. Geol. Surv., Prof. Paper 91, 1916, p. 262, pl. lviii, f. 7.
Wilcox (Holly Springs): Holly Springs and Vaughans, Benton County, Miss.
Lagrange: Wickliffe, Ballard County, Ky., and Puryear, Henry County, Tenn.

Anacardites minor Berry, U. S. Geol. Surv., Prof. Paper 91, 1916, p. 262, pl. lvii, f. 4.
Lagrange: Puryear, Henry County, Tenn.

Anacardites puryearensis Berry, U. S. Geol. Surv., Prof. Paper 91, 1916, p. 261, pl. lvii, f. 6.
Lagrange: Puryear, Henry County, Tenn.

Anacardites serratus Berry, U. S. Geol. Surv., Prof. Paper 91, 1916, p. 263, pl. lviii, f. 8.
Lagrange: Near Grand Junction, Fayette County, Tenn.

ANDROMEDA Linné, Sp. pl., 1753, p. 393.

Andromeda acuminata Lesquereux = *Andromeda dakotana.*

Andromeda affinis Lesquereux, U. S. Geol. and Geog. Surv. Terr., Ann. Rept. 1874 [1876], p. 348; Rept. U. S. Geol. Surv. Terr., vol. 8 (Cret. and Tert. Fl.), 1883, p. 60, pl. ii, f. 5.
Livingston: Bozeman coal field, Mont.
—— Las Animas, Colo.

Andromeda angustifolia Berry, Bull. Torr. Bot. Club, vol. 41, 1914, p. 298 [name].
Magothy: Maryland.

Andromeda cookii Berry, Bull. Torr. Bot. Club, vol. 36, 1909, p. 261; idem, vol. 37, 1910, p. 29; Geol. Surv. N. J., Bull. 3, 1911, p. 206; Md. Geol. Surv., Upper Cret., 1916, p. 887, pl. lxxxix, f. 3.
Andromeda flexuosa Newberry [homonym, *A. flexuosa* Moon, 1849], U. S. Geol. Surv., Mon. 26, 1896, p. 121, pl. xxxiv, f. 1–5.—Hollick, N. Y. Bot. Gard., Bull., vol. 3, 1904, p. 416, pl. lxxix, f. 2; U. S. Geol. Surv., Mon. 50, 1906, p. 101, pl. xxxix, f. 6.
Magothy: Grove Point and Round Bay, Md.; Glen Cove, Long Island, N. Y.
Raritan: Sayreville, Woodbridge, Milltown, N. J.

Andromeda crassa Lesquereux, U. S. Nat. Mus., Proc., vol. 11, 1888, p. 16.—Knowlton, U. S. Geol. Surv., Bull. 204, 1902, p. 83, pl xvi, f. 3.
Mascall: Van Horn's ranch, John Day Basin, Oreg.

Andromeda cretacea Lesquereux, U. S. Geol. Surv., Mon. 17, 1891 [1892], p. 117, pl. xvii, f. 17, 18; pl. xxiv, f. 5.—Berry, Bull. Torr. Bot. Club, vol. 37, 1910, p. 504; U. S. Geol. Surv., Prof. Paper 84, 1914, p. 120, pl. xxiv, f. 2; U. S. Geol. Surv., Prof. Paper 112, 1919, p. 132.

Dakota: *Ellsworth County, Kans.
Eutaw: McBrides Ford, Columbus, Ga.

Andromeda dakotana Knowlton and Cockerell, n. name.

Andromeda acuminata Lesquereux U. S. Geol. and Geog. Surv. Terr., Bull., vol. 1, 1875 [1876], p. 393. [Homonym, Aiton, 18—.]

Dakota: Kansas.

Andromeda delicatula Lesquereux, Rept. U. S. Geol. Surv. Terr., vol. 8 (Cret. and Tert. Fl.), 1883, p. 175, pl. xxxiv, f. 10, 11.—Penhallow, Rept. Tert. Pl. Brit. Col., 1908, p. 38.

Green River: Uinta County (formerly wrongly called Randolph County), Wyo.
Eocene: Tranquille River, British Columbia
Fort Union: Bull Mountains, Mont.

Andromeda delicatula Lesquereux. Hollick, 1899 = *Mespilodaphne coushatta.*

Andromeda dubia Lesquereux = *Diospyros brachysepala.*

Andromeda eolignitica Hollick. Veatch, U. S. Geol. Surv., Prof. Paper 46, 1906, pl. xvii, f. 2 = *Rhamnus coushatta.*

Andromeda eolignitica Hollick = *Mespilodaphne eolignitica.*

Andromeda euphorbiophylloides Berry, U. S. Geol. Surv., Prof. Paper 84, 1914, p. 59, pl. xiv, f. 7.

Black Creek (Middendorf): Middendorf, S. C.

Andromeda flexuosa Newberry = *Andromeda cookii.*

Andromeda grandifolia Berry, Bull. Torr. Bot. Club, vol. 34, 1907, p. 204, pl. 15, f. 3; idem, vol. 37, 1910, p. 28; Geol. Surv. N. J., Bull. 3, 1911, p. 205, U. S. Geol. Surv., Prof. Paper 84, 1914, p. 59, pl. xiv, f. 10; Md. Geol. Surv. Upper Cret., 1916, p. 889; U. S. Geol. Surv., Prof. Paper 112, 1919, p. 131, pl. xxvii, f. 7.

Andromeda latifolia Newberry, U. S. Geol. Surv., Mon. 26, 1895 [1896], p. 120, pl. xxxiii, f. 6–8, 10; pl. xxxvi, f. 10.—Hollick, N. Y. Bot. Gard., Bull., vol. 3, 1904, p. 416, pl. 79, f. 3; U. S. Geol. Surv., Mon. 50, 1906, p. 100, pl. xxxix, f. 1.

Magothy: Grove Point, Md.
Raritan: Oak Neck, Long Island, N. Y.; Sayreville, Woodbridge, and Milltown, N. J.
Black Creek: Court House Bluff, Cape Fear River, N. C.; Rocky Point, S. C.
Tuscaloosa: Shirley Mill, Glen Allen, etc., Ala.

Andromeda grayana Heer, Neue Denkschr., Mém. 3, vol. 21, 1865, p. 7, pl. i, f. 7–9; Fl. foss. arct., vol. 2, Abt. 2, 1869, p. 34, pl. viii, f. 5.—Lesquereux, U. S. Geol. and Geog. Surv. Terr., Ann. Rept. 1871 [1872], p. 271; idem, Suppl., p. 14; Rept. U. S. Geol. Surv. Terr., vol. 7 [Tert. Fl.], 1878, p. 234, pl. xl, f. 4.—Knowlton, U. S. Nat. Mus., Proc., vol. 17, 1894, p. 225; Geol. Soc. Am., Bull., vol. 5, 1893, p. 584; Harriman Alaska Exped., vol. 4, 1904, p. 157; U. S. Geol. Surv., Bull. 105, 1893, p. 56; U. S. Geol. Surv., Mon. 32, pt. 2, 1899, p. 661.

Kenai: Port Graham, and Kukak Bay, Alaska.

—— Burrard Inlet, British Columbia.

Livingston: Mount Everts, Yellowstone National Park; North Fork Mission Creek, and Livingston area, Mont.

Andromeda? lanceolata Knowlton, U. S. Geol. Surv., Prof. Paper 101, 1917, p. 344, pl. cix, f. 3; pl. cx, f. 1.

Raton: Turner mine, Wootton, Colo.

Andromeda latifolia Newberry = ***Andromeda grandifolia***.

Andromeda latifolia Newberry, U. S. Geol. Surv., Mon. 26, 1902, pl. xxxiii, f. 9 = *Persoonia lesquereuxii*.

Andromeda linearifolia Lesquereux, U. S. Nat. Mus., Proc., vol. 10, 1887, p. 42, pl. iii, f. 2, 3.

Denver: Silver Cliff, Colo.

Andromeda linifolia Lesquereux, U. S. Geol. Surv., Mon. 17, 1891 [1892], p. 118, pl. lii, f. 5.

Dakota: Ellsworth County, Kans.; 2 miles south of Green River, Utah.

Andromeda novaecaesareae Hollick, in Newberry, U. S. Geol. Surv., Mon. 26, 1895 [1896], p. 121, pl. xlii, f. 9–12, 28–31.—Berry, Bull. Torr. Bot. Club, vol. 33, 1906, p. 181; idem, vol. 34, 1907, p. 204; idem, vol. 37, 1910, pp. 29, 505; idem, vol. 39, 1912, p. 405; Geol. Surv. N. J., Bull. 3, 1911, p. 204; U. S. Geol. Surv., Prof. Paper 84, 1914, pp. 58, 120, pl. xiv, f. 5, 6; pl. xxiv, f. 1; Md. Geol. Surv., Upper Cret., 1916, p. 885, pl. lxxxix, f. 1, 2; Bull. Torr. Bot. Club, vol. 43, 1916, p. 301; U. S. Geol. Surv., Prof. Paper 112, 1919, p. 130, pl. xxx, f. 1, 2.

Raritan: South Amboy, N. J.

Magothy: New Jersey; Grove Point and Round Bay, Md.

Black Creek: Blackmans Bluff, Neuse River, N. C.

Black Creek (Middendorf): Middendorf, S. C.

Ripley (Cusseta): Buena Vista, Ga.

Ripley: Coffee Bluff, Hardin County, Tenn.

Woodbine, Arthurs Bluff, Tex.

Tuscaloosa: Shirleys Mill and Glen Allen, Ala.

Bingen: Mine Creek, near Nashville, Ark.

Andromeda parlatorii Heer, Phyll. crét. d. Nebr., 1866, p. 18, pl. i, f. 5.—Lesquereux, U. S. Geol. Surv. Terr., Rept., vol. 6 (Cret. Fl.), 1874, p. 88, pl. xxiii, f. 6, 7; pl. xxviii, f. 15; U. S. Geol. Surv., Mon. 17, 1891 [1892], p. 115, pl. xix, f. 1; pl. lii, f. 6.—Hollick, N. Y. Acad. Sci., Annals, vol. 11, 1898, p. 420, pl. xxxvii, f. 7.—Berry, N. Y. Bot. Gard., Bull., vol. 3, 1903, p. 97, pl. l, f. 1–4, Bull. Torr. Bot. Club, vol. 31, 1904, p. 79, pl. i, f. 1, 2; idem, vol. 33, 1906, p. 181; idem, vol. 34, 1907, p. 203, pl. xv, f. 2.—Hollick, U. S. Geol. Surv., Mon. 50, 1906, p. 101, pl. xxxix, f. 2–5.—Berry, Bull. Torr. Bot. Club, vol. 37, 1910, p. 29, Geol. Surv. N. J., Bull. 3, 1911, p. 206; U. S. Geol. Surv., Prof. Paper 84, 1914, p. 60; Md. Geol. Surv. Upper Cret., 1916, p. 888, pl. lxxxix, f. 4; U. S. Geol. Surv., Prof. Paper 112, 1919, p. 131; Bull. Torr. Bot. Club, vol. 44, 1917, p. 187.

Prunus? parlatorii Lesquereux, Am. Jour. Sci., 2d ser., vol. 46, 1868, p. 102.

Dakota: Kansas; Nebraska; New Olm, Minn.

Dakota (?): Upper Kanab, Utah.

Raritan: Tottenville, Staten Island, N. Y.; Milltown and Hylton pits, N. J.

Magothy: Cliffwood, N. J.; Deep Cut, Del.; Grove Point, Md.; Nashaquitsa and Gay Head, Marthas Vineyard, Mass.; Glen Cove, Long Island, N. Y.

Black Creek: Court House Bluff, Cape Fear River, N. C.; Langley and Rocky Point, S. C.

Tuscaloosa: Shirleys Mill, Glen Allen, and Cottondale, Ala.

Eutaw: Near Havana, Ala.

Bingen: Mine Creek near Nashville, Ark.

Andromeda parlatorii longifolia Lesquereux, U. S. Geol. Surv., Mon. 17, 1891 [1892], p. 116, pl. lxiv, f. 19.

Dakota: Ellsworth County, Kans.

Andromeda pfaffiana Heer, Fl. foss. arct., vol. 6, Abt. 2, 1882, p. 79, pl. xxv, f. 6; pl. xxxviii, f. 5–7; pl. xliv, f. 12.—Lesquereux, U. S. Geol. Surv., Mon. 17, 1891 [1892], p. 116, pl. xviii, f. 7, 8; pl. lii, f. 7.—Knowlton, in Woodruff, U. S. Geol. Surv., Bull. 452, 1911, p. 20.

Dakota: Near Delphos, Kans.; near Lander, Wyo.

Andromeda protogaea Unger, Denkschr. Wien. Akad., vol. 2, 1850, p. 173 [43], pl. xliv [xxiii], f. 1–9.—Lesquereux, U. S. Nat. Mus., Proc., vol. 11, 1888, p. 20.

Eocene?: Cherry Creek, Oreg.

Andromeda reticulata? Ettingshausen. Lesquereux, U. S. Geol. and Geog. Surv. Terr., Ann. Rept. 1871 [1872], p. 298. [Not afterward recognized by Lesquereux. *A. reticulata* Ettingshausen is a homonym.]

Livingston: 6 miles above Spring Canyon, Mont.

Andromeda rhomboidalis Lesquereux = *Cotinus fraterna*.

Andromeda scripta Knowlton, U. S. Geol. Surv., Prof. Paper 101, 1917, p. 344, pl. cxii, f. 1, 2.
Raton: Rugby, Wichita, and near Walsenburg, Colo.

Andromeda scudderiana Cockerell = *Cotinus fraterna*.

Andromeda snowii Lesquereux, U. S. Geol. Surv., Mon. 17, 1891 [1892], p. 117, pl. xvii, f. 16.—Berry, Bull. Torr. Bot. Club, vol. 39, 1912, p. 405.
Dakota: Ellsworth County, Kans.
Woodbine: Arthurs Bluff, Tex.

Andromeda tenuinervis Lesquereux, U. S. Geol. Surv., Mon. 17, 1891 [1892], p. 116, pl. xxxviii, f. 7.—Hollick, U. S. Geol. Surv., Mon. 50, 1906, p. 102, pl. xxxix, f. 7.
Rhamnus pfaffiana Heer. Hollick, N. Y. Acad. Sci., Trans., vol. 11, 1892, p. 103, pl. iv, f. 2.
Dakota: Ellsworth County, Kans.
Raritan: Tottenville, Staten Island, N. Y.

Andromeda vaccinifolia affinis Lesquereux = *Cassia glenni*.

Andromeda wardiana Lesquereux, U. S. Geol. Surv., Mon. 17, 1891 [1892], p. 119, pl. lxiv, f. 17.—Berry, Bull. Torr. Bot. Club, vol. 37, 1910, p. 504; U. S. Geol. Surv., Prof. Paper 84, 1914, p. 120, pl. xxiv, f. 3; Bull. Torr. Bot. Club, vol. 43, 1916, p. 302; U. S. Geol. Surv., Prof. Paper 112, 1919, p. 132, pl. xxvii, f. 6.
Dakota: Ellsworth County, Kans.
Tuscaloosa: Near Iuka, Miss.; Shirleys Mill, Ala.
Eutaw: McBrides Ford, Columbus, Ga.; Coffee Bluff, Hardin County, Tenn.

Andromeda sp., Lesquereux, U. S. Geol. and Geog. Surv. Terr., Ann. Rept. 1871 [1872], Suppl., p. 9. [Not afterward mentioned.]
Mesaverde: Point of Rocks, Wyo.

Andromeda sp., Lesquereux, Rept. on clays in N. J., 1878, p. 28.
Raritan?: Washington, N. J.

ANDROVETTIA Hollick and Jeffrey, N. Y. Bot. Gard., Mem., vol. 3, 1909, p. 22. [Type, *A. statenensis*.]

Androvettia carolinensis Berry, Bull. Torr. Bot. Club, vol. 37, 1910, p. 183, pl. xix, f. 1–6; U. S. Geol. Surv., Prof. Paper 112, 1919, p. 62, pl. vii, f. 1–10.
Black Creek: 3 miles below Dunbars Bridge, Tar River, N. C.
Tuscaloosa: Near Iuka, Miss.

Androvettia elegans Berry, U. S. Geol. Surv., Prof. Paper 84, 1914, p. 103, pl. xviii, f. 1–10; idem, Prof. Paper 112, 1919, p. 64.
Eutaw: McBrides Ford, Chattahoochee River, Ga.

Androvettia statenensis Hollick and Jeffrey, N. Y. Bot. Gard., Mem., vol. 3, 1909, p. 22, pl. iii, f. 1–5; pl. vii, f. 1–8; pl. viii, f. 1, 2; pl. xxviii, f. 5–8; pl. xxix, f. 1–6.
Raritan: Kreischerville, Staten Island, N. Y.

Androvettia sp., Berry, Bull. Torr. Bot. Club, vol. 37, 1910, p. 504.
Tuscaloosa: McBrides Ford, near Columbus, Ga.

ANEMIA Swartz, Syn. Fil., 1806, p. 155.

Anemia elongata (Newberry) Knowlton, U. S. Geol. Surv., Prof. Paper 130, 19—, p. —, pl. ii, f. 2 [in preparation].

Sphenopteris (*Asplenium*) *elongatum* Newberry, Boston Jour. Nat. Hist., vol. 7, 1863, p. 511.

Anemia subcretacea (Saporta) Gardner and Ettingshausen, Monog. British Eocene Fl., vol. 1, pt. ii, 1880, p. 45, pl. viii; pl. ix.—Knowlton, U. S. Geol. Surv., Bull. 152, 1898, p. 34; U. S. Geol. Surv., Mon. 32, pt. 2, 1899, p. 657.

Anemia perplexa Hollick, in Newberry, U. S. Geol. Surv., Mon. 35, 1898, p. 3, pl. xv, f. 1, 1a; pl. xvi, f. 3; pl. lxiii, f. 1–4.

Gymnogramma haydenii Lesquereux, U. S. Geol. and Geog. Surv. Terr., Ann. Rept. 1871 [1872], p. 295; Rept. U. S. Geol. Surv. Terr., vol. 7 (Tert. Fl.), 1878, p. 59, pl. v, f. 1–3.

Anemia haydenii (Lesquereux) Cockerell, Torreya, vol. 9, 1909, p. 142.

Laramie: Erie, Colo. (?)
"Laramie": ? San Juan Basin, N. Mex.; near Thompsons, Utah.
Livingston: ? Mount Everts, Yellowstone National Park; Livingston, Mont.
Mesaverde: ? Point of Rocks, Wyo.

Anemia eocenica Berry, U. S. Geol. Surv., Prof. Paper 91, 1916, p. 164, pl. ix, f. 7; pl. x, f. 2; pl. xi, f. 1, 2.
Lagrange: Puryear, Henry County, Tenn.
Wilcox: Near Boydsville, Clay County, Ark., and Mansfield, La.

Anemia fremonti Knowlton, U. S. Geol. Surv., Prof. Paper 108, 1917, pp. 84–85, pl. xxxi, f. 6; pl. xxxii, f. 1–3.
Frontier: About 1 mile east of Cumberland, Wyo.

Anemia? gracillima (Lesquereux) Cockerell, Am. Mus. Nat. Hist., Bull., vol. 24, 1908, p. 76.

Adiantites gracillimus Lesquereux, Rept. U. S. Geol. Surv. Terr., vol. 8 (Cret. and Tert. Fl.), 1883, p. 137, pl. xxi, f. 8.

Miocene: Florissant, Colo.

Anemia haydenii (Lesquereux) Cockerell = *Anemia elongata*.

Anemia hesperia Knowlton, U. S. Geol. Surv., Prof. Paper 98, 1916, p. 332, pl. lxxxiv, f. 3.
Fruitland: 10 miles south of Jewett, San Juan County, N. Mex.

Anemia occidentalis Knowlton, U. S. Geol. Surv., Prof. Paper 101, 1917, p. 285, pl. liv, f. 2.
Raton: Near Trinidad, Riley Canyon, and Cokedale, Colo.; Yankee, N. Mex.

Anemia perplexa Hollick, in Newberry = ***Anemia elongata.***

Anemia robusta Hollick, Torreya, vol. 2, 1902, p. 145, pl. iii, f. 1.—Knowlton, U. S. Geol. Surv., Prof. Paper 101, 1917, p. 248.
Vermejo: Florence, Colo.

Anemia stricta Newberry = ***Dicksonia groenlandica.***

Anemia subcretacea (Saporta) Gardner and Ettingshausen (of American authors) = ***Anemia elongata.***

Anemia supercretacea Hollick, Torreya, vol. 2, 1902, p. 145, pl. 3, f. 6, 7.—Cockerell, Torreya, vol. 9, 1909, p. 142.—Knowlton, U. S. Geol. Surv., Prof. Paper 101, 1917, p. 248.
Vermejo: *Florence, Colo.; ? near Gallup, N. Mex.
Laramie: Marshall, Colo.

Anemia sp., Knowlton, Wash. Geol. Surv., Ann. Rept., vol. 1, 1902, p. 33.
Eocene: Skykomish, Snohomish County, Wash.

Anemia sp., Knowlton, U. S. Geol. Surv., Prof. Paper 98, 1916, p. 87, pl. xv, f. 6, 7.
Fox Hills: Wildcat Mound, near Milliken, Colo.

Anemia sp., Knowlton, U. S. Geol. Surv., Prof. Paper 98, 1916, p. 333, pl. lxxxiv, f. 4.
Fruitland: 30 miles south of Farmington, San Juan County, N. Mex.

Anemia sp., Knowlton, U. S. Geol. Surv., Prof. Paper 130, 19—, p. —, pl. ii, f. 1 [in preparation].
Laramie: 4 miles north of Colorado Springs, Colo.

ANGIOPTERIDIUM Schimper, Pal. vég., vol. 1, 1869, p. 602.

Angiopteridium auriculatum Fontaine = ***Taeniopteris auriculata.***

Angiopteridium californicum Fontaine, Am. Jour. Sci., 4th ser., vol. 2, 1896, p. 274; in Ward, U. S. Geol. Surv., Twentieth Ann. Rept., pt. ii, 1900, p. 351, pl. lv, f. 2–5.
Jurassic: Oroville, Calif.

Angiopteridium canmorense Dawson, Roy. Soc. Canada, Trans., vol. 10, 1892, sec. 4, p. 83, f. 2.—Fontaine, in Diller and Stanton, Geol. Soc. Am., Bull., vol. 5, 1894, p. 450; in Stanton, U. S. Geol. Surv., Bull. 133, 1895 [1896], p. 15; in Ward, U. S. Geol. Surv., Mon. 48, 1905 [1906], p. 239, pl. lxvi, f. 1–4.

Angiopteridium nervosum Fontaine. Fontaine, in Diller and Stanton, op. cit., p. 450; in Stanton, op. cit., p. 15.

Horsetown: ? Elder Creek, Tehama County, Calif.
Knoxville: ? Tehama County, Calif.
Kootenai: Canmore, British Columbia.

Angiopteridium densinerve Fontaine = ***Taeniopteris nervosa.***

Angiopteridium dentatum Fontaine, U. S. Geol. Surv., Mon. 15, 1889, p. 117, pl. xxx, f. 6, 7.
Patuxent: Fredericksburg, Va.

Angiopteridium ellipticum Fontaine, U. S. Geol. Surv., Mon. 15, 1889, p. 114, pl. xxix, f. 3.
Patuxent: Fredericksburg, Va.

Angiopteridium nervosum Fontaine = *Taeniopteris nervosa.*

Angiopteridium nervosum Fontaine. Fontaine, in Diller and Stanton, Geol. Soc. Am., Bull., vol. 5, 1894, p. 450; in Stanton, U. S. Geol. Surv., Bull. 133, 1895 [1896], p. 15 = *Angiopteridium canmorense.*

Angiopteridium nervosum Fontaine. Fontaine, in Stanton, U. S. Geol. Surv., Bull. 133, 1895 [1896], p. 22 = *Taeniopteris nervosa.*

Angiopteridium oregonense Fontaine = *Sagenopteris oregonensis.*

Angiopteridium ovatum Fontaine, U. S. Geol. Surv., Mon. 15, 1889, p. 115, pl. xxix, f. 6, 7.
Patuxent: Fredericksburg, Va.

Angiopteridium pachyphyllum Fontaine = *Taeniopteris nervosum.*

Angiopteridium strictinerve Fontaine = [in part] *Taeniopteris nervosa.*

Angiopteridium strictinerve Fontaine. Fontaine, in Diller and Stanton, also in Stanton = *Nageiopsis longifolia.*

Angiopteridium strictinerve Fontaine [in part] = *Nilsonia oregonensis.*

Angiopteridium strictinerve Fontaine [in part] = *Taeniopteris shastensis.*

Angiopteridium strictinerve latifolium Fontaine = *Taeniopteris nervosa.*

Angiosperm, ament of ?, Fontaine, U. S. Geol. Surv., Mon. 15, 1889, p. 272, pl. cxxxv, f. 16.
Patuxent: Near Dutch Gap Canal, Va.

ANISOPHYLLUM Lesquereux, Rept. U. S. Geol. Surv. Terr., vol. 6 (Cret. Fl.), 1874, p. 98. [Type, *Quercus semialatus.*]

Anisophyllum semialatum (Lesquereux) Lesquereux, Rept. U. S. Geol. Surv. Terr., vol. 6 (Cret. Fl.), 1874, p. 98, pl. vi, f. 1–5.
Quercus semialatus Lesquereux, Am. Jour. Sci., 2d ser., vol. 46, 1868, p. 96.
Dakota: Beatrice, Nebr.

Anisophyllum sp., Dawson, Roy. Soc. Canada, Trans., vol. 1, sec. 4, 1882 [1883], p. 28, pl. viii, f. 34.
Upper Cretaceous: Baynes Sound, British Columbia.

ANOMASPIS Hollick and Jeffrey, N. Y. Bot. Gard., Mem., vol. 3, 1909, p. 49. [Type, *A. tuberculata.*]

Anomaspis hispida Hollick and Jeffrey, N. Y. Bot. Gard., Mem., vol. 3, 1909, p. 50, pl. x, f. 4, 8, 9.
Raritan: Kreischerville, Staten Island, N. Y.

Anomaspis tuberculata Hollick and Jeffrey, N. Y. Bot. Gard., Mem., vol. 3, 1909, p. 49, pl. x, f. 5, 6; pl. xxv, f. 5; pl. xxvi, f. 1.
Raritan: Kreischerville, Staten Island, N. Y.

ANOMALOFILICITES Hollick, N. Y. Bot. Gard., Mem., vol. 6, 1916, p. 474. [Type, *A. monstrosus*.]

Anomalofilicites monstrosus Hollick, N. Y. Bot. Gard., Mem., vol. 6, 1916, p. 474, pl. xxxi, f. 1–3.
Fort Union: Dawson County, Mont.

ANOMOPHYLLITES Watelet, Descr. pl. foss. d. bassin d. Paris, 1866, p. 99.

Anomophyllites bridgetonensis Hollick, Bull. Torr. Bot. Club, vol. 24, 1897, p. 329, pls. 311–313.
Pleistocene: Bridgeton, N. J.

ANOMOZAMITES Schimper, Traité de paléontologie végétale, vol. 2, 1870, p. 140.

Anomozamites acutilobus? Heer, Fl. foss. arct., vol. 4, 1876, Abt. 2, p. 102, pl. xxiii, f. 1a; pl. xxiv, f. 1–3; pl. xxv, f. 9; pl. xxvii, f. 3b.—Dawson, Roy. Soc. Canada, Trans., vol. 3, sec. 4, 1885 [1886], p. 7, pl. i, f. 7.
Kootenai: Near Canmore, British Columbia.

Anomozamites angustifolius Fontaine = *Taeniopteris nervosa*.

Anomozamites? egyptiacus Fontaine, in Ward, U. S. Geol. Surv., Twentieth Ann. Rept., pt. 2, 1900, p. 290, pl. xxxix, f. 5.
Triassic (Newark): North Carolina.

Anomozamites princeps (Oldham and Morris) Schimper? Traité de paléontologie végétale, vol. 2, 1870, p. 142.—Fontaine, in Ward, U. S. Geol. Surv., Twentieth Ann. Rept., pt. 2, 1900, p. 242, pl. xxvi, f. 1.
Pterophyllum princeps Oldham and Morris, Mem. Geol. Surv. India, Pal. Indica, vol. 1, 1862, p. 23.
Triassic (Newark): York County, Pa.

Anomozamites schmidtii Heer, Fl. foss. arct., vol. 4, pt. 2, 1876, p. 100, pl. xxiii, f. 2, 3; pl. xxiv, f. 4–7.—Knowlton, in Atwood, U. S. Geol. Surv., Bull. 467, 1911, p. 45.
Chignik (Upper Cretaceous): Chignik Lagoon, Alaska.

Anomozamites virginicus Fontaine = *Taeniopteris nervosa*.

Anomozamites sp., Dawson, Roy. Soc. Canada, Trans., vol. 10, sec. 4, 1892 [1893], p. 91.
Kootenai: Martin Brook, British Columbia.

ANONA Linné, Sp. pl., 1753, p. 536.

Anona ampla Berry, U. S. Geol. Surv., Prof. Paper 91, 1916, p. 217, pl. xxxix, f. 1; pl. xl, f. 1; pl. xli, f. 3.
Lagrange: Puryear, Henry County, Tenn.
Wilcox: Hardys Mill, Greene County, Ark.; Naborton, De Soto Parish, La.

Anona coloradensis Knowlton, U. S. Geol. Surv., Prof. Paper 130, 19—, p. —, pl. xviii, f. 4 [in preparation].
Laramie: Cowan station, near Denver, Colo.

Anona cretacea Lesquereux, Rept. U. S. Geol. Surv. Terr., vol. 8 (Cret. and Tert. Fl.), 1883, p. 77.
Dakota: Glasco, Kans.

Anona eolignitica Berry, U. S. Geol. Surv., Prof. Paper 91, 1916, p. 217, pl. xlii, f. 2–4.
Lagrange: Puryear, Henry County, Tenn.
Wilcox: Near Naborton, De Soto Parish, La.

Anona glabra Linné. Berry, Jour. Geol., vol. 25, 1917, p. 662.
Pleistocene: Vero, Fla.

Anona robusta Lesquereux, Rept. U. S. Geol. Surv. Terr., vol. 8 (Cret. and Tert. Fl.), 1883, p. 124, pl. xx, f. 4.—Knowlton, U. S. Geol. Surv., Prof. Paper 130, 19—, p. —, pl. xvii, f. 7 [in preparation].
Laramie: Golden, Colo.

Anona spoliata Cockerell, Am. Jour. Sci., 4th ser., vol. 26, 1908, p. 542, f. 7.—Knowlton, U. S. Nat. Mus., Proc., vol. 51, 1916, p. 271.
Miocene: *Florissant, Colo.

Anona wilcoxiana Berry, U. S. Geol. Surv., Prof. Paper 91, 1916, p. 216, pl. xli, f. 1, 2.
Lagrange: Puryear, Henry County, and near Trenton, Gibson County, Tenn.
Wilcox: Near Naborton, De Soto Parish, La.

ANTHOLITHES Brongniart, Mém. Mus. hist. nat., Paris, vol. 8 (Classif. vég. foss.), 1822, p. 320.

Antholithes amissus Heer, Fl. foss. arct., vol. 1, 1868, p. 139, pl. xxiii, f. 12.—Dawson, Roy. Soc. Canada, Trans., vol. 7, 1889, p. 69.—Penhallow, Rept. Tert. Pl. Brit. Col., 1908, p. 38.
Eocene: Mackenzie River, Northwest Territory.

Antholithes amoenus Lesquereux, Rept. U. S. Geol. Surv. Terr., vol. 8 (Cret. and Tert. Fl.), 1883, p. 203, pl. xxxiv, f. 13–15.
Miocene: Florissant, Colo.

Antholithes gaudium-rosae Ward, U. S. Geol. Surv., Mon. 48, 1905 [1906], p. 492.
Antholithus gaudium-rosae Ward, U. S. Geol. Surv., Fifteenth Ann. Rept., 1895, p. 355, pl. iii, f. 7.
Patapsco: Mount Vernon, Va.

Antholithes horridus (Dawson) Dawson, Roy. Soc. Canada, Trans., vol. 3, sec. 4, 1885 [1886], p. 7.
Carpolithes horridus Dawson, Roy. Soc. Canada, Trans., vol. 1, sec. 4, 1882 [1883], p. 21, pl. i, f. 3, 3a, 3b.
Cretaceous: Pine River, Northwest Territory.
Kootenai: North Fork of Old Man River, British Columbia.

Antholithes improbus Lesquereux, Rept. U. S. Geol. Surv. Terr., vol. 8 (Cret. and Tert. Fl.), 1883, p. 204, pl. xl, f. 20, 21.
Green River: Uinta (formerly erroneously called Randolph) County, Wyo.

Antholithes obtusilobus Lesquereux, U. S. Geol. Surv. Terr., vol. 8 (Cret. and Tert. Fl.), 1883, p. 203, pl. xxxii, f. 20.
Miocene: Florissant, Colo.

Antholithes pediloides Cockerell, Bot. Gaz., vol. 59, 1915, p. 332, text f. 1.
Miocene: Florissant, Colo.

Antholithes sp., Dawson, Roy. Soc. Canada, Trans., vol. 8, sec. 4, 1890, p. 82, text f. 11.—Penhallow, Rept. Tert. Pl. Brit. Col., 1908, p. 38.
Eocene: Similkameen River, British Columbia.

Antholithus arundites Berry, U. S. Geol. Surv., Prof. Paper 91, 1916, p. 350, pl. cxii, f. 6.
Lagrange: Puryear, Tenn.

Antholithus gaudium-rosae Ward = *Antholithes gaudium-rosae.*

Antholithus marshallensis Berry, U. S. Geol. Surv., Prof. Paper 91, 1916, p. 350, pl. cxii, f. 1.
Wilcox (Holly Springs): Early Grove, Miss.

APEIBOPSIS Heer, Fl. tert. Helv., vol. 3, 1859, p. 37.

Apeibopsis cyclophylla Lesquereux, U. S. Geol. Surv., Mon. 17, 1891 [1892], p. 180, pl. xxv, f. 6.
Dakota: Near Delphos, Kans.

Apeibopsis? discolor (Lesquereux) Lesquereux, Rept. U. S. Geol. Surv. Terr., vol. 7 (Tert. Fl.), 1878, p. 259, pl. xlvi, f. 4–7.
Rhamnus discolor Lesquereux [part], U. S. Geol. and Geog. Surv. Terr, Ann Rept. 1872 [1873], p. 398.
Post-Laramie: Black Buttes, Wyo.
Hanna: Carbon, Wyo.

Apeibopsis gaudini Lesquereux, Am. Jour. Sci., vol. 32, 1861, p. 358; Geol. Vt., vol. 2, 1861, p. 715, f. 139, 140.—Perkins, Rept. Vt. State Geol., 1903–4 [1904], p. 201, pl. lxxx, f. 120, 128; Geol. Soc. Am., Bull., vol. 16, 1905, p. 508, pl. 86, f. 7; Rept. Vt. State Geol., 1905–6 [1906], p. 221, pl. lii, f. 7; pl. lvii, f. 16, 17.
Miocene: Brandon, Vt.

Apeibopsis heerii Lesquereux, Am. Jour. Sci., 2d ser., vol. 32, 1861, p. 358; Geol. Vt., vol. 2, 1861, p. 715, f. 131, 132, 133.—Perkins, Rept. Vt. State Geol., 1903–4 [1904], p. 201, pl. lxxx, f. 118, 121, 124; idem, 1905-6 [1906], p. 221, pl. lviii, f. 7.
Miocene: Brandon, Vt.

Apeibopsis? laramiensis Knowlton, U. S. Geol. Surv., Prof. Paper 130, 19—, p. —, pl. vii, f. 4 [in preparation].
Laramie: Lyden Gulch, 6½ miles north of Golden, Colo.

Apeibopsis? neomexicana Knowlton, U. S. Geol. Surv., Prof. Paper 101, 1917, p. 336, pl. ci, f. 3.
Raton: Near Yankee, N. Mex.

Apeibopsis parva Perkins, Rept. Vt. State Geol., 1903-4 [1904], p. 202, pl. lxxx, f. 148, 152.
Miocene: Brandon, Vt.

APOCYNOPHYLLUM Unger, Gen. et sp. pl. foss., 1850, p. 433.

Apocynophyllum constrictum Berry, U. S. Geol. Surv., Prof. Paper 91, 1916, p. 344, pl. ciii, f. 4.
Wilcox: Benton, Ark.

Apocynophyllum lesquereuxii Ettingshausen, Denks. d. K. Akad. d. W., Math.-naturw. Classe, vol. 47, 1883, p. 132 [32]; Geol. Surv. N. S. W., Mem. Palaeont. No. 2 (Tert. Fl. Australia), 1888, p. 51.—Knowlton, U. S. Geol. Surv., Prof. Paper 101, 1917, p. 345.
Quercus neriifolia Al. Braun. Lesquereux, Rept. U. S. Geol. Surv. Terr., vol. 7 (Tert. Fl.), 1878, p. 150, pl. xix, f. 5.
Raton: Fishers Peak, Colo.; 12 miles east of Raton, N. Mex.

Apocynophyllum linifolium Knowlton, U. S. Geol. Surv., Prof. Paper 101, 1917, p. 346, pl. civ, f. 3.
Raton: Green Canyon mine, Aguilar, Colo.

Apocynophyllum mississippiensis Berry, U. S. Geol. Surv., Prof. Paper 91, 1916, p. 342, f. 6.
Wilcox (Grenada): Grenada, Miss.
Wilcox: Mansfield, La.

Apocynophyllum pealii Ettingshausen, Denks. d. K. Akad. d. W., Math.-naturw. Classe, vol. 47, 1883, p. 132 [32]; Geol. Surv. N. S. W., Mem. Palaeont. No. 2 (Tert. Fl. Australia), 1888, p. 51.
Quercus neriifolia Al. Braun. Lesquereux, Rept. U. S. Geol. Surv. Terr., vol. 7 (Tert. Fl.), 1878, p. 150, pl. xix, f. 4.
Miocene: Near Florissant, Colo.

Apocynophyllum sapindifolium Hollick, Geol. Surv. La., Special Rept. 5, 1899, p. 288, pl. xlvi, f. 3.—Berry, U. S. Geol. Surv., Prof. Paper 91, 1916, p. 344, pl. cii, f. 1; pl. cviii, f. 5
Wilcox: Coushatta, La.
Wilcox (Grenada): Grenada, Miss.
Lagrange: Puryear, Tenn.

Apocynophyllum scudderi Lesquereux, Rept. U. S. Geol. Surv. Terr., vol. 8 (Cret. and Tert. Fl.), 1883, p. 172, pl. xlv-A, f. 1-5.
Green River: Alkali station, Wyo.

Apocynophyllum sordidum Lesquereux, U. S. Geol. Surv., Mon. 17, 1891 [1892], p. 109, pl. lxiv, f. 11.
Dakota: Ellsworth County, Kans.

Apocynophyllum tabellarum (Lesquereux) Berry, U. S. Geol. Surv., Prof. Paper 91, 1916, p. 343, pl. cii, f. 2–5; pl. ciii, f. 5.
Salix tabellaris Lesquereux, Am. Philos. Soc., Trans., vol. 13, 1869, p., 414, pl. xvii, f. 4.
Wilcox (Holly Springs): Oxford and Vaughns, Miss.
Wilcox: Boydville and Naborton, La., and Old Port Caddo Landing, Harrison County, Tex.
Lagrange: Puryear, Tenn.

Apocynophyllum? taenifolium Knowlton, U. S. Geol. Surv. Prof. Paper 130, 19—, p. —, pl. xvi, f. 2 [in preparation].
Laramie: Lyden Gulch, 6½ miles north of Golden, Colo.

Apocynophyllum wilcoxensis Berry, U. S. Geol. Surv., Prof. Paper 91, 1916, p. 342, pl. ciii, f. 2, 3; pl. cviii, f. 4.—Knowlton, U. S. Geol. Surv., Prof. Paper 101, 1917, p. 345, pl. ciii, f. 3; pl. cv, f. 1, 2; pl. cvi, f. 1.
Wilcox (Holly Springs): Oxford, Miss.
Wilcox: Mansfield and Naborton, La.
Lagrange: Puryear, Tenn.
Raton: Primero, Cokedale, and near Trinidad, Colo.

ARACEAEITES Fritel, Mém. Soc. géol. France, vol. 16, No. 40, 1910, p. 28.

Araceaeites friteli Berry, U. S. Geol. Surv., Prof. Paper 91, 1916, p. 175, pl. cxiv, f. 3, 4.
Wilcox: Naborton, De Soto Parish, La.

ARALIA (Tournefort) Linné, Sp. pl., 1753, p. 273.

Aralia acerifolia Lesquereux = *Aralia dakotana.*

Aralia angustiloba Lesquereux, Mus. Comp. Zool., Mem., vol. 6, No. 2, 1878, p. 22, pl. v, f. 4, 5.—Knowlton, in Lindgren, Jour. Geol., vol. 4, 1896, p. 890; U. S. Geol. Surv., Prof. Paper 73, 1911, p. 59.
Miocene (auriferous gravels): Chalk Bluffs, Nevada County; Independence Hill, Placer County, Calif.
Green River(?): Pine Bluff, Wyo.
?Eocene: Southwest Oregon.

Aralia berberidifolia Lesquereux, U. S. Geol. Surv., Mon. 17, 1891 [1892], p. 135, pl. xvi, f. 11.
Dakota: Ellsworth County, Kans.

Aralia brittoniana Berry, N. Y. Bot. Gard., Bull., vol. 3, 1903, p. 96, pl. 45, f. 3; Geol. Surv. N. J., Ann. Rept. 1905 [1906], p. 138
Magothy: Cliffwood, N. J.

Aralia coloradensis Knowlton, U. S. Geol. Surv., Prof. Paper 101, 1917, p. 341, pl. cvii, f. 2..
Raton: Riley Canyon, Cokedale, Colo.

Aralia concainna Newberry = *Aralia wellingtoniana.*

Aralia concreta Lesquereux, U. S. Geol. and Geog. Surv. Terr., Bull., vol. 1, 1875 [1876], p. 394; idem, Ann. Rept. 1874 [1876], p. 349, pl. iv, f. 2–4.
Aralia semiorbiculata Lesquereux, U. S. Geol. and Geog. Surv. Terr., Bull., vol. 1, 1875 [1878], p. 395.
Dakota: Kansas.

Aralia coriacea Velenovsky, Fl. böhm. Kreideform., pt. 3, 1884, p. 11, pl. i [xvi], f. 1–9; pl. ii [xvii], f. 2.—Hollick, Geol. Soc. Am., Bull., vol. 7, 1895, p. 13; N. Y. Bot. Gard., Bull., vol. 3, 1904, p. 415, pl. lxxiii, f. 3; U. S. Geol. Surv., Mon. 50, 1906, p. 99, pl. xxxviii f. 5, 6; N. Y. Bot. Gard., Bull. 8, 1912, p. 167, pl. clxvii, f. 6.
Magothy: Glen Cove, Long Island, N. Y.; Gay Head, Marthas Vineyard, Mass.

Aralia cottondalensis Berry, U. S. Geol. Surv., Prof. Paper 112, 1919, p. 128, pl. xxvi, f. 1–3.
Aralia wellingtoniana Lesquereux. Ward, in Smith, E. A., On the geology of the Coastal Plain of Alabama, 1894, p. 348
Tuscaloosa: Cottondale, Ala.

Aralia dakotana Knowlton and Cockerell, n. name.
Aralia acerifolia Lesquereux, Rept. U. S. Geol. Surv. Terr., vol. 8 (Cret. and Tert. Fl.), 1883, pp. 232, 265, pl. xlix, f. 5; pl. xlv-B, f. 1.—Penhallow, Rept. Tert. Pl. Brit. Col., 1908, p. 38.—Berry, U. S. Geol. Surv., Prof. Paper 91, 1916, p. 328. [Homonym, Willdenow, 18—].
Fort Union: Badlands, N. Dak.
Paskapoo (Fort Union): Quilchena, British Columbia.
Wilcox (Grenada): Grenada, Miss.

Aralia digitata Ward = *Aralia wardiana.*

Aralia dissecta Lesquereux, Rept. U. S. Geol. Surv. Terr., vol. 8 (Cret. and Tert. Fl.), 1883, p. 176, pl. xxxv, f.
Miocene: Florissant, Colo.

?*Aralia dubia* Fontaine = *Sapindopsis magnifolia.*

Aralia eutawensis Berry, U. S. Geol. Surv., Prof. Paper 84, 1914, p. 119, pl. xx, f. 7; idem, Prof. Paper 112, 1919, p. 128.
Eutaw: McBrides Ford, Ga.

?*Aralia fontainei* Knowlton = *Sapindopsis magnifolia.*

Aralia formosa Heer, Kreidefl. v. Moletein, 1869, p. 18, pl. viii, f. 3.—Lesquereux, Rept. U. S. Geol. Surv. Terr., vol. 8 (Cret. and Tert. Fl.), 1883, p. 60, pl. xi, f. 3, 4.—Newberry, U. S. Geol. Surv., Mon. 35, 1895 [1896], p. 116, pl. xxii, f. 8.—Berry, Geol. Surv. N. J., Bull. 3, 1911, p. 202.
Dakota: Morrison, Colo.
Raritan: South Amboy, N. J.

Aralia? gracilis (Lesquereux) Lesquereux, Rept. U. S. Geol. Surv. Terr., vol. 7 (Tert. Fl.), 1878, p. 236, pl. xxxix, f.

Liquidambar gracilis Lesquereux, U. S. Geol. and Geog. Surv. Terr., Ann. Rept. 1871 [1872], p. 287.

Green River(?): Bridgers Pass, Wyo.

Aralia groenlandica Heer, Fl. foss. arct., vol. 6, Abt. 2, 1882, p. 84, pl. xxxviii, f. 3; pl. xxxix, f. 1; pl. xlvi, f. 16, 17.—Lesquereux, U. S. Geol. Surv., Mon. 17, 1892, p. 134, pl. liv, f. 1–3.—Newberry, U. S. Geol. Surv., Mon. 26, 1895 [1896], p. 116, pl. xxviii, f. 4.—Berry, N. Y. Bot. Gard., Bull., vol. 3, 1903, p. 95, pl. xlv, f. 4; Geol. Surv. N. J., Ann. Rept., 1905 [1906], p. 138.—Hollick, U. S. Geol. Surv., Mon. 50, 1906, p. 98, pl. xxxvii, f. 3–6.—Berry, Geol. Surv. N. J., Bull. 3, 1911, p. 199; Bull. Torr. Bot. Club, vol. 38, 1911, p. 408; Md. Geol. Surv., Upper Cret., 1916, p. 875.

Sterculia krejcii Velenovsky. Hollick, Geol. Soc. Am., Bull., vol. 7, 1895, p. 13.

Dakota: Near Fort Harker, Kans.

Magothy: Cliffwood, N. J.; Nashaquitsa and Gay Head, Marthas Vineyard, Mass.; Sullivans Cove, Severn River, Md.

Raritan: Woodbridge, N. J.

Aralia jorgenseni Heer? Fl. foss. arct., vol. 7, 1883, p. 116, pl. ci, f. 1.—Berry, U. S. Geol. Surv., Prof. Paper 91, 1916, p. 328.

Wilcox (Grenada): Grenada, Miss.

Aralia lasseniana Lesquereux, U. S. Nat. Mus., Proc., vol. 11, 1888, p. 28, pl. xiv, f. 5. [Doubtful status.]

Eocene: Lassen County, Calif.

Aralia looziana Saporta and Marion, Mém. cour. Acad. roy. de Belgique, vol. 41, 1878, p. 77, pl. xiii, f. 1–3.—Ward, U. S. Geol. Surv., Sixth Ann. Rept., 1884–85 [1886], p. 554, pl. xlviii, f. 10–12; U. S. Geol. Surv., Bull. 37, 1887, p. 61, pl. xxvii, f.

Fort Union: Clear Creek, near Glendive, Mont.

Aralia macrophylla Newberry = *Aralia wyomingensis.*

Aralia masoni Lesquereux, U. S. Geol. Surv., Mon. 17, 1891 [1892], p. 133, pl. xv, f. 4.

Dakota: Near Delphos, Kans.

Aralia mattewanensis Berry, N. Y. Bot. Gard., Bull., vol. 3, 1903, p. 95, pl. 43, f. 2; pl. 46, f. 6; Geol. Surv. N. J., Ann. Rept. 1905 [1906], p. 138.

Magothy: Cliffwood, N. J.

Aralia nassauensis Hollick, Bull. Torr. Bot. Club, vol. 21, 1894, p. 55, pl. 174, f. 3, 7; U. S. Geol. Surv., Mon. 50, 1906, p. 99, pl. xxxviii, f. 1, 2.

Magothy: Brooklyn, N. Y.

Aralia newberryi Berry, Bull. Torr. Bot. Club, vol. 34, 1907, p. 201, pl. xv, f. 1; Geol. Surv. N. J., Bull. 3, 1911, p. 197.

Aralia palmata Newberry, U. S. Geol. Surv., Mon. 26, 1895 [1896], p. 117, pl. xxxix, f. 6, 7.—Berry, N. Y. Bot. Gard., Bull., vol. 3, 1903, p. 93, pl. xliv; Bull. Torr. Bot. Club, vol. 31, 1904, p. 79, pl. iv, f. 12; Geol. Surv. N. J., Ann. Rept. 1905 [1906], p. 139.—Hollick, U. S. Geol. Surv., Mon. 50, 1906, p. 98, pl. xxxviii, f. 4.

Aralia rotundiloba Newberry? Hollick, N. Y. Acad. Sci., Annals, vol. 11, 1898, p. 421, pl. xxxviii, f. 2.

Aralia sp., Hollick, N. Y. State Mus., Fifty-fifth Ann. Rept. 1901 [1903], p. 250.

Raritan: Tottenville, Staten Island, N. Y.; Woodbridge, N. J.

Magothy: Cliffwood, N. J.

Black Creek: Court House Bluff, Cape Fear River, N. C.

Aralia notata Lesquereux, Rept. U. S. Geol. Surv. Terr., vol. 7 (Tert. Fl.), 1878, p. 237, pl. xxxix, f. 2–4.—Ward, U. S. Geol. Surv., Sixth Ann. Rept., 1884–85 [1886], p. 553, pl. xlviii, f. 8; U. S. Geol. Surv., Bull. 37, 1887, p. 60, pl. xxvii, f. 1.—Knowlton, U. S. Geol. Surv., Mon. 32, pt. 2, 1899, p. 745, pl. c, f. 1; Bull. Torr. Bot. Club, vol. 29, 1902, p. 706.—Penhallow, Roy. Soc. Canada, Trans., 2d ser., vol. 8, 1902, p. 70; Rept. Tert. Pl. Brit. Col., 1908, p. 38.—Lesquereux, U. S. Nat. Mus., Proc., vol. 11, 1888, p. 40, pl. xvii, f. 1.—Knowlton, Washington Acad. Sci., Proc., vol. 11, 1909, pp. 194, 196, 198.—Berry, U. S. Geol. Surv., Prof. Paper 91, 1916, p. 327, pl. xcvii, f. 4.

Platanus dubia Lesquereux. Schimper, Pal. vég., 3, 1874, p. 35.—U. S. Geol. and Geog. Surv. Terr., Ann. Rept. 1873 [1874], p. 406. [Homonym *P. dubia* (Ettingshausen).]

Aralia sp. Knowlton, U. S. Geol. Surv., Bull. 204, 1902, p. 81.

Fort Union: Burleigh County, N. Dak.; Milk River, Fort Peck Indian Reservation, Bull Mountains, Fish Creek, Clear Creek near Glendive, Yellowstone River, etc., Mont.; Elk Creek, Yellowstone National Park; Owl Creek, Wyo.

Paskapoo (Fort Union): Quilchena, Horsefly, and Tulameen rivers, British Columbia.

Denver: Mount Brass, Middle Park, and Golden, Colo.

Lance: Near Melville, 25 miles northwest of Red Lodge, bluffs opposite Glendive, Mont.

Wilcox: Hardys Mill, near Gainsville, Ark.; near Mansfield, La.

Aralia notata Lesquereux. Lesquereux, U. S. Nat. Mus., Proc., vol. 11, 1888, p. 20 = *Aralia* sp? Knowlton.

Aralia palmata Newberry = *Aralia newberryi*.

Aralia patens Newberry (MS.). Hollick, Bull. Torr. Bot. Club, vol. 21, 1894, p. 54, pl. 174, f. 4.—Newberry, U. S. Geol. Surv., Mon. 26, 1895 [1896], p. 117, pl. xxviii, f. 3.—Hollick, U. S. Geol. Surv., Mon. 50, 1906, p. 98, pl. xxxviii, f. 3.—Berry, Geol. Surv. N. J., Bull. 3, 1911, p. 201.

Magothy: ? Glen Cove, Long Island, N. Y.

Raritan: Woodbridge, N. J.

Aralia polymorpha Newberry, U. S. Geol. Surv., Mon. 26, 1895 [1896], p. 118, pl. xxix, f. 1–5.
Raritan: Woodbridge, N. J.

Aralia pungens Lesquereux = *Artocarpus pungens*.

Aralia pungens Lesquereux. Lesquereux, U. S. Nat. Mus., Proc., vol. 11, 1888, p. 16 = *Artocarpus californica*.

Aralia quinquepartita Lesquereux, U. S. Geol. and Geog. Surv. Terr., Bull., vol. 1, 1875 [1876], p. 394; idem, Ann. Rept., 1871 [1872], p. 302.—Newberry, U. S. Geol. Surv., Mon. 35, 1895 [1896], p. 115, pl. xl, f. 1, 2.—Berry, Geol. Surv. N. J., Bull. 3, 1911, p. 198, pl. xx, f. 8; pl. xxiv, f. 5.
Dakota: Near Fort Harker, Kans.
Dakota(?): Drummond, Mont.
Raritan: Woodbridge, N. J.

Aralia radiata Lesquereux, Rept. U. S. Geol. Surv. Terr., vol. 8 (Cret and Tert. Fl.), 1883, p. 64, pl. vii, f. 2, 3.
Dakota: Kansas.

Aralia ravniana Heer, Fl. foss. arct., vol. 6, 1882, Abt. 5, p. 84, pl. xxxviii, f. 1–2.—Berry, N. Y. Bot. Gard., Bull., vol. 3, 1903, p. 92, pl. xlvi, f. 7; pl. liii, f. 2; pl. lvii, f. 1; Bull. Torr. Bot. Club, vol. 31, 1904, p. 79.—Hollick, U. S. Geol. Surv., Mon. 50, 1906, p. 99, pl. xxxvii, f. 1, 2.—Berry, Bull. Torr. Bot. Club, vol. 37, 1910, p. 27; Md. Geol. Surv., Upper Cret., 1916, p. 876, pl. lxxxii, f. 4, pl. lxxxiii, f. 1–4.
Sterculia snowii Lesquereux? Hollick, N. Y. Acad. Sci., Annals, vol. 11, 1898, p. 422, pl. xxxviii, f. 4.
Magothy: Cliffwood, N. J.; Gay Head, Marthas Vineyard, Mass.; Grove Point, Md.
Raritan: Tottenville, Staten Island, N. Y.

Aralia rotundata Dawson, Roy. Soc. Canada, Trans., vol. 3, sec. 4, 1885 [1886], p. 14, pl. iv, f. 5.
Mill Creek: Mill Creek, British Columbia.

Aralia rotundiloba Newberry, U. S. Geol. Surv., Mon. 26, 1895 [1896], p. 118, pl. xxviii, f. 5.—Berry, Geol. Surv. N. J., Bull. 3, 1911, p. 200.
Raritan: Woodbridge, N. J.

Aralia rotundiloba Newberry? Hollick, N. Y. Acad. Sci., Annals, vol. 11, 1898, p. 421, pl. xxxviii, f. 21 = *Aralia newberryi*.

Aralia saportana Lesquereux, U. S. Geol. and Geog. Surv. Terr., Bull., vol. 1, 1875 [1876], p. 394; idem, Ann. Rept., 1874 [1876], p. 350, pl. i, f. 2, 2a; Rept. U. S. Geol. Surv. Terr., vol. 8 (Cret. and Tert. Fl.), 1883, p. 61, pl. viii, f. 1, 2; pl. ix, f. 1, 2.
Dakota: Kansas; Lander, Wyo.

Aralia saportana deformata Lesquereux, U. S. Geol. Surv., Mon. 17, 1891 [1892], p. 131, pl. xxiii, f. 1, 2.
Dakota: Ellsworth County, Kans.

Aralia semiorbiculata Lesquereux = *Aralia concreta.*

Aralia? serrata Knowlton, U. S. Geol. Surv., Prof. Paper 101, 1917, p. 341, pl. cviii, f. 4.
Raton: Riley Canyon, Cokedale, Colo.; Salyers Creek, N. Mex.

Aralia serrulata Knowlton, U. S. Geol. Surv., Mon. 32, pt. 2, 1899, p. 747, pl. ci, f. 3.
Fort Union: Near mouth of Elk Creek, Yellowstone National Park.

Aralia subemarginata Lesquereux, Rept. U. S. Geol. Surv. Terr., vol. 8 (Cret. and Tert. Fl.), 1883, p. 63; U. S. Geol. Surv., Mon. 17, 1892, p. 133, pl. xv, f. 3.
Dakota: Near Fort Harker, Kans.

Aralia tenuinervis Lesquereux, Rept. U. S. Geol. Surv. Terr., vol. 8 (Cret. and Tert. Fl.), 1883, p. 63, pl. vii, f. 4.
Dakota: Kansas.

Aralia towneri Lesquereux, U. S. Geol. and Geog. Surv. Terr., Bull., vol. 1, 1875 [1876], p. 395; idem, Ann. Rept. 1874 [1876], p. 349, pl. iv, f. 1; Rept. U. S. Geol. Surv. Terr., vol. 8 (Cret. and Tert. Fl.), 1883, p. 62, pl. vi, f. 4; U. S. Geol. Surv., Mon. 17, 1892, p. 132, pl. xxiii, f. 3, 4; pl. xxxi, f. 1.—Berry, N. Y. Bot. Gard., Bull., vol. 3, 1903, p. 92; Geol. Surv. N. J., Ann. Rept. 1905 [1906], p. 138.
Dakota: Ellsworth County, Kans.
Magothy: Cliffwood, N. J.

Aralia towneri Lesquereux. Ward, Jour. Geol., vol. 2, 1894, p. 261 = *Sassafras mudgii.*

Aralia transversinervia Saporta and Marion. Hollick, Bull. Torr. Bot. Club, vol. 21, 1894, p. 54, pl. clxxvi, f. 1 = *Ficus fracta.*

Aralia triloba Newberry, N. Y. Lyc. Nat. Hist., Ann., vol. 9, 1868, p. 58.—[Lesquereux], U. S. Geol. and Geog. Surv. Terr., Ill. Cret. and Tert. Fl., 1878, pl. xxv, f. 4, 5.—Newberry, U. S. Geol. Surv., Mon. 35, 1898, p. 121, pl. xl, f. 4, 5.
Fort Union: Fort Clarke, N. Dak.

Aralia tripartita Lesquereux = *Sterculia tripartita.*

Aralia veatchii Knowlton, U. S. Geol. Surv., Prof. Paper 108, 1917, pp. 92–93, pl. xxxvi, f. 4; pl. xxxvii, f. 4; pl. xxxviii, f. 1; pl. xxxix.
Aralia sp. cf. *A. saportana* Lesquereux. Knowlton, in Veatch, Am. Jour. Sci., 4th ser., vol. 21, 1906, p. 459.
Frontier: About 1 mile east of Cumberland, Wyo.

Aralia? vernonensis Fontaine, in Ward, U. S. Geol. Surv., Mon. 48, 1905 [1906], p. 492, pl. cvii, f. 6.
Patapsco: Mount Vernon, Va.

Aralia wardiana Knowlton and Cockerell, n. name.
Aralia digitata Ward, U. S. Geol. Surv., Sixth Ann. Rept., 1884-85 [1886], p. 554, pl. xlix, f. 2-5; U. S. Geol. Surv., Bull. 37, 1887, p. 62, pl. xxvii, f. 3-5.—Knowlton, U. S. Geol. Surv., Bull. 204, 1902, p. 81.—Lesquereux, U. S. Nat. Mus., Proc., vol. 11, 1888, p. 20, pl. xi, f. 4. [Homonym, Roxburg, 1814.]
Fort Union: Clear Creek, Glendive, and Roundup, Mont.
Clarno (lower part): Cherry Creek, Crook County, Oreg.

Aralia washingtoniana Berry, Bull. Torr. Bot. Club, vol. 37, 1910, p. 27, pl. viii, f. 4; Md. Geol. Surv. Upper Cret., 1916, p. 878, pl. lxxxii, f. 3.
Raritan: East Washington Heights, D. C.

Aralia wellingtoniana Lesquereux, U. S. Geol. Surv., Mon. 17, 1891 [1892], p. 131, pl. xxi, f. 1; pl. xxii, f. 2, 3.—Newberry, U. S. Geol. Surv., Mon. 35, 1895 [1896], p. 114, pl. xxvi, f. 1.—Berry, Geol. Surv. N. J., Bull. 3, 1911, p. 202, pl. xxv, f. 7; Bull. Torr. Bot. Club, vol. 39, 1912, p. 402.
Aralia concinna Newberry, U. S. Geol. Surv., Mon. 16, 1895 [1896], p. 114.
Dakota: Carneiro, Ellsworth County, Kans.
Raritan: Woodbridge, N. J.
Woodbine: Arthurs Bluff, Tex.

Aralia wellingtoniana Lesquereux. Ward, in Smith, E. A., On the geology of the Coastal Plain of Alabama, 1894, p. 348 = *Aralia cottondalensis*.

Aralia wellingtoniana vaughanii Knowlton, in Hill, U. S. Geol. Surv., Twenty-first Ann. Rept., pt. 7, 1899-1900 [1901], p. 317.
Woodbine: Arthurs Bluff, Lamar County, Tex.

Aralia westoni Dawson, Roy. Soc. Canada, Trans., vol. 3, sec. 4, 1885 [1886], p. 14, pl. iv, f. 6.
Mill Creek: Mill Creek, British Columbia.

Aralia whitneyi Lesquereux, Mus. Comp. Zool., Mem., vol. 6, No. 2, 1878, p. 20, pl. v, f. 1; U. S. Nat. Mus., Proc., vol. 11, 1888, p. 16.—Knowlton, in Lindgren, Jour. Geol., vol. 4, 1896, p. 890.—Knowlton, U. S. Geol. Surv., Bull. 204, 1902, p. 82; U. S. Geol. Surv., Mon. 32, pt. 2, 1899, p. 748, pl. xcix, f. 3.—Knowlton, in Lindgren, U. S. Geol. Surv., Prof. Paper 73, 1911, p. 59.
Miocene: Fossil Forest, Yellowstone National Park.
Miocene (auriferous gravels): Chalk Bluffs, Nevada County; Independence Hill, near Susanville, Placer County, Calif.
Mascall: ?Van Horn's ranch, Grant County, Oreg.
Fort Union ?: Owl Creek, Wyo. ?
Eocene: Southwestern Oregon.

Aralia wrightii Knowlton, U. S. Geol. Surv., Mon. 32, pt. 2, 1899, p. 744, pl. ci, f. 1.
Miocene: Fossil Forest Ridge, Yellowstone National Park.

Aralia wyomingensis Knowlton and Cockerell, n. name.
Aralia macrophylla Newberry, U S. Nat. Mus., Proc., vol. 5, 1882 [1883], p. 513; U. S. Geol. Surv., Mon. 35, 1898, p. 121, pl. lxvii, f. 1; pl. xlviii, f. 1. [Homonym, Lindley, 1844.]
Green River: Green River, Wyo.

Aralia zaddachi? Heer, Mioc. Balt. Fl., 1869, p. 89, pl. xv, f. 1b.—Lesquereux, Mus. Comp. Zool., Mem., vol. 6, No. 2, 1878, p. 21, pl. v, f. 2, 3.—Knowlton, in Lindgren, U. S. Geol. Surv., Prof. Paper 73, 1911, p. 59.
Miocene (auriferous gravels): Table Mountain, Tuolumne County, Calif.

Aralia sp., Berry, Bull. Torr. Bot. Club, vol. 37, 1910, p. 504.
Tuscaloosa: McBrides Ford, near Columbus, Ga.

Aralia sp., Dawson, Roy. Soc. Canada, Trans., vol. 3, sec. 4, 1885 [1886], p. 14.
Mill Creek: North Fork of Old Man River, British Columbia.

Aralia sp., Hollick, N. Y. State Mus., 1901 [1903], p. r. 50=*Aralia newberryi.*

Aralia sp., Knowlton, Geol. Soc. Am., Bull., vol. 8, 1897, p. 133; Washington Acad. Sci., Proc., vol. 11, 1909, p. 207.
Lance: Lance Creek, Converse County, Wyo.

Aralia sp., Knowlton, Geol. Soc. Am., Bull., vol. 8, 1897, p. 133.
Lance: Lance Creek, Converse County, Wyo.

Aralia sp., Knowlton, U. S. Geol. Surv., Mon. 32, pt. 2, 1899, p. 749.
Miocene: Yellowstone National Park.

Aralia sp., Knowlton, U. S. Geol. Surv., Bull. 204, 1902, p. 82, pl. xv, f. 4.
Aralia sp., Knowlton, in Merriam, Univ. Calif., Bull. Dept. Geol., vol. 2, 1901, p. 289.
Clarno (lower part): Near Clarnos Ferry, John Day Basin, Oreg.

Aralia? sp., Knowlton, U. S. Geol. Surv., Bull. 204, 1902, p. 82, pl. xvi, f. 8.
Aralia? sp., Knowlton, in Merriam, Univ. Calif., Bull. Dept. Geol., vol. 2, 1901, p. 289.
Clarno (lower part): Near Clarnos Ferry, John Day Basin, Oreg.

Aralia sp.? Knowlton, U. S. Geol. Surv., Bull. 204, 1902, p. 81.
Aralia notata Lesquereux. Lesquereux, U. S. Nat. Mus., Proc., vol. 11, 1888, p. 20.
Clarno (lower part): Cherry Creek, Crook County, Oreg.

Aralia sp., Knowlton, Washington Acad. Sci., Proc., vol. 11, 1909, p. 191.
Lance: Forsyth, Mont.; Converse County, Wyo.

Aralia cf. *A. saportana* Lesquereux. Knowlton, in Veatch, Am. Jour. Sci., 4th ser., vol. 21, 1906, p. 459 = *Aralia veatchii.*

ARALIAEPHYLLUM Fontaine, U. S. Geol. Surv., Mon. 15, 1889 [1890], p. 316. [Type, *A. magnifolium.*]

Araliaephyllum aceroides Fontaine = *Araliaephyllum magnifolium.*

Araliaephyllum acutilobum Fontaine = *Araliaephyllum crassinerve.*

Araliaephyllum crassinerve (Fontaine) Berry, Md. Geol. Surv., Lower Cret., 1912, p. 490, pl. xcv, f. 1–3.

Platanophyllum crassinerve Fontaine, U. S. Geol. Surv., Mon. 15, 1889 [1890], p. 316, pl. clviii, f. 5.

Araliaephyllum obtusilobum Fontaine, U. S. Geol. Surv., Mon. 15, 1889 [1890], p. 317, pl. clxiii, f. 1, 4; pl. clxiv, f. 3.

Araliaephyllum acutilobum Fontaine, U. S. Geol. Surv., Mon. 15, 1889 [1890], p. 318, pl. clxiii, f. 2.

Patapsco: Deep Bottom, Brooke, Va.; near Glymont and Stump Neck, Md.

Araliaephyllum magnifolium Fontaine, U. S. Geol. Surv., Mon. 15, 1889 [1890], p. 318, pl. clix, f. 9, 10.—Berry, Md. Geol. Surv., Lower Cret., 1912, p. 491, pl. xcvi, f. 1–5.

Araliaephyllum aceroides Fontaine, U. S. Geol. Surv., Mon. 15, 1889 [1890], p. 319, pl. clvi, f. 11; pl. clxii, f. 2.

Aceriphyllum aralioides Fontaine, U. S. Geol. Surv., Mon. 15, 1889 [1890], p. 321, pl. clxiii, f. 8.

Hederaephyllum angulatum Fontaine, U. S. Geol. Surv., Mon. 15, 1889 [1890], p. 324, pl. clxii, f. 1.

Patapsco: Near Brooke, 72d milepost, Va.; Stump Neck, Md.

Araliaephyllum obtusilobum Fontaine = *Araliaephyllum crassinerve.*

ARALINIUM Platen, Sitzungsb. Naturf. Ges., Leipzig, vol. 34, 1907 [1908], p. 59. [Type, *A. excellens.*]

Aralinium excellens Platen, Sitzungsb. Naturf. Ges., Leipzig, vol. 34, 1907 [1908], p. 59.

Miocene (auriferous gravels): Nevada County, Calif.

Aralinium lindgreni Platen, Sitzungsb. Naturf. Ges., Leipzig, vol. 34, 1907 [1908], p. 64.

Auriferous gravels (?): California.

Aralinium multiradiatum Platen, Sitzungsb. Naturf. Ges., Leipzig, vol. 34, 1907 [1908], p. 63.

Miocene (auriferous gravels): Nevada County, Calif.

Aralinium parenchymaticum Platen, Sitzungsb. Naturf. Ges., Leipzig, vol. 34, 1907 [1908], p. 62.

Miocene (auriferous gravels): Nevada County, Calif.

Araliopsis (Lesquereux) Berry [preoccupied by Engler, 1895, for African genus of Rutaceae], Bull. Torr. Bot. Club, vol. 38, 1911, p. 413 = *Araliopsoides.*

Araliopsis breviloba Berry = *Araliopsoides breviloba.*

Araliopsis cretacea dentata (Lesquereux) Berry = *Araliopsoides cretacea dentata.*

Araliopsis cretacea salisburiaefolia (Lesquereux) Berry = *Araliopsoides cretacea salisburiaefolia.*

Araliopsis sp., Lesquereux, Rept. on clay in New Jersey, 1878, p. 29.
Raritan ?: Sayreville, N. J.

ARALIOPSOIDES Berry, Md. Geol. Surv., Upper Cret., 1916, p. 878. [Type, *Sassafras cretaceum.*]

Araliopsoides breviloba (Berry) Berry, Md. Geol. Surv., Upper Cret., 1916, p. 878, pl. lxxxvi, f. 2.
Araliopsis breviloba Berry, Bull. Torr. Bot. Club, vol. 38, 1911, p. 417.
Raritan: Bull Mountain, Cecil County, Md.

Araliopsoides cretacea (Newberry) Berry, Md. Geol. Surv., Upper Cret., 1916, p. 879, pl. lxxiv, f. 3; pl. lxxxiv, f. 1, 2; pl. lxxxv, f. 1–5; pl. lxxxviii, f. 1–3; Bull. Torr. Bot. Club, vol. 38, 1911, p. 413.
Sassafras (Araliopsis) cretaceum Newberry, N. Y. Lyc. Nat. Hist., Annals, vol. 9, 1868, p. 14.—[Lesquereux], U. S. Geol. and Geog. Surv. Terr., Ill. Cret. and Tert. Pl., pl. vi, f. 1–4.—Lesquereux, U. S. Geol. Surv. Terr., Rept., vol. 6 (Cret. Fl.), 1874, p. 80, pl. xi, f. 1, 2; pl. xii, f. 2; U. S. Geol. Surv., Mon. 17, 1891 [1892], p. 102.—Newberry, idem, Mon. 35, 1898, p. 98, pl. vii, f. 1–4; pl. vii, f. 1–3; pl. viii, f. 1, 2.—?Hollick, idem, Mon. 50, 1906, p. 77, pl. xxx, f. 10.—Penhallow, Roy. Soc. Canada, Trans., 3d ser., vol. 1, 1907, sec. 4, p. 310.—Berry, Bull. Torr. Bot. Club, vol. 37, 1910, p. 22.
Dakota: *Blackbird Hill, Nebr.; Smoky Hill, Fort Harker, etc., Kans.
Magothy: Nashaquitsa, Marthas Vineyard, Mass.; Brightseat and Overlook Inn, Md.; Pennsylvania Avenue extended, D. C.
—— ? International Boundary between Pasayten and Skagit rivers, British Columbia. [Misidentification.]
Raritan: Bull Mountain and Shannon Hill, Cecil County, Md.; East Washington Heights, D. C.

Araliopsoides cretacea dentata (Lesquereux) Berry, Md. Geol. Surv., Upper Cret., 1916, p. 882, pl. lxxxvii, f. 1; Bull. Torr. Bot. Club, vol. 38, 1911, p. 417.
Sassafras (Araliopsis) cretaceum dentatum Lesquereux, U. S. Geol. and Geog. Surv. Terr., Ann. Rept. 1874 [1876], p. 344; Rept. U. S. Geol. Surv. Terr., vol. 6 (Cret. Fl.), 1874, pl. xi, f. 1, 2.
Dakota: Kansas.
Raritan: Bull Mountain, Cecil County, Md.

Araliopsoides cretacea salisburiaefolia (Lesquereux) Berry, Md. Geol. Surv., Upper Cret., 1916, p. 883, pl. lxxxvi, f. 1; pl. lxxxvii, f. 2, 3.
Araliopsis cretacea salisburiaefolia (Lesquereux) Berry, Bull. Torr. Bot. Club, vol. 38, 1911, p. 416.

Araliopsoides cretacea salisburiaefolia—Continued.

Cissites salisburiaefolius (Lesquereux) Lesquereux, Rept. U. S. Geol. Surv. Terr., vol. 8 (Cret. and Tert. Fl.), 1883, p. 66; U. S. Geol. Surv., Mon. 17. 1892, p. 164.—Ward, U. S. Geol. Surv., Nineteenth Ann. Rept., pt. 2, 1899, p. 707, pl. clxxi, f. 5.

Populites salisburyiaefolius Lesquereux, Am. Jour. Sci., 2d ser., vol. 46, 1868, p. 94.

Sassafras obtusus Lesquereux, U. S. Geol. and Geog. Surv. Terr., Ann. Rept. 1872 [1873], p. 424.

Cissites obtusum Lesquereux, U. S. Geol. and Geog. Surv. Terr., Ann. Rept. 1874 [1876], p. 354.

Sassafras (Araliopsis) cretaceum obtusum Lesquereux, Rept. U. S. Geol. Surv. Terr., vol. 6 (Cret. Fl.), 1874, v. 80, pl. xii, f. 3; pl. xiii, f. 1; U. S. Geol. Surv., Mon. 17, 1891 [1892], p. 102.

Sassafras (Araliopsis) obtusum Lesquereux, U. S. Geol. Surv. Terr., Rept., vol. 8 (Cret. and Tert. Fl.), 1883, p. 56.

Dakota: *Salina River and Belvidere, Kans.; Nebraska; Minnekahta Creek near Hot Springs, S. Dak.

Raritan: Bull Mountain, Cecil County, Md.

ARAUCARIA Jussieu, Gen., 1789, p. 413.

Araucaria bladensis Berry, Bull. Torr. Bot. Club, vol .35, 1908, p. 255, pls. xii, xiii, xiv, f. 1–3; idem, vol. 37, 1910, p. 505; idem, vol. 38, 1911, p. 405; U. S. Geol. Surv., Prof. Paper 84, 1914, pp. 19, 105, pl. iii, f. 6, 7; pl. xix, f. 1, 2; Md. Geol. Surv., Upper Cret., 1916, p. 777, pl. liv, f. 1; U. S. Geol. Surv., Prof. Paper 112, 1919, p. 61, pls. ix, x.

Black Creek: Parker Landing, Tar River; 95½, 92, and 87 miles above Newbern, Neuse River; Big Bend, etc., Black River; Cape Fear River; all North Carolina; 3 or 4 miles northeast of Florence, S. C.

Eutaw: Chimney Bluff below Columbus, Ga., near Havana, Hale County, Ala.; western Georgia.

Ripley (Cusseta): Buena Vista, Ga.

Magothy: Grove Point, Md.

Araucaria clarkii Berry, Bull. Torr. Bot. Club, vol. 37, 1910, p. 182.

Black Creek: Court House Bluff, Cape Fear River, N. C.

Araucaria darlingtonensis Berry, U. S. Geol. Surv., Prof. Paper 84, 1914, p. 20, pl. iii, f. 1; Bull. Torr. Bot. Club, vol. 44, 1917, p. 174.

Black Creek: Near Darlington, S. C.

Bingen: Near Maxwell Spur, Pike County, Ark.

Araucaria hatcheri Wieland, Geol. Surv. S. Dak., Bull. 4, Rept. for 1908, p. 80 [6], f. 2.

Lance: Converse County, Wyo.

Araucaria hespera Wieland, Geol. Surv. S. Dak., Bull. 4, Rept. for 1908, p. 78 [4], f. 1.

Upper Cretaceous?: Grand Valley near mouth of Cottonwood Creek, S. Dak.

Araucaria jeffreyi Berry, Bull. Torr. Bot. Club, vol. 35, 1908, p. 258, pl. xvi; idem, vol. 37, 1910, p. 505; U. S. Geol. Surv. Prof. Paper 84, 1914, pp. 20, 105; idem, Prof. Paper 112, 1919, p. 62.
Black Creek: Black, Neuse, and Tar rivers, N. C.
Black Creek (Middendorf): Langley, S. C.
Eutaw: Chimney Bluff below Columbus, Ga.
Ripley (Cusseta): Byron, Ga.

Araucaria marylandica Berry, Md. Geol. Surv., Upper Cret., 1916, p. 779, pl. liv, f. 2.
Magothy: Little Rounde Bay, Md.

Araucaria obtusifolia Fontaine, U. S. Geol. Surv., Mon. 15, 1889, p. 249, pl. lxxxv, f. 13.
Patuxent: Trents Reach, Va.

Araucaria peregrina Lindley and Hutton = *Pagiophyllum peregrinum*.

Araucaria podocarpoides Fontaine, U. S. Geol. Surv., Mon. 15, 1889, p. 249, pl. lxxxvi, f. 4.
Patapsco: Near Brooke, Va.

Araucaria spatulata Newberry, N. Y. Lyc. Nat. Hist., Ann., vol. 9, 1868, p. 10.—[Lesquereux], U. S. Geol. and Geog. Surv. Terr., Ill. Cret. and Tert. Pl., 1878, pl. ii, f. 5, 5a.—Lesquereux, Rept. U. S. Geol. Surv. Terr., vol. 8, 1883 (Cret. and Tert. Fl.), p. 30; U. S. Geol. Surv., Mon. 17, 1892, p. 32.—Newberry, idem, Mon. 35, 1898, p. 17, pl. i, f. 5, 5a.
Dakota: Sage Creek, Nebr.

Araucaria zamioides Fontaine, U. S. Geol. Surv., Mon. 15, 1889, p. 250, pl. cxxi, f. 1.
Patuxent: Near Potomac Run, Va.

ARAUCARIOPITYS Jeffrey, Bot. Gaz., vol. 44, 1907, p. 435. [Type, *A. americana*.]

Araucariopitys americana Jeffrey, Bot. Gaz., vol. 44, 1907, p. 435, pl. xxviii–xxx.—Hollick and Jeffrey, N. Y. Bot. Gard., Mem., vol. 3, 1909, p. 54.
Raritan: Kreischerville, Staten Island, N. Y.

ARAUCARIOXYLON Kraus, in Schimper, Pal. vég., vol. 2, 1872, p. 380.

Araucarioxylon arizonicum Knowlton, U. S. Nat. Mus., Proc., vol. 11, 1888, p. 1, pl. i, f. 1–5; idem, vol. 13, 1890, p. 285.—Penhallow, Man. N. A. Gym., 1907, p. 207.—Knowlton, Am. Forestry, vol. 19, 1913, p. 210.
Araucarites möllhausianus? Göppert, in Möllhausen, Tagebuch einer Reise v. Mississippi, nach d. Küsten d. Südsee, 1858, p. 429 [name].
Triassic: Fort Wingate and Abiquiu, N. Mex.; Chalcedony Park, Ariz.; Cedar City, Utah.

Araucarioxylon arizonicum neogaeum Platen, Sitzungsb. Naturf. Ges., Leipzig, vol. 34, 1907 [1908], p. 102.
Tertiary: Comstock mines, Nevada.

Araucarioxylon hoppertonae Knowlton, in Ward, U. S. Geol. Surv., Nineteenth Ann. Rept., pt. 2, 1899, p. 644, pls. clviii, clix.—Penhallow, Man. N. A. Gym., 1907, p. 208.
Araucarioxylon sp. Knowlton, in Ward, Jour. Geol., vol. 2, 1894, pp. 260, 261.
Lakota: 2 miles southwest of Minnekahta station, S. Dak.

Araucarioxylon noveboracense Hollick and Jeffrey, N. Y. Bot. Gard., Mem., vol. 3, 1909, p. 58, pl. xxi, f. 1–3, 5, 6.
Wood of *Araucaria* or an allied genus. Jeffrey, Annals Bot., vol. 20, p. 388, pl. xxviii, f. 14, 15.
Raritan: Kreischerville, Staten Island, N. Y.

Araucarioxylon? obscurum Knowlton, in Ward, U. S. Geol. Surv., Twentieth Ann. Rept., pt. 2, 1900, p. 418, pl. clxxviii.
Morrison: Freezeout Hills, Carbon County, Wyo.

Araucarioxylon prosseri Penhallow, Roy. Soc. Canada, Trans., 2d ser., vol. 6, 1900, sec. 4, p. 77; Man. N. A. Gym., 1907, p. 207.
——: Sun City, Medicine Lodge River, Baker County, Kans.

Araucarioxylon virginianum Knowlton, U. S. Nat. Mus., Proc., vol. 11, 1888, p. 4 [name]; Am. Geol., vol. 3, 1889, p. 106; U. S. Geol. Surv., Bull. 56, 1889, p. 50, pl. vii, f. 2–5; in Shaler and Woodworth, U. S. Geol. Surv., Nineteenth Ann. Rept., pt. 2, 1899, p. 516, pl. lii, f. 7–10; in Ward, idem, Twentieth Ann. Rept., pt. 2, 1900, p. 274, pl. xxxvii, f. 1–6.—Penhallow, Man. N. A. Gym., 1907, p. 206.—Wherry, Acad. Nat. Sci. Philadelphia., Proc., 1912, p. 368, pl. iii, f. 1–3.
Triassic (Newark): Taylorsville, Va.; Lockville, N. C.; Bucks, Montgomery, Chester, Lancaster, and York counties, Pa.

Araucarioxylon woodworthi Knowlton, in Shaler and Woodworth, U. S. Geol. Surv., Nineteenth Ann. Rept., pt. 2, 1899, p. 517, pl. lii, f. 1–6; in Ward, idem, Twentieth Ann. Rept., pt. 2, 1900, p. 273, pl. xxxvii, f. 7–9.—Penhallow, Man. N. A. Gym., 1907, p. 206.
Triassic (Newark): Road between Walnut Grove and Germantown, and near Lockville, N. C.

Araucarioxylon sp. Holden, Bot. Gaz., vol. 58, 1914, p. 168, pl. xii, f. 1–6.
Magothy: Cliffwood, N. J.

Araucarioxylon sp. Hollick and Jeffrey, Am. Nat., vol. 40, 1906, p. 203, pl. v, f. 5, 6.
Raritan: Kreischerville, Staten Island, N. Y.

Araucarioxylon sp. Knowlton = *Araucarioxylon hoppertonae.*

Araucarioxylon sp. Platen, Sitzungsb. Naturf. Ges., Leipzig, vol. 34, 1907 [1908], p. 108.
Triassic: Chalcedony Park, Adamana, Ariz.

ARAUCARITES Presl, in Sternberg, Flora d. Vorwelt, fasc. 2, 1833, p. 203.

Araucarites aquiensis Fontaine, U. S. Geol. Surv., Mon. 15, 1889, p. 264, pl. cxxxiii, f. 8–12; in Ward, idem, Mon. 48, 1905 [1906], pp. 489, 514, 528.—Berry, Md. Geol. Surv., Lower Cret., 1912, p. 398.
Patapsco: Near Brooke, Aquia Creek, Widewater, Mount Vernon, and Chinkapin Hollow, Va.; Fort Foote, Md.

Araucarites carolinensis Fontaine, U. S. Geol. Surv., Mon. 6, 1883, p. 119, pl. xlix, f. 8 = *Lepacyclotes circularis.*

Araucarites carolinensis Fontaine, U. S. Geol. Surv., Mon. 6, 1883, pp. 118, 119, pl. lii, f. 4, 4a = *Lepacyclotes ellipticus.*

Araucarites chiquito Ward, U. S. Geol. Surv., Twentieth Ann. Rept., pt. 2, 1900, p. 322.
Triassic: Little Colorado River, Ariz.

Araucarites? cuneatus Ward, U. S. Geol. Surv., Nineteenth Ann. Rept., pt. 2, 1899, p. 670, pl. clxiii, f. 10.
Lakota: Barr's tunnel, Crook County, Wyo.

Araucarites möllhausianus Göppert, in Möllhausen, Tagebuch einer Reise vom Mississippi nach d. Küsten d. Südsee, 1858, p. 492 [name] = *Araucarioxylon arizonicum* Knowlton.

Araucarites ovatus Hollick, N. Y. Acad. Sci., Trans., vol. 16, 1897, p. 128, pl. xii, f. 3a, 4.—Berry, N. Y. Bot. Gard., Bull., vol. 3, 1903, p. 61; Geol. Surv. N. J., Ann. Rept. 1905 [1906], p. 138.
Magothy: Cliffwood, N. J.

Araucarites patapscoensis Berry, Md. Geol. Surv., Lower Cret., 1912, p. 399, pl. lxxvii, f. 5, 5a.
Patapsco: *Widewater, Va.

Araucarites? pennsylvanicus Fontaine, in Ward, U. S. Geol. Surv., Twentieth Ann. Rept., pt. 2, 1900, p. 253, pl. xxxiv, f. 1, 2.
Triassic (Newark): York County, Pa.

Araucarites reichenbachi Geinitz = *Geinitzia reichenbachi* (according to Hollick and Jeffrey).

Araucarites virginicus Fontaine, U. S. Geol. Surv., Mon. 15, 1889 [1890], p. 263, pl. cxxxiv, f. 7.
Patuxent: Trents Reach, Va.

Araucarites virginicus Fontaine. Fontaine, in Ward = *Pinus vernonensis.*

Araucarites wardi Hill = Animal remains (Bryozoa?) cf. Ward, U. S. Geol. Surv., Mon. 48, pp. 336 et seq.

Araucarites wyomingensis Fontaine, in Ward, U. S. Geol. Surv., Nineteenth Ann. Rept., pt. 2, 1899, p. 669, pl. clxiii, f. 1–9.
Lakota: South Fork Hay Creek, and Cottle's ranch, Crook County, Wyo.

Araucarites yorkensis Fontaine, in Ward, U. S. Geol. Surv., Twentieth Ann. Rept., pt. 2, 1900, p. 254, pl. xxxiv, f. 3.
Triassic (Newark): York County, Pa.

Araucarites zeilleri Berry, Bull. Torr. Bot. Club, vol. 35, 1908, p. 252, pl. xi, f. 2.
Magothy: Cliffwood, N. J.

Araucarites? sp., Fontaine, in Ward, U. S. Geol. Surv., Mon. 48, 1905 [1906], p. 131, pl. xxxv, f. 9.
Jurassic: Douglas County, Oreg.

ARBUTUS Linné, Sp. pl., 1753, p. 395.

Arbutus menziesii Pursh. Hannibal, Bull. Torr. Bot. Club, vol. 38, 1911, p. 339, pl. xv, f. 1.
Pliocene: Near Portola, Santa Cruz Mountains, Calif.

ARCTOSTAPHYLOS Adanson, Fam. Pl., vol. 2, 1763, p. 165.

Arctostaphylos elliptica Knowlton, U. S. Geol. Surv., Mon. 32, pt. 2, 1899, p. 750, pl. xcvii, f. 2.
Fort Union: Below mouth of Elk Creek, Yellowstone National Park.

Arctostaphylos manzanita Parry. Hannibal, Bull. Torr. Bot. Club, vol. 38, 1911, p. 339, pl. xv, f. 5.
Pliocene: Near Portola and in Calabasas Canyon, Santa Cruz Mountains, Calif.

ARISAEMA Martius, Flora, vol. 14, 1831, p. 459.

Arisaema cretacea Lesquereux, U. S. Geol. Surv., Mon. 17, 1891 [1892], p. 38, pl. xlvi, f. 1.—Berry, N. Y. Bot. Gard., Bull., vol. 3, 1903, p. 66, pl. 46, f. 4; Geol. Surv. N. J., Ann. Rept. 1905 [1906], p. 138.
Arisaema? dubia Hollick, N. Y. Acad. Sci., Trans., vol. 16, 1897, p. 130, pl. xii, f. 6.
Dakota: Near Fort Harker, Kans.
Magothy: Cliffwood, N. J.

Arisaema? dubia Hollick = *Arisaema cretacea.*

Arisaema? mattewanense Hollick, N. Y. Acad. Sci., Trans., vol. 16, 1897, p. 130, pl. xii, f. 7.—Berry, N. Y. Bot. Gard., Bull., vol. 3, 1903, p. 67; Geol. Surv. N. J., Ann. Rept. 1905 [1906], p. 138.
Magothy: Cliffwood, N. J.

ARISAEMITES Knowlton, in Lindgren, Jour. Geol., vol. 4, 1896, p. 889 [name].

Arisaemites sp., Knowlton, in Lindgren, Jour. Geol., vol. 4, 1896, p. 889 [name].

Miocene (auriferous gravels): Independence Hill, Placer County, Calif.

ARISTOLOCHIA (Tournefort) Linné, Sp. pl., 1753, p. 960.

Aristolochia brittoni Knowlton, U. S. Geol. Surv. Prof. Paper 130, 19—, p.—, pl. xxiii, f. 3-5 [in preparation].

Laramie: Coal Creek and Marshall, Boulder County, Colo.

Aristolochia cordifolia Newberry = *Aristolochia crassifolia.*

Aristolochia crassifolia (Newberry) Cockerell, Am. Mus. Nat. Hist., Bull., vol. 24, 1908, p. 90.

Catalpa crassifolia Newberry, N. Y. Lyc. Nat. Hist., Ann., vol. 9, 1868, p. 56.—[Lesquereux], U. S. Geol. and Geog. Surv. Terr., Ill. Cret. and Tert. Pl., 1878, pl. xxii.—Penhallow, Rept. Tert. Pl. Brit. Col., 1908, p. 44.

Aristolochia cordifolia Newberry, N. Y. Lyc. Nat. Hist., Ann., vol. 9, 1868, p. 74.—[Lesquereux], U. S. Geol. and Geog. Surv. Terr., Ill. Cret. and Tert. Pl., 1878, pl. xxv, f. 7.—Newberry, U. S. Geol. Surv., Mon. 35, 1898, p. 90, pl. xxxix; pl. xl, f. 7; pl. lx, f. 4.

Fort Union: Fish Creek, Tongue River, and Banks of Yellowstone River, Mont.; Morristown, S. Dak.; Rock Springs and Buffalo, Wyo.

Paskapoo (Fort Union): Red Deer River and Porcupine Creek, Alberta.

Aristolochia curvata Lesquereux = *Aristolochites curvatus.*

Aristolochia dentata (Heer) Lesquereux = *Aristolochites dentata.*

Aristolochia? elongata Knowlton, U. S. Geol. Surv., Prof. Paper 101, 1917, p. 307, pl. lxxviii, f. 3.

Raton: Near Berwind, Colo.

Aristolochia mortua Cockerell, Am. Mus. Nat. Hist., Bull., vol. 24, 1908, p. 90, pl. viii, f. 25.

Miocene: Florissant, Colo.

Aristolochia obscura Lesquereux, Am. Jour. Sci., 2d ser., vol. 32, 1861, p. 359; Geol. Vt., vol. 2, 1861, p. 715, f. 137, 138, 141.—Perkins, Rept. Vt. State Geol. 1903-4 [1904], p. 202, pl. lxxx, f. 123, 125.

Miocene: Brandon, Vt.

Aristolochia oeningensis Heer, Fl. tert. Helv., vol. 2, 1856, p. 104, pl. c, f. 113.—Lesquereux, Am. Jour. Sci., 2d ser., vol. 32, 1861, p. 359; Geol. Vt., vol. 2, 1861, p. 715, f. 134.

Miocene: Brandon, Vt.

Aristolochia wilcoxiana Berry, U. S. Geol. Surv., Prof. Paper 91, 1916, p. 211, pl. xxxviii, f. 3.

Lagrange: Near Grand Junction, Fayette County, Tenn.

Aristolochia williardiana Knowlton, U. S. Nat. Mus., Proc., vol. 51, 1916, p. 268, pl. xxii, f. 1, 2.
Miocene: *Florissant, Colo.

ARISTOLOCHIAEPHYLLUM Fontaine, U. S. Geol. Surv., Mon. 15, 1889 [1890], p. 322. [Type, *A. crassinerve*.]

Aristolochiaephyllum? cellulare Ward, U. S. Geol. Surv., Mon. 48, 1905 [1906], p. 492, pl. cviii, f. 5.—Berry, Md. Geol. Surv., Lower Cret., 1912, p. 507.
Patapsco: Mount Vernon, Va.

Aristolochiaephyllum crassinerve Fontaine, U. S. Geol. Surv., Mon. 15, 1889 [1890], p. 322, pl. clx, f. 3-6; in Ward, U. S. Geol. Surv., Mon. 48, 1905 [1906], pp. 481, 504, 528, pl. cix, f. 1.—Berry, Md. Geol. Surv., Lower Cret., 1912, p. 507, pl. xcvii.
Patapsco: Near Brooke, Hell Hole, and near 72d milepost, Va.; Fort Foote, Md.

ARISTOLOCHITES Heer, Mém. Soc. helv. d. sci. nat. (Phyll. crét. d. Nebr.), 1866, p. 18. [Type, *A. dentata*.]

Aristolochites acutus Perkins, Rept. Vt. State Geol., 1903-4 [1904], p. 205, pl. lxxxi, f. 159.
Miocene: Brandon, Vt.

Aristolochites apicalis Perkins, Rept. Vt. State Geol., 1903-4 [1904], p. 203, pl. lxxx, f. 144.
Miocene: Brandon, Vt.

Aristolochites brandonianus Perkins, Rept. Vt. State Geol., 1903-4 [1904], p. 203, pl. lxxx, f. 138.
Miocene: Brandon, Vt.

Aristolochites coniodeus Perkins, Rept. Vt. State Geol., 1903-4 [1904], p. 204, pl. lxxxi, f. 154.
Miocene: Brandon, Vt.

Aristolochites crassicostatus Perkins, Rept. Vt. State Geol., 1903-4 [1904], p. 205, pl. lxxxi, f. 161, 162, 165.
Miocene: Brandon, Vt.

Aristolochites cuneatus Perkins, Rept. Vt. State Geol., 1903-4 [1904], p. 206, pl. lxxxi, f. 163.
Miocene: Brandon, Vt.

Aristolochites curvatus (Lesquereux) Perkins, Rept. State Geol., 1903-4 [1904], p. 205, pl. lxxxi, f. 158.
Aristolochia curvata Lesquereux, Am. Jour. Sci., 2d ser., vol. 32, 1861, p. 359; Geol. Vt., vol. 2, 1861, p. 715, f. 135, 136.
Miocene: Brandon, Vt.

Aristolochites dentata Heer, Mém. Soc. helv. d. sci. nat. (Phyll. crét. d. Nebr.), 1866, p. 18, pl. ii, f. 1, 2.—Lesquereux, U. S. Geol. Surv. Terr., vol. 6 (Cret. Fl.), 1874, p. 87, pl. xxx, f. 6.

Aristolochia dentata Heer, Lesquereux, Rept. U. S. Geol. Surv. Terr., vol. 8 (Cret. and Tert. Fl.), 1883, p. 59.

Dakota: Nebraska.

Aristolochites dubius Perkins, Rept. Vt. State Geol., 1903–4 [1904], p. 204, pl. lxxx, f. 153.
Miocene: Brandon, Vt.

Aristolochites elegans Perkins, Rept. Vt. State Geol., 1903–4 [1904], p. 203, pl. lxxx, f. 145; Geol. Soc. Am., Bull., vol. 16, 1905, p. 508, pl. lxxxvi, f. 6; Rept. Vt. State Geol., 1905–6 [1906], pl. lii, f. 6.
Miocene: Brandon, Vt.

Aristolochites excavatus Perkins, Rept. Vt. State Geol., 1903–4 [1904], p. 205, pl. lxxxi, f. 160.
Miocene: Brandon, Vt.

Aristolochites globosus Perkins, Rept. Vt. State Geol., 1903–4 [1904], p. 206, pl. lxxx, f. 129, 151.
Miocene: Brandon, Vt.

Aristolochites infundibuliformis Lesquereux, U. S. Geol. and Geog. Surv. Terr., Bull., vol. 1, 1875 [1876], p. 393. [Apparently not afterward referred to.]

Aristolochites irregularis Perkins, Rept. Vt. State Geol., 1903–4 [1904], p. 204, pl. lxxx, f. 149.
Miocene: Brandon, Vt.

Aristolochites latisulcatus Perkins, Rept. Vt. State Geol., 1903–4 [1904], p. 204, pl. lxxxi, f. 155.
Miocene: Brandon, Vt.

Aristolochites majus Perkins, Rept. Vt. State Geol., 1903–4 [1904], p. 206, pl. lxxx, f. 134; pl. lxxxi, f. 164; Geol. Soc. Am., Bull., vol. 16, 1905, p. 508, pl. lxxxvi, f. 4; Rept. Vt. State Geol., 1905–6 [1906], pl. lii, f. 4.
Miocene: Brandon, Vt.

Aristolochites ovoides Perkins, Rept. Vt. State Geol., 1903–4 [1904] p. 230, pl. lxxx, f. 136, 137.
Miocene: Brandon, Vt.

Aristolochites rugosus Perkins, Rept. Vt. State Geol., 1903–4 [1904], p. 203, pl. lxxx, f. 115.
Miocene: Brandon, Vt.

Aristolochites sulcatus Perkins, Rept. Vt. State Geol., 1903–4 [1904], p. 204, pl. lxxxi, f. 156, 157; Geol. Soc. Am., Bull., vol. 16, 1905, p. 508, pl. 86, f. 5; Rept. Vt. State Geol., 1905–6 [1906], pl. lii, f. 5.
Miocene: Brandon, Vt.

Aristolochites sp., Knowlton, Bull. Torr. Bot. Club, vol. 29, 1902, pl. 25, f. 6, 13.—Perkins, Rept. Vt. State Geol., 1903–4 [1904]. p. 185, text f. 6.
Miocene: Brandon, Vt.

ARTHRODENDRON Ulrich, Harriman Alaska Exped., vol. 4, 1904, p. 138. [Type, *A. diffusum*.]

Arthrodendron diffusum Ulrich, Harriman Alaska Exped., vol. 4, 1904, p. 138, pl. xiv, f. 1–3.
Yakutat: Pogibshi Island, opposite Kadiak, Alaska.

Arthrotaxopsis of authors = *Athrotaxopsis*.

ARTOCARPIDIUM Unger, Denkschr. Wien. Akad. (Foss. Fl. v. Sotzka), vol. 2, 1850, p. 254.

Artocarpidium cretaceum Ettingshausen, Kreidefl. v. Niedersch., 1867, p. 251, pl. ii, f. 4.—Lesquereux, U. S. Geol. Surv., Mon. 17, 1892, p. 86, pl. l, f. 7.
Dakota: Kansas.

Artocarpidium olmediaefolium? Unger = *Celastrinites artocarpidioides*.

ARTOCARPOIDES Saporta, Ann. sci. nat., 5th ser., Bot., vol. 3 (Études, vol. 2, pt. 1), 1865, p. 46 (42).

Artocarpoides wilcoxensis Berry, U. S. Geol. Surv., Prof. Paper 91, 1916, p. 194, pl. cix, f. 5.
Lagrange: Puryear, Henry County, Tenn.

ARTOCARPOPHYLLUM Dawson, Roy. Soc. Canada, Trans., vol. 11, 1893 [1894], sec. 4, p. 60. [Type, *A. occidentale*.]

NOTE.—The name *Artocarpophyllum* was first used by Crié (Palaeont. Abh., vol. 5, No. 2, 1889, p. 90), but as it is a nomen nudum, Dawson's name must stand.

Artocarpophyllum occidentale Dawson, Roy. Soc. Canada, Trans., vol. 11, sec. 4, 1893 [1894], p. 60, pl. xii, f. 51; pl. xiii, f. 52.
Upper Cretaceous: Vancouver Island, British Columbia.

ARTOCARPUS Forster, Char. gen., 1776, p. 101.

Artocarpus californica Knowlton, Science, vol. 21, 1893, p. 24; ?U. S. Geol. Surv., Bull. 204, 1902, p. 56.
Aralia pungens Lesquereux. Lesquereux, U. S. Nat. Mus., Proc., vol. 11, 1888, p. 16.
Myrica (*Aralia*) *lessigii?* Lesquereux. Lesquereux, idem, p. 16.
Miocene (auriferous gravels): Independence Hill, Placer County, Calif.
Mascall: ?Van Horn's ranch, Grant County, Oreg.

Artocarpus dissecta Knowlton, U. S. Geol. Surv., Prof. Paper 101, 1917, p. 267, pl. xlii, f. 6.
Vermejo: Walsenburg, Colo.

Artocarpus dubia Hollick, Geol. Surv. La., Special Rept. 5, 1899, p. 281, pl. xxxviii, f. 3.—Berry, U. S. Geol. Surv., Prof. Paper 91, 1916, p. 196, pl. xxix, f. 2; pl. cxiii, f. 1, 2.
Wilcox: *Coushatta, Naborton and Shreveport, La.

Artocarpus lessigiana (Lesquereux) Knowlton, Science, vol. 21, 1893, p. 24; Hollick, Geol. Surv. La., Special Rept. 5, 1899, p. 281, pl. xxxvii.—Berry, U. S. Geol. Surv., Prof. Paper 91, 1916, p. 194, pl. xxvi, f. 1.
Myrica? lessigiana Lesquereux, U. S. Geol. and Geog. Surv. Terr., Bull., vol. 1, 1875 [1876], p. 386; idem, Ann. Rept. 1874 [1876], p. 312.
Myrica? lessigii Lesquereux, Rept. U. S. Geol. Surv. Terr., vol. 7 (Tert. Fl.), 1878, p. 136, pl. lxiv, f. 1.
Laramie: *Coal Creek, Boulder County; Cowan station, near Denver, and Morrison, Colo.

Artocarpus liriodendroides Knowlton, U S. Geol. Surv., Prof. Paper 130, 19—, p. —, pl. xxi, f. 3 [in preparation].
Laramie: Coal Creek, Boulder County, Colo.

Artocarpus pungens (Lesquereux) Hollick, Geol. Surv. La., Special Rept. 5, 1899, p. 281, pl. xxxviii, f. 1, 2.—Berry, U. S. Geol. Surv., Prof. Paper 91, 1916, p. 195, pl. xxv, f. 1; pl. xxvii, f. 1; pl. xxix, f. 1.
Aralia pungens Lesquereux, Rept. U. S. Geol. Surv. Terr., vol. 8 (Cret. and Tert. Fl.), 1883, p. 123, pl. xix, f. 3, 4.
Denver: Golden, Colo.
Wilcox (Grenada): Grenada, Miss.
Wilcox: Benton, Ark.; Coushatta and Naborton, La.

Artocarpus? quercoides Knowlton, U. S. Geol. Surv., Mon. 32, pt. 2, 1899, p. 716, pl. xcii, f. 1.
Fort Union: Near mouth of Elk Creek, Yellowstone National Park.

Artocarpus similis Knowlton, U. S. Geol. Surv., Prof. Paper 101, 1917, p. 306, pl. lxxvii; pl. lxxviii, f. 1, 2.
Raton: *Wet Canyon, near Weston, Colo.; also Wootton, Walsenburg, Strong, and Cokedale, Colo.

ARUNDINARIA Richard, in Michaux, Fl. Bor. Am., vol. 1, 1803, p. 73.

Arundinaria macrosperma Michaux. Berry, Am. Jour. Sci., 4th ser., vol. 29, 1910, p. 392.
Pleistocene: Black Warrior River, 356 and 311½ miles above Mobile, and Abercrombies Landing, Chattahoochee River, Ala.

Arundinaria sp. (probably *A. macrosperma* Michaux). Berry, Torreya, vol. 14, 1914, p. 161.
Pleistocene: Monroeville, Monroe County, Ala.

ARUNDO (Tournefort) Linné, Sp. pl., 1753, p. 81.

Arundo cretaceus Lesquereux = *Phragmites cretaceus*.

Arundo göpperti? Münster, in Lesquereux, U. S. Geol. and Geog. Surv. Terr., Ann. Rept. 1871 [1872], Suppl., p. 5; Rept. U. S. Geol. Surv. Terr., vol. 7 (Tert. Fl.), 1878, p. 86, pl. viii, f. 3–5.
Green River: Green River, Wyo.
Denver: Silver Cliff, Colo.

Arundo grönlandica Heer? Berry, U. S. Geol. Surv., Prof. Paper 84, 1914, p. 28, pl. iv, f. 7.
Black Creek (Middendorf): Rocky Point, S. C.

Arundo? obtusa Lesquereux, U. S. Geol. and Geog. Surv. Terr., Bull., vol. 1, 1875 [1876], p. 385; idem, Ann. Rept., 1874 [1875], p. 311; Rept. U. S. Geol. Surv. Terr., vol. 7 (Tert. Fl.), 1878, p. 87, pl. viii, f. 9, 9c.
Denver: Golden, Colo.

Arundo pseudogöpperti Berry, U. S. Geol. Surv., Prof. Paper 84, 1914, p. 134, pl. xxiv, f. 7.
Barnwell (Twiggs): Phinizy Gully, Columbia County, Ga.

Arundo reperta Lesquereux, U. S. Geol. and Geog. Surv. Terr., Bull., vol. 1, 1875 [1876], p. 384; idem, Ann. Repor. 1874 [1876], p. 311.
Green River: Green River, Wyo.

Arundo sp., Dawson, Geol. Surv. Canada, vol. 8, 1894, p. 36, 37c.—Penhallow, Rept. Tert. Pl. Brit. Col., 1908, p. 38.
Eocene: Omineca River, British Columbia.

ASIMINA Adanson, Fam. Pl., vol. 2, 1763, p. 365.

Asimina eocenica Lesquereux, U. S. Geol. and Geog. Surv. Terr., Ann. Rept. 1872 [1873], p. 387; Rept. U. S. Geol. Surv. Terr., vol. 7 (Tert. Fl.), 1878, p. 251, pl. xliii, f. 5–8.—Berry, U. S. Geol. Surv., Prof. Paper 91, 1916, p. 14.
Denver: Golden, Colo.
Hanna: Carbon, Wyo.
Mesaverde: ?Rock Springs area, Wyo.
Fort Union: Wind River Basin, Wyo.
Midway(?): Earle, Tex.

Asimina leiocarpa Lesquereux, Am. Philos. Soc., Trans., vol. 13, 1869, p. 422, pl. xv, f. 8.—Berry, U. S. Geol. Surv., Prof. Paper 91, 1916, p. 218.
Wilcox ("Eolignitic"): Oxford, Miss.
Wilcox (Ackerman): Hurleys, Benton County, Miss.

Asimina triloba (Linné) Dunal. Penhallow, Geol. Soc. Am., Bull., vol. 1, 1890, p. 323.—Hollick, Bull. Torr. Bot. Club, vol. 19, 1892, p. 331.—Penhallow, Brit. Assoc. Adv. Sci., Bradford meeting, 1900, sec. C, p. 328 [c 1].
Pleistocene: Don Valley, Toronto, Canada; Bridgeton, N. J.

ASPIDIOPHYLLUM Lesquereux, U. S. Geol. and Geog. Surv. Terr., Ann. Rept. 1874 [1876], p. 361. [Type, *A. trilobatum*.]

Aspidiophyllum dentatum Lesquereux, Rept. U. S. Geol. Surv. Terr., vol. 8 (Cret. and Tert. Fl.), 1883, p. 88; U. S. Geol. Surv., Mon. 17, 1892, p. 212, pl. xxxix, f. 1.
Dakota: Ellsworth County?, Kans.

Aspidiophyllum platanifolium Lesquereux, Rept. U. S. Geol. Surv. Terr., vol. 8 (Cret and Tert. Fl.), 1883, p. 88, pl. ii, f. 4.
Dakota: Clay County, Kans.

Aspidiophyllum trilobatum Lesquereux, U. S. Geol. and Geog. Surv. Terr., Ann. Rept. 1874 [1876], p. 361, pl. ii, f. 1, 2; Rept. U. S. Geol. Surv. Terr., vol. 8 (Cret. and Tert. Fl.), 1883, p. 87, pl. xii, f. 1; pl. xiii, f. 1–5; pl. xiv, f. 1.—Berry, Bull. Torr. Bot. Club, vol. 38, 1911, p. 410; Md. Geol. Surv., Upper Cret., 1916, p. 826, pl. lx, f. 1, 2; pl. lxi, f. 1, 2.
Dakota: Kansas.
Raritan: Shannon Hill, Cecil and Anne Arundel counties, Md.; East Washington Heights, D. C.

ASPIDIUM Swartz, Jour. Bot., vol. 2, 1800, p. 4.

Aspidium angustipinnatum Fontaine = *Cladophlebis albertsii*.

Aspidium angustipinnatum montanense Fontaine = *Cladophlebis albertsii montanense*.

Aspidium cystopteroides Fontaine = *Dryopterites cystopteroides*.

Aspidium dentatum Fontaine = *Dryopterites dentata*.

Aspidium dunkeri (Schimper) Fontaine = *Cladophlebis ungeri*.

Aspidium ellipticum Fontaine = *Dryopterites elliptica*.

Aspidium (Lastrea) fischeri? Heer. Lesquereux, 1870 [1872] = *Dryopteris laramiensis*.

Aspidium fredericksburgense Fontaine = *Cladophlebis distans*.

Aspidium goldianum Lesquereux = *Dryopteris lesquereuxii*.

Aspidium heterophyllum Fontaine = *Cladophlebis parva*.

Aspidium heterophyllum Fontaine. Fontaine, in Diller and Stanton, Geol. Soc. Am., Bull., vol. 5, 1894, p. 450 in part; in Stanton, U. S. Geol. Surv., Bull. 133, 1895, p. 15 in part = *Cladophlebis parva*, part *Gleichenia nordenskiöldi*.

Aspidium heterophyllum Fontaine = *Cladophlebis parva*.

Aspidium kennerleyi Newberry = *Dryopteris kennerleyi*.

Aspidium lakesii (Lesquereux) Knowlton = *Dryopteris arguta*.

Aspidium macrocarpum Fontaine = *Dryopterites macrocarpa*.

Aspidium microcarpum Fontaine = *Dryopterites dentata*.

Aspidium monocarpum Fontaine = *Dicksonia oregonensis*.

Aspidium montanense Fontaine = *Dryopteris montanense.*

Aspidium oblongifolium Fontaine = *Dryopterites virginica.*

Aspidium oerstedi? Herr, Fontaine = *Cladophlebis albertsii.*

Aspidium oerstedi Heer. Lesquereux, U. S. Nat. Mus., Proc., vol. 11, 1888, p. 32 in part = *Onychiopsis psilotoides, Cladophlebis alata,* and *Cladophlebis huttoni.*

Aspidium parvifolium Fontaine = *Cladophlebis ungeri.*

Aspidium pinnatifidum Fontaine = *Dryopterites pinnatifida.*

Aspidium (Lastrea) pulchellum? Heer. Lesquereux, 1870 [1872] = *Dryopteris laramiensis.*

Aspidium virginicum Fontaine = *Dryopterites virginica.*

ASPLENIOPTERIS Fontaine, U. S. Geol. Surv., Mon. 15, 1889, p. 117. [Type, *A. pinnatifida.*]

Aspleniopteris adiantifolia Fontaine, U. S. Geol. Surv., Mon. 15, 1889, p. 118, pl. xvi, f. 6.—Berry, Md. Geol. Surv., Lower Cret., 1911, p. 267.
Patuxent: Fredericksburg, Va.

Aspleniopteris pinnatifida Fontaine, U. S. Geol. Surv., Mon. 15, 1889, p. 118, pl. xxii, f. 1–3, 6, 7.—Berry, Md. Geol. Surv., Lower Cret., 1911, p. 265, pl. xxviii, f. 5, 6.
Patuxent: Fredericksburg, Va.

Aspleniopteris pinnatifida? Fontaine. Fontaine, in Diller and Stanton, Geol. Soc. Am., Bull., vol. 5, 1894, p. 450; in Stanton, U. S. Geol. Surv., Bull. 133, 1895, p. 15 = *Dicksonia pachyphylla.*

ASPLENITES Göppert, Syst. Fil. Foss.: Nova Acta Acad. Caes. Leop. Nat. Cur., vol. 17, 1836, p. 277.

Asplenites rösserti (Presl) Schenk, var. Schenk. Fontaine, U. S. Geol. Surv., Mon. 6, 1883, p. 105, pl. liv, f. 9.—Ward, U. S. Geol. Surv., Twentieth Ann. Rept., pt. 2, 1900, p. 423.
Fern, undetermined, Emmons, American geology, pt. 6, 1857, p. 104, pl. iv, f. 3.
Triassic (Newark): North Carolina.

Asplenites sp., Dawson, Roy. Soc. Canada, Trans., 2d ser., vol. 1, sec. 4, 1895–96, p. 142, f. 4.—Penhallow, Rept. Tert. Pl. Brit. Col., 1908, p. 39.
Eocene: Stanley Park, Vancouver, British Columbia.

ASPLENIUM Linné, Sp. pl., 1753, p. 1078.

Asplenium albertum Dawson, Roy. Soc. Canada, Trans., vol. 3, sec. 4, 1885 [1886], p. 11, pl. iii, f. 6.
Mill Creek: Mill Creek, British Columbia.

Asplenium argutulum Heer = *Cladophlebis argutula.*

Asplenium cecilensis Berry, Bull. Torr. Bot. Club, vol. 38, 1911, p. 403, pl. xviii, f. 4, 5; Md. Geol. Surv., Upper Cret., p. 766, pl. l, f. 3, 4.

Magothy: *Grove Point, Md.

Asplenium? coloradense Knowlton, U. S. Geol. Surv., Prof. Paper 101, 1917, p. 245, pl. xxx, f. 1, 2.

Asplenium magnum Knowlton. Hollick, Torreya, vol. 2, 1902, p. 146, pl. iv, f. 1, 2.

Vermejo: Rockvale and Walsenburg, Colo.

Asplenium crossii Knowlton, U. S. Geol. Surv., Bull. 152, 1898, p. 44.

Diplazium mülleri? Heer. Lesquereux, U. S. Geol. and Geog. Surv. Terr., Ann. Rept. 1873 [1874], p. 393; Rept. U. S. Geol. Surv. Terr., vol. 7 (Tert. Fl.), 1878, p. 55, pl. iv, f. 10, 10a. [Not *Diplazium mülleri* Heer.]

Denver: Golden, Colo.

Asplenium dicksonianum Heer, Fl. foss. arct., vol. 3, pt. 2, 1874, p. 31, pl. i, f. 1–5.—Lesquereux, U. S. Geol. Surv., Mon. 17, 1892, p. 24, pl. i, f. 1, 1a.—Newberry, U. S. Geol. Surv., Mon. 26, 1895 [1896], p. 39, pl. i, f. 6, 7; pl. ii, f. 1–8; pl. iii, f. 3.—Dawson, Roy. Soc. Canada, Trans., vol. 3, 1885 [1886], p. 24.—Fontaine, in Ward, U. S. Geol. Surv., Nineteenth Ann. Rept., pt. 2, 1899, pp. 664, 704, pl. clxii, f. 6–8; pl. clxx, f. 1.—Berry, Geol. Surv. N. J., Bull. 3, 1911, p. 68, pl. v, f. 3, 4; Md. Geol. Surv., Upper Cret., 1916, p. 767; U. S. Geol. Surv., Prof. Paper 112, 1919, p. 53.

Dakota: Ellsworth County, Kans.; Minnekahta station, 4 miles below Hot Springs, S. Dak.

Raritan: Woodbridge and Milltown, N. J.

Kootenai: Near Canmore, Coal Creek, and North Fork of Old Man River, British Columbia.

Lakota: Red Canyon, Minnekahta region, S. Dak.

Patapsco: Fort Foote and Shannon Hill, Md.; Overlook Inn, D. C.

Tuscaloosa: Shirleys Mill, Ala.

Asplenium diforme Sternburg = *Comptonia diforme.*

Asplenium distans Heer. Dawson = *Cladophlebis virginiensis.*

Asplenium dubium Fontaine, U. S. Geol. Surv., Mon. 15, 1889, p. 109, pl. x, f. 9.

Patuxent: Fredericksburg, Va.

Asplenium eolignitica Berry, U. S. Geol. Surv., Prof. Paper 91, 1916, p. 167, pl. xi, f. 3.

Wilcox (Ackerman): Kemper County, Miss.

Wilcox: Hardys Mill, Greene County, Ark., and Old Port Caddo Landing, Harrison County, Tex.

Asplenium erosum (Lesquereux) Knowlton = *Allantodiopsis erosa.*

Asplenium foersteri Debey and Ettingshausen, Denkschr. Wien Akad. (Die urwelt. Acrobryen), vol. 17, 1859, p. 193 [13], pl. ii, f. 4–7, 11.—Newberry, U. S. Geol. Surv., Mon. 26, 1895 [1896], p. 41, pl. iv, f. 1–11.—Berry, Geol. Surv. N. J., Bull. 3, 1911, p. 69, pl. v, f. 1, 2.

Raritan: Woodbridge, South Amboy, N. J.; Severn River, Md.

Asplenium foersteri Heer. Lesquereux, U. S. Nat. Mus., Proc., vol. 11, 1888, p. 32 = *Cladophlebis huttoni*.

Asplenium haguei Knowlton, U. S. Geol. Surv., Mon. 32, pt. 2, 1899, p. 655, pl. lxxvii, f. 1, 2.

"Laramie?" Wolverine Creek, Yellowstone National Park.

Asplenium hurleyensis Berry, U. S. Geol. Surv., Prof. Paper 91, 1916, p. 168, pl. x, f. 1.

Wilcox (Ackerman): Hurleys, Benton County, Miss.

Asplenium iddingsi Knowlton, U. S. Geol. Surv., Mon. 32, pt. 2, 1899, p. 666, pl. lxxix, f. 2, 3; pl. lxxx, f. 9, 10.

Fort Union: Yellowstone National Park.

Asplenium jerseyensis Berry, Geol. Surv. N. J., Bull. 3, 1911, p. 71, text f. 5.

Raritan: *Milltown, N. J.

Asplenium magnum Knowlton, U. S. Geol. Surv., Mon. 32, pt. 2, 1899, p. 667, pl. lxxix, f. 5–8, 8a.

Fort Union: Near mouth of Elk Creek, Yellowstone National Park.

Asplenium magnum Knowlton. Hollick, Torreya, vol. 2, 1902, p. 146, pl. iv, f. 1, 2 = *Asplenium coloradense*.

Asplenium magnum intermedium Duror, Jour. Geol., vol. 24, 1916, p. 571, f. 3.

Swauk: Skykomish Basin, Wash.

Asplenium martini Knowlton, U. S. Geol. Surv., Prof. Paper 130, 19—, p. —, pl. ii, f. 6 [in preparation].

Laramie: Reliance mine, Erie, Colo.

Asplenium martinianum Dawson, Roy. Soc. Canada, Trans., vol. 3, 1885 [1886], p. 5, pl. i, f. 1.

Kootenai: Martin Creek, Anthracite, and North Fork of Old Man River, British Columbia.

Asplenium neomexicanum Knowlton, U. S. Geol. Surv., Prof. Paper 98, 1916, p. 331, pl. lxxxiv, f. 5–9.

Asplenium sp., Knowlton, U. S. Geol. Surv., Bull. 163, 1900, p. 20, pl. 111, f. 11.

Kirtland: Near Pina Veta China, San Juan County, N. Mex.
Mesaverde: North fork of Dutton Creek, Laramie Plains, Wyo.

Asplenium niobrara Dawson, Roy. Soc. Canada, Trans., vol. 1, 1882 [1883], sec. 4, p. 20, pl. i, f. 1.
"Peace River": Northwest Territory.

Asplenium occidentale Knowlton, U. S. Geol. Surv., Prof. Paper 108, 1917, p. 84, pl. xxxi, f. 2–5.
Frontier: About 1 mile east of Cumberland, Wyo.

Asplenium? primero Knowlton, U. S. Geol. Surv., Prof. Paper 101, 1917, p. 285, pl. liv, f. 4.
Raton: Primero, Colo.

Asplenium raritanense Berry, Bull. Torr. Bot. Club, vol. 36, 1909, p. 246, pl. xviii, f. 1; Geol. Surv. N. J., Bull. 3, 1911, p. 70, text f. 4.
Raritan: Hylton Pits, N. J.

Asplenium remotidens Knowlton, U. S. Geol. Surv., Mon. 32, pt. 2, 1899, p. 669, pl. lxxx, f. 7.
Fort Union: Near mouth of Elk Creek, Yellowstone National Park.

Asplenium spectabile Heer = *Cladophlebis spectabilis.*

Asplenium subsimplex (Lesquereux) Knowlton = *Allantodiopsis erosa.*

Asplenium tenellum Knowlton, U. S. Geol. Surv., Bull. 163, 1900, p. 19, pl. iii, f. 1, 2.
Mesaverde: North fork of Dutton Creek, Laramie Plains, Wyo.

Asplenium tenerum Lesquereux = *Dennstaedtia americana.*

Asplenium whitbiense tenue var. *a* Heer = *Cladophlebis whitbiensis tenuis.*

Asplenium wyomingense Knowlton, U. S. Geol. Surv., Bull. 163, 1900, p. 19, pl. iii, f. 12.
Asplenium? n. sp., Knowlton, Geol. Soc. Am., Bull., vol. 8, 1897, p. 140.
Mesaverde: Dutton Creek, Laramie Plains, Wyo.

Asplenium sp., Knowlton, Geol. Soc. Am., Bull., vol. 8, 1897, p. 133.
Lance: Lance Creek, Converse County, Wyo.

Asplenium? sp., Knowlton, Geol. Soc. Am., Bull., vol. 8, 1897, p. 140 = *Asplenium wyomingense.*

Asplenium sp., Knowlton, U. S. Geol. Surv., Bull. 163, 1900, p. 20, pl. iii, f. 11 = *Asplenium neomexicanum.*

Asplenium sp., Knowlton, U. S. Geol. Surv., Prof. Paper 101, 1917, p. 246. pl. xxx, f. 4.
Vermejo: Coal Creek near Rockvale, Colo.

ASTER (Tournefort) Linné, Sp. pl., 1753, p. 872.

Aster florissantia Cockerell, Am. Mus. Nat. Hist., Bull., vol. 24, 1908, p. 108, pl. ix, f. 35.
Miocene: Florissant, Colo.

ASTEROCARPUS Goeppert, Syst. Fil. Foss., 1836, p. 188.

Asterocarpus falcatus (Emmons) Fontaine, in Ward, U. S. Geol. Surv., Twentieth Ann. Rept., pt. 2, 1900, p. 237, pl. xxii, f. 3; p. 282, pl. xxxviii, f. 5, 6.

Pecopteris falcatus Emmons, Geol. Rept. Midland Counties, N. C., 1856, p. 327, pl. iv, f. 9; American geology, pt. 6, 1857, p. 100, pl. iv, f. 9.

Pecopteris falcatus variabilis Emmons, American geology, pt. 6, 1857, p. 100, pl. iv, f. 5.

Pecopteris carolinensis Emmons, Geol. Rept. Midland Counties, N. C., 1856, p. 327, pl. iv, f. 1, 2; American geology, pt. 6, 1857, p. 100, text f. 68, pl. iv, f, 1, 2.

Asterocarpus virginiensis Fontaine, U. S. Geol. Surv., Mon. 6, 1883, p. 41, pl. xix, f. 2, 2a, 3–5; pl. xx, f. 1, 2; pl. xxi, f. 1, 2; pl. xxii, f. 1–3; pl. xxiii, f. 1–4; pl. xxiv, f. 1–2.

Laccopteris emmonsi Fontaine, U. S. Geol. Surv., Mon. 6, 1883, p. 102, pl. xlviii, f. 6, 7.

Laccopteris carolinensis (Emmons) Fontaine, op. cit., p. 102, pl. xlix, f. 11, 12.

Triassic (Newark): North Carolina; York County, Pa.; Carbon Hill, Clover Hill, Midlothian, and Manakin, Va.

Asterocarpus falcatus obtusilobus (Fontaine) Knowlton, n. comb.

Asterocarpus virginiensis obtusiloba Fontaine, U. S. Geol. Surv., Mon. 6, 1883, p. 45, pl. xxi, f. 3, 4; pl. xxiv, f. 3–5a; pl. xxv, f. 1, 1a.

Asterocarpus falcatus obtusifolius (Fontaine) Ward, U. S. Geol. Surv., Twentieth Ann. Rept., pt. 2, 1900, p. 423.

Triassic (Newark): Clover Hill, Va.

Asterocarpus falcatus obtusifolius (Fontaine) Ward = *Asterocarpus falcatus obtusilobus.*

Asterocarpus lanceolatus Göppert = *Laccopteris lanceolata.*

Asterocarpus penticarpus Fontaine, U. S. Geol. Surv., Mon. 6, 1883, p. 48, pl. xxvi, f. 2, 2a.

Triassic (Newark): Clover Hill, Va.

Asterocarpus platyrachis Fontaine, U. S. Geol. Surv., Mon. 6, 1883, pp. 46, 104, pl. xxv, f. 2–6; pl. xxvi, f. 1; pl. xlix, f. 2.

Neuropteris sp., Emmons, American geology, pt. 6, 1857, p. 102, f. 71.

Triassic (Newark): Clover Hill, Va.; North Carolina.

Asterocarpus virginiensis Fontaine = *Asterocarpus falcatus.*

Asterocarpus virginiensis obtusiloba Fontaine = *Asterocarpus falcatus obtusilobus.*

ATHROTAXIS Don, Linn. Trans., vol. 18, pt. 2, p. 171, 1839.

Athrotaxis? eolignitica Berry, U. S. Geol. Surv., Prof. Paper 91, 1916, p. 173, pl. xv, f. 1, 2.

Lagrange: Puryear, Henry County, Tenn.

ATHROTAXOPSIS Fontaine, U. S. Geol. Surv., Mon. 15, 1889 [1890], p. 239. [Type, *A. grandis.*]

Athrotaxopsis expansa Fontaine, U. S. Geol. Surv., Mon. 15, 1890, pt. 1, p. 241, pl. cxiii, f. 5, 6; pl. cxv, f. 2; pl. cxvii, f. 6 [not pl. cxxxv, f. 15, 18, 22, which = *Sphenolepis kurriana*].—Fontaine, in Ward, U. S. Geol. Surv., Mon. 48, 1906, pp. 504, 520, 546, 571 [not pp. 533, 535, 538, pl. cix, figs. 12, 13, which = *Sequoia ambigua*, and p. 547, which = *Widdringtonites ramosus*].—Berry, U. S. Nat. Mus., Proc., vol. 40, 1911, p. 297; Md. Geol. Surv., Lower Cret., 1912, p. 439, pl. lxxiv, f. 1.

Taxodium (Glyptostrobus) expansum Fontaine, U. S. Geol. Surv., Mon. 15, 1890, p. 252, pl. cxxiii, f. 1.

Glyptostrobus expansus (Fontaine) Ward, U. S. Geol. Surv., Mon. 48, 1906, p. 543.

Sphenolepidium sternbergianum densifolium Fontaine, in Ward, U. S. Geol. Surv., Mon. 48, 1906, p. 524, pl. cxii, f. 11 [not f. 1, 10].

Glyptostrobus brookensis Fontaine, in Ward, U. S. Geol. Surv., Mon. 48, 1906, pp. 483, 486, 520. [Not other citations.]

Patuxent: Potomac Run, telegraph station near Lorton, Trents Reach, Dutch Gap, Cockpit Point, Va.; Springfield ?, Md.

Arundel: Langdon, D. C.; Bay View and Tip Top ?, Md.

Athrotaxopsis expansa Fontaine [in part] = *Sequoia ambigua.*

Athrotaxopsis expansa Fontaine. Fontaine, in Ward, Mon. 48, 1905 [1906], p. 547 = *Widdringtonites ramosus.*

Athrotaxopsis expansa Fontaine [in part] = *Sphenolepis kurriana.*

Athrotaxopsis grandis Fontaine [in part], U. S. Geol. Surv., Mon. 15, 1889, p. 240, pl. cxiv, f. 1–3; pl. cxvi, f. 1–4; pl. cxxxv, f. 10; in Ward, U. S. Geol. Surv., Mon. 48, 1906, p. 546.—Berry, U. S. Nat. Mus., Proc., vol. 40, 1911, p. 296; Md. Geol. Surv., Lower Cret., 1912, p. 441, pl. lxxvi; pl. lxxvii, f. 6.

Athrotaxopsis tenuicaulis Fontaine, U. S. Geol. Surv., Mon. 15, 1889, p. 241, pl. cxiv, f. 4, 5; pl. cxv, f. 6; pl. cxvii, f. 2; in Ward, U. S. Geol. Surv., Nineteenth Ann. Rept., pt. 2, 1899, p. 674, pl. clxiv; U. S. Geol. Surv., Mon. 48, 1906, pp. 516, 520, 538, 546, 571.

Athrotaxopsis pachyphylla Fontaine, U. S. Geol. Surv., Mon. 15, 1889, p. 242, pl. cxv, f. 1, 3; pl. cxvii, f. 1, 3–5.

Patuxent: Dutch Gap and Fredericksburg, Va.; Sixteenth Street, D. C.; Springfield, Md.

Arundel: Tip Top, Arlington, and Bay View, Md.; Langdon, D. C.

Fuson: Pine Creek, Crook County, Wyo.

Athrotaxopsis pachyphylla Fontaine = *Athrotaxopsis grandis.*

Athrotaxopsis tenuicaulis Fontaine = *Athrotaxopsis grandis.*

AVICENNIA Linné, Sp. pl., 1753, p. 110.

Avicennia eocenica Berry, U. S. Geol. Surv., Prof. Paper 91, 1916, p. 347, pl. civ, f. 6.

Lagrange: Puryear, Tenn.

Avicennia nitidaformis Berry, U. S. Geol. Surv., Prof. Paper 91, 1916, p. 347, pl. cvii, f. 4.

Wilcox (Holly Springs): Early Grove, Miss.

AZOLLOPHYLLUM Penhallow, Roy. Soc. Canada, Trans., vol. 8, 1890 [1891], sec. 4, p. 77. [Type, *A. primaevum.*]

Azollophyllum primaevum Penhallow, Roy. Soc. Canada, Trans., vol. 8, 1890 [1891], sec. 4, p. 77, f. 2; Rept. Tert. Pl. Brit. Col., 1908, p. 39.

Oligocene?: Stump Lake, Similkameen Valley, British Columbia.

BAIERA Braun, Flora, 1843, p. 33.

Baiera brevifolia Newberry = *Ginkgo siberica.*

Baiera foliosa Fontaine, U. S. Geol. Surv., Mon. 15, 1889, p. 213, pl. xciv, f. 13.—Berry, Md. Geol. Surv., Lower Cret., 1912, p. 372, pl. lix.

Patuxent: Dutch Gap Canal and Sailor's Tavern, Va.

Baiera gracilis (Bean) Bunbury, Quart. Jour. Geol. Soc. London, vol. 7, 1851, p. 182, pl. xii, f. 3.—Fontaine, in Ward, U. S. Geol. Surv., Mon. 48, 1905 [1906], p. 168, pl. xliv, f. 2.—Schrader, Geol. Soc. Am., Bull., vol. 13, 1902, p. 245.

Schizopteris gracilis Bean in Morris, Cat. Brit. Foss., 1843, p. 20.

Jurassic: Cape Lisburne, Alaska.

Baiera grandis Heer? Hollick, U. S. Geol. Surv., Mon. 50, 1906, p. 36, pl. ii, f. 44–46.

Magothy: Gay Head, Marthas Vineyard, Mass.

Baiera incurvata Heer. Fl. foss. arct., vol. 6, Abt. 2, 1882, p. 45, pl. xiii, f. 6.—Newberry, U. S. Geol. Surv., Mon. 26, 1895 [1896], p. 60, pl. x, f. 6.—Berry, Geol. Surv. N. J., Bull. 3, 1911, p. 100.

Raritan: Woodbridge, N. J.

Baiera longifolia (Pomel) Heer, Fl. foss. arct., vol. 4, Abt. 1, 1877, p. 39, pl. viii, f. 6.—Dawson, Roy. Soc. Canada, Trans., vol. 3, sec. 4, 1885 [1886], p. 9, pl. ii, f. 5.

Dicropteris longifolia Pomel, Ber. d. deutsch. Naturf. Gesell., 1847, p. 339.

Kootenai: Martin Creek, British Columbia.

Baiera münsteriana (Presl) Heer, in Saporta, Pal. franç. (Pl. jurass.), vol. 3, 1878, p. 272.—Fontaine, in Ward, U. S. Geol. Surv., Twentieth Ann. Rept., pt. 2, 1900, p. 249, pl. xxxi, f. 1, 2.

Sphaerococites münsterianus Presl, Flora d. Vorwelt, vol. 2 (Versuch), 1838, pts. 7 and 8, p. 105, pl. xxviii, f. 3.

Baiera münsteriana Unger. Newberry, N. Y. Acad. Sci., Trans., vol. 6, 1886–87 [1887], p. 126; U. S. Geol. Surv., Mon. 14, 1888, p. 84, pl. xxii, f. 1.

Triassic (Newark): Durham, Conn.; North Carolina; York County, Pa.

Baiera münsteriana Unger. Newberry = *Baiera münsteriana* (Presl) Heer.

Baiera multifida Fontaine, U. S. Geol. Surv., Mon. 6, 1883, pp. 87, 118, pl. xlv, f. 3; pl. xlvi, f. 1–3; pl. xlvii, f. 1, 2; pl. liii, f. 1; in Ward, U. S. Geol. Surv., Twentieth Ann. Rept., pt. 2, 1900, pp. 304, 361, pl. xliii, f. 4; pl. lxv, f. 1, 2.

?*Noeggerathia striata* Emmons, Am. Geol., pt. vi, 1857, p. 127, f. 96.

Triassic (Newark): Clover Hill and Carbon Hill, Va.; North Carolina?.

Jurassic: Oroville, Calif.?

Baiera palmata Heer. Lesquereux, U. S. Nat. Mus., Proc., vol. 11, 1888, p. 31, pl. xvi, f. 4, 5. = *Zamites megaphyllus.*

BAIEROPSIS Fontaine, U. S. Geol. Surv., Mon. 15, 1889, p. 205, [Type, *B. expansa*.]

Baieropsis adiantifolia Fontaine = *Acrostichopteris adiantifolia.*

Baieropsis adiantifolia minor Fontaine = *Acrostichopteris parvifolia.*

Baieropsis denticulata Fontaine = *Acrostichopteris cyclopteroides.*

Baieropsis denticulata angustifolia Fontaine = *Acrostichopteris longipennis.*

Baieropsis expansa Fontaine, U. S. Geol. Surv., Mon. 15, 1889 [1890], p. 207, pl. lxxxix, f. 1 = *Schizaeopsis americana.*

Baieropsis expansa Fontaine, U. S. Geol. Surv., Mon. 15, 1889 [1890], p. 207, pl. lxxxix, f. 3; pl. xci, f. 2; pl. xcii, f. 5 = *Acrostichopteris expansa.*

Baieropsis foliosa Fontaine = *Acrostichopteris longipennis.*

Baieropsis longifolia Fontaine = *Acrostichopteris pluripartita.*

Baieropsis macrophylla Fontaine = *Schizaeopsis americana.*

Baieropsis pluripartita Fontaine = *Acrostichopteris pluripartita.*

Baieropsis pluripartita minor Fontaine = *Acrostichopteris pluripartita.*

Baieropsis sp., Dawson, Roy. Soc. Canada, Trans., vol. 10, sec. 4, 1892, p. 87.

Kootenai: Anthracite, British Columbia.

BAMBUSIUM Unger, Syn. pl. foss., 1845, p. 166.

Bambusium carolinense Fontaine, U. S. Geol. Surv., Mon. 6, 1883, p. 120, pl. lii, fig. 1.

Triassic (Newark): Virginia.

Bambusium ? sp., Fontaine = *Loperia simplex.*

BANISTERIA Linné, Sp. pl., vol. 1, 1753, p. 427.

Banisteria fructuosa Berry, U. S. Geol. Surv., Prof. Paper 91, 1916, p. 257, pl. lvi, f. 8, 9.

Lagrange: Puryear, Henry County, Tenn.

Banisteria pseudolaurifolia Berry, U. S. Geol. Surv., Prof. Paper 91, 1916, p. 255, pl. lvi, f. 6, 7; pl. cx, f. 1, 2.

Lagrange: Puryear, Henry County, Tenn.; Wickliffe, Ballard County, Ky.

Banisteria repandifolia Berry, U. S. Geol. Surv., Prof. Paper 91, 1916, p. 256, pl. lvi, f. 3, 4.

Lagrange: Puryear, Henry County, Tenn.

Banisteria wilcoxiana Berry, U. S. Geol. Surv., Prof. Paper 91, 1916, p. 257, pl. lvi, f. 5.

Sapindus dubius Lesquereux. Lesquereux, U. S. Nat. Mus., Proc., vol. 11, 1888, p. 13 [part].

Lagrange: Wickliffe, Ballard County, Ky.; Grand Junction, Fayette County, Tenn.

BANKSIA Forster, Char. gen., 1776, p. 7.

Banksia helvetica Heer. Lesquereux = *Bumelia pseudotenax*.

Banksia puryearensis Berry, U. S. Geol. Surv., Prof. Paper 91, 1916, p. 211, pl. xxxvi, f. 4.

Lagrange: Puryear, Henry County, Tenn.

Banksia pusilla Velenovsky, Fl. boehm. Kreidef., Theil 2, 1883, p. 32 [7], pl. ix [i], f. 14–17.—Hollick, N. Y. Acad. Sci., Trans., vol. 16, 1897, p. 132, pl. xiii, f. 7.—Berry, N. Y. Bot. Gard., Bull., vol. 3, 1903, p. 75; Geol. Surv. N. J., Ann. Rept. 1905 [1906], p. 138.

Magothy: Cliffwood, N. J.

Banksia saffordi (Lesquereux) Berry, U. S. Geol. Surv., Prof. Paper 91, 1916, p. 208, pl. xxxvi, f. 5, 6.

Quercus saffordi Lesquereux, Am. Jour. Sci., 2d ser., vol. 27, 1859, p. 364; 2d Rept. Geol. Recon. Ark., 1860, p. 319, pl. vi, f. 3; Geol. Tenn., 1869, p. 427, pl. K, f. 2a–2c; U. S. Nat. Mus., Proc., vol. 11, 1888, p. 13, pl. v, f. 1–3.—Knowlton, in Glenn, U. S. Geol. Surv., Water-Supply Paper 164, 1906, p. 38.

Wilcox (Grenada): Grenada, Grenada County, Miss.

Lagrange: Sommerville and near Grand Junction, Fayette County, and Puryear, Henry County, Tenn.; Wickliffe, Ballard County, and Boaz, Graves County, Ky.

Banksia tenuifolia Berry, U. S. Geol. Surv., Prof. Paper 91, 1916, p. 210, pl. xxxvi, f. 1–3.

Wilcox: Near Boydsville, Clay County, Ark.

Lagrange: Puryear, Henry County, Tenn.; Wickliffe, Ballard County, Ky.

BANKSITES Saporta, in Heer, Recherches clim. vég. pays tert., 1861, p. 138.

Banksites lineatus Lesquereux, Rept. U. S. Geol. Surv. Terr., vol. 8 (Cret and Tert. Fl.), 1883, p. 165, pl. xxxii, f. 21.—Knowlton, U. S. Nat. Mus., Proc., vol. 51, 1916, p. 268.

Miocene: *Florissant, Colo.

Banksites saportanus Velenovsky. Beitr. Pal. Österr.-Ungarns u. d. Orients (Böhm. Kreidef.), vol. 3, no. 1, pt. 2, 1883, p. 7 (32), pl. i (ix), f. 18–20.—Hollick, U. S. Geol. Surv., Mon. 50, 1906, p. 60, pl. viii, f. 20, 21.

Celastrophyllum benedeni Saporta and Marion. Hollick, Bull. Torr. Bot. Club, vol. 21, 1894, p. 58, pl. clxxvii, f. 3.

Magothy: Gay Head, Marthas Vineyard, Mass.

BAUHINIA Linné, Sp. pl., 1753, p. 374.

Bauhinia alabamensis Berry, Am. Jour. Sci., 4th ser., vol. 29, 1910, p. 256, text f. 1; U. S. Geol. Surv., Prof. Paper 112, 1919, p. 99, pl. xxiii, f. 8.

Eutaw: Near Havana, Hale County, Ala.

Bauhinia cretacea Newberry, Bull. Torr. Bot. Club, vol. 13, 1886, p. 77, pl. lvi; U. S. Geol. Surv., Mon. 26, 1895 [1896], p. 91, pl. xliii, f. 1–4; pl. xliv, f. 1–3.—Berry, Geol. Surv. N. J., Bull. 3, 1911, p. 162, pl. xix, f. 3; U. S. Geol. Surv., Prof. Paper 112, 1919, p. 97.

Raritan: Woodbridge, N. J.
Tuscaloosa: Shirleys Mill and Cottondale, Ala.
Eutaw: Near Havana, Ga.

Bauhinia? gigantea Newberry, U. S. Geol. Surv., Mon. 26, 1895 [1896], p. 93, pl. xx, f. 1.—Berry, Geol. Surv. N. J., Bull. 3, 1911, p. 164.

Raritan: Woodbridge, N. J.

Bauhinia marylandica Berry, Torreya, vol. 8, 1908, p. 219, text f. 3; Md. Geol. Surv., Upper Cret., 1916, p. 846, pl. lxxv, f. 5–7; U. S. Geol. Surv., Prof. Paper 112, 1919, p. 98.

Magothy: Grove Point and Round Bay, Md.
Tuscaloosa: Shirleys Mill and Cottondale, Ala.

Bauhinia pseudocotyledon Cockerell, Torreya, vol. 9, 1909, p. 184, text f.

Miocene: Florissant, Colo.

Bauhinia ripleyensis Berry, Bull. Torr. Bot. Club, vol. 43, 1916, p. 294, pl. xvi, f. 1; U. S. Geol. Surv., Prof. Paper 112, 1919, p. 100, pl. xxiii, f. 7.

Ripley: Cowikee Creek, Barbour County, Ala.; Selmer, McNairy County, Tenn.

Bauhinia sp., Knowlton, Washington Acad. Sci., Proc., vol. 11, 1909, p. 211.

Lance: Near Ranchester, Wyo.

BENNETTITES Carruthers, Linn. Soc. London, Trans., vol. 26, 1868, p. 681.

Bennettites dacotensis McBride = *Cycadeoidea dacotensis* and *Cycadeoidea mcbridei.*

Bennettites sp. Carruthers = *Cycadeoidea marylandica.*

BENZOIN Fabricius, Enum. Pl. Hort. Helmst., 1763.

Benzoin antiquum? Heer, Fl. tert. Helv., vol. 2, 1855, p. 81, pl. xc, f. 1-8.—Lesquereux, U. S. Geol. and Geog. Surv. Terr., Ann. Rept. 1872 [1873], p. 379. [Not later mentioned by Lesquereux.]

——: Golden, Colo.

Benzoin masoni (Lesquereux) Knowlton, U. S. Geol. Surv., Bull. 152, 1898, p. 47.

Lindera masoni Lesquereux, U. S. Geol. Surv., Mon. 17, 1891 [1892], p. 96, pl. xviii, f. 9, 10.

Dakota: Near Brookville, Kans.

Benzoin cf. **B. melissaefolium** (Walter) Nees. Berry, Jour. Geol., vol. 25, 1917, p. 662.

Pleistocene: Vero, Fla.

Benzoin venustum (Lesquereux) Knowlton, U. S. Geol. Surv., Bull. 152, 1898, p. 47.—Berry, Bull. Torr. Bot. Club, vol. 39, 1912, p. 399.

Lindera venusta Lesquereux, U. S. Geol. Surv., Mon. 17, 1891 [1892], p. 95, pl. xvi, f. 1, 2.

Dakota: Ellsworth County, Kans.

Woodbine: Arthurs Bluff, Tex.

BERBERIS (Tournefort) Linné, Sp. pl., 1753, p. 330.

Berberis? gigantea Knowlton, U. S. Geol. Surv., Bull. 204, 1902, p. 57, pl. xi, f. 1.

Mascall: Van Horn's ranch, John Day Basin, Oreg.

Berberis simplex Newberry = *Odostemon simplex*.

Berberis sp., Berry, Jour. Geol., vol. 15, 1907, p. 343 = *Vaccinium arboreum*.

BERCHEMIA Necker, Elem., vol. 2, 1790, p. 122.

Berchemia multinervis (Al. Braun) Heer. Lesquereux, Rept. U. S. Geol. Surv. Terr., vol. 7 (Tert. Fl.), 1878, p. 277, pl. lii, f. 9, 10; U. S. Geol. and Geog. Surv. Terr., Ann. Rept. 1876, p. 517, 1878.—Ward, U. S. Geol. Surv., Sixth Ann. Rept., 1884-85 [1886], p. 554, pl. li, f. 12; idem, Bull. 37, 1887, p. 73, pl. xxxiii, f. 1.—Knowlton, Washington Acad. Sci., Proc., vol. 11, 1909, p. 213; U. S. Geol. Surv., Prof. Paper 101, 1917, p. 333, pl. ci, f. 5.

Berchemia multinervis (Al. Braun) Heer. Ward, U. S. Geol. Surv., Sixth Ann. Rept., 1884-85 [1886], p. 554, pl. li, f. 13; idem, Bull. 37, 1887, p. 73, pl. xxxiii, f. 2—*Rhamnus goldianus*.

Berchemia parvifolia Lesquereux, Am. Jour. Sci., 2d ser., vol. 45, 1868, p. 207; U. S. Geol. and Geog. Surv. Terr., Ann. Rept., 1869, p. 96 [reprint, 1873, p. 196]; idem, 1870 [1872], p. 382; idem, 1871 [1872], Suppl. p. 15; idem, 1873 [1874], p. 382.

Denver: Golden, Colo.

Dawson: Near Sedalia, Colo.

Raton: Raton Pass, and near Yankee, N. Mex.

Lance: Near Ilo post office and near Kirby, Wyo.

Fort Union: Bull Mountains, Mont.

Birchemia parvifolia Lesquereux = *Berchemia multinervis.*

Berchemia priscaformis Berry, U. S. Geol. Surv., Prof. Paper 98, 1916, p. 69, pl. xii, f. 11, 12.
Calvert: Good Hope Hill, D. C.

BETULA (Tournefort) Linné, Sp. pl., 1753, p. 982.

Betula aequalis Lesquereux, Mus. Comp. Zool., Mem., vol. 6, No. 2, 1878, p. 2, pl. i, f. 2-4.—Knowlton, in Lindgren, Jour. Geol., vol. 4, 1896, p. 889.
Miocene (auriferous gravels): Chalk Bluffs, Nevada County, and Independence Hill, Placer County, Calif.

Betula alaskana Lesquereux, U. S. Nat. Mus., Proc., vol. 5, 1882 [1883], p. 446, pl. vi, f. 14; Rept. U. S. Geol. Surv. Terr., vol. 8 (Cret and Tert. Fl.), 1883, p. 258.—Knowlton, U. S. Nat. Mus., Proc., vol. 17, 1894, p. 221; Geol. Soc. Am., Bull., vol. 5, 1893, p. 583.
Kenai: Chignik Bay, Alaska.

Betula angustifolia Newberry, U. S. Nat. Mus., Proc., vol. 5, 1882 [1883], p. 508; U. S. Geol. Surv., Mon. 35, 1898, p. 63, pl. xlvi, f. 5; pl. xlvii, f. 5.—Knowlton, U. S. Geol. Surv., Bull. 204, 1902, p. 41.—Penhallow, Rept. Tert. Pl. Brit. Col., 1908, p. 39.
Clarno (upper part): Bridge Creek, Oreg.
Eocene: Quilchena, British Columbia.

Betula basiserrata Ward, U. S. Geol. Surv., Sixth Ann. Rept., 1884-85 [1886], p. 551, pl. xl, f. 3; U. S. Geol. Surv., Bull. 37, 1887, p. 32, pl. xiv, f. 4.
Fort Union: Sevenmile Creek, Glendive, Mont.

Betula beatriciana Lesquereux, Am. Jour. Sci., 2d ser., vol. 46, 1868, p. 95; U. S. Geol. Surv., Mon. 17, 1892 [1893], p. 59, pl. iii, f. 16.
Dakota: *Beatrice, Nebr.; Ellsworth County, Kans.

Betula bendirei Knowlton, U. S. Geol. Surv., Bull. 204, 1902, p. 40, pl. iv, f. 2.
Clarno (upper part): Bridge Creek, Oreg.

Betula caudata Göppert. Lesquereux = *Betula stevensonii.*

Betula coryloides Ward, U. S. Geol. Surv., Sixth Ann. Rept., 1884-85 [1886], p. 551, pl. xl, f. 2; U. S. Geol. Surv., Bull. 37, 1887, p. 31, pl. xiv, f. 3.
Fort Union: Sevenmile Creek, Glendive, Mont.

Betula? dayana Knowlton, U. S. Geol. Surv., Bull. 204, 1902, p. 41, pl. iv, f. 4.
Mascall: Van Horn's ranch, John Day Basin, Oreg.

Betula deltoides Knowlton, U. S. Nat. Mus., Proc., vol. 51, 1916, p. 264, pl. xix, f. 3.
Miocene *Florissant, Colo.

Betula elliptica Saporta, Climat et vég. tert., 1861, p. 160.—Lesquereux, Rept. U. S. Geol. Surv. Terr., vol. 8 (Cret. and Tert. Fl.), 1883, p. 242, pl. li, f. 6.
Miocene: South of Mount Diablo, Calif. [by error "John Day Valley, Oregon"].

Betula fallax Lesquereux, Mus. Comp. Zool., Bull., vol. 16, 1888, p. 45.
Denver: Golden, Colo.

Betula florissanti Lesquereux, Rept. U. S. Geol. Surv., Terr., vol. 8 (Cret. and Tert. Fl.), 1883, p. 150, pl. xxvii, f. 11.
Miocene: Florissant, Colo.

Betula göpperti Lesquereux, Tert. Fl., pl. xvii, f. 21, 22=*Betula stevensonii.*

Betula göpperti Lesquereux, Tert. Fl., pl. xvii, f. 23, 23a=*Alnus americana.*

Betula gracilis? Ludwig, Palaeontographica, vol. 8, 1861, p. 99, pl. xxxii, f. 3-6.—Lesquereux, U. S. Geol. and Geog. Surv. Terr., Ann. Rept. 1873 [1874], p. 397; Rept. U. S. Geol. Surv. Terr., vol. 7 (Tert. Fl.), 1878, p. 138, pl. xvii, f. 20.
Denver: Golden, Colo.

Betula grandifolia Ettingshausen, Denkschr. Wien. Akad. (Foss. Fl. v. Bilin), vol. 26, 1867, p. 123, pl. xiv, f. 23, 25.—Heer, Fl. foss. arct., vol. 2, Abt. 2 (Fl. foss. Alask.), 1868, p. 29, pl. v, f. 8.—Knowlton, U. S. Nat. Mus., Proc., vol. 17, 1894, p. 221; Geol. Soc. Am., Bull., vol. 5, 1893, p. 583.
Kenai: Port Graham and Hess Creek, Alaska.

Betula heterodonta Newberry, U. S. Nat. Mus., Proc., vol. 5, 1882 [1883], p. 508; U. S. Geol. Surv., Mon. 35, 1898, p. 64, pl. xliv, f. 1-4; pl. xlv, f. 1, 6.—Knowlton, U. S. Geol. Surv., Bull. 204, 1902, p. 40.—Penhallow, Rept. Tert. Pl. Brit. Col., 1908, p. 39.
Clarno (upper part): Bridge Creek and Trout Creek, Harney County, Oreg.
Oligocene: Quilchena, British Columbia.

Betula iddingsi Knowlton, U. S. Geol. Surv., Mon. 32, pt. 2, 1899, p. 698, pl. lxxxvi, f. 4, 5.
Fort Union: Near mouth of Elk Creek, Yellowstone National Park.

Betula lutea Michaux f. Penhallow, Roy. Soc. Canada, Trans., 2d ser., vol. 2, 1896, pp. 66, 70; Brit. Assoc. Adv. Sci., Bradford meeting, 1900, p. 335.
Pleistocene: Greens Creek and Besserers Wharf, Ottawa River, Canada.

Betula macclintockia Cramer, in Heer, Fl. foss. arct., vol. 1, 1868, p. 174, pl. xxxiv, f. 4a, b; pl. xxxix, f. 1-9.
Tertiary: Banks Land.

Betula macrophylla Göppert, Fl. tert. Helv., vol. 3, 1859, p. 311.—Heer, Fl. foss. arct., vol. 6, Abt. 3, 1880, p. 14, pl. ii, f. 3-5.—Penhallow, Rept. Tert. Pl. Brit. Col., 1908, p. 39.
Miocene: Tranquille River, British Columbia; Mackenzie River valley.

Betula nigra Linné. Knowlton, Am. Geol., vol. 18, 1896, p. 371.—Berry, Jour. Geol., vol. 15, 1907, p. 341; Am. Nat., vol. 41, 1907, p. 692, pl. ii, f. 2-4; idem, vol. 43, 1909, p. 435; Am. Jour. Sci., 4th ser., vol. 29, 1910, p. 393; Torreya, vol. 14, 1914, p. 162; U. S. Nat. Mus., Proc., vol. 48, 1915, p. 298.
Pleistocene: Morgantown, W. Va.; Neuse and Roanoke rivers, N. C.; Abercrombies Landing, Chattahoochee River, Black Warrior River above Mobile, Ala.; Rappahannock River 1½ miles below Port Royal, Va.

Betula parce-dentata Lesquereux, Rept. U. S. Geol. Surv. Terr., vol. 8 (Cret. and Tert. Fl.), 1883, p. 242, pl. l, f. 12.
Miocene: South of Mount Diablo, Calif. [by error John Day Valley, Oreg.].

Betula perantiqua Dawson, Roy. Soc. Canada, Trans., vol. 1, sec, 4, 1882 [1883], p. 27, pl. vii, f. 27.
Upper Cretaceous: Baynes Sound, Port McNeal, Vancouver Island, British Columbia.

Betula prenigra Berry, U. S. Geol. Surv., Prof. Paper 98, 1916, p. 198, pl. xlv, f. 14, 15.
Citronelle: Lambert, Ala.

Betula prisca Ettingshausen, Foss. Fl. v. Wien, 1851, p. 11, pl. i, f. 15-17.—Heer, Fl. foss. arct., vol. 2, Abt. 2, 1869, p. 28, pl. v, f. 3-6.—Ward, U. S. Geol. Surv., Sixth Ann. Rept., 1884-85 [1886], p. 551, pl. xl, f. 1; U. S. Geol. Surv., Bull. 37, 1887, p. 31, pl. xiv, f. 2. —Knowlton, U. S. Nat. Mus., Proc., vol. 17, 1894, p. 221; Geol. Soc. Am., Bull., vol. 5, 1893, p. 583.—Penhallow, Rept. Tert. Pl. Brit. Col., 1908, p. 40.
Fort Union: Seven Mile Creek, near Glendive, Mont.
Eocene: Quesnel, British Columbia.
Kenai: Port Graham, Alaska.

Betula pseudo-fontinalis Berry, Jour. Geol., vol. 15, 1907, p. 341.
Pleistocene: Neuse River, N. C.

Betula schimperi Lesquereux, Mus. Comp. Zool., Bull., vol. 16, 1888, p. 45.
Denver: Golden, Colo.

Betula stevensonii Lesquereux, U. S. Geol. and Geog. Surv. Terr., Ann. Rept. 1871 [1872], p. 293; Rept. U. S. Geol. Surv. Terr., vol. 7 (Tert. Fl.), 1878, p. 139, pl. xviii, f. 1–5.—Penhallow, Tert. Pl. Brit. Col., 1908, p. 40.

Betula göpperti Lesquereux, Rept. U. S. Geol. Surv. Terr., vol. 7 (Tert. Fl.), 1878, p. 138, pl. xvii, f. 21–23.

Betula? caudata Göppert. Lesquereux, U. S. Geol. and Geog. Surv. Terr., Rept. 1871 [1872], p. 293.

Evanston: Evanston, Wyo.

——: Carbon, Wyo.

Oligocene?: Similkameen River, British Columbia.

Betula truncata Lesquereux, Rept. U. S. Geol. Surv. Terr., vol. 8 (Cret. and Tert. Fl.), 1883, p. 150, pl. xxviii, f. 7, 8.—Knowlton, U. S. Nat. Mus., Proc., vol. 51, 1916, p. 262.

Miocene: Florissant, Colo.

Betula vogdesii Lesquereux, U. S. Geol. and Geog. Surv. Terr., Bull., vol. 1, 1875 [1876], p. 386; U. S. Geol. and Geog. Surv., Terr., Ann. Rept. 1874 [1876], p. 312; Rept. U. S. Geol. Surv. Terr., vol. 7 (Tert. Fl.), 1878, p. 137, pl. xvii, f. 18, 19.

——: Fort Fetterman, Wyo.

Betula sp., Berry, U. S. Nat. Mus., Proc., vol. 48, 1915, p. 298.

Pleistocene: Hickman, Ky.

Betula sp., Dawson, Geol. Surv. Canada, Rept. progress 1872–73 [1873], appendix i, p. 68.

Upper Cretaceous: Vancouver Island, British Columbia.

Betula sp., Dawson, Roy. Soc. Canada, Trans., vol. 1, 1882 [1883], sec. 4, p. 22.

——: Peace River, Northwest Territory.

Betula sp., Dawson, Roy. Soc. Canada, Trans., vol. 5, sec. 4, 1887, p. 33 [wood].

Belly River: Ribestone Creek, Canada.

Pierre: Swift Current, Canada.

Betula sp., Dawson, Roy. Soc. Canada, Trans., vol. 5, 1887, sec. 4, p. 34 [fossil wood].

Paskapoo (Fort Union): Souris River, British Columbia.

Betula sp., Knowlton, Washington Acad. Sci., Proc., vol. 11, 1909, p. 189.

Lance: 9 miles west of Miles City, Mont.

Betula sp. (branch) Knowlton, Harriman Alaska Exped., vol. 4, 1904, p. 153, pl. xxiv, f. 2.

Kenai: Kukak Bay, Alaska.

Betula sp., Knowlton, U. S. Geol. Surv., Bull. 316, pt. 2, 1907, p. 202.

Fort Union: Signal Butte, Miles City, Mont.

Betula sp.? Newberry, U. S. Geol. Surv., Mon. 35, 1898, p. 65, pl. lvii, f. 4.
Clarno (upper part): Bridge Creek, Oreg.?

Betula sp., Penhallow, Roy. Soc. Canada, Trans., 3d ser., vol. 1, sec. 4, 1907, p. 301.
Eocene: Kettle River near Midway, British Columbia.

Betula sp., Penhallow, Roy. Soc. Canada, Trans., 3d ser., vol. 1, sec. 4, 1907, p. 294.
Eocene: Kettle River near Midway, British Columbia.

BETULITES Göppert, De floribus in statu fossili, Breslau, 1837, p. 22.

Betulites denticulata Heer, Phyll. crét. d. Nebr., 1866, p. 15, pl. iv, f. 5, 6.
Dakota: Blackbird Hill, Nebr.

Betulites? hatcheri Knowlton, U. S. Geol. Surv., Bull. 257, p. 141, pl. xviii, f. 3.
Judith River: Willow Creek, Fergus County, Mont.

Betulites populifolius Lesquereux, U. S. Geol. Surv., Mon. 17, 1891 [1892], p. 64, pl. vi, f. 1, 2.—Berry, Bull. Torr. Bot. Club, vol. 32, 1905, p. 46, pl. ii, f. 6; Geol. Surv. N. J., Ann. Rept. 1905 [1906], p. 138.
Dakota: Ellsworth County, Kans.
Magothy: Cliffwood, N. J.

Betulites rugosus Lesquereux, U. S. Geol. Surv., Mon. 17, 1891 [1892], p. 65, pl. vi, f. 3–5.
Dakota: Ellsworth County, Kans.

Betulites snowii Lesquereux, U. S. Geol. Surv., Mon. 17, 1891 [1892], p. 64, pl. v, f. 1–4.
Dakota: Ellsworth County, Kans.

Betulites westii Lesquereux, U. S. Geol. Surv., Mon. 17, 1891 [1892], p. 60, pl. iv, f. 1–22; pl. v, f. 5–14.
Dakota: Ellsworth County, Kans.

Betulites westii crassus Lesquereux, U. S. Geol. Surv., Mon. 17, 1891 [1892], p. 63, pl. v, f. 15–17.
Dakota: Ellsworth County, Kans.

Betulites westii cuneatus Lesquereux, U. S. Geol. Surv., Mon. 17, 1891 [1892], p. 62, pl. v, f. 8.
Dakota: Ellsworth County, Kans.

Betulites westii grewiopsideus Lesquereux, U. S. Geol. Surv., Mon. 17, 1891 [1892], p. 63, pl. lxiv, f. 10.
Dakota: Ellsworth County, Kans.

Betulites westii inaequilateralis Lesquereux, U. S. Geol. Surv., Mon. 17, 1891 [1892], p. 62, pl. v, f. 10–13.
Dakota: Ellsworth County, Kans.

Betulites westii lanceolatus Lesquereux, U. S. Geol. Surv., Mon. 17, 1891 [1892], p. 62, pl. v, f. 14.
Dakota: Ellsworth County, Kans.

Betulites westii latifolius Lesquereux, U. S. Geol. Surv., Mon. 17, 1891 [1892], p. 61, pl. iv, f. 9–11.
Dakota: Ellsworth County, Kans.

Betulites westii multinervis Lesquereux, U. S. Geol. Surv., Mon. 17, 1891 [1892], p. 62, pl. iv, f. 20–22.
Dakota: Ellsworth County, Kans.

Betulites westii oblongus Lesquereux, U. S. Geol. Surv., Mon. 17, 1891 [1892], p. 61, pl. iv, f. 17–19.
Dakota: Ellsworth County, Kans.

Betulites westii obtusa[us] Lesquereux, according to Snow, Kansas Acad. Sci., Trans., vol. 11, 1887–88 [1889], p. 35, f. 1.—Lesquereux, U. S. Geol. Surv., Mon. 17, 1891 [1892], p. 61, pl. iv, f. 5–8.
Dakota: Ellsworth County, Kans.

Betulites westii populoides Lesquereux, U. S. Geol. Surv., Mon. 17, 1891 [1892], p. 63.
Dakota: Ellsworth County, Kans.

Betulites westii quadratifolius Lesquereux, U. S. Geol. Surv., Mon. 17, 1891 [1892], p. 62, pl. v, f. 9.
Dakota: Ellsworth County, Kans.

Betulites westii reniformis Lesquereux, U. S. Geol. Surv., Mon. 17, 1891 [1892], p. 62, pl. v, f. 5.
Dakota: Ellsworth County, Kans.

Betulites westii rhomboidalis Lesquereux, U. S. Geol. Surv., Mon. 17, 1891 [1892], p. 62, pl. v, f. 6, 7.
Dakota: Ellsworth County, Kans.

Betulites westii rotundatus Lesquereux, U. S. Geol. Surv., Mon. 17, 1891 [1892], p. 61, pl. iv, f. 12–16.
Dakota: Ellsworth County, Kans.

Betulites westii subintegrifolius Lesquereux, U. S. Geol. Surv., Mon. 17, 1891 [1892], p. 61, pl. iv, f. 1–4.
Dakota: Ellsworth County, Kans.

Betulites, stipules of, Lesquereux, U. S. Geol. Surv., Mon. 17, 1891 [1892], p. 65, pl. v, f. 18.
Dakota: Kansas.

BETULOXYLON Kaiser, Botanisches Centralblatt, vol. 1, 1880, p. 511.

Betuloxylon? sp., Jeffrey and Chrysler, Rept. Vt. State Geol., 1905–6 [1906], p. 200, pl. l, f. 7–11.
Miocene: Brandon, Vt.

BICARPELLITES Perkins, Rept. Vt. State Geol., 1903–4 [1904], p. 90. [Type, *Carpolithes grayana.*]

Bicarpellites abbreviatus Perkins, Rept. Vt. State Geol., 1905–6 [1906], p. 209, pl. lvi, f. 1.
Miocene: Brandon, Vt.

Bicarpellites attenuatus Perkins, Rept. Vt. State Geol., 1905–6 [1906], p. 210, pl. liv, f. 9.
Miocene: Brandon, Vt.

Bicarpellites bicarinatus Perkins, Rept. Vt. State Geol., 1905–6 [1906], p. 210, pl. liv, f. 10.
Miocene: Brandon, Vt.

Bicarpellites brevis Perkins, Rept. Vt. State Geol., 1905–6 [1906], p. 213, pl. lv, f. 13.
Miocene: Brandon, Vt.

Bicarpellites carinatus Perkins, Rept. Vt. State Geol., 1905–6 [1906], p. 210, pl. liv, f. 11.
Miocene: Brandon, Vt.

Bicarpellites crassus Perkins, Rept. Vt. State Geol., 1905–6 [1906], p. 211, pl. lv, f. 2, 3.
Miocene: Brandon, Vt.

Bicarpellites crateriformis Perkins, Rept. Vt. State Geol., 1905–6 [1906], p. 211, pl. lv, f. 1.
Miocene: Brandon, Vt.

Bicarpellites grayana (Lesquereux) Perkins, Rept. Vt. State Geol., 1903–4 [1904], p. 190, lxxviii, f. 69.
Carpolithes grayana Lesquereux, Am. Jour. Sci., 2d ser., vol. 32, 1861, p. 356; Geol. Vt., vol. 2, 1861, p. 714, f. 122.
Miocene: Brandon, Vt.

Bicarpellites inequalis Perkins, Rept. Vt. State Geol., 1905–6 [1906], p. 211, pl. lv, f. 4.
Miocene: Brandon, Vt.

Bicarpellites knowltoni Perkins, Rept. Vt. State Geol., 1903–4 [1904], p. 191, pl. lxxviii, f. 71, 72; Geol. Soc. Am., Bull., vol. 16, 1905, p. 510, pl. 86, f. 13; Rept. Vt. State Geol., 1905–6 [1906], pl. lii, f. 13.
Miocene: Brandon, Vt.

Bicarpellites lanceolatus Perkins, Rept. Vt. State Geol., 1905–6 [1906], p. 211, pl. lv, f. 5.
Miocene: Brandon, Vt.

Bicarpellites latus Perkins, Rept. Vt. State Geol., 1905–6 [1906], p. 212, pl. lv, f. 6, 7.
Miocene: Brandon, Vt.

Bicarpellites major Perkins, Rept. Vt. State Geol., 1905–6 [1906], p. 212, pl. lv, f. 8, 11, 12.
Miocene: Brandon, Vt.

Bicarpellites medius Perkins, Rept. Vt. State Geol., 1905–6 [1906], p. 212, pl. lv, f. 9, 10.
Miocene: Brandon, Vt.

Bicarpellites minimus Perkins, Rept. Vt. State Geol., 1903–4 [1904], p. 192, pl. lxxviii, f. 79.
Miocene: Brandon, Vt.

Bicarpellites obesus Perkins, Rept. Vt. State Geol., 1903–4 [1904], p. 191, pl. lxxviii, f. 75.
Miocene: Brandon, Vt.

Bicarpellites ovatus Perkins, Rept. Vt. State Geol., 1905–6 [1906], p. 213, pl. lv, f. 14.
Miocene: Brandon, Vt.

Bicarpellites papillosus Perkins, Rept. Vt. State Geol., 1905–6 [1906], p. 213, pl. lv, f. 15.
Miocene: Brandon, Vt.

Bicarpellites parvus Perkins, Rept. Vt. State Geol., 1905–6 [1906], p. 214, pl. lv, f. 16.
Miocene: Brandon, Vt.

Bicarpellites quadrangulatus Perkins, Rept. Vt. State Geol., 1905–6 [1906], p. 214, pl. lvi, f. 2.
Miocene: Brandon, Vt.

Bicarpellites rotundus Perkins, Rept. Vt. State Geol., 1903–4 [1904], p. 191, pl. lxxviii, f. 75.
Miocene: Brandon, Vt.

Bicarpellites rugosus Perkins, Rept. Vt. State Geol., 1903–4 [1904], p. 191, pl. lxxviii, f. 70; Geol. Soc. Am., Bull., vol. 16, 1905, p. 510, pl. lxxxvi, f. 14; Rept. Vt. State Geol., 1905–6 [1906], pl. lii, f. 14.
Miocene: Brandon, Vt.

Bicarpellites solidus Perkins, Rept. Vt. State Geol., 1905–6 [1906], p. 215, pl. lvi, f. 4.
Miocene: Brandon, Vt.

Bicarpellites sulcatus Perkins, Rept. Vt. State Geol., 1905–6 [1906], p. 215, pl. lvi, f. 5, 6.
Miocene: Brandon, Vt.

Bicarpellites vermontanus (Lesquereux) Perkins, Rept. Vt. State Geol., 1903–4 [1904], p. 192, pl. lxxviii, f. 88, 89.

Carya vermontana Lesquereux, Am. Jour. Sci., 2d ser., vol. 32, 1861, p. 357 Geol. Vt., vol. 1, 1861, p. 230, f. 130; vol. 2, 1861, p. 713.

Hicorea vermontana (Lesquereux) Knowlton, U. S. Geol. Surv., Bull. 152, 1898, p. 118.

Miocene: Brandon, Vt.

BLECHNUM Linné, Sp. pl., 1753, p. 1077.

Blechnum göpperti Ettingshausen, Farnkräuter, 1865, p. 153.—Lesquereux, U. S. Geol. and Geog. Surv. Terr., Ann. Rept. 1871 [1872], p. 283. [Not afterward recognized by Lesquereux.]
Green River?: Henrys Fork, Utah.

BOMBACITES Berry, U. S. Geol. Surv., Prof. Paper 91, 1916, p. 289. [Type, *B. formosus*.]

Bombacites formosus Berry, U. S. Geol. Surv., Prof. Paper 91, 1916, p. 289, pl. lxxv, f. 1.
Lagrange: Puryear, Tenn.

Bombacites wilcoxianus Berry, U. S. Geol. Surv., Prof. Paper 91, 1916, p. 291, pl. lxxv, f. 2.
Lagrange: Puryear, Tenn.
Wilcox: Near Naborton, La.

BOMBAX Linné, Sp. pl., 1753, p. 511.

Bombax virginiensis Fontaine, U. S. Geol. Surv., Mon. 15, 1889, p. 310, pl. cli, f. 4.
Patapsco: Near Brooke, Va.

BRACHYOXYLON Hollick and Jeffrey, N. Y. Bot. Gard., Mem., vol. 3, 1909, p. 54. [Type, *B. notabile*.]

Brachyoxylon notabile Hollick and Jeffrey, N. Y. Bot. Gard., Mem., vol. 3, 1909, p. 54, pl. xiii, f. 1–6; pl. xiv, f. 2.
Brachyphyllum macrocarpum Newberry. Hollick, and Jeffrey, Am. Nat., vol. 40, 1906, pp. 203, 214, pl. v, f. 1–4.
Araucarioxylon sp., Hollick and Jeffrey, Am. Nat., vol. 40, 1906, p. 203, pl. v, f. 5, 6.
Brachyphyllum, wood of, Jeffrey, Annals Bot., vol. 20, 1906, pp. 384–386, pl. xxvii, f. 1–12.
Raritan: Kreischerville, Staten Island, N. Y.

Brachyoxylon pennsylvanicum (Wherry) Wherry, Acad. Nat. Sci. Philadelphia, Proc., 1912, p. 370, pl. iv, fig. 1–5.
Cedroxylon? pennsylvanicum Wherry, Acad. Nat. Sci. Phila., Proc., 1912, p. 370 [name].
Triassic (Newark): Pennsylvania.

Brachyoxylon sp. Holden, Bot. Gaz., vol. 58, 1914, p. 171, pl. xiii, fig. 16; pl. xiv, f. 17–19.
Magothy: Cliffwood, N. J.

BRACHYPHYLLUM Brongniart, Prod. d'une hist. vég. foss., Paris, 1828, p. 109.

Brachyphyllum crassicaule Fontaine, U. S. Geol. Surv., Mon. 15, 1889 [1890], p. 221, pl. c, f. 4; pl. cix, f. 1–7; pl. cx, f. 1–3; pl. cxi, f. 6, 7; pl. cxii, f. 6–8; pl. clxviii, f. 9; in Ward, U. S. Geol. Surv., Mon. 48, 1906, pp. 529, 557, pl. cxiii, f. 6.—Berry, U. S. Nat. Mus., Proc., vol. 40, 1911, p. 305; Md. Geol. Surv., Lower Cret., 1912, p. 393, pl. lxiv, f. 1–6.

Brachyphyllum crassicaule—Continued.

Patuxent: Trents Reach and Dutch Gap, Va.

Patapsco: Near Brooke, near Widewater, Dumfries Landing, Va.; Fort Foote, Federal Hill (Baltimore), near Glymont and Stump Neck, Md.

Brachyphyllum crassum Lesquereux = *Brachyphyllum macrocarpum*.

Brachyphyllum macrocarpum Newberry, U. S. Geol. Surv., Mon. 26, 1895 [1896], p. 51 [in footnote].—Hollick, N. Y. Bot. Gard., Bull., vol. 3, 1904, p. 406, pl. lxx, f. 4, 5.—Berry, Bull. Torr. Bot. Club, vol. 32, 1905, p. 44, pl. 2, f. 9; Geol. Surv. N. J., Ann. Rept. 1905 [1906], pp. 138, 139; Bull. Torr. Bot. Club, vol. 33, 1906, p. 166, pl. ix, f. 1–3.—Hollick, U. S. Geol. Surv., Mon. 50, 1906, p. 44, pl. iii, f. 9, 10.—Hollick and Jeffrey, Am. Nat., vol. 40, 1906, p. 201, pl. i, f. 15–17; pl. iii, f. 1–5; pl. iv, f. 1–6; pl. v, f. 1–4; N. Y. Bot. Gard., Mem., vol. 3, 1909, p. 33, pl. iv, f. 12–14; pl. ix, f. 7, 8; pl. xi, f. 1, 2, 4, 5; pl. xii, f. 1–6; pl. xiii, f. 1.—Knowlton, U. S. Geol. Surv., Bull. 163, 1900, p. 29, pl. iv, f. 5–6.—Berry, Bull. Torr. Bot. Club, vol. 37, 1910, pp. 20, 183, 504; Geol. Surv. N. J., Bull. 3, 1911, p. 81, pl. vii; U. S. Geol. Surv., Prof. Paper 84, 1914, p. 21, pl. iii, f. 2; Md. Geol. Surv., Upper Cret., 1916, p. 781, pl. liv, f. 4, 5.—Knowlton, U. S. Geol. Surv., Prof. Paper 101, 1917, p. 249, pl. xxxi, f. 4.—Berry, U. S. Geol. Surv., Prof. Paper 112, 1919, p. 59, pl. v, f. 9.

Brachyphyllum crassum (Lesquereux) Lesquereux, U. S. Nat. Mus., Proc., vol. 10, 1887, p. 34; U. S. Geol. Surv., Mon. 17, 1891 [1892], p. 32, pl. ii, f. 5.—Newberry, U. S. Geol. Surv., Mon. 26, 1895 [1896], p. 51, pl. vii, f. 1–7. [Homonym, Tenison-Woods, 1883.]

Thuites crassus Lesquereux, Rept. U. S. Geol. Surv., Terr., vol. 8 (Cret. and Tert. Fl.), 1883, p. 32.

Brachyphyllum sp., Knowlton, Geol. Soc. Am., Bull., vol. 8, 1897, p. 137.

Dakota: Kansas.

Magothy: Cliffwood, N. J.; Deep Cut, Del.; Grove Point, Md.

Raritan: Woodbridge and South Amboy, N. J.; Little Neck, Northport Harbor, Long Island, N. Y.; Kreischerville, Staten Island, N. Y.

Mesaverde: Harper station and North Fork Dutton Creek, Wyo.

———: ?Upper Kanab Valley, Utah.

Black Creek: Court House Bluff, Cape Fear River, N. C.

Black Creek (Middendorf): Middendorf, S. C.

Vermejo: Cimarron, N. Mex.; Trinidad, Colo.

Tuscaloosa: Shirley Mill and Whites Bluff, Ala.

Eutaw: Near Havana, Houston County, Ga.; McBrides Ferry, Ala.

Brachyphyllum macrocarpum Newberry. Hollick and Jeffrey, Am. Nat., vol. 40, 1906, p. 203, pl. v, f. 1–4 = *Brachyoxylon notabile*.

Brachyphyllum macrocarpum formosum Berry, Bull. Torr. Bot. Club, vol. 39, p. 392, pl. xxx; U. S. Geol. Surv., Prof. Paper 84, 1914, p. 106, Md. Geol. Surv., Upper Cret., 1916, p. 783, pl. liii, f. 1.
Woodbine: Arthurs Bluff, Lamar County, Tex.
Magothy: Round Bay, Md.
Eutaw: McBrides Ford, Chattahoochee River, Ga.

Brachyphyllum mamillaria Brongniart, Dist. sci. nat., vol. 57 (Prodrome), 1828, p. 113 (109) [name].—Fontaine, in Ward, U. S. Geol. Surv., Mon. 48, 1905 [1906], p. 130, pl. xxxv, f. 4–8.
Jurassic: Douglas County, Oreg.

Brachyphyllum muensteri Schenk = *Cheirolepis muensteri.*

Brachyphyllum parceramosum Fontaine, U. S. Geol. Surv., Mon. 15, 1890, p. 223, pl. cx, fig. 4; in Ward, U. S. Geol. Surv., Mon. 48, 1906, pp. 517, 538.—Berry, U. S. Nat. Mus., Proc., vol. 40, 1911, p. 306; Md. Geol. Surv., Lower Cret., 1912, p. 395, pl. lxv, f. 4, 5.
Brachyphyllum texense Fontaine, U. S. Nat. Mus., Proc., vol. 16, 1893, p. 209, pl. xxxviii, figs. 3–5; pl. xxxix, figs. 1, 1a.
Patuxent: Telegraph station near Lorton, Va.; New Reservoir, D. C.
Arundel: Arlington, Md.
Trinity: Glen Rose, Tex.

Brachyphyllum? storrsii Ward, U. S. Geol. Surv., Mon. 48, 1905 [1906], p. 176, pl. xlv, f. 6. [Probably not a plant.]
Brachyphyllum? sp., Fontaine, in Diller, Am. Jour. Sci., 4th ser., vol. 15, 1903, p. 352.
Jurassic ?: Trinity County, Calif.

Brachyphyllum texense Fontaine = *Brachyphyllum parceramosum.*

Brachyphyllum yorkense Fontaine, in Ward, U. S. Geol. Surv., Twentieth Ann. Rept., pt. 2, 1900, p. 251, pl. xxxi, f. 6–9.
Triassic (Newark): York County, Pa.

Brachyphyllum sp. (cone), Fontaine, U. S. Geol. Surv., Mon. 15, 1889 [1890], p. 223, pl. cxxxv, f. 8.
Patuxent: Trents Reach, Va.

Brachyphyllum sp. (cone), Fontaine, U. S. Geol. Surv., Mon. 15, 1889 [1890], p. 224, pl. clxviii, f. 2.
Patuxent: Trents Reach, Va.

Brachyphyllum sp. (cone) Fontaine, U. S. Geol. Surv., Mon. 15, 1889 [1890], p. 224, pl. cxxxv, f. 9.
Patuxent: Trents Reach, Va.

Brachyphyllum, wood of, Jeffrey, Annals Bot., vol. 20, 1906, pp. 384–386, pl. xxvii, f. 1–12.
Raritan: Kreischerville, Staten Island, N. Y.

Brachyphyllum sp. ? cone of, Hollick and Jeffrey, N. Y. Bot. Gard., Mon., vol. 3, 1909, p. 37, pl. ix, f. 5, 6; pl. xi, f. 3; pl. xiv, f. 3.
Raritan: Kreischerville, Staten Island, N. Y.

Brachyphyllum? sp. Fontaine, in Diller, Am. Jour. Sci., 4th ser., vol. 15, 1903, p. 352 = *Brachyphyllum? storrsii.*

Brachyphyllum sp., Knowlton, Geol. Soc. Am., Bull., vol. 8, 1896, pp. 137, 140 = *Brachyphyllum macrocarpum.*

Brachyphyllum sp., Newberry, in Macomb, Rept. Expl. Exped., 1876, p. 69, pl. iv, f. 4.
Triassic: Abiquiu, N. Mex.

BRANDONIA Perkins, Rept. Vt. State Geol., 1903–4 [1904], p. 192. [Type, *B. globulus.*]

Brandonia globulus Perkins, Rept. Vt. State Geol., 1903–4 [1904], p. 192, pl. lxxviii, f. 73, 74; Geol. Soc. Am., Bull., vol. 16, 1905, p. 513, pl. 87, f. 23; Rept. Vt. State Geol., 1905–6 [1906], pl. liii, f. 23.
Miocene: Brandon, Vt.

BRASENIA Schreber, Gen. pl., 1789, p. 372.

Brasenia antiqua Newberry, U. S. Nat. Mus., Proc., vol. 5, 1882 [1883], p. 514; U. S. Geol. Surv., Mon. 35, 1898, p. 93, pl. lxviii, f. 7.
Green River: Green River, Wyo.

Brasenia antiqua Dawson = *Nelumbo dawsoni.*

Brasenia peltata Pursh = *Brasenia purpurea.*

Brasenia purpurea (Michaux) Caspary.
Brasenia peltata Pursh. Penhallow, Geol. Soc. Am., Bull., vol. 1, 1890, p. 326; Roy. Soc. Canada, Trans., 2d ser., vol. 2, 1896, p. 70; Brit. Assoc. Adv. Sci., Bradford meeting, 1900, sec. C, p. 335 [8].—Berry, Jour. Geol., vol. 25, 1917, p. 662.
Pleistocene: Greens Creek and Besserers Wharf, Ottawa River, Canada; Vero, Fla.

BROMELIA Linné, Sp. pl., 1753, p. 285.

Bromelia? tenuifolia Lesquereux, U. S. Geol. Surv., Mon. 17, 1891 [1892], p. 41, pl. i, f. 13.
Dakota: Ellsworth County, Kans.

BROMUS Linné, Sp. pl., 1753, p. 76.

Bromus ciliatus Linné. Penhallow, Geol. Soc. Am., Bull., vol. 1, 1890, p. 332; Roy. Soc. Canada, Trans., 2d ser., vol. 2, 1896, p. 70; Brit. Assoc. Adv. Sci., Bradford meeting, 1900, sec. C, p. 335 [8].
Pleistocene: Greens Creek, Ottawa River, Canada.

BRUNSWICKIA Wherry, U. S. Nat. Mus., Proc., vol. 51, 1916, p. 329. [Type, *B. dubia.*]

Brunswickia dubia Wherry, U. S. Nat. Mus., Proc., vol. 51, 1916, p. 329, pl. xxx.
Triassic (Newark): Near Sellersville station, Bucks County, and Coopersburg, Lehigh County, Pa.

BRYUM (Dillenius) Linné, Sp. pl., 1753, p. 1115.

Bryum sp., Hinde, Canadian Jour. [Can. Institute], vol. 15, 1878, p. 399.
Pleistocene: Scarboro Heights, Ontario, Canada.

BUETTNERIA Löfling, It., 1758, p. 313.

Buettneria? perplexans Cockerell, Am. Mus. Nat. Hist., Bull., vol. 24, 1908, p. 104, pl. x, f. 39.
Miocene: Florissant, Colo.

BUMELIA Swartz, Prodr. Veg. Ind. Occ., 1788, p. 49.

Bumelia americana (Lesquereux) Berry, U. S. Geol. Surv., Prof. Paper 91, 1916, p. 337, pl. c, f. 6.
Sapotacites americanus Lesquereux, in Safford, Geol. Tenn., 1869, p. 428, pl. K, f. 8.
Lagrange: Puryear, Tenn.; * Somerville, Tenn.

Bumelia florissanti Lesquereux, Rept. U. S. Geol. Surv. Terr., vol. 8 (Cret. and Tert. Fl.), 1883, p. 174, pl. xxxiv, f. 4.
Miocene: Florissant, Colo.

Bumelia grenadensis Berry, U. S. Geol. Surv., Prof. Paper 91, 1916, p. 338, pl. cviii, f. 3.
Wilcox (Grenada): Grenada, Miss.

Bumelia hurleyensis Berry, U. S. Geol. Surv., Prof. Paper 91, 1916, p. 338, pl. cviii, f. 2.
Wilcox (Ackerman): Hurleys, Benton County, Miss.

Bumelia marcouana (Heer) Lesquereux = *Liriodendron marcouanum*.

Bumelia oklahomensis Berry, U. S. Nat. Mus., Proc., vol. 54, 1918, p. 634, pl. xciv, f. 1.
Miocene: Beaver Creek, 10 miles east of Beaver City, Okla.

Bumelia praenuntia Berry, Bull. Torr. Bot. Club, vol. 41, 1914, p. 298 [name]; Md. Geol. Surv., Upper Cret., 1916, p. 893, pl. xc, f. 1.
Magothy: Grove Point, Md.

Bumelia preangustifolia Berry, U. S. Geol. Surv., Prof. Paper 98, 1916, p. 204, pl. xlvii, f. 11.
Citronelle: Lambert, Ala.

Bumelia pseudohorrida Berry, U. S. Geol. Surv., Prof. Paper 91, 1916, p. 336, pl. c, f. 1.
Lagrange: Puryear, Tenn.

Bumelia pseudo-lanuginosa Hollick, Md. Geol. Surv., Pliocene and Pleistocene, 1906, p. 236, pl. lxxi, f. 18, 19.
Pleistocene (Sunderland): Head of Island Creek, Calvert County, Md.

Bumelia pseudotenax Berry, U. S. Geol. Surv., Prof. Paper 91, 1916, p. 337, pl. c, f. 2.
Banksia helvetica Herr. Lesquereux, Am. Philos. Soc., Trans., vol. 13, 1869, p. 418, pl. xvi, f. 6.

Bumelia pseudotenax—Continued.
Wilcox (Ackerman): Hurleys, Miss.
Wilcox (Holly Springs): Early Grove, Miss.
Wilcox: Calaveras Creek, Wilson County, Tex.

Bumelia? rhomboidea Lesquereux, U. S. Geol. Surv., Mon. 17, 1891 [1892], p. 113, pl. li, f. 10.
Dakota: Kansas.

Bumelia wilcoxiana Berry, U. S. Geol. Surv., Prof. Paper 91, 1916, p. 336, pl. c, f. 4, 5; pl. cvii, f. 3.
Wilcox (Holly Springs): Early Grove, Miss.
Lagrange: Puryear, Tenn.

CABOMBA Aublet, Pl. Guian., vol. 1, 1775, p. 321.

Cabomba? gracilis Newberry, U. S. Nat. Mus., Proc., vol. 5, 1882 [1883], p. 514.—[Lesquereux], U. S. Geol. and Geog. Surv. Terr., Ill. Cret. and Tert. Pl., 1878, pl. vii, f. 1, under "aquatic rootlets of *Equisetum*," pl. viii, f. 2, under "*Equisetum*."—Newberry, U. S. Geol. Surv., Mon. 35, 1898, p. 91, pl. xxii, f. 1; pl. xxiii, f. 1.
Fort Union: Fort Union, N. Dak.

Cabomba grandis Newberry = *Cabomba inermis*.

Cabomba inermis (Newberry) Hollick, in Newberry, U. S. Geol. Surv., Mon. 35, 1898, p. 92, pl. xxii, f. 2; pl. xxiii, f. 2.
Psilotum inerme Newberry, N. Y. Lyc. Nat. Hist., Ann., vol. 9, 1868, p. 38.—[Lesquereux], U. S. Geol. and Geog. Surv. Terr., Ill. Cret. and Tert. Pl., 1878, pl. vii, f. 2, under "aquatic rootlets of *Equisetum*," pl. viii, f. 3.
Psilotitis inermis (Newberry) Schimper, Pal. vég., vol. 3, 1874, p. 547.—Knowlton, U. S. Geol. Surv., Bull. 152, 1898, p. 186.
Cabomba grandis Newberry, U. S. Nat. Mus., Proc., vol. 5, 1883, p. 514. Cf. Newberry, U. S. Geol. Surv., Mon. 35, 1898, footnote, p. 92.
Fort Union: Fort Union, N. Dak.

CAENOMYCES Berry, U. S. Geol. Surv., Prof. Paper 91, 1916, p. 162.

Caenomyces annulata Berry, U. S. Geol. Surv., Prof. Paper 91, 1916, p. 163, pl. xlv, f. 17*b*.
Lagrange: Near Grand Junction, Fayette County, Tenn.

Caenomyces cassiae Berry, U. S. Geol. Surv., Prof. Paper 91, 1916, p. 163, pl. ix, f. 1.
Lagrange: Near Grand Junction, Fayette County, Tenn.

Caenomyces laurinea Berry, U. S. Geol. Surv., Prof. Paper 91, 1916, p. 162, pl. lxxxviii, f. 4.
Wilcox (Holly Springs): Lafayette County, Miss.

Caenomyces pestalozzites Berry, U. S. Geol. Surv., Prof. Paper 91, 1916, p. 162, pl. ix, f. 2, 3.
Wilcox (Holly Springs): Lafayette and Grenada counties, Miss.

Caenomyces sapotae Berry, U. S. Geol. Surv., Prof. Paper 91, 1916, p. 162, pl. xcix, f. 4.
Wilcox (Holly Springs): Lafayette County, Miss.

CAESALPINIA Linné, Sp. pl., 1753, p. 380.

Caesalpinia citronellensis Berry, U. S. Geol. Surv., Prof. Paper 98, 1916, p. 202, pl. xlvii, f. 6.
Citronelle: Lambert, Ala.

Caesalpinia cookiana Hollick, in Newberry, U. S. Geol. Surv., Mon. 26, 1895 [1896], p. 94, pl. xlii, f. 49, 50.—Berry, Geol. Surv. N. J., Bull. 3, 1911, p. 161.
Raritan: New Jersey, but no locality given by Newberry.

Caesalpinia linearis Lesquereux = *Mimosites linearis.*

Caesalpinia middendorfensis Berry, U. S. Geol. Surv., Prof. Paper 84, 1914, p. 46, pl. x, f. 7.
Black Creek (Middendorf): Middendorf, S. C.

Caesalpinia raritanensis Berry, Bull. Torr. Bot. Club, vol. 36, 1909, p. 257; Geol. Surv. N. J., Bull. 3, 1911, p. 161, pl. xx, f. 3.
Raritan: South Amboy, N. J.

Caesalpinia wilcoxiana Berry, U. S. Geol. Surv., Prof. Paper 91, 1916, p. 235, pl. l, f. 9–12.
Wilcox (Holly Springs): Early Grove and Holly Springs, Marshall County, Miss.
Lagrange: Puryear, Henry County, and Pinson, Madison County, Tenn.

CAESALPINITES Saporta, Ann. sci. nat., 4th ser., Bot., vol. 17 (Études, vol. 1, pt. 2), 1862, p. 288 (135).

Caesalpinites aculeatafolia Berry, U. S. Geol. Surv., Prof. Paper 91, 1916, p. 237, pl. l, f. 15.
Lagrange: Puryear, Henry County, Tenn.

Caesalpinites bentonensis Berry, U. S. Geol. Surv., Prof. Paper 91, 1916, p. 237, pl. l, f. 14.
Wilcox: Benton, Saline County, Ark.

Caesalpinites mississippiensis Berry, U. S. Geol. Surv., Prof. Paper 91, 1916, p. 237, pl. l, f. 16.
Wilcox (Holly Springs): Holly Springs, Marshall County, Miss.

Caesalpinites pinsonensis Berry, U. S. Geol. Surv., Prof. Paper 91, 1916, p. 236, pl. l, f. 13.
Lagrange: Pinson, Marshall County, Tenn.

Calamites arenaceus Brongniart = *Equisetum rogersii.*

Calamites disjunctus Emmons = *Equisetum rogersii.*

Calamites planicostatus Rogers = *Schizoneura planicostata.*

Calamites punctatus Emmons = *Sphenozamites rogersianus.*

Calamites rogersii Bunbury = *Equisetum rogersii.*

Calamopsis danai Lesquereux = *Chamaederea danai.*

Callicoma microphylla Ettingshausen. Lesquereux = *Myrica drymeia.*

CALLISTEMOPHYLLUM Ettingshausen, Abhand. K. k. geol. Reichsanst., vol. 2, pt. 3, No. 2, 1853, p. 83.

Callistemophyllum heerii Ettingshausen, Sitzb. Wien. Akad., vol. 55, Abt. 1, 1867, p. 261, pl. iii, f. 9.—Lesquereux, U. S. Geol. Surv., Mon. 17, 1891 [1892], p. 138, pl. xxxviii, f. 8.
Dakota: Ellsworth County, Kans.

Callistemophyllum latum Dawson, Roy. Soc. Canada, Trans., vol. 7, sec. 4, 1889, p. 72, pl. x, f. 8.—Penhallow, Rept. Tert. Pl. Brit. Col., 1908, p. 40.
Eocene?: Mackenzie River, Northwest Territory.

CALLITRIS Ventenat, Dec. Nov. Gen., 1808, p.

Callitris? sp., Fontaine, U. S. Geol. Surv., Mon. 15, 1889, p. 272, pl. clxix, f. 8.
Patuxent: Dutch Gap Canal, Va.

CALYCITHES [*Calycites* of authors] Massalongo, Schizzo Geogn., Verona, 1850, p. 72.

Calycites alatus Hollick, N. Y. Bot. Gard., Bull., vol. 3, 1904, p. 417, pl. 72, f. 8; U. S. Geol. Surv., Mon. 50, 1906, p. 109, pl. v, f. 24.
Tricalycites papyraceus Newberry. Hollick, N. Y. Acad. Sci., Trans., vol. 15, 1895, p. 6.
Magothy: Montauk Point, Long Island, N. Y.

Calycites davillaformis Berry, U. S. Geol. Surv., Prof. Paper 91, 1916, p. 350, pl. civ, f. 7.
Lagrange: Puryear, Tenn.

Calycites diospyriformis Newberry, U. S. Geol. Surv., Mon. 26, 1895 [1896], p. 132, pl. xlvi, f. 39–41.—Berry, Geol. Surv. N. J., Bull. 3, 1911, p. 213, pl. xxix, f. 2–4.
Raritan: Woodbridge, N. J.

Calycites hexaphylla Lesquereux = *Diospyros hexaphylla.*

Calycites middendorfensis Berry, U. S. Geol. Surv., Prof. Paper 84, 1914, p. 63, pl. x, f. 4.
Black Creek (Middendorf): Middendorf, S. C.

Calycites obovatus Hollick, U. S. Geol. Surv., Mon. 50, 1906, p. 109, pl. v, f. 23.
Magothy: Nashaquitsa, Marthas Vineyard, Mass.

Calycites ostryaformis Berry, U. S. Geol. Surv., Prof. Paper 91, 1916, p. 351, pl. civ, f. 4, 5.
Wilcox: Calaveras Creek, Wilson County, Tex.

116234°—19—Bull. 696——9

Calycites parvus Newberry, U. S. Geol. Surv., Mon. 26, 1895 [1896], p. 131, pl. xlvi, f. 28, 29.—Berry, Geol. Surv. N. J., Bull. 3, 1911, p. 214.
  Raritan: Woodbridge, N. J.

Calycites polysepala Newberry, N. Y. Lyc. Nat. Hist., Ann., vol. 9, 1868, p. 31 [name].—[Lesquereux], U. S. Geol. and Geog. Surv. Terr., Ill. Cret. and Tert. Pl., 1878, pl. xxv, f. 3.—Newberry, U. S. Geol. Surv., Mon. 35, 1898, p. 139, pl. xl, f. 3.
  Fort Union: *Fort Union, N. Dak., near mouth of Yellowstone River; near Jordan and near Hell Creek, Mont.
  Hanna: Carbon County, Wyo.

Calycites sexpartitus Berry, U. S. Geol. Surv., Prof. Paper 112, 1919, p. 139, pl. xxix, f. 8.
  Tuscaloosa: Snow place, Tuscaloosa County, Ala.

Calycites sp., Lesquereux, U. S. Geol. Surv., Mon. 17. 1891 [1892], p. 221, pl. xxii, f. 8.
  Dakota: Kansas.

CALYPTRANTHES Swartz, Prodr., 1788, p. 5.

Calyptranthes eocenica Berry, U. S. Geol. Surv., Prof. Paper 91, 1916, p. 319, pl. xc, f. 5.
  Lagrange: Puryear, Tenn.

CANAVALIA Adanson, Pl. Fam., II, 1763, p. 325.

Canavalia acuminata Berry, U. S. Geol. Surv., Prof. Paper 91, 1916, p. 249, pl. cx, f. 4, 5.
  Wilcox (Holly Springs): Early Grove, Marshall County, Miss.
  Lagrange: Puryear, Henry County, Tenn.

Canavalia eocenica Berry, U. S. Geol. Surv., Prof. Paper 91, 1916, p. 248, pl. liii, f. 3–6.
  Wilcox (Grenada): Grenada, Grenada County, Miss.
  Wilcox (Holly Springs): Holly Springs, Miss.
  Lagrange: Puryear, Henry County, Tenn.

CANCELLOPHYCUS Saporta, Paléont. franç. (Pl. jurass., vol. 1), 1872, p. 126.

Cancellophycus rhombicum Ulrich, Harriman Alaska Exped., vol. 4, 1904, p. 139, pl. xx, f. 1.
  Yakutat: Pogibski Island, opposite Kodiak, Alaska.

CANNA Linné, Sp. pl., 1753, p. 1.

Canna eocenica Berry, U. S. Geol. Surv., Prof. Paper 91, 1916, p. 181, pl. xv, f. 7, 8.
  Wilcox (Holly Springs): Oxford, Miss.
  Wilcox (Grenada): Grenada, Miss.
  Wilcox: Old Port Caddo Landing, Harrison County, Tex.

Canna? magnifolia Knowlton, U. S. Geol. Surv., Prof. Paper 101, 1917, p. 254, pl. xxxvi, f. 3.
Vermejo: *Rockland mine, Walsenburg, Colo.; also Rockvale, Colo.

Canna? sp., Knowlton, U. S. Geol. Surv., Prof. Paper 101, 1918, p. 255.
Vermejo: Rockvale, Colo.

CAPPARITES Berry, U. S. Geol. Surv., Prof. Paper 112, 1919, p. 95. [Type *C. cynophylloides*.]

Capparites cynophylloides Berry, U. S. Geol. Surv., Prof. Paper 112, 1919, p. 95, pl. xxii, f. 1.
Tuscaloosa: Shirleys Mill, Ala.

Capparites orbicularis Berry, U. S. Geol. Surv., Prof. Paper 112, 1919, p. 96, pl. xxii, f. 2, 3.
Tuscaloosa: Shirleys Mill, Ala.

CARAPA Aublet, Guian. II. Suppl., 1775, p. 32.

Carapa eolignitica Berry, U. S. Geol. Surv., Prof. Paper 91, 1916, p. 253, pl. lv, f. 4; pl. lx, f. 4.—Knowlton, U. S. Geol. Surv., Prof. Paper, 101, 1918, p. 327, pl. lxix, f. 2.
Lagrange: Puryear, Henry County, Tenn., and Wickliffe, Ballard County, Ky.
Raton: Near Rugby, Colo. [Identification ?]

Carapa xylocarpoides Berry, Am. Jour. Sci., 4th ser., vol. 43, 1917, p. 298, f. 1a (p. 299).
Claiborne: Fort Gaines, Clay County, Ga.

Carduus florissantensis Cockerell = *Lomatites hakeaefolia*.

CAREX (Ruppius) Linné, Sp. pl., 1753, p. 972.

Carex aquatilis Wahlenberg. Macoun, in Coleman, Jour. Geol., vol. 3, 1895, p. 626.—Penhallow, Roy. Soc. Canada, Trans., 2d ser., vol. 2, sec. 4, 1896, p. 70; Brit. Assoc. Adv. Sci., Bradford meeting, 1900, p. 335.
Pleistocene: Scarborough Heights, Ontario, Canada.

Carex berthoudi Lesquereux, U. S. Geol. and Geog. Surv. Terr., Ann. Rept. 1872 [1873], p. 377; Rept. U. S. Geol. Surv. Terr., vol. 7 (Tert. Fl.), 1878, p. 92, pl. ix, f. 3, 4.
Denver: Golden, Colo.

Carex burrardiana Penhallow, in Dawson, Roy. Soc. Canada, Trans., 2d ser., vol. 1, sec. 4, 1895–96, p. 145, pl. iv, f. 6.—Penhallow, Rept. Tert. Pl. Brit. Col., 1908, p. 40.
Eocene: Burrard Inlet, Vancouver, British Columbia.

Carex clarkii Berry, Am. Nat., vol. 39, 1905, p. 347, text f. 1; Bull. Torr. Bot. Club, vol. 33, 1906, p. 169; Geol. Surv. N. J., Ann. Rept., 1905 [1906], p. 138; Johns Hopkins Univ. Circ., 1907, No. 7, p. 81; U. S. Geol. Surv., Prof. Paper 84, 1914, p. 29; Md. Geol. Surv., Upper Cret., 1911, p. 808.

Carex clarkii—Continued.
Magothy: Morgan, Cliffwood, and Kinkora, N. J.; Deep Cut, Del.; Grove Point, Md.
Black Creek (Middendorf): Rocky Point, S. C.

Carex magellanica Lamarck. Penhallow, Geol. Soc. Am., Bull., vol. 1, 1890, p. 325; Roy. Soc. Canada, Trans., 2d ser., vol. 2, sec. 4, 1896, p. 70; Brit. Assoc. Adv. Sci., Bradford meeting, 1900, p. 335[8].
Pleistocene: Greens Creek, Ottawa River, Canada.

Carex servata Heer, Fl. foss. arct., vol. 2, Abt. 2 (Fl. foss. alask.), 1869, p. 24, pl. i, f. 13, 13c, d.—Knowlton, U. S. Nat. Mus., Proc., vol. 17, 1894, p. 216; Geol. Soc. Am., Bull., vol. 5, 1893, p. 580.
Kenai: Port Graham and Herendeen Bay, Alaska.

Carex utriculata Boott. Macoun, in Coleman, Jour. Geol., vol. 3, 1895, p. 626.—Penhallow, Roy. Soc. Canada, Trans., 2d ser., vol. 2, sec. 4, 1896, p. 70; Brit. Assoc. Adv. Sci., Bradford meeting, 1900, p. 335[8].
Pleistocene: Scarborough Heights, Ontario, Canada.

Carex vancouverensis Penhallow, in Dawson, Roy. Soc. Canada Trans., 2d ser., vol. 1, sec. 4, 1895–96, p. 144, pl. iv, f. 5.—Penhallow, Rept. Tert. Pl. Brit. Col., 1908, p. 41.
Eocene: Burrard Inlet, Vancouver, British Columbia.

Carex sp., Berry, Jour. Geol., vol. 25, 1917, p. 662.
Pleistocene: Vero, Fla.

Carex sp., Lesquereux, U. S. Nat. Mus., Proc., vol. 10, 1867, p. 36.—Knowlton, idem, vol. 17, 1894, p. 216; Geol. Soc. Am., Bull., vol.5, 1893, p. 580.
Kenai ?: Sitka, Alaska.

Carex sp., Penhallow, Rept. Tert. Pl. Brit. Col., 1908, p. 41.
Eocene ?: Quilchena and Horsefly River, British Columbia.

CARPINOXYLON Vater, Zeitschr. Deutsch. geol. Ges., vol. 36, 1884, p. 848.

Carpinoxylon pfefferi Platen, Sitzungsb. Naturf. Ges., Leipzig, vol. 34, 1907 [1908], p. 40, pl. ii, f. 1, 2.
—— ?: California.

CARPINUS (Tournefort) Linné, Sp. pl., 1753, p. 998.

Carpinus attenuata Lesquereux, Rept. U. S. Geol. Surv. Terr., vol. 8 (Cret. and Tert. Fl.), 1883, p. 152, pl. xxvii, f. 10.—Knowlton, U. S. Nat. Mus., Proc., vol. 51, 1916, p. 263.
Miocene: *Florissant, Colo.

Carpinus betuloides Unger, Chloris Protogaea, 1845, pt. 6, p. lxxix [name]; Syn. Pl. Foss., 1845, p. 220.—Knowlton, U. S. Geol. Surv., Bull. 204, 1900, p. 38.
Clarno (upper part): Bridge Creek, Oreg.

Carpinus caroliniana Walter. Berry, Am. Nat., vol. 41, 1907, p. 692, pl. i, f. 8, 9; Jour. Geol., vol. 15, 1907, p. 340; Am. Jour. Sci., 4th ser., vol. 29, 1910, p. 395.
  Pleistocene: Abercrombies Landing, Chattahoochee River, Ala.; Neuse River, N. C.

Carpinus elkoana (Lesquereux) Cockerell, Univ. Colorado Studies, vol. 3, 1906, p. 174.
  Quercus elkoana Lesquereux, U. S. Geol. and Geog. Surv. Terr., Ann. Rept. 1873 [1874], p. 413.
  Miocene: Elko station, Nev.

Carpinus fraterna Lesquereux, Rept. U. S. Geol. Surv. Terr., vol. 8 (Cret. and Tert. Fl.), 1883, p. 152, pl. xxvii, f. 12–14.—Knowlton, U. S. Nat. Mus., Proc., vol. 51, 1916, p. 263.
  Miocene: *Florissant, Colo.

Carpinus grandis Unger, Synop. Pl. Foss., 1845, p. 220.—Lesquereux, Am. Jour. Sci., vol. 32, 1861, p. 360; Geol. Vt., vol. 2, 1861, p. 716, f. 151?—Newberry, Boston Jour. Nat. Hist., vol. 7, 1863, p. 519.—Heer, Fl. foss. arct., vol. 2, Abt. 2, 1869, p. 29, pl. ii, f. 12.—Lesquereux, U. S. Geol. and Geog. Surv. Terr., Ann. Rept. 1874 [1876], p. 313; Rept. U. S. Geol. Surv. Terr., vol. 7 (Tert. Fl.), 1878, p. 143, pl. xix, f. 9; pl. lxiv, f. 8–10.—Knowlton, U. S. Nat. Mus., Proc., vol. 17, 1894, p. 220; Geol. Soc. Am., Bull., vol. 5, 1893, p. 582.—Newberry, U. S. Geol. Surv., Mon. 35, 1898, p. 59, pl. liv, f. 3 in part; pl. lv, f. 6.—Knowlton, U. S. Geol. Surv., Bull. 204, 1902, p. 38.—Penhallow, Rept. Tert. Pl. Brit. Col., 1908, p. 41.—Berry, Jour. Geol., vol. 17, 1909, p. 23.
  Mascall: John Day Basin, Oreg.
  Miocene?: Brandon, Vt.
  Eocene?: Quilchena, Similkameen River, Tulameen River, Stump Lake, and Tranquille River, British Columbia.
  Miocene: Florissant, Colo.
  Calvert: Richmond, Va.
  Kenai: Cook Inlet and Port Graham, Alaska.
  Eocene: Birch Bay, Wash.

Carpinus pseudo-caroliniana Hollick, Md. Geol. Surv., Pliocene and Pleistocene, 1906, p. 225, pl. lxxi, f. 10.
  Pleistocene (Sunderland): Head of Island Creek, Calvert County, Md.

Carpinus pyramidalis (Göppert) Heer. Lesquereux, U. S. Nat. Mus., Proc., vol. 11, 1888, p. 18 = *Carpinus grandis*. [Cf. Knowlton, U. S. Geol. Surv., Bull. 204, 1902, p. 38.]

CARPITES Schimper, Pal. vég., vol. 3, 1874, p. 421.

Carpites alatus Knowlton, U. S. Geol. Surv., Bull. 257, p. 150, pl. xvi, f. 2.
Judith River: *Willow Creek, Fergus County, and north of Musselshell River, Mont.

Carpites baueri Knowlton, U. S. Geol. Surv., Prof. Paper 98, 1916, p. 342, pl. lxxxviii, f. 2.
Fruitland: Coal Creek, 35 miles south of Farmington, San Juan County, N. Mex.

Carpites bursaeformis Lesquereux, Rept. U. S. Geol. Surv. Terr., vol. 7 (Tert. Fl.), 1878, p. 306, pl. lx, f. 30.
Post-Laramie: Black Buttes, Wyo.

Carpites cocculoides (Heer) Lesquereux, Rept. U. S. Geol. Surv. Terr., vol. 7 (Tert. Fl.), 1878, p. 307, pl. lx, f. 32–35.
Carpolithes cocculoides Heer, Fl. foss. arct., vol. 2, 1869, p. 484, pl. lii, f. 9.
Hanna: Carbon, Wyo.

Carpites cocculoides major Lesquereux, Rept. U. S. Geol. Surv. Terr., vol. 7 (Tert. Fl.), 1878, p. 307, pl. lx, f. 38, 39.
Hanna: Carbon, Wyo.

Carpites coffeaeformis Lesquereux, Rept. U. S. Geol. Surv. Terr., vol. 7 (Tert. Fl.), 1878, p. 303, pl. lx, f. 6, 7.—Knowlton, U. S. Geol. Surv., Prof. Paper 101, 1918, p. 348.
Denver: *Golden, Colo.
Raton: Near Brilliant, N. Mex.

Carpites coniger Lesquereux, U. S. Geol. Surv., Mon. 17, 1891 [1892], p. 221, pl. xxxviii, f. 17.
Dakota: Ellsworth County, Kans.

Carpites cordiformis Lesquereux, U. S. Geol. Surv., Mon. 17, 1891 [1892], p. 220, pl. xxii, f. 9.
Dakota: Kansas.

Carpites costatus Lesquereux, Rept. U. S. Geol. Surv. Terr., vol. 7 (Tert. Fl.), 1878, p. 303, pl. lx, f. 5.
Denver: Golden, Colo.

Carpites dakotensis Knowlton, n. sp.
Nordenskiöldia borealis Herr. Lesquereux, U. S. Geol. Surv., Mon. 17, 1892, p. 219, pl. xliv, f. 6.
Dakota: Kansas.

Carpites fragariaeformis Lesquereux, U. S. Nat. Mus., Proc., vol. 11, 1888, p. 16.
Mascall: John Day Basin, Oreg.

Carpites gemmaceus Lesquereux, Rept. U. S. Geol. Surv. Terr., vol. 8 (Cret. and Tert. Fl.), 1883, p. 204, pl. xl, f. 19.
Miocene: Florissant, Colo.

Carpites glumaeformis Lesquereux, Rept. U. S. Geol. Surv. Terr., vol. 7 (Tert. Fl.), 1878, p. 304, pl. xxxv, f. 4d; pl. lx, f. 14–17.
Post-Laramie: Black Buttes, Wyo.

Carpites inequalis Perkins, Rept. Vt. State Geol., 1903–4 [1904], p. 193, pl. lxxviii, f. 85.
Miocene: Brandon, Vt.

Carpites judithae Knowlton, U. S. Geol. Surv., Bull. 257, p. 149, pl. xvi, f. 4.
Judith River: *Willow Creek, Fergus County, and 7 miles south of Chinook, Mont.

Carpites lakesii Knowlton, U. S. Geol. Surv., Prof. Paper 130, 19—, p. —, pl. xix, f. 6–8 [in preparation].
Laramie: Near Golden, Colo.

Carpites laurineus Lesquereux, Rept. U. S. Geol. Surv. Terr., vol. 7 (Tert. Fl.), 1878, p. 304, pl. lx, f. 20, 21.
Denver: Golden, Colo.

Carpites lesquereuxiana Knowlton, U. S. Geol. Surv., Prof. Paper 130, 19—, p. — [in preparation].
Laramie: Coal Creek, Boulder County, Colo.

Carpites ligatus Lesquereux, Rept. U. S. Geol. Surv. Terr., vol. 7 (Tert. Fl.), 1878, p. 307, pl. lx, f. 36, 36a.
Raton?: Placer Mountain, N. Mex.

Carpites lineatus (Newberry) Lesquereux, Rept. U. S. Geol. Surv. Terr., vol. 7 (Tert. Fl.), 1878, p. 302, pl. lx, f. 1–1d.
Carpolithus lineatus Newberry, N. Y. Lyc. Nat. Hist., Annals, vol. 9, 1868, p. 31.
Evanston: Evanston, Wyo.

Carpites liriophylli? Lesquereux, Rept. U. S. Geol. Surv. Terr., vol. 8 (Cret. and Tert. Fl.), 1883, p. 77, pl. xi, f. 5.—Berry, Bull. Torr. Bot. Club, vol. 33, 1906, p. 174; Md. Geol. Surv., Upper Cret., 1916, p. 839, pl. lxx, f. 4, 5.
Dakota: *Morrison, Colo.
Magothy: Deep Cut, Del.

Carpites milioides Lesquereux, Rept. U. S. Geol. Surv. Terr., vol. 8 (Cret. and Tert. Fl.), 1883, p. 204, pl. xl, f. 18.
Miocene: Florissant, Colo.

Carpites minutulus Lesquereux, Rept. U. S. Geol. Surv. Terr., vol. 7 (Tert. Fl.), 1878, p. 305, pl. lx, f. 25.—Berry, Bull. Torr. Bot. Club, vol. 32, 1905, p. 47.
Denver: *Golden, Colo.
Magothy: Cliffwood, N. J.

Carpites mitratus Lesquereux, Rept. U. S. Geol. Surv. Terr., vol. 7 (Tert. Fl.), 1878, p. 304, pl. lx, f. 18, 19.
Post-Laramie: Black Buttes, Wyo.

Carpites myricarum Lesquereux = *Ginkgo* sp. Knowlton.

Carpites obovatus Lesquereux, U. S. Geol. Surv., Mon. 17, 1891 [1892], p. 221, pl. lxii, f. 5.
Dakota: Ellsworth County, Kans.

Carpites ovalis Perkins, Rept. Vt. State Geol., 1903–4 [1904], p. 194, pl. lxxviii, f. 86.
Miocene: Brandon, Vt.

Carpites oviformis Lesquereux, Rept. U. S. Geol. Surv. Terr., vol. 7 (Tert. Fl.), 1878, p. 302, pl. xxx, f. 6a.
Denver: Golden, Colo.

Carpites pealei Lesquereux, Rept. U. S. Geol. Surv. Terr., vol. 7 (Tert. Fl.), 1878, p. 306, pl. lx, f. 31.
Miocene: Florissant, Colo.

Carpites pedunculatus Knowlton, U. S. Geol. Surv., Mon. 32, pt. 2, 1899, p. 755, pl. ciii, f. 3.
Fort Union: Elk Creek, Yellowstone River, Yellowstone National Park.

Carpites pruni Knowlton, U. S. Geol. Surv., Bull. 257, p. 149, pl. xv, f. 6, 6a.
Judith River: Willow Creek, Fergus County, Mont.

Carpites rhomboidalis Lesquereux, Rept. U. S. Geol. Surv. Terr., vol. 7 (Tert. Fl.), 1878, p. 306, pl. lx, f. 28, 29.—Knowlton, U. S. Geol. Surv., Prof. Paper 130, 19—, p. — [in preparation].
Laramie: *Golden, Colo.

Carpites spiralis (Lesquereux) Lesquereux, Rept. U. S. Geol. Surv. Terr., vol. 7 (Tert. Fl.), 1878, p. 306, pl. lx, f. 27.
Carpolithes spiralis Lesquereux, U. S. Geol. and Geog. Surv. Terr., Ann. Rept. 1871 [1872], Suppl. p. 16.
Raton ?: Placer Mountain, N. Mex.

Carpites tiliaceus (Heer) Lesquereux, U. S. Geol. Surv., Mon. 17, 1891 [1892], p. 221, pl. xxii, f. 6, 7.
Carpolithes tiliaceus Heer, Mioc. Balt. Fl., 1869, p. 101, pl. xxx, f. 42, 43.
Dakota: Kansas.

Carpites triangulosus Lesquereux, Rept. U. S. Geol. Surv. Terr., vol. 7 (Tert. Fl.), 1878, p. 302, pl. lx, f. 4; pl. lxii, f. 19, 20.
Denver: Golden, Colo.

Carpites trigonus Perkins, Rept. Vt. State Geol., 1903–4 [1904], p. 194, pl. lxxviii, f. 87.
Miocene: Brandon, Vt.

Carpites utahensis Lesquereux, Rept. U. S. Geol. Surv. Terr., vol. 7 (Tert. Fl.), 1878, p. 305, pl. lx, f. 22.
Evanston: Evanston, Wyo.

Carpites valvatus Lesquereux, Rept. U. S. Geol. Surv. Terr., vol. 7 (Tert. Fl.), 1878, p. 307, pl. lx, f. 37.
——: Fort Steele, Wyo.

Carpites verrucosus Lesquereux, Rept. U. S. Geol. Surv. Terr., vol. 7 (Tert. Fl.), 1878, p. 305, pl. lx, f. 23.
Post-Laramie: Black Buttes, Wyo.

Carpites viburni Lesquereux, U. S. Geol. and Geog. Surv. Terr., Bull., vol. 1, 1875 [1876], p. 382; idem, Ann. Rept., 1874 [1876], p. 308; Rept. U. S. Geol. Surv. Terr., vol. 7 (Tert. Fl.), 1878, p. 305, pl. lx, f. 26, 26a.
Post-Laramie: Black Buttes, Wyo.

Carpites? sp., Lesquereux, U. S. Geol. Surv., Mon. 17, 1891 [1892], p. 221.
Carpolithes? Lesquereux, Rept. U. S. Geol. Surv. Terr., vol. 6 (Cret. Fl.), 1874, p. 114, pl. xxvii, f. 5; pl. xxx, f. 11.
Dakota: Kansas?

Carpites sp., Knowlton, in Spurr, U. S. Geol. Surv., Eighteenth Ann. Rept., pt. 3, 1898, p. 192.
Kenai: Miller's mine, Yukon River, Alaska.

Carpites sp., Knowlton, Washington Acad. Sci., Proc., vol. 11, 1909, p. 207.
Lance: Converse County, Wyo.

Carpites sp., Knowlton, Washington Acad. Sci., Proc., vol. 11, 1909, p. 207.
Lance: Converse County, Wyo.

CARPOLITHES Schlotheim, Petrefactenkunde, 1820, p. 418.

Carpolithes arachioides Lesquereux = *Leguminosites? arachioides.*

Carpolithes brandoniana Lesquereux, Am. Jour. Sci., 2d ser., vol. 32, 1861, p. 356; Geol. Vt., vol. 1, 1861, p. 229, f. 111; vol. 2, 1861, p. 713.—Knowlton, Bull. Torr. Bot. Club, vol. 29, 1902, p. 640, pl. 125, f. 1, 2, 11, 12.—Perkins, Rept. Vt. State Geol., 1903–4 [1904], p. 175, pl. lxxv, f. 10, 11, 20.
Miocene: Brandon, Vt.

Carpolithes brandoniana elongata Lesquereux = *Glossocarpellites elongatus.*

Carpolithes brandoniana obtusa Lesquereux = *Glossocarpellites obtusus.*

Carpolithes bucklandii? Williamson, MS. Lindley and Hutton, Foss. Fl. Great Brit., vol. 3, 1836, p. 103, pl. clxxxix, f. 3, 5.—Fontaine, in Ward, U. S. Geol. Surv., Mon. 48, 1905 [1906], p. 138, pl. xxxvii, f. 9.
Jurassic: Douglas County, Oreg.

Carpolithes bursaeformis Lesquereux = *Prunoides bursaeformis.*

Carpolithes cocculoides? Heer, Fl. foss. arct., vol. 2, Abt. 4, 1869, p. 484, pl. lii, f. 9, 9b.—Lesquereux, U. S. Geol. and Geog. Surv. Terr., Ann. Rept. 1871 [1872], p. 290. [Not afterward recognized by Lesquereux.]
——: Carbon, Wyo.

Carpolithes compositus Lesquereux = *Palmocarpon compositum*.

Carpolithes corrugatus (Lesquereux) Cockerell, Am. Jour. Sci., 4th ser., vol. 27, 1909, p. 447.
Palmocarpon corrugatum Lesquereux, Rept. U. S. Geol. Surv. Terr., vol. 7 (Tert. Fl.), 1878, p. 121, pl. xi, f. 10, 11.
Denver: Golden, Colo.

Carpolithes dentatus Penhallow, in Dawson, Roy. Soc. Canada, Trans., vol. 8, 1890, p. 89, text f. 26; Rept. Tert. Pl. Brit. Col., 1908, p. 42.
Eocene?: Stump Lake, Quilchena, and Horsefly River, British Columbia.
Eocene?: Tranquille River, British Columbia.

Carpolithes elongatus (Lesquereux) Perkins = *Glossocarpellites elongatus*.

Carpolithes emarginatus Perkins, Rept. Vt. State Geol., 1903–4 [1904], p. 177, lxxv, f. 4.
Miocene: Brandon, Vt.

Carpolithes falcatus Lesquereux, U. S. Geol. and Geog. Surv. Terr., Ann. Rept. 1872 [1873], p. 398. [Not afterward recognized by the author.]
——: Black Buttes, Wyo.

Carpolithes fissilis Lesquereux = *Tricarpellites fissilis*.

Carpolithes grandis Perkins = *Glossocarpellites grandis*.

Carpolithes grayana Lesquereux = *Bicarpellites grayana*.

Carpolithes hitchcockii Perkins, Rept. Vt. State Geol., 1903–4 [1904], p. 179, pl. lxxv, f. 19.
Miocene: Brandon, Vt.

Carpolithes horridus Dawson = *Antholithes horridus*.

Carpolithes irregularis Lesquereux, Am. Jour. Sci., 2d ser., vol. 32, 1861, p. 356; Geol. Vt., vol. 2, 1861, p. 714, f. 120, 121, 123, 125, 128.
Miocene: Brandon, Vt.

Carpolithes lescurii Hitchcock = *Nyssa lescurii*.

Carpolithes lineatus Newberry, N. Y. Lyc. Nat. Hist., Ann., vol. 9, 1868, p. 31 [name].—[Lesquereux], U. S. Geol. and Geog. Surv. Terr., Ill. Cret. and Tert. Fl., 1878, pl. xxv, f. 1.—Newberry, U. S. Geol. Surv., Mon. 35, 1898, p. 138, pl. xl, f. 1.—Knowlton, Wash. Acad. Sci., Proc., vol. 11, 1909, p. 198.
Fort Union: Fort Union, N. Dak.
Lance: 9 miles above Glendive, Mont.

Carpolithes macrophyllus Cockerell, Torreya, vol. 11, 1911, p. 235, text f. 1.—Knowlton, U. S. Nat. Mus., Proc., vol. 51, 1916, p. 289, pl. xxvii, f. 7.
Miocene: *Florissant, Colo.

Carpolithes (Zamites) meridionalis Dawson, Roy. Soc. Canada, Trans., vol. 11, 1893 [1894], sec. 4, p. 64, pl. x, f. 42.
Upper Cretaceous: Port McNeill, Vancouver Island.

Carpolithes mexicanus Lesquereux = *Palmocarpon mexicanum.*

Carpolithes mucronatus Perkins, Rept. Vt. State Geol., 1903–4 [1904], p. 179, pl. lxxv, f. 15.
Miocene: Brandon, Vt.

Carpolithes obtusus (Lesquereux) Perkins = *Glossocarpellites obtusus.*

Carpolithes osseus Lesquereux, U. S. Geol. and Geog. Surv. Terr., Ann. Rept. 1872 [1873], p. 404 [not afterward recognized by its author].—Knowlton, U. S. Geol. Surv., Mon. 32, pt. 2, 1899, p. 754.
Fort Union: Elk Creek, Yellowstone National Park.

Carpolithes ovatus Perkins, Rept. Vt. State Geol., 1903–4 [1904], p. 178, pl. lxxv, f. 9.
Miocene: Brandon, Vt.

Carpolithes palmarum Lesquereux = *Palmocarpon palmarum.*

Carpolithes parvus Perkins = *Glossocarpellites parvus.*

Carpolithes pruniformis Lesquereux. Perkins, Rept. Vt. State Geol., 1903 [1904], p. 208 [synonymy] = *Prunoides bursaeformis.*

Carpolithes seminulum Heer, Fl. foss. arct., vol. 1, 1868, p. 139, pl. xxiii, f. 1.—Penhallow, Rept. Tert. Pl. Brit. Col., 1908, p. 43.
Eocene: Mackenzie River basin, Northwest Territory.

Carpolithes simplex Perkins, Rept. Vt. State Geol., 1903–4 [1904], p. 178, pl. lxxv, f. 12, 16.
Miocene: Brandon, Vt.

Carpolithes solidus Perkins, Rept. Vt. State Geol., 1903–4 [1904], p. 179, pl. lxxv, f. 17, 18.
Miocene: Brandon, Vt.

Carpolithes spinosus Newberry, U. S. Nat. Mus., Proc., vol. 5, 1882 [1883], p. 514; U. S. Geol. Surv., Mon. 35, 1898, p. 138, pl. lxviii, f. 2, 3.—Knowlton, U. S. Geol. Surv., Prof. Paper 101, 1917, p. 349.
Vermejo or Raton?: North branch of Purgatoire River, Colo.

Carpolithes spiralis Lesquereux = *Carpites spiralis.*

Carpolithes tiliaceus Herr = *Carpites tiliaceus.*

Carpolithes venosus? Sternberg. Lesquereux, Am. Jour. Sci., 2d ser., vol. 32, 1861, p. 361 = *Staphidoides perkinsi.*

Carpolithes vermontanus Perkins, Rept. Vt. State Geol., 1903–4 [1904], p. 179, pl. lxxv, f. 13.
Miocene: Brandon, Vt.

Carpolithes sp., Dawson, Brit. N. A. Bound. Com. (Rept. Geol. and Rec. Vic. 49th Parallel), 1875, App. A, p. 330.—Penhallow, Rept. Tert. Pl. Brit. Col., 1908, p. 41, f. 5.
Paskapoo (Fort Union): Badlands west of Woody Mountain; Mackenzie River, Porcupine Creek, Great Valley, and Saskatchewan.

Carpolithes sp., Dawson, Roy. Soc. Canada, Trans., vol. 1, 1882 [1883], p. 29.
Upper Cretaceous: Baynes Sound, Vancouver Island.

Carpolithes sp., Dawson, Roy. Soc. Canada, Trans., vol. 10, 1892, p. 90, text f. 15.
Kootenai: Anthracite, British Columbia.

Carpolithes sp., Lesquereux, U. S. Geol. and Geog. Surv. Terr., Ann. Rept., 1873 [1874], p. 418.
Miocene: Florissant, Colo.

Carpolithes sp., Penhallow, Rept. Tert. Pl. Brit. Col., 1908, p. 42, f. 6.
Eocene?: Tranquille River, British Columbia.

Carpolithus of authors = *Carpolithes* Schlotheim.

Carpolithus agglomeratus Fontaine, U. S. Geol. Surv., Mon. 15, 1889, p. 267, pl. cxxxiv, f. 5.
Patuxent: Fredericksburg, Va.

Carpolithus barrensis Ward, U. S. Geol. Surv., Nineteenth Ann. Rept., pt. 2, 1899, p. 692, pl. clxix, f. 14, 15.
Lakota: Barr's tunnel, Crook County, Wyo.

Carpolithus brookensis Fontaine, U. S. Geol. Surv., Mon. 15, 1889, p. 268, pl. cxxxv, f. 2, 4; pl. cxxxvi, f. 6; pl. clxvii, f. 6.
Patuxent: Dutch Gap Canal, Va.
Patapsco: Brooke, Va.

Carpolithus cliffwoodensis Berry, N. Y. Bot. Gard., Bull., vol. 3, 1903, p. 100, pl. xlviii, f. 6; Geol. Surv. N. J., Ann. Rept. 1905 [1906], p. 138.
Magothy: Cliffwood, N. J.

Carpolithus conjugatus Fontaine, U. S. Geol. Surv., Mon. 15, 1889, p. 267, pl. cxxxiv, f. 9.
Patuxent: Fredericksburg, Va.

Carpolithus curvatus Fontaine, U. S. Geol. Surv., Mon. 15, 1889, p. 269, pl. cxxxv, f. 17.
Patuxent: Fredericksburg, Va.

Carpolithus dictyolomoides Berry, U. S. Geol. Surv., Prof. Paper 91, 1916, p. 353, pl. cxi, f. 2, 3.
Lagrange: Puryear, Tenn.

Carpolithus douglasensis Fontaine, in Ward, U. S. Geol. Surv., Mon. 48, 1905 [1906], p. 139, pl. xxxvii, f. 13.
Jurassic: Douglas County, Oreg.

Carpolithus drupaeformis Hollick, N. Y. Acad. Sci., Trans., vol. 16, 1897, p. 134, pl. xi, f. 4, 4a.—Berry, N. Y. Bot. Gard., Bull., vol. 3, 1903, p. 101.
Magothy: Cliffwood, N. J.

Carpolithus dubius Berry = *Carpolithus juglandiformis*.

Carpolithus elongatus Fontaine, in Ward, U. S. Geol. Surv., Mon. 48, 1905 [1906], p. 139, pl. xxxvii, f. 12.
Jurassic: Douglas County, Oreg.

Carpolithus euonymoides Hollick, U. S. Geol. Surv., Mon. 50, 1906, p. 110, pl. vii, f. 2.
Carpolithus sp., Hollick, N. Y. Acad. Sci., Trans., vol. 12, 1892, p. 38, pl. i, f. 4.
Raritan: Tottenville, Staten Island, N. Y.

Carpolithus fasciculatus Fontaine, U. S. Geol. Surv., Mon. 15, 1889, p. 265, pl. cxxxiv, f. 1; in Ward, U. S. Geol. Surv., Nineteenth Ann. Rept., pt. 2, 1899, p. 691, pl. clxix, f. 11, 12.
Patuxent: *Fredericksburg, Va.
Lakota: South Fork of Hay Creek, Crook County, Wyo.

Carpolithus floribundus Newberry, U. S. Geol. Surv., Mon. 26, 1895 [1896], p. 133, pl. xlvi, f. 17–21.—Hollick, idem, Mon. 50, 1906, p. 110, pl. vii, f. 20, 21.—Berry, Geol. Surv. N. J., Bull. 3, 1911, p. 216; U. S. Geol. Surv., Prof. Paper 112, 1919, p. 139, pl. xxix, f. 1.
Raritan: Woodbridge, N. J.
Magothy: Gay Head, Marthas Vineyard, Mass.
Tuscaloosa: Shirleys Mill, Ala.

Carpolithus foenarius Ward, U. S. Geol. Surv., Nineteenth Ann. Rept., pt. 2, 1899, p. 693, pl. clxix, f. 17, 18.
Lakota: Hay Creek coal field, Crook County, Wyo.

Carpolithus geminatus Fontaine, U. S. Geol. Surv., Mon. 15, 1889, p. 267, pl. cxxxiv, f. 10.
Patuxent: Dutch Gap Canal, Va.

Carpolithus grenadensis Berry, U. S. Geol. Surv., Prof. Paper 91, 1916, p. 352, pl. cxii, f. 15.
Wilcox (Grenada): Grenada, Miss.

Carpolithus harveyi Fontaine, U. S. Nat. Mus., Proc., vol. 16, 1893, p. 278, pl. xliii, f. 3.
Trinity: Glen Rose, Tex.

Carpolithus henryensis Berry, U. S. Geol. Surv., Prof. Paper 91, 1916, p. 352, pl. cxii, f. 16.
Lagrange: Puryear, Tenn.

Carpolithus hirsutus Newberry, U. S. Geol. Surv., Mon. 26, 1895 [1896], p. 134, pl. clvi, f. 14, 14a.—Hollick, idem, Mon. 50, 1906, p. 110, pl. vii, f. 3–8.—Berry, Geol. Surv. N. J., Bull. 3, 1911, p. 216.

Carpolithus spinosus Newberry. Hollick, Geol. Soc. Am., Bull., vol. 7, 1895, p. 13.

Raritan: Woodbridge, N. J.
Magothy: Gay Head, Marthas Vineyard, Mass.

Carpolithus hyoseritiformis Berry, U. S. Geol. Surv., Prof. Paper 91, 1916, p. 353, pl. cxii, f. 4.
Lagrange: Puryear, Tenn.

Carpolithus juglandiformis Berry, N. Y. Bot. Gard., Bull., vol. 3, 1903, p. 100, pl. xlvi, f. 8; Bull. Torr. Bot. Club, vol. 31, 1904, p. 81.

Carpolithus dubius Berry, N. Y. Bot. Gard., Bull., vol. 3, 1903, p. 100, pl. xlviii, f. 7.

Magothy: Cliffwood, N. J.

Carpolithus latus Fontaine, U. S. Geol. Surv., Mon. 15, 1889, p. 269, pl. cxxxv, f. 1, 3.
Patuxent: Dutch Gap Canal, Va.

Carpolithus mattewanensis Berry, Bull. Torr. Bot. Club, vol. 32, 1905, p. 48, pl. ii, f. 7; Geol. Surv. N. J., Ann. Rept. 1905 [1906], p. 138.
Magothy: Cliffwood, N. J.

Carpolithus montium-nigrorum Ward, U. S. Geol. Surv., Nineteenth Ann. Rept., pt. 2, 1899, p. 692, pl. clxix, f. 13.
Lakota: South Fork Hay Creek, Crook County, Wyo.

Carpolithus mucronatus Fontaine, U. S. Geol. Surv., Mon. 15, 1889, p. 270, pl. cxxxvi, f. 15.
Patapsco: Near Brooke, Va.

Carpolithus najasoides Berry, in Cooke and Shearer, U. S. Geol. Surv., Prof. Paper 120, 1918, p. 55 [name].
Barnwell (Twiggs): Twiggs County, Ga.

Carpolithus obovatus Fontaine, U. S. Nat. Mus., Proc., vol. 16, 1873, p. 278, pl. xliii, f. 5.
Trinity: Glen Rose, Tex.

Carpolithus olallensis Ward, U. S. Geol. Surv., Mon. 48, 1905 [1906], p. 137, pl. xxxvii, f. 7, 8.
Jurassic: Douglas County, Oreg.

Carpolithus oregonensis Fontaine, in Ward, U. S. Geol. Surv., Mon. 48, 1905 [1906], p. 139, pl. xxxvii, f. 10, 11.
Jurassic: Douglas County, Oreg.

Carpolithus ostryaeformis Berry, Bull. Torr. Bot. Club, vol. 32, 1905, p. 47.

Carpolithus virginiensis Fontaine? Berry, N. Y. Bot. Gard., Bull., vol. 3, 1903, p. 100, pl. 48, f. 5.

Magothy: Cliffwood, N. J.

Carpolithus ovaeformis Newberry, U. S. Geol. Surv., Mon. 26, 1895 [1896], p. 134, pl. xlvi, f. 15, 16.—Berry, Geol. Surv. N. J., Bull. 3, 1911, p. 217.

Raritan: Woodbridge, N. J.

Carpolithus pilocarpoides Berry, U. S. Geol. Surv., Prof. Paper 91, 1916, p. 352, pl. cxii, f. 11.

Wilcox (Grenada): Grenada, Miss.

Carpolithus prangosoides Berry, U. S. Geol. Surv., Prof. Paper 91, 1916, p. 351, pl. civ, f. 9.

Lagrange: Puryear, Tenn.

Carpolithus proteoides Berry, U. S. Geol. Surv., Prof. Paper 91, 1916, p. 353, pl. cxii, f. 2.

Lagrange: Puryear, Tenn.

Carpolithus pruniformis Newberry, U. S. Geol. Surv., Mon. 26, 1895 [1896], p. 133, pl. xlvi, f. 42.—Berry, Geol. Surv. N. J., Bull. 3, 1911, p. 215.

Raritan: Woodbridge, N. J.

Carpolithus puryearensis Berry, U. S. Geol. Surv., Prof. Paper 91, 1916, p. 351, pl. civ, f. 8.

Lagrange: Puryear, Tenn.

Carpolithus septloculus Berry, Maryland Geol. Surv., Upper Cret., 1916, p. 900, pl. lxxxiv, f. 3.

Magothy: Deep Cut, Del.

Carpolithus sessilis Fontaine, U. S. Geol. Surv., Mon. 15, 1889, p. 269, pl. cxxxvi, f. 9.

Patuxent: Near Dutch Gap Canal, Va.

Carpolithus sophorites Berry, U. S. Geol. Surv., Prof. Paper 91, 1916, p. 352, pl. cxii, f. 7.

Wilcox (Grenada): Grenada, Miss.

Carpolithus spinosus Newberry. Hollick, Geol. Soc. Am., Bull., vol. 7, 1895, p. 13 = *Carpolithus hirsutus*.

Carpolithus storrsii Fontaine, Am. Jour. Sci., 4th ser., vol. 2, 1896, p. 274; in Ward, U. S. Geol. Surv., Twentieth Ann. Rept., pt. 2, 1900, p. 363, pl. lxv, f. 4–6.

Jurassic: Oroville, Calif.

Carpolithus tennesseensis Berry, U. S. Geol. Surv., Prof. Paper 91, 1916, p. 352, pl. cxii, f. 17.

Lagrange: Puryear, Tenn.

Carpolithus ternatus Fontaine, U. S. Geol. Surv., Mon. 15, 1889, p. 265, pl. cxxxiv, f. 2–4, 6, 8.
Patuxent: Fredericksburg, Va.

Carpolithus tuscaloosensis Berry, U. S. Geol. Surv., Prof. Paper 112, 1919, p. 139.
Tuscaloosa: Snow place, near Northport, and near Jones Ferry, all Tuscaloosa County, Ala.

Carpolithus vaccinioides Hollick, U. S. Geol. Surv., Mon. 50, 1906, p. 110, pl. vii, f. 19, 19a.
Carpolithus sp., Hollick, N. Y. Acad. Sci., Trans., vol. 12, 1892, p. 38, pl. i, f. 16, 16a.
Raritan: Kreischerville, Staten Island, N. Y.

Carpolithus virginiensis Fontaine, U. S. Geol. Surv., Mon. 15, 1889, p. 266, pl. cxxxiv, f. 11–14; pl. cxxxv, f. 1–5; pl. clxviii, f. 7; in Ward, U. S. Geol. Surv., Nineteenth Ann. Rept., pt. 2, 1899, p. 693, pl. clxix, f. 16.
Patuxent: Telegraph station and near Dutch Gap Canal, Va.
Lakota: North side of Pine Creek, Crook County, Wyo.
Kootenai: Great Falls, Mont.

Carpolithus virginiensis Fontaine? Berry, N. Y. Bot. Gard., Bull., vol. 3, 1903, p. 100, pl. xlviii, f. 5 = *Carpolithus ostryaeformis.*

Carpolithus woodbridgensis Newberry, U. S. Geol. Surv., Mon. 26, 1895 [1896], p. 133, pl. xlvi, f. 22.—Berry, Geol. Surv. N. J., Bull. 3, 1911, p. 217.
Raritan: Woodbridge, N. J.

Carpolithus sp. Hollick, N. Y. Acad. Sci., Trans., vol. 12, 1892, p. 38, pl. i, f. 16 = *Carpolithus vaccinioides.*

Carpolithus sp. Hollick, N. Y. Acad. Sci., Trans., vol. 12, 1892, p. 38, pl. i, f. 4 = *Carpolithus euonymoides.*

Carpolithus sp., Hollick, N. Y. Acad. Sci., Trans., vol. 12, 1892, p. 38, pl. i, f. 8; U. S. Geol. Surv., Mon. 50, 1906, p. 111, pl. vii, f. 9, 10.
Magothy: Gay Head, Marthas Vineyard, Mass.
Raritan: Kreischerville, Staten Island, N. Y.

Carpolithus sp. Hollick, N. Y. Acad. Sci., Trans., vol. 12, 1892, p. 38, pl. i, f. 6; U. S. Geol. Surv., Mon. 50, 1906, p. 111, pl. vii, f. 11.
Raritan: Green Ridge, Staten Island, N. Y.

Carpolithus sp. Hollick, N. Y. Acad. Sci., Trans., vol. 12, 1892, p. 38, pl. i, f. 11; U. S. Geol. Surv., Mon. 50, 1906, p. 111, pl. vii, f. 12.
Raritan: Kreischerville, Staten Island, N. Y.

Carpolithus sp. Hollick, N. Y. Acad. Sci., Trans., vol. 12, 1892, p. 38, pt. 1, f. 12; U. S. Geol. Surv., Mon. 50, 1906, p. 211, pl. vii, f. 13.
Raritan: Kreischerville, Staten Island, N. Y.

Carpolithus sp. Hollick, N. Y. Acad. Sci., Trans., vol. 12, 1892, p. 39, pl. i, f. 15; U. S. Geol. Surv., Mon. 50, 1906, p. 111, pl. vii, f. 14.
Raritan: Kreischerville, Staten Island, N. Y.

Carpolithus sp. Hollick, N. Y. Acad. Sci., Trans., vol. 12, 1892, p. 39, pl. i, f. 14; U. S. Geol. Surv., Mon. 50, 1906, p. 111, pl. vii, f. 15.
Raritan: Kreischerville, Staten Island, N. Y.

Carya Nuttall, 1818 = *Hicoria* Rafinesque, 1808.

Carya alba Nuttall = *Hicoria alba.*

Carya antiquorum Newberry = *Hicoria antiquora.*

Carya bilinica (Unger) Ettingshausen = *Hicoria juglandiformis.*

Carya bruckmani Heer = *Hicoria princetonia.*

Carya elaenoides (Unger) Heer = *Hicoria elaenoides.*

Carya globulosa Lesquereux, in Knowlton = *Cucumites lesquereuxii.*

Carya heerii (Ettingshausen) Heer, Fl. tert. Helv., vol. 3, 1859, p. 93.—Lesquereux, U. S. Geol. and Geog. Surv. Terr., Ann. Rept., 1871 [1872], p. 289. [Not afterward recognized by Lesquereux.]
Green River: Green River, Wyo. ?

Carya olivaeformis Nuttall = *Hicoria pecan.*

Carya porcina Nuttall. Meehan, in Mercer, Acad. Nat. Sci. Philadelphia, Jour., 2d ser., vol. 11, 1899, pp. 277, 281, text f. 8 = *Hicoria glabra.*

Carya rostrata (Göppert) Schimper = *Hicoria rostrata.*

Carya vermontana Lesquereux = *Bicarpellites vermontanus.*

Carya verrucosa Lesquereux = *Hicoria verrucosa.*

Carya sp. Dawson = *Hicoria* sp.

CASSIA Linné, Sp. pl., 1753, p. 376.

Cassia bentonensis Berry, U. S. Geol. Surv., Prof. Paper 91, 1916, p. 229, pl. l, f. 1.
Wilcox: Benton, Saline County, Ark.; Calaveras Creek, Wilson County, Tex.

Cassia concinna? Heer. Lesquereux = *Cassia evanstonensis.*

Cassia emarginata Berry, U. S. Geol. Surv., Prof. Paper 91, 1916, p. 223, pl. xlv, f. 17b; pl. xlviii, f. 5.
Wilcox (Holly Springs): Holly Springs, Marshall County, Miss.
Lagrange: Grand Junction, Fayette County, Tenn.

Cassia eolignitica Berry, U. S. Geol. Surv., Prof. Paper 91, 1916, p. 229, pl. xlviii, f. 2–4.
Wilcox (Holly Springs): Early Grove, Marshall County, Miss.
Wilcox: Mansfield, La.
Lagrange: Puryear, Henry County, and Grand Junction, Fayette County, Tenn.

Cassia evanstonensis Knowlton and Cockerell, n. name.
Cassia concinna? Heer. Lesquereux, U. S. Geol. and Geog. Surv. Terr., Ann. Rept. 1872 [1873], p. 402; Rept. U. S. Geol. Surv. Terr., vol. 7 (Tert. Fl.), 1878, p. 299, pl. lix, f. 8, 8a.
Evanston: Evanston, Wyo.

Cassia fayettensis Berry, U. S. Geol. Surv., Prof. Paper 91, 1916, p. 232, pl. xlix, f. 5–8.
Wilcox (Holly Springs): Holly Springs, Marshall County, and Vaughans, Benton County, Miss.
Lagrange: Puryear, Henry County, Grand Junction, Fayette County, Tenn., and Wickliffe, Ballard County, Ky.

Cassia fischeri Herr. Lesquereux = *Cytisus florissantinus*.

Cassia fisheriana Knowlton, U. S. Geol. Surv., Prof. Paper 101, 1918, p. 327, pl. lxxviii, f. 4.
Raton: Fishers Peak mine, Trinidad, Colo.

Cassia georgiana Berry, in Cooke and Shearer, U. S. Geol. Surv., Prof. Paper 120, 1918, p. 55 [name].
Barnwell (Twiggs): Twiggs County, Ga.

Cassia glenni Berry, U S. Geol. Surv., Prof. Paper 91, 1916, p. 233, pl. xlv, f. 15, 16, 17a, 18; pl. lii, f. 6.
Andromeda vaccinifolia affinis Lesquereux, Am. Jour. Sci., vol. 27, 1859, p. 364; in Safford, Geol. Tenn., 1869, p. 428, pl. K, f. 4a, 4b.
Wilcox (Holly Springs): Holly Springs and Early Grove, Marshall County, Miss.
Lagrange: Somerville and Grand Junction, Fayette County, Tenn.; Trenton, Gibson County, Tenn.; and Wickliffe, Ballard County, Ky.

Cassia glenni major Berry, U. S. Geol. Surv., Prof. Paper 91, 1916, p. 234, pl. cxi, f. 4.
Lagrange: Puryear, Henry County, Tenn.

Cassia idahoensis Knowlton, n. name.
Cassia obtusa Knowlton, U. S. Geol. Surv., Eighteenth Ann. Rept., pt. 3, 1898, p. 731, pl. c, f. 4, 5. [Homonym, *C. obtusa* Clos., C. Gay, Fl. Chile, vol. 2, 1845–1852, p. 235.]
Payette: Marsh, Idaho.

Cassia insularis Hollick, N. Y. Bot. Gard., Bull., vol. 8, 1912, p. 164, pl. clxvii, f. 3.
Magothy: Glen Cove, N. Y.

Cassia? laramiensis Knowlton, U. S. Geol. Surv., Prof. Paper 130, 19—, p. —, pl. xix, f. 3 [in preparation].
Laramie: Marshall, Colo.

Cassia lowii Berry, U. S. Geol. Surv., Prof. Paper 91, 1916, p. 234, pl. lii, f. 7–9.
Wilcox (Grenada): Grenada, Grenada County, Miss.

Cassia marshallensis Berry, U. S. Geol. Surv., Prof. Paper 91, 1916, p. 232, pl. l, f. 6, 7.
Wilcox (Holly Springs): Early Grove, Marshall County, Miss.
Lagrange: Grand Junction, Fayette County, Tenn., and Wickliffe, Ballard County, Ky.

Cassia mississippiensis Berry, U. S. Geol. Surv., Prof. Paper 91, 1916, p. 235, pl. li, f. 10, 11.
Wilcox (Grenada): Grenada, Grenada County, Miss.
Lagrange: Grand Junction, Fayette County, Tenn.

Cassia obtusa Knowlton = *Cassia idahoensis.*

Cassia phaseolites? Unger. Lesquereux, U. S. Nat. Mus., Proc., vol. 11, 1888, p. 16 = *Salix engelhardti.*

Cassia podogonioides Ettingshausen = *Podogonium americanum.*

Cassia polita Lesquereux, U. S. Geol. Surv., Mon. 17, 1891 [1892], p. 146.
Dakota: Kansas.

Cassia problematica Lesquereux, U. S. Geol. Surv., Mon. 17, 1891 [1892], p. 146, pl. xxxviii, f. 3.
Dakota: Ellsworth County, Kans.

Cassia puryearensis Berry, U. S. Geol. Surv., Prof. Paper 91, 1916, p. 230, pl. li, f. 13, 14.
Lagrange: Puryear, Henry County, Tenn.

Cassia richardsoni Knowlton, U. S. Geol. Surv., Prof. Paper 101, 1918, p. 327, pl. xciii, f. 2.
Raton: Berwind, Colo.

Cassia sapindoides Knowlton, U. S. Geol. Surv., Prof. Paper 101, 1918, p. 327, pl. xcvi, f. 1.
Raton: Wootton, Colo.

Cassia tennesseensis Berry, U. S. Geol. Surv., Prof. Paper 91, 1916, p. 228, pl. xlix, f. 3, 4.
Wilcox (Holly Springs): Early Grove, Marshall County, Miss.
Lagrange: Grand Junction, Fayette County, Tenn.

Cassia toraformis Berry, U. S. Geol. Surv., Prof. Paper 98, 1916, p. 68, pl. xii, f. 6, 7.
Calvert: Good Hope Hill, D. C.

Cassia vaughani Berry, U. S. Geol. Surv., Prof. Paper 112, 1919, p. 100, pl. xxii, f. 8.
Tuscaloosa: Glen Allen, Ala.

Cassia wilcoxiana Berry, U. S. Geol. Surv., Prof. Paper 91, 1916, p. 230, pl. i, f. 2–5.
Wilcox (Holly Springs): Holly Springs, Marshall County, Miss.

Cassia sp., Hollick, U. S. Geol. Surv., Mon. 50, 1906, p. 84, pl. xxxii, f. 13.

Magothy: Gay Head, Marthas Vineyard, Mass.

Cassia sp. ? Hollick, Maryland Geol. Surv., Pliocene and Pleistocene, 1906 p. 233, pl. lxxi, f. 20.

Pleistocene (Sunderland): Head of Island Creek, Calvert County, Md.

Cassia sp. ? (fruit) Newberry, U. S. Geol. Surv., Mon. 35, 1898, p. 113, pl. xlvi, f. 10.—Knowlton, U. S. Geol. Surv., Bull. 204, 1902, p. 69.

Clarno (upper part): Bridge Creek, Grant County, Oreg.

CASTALIA Salisbury, in König and Sims, Ann. Bot., vol. 2, 1805, p. 71.

Castalia? duttoniana Knowlton, U. S. Geol. Surv., Bull. 163, p. 55, pl. xiii, f. 7.

Castalia sp. Knowlton, in Stanton and Knowlton, Geol. Soc. Am., Bull., vol. 8, 1897, p. 140.

Mesaverde: North Fork Dutton Creek, Laramie Plains, Wyo.

Castalia leei Knowlton, U. S. Geol. Surv., Prof. Paper 101, 1917, p. 307, pl. lxxix, f. 2.

Raton: Near Walsenburg, Colo.

Castalia stantoni Knowlton, U. S. Geol. Surv., Bull. 257, p. 147, pl. xix, f. 4.

Judith River: *Willow Creek, Fergus County, and Bull Mountain field, Mont.

Castalia sp., Knowlton, Geol. Soc. Am., Bull., vol. 8, 1897, p. 133.

Lance: Lance Creek, Converse County, Wyo.

Castalia sp. Knowlton, in Stanton and Knowlton, Geol. Soc. Am., Bull., vol. 8, 1897, p. 140 = ***Castalia? duttoniana.***

Castalia sp. Knowlton = *Nelumbo?* sp.

CASTANEA Adanson, Fam. Pl., vol. 2, 1763, p. 375.

Castanea atavia Unger. Lesquereux, Rept. U. S. Geol. Surv. Terr., vol. 8 (Cret. and Tert. Fl.), pl. lii, f. 2 = ***Quercus horniana.***

Castanea castaneaefolia (Unger) Knowlton, U. S. Geol. Surv., Bull. 152, 1898, p. 60.—Penhallow, Roy. Soc. Canada, Trans., 2d ser., vol. 8, 1902, p. 69; Rept. Tert. Pl. Brit. Col., 1908, p. 43.

Fagus castaneaefolia Unger, Chlor. Prot., 1847, p. 104, pl. xxviii, f. 1.

Castanea ungeri Heer, Philos. Trans., Lond., vol. 159, 1869, p. 470, pl. xlv, f. 1-3; pl. xlvi, f. 8; Fl. foss. arct., vol. 2, Abt. 2, 1869, p. 32, pl. vii, f. 1-3.—Lesquereux, Rept. U. S. Geol. Surv. Terr., vol. 8 (Cret. and Tert. Fl.), 1883, p. 246, pl. lii, f. 1, 3-7.—Knowlton, in Lindgren, Jour. Geol., vol. 4, 1896, p. 890.—Knowlton, U. S. Nat. Mus., Proc., vol. 17, 1894, p. 219; Geol. Soc. Am., Bull., vol. 5, 1893, p. 581.

Mascall: John Day Basin, Oreg.

?Miocene (auriferous gravels): Corral Hollow, Alameda County, Calif.

Castanea claibornensis Berry, U. S. Geol. Surv., Prof. Paper 84, 1914, p. 138, pl. xxviii, f. 1, 2.
Barnwell (Twiggs): Grovetown, Ga.

Castanea dolicophylla Cockerell, Am. Mus. Nat. Hist., Bull., vol. 24, 1908, p. 87; Am. Jour. Sci., 4th ser., vol. 26, 1908, p. 68 text, f. 2 ?; Pop. Sci. Mo., vol. 73, 1908, p. 123, f. in text.
Miocene: Florissant, Colo.

Castanea intermedia Lesquereux, U. S. Geol. and Geog. Surv. Terr., Bull., vol. 1, 1875 [1876], p. 386; U. S. Geol. and Geog. Surv. Terr., Ann. Rept. 1874 [1876], p. 313; Rept. U. S. Geol. Surv. Terr., vol. 7 (Tert. Fl.), 1878, p. 164, pl. xxi, f. 7.—Penhallow, Rept. Tert. Pl. Brit. Col., 1908, p. 43.—Knowlton, U. S. Geol. Surv., Prof. Paper 101, 1917, p. 297, pl. lxviii, f. 2.
Denver: Middle Park, Colo.
Eocene: Quilchena, British Columbia.
Raton: Wootton, Colo.

Castanea nana? Elliott. Lesquereux, Am. Jour. Sci., 2d ser., vol. 27, 1859, p. 365 = *Castanea pumila*.

Castanea pulchella Knowlton, U. S. Geol. Surv., Mon. 32, pt. 2, 1899, p. 702, pl. lxxxvi, f. 6–8; pl. lxxxvii, f. 1–3.
Fort Union: Elk Creek, Yellowstone National Park.
Miocene: Fossil Forest Ridge, Yellowstone National Park.

Castanea pumila (Linné) Miller. Knowlton, Am. Geol., vol. 18, 1896, p. 371.
Castanea nana? Elliott. Lesquereux, Am. Jour. Sci., 2d ser., vol. 27, 1859, p. 365.
Pleistocene: Banks of Mississippi River near Columbus, Ky.; Morgantown, W. Va.

Castanea ungeri Heer = *Castanea castaneaefolia*.

Castanea sp. Dawson, Roy. Soc. Canada, Trans., vol. 4, sec. 4, 1886 [1887], p. 27, pl. i, f. 8.—Penhallow, Rept. Tert. Pl. Brit. Col., 1908, p. 43.
Eocene ?: Souris River and Porcupine Creek, British Columbia.

Castanea sp. Knowlton, in Lindgren, Jour. Geol., vol. 4, 1896, p. 890.
Miocene (auriferous gravels): Independence Hill, Placer County, Calif.

Castanea sp., Hollick, Bull. Torr. Bot. Club, vol. 19, 1892, p. 332.
Pleistocene: Bridgeton, N. J.

CASTANOPSIS Spach, Hist. Veg. Phan., vol. 11, 1842, p. 185.

Castanopsis chrysophylloides Lesquereux, Mus. Comp. Zool., Mem., vol. 6, No. 2, 1878, p. 9, pl. ii, f. 10.—Knowlton, in Lindgren, Jour. Geol., vol. 4, 1896, p. 880.
Miocene (auriferous gravels): Chalk Bluffs, Nevada County, and Independence Hill, Placer County, Calif.

CASUARINA (Rumphius) Adanson, Fam., vol. 2, 1763, p. 481.

Casuarina covillei Ward, U. S. Geol. Surv., Fifteenth Ann. Rept., 1895, p. 353, pl. iii, f. 2.
Patapsco: Mount Vernon, Va.

CATALPA Scopoli, Introd., 1771, p. 170.

Catalpa crassifolia Newberry = *Aristolochia crassifolia.*

CAUDEX Lesquereux, Rept. U. S. Geol. Surv. Terr., vol. 8 (Cret. and Tert. Fl.), 1883, p. 94. [Type, *Caulinites spinosa.*]

Caudex spinosus (Lesquereux) Lesquereux, Rept. U. S. Geol. Surv. Terr., vol. 8 (Cret. and Tert. Fl.), 1883, p. 91.
Caulinites spinosa Lesquereux, U. S. Geol. and Geog. Surv. Terr., Ann. Rept. 1872 [1873], p. 422.
Dakota: Kansas.

CAULERPITES (Brongniart) Sternberg, Vers., vol. 2, 1833, p. 20.

Caulerpites fastigiatus Sternberg = *Sequoia fastigiata.*

Caulerpites incrassatus (Lesquereux) Lesquereux, Rept. U. S. Geol. Surv. Terr., vol. 7 (Tert. Fl.), 1878, p. 40, pl. i, f. 11, 12.—Knowlton, U. S. Geol. Surv., Prof. Paper 101, 1919, p. 243.
Delesseria incrassata Lesquereux, U. S. Geol. and Geog. Surv. Terr., Ann. Rept. 1872 [1873], p. 374.
Trinidad: Trinidad, Colo.

Caulerpites lingulatus (Lesquereux) Knowlton, U. S. Geol. Surv., Bull. 152, 1898, p. 61.
Delesseria lingulata Lesquereux, U. S. Geol. and Geog. Surv. Terr., Ann. Rept. 1872 [1873], p. 374; Rept. U. S. Geol. Surv. Terr., vol. 7 (Tert. Fl.), 1878, p. 41.
Trinidad: Raton Mountains, N. Mex.

"Caulinia laevis (Göppert) Göppert," Schles. Gesell. f. Vaterländ. Kultur, 1861, p. 202. [Probably = *Phragmites alaskana* Heer.]
Kenai: Ninilchik, east shore of Cook Inlet, Alaska.

CAULINITES Brongniart, Prod. hist. vég. foss., 1828, p. 115.

Caulinites beckeri Lesquereux, U. S. Nat. Mus., Proc., vol. 10, 1887, p. 36, pl. i, f. 3; pl. ii, f. 2–4.
Pleistocene: Clear Lake, Calif.

Caulinites fecundus Lesquereux = *Onoclea fecunda.*

Caulinites inquirendus Hollick = *Onoclea inquirenda.*

Caulinites sparganioides Lesquereux, U. S. Geol. and Geog. Surv. Terr., Ann. Rept. 1872 [1873], p. 391; Rept. U. S. Geol. Surv. Terr., vol. 7 (Tert. Fl.), 1878, p. 99, pl. xiv, f. 4–11.
Livingston ?: Bozeman coal field, Mont.
Post-Laramie: Black Buttes, Wyo.

Caulinites spinosa Lesquereux = *Caudex spinosus.*

Caulinites sp. Berry, U. S. Nat. Mus., Proc., vol. 54, 1819, p. 629.
Miocene: Beaver Creek, 10 miles east of Beaver City, Okla.

CEANOTHUS Linné, Sp. pl., 1753, p. 195.

Ceanothus! americanus? Linné. Lesquereux, Am. Jour. Sci., 2d ser., vol. 27, 1859, p. 365.

Pleistocene: Banks of Mississippi River near Columbus, Ky.

Ceanothus cinnamomoides Lesquereux = *Zizyphus cinnamomoides*.

Ceanothus constrictus Hollick, U. S. Geol. Surv., Mon. 50, 1906, p. 93, pl. xxxiv, f. 15–17.

Magothy: Glen Cove, Long Island, N. Y.: Gay Head, Marthas Vineyard, Mass.

Ceanothus cretaceus Dawson, Roy. Soc. Canada, Trans., vol. i, sec. 4, 1882 [1883], p. 28, pl. viii, f. 33.

Upper Cretaceous: Baynes Sound, Fort McNeill, Vancouver Island.

Ceanothus eriensis Knowlton, U. S. Geol. Surv., Prof. Paper 130, 19—, p. —, pl. xxvi, f. 3–6 [in preparation].

Laramie: Erie, Colo.

Ceanothus fibrillosus Lesquereux = *Zizyphus fibrillosus*.

Ceanothus meigsii Lesquereux = *Zizyphus meigsii*.

Ceanothus ovatifolius Knowlton, U. S. Geol. Surv., Prof. Paper 130, 19—, pl —, pl. xxv, f. 3 [in preparation].

Laramie: Morrison, Colo.

Ceanothus? sp. Dawson, Roy. Soc. Canada, Trans., vol. 8, sec. 4, 1890 [1891], p. 84, text f. 14.—Penhallow, Rept. Tert. Pl. Brit. Col., 1908, p. 44.

Oligocene: North Fork of Similkameen River, British Columbia.

CEBATHA Forskål, Fl. Aegypt., 1775, p. 171.

Cebatha carolina (Linné) Britton. Berry, U. S. Nat. Mus., Proc., vol. 48, 1915, p. 300, pl. xii, f. 3–5.

Menispermum canadense Linné. Knowlton, in Glenn, U. S. Geol. Surv., Water-Supply Paper 164, 1906, p. 38.

Pleistocene: Hickman and Columbus, Ky.

Cebatha haydenianus (Ward) Knowlton = *Cocculus haydenianus*.

Cedar cone Bibbins, 1895 = *Cedrus leei*.

CEDRELA P. Browne, Jamaic., 1756, p. 158.

Cedrela mississippiensis Berry, U. S. Geol. Surv., Prof. Paper 91, 1916, p. 254, pl. lv, f. 5.

Wilcox (Holly Springs): Early Grove, Marshall County, Miss.

Cedrela odoratifolia Berry, U. S. Geol. Surv., Prof. Paper 91, 1916, p. 255, pl. lvii, f. 7.

Lagrange: Puryear, Henry County, Tenn.

Cedrela puryearensis Berry, U. S. Geol. Surv., Prof. Paper 91, 1916, p. 254, pl. lvi, f. 2.

Lagrange: Puryear, Henry County, Tenn.

Cedrela wilcoxiana Berry, U. S. Geol. Surv., Prof. Paper 91, 1916, p. 253, pl. lvi, f. 1.
Wilcox (Holly Springs): Holly Springs, Marshall County, Miss.
Lagrange: Puryear, Henry County, Tenn.

CEDROXYLON Kraus, in Schimper, Pal. vég., vol. 2, 1870, p. 370.

Cedroxylon? pennsylvanianum Wherry = *Brachyoxylon pennsylvanianum.*

Cedroxylon sp. Dawson, Brit. N. A. Bound. Com. (Rept. Geol. and Rec. Vic. 49th Parallel), 1875, App. A, p. 331, pl. xv, f. 1 = *Sequoia langsdorfii.*

CEDRUS Miller, Gard. Dict., ed. 3, 1737.

Cedrus leei (Fontaine) Berry, Maryland Geol. Surv., Lower Cret., 1912, p. 411, pl. lxxvii, f. 4, 4a.
Cedar cone, Bibbins, Johns Hopkins Univ. Cir., vol. 15, 1895, p. 19, f. F.
Pinites leei Fontaine, in Ward, U. S. Geol. Surv., Mon. 48, 1905 [1906], p. 570, pl. cxix, f. 6, 7.
Arundel: Union tunnel, Baltimore, Md.

CELASTRINITES Saporta, Ann. Sci. Nat., Bot., 5th ser., vol. 3, 1865, p. 52.

Celastrinites alatus Knowlton, U. S. Geol. Surv., Prof. Paper 130, 19—, p. —, pl. xxv, f. 4, 5; pl. xxvi, f. 1, 2 [in preparation].
Laramie: Erie and near Golden, Colo.

Celastrinites ambiguus (Lesquereux) Knowlton, U. S. Geol. Surv., Bull. 152, 1898, p. 62.
Myrica ambigua Lesquereux, U. S. Geol. and Geog. Surv. Terr., Ann. Rept. 1871 [1872], p. 297.
Celastrinites laevigatus Lesquereux, Rept. U. S. Geol. Surv. Terr., vol. 7 (Tert. Fl.), 1878, p. 269, pl. xvii, f. 16, 16a.
Livingston: Bozeman coal field, Mont.

Celastrinites artocarpidioides Lesquereux, Rept. U. S. Geol. Surv. Terr., vol. 7 (Tert. Fl.), 1878, p. 268, pl. xxxv, f. 3.
Artocarpidium olmediaefolium Unger. Lesquereux, U. S. Geol. and Geog. Surv. Terr., Ann. Rept. 1873 [1874], p. 400.
Denver: Golden, Colo.

Celastrinites cowanensis Knowlton, U. S. Geol. Surv., Prof. Paper 130, 19—, p. —, pl. xvi, f. 6 [in preparation].
Laramie: Cowan station near Denver, Colo.

Celastrinites elegans Lesquereux, Rept. U. S. Geol. Surv., Terr., vol. 8 (Cret. and Tert. Fl.), 1883, p. 185, pl. xxxi, f. 9, 10.—Knowlton, U. S. Nat. Mus., Proc., vol. 51, 1916, p. 281, pl. xxi, f. 7.
Celastrus greithianus Heer, Lesquereux, Rept. U. S. Geol. Surv. Terr., vol. 8 (Cret. and Tert. Fl.), 1883, p. 184.
Miocene. *Florissant, Colo.

Celastrinites eriensis Knowlton, U. S. Geol. Surv., Prof. Paper 130, 19—, p. —, pl. xxvi, f. 2 [in preparation].
Laramie: Erie, Colo.

Celastrinites laevigatus Lesquereux = *Celastrinites ambiguus.*

CELASTROPHYLLUM Göppert, Neues Jahrb., 1853, p. 435 [name]; Tertiärfl. d. Insel Java, 1854, p. 52.

Celastrophyllum acutidens Fontaine, U. S. Geol. Surv., Mon. 15, 1889, p. 305, pl. clvi, f. 8; in Ward, U. S. Geol. Surv., Mon. 48, 1905 [1906], p. 529, pl. cxiii, f. 7, 8.—Berry, Md. Geol. Surv., Lower Cret., 1912, p. 478, pl. lxxxix.

Celastrophyllum obtusidens Fontaine, U. S. Geol. Surv., Mon. 15, 1889 [1890], p. 306, pl. clvi, f. 5.

Myrica brookensis Fontaine, U. S. Geol. Surv., Mon. 15, 1889 [1890], p. 310, pl. cl, f. 11; pl. clvi, f. 10; in Ward, idem, Mon. 48, 1905 [1906], p. 573, pl. cviii, f. 8.

Celastrophyllum pulchrum Ward, U. S. Geol. Surv., Nineteenth Ann. Rept., pt. 2, 1899, p. 706, pl. clxxi, f. 3, 4.

Patapsco: Fort Foote, Md.; White House Bluff, Brooke, Widewater, Chinkapin Hollow, Va.

Dakota: Evans quarry, 5 miles east of Hot Spring, S. Dak.

Celastrophyllum alabamensis Berry, U. S. Geol. Surv., Prof. Paper 112, 1919, p. 110, pl. xxiv, f. 10–12.

Tuscaloosa: Near Centerville, Bibb County, Ala.

Celastrophyllum albaedomus Ward, U. S. Geol. Surv., Mon. 48, 1905 [1906], p. 489, pl. cviii, f. 4.—Berry, Md. Geol. Surv., Lower Cret., 1912, p. 480, pl. xc, f. 4.

Patapsco: White House Bluff, Va.

Celastrophyllum angustifolium Newberry, U. S. Geol. Surv., Mon. 26, 1895 [1896], p. 100, pl. xiv, f. 8–17.

Raritan: Woodbridge, N. J.

Celastrophyllum arcinerve Fontaine, U. S. Geol. Surv., Mon. 15, 1889, p. 304, pl. cxliv, f. 3.

Patuxent: Fredericksburg, Va.

Celastrophyllum benedeni Saporta and Marion. Hollick, Bull. Torr. Bot. Club, vol. 21, 1894, p. 58, pl. clxxvii, f. 3 = *Banksites saportanus.*

Celastrophyllum benedeni Saporta and Marion. Hollick, Bull. Torr. Bot. Club, vol. 21, 1894, p. 58, pl. clxxvii, f. 4 = *Elaeodendron* sp.

Celastrophyllum brittonianum Hollick, in Newberry, U. S. Geol. Surv., Mon. 26, 1895 [1896], p. 105, pl. xlii, f. 37, 38, 46, 47; Ward, U. S. Geol. Surv., Fifteenth Ann. Rept., 1895, pp. 349, 358, 377, 378, 379.—Fontaine, in Ward, U. S. Geol. Surv., Mon. 48, 1905 [1906], p. 493, pl. cvii, f. 7.—Berry, Geol. Surv. N. J., Bull. 3, 1911, p. 180; Md. Geol. Surv., Lower Cret., 1912, p. 479, pl. cx, f. 3; U. S. Geol. Surv., Prof. Paper 112, 1919, p. 110, pl. xxiv, f. 4.

Patapsco: Mount Vernon, Va.

Raritan: New Jersey.

Tuscaloosa: Shirleys Mill, Ala.

Celastrophyllum brookense Fontaine = *Celastrophyllum parvifolium.*

Celastrophyllum carolinensis Berry, U. S. Geol. Surv., Prof. Paper 84, 1914, p. 51, pl. xiii, f. 1–5; idem, Prof. Paper 112, 1919, p. 109, pl. xxiv, f. 6, 7.
Black Creek (Middendorf): Langley, S. C.
Tuscaloosa: Snow place, Tuscaloosa County, Ala.

Celastrophyllum crassipes Lesquereux, U. S. Geol. Surv., Mon. 17, 1891 [1892], p. 174, pl. lvii, f. 6, 7.—Hollick, N. Y. Bot. Gard., Bull., vol. 8, 1912, p. 165, pl. clxiv, f. 6.
Dakota: Kansas.
Magothy: ?Roslyn, N. Y.

Celastrophyllum crenatum Heer, Fl. foss. arct., vol. 7, 1883, p. 41, pl. lxii, f. 21.—Newberry, U. S. Geol. Surv., Mon. 26, 1895 [1896], p. 99, pl. xlviii, f. 1–19.—Berry, Bull. Torr. Bot. Club, vol. 34, 1907, p. 197, pl. xiii, f. 5; Geol. Surv. N. J., Bull. 3, 1911, p. 178, pl. xxii, f. 9; pl. xxiii, f. 2; U. S. Geol. Surv., Prof. Paper 84, 1914, p. 50; Md. Geol. Surv., Upper Cret., 1916, p. 852; U. S. Geol. Surv., Prof. Paper 112, 1919, p. 109.
Raritan: South Amboy and Sayreville, N. J.
Black Creek: Court House Bluff, Cape Fear River, N. C.
Black Creek (Middendorf): Middendorf, S. C.
Tuscaloosa: Cottondale and Snow place, Tuscaloosa County, Ala.

Celastrophyllum crenatum ellipticum Berry, U. S. Geol. Surv., Prof. Paper 112, 1919, p. 110, pl. xxiv, f. 1, 2.
Tuscaloosa: Shirleys Mill, Ala.

Celastrophyllum cretaceum Lesquereux, U. S. Geol. Surv., Mon. 17, 1891 [1892], p. 173, pl. xxxviii, f. 12–14.—Newberry, U. S. Geol. Surv., Mon. 26, 1895 [1896], p. 100, pl. xlii, f. 13.—Berry, Geol. Surv. N. J., Bull. 3, 1911, p. 177.
Dakota: Ellsworth County, Kans.
Raritan: New Jersey.

Celastrophyllum decurrens Lesquereux, U. S. Geol. Surv., Mon. 17, 1891 [1892], p. 172, pl. xxxvi, f. 1.—Berry, Geol. Surv. N. J., Bull. 3, 1911, p. 176, pl. xxii, f. 8; U. S. Geol. Surv., Prof. Paper 112, 1919, p. 106.
Dakota: Ellsworth County, Kans.
Raritan: South Amboy, Milltown, N. J.
Tuscaloosa: Shirleys Mill, Ala.

Celastrophyllum decurrens Lesquereux? Hollick, Bull. Torr. Bot. Club, vol. 21, 1894, p. 59, pl. clxxix, f. 1 = *Liriodendropsis spectabilis.*

Celastrophyllum denticulatum Fontaine, U. S. Geol. Surv., Mon. 15, 1889 [1890], p. 306, pl. clxix, f. 10; pl. clxxii, f. 7.—Berry, Md. Geol. Surv., Lower Cret., 1912, p. 475, pl. xc, f. 1, 2.
Patapsco: Baltimore, Md.

Celastrophyllum elegans Berry, N. Y. Bot. Gard., Bull., vol. 3, 1903, p. 84, pl. 43, f. 6; Bull. Torr. Bot. Club, vol. 32, 1905, p. 46, pl. ii, f. 1; Geol. Surv. N. J., Ann. Rept. 1905 [1906], p. 138; U. S. Geol. Surv., Prof. Paper 84, 1914, p. 50, pl. xiv, f. 11.

Magothy: Cliffwood, N. J.

Black Creek (Middendorf): Rocky Point, S. C.

Celastrophyllum ensifolium (Lesquereux) Lesquereux, Rept. U. S. Geol. Surv. Terr., vol. 7 (Cret. Fl.), 1874, p. 108, pl. xxi, f. 2, 3.

Magnolia ensifolia Lesquereux, U. S. Geol. and Geog. Surv. Terr., Ann. Rept. 1871 [1872], p. 302.

Dakota: Southern Kansas; near Lander, Wyo.

Celastrophyllum grandifolium Newberry, U. S. Geol. Surv., Mon. 26, 1895 [1896], p. 104, pl. xix, f. 8; pl. xxi, f. 1–4.—? Hollick, U. S. Geol. Surv., Mon. 50, 1906, p. 88, pl. xxxiii, f. 8.—Berry, Geol. Surv. N. J., Bull. 3, 1911, p. 179, pl. xxiii, f. 1; U. S. Geol. Surv., Prof. Paper 112, 1919, p. 108, pl. xxiv, f. 5.

Raritan: Milltown, N. J.

Magothy: Nashaquitsa, Marthas Vineyard, Mass.

Tuscaloosa: Shirleys Mill and Cottondale, Ala.

Celastrophyllum gymindaefolium Berry, U. S. Geol. Surv., Prof. Paper 112, 1919, p. 111, pl. xxiv, f. 9.

Tuscaloosa: Shirleys Mill, Ala.

Celastrophyllum hunteri Ward, U. S. Geol. Surv., Fifteenth Ann. Rept. 1895, p. 358, pl. iv, f. 9.—Fontaine, in Ward, U. S. Geol. Surv., Mon. 48, 1905 [1906], p. 494, pl. cviii, f. 6.—Berry, Md. Geol. Surv., Lower Cret., 1912, p. 480, pl. xc, f. 5, 10, 11.

Celastrophyllum? saliciforme Ward, U. S. Geol. Surv., Mon. 48, 1905 [1906], p. 494, pl. cviii, f. 7.

Patapsco: Mount Vernon, Va.

Celastrophyllum latifolium Fontaine, U. S. Geol. Surv., Mon. 15, 1889, p. 306, pl. clxxii, f. 3, 6; pl. clxxiii, f. 13; in Ward, U. S. Geol. Surv., Mon. 48, 1905 [1906], p. 559, pl. cxvi, f. 6.—Berry, Md. Geol. Surv., Lower Cret., 1912, p. 477, pl. xc, f. 6–9.

Celastrophyllum tenuinerve Fontaine, U. S. Geol. Surv., Mon. 15, 1889 [1890], p. 306, pl. clxxii, f. 2.

Celastrophyllum obovatum Fontaine, U. S. Geol. Surv., Mon. 15, 1889 [1890], p. 307, pl. clxxii, f. 9, 10; in Ward, idem, Mon. 48, 1905 [1906], pp. 550, 560, pl. cxv, f. 6; pl. cxvii, f. 2, 3.

Proteaephyllum uhleri Fontaine, in Ward, U. S. Geol. Surv., Mon. 48, 1905 [1906], p. 564, pl. cxviii, f. 5.

Patapsco: Federal Hill (Baltimore), and Vinegar Hill, Md.

Celastrophyllum marylandicum Fontaine = *Plantaginopsis marylandica*.

Celastrophyllum minus Hollick, in Newberry, U. S. Geol. Surv., Mon. 26, 1895 [1896], p. 105, pl. xlii, f. 51, 52.—Berry, Geol. Surv. N. J., Bull. 3, 1911, p. 173, pl. xxii, f. 3.

Raritan: Milltown, N. J.

Celastrophyllum myrsinoides Lesquereux, U. S. Geol. Surv., Mon. 17, 1891 [1892], p. 174, pl. lvii, f. 8, 9.
Dakota: Kansas.

Celastrophyllum newberryanum Hollick, in Newberry, U. S. Geol. Surv., Mon. 26, 1895 [1896], p. 101, pl. xlix, f. 1–27.—Hollick, N. Y. Acad. Sci., Trans., vol. 16, 1897, p. 133, pl. xiv, f. 1.—Berry, Bull. Torr. Bot. Club, vol. 31, 1904, p. 78; N. Y. Bot. Gard., Bull., vol. 3, 1903, p. 85; Geol. Surv. N. J., Ann. Rept. 1905 [1906], p. 138; Geol. Surv. N. J., Bull. 3, 1911, p. 174, pl. xxii, f. 5–7; U. S. Geol. Surv., Prof. Paper 112, 1919, p. 108.
Magothy: Cliffwood, N. J.
Raritan: South Amboy and Sayreville, N. J.
Tuscaloosa: Shirleys Mill, Ala.

Celastrophyllum obliquum Knowlton, in Lesquereux, U. S. Geol. Surv., Mon. 17, 1891 [1892], p. 173, pl. lvii, f. 5.
Dakota: Near Fort Harker, Kans.

Celastrophyllum obovatum Fontaine = *Celastrophyllum latifolium.*

Celastrophyllum obtusidens Fontaine = *Celastrophyllum acutidens.*

Celastrophyllum parvifolium (Fontaine) Berry, Md. Geol. Surv., Lower Cret., 1912, p. 476.
Saliciphyllum parvifolium Fontaine, U. S. Geol. Surv., Mon. 15, 1889 [1890], p. 303, pl. clxxii, f. 5.
Saliciphyllum ellipticum Fontaine, U. S. Geol. Surv., Mon. 15, 1889 [1890], p. 303 [part].
Celastrophyllum brookense Fontaine, U. S. Geol. Surv., Mon. 15, 1889 [1890], p. 305, pl. clviii, f. 8; pl. clix, f. 7; Fontaine, in Ward, U. S. Geol. Surv., Mon. 48, 1905 [1906], p. 505, pl. cx, f. 10.
Patapsco: Wellhams and Federal Hill (Baltimore), Md.; Hell Hole and 72d milepost, Va.

Celastrophyllum praecrassipes Berry, U. S. Geol. Surv., Prof. Paper 112, 1919, p. 111, pl. xxiv, f. 13.
Tuscaloosa: Shirleys Mill, Ala.

Celastrophyllum proteoides Fontaine = *Thinnfeldia granulata.*

Celastrophyllum pulchrum Ward = *Celastrophyllum acutidens.*

Celastrophyllum robustum Newberry, U. S. Geol. Surv., Mon. 26, 1895 [1896], p. 103, pl. xlii, f. 41, 42.
Raritan?: New Jersey.

Celastrophyllum? saliciforme Ward = *Celastrophyllum hunteri.*

Celastrophyllum shirleyensis Berry, U. S. Geol. Surv., Prof. Paper 112, 1919, p. 107, pl. xxiv, f. 3, 8.
Tuscaloosa: Shirleys Mill, Ala.

Celastrophyllum spatulatum Newberry, U. S. Geol. Surv., Mon. 26, 1895 [1896], p. 103, pl. xlii, f. 43–45.—Berry, Geol. Surv. N. J., Bull. 3, 1911, p. 178, pl. xxii, f. 4.
Raritan: South Amboy, N. J.

Celastrophyllum tenuinerve Fontaine = *Celastrophyllum latifolium*.

Celastrophyllum undulatum Newberry, U. S. Geol. Surv., Mon. 26, 1895 [1896], p. 102, pl. xxxviii, f. 1–3.—Berry, Bull. Torr. Bot. Club, vol. 37, 1910, p. 198; Geol. Surv. N. J., Bull. 3, 1911, p. 175; Md. Geol. Surv., Upper Cret., 1916, p. 853; U. S. Geol. Surv., Prof. Paper 112, 1919, p. 107.
  Raritan: Woodbridge and Sayreville, N. J.
  Tuscaloosa: Cottondale, Ala.
  Black Creek: Court House Bluff, Cape Fear River, N. C.
  Magothy: Round Bay, Md.

Celastrophyllum sp. Hollick, in Ries, School of Mines Quart., vol. 15, 1874, p. 355.
  Raritan: Northport, Long Island, N. Y.

CELASTRUS Linné, Sp. pl., 1753, p. 196.

Celastrus alnifolius Ward = *Celastrus montanensis*.

Celastrus arctica Heer, Fl. foss. arct., vol. 7, 1883, p. 40, pl. lxi, f. 5d, 5e.—Newberry, U. S. Geol. Surv., Mon. 26, 1895 [1896], p. 98, pl. xiii, f. 8–18.—Hollick, N. Y. Bot. Gard., Bull., vol. 3, 1908, p. 408, pl. lxx, f. 12, 13; U. S. Geol. Surv., Mon. 50, 1906, p. 88. pl. xxxiii, f. 9–11.—Berry, Bull. Torr. Bot. Club, vol. 38, 1911, p. 407; Geol. Surv. N. J., Bull. 3, 1911, p. 172, pl. xxv, f. 1–5; Md. Geol. Surv., Upper Cret., 1916, p. 850, pl. lxxvii, f. 7.
  Raritan: South Amboy, N. J.
  Magothy: Little Neck, Northport Harbor, Long Island, N. Y.; ?Black Rock Point, Block Island, R. I.; Little Round Bay, Severn River, Md.

Celastrus borealis Heer, Fl. foss. arct., vol. 2, Abt. 2 (Fl. foss. alask.), 1869, p. 37, pl. x, f. 4.—Knowlton, U. S. Nat. Mus., Proc., vol. 17, 1894, p. 229; Geol. Soc. Am., Bull., vol. 5, 1893, p. 586.
  Kenai: Port Graham, Alaska.

Celastrus bruckmanni Al. Braun, Stizenb. Verzeich., 1851, p. 87.–Heer, Fl. foss. arct., vol. 6, Abt. 1, 1880, p. 14, pl. vi, f. 5.—Berry, Jour. Geol., vol. 17, 1909, p. 28, text f. 8.
  Calvert: Richmond, Va.

Celastrus bruckmannifolia Berry, U. S. Geol. Surv., Prof. Paper 91, 1916, p. 265, pl. lxi, f. 1.
  Wilcox (Holly Springs): Holly Springs, Marshall County, Miss.

Celastrus confluens Knowlton, U. S. Geol. Surv., Bull. 204, 1902, p. 71, pl. ii, f. 1–3.
  Mascall: Van Horn's ranch, John Day Basin, Oreg.

Celastrus culveri Knowlton, U. S. Geol. Surv., Mon. 32, pt. 2, 1899, p. 732, pl. xcvii, f. 4.
  Fort Union: Elk Creek, Yellowstone National Park.

Celastrus curvinervis Ward, U. S. Geol. Surv., Sixth Ann. Rept., 1884–85 [1886], p. 555, pl. liii, f. 9, 10; U. S. Geol. Surv., Bull. 37, 1887, p. 82, pl. xxxvi, f. 3, 4.—Knowlton, Washington Acad. Sci., Proc., vol. 11, 1909, p. 189.

Fort Union: Burns's ranch, near Glendive, and Bull Mountain field, Mont.; near Moreau River, S. Dak.

Lance: West of Miles City, Mont.

Celastrus dignatus Knowlton, U. S. Geol. Surv., Bull. 204, 1902, p. 71, pl. xi, f. 5.

Mascall: Van Horn's ranch, John Day Basin, Oreg.

Celastrus ellipticus Knowlton, U. S. Geol. Surv., Mon. 32, pt. 2, 1899, p. 734, pl. xcvii, f. 3.

Fort Union: Elk Creek, Yellowstone National Park.

Celastrus eolignitica Berry, U. S. Geol. Surv., Prof. Paper 91, 1916, p. 266, pl. lix, f. 8–11.

Wilcox (Holly Springs): Early Grove, Marshall County, Miss.

Lagrange: Near Grand Junction, Fayette County, Tenn.

Celastrus ferrugineus Ward, U. S. Geol. Surv., Sixth Ann. Rept., 1884–85 [1886], p. 555, pl. lii, f. 11–14; U. S. Geol. Surv., Bull. 37, 1887, p. 78, pl. xxxiv, f. 1–4.—Knowlton, Washington Acad. Sci., Proc., vol. 11, 1909, p. 215.

Fort Union: *Burns's ranch and Iron Bluff, near Glendive and Bull Mountains, Mont.; Minturn, 8 miles northeast of Parkman, and Table Mountain, near Black Buttes, Wyo.

Lance: Near Kirby, Wyo.

Celastrus fraxinifolius Lesquereux, Rept. U. S. Geol. Surv. Terr., vol. 8 (Cret. and Tert. Fl.), 1883, p. 184, pl. xxxiii, f. 2–4; pl. xl, f. 10.—Knowlton, U. S. Nat. Mus., Proc., vol. 51, 1916, p. 280.

Miocene: *Florissant, Colo.

Celastrus gaudini Lesquereux, Mus. Comp. Zool., Bull., vol. 16, 1888, p. 54.

Denver: Golden, Colo.

Celastrus greithianus Heer. Lesquereux, Rept. U. S. Geol. Surv. Terr., vol. 8 (Cret. and Tert. Fl.), 1883, p. 184 = *Celastrinites elegans*.

Celastrus grewiopsis Ward, U. S. Geol. Surv., Sixth Ann. Rept., 1884–85 [1886], p. 555, pl. liii, f. 8; U. S. Geol. Surv., Bull. 37, 1887, p. 81, pl. xxxvi, f. 2.

Fort Union: Burns's ranch, Glendive, Mont.

Celastrus haddeni Knowlton, U. S. Geol. Surv., Prof. Paper 101, 1918, p. 271, pl. xlv, f. 2.

Vermejo: Coal Creek, Rockvale, Colo.

Celastrus? hesperius Knowlton, U. S. Geol. Surv., Prof. Paper 101, 1918, p. 271, pl. xliv, f. 5.

Vermejo: Coal Creek, Rockvale, Colo.

Celastrus inaequalis Knowlton, U. S. Geol. Surv., Mon. 32, pt. 2, 1899, p. 733, pl. xcviii, f. 3.
Fort Union: Elk Creek, Yellowstone National Park.

Celastrus lacoei Lesquereux, Rept. U. S. Geol. Surv. Terr., vol. 8 (Cret. and Tert. Fl.), 1883, p. 184.—Knowlton, U. S. Nat. Mus., Proc., vol. 51, 1916, p. 281, pl. xxiv, f. 6.
Miocene: *Florissant, Colo.

Celastrus lindgreni Knowlton, U. S. Geol. Surv., Eighteenth Ann. Rept., pt. 3, 1898, p. 732, pl. xcix, f. 13; pl. c, f. 6.
Payette: Idaho City, Idaho.

Celastrus minor Berry, U. S. Geol. Surv., Prof. Paper 91, 1916, p. 266, pl. lxi, f. 3, 4.
Wilcox (Holly Springs): Holly Springs, Marshall County, Miss.

Celastrus montanensis Knowlton and Cockerell, n. name.
Celastrus alnifolius Ward, U. S. Geol. Surv., Sixth Ann. Rept. 1884–85 [1886], p. 555, pl. liii, f. 1, 2; U. S. Geol. Surv., Bull. 37, 1887, p. 80, pl. xxxv. f. 1, 2.—Knowlton, Washington Acad. Sci., Proc., vol. 11, 1909, p. 189; Jour. Geol., vol. 19, 1911, p. 369. [Homonym, Don, 1825.]
Fort Union: Burns's ranch, near Glendive and Bull Mountain field, Mont.; ?McCord coal bank, Cannonball River, N. Dak.
Lance: Cannonball River, N. Dak.; west of Miles City, Mont.

Celastrus ovatus Ward = *Celastrus wardii*.

Celastrus pterospermoides Ward, U. S. Geol. Surv., Sixth Ann. Rept., 1884–85 [1886], p. 555, pl. liii, f. 3–6; U. S. Geol. Surv., Bull. 37, 1887, p. 80, pl. xxxv, f. 3–6.—Knowlton, Washington Acad. Sci., Proc., vol. 11, 1909, p. 189.
Fort Union: *Burns's ranch and Iron Bluff, near Glendive, Bull Mountains, and 6 miles west of Jordan, Mont.; near Kirby and Rock Springs area, Wyo.
Lance: West of Miles City, Mont.

Celastrus serratus Knowlton, U. S. Geol. Surv., Prof. Paper 101, 1918, p. 329, pl. xcviii, f. 3; pl. xcix, f. 4; pl. c, f. 1.
Raton: Primero and Woolton, Colo.

Celastrus taurinensis Ward, U. S. Geol. Surv., Sixth Ann. Rept., 1884–85 [1886], p. 555, pl. lii, f. 15, 16; U. S. Geol. Surv., Bull. 37, 1887, p. 79, pl. xxxiv, f. 5, 6.—?Hollick, Geol. Surv. La., Special Rept. 5, 1899, p. 285, pl. xlvi, f. 1.—Knowlton, Washington Acad. Sci., Proc., vol. 11, 1909, p. 213.—Berry, U. S. Geol. Surv., Prof. Paper 91, 1916, p. 267, pl. lx, f. 1–3.
Fort Union: *Bull Mountains and Burns's ranch, near Glendive, Mont.; Ilo Post Office, Big Horn County, Wyo.
Wilcox: Coushatta, La.
Lance: Near Ilo Post Office, Big Horn Basin, Wyo.

Celastrus veatchi Hollick, Geol. Surv. La., Special Rept. 5, 1899, p. 285, pl. xliii, f. 4, 5.—Berry, U. S. Geol. Surv., Prof. Paper 91, 1916, p. 267, pl. lxi, f. 2.

Wilcox: *Coushatta, La.

Lagrange: Near Trenton, Gibson County, Tenn.

Wilcox: (Holly Springs): Early Grove, Marshall County, Miss.

Celastrus wardii Knowlton and Cockerell, n. name.

Celastrus ovatus Ward, U. S. Geol. Surv., Sixth Ann. Rept., 1884–85 [1886], p. 555, pl. liii, f. 7; U. S. Geol. Surv., Bull. 37, 1887, p. 71, pl. xxxvi, f. 1. —Knowlton, Washington Acad. Sci., Proc., vol. 11, 1909, p. 190. [Homonym, Hill, 1865.]

Fort Union: *Iron Bluff, Glendive, and Miles City, Mont.

Lance: Signal Butte, Miles City, Mont.

Celastrus? sp. Hollick, N. Y. Acad. Sci., Trans., vol. 12, 1892, p. 38, pl. i, f. 4.

Raritan: Tottenville, Staten Island, N. Y.

Celastrus sp. Knowlton, Geol. Soc. Am., Bull., vol. 8, 1897, p. 130.

Mesaverde: Harper, Wyo.

Celastrus sp. Knowlton, Washington Geol. Surv., Ann. Rept., vol. 1, 1902, p. 33.

Eocene: Skykomish, Snohomish County, Wash.

Celastrus? sp. Knowlton, U. S. Geol. Surv., Prof. Paper 101, 1918, p. 271, pl. xlvii, f. 6.

Vermejo: Vermejo Park, N. Mex.

Celastrus sp. Knowlton, U. S. Geol. Surv., Prof. Paper 101, 1917, p. 329, pl. xcviii, f. 1.

Raton: Near Yankee, N. Mex.

CELTIS (Tournefort) Linné, Sp. pl., 1753, p. 1043.

Celtis brevifolia Lesquereux = *Ficus schimperi.*

Celtis lingualis Knowlton and Cockerell, n. name.

Celtis rugosa Newberry, U. S. Nat. Mus., Proc., vol. 5, 1882 [1883], p. 510. [Homonym, Willdenow, 1897.]

Fort Union: Tongue River, Wyo.

Celtis mccoshii Lesquereux, Rept. U. S. Geol. Surv. Terr., vol. 8 (Cret. and Tert. Fl.), 1883, p. 163, pl. xxxviii, f. 7, 8.—Knowlton, U. S. Nat. Mus., Proc., vol. 51, 1916, p. 266.

Miocene: Florissant, Colo.; Uinta County (formerly wrongly called Randolph County), Wyo.

Celtis newberryi Knowlton and Cockerell, n. name.

Celtis parvifolia Newberry, U. S. Nat. Mus., Proc., vol. 5, 1882 [1883], p. 510; U. S. Geol. Surv., Mon. 35, 1898, p. 84, pl. liii, f. 6. [Homonym, Rich, 1853.]

Fort Union: Tongue River, Wyo.

Celtis? ovata Lesquereux = *Ampelophyllites ovatus.*

Celtis parvifolia Newberry = *Celtis newberryi.*

Celtis pseudo-crassifolia Hollick, Md. Geol. Surv., Pliocene and Pleistocene, 1906, p. 230, pl. lxxi, f. 9.
  Pleistocene (Sunderland): Head of Island Creek, Calvert County, Md.

Celtis rugosa Newberry = *Celtis lingualis.*

CEPHALANTHUS Linné, Sp. pl., 1753, p. 95.

Cephalanthus occidentalis Linné. Hannibal, Bull. Torr. Bot. Club, vol. 38, 1911, p. 339.
  Pliocene: Near Portola and in Calabasas Canyon, Santa Cruz Mountains, Calif.

CEPHALOTAXOPSIS Fontaine, U. S. Geol. Surv., Mon. 15, 1889, p. 235. [Type, *C. magnifolia.*]

Cephalotaxopsis brevifolia Fontaine, U. S. Geol. Surv., Mon. 15, 1889, p. 238, pl. cv, f. 3; pl. cvi, f. 5; pl. cvii, f. 5.—Berry, U. S. Nat. Mus., Proc., vol. 40, 1911, p. 300; Md. Geol. Surv., Lower Cret., 1912, p. 379, pl. lx, f. 2.
  Cephalotaxopsis microphylla Fontaine, U. S. Geol. Surv., Mon. 15, 1889 [1890], p. 238, pl. cviii, f. 5; pl. cix, f. 9.
  Patuxent: Fredericksburg, near Dutch Gap Canal and near Potomac Run, Va.

Cephalotaxopsis carolinensis Fontaine, in Ward, U. S. Geol. Surv., Twentieth Ann. Rept., pt. 2, 1900, p. 204, pl. xlviii, f. 5.
  Triassic (Newark): North Carolina.

Cephalotaxopsis magnifolia Fontaine, U. S. Geol. Surv., Mon. 15, 1889 [1890], p. 336, pl. civ, f. 4, 5; pl. cv, f. 1, 2, 4; pl. cvi, f. 1, 3; pl. cvii, f. 1, 2, 4; pl. cviii, f. 1, 3, 4; in Ward, U. S. Geol. Surv., Nineteenth Ann. Rept., pt. 2, 1899, p. 686, pl. clxii, f. 16; pl.clxix, f. 3, 4.—Berry, U. S. Nat. Mus., Proc., vol. 40, 1911, p. 299; Md. Geol. Surv., Lower Cret., 1912, p. 377, pl. lx, f. 1.
  ?*Cephalotaxopsis ramosa* Fontaine. Fontaine, in Ward, U. S. Geol. Surv., Mon. 48, 1905 [1906], p. 258, pl. lxviii, f. 5-7? [not p. 311, pl. lxxiii, f. 8, which=*Oleandra graminaefolia*, or p. 547, which=*Nageiopsis angustifolia*].
  Cephalotaxopsis? rhytidodes Ward, U. S. Geol. Surv., Mon. 48, 1905 [1906], p. 258, pl. lxviii, f. 8.
  Cephalotaxopsis sp.? Fontaine, in Diller and Stanton, Geol. Soc. Am., Bull., vol. 5, 1894, p. 450.
  Patuxent: Fredericksburg, Dutch Gap, and near Potomac Run, Va.
  Fuson: Pine Creek, Cook County, Wyo.
  Lakota: South Fork Hay Creek, Crook County, Wyo.

Cephalotaxopsis magnifolia? Fontaine. Fontaine, in Diller and Stanton, Geol. Soc. Am., Bull., vol. 5, 1894, p. 450 = *Cephalotaxopsis ramosa.*

Cephalotaxopsis microphylla Fontaine = *Cephalotaxopsis brevifolia.*

Cephalotaxopsis ramosa Fontaine, U. S. Geol. Surv., Mon. 15, 1889, p. 237, pl. civ, f. 2, 3; pl. cvi, f. 2, 4; pl. cvii, f. 3; pl. cviii, f. 2; in Ward, U. S. Geol. Surv., Mon. 48, 1905 [1906], p. 258, pl. lxviii, f. 5–7.

Cephalotaxopsis magnifolia? Fontaine. Fontaine, in Diller and Stanton, Geol. Soc. Am., Bull., vol. 5, 1894, p. 450.

Patuxent: Fredericksburg and near Potomac Run, Va.

Horsetown: Near Ono, Shasta County, and near Lowry, Tehama County, Calif.

Knoxville: Near Paskenta, Tehama County, Calif.

Cephalotaxopsis ramosa Fontaine. Fontaine, in Ward, U. S. Geol. Surv., Mon. 48, 1905 [1906], p. 311, pl. lxxiii, f. 8 = *Oleandra gramineafolia.*

Cephalotaxopsis ramosa Fontaine [part] = *Cephalotaxopsis magnifolia.*

Cephalotaxopsis ramosa Fontaine? Fontaine, in Ward, Mon. 48, 1905 [1906], p. 547 = *Nageiopsis angustifolia.*

?*Cephalotaxopsis? rhytidodes* Ward = *Cephalotaxopsis magnifolia.*

Cephalotaxopsis sp.? Fontaine, in Diller and Stanton, Geol. Soc. Am., Bull., vol. 5, 1894, p. 450 = *Cephalotaxopsis? magnifolia.*

Cephalotaxopsis sp. Dawson, Roy. Soc. Canada, Trans., vol. 10, 1892 [1893], p. 89.

Kootenai: Anthracite, British Columbia.

CEPHALOTAXOSPERMUM Berry, Bull. Torr. Bot. Club, vol. 37, 1910, p. 186. [Type, *C. carolinianum.*]

Cephalotaxospermum carolinianum Berry, Bull. Torr. Bot. Club, vol. 37, 1910, p. 187; U. S. Geol. Surv., Prof. Paper 84, 1914, p. 18, pl. iii, f. 4; idem, Prof. Paper 112, 1919, p. 56.

Black Creek: Black River, Sampson County, and Parker Landing, Tar River(?), N. C.; near Florence, S. C.

Eutaw: Near Havana, Hale County, Ala.

CEPHALOTAXUS Sieboldt and Zuccarini, Fl. Japon., 1842, p. 130.

Cephalotaxus? coloradensis Knowlton, U. S. Geol. Surv., Prof. Paper 98–H, 1916, p. 89, pl. xv, f. 4.

Fox Hills: About 1 mile east of Milliken, Colo.

Ceratopetalum americanum Ettingshausen = *Myrica drymeja.*

CERCIDOXYLON Platen, Sitzungsb. Naturf. Ges., Leipzig, vol. 34, 1907 [1908], p. 138. [Type, *C. zirkeli.*]

Cercidoxylon zirkeli Platen, Sitzungsb. Naturf. Ges., Leipzig, vol. 34, 1907 [1908], p. 139, pl. ii, f. 5, 6.

Pliocene: Running Water River, Nebr.

CERCIS Linné, Sp. pl., 1753, p. 374.

Cercis borealis Newberry, U. S. Nat. Mus., Proc., vol. 5, 1882 [1883], p. 510.

Fort Union: Valley of Yellowstone River, Mont.

Cercis canadensis Linné. Penhallow, Am. Nat., vol. 41, 1907, p. 446.—Berry, Torreya, vol. 9, 1909, p. 72, text f. 2.
Pleistocene: Don Valley, Toronto, Canada; Neuse River, N. C.

Cercis eocenica Lesquereux, U. S. Geol. and Geog. Surv. Terr., Ann. Rept. 1872 [1873], p. 384. [Not afterward recognized by its author.]
Laramie: Erie, Colo.

Cercis? nevadensis Knowlton, U. S. Geol. Surv., Twenty-first Ann. Rept., pt. 2, 1901, p. 217, pl. xxx, f. 23.
Esmeralda: 4.5 kilometers northeast of Emigrant Peak, Esmeralda County, Nev.

Cercis parvifolia Lesquereux, Rept. U. S. Geol. Surv. Terr., vol. 8 (Cret. and Tert. Fl.), 1883, p. 201, pl. xxxi, f. 5–7.—Penhallow, Roy. Soc. Canada, Trans., 2d ser., vol. 8, sec. 4, 1902, p. 63; Rept. Tert. Pl. Brit. Col., 1908, p. 44.—Knowlton, U. S. Nat. Mus., Proc., vol. 51, 1916, p. 276, pl. xxv, f. 1, 2.
Miocene: Florissant, Colo.
Paskapoo (Fort Union): Red Deer River, Alberta.

Cercis truncata Lesquereux, Rept. U. S. Geol. Surv. Terr., vol. 8 (Cret. and Tert. Fl.), 1883, p. 237.
Fort Union: Badlands, N. Dak.

Cercis wilcoxiana Berry, U. S. Geol. Surv., Prof. Paper 91, 1916, p. 228, pl. xlix, f. 1.
Wilcox (Holly Springs): Vaughns, Benton County, Miss.
Lagrange: Grand Junction, Fayette County, Tenn.

CERCOCARPUS Humboldt, Bonpland, and Kunth, Nov. gen. et sp., vol. 6, 1823, p. 232.

Cercocarpus antiquus Lesquereux, Mus. Comp. Zool., Mem., vol. 6, No. 2, 1878, p. 37, pl. x, f. 6–11; Rept. U. S. Geol. Surv. Terr., vol. 8 (Cret. and Tert. Fl.), 1883, p. 265, pl. xlv–B, f. 2.—Knowlton, in Lindgren, Jour. Geol., vol. 4, 1896, p. 890.
Miocene (auriferous gravels): Table Mountain, Tuolumne County, and Independence Hill, Placer County, Calif.

Cercocarpus betulaefolius Nuttall. Hannibal, Bull. Torr. Bot. Club, vol. 38, 1911, p. 337.
Pliocene: Calabasas Canyon, Santa Cruz Mountains, Calif.

Cercocarpus orestesi Knowlton, U. S. Geol. Surv., Prof. Paper 101, 1918, p. 325, pl. xcv, f. 2.
Raton: Raton Valley, N. Mex.

CHAETOCHLOA Scribner, Bull. U. S. Div. Agrost., vol. 4, p. 38, 1897.

Chaetochloa sp., Berry, Torreya, vol. 14, 1914, p. 160.
Pleistocene: Seven Springs, Neuse River, Wayne County, N. C.

CHAMAECYPARIS Spach, Hist. vég., Phan., vol. 11, 1842, p. 329.

Chamaecyparis nutkaensis Spach. Knowlton, Jour. Geol., vol. 3, 1895, p. 531.

Interglacial: Muir Glacier, Alaska.

Chamaecyparis sphaeroidea Spach = *Chamaecyparis thyoides.*

Chamaecyparis thyoides (Linné), Britton, Sterns, and Poggenberg. Knowlton, U. S. Geol. Surv., Bull. 152, 1898, p. 67.—Penhallow, Man. N. A. Gym., 1907, p. 231.

Cupressus thyoides Linné. Penhallow, Roy. Soc. Canada Trans., 2d ser., vol. 2, sec. 4, 1896, p. 73.

Chamaecyparis sphaeroidea Spach. Penhallow, Brit. Assoc. Adv. Sci., Bradford meeting, 1900, p. 335 [8].

Pleistocene: New Jersey; Don Valley, Toronto, Canada.

CHAMAEDOREA Willdenow, Act. Ac. Berol., 1804, p. 40.

Chamaedorea danai (Lesquereux) Berry, U. S. Geol. Surv., Prof. Paper 91, 1916, p. 179, pl. xii, f. 4; pl. xiii, f. 1–3.

Cycas Hilgard, Rept. on geology and agriculture of Mississippi, 1860, pp. 108, 117.

Calamopsis danai Lesquereux, in Dana, Manual of geology, 1st ed., 1866, p. 513, f. 795; Am. Philos. Soc., Trans., vol. 13, 1869, p. 411, pl. xiv, f. 1–3.

Wilcox ("Eolignitic"): *Mississippi.

Wilcox (Ackerman): Colemans Mill, Choctaw County, and Bendon?, Saline County, Ark..

CHARA (Vaillant) Linné, Sp. pl., 1753, p. 1156.

Chara compressa Knowlton, Bot. Gaz., vol. 13, 1888, p. 156, text f. 1, 2.

Eocene: Wales, Utah.

Chara estanciana Hannibal, Science, new ser., vol. 48, 1918, p. 578.

"Jura-Cretaceous?": Near Arkansas River, opposite old Lewisburg, Ark. [sec. 6, T. 5 N., R. 16 W.].

Chara? glomerata Lesquereux, Rept. U. S. Geol. Surv. Terr., vol. 8 (Cret. and Tert Fl.), 1883, p. 135, pl. xxi, f. 12.

Miocene: Florissant, Colo.

Chara peritula Cockerell, Am. Mus. Nat. Hist., Bull., vol. 24, 1908, p. 75.

Miocene: Florissant, Colo.

Chara springerae Knowlton, Torreya, vol. 2, 1902, p. 72, f.

Pleistocene: Arroyo Pecos, Las Vegas, N. Mex.

Chara stantoni Knowlton, Bot. Gaz., vol. 18, 1893, p. 141, text f. 1–3.

Bear River: Cokeville, Wyo.

Chara sp. Hinde, Canadian Jour. [Canadian Inst.], vol. 15, 1878, p. 399.

Pleistocene: Scarborough Heights, Ontario.

Cheilanthites göpperti Dunker = *Ruffordia göpperti.*

CHEIROLEPIS Schimper, Pal. vég., vol. 2, 1870, p. 247. [Type, *Brachyphyllum muensteri*.]

Cheirolepis latus Brown, Acad. Nat. Sci. Philadelphia, Proc., vol. 63, 1911, p. 20, pl. v.
Triassic (Newark): Bucks County, Pa.

Cheirolepis muensteri (Schenk) Schimper, Traité pal. vég., vol. 2, 1870, p. 248.—Fontaine, U. S. Geol. Surv., Mon. 6, 1883, pp. 88, 99, pl. xlvii, f. 6, 7; pl. xlix, f. 10.—Newberry, idem, Mon. 14, 1888, p. 90, pl. xxii, f. 4, 4a.—Fontaine, in Ward, U. S. Geol. Surv., Twentieth Ann. Rept., pt. 2, 1900, p. 252, pl. xxxiii, f. 1, 2.—Brown, Acad. Nat. Sci. Philadelphia, Proc., vol. 63, 1911, p. 20.—Knowlton, in Stone, U. S. Geol. Surv., Bull. 471, 1912, p. 143.
Brachyphyllum muensteri Schenk, Fl. d. Grenzschichten des Keupers u. Lias Frankens, p. 187, pl. xliii, f. 1-3, 4-12.
Walchea angustifolia Emmons, American geology, pt. 6, 1857, p. 35, pl. iii, f. 3.
Triassic (Newark): Cumberland coal field, Va.; Milford, N. J.; Dan River, N. C.; York and Bucks counties, Pa.

Cheirolepis muensteri Fontaine, U. S. Geol. Surv., Mon. 6, 1883, p. 108, pl. liii, f. 3 = *Palissya brevifolia*.

Cheirolepis muensteri (Schenk) Schimper. Fontaine, U. S. Geol. Surv., Mon. 6, 1883, p. 108, pl. l, f. 3 = *Palissya diffusa*.

CHIONANTHUS Linné, Sp. pl., 1753, p. 8.

Chionanthus membranaceus Knowlton, U. S. Geol. Surv., Prof. Paper 101, 1918, p. 345, pl. cviii, f. 2.
Raton: Aguilar, Colo.

CHIROPTERIS Kurr, in Bronn, Neues Jahrb., 1858, p. 143.

Chiropteris spatulata Newberry = *Sagenopteris elliptica*.

Chiropteris williamsii Newberry, Am. Jour. Sci., 3d ser., vol. 41, 1891, p. 198, pl. xiv, f. 10, 11.
Kootenai: Great Falls, Mont.

CHONDRITES Sternberg, Vers., vol. 2, 1833, p. 25.

Chondrites alpestris Heer, Fl. foss. Helv., 1877, p. 109, pl. xlii, f. 13-16.—Ulrich, Harriman Alaska Exped., vol. 4, 1904, p. 136, pl. xviii, f. 4.
Yakutat: Pogibshi Island, opposite Kodiak, Alaska.

Chondrites bulbosus Lesquereux, U. S. Geol. and Geog. Surv. Terr., Ann. Rept., 1872 [1873], p. 373; Rept. U. S. Geol. Surv. Terr., vol. 7 (Tert. Fl.), 1878, p. 42, pl. i, f. 14.—Knowlton, U. S. Geol. Surv., Prof. Paper 101, 1917, p. 243.
Trinidad: Raton Mountains, N. Mex.

Chondrites divaricatus Fischer-Ooster, Foss. fucoiden, 1858, p. 45.—Heer, Fl. foss. Helv., 1877, p. 107, pl. xli, f. 6, 7; pl. xlii, f. 11, 12.—Ulrich, Harriman Alaska Exped., 1904, vol. 4, p. 136, pl. xvi, f. 1, 2.
Yakutat: Pogibshi Island, opposite Kodiak, Alaska.

Chondrites filiciformis Lesquereux, U. S. Nat. Mus., Proc., vol. 11, 1888, p. 32, pl. xvi, f. 1.—Fontaine, in Ward, U. S. Geol. Surv., Mon. 48, 1905 [1906], p. 154.—Knowlton, U. S. Nat. Mus., vol. 17, 1894 p. 211; Geol. Soc. Am., Bull., vol. 5, 1893, p. 577.
Jurassic: Cape Lisburne, Alaska.

Chondrites flexuosus Newberry, U. S. Geol. Surv., Mon. 26, 1895 [1896], p. 34, pl. i, f. 1, 4.—Berry, N. Y. Bot. Gard., Bull., vol. 3, 1903, p. 100; Geol. Surv. N. J., Ann. Rept., 1905 [1906], p. 138; idem Bull. 3, 1911, p. 63.
Magothy: Cliffwood, N. J.
Raritan: Woodbridge and Sayreville, N. J.

Chondrites gracilis Emmons, Geol. Rept. Midland Counties, N. C., 1856, p. 288, pl. ii, f. 4.
Triassic (Newark): Near Egypt, N. C.

Chondrites heerii Eichwald, Geognost.-Palaeontolog. Bemerk. ü. Halbinsel. Mangischlak und Aleut. Inseln, 1871, p. 111, pl. iv, f. 1.—Knowlton, Geol. Soc. Am., Bull., vol. 5, 1893, p. 577; U. S. Nat. Mus., Proc., vol. 17, 1894, p. 211; U. S. Geol. Surv., Seventeenth Ann. Rept., pt. 1, 1896, p. 876.
Chondrites sp. Heer, Fl. foss. arct., vol. 2, Abt. 2 (Fl. foss. alask.), 1869, p. 21, pl. x, f. 5.
Triassic?: Kachemak Bay, Woody Island, Kadiak, Alaska.

Chondrites interruptus Emmons, Geol. Rept. Midland Counties N. C., 1856, p. 288.
Triassic (Newark): Near Egypt, N. C.

Chondrites ramosus Emmons, Geol. Rept. Midland Counties N. C., 1856, p. 289.
Triassic (Newark): Egypt, N. C.

Chondrites subsimplex Lesquereux, U. S. Geol. and Geog. Surv. Terr., Ann. Rept., 1872 [1873], p. 373; Rept. U. S. Geol. Surv. Terr., vol. 7 (Tert. Fl.), 1878, p. 41, pl. i, f. 13.
Trinidad: Trinidad, Colo.

Chondrites sp. Heer = *Chondrites heerii.*

CHONDROPHYLLUM Necker, Elem. Bot., vol. 3, 1790, p. 347.

Chondrophyllum nordenskiöldi Heer. Berry = *Pistia nordenskiöldi.*

Chondrophyllum nordenskiöldi? Heer. Lesquereux = *Hedera ovalis.*

Chondrophyllum obovatum Newberry, U. S. Geol. Surv., Mon. 26, 1895 [1896], p. 118, pl. xlii, f. 26, 27.—Berry, Geol. Surv. N. J., Bull. 3, 1911, p. 218
Raritan: Woodbridge, N. J.

Chondrophyllum orbiculatum Heer, Fl. foss. arct., vol. 3, Abt. 2, 1874, p. 115, pl. xxxi, f. 3c; pl. xxxii, f. 13.—Hollick, N. Y. Acad. Sci., Trans., vol. 12, 1892, p. 35, pl. ii, f. 2b; U. S. Geol. Surv., Mon. 50, 1906, p. 100, pl. xxxviii, f. 8a.
Raritan: Tottenville, Staten Island, N. Y.

Chondrophyllum orbiculatum Heer. Lesquereux = ***Hedera orbiculata***.

Chondrophyllum reticulatum Hollick, in Newberry, U. S. Geol. Surv., Mon. 27, 1895 [1896], p. 119, pl. xli, f. 6, 7; Geol. Surv. N. J., Bull., vol. 3, 1911, p. 218.
Raritan: New Jersey.

CHRYSOBALANUS Linné, Sp. pl., 1753, p. 513.

Chrysobalanus eocenica Berry, U. S. Geol. Surv., Prof. Paper 91, 1916, p. 220, pl. xliv, f. 3, 4; pl. cxii, f. 8–10.
Wilcox (Grenada): Grenada, Grenada County, Miss.
Lagrange: Puryear, Henry County, Tenn.

Chrysobalanus inaequalis (Lesquereux) Berry, U. S. Geol. Surv., Prof. Paper 91, 1916, p. 220, pl. xliv, f. 8–10.
Elaeagnus inaequalis Lesquereux, Am. Jour. Sci., vol. 27, 1859, p. 364, in Safford's Geol. Tenn., 1869, p. 428, pl. K, f. 7.
Wilcox (Grenada): Grenada, Grenada County, Miss.
Lagrange: Summerville, Fayette County, and Puryear, Henry County, Tenn.

Chrysobalanus pollardiana Knowlton, U. S. Geol. Surv., Twenty-first Ann. Rept., pt. 2, 1901, p. 216, pl. xxx, f. 19.
Esmeralda: 4.5 kilometers northeast of Emigrant Peak, Esmeralda County, Nev.

CHRYSOPHYLLUM Linné, Sp. pl., 1753, p. 852.

Chrysophyllum ficifolia Berry, U. S. Geol. Surv., Prof. Paper 91, 1916, p. 335, pl. c, f. 7.
Wilcox (Grenada): Grenada, Miss.
Lagrange: Pinson, Tenn.

CINCHONIDIUM Unger, Gen. et sp. pl. foss., 1850, p. 430.

Cinchonidium copeanum (Lesquereux) Ettingshausen, Denks. d. k. Akad. d. W., Math.-Naturw. Classe, vol. 47, 1883, p. 130 [30]; Geol. Surv. N. S. W., Mem. Palaeont. No. 2 (Tert. Fl. Australia), 1888, p. 49.
Diospyros copeana Lesquereux, U. S. Geol. and Geog. Surv. Terr., Ann. Rept. 1873 [1874], p. 414; Rept. U. S. Geol. Surv. Terr., vol. 7 (Tert. Fl.), 1878, p. 232, pl. xl, f. 11.
Miocene: Elko station, Nev.

Cinchonidium ovale Lesquereux, Rept. U. S. Geol. Surv. Terr., vol. 8 (Cret. and Tert. Fl.), 1883, p. 229, pl. xlviii, f. 8–10b.
Fort Union: Badlands, N. Dak.

Cinchonidium? turneri Knowlton, U. S. Geol. Surv., Twenty-first Ann. Rept., pt. 2, 1901, p. 218, pl. xxx, f. 9–11.
Esmeralda: 3.8 and 4.5 kilometers northeast of Emigrant Peak, Esmeralda County, Nev.

CINNAMOMUM (Burmann) R. Brown, Prodr., I, 1810, p. 62.

Cinnamomum affine Lesquereux, Am. Jour. Sci., 2d ser., vol. 45, 1868, p. 206; U. S. Geol. and Geog. Surv. Terr., Ann. Rept. 1869, p. 169, reprint 1873; idem, 1872 [1873], p. 383; idem, 1873 [1874], p. 401; Rept. U. S. Geol. Surv. Terr., vol. 7 (Tert. Fl.), 1878, p. 219, pl. xxxvii, f. 1–4, 7 [not f. 5, which = *Ficus trinervis* Knowlton].—Knowlton, U. S. Geol. Surv., Bull. 163, 1900, p. 59, pl. xiv, f. 2.—Cockerell, Torreya, vol. 9, 1909, p. 142.—Berry, U. S. Geol. Surv., Prof. Paper 91, 1916, pp. 13, 299, pl. iii, f. 2.—Knowlton, idem, Prof. Paper 130, 19—, p. —, pl. viii, f. 4; pl. xvii, f. 6 [in preparation].

Cinnamomum polymorphum Al. Braun. Lesquereux, Rept. U. S. Geol. Surv. Terr., vol. 7 (Tert. Fl.), 1878, p. 221, pl. xxxvii, f. 6 [not f. 10].

Cinnamomum rossmässleri Heer. Lesquereux, U. S. Geol. and Geog. Surv. Terr., Ann. Rept. 1872 [1873], p. 379.

[NOTE.—*Cinnamomum affine* has been recorded from many localities and horizons, but as it is not now accepted as occurring in beds younger than Laramie, these localities are not listed.]

Laramie: *Marshall, Golden, and Cowan station, near Denver, Colo.

Montana(?): Coalville, Utah.

Midway(?): Earle, Bexar County, Tex.

Cinnamomum affine Lesquereux, Tert. Fl., pl. xxxvii, f. 5 = *Ficus trinervis*.

Cinnamomum affine Lesquereux. Ward, U. S. Geol. Surv., Sixth Ann. Rept., 1884–85 [1886], p. 553, pl. xlvii, f. 1–3; idem, Bull. 37, 1887, p. 50, pl. xxiv, f. 3–5 = *Ficus trinervis*.

Cinnamomum angustum Berry, in Cooke and Shearer, U. S. Geol. Surv., Prof. Paper 120, 1918, p. 55 [name].

Barnwell (Twiggs): Twiggs County, Ga.

Cinnamomum bendirei Knowlton, U. S. Geol. Surv., Bull. 204, 1902, p. 59, pl. x, f. 4.

Clarno (upper part): Bridge Creek, Oreg.

Cinnamomum buchi Heer, Fl. tert. Helv., vol. 2, 1856, p. 90, pl. 95, f. 1–8.—Hollick, Rept. Geol. Surv. La., 1899 [1900], p. 283, pl. xliii, f. 1.—Berry, U. S. Geol. Surv., Prof. Paper 91, 1916, p. 299, pl. lxxix, f. 10.

Wilcox: Near Coushatta, La.

Cinnamomum canadense Dawson, Roy. Soc. Canada, Trans., vol. 3, sec. 4, 1885 [1886], p. 13, pl. iv, f. 7.

Mill Creek: Mill Creek, British Columbia.

Cinnamomum corrugatum Perkins, Rept. Vt. State Geol., 1903–4 [1904], p. 200, pl. lxxix, f. 105.

Miocene: Brandon, Vt.

Cinnamomum crassipes Lesquereux, Am. Jour. Sci., 2d ser., vol. 27, 1859, p. 361.

——: Bellingham Bay, Wash.

Cinnamomum crassipetiolatum Hollick, U. S. Geol. Surv., Mon. 50, 1906, p. 74, pl. xxx, f. 3, 4.

Magothy: Glen Cove, Long Island, N. Y.

Cinnamomum dilleri Knowlton, U. S. Geol. Surv., Twentieth Ann. Rept., pt. 3, 1900, p. 47, pl. iv, f. 1; U. S. Geol. Surv., Bull. 204, 1902, p. 59.

Eocene: Comstock, Douglas County, Oreg.

Clarno (lower part): Cherry Creek, Crook County, Oreg.

Cinnamomum ellipsoideum Saporta and Marion, Mém. Acad. roy. Belgique, vol. 41, 1878, p. 61, pl. ix, f. 7–9.—Lesquereux, U. S. Geol. Surv., Mon. 17, 1891 [1892], p. 105, pl. li, f. 8, 9.

Dakota: Near Glencoe, Kans.

Cinnamomum ellipticum Knowlton, U. S. Geol. Surv., Bull. 105, 1893, p. 54.

Cinnamomum polymorphum Al. Braun. Lesquereux, Rept. U. S. Geol. Surv. Terr., vol. 7 (Tert. Fl.), 1878, p. 221, pl. xxxvii, f. 10 [not f. 6, which= *C. affine*, q. v.].

Livingston: Bozeman coal field, Livingston area, and Fish Creek, Mont.

Cinnamomum? ficifolium Knowlton, U. S. Geol. Surv., Prof. Paper 101, 1917, p. 318, pl. xc, f. 3.

Raton: Apishapa Canyon, near Aboton, Colo.

Cinnamomum heerii (Lesquereux) Lesquereux Am. Jour. Sci., 2d ser., vol. 27, 1859, p. 361; Am. Philos. Soc., Trans., vol. 13, 1869, p. 431, pl. xxiii, f. 12; Rept. U. S. Geol. Surv. Terr., vol. 6, 1874, p. 84, pl. xxviii, f. 11; idem, vol. 8 (Cret. and Tert. Fl.), 1883, p. 54; U. S. Geol. Surv. Mon. 17, 1891 [1892], p. 105, pl. xv, f. 1.—Newberry, U. S. Geol. Surv., Mon. 35, 1898, p. 100, pl. xvii, f. 1-3.—Berry, Geol. Surv. N. J., Ann. Rept. 1905 [1906], p. 139; Bull. Torr. Bot. Club, vol. 33, 1906, p. 179.—Hollick, U. S. Geol. Surv., Mon. 50, 1906, p. 75, pl. xxx, f. 5, 6.—Berry, Bull. Torr. Bot. Club, vol. 34, 1907, p. 200, pl. xiii, f. 2, 3; idem, vol. 37, 1910, p. 504; U. S. Geol. Surv., Prof. Paper 84, 1914, p. 118, pl. xxi, f. 8; Bull. Torr. Bot. Club, vol. 43, 1916, p. 298; U. S. Geol. Surv., Prof. Paper 112, 1919, p. 118.

Daphnogene heerii Lesquereux, U. S. Geol. and Geog. Surv. Terr., Ann. Rept. 1874 [1876], p. 343.

Dakota: Near Glencoe, Kans.

Raritan: South Amboy, N. J.

Magothy: Cliffwood, N. J.; Gay Head, Marthas Vineyard, Mass.

Black Creek: Court House Bluff, Cape Fear River, N. C.

——: Orcas Island, Wash.

Eutaw: McBrides Ford, Ga.; Coffee Bluff, Hardin County, Tenn.

Cinnamomum hesperium Knowlton, U. S. Geol. Surv., Prof. Paper 108, 1917, p. 88, pl. xxxviii, f. 2.

Frontier: About 1½ miles south of Cumberland, Wyo.

Cinnamomum intermedium Newberry = *Cinnamomum newberryi.*

Cinnamomum lanceolatum (Unger) Heer. Ward, U. S. Geol. Surv., Sixth Ann. Rept., 1884–85 [1886], p. 553, pl. xlvi, f. 12; idem, Bull. 37, 1887, p. 49, pl. xxiv, f. 2 = *Cinnamomum wardii.*

Cinnamomum laramiense Knowlton, U. S. Geol. Surv., Prof. Paper 130, 19—, p. —, pl. xxii, f. 3 [in preparation].

Laramie: Morrison and Cowan station, near Denver, Colo.

Cinnamomum lignitum Perkins, Rept. Vt. State Geol., 1903–4 [1904], p. 200, pl. lxxix, f. 106; Geol. Soc. Am., Bull., vol. 16, 1905, p. 514, pl. lxxxvii, f. 24; Rept. Vt. State Geol., 1905–6 [1906], pl. liii, f. 24.

Miocene: Brandon, Vt.

Cinnamomum linifolium Knowlton, U. S. Geol. Surv., Prof. Paper 101, 1917, p. 319, pl. lxxxviii, f. 3–7.

Raton: Near Starkville, Colo., and Dawson, N. Mex.

Cinnamomum marioni Lesquereux, U. S. Geol. Surv., Mon. 17, 1891 [1892], p. 106, pl. li, f. 6, 7.

Dakota: Near Fort Harker, Kans.

Cinnamomum membranaceum (Lesquereux) Hollick, U. S. Geol. Surv., Mon. 50, 1906, p. 75, pl. xxix, f. 5, 6.—Berry, Bull. Torr. Bot. Club, vol. 39, 1912, p. 401.

Paliurus membranaceus Lesquereux, Am. Jour. Sci., 2d ser., vol. 46, 1868, p. 101; Rept. U. S. Geol. Surv. Terr., vol. 6 (Cret. Fl.), 1874, p. 108, pl. xx, f. 6; U. S. Geol. Surv., Mon. 17, 1891 [1892], p. 167, pl. xxxv, f. 5.

Magothy: Gay Head, Marthas Vineyard, Mass.

Woodbine: Arthurs Bluff, Tex.

Dakota: Decatur, Nebr.; Pipe Creek, Kans.

Cinnamomum middendorfensis Berry, U. S. Geol. Surv., Prof. Paper 84, 1914, p. 55, pl. viii, f. 14; pl. ix, f. 1.

Black Creek (Middendorf): Middendorf, S. C.

Cinnamomum mississippiensis Lesquereux, in Dana, Manual of geology, 1st ed., 1866, p. 513, f. 794; Am. Philos. Soc., Trans., vol. 13, 1869, p. 418, pl. xix, f. 2.—Knowlton, Am. Geol., vol. 16, 1895, p. 308.—Berry, U. S. Geol. Surv., Prof. Paper 91, 1916, p. 298, pl. xxxvii, f. 2.—Knowlton, U. S. Geol. Surv., Prof. Paper 101, 1917, p. 320, pl. lxxxix, f. 2.

Wilcox (Ackerman): Raglands Branch, Lafayette County, Miss.

Wilcox: Colemans Mill near New Prospect, Choctaw County, Miss.

Raton: Near Raton and Yankee, N. Mex.; Walsenburg, Abeton, Dean, Trinidad, etc., Colo.

Cinnamomum newberryi Berry, Bull. Torr. Bot. Club, vol. 38, 1911, p. 423; Geol. Surv. N. J., Bull. 3, 1911, p. 150, pl. xvi, f. 3; U. S. Geol. Surv., Prof. Paper 84, 1914, pp. 57, 117, pl. ix, f. 12, 13; pl. xxi, f. 9–11; Md. Geol. Surv., Upper Cret., 1916, p. 860, pl. lxxi, f. 6; U. S. Geol. Surv., Prof. Paper 112, 1919, p. 118, pl. xxi, f. 6–9; Bull. Torr. Bot. Club, vol. 44, 1917, p. 187.

Cinnamomum sezannense Watelet. Lesquereux, U. S. Geol. Surv., Mon. 17, 1891 [1892], p. 107, pl. xii, f. 6, 7.—Dawson, Roy. Soc. Canada, Trans., 1st ser., vol. 11, 1894, p. 64, pl. xiii, f. 58.—Hollick, Bull. Torr. Bot. Club, vol. 21, 1894, p. 53, pl. clxxx, f. 5, 7.—Penhallow, Roy. Soc. Canada, Trans., 2d ser., vol. 8, 1902, p. 46.—Hollick, N. Y. State Mus., Fifty-fifth Ann. Rept., for 1901 [1903], p. r50.

Cinnamomum intermedium Newberry, U. S. Geol. Surv., Mon. 26, 1895 [1896], p. 89, pl. xxix, f. 1–8, 10.—Berry, N. J. Geol. Surv., Ann. Rept. 1905 [1906], p. 139, pl. xx, f. 2–6; Bull. Torr. Bot. Club, vol. 33, 1906, p. 179, pl. vii, f. 3, 4.—Hollick, U. S. Geol. Surv., Mon. 50, 1906 [1907], p. 74, pl. xxix, f. 7; pl. xxx, f. 1, 2.—Berry, Bull. Torr. Bot. Club, vol. 37, 1910, p. 27.

Raritan: Woodbridge, Sayreville, and South Amboy, N. J.; East Washington Heights, D. C.

Magothy: Cliffwood, N. J.; Deep Cut, Del.; Grove Point and Round Bay, Md.; Glen Cove, Sea Cliff, and Manhasset Neck, Long Island, N. Y.

Eutaw: McBrides Ford, Columbus, Ga.

Tuscaloosa: Shirleys Mill, Glen Allen, and Cottondale, Ala.

Dakota: Ellsworth County, Kans.

Upper Cretaceous: Port McNeil, Vancouver Island, British Columbia.

Black Creek (Middendorf): Rocky Point and below Columbia, S. C.

Bingen: Mine Creek near Nashville, Ark.

Cinnamomum novae-angliae Lesquereux, Am. Jour. Sci., 2d ser., vol. 32, 1861, p. 360, f. 16; Geol. Vt., vol. 2, 1861, p. 716, f. 148.—Perkins, Rept. Vt. State Geol., 1903–4 [1904], p. 200, pl. lxxviii, f. 82; pl. lxxix, f. 108.

Miocene: Brandon, Vt.

Cinnamomum oblongatum Berry, U. S. Geol. Surv., Prof. Paper 91, 1916, p. 297, pl. lxxix, f. 1, 2; pl. lxxxiii, f. 6.

Lagrange: Puryear, Tenn.

Wilcox: De Soto Parish, La.

Cinnamomum obovatus Berry, U. S. Geol. Surv., Prof. Paper 91, 1916, p. 296, pl. xxix, f. 3.

Wilcox (Holly Springs): Holly Springs, Miss.

Cinnamomum ovoides Perkins, Rept. Vt. State Geol., 1903–4 [1904], p. 199, pl. lxxix, f. 104, 107.

Miocene: Brandon, Vt.

Cinnamomum polymorphum Al. Braun. Lesquereux, Rept. U. S. Geol. Surv. Terr., vol. 7 (Tert. Fl.), 1878, p. 221, pl. xxxvii, f. 6 = *Cinnamomum affine.*

Cinnamomum polymorphum Al. Braun. Lesquereux, Rept. U. S. Geol. Surv., Terr., vol. 7 (Tert. Fl.), 1878, p. 221, pl. xxxvii, f. 10 = *Cinnamomum ellipticum.*

Cinnamomum postnewberryi Berry, U. S. Geol. Surv., Prof. Paper 91, 1916, p. 298, pl. lxxix, f. 9.

Cinnamomum scheuchzeri Heer. Hollick, La. Geol. Surv., Special Rept. 5, 1899, p. 283, pl. xli, f. 4.

Daphnogene kanei Heer. Hollick, idem, p. 284, pl. xli, f. 3.

Wilcox: Caddo Parish, La., and Hardys Mill, Green County, Ark.

Cinnamomum scheuchzeri Heer, Fl. Tert. Helv., vol. 2, 1856, p. 85, pl. xci, f. 4–22; pl. xcii, pl. xciii, f. 1, 5.—Lesquereux, Rept. U. S. Geol. Surv. Terr., vol. 6 (Cret. Fl.), 1874, p. 83, pl. xxx, f. 2, 3; idem, vol. 7 (Tert. Fl.), 1878, p. 220, pl. xxxviii, f. 8; U. S. Geol. Surv., Mon. 17, 1891 [1892], p. 104, pl. xi, f. 1.

Daphnogene cretacea Lesquereux, U. S. Geol. and Geog. Surv. Terr., Ann. Rept., 1874 [1876], p. 343.

Dakota: Kansas; New Ulm, Minn.

Cinnamomum scheuchzeri Heer, Hollick = *Cinnamomum postnewberryi.*

Cinnamomum sezannense Watelet = *Cinnamomum newberryi.*

Cinnamomum sezannense Watelet. Hollick, 1899 = *Oreodaphne obtusifolia.*

Cinnamomum spectabile Heer, Fl. tert. Helv., vol. 2, 1856, p. 91, pl. xcvi, f. 1–8.—Knowlton, U. S. Geol. Surv., Mon. 32, pt. 2, 1899, p. 727, pl. xciv, f. 6.

Miocene: Tower Creek, Yellowstone National Park.

Cinnamomum? stantoni Knowlton, U. S. Geol. Surv., Bull. 162, 1900, p. 60, pl. xv, f. 1.

Montana: Coalville, Utah.

Cinnamomum vera Berry, U. S. Geol. Surv., Prof. Paper 91, 1916, p. 297, pl. lxxix, f. 3–5; pl. lxxxvii, f. 4.

Wilcox (Holly Springs): Oxford and Holly Springs, Miss.

Lagrange: Puryear, Tenn.

Cinnamomum wardii Knowlton, U. S. Geol. Surv., Bull. 152, 1898, p. 69.

Cinnamomum lanceolatum (Unger) Heer. Ward, U. S. Geol. Surv., Sixth Ann. Rept., 1884–85 [1886], p. 553, pl. xlvi, f. 12; idem, Bull. 37, 1887, p. 49, pl. xxiv, f. 2.

Laramie (?): Hodges Pass, Wyo.

Cinnamomum? sp. Dawson, Roy. Soc. Canada, Trans., vol. 8, sec. 4, 1890, p. 84, text f. 15.

Miocene?: North Fork of Similkameen River, British Columbia.

Cinnamomum sp. Hollick, Bull. Torr. Bot. Club, vol. 19, 1892, p. 332.
Pleistocene: Bridgeton, N. J.

Cinnamomum sp. Hollick, U. S. Geol. Surv., Mon. 50, 1906, p. 75, pl. xxx, f. 7.
Magothy: Gay Head, Marthas Vineyard, Mass.

Cinnamomum sp. Knowlton, Geol. Soc. Am., Bull., vol. 8, 1897, p. 145.
Post-Laramie: Black Buttes, Wyo.

Cinnamomum sp. Knowlton, Wash. Geol. Surv., Ann. Rept., vol. 1, 1902, p. 33.
Miocene: Black River Junction, King County, Wash.

Cinnamomum sp. Knowlton, Wash. Geol. Surv., Ann. Rept., vol. 1, 1902, p. 33.
Eocene: Coal Creek near Keese, Whatcom County, Wash.

Cinnamomum sp. Knowlton, Wash. Geol. Surv., Ann. Rept., vol. 1, 1902, p. 33.
Eocene: Coal Creek near Keese, Whatcom County, Wash.

Cinnamomum sp. Knowlton, Wash. Geol. Surv., Ann. Rept., vol. 1, 1902, p. 32.
Eocene: Cokedale Skagit County, Wash.

Cinnamomum sp. Knowlton, Wash. Geol. Surv., Ann. Rept., vol. 1, 1902, p. 33.
Eocene: Republic, Ferry County, Wash.

Cinnamomum sp. Knowlton, in Washburne, U. S. Geol. Surv., Bull. 590, 1914, p. 36.
Eocene?: Eugene, Oreg.

Cinnamomum? sp. Knowlton, U. S. Geol. Surv., Prof. Paper 108, 1917, p. 89, pl. xxxv, f. 2.
Frontier: About one-half mile east of Cumberland, Wyo.

CISSITES Heer, Mém. Soc. helvétique d. sci. nat. (Phyll. crét. d. Nebraska), Zürich, 1866, p. 19. [Type, *C. insignis*.]

Cissites acerifolius Lesquereux, U. S. Geol. Surv., Mon. 17, 1891 [1892], p. 163, pl. lviii, f. 1.
Dakota: Near Fort Harker, Kans.

Cissites acuminatus Lesquereux, U. S. Geol. and Geog. Surv. Terr., Bull., vol. 1, 1875 [1876], p. 396; idem, Ann. Rept., 1874 [1876], p. 353, pl. viii, f. 1.
Dakota: Near Fort Harker, Kans.

Cissites acutiloba Hollick, Bull. Torr. Bot. Club, vol. 22, 1895, p. 227, pl. ccxxxvii, f. 3.
Dakota: Fort Harker, Kans.

Cissites affinis (Lesquereux) Lesquereux, U. S. Geol. and Geog. Surv. Terr., Ann. Rept. 1874 [1876], p. 352; Rept. U. S. Geol. Surv. Terr., vol. 8 (Cret. and Tert. Fl.), 1883, p. 67.

Populites affinis Lesquereux, U. S. Geol. and Geog. Surv. Terr., Ann. Rept., 1872 [1873], p. 423.

Platanus affinis Lesquereux, Rept. U. S. Geol. Surv. Terr., vol. 6 (Cret. Fl.), 1874, p. 71, pl. iv, f. 4.—Knowlton, in Woodruff, U. S. Geol. Surv., Bull. 452, 1911, p. 20.

Dakota: Salina Valley, Kans.; Lander, Wyo.

Mill Creek: Mill Creek and North Fork of Old Man River, British Columbia.

Cissites affinis ampla (Dawson) Knowlton, U. S. Geol. Surv., Bull, 152, 1898, p. 70.

Platanus affinis ampla Dawson, Roy. Soc. Canada, Trans., vol. 3, sec. 4, 1885 [1886], p. 12.

Mill Creek: Mill Creek, British Columbia.

Cissites alatus Lesquereux, U. S. Geol. Surv., Mon. 17, 1891 [1892], p. 160, pl. xxiii, f. 6.

Dakota: Near Delphos, Kans.

Cissites brownii Lesquereux, U. S. Geol. Surv., Mon. 17, 1891 [1892], p. 162, pl. xviii, f. 11.

Dakota: Kansas and Minnesota.

Cissites crispus Velenovsky, Fl. böhm. Kreidef., 1885, pt. 4, p. 12 [73], pl. iv [xxvii], f. 6.—Berry, Bull. Torr. Bot. Club, vol. 43, 1916, p. 296; U. S. Geol. Surv., Prof. Paper 112, 1919, p. 115.

Ripley: Near Selmer, McNairy County, Tenn.

Cissites crispus Velenovsky. Newberry, 1896 = *Cissites newberryi.*

Cissites cyclophylla Lesquereux = *Populites cyclophyllus.*

Cissites dentato-lobatus Lesquereux, U. S. Geol. Surv., Mon. 17, 1891 [1892], p. 164, pl. lxvi, f. 4.

Dakota: Ellsworth County, Kans.

Cissites formosus Heer, Fl. foss. arct., vol. 6, Abt. 2, 1882, p. 85, pl. xxi, f. 5–8.—Lesquereux, U. S. Geol. Surv., Mon. 17, 1891 [1892], p. 161, pl. xxi, f. 5.—Newberry, U. S. Geol. Surv., Mon. 26, 1895 [1896], p. 107, pl. xlvii, f. 1–8.—Hollick, U. S. Geol. Surv., Mon. 50, 1906, p. 94, pl. xxxvii, f. 7.—Berry, Geol. Surv. N. J., Bull. 3, 1911, p. 185; U. S. Geol. Surv., Prof. Paper 112, 1919, p. 115.

Dakota: Ellsworth County, Kans.

Raritan: Sayreville, Woodbridge, South Amboy, N. J.; Dorosis Island, Glen Cove, Long Island, N. Y.

Tuscaloosa: Glen Allen, Ala.

Cissites formosus magothiensis Berry, Bull. Torr. Bot. Club, vol. 37, 1910, p. 25: Md. Geol. Surv., Upper Cret., 1916, p. 855, pl. lxxviii, f. 4.
Magothy: Grove Point, Md.

Cissites harkerianus (Lesquereux) Lesquereux, U. S. Geol. and Geog. Surv. Terr., Bull., vol. 1, 1875 [1876], p. 397; idem, Ann. Rept. 1874 [1876], p. 352, pl. vii, f. 1, 2.
Sassafras harkeriana Lesquereux, U. S. Geol. and Geog. Surv. Terr., Ann. Rept., 1872 [1873], p. 425.
Sassafras (*Araliopsis*) *harkerianum* Lesquereux, Rept. U. S. Geol. Surv. Terr., vol. 6 (Cret. Fl.), 1874, p. 81, pl. xi, f. 3, 4: pl. xxvii, f. 2.
Dakota: Kansas.

Cissites heerii Lesquereux, U. S. Geol. and Geog. Surv. Terr., Bull., vol. 1, 1875 [1876], p. 396; idem, Ann. Rept. 1874 [1876], p. 353, pl. vi, f. 3.
Dakota: Fort Harker, Kans.

Cissites ingens Lesquereux, U. S. Geol. Surv., Mon. 17, 1891 [1892], p. 159, pl. xix, f. 2, 2a.—Ward, U. S. Geol. Surv., Nineteenth Ann. Rept., pt. 2, 1899, p. 708, pl. clxxii, f. 1, 2.
Dakota: Ellsworth County, Kans.; Minnekahta Creek, 4 miles below Hot Springs, S. Dak.

Cissites ingens parvifolia Lesquereux. U. S. Geol. Surv., Mon. 17, 1891 [1892], p. 160, pl. lvii, f. 3, 4.
Dakota: Near Fort Harker, Kans.

Cissites insignis Heer, Mém. Soc. helvétique sci. nat. (Phyll. crét. d. Nebraska), 1866, p. 19, pl. ii, f. 3, 4.
Dakota: Tekamah, Nebr.

Cissites microphyllus Lesquereux, U. S. Nat. Mus., Proc., vol. 10, 1887, p. 44, pl. iii, f. 11.
Pleistocene: Clear Lake, Calif.

Cissites newberryi Berry, Md. Geol. Surv., Upper Cret., 1916, p. 856.
Cissites crispus Newberry, U. S. Geol. Surv., Mon. 26, 1895 [1896], p. 108, pl. xlii, f. 20–23.—Berry, Bull. Torr. Bot. Club, vol. 33, 1906, p. 177; Geol. Surv. N. J., Bull. 3, 1911, p. 186.
Raritan: Woodbridge and South Amboy, N. J.
Magothy: Cliffwood, N. J.; Deep Cut, Del.
Ripley: ————, Tenn.

Cissites obtusilobus Lesquereux, U. S. Geol. Surv., Mon. 17, 1891 [1892], p. 161, pl. xxxiii, f. 5.
Dakota: Ellsworth County, Kans.

Cissites obtusum Lesquereux = *Araliopsis cretacea salisburiaefolia*.

Cissites panduratus Knowlton, U. S. Geol. Surv., Prof. Paper 101, 1918, p. 274, pl. xlix, f. 10.
Vermejo: Walsenburg, Colo.

Cissites parvifolius (Fontaine) Berry, Md. Geol. Surv., Lower Cret., 1912, p. 482, pls. xci, xcii.

Vitiphyllum parvifolium Fontaine, U. S. Geol. Surv., Mon. 15, 1889 [1890], p. 309, pl. clxxii, f. 11, 12; in Ward, idem, Mon. 48, 1905 [1906], p. 558.

Vitiphyllum multifidum Fontaine, U. S. Geol. Surv., Mon. 15, 1889 [1890], p. 309, pl. clxxiii, f. 1–9.—Ward, U. S. Geol. Surv., Sixteenth Ann. Rept., pt. 1, 1896, p. 539, pl. cvii, f. 2–5.—Fontaine, in Ward, idem, Mon. 48, 1905 [1906], pp. 553, 565, pl. cxix, f. 2–5.

Patapsco: Federal Hill (Baltimore), Wellhams?, and Vinegar Hill?, Md.

Cissites platanoidea Hollick, Bull. Torr. Bot. Club, vol. 22, 1895, p. 226, pl. ccxxxvii, f. 2.

Dakota: Fort Harker, Kans.

Cissites populoides Lesquereux, U. S. Geol. Surv., Mon. 17, 1891 [1892], p. 162, pl. xviii, f. 12–14.

Dakota: Near Delphos, Kans.

Cissites salisburiaefolius Lesquereux = *Araliopsis cretacea salisburiaefolia.*

CISSUS Linné, Sp. pl., 1753, p. 117.

Cissus browniana Lesquereux, in Winchell, Geol. and Nat. Hist. Surv. Minnesota, Thirteenth Ann. Rept., 1884 [1885], p. 77 [name].—Lesquereux, Geol. and Nat. Hist. Surv. Minnesota, Final Rept., vol. 3, pt. 1, 1893, p. 17, pl. A, f. 8.

Dakota: Near New Ulm, Minn.

Cissus coloradensis Knowlton and Cockerell, n. name.

Cissus laevigata Lesquereux, U. S. Geol. and Geog. Surv. Terr., Ann. Rept. 1872 [1873], p. 380; Rept. U. S. Geol. Surv. Terr., vol. 7 (Tert. Fl.), 1878, p. 238, pl. xl, f. 12, 13.—Knowlton, U. S. Geol. Surv., Prof. Paper 101, 1917, p. 340, pl. ciii, f. 1. [Homonym, Blume, 1825–26.]

Denver: Golden, Colo.

Dawson: Sedalia, Colo.

Raton: Abeton, Colo.

Cissus corylifolia Lesquereux, Mus. Comp. Zool., Bull., vol. 16, 1888, p. 52.

Denver: Golden, Colo.

Cissus duplicato-serrata Lesquereux, Mus. Comp. Zool., Bull., vol. 16, 1888, p. 52.

Denver: Golden, Colo.

Cissus grossedentata Knowlton, U. S. Geol. Surv., Prof. Paper 101, 1918, p. 340, pl. civ, f. 1.

Raton: Morley, Colo.

Cissus haguei Knowlton, U. S. Geol. Surv., Mon. 32, pt. 2, 1899, p. 741, pl. ci, f. 2.

Miocene: Fossil Forest Ridge, Yellowstone National Park.

Cissus laevigata Lesquereux = *Cissus coloradensis.*

Cissus lesquereuxii Knowlton, n. name.

Cissus tricuspidata (Heer) Lesquereux, Rept. U. S. Geol. Surv. Terr., vol. 7 (Tert. Fl.), 1878, p. 240, pl. xli, f. 4–7. [Homonym, Sieboldt and Zuccarini, 1846.]

Vitis (Cissus) tricuspidata Heer, Mioc. Balt. Fl., 1869, p. 91, pl. xxviii, f. 18, 19.—Lesquereux, U. S. Geol. and Geog. Surv. Terr., Ann. Rept., 1872 [1873], p. 296.

Post-Laramie: Black Buttes, Wyo.

Cissus lobato-crenata Lesquereux, U. S. Geol. and Geog. Surv. Terr., Ann. Rept. 1872 [1873], p. 396; Rept. U. S. Geol. Surv. Terr., vol. 7 (Tert. Fl.), 1878, p. 240, pl. xli, f. 1–3.

Denver: Mount Bross, Middle Park, and Sedalia, Colo.

Post-Laramie: Black Buttes, Wyo.

Cissus parrottiaefolia Lesquereux, U. S. Geol. and Geog. Surv. Terr., Bull., vol. 1, 1876, p. 388; idem, Ann. Rept. 1874 [1876], p. 314; Rept. U. S. Geol. Surv. Terr., vol. 7 (Tert. Fl.), 1878, p. 239, pl. xl, f. 15–17; pl. xlii, f. 1.

Green River: Green River, Wyo.

Lance?: Medicine Bow, Wyo.

Cissus tricuspidata (Heer) Lesquereux = *Cissus lesquereuxii.*

CITHAREXYLON Linné, Sp. pl., 1753, p. 156.

Citharexylon eoligniticum Berry, U. S. Geol. Surv., Prof. Paper 91, 1916 p. 346, pl. cvi, f. 10.

Wilcox (Holly Springs): Holly Springs, Miss.

Wilcox (Grenada): Grenada, Miss.

CITROPHYLLUM Berry, Bull. Torr. Bot. Club, vol. 36, 1909, p. 258. [Type, *Ficus aligera* Lesquereux.]

Citrophyllum aligerum (Lesquereux) Berry, Bull. Torr. Bot. Club, vol. 36, 1909, p. 258, pl. xviiia, f. 1–8; U. S. Geol. Surv., Prof. Paper 84, 1914, p. 47; Geol. Surv. N. J., Bull. 3, 1911, p. 169, pl. xxi, f. 1–8; U. S. Geol. Surv., Prof. Paper 112, 1919, p. 104.

Ficus aligera Lesquereux, U. S. Geol. Surv., Mon. 17, 1891 [1892], p. 84, pl. x, f. 3–6.—Berry, Bull. Torr. Bot. Club, vol. 33, 1906, p. 172; Geol. Surv. N. J., Ann. Rept. 1905 [1906], p. 139.

Tuscaloosa: Cottondale and Shirleys Mill, Ala.

Black Creek (Middendorf): Middendorf and Miles Mill, S. C.

Raritan: South Amboy, N. J.

Dakota: *Pipe Creek, Cloud County, Kans.

Citrophyllum wilcoxianum Berry, U. S. Geol. Surv., Prof. Paper 91, 1916, p. 252, pl. lv, f. 3.

Lagrange: Puryear, Henry County, Tenn.

CLADOPHLEBIS Brongniart, Tableau, 1849, p. 25.

Cladophlebis acuta Fontaine = *Cladophlebis virginiensis.*

Cladophlebis acuta angustifolia Fontaine = *Cladophlebis virginiensis.*

Cladophlebis acutiloba (Heer) Fontaine, in Ward, U. S. Geol. Surv., Mon. 48, 1905 [1906], p. 72, pl. xi, f. 11, 12.

Dicksonia acutiloba Heer, Fl. foss. arct., vol. 4, Abt. 2, 1876, p. 92, pl. xviii, f. 4.

Jurassic: Douglas County, Oreg.

Cladophlebis alabamensis Berry, U. S. Geol. Surv., Prof. Paper 112, 1919, p. 50, pl. v, f. 8.

Tuscaloosa: Snow place, Tuscaloosa County, Ala.

Cladophlebis alata Fontaine = *Cladophlebis browniana*.

Cladophlebis alata Fontaine. Fontaine, in Ward, U. S. Geol. Surv., Mon. 48, 1905, p. 158, pl. xxxix, f. 9–11; pl. xl, f. 1–9.—Knowlton, U. S. Geol. Surv., Prof. Paper 85, 1914, p. 49, pl. v, f. 3, 4; pl. vi, f. 4.

Aspidium oerstedi Heer. Lesquereux, U. S. Nat. Mus., Proc., vol. 11, 1888, p. 32 (in part).

?*Pinus staratschini* Heer. Lesquereux, idem, p. 32.

Jurassic: Cape Lisburne, Alaska.

Cladophlebis albertsii (Dunker) Brongniart, Tableau, 1849, p. 107.—Berry, U. S. Nat. Mus., Proc., vol. 41, 1911, p. 310; Md. Geol. Surv., Lower Cret., 1911, p. 252, pl. xxxii, f. 3, 4.

Neuropteris albertsii Dunker, Mon. Norddeutsch. Wealdenbildung. 1846, p. 8, pl. vii, f. 6, 6a.

Cladophlebis inclinata Fontaine. U. S. Geol. Surv., Mon. 15, 1889 [1890], p. 76, pl. x, f. 3, 4; pl. xx, f. 8.

Cladophlebis denticulata Fontaine, idem, p. 71, pl. iv, f. 2; pl. vii, f. 7.

Cladophlebis sp. Fontaine, idem, p. 76, pl. x, f. 5, 8; pl. xx, f. 7.

Cladophlebis pachyphylla Fontaine, idem, p. 80, pl. xxv, f. 9.

Cladophlebis sp. Fontaine, idem. p. 77, pl. xv, f. 6; pl. xix, f. 3.

Aspidium angustipinnatum Fontaine, idem. p. 98, pl. xvi, f. 1, 3, 8; pl. xvii, f. 1; pl. xix, f. 10.

Dryopteris angustipinnata (Fontaine) Knowlton, U. S. Geol. Surv., Bull. 152, 1898, p. 91.—Fontaine, in Ward, U. S. Geol. Surv., Mon. 48, 1906. p. 540. 544, 548; pl. cxiv, f. 6.

Aspidium oerstedi? Heer. Fontaine, U. S. Geol. Surv., Mon. 15, 1890. p. 99, pl. xix, f. 4.

Dryopteris oerstedi? (Heer) Knowlton, U. S. Geol. Surv., Bull. 152. 1898. p. 92.

Patuxent: Potomac Run, telegraph station near Lorton, Dutch Gap, Trents Reach, Fredericksburg, Va.

Arundel: Arlington, Hanover, Bay View, Md.

Cladophlebis albertsii montanense (Fontaine) Knowlton, n. comb.

Dryopteris angustipinnata montanense (Fontaine) Knowlton, U. S. Geol. Surv., Bull. 152, 1898, p. 91.

Aspidium angustipinnatum montanense Fontaine, U. S. Nat. Mus., Proc., vol. 15, 1892, p. 491, pl. lxxxiv, f. 1, 1a.

Kootenai: Great Falls, Mont.

Cladophlebis angustifolia Newberry, Am. Jour. Sci., 3d ser., vol. 41, 1891, p. 200, pl. xiv, f. 8.

Kootenai: Great Falls, Mont.

Cladophlebis argutula (Heer) Fontaine, Am. Jour. Sci., 4th ser., vol. 2, 1896, p. 274.—Fontaine, in Ward, U. S. Geol. Surv., Twentieth Ann. Rept., pt. 2, 1900, p. 345, pl. l, f. 1–6.

Asplenium argutulum Heer, Fl. foss. arct., vol. 4, pt. 2, 1876, pp. 24, 41, 96, 118, 120, pl. iii, f. 7, 7b, 7c, 7d: pl. xix, f. 1, 1b, 2, 3, 3b, 3c, 4.

Jurassic: Oroville, Calif.

Cladophlebis auriculata Fontaine, U. S. Geol. Surv., Mon. 6, 1883, p. 50, pl. xxvi, f. 6, 7.

Triassic (Newark): Carbon Hill, Va.

Cladophlebis brevipennis Fontaine = *Cladophlebis rotundata.*

Cladophlebis columbiana Dawson, Roy. Soc. Canada, Trans., vol. 11, 1893 [1894], sec. 4, p. 55, pl. v, f. 4, 5.

Upper Cretaceous: Vancouver Island, British Columbia.

Cladophlebis browniana (Dunker) Seward, Wealden Fl., pt. 1, 1894, p. 99, pl. vii, f. 4.—Knowlton, Smithsonian Misc. Coll., vol. 50, 1907, p. 108, pl. xi, f. 1; in Diller, Geol. Soc. Am., Bull., vol. 19, 1908, p. 386.—Fontaine, in Ward, U. S. Geol. Surv., Mon. 48, 1906, pp. 272, 510, 517, 538, 544, 547, 557, 572.—Berry, U. S. Nat. Mus., Proc., vol. 41, 1911, p. 312; Md. Geol. Surv., Lower Cret., 1911, p. 243, pl. xxix, f. 1, 2.

Pecopteris browniana Dunker, Mon. Norddeutsch. Wealdenbildung., 1846, p. 5, pl. viii, f. 7.—Fontaine, U. S. Nat. Mus., Proc., vol. 15, 1892, p. 492.

Cladophlebis inaequiloba Fontaine, U. S. Geol. Surv., Mon. 15, 1889 [1890], p. 80, pl. xxv, f. 8; in Ward, idem, Mon. 48, 1906, p. 510.

Cladophlebis petiolata Fontaine, op. cit. (Mon. 15), p. 80, pl. xxii, f. 8.

Cladophlebis oblongifolia Fontaine, op. cit. (Mon. 15), p. 74 (part), pl. vii, f. 5 [not f. 3, 4, which = *C. virginiensis*].

Cladophlebis crenata Fontaine, op. cit. (Mon. 15), p. 75, pl. ix, f. 7–9; pl. x, f. 1, 2; pl. xiii, f. 1–3; pl. xix, f. 7; pl. xx, f. 6; in Ward, op. cit. (Mon. 48), 1906, p. 547.

Cladophlebis alata Fontaine, op. cit. (Mon. 15), p. 77, pl. xix, f. 5; in Ward, op. cit. (Mon. 48), pp. 229, 480, 510, 544, 557, pl. lxv, f. 17–21.

Cladophlebis inclinata Fontaine, in Diller and Stanton, Geol. Soc. Am., Bull., vol. 5, 1894, p. 450.

Cladophlebis sp., Fontaine, op. cit. (Mon. 15), p. 78, pl. xix, f. 2.

Pecopteris strictinervis Fontaine, op. cit. (Mon. 15), p. 84, pl. xiii, f. 6–8; pl. xix, f. 9; pl. xx, f. 3; pl. xxii, f. 13; pl. clxx, f. 5, 6.

Pecopteris ovodentata Fontaine, op. cit. (Mon. 15), p. 85, pl. xv, f. 8; pl. xxii, f. 12; pl. xxiii, f. 1.

Pecopteris microdonta Fontaine, op. cit. (Mon. 15), p. 85, pl. xix, f. 8; pl. xx, f. 5, 11.

Pecopteris virginiensis Fontaine, op. cit. (Mon. 15), p. 82, pl. viii, f. 1–7; pl. ix, f. 1–6; pl. xxiv, f. 2; pl. clxix, f. 3; in Ward, op. cit (Mon. 48), pp. 480, 538, 552, pl. cxvi, f. 3, 4.

Pecopteris constricta Fontaine, op. cit. (Mon. 15), p. 86, pl. xx, f. 1, 2, 4; in Ward, op. cit. (Mon. 48), p. 519.

Pecopteris socialis Heer. Fontaine, op. cit. (Mon. 15), p. 87, pl. xxi, f. 7.

Pecopteris angustipennis Fontaine, op. cit. (Mon. 15), p. 87, pl. xxi, f. 10.

?*Pinus staratschini* Heer. Lesquereux, U. S. Nat. Mus., Proc., vol. 11, 1888, p. 32.

Cladophlebis browniana—Continued.

Patuxent: Fredericksburg, Dutch Gap, Alum Rock, telegraph station near Lorton, Potomac Run, Va.; New Reservoir, Ivy City, D. C.; Broad Creek?, Md.

Arundel: Arlington, Hanover, Howard Brown estate, Md.

Patapsco: Brooke and vicinity, Chinkapin Hollow, Va.; Federal Hill (Baltimore), Vinegar Hill, Md.

Kootenai: Great Falls and Skull Butte, Mont.; Anthracite, Green Hills, and Morrisey, British Columbia.

Horsetown: Horsetown and near Ono, Calif.

Knoxville: Tehama County, Calif.

Cladophlebis constricta Fontaine, U. S. Geol. Surv., Mon. 15, 1889 [1890], p. 68, pl. ii, f. 11; pl. iii, f. 2; pl. vi, f. 5, 6, 8–14; pl. xxi, f. 9, 13; pl. clxix, f. 2; in Ward, U. S. Geol. Surv., Mon. 48, 1906, pp. 280, 297, 504, 528, 547, pl. lxxvii, f. 26.—Penhallow, Summ. Geol. Surv. Canada, 1904 (1905), p. 9.—Knowlton, Smithsonian Misc. Coll., vol. 50, 1907, p. 109.—Berry, U. S. Nat. Mus., Proc., vol. 41, 1911, p. 314; Md. Geol. Surv., Lower Cret., 1911, p. 246, pl. xxix, f. 3.

Cladophlebis latifolia Fontaine, U. S. Geol. Surv., Mon. 15, 1889 [1890], p. 69, pl. iii, f. 1; pl. vi, f. 4.

Cladophlebis virginiensis Fontaine. Fontaine, in Ward, U. S. Geol. Surv., Mon. 48, 1906, p. 512, pl. cxi, f. 7.

Patuxent: Fredericksburg, Va.

Patapsco: Hell Hole?, Brooke, Deep Bottom, Va.; Federal Hill (Baltimore), Vinegar Hill, Fort Foote?, Md.

Kootenai: Great Falls and Geyser, Mont.

Cladophlebis crenata Fontaine = *Cladophlebis browniana.*

Cladophlebis densifolia Fontaine, Am. Jour. Sci., 4th ser., vol. 2, 1896, p. 274; in Ward, U. S. Geol. Surv., Twentieth Ann. Rept., pt. 2, 1900, p. 347, pl. li.

Jurassic: Oroville, Calif.

Cladophlebis denticulata (Brongniart) Nathorst, Bidrag till Sveriges Foss. Fl. Växter fr. Rät. Form. vid Pålsjö, 1876, p. 19.—Fontaine, in Ward, U. S. Geol. Surv., Mon. 48, 1905 [1906], p. 68, pl. xi, f. 1–7.

Pecopteris denticulata Brongniart, Prodrome, 1828, pp. 57, 198 [name].

Jurassic: Douglas County, Oreg.

Cladophlebis denticulata Fontaine = *Cladophlebis albertsii.*

Cladophlebis distans Fontaine, U. S. Geol. Surv., Mon. 15, 1889 [1890], p. 77, pl. xiii, f. 4, 5; in Ward, U. S. Geol. Surv., Mon. 48, 1906, pp. 280, 572.—Berry, U. S. Nat. Mus., Proc., vol. 41, 1911, p. 315; Md. Geol. Surv., Lower Cret., 1911, p. 258, pl. xxxii, f. 5, 6.

Aspidium fredericksburgense Fontaine, U. S. Geol. Surv., Mon. 15, 1889 [1890], p. 94, pl. xi, f. 1–6; pl. xii, f. 1–6; pl. xvi, f. 9; pl. xix, f. 6, 7.—Penhallow, Roy. Soc. Canada, Trans., vol. 1, 1908, sec. 4, p. 307.

Cladophlebis distans—Continued.

Dryopteris fredericksburgensis (Fontaine) Knowlton, U. S. Geol. Surv., Bull. 152, 1898, p. 92.—Fontaine, in Ward, U. S. Geol. Surv., Mon. 48, 1906, pp. 280, 512, 538, 548, pl. cxii, f. 2.

Kootenai: Great Falls, Mont.

Patuxent: Fredericksburg, Dutch Gap, telegraph station near Lorton, Va.; Broad Creek, Md.

Arundel: Arlington, Md.

Patapsco: Chinkapin Hollow, Va.

Cladophlebis dunkeri (Schimper) Seward. Knowlton, U. S. Geol. Surv., Bull. 152, 1900, p. 73 = *Cladophlebis ungeri*.

Cladophlebis falcata Fontaine = *Cladophlebis virginiensis*.

Cladophlebis falcata montanensis Fontaine = *Cladophlebis virginiensis*.

Cladophlebis fisheri Knowlton, Smithsonian Misc. Coll., vol. 50, 1907, p. 109, pl. xi, f. 2, 2a.

Kootenai: Six miles southwest of Geyser, Fergus County, Mont.

Cladophlebis haiburnensis (Lindley and Hutton) Brongniart, Tableau, 1849, p. 105.—Fontaine, in Ward, U. S. Geol. Surv., Mon. 48, 1905 [1906], p. 71, pl. xi, f. 8–10.

Pecopteris haiburnensis Lindley and Hutton, Foss. Fl. Gt. Brit., vol. 3, 1836, pl. clxxxviii.

Jurassic: Douglas County, Oreg.

Cladophlebis heterophylla Fontaine, U. S. Nat. Mus., Proc., vol. 15, 1892, p. 493, pl. lxxxiv, f. 2; in Ward, U. S. Geol. Surv., Mon. 48, 1905 [1906], p. 294, pl. lxxi, f. 21–25.—Knowlton, Smithsonian Misc. Coll., vol. 50, 1907, p. 108.

Kootenai: Great Falls, Skull Butte, Fergus County, and 6 miles southwest of Geyser, Cascade County, Mont.

Cloverly: No Wood Creek, Big Horn Basin, Wyo.

Knoxville: California.

Cladophlebis hirta? Möller, Bidrag till Bornholm fossila flora, Pteridophyter: Kong. Fysiografiska Sällskapots, Handl., vol. 13, 1902, p. 30, pl. ii, f. 23, 24; pl. iii, f. 2.—Knowlton, U. S. Nat. Mus., Proc., vol. 51, 1916, p. 454, pl. 81, f. 3.

Lower Jurassic: Upper Matanuska Valley, Alaska.

Cladophlebis huttoni (Dunker) Fontaine, in Ward, U. S. Geol. Surv., Mon. 48, 1905 [1906], p. 161, pl. xli–xliii.—Knowlton, U. S. Geol. Surv., Prof. Paper 85, 1914, p. 48, pl. vi, f. 3.

Neuropteris huttoni Dunker, Monogr. d. Norddeutsch. Wealdenbildung, 1846, p. 9, pl. viii, f. 1.

Aspidium oerstedi Heer. Lesquereux, U. S. Nat. Mus., Proc., vol. 11, 1888, p. 32, in part.

Asplenium foersteri Debey and Ettingshausen. Lesquereux, U. S. Nat. Mus., Proc., vol. 11, 1888, p. 32.

Jurassic: Cape Lisburne, Alaska.

Cladophlebis inaequiloba Fontaine, U. S. Geol. Surv., Mon. 15, 1889, p. 80, pl. xxv, f. 8.

Patuxent: Fredericksburg, Va.

Cladophlebis inclinata Fontaine [part] = *Cladophlebis albertsii.*

Cladophlebis inclinata Fontaine [part] = *Cladophlebis browniana.*

Cladophlebis inclinata Fontaine? Fontaine, in Diller and Stanton, Geol. Soc. Am., Bull., vol. 5, 1894, p. 450, in part; in Stanton, U. S. Geol. Surv., Bull. 133, 1895 [1896], p. 15, in part = *Cladophlebis alata* and *Cladophlebis parva.*

Cladophlebis indica (Oldham and Morris) Fontaine?, Am. Jour. Sci., 4th ser., vol. 2, 1896, p. 274.—Fontaine, in Ward, U. S. Geol. Surv., Twentieth Ann. Rept., pt. 2, 1900, p. 348, pl. lii, f. 1.

Pecopteris (Alethopteris) indica Oldham and Morris, Palaeontologica Indica, ser. 2, vol. 1, pt. 1, 1862, p. 47. pl. xxvii.

Jurassic: Oroville, Calif.?

Cladophlebis latifolia Fontaine = *Cladophlebis constricta.*

Cladophlebis microphylla Fontaine, U. S. Geol. Surv., Mon. 6, 1883, p. 51, pl. xxvii, f. 2, 2a.

Triassic (Newark): Clover Hill, Va.

Cladophlebis oblongifolia Fontaine [part] = *Cladophlebis browniana.*

Cladophlebis oblongifolia Fontaine [part] = *Cladophlebis virginiensis.*

Cladophlebis obtusifolia (Andrae) Ward, U. S. Geol. Surv., Twentieth Ann. Rept., pt. 2, 1898–99 [1900], p. 424.

Cladophlebis obtusiloba (Andrae) Fontaine, U. S. Geol. Surv., Mon. 6, 1883, p. 106, pl. xlix, f. 7. [Fontaine's error for *obtusifolia.*]

Sphenopteris obtusifolia Andrae, Abh. K. k. geol. Reichsanst., vol. 2, pt. 3, vol. 4, 1855, p. 32, pl. vi, f. 9.

Triassic (Newark): North Carolina.

Cladophlebis obtusiloba (Andrae) Fontaine = *Cladophlebis obtusifolia.*

Cladophlebis ovata Fontaine, U. S. Geol. Surv., Mon. 6, 1883, p. 50, pl. xxvi, f. 5, 5a; pl. xxvii, f. 3.

Triassic (Newark): Clover Hill, Va.

Cladophlebis pachyphylla Fontaine = *Cladophlebis albertsii.*

Cladophlebis parva Fontaine, U. S. Geol. Surv., Mon. 15, 1889 [1890], p. 73, pl. iv, f. 7; pl. vi, f. 1–3; in Ward, U. S. Geol. Surv., Nineteenth Ann. Rept., pt. 2, 1899, p. 657, pl. clx, f. 18; U. S. Geol. Surv., Mon. 48, 1905 [1906], pp. 225, 280, 510, 538, pl. lxv, f. 5–8.—Knowlton, in Diller, Geol. Soc. Am., Bull., vol. 19, 1908, p. 386.—Berry, U. S. Nat. Mus., Proc., vol. 41, 1911, p. 316; Md. Geol. Surv., Lower Cret., 1911, p. 250, pl. xxviii, f. 1, 2; pl. xxx; pl. xxxi.

Cladophlebis inclinata Fontaine. Fontaine, in Diller and Stanton, Geol. Soc. Am., Bull., vol. 5, 1894, p. 450; in Stanton, U. S. Geol. Surv., Bull. 133, 1896, p. 15.

Cladophlebis parva—Continued.

Cladophlebis sp. Fontaine, U. S. Geol. Surv., Mon. 15, 1890, p. 81, pl. xxvi, f. 15.

Aspidium heterophyllum Fontaine, U. S. Geol. Surv., Mon. 15, 1889 [1890], p. 96, pl. xiv, f. 1–5; pl. xv, f. 1–5; in Diller and Stanton, Geol. Soc. Am., Bull., vol. 5, 1894, p. 450; in Stanton, U. S. Geol. Surv., Bull. 133, 1896, p. 15.

Dryopteris heterophylla (Fontaine) Knowlton, U. S. Geol. Surv., Bull. 152, 1898, p. 92.—Fontaine, in Ward, U. S. Geol. Surv., Mon. 48, 1906, pp. 483, 550, pl. cxv, f. 7, 8.

Patuxent: Fredericksburg, Cockpit Point, Potomac Run, Va.
Arundel: Arlington ?, Md.
Patapsco: Vinegar Hill, Md.
Lakota: Hay Creek, Crook County, Wyo.
Knoxville: Tehama County, Calif.

Cladophlebis pecopteroides Fontaine, in Ward, U. S. Geol. Surv., Mon. 48, 1905 [1906], p. 73, pl. xi, f. 13–15.
Jurassic: Douglas County, Oreg.

Cladophlebis petiolata Fontaine = *Cladophlebis browniana*.

Cladophlebis pseudowhitbiensis Fontaine, U. S. Geol. Surv., Mon. 6, 1883, p. 52, pl. xxvii, f. 4, 4a.
Triassic (Newark): Clover Hill, Va.

Cladophlebis rarinervis Fontaine. Ward, U. S. Geol. Surv., Twentieth Ann. Rept., pt. 2, 1898–99 [1900], p. 424. Error for *Thyrsopteris rarinervis*.

Cladophlebis reticulata Fontaine, in Ward, U. S. Geol. Surv., Twentieth Ann. Rept., pt. 2, 1900, p. 235, pl. xxi, f. 1–5.
Triassic (Newark): York County, Pa.

Cladophlebis rotundata Fontaine, U. S. Geol. Surv., Mon. 15, 1889 [1890], p. 78, pl. xx, f. 9, 10.—?Penhallow, Summ. Geol. Surv. Canada, 1904 [1905], p. 9.—Fontaine, in Ward, U. S. Geol. Surv., Mon. 48, 1905 [1906], pp. 491, 510.—Berry, U. S. Nat. Mus., Proc., vol. 41, 1911, p. 317; Md. Geol. Surv., Lower Cret., 1911, p. 247.

Cladophlebis brevipennis Fontaine, U. S. Geol. Surv., Mon. 15, 1889 [1890], p. 81, pl. xxxvi, f. 1.

Patuxent: Fredericksburg, Dutch Gap, Va.
Patapsco: Mount Vernon, Chinkapin Hollow, Hell Hole, Brooke, Deep Bottom, Va.; Federal Hill (Baltimore), Vinegar Hill, Fort Foote ?, Md.

Cladophlebis rotundiloba Fontaine, U. S. Geol. Surv., Mon. 6, 1883, p. 52, pl. xxvii, f. 1, 1a.
Triassic (Newark): Hanover Junction, Va.

Cladophlebis skagitensis Penhallow, Roy. Soc. Canada, Trans., 3d ser., vol. 1, sec. 4, 1907, p. 306, text f. 2.
Lower Cretaceous: International boundary between Pasayten and Skagit rivers, British Columbia.

Cladophlebis socialis (Heer) Berry, Bull. Torr. Bot. Club, vol. 38, 1911, p. 409; Md. Geol. Surv., Upper Cret., 1916, p. 765.

Pecopteris socialis Heer, Fl. foss. arct., vol. 6, Abt. 2, 1882, pl. vii, f. 4; pl. viii, f. 15; pl. xxxii, f. 9.

Raritan: *Shannon Hill, Cecil County, Md.

Cladophlebis spectabilis (Heer) Fontaine, Am. Jour. Sci., 4th ser., vol. 2, 1896, p. 274; in Ward, U. S. Geol. Surv., Twentieth Ann. Rept., pt. 2, 1900, p. 345, pl. xlix, f. 4, 5.

Asplenium spectabile Heer, Fl. foss. arct., vol. 4, Abt. 2, 1876, pp. 96, 120, pl. xxi, f. 1, 2a, 2c, 2d.

Jurassic: Oroville, Calif.

Cladophlebis sphenopteroides Fontaine = *Ruffordia göpperti.*

Cladophlebis subfalcata Fontaine, U. S. Geol. Surv., Mon. 6, 1883, p. 49, pl. xxix, f. 5, 5a.

Triassic (Newark): Manakin, Va.

Cladophlebis ungeri (Dunker) Ward, in Fontaine, in Ward, U. S. Geol. Surv., Mon. 48, 1905 [1906], pp. 228, 510, 538, pl. lxv, f. 15, 16.—Knowlton, in Diller, Geol. Soc. Am., Bull., vol. 19, 1908, p. 386.—Berry, U. S. Nat. Mus., Proc., vol. 41, 1911, p. 318; Md. Geol. Surv., Lower Cret., 1911, p. 255, pl. xxxiii, f. 1, 2.

Pecopteris ungeri Dunker, Mon. Norddeutsch. Wealdenbildung, 1846, p. 6, pl. ix, f. 10.

Aspidium dunkeri (Schimper) Fontaine, U. S. Geol. Surv., Mon. 15, 1889 [1890], p. 101, pl. xxii, f. 9–9b; pl. xxv, f. 11, 12; pl. xxvi, f. 2, 8, 9, 18; pl. liv, f. 3. 9; Diller and Stanton, Geol. Soc. Am., Bull., vol. 5, 1894, p. 450; in Stanton, U. S. Geol. Surv., Bull. 133, 1896, p. 15.

Aspidium parvifolium Fontaine, U. S. Geol. Surv., Mon. 15, 1889 [1890], p. 100, pl. xxi, f. 6; pl. xxiv, f. 8; pl. xxv, f. 10; pl. xxvi, f. 1, 14, 16, 17.

Dryopteris parvifolia (Fontaine) Knowlton, U. S. Geol. Surv., Bull. 152, 1898, p. 92; Fontaine, in Ward, U. S. Geol. Surv., Mon. 48, 1906, p. 486.

Pecopteris brevipennis Fontaine, U. S. Geol. Surv., Mon. 15, 1889 [1890], p. 86, pl. xxi, f. 1–3; in Ward, U. S. Geol. Surv., Mon. 48, 1906, p. 510.

Pecopteris pachyphylla Fontaine, U. S. Geol. Surv., Mon. 15, 1889 [1890], p. 88, pl. xxvi, f. 4, 5.

Cladophlebis dunkeri (Schimper) Seward. Knowlton, U. S. Geol. Surv., Bull. 152, 1898, p. 73.

Patuxent: Woodbridge, Fredericksburg, Dutch Gap, telegraph station near Lorton, Va.

Patapsco: Chinkapin Hollow, Va.

Knoxville: Elder Creek near Lowry, Tehama County, Calif.

Cladophlebis vaccensis Ward, U. S. Geol. Surv., Mon. 48, 1905 [1906], pp. 66, 149, 157, pl. x, f. 8–12; pl. xxxviii, f. 5, 6; pl. xxxix, f. 7, 8.

Cladophlebis whitbiensis tenuis var. *a* Heer. Fontaine, Am. Jour. Sci., 4th ser., vol. 2, 1896, p. 274; in Ward, U. S. Geol. Surv., Twentieth Ann. Rept., pt. 2, 1898–99 [1900], p. 346, pl. l, f. 7.

?*Pecopteris denticulata* Heer. Lesquereux, U. S. Nat. Mus., Proc., vol. 11, 1888, p. 32.

Jurassic: Oroville, Calif.; Douglas and Curry counties, Oreg.; Cape Lisburne, Alaska.

Cladophlebis virginiensis Fontaine, U. S. Geol. Surv., Mon. 15, 1889 [1890], p. 70, pl. iii, f. 3–8; pl. iv, f. 1, 3–6.—Berry, U. S. Nat. Mus., Proc., vol. 41, 1911, p. 320; Md. Geol. Surv., Lower Cret., 1911, p. 248, pl. xxix, f. 4–6.

Cladophlebis falcata Fontaine, U. S. Geol. Surv., Mon. 15, 1889 [1890], p. 72, pl. iv, f. 8; pl. v, f. 1–6; pl. vi, f. 7; pl. vii, f. 1, 2; in Ward, U. S. Geol. Surv., Mon. 48, 1905 [1906], pp. 227, 280, 511, 548, pl. lxv, f. 12–14; pl. cxi, f. 6.—Knowlton, in Diller, Geol. Soc. Am., Bull., vol. 19, 1908, p. 386.

Cladophlebis acuta Fontaine, U. S. Geol. Surv., Mon. 15, 1889 [1890], p. 75, pl. v, f. 7; pl. vii, f. 6; pl. x, f. 6, 7; pl. xi, f. 7, 8; pl. clxvi, f. 5; in Ward, U. S. Geol. Surv., Mon. 48, 1906, p. 538, pl. cxiv, f. 3, 4.

Cladophlebis falcata montanensis Fontaine, in Ward, U. S. Geol. Surv., Mon. 48, 1905 [1906], p. 291, pl. lxxi, f. 14–20.

Cladophlebis acuta angustifolia Fontaine, in Ward, U. S. Geol. Surv., Mon. 48, 1905 [1906], p. 539, pl. cxiv, f. 5.

Thinnfeldia variabilis Fontaine, in Diller and Stanton, Geol. Soc. Am., Bull., vol. 5, 1894, p. 450; in Stanton, U. S. Geol. Surv., Bull. 133, 1896, p. 15.

Asplenium distans Heer. Dawson, Roy. Soc. Canada, Trans., vol. 3, 1886, sec. 4, p. 5, pl. iii, f. 7.

Thinnfeldia montanensis Fontaine, in Weed and Pirsson, U. S. Geol. Surv., Eighteenth Ann. Rept., 1896–97 [1898], pt. 3, p. 481.

Cladophlebis oblongifolia Fontaine, U. S. Geol. Surv., Mon. 15, 1889 [1890], p. 74 (part), pl. vii, f. 3, 4.

Patuxent: Fredericksburg, Dutch Gap, Potomac Run, Va.
Arundel: Arlington, Md.
Patapsco: Vinegar Hill, Md.; Chinkapin Hollow, Va.
Kootenai: Canmore, British Columbia.

Cladophlebis virginiensis Fontaine. Fontaine, in Ward = *Cladophlebis constricta.*

Cladophlebis whitbiensis tenuis var. *a* Heer?. Fontaine, Am. Jour. Sci., 4th ser., vol. 2, 1896, p. 274; in Ward, U. S. Geol. Surv., Twentieth Ann. Rept., pt. 2, 1900, p. 346, pl. l, f. 7 = *Cladophlebis vaccensis.*

Cladophlebis wyomingensis Fontaine, in Ward, U. S. Geol. Surv., Nineteenth Ann. Rept., pt. 2, 1899, p. 656, pl. clx, f. 16, 17.
Fuson: Oak Creek, Crook County, Wyo.

Cladophlebis? Fontaine, U. S. Geol. Surv., Mon. 15, 1889 [1890], p. 76, pl. x, f. 5, 8; pl. xx, f. 7 = *Cladophlebis albertsii.*

Cladophlebis sp. Fontaine, U. S. Geol. Surv., Mon. 15, 1889 [1890], p. 78, pl. xix, f. 2 = *Cladophlebis browniana.*

Cladophlebis sp. Fontaine, U. S. Geol. Surv., Mon. 15, 1889 [1890], p. 77, pl. xv, f. 6; pl. xix, f. 3 = *Cladophlebis albertsii.*

Cladophlebis sp. Fontaine, U. S. Geol. Surv., Mon. 15, 1889 [1890], p. 77, pl. x, f. 5, 8; pl. xx, f. 7 = *Cladophlebis albertsii.*

Cladophlebis sp. Fontaine, U. S. Geol. Surv., Mon. 15, 1889 [1890], p. 77, pl. xv, f. 6; pl. xix, f. 3.
Patuxent: Potomac Run, Va.
Patapsco: Near Brooke, Va.

Cladophlebis sp. Fontaine, U. S. Geol. Surv., Mon. 15, 1889 [1890], p. 78, pl. xix, f. 2.
Patuxent: Telegraph station near Lorton, Va.

Cladophlebis sp. Fontaine, U. S. Geol. Surv., Mon. 15, 1889 [1890], p. 81, pl. xxvi, f. 15 = *Cladophlebis parva*.

Cladophlebis sp. Dawson, Roy. Soc. Canada, Trans., vol. 10, sec. 4, 1892, p. 85.
Kootenai: Anthracite, British Columbia.

CLADOSPORITES Felix, Zeitschr. Deutsch. geol. Ges., vol. 46, 1894, p. 276.

Cladosporites fasciculatus Berry, Mycologia, vol. 8, 1916, p. 77, pl. clxxxii, f. 1, 2.
Eocene: Westmoreland Bluff, Trinity River, Tex.

Cladosporites ligni-perditor Whitford, Nebr. Univ. Studies, vol. 14, 1914, p. 182 (2), pls. i–ii.
Pliocene: 20 miles south of Agate, Nebr.

Cladosporites oligocaenicum Berry, Mycologia, vol. 8, 1916, p. 79, pl. clxxxi, f. 1–8.
Lower Oligocene: Bayou Pierre, Miss.

Clathropodium mirabile (Lesquereux) Ward = *Cycadeoidea mirabilis*.

CLATHROPTERIS Brongniart, Prodrome, 1828, p. 70 [62].

Clathropteris platyphylla (Göppert) Brongniart, Tableau, 1849, p. 81 [32].—Newberry, N. Y. Acad. Sci., Trans., vol. 6, 1886 [1887], p. 126; U. S. Geol. Surv., Mon. 14, 1888, p. 94, pl. xxii, f. 6.
Triassic (Newark): Sunderland, Westfield, and Durham, Conn.; Newark and Milford, N. J.; Easthampton, Mass.

Clathropteris platyphylla expansa Saporta, Pal. franç. (Pl. jurass.), vol. 1, 1873, p. 337, pl. xxvi, f. 1; etc.—Fontaine, U. S. Geol. Surv., Mon. 6, 1883, p. 54, pl. xxxi, f. 3, 4: pl. xxxii, f. 1; pl. xxxiii, f. 1; pl. xxxiv, f. 1; pl. xxxv, f. 2, 2a.
Triassic (Newark): Clover Hill, Va.

Clathropteris rectiusculus Hitchcock, Am. Jour. Sci., 2d ser., vol. 20, 1855, p. 24, text f. 2.
Triassic (Newark): Mount Tom, Easthampton, Mass.

CLETHRA Linné, Sp. pl., 1753, p. 396.

Clethra alnifolia Linné. Hollick, Bull. Torr. Bot. Club, vol. 19, 1892, p. 332.—Penhallow, Roy. Soc. Canada, Trans., 2d ser., vol. 10, sec. 4, 1904, p. 73.
Pleistocene: Bridgeton, N. J.; Don Valley, Toronto, Canada.

CLINTONIA Rafinesque, Jour. Phys., vol. 89, 1819, p. 102.

Clintonia oblongifolia Penhallow, Roy. Soc. Canada, Trans., 2d ser., vol. 8, 1902, p. 55, f. 6; Rept. Tert. Pl. Brit. Col., 1908, p. 45.
Paskapoo (Fort Union): Red Deer River, Alberta.

COCCOLOBA Linné, Sp. pl., 1764, p. 523.

Coccoloba laevigata Lesquereux, U. S. Geol. and Geog. Surv. Terr., Ann. Rept. 1872 [1873], p. 387; Rept. U. S. Geol. Surv. Terr., vol. 7 (Tert. Fl.), 1878, p. 208, pl. xxxv, f. 7.
Hanna: Carbon, Wyo.

COCCOLOBIS P. Browne, Jamaic., 1756, p. 209.

Coccolobis eolignitica, Berry, U. S. Geol. Surv., Prof. Paper 91, 1916, p. 212, pl. xxxviii, f. 4.
Lagrange: Puryear, Henry County, Tenn.

Coccolobis uviferafolia Berry, U. S. Geol. Surv., Prof. Paper 91, 1916, p. 212, pl. lxxxvii, f. 5.
Lagrange: Puryear, Henry County, Tenn.

COCCOLOBITES Berry, Maryland Geol. Surv., Upper Cret., 1916, p. 830. [Type, *C. cretaceus*.]

Coccolobites cretaceus Berry, Maryland Geol. Surv., Upper Cret., 1916, p. 830, pl. lxviii, f. 1.
Magothy: Grove Point, Md.

COCCULITES Heer, Fl. foss. arct., vol. 3 (Mioc. Fl. Arct. Zone, pt. 4), 1874, p. 21.

Cocculites imperfectus Hollick, U. S. Geol. Surv., Mon. 50, 1906, p. 63, pl. xii, f. 14.
Magothy: Gay Head, Marthas Vineyard, Mass.

Cocculites inquirendus Hollick, U. S. Geol. Surv., Mon. 50, 1906, p. 63, pl. xii, f. 13.
Magothy: Gay Head, Marthas Vineyard, Mass.

COCCULUS (Bauhin) Linné, Sp. pl., 1753, p. 341.

Cocculus cinnamomeus Velenovsky, Fl. böhm. Kreideform., pt. 4, 1885, p. 4 [65], pl. viii [xxxi], f. 16–21.—Hollick, U. S. Geol. Surv., Mon. 50, 1906, p. 62, pl. xii, f. 10–12.—Berry, U. S. Geol. Surv., Prof. Paper 112, 1919, p. 93, pl. xvii, f. 1.
Magothy: Gay Head, Marthas Vineyard, Mass.
Tuscaloosa: Cottondale, Ala.

Cocculus haydenianus Ward, U. S. Geol. Surv., Sixth Ann. Rept., 1884–85 [1886], p. 556, pl. lix, f. 1–5; U. S. Geol. Surv., Bull. 37, 1887, p. 100, pl. xlvii, f. 1–4; pl. xlviii, f. 1.—Knowlton, Washington Acad. Sci., Proc., vol. 11, 1909, pp. 189, 198, 200, 213, 215.
Cebatha haydenianus (Ward) Knowlton, U. S. Geol. Surv., Bull. 152, 1898, p. 62.
Fort Union: Burns's ranch and Iron Bluff, Glendive, Hell Creek region, Musselshell River area, and Fish Creek, Mont.; Dawson County, N. Dak.
Lance: Glendive, Mont.

Cocculus minutus Hollick, N. Y. Bot. Gard., Bull., vol. 3, 1904, p. 407, pl. 70, f. 6; U. S. Geol. Surv., Mon. 50, 1906, p. 62, pl. xii, f. 9.
Magothy: Northport, Long Island, N. Y.

Cocculus polycarpaefolius Berry, U. S. Geol. Surv., Prof. Paper 112 1919, p. 94, pl. xvii, f. 2, 3.
Tuscaloosa: Glen Allen, Ala.

Cocculus problematicus Berry, U. S. Geol. Surv., Prof. Paper 112, 1919, p. 94, pl. xvii, f. 4.
Tuscaloosa: Glen Allen, Ala.

COLUTEA (Tournefort) Linné, Sp. pl., 1753, p. 723.

Colutea boweniana Lesquereux, Rept. U. S. Geol. Surv. Terr., vol. 8 (Cret. and Tert. Fl.), 1883, p. 255, pl. lvii, f. 4.
Miocene (auriferous gravels): Bowen Claim, Placer County, Calif.

Colutea obovata Berry, Bull. Torr. Bot. Club, vol. 33, 1906, p. 175, pl. viii, f. 5, 6; Md. Geol. Surv., Upper Cret., 1916, p. 844, pl. lxxvi, f. 1, 2; U. S. Geol. Surv., Prof. Paper 112, 1919, p. 101, pl. xxiii, f. 3.
Magothy: Grove Point, Md.
Tuscaloosa: Shirleys Mill, Ala.

Colutea primordialis Heer, Fl. foss. arct., vol. 6, Abt. 2, 1882, p. 99, pl. xxvii, f. 7–11; pl. lxiii, f. 7, 8.—Lesquereux, U. S. Geol. Surv., Mon. 17, 1891 [1892], p. 148, pl. xiii, f. 8, 9.—Newberry, U. S. Geol. Surv., Mon. 26, 1895 [1896], p. 97, pl. xix, f. 4, 5.—Hollick, U. S. Geol. Surv., Mon. 50, 1906, p. 84, pl. xxxii, f. 14, 15.—Berry, Bull. Torr. Bot. Club, vol. 37, 1910, p. 24; idem, vol. 38, 1911, p. 407; idem, vol. 39, 1912, p. 396; Geol. Surv. N. J., Bull. 3, 1911, p. 156, pl. xx, f. 4; Md. Geol. Surv., Upper Cret., 1916, p. 845, pl. lxxv, f. 3; Bull. Torr. Bot. Club, vol. 44, 1917, p. 184.
Dakota: Near Delphos, Kans.
Raritan: Woodbridge, South Amboy, N. J.; ?Eatons Neck, L. I.
Magothy: Gay Head, Marthas Vineyard, Mass.; Grove Point, Md.
Woodbine: Arthurs Bluff, Tex

Colutea speciosa Knowlton, U. S. Geol. Surv., Prof. Paper 101, 1918, p. 270, pl. xliv, f. 4.
Vermejo: Vermejo Park, N. Mex.

COMBRETANTHITES Berry, U. S. Nat. Mus., Proc., vol. 45, 1913, p. 261. [Type, *C. eocenica*.]

Combretanthites eocenica Berry, U. S. Nat. Mus., Proc., vol. 45, 1913, p. 262, text f. 1, pl. xxi, f. 1–4; U. S. Geol. Surv., Prof. Paper 91, 1916, p. 322, pl. xcvi, f. 1–5.
Lagrange: Near Grand Junction, Fayette County, Tenn.

COMBRETUM Linné, Sp. pl., 1753, p. —.

Combretum obovalis Berry, U. S. Geol. Surv., Prof. Paper 91, 1916, p. 322, pl. xci, f. 6, 7.
Lagrange: Puryear, Tenn.
Wilcox: Mansfield, La.

Combretum ovalis (Lesquereux) Berry, U. S. Geol. Surv., Prof. Paper 91, 1916, p. 321, pl. xciii, f. 1; pl. xciv, f. 1.
Magnolia ovalis Lesquereux, Am. Philos. Soc., Trans., vol. 13, 1869, p. 422, pl. xxi, f. 3, 4.
Magnolia cordifolia Lesquereux, idem, p. 422, pl. xxii, f. 1, 2.
Wilcox (Ackerman): Hurleys, Miss. Old Port Caddo Landing, Harrison County, Tex.
Lagrange: Grand Junction, Tenn.

Combretum wilcoxensis Berry, U. S. Geol. Surv., Prof. Paper 91, 1916, p. 321, pl. lxxxix, f. 1, 2.
Lagrange: Puryear, Tenn.

COMEPHYLLUM Emmons, American geology, pt. 6, 1857, p. 128.

Comephyllum cristatum Emmons, American geology, pt. 6, 1857, p. 128, f. 97.—Fontaine, in Ward, U. S. Geol. Surv., Twentieth Ann. Rept., pt. 2, 1900, p. 311, pl. xlvii, f. 3.
Triassic (Newark): North Carolina.

COMPTONIA Banks. Gaertner, Fr. and Sem., vol. 2, 1791, p. 58, pl. xc.

Comptonia acutiloba (Lesquereux) Cockerell, Univ. Colorado Studies, vol. 3, 1906, p. 173; Am. Mus. Nat. Hist., Bull., vol. 24, 1908, p. 81.
Myrica latiloba acutiloba Lesquereux, U. S. Geol. and Geog. Surv. Terr., Ann. Rept. 1873 [1874], p. 412; Rept. U. S. Geol. Surv. Terr., vol. 7 (Tert. Fl.), 1878, p. 134, pl. xvii, f. 13.
Miocene: Florissant, Colo.

Comptonia columbiana Dawson = *Comptonia diforme*.

Comptonia cuspidata Lesquereux, U. S. Nat. Mus., Proc., vol. 5, 1882 [1883], p. 445, pl. vi, f. 10–12; Rept. U. S. Geol. Surv. Terr., vol. 8 (Cret. and Tert. Fl.), 1883, p. 258.—Berry, Am. Nat., vol. 40, 1906, p. 503.
Myrica (*Comptonia*) *cuspidata* (Lesquereux) Dawson. Knowlton, U. S. Nat. Mus., Proc., vol. 17, 1894, p. 221; Geol. Soc. Am., Bull., vol. 5, 1893, p. 583; U. S. Geol. Surv., Bull. 152, 1898, p. 146.
Kenai: Coal Harbor, Unga Island, Alaska.

Comptonia diforme (Sternberg) Berry, Am. Nat., vol. 40, 1896, p. 495.—Penhallow, Rept. Tert. Pl. Brit. Col., 1908, p. 45.
Asplenium diforme Sternberg, Fl. d. Vorwelt, vol. 2, 1822, pp. 29, 33; pl. xxiv, f. 1.
Comptonia columbiana Dawson, Roy. Soc. Canada, Trans., vol. 8, 1890, sec. 4, p. 81, text f. 10.
Oligocene: Kamloops Lake and Tulameen River, British Columbia.

Comptonia dryandroides Unger, Foss. Fl. v. Sotzka, 1850, p. 31 (161), pl. vi (xxvii), f. 1.—Berry, Am. Nat., vol. 40, 1906, p. 502.—Penhallow, Rept. Tert. Pl. Brit. Col., 1908, p. 46.

Myrica (*Comptonia*) *cuspidata* (Lesquereux) Dawson, Roy. Soc. Canada, Trans., vol. 8, 1890, sec. 4, p. 80, f. 9.

Eocene ?: Similkameen River and Quilchena, British Columbia.

Comptonia insignis (Lesquereux) Cockerell, Univ. Colorado Studies, vol. 3, 1906, p. 173.—Berry, Am. Nat., vol. 40, 1906, p. 499.—Cockerell, Am. Mus. Nat. Hist., Bull., vol. 24, 1908, p. 81.—Knowlton, U. S. Nat. Mus., Proc., vol. 51, 1916, p. 260.

Myrica insignis Lesquereux, U. S. Geol. and Geog. Surv. Terr., Ann. Rept. 1874 [1876], p. 312; Rept. U. S. Geol. Surv. Terr., vol. 7 (Tert. Fl.), 1878, p. 135, pl. lxv, f. 7, 8.

Myrica alkalina Lesquereux, Rept. U. S. Geol. Surv. Terr., vol. 8 (Cret. and Tert. Fl.), 1883, p. 149, pl. xlv–A, f. 10–15.

Miocene: Florissant, Colo.; ?Alkali station, Wyo.

Comptonia microphylla (Heer) Berry, Am. Nat., vol. 40, 1906, p. 508, pl. iv, f. 1, 3, 4; Geol. Surv. N. J., Bull. 3, 1911, p. 108.

Rhus microphylla Heer, Fl. foss. arct., vol. 3, Abt. 2, 1874, p. 117, pl. xxxii, f. 18.

Myrica (*Comptonia*) *parvula* Heer, idem, vol. 7, 1883, p. 20, pl. lv, f. 1–3.—Newberry, U. S. Geol. Surv., Mon. 26, 1895 [1896], p. 63, pl. xix, f. 6.

Raritan: Sayreville, N. J.

Comptonia partita (Lesquereux) Berry, Am. Nat., vol. 40, 1906, p. 512.

Myrica partita Lesquereux, U. S. Geol. and Geog. Surv. Terr., Ann. Rept. 1873 [1874], p. 412; Rept. U. S. Geol. Surv. Terr., vol. 7 (Tert. Fl.), 1878, p. 134, pl. xvii, f. 14.—Dawson, Geol. Surv. Canada, 1877–78, p. 27.—Penhallow, Rept. Tert. Pl. Brit. Col., 1908, p. 46.

Miocene: Elko station, Nev.

Eocene ?: Tulameen River, British Columbia.

Comptonia praemissa Lesquereux, U. S. Nat. Mus., Proc., vol. 5, 1882 [1883], p. 445, pl. vi, f. 13.—Berry, Am. Nat., vol. 40, 1906, p. 504.

Myrica (*Comptonia*) *praemissa* (Lesquereux) Knowlton, U. S. Nat. Mus., Proc., vol. 17, 1894, p. 222; U. S. Geol. Surv., Seventeenth Ann. Rept., pt. 1, 1896, p. 885; U. S. Geol. Surv., Bull. 152, 1898, p. 148.

Kenai: Coal Harbor, Unga Island, Alaska.

Comptonia quilchenensis Penhallow, Rept. Tert. Pl. Brit. Col., 1908, p. 46, text f. 13.

Eocene ?: Quilchena, British Columbia.

Cone of *Pinus?* sp., Fontaine, U. S. Geol. Surv., Mon. 15, 1889, p. 272, pl. clxx, f. 4.

Patapsco: Baltimore, Md.

Cone (undetermined) Berry, Bull. Torr. Bot. Club, vol. 32, 1905, p. 45, pl. i, f. 1.

Magothy. Cliffwood, N. J.

CONFERVITES Brongniart, Hist. vég. foss., vol. 1, 1828, p. 35.

Confervites dubius Berry, Am. Nat., vol. 37, 1903, p. 677, f. 9; Geol. Surv. N. J., Ann. Rept. 1905 [1906], p. 138.

Magothy: Cliffwood, N. J.

Conifer (*a*), ament of, Fontaine, U. S. Geol. Surv., Mon. 15, 1889 [1890], p. 225, pl. cxxxvi, f. 2.

Patapsco: Near Brooke, Va.

Conifer (*b*), ament of, Fontaine, U. S. Geol. Surv., Mon. 15, 1889 [1890], p. 225, pl. cxxxvi, f. 3.

Patapsco: Near Brooke, Va.

Conifer (*c*), ament of, Fontaine, U. S. Geol. Surv., Mon. 15, 1889 [1890], p. 226, pl. cxxxvi, f. 4.

Patapsco: Near Brooke, Va.

Conifer (*d*), ament of, Fontaine, U. S. Geol. Surv., Mon. 15, 1889 [1890], p. 226, pl. cxxxvi, f. 5.

Patapsco: Near Cherry Hill station, Va.

Conifer (*e*), ament of, Fontaine, U. S. Geol. Surv., Mon. 15, 1889 [1890], p. 226, pl. cxxxvi, f. 7.

Patapsco: Near Brooke, Va.

Conifer (*f*), ament of, Fontaine, U. S. Geol. Surv., Mon. 15, 1889 [1890], p. 227, pl. cxxxvi, f. 8.

Patapsco: Near Brooke, Va.

Conifer, male ament of, Fontaine, in Ward, U. S. Geol. Surv., Nineteenth Ann. Rept., pt. 2, 1899, p. 687, pl. clxix, f. 5.

Lakota: South Fork of Hay Creek, Crook County, Wyo.

Conifer, male ament of, Fontaine, in Ward, U. S. Geol. Surv., Mon. 48, 1905 [1906], p. 135, pl. xxxvi, f. 13.

Jurassic: Douglas County, Oreg.

Coniferous stem, Berry, Bull. Torr. Bot. Club, vol. 31, 1904, p. 81, pl. v, f. 5.

Magothy: Cliffwood, N. J.

Coniferous wood in stomach of *Mastodon*, Asa Gray, Bost. Soc. Nat. Hist., Proc., vol. 2, 1846, p. 93.

Coniferous leaves, Knowlton, Smithsonian Misc. Coll., vol. 50, 1907, p. 127, pl. xii, f. 5, 6.

Kootenai: Skull Butte, near Stanford, Fergus County, Mont.

CONIOPTERIS Brongniart, Dict. univ. hist. nat., vol. 13 (Tableau), 1849, p. 75 (26).

Coniopteris burejensis (Zalessky) Seward, Jurassic plants from Amurland, Com. géol., Mém., new ser., pt. 81, 1912, p. 22, pl. i, f. 1–5; pl. iii, f. 18–21.—Knowlton, U. S. Geol. Surv., Prof. Paper 85, 1914, p. 46, pl. v, f. 1.

Dicksonia burejensis Zalessky. Knowlton, in Collier, U. S. Geol. Surv., Bull. 278, 1906, p. 29.

Dicksonia n. sp., Knowlton, in Collier, op. cit., p. 29.

Cladophlebis vaccensis Ward. Fontaine, in Ward, U. S. Geol. Surv., Mon. 48, 1905 [1906], p. 157, pl. xxxix, f. 7, 8.

Jurassic: Cape Lisburne, Alaska.

Coniopteris hymenophylloides (Brongniart) Seward, Jurassic Flora of Yorkshire Coast, 1900, p. 98, pl. xvi, f. 4–6; pl. xvii, f. 3, 6–8, pl. xx, f. 1, 2; pl. xxi, f. 1–3, 4, 4a.—Fontaine, in Ward, U. S. Geol. Surv., Mon. 48, 1905 [1906], p. 59, pl. viii, f. 1–3.—Knowlton, U. S. Geol. Surv., Prof. Paper 85, 1914, p. 47, pl. v, f. 2.

Sphenopteris hymenophylloides Brongniart, Prodrome, 1828, pp. 51, 198 [name]; Hist. vég. foss., 1829, p. 189, pl. lvi, f. 4, 4b.

Adiantites amurensis Heer, Fl. foss. arct., vol. 4, 1876, Abt. 2, p. 94, pl. xxi, f. 6, 6a, b, c, d.

Jurassic: Douglas County, Oreg.

Jurassic (Corwin): Cape Lisburne region, Alaska.

CONIOSPERMITES Ettingshausen, Sitzb. Wien. Akad., vol. 55, Abt. 1, 1867, p. 254.

Coniospermites ellipticus Fontaine, U. S. Geol. Surv., Mon. 15, 1889 [1890], p. 279, pl. cxxxviii, f. 14.

Patuxent: Near Potomac Run, Va.

Conites gibbus Reuss = *Microzamia gibba.*

CONOCARPITES Berry, U. S. Geol. Surv., Prof. Paper 112, 1919, p. 127. [Type, *C. formosus*.]

Conocarpites formosus Berry, U. S. Geol. Surv., Prof. Paper 112, 1919, p. 127, pl. xxviii, f. 9.

Tuscaloosa: Glen Allen, Ala.

CONOCARPUS Linné, Sp. pl., 1753, p. 176.

Conocarpus eocenica Berry, U. S. Geol. Surv., Prof. Paper 84, 1914, p. 147, pl. xxix, f. 4–7.

Barnwell (Twiggs): Phinizy Gully, Columbia County, and near Macon, Ga.

Conocarpus eoligniticus Berry, U. S. Geol. Surv., Prof. Paper 91, 1916, p. 325, pl. xcv, f. 1, 2.

Lagrange: Puryear, Tenn.

Wilcox: Naborton, La.

CORDIA Linné, Sp. pl., 1753, p. 190.

Cordia apiculata (Hollick) Berry, Md. Geol. Surv., Upper Cret., 1916, p. 897, pl. xc, f. 6; U. S. Geol. Surv., Prof. Paper 112, 1919, p. 137, pl. xxx, f. 7, 8; Bull. Torr. Bot. Club, vol. 44, 1917, p. 189.

Populus? apiculata Newberry MS. Hollick, N. Y. Acad. Sci., Trans., vol. 12, 1892, p. 31, pl. iii, f. 2.—Newberry, U. S. Geol. Surv., Mon. 26, 1895 [1896], p. 65, pl. xv, f. 3, 4.—Berry, Bull. Torr. Bot. Club, vol. 33, 1906, p. 172; Geol. Surv. N. J., Bull. 3, 1911, p. 111, pl. xi, f. 4.—Hollick, U. S. Geol. Surv., Mon. 50, 1907, p. 49, pl. vii, f. 28, 29.—Berry, Geol. Surv. N. J., Bull. 3, 1911, p. 111, pl. xi, f. 4.

Magothy: Glen Cove, Long Island, N. Y.; Deep Cut, Del.

Raritan: Woodbridge, N. J.; Arrochar, Staten Island, N. Y.

Tuscaloosa: Cottondale, Glen Allen, and Shirleys Mill, Ala.

Bingen: Near Maxwell Spur, Pike County, Ark.

Cordia eocenica Berry, U. S. Geol. Surv., Prof. Paper 91, 1916, p. 345, pl. cvi, f. 11, 12.
Lagrange: Puryear, Tenn.

Cordia? lowii Berry, U. S. Geol. Surv., Prof. Paper 91, 1916, p. 346, pl. cvii, f. 1.
Wilcox (Ackerman): Hurleys, Miss.

Cordia? tiliaefolia Al. Braun = *Ficus tiliaefolia.*

CORNOPHYLLUM Newberry, U. S. Geol. Surv., Mon. 26, 1895 [1896], p. 119. [Type, *C. vetustum.*]

Cornophyllum obtusatum Berry, U. S. Geol. Surv., Prof. Paper 112, 1919, p. 129, pl. xxvi, f. 7, 8.
Tuscaloosa: Cottondale, Ala.

Cornophyllum vetustum Newberry, U. S. Geol. Surv., Mon. 26, 1895 [1896], p. 119, pl. xix, f. 10.—Berry, Bull. Torr. Bot. Club, vol. 39, 1912, p. 404; Geol. Surv. N. J., Bull. 3, 1911, p. 196; U. S. Geol. Surv., Prof. Paper 112, 1919, p. 129.
Raritan: Woodbridge, Milltown, N. J.
Woodbine: Arthurs Bluff, Tex.
Tuscaloosa: Glen Allen, Ala.

CORNUS (Tournefort) Linné, Sp. pl., 1753, p. 117.

Cornus acuminata Newberry = *Cornus newberryi.*

Cornus cecilensis Berry, Bull. Torr. Bot. Club, vol. 38, 1911, p. 408, pl. xix, f. 4; Md. Geol. Surv., Upper Cret., 1916, p. 884, pl. lxxxii, f. 2.
Magothy: Grove Point, Md.

Cornus emmonsii Ward, U. S. Geol. Surv. Sixth Ann. Rept., 1884–85 [1886], p. 553, pl. xlvii, f. 3 [not f. 2, which = *C. impressa*]; idem, Bull. 37, 1887, p. 55, pl. xxvi, f. 3 [not f. 2].
Cornus impressa Lesquereux. Knowlton, U. S. Geol. Surv., Bull. 163, 1900, p. 68.
Mesaverde: Point of Rocks, Wyo.

Cornus emmonsii Ward, U. S. Geol. Surv., Sixth Ann. Rept., 1884–85 [1886], p. 553, pl. xlviii, f. 2 = *Cornus impressa.*

?**Cornus ferox** Unger, Gen. et sp. pl. foss., 1850, p. 441.—Lesquereux, U. S. Nat. Mus., Proc., vol. 11, 1888, p. 21; Knowlton, U. S. Geol. Surv., Bull. 204, 1902, p. 83.
Clarno (lower part): Cherry Creek, Crook County, Oreg.

Cornus forchhammeri Heer, Fl. foss. arct., vol. 6, pt. 5, 1882, p. 85, pl. xliv, f. 13.—Berry, Bull. Torr. Bot. Club, vol. 37, 1910, p. 27; Md. Geol. Surv., Upper Cret., 1916, p. 885, pl. lxxxii, f. 1.
Magothy: Grove Point, Md.

Cornus fosteri Ward, U. S. Geol. Surv., Sixth Ann. Rept., 1884–85 [1886], p. 553, pl. xlvii, f. 8; U. S. Geol. Surv., Bull. 37, 1887, p. 54, pl. xxv, f. 5.

Fort Union: Sevenmile Creek near Glendive, Mont.

Cornus glabrata Bentham. Hannibal, Bull. Torr. Bot. Club, vol. 38, 1911, p. 338, pl. xv, f. 4.

Pliocene: Calabasas Canyon, Santa Cruz Mountains, Calif.

Cornus holmesii Lesquereux, U. S. Geol. and Geog. Surv. Terr., Ann. Rept. 1873 [1874], p. 402. [Not afterward recognized.]

Denver: Coal Creek, east of Denver, Colo.

Cornus hyperborea Heer. Lesquereux, U. S. Nat. Mus., Proc., vol. 11, 1888, p. 29, pl. xv, f. 3 = *Magnolia inglefieldi.*

Cornus impressa Lesquereux, U. S. Geol. and Geog. Surv. Terr., Ann. Rept. 1873 [1874], p. 408; Rept. U. S. Geol. Surv. Terr., vol. 7 (Tert. Fl.), 1878, p. 243, pl. xlii, f. 3.

Cornus emmonsii Ward, U. S. Geol. Surv., Sixth Ann. Rept., 1884–85 [1886], p. 553, pl. xlviii, f. 2 [not f. 3]; idem, Bull. 37, 1887, p. 55, pl. xxvi, f. 2 [not f. 3].

Denver: Denver and Mount Bross, Middle Park, Colo.

Cornus impressa Lesquereux. Knowlton, U. S. Geol. Surv., Bull. 163, 1900, p. 68 = *Cornus emmonsii.*

Cornus incompletus Lesquereux, Am. Jour. Sci., 2d ser., vol. 45, 1868, p. 206. [Not afterward referred to by its author.]

——: Marshalls, Colo.

Cornus kelloggii Lesquereux, Mus. Comp. Zool., Mem., vol. 6, No. 2, 1878, p. 23, pl. vi, f. 3.—Knowlton, in Lindgren, U. S. Geol. Surv., Prof. Paper 73, 1911, p. 59.

Miocene (auriferous gravels): Chalk Bluff, Nevada County, Calif.

Cornus neomexicana Knowlton, U. S. Geol. Surv., Prof. Paper 101, 1917, p. 342, pl. cix, f. 1.

Raton: Salyers Creek, N. Mex.

Cornus newberryi Hollick, in Knowlton, U. S. Geol. Surv., Bull. 152, 1898, p. 76.—Newberry, U. S. Geol. Surv., Mon. 35, 1898, p. 124, pl. xxxvii, f. 2–4.—Knowlton, U. S. Geol. Surv., Mon. 32, pt. 2, 1899, p. 749, pl. ciii, f. 6.—Penhallow, Rept. Tert. Pl. Brit. Col., 1908, p. 47.—Knowlton, Washington Acad. Sci., Proc., vol. 11, 1909, p. 189; Jour. Geol., vol. 19, 1911, p. 370.

Cornus acuminata Newberry, N. Y. Lyc. Nat. Hist., Annals, vol. 9, 1868. p. 71.—[Lesquereux], U. S. Geol. and Geog. Surv. Terr., Ill. Cret. and Tert. Pl., 1878, pl. xx, f. 2–4. [Homonym, Weber, 1852.]

Fort Union: Banks of Yellowstone River, below mouth of Elk Creek, Yellowstone National Park; Craig Creek, Stinking Water Valley.

Paskapoo (Fort Union): Quilchena, British Columbia.

Lance: Rattlesnake Butte, Cheyenne Indian Reservation, S. Dak.; bluffs opposite Miles City, Mont.

Cornus obesus Dawson, Roy. Soc. Canada, Trans., vol. 11, sec. 4, 1893 [1894], p. 62, pl. ix, f. 30.
  Upper Cretaceous: Vancouver Colliery, Vancouver Island, British Columbia.

Cornus orbifera Heer, Gartenflora, vol. 2, 1853, p. 293, pl. lxvi, f. 9.—Lesquereux, U. S. Nat. Mus., Proc., vol. 5, 1882 [1883], p. 448, pl. x, f. 6; Rept. U. S. Geol. Surv. Terr., vol. 8 (Cret. and Tert. Fl.), 1883, p. 262.—Knowlton, U. S. Nat. Mus., Proc., vol. 17, 1894, p. 226; Geol. Soc. Am., Bull., vol. 5, 1893, p. 585.
  Kenai: Cook Inlet, Alaska.

Cornus orbifera Heer. Lesquereux, U. S. Geol. and Geog. Surv. Terr., Ann. Rept. 1873 [1874], p. 402 = *Cornus suborbifera.*

Cornus ovalis Lesquereux, Mus. Comp. Zool., Mem., vol. 6, No. 2, 1878, p. 23, pl. vi, f. 1, 2.—Knowlton, in Lindgren, Jour. Geol., vol. 4, 1896, p. 890; U. S. Geol. Surv., Prof. Paper, 73, 1901, p. 59.
  Miocene (auriferous gravels): Table Mountain, Tuolumne County, Independence Hill, Placer County, Calif.

Cornus platyphylloid Lesquereux = *Eorhamnidium platyphylloides.*

Cornus praecox Lesquereux, U. S. Geol. Surv., Mon. 17, 1891 [1892], p. 125, pl. xxiii, f. 5.
  Dakota: Ellsworth County, Kans.

Cornus praeimpressa Knowlton, U. S. Geol. Surv., Prof. Paper 130, 19—, p. —, pl. xiv, f. 5; pl. xix, f. 2a [in preparation].
  Laramie: Cowan station, near Denver, Colo.

Cornus rhamnifolia O. Weber, Palaeontogr., vol. 2, 1852, p. 152, pl. iv, f. 8.—Lesquereux, U. S. Geol. and Geog. Surv. Terr., Ann. Rept. 1871 [1872], Suppl., p. 9; Rept. U. S. Geol. Surv. Terr., vol. 7 (Tert. Fl.), 1878, p. 244, pl. lxii, f. 6.—Penhallow, Roy. Soc. Canada, Trans., 2d ser., vol. 8, 1902, p. 62.
  Paskapoo (Fort Union): Red Deer River, Alberta.
  Livingston: Bozeman coal field, Mont.
  ——: Near Point of Rocks, Wyo.

Cornus studeri? Heer, Fl. tert. Helv., vol. 3, 1859, p. 27, pl. cv, f. 18–21.—Lesquereux, U. S. Geol. and Geog. Surv. Terr., Ann. Rept., 1871 [1872], Suppl., p. 9; Rept. U. S. Geol. Surv. Terr., vol. 7 (Tert. Fl.), 1878, p. 244, pl. xlii, f. 4, 5.—Ward, U. S. Geol. Surv., Sixth Ann. Rept., 1884–85 [1886], p. 553, pl. xlviii, f. 1; idem, Bull. 37, 1887, p. 55, pl. xxvi, f. 1.—Knowlton, U. S. Geol. Surv., Bull. 163, 1900, p. 68, pl. xv, f. 3 [not this species].—Berry, U. S. Geol. Surv., Prof. Paper 91, 1916, p. 331, pl. lxviii, f. 3.—Knowlton, U. S. Geol. Surv., Prof. Paper 101, 1918, p. 342, pl. cix, f. 2.
  Denver: Golden, Colo.
  Raton: Aguilar, Colo.
  Mesaverde (?): Point of Rocks, Wyo.
  Wilcox: Coushatta and Naborton, La.

Cornus suborbifera Lesquereux, Rept. U. S. Geol. Surv. Terr., vol. 7 (Tert. Fl.), 1878, p. 243, pl. xlii, f. 2.—?Penhallow, Rept. Tert. Pl. Brit. Col., 1908, p. 47.—Knowlton, U. S. Geol. Surv., Prof. Paper 130, 19—, p. —, pl. xv, f. 2, 2a [in preparation].

Cornus orbifera Heer. Lesquereux, U. S. Geol. and Geog. Surv. Terr., Ann. Rept. 1873 [1874], p. 402.

Laramie: Golden, Colo.

?Eocene: Tulameen River, British Columbia.

Cornus wrightii Knowlton, U. S. Geol. Surv., Mon. 32, pt. 2, 1899, p. 749, pl. ciii, f. 4, 5.

Miocene: Fossil Forest, Yellowstone National Park.

Cornus sp., Knowlton, U. S. Geol. Surv., Prof. Paper 130, 19—, p. —, pl. xiv, f. 4 [in preparation].

Laramie: Cowan station, near Denver, Colo.

Cornus sp. Penhallow, Roy. Soc. Canada, Trans., 2d ser., vol. 2, sec. 4, 1896, p. 70.

Pleistocene: Greenville, N. J.

Cornus sp. Knowlton, in Lindgren, Jour. Geol., vol. 4, 1896, p. 890.

Miocene (auriferous gravels): Independence Hill, Placer County, Calif.

CORYLUS (Tournefort) Linné, Sp. pl., 1753, p. 998.

Corylus americana Walter. Newberry, N. Y. Lyc. Nat. Hist., Annals, vol. 9, 1868, p. 59.—[Lesquereux], U. S. Geol. and Geog. Surv. Terr., Ill. Cret. and Tert. Pl., 1878, pl. xiv, f. 8–10.—Ward, U. S. Geol. Surv., Sixth Ann. Rept., 1884–85 [1886], p. 551, pl. xxxviii, f. 1–5; idem, Bull. 37, 1887, p. 28, pl. xi, f. 3–5; pl. xii, f. 1, 2.—Penhallow, Roy. Soc. Canada, Trans., 2d ser., vol. 8, sec. 4, 1902, p. 58; Rept. Tert. Pl. Brit. Col., 1908, p. 47.—Knowlton, Washington Acad. Sci., Proc., vol. 11, 1909, pp. 189, 190, 198, 213.—Meehan, in Mercer, Acad. Nat. Sci. Philadelphia, Jour., 2d ser., vol. 11, 1899, p. 281.

Corylus americana fossilis Newberry, U. S. Geol. Surv., Mon. 35, 1898, p. 60, pl. xxix, f. 8–10.

Pleistocene: Port Kennedy, Pa.

Fort Union: Fort Union, N. Dak.; Sevenmile Creek near Glendive, Mont.

Paskapoo (Fort Union): Red Deer River, Alberta; Porcupine Creek, Saskatchewan; Quilchena, Horsefly River, and Tranquille River, British Columbia.

Lance: Bluffs opposite Miles City, Signal Butte near Miles City, and 9 miles above Glendive, Mont.; Ilo, Big Horn Basin, Wyo.

Corylus fosteri Ward, U. S. Geol. Surv., Sixth Ann. Rept., 1884–85 [1886], p. 553, pl. xlvii, f. 8; idem, Bull. 37, 1887, p. 29, pl. xiii, f. 5, 6.

Fort Union: Clear Creek, Mont.

Corylus grandifolia Newberry = *Corylus macquarrii.*

Corylus harrimani Knowlton, Harriman Alaska Exped., vol. 4, 1904, p. 154, pl. xxiii, f. 1.

Kenai: Kukak Bay, Alaska.

Corylus macquarrii (Forbes) Heer, Urwelt d. Schweiz, 1865, p. 321; Fl. foss. arct., vol. 2, Abt. 2, 1869, p. 29, pl. iii, f. 9; pl. iv, f. 1–5, 8.—Eichwald, Geognost.-Palaeontolog. Bemerk. ü. Halbinsel Mangeschlak u. Aleutischen Inseln, 1871, p. 112, pl. iv, f. 6.—Ward, U. S. Geol. Surv., Sixth Ann. Rept., 1884–85 [1886], p. 551, pl. xxxix, f. 7; idem, Bull. 37, 1887, p. 30, pl. xiii, f. 7.—Knowlton, U. S. Nat. Mus., Proc., vol. 17, 1894, p. 219, pl. ix, f. 4; Geol. Soc. Am., Bull., vol. 5, 1893, p. 582.—Newberry, U. S. Geol. Surv., Mon. 35, 1898, p. 61, pl. xxxii, f. 5; pl. xlviii, f. 4.—Knowlton, in Spurr, U. S. Geol. Surv., Eighteenth Ann. Rept., pt. 3, 1898, pp. 185, 192, 195.—Knowlton, U. S. Geol. Surv., Bull. 204, 1902, p. 38; U. S. Geol. Surv., Mon. 32, pt. 2, 1899, p. 699, pl. lxxxviii, f. 3.—Penhallow, Roy. Soc. Canada, Trans., 2d ser., vol. 8, sec. 4, 1902, p. 59.—Knowlton, Harriman Alaska Exped., vol. 4, 1904, p. 153.—Penhallow, Rept. Tert. Pl. Brit. Col., 1908, p. 48.

Alnites macquarrii Forbes, Quart. Jour. Geol. Soc. London, vol. 7, 1851, p. 103, pl. iv, f. 3.

Corylus grandifolia Newberry, N. Y. Lyc. Nat. Hist., Annals, vol. 9, 1868, p. 59.—[Lesquereux], U. S. Geol. and Geog. Surv. Terr., Ill. Cret. and Tert. pl., 1878, pl. xv, f. 5.

Alnus pseudo-glutinosa Göppert. Göppert, Schlesisch. Gesell. f. vaterländ. Kultur, 1861, p. 202.

Fort Union: Fort Union, N. Dak.; Sevenmile Creek, Glendive, Mont.

Paskapoo (Fort Union): Red Deer River, Alberta; Porcupine Creek, Saskatchewan; Horsefly River, British Columbia; Mackenzie River, Northwest Territory.

Clarno (upper part): Bridge Creek, Oreg.

Kenai: Port Graham; Ninilchik, Kuiu Island, Unga Island, Millers Creek, Yukon River, 25 miles below Mission Creek; above Mynook Creek; Herendeen Bay; Kukak Bay, Alaska.

Miocene (intermediate flora): Fossil Forest, Yellowstone National Park.

Corylus macquarrii macrophylla Heer, Fl. foss. arct., vol. 1, 1868, p. 105; idem, vol. 2, Abt. 2 (Fl. foss. alask.), 1869, p. 30, pl. iv, f. 6, 7.—Knowlton, U. S. Nat. Mus., Proc., vol. 17, 1894, p. 220; Geol. Soc. Am., Bull., vol. 5, 1893, p. 58.

Kenai: Port Graham, Alaska.

Corylus orbiculata Newberry, N. Y. Lyc. Nat. Hist., Annals, vol. 9, 1868, p. 58.—[Lesquereux], U. S. Geol. and Geog. Surv. Terr., Ill. Cret. and Tert. Pl., 1878, pl. xv, f. 4.—Newberry, U. S. Geol. Surv., Mon. 35, 1898, p. 62, pl. xxxii, f. 4.

Fort Union: Fort Union, N. Dak.

Corylus? palachei Knowlton, Harriman Alaska Exped., vol. 4, 1904, p. 154, pl. xxii, f. 2; pl. xxviii, f. 1.
Kenai: Kukak Bay, Alaska.

Corylus rostrata Aiton. Newberry, N. Y. Lyc. Nat. Hist., Annals, vol. 9, 1868, p. 60.—Ward, U. S. Geol. Surv., Sixth Ann. Rept., 1884–85 [1886], p. 551, pl. xxxix, f. 1–4; idem, Bull. 37, 1887, p. 29, pl. xiii, f. 1–4.—Penhallow, Rept. Tert. Pl. Brit. Col., 1898, p. 48.—Knowlton, Washington Acad. Sci., Proc., vol. 11, 1909, pp. 190, 197; in Calvert, U. S. Geol. Surv., Bull. 471–D, 1912, p. 16.
Corylus rostrata fossilis Newberry, U. S. Geol. Surv., Mon. 35, 1898, p. 63, pl. xxxii, f. 1–3.
Fort Union: Fort Union, N. Dak.; Sevenmile Creek, Glendive, Mont.
Paskapoo (Fort Union): Porcupine Creek and Great Valley, Saskatchewan.
Lance: Signal Butte, Miles City, mouth of Cedar Creek, Glendive and opposite Glendive, Mont.

Corylus rostrata fossilis Newberry = *Corylus rostrata*.

Corylus sp. Knowlton, in Lindgren, Jour. Geol., vol. 4, 1896, p. 890.
Miocene (auriferous gravels): Independence Hill, Placer County, Calif.

COTINUS Adanson, Fam. Pl., vol. 2, 1763, p. 345.

Cotinus fraterna (Lesquereux) Cockerell, Torreya, vol. 5, 1905, p. 12; Univ. Colorado Studies, vol. 3, 1906, p. 170; Am. Mus. Nat. Hist., Bull., vol. 24, 1908, p. 99.—Knowlton, U. S. Nat. Mus., Proc., vol. 51, 1916, p. 279, pl. xxiv, f. 1.
Rhus fraterna Lesquereux, Rept. U. S. Geol. Surv. Terr., vol. 8 (Cret. and Tert. Fl.), 1883, p. 192, pl. xli, f. 1, 2.
Andromeda rhomboidalis Lesquereux, idem, p. 176.
Andromeda scudderiana Cockerell, Am. Mus. Nat. Hist., Bull., vol. 24, 1908, p. 105.
Miocene: *Florissant, Colo.

CRATAEGUS (Tournefort) Linné, Sp. pl., 1753, p. 475.

Crataegus acerifolia Lesquereux = *Sorbus diversifolia*.

Crataegus aceroides Lesquereux, U. S. Geol. Surv., Mon. 17, 1891 [1892], p. 143, pl. liv, f. 8; pl. lv, f. 1.
Dakota: Ellsworth County, Kans.

Crataegus? aequidentata Lesquereux, Rept. U. S. Geol. Surv. Terr., vol. 7 (Tert. Fl.), 1878, p. 297, pl. lviii, f. 4, 4a.
Hanna: Carbon, Wyo.

Crataegus antiqua Heer, Fl. foss. arct., vol. 1, 1868, p. 125, pl. l, f. 1, 2.—Lesquereux, Mus. Comp. Zool., Bull., vol. 16, 1888, p. 57.
Denver: Golden, Colo.

Crataegus atavina Heer, Fl. foss. arct., vol. 7, 1883, p. 43, pl. lxiv, f. 11.—Lesquereux, Geol. and Nat. Hist. Surv., Minn., vol. 3, 1893, p. 20, pl. B, f. 8.
Dakota: New Ulm, Minn.

Crataegus betulaefolia Lesquereux, Mus. Comp. Zool., Bull., vol. 3, 1888, p. 56.
Denver: Golden, Colo.

Crataegus coccineaefolia Berry, Jour. Geol., vol. 15, 1907, p. 345.
Pleistocene: Neuse River, N. C.

Crataegus crus-galli? Linné. Meehan, in Mercer, Acad. Nat. Sci. Philadelphia, Jour., 2d ser., vol. 11, 1899, p. 281.
Pleistocene: Port Kennedy, Pa.

Crataegus engelhardti Lesquereux, Mus. Comp. Zool., Bull., vol. 3, 1888, p. 56.
Denver: Golden, Colo.

Crataegus flavescens Newberry = *Crataegus newberryi*.

Crataegus holmesii Lesquereux, U. S. Nat. Mus., Proc., vol. 10, 1887, p. 43, pl. iii, f. 7–9.
Denver: Silver Cliff, Colo.

Crataegus imparilis Knowlton, U. S. Geol. Surv., Bull. 204, 1902, p. 66, pl. x, f. 2.
Mascall: Near Van Horn's ranch, John Day Basin, Oreg.

Crataegus lacoei Lesquereux, U. S. Geol. Surv., Mon. 17, 1891 [1892], p. 143, pl. lxvi, f. 2; pl. lxiv, f. 14.
Dakota: Ellsworth County, Kans.

Crataegus laurenciana Lesquereux, U. S. Geol. Surv., Mon. 17, 1891 [1892], p. 142, pl. xxxviii, f. 1.
Dakota: Ellsworth County, Kans.

Crataegus lesquereuxii Cockerell = *Sorbus diversifolia*.

Crataegus marcouiana Lesquereux, U. S. Nat. Mus., Proc., vol. 11, 1888, p. 36, pl. xiv, f. 2; pl. xv, f. 1, 2.
——: "Fossil Point, Plane Table Sheet," = Wingate, Ariz. ?

Crataegus marcouiana subintegrifolia Lesquereux, U. S. Nat. Mus., Proc., vol. 11, 1888, p. 36, pl. xiv, f. 2.
——: "Fossil Point, Plane Table Sheet" = Wingate, Ariz. ?

Crataegus monmouthensis Berry, Johns Hopkins Univ. Circ., new ser., No. 7, 1907, p. 675 (67).
Magothy: Cliffwood, N. J.

Crataegus myricoides Lesquereux, Mus. Comp. Zool., Bull., vol. 16, 1888, p. 56.
Denver: Golden, Colo.

Crataegus newberryi Cockerell, Univ. Colorado Studies, vol. 5, 1907 [1908], p. 43; Am. Mus. Nat. Hist., Bull., vol. 24, 1908, p. 95.

Crataegus flavescens Newberry, U. S. Nat. Mus., Proc., vol. 5, 1882 [1883], p. 507; U. S. Geol. Surv., Mon. 35, 1898, p. 112, pl. xlviii, f. 1.—Knowlton, U. S. Geol. Surv., Bull. 152, 1898, p. 77; idem, Bull. 204, 1902, p. 66, pl. x, f. 1; in Merriam, Univ. Calif. Bull. Dept. Geol., vol. 2, 1901, p. 289. [Homonym, Bosc, 1825.]

Myrica diversifolia Lesquereux, Rept. U. S. Geol. Surv. Terr., vol. 8 (Cret. and Tert. Fl.), 1883, p. 241, pl. l, f. 10.

Clarno (upper part): Bridge Creek, Grant County, Oreg.

Crataegus punctata Jacques. Penhallow, Brit. Assoc. Adv. Sci., Bradford meeting, 1900, p. 335.

Pleistocene: Don Valley, Toronto, Canada.

Crataegus spathulatoides Berry, Jour. Geol., vol. 15, 1907, p. 344.

Pleistocene: Neuse River, N. C.

Crataegus tenuinervis Lesquereux, U. S. Geol. Surv., Mon. 17, 1891 [1892], p. 142, pl. liv, f. 5–7.

Dakota: Near Fort Harker, Kans.

Crataegus tranquillensis Penhallow, Rept. Tert. Pl. Brit. Col., 1908, p. 49, text f. 8.

Oligocene: Tranquille and Horsefly rivers, British Columbia.

Crataegus tulameensis Penhallow, Rept. Tert. Pl. Brit. Col., 1908, p. 48, text f. 7.

Oligocene: Tulameen River, British Columbia.

Crataegus sp. Berry, Am. Jour. Sci., 4th ser., vol. 34, 1912, p. 222, f. 3.

Pleistocene: Buena Vista, Va.

Crataegus sp. Knowlton, U. S. Nat. Mus., Proc., vol. 51, 1916, p. 275, pl. xxi, f. 4.

Miocene: Florissant, Colo.

CREDNERIA Zenker, Beiträge z. Naturgeschichte d. Urwelt, 1833, p. 15.

Credneria? daturaefolia Ward, U. S. Geol. Surv., Sixth Ann. Rept., 1884–85 [1886], p. 556, pl. lvii, f. 1–5; pl. lviii, f. 1–5; U. S. Geol. Surv., Bull. 37, 1887, p. 97, pls. xliii–xlv.—Knowlton, Washington Acad. Sci., Proc., vol. 11, 1909, p. 194.

Populus? daturaefolia (Ward) Cockerell, Am. Mus. Nat. Hist., Bull., vol. 24, 1908, p. 83.

Fort Union: Sevenmile Creek near Glendive, Mont.

Lance: Near Melville, Mont.

Credneria leconteana Lesquereux = *Protophyllum leconteanum.*

Credneria macrophylla Heer, Neue Denkschr. Schw. Gesellsch. Nat., vol. 23, pl. 2 (Kreide-Fl. Moletin), 1869, p. 16, pl. iv.—Hollick, N. Y. Bot. Gard., Bull., vol. 8, 1912, p. 168, pl. clxx, f. 4.

Magothy: Roslyn, N. Y.

Credneria? microphylla Lesquereux, U. S. Geol. and Geog. Surv. Terr., Bull., vol. 1, 1875 [1876], 397.
Dakota: Kansas.

Credneria? pachyphylla Knowlton, U. S. Geol. Surv., Mon. 32, pt. 2, 1899, p. 742, pl. ci, f. 6.
Fort Union: Elk Creek, Yellowstone National Park.

Credneria protophylloides Knowlton, U. S. Geol. Surv., Prof. Paper 101, 1918, p. 267, pl. xlvi.
Vermejo: Santa Clara Canyon, Colo.

CROTON Linné, Sp. pl., 1753, p. 1004.

Croton? furcatulum Cockerell, Torreya, vol. 9, 1909, p. 18, text f.
Miocene: Florissant, Colo.

CROTONOPHYLLUM Velenovsky, Abh. K. böhm. Ges. Wiss., Prag, 7th ser., vol. 3, No. 3, 1889, p. 20.

Crotonophyllum appendiculatum Berry, U. S. Geol. Surv., Prof. Paper 91, 1916, p. 259, pl. lviii, f. 1.
Lagrange: Puryear, Henry County, Tenn.

Crotonophyllum cretaceum Velenovsky, Květena českého cenomanu, 1889, p. 20, pl. v, f. 4–11.—Berry, Md. Geol. Surv., Upper Cret., 1916, p. 847, pl. lxxvi, f. 7, 8.
Magothy: Grove Point, Md.

Crotonophyllum eocenicum Berry, U. S. Geol. Surv., Prof. Paper 91, 1916, p. 258, pl. lviii, f. 2.
Lagrange: Puryear, Henry County, Tenn.

Crotonophyllum panduraeformis Berry, U. S. Geol. Surv., Prof. Paper 84, 1914, p. 48, pl. vii, f. 5–10; idem, Prof. Paper 112, 1919, p. 105.
Black Creek (Middendorf): Middendorf, Langley, Miles Mill, and Rocky Point, S. C.
Tuscaloosa: Shirleys Mill, Ala.

CRYPTOCARYA R. Brown, Prodr., vol. 1, 1810, p. 402.

Cryptocarya eolignitica Hollick, Rept. Geol. Surv. La., 1899 [1900], p. 283, pl. xlii, f. 1.—Berry, U. S. Geol. Surv., Prof. Paper 91, 1916, p. 312, pl. lxxxviii, f. 6.
Wilcox: Near Coushatta, La.

CTENIS Lindley and Hutton, Foss. Fl. Great Britain, vol. 2, 1834, p. 63.

Ctenis auriculata Fontaine, Am. Jour. Sci., 4th ser., vol. 2, 1896, p. 274; in Ward, U. S. Geol. Surv., Twentieth Ann. Rept., pt. 2, 1900, p. 356, pl. lviii, f. 1–3; U. S. Geol. Surv., Mon. 48, 1905 [1906], p. 117, pl. xxix, f. 1.
Jurassic: Oroville, Calif.; Douglas County, Oreg.

Ctenis grandifolia Fontaine, Am. Jour. Sci., 4th ser., vol. 2, 1896, p. 274; in Ward, U. S. Geol. Surv., Twentieth Ann. Rept., pt. 2, 1900, p. 354, pl. liii, f. 2; pl. lvi, f. 6, 7; pl. lvii; U. S. Geol. Surv., Mon. 48, 1905 [1906], p. 116, pl. xxviii, f. 2–8.

Jurassic: Oroville, Calif.; Douglas County, Oreg.

Ctenis imbricata Fontaine, U. S. Geol. Surv., Mon. 15, 1889 [1890], p. 177, pl. cxxxviii, f. 10–12.

Patuxent: Fredericksburg, Va.

Ctenis leckenbyi Bean = *Ptilozamites leckenbyi*.

Ctenis orovillensis Fontaine, Am. Jour. Sci., 4th ser., vol. 2, 1896, p. 274 [name]; in Ward, U. S. Geol. Surv., Twentieth Ann. Rept., pt. 2, 1900, p. 357, pl. lviii, f. 4; U. S. Geol. Surv., Mon. 48, 1905 [1906], p. 115, pl. xxvii, f. 1–5; pl. xxviii, f. 1.

Jurassic: Oroville, Calif.; Douglas County, Oreg.

Ctenis sulcicaulis (Phillips) Ward, U. S. Geol. Surv., Mon. 48, 1905 [1906], p. 113, pl. xxv, f. 9; pl. xxvi; p. 149, pl. xxxviii, f. 7, 8.

Cycadites sulcicaulis Phillips, Geol. Yorkshire, 1829, p. 128.

Jurassic: Douglas County, Oreg.; Curry County, Oreg.

Ctenis wardii Fontaine, Am. Jour. Sci., 4th ser., vol. 2, 1896, p. 274; in Ward, U. S. Geol. Surv., Twentieth Ann. Rept., pt. 2, 1900, p. 357, pl. lix; pl. lx; pl. lxvii, f. 5.

Jurassic: Oroville, Calif.

CTENOPHYLLUM Schimper, Pal. vég., vol. 2, 1870, p. 143.

Ctenophyllum angustifolium Fontaine, Am. Jour. Sci., 4th ser., vol. 2, 1896, p. 274; in Ward, U. S. Geol. Surv., Twentieth Ann. Rept., pt. 2, 1900, p. 360, pl. lxiii, f. 2, 3; U. S. Geol. Surv., Mon. 48, 1905 [1906], p. 105, pl. xxii.—Knowlton, U. S. Nat. Mus., Proc., vol. 51, 1916, p. 458, pl. 80, f. 2.

Jurassic: Oroville, Calif.; Douglas County, Oreg.

Lower Jurassic: Upper Matanuska Valley, Alaska.

Ctenophyllum braunianum abbreviatum (F. Braun) Schimper. Fontaine, in Ward, U. S. Geol. Surv., Twentieth Ann. Rept., pt. 2, 1900, p. 293, pl. xxxix, f. 8, 9.

Pterozamites abbreviatus F. Braun, in Münster, Beiträge zur Petrefactenkunde, vol. 2, pt. 6, 1843, p. 30.

Zamites obtusifolius Rogers, Trans. Assoc. Am. Geol. and Nat., Philadelphia, 1843, p. 312, pl. xiv.—Emmons, American geology, pt. 6, 1856, p. 331.

Pterozamites obtusifolius (Rogers) Emmons, op. cit., p. 118, f. 85.

Pterozamites gracilis Emmons, op. cit., p. 118, f. 86.

Triassic (Newark): North Carolina.

Ctenophyllum braunianum angustum (F. Braun) Schimper. Fontaine, in Ward, U. S. Geol. Surv., Twentieth Ann. Rept., pt. 2, 1900, pl. xxxix, f. 6, 7.

Pterozamites decussatus Emmons, Geol. Rept. Midland Counties, N. C., 1856, p. 330, pl. iii, f. 1; American geology, pt. 6, 1857, p. 117, pl. iii, f. 1.

Ctenophyllum braunianum angustum—Continued.

Zamites graminioides Emmons, op. cit. (N. C.). p. 330 (*Dioonites graminioides*, p. 349), pl. iv, f. 11.

Pterozamites sp. Emmons. op cit. (N. C.). p. 349, pl. iii, f. 8.

Dionites linearis (*Zamites graminioides*) Emmons, American geology, p. 121, pl. iv, f. 11.

Pterophyllum decussatum (Emmons) Fontaine, U. S. Geol. Surv., Mon. 6, 1883, p. 111, pl. li, f. 2.

Pterophyllum spatulatum (Emmons) Fontaine, U. S. Geol. Surv., Mon. 6, 1883, p. 114, pl. liii, f. 6.

Pterozamites spatulatus Emmons, American geology, pt. 6, 1857, p. 120, f. 88.

Triassic (Newark): Ellingtons and Haw River, N. C.; Cumberland area, Va.

Ctenophyllum braunianum var. a Göppert. Fontaine, U. S. Geol. Surv., Mon. 6, 1883, p. 69, pl. xxxiv, f. 2–4a; pl. xxxv, f. 1; pl. xxxvii, f. 1, 2; pl. xxxviii, f. 1–2a.

Triassic (Newark): Richmond, Va., coal field.

Ctenophyllum densifolium Fontaine, Am. Jour. Sci., 4th ser., vol. 2, 1896, p. 274; in Ward, U. S. Geol. Surv., Twentieth Ann. Rept., pt. 2, 1900, p. 358, pl. lxi.

Jurassic: Oroville, Calif.

Ctenophyllum emmonsi Fontaine = *Ctenophyllum robustum*.

Ctenophyllum giganteum Fontaine, U. S. Geol. Surv., Mon. 6, 1883, p. 76, pl. xxxix, f. 5.

Triassic (Newark): Clover Hill, Va.

Ctenophyllum grandifolium Fontaine, U. S. Geol. Surv., Mon. 6, 1883, p. 73, pl. xxxix, f. 1–3; pl. xl; pl. xli; pl. xlii, f. 1; in Ward, U. S. Geol. Surv., Twentieth Ann. Rept., pt. 2, 1900, p. 243, pl. xxvii.—Knowlton, in Stone, U. S. Geol. Surv., Bull. 471, 1916, p. 143.

Triassic (Newark): Clover Hill, Va.; York County, Pa.; Dan River, N. C.

Ctenophyllum grandifolium Fontaine. Fontaine, in Turner, Jour. Geol., vol. 3, 1895, p. 395 = *Ctenophyllum grandifolium storrsii*.

Ctenophyllum grandifolium storrsii Fontaine, Am. Jour. Sci., 4th ser., vol. 2, 1896, p. 274; in Ward, U. S. Geol. Surv., Twentieth Ann. Rept., pt. 2, 1900, p. 359, pl. liii, f. 3; pl. lxii; pl. lxiii, f. 1; pl. lxvi, f. 3.

Ctenophyllum grandifolium Fontaine. Fontaine, in Turner, Jour. Geol., vol. 3, 1895, p. 395.

Jurassic: Oroville, Calif.

Ctenophyllum latifolium Fontaine = *Ctenopsis latifolia*.

Ctenophyllum lineare (Emmons) Fontaine, U. S. Geol. Surv., Mon. 6, 1883, p. 114, pl. liv, f. 2; in Ward, U. S. Geol. Surv., Twentieth Ann. Rept., pt. 2, 1900, p. 293.

Pterozamites linearis Emmons, American geology, pt. 6, 1857, p. 120, f. 87.

Triassic (Newark): Haw River, N. C.

Ctenophyllum pachynerve Fontaine, in Ward, U. S. Geol. Surv., Mon. 48, 1905 [1906], p. 106, pl. xxiii, f. 1–4.
Jurassic: Douglas County, Oreg.

Ctenophyllum robustum (Emmons) Fontaine, U. S. Geol. Surv., Mon. 6, 1883, p. 116, pl. liv, f. 6, 7; in Ward, U. S. Geol. Surv., Twentieth Ann. Rept., pt. 2, 1900, p. 294, pl. xxxix, f. 10.
Pterophyllum robustum Emmons, American geology, pt. 6, 1857, p. 122, f. 91.
Pterophyllum robustum var.? Emmons, op. cit., p. 123, f. 92.
Pterozamites obtusus Emmons, op. cit., p. 119, f. 86a.
Ctenophyllum emmonsi Fontaine, U. S. Geol. Surv., Mon. 6, 1883, p. 113, pl. liv, f. 1.
Triassic (Newark): Ellingtons, N. C.

Ctenophyllum taxinum (Lindley and Hutton) Fontaine, U. S. Geol. Surv., Mon. 6, 1883, p. 67, pl. xxxiii, f. 3–4a.
Zamia taxina Lindley and Hutton, Foss. Fl. Gt. Brit., vol. 3, 1835, p. 65, pl. clxxv.
Triassic (Newark): Midlothian, Va.

Ctenophyllum truncatum Fontaine, U. S. Geol. Surv., Mon. 6, 1883, p. 68, pl. xxxviii, f. 3–5.
Triassic (Newark): Clover Hill, Va.

Ctenophyllum wannerianum Fontaine, in Ward, U. S. Geol. Surv., Twentieth Ann. Rept., pt. 2, 1900, p. 243, pl. xxviii, f. 1.
Triassic (Newark): York County, Pa.

Ctenophyllum wardii Fontaine, Am. Jour. Sci., 4th ser., vol. 2, 1896, p. 274; in Ward, U. S. Geol. Surv., Twentieth Ann. Rept., pt. 2, 1900, p. 357, pl. lix; pl. lx; pl. lxvii, f. 5; U. S. Geol. Surv., Mon. 48, 1905 [1906], p. 107, pl. xxiii, f. 5–12.
Jurassic: Oroville, Calif.; Douglas County, Oreg.

Ctenophyllum? sp. Fontaine, U. S. Nat. Mus., Proc., vol. 13, 1890, p. 285.
Triassic: Abiquiu, N. Mex.

Ctenophyllum? sp. Fontaine n. sp.?, in Ward, U. S. Geol. Surv., Mon. 48, 1905 [1906], p. 149, pl. xxxviii, f. 9, 10.
Jurassic: Curry County, Oreg.

CTENOPSIS Berry, Md. Geol. Surv., Lower Cret., 1911, p. 347. [Type, *Ctenophyllum latifolium* Fontaine.]

Ctenopsis latifolia (Fontaine) Berry, Md. Geol. Surv., Lower Cret., 1911, p. 349, pl. lv, f. 1, 2.
Ctenophyllum latifolium Fontaine, U. S. Geol. Surv., Mon. 15, 1889 [1890], p. 175, pl. lxviii, f. 2, 3; in Ward, U. S. Geol. Surv., Mon. 48, 1905 [1906], p. 255, pl. lxvii, f. 10.
Podozamites grandifolius Fontaine, U. S. Geol. Surv., Mon. 15, 1889 [1890], p. 180, pl. lxxxii, f. 2; pl. lxxxiii, f. 5 [not other references].
Patuxent: Fredericksburg and Potomac Run, Va.
Horsetown (?): Near Ono, Shasta County, Calif.

CTENOPTERIS Saporta, Pal. France, vol. 1, 1873, p. 351.

Ctenopteris angustifolia Fontaine, U. S. Geol. Surv., Mon. 15, 1889 [1890], p. 159, pl. lxv, f. 2; pl. lxvii, f. 4; in Ward, U. S. Geol. Surv., Mon. 48, 1905 [1906], p. 510.—Berry, Md. Geol. Surv., Lower Cret., 1911, p. 353.

Patuxent: *Potomac Run and Chinkapin Hollow, Va.

Ctenopteris columbiensis Penhallow, Roy. Soc. Canada, Trans., 2d ser., vol. 8, 1902, sec. 4, p. 36, f. 1.

Upper Cretaceous: Port McNeil, Vancouver Island.

Ctenopteris insignis Fontaine, U. S. Geol. Surv., Mon. 15, 1889 [1890], p. 156, pl. lxi, f. 4, 5; pl. lxii, f. 1; pl. lxiii, f. 1, 2; in Ward, U. S. Geol. Surv., Mon. 48, 1905 [1906], p. 521, pl. cxii, f. 7.—Berry, Md. Geol. Surv., Lower Cret., 1911, p. 352.

Ctenopteris virginiensis Fontaine, U. S. Geol. Surv., Mon. 15, 1889 [1890], p. 157, pl. lxii, f. 4; pl. lxv, f. 1; pl. lxvi, f. 4.

Ctenopteris minor Fontaine, idem, p. 157, pl. lxvii, f. 3.

Zamiopsis insignis Fontaine. Fontaine, in Ward, U. S. Geol. Surv., Mon. 48, 1905 [1906], p. 525, pl. cxiii, f. 4, 5.

Patuxent: *Fredericksburg, Potomac Run, Va.

Arundel: Langdon, D. C.

Ctenopteris integrifolia Fontaine = *Scleropteris elliptica.*

Ctenopteris longifolia Fontaine, U. S. Geol. Surv., Mon. 15, 1889 [1890], p. 159, pl. lxvii, f. 5.—Berry, Md. Geol. Surv., Lower Cret., 1911, p. 354.

Patuxent: *Fredericksburg, Va.

Ctenopteris minor Fontaine = *Ctenopteris insignis.*

Ctenopteris virginiensis Fontaine = *Ctenopteris insignis.*

CUCUMITES Bowerbank, Hist. Foss. Fruits and Seeds, London Clays, 1840, p. 90.

Cucumites globulosus (Knowlton) Cockerell = *Cucumites lesquereuxii.*

Cucumites lesquereuxii Knowlton, Bull. Torr. Bot. Club, vol. 29, 1902, p. 641, pl. xxv, f. 3–5.—Perkins, Vt. State Geol., Rept. 1903–4 [1904], p. 211, f. 3–5 (p. 185).

Carya globulosa Lesquereux, in Knowlton, Bull. Torr. Bot. Club, vol. 29, 1902, p. 640.

Cucumites globulosus (Lesquereux) Cockerell, Science, vol. 17, 1903, p. 189.

Miocene: Brandon, Vt.

CUNNINGHAMITES Presl, in Sternberg, Vers., vol. 2, 1833, p. 203.

Cunninghamites elegans (Corda) Endlicher, Synop. Conif., 1847, p. 270.—Newberry, U. S. Geol. Surv., Mon. 26, 1895 [1896], p. 48, pl. v, f. 1–7.—Hollick, N. Y. Bot. Gard., Bull., vol. 2, 1902, p. 402, pl. xli, f. 11.—Berry, N. Y. Bot. Gard., Bull., vol. 3, 1903, p. 64; Bull. Torr. Bot. Club, vol. 31, 1904, p. 70, pl. iii, f. 7, 8; Geol. Surv. N. J., Ann. Rept. 1905 [1906], p. 138.—Knowlton, U. S. Geol.

Cunninghamites elegans —Continued.
Surv., Bull. 257, 1905, p. 135, pl. xv, f. 1.—Hollick, U. S. Geol. Surv., Mon. 50, 1906, p. 41, pl. iii, f. 1.—Berry, Bull. Torr. Bot. Club, vol. 37, 1910, pp. 186, 505, pl. xx, f. 1–4; U. S. Geol. Surv., Prof. Paper 84, 1914, pp. 24, 106; idem, Prof. Paper 112, 1919, p. 58.
Raritan: South Amboy and Keyport, N. J.
Magothy: Cliffwood, N. J.; Chappaquiddick Island, Mass.
Judith River: Willow Creek, Fergus County, Mont.
Black Creek: Black River, N. C.; Rocky Point, S. C.
Ripley (Cusseta): Byron, Ga.

Cunninghamites recurvatus? Hosius and Von der Marck. Knowlton, U. S. Geol. Surv., Bull. 257, p. 136, pl. xvi, f. 6.
Judith River: Willow Creek, Fergus County, Mont.

Cunninghamites sphenolepis F. Braun = *Palissya sphenolepis.*

Cunninghamites squamosus Heer, Kreide Fl. v. Quedlinburg, 1872, p. 9, pl. i, f. 5–7.—Hollick, N. Y. Acad. Sci., Trans., vol. 16, 1897, p. 129, pl. xi, f. 3.—Berry, N. Y. Bot. Gard., Bull., vol. 3, 1903, p. 64, pl. xlviii, f. 14, 19; Geol. Surv. N. J., Ann. Rept. 1905 [1906], p. 138; Bull. Torr. Bot. Club, vol. 33, 1906, p. 165.
Magothy: Morgan and Cliffwood, N. J.

Cunninghamites? sp. Knowlton, U. S. Geol. Surv., Bull. 162, 1900, p. 29, pl. v, f. 3.
Mesaverde: North Fork Dutton Creek, Laramie Plains, Wyo.

CUPANITES Schimper, Pal. vég., vol. 3, 1874, p. 170.

Cupanites eoligniticus Berry, U. S. Geol. Surv., Prof. Paper 91, 1916, p. 269, pl. lxiv, f. 8, 9; pl. lxv, f. 1–3.
Lagrange: Puryear, Henry County, Tenn.
Wilcox: Near Naborton, De Soto Parish, La.

Cupanites loughridgii Berry, U. S. Geol. Surv., Prof. Paper 91, 1916, p. 269, pl. lxv, f. 4.
Myrica copeana Lesquereux. Lesquereux, U. S. Nat. Mus., Proc., vol. 11, 1888, p. 12.
Lagrange: Wickliffe, Ballard County, Ky.

Cupanites nigricans Berry, in Cooke and Shearer, U. S. Geol. Surv., Prof. Paper 120, 1918, p. 55 [name].
Barnwell (Twiggs): Twiggs County, Ga.

CUPRESSINOXYLON Göppert, Monogr. d. foss. Conif., 1850, p. 196.

Cupressinoxylon arkansanum Knowlton, Geol. Surv. Ark., Ann. Rept. 1889, vol. 2, 1889, p. 253, pl. ix, f. 1, 2.—Penhallow, Man. N. A. Gym., 1907, p. 240.
Wilcox?: Near Wittsburg, Cross County, Ark.

Cupressinoxylon? bibbinsi Knowlton, Science, new ser., vol. 3, No. 68, 1896, p. 584, text f. 1–4.—Berry, Md. Geol. Surv., Upper Cret., 1916, p. 791.

Magothy: Cape Sable, Md.

Cupressinoxylon calli Knowlton, Geol. Surv. Ark., Ann. Rept. 1889, vol. 2, 1889, p. 254, pl. ix, f. 3–7.—Penhallow, Man. N. A. Gym., 1907, p. 244.—Berry, U. S. Geol. Surv., Prof. Paper 91, 1916, p. 174, pl. xvi, f. 1–5.

Wilcox: 5½ miles northwest of Gainesville, Green County, Ark.

Cupressinoxylon cheyennense Penhallow, Roy. Soc. Canada, Trans., 2d ser., vol. 6, 1900 [1901], sec. 4, p. 76; Man. N. A. Gym., 1907, p. 238.

Cheyenne: East of Stokes Hill, Kiowa and Baker county line, Kans.

Cupressinoxylon coloradense Knowlton, U. S. Geol. Surv., Prof. Paper 101, 1918, p. 252, pl. xxxiii, f. 1–4; pl. xxxiv, f. 1; pl. xxxv, f. 1–6.

Vermejo: Near Trinidad and Newland Creek, Canon City field, Colo.

Cupressinoxylon columbianum Knowlton, Am. Geol., vol. 3, 1889, p. 106 [name]; U. S. Geol. Surv., Bull. 56, 1889, p. 49, pl. iv, f. 5; pl. v, f. 3, 4; pl. vi, f. 1–5; pl. vii, f. 1.—Penhallow, Man. N. A. Gym., 1907, p. 241.

Patuxent: Dutch Gap Canal, James River, Va.

Cupressinoxylon comanchense Penhallow, Roy. Soc. Canada, Trans., 2d ser., vol. 6, 1900 [1901], sec. 4, p. 77; Man. N. A. Gym., 1907, p. 239.

Comanche: Northwest of Ashland, Clark County, Kans.

Cupress[in]oxylon dawsoni Penhallow, Roy. Soc. Canada, Trans., 2d ser., vol. 9, 1903, sec. 4, p. 46, f. 9–11; idem, vol. 10, 1904, sec. 4, p. 60; idem, 3d ser., vol. 1, 1907, sec. 4, p. 95; Man. N. A. Gym., 1907, p. 240; Rept. Tert. Pl. Brit. Col., 1908, p. 50, text f. 9, 10.

Cupressoxylon sp. *b*? Dawson, Brit. N. A. Bound. Com. (Rept. Geol. and Res. Vic. 49th Parallel), 1875, App. A, p. 331.

Cupressoxylon sp. *e*? Dawson, op. cit., p. 331, pl. xv, f. 5.

Yegua: Sommerville, Burleson County, Tex.

Paskapoo (Fort Union): Porcupine Creek and Great Valley, S. Dak.

Cretaceous: Medicine Hat, Alberta.

Cupressinoxylon distichum Mercklin, Palaeodendr. Ross., 1856, p. 64, pl. xix, f. 1–6.—Platen, Sitzungsb. Naturf. Ges., Leipzig, vol. 34, 1907 [1908], p. 38.

——?: California.

Cupressinoxylon dubium Cramer, in Heer, Fl. foss. arct., vol. 1, 1868, p. 173, pl. xxxiv, f. 3; pl. xxxviii, f. 1–6.
Miocene: Banks Land, Baring Island.

Cupressinoxylon elongatum Knowlton, U. S. Nat. Mus., Proc., vol. 11, 1888, p. 7, pl. iii, f. 1–4.—Penhallow, Man. N. A. Gym., 1907, p. 242.
Fort Union: Near Glendive, Mont.

Cupressinoxylon erraticum Mercklin, Palaeodendr. Ross., St. Petersburg, 1856, p. 60, pl. xiv.—Felix, Zeitschr. d. Deutsch. geol. Gesell., vol. 38, 1886, p. 484.—Knowlton, U. S. Nat. Mus., Proc., vol. 17, 1894, p. 215; Geol. Soc. Am., Bull., vol. 5, 1893, p. 579.
——: Copper Island, Alaska.

Cupressinoxylon eutreron Felix, Zeitschr. Deutsch. d. geol. Gesell., Jahr. 1896, p. 255.
Miocene: Yancey's, Yellowstone National Park.

Cupressinoxylon glasgowi Knowlton, U. S. Nat. Mus., Proc., vol. 11, 1888, p. 6, pl. ii, f. 1–5.—Penhallow, Man. N. A. Gym., 1907, p. 242.
Tertiary: Emmet County, Iowa.

Cupress[in]oxylon macrocarpoides Penhallow, Roy. Soc. Canada, Trans., 2d ser., vol. 10, 1904, sec. 4, p. 59; idem, 3d ser., vol. 1, 1907, sec. 4, p. 296; Man. N. A. Gym., 1907, p. 238; Rept. Tert. Pl. Brit. Col., 1908, p. 51; Ottawa Nat., vol. 22, 1908, p. 82.
——: Twentymile Creek near Medicine Hat, Alberta.
Oligocene: Kettle River, British Columbia.
Oligocene?: Red Deer River, Alberta.

Cupressinoxylon mcgeei Knowlton = *Podocarpoxylon mcgeei.*

Cupress[in]oxylon pannonicum (Unger) Felix, Jahrb. K. ung. geol. Anst., vol. 7, 1884, p. 36.—Platen, Sitzungsb. Naturf. Ges., Leipzig, vol. 34, 1907 [1908], p. 95.
——?: Nevada [no locality].

Cupressinoxylon polyommatum Cramer, in Heer, Foss. fl. arct., vol. 1, 1868, p. 172, pl. xxxiv, f. 2a, 6; pl. xxv, f. 2, 3; pl. xxxvii, f. 1–7.
Miocene: Banks Land.

Cupressinoxylon pulchellum Knowlton, Am. Geol., vol. 3, 1889, p. 106 [name]; U. S. Geol. Surv., Bull. 56, 1889, p. 45, pl. ii, f. 1–4.—Penhallow, Man. N. A. Gym., 1907, p. 239.
Patuxent: Spring Hill, Appomattox River, Va.

Cupressinoxylon pulchrum Cramer, in Heer, Fl. foss. arct., vol. 1, 1868, p. 171, pl. xxxiv, f. 1; pl. xxxvi, f. 6–8.
Miocene: Ballast Bay, Banks Land.

Cupressinoxylon taxodioides Conwentz, Schriften Naturf. Gesell. Danzig, vol. 4, 1878, p. 124; Neues Jahrb., 1878, p. 812, pls. xii, xiv.—Platen, Sitzsungsb. Naturf. Ges., Leipzig, vol. 34, 1907 [1908], p. 4, pl. i, f. 1–4.

Tertiary: *Calistoga, Calif.

——?: Nevada [no locality].

Cupressinoxylon? vermejoense Knowlton, U. S. Geol. Surv., Prof. Paper 101, 1918, p. 253, pl. xxxiv, f. 2, 3; pl. xxxv, f. 7.

Vermejo: Canon City field, Colo.

Cupressinoxylon wardi Knowlton, Am. Geol., vol. 3, 1889, p. 106 [name]; U. S. Geol. Surv., Bull. 56, 1889, p. 48, pl. iv, f. 1–4; pl. v, f. 1, 2.—Penhallow, Man. N. A. Gym., 1907, p. 241.—Berry, Md. Geol. Surv., Lower Cret., 1912, p. 415, pl. lxviii, f. 1–6; pl. lxix, f. 7, 8.

Patuxent: *Near Montello, D. C.; Neabsco Creek, Va.; Clifton (Baltimore), Md.

Cupressinoxylon sp., Berry, U. S. Geol. Surv., Prof. Paper 112, 1919, p. 70.

Eutaw: Coffee Bluff, Hardin County, Tenn.

Cupressoxylon sp. *a* Dawson, Brit. N. A. Bound. Com. (Rept. Geol. and Rec. Vic. 49th parallel), 1875, App. A, p. 331, pl. xv, f. 3 = *Sequoia burgessii.*

Cupressoxylon sp. *b? e?* Dawson, op. cit., p. 331, pl. xv, f. 5 = *Cupress*[*in*]*oxylon dawsoni.*

Cupressoxylon sp. *c* Dawson, op. cit., p. 331, pl. xv, f. 4 = *Taxodium distichum.*

Cupress[in]oxylon sp. *d*, Dawson, op. cit., p. 331.

Paskapoo (?): Vicinity of 49th parallel, Alberta.

Cupressoxylon sp. *f*, Dawson, op. cit., p. 331 = *Pseudotsuga miocena.*

Cupressinoxylon sp. Knowlton, in Turner, Geol. Soc. Am., Bull., vol. 2, 1891, p. 394.

Knoxville: Eagle Point, Calif.

Cupress[in]oxylon sp., Dawson, Geol. Surv. Canada, Rept. Progress, 1872–73 [1873], App. i, p. 67.—Penhallow, Rept. Tert. Pl. Brit. Col., 1908, p. 49.

Cretaceous ?: Edmonton, Alberta.

Cupressinoxylon sp., Hollick and Jeffrey, N. Y. Bot. Gard., Mem., vol. 3, 1909, p. 65, pl. xx, f. 2.

Raritan: Kreischerville, Staten Island, N. Y.

CUPRESSITES Brongniart, Dict. sci. nat., vol. 57 (Prodrome), 1828, p. 112 (109).

Cupressites cookii Newberry, Lyc. Nat. Hist. N. Y., Annals, vol. 9, 1868, p. 9; U. S. Geol. Surv., Mon. 26, 1895 [1896], p. 61 [name].
Raritan (?): New Jersey.

CUPRESSUS Linné, Sp. pl., 1753, p. 1002.

Cupressus macrocarpa Hartweg. Knowlton, Washington Acad. Sci., Jour., vol. 6, 1916, p. 85.
Pleistocene: Rancho La Brea, Los Angeles, Calif.

Cupressus thyoides Linné = *Chamaecyparis thyoides.*

CYATHEITES Göppert, Syst. Fil. Foss., 1836, p. 319.

Cyatheites? nebraskana (Heer) Knowlton, U. S. Geol. Surv., Bull. 152, 1898, p. 81.
Pecopteris nebraskana Heer. Saporta, Fl. Foss. d. Sézanne, 1868, p. 332, text f. 8.—Lesquereux, Rept. U. S. Geol. Surv. Terr., vol. 6 (Cret. Fl.), 1874, p. 46, pl. xxix, f. 5, 5a.
Dakota: Kansas.

CYCADELLA Ward, Washington Acad. Sci., Proc., vol. 1, 1900, p. 263. [Type, *C. knowltoniana.*]

Cycadella beecheriana Ward, Washington Acad. Sci., Proc., vol. 1, 1900, p. 265, pl. xvi; U. S. Geol. Surv., Twentieth Ann. Rept., pt. 2, 1900, p. 394, pls. lxxvii, lxxviii; U. S. Geol. Surv., Mon. 48, 1905 [1906], p. 183.
Morrison: Freezeout Hills, Carbon County, Wyo.

Cycadella carbonensis Ward, Washington Acad. Sci., Proc., vol. 1, 1900, p. 282; U. S. Geol. Surv., Twentieth Ann. Rept., pt. 2, 1900, p. 415, pls. clxx–clxxi.
Morrison: Freezeout Hills, Carbon County, Wyo.

Cyadella cirrata Ward, Washington Acad. Sci., Proc., vol. 1, 1900, p. 272; U. S. Geol. Surv., Twentieth Ann. Rept., pt. 2, 1900, p. 403, pls. cxxiii–cxxix.
Morrison: Freezeout Hills, Carbon County, Wyo.

Cycadella compressa Ward, Washington Acad. Sci., Proc., vol. 1, 1900, p. 269; U. S. Geol. Surv., Twentieth Ann. Rept., pt. 2, 1900, p. 398, pls. xcvi, pl. xcvii.
Morrison: Freezeout Hills, Carbon County, Wyo.

Cycadella concinna Ward, Washington Acad. Sci., Proc., vol. 1, 1900, p. 280; U. S. Geol. Surv., Twentieth Ann. Rept., pt. 2, 1900, p. 412, pl. clxii.
Morrison: Freezeout Hills, Carbon County, Wyo.

Cycadella contracta Ward, Washington Acad. Sci., Proc., vol. 1, 1900, p. 277; U. S. Geol. Surv., Twentieth Ann. Rept., pt. 2, 1900, p. 409, pls. cxlviii–cliii.
Morrison: Freezeout Hills, Carbon County, Wyo.

Cycadella crepidaria Ward, Washington Acad. Sci., Proc., vol. 1, 1900, p. 280; U. S. Geol. Surv., Twentieth Ann. Rept., pt. 2, 1900, p. 413, pls. clxiii–clxiv.
Morrison: Freezeout Hills, Carbon County, Wyo.

Cycadella exogena Ward, Washington Acad. Sci., Proc., vol. 1, 1900, p. 273; U. S. Geol. Surv., Twentieth Ann. Rept., pt. 2, 1900, p. 404, pls. cxxx–cxxxvii.
Morrison: Freezeout Hills, Carbon County, Wyo.

Cycadella ferruginea Ward, Washington Acad. Sci., Proc., vol. 1, 1900, p. 276; U. S. Geol. Surv., Twentieth Ann. Rept., pt. 2, 1900, p. 408, pls. cxlv–cxlvii.
Morrison: Freezeout Hills, Carbon County, Wyo.

Cycadella gelida Ward, Washington Acad. Sci., Proc., vol. 1, 1900, p. 281; U. S. Geol. Surv., Twentieth Ann. Rept., pt. 2, 1900, p. 414, pls. clxv–clxix.
Morrison: Freezeout Hills, Carbon County, Wyo.

Cycadella gravis Ward, Washington Acad. Sci., Proc., vol. 1, 1900, p. 277; U. S. Geol. Surv., Twentieth Ann. Rept., pt. 2, 1900, p. 410, pl. cliv.
Morrison: Freezeout Hills, Carbon County, Wyo.

Cycadella jejuna Ward, Washington Acad. Sci., Proc., vol. 1, 1900, p. 279; U. S. Geol. Surv., Twentieth Ann. Rept., pt. 2, 1900, p. 412, pls. clviii–clxi.
Morrison: Freezeout Hills, Carbon County, Wyo.

Cycadella jurassica Ward, Washington Acad. Sci., Proc., vol. 1, 1900, p. 270; U. S. Geol. Surv., Twentieth Ann. Rept., pt. 2, 1900, p. 399, pls. xcviii–cxii.
Morrison: Freezeout Hills, Carbon County, Wyo.

Cycadella knightii Ward, Washington Acad. Sci., Proc., vol. 1, 1900, p. 283, pl. xxi; U. S. Geol. Surv., Twentieth Ann. Rept., pt. 2, 1900, p. 416, pls. clxxii–clxxvii.
Morrison: Freezeout Hills, Carbon County, Wyo.

Cycadella knowltoniana Ward, Washington Acad. Sci., Proc., vol. 1, 1900, p. 267, pls. xviii–xx; U. S. Geol. Surv., Twentieth Ann. Rept., pt. 2, 1900, p. 396, pl. lxx, f. 1–3; pls. xci–xcv.
Morrison: Freezeout Hills, Carbon County, Wyo.

Cycadella nodosa Ward, Washington Acad. Sci., Proc., vol. 1, 1900, p. 271; U. S. Geol. Surv., Twentieth Ann. Rept., pt. 2, 1900, p. 401, pls. cxiii–cxxii.
Morrison: Freezeout Hills, Carbon County, Wyo.

Cycadella ramentosa Ward, Washington Acad. Sci., Proc., vol. 1, 1900, p. 275; U. S. Geol. Surv., Twentieth Ann. Rept., pt. 2, 1900, p. 406, pl. lxx, f. 4, 5; pls. cxxxviii–cxliv.
Morrison: Freezeout Hills, Carbon County, Wyo.

Cycadella reedii Ward, Washington Acad. Sci., Proc., vol. 1, 1900, p. 264, pl. xv; U. S. Geol. Surv., Twentieth Ann. Rept., pt. 2, 1900, p. 392, pl. lxx; U. S. Geol. Surv., Mon. 48, 1905 [1906], p. 182, pl. xlvii, f. 3.—Wieland, Am. fossil cycads, vol. 2, 1916, pl. xxviii, f. 6.
Morrison: Freezeout Hills, Carbon County, Wyo.

Cycadella utopiensis (Ward) Wieland, in Ward, U. S. Geol. Surv., Mon. 48, 1905 [1906], p. 204, pl. lxiii, f. 2.
Cycadeoidea utopiensis Ward, Am. Jour. Sci., 4th ser., vol. 10, 1900, p. 338, pl. iii, upper figure.
Morrison: "50 miles west of Hot Springs [S. Dak.], in Wyoming." Exact locality thought to be near head of Skull Creek, 4 miles south of Inyankara Mountain, Crook County, Wyo.

Cycadella verrucosa Ward, Washington Acad. Sci., Proc., vol. 1, 1900, p. 278; U. S. Geol. Surv., Twentieth Ann. Rept., pt. 2, 1900, p. 410, pls. clv–clvii.
Morrison: Freezeout Hills, Carbon County, Wyo.

Cycadella wyomingensis Ward, Washington Acad. Sci., Proc., vol. 1, 1900, p. 266, pl. xvii, U. S. Geol. Surv., Twentieth Ann. Rept., pt. 2, 1900, p. 395, pls. lxxix–xc.—Wieland, Am. fossil cycads, vol. 2, 1916, pl. xxviii, f. 5.
Morrison: Freezeout Hills, Carbon County, Wyo.

Cycadeocarpus (*Dioonites*) *columbianus* Dawson = *Dioonites columbianus.*

CYCADEOIDEA Buckland, Geol. Soc. London, Proc., vol. 1, 1827, p. 80.

Cycadeoidea aspera Ward, U. S. Nat. Mus., Proc., vol. 21, 1898, p. 213; U. S. Geol. Surv., Nineteenth Ann. Rept., pt. 2, 1899, p. 624, pl. cxvii.
Lakota: Blackhawk region, S. Dak.

Cycadeoidea bibbinsi[1] Ward, Biol. Soc. Washington, Proc., vol. 11, 1897, p. 15; U. S. Geol. Surv., Mon. 48, 1905 [1906], p. 456, pl. lxxxiii, f. 3; pl. lxxxiv, f. 3; pl. lxxxv; pl. lxxxvii, f. I, 2, 4, 6, 7; II, 1, 2, 4, 8, 9, 10, 11; III, 5, 7; IV, 2, 3, 6, 8, 11; V, 12, 15; pl. lxxxix, f. I, 5; II, 2, 5, 6; pls. ci–civ.—Berry, Md. Geol. Surv., Lower Cret, 1911, p. 327, pls. xlviii.—Wieland, Am. fossil cycads, vol. 2, 1916, pl. ii.
Patuxent?: Maryland.

[1] The Maryland cycad trunks (*Cycadeoidea*)—over 100 in number—came from a narrow belt of the Potomac group, mainly between Baltimore and the District of Columbia. With possibly a single exception none were found in place, all having been redeposited in beds ranging in age from Patuxent to Raritan. It is to be presumed, however, that they were laid down originally in the Patuxent, and in the following pages the age is recorded as Patuxent?, and the locality as Maryland, as it is neither possible nor perhaps, necessary in the present connection to give the precise localities. Cf. Ward, U. S. Geol. Surv., Mon. 48 1905 [1906], pl. lxxx.

Cycadeoidea cicatricula Ward, U. S. Nat. Mus., Proc., vol. 21, 1898, p. 203; U. S. Geol. Surv., Nineteenth Ann. Rept., pt. 2, 1899, p. 609, pls. xxxiii, xxxiv.—Wieland, Am. fossil cycads, vol. 2, 1916, p. 48.

Cycadeoidea rhombica Ward, Am. Jour. Sci., 4th ser., vol. 10, 1900, p. 336, pl. ii, upper figure.—Wieland, Am. fossil cycads, vol. 1, 1906, pl. v, f. 4.

Lakota: Blackhawk, S. Dak.

Cycadeoidea clarkiana Ward, U. S. Geol. Surv., Mon. 48, 1905 [1906], p. 472, pl. lxxxix, f. I, 2, 4; pl. cvi.—Berry, Md. Geol. Surv., Lower Cret., 1911, p. 328, pl. xlix.

Patuxent?: Maryland.

Cycadeoidea colei Ward, U. S. Nat. Mus., Proc., vol. 21, 1898, p. 211; U. S. Geol. Surv., Nineteenth Ann. Rept., pt. 2, 1899, p. 619, pls. cx–cxii.

Lakota: Minnekahta region, S. Dak.

Cycadeoidea colossalis Ward, U. S. Nat. Mus., Proc., vol. 21, 1898, p. 197; U. S. Geol. Surv., Nineteenth Ann. Rept., pt. 2, 1899, p. 603, pls. lxvii–lxxii.—Wieland, Am. fossil cycads, vol. 1, 1906, pp. 7, 52, 98, 99, 104, pl. v, f. 3; Am. Jour. Sci., 4th ser., vol. 38, 1914, p. 121; Am. fossil cycads, vol. 2, 1916, p. 67, f. 23–25, 27, 28, pl. lvii.

Lakota: Minnekahta region, S. Dak.

Cycadeoidea dacotensis (McBride) Ward, emend, U. S. Geol. Surv., Nineteenth Ann. Rept., pt. 2, 1899, p. 602, pls. lxii–lxvi.—Wieland, Am. Jour. Sci., 4th ser., vol. 11, 1901, pp. 428–432, f. 1–3; Yale Sci. Monthly, vol. 6, 1900, p. 7, f. 8.—Ward, Am. Jour. Sci., 4th ser., vol. 10, 1900, p. 332.—Wieland, Am. fossil cycads, vol. 1, 1906, pp. 133, 144–162, 176–183, 184, f. 14, 28, 30, 34, 70–75, 80–82, 94–100, pl. v, f. 2; pl. vi, f. 8; pl. xx, f. 1–4; pl. xxxiv–xlii; Am. Jour. Sci., 4th ser., vol. 25, 1908, p. 94; idem, vol. 33, 1909, pp. 74, 75; idem, vol. 38, 1914, pp. 121, 122; Am. fossil cycads, vol. 2, 1916, p. 52, f. 1, 4, 15–26, 47; pls. xiv, xv, xvi, xvii, xviii, xix, xx, f. 1–3a; pls. lv, lvi, lviii, f. 1.

Bennettites dacotensis McBride, in part, Am. Geol., vol. 12, 1893, p. 249, pl. xi, f. 1 [not f. 2]; State Univ. Iowa, Bull. Lab. Nat. Hist., vol. 2, 1893, p. 391, pl. xii, f. 1 [not f. 2].

Cycadeoidea dacotensis (McBride) Ward, in part, Biol. Soc. Washington, Proc., vol. 9, 1894, p. 86.

Lakota: Minnekahta and Blackhawk regions, and 2 miles west of Sturgis, S. Dak.

Cycadeoidea dartoni Wieland, Am. fossil cycads, vol. 2, 1916, p. 95, f. 36, 37, 46, 50A, 53A; pl. xxxiii, f. 3; pl. xxxiv, f. 1, 2; pl. xxxv, f. 6; pl. xxxvi, f. 2, 3; pls. xxxviii, l, lii, f. 2; pl. liii, f. 1, 2.

Lakota: 4 miles west of Hermosa, S. Dak.

Cycadeoidea emmonsi (Fontaine) Ward, Biol. Soc. Washington, Proc., vol. 9, 1894, p. 86; U. S. Geol. Surv., Twentieth Ann. Rept. pt. 2, 1900, p. 302, pl. xliii, f. 3.
Zamiostrobus emmonsi Fontaine, U. S. Geol. Surv., Mon. 6, 1883, p. 117, pl. lii, f. 5.
Triassic (Newark): North Carolina.

Cycadeoidea excelsa Ward, U. S. Nat. Mus., Proc., vol. 21, 1898, p. 225; U. S. Geol. Surv., Nineteenth Ann. Rept., pt. 2, 1899, p. 637, pls. cliii–clv.—Wieland, Am. fossil cycads, vol. 2, 1916, p. 88.
Lakota: 2 miles below Hot Springs, S. Dak.

Cycadeoidea fisherae Ward, U. S. Geol. Surv., Mon. 48, 1905 [1906], p. 470, pl. lxxxvii, f. III, 9; pl. cv.—Berry, Md. Geol. Surv., Lower Cret., 1911, p. 329, pl. l.
Patuxent? Maryland.

Cycadeoidea fontaineana Ward, Biol. Soc. Washington, Proc., vol. 11, 1897, p. 13; U. S. Geol. Surv., Mon. 48, 1905 [1906], p. 439, pl. lxxxvi; pl. lxxxvii, f. I, 1; III, 2, 6, 8; IV, 1, 7, 9; V, 1, 10, 11, 13, 14, 16, 18, 21; pl. lxxxix, f. I, 1; III, 1, 3, 6, 7, 11; pls. xcv–xcviii.—Berry, Md. Geol. Surv., Lower Cret., 1911, p. 324, pl. xlv.
Patuxent?: Maryland.

Cycadeoidea formosa Ward, U. S. Nat. Mus., Proc., vol. 21, 1898, p. 222; U. S. Geol. Surv., Nineteenth Ann. Rept., pt. 2, 1899, p. 634, pls. cxliv–cxlvi.—Wieland, Am. fossil cycads, vol. 1, 1906, pls. vi, xiv; idem, vol. 2, 1916, p. 47.
Lakota: Blackhawk region, S. Dak.

Cycadeoidea furcata Ward, U. S. Nat. Mus., Proc., vol. 21, 1898, p. 210; U. S. Geol. Surv., Nineteenth Ann. Rept., pt. 2, 1899, p. 618, pls. cvi–cix.
Lakota: Minnekahta region, S. Dak.

Cycadeoidea goucheriana Ward, Biol. Soc. Washington, Proc., vol. 11, 1897, p. 14; U. S. Geol. Surv., Mon. 48, 1905 [1906], p. 451, pl. lxxxvii, f. I, 3; pl. lxxxix, f. I, 3; pl. xcix.—Berry, Md. Geol. Surv., Lower Cret., 1911, p. 325, pl. xlvi.
Patuxent (?): Maryland.

Cycadeoidea heliochorea Ward, Am. Jour. Sci., 4th ser., vol. 10, 1900, p. 337, pl. iv, f. 722.
Lakota: 10 miles west of Sundance, Wyo.

Cycadeoidea ingens Ward, U. S. Nat. Mus., Proc., vol. 21, 1898, p. 221.—Wieland, Am. Jour. Sci., 4th ser., vol. 7, 1899, pp. 219–226, pls. ii–iv; idem, pp. 305–308, pl. vii.—Ward, U. S. Geol. Surv., Nineteenth Ann. Rept., pt. 2, 1899, p. 632, pls. cxxxiii–cxliii; Am. Jour. Sci., 4th ser., vol. 10, 1900, p. 352.—Wieland, Am. Jour. Sci., 4th ser., vol. 11, 1900, p. 423.—Am. fossil cycads, vol. 1, 1906, pls. i–ivA, text f. 19, 44, 45, 49, 51, 54; Am. Jour. Sci., 4th ser., vol. 38, 1914, p. 454; Am. fossil cycads, vol. 2, 1916, p. 42, text f. 12, 13, pl. v, f. 5, 6; pl. vii, f. 2; pls. ix, x, xii; pl. lviii, f. 5.
Lakota: Blackhawk region, S. Dak.

Cycadeoidea insoleta Ward, U. S. Nat. Mus., Proc., vol. 21, 1898, p. 214; U. S. Geol. Surv., Nineteenth Ann. Rept., pt. 2, 1899, p. 625, pls. cxviii, cxix.
Lakota: Minnekahta region, S. Dak.

Cycadeoidea jenneyana Ward, Biol. Soc. Washington, Proc., vol. 9, 1894, p. 87; U. S. Nat. Mus., Proc., vol. 21, 1898, p. 216; U. S. Geol. Surv., Nineteenth Ann. Rept., pt. 2, 1899, p. 627, pls. cxxi–cxxxii.—Wieland, Am. fossil cycads, vol. 1, 1906, pp. 78–80, 131, 132; pl. xiv, f. 65; idem, vol. 2, 1916, p. 36, f. 9–11, pl. vi, f. 1–3; pl. vii, f. 1; pl. viii; pls. x, xi.
Lakota: Minnekahta and Blackhawk regions, S. Dak.

Cycadeoidea marshiana Ward, U. S. Nat. Mus., Proc., vol. 21, 1898, p. 208; U. S. Geol. Surv., Nineteenth Ann. Rept., pt. 2, 1899, p. 616, pls. ci–cv.—Wieland, Yale Sci. Monthly, vol. 6, 1900, pp. 1–11, f. 7, 9, 13; Am. Jour. Sci., 4th ser., vol. 10, 1900, p. 352; Am. fossil cycads, vol. 1, 1906, pp. 135, 136, f. 67, 67a, pl. v, f. 1; pls. vii, viii, ix, f. 1; pls. xii, xxxiii; Am. Jour. Sci., 4th ser., vol. 33, 1912, p. 75–87, f. 1–9; Am. fossil cycads, vol. 2, 1916, p. 56, f. 16–19; pls. xxii–xxiv, xxv, f. 1–3, 5; pls. xxvi, xxvii.
Lakota: Minnekahta region, S. Dak.

Cycadeoidea marylandica (Fontaine) Capillini and Solms-Laubach, Mem. Real. Accad. Sci. Ist. Bologna, 5th ser., vol. 2, 1892, pp. 179, 180, 186.—Ward, Biol. Soc. Washington, Proc., vol. 9, 1894, p. 86; idem, vol. 11, 1897, p. 9; U. S. Geol. Surv., Mon. 48, 1905 [1906], p. 416, pls. lxxxi, lxxxii; pl. lxxxiii, f. 1, 2, 4; pl. lxxiv—f. 1, 2; pl. lxxxvii, f. II, 3, 5, 6, 7; III, 1, 4; IV, 4, 5, 12–14; V, 2, 3–7, 17; pl. lxxxviii; pl. lxxxix, f. II, 3, 4, 7; III, 2, 5, 8; pls. xc–xcii.—Wieland, Am. fossil cycads, vol. 1, 1906, f. 1.—Berry, Md. Geol. Surv., Lower Cret., 1911, p. 320, pl. xlii.—Wieland, Am. fossil cycads, vol. 2, 1916, pl. i, f. 3, 4.

Cycas sp., Tyson, First Rept. State Agr. Chem. Maryland, 1860, p. 42.
Bennettites sp. Carruthers, Linn. Soc. London, Trans., vol. 26, 1870, p. 708.
Cycadoidea sp. Fontaine, Am. Jour. Sci., 3d ser., vol. 17, 1879, p. 157.
Tysonia marylandica Fontaine, U. S. Geol. Surv., Mon. 15, 1889 [1890], p. 193, pl. clxxiv–clxxx.

Patuxent: Link Gulch, near Arbutus, Baltimore County, Md.

Cycadeoidea mcbridei Ward, U. S. Nat. Mus., Proc., vol. 21, 1898, p. 205; U. S. Nat. Mus., Nineteenth Ann. Rept., pt. 2, 1899, p. 612, pls. xci–c.—Wieland, Am. fossil cycads, vol. 1, 1906, p. 65, f. 33, pl. vi; idem, vol. 2, 1916, p. 65, f. 22.

Bennettites dacotensis McBride, in part, Am. Geol., vol. 12, 1893, p. 249, pl. xi, f. 2 [not f. 1]; State Univ. Iowa, Bull. Lab. Nat. Hist., vol. 2, 1893, p. 391, pl. xii, f. 2 [not f. 1].

Lakota: Minnekahta region, S. Dak.

Cycadeoidea mcgeeana Ward, Biol. Soc. Washington, Proc., vol. 11, 1897, p. 12; U. S. Geol. Surv., Mon. 48, 1905 [1906], p. 434, pl. lxxxvii, f. III, 3, 10; IV, 15; V, 8, 9, 19, 20; pl. lxxxix, f. II, 1, 8; III, 4, 9, 10; pl. xciv.—Berry, Md. Geol. Surv., Lower Cret., 1911, p. 323, pl. xliv.

Patuxent ?: Maryland.

Cycadeoidea minima Ward, Am. Jour. Sci., 4th ser., vol. 10, 1900, p. 341, pl. ii, f. 53.

Lakota: Minnekahta region, S. Dak.

Cycadeoidea minnekahtensis Ward, U. S. Nat. Mus., Proc., vol. 21, 1898, p. 200; U. S. Geol. Surv., Nineteenth Ann. Rept., pt. 2, 1899, p. 606, pls. lxxvi–lxxix.

Lakota: Minnekahta region, S. Dak.

Cycadeoidea mirabilis (Lesquereux) Ward, Biol. Soc. Washington, Proc., vol. 9, 1894, p. 86.—Wieland, Am. fossil cycads, vol. 2, 1916, p. 109; pl. i, f. 1, 2.—Knowlton, U. S. Geol. Surv., Prof. Paper 130, 19—, p. — [in preparation].

Zamiostrobus? mirabilis Lesquereux, U. S. Geol. and Geog. Surv. Terr., Bull., vol. 1, 1875 [1876], p. 383; Rept. U. S. Geol. Surv Terr., vol. 7 (Tert. Fl.), 1878, p. 70, pl. lxiii, f. 1.

Nelumbium James, Science, vol. 3, 1884, p. 434.

Clathropodium mirabile (Lesquereux) Ward, Science, vol. 3, 1884, p. 532.

Cycadeoidea zamiostrobus Solms, Mem. Real. Accad. Sci. Ist. Bologna, 5th ser., vol. 2, 1892, p. 210.

Laramie ?: Golden, Colo.

Cycadeoidea munita Cragin, Bull. Washburn College Lab. Nat. Hist., Topeka, Kans., vol. 2, 1889, p. 65.—Ward, U. S. Geol. Surv., Nineteenth Ann. Rept., pt. 2, 1899, p. 541.

Cheyenne: Belvidere, Kans.

Cycadeoidea nana Ward, U. S. Nat. Mus., Proc., vol. 21, 1898, p. 227; U. S. Geol. Surv., Nineteenth Ann. Rept., pt. 2, 1899, p. 639, pls. clvi–clvii.—Wieland, Am. Jour. Sci., 4th ser., vol. 33, 1912, p. 87–89, f. 10; Am. fossil cycads, vol. 2, 1916, p. 76; pl. xxx, f. 2–4; pl. xxxvii.

Lakota: Minnekahta region, S. Dak.

Cycadeoidea nigra Ward, U. S. Geol. Surv., Twentieth Ann. Rept., pt. 2, 1900, p. 378, pl. lxviii; pl. lxix.—Wieland, Am. fossil cycads, vol. 1, 1906, pp. 7, 12, 22, 52, 60–64, f. 31, 33; idem, vol. 2, 1916, p. 111.

Jurassic (?): Boulder, Colo.

Cycadeoidea occidentalis Ward, U. S. Nat. Mus., Proc., vol. 21, 1898, p. 215; U. S. Geol. Surv., Nineteenth Ann. Rept., pt. 2, 1899, p. 626, pl. cxx.

Lakota: Minnekahta region, S. Dak.

Cycadeoidea paynei Ward, U. S. Nat. Mus., Proc., vol. 21, 1898, p. 212; U. S. Geol. Surv., Nineteenth Ann. Rept., pt. 2, 1899, p. 620, pls. cxiii–cxv; Am. Jour. Sci., vol. 10, 1900, p. 333.—Wieland, Am. fossil cycads, vol. 1, 1906, p. 131, 163, 164, 167; idem, vol. 2, 1916, p. 85, f. 35; pl. xxv, f. 4; pl. xxx, f. 1; pl. xxxiii, f. 4; pl. xxxv, f. 3, 4; pl. xxxv, f. 3–5; pl. xxxvi, f. 4, 5.
Lakota: Minnekahta region, S. Dak.

Cycadeoidea paynei Ward [in part] = *Cycadeoidea wielandi*.

Cycadeoidea protea Ward, Am. Jour. Sci., 4th ser., vol. 10, 1900, p. 343, pl. iv, f. 457.—Wieland, Am. fossil cycads, vol. 2, 1916, p. 87.
Lakota: Minnekahta region, S. Dak.

Cycadeoidea pulcherrima Ward, U. S. Nat. Mus., Proc., vol. 21, 1898, p. 201; U. S. Geol. Surv., Nineteenth Ann. Rept., pt. 2, 1899, p. 608, pls. lxxx–lxxxii.—Wieland, Am. fossil cycads, vol. 2, 1916, pl. xxviii, f. 1–4.
Lakota: Minnekahta region, S. Dak.

Cycadeoidea reticulata Ward, Am. Jour. Sci., 4th ser., vol. 10, 1900, p. 340, pl. iv, f. 342.
Lakota: Minnekahta region, S. Dak.

Cycadeoidea rhombica Ward, Am. Jour. Sci., 4th ser., vol. 10, 1900, p. 336, pl. ii (upper fig.).—Wieland, Am. fossil cycads, vol. 1, 1906, p. 262, pl. v, f. 4; idem, vol. 2, 1916, p. 252, pl. v, f. 3, 4.
Lakota: Blackhawk region, S. Dak.

Cycadeoidea stantoni Ward, U. S. Geol. Surv., Mon. 48, 1905 [1906], p. 276, pl. lxx.—Wieland, Am. fossil cycads, vol. 1, 1906, p. 185; idem, vol. 2, 1916, p. 111.
Chico (?): Grapevine Creek, 6 miles west of Sites, Colusa County, Calif.

Cycadeoidea stillwelli Ward, U. S. Nat. Mus., Proc., vol. 21, 1898, p. 223; U. S. Geol. Surv., Nineteenth Ann. Rept., pt. 2, 1899, p. 635, pls. cxlvii–clii.—Wieland, Am. fossil cycads, vol. 1, 1906, pl. vi, f. 4, text f. 35; idem, vol. 2, 1916, p. 48, text f. 73C; pl. v, f. 1, 2.
Lakota: Minnekahta and Blackhawk regions, S. Dak.

Cycadeoidea superba Ward, Am. Jour. Sci., 4th ser., vol. 10, 1900, p. 334; pl. iii, f. 137, 146.
Lakota: Minnekahta region, S. Dak.

Cycadeoidea turrita Ward, U. S. Nat. Mus., Proc., vol. 21, 1898, p. 203; U. S. Geol. Surv., Nineteenth Ann. Rept., pt. 2, 1899, p. 610, pls. xxxv–xc.
Lakota: Minnekahta region, S. Dak.

Cycadeoidea tysoniana Ward, Biol. Soc. Washington, Proc., vol. 11, 1897, p. 11; U. S. Geol. Surv., Mon. 48, 1905 [1906], p. 432, pl. lxxxvii, f. I, 5; f. V, 4; pl. xciii.—Berry, Md. Geol. Surv., Lower Cret., 1911, p. 323, pl. xliii.
Patuxent?: Maryland.

Cycadeoidea uddeni Wieland, Am. fossil cycads, vol. 2, 1916, p. 113, pl. iii, f. 1–3.
Upson: Near Paloma, Tex.

Cycadeoidea uhleri Ward, Biol. Soc. Washington, Proc., vol. 11, 1897, p. 15; U. S. Geol. Surv., Mon. 48, 1905 [1906], p. 454, pl. lxxxvii, f. IV, 10; pl. c.—Berry, Md. Geol. Surv., Lower Cret., 1911, p. 326.
Patuxent?: Maryland.

Cycadeoidea utopiensis Ward = *Cycadella utopiensis.*

Cycadeoidea wellsii Ward, U. S. Nat. Mus., Proc., vol. 21, 1898, p. 199; U. S. Geol. Surv., Nineteenth Ann. Rept., pt. 2, 1899, p. 605, pls. lxxiii–lxxv.
Lakota: Minnekahta region, S. Dak.

Cycadeoidea wielandi Ward, U. S. Geol. Surv., Nineteenth Ann. Rept., pt. 2, 1899, p. 621, pl. cxvi.—Wieland, Am. Jour. Sci., 4th ser., vol. 7, 1899, pp. 384–388, pl. viii, f. 1, 3; pl. ix, f. 8–10; pl. x, f. 13–16; Yale Sci. Monthly, vol. 6, 1900, p. 218, pl. i, f. 6.—Ward, Am. Jour. Sci., vol. 10, 1909, p. 33.—Wieland, Am. fossil cycads, vol. 1, 1906, pp. 108–126, f. 23, 36–42, 58, 61, pl. vi, f. 1, 6; pl. xxi, xxii, xxiii, f. 1, 7; pl. xxiv, f. 1, 3, 5, 6; pl. xxx, f. 4; pl. xxx, f. 1; Am. Jour. Sci., vol. 22, 1910, p. 174, f. 3, 4; Am. fossil cycads, vol. 2, 1916, p. 77, f. 29–34, 51; pl. xx, f. 4, 5; pl. xxix, f. 1; pls. xxxi, xxxii, f: 1–3; pls. li, lii, f. 1.
Cycadeoidea paynei Ward [in part], U. S. Nat. Mus., Proc., vol. 21, 1898, pp. 212, 213.
Lakota: Minnekahta region, S. Dak.

Cycadeoidea zamiostrobus Solms = *Cycadeoidea mirabilis.*

Cycadoidea sp. Fontaine = *Cycadeoidea marylandica.*

CYCADEOMYELON Saporta, Pal. franç. (Pl. jurass., vol. 2), 1875, p. 331.

Cycadeomyelon yorkense Fontaine, in Ward, U. S. Geol. Surv., Twentieth Ann. Rept., pt. 2, 1900, p. 248, pl. xxx.
Palissya? sp., Newberry, U. S. Geol. Surv., Mon. 14, 1888, p. 94, pl. xxvi, f. 1, 2.
Triassic (Newark): Newark, N. J.; York County, Pa.

CYCADEOSPERMUM Saporta, Pal. franç. (Pl. jurass., vol. 2), 1875, p. 235.

Cycadeospermum acutum Fontaine, U. S. Geol. Surv., Mon. 15, 1889 [1890], p. 270, pl. cxxxv, f. 12, in Ward, idem, vol. 48, 1905 [1906], pp. 480, 535.—Berry, Md. Geol. Surv., Lower Cret., 1912, p. 368.
Patuxent: Potomac Run, Va.
Arundel: Muirkirk ?, Md.

Cycadeospermum angustum Fontaine, U. S. Geol. Surv., Mon. 15, 1889 [1890], p. 271, pl. cxxxv, f. 20.
Patuxent: Potomac Run, Va.

Cycadeospermum californicum Fontaine, in Ward, U. S. Geol. Surv., Mon. 48, 1905 [1906], p. 257, pl. lxviii, f. 4.
Knoxville: Near Lowrey, Tehama County, Calif.

Cycadeospermum columnare Lesquereux, U. S. Geol. Surv., Mon. 17, 1891 [1892], p. 31, pl. xliv, f. 7, 8.
Dakota: Ellsworth County, Kans.

Cycadeospermum ellipticum Fontaine = *Cycadeospermum spatulatum*.

Cycadeospermum lineatum Lesquereux, U. S. Geol. Surv., Mon. 17, 1891 [1892], p. 30, pl. i, f. 14.
Dakota: Near Delphos, Kans.

Cycadeospermum marylandicum Berry, Md. Geol. Surv., Lower Cret., 1912, p. 367, pl. lxxvii, f. 8.
Arundel: *Muirkirk, Md.

Cycadeospermum montanense Fontaine, in Ward, U. S. Geol. Surv., Mon. 48, 1905 [1906], p. 310, pl. lxxiii, f. 7.
Kootenai: Near Geyser, Cascade County, Mont.

Cycadeospermum obovatum Fontaine, U. S. Geol. Surv., Mon. 15, 1889 [1890], p. 270, pl. cxxxv, f. 13; in Ward, U. S. Geol. Surv., Mon. 48, 1905 [1906], p. 485, pl. cvii, f. 5.—Berry, Md. Geol. Surv., Lower Cret., 1912, p. 368.
Arundel: Langdon, D. C.; German's iron mine ?, Md.
Patapsco: Fort Foote, Md.

Cycadeospermum oregonense Fontaine, in Ward, U. S. Geol. Surv., Mon. 48, 1905 [1906], p. 118, pl. xxix, f. 4.
Jurassic: Douglas County, Oreg.

Cycadeospermum ovatum Fontaine, in Ward, U. S. Geol. Surv., Mon. 48, 1905 [1906], p. 118, pl. xxix, f. 5.
Jurassic: Douglas County, Oreg.

Cycadeospermum rotundatum Fontaine, U. S. Geol. Surv., Mon. 15, 1889, p. 271, pl. cxxxvi, f. 12; U. S. Nat. Mus., Proc., vol. 16, 1893, p. 279, pl. xliii, f. 6; in Ward, U. S. Geol. Surv., Nineteenth Ann. Rept., pt. 2, 1899, p. 667, pl. clxii, f. 19; U. S. Geol. Surv., Mon. 48, 1905 [1906], pp. 280, 537.—Berry, Md. Geol. Surv., 1912, p. 369.
Patuxent: Kankeys, Va.
Arundel: Contee, Md.
Trinity: Glen Rose, Tex.
Fuson: Oak Creek, Crook County, Wyo.

Cycadeospermum spatulatum Fontaine, U. S. Geol. Surv., Mon. 15, 1889 [1890], p. 271; pl. cxxxv, f. 11, 21.—Berry, Md. Geol. Surv., Lower Cret., 1912, p. 370.

Cycadeospermum ellipticum Fontaine, U. S. Geol. Surv., Mon. 15, 1889 [1890], p. 271, pl. cxxxv, f. 19; in Ward, idem, vol. 48, 1905 [1906], p. 520.

Patuxent: Near Potomac Run, Dutch Gap, and Fredericksburg, Va.
Arundel: Langdon, D. C.

Cycadeospermum wanneri Fontaine, in Ward, U. S. Geol. Surv., Twentieth Ann. Rept., pt. 2, 1900, p. 247, pl. xxix, f. 10.

Triassic (Newark): York County, Pa.

CYCADINOCARPUS Schimper, Pal. vég., vol. 2, 1870, p. 208.

Cycadinocarpus chapini Newberry, U. S. Geol. Surv., Mon. 14, 1888, p. 92, pl. xxiv, f. 4.—Chapin, Scientific Assoc. Meriden, Conn., Proc. and Trans., vol. 4, 1889–90, p. 62, text f.

Triassic (Newark): Durham, Conn.

Cycadinocarpus circularis Newberry, U. S. Geol. Surv., Mon. 26, 1895 [1896], p. 46, pl. xlvi, f. 1–4.—Berry, Bull. Torr. Bot. Club, vol. 37, 1910, p. 182; Geol. Surv. N. J., Bull. 3, 1911, p. 79; U. S. Geol. Surv., Prof. Paper 112, 1919, p. 56.

Raritan: Woodbridge, N. J.
Black Creek: Black River, N. C.
Tuscaloosa: Glen Allen, Ala.

CYCADITES Buckland, Geol. and Mineral., vol. 1, 1836, p. 496.

Cycadites acutus Emmons, Geol. Rept. Midland Counties N. C., 1856, p. 330; American geology, pt. 6, 1857, p. 114, f. 81.—Fontaine, U. S. Geol. Surv., Mon. 6, 1883, p. 109, pl. li, f. 3; in Ward, U. S. Geol. Surv., Twentieth Ann. Rept., pt. 2, 1900, p. 300.

Triassic (Newark): North Carolina.

Cycadites althausii Dunker = *Matonidium althausii.*

Cycadites longifolius Emmons = *Podozamites longifolius.*

Cycadites pungens Lesquereux, U. S. Geol. Surv., Mon. 17, 1891 [1892], p. 30, pl. ii, f. 6.

Dakota: Kansas.

Cycadites sulcicaulis Phillips = *Ctenis sulcicaulis.*

Cycadites tenuinervis Fontaine, U. S. Geol. Surv., Mon. 6, 1883, p. 84, pl. xliv, f. 4–6; in Ward, U. S. Geol. Surv., Twentieth Ann. Rept., pt. 2, 1900, p. 300, pl. xliii, f. 1.

Triassic (Newark): Hanover County, Va.; North Carolina.

Cycadites unjiga Dawson, Roy. Soc. Canada, Trans., vol. 1, 1882 [1883], sec. 4, p. 20, pl. i, f. 2, 2a.—Penhallow, Roy. Soc. Canada, Trans., 3d ser., vol. 1, 1907, sec. 4, p. 308.

——: Pine River Forks, Table Mountain, and Peace River, Northwest Territory.

Lower Cretaceous: International boundary between Pasayten and Skagit rivers, British Columbia.

Cycadites? sp. Fontaine, U. S. Nat. Mus., Proc., vol. 13, 1890, p. 284.
Triassic: Abiquiu, N. Mex.

Cycadites sp. Penhallow, Roy. Soc. Canada, Trans., 2d ser., vol. 8, 1902, sec. 4, p. 41.
——: Alliford Bay, Queen Charlotte Islands?

CYCADOSPADIX Schimper, Pal. Vég., vol. 2, 1870, p. 207.

Cycadospadix? sp. Cockerell, Washington Acad. Sci. Jour., vol. 6, 1916, p. 110, f. 1.
Lower Cretaceous: Near Hovenweep Canyon, latitude 37° 53′ N., longitude 108° 57′ W., southwestern Colorado.

CYCAS Linné, Sp. pl., 1753, p. 1188.

Cycas sp. Tyson = *Cycadeoidea marylandica.*

CYCLOPITYS Schmalhausen, Mém. Acad. imp. sci. St. Petersburg, 7th ser., vol. 27, No. 4, 1879, p. 41.

Cyclopitys oregonensis Fontaine, in Ward, U. S. Geol. Surv., Mon. 48, 1905 [1906], p. 132, pl. xxxvi, f. 1, 2.
Jurassic: Douglas County, Oreg.

CYCLOPTERIS Brongniart, Dict. sci. nat., vol. 57 (Prodrome), 1828, p. 61.

Cyclopteris digitata Brongniart = *Ginkgo digitata.*

Cyclopteris huttoni Sternberg = *Ginkgo huttoni.*

Cyclopteris mantelli Dunker = *Sagenopteris mantelli.*

Cyclopteris moquensis Newberry, in Ives, Rept. on Colorado River of the West, explored in 1857 and 1858, pt. 3, 1861, p. 129, pl. iii, f. 1, 2.—Fontaine, in Ward, U. S. Geol. Surv., Twentieth Ann. Rept., pt. 2, 1900, p. 337.
Triassic (?): Moqui villages, Ariz.

Cyclopteris obscurus Emmons = *Sagenopteris emmonsi.*

Cyclopteris sp. Emmons, Geol. Rept. Midland Counties N. C., 1856, p. 329, pl. iv, f. 10 = *Sagenopteris rhoifolia* (Fontaine, U. S. Geol. Surv., Mon. 6, 1883, p. 104, pl. xlix, f. 5) = *Sagenopteris emmonsi.*

CYPARISSIDIUM Heer, Kgl. Sv. Vet. Akad., Handl., vol. 12, No. 6 (Fl. foss. arct., vol. 3, pt. 2), 1874, p. 74. [Type, *Widdringtonites gracilis.*]

Cyparissidium gracile? (Heer) Heer, Fl. foss. arct., vol. 3, 1874, p. 74, pl. xvii, f. 5b, 5c; pl. xix, f. 1–10; pl. xx, f. 1d, 1e; pl. xxi, f. 9b, 9d.—Hollick, U. S. Geol. Surv., Mon. 50, 1906, p. 46, pl. iii, f. 2.
Widdringtonites gracilis Heer, Fl. foss. arct., vol. 1, 1868, p. 83, pl. xlii, f. 1e, 1ee, 1f, 1g, 3c.
"*Sequoia reichenbachi* Geinitz?" Hollick, N. Y. Acad. Sci., Trans., vol. 12, 1892, p. 30, pl. i, f. 18.
Raritan: Tottenville, Staten Island, N. Y.

CYPERACITES Schimper, Pal. vég., vol. 2, 1870, p. 412. [*Cyperites* Lindley and Hutton, Foss. Fl. Gt. Brit., vol. 1, 1831–1833, p. 124 = *Lepidodendron*. *Cyperites* Heer, Fl. tert. Helv., vol. 1, 1855, p. 75 = *Cyperacites* Schimper.]

Cyperacites angustior (Al. Braun) Schimper, Pal. vég., vol. 2, 1870, p. 414.—Knowlton, U. S. Geol. Surv., Mon. 32, pt. 2, 1899, p. 684.

Cyperites angustior Al. Braun, in Stizenberger, Versteinerungen, 1851, p. 75.—Lesquereux, U. S. Geol. and Geog. Surv. Terr., Ann. Rept. 1872 [1873], p. 403.

Fort Union: Elk Creek, Yellowstone National Park.

Cyperacites borealis (Heer) Heer, Fl. foss. arct., vol. 1, 1868, p. 65.

Cyperites borealis Heer, Fl. foss. arct., vol. 1, 1868, p. 96, pl. xlv, f. 3.—Lesquereux, U. S. Nat. Mus., Proc., vol. 10, 1887, p. 36.

Pliocene ?: Applegarth Canyon.

Cyperacites canaliculatus (Heer) Schimper, Pal. vég., vol. 2, 1870, p. 415.

Cyperites canaliculitus Heer, Fl. tert. Helv., vol. 1, 1855, p. 77, pl. xxviii, f. 5; U. S. Nat. Mus., Proc., vol. 10, 1887, p. 36.

——: Near Strahlenberg, Utah.

Cyperacites giganteus Knowlton, U. S. Geol. Surv., Mon. 32, pt. 2, 1899, p. 684, pl. lxxxii, f. 10.

Fort Union: Crescent Hill, Yellowstone National Park.

Cyperacites haydenii (Lesquereux) Knowlton, U. S. Geol. Surv., Bull. 152, 1898, p. 83.—Penhallow, Roy. Soc. Canada, Trans., 3d ser., vol. 1, 1907, sec. 4, p. 291.

Cyperites haydenii Lesquereux, Rept. U. S. Geol. Surv. Terr., vol. 8 (Cret. and Tert. Fl.), 1883, p. 140, pl. xxiii, f. 1–3a.—Penhallow, Rept. Tert. Pl. Brit. Col., 1908, p. 52.

Green River: Uinta [not Randolph] County, Wyo.

Oligocene: Kettle River near Midway, British Columbia.

Cyperacites? hillsii Knowlton, U. S. Geol. Surv., Prof. Paper 130, 19—, p. —, pl. xx, f. 6 [in preparation].

Laramie: Erie and Coal Creek, Colo.

Cyperacites paucinervis (Heer) Schimper, Pal. vég., vol. 2, 1870, p. 414.

Cyperites paucinervis Heer, Fl. tert. Helv., vol. 1, 1855, p. 79, pl. xxix, f. 4.—Dawson, Roy. Soc. Canada, Trans., 2d ser., vol. 1, 1895 [1896], p. 144.—Penhallow, Rept. Tert. Pl. Brit. Col., 1908, p. 52.

Eocene: Stanley Park, Vancouver, British Columbia.

Cyperacites? tessellatus Knowlton, U. S. Geol. Surv., Prof. Paper 130, 19—, p. —, pl. iii, f. 1, 2 [in preparation].

Laramie: Popes Bluff, Pikeview, Colo.

Cyperacites sp. Berry, U. S. Nat. Mus., Proc., vol. 54, 1918, p. 629.

Miocene: Beaver Creek, 10 miles east of Beaver City, Okla.

Cyperacites? sp. (Dawson) Knowlton, U. S. Geol. Surv., Bull. 152, 1898, p. 83.

Cyperites? sp., Dawson, Roy. Soc. Canada, Trans., vol. 8, 1890, sec. 4, p. 90, text f. 31.

Oligocene: Similkameen River, British Columbia.

Cyperacites sp. (Dawson) Knowlton, U. S. Geol. Surv., Bull. 152, p. 83.

Cyperites sp., Dawson, Roy. Soc. Canada, Trans., vol. 10, 1892, sec. 4, p. 91, text f. 16.

Kootenai: Anthracite, British Columbia.

Cyperacites? sp., Knowlton, U. S. Geol. Surv., Mon. 32, pt. 2, 1899, p. 685, pl. lxxxiii, f. 6.

Fort Union: Crescent Hill, Yellowstone National Park.

Cyperacites sp., Knowlton, U. S. Geol. Surv., Bull. 257, 1905, p. 130.

Judith River: Willow Creek, Fergus County, Mont.

Cyperacites sp., Knowlton, U. S. Geol. Surv., Bull. 162, 1900, p. 32, pl. v, f. 8.

Mesaverde: Point of Rocks, Wyo.

Cyperacites sp., Knowlton, Geol. Soc. Am., Bull., vol. 8, 1897, p. 145.

Post-Laramie: Black Buttes, Wyo.

Cyperacites? sp., Knowlton, U. S. Geol. Surv., Mon. 32, pt. 2, 1899, p. 684, pl. lxxxiii, f. 4.

Fort Union: Yellowstone National Park.

Cyperacites sp., Knowlton, U. S. Geol. Surv., Bull. 204, 1902, p. 28, pl. i, f. 9..

Mascall: Van Horn's ranch, John Day Basin, Oreg.

Cyperacites sp., Knowlton, Washington Acad. Sci., Proc., vol. 11, 1909, p. 207.

Lance: Converse County, Wyo.

Cyperacites? sp., Knowlton, U. S. Geol. Surv., Prof. Paper 130, 19—, p. —, pl. iv, f. 1 [in preparation].

Laramie: Leyden Gulch, 6½ miles north of Golden, Colo.

Cyperacites sp. Hollick, Geol. Surv. La., Special Rept. 5, 1899, p. 279, pl. xxxii, f. 3, 4.

Wilcox: Shreveport, La.

Cyperacites sp. Hollick, Bull. Torr. Bot. Club, vol. 21, 1894, p. 63, pl. clxxx, f. 3.—Berry, U. S. Geol. Surv., Prof. Paper 112, 1919, p. 71.

Tuscaloosa: Northport, Ala.

Cyperacites sp. (Hollick) Hollick, Rept. Geol. Surv. La., 1899 [1900], p. 279, pl. xxxii, f. 3, 4.—Berry, U. S. Geol. Surv., Prof. Paper 91, 1916, p. 174.

Wilcox: Near Shreveport, La.

Cyperacites sp. Hollick, U. S. Geol. Surv., Mon. 50, 1906, p. 48, pl. vi, f. 7, 8.

Magothy: Glen Cove, Long Island, N. Y.

Cyperacites sp. (Penhallow) Knowlton, n. comb.

Cyperites sp. Penhallow, Rept. Tert. Pl. Brit. Col., 1908, p. 51.

Oligocene: Tulameen, Horsefly, Similkameen, Tranquille, and Kettle rivers, British Columbia.

Cyperacites sp. Penhallow, Roy. Soc. Canada, Trans., 3d ser., vol. 1, sec. 4, 1907, p. 293.
Oligocene: Kettle River near Midway, British Columbia.

Cyperites of authors = *Cyperacites* Schimper.

Cyperites angustior Al. Braun = *Cyperacites angustior.*

Cyperites borealis Heer = *Cyperacites borealis.*

Cyperites canaliculatus Heer = *Cyperacites canaliculatus.*

Cyperites deucalionis Heer, Fl. tert. Helv., vol. 1, 1855, p. 78, pl. xxix, f. 1; pl. xxvi, f. 13b; pl. xxx, f. 3h.—Lesquereux, U. S. Geol. and Geog. Surv. Terr., Ann. Rept. 1871 [1872], p. 285. [Not afterward referred to by Lesquereux.]
Green River?: Barrel Springs, Wyo.

Cyperites haydenii Lesquereux = *Cyperacites haydenii.*

Cyperites paucinervis Heer = *Cyperacites paucinervis.*

Cyperites sp. Dawson = *Cyperacites* sp.

Cyperites? sp. Dawson = *Cyperacites* sp.

CYPERUS (Tournefort) Linné, Sp. pl., 1753, p. 44.

Cyperus! braunianus? Heer, Fl. tert. Helv., vol. 1, 1855, p. 72, pl. xxii, f. 6; pl. xxvii, f. 4–7.—Lesquereux, U. S. Geol. and Geog. Surv. Terr., Ann. Rept. 1871 [1872], p. 285. [Not afterward mentioned by Lesquereux.]
Green River?: Barrel Springs, Wyo.

Cyperus chavanensis "Heer" = *Cyperus chavannesi.*

Cyperus chavannesi Heer, Fl. tert. Helv., vol. 1, 1855, p. 72, pl. xxii, f. 7; pl. xxviii, f. 1.
Cyperus "*chavanensis*" Heer. Lesquereux, U. S. Geol. and Geog. Surv. Terr., Ann. Rept. 1871 [1872], p. 291; Rept. U. S. Geol. Surv. Terr., vol. 7 (Tert. Fl.), 1878, p. 92, pl. ix, f. 1, 2.
Green River: Green River, Wyo.
Evanston: Evanston, Wyo.

Cyperus sp. Knowlton, Am. Geol., vol. 18, 1896, p. 371.
Pleistocene: Morgantown, W. Va.

Cyperus sp., Knowlton, Geol. Soc. Am., Bull., vol. 8, 1897, p. 133.
Lance: Lance Creek, Converse County, Wyo.

Cyperus sp. Penhallow, Am. Nat., vol. 41, 1907, p. 446.
Pleistocene: Don Valley, Toronto, Canada.

CYTISUS (Tournefort) Linné, Sp. pl., 1753, p. 739.

Cytisus florissantianus Lesquereux, Rept. U. S. Geol. Surv. Terr., vol. 8 (Cret. and Tert. Fl.), 1883, p. 200, pl. xxxix, f. 14.—Knowlton, U. S. Nat. Mus., Proc., vol. 51, 1916, p. 277, pl. xxi, f. 5; pl. xxiii, f. 3; pl. xxiv, f. 4.
Cassia fischeri Heer. Lesquereux, Rept. U. S. Geol. Surv. Terr., vol. 8 (Cret. and Tert. Fl.), 1883, p. 202.
Miocene: *Florissant, Colo.

Cytisus modestus Lesquereux = *Ptelea modesta.*

CZEKANOWSKIA Heer, Fl. foss. arct., vol. 4, Abt. 2, 1876, p. 70.

Czekanowskia capillaris Newberry, U. S. Geol. Surv., Mon. 26, 1895 [1896], p. 61, pl. ix, f. 14–16.—Hollick and Jeffrey, N. Y. Bot. Gard., Mem., vol. 3, 1909, p. 63, pl. vi, f. 1–3.—Berry, Geol. Surv. N. J., Bull. 3, 1911, p. 101; Md. Geol. Surv., Upper Cret., 1916, p. 804.

Raritan: Woodbridge, N. J.; Kreischerville, Staten Island, N. Y.; Forked Creek, Severn River, Md.

Czekanowskia dichotoma? (Heer) Heer, Fl. foss. arct., vol. 6, Abt. 2, 1882, p. 14, pl. ii, f. 12b, 12c; pl. iii, f. 1.—Hollick, U. S. Geol. Surv., Mon. 50, 1906, p. 36, pl. v, f. 7.

Sclerophyllina dichotoma Heer, Fl. foss. arct., vol. 1, 1868, p. 82, pl. xliv, f. 6.—Hollick, N. Y. Bot. Gard., Bull., vol. 2, 1902, p. 404, pl. xli, f. 10.

Magothy: Chappaquiddick, Marthas Vineyard, Mass.
Raritan: Woodbridge, N. J.
Tuscaloosa: Alabama.

Czekanowskia nervosa Heer, Contr. Fl. foss. du Portugal, 1881, p. 18, pl. xvii, f. 5–7a, 8–11.—Fontaine, in Ward, U. S. Geol. Surv., Nineteenth Ann. Rept., pt. 2, 1899, p. 685, pl. clxix, f. 1, 2.

Lakota: Barrett and Webster's ranch, Crook County, Wyo.

DACTYOLEPIS Hollick and Jeffrey, N. Y. Bot. Gard., Mem., vol. 3, 1909, p. 52. [Type, *D. cryptomerioides*.]

Dactyolepis cryptomerioides Hollick and Jeffrey, N. Y. Bot. Gard., Mem., vol. 3, 1909, p. 52, pl. x, f. 12, 13.

Raritan: Kreischerville, Staten Island, N. Y.

DALBERGIA Linné f., Suppl., 1781, p. 52.

Dalbergia apiculata Newberry, U. S. Geol. Surv., Mon. 26, 1895 [1896], p. 90, pl. xlii, f. 17–19.—Berry, Geol. Surv. N. J., Bull. 3, 1911, p. 165.

Raritan: Woodbridge, N. J.

Dalbergia calvertensis Berry, Jour. Geol., vol. 17, 1909, p. 28, text f. 7.

Calvert: Richmond, Va.

Dalbergia? coloradensis Knowlton, U. S. Nat. Mus., Proc., vol. 51, 1916, p. 278, pl. xix, f. 4.

Miocene: *Florissant, Colo.

Dalbergia cuneifolia Heer, Fl. tert. Helv., vol. 3, 1859, p. 104, pl. cxxxiii, f. 20.—Lesquereux, Rept. U. S. Geol. Surv. Terr., vol. 8 (Cret. and Tert. Fl.), 1883, p. 200, pl. xxxiv, f. 6, 7.

Miocene: Florissant, Colo.

Dalbergia eocenica Berry, U. S. Geol. Surv., Prof. Paper 91, 1916, p. 245, pl. liii, f. 1, 2.
Lagrange: Puryear, Henry County, Tenn.

Dalbergia hyperborea Heer, Fl. foss. arct., vol. 6, Abt. 2, 1882, p. 102, pl. xxvi, f. 4a.—Hollick, N. Y. Acad. Sci., Trans., vol. 11, 1892, p. 7, pl. iv, f. 7; U. S. Geol. Surv., Mon. 50, 1906, p. 85, pl. xxxii, f. 10.
Raritan: Tottenville, Staten Island, N. Y.

Dalbergia irregularis Hollick, U. S. Geol. Surv., Mon. 50, 1906, p. 85, pl. xxxii, f. 11.
Magothy: Gay Head, Marthas Vineyard, Mass.

Dalbergia lesquereuxii Ettingshausen, Denks. K. Akad. W., Math.-naturw. Classe, vol. 47, 1883, p. 142 [42]; Geol. Surv. N. S. W., Mem. Palaeont. No. 2 (Tert. Fl. Australia), 1888, p. 68.
Vaccinium reticulatum? Al. Braun. Lesquereux, Rept. U. S. Geol Surv. Terr., vol. 7 (Tert. Fl.), 1878, p. 235, pl. lix, f. 6; Rept. U. S. Geol. Surv. Terr., vol. 8 (Cret. and Tert. Fl.), 1883, p. 261.
Vaccinium retigerum Cockerell, Am. Mus. Nat. Hist., Bull., vol. 24, 1908, p. 105.
Miocene: Florissant, Colo.

Dalbergia minor Hollick, U. S. Geol. Surv., Mon. 50, 1906, p. 85, pl. xxxii, f. 12.
Magothy: Gay Head, Marthas Vineyard, Mass.

Dalbergia? minuta Knowlton, U. S. Nat. Mus., Proc., vol. 51, 1916, p. 277, pl. xxiv, f. 3.
Miocene: *Florissant, Colo.

Dalbergia monospermoides Berry, U. S. Geol. Surv., Prof. Paper 91, 1916, p. 246, pl. liv, f. 3.
Lagrange: Puryear, Henry County, Tenn.

Dalbergia rinkiana Heer. Hollick, N. Y. Acad. Sci., Trans., vol. 12, 1893, p. 236, pl. vi, f. 5 = *Hymenaea dakotana*.

Dalbergia severnensis Berry, Bull. Torr. Bot. Club, vol. 38, 1911, p. 407, pl. xix, f. 2; Md. Geol. Surv., Upper Cret., 1916, p. 847, pl. lxxvi, f. 3.
Magothy: Little Round Bay, Severn River, Md.

Dalbergia tennesseensis Berry, U. S. Geol. Surv., Prof. Paper 91, 1916, p. 246, pl. liv, f. 4.
Lagrange: Puryear, Henry County, Tenn.

Dalbergia wilcoxiana Berry, U. S. Geol. Surv., Prof. Paper 91, 1916, p. 246, pl. liii, f. 7; pl. liv, f. 1, 2.
Lagrange: Puryear, Henry County, Tenn.

DALBERGITES Berry, U. S. Geol. Surv., Prof. Paper 91, 1916, p. 247. [Type, *D. ellipticifolius*.]

Dalbergites ellipticifolius Berry, U. S. Geol. Surv., Prof. Paper 91, 1916, p. 247, pl. liv, f. 10.

Wilcox (Grenada): Grenada, Grenada County, Miss.

Dalbergites ovatus Berry, U. S. Geol. Surv., Prof. Paper 91, 1916, p. 247, pl. liv, f. 11.

Wilcox (Grenada): Grenada, Grenada County, Miss.

DAMMARA Gaertner, Fruct., vol. 2, 1791, p. 100.

Dammara acicularis Knowlton, U. S. Geol. Surv., Bull. 257, p. 134, pl. xv, f. 2–5.

Judith River: *Willow Creek, Fergus County, and 6½ miles east of Billings, Mont.

Dammara borealis Heer, Fl. foss. arct., vol. 6, Abt. 2, 1882, p. 54, pl. xxxvii, f. 5.—Hollick, N. Y. Acad. Sci., Trans., vol. 12, 1892, p. 31, pl. i, f. 17.—Newberry, U. S. Geol. Surv., Mon. 26, 1895 [1896], p. 46, pl. x, f. 8.—Hollick, N. Y. Bot. Gard., Bull., vol. 2, 1902, p. 402, pl. xli, f. 6; N. Y. State Mus., Fifty-fifth Ann. Rept., 1901 [1903], p. r49; U. S. Geol. Surv., Mon. 50, 1906, p. 37, pl. ii, f. 2–11, 12–26 in part.—Hollick and Jeffrey, Am. Nat., vol. 40, 1906, p. 197, pl. i, f. 2.— Berry, Bull. Torr. Bot. Club, vol. 37, 1910, p. 185; Geol. Surv. N. J., Bull. 3, 1911, p. 80; U. S. Geol. Surv., Prof. Paper 112, 1919, p. 58.

Dammara microlepis Heer. Hollick, N. Y. Bot. Gard., Bull., vol. 3, 1904, p. 410, pl. lxxi, f. 9, 10.

Eucalyptus geinitzi Heer. White, Am. Jour. Sci., 3d ser., vol. 39, 1890, p. 98, pl. ii, f. 9, 10.

Dammara cliffwoodensis Hollick. Berry, N. Y. Bot. Gard., Bull., vol. 3, 1903, p. 61, pl. xlviii, f. 8–11; Bull. Torr. Bot. Club, vol. 31, 1904, p. 69, pl. i, f. 11.

Magothy: Chappaquiddick Island and Gay Head, Marthas Vineyard, Mass.; Glen Cove, Long Island, N. Y.

Raritan: Woodbridge, N. J.; Tottenville, Staten Island, N. Y.

Black Creek: Court House Bluff, Cape Fear River, N. C.

Tuscaloosa: Snow place and Shirleys Mill, Ala.

Dammara? cliffwoodensis Hollick, N. Y. Acad. Sci., Trans., vol. 16, 1897, p. 128, pl. xi, f. 5–7; Geol. Surv. N. J., Ann. Rept. 1905 [1906], p. 138.—Hollick and Jeffrey, Am. Nat., vol. 40, 1906, p. 198, pl. i, f. 3.—Berry, Bull. Torr. Bot. Club, vol. 38, 1911, p. 400; Md. Geol. Surv., Upper Cret., 1916, p. 776, pl. liv, f. 3.

Magothy: Cliffwood, N. J.; Little Round Bay, Md.

Matawan: Millersville, Md.

Dammara cliffwoodensis Hollick. Berry, N. Y. Bot. Gard., Bull., vol. 3, 1903, p. 61, pl. xlviii, f. 8–11; Bull. Torr. Bot. Club, vol. 31, 1904, p. 69, pl. i, f. 11 = *Dammara borealis*.

Dammara microlepis Heer. Hollick, N. Y. Bot. Gard., Bull., vol. 3, 1904, p. 410 = *Dammara borealis*.

Dammara microlepis Heer (?). Hollick, N. Y. Acad. Sci., Ann., vol. 11, 1898, p. 57, pl. iii, f. 9a, b = *Protodammara speciosa.*

Dammara minor Hollick = *Protodammara speciosa.*

Dammara northportensis Hollick, N. Y. Bot. Gard., Bull., vol. 3, 1904, p. 405, pl. lxx, f. 1, 2; U. S. Geol. Surv., Mon. 50, 1906, p. 39, pl. ii, f. 33, 34.—Hollick and Jeffrey, Am. Nat., vol. 40, 1906, p. 198, pl. i, f. 4.

Magothy: Little Neck, Northport Harbor, Long Island, N. Y.

Dammara? sp., Knowlton, Washington Acad. Sci., Proc., vol. 11, 1909, p. 204.

Lance: Weston County, Wyo.

Dammara sp. Knowlton, U. S. Geol. Surv. Prof. Paper 130, 19—, p. —, pl. ii, f. 4 [in preparation].

Laramie: Marshall, Colo.

DAMMARITES Presl, in Sternberg, Vers., vol. 2, 1833, p. 203.

Dammarites caudatus (Lesquereux) Lesquereux, U. S. Geol. Surv., Mon. 17, 1891 [1892], p. 32, pl. i, f. 9, 10.

Podozamites caudatus Lesquereux, Rept. U. S. Geol. Surv. Terr., vol. 8 (Cret. and Tert. Fl.), 1883, p. 29.

Podozamites praelongus Lesquereux, Rept. U. S. Geol. Surv. Terr., vol. 8 (Cret. and Tert. Fl.), 1883, p. 29.

Dakota: Ellsworth County, Kans.

Dammarites dubius Dawson, Roy. Soc. Canada, Trans., vol. 11, 1893 [1894], sec. 4, p. 56, pl. vi, f. 8.

Upper Cretaceous: Nanaimo, British Columbia.

Dammarites emarginatus (Lesquereux) Lesquereux, U. S. Geol. Surv., Mon. 17, 1891 [1892], p. 33, pl. i, f. 11.

Podozamites emarginatus Lesquereux, Rept. U. S. Geol. Surv. Terr., vol. 8 (Cret. and Tert. Fl.), 1883, p. 29.

Dakota: Near Glascoe, Kans.

DANAEOPSIS Heer, Fl. foss. Helv., 1876, p. 71.

Danaeopsis storrsii Fontaine, in Ward, U. S. Geol. Surv., Mon. 48, 1905 [1906], p. 87, pl. xv, f. 6–9.

Jurassic: Douglas County, Oreg.

Danaeopsis? sp., Fontaine, in Ward, U. S. Geol. Surv., Twentieth Ann. Rept., pt. 2, 1900, p. 284, pl. xxxviii, f. 7.

Triassic (Newark): North Carolina.

DAPHNOGENE Unger, Synop. pl. foss., 1845, p. 227.

Daphnogene anglica? Heer, Fl. tert. Helv., vol. 3, 1859, p. 315.—Lesquereux, U. S. Geol. and Geog. Surv. Terr., Ann. Rept. 1873 [1874], p. 401; Rept. U. S. Geol. Surv. Terr., vol. 7 (Tert. Fl.), 1878, p. 222, pl. xxxvii, f. 9.

Denver: Golden, Colo.

Daphnogenea cretacea Lesquereux = *Cinnamomum scheuchzeri.*

Daphnogene elegans Watelet. Ward, U. S. Geol. Surv., Sixth Ann. Rept., 1884–85 [1886], p. 553, pl. xlvii, f. 4; idem, Bull. 37, 1887, p. 51, pl. xxv, f. 1.
Post-Laramie: Black Buttes, Wyo.

Daphnogene heerii Lesquereux = *Cinnamomum heerii.*

Daphnogene kanei Heer. Hollick = *Cinnamomum postnewberryi.*

DAPHNOPHYLLUM Heer, Neue Denkschr. Schw. Gesell., vol. 23, 1869, p. 17.

Daphnophyllum angustifolium Lesquereux, U. S. Geol. Surv., Mon. 17, 1891 [1892], p. 98, pl. xxxvi, f. 8.
Dakota: Ellsworth County, Kans.

Daphnophyllum dakotense Lesquereux, U. S. Geol. Surv., Mon. 17, 1891 [1892], p. 99, pl. li, f. 1–4; pl. lii, f. 1.
Dakota: Probably Ellsworth County, Kans.

Daphnophyllum sp. Lesquereux, Rept. on Clay in N. J., 1878, p. 29 = ?
Raritan: Sayreville, N. J.

DAVALLIA Smith, Mem. Act. Turin, vol. 5, 1793, p. 414.

Da[e]vallia? montana Knowlton, U. S. Geol. Surv., Mon. 32, pt. 2, 1899, p. 671, pl. lxxix, f. 4.
Miocene: Fossil Forest, Yellowstone National Park.

Davallia (Stenoloma) tenuifolia Swartz. Dawson, Brit. N. A. Bound. Com. (Rept. Geol. and Res. Vic. 49th Parallel), 1875, App. A, p. 329, pl. xvi, f. 1, 1a, 2, 2a = *Dennstaedtia americana.*

DAVALLITES Dawson, Roy. Soc. Canada, Trans., vol. 1, 1882 [1883], p. 25. [Type, *D. richardsoni.*]

Davallites richardsoni Dawson, Roy. Soc. Canada, Trans., vol. 1, 1882 [1883], sec. 4, p. 25, pl. v, f. 18–18b.
Upper Cretaceous: Protection Island, Vancouver Island, British Columbia.

DELESSERIA Lamarck, Ann. Mus., vol. 20, 1813, p. 122.

Delesseria fulva Lesquereux, U. S. Geol. and Geog. Surv. Terr., Ann. Rept. 1872 [1873], p. 376; Rept. U. S. Geol. Surv. Terr., vol. 7 (Tert. Fl.), 1878, p. 39, pl. i, f. 10.—Knowlton, U. S. Geol. Surv., Prof. Paper 130, 19—, p. —, pl. i, f. 4 [in preparation].
Laramie: Golden, Colo.

Delesseria incrassata Lesquereux = *Caulerpites incrassatus.*

Delesseria lingulata Lesquereux = *Caulerpites lingulatus.*

DENDRIUM Desvaux, Jour. Bot., 2d ser., vol. 1, 1813, pp. 3–6. [*Leiophyllum* Persoon.]

Dendrium pleistocenicum Berry, Jour. Geol., vol. 15, 1907, p. 346; Am. Nat., vol. 43, 1909, p. 436.
Pleistocene: Neuse River, N. C.; Rappahannock River, 1½ miles below Port Royal, Va.

DENDROPHYCUS Lesquereux, Second Geol. Surv. Pa., Rept. Prog. P. (Coal Fl., vol. 3), 1884, p. 699.

Dendrophycus shoemakeri Ward, U. S. Geol. Surv., Twentieth Ann. Rept., pt. 2, 1900, p. 256, pl. xxxv, f. 2; pl. xxxvi.
Triassic (Newark): Seneca, Md.

Dendrophycus triassicus Newberry, N. Y. Acad. Sci., Trans., vol. 6, 1886–87 [1887], p. 126 [name]; U. S. Geol. Surv., Mon. 14, 1888, p. 82, pl. xxi, f. 1.
Triassic (Newark): Portland, Conn.

DENNSTAEDTIA Bernhardi, Schrader's Jour., 1800, vol. 2, p. 124.

Dennstaedtia americana Knowlton, Smithsonian Misc. Coll. (quart. issue), vol. 53, 1910, p. 492, pl. lxiii, f. 4; pl. lxiv, f. 3–5.
Davalia (Stenoloma) tenuifolia Swartz. Dawson, Brit. N. A. Boundary Com. (Rept. Geol. and Res. Vicinity 49th Parallel), 1875, App. A, p. 329, pl. xvi, f. 1, 1a, 2, 2a; Roy. Soc. Canada, Trans., vol. 4, 1886 [1887], p. 21, pl. i, f. 1, 1a, 1b.—Penhallow, Rept. Tert. Pl. Brit. Col., 1908, p. 52.
Asplenium tenerum Lesquereux, Rept. U. S. Geol. Surv. Terr., vol. 8 (Cret. and Tert. Fl.), 1883, p. 221, pl. xlviA, f. 1, 2. [Not *Asplenium tenerum* Forster, 1786.]
Sphenopteris blomstrandi Heer, Fl. foss. arct., vol. 1, 1868, p. 155, pl. xxix, f. 1a–e, 2a, b, 3, 4a–c, 9a, b.—Penhallow, Roy. Soc. Canada, Trans., 2d ser., vol. 8, 1902, p. 48; Rept. Tert. Pl. Brit. Col., 1908, p. 90.
Fort Union: Gilmore station, N. Dak.?; Black Butte, Custer Trail ranch near Medora, and Sentinel Butte, N. Dak.; 35 miles southeast of Rock Springs, Wyo.
Paskapoo (Fort Union): Red Deer River, Alberta; Porcupine Creek, Saskatchewan.

Dennstaedtia? fremonti (Hall) Knowlton, U. S. Geol. Surv., Prof. Paper 108, 1917, p. 83, pl. xxx, f. 5; pl. xxxi, f. 1.
Sphenopteris fremonti Hall, in Frémont, Rept. Expl. Exped. to Rocky Mountains in 1843, etc., App. B, 1845, p. 304, pl. ii, f. 3, 3a.
Frontier: 1 mile east of Cumberland, Wyo.

DERMATOPHYLLITES Göppert and Berendt, in Berendt, Bernstein reste Vorwelt, vol. 1, 1845, pt. 1, p. 76.

Dermatophyllites acutus Heer, Fl. foss. arct., vol. 6, Abt. 2, 1882, p. 80, pl. xlii, f. 7; Berry, U. S. Geol. Surv., Prof. Paper 112, 1919, p. 132, pl. xxvii, f. 8.
Tuscaloosa: Shirleys Mill, Ala.

DEWALQUEA Saporta and Marion, Mém. cour. et d. sav. étrangers d. l'Acad., vol. 37, 1874, p. 55.

Dewalquea dakotensis Lesquereux, U. S. Geol. Surv., Mon. 17, 1891 [1892], p. 211, pl. lix, f. 5, 6.
Dakota: Kansas.

Dewalquea grönlandica Heer? Fl. foss. arct., vol. 6, Abt. 2, 1882, p. 87, pl. xxix, f. 18, 19; pl. xliv, f. 11.—Newberry, U. S. Geol. Surv., Mon. 26, 1895 [1896], p. 129, pl. xli, f. 2, 3—Hollick, N. Y. Acad. Sci., Ann., vol. 11, 1898, p. 423, pl. xxxvi, f. 7.—Berry, N. Y. Bot. Gard., Bull., vol. 3, 1903, p. 98, pl. lvii, f. 3; Geol. Surv. N. J., Ann. Rept. 1905 [1906], p. 138.—Hollick, U. S. Geol. Surv., Mon. 50, 1906, p. 106, pl. viii, f. 25.—Berry, Bull. Torr. Bot. Club, vol. 34, 1907, p. 194.

Magothy: Cliffwood, N. J.

Raritan: New Jersey (no locality given); Tottenville Staten Island, N. Y.

Black Creek: Blackmans Bluff, Neuse River, N. C.

Dewalquea grönlandica Heer. Newberry, U. S. Geol. Surv., Mon. 26, 1895, pl. xli, f. 12 [not f. 2, 3] = *Salix lesquereuxii*.

Dewalquea haldemiana (Debey) Saporta and Marion. Lesquereux, U. S. Nat. Mus., Proc., vol. 10, 1887, p. 46.

Cretaceous: Upper Kanab Valley, Utah.

Dewalquea haldemiana (Debey) Saporta and Marion. Hollick, N. Y. Acad. Sci., Trans., vol. 12, 1892, p. 36, pl. ii, f. 2a, 10 = *Salix flexuosa*.

Dewalquea insigniformis Berry, Bull. Torr. Bot. Club, vol. 44, 1917, p. 179, pl. vii, f. 6, 7.

Bingen: Near Maxwell Spur and near Murfreesboro, Pike County, Ark.

Dewalquea insignis Hosius and Von der Marck, Palaeontogr., vol. 26, 1880, p. 172 [48 of reprint], pls. xxxii, xxxiii, xxxiv.—Hollick, N. Y. Acad. Sci., Trans., vol. 12, 1892, p. 36, pl. i, f. 9; U. S. Geol. Surv., Mon. 50, 1906, p. 106, pl. viii, f. 24.

Raritan: Tottenville, Staten Island, N. Y.

Dewalquea primordialis Lesquereux, in Winchell, Geol. and Nat. Hist. Surv. Minn., Thirteenth Ann. Rept., 1884 [1885], p. 77 [name].—Lesquereux, Geol. and Nat. Hist. Surv. Minn., Final Rept., vol. 3, pt. 1, 1893, p. 18, pl. A, f. 10.

Dakota: Near New Ulm, Minn.

Dewalquea pulchella Knowlton, U. S. Geol. Surv., Prof. Paper 108, 1917, p. 90, pl. xxxvi, f. 1–3; pl. xxxviii, f. 1–3.

Frontier: About 1 mile east of Cumberland, Wyo.

Dewalquea smithi Berry, Torreya, vol. 10, 1910, p. 36, text f. 1; U. S. Geol. Surv., Prof. Paper 84, 1914, p. 41, pl. viii, f. 3–9, text f. 1; Bull. Torr. Bot. Club, vol. 43, 1916, p. 293; U. S. Geol. Surv., Prof. Paper 112, 1919, p. 86, pl. xiv, f. 1; pl. xvi, f. 2, 3.

Tuscaloosa: Shirleys Mill and White Bluffs on Black Warrior River, near Mobile, Ala.

Black Creek (Middendorf): Langley and Miles Mill, S. C.

Eutaw: Coffee Bluff, Hardin County, Tenn.

Dewalquea trifoliata Newberry, U. S. Geol. Surv., Mon. 26, 1895 [1896], p. 129, pl. xxii, f. 4–7.—Berry, Geol. Surv. N. J., Bull. 3, 1911, p. 128.
Raritan: Woodbridge, N. J.

Dewalquea near *D. insignis* Hosius and Von der Marck. Knowlton, in Veatch, Am. Jour. Sci., 4th ser., vol. 21, 1906, p. 459 = ***Dewalquea pulchella.***

DICHOTOZAMITES Berry, Md. Geol. Surv., Lower Cret., 1912, p. 364. [Type, *Sequoia cycadopsis.*]

Dichotozamites cycadopsis (Fontaine) Berry, Md. Geol. Surv., Lower Cret., 1912, p. 365, pl. lxxvii, f. 2, 3.
Sequoia cycadopsis Fontaine, U. S. Geol. Surv., Mon. 15, 1889 [1890], p. 243, pl. cxii, f. 9–11; pl. cxiii, f. 1, 2; in Ward, U. S. Geol. Surv., Mon. 48, 1905 [1906], pp. 489, 553, pl. cix, f. 11.
Patuxent: Mount Vernon and near Brooke, Va.; Fort Foote, Md.

DICKSONIA L'Héritier, Sertum Angl., 1788, p. 30.

Dicksonia acutiloba Heer = ***Cladophlebis acutiloba.***

Dicksonia burejensis Zalessky = ***Coniopteris burejensis.***

Dicksonia groenlandica Heer, Fl. foss. arct., vol. 6, Abt. 2, 1882, p. 23, pl. xxxv, f. 8, 9; idem, vol. 7, 1883, p. 2, pl. xlviii, f. 1–3.—Berry, Bull. Torr. Bot. Club, vol. 36, 1909, p. 245; Geol. Surv. N. J., Bull. 3, 1911, p. 66; U. S. Geol. Surv., Prof. Paper 112, 1919, p. 53; Bull. Torr. Bot. Club, vol. 44, 1917, p. 171.
Anemia stricta Newberry, U. S. Geol. Surv., Mon. 26, 1895 [1896], p. 38, pl. iii, f. 1, 2.
Raritan: Woodbridge, N. J.
Tuscaloosa: Snow place and Shirleys Mill, Ala.
Bingen: Kaolin cuts, Pike County, and Mine Creek, Howard County, Ark.

Dicksonia montanensis Fontaine, in Ward, U. S. Geol. Surv., Mon. 48, 1905 [1906], p. 286, pl. lxxi, f. 1–4.
Kootenai: Near Geyser, Cascade County, Mont.

Dicksonia munda Dawson, Roy. Soc. Canada, Trans., vol. 3, 1885 [1886], sec. 4, p. 11, pl. iii, f. 5, 5a.
Mill Creek: Mill Creek, British Columbia.

Dicksonia oregonensis Fontaine, in Ward, U. S. Geol. Surv., Mon. 48, 1905 [1906], pp. 55, 148, pl. vi, f. 3–9; pl. vii; pl. xxxviii, f. 1, 2.
Dryopteris monocarpa (Fontaine) Knowlton. Ward, U. S. Geol. Surv., Twentieth Ann. Rept. 1900, pt. 2, p. 369.
Jurassic: Douglas and Curry counties, Oreg.

Dicksonia pachyphylla Fontaine, in Ward, U. S. Geol. Surv., Mon. 48, 1905 [1906], pp. 224, 288, pl. lxv, f. 1; pl. lxxi, f. 5–11.

Aspleniopteris pinnatifida? Fontaine. Fontaine, in Diller and Stanton, Geol. Soc. Am., Bull., vol. 5, 1894, p. 450; in Stanton, U. S. Geol. Surv., Bull. 133, 1895, p. 15.

Knoxville: Elder Creek, Tehama County, Calif.
Kootenai: Near Geyser, Cascade County, Mont.
Cloverly: No Wood Creek, Big Horn Basin, Wyo.

Dicksonia saportana Heer, Fl. foss. arct., vol. 4, Abt. 2, 1876, pp. 89, 90, pl. xvii, f. 1, 2; pl. xviii, f. 1–3.—Fontaine, in Ward, U. S. Geol. Surv., Twentieth Ann. Rept., pt. 2, 1900, p. 340; U. S. Geol. Surv., Mon. 48, 1905 [1906], p. 155, pl. xxxix, f. 1, 2.

Jurassic: Cape Lisburne, Alaska.
Mariposa (?): Near Princeton, Calif.

Dicksonia sp., Dawson, Roy. Soc. Canada, Trans., vol. 3, 1885 [1886], sec. 4, p. 5.

Kootenai: Martin Creek, British Columbia.

Dicksonia n. sp., Knowlton, in Collier = *Coniopteris burejensis*.

DICKSONIOPSIS Berry, Md. Geol. Surv., Lower Cret., 1911, p. 237. [Type, *Scleropteris vernonensis* Ward.]

Dicksoniopsis vernonensis (Ward) Berry, Md. Geol. Surv., Lower Cret., 1911, p. 237, pl. xxviii, f. 3, 4.

Scleropteris vernonensis Ward, U. S. Geol. Surv., Fifteenth Ann. Rept., 1895, p. 349, pl. ii, f. 1–3.—Fontaine, in Ward, U. S. Geol. Surv., Mon. 48, 1905 [1906], p. 501, pl. cvii, f. 10.

Dryopteris virginica (Fontaine) Knowlton. Fontaine, in Ward, U. S. Geol. Surv., Mon. 48, 1905 [1906], p. 491.

Dryopteris parvifolia (Fontaine) Knowlton. Fontaine, in Ward, U. S. Geol. Surv., Mon. 48, 1905 [1906], p. 541, pl. cxiv, f. 7. [Not other citations of this species.]

Arundel: Arlington, Md.
Patapsco: Mount Vernon, White House Bluff, Va.

DICOTYLEDON? (male ament of ?) Fontaine, in Ward, U. S. Geol. Surv., Nineteenth Ann. Rept., pt. 2, 1899, p. 691, pl. clxix, f. 10.

Lakota: South Fork of Hay Creek, Crook County, Wyo.

DICRANOPTERIS Schenk, Foss. Fl. Grenzsch. Keup. und Lias Frankens, 1867, p. 145. [Preoccupied by the living *Dicranopteris* Bernhardi; Schrader, Neues Jour., vol. 1, pt. 2, 1806, pp. 26, 38.]

Dicranopteris sp., Fontaine, U. S. Geol. Surv., Mon. 6, 1883, p. 63, pl. xxx, f. 6.

Triassic (Newark): Clover Hill, Va.

Dicropteris longifolia Pomel = *Baiera longifolia*.

DICTYOPHYLLUM Lindley and Hutton, Foss. Fl. Great Britain, vol. 2, 1834, p. 65.

Dictyophyllum nilssoni (Brongniart) Göppert, Gattungen d. fossilen Pflanzen, pts. 5 and 6, 1846, p. 119.—Knowlton, U. S. Nat. Mus., Proc., vol. 51, 1916, p. 454, pl. lxxxii; f. 1–4.
Lower Jurassic: Upper Matanuska Valley, Alaska.

DIDYMOSORUS Debey, Amtl. Ber. 25, Versamml. Deutsch. Naturf. Aertzte, 1849, p. 299 (31).

Didymosorus? bindrabunensis acutifolius Fontaine, in Ward, U. S. Geol. Surv., Twentieth Ann. Rept., pt. 2, 1900, p. 353, pl. lvi, f. 2, 3.
Jurassic: Oroville, Calif.

DIDYMOSPHAERIA (Fuckel) Rostrock, Eclogae, III, 1810, p. 49.

Didymosphaeria betheli Cockerell, Am. Mus. Nat. Hist., Bull., vol. 24, 1908, p. 75, text f. 1.
Miocene: Florissant, Colo.

DILLENITES Berry, U. S. Geol. Surv., Prof. Paper 91, 1916, p. 291. [Type, *Quercus microdentata*.]

Dillenites microdentatus (Hollick) Berry, U. S. Geol. Surv., Prof. Paper 91, 1916, p. 291, pl. lxxv, f. 3; pl. lxxvii, f. 1; pl. cxiv, f. 5.
Quercus microdentata Hollick, La. Geol. Surv., Special Rept. 5, 1899, p. 280, pl. xxxiv.
Wilcox: *Coushatta, near Mansfield and near Naborton, La.

Dillenites ovatus Berry, U. S. Geol. Surv., Prof. Paper 91, 1916, p. 292, pl. lxviii, f. 2.
Ulmus tenuinervis Lesquereux. Hollick, La. Geol. Surv., Special Rept. 5, 1899, p. 280, pl. xxxii, f. 6.
Wilcox: Coushatta, La.
Wilcox (Ackerman): Hurleys, Miss.

Dillenites serratus Berry, U. S. Geol. Surv., Prof. Paper 91, 1916, p. 293, pl. lxxv, f. 6.
Wilcox (Holly Springs): Holly Springs, Miss.

Dillenites tetracerafolia Berry, U. S. Geol. Surv., Prof. Paper 91, 1916, p. 293, pl. lxxv, f. 4, 5.
Wilcox (Grenada): Grenada, Miss.
Lagrange: Puryear, Tenn.

Dillenites texensis Berry, U. S. Geol. Surv., Prof. Paper 91, 1916, p. 294, pl. lxviii, f. 5.
Wilcox: Pope Bend of Colorado River, Bastrop County, Tex.
Wilcox (Grenada): Grenada, Miss.

DIOONITES Miquel, Tijdschr. v. Wisen. Nat.-Wetens., vol. 4, 1851, p. 211 [11 of reprint].

Dioonites borealis Dawson, Roy. Soc. Canada, Trans., vol. 1, 1882 [1883], sec. 4, p. 24, pl. iii, f. 37; idem, vol. 3, 1885, p. 6, pl. i, f. 2.
——: Willow Creek, Northwest Territory.
Kootenai: Martin Creek and Canmore, British Columbia.

Dioonites buchianus (Ettingshausen) Bornemann, Org. Reste d. Lettenkohlengr. Thüringens, 1856, p. 57.—Fontaine, U. S. Geol. Surv., Mon. 15, 1889 [1890], p. 182, pl. lxviii, f. 1; pl. lxix, f. 1, 3; pl. lxx, f. 2, 3; pl. lxxi, f. 1; pl. lxxii, f. 1, 2; pl. lxxiii, f. 1–3; pl. lxxiv, f. 1–3; U. S. Nat. Mus., Proc., vol. 16, 1893, p. 264, pl. xxxvi, f. 5; in Ward, U. S. Geol. Surv., Mon. 48, 1905 [1906], pp. 244, 479, pl. lxvi, f. 16, 17; pl. cviii, f. 1.—Knowlton, in Diller, Geol. Soc. Am., Bull., vol. 19, 1908, p. 386.—Knowlton, Am. Jour. Sci., vol. 30, 1910, p. 42.—Berry, Md. Geol. Surv., Lower Cret., 1911, p. 332, pls. li, lii.

Pterophyllum buchianum Ettingshausen, Abh. K. k. geol. Reichsanst., vol. 1, Abt. 3, No. 2, 1852, p. 21, pl. i, f. 1.

Patuxent: Cockpit Point, telegraph station near Lorton, Kankeys, Trents Reach, Dutch Gap and vicinity, Va.; New Reservoir ?, D. C.

Arundel: Arlington ?, Md.

Trinity: Glen Rose, Tex.

Horsetown: Shasta County, Calif.

Dioonites buchianus abietinus (Göppert) Ward, U. S. Geol. Surv., Mon. 48, 1905 [1906], pp. 250, 486, pl. lxvii, f. 1–3; pl. cviii, f. 2.

Pterophyllum abietinum Göppert, in Dunker, Monogr. d. Norddeutsch. Wealdenbildung, 1846, p. 15, pl. vii, f. 2.

Dioonites buchianus angustifolius Fontaine, U. S. Geol. Surv., Mon. 15, 1889 [1890], p. 185, pl. lxvii, f. 6; pl. lxviii, f. 2; U. S. Nat. Mus., Proc., vol. 16, 1893, p. 265, pl. xxxvi, f. 6; in Diller and Stanton, Geol. Soc. Am., Bull., vol. 5, 1904, p. 450.

Horsetown: Riddles, Oreg.; Shasta County, Calif.

Knoxville: Tehama County, Calif.

Patuxent: Dutch Gap Canal, Kankeys, Occoquan, and Lorton station, Va.

Dioonites buchianus angustifolius Fontaine = *Dioonites buchianus abietinus*.

Dioonites buchianus obtusifolius Fontaine, U. S. Geol. Surv., Mon. 15, 1889 [1890], p. 184, pl. clxviii, f. 3.

Patuxent: Dutch Gap Canal, Va.

Dioonites buchianus rarinervis Fontaine, U. S. Nat. Mus., Proc., vol. 16, 1893, p. 264, pl. xxxvi, f. 3, 4.—?Fontaine, in Diller and Stanton, Geol. Soc. Am., Bull., vol. 5, 1894, p. 450.—?Fontaine, in Ward, U. S. Geol. Surv., Mon. 48, 1905 [1906], p. 251, pl. lxvii, f. 4.

Trinity: Glen Rose, Tex.

Knoxville and Horsetown: Elder Creek, Tehama County, Calif.

Dioonites carnallianus (Göppert) Bornemann, Ueber organische Reste der Lettenkohlengruppe Thüringens, 1856, p. 56.—Fontaine, in Ward, U. S. Geol. Surv., Twentieth Ann. Rept., pt. 2, 1900, p. 244, pl. xxviii, f. 2.

Pterophyllum carnallianum Göppert, Uebersicht Schles. Ges., 1843, p. 130, pl. i, f. 4.

Triassic (Newark): York County, Pa.

Dioonites columbianus (Dawson) Dawson, Roy. Soc. Canada, Trans., vol. 10, 1892 [1893], sec. 4, p. 91.

Cycadeocarpus (*Dioonites*) *columbianus* Dawson, Geol. Surv. Canada, Rept. Prog. 1872–73 [1873], App. 1, p. 69, pl. i, f. 1–12; Am. Jour. Sci., 3d ser., vol. 7, 1874, p. 49.

Lower Cretaceous: Skidegate Channel, Queen Charlotte Island.

Dioonites dunkerianus (Göppert) Miquel, Tijdschr. v. Wis.- en Nat.-Watens, vol. 4, 1851, p. 212.—Fontaine, U. S. Nat. Mus., Proc., vol. 16, 1893, p. 265, pl. xxxvi, f. 12; pl. xxxvii, f. 1; in Diller and Stanton, Geol. Soc. Am., Bull., vol. 5, 1894, p. 450; in Ward, U. S. Geol. Surv., Mon. 48, 1905 [1906], p. 243, pl. lxvi, f. 15.—Knowlton, Am. Jour. Sci., 4th ser., vol. 30, 1910, p. 42.

Pterophyllum dunkerianum Göppert, Uebersicht d. Arbeiten d. Schles. Ges. f. Vaterl. Kultur, 1843, p. 134.

Trinity: Glen Rose, Tex.

Horsetown: Near Horsetown, Shasta County, Calif.

Dioonites graminioides Emmons = *Ctenophyllum braunianum angustum.*

Dionites linearis Emmons = *Ctenophyllum braunianum angustum.*

Dioonites longifolius (Emmons) Fontaine = *Podozamites longifolius.*

DIOSCOREA Linné, Sp. pl., 1753, p. 1462.

Dioscorea? cretacea Lesquereux, Rept. U. S. Geol. Surv. Terr., vol. 6 (Cret. Fl.), 1874, p. 56, pl. xxviii, f. 10.

Dakota: Western Kansas.

DIOSPYROS Linné, Sp. pl., 1753, p. 1057.

Diospyros alaskana Schimper, Pal. vég., vol. 2, 1872, p. 949.—Dawson, Geol. Surv. Canada, 1875–76, p. 78.—Knowlton, U. S. Nat. Mus., Proc., vol. 17, 1874, p. 224; Geol. Soc. Am., Bull., vol. 5, 1893, p. 584; U. S. Geol. Surv., Bull. 204, 1902, p. 83.—Penhallow, Rept. Tert. Pl. Brit. Col., 1908, p. 52.

Diospyros lancifolia Lesquereux, Am. Jour. Sci., 2d ser., vol. 27, 1859, p. 361; U. S. Geol. and Geog. Surv. Terr., Ann. Rept. 1871 [1872], p. 293. [Homonym, Al. Braun, 1850.]

Kenai: Ninilchik, Alaska.

——?: Evanston, Wyo.

Clarno (lower part): Cherry Creek, Crook County, Oreg.

Eocene: Blackwater River, British Columbia.

Diospyros ambigua Lesquereux = *Diospyros nebraskana.*

Diospyros amboyensis Berry, Bull. Torr. Bot. Club, vol. 36, 1909, p. 262; Geol. Surv. N. J., Bull. 3, 1911, p. 212; U. S. Geol. Surv., Prof. Paper 112, 1919, p. 134, pl. xxvii, f. 5.

Phyllites ellipticus Newberry, U. S. Geol. Surv., Mon. 26, 1896, p. 130, pl. xxiv, f. 9.

Tuscaloosa: Shirleys Mill, Ala.

Raritan: Woodbridge, N. J.

Diospyros anceps Heer, Fl. tert. Helv., vol. 3, 1859, p. 12, pl. cii, f. 15–18.—Lesquereux, U. S. Nat. Mus., Proc., vol. 5, 1882 [1883], p. 448, pl. x, f. 1, 2; Rept. U. S. Geol. Surv. Terr., vol. 8 (Cret. and Tert. Fl.), 1883, p. 261.—Knowlton, U. S. Nat. Mus., Proc., vol. 17, 1894, p. 224; Geol. Soc. Am., Bull., vol. 5, 1893, p. 582.
  Kenai: Cook Inlet, Alaska.

Diospyros anceps Lesquereux = *Diospyros nebraskana.*

Diospyros apiculata Lesquereux, U. S. Geol. Surv., Mon. 17, 1891 [1892], p. 110, pl. xiv, f. 3.—Hollick, Geol. Soc. Am., Bull., vol. 7, 1895, p. 13; ?U. S. Geol. Surv., Mon. 50, 1906, p. 103, pl. lx, f. 4–6.
  Rhamnus pfaffiana Heer. Hollick, N. Y. Acad. Sci., Trans., vol. 11, 1892, p. 103, pl. iv, f. 3.
  Dakota: Near Delphos, Kans.
  Raritan: Princes Bay, Staten Island, N. Y.
  Magothy: Gay Head, Marthas Vineyard, Mass.; Glen Cove, Long Island, N. Y.

Diospyros berryana Knowlton, U. S. Geol. Surv., Prof. Paper 130, 19—, p. —, pl. xvii, f. 5 [in preparation].
  Laramie: 2 miles east of Lafayette, Colo.

Diospyros brachysepala Al. Braun, Die Tertiär Flora von Oeningen; Neues Jahrb., 1845, p. 170.—Lesquereux, U. S. Geol. and Geog. Surv. Terr., Ann. Rept. 1872 [1873], p. 394; idem, 1873 [1874], p. 401; idem, 1874 [1876], p. 306; idem, 1876 [1878], p. 511; idem, Bull., vol. 1, 1875, p. 367; Rept. U. S. Geol. Surv. Terr., vol. 7 (Tert. Fl.), 1878, p. 262, pl. xl, f. 7–10; pl. lxiii, f. 6; idem, vol. 8 (Cret. and Tert. Fl.), 1883, p. 174, pl. xxxiv, f. 1, 2; U. S. Nat. Mus., Proc., vol. 10, 1887, p. 41.—Ward, U. S. Geol. Surv., Sixth Ann. Rept., 1884–85 [1886], p. 556, pl. lx, f. 4, 5; idem, Bull. 37, 1887, p. 104, pl. xlix, f. 1, 2.—Knowlton, U. S. Geol. Surv., Mon. 32, pt. 2, 1899, p. 751; idem, Bull. 160, 1900, p. 74, pl. xviii, f. 3.—Cockerell, Univ. Colorado Studies, vol. 3, 1906, p. 169.—Berry, U. S. Geol. Surv., Prof. Paper 91, 1916, p. 333.—Knowlton, U. S. Nat. Mus., Proc., vol. 51, 1916, p. 285.—Berry, U. S. Nat. Mus., Proc., vol. 54, 1918, p. 635, pl. xcv, f. 3.
  Diospyros princetonia Cockerell, Am. Mus. Nat. Hist., Bull., vol. 24, 1906, p. 105, pl. x, f. 36.
  Andromeda dubia Lesquereux, Am. Jour. Sci., vol. 27, 1859, p. 364; Geol. Tenn., 1869, p. 428, pl. K, f. 5.
  Rhamnus inaequalis Lesquereux, U. S. Geol. and Geog. Surv. Terr., Ann. Rept. 1873 [1874], p. 405; Rept. U. S. Geol. Surv. Terr., vol. 7 (Tert. Fl.), 1878, p. 279, pl. liii, f. 16.
  Mesaverde (?): Point of Rocks, Wyo.
  Post-Laramie: Black Buttes, Wyo.
  Denver: Golden, Sand Creek, and Sedalia, Colo.
  Fort Union (?): Near Glendive, Mont.
  Miocene (intermediate): Yellowstone National Park.

Diospyros brachysepala—Continued.
Miocene: ?Florissant, Colo. [Identification?]
Lagrange: Puryear and Somerville, Tenn.
Wilcox: Calaveras Creek, Wilson County, Tex.
Miocene: Beaver Creek, 10 miles east of Beaver City, Okla.

Diospyros? celastroides Lesquereux, U. S. Geol. Surv., Mon. 17, 1891 [1892], p. 113, pl. xx, f. 7.
Dakota: Ellsworth County, Kans.

Diospyros copeana Lesquereux = *Cinchonidium copeanum.*

Diospyros cuspidata Kirchner = *Populus crassa.*

Diospyros elliptica Knowlton, U. S. Geol. Surv., Bull. 204, 1902, p. 83, pl. xvi, f. 5; in Smith, G. O., U. S. Geol. Surv., Folio 86 (Ellensburg, Wash.), 1903, p. 3.
Mascall: Near Van Horn's ranch, John Day Basin, Oreg.
Ellensburg: Ellensburg and Kelly Hollow, Wenas Valley, Wash.

Diospyros eminens Dawson, Roy. Soc. Canada, Trans., vol. 11, 1893 [1894], sec. 4, p. 62, pl. x, f. 40.
Upper Cretaceous: Port McNeill, Vancouver Island, British Columbia.

Diospyros? ficoidea Lesquereux, U. S. Geol. and Geog. Surv. Terr., Bull., vol. 1, 1875 [1876], p. 387; idem, Ann. Rept., 1874 [1876], p. 314; Rept. U. S. Geol. Surv. Terr., vol. 7 (Tert. Fl.), 1878, p. 231, pl. xl, f. 5, 6.—Ward, U. S. Geol. Surv., Sixth Ann. Rept., 1884–85 [1886], p. 556, pl. lx, f. 6, 7; idem, Bull. 37, 1887, p. 105, pl. xlix, f. 3, 4.
Adaville: Hodges Pass, Wyo.
Fort Union: Near Glendive, Mont.

Diospyros haguei Knowlton, U. S. Geol. Surv., Mon. 32, pt. 2, 1899, p. 752, pl. c, f. 3; Washington Acad. Sci., Proc., vol. 11, 1909, p. 189.
Fort Union: Below mouth of Elk Creek, Yellowstone National Park.
Lance: Near Miles City, Mont.

Diospyros hexaphylla (Lesquereux), Knowlton, n. comb.
Calycites hexaphylla Lesquereux, U. S. Geol. and Geog. Surv. Terr., Ann. Rept. 1872 [1873], p. 402.
Diospyros wodani Unger. Lesquereux, Rept. U. S Geol. Surv. Terr., vol. 7 (Tert. Fl.), 1878, p. 233, pl. lix, f. 13.
Evanston: Evanston, Wyo.

Diospyros judithae Knowlton, U. S. Geol. Surv., Bull. 257, p. 146, pl. xviii, f. 4, 5; pl. xix, f. 3.
Judith River: Willow Creek, Fergus County, Mont.

Diospyros lamarensis Knowlton, U. S. Geol. Surv., Mon. 32, pt. 2, 1899, p. 751, pl. xcv, f. 5, 6; pl. xcvi, f. 4.
Miocene: Lamar River, Yellowstone National Park.

Diospyros lancifolia Lesquereux = *Diospyros alaskana*.

Diospyros? leei Knowlton, U. S. Geol. Surv., Prof. Paper 101, 1917, p. 275, pl. xlviii, f. 2.

Vermejo: Vermejo Park, N. Mex.

Diospyros lesquereuxi Knowlton and Cockerell, n. name.

Diospyros rotundifolia Lesquereux, Rept. U. S. Geol. Surv. Terr., vol. 6 (Cret. Fl.), 1874, p. 111, pl. xxx, f. 1; U. S. Geol. Surv., Mon. 17, 1891 [1892], p. 112, pl. xvii, f. 8-11.—Berry, Geol. Surv. N. J., Ann. Rept. 1905 [1906], p. 139; Bull. Torr. Bot. Club, vol. 33, 1906, p. 181; Md. Geol. Surv., Upper Cret., 1916, p. 895, pl. xc, f. 3; U. S. Geol. Surv., Prof. Paper 112, 1919, p. 135, pl. xxvii, f. 4; pl. xxx, f. 415. [Homonym, Hiern, 1873.]

Dakota: Western Kansas.

Magothy: Cliffwood, N. J.; Glen Cove, Long Island, N. Y.; Deep Cut, Del.; Grove Point, Md.

Tuscaloosa: Shirleys Mill and Glen Allen, Ala.

Diospyros nebraskana Knowlton and Cockerell, n. name.

Diospyros ambigua Lesquereux, Rept. U. S. Geol. Surv. Terr., vol. 8 (Cret. and Tert. Fl.), 1883, p. 60; U. S. Geol. Surv., Mon. 17, 1891, p. 110. [Homonym, Ventenat, Jard. Malm., 1803-4, pl. xvii.]

Quercus anceps Lesquereux, Am. Jour. Sci., vol. 46, 1868, p. 96.

Diospyros anceps (Lesquereux) Lesquereux, Rept. U. S. Geol. Surv. Terr., vol. 6 (Cret. Fl.), 1874, p. 89, pl. vi, f. 6. [Homonym, Heer, 1859.]

Dakota: Nebraska.

Diospyros nitida Dawson, Roy. Soc. Canada, Trans., vol. 1, 1882 [1883], sec. 4, p. 22, pl. iii, f. 10.

Cretaceous: Peace River, Northwest Territory.

Diospyros? obtusata Ward, U. S. Geol. Surv., Sixth Ann. Rept., 1884-85 [1886], p. 556, pl. lx, f. 8; U. S. Geol. Surv., Bull. 37, 1887, p. 105, pl. xlix, f. 5.

Fort Union: Sevenmile Creek, Glendive, Mont.

Diospyros primaeva Heer, Phyll. crét. Nebr., 1866, p. 19, pl. i, f. 6, 7; Fl. foss. arct., vol. 6, Abt. 2, 1882, p. 80, pl. xviii, f. 11; idem, vol. 7, 1883, p. 31, pl. lx, f. 5a, b, c.—Lesquereux, U. S. Geol. Surv., Mon. 17, 1891 [1892], p. 109, pl. xx, f. 1-3.—Smith, Geol. Coastal Plain Ala., 1894, p. 348.—Newberry, U. S. Geol. Surv., Mon. 26, 1895 [1896], p. 124, pl. xxx, f. 1-5.—Knowlton, U. S. Geol. Surv., Twenty-first Ann. Rept., pt. 7, 1901, p. 317, pl. xxxix, f. 3.—Berry, Bull. Torr. Bot. Club, vol. 32, 1905, p. 46, pl. ii, f. 2; idem, vol. 34, 1907, p. 204; U. S. Geol. Surv., Prof. Paper 84, 1914, p. 61, pl. xi, f. 3; pl. xiv, f. 12, 13; Berry, Geol. Surv. N. J., Bull. 3, 1911, p. 211; Md. Geol. Surv., Upper Cret., 1916, p. 894, pl. xc, f. 4; Bull. Torr. Bot. Club, vol. 43, 1916, p. 303; U. S. Geol. Surv., Prof. Paper 112, 1919, p. 134, pl. xxx, f. 3.

Dakota: Ellsworth County, Kans.; Nebraska.

Magothy: Cliffwood, N. J.; Bodkin Point, Md.

Diospyros primaeva—Continued.
Black Creek: Court House Bluff, Cape Fear River, N. C.
Black Creek (Middendorf): Middendorf, Langley, and Rocky Point, S. C.
Raritan: South Amboy, N. J.; Bull Mountain, Cecil County, Md.
Woodbine: Arthurs Bluff, Tex.
Eutaw: Coffee Bluff, Hardin County, Tenn.; Havana, Ala.
Tuscaloosa: Shirleys Mill, Glen Allen, and Cottondale, Ala.

Diospyros primaeva Newberry (Lesquereux), Ill. Cret. and Tert. Fl., pl. iii, f. 8 = *Phyllites vanonae.*

Diospyros primaeva Heer. Hollick, N. Y. Acad. Sci., Trans., vol. 12, 1893, p. 236, pl. vii, f. 5; N. Y. State Mus., Fifty-fifth Ann. Rept., 1901 [1903], p. r51 = *Diospyros pseudoanceps.*

Diospyros princetonia Cockerell = *Diospyros brachysepala.*

Diospyros prodromus Heer. Hollick, U. S. Geol. Surv., Mon. 50, 1906, p. 104, pl. xl, f. 12.
Magothy: Glen Cove, Long Island, N. Y.

Diospyros provecta Velenovsky, Fl. böhm. Kreidef., pt. 3, 1884, p. 2 [49], pl. viii [xxiii], f. 1–5, 10.—Hollick, U. S. Geol. Surv., Mon. 50, 1906, p. 104, pl. xl, f. 7–10.
Rhamnus pfaffiana Heer. Hollick, N. Y. Acad. Sci., Trans., vol. 11, 1892, p. 103, pl. iv, f. 1.
Diospyros steenstrupi Heer. Hollick, idem, vol. 12, 1893, p. 34, pl. iii, f. 8.
Myrsine elongata Newberry. Hollick, N. Y. Bot. Gard., Bull., vol. 2, 1902, p. 405, pl. xli, f. 2.
Magothy: Gay Head and Chappaquiddick Island, Marthas Vineyard, Mass.
Raritan: Tottenville, Staten Island, N. Y.

Diospyros pseudoanceps Lesquereux, U. S. Geol. Surv., Mon. 17, 1891 [1892], p. 111, pl. xxii, f. 1; Geol. and Nat. Hist. Surv. Minn., vol. 3, 1893, p. 17, pl. B, f. 6.—Hollick, U. S. Geol. Surv., Mon. 50, 1906, p. 104, pl. xl, f. 3.
Diospyros primaeva Heer. Hollick, N. Y. Acad. Sci., Trans., vol. 12, 1893, p. 236, pl. vii, f. 5; N. Y. State Mus., Fifty-fifth Ann. Rept., 1901 [1903], p. r51.
Dakota: Kansas; New Ulm, Minn.
Magothy: Glen Cove, Long Island, N. Y.

Diospyros raritanensis Berry, Jour. Geol., vol. 18, 1910, p. 254 [name].
Raritan: New Jersey.

Diospyros rotundifolia Lesquereux = *Diospyros lesquereuxi.*

Diospyros rotundifolia Lesquereux. Hollick, Bull. Torr. Bot. Club, vol. 21, 1894, p. 53 = *Myrsine borealis.*

Diospyros steenstrupi? Heer, Fl. foss. arct., vol. 7, 1883, p. 32, pl. lxiv, f. 1; U. S. Geol. Surv., Mon. 17, 1891 [1892], p. 111, pl. xvi, f. 9.
Dakota: Kansas.

Diospyros steenstrupi Heer. Hollick, N. Y. Acad. Sci., Trans., vol. 12, 1893, p. 34, pl. iii, f. 8 = *Diospyros provecta*.

Diospyros stenosepala Heer, Fl. foss. arct., vol. 2, Abt. 2, 1869, p. 35, pl. viii, f. 7, 8.—Lesquereux, U. S. Geol. and Geog. Surv. Terr., Ann. Rept. 1871 [1872], p. 296.—Knowlton, U. S. Nat. Mus., Proc., vol. 17, 1894, p. 224; Geol. Soc. Am., Bull., vol. 5, 1893, p. 584; U. S. Geol. Surv., Mon. 32, pt. 2, 1899, p. 662.
Kenai: Ninilchik, Alaska.
"Laramie": Snake River below Yellowstone National Park.

Diospyros vancouverensis Dawson, Roy. Soc. Canada, Trans., vol. 1, 1882 [1883], sec. 4, p. 28, pl. viii, f. 32.
Upper Cretaceous: Nanaimo; Port McNeill, Vancouver Island.

Diospyros vera Berry, Bull. Torr. Bot. Club, vol. 38, 1911, p. 418, pl. xix, f. 5; Md. Geol. Surv., Upper Cret., 1916, p. 896, pl. xc, f. 5.
Raritan: East Washington Heights, D. C.

Diospyros virginiana turneri Lesquereux, U. S. Nat. Mus., Proc., vol. 11, 1888, p. 35.
Miocene: Contra Costa, Calif.

Diospyros wilcoxiana Berry, U. S. Geol. Surv., Prof. Paper 91, 1916, p. 334, pl. ci, f. 1, 2.
Lagrange: Puryear, Tenn.

Diospyros wodani Unger. Lesquereux = *Diospyros hexaphylla*.

Diospyros sp. Dawson, Brit. N. A. Bound. Com. (Rept. Geol. and Res. Vic. 49th Parallel), 1875, App. A, p. 330.
Paskapoo (Fort Union): Porcupine Creek, Alberta.

Diospyros (calyx) Dawson, Roy. Soc. Canada, Trans., vol. 11, 1893 [1894], sec. 4, p. 62, pl. x, f. 41.
Upper Cretaceous: Port McNeill, Vancouver Island.

Diplazium mülleri? Heer. Lesquereux = *Asplenium crossii*.

DISTICHIUM Bruchman, Schimper, and Gümbel, Bryol. Europe. Foss., 1846, pp. 18–20.

Distichium capillaceum Bruchman and Schimper. Penhallow, Roy. Soc. Canada, Trans., 2d ser., vol. 2, 1896, sec. 4, p. 62; Brit. Assoc. Adv. Sci., Bradford meeting, 1900, p. C335 [7].
Pleistocene: Long Portage, Missinaibi River.

DODONAEA (Plumier) Linné, Mantissa plantarum, vol. 2, 1771, p. 149.

Dodonaea knowltoni Berry, U. S. Geol. Surv., Prof. Paper 91, 1916, p. 271, pl. lxiv, f. 3.
Lagrange: Puryear, Henry County, Tenn.

Dodonaea viscosoides Berry, U. S. Geol. Surv., Prof. Paper 84, 1914, p. 142, pl. xxviii, f. 4–8.
Barnwell (Twiggs): Phinizy Gully, Columbia County, Grovetown, and near Macon, Ga.

Dodonaea wilcoxiana Berry, U. S. Geol. Surv., Prof. Paper 91, 1916, p. 270, pl. xxxviii, f. 2.
Wilcox (Holly Springs): Holly Springs, Marshall County, Miss.

Dodonaea sp. [seeds of] Lesquereux, Rept. U. S. Geol. Surv. Terr., vol. 8 (Cret. and Tert. Fl.), 1883, p. 182, pl. xxxvi, f. 5.—Knowlton, U. S. Nat. Mus., Proc., vol. 51, 1916, p. 276.
Miocene: Florissant, Colo.

DOLICHITES Unger, Steiermärkische Zeitschr., Gratz (n. F.), vol. 5, 1838 [1839], p. 104 (30).

Dolichites deusseni Berry, U. S. Geol. Surv., Prof. Paper 91, 1916, p. 14, pl. iii, f. 3.
Midway?: Earle, Tex.

DOMBEYOPSIS Unger, Steiermärkische Zeitschr., Gratz, vol. 9, 1 heft, 1848, p. 47(?) [24 of reprint]; Gen. et sp. pl. foss., 1850, p. 447.

Dombeyopsis aequifolia Göppert. Lesquereux, U. S. Geol. and Geog. Surv. Terr., Ann. Rept. 1871 [1872], Suppl., p. 10 = *Ficus tiliaefolia.*

Dombeyopsis grandifolia? Unger, Gen. et sp. pl. foss., 1850, p. 447.—Lesquereux, U. S. Geol. and Geog. Surv. Terr., Ann. Rept. 1873 [1874], p. 404; Rept. U. S. Geol. Surv. Terr., vol. 7 (Tert. Fl.) 1878, p. 255, pl. xlvii, f. 6.
Denver: Golden, Colo.

Dombeyopsis islandica Heer, Fl. tert. Helv., vol. 3, 1859, p. 319; Fl. foss. arct., vol. 1, 1868, p. 151, pl. xxvii, f. 10.—Dawson, Geol. Surv. Canada, 1875–76.—Penhallow, Rept. Tert. Pl. Brit. Col., 1908, p. 52.
Oligocene: Quesnel, British Columbia.

Dombeyopsis magnifolia Knowlton, U. S. Geol. Surv., Prof. Paper 101, 1917, p. 337, pl. cii, f. 2.
Raton: Ute Park, N. Mex.

Dombeyopsis obtusa Lesquereux, U. S. Geol. and Geog. Surv. Terr., Ann. Rept. 1872 [1873], p. 375; Rept. U. S. Geol. Surv. Terr., vol. 7 (Tert. Fl.), 1878, p. 255, pl. xlvii, f. 4, 5.—Knowlton, U. S. Geol. Surv., Prof. Paper 130, 19—, p. —, —, pl. xiii, f. 4; pl. xx, f. 12; pl. xxvii, f. 1–4 [in preparation].
Laramie: *Gehrung's coal mine, Colorado Springs, near Golden, Coal Creek, and Marshall, Colo.
Denver: Golden and Sedalia, Colo.
Dawson: Pulpit Rock, Colorado Springs, Colo.

Dombeyopsis obtusiloba Lesquereux, var. ? = *Menispermites obtusiloba.*

Dombeyopsis occidentalis Lesquereux = *Ficus occidentalis.*

Dombeyopsis ovata Knowlton, U. S. Geol. Surv., Prof. Paper 130, 19—, p. —, pl. xxiv, f. 1-3, 9 [in preparation].
Laramie: Erie, Colo.

Dombeyopsis platanoides Lesquereux, U. S. Geol. Surv. Terr., vol. 7 (Tert. Fl.), 1878, p. 254, pl. xlvii, f. 1, 2.—Knowlton, U. S. Geol. Surv., Bull. 105, 1893, p. 57; U. S. Geol. Surv., Mon. 32, pt. 2, 1899, p. 661, pl. lxxviii, f. 1.
Quercus platania var. *rotundifolia* Lesquereux, U. S. Geol. and Geog. Surv. Terr., Ann. Rept. 1872 [1873], p. 405.
——: Mount Everts, Yellowstone National Park.
Livingston: *Spring Canyon near Fort Ellis, Mont.

Dombeyopsis? sinuata Knowlton, U. S. Geol. Surv., Prof. Paper 130, 19—, p. —, pl. xxv, f. 1, 2 [in preparation].
Laramie: Coal Creek, Boulder County, Colo.

Dombeyopsis tiliaefolia (Al. Braun) Unger = *Ficus tiliaefolia.*

Dombeyopsis trivialis Lesquereux, U. S. Geol. and Geog. Surv. Terr., Ann. Rept. 1872 [1873], p. 380; Rept. U. S. Geol. Surv. Terr., vol. 7 (Tert. Fl.), 1878, p. 255, pl. xlvii, f. 3.—Knowlton, U. S. Geol. Surv., Prof. Paper 130, 19—, p. —, pl. xii, f. 3; pl. xiv, f. 3 [in preparation].
Laramie: *Golden and 4 miles north of Colorado Springs, Colo.

Dombeyopsis sp. Knowlton, Jour. Geol., vol. 19, 1911, p. 370.
Lance: Sec. 7, T. 17 N., R. 24 E., S. Dak.

DORYANTHES Berry, Bull. Torr. Bot. Club, vol. 38, 1911, p. 406. [Type, *D. cretacea.*]

Doryanthes cretacea Berry, Bull. Torr. Bot. Club, vol. 38, 1911, p. 406; U. S. Geol. Surv., Prof. Paper 84, 1914, p. 108, pl. xvii, f. 3; Md. Geol. Surv., Upper Cret., 1916, p. 806, pl. lvi, f. 6; U. S. Geol. Surv., Prof. Paper 112, 1919, p. 70, pl. xiii, f. 1.
Magothy: Cliffwood, N. J.; Severn River, Md.
Ripley (Cusseta): Buena Vista, Ga.
Eutaw: Near Havana, Hale County, Ala.

DORSTENIA (Plumier) Linné, Sp. pl. 1753, p. 176.

Dorstenia? sp. Penhallow, Roy. Soc. Canada, Trans., 3d ser., vol. 1, sec. 4, 1907, p. 310, text f. 5.
Lower Cretaceous?: International boundary between Pasayten and Skagit rivers, British Columbia.

DROSERA Linné, Sp. pl., 1753, p. 281.

Drosera rotundifolia Linné. Dawson, Can. Nat., new ser., vol. 3, 1868, p. 70; Geol. Hist. Pl., 1888, p. 228.—Penhallow, Roy. Soc. Canada, Trans., 2d ser., vol. 2, 1896, sec. 4, p. 70; Brit. Assoc. Adv. Sci., Bradford meeting, 1900, p. 335.
Pleistocene: Greens Creek, Ottawa River, Canada.

DRUPA Lesquereux, Am. Jour. Sci., 2d ser., vol. 32, 1861, p. 360. [Type, *D. rhabdosperma.*]

Drupa rhabdosperma Lesquereux, Am. Jour. Sci., 2d ser., vol. 32, 1861, p. 360; Geol. Vt., vol. 2, 1861, p. 716, f. 150.—Perkins, Rept. Vt. State Geol., 1903–4 [1904], p. 210, pl. lxxxi, f. 168–170; Geol. Soc. Am., Bull., vol. 16, 1905, p. 514, pl. lxxxv. ii, f. 25; Rept. Vt. State Geol. 1905–6 [1906], p. 228, pl. liii, f. 25.
  Miocene: Brandon, Vt.

Dryandra vindobonensis Ettingshausen = *Myrica vindobonensis.*

DRYANDROIDES Unger, Gen. sp. pl. foss., 1850, p. 428.

Dryandroides cleburni (Lesquereux) Ettingshausen, Denks. K. Akad. d. W., Math.-naturw. Classe, vol. 47, 1883, p. 130 [30]; Geol. Surv. N. S. W., Mem. Palaeont. No. 2 (Tert. Fl. Australia), 1888, p. 48.
    Quercus cleburni Lesquereux, U. S. Geol. and Geog. Surv. Terr., Ann. Rept. 1873 [1874], p. 399; Rept. U. S. Geol. Surv. Terr., vol. 7 (Tert. Fl.), 1878, p. 154, pl. xx, f. 2.
  Post-Laramie: Black Buttes, Wyo.

Dryandroides lanceolata Knowlton, U. S. Geol. Surv., Prof. Paper 108, 1917, p. 89, pl. xxxiv, f. 7.
  Frontier: About 1½ miles south of Cumberland, Wyo.

Dryandroides lignitum (Unger) Ettingshausen. Lesquereux, U. S. Nat. Mus., Proc., vol. 10, 1887, p. 41 = *Myrica lignitum.*

Dryandroides quercinea Velenovsky, Fl. böhm. Kreideform., pt. 2, 1883, p. 8 (33), pl. ii (x), f. 8a–15.—Hollick, U. S. Geol. Surv., Mon. 50, 1906, p. 60, pl. viii, f. 18, 19.
  Magothy: Gay Head, Marthas Vineyard, Mass.

Dryandroides undulata Heer = *Myrica undulata.*

DRYOPHYLLUM Debey, Sur les feuilles querciformes d'Aix-la-Chapelle, 1881, p. 7.

Dryophyllum amplum Berry, U. S. Geol. Surv., Prof. Paper 91, 1916, p. 193, pl. cxvii, f. 1–4.
  Wilcox: Naborton, De Soto Parish, La.

Dryophyllum anomalum Berry, U. S. Geol. Surv., Prof. Paper 91, 1916, p. 189, pl. xxiv, f. 2, 3.
  Lagrange: Puryear, Henry County, Tenn.

Dryophyllum aquamarum Ward, U. S. Geol. Surv., Sixth Ann. Rept., 1884–85 [1886], p. 551, pl. xxxvii, f. 3–5; U. S. Geol. Surv., Bull. 37, 1887, p. 26, pl. x, f. 2–4.—Knowlton, U. S. Geol. Surv., Prof. Paper 101, 1918, p. 299, pl. lxx, f. 2.
  Post-Laramie: Black Buttes, Wyo.
  ?Raton: Wootton, Colo.

Dryophyllum basidentatum Ward = *Quercus carbonensis.*

Dryophyllum bruneri Ward, U. S. Geol. Surv., Sixth Ann. Rept., 1884–85 [1886], p. 551, pl. xxxvi, f. 6–9; idem, Bull. 37, 1887, p. 27, pl. x, f. 5–8; idem, Prof. Paper 101, 1918, p. 259, pl. liii, f. 5.
?Adaville: Hodges Pass, Wyo.
Mesaverde: Point of Rocks, Wyo.
Trinidad: Canon City field, Colo.

Dryophyllum (Quercus) crenatum Lesquereux, U. S. Geol. and Geog. Surv. Terr., Bull., vol. 1, 1875 [1876], p. 371; idem, Ann. Rept. 1874 [1876], p. 301; Rept. U. S. Geol. Surv. Terr., vol. 7 (Tert. Fl.), 1878, p. 162, pl. lxii, f. 10, 11.—Knowlton, U. S. Geol. Surv., Bull. 163, 1900, p. 42.
Mesaverde: Point of Rocks, Wyo.

Dryophyllum elongatum Dawson = *Quercus vancouveriana.*

Dryophyllum falcatum Ward, U. S. Geol. Surv., Sixth Ann. Rept., 1884–85 [1886], p. 551, pl. xxxvii, f. 10; U. S. Geol. Surv., Bull. 37, 1887, p. 27, pl. xi, f. 1.—Knowlton, U. S. Geol. Surv., Bull. 163, 1900, p. 42, pl. viii, f. 1.
Adaville: Hodges Pass, Wyo.
Mesaverde: Point of Rocks, Wyo.

Dryophyllum gracile Debey, Feuilles querciformes d'Aix-la-Chapelle: Compte rendu Congrès bot. et hort. Bruxelles 10, 1887, f. 10, 11.—Berry, Bull. Torr. Bot. Club, vol. 43, 1916, p. 290, pl. xvi, f. 6; U. S. Geol. Surv., Prof. Paper 112, 1919, p. 75, pl. xxxii, f. 2.
Ripley: Near Cypress and near Selmer, McNairy County, Tenn.

Dryophyllum (Quercus) holmesii Lesquereux = *Quercus holmesii.*

Dryophyllum (Quercus) latifolium Lesquereux = *Quercus dryophyllopsis.*

Dryophyllum longipetiolatum Knowlton, U. S. Geol. Surv., Mon. 32, pt. 2, 1899, p. 710, pl. lxxxviii, f. 6, 7.
Miocene: Yellowstone National Park.

Dryophyllum moorii (Lesquereux) Berry, U. S. Geol. Surv., Prof. Paper 91, 1916, p. 190, pl. xxii, f. 1; pl. xxiii, f. 1–3.—Knowlton, U. S. Geol. Surv., Prof. Paper 101, 1917, p. 299, pl. lxx, f. 1.

Quercus moorii Lesquereux, Am. Philos. Soc., Trans., vol. 13, 1869, p. 415, pl. xvi, f. 1–3; U. S. Nat. Mus., Proc., vol. 11, 1888, p. 31.—Knowlton, in Glenn, U. S. Geol. Surv., Water-Supply Paper 164, 1906, p. 38; in Lindgren, U. S. Geol. Surv., Prof. Paper 73, 1911, pp. 60, 61.

Wilcox (Ackerman): Hurleys, Benton County, Miss.
Wilcox: Old Port Caddo Landing, Harrison County, Tex.
Lagrange: Wickliffe, Ballard County, Ky.
Raton: Near Dawson, N. Mex.

Dryophyllum neillianum Dawson = *Quercus vancouveriana.*

Dryophyllum occidentale Dawson = *Quercus vancouveriana.*

Dryophyllum primordiale Lesquereux = *Quercus primordialis.*

Dryophyllum puryearensis Berry, U. S. Geol. Surv., Prof. Paper 91, 1916, p. 192, pl. xxi, f. 2, 3.

Wilcox (Grenada): Grenada, Grenada County, Miss.

Lagrange: Puryear, Henry County, Tenn.; Wickliffe, Ballard County, Ky.

Dryophyllum (*Quercus*) *salicifolium* Lesquereux = *Quercus holmesii.*

Dryophyllum stanleyanum Dawson, Roy. Soc. Canada, Trans., 2d ser., vol. 1, 1895 [1896], sec. 4, p. 147, pl. vii, f. 13.—Penhallow, Rept. Tert. Pl. Brit. Col., 1908, p. 52.

Eocene: Stanley Park, Vancouver, British Columbia.

?Oligocene: Quilchena, British Columbia.

Dryophyllum subfalcatum Lesquereux, U. S. Geol. and Geog. Surv. Terr., Bull., vol. 1, 1875 [1876], p. 379; idem, Ann. Rept. 1874 [1876], p. 301; Rept. U. S. Geol. Surv. Terr., vol. 7 (Tert. Fl.), 1878, p. 163, pl. lxiii, f. 10.—Knowlton, U. S. Geol. Surv., Bull. 163, 1900, p. 41.

Quercus gracilis Newberry, U. S. Nat. Mus., Proc., vol. 5, 1882 [1883], p. 504; U. S. Geol. Surv., Mon. 35, 1898, p. 75, pl. lxvii, f. 4.

Mesaverde: Point of Rocks, Wyo.

Dryophyllum tennesseensis Berry, U. S. Geol. Surv., Prof. Paper 91, 1916, p. 191, pl. xix, f. 6; pl. xx, f. 1–3; pl. xxi, f. 1, 4, 5; pl. xxii, f. 2.—Knowlton, U. S. Geol. Surv., Prof. Paper 101, 1919, p. 299, pl. lxix, f. 3–5.

Quercus ! crassinervis ? Göppert. Lesquereux, Am. Jour. Sci., vol. 27, 1859, p. 364; Geol. Tenn., 1869, p. 427, pl. K, f. 1.

Quercus cf. *Q. cuspidata* (Rossmässler) Unger. Lesquereux, U. S. Nat. Mus., Proc., vol. 11, 1888, p. 12.

Wilcox (Holly Springs): Oxford and Holly Springs, Miss.

Wilcox (Grenada): Grenada, Grenada County, Miss.

Wilcox: Naborton, De Soto Parish, La.

Lagrange: *Somerville, Fayette County, and Puryear, Henry County, Tenn.; Graves County, Ky.; Wickliffe, Ballard County, Ky.

Raton: Rugby, Colo.

Dryophyllum sp. Dawson = *Quercus vancouveriana.*

DRYOPTERIS Adanson, Fam. Pl., vol. 2, 1763, p. 20.

Dryopteris angustipinnata (Fontaine) Knowlton = *Cladophlebis albertsii.*

Dryopteris angustipinnata montanense Fontaine (Knowlton) = *Cladophlebis albertsii montanense.*

Dryopteris arguta (Lesquereux) Knowlton, U. S. Geol. Surv., Bull. 152, 1898, p. 91.

Lathraea arguta Lesquereux, Am. Jour. Sci., 2d ser., vol. 45, 1868, p. 207.

Sphenopteris eocenica Ettingshausen. Lesquereux, U. S. Geol. and Geog. Surv. Terr., Ann. Rept. 1872 [1873], p. 376.

Sphenopteris membranacea Lesquereux, U. S. Geol. and Geog. Surv. Terr., Ann. Rept. 1873 [1874], p. 394; Rept. U. S. Geol. Surv. Terr., vol. 7 (Tert. Fl.), 1878, p. 50, pl. ii, f. 2–3a.

Sphenopteris lakesii Lesquereux, Rept. U. S. Geol. Surv. Terr., vol. 7 (Tert. Fl.), 1878, p. 49, pl. ii, f. 1, 1a.

Aspidium lakesii (Lesquereux) Knowlton, Biol. Soc. Washington, Proc., vol. 7, 1892, p. 154; U. S. Geol. Surv., Bull. 105, 1893, p. 45, pl. vi, f. 1–4.

Denver: Golden, Colo.

Dryopteris? carbonensis Knowlton, U. S. Geol. Surv., Prof. Paper 130, 19—, p. —, pl. xx, f. 3–5 [in preparation].

Laramie: Mount Carbon, Morrison, Colo.

Dryopteris? civica (Dawson) Knowlton, n. comb.

Neuropteris civica Dawson, Roy. Soc. Canada, Trans., 2d ser., vol. 1, 1895–96, p. 141, pl. iv, f. 2.—Penhallow, Rept. Tert. Pl. Brit. Col., 1908, p. 64.

Eocene: Burrard Inlet, Vancouver, British Columbia.

Dryopteris? cladophleboides Knowlton, U. S. Geol. Surv., Prof. Paper 101, 1918, p. 284, pl. liv, f. 1.

Raton: Tercio, Colo.

Dryopteris coloradensis Knowlton, U. S. Geol. Surv., Prof. Paper 108, 1917, p. 83, pl. xxx, f. 3, 4.

Frontier: About 1 mile east of Cumberland, Wyo.

Dryopteris cystopteroides (Fontaine) Knowlton = *Dryopterites cystopteroides.*

Dryopteris dentata (Fontaine) Knowlton = *Dryopterites dentata.*

Dryopteris elliptica (Fontaine) Knowlton = *Dryopterites elliptica.*

Dryopteris fredericksburgensis (Fontaine) Knowlton = *Cladophlebis distans.*

Dryopteris georgii Knowlton, U. S. Geol. Surv., Prof. Paper 130, 19—, p. —, pl. i, f. 6, 7 [in preparation].

Laramie: Columbia mine, near Louisville Junction, Colo.

Dryopteris? gleichenoides Knowlton, U. S. Geol. Surv., Twenty-first Ann. Rept., pt. 2, 1901, p. 211, pl. xxx, f. 5–7.

Esmeralda: 3.8 kilometers northeast of Emigrant Peak, Esmeralda County, Nev.

Dryopteris heterophylla (Fontaine) Knowlton = *Cladophlebis parva.*

Dryopteris idahoensis Knowlton, U. S. Geol. Surv., Eighteenth Ann. Rept., pt. 3, 1898, p. 721, pl. xcix, f. 1, 2.

Payette: Marsh, Idaho.

Dryopteris kennerlyi (Newberry) Knowlton, U. S. Geol. Surv., Bull. 152, 1898, p. 92.

Aspidium kennerlyi Newberry, Boston Jour. Nat. Hist., vol. 7, 1863, p. 513; U. S. Geol. Surv., Mon. 35, 1898, p. 11, pl. xvi, f. 4, 5.

Upper Cretaceous: Nanaimo, Vancouver Island, British Columbia.

Dryopteris? kootaniensis Knowlton, Smithsonian Misc. Coll., vol. 50, 1907, p. 111, pl. xi, f. 4, 4a.

Kootenai: 6 miles southwest of Geyser, Cascade County, Mont.

Dryopteris laramiensis Knowlton, U. S. Geol. Surv., Prof. Paper 130, 19—, p. —, pl. i, f. 5 [in preparation].

Lastrea (Goniopteris) intermedia Lesquereux, Rept. U. S. Geol. Surv. Terr., vol. 7 (Tert. Fl.), 1878, p. 56, pl. iv, f. 14. [*intermedia* not available in *Dryopteris* on account of *D. intermedia* (Gray), 1848.]

Aspidium (Lastrea) pulchellum? Heer, or *A. fischeri?* Heer. Lesquereux, U. S. Geol. and Geog. Surv. Terr., Ann. Rept. 1871 [1872], p. 384.

Laramie: Golden, Colo.

Dryopteris lesquereuxii Knowlton, U. S. Geol. Surv., Prof. Paper 130, 19—, p. —.

Aspidium goldianum Lesquereux, U. S. Geol. and Geog. Surv. Terr., Ann. Rept. 1873 [1874], p. 393. [Homonym, *Dryopteris goldiana* (Hooker) Gray, 1848.]

Lastrea (Goniopteris) goldianum (Lesquereux) Lesquereux, Rept. U. S. Geol. Surv. Terr., vol. 7 (Tert. Fl), 1878, p. 56, pl. iv, f. 13.

Laramie: Golden, Colo.

Dryopteris lloydii Knowlton, Torreya, vol. 15, 1915, p. 68, f. 1–5.

Judith River: Near mouth of Judith River, Fergus County, Mont.

Dryopteris macrocarpa (Fontaine) Knowlton = *Dryopterites macrocarpa.*

Dryopteris microcarpa (Fontaine) Knowlton = *Dryopterites dentata.*

Dryopteris monocarpa (Fontaine) Knowlton. Ward, U. S. Geol. Surv., Twentieth Ann. Rept., pt. 2, 1900, p. 369 = *Dicksonia oregonensis.*

Dryopteris montanensis (Fontaine) Knowlton, U. S. Geol. Surv., Bull. 152, 1898, p. 92; Smithsonian Misc. Coll., vol. 50, 1907, p. 111.

Aspidium montanense Fontaine, U. S. Nat. Mus., Proc., vol. 15, 1892, p. 490, pl. lxxxii, f. 1–3; pl. lxxxiii, f. 1–3a.

Kootenai: Great Falls, Mont.

Dryopteris nigricans (Lesquereux) Knowlton, n. comb.

Sphenopteris nigricans Lesquereux, U. S. Geol. and Geog. Surv. Terr., Ann. Rept. 1873 [1874], p. 394; Rept. U. S. Geol. Surv. Terr., vol. 7 (Tert. Fl.), 1878, p. 51, pl. ii, f. 4–5a.

Denver: Golden, Colo.

Post-Laramie: Black Buttes, Wyo.

Dryopteris oblongifolia (Fontaine) Knowlton = *Dryopterites virginica.*

Dryopteris oerstedi? (Heer) Knowlton = *Cladophlebis albertsii.*

Dryopteris parvifolia (Fontaine) Knowlton = *Cladophlebis ungeri.*

Dryopteris parvifolia (Fontaine) Knowlton. Fontaine, in Ward = *Dicksoniopsis vernonensis.*

Dryopteris pinnatifida (Fontaine) Knowlton = *Dryopterites pinnatifida.*

Dryopteris polypodioides (Ettingshausen) Knowlton, U. S. Geol. Surv., Bull. 152, 1898, p. 93.

Goniopteris polypodioides Ettingshausen. Lesquereux, U. S. Geol. and Geog. Surv. Terr., Ann. Rept. 1873 [1874], p. 394.

Lastrea (Goniopteris) polypodioides? (Ettingshausen) Lesquereux, Rept. U. S. Geol. Surv. Terr., vol. 7 (Tert. Fl.), 1878, p. 57, pl. iv, f. 11, 12.

Denver: Sand Creek, Colo.

Dryopteris scansa Cockerell, Am. Mus. Nat. Hist., Bull., vol. 24, 1908, p. 76, pl. vi, f. 9; pl. x, f. 38.—Knowlton, U. S. Nat. Mus., Proc., vol. 51, 1916, p. 246.

Miocene: *Florissant, Colo.

Dryopteris stephensoni Berry, U. S. Geol. Surv., Prof. Paper 84, 1914, p. 103, pl. xvii, f. 1, 2; idem, Prof. Paper 112, 1919, p. 52.

Ripley: Near Byron, Houston County, Ga.

Tuscaloosa: Snow place, Tuscaloosa County, Ala.

Dryopteris virginica (Fontaine) Knowlton = *Dryopterites virginica.*

Dryopteris virginica (Fontaine) Knowlton. Fontaine, in Ward = *Dicksoniopsis vernonensis.*

Dryopteris weedii Knowlton, U. S. Geol. Surv., Mon. 32, pt. 2, 1899, p. 669, pl. lxxx, f. 8; pl. lxxxi, f. 2.

Fort Union: Near mouth of Elk Creek, Yellowstone National Park.

Dryopteris xantholithensis Knowlton, U. S. Geol. Surv., Mon. 32, pt. 2, 1899, p. 671, pl. lxxxi, f. 1.

Miocene: Fossil Forest Ridge, Yellowstone National Park.

Dryopteris sp. Berry, Bull. Torr. Bot. Club, vol. 37, 1910, p. 505.

Ripley (Cusseta): Near Byron, Ga.

Dryopteris sp. (Knowlton) Knowlton, U. S. Geol. Surv., Bull. 152, p. 93.

Aspidium sp. Knowlton, Geol. Soc. Am., Bull., vol. 8, 1897, p. 140.

——: North fork of Dutton Creek, Laramie Plains, Wyo.

Dryopteris sp. Knowlton, Washington Acad. Sci., Proc., vol. 11, 1909, p. 191.

Lance: Forsyth, Mont.

Dryopteris? sp. Knowlton, U. S. Geol. Surv., Prof. Paper 101, 1918, p. 284.

Raton: Yankee, N. Mex.

DRYOPTERITES Berry, Md. Geol. Surv., Lower Cret., 1911, p. 259. [Type, *Aspidium macrocarpum* Fontaine.]

Dryopterites cystopteroides (Fontaine) Berry, Md. Geol. Surv., Lower Cret., 1911, p. 262.

Aspidium cystopteroides Fontaine, U. S. Geol. Surv., Mon. 15, 1889 [1890], p. 99, pl. xvi, f. 2.

Dryopteris cystopteroides (Fontaine) Knowlton, U. S. Geol. Surv., Bull. 152, 1898, p. 92.

Patuxent: Fredericksburg, Va.

Dryopterites dentata (Fontaine) Berry, Md. Geol. Surv., Lower Cret., 1911, p. 263.

Aspidium dentatum Fontaine, U. S. Geol. Surv., Mon. 15, 1889 [1890], p. 102, pl. xxv, f. 6, 7, 14, 15.

Dryopteris dentata (Fontaine) Knowlton, U. S. Geol. Surv., Bull. 152, 1898, p. 91.

Aspidium microcarpum Fontaine, U. S. Geol. Surv., Mon. 15, 1889 [1890], p. 103, pl. lix, f. 2, 12; pl. lx, f. 6, 7.

Dryopteris microcarpa (Fontaine) Knowlton, U. S. Geol. Surv., Bull. 152, 1898, p. 92.

Patuxent: Dutch Gap, Potomac Run, and telegraph station near Lorton, Va.

Dryopterites elliptica (Fontaine) Berry, Md. Geol. Surv., Lower Cret., 1911, p. 262.

Aspidium ellipticum Fontaine, U. S. Geol. Surv., Mon. 15, 1890, p. 95, pl. xiii, f. 9, 10.

Dryopteris elliptica (Fontaine) Knowlton, U. S. Geol. Surv., Bull. 152, 1898, p. 92.

Patuxent: Potomac Run, Va.

Patapsco: Near Brooke, Va.

Dryopterites macrocarpa (Fontaine) Berry, Md. Geol. Surv., Lower Cret., 1911, p. 261.

Aspidium macrocarpum Fontaine, U. S. Geol. Surv., Mon. 15, 1890, p. 103, pl. xvii, f. 2.

Dryopteris macrocarpa (Fontaine) Knowlton, U. S. Geol. Surv., Bull. 152, 1898, p. 92.

Patuxent: Dutch Gap, Va.

Dryopterites pinnatifida (Fontaine) Berry, Md. Geol. Surv., Lower Cret., 1911, p. 261.

Aspidium pinnatifidum Fontaine, U. S. Geol. Surv., Mon. 15, 1890, p. 101, pl. xxi, f. 15.

Dryopteris pinnatifida (Fontaine) Knowlton, U. S. Geol. Surv., Bull. 152, 1898, p. 92.

Patapsco: Near Brooke, Va.

Dryopterites stephensoni Berry, U. S. Geol. Surv., Prof. Paper 84, 1914, p. 103, pl. xvii, f. 1, 2.

Ripley (Cusseta): Byron, Ga.

Dryopterites virginica (Fontaine) Berry, Md. Geol. Surv., Lower Cret., 1911, p. 264.

Aspidium virginicum Fontaine, U. S. Geol. Surv., Mon. 15, 1889 [1890], p. 97, pl. xv, f. 14.

Dryopteris virginica (Fontaine) Knowlton, U. S. Geol. Surv., Bull. 152, 1898, p. 93.

Aspidium oblongifolium Fontaine, U. S. Geol. Surv., Mon. 15, 1889 [1890], p. 100, pl. xxi, f. 5.

Dryopteris oblongifolia (Fontaine) Knowlton, U. S. Geol. Surv., Bull. 152, 1898, p. 92.

Patuxent: Potomac Run, Va.
Patapsco: Near Brooke?, Va.

DRYPETES Vahl, Eclog., fasc. 3, 1815, p. 49.

Drypetes praelaterifolia Berry, U. S. Geol. Surv., Prof. Paper 91, 1916, p. 258, pl. lviii, f. 3.

Wilcox (Holly Springs): Holly Springs, Marshall County, Miss.

Drypetes prekeyensis Berry, U. S. Geol. Surv., Prof. Paper 91, 1916, p. 258, pl. lviii, f. 4.

Lagrange: Puryear, Henry County, Tenn.

DYCTUOCAULUS Emmons, Geol. Rept. Midland Counties N. C., 1856, p. 293.

Dyctuocaulus striatus Emmons, Geol. Rept. Midland Counties N. C., 1856, p. 293, pl. 1, f. 3; American geology, pt. 6, 1857, p. 38, pl. vi, f. 3.—Ward, U. S. Geol. Surv., Twentieth Ann. Rept., pt. 2, 1900, p. 426.

Triassic (Newark): Farmville, N. C.; Pennsylvania?.

EBENOXYLON Felix, Stud. foss. Hölz., 1882, p. 71.

Ebenoxylon boreale Platen, Sitzungsb. Naturf. Ges., Leipzig, vol. 34, 1907 [1908], p. 147.

Kenai?: Alaska [no locality].

Ebenoxylon speciosum Platen, Sitzungsb. Naturf. Ges., Leipzig, vol. 34, 1907 [1908], p. 68.

Miocene (auriferous gravels): Nevada County, Calif.

ECHITONIUM Unger, Syn. pl. foss., 1845, p. 230.

Echitonium lanceolatum Ettingshausen, Beiträge z. Kenntniss d. Tertiarfl. Australiens. K. Akad. Wiss. Wien Denkschr., vol. 47, 1883, p. 134.—Berry, U. S. Geol. Surv., Prof. Paper 91, 1916, p. 345, pl. ciii, f. 1.

Lagrange: Puryear, Tenn.

Echitonium sophiae O. Weber, Palaeontogr., vol. 2, 1852, p. 187, pl. xx, f. 17.—Lesquereux, Am. Jour. Sci., 2d ser., vol. 45, 1868, p. 206; U. S. Nat. Mus., Proc., vol. 10, 1887, p. 41.

——: White River, Utah.

Elaeagnus inaequalis Lesquereux = *Chrysobalanus inaequalis.*

ELAEODENDRON Jacques f., Nova Acta Helv., vol. 1, 1787, p. 36.

Elaeodendron helveticum Heer, Fl. tert. Helv., vol. 3, 1859, p. 71, pl. cxxii, f. 5.—Lesquereux, U. S. Nat. Mus., Proc., vol. 5, 1882 [1883], p. 449; Rept. U. S. Geol. Surv. Terr., vol. 8 (Cret. and Tert. Fl.), 1883, p. 263.—Knowlton, U. S. Nat. Mus., Proc., vol. 17, 1894, p. 228; Geol. Soc. Am., Bull., vol. 5, 1893, p. 586.
Kenai: Unga Island, Alaska.

Elaeodendron marylandicum Berry, Bull. Torr. Bot. Club, vol. 37, 1910, p. 24, pl. viii, f. 1; Md. Geol. Surv., Upper Cret., 1916, p. 849, pl. lxxvii, f. 3–6.
Magothy: Grove Point, Md.

Elaeodendron polymorphum Ward, U. S. Geol. Surv., Sixth Ann. Rept., 1884–85 [1886], p. 555, pl. liv, f. 6–12; U. S. Geol. Surv., Bull. 37, 1887, p. 84, pl. xxxviii, f. 1–7.—Knowlton, U. S. Geol. Surv., Mon. 32, pt. 2, 1899, p. 734, pl. xcvii, f. 1; Washington Acad. Sci., Proc., vol. 11, 1909, pp. 189, 198.
Fort Union: Burns's ranch, Glendive, Mont.; Yancey fossil forest, Yellowstone National Park.
Lance: West of Miles City and mouth of Cedar Creek, near Glendive, Mont.

Elaeodendron serrulatum Ward, U. S. Geol. Surv., Sixth Ann. Rept., 1884–85 [1886], p. 555, pl. liv, f. 3–5; U. S. Geol. Surv., Bull. 37, 1887, p. 83, pl. xxxvii, f. 3–5.
Fort Union: Burns's ranch and Sevenmile Creek, Glendive, Mont.

Elaeodendron speciosum Lesquereux, U. S. Geol. Surv., Mon. 17, 1891 [1892], p. 175, pl. xxxvi, f. 2, 3.
Dakota: Ellsworth County, Kans.

Elaeodendron strictum Hollick, U. S. Geol. Surv., Mon. 50, 1906, p. 89, pl. xxxiii, f. 6.
Magothy: Gay Head, Marthas Vineyard, Mass.

Elaeodendron sp., Hollick, U. S. Geol. Surv., Mon. 50, 1906, p. 89, pl. xxxiii, f. 7.
Celastrophyllum benedini Saporta and Marion. Hollick, Bull. Torr. Bot. Club, vol. 21, 1894, p. 58, pl. 177, f. 4.
Magothy: Glen Cove, Long Island, N. Y.

ELAEODENDROXYLON Platen, Sitzungsb. Naturf. Ges., Leipzig, vol. 34, 1907 [1908], p. 120. [Type, *E. polymorphum*.]

Elaeodendroxylon polymorphum Platen, Sitzungsb. Naturf. Ges., Leipzig, vol. 34, 1907 [1908], p. 120.
Miocene: Amethyst Mountain, Yellowstone National Park.

ELATIDES Heer, Mém. Acad. imp. sci. St.-Pétersbourg, vol. 22, No. 12, 1876, p. 77.

Elatides curvifolia (Dunker) Nathorst, Kongl. Svenska Vetenskaps.-Akad. Handl., vol. 30, 1897, p. 35, pl. 1, f. 25–27; pl. ii, f. 3–6; pl. iv, f. 1–18; pl. vi, f. 6–8.—Knowlton, U. S. Geol. Surv., Prof. Paper 85, 1914, p. 53, pl. viii, f. 1.
Jurassic: Cape Lisburne, Alaska.

Elodea canadensis Michaux = *Philotria canadensis*.

EMBOTHRIOPSIS Hollick, N. Y. Bot. Gard., Bull., vol. 8, 1912, p. 159. [Type, *E. presagita*.]

Embothriopsis presagita Hollick, N. Y. Bot. Gard., Bull., vol. 8, 1912, p. 159, pl. 165, fig. 1.
Magothy: Glen Cove, N. Y.

EMBOTHRITES Unger, Gen. et sp. pl. foss, 1850, p. 428.

Embothrites? daphneoides (Lesquereux) Lesquereux, Rept. U. S. Geol. Surv. Terr., vol. 8 (Cret. and Tert. Fl.), 1883, p. 51.
Embothrium? daphneoides Lesquereux, Rept. U. S. Geol. Surv. Terr., vol. 6 (Cret. Fl.), 1874, p. 87, pl. xxx, f. 10.
Dakota: Kansas.

Embothrium? daphneoides Lesquereux = *Embothrites? daphneoides*.

ENCEPHALARTOPSIS Fontaine, U. S. Geol. Surv., Mon. 15, 1889 [1890], p. 174. [Type, *E. nervosa*.]

Encephalartopsis nervosa Fontaine, U. S. Geol. Surv., Mon. 15, 1889 [1890], p. 174, pl. lxx, f. 4; pl. lxxi, f. 3, 4; pl. lxxii, f. 3, 4.
Patuxent: Fredericksburg, Va.

Encephalartopsis? oregonensis Fontaine, in Ward, U. S. Geol. Surv., Mon. 48, 1905 [1906], p. 117, pl. xxix, f. 2, 3.
Jurassic: Douglas County, Oreg.

ENCEPHALARTOS Lehman, Pugill., vol. 6, 1834, p. 3.

Encephalartos cretaceus Lesquereux, U. S. Geol. Surv., Mon. 17, 1891 [1892], p. 29, pl. i, f. 12.
Dakota: Ellsworth County, Kans.

ENGELHARDTIA Leschin, in Blume, Bijdr., 1825, p. 528.

Engelhardtia ettingshauseni Berry, U. S. Geol. Surv., Prof. Paper 91, 1916, p. 185, pl. xix, f. 1, 3, 5.
Wilcox (Grenada): Grenada, Grenada County, Miss.
Wilcox (Holly Springs): Early Grove and Holly Springs, Marshall County, Miss.
Lagrange: Puryear, Henry County, Tenn.

Engelhardtia mississippiensis Berry, Am. Jour. Sci., 4th ser., vol. 31, 1911, p. 494, text f. 2; Plant World, vol. 15, 1912, p. 236, text f. 3, U. S. Geol. Surv., Prof. Paper 91, 1916, p. 183, pl. xvii, f. 1.
Wilcox (Holly Springs): Early Grove, Marshall County, Miss.

Engelhardtia oxyptera Saporta, Am. sci. nat., 5th ser., Bot., vol. 4 (Études), 1865, p. 344, pl. xii, f. 2.—Lesquereux, Rept. U. S. Geol. Surv. Terr., vol. 8 (Cret. and Tert. Fl.), 1883, p. 192.
Miocene: Florissant, Colo.

Engelhardtia puryearensis Berry, U. S. Geol. Surv., Prof. Paper 91, 1916, p. 185, pl. xvii, f. 6, 7.
Lagrange: Puryear, Henry County, Tenn.

EOACHRAS Berry, Am. Jour. Sci., 4th ser., vol. 39, 1915, p. 210. [Type, *E. eocenica*.]

Eoachras eocenica Berry, Am. Jour. Sci., 4th ser., vol. 39, 1915, p. 210, pl. 1, f. 1–3.
Claiborne (upper part): Near Lexington, Holmes County, Miss.

EORHAMNIDIUM Berry, U. S. Geol. Surv., Prof. Paper 112, 1919, p. 113. [Type, *E. cretaceum*.]

Eorhamnidium cretaceum Berry, U. S. Geol. Surv., Prof. Paper 112, 1919, p. 114, pl. xxviii, f. 10.
Tuscaloosa: Shirleys Mill and Cottondale, Ala.

Eorhamnidium platyphylloides (Lesquereux) Berry, U. S. Geol. Surv., Prof. Paper 112, 1919, p. 114.
Cornus platyphylloides Lesquereux, U. S. Geol. Surv., Mon. 17, 1891 [1892], p. 126, pl. lxiv, f. 15.
Dakota: Ellsworth County, Kans.
Tuscaloosa: Cottondale, Ala.

EPHEDRITES Goeppert and Berger, in Berendt, Bernst. Org. Vorw., vol. 1, pt. 1, 1845, p. 105.

Ephedrites? vernonensis Fontaine, in Ward, U. S. Geol. Surv., Mon. 48, 1905 [1906], p. 495, pl. cvii, f. 8.
Patapsco: Mount Vernon, Va.

Equisetites burchardti Dunker = *Equisetum burchardti*.

Equisetites phillipsii Dunker = *Equisetum phillipsii*.

EQUISETUM (Tournefort) Linné, Sp. pl., 1753, p. 1061.

Equisetum abiquiense Fontaine, U. S. Nat. Mus., Proc., vol. 13, 1890, p. 283, pl. xxii, f. 1.
Triassic: Abiquiu, N. Mex.

Equisetum arcticum Heer, Öfv. Sv. Vet.-Akad. Förh., vol. 23, 1866, p. 150.—Penhallow, Roy. Soc. Canada, Trans., 2d ser., vol. 8, 1902, p. 49; Rept. Tert. Pl. Brit. Col., 1908, p. 53.
Paskapoo (Fort Union): Red Deer River, Alberta.

Equisetum arundiniforme Rogers = *Equisetum rogersii*.

Equisetum arvense Linné. Knowlton, Am. Geol., vol. 18, 1896, p. 371.
Pleistocene: Morgantown, W. Va.

Equisetum burchardti (Dunker) Brongniart, Tableau, 1849, p. 107.—Berry, Md. Geol. Surv., Lower Cret., 1911, p. 316, pl. xli, f. 3–6.—Cockerell, Washington Acad. Sci. Jour., vol. 6, 1916, p. 110.

Equisetites burchardti Dunker, Mon. Norddeutsch. Wealdenbild., 1846, p. 2, pl. v, f. 7.

Equisetum virginicum Fontaine, U. S. Geol. Surv., Mon. 15, 1889 [1890], p. 63, pl. i, f. 1–6, 8; pl. ii, f. 1–3, 6, 7, 9; in Ward, U. S. Geol. Surv., Nineteenth Ann. Rept., pt. 2, 1899, p. 650, pl. clx, f. 1; U. S. Geol. Surv., Mon. 48, 1905 [1906], pp. 483, 486, 519.

Equisetum marylandicum Fontaine, U. S. Geol. Surv., Mon. 15, 1889 [1890], p. 65, pl. ii, f. 10; in Ward, U. S. Geol. Surv., Mon. 48, 1905 [1906], pp. 517, 557, pl. cix, f. 10.

Equisetum sp., rhizome of, Fontaine, U. S. Geol. Surv., Mon. 15, 1889 [1890], p. 66, pl. clxx, f. 8.

Patuxent: Terra Cotta and New Reservoir, D. C.; Dutch Gap, Cockpit Point, telegraph station near Lorton, Va.

Patapsco: Federal Hill (Baltimore), Md.

Fuson: Pine Creek, Crook County, Wyo.

Lakota: Hay Creek coal field, Crook County, Wyo.

Lower Cretaceous: Near Hovenweep Canyon, latitude 37° 53′ N., longitude 108° 57′ W., southwestern Colorado.

Equisetum canaliculatum Knowlton, U. S. Geol. Surv., Mon. 32, pt. 2, 1899, p. 675, pl. lxxxi, f. 6, 7.

Fort Union: Yellowstone National Park.

Miocene: Yellowstone National Park.

Equisetum collieri Knowlton, U. S. Geol. Surv., Prof. Paper 85, 1914, p. 50, pl. v, f. 5.

Jurassic: Cape Lisburne, Alaska.

Equisetum columnaroides Emmons = *Equisetum rogersii*.

Equisetum deciduum Knowlton, U. S. Geol. Surv., Mon. 32, pt. 2, 1899, p. 676, pl. lxxxi, f. 5.

Fort Union: Elk Creek, Yellowstone National Park.

Equisetum florissantense Cockerell, Torreya, vol. 15, 1915, p. 266, text f. 1.

Miocene: Florissant, Colo.

Equisetum fluviatile Linné. Penhallow, Geol. Soc. Am., Bull., vol. 1, 1890, p. 327; Roy. Soc. Canada, Trans., 2d ser., vol. 2, 1896, pp. 66, 71.

Equisetum limosum Linné. Penhallow, Brit. Assoc. Adv. Sci., Bradford meeting, 1900, p. 335.

Pleistocene: Greens Creek, Toronto, Canada.

Equisetum globulosum Lesquereux, U. S. Nat. Mus., Proc., vol. 5, 1882 [1883], p. 444; Rept. U. S. Geol. Surv. Terr., vol. 8 (Cret. and Tert. Fl.), 1883, p. 222, pl. xlviii, f. 3.—Knowlton, Harriman Alaska Exped., vol. 4, 1904, p. 149; Geol. Soc. Am., Bull., vol. 5, 1893, p. 577; U. S. Nat. Mus., Proc., vol. 17, 1894, p. 212.

Kenai: Chignik Bay ?; Kukak Bay, Alaska.

Fort Union: Badlands, N. Dak.

Equisetum haguei Knowlton, U. S. Geol. Surv., Mon. 32, pt. 2, 1899, p. 674, pl. lxxxi, f. 3, 4.
Miocene: Lost Creek, Yellowstone National Park.

Equisetum haydenii Lesquereux, U. S. Geol. and Geog. Surv. Terr., Ann. Rept. 1871 [1872], p. 284; Rept. U. S. Geol. Surv. Terr., vol. 7 (Tert. Fl.), 1878, p. 67, pl. vi, f. 2–4.
Green River ?: Barrell Springs, Wyo.
Hanna: Carbon, Wyo.

Equisetum hornii Lesquereux = *Equisetum oregonense.*

Equisetum knowltoni Fontaine, U. S. Nat. Mus., Proc., vol. 13, 1890, p. 283, pl. xxiii, f. 2; pl. xxiv, f. 3, 4.
Triassic: Abiquiu, N. Mex.

Equisetum laevigatum Lesquereux = *Equisetum perlaevigatum.*

Equisetum lesquereuxii Knowlton, U. S. Geol. Surv., Bull. 152, 1898, p. 94; U. S. Geol. Surv., Mon. 32, pt. 2, 1899, p. 675.
Equisetum limosum? Linné. Lesquereux, U. S. Geol. and Geog. Surv. Terr., Ann. Rept. 1871 [1872], p. 299; Rept. U. S. Geol. Surv. Terr., vol. 7 (Tert. Fl.), 1878, p. 69, pl. vi, f. 5. [Not the living *E. limosum!*]
Miocene ?: "Near Yellowstone Lake, among basaltic rocks."

Equisetum limosum? Linné. Lesquereux, U. S. Geol. and Geog. Surv. Terr., Ann. Rept. 1871 [1872], p. 299 = *Equisetum lesquereuxii.*

Equisetum limosum Linné = *Equisetum fluviatile.*

Equisetum lyelli Mantell, Geol. S. E. England, 1833, p. 217.—Fontaine, U. S. Geol. Surv., Mon. 15, 1889, p. 65, pl. i, f. 7; pl. ii, f. 4, 5; U. S. Nat. Mus., Proc., vol. 15, 1892, p. 489.—Dawson, Roy. Soc. Canada, Trans., vol. 10, 1892 [1893], sec. 4, p. 83, f. 1.—Fontaine, in Weed and Pirsson, U. S. Geol. Surv., Eighteenth Ann. Rept., pt. 3, 1898, p. 481; in Ward, U. S. Geol. Surv., Mon. 48, 1905 [1906], p. 301, pl. lxxii, f. 12–14.—Berry, Md. Geol. Surv., Lower Cret., 1911, p. 311, pl. xli, f. 7, 8.
Patuxent: Springfield ?, Md.; Fredericksburg, Chinkapin Hollow ?, Dutch Gap, Va.
Kootenai: Great Falls, Geyser, and Giltedge, Mont.
Cloverly: No Wood Creek, Big Horn County, Wyo.

Equisetum marylandicum Fontaine = *Equisetum burchardti.*

Equisetum meriani? Brongniart, Hist. vég. foss., vol. 1, 1828, p. 115, pl. xii, f. 13.—Newberry, U. S. Geol. Surv., Mon. 14, 1888, p. 86. [Probably referable to *Equisetum rogersii.*]
Triassic (Newark): Newark, N. J.

Equisetum montanense Fontaine = *Equisetum phillipsii.*

Equisetum muensteri (Sternberg) Brongniart ? Fontaine, in Ward, U. S. Geol. Surv., Twentieth Ann. Rept., pt. 2, 1900, p. 333.
Jurassic (Foreman): Taylorsville, Calif.

Equisetum newberryi Knowlton and Cockerell, n. name.

Equisetum robustum Newberry, Boston Jour. Nat. Hist., vol. 7, 1863, p. 513; U. S. Geol. Surv., Mon. 35, 1898, p. 15, pl. xvi, f. 1, 2. [Homonym, Al. Braun, 1844.]

Puget: Bellingham Bay, Wash.

Equisetum nodosum Lesquereux = *Phragmites cretaceus*.

Equisetum oregonense Newberry, U. S. Nat. Mus., Proc., vol. 5, 1882 [1883], p. 503; U. S. Geol. Surv., Mon. 35, 1898, p. 14, pl. lxv, f. 7.—Knowlton, U. S. Geol. Surv., Bull. 204, 1902, p. 22.

Equisetum hornii Lesquereux, U. S. Nat. Mus., Proc., vol. 11, 1888, p. 23.

Clarno (lower part): Cherry Creek, Currant Creek, and 3 miles above Clarnos Ferry, Oreg.

Equisetum parlatorii (Heer) Schimper, Pal. vég., vol. 1, 1869, p. 261.—Penhallow, Rept. Tert. Pl. Brit. Col., 1908, p. 54.

Physagenia parlatorii Heer, Fl. tert. Helv., vol. 1, 1854, p. 109, pl. xlii, f. 2–17.—Dawson, Brit. N. A. Bound Com. (Rept. Geol. and Res. Vic. 49th Parallel), 1875, App. A, p. 329, pl. xvi, f. 3, 4.

Paskapoo (Fort Union): Great Valley, Saskatchewan.

Equisetum perlaevigatum Cockerell, West Am. Scientist, vol. 6, 1889, p. 154.—Knowlton, U. S. Geol. Surv., Prof. Paper 130, 19—, p. —, pl. i, f. 8, 9 [in preparation].

Equisetum laevigatum Lesquereux, U. S. Geol. and Geog. Surv. Terr., Ann. Rept. 1873 [1874], p. 395; Rept. U. S. Geol. Surv. Terr., vol. 7 (Tert. Fl.), 1878, p. 68, pl. vi, f. 6, 7. [Homonym, Al. Braun, 1867.]

Laramie: Golden, Colo.

Equisetum phillipsii (Dunker) Brongniart, Tableau, 1849, p. 107.—Fontaine, in Ward, U. S. Geol. Surv., Mon. 48, 1905, p. 298, pl. lxxii, f. 1–11.—Knowlton, Smithsonian Misc. Coll., vol. 50, 1907, p. 120.

Equisetites phillipsii Dunker, 1843.

Equisetum montanense Fontaine, in Weed and Pirsson, U. S. Geol. Surv., Eighteenth Ann. Rept., pt. 3, 1898, p. 481 [name].

Kootenai: Great Falls, Grafton, and Geyser, Mont.

Equisetum robustum Newberry = *Equisetum newberryi*.

Equisetum rogersii (Bunbury) Schimper, Pal. vég., vol. 1, 1869, p. 276.—Fontaine, U. S. Geol. Surv., Mon. 6, 1883, p. 10, pl. i, f. 2; pl. ii, f. 1, 2.—Newberry, idem, Mon. 14, 1888, p. 85, pl. xxii, f. 5, 5a.—Fontaine, in Ward, U. S. Geol. Surv., Twentieth Ann. Rept., pt. 2, 1900, pp. 241, 288, pl. xxv, f. 11, 12.—Holden, Ann. Bot., vol. 27, 1913, p. 253.

Calamites rogersii Bunbury, Quart. Jour. Geol. Soc., 1851, Proc., p. 190.

Calamites arenaceus Emmons, Geol. Rept. Midland Counties N. C., 1856, p. 334; American geology, pt. 6, 1857, p. 109, f. 77, 78.

Calamites disjunctus Emmons, Geol. Rept. Midland Counties N. C., 1856, p. 334, pl. iv, f. 7; American geology, pt. 2, 1857, p. 110, pl. iv, f. 4.

Equisetum columnaroides Emmons, Geol. Rept. Midland Counties N. C., 1856, p. 290, pl. ii, f. 3; American geology, pt. 2, 1857, p. 35, pl. vi, f. 3.

Equisetum rogersii—Continued.

Equisetum arundiniforme Rogers, Rept. Assoc. Am. Geol. and Nat., 1843, p. 303.

Phytolithus striaticulmis Martin, Petrefacta Derbeyensis, 1809, p. 27, pl. xxv.—Fontaine, in Ward, U. S. Geol. Surv., Twentieth Ann. Rept., pt. 2, 1900, p. 258.

Triassic (Newark): Richmond coal field, Va.; North Carolina; York County, Pa.; Milford, N. J.; Martins Head, New Brunswick.

Equisetum scirpoides Michaux. Dawson, Can. Nat., new ser., vol. 3, 1868, p. 73; Geol. Hist. Pl., 1888, p. 230.

Pleistocene: Greens Creek, Ottawa River, Canada.

Equisetum similkamense Dawson, Geol. Surv. Canada, Rept. Progress, 1877–78 [1879], p. 186B; Roy. Soc. Canada, Trans., vol. 8, 1890, sec. 4, p. 76, f. 1–1d.—Penhallow, Rept. Tert. Pl. Brit. Col., 1908, p. 54.

Oligocene: Ninemile Creek, Similkameen River, British Columbia.

Equisetum sylvaticum? Linné. Penhallow, Geol. Soc. Am., Bull., vol. 1, 1890, p. 327; Roy. Soc. Canada, Trans., 2d ser., vol. 2, 1896, p. 71; Brit. Assoc. Adv. Sci., Bradford meeting, 1900, p. 335.

Pleistocene: Greens Creek, Ottawa River, Canada.

Equisetum texense Fontaine, U. S. Nat. Mus., Proc., vol. 16, 1893, p. 263, pl. xxxvi, f. 1; in Diller and Stanton, Geol. Soc. Am., Bull., vol. 5, 1894, p. 450; in Ward, U. S. Geol. Surv., Mon. 48, 1905 [1906], p. 243, pl. lxvi, f. 14.

Trinity: Glen Rose, Tex.

Knoxville: Tehama County, Calif.

Equisetum virginicum Fontaine = *Equisetum burchardti*.

Equisetum wyomingense Lesquereux, U. S. Geol. and Geog. Surv. Terr., Ann. Rept. 1873 [1874], p. 409; Rept. U. S. Geol. Surv. Terr., vol. 7 (Tert. Fl.), 1878, p. 69, pl. vi, f. 8–11.—Newberry, U. S. Geol. Surv., Mon. 35, 1898, p. 15, pl. lxv, f. 8.

Green River: Green River, Wyo.

Equisetum? sp. Berry, U. S. Geol. Surv., Prof. Paper 112, 1919, p. 49.

Tuscaloosa: Snow place, Tuscaloosa County, Ala.

Equisetum sp., Dawson, Brit. N. A. Bound. Com. (Rept. Geol. and Res. Vic. 49th Parallel), 1875, App. A, p. 329.

Paskapoo (Fort Union): Porcupine Creek, British Columbia.

Equisetum sp. Dawson, Roy. Soc. Canada, Trans., vol. 4, 1886 [1887], sec. 4, p. 22, pl. i, f. 2.—Penhallow, Rept. Tert. Pl. Brit. Col., 1908, p. 53.

Paskapoo (Fort Union): Porcupine Creek and Great Valley, British Columbia.

Equisetum sp., rhizome of, Fontaine = *Equisetum burchardti.*

Equisetum sp. Fontaine, U. S. Geol. Surv., Mon. 15, 1889 [1890], p. 65, pl. ii, f. 8.
Patuxent: Fredericksburg, Va.

Equisetum? sp. Fontaine, in Ward, U. S. Geol. Surv., Mon. 48, 1905 [1906], p. 88, pl. xv, f. 10.
Jurassic: Douglas County, Oreg.

Equisetum sp. Knowlton, U. S. Geol. Surv., Eighteenth Ann. Rept., pl. 3, 1888, p. 722.
Payette: Idaho City, Idaho.

Equisetum sp. Knowlton, Geol. Soc. Am., Bull., vol. 8, 1897, p. 133.
Lance: Lance Creek, Converse County, Wyo.

Equisetum sp. Knowlton, U. S. Geol. Surv., Bull. 204, 1902, p. 23, pl. i, f. 1.
Mascall: Near Van Horn's ranch, John Day Basin, Oreg.

Equisetum? sp. Knowlton, in Veatch, Am. Jour. Sci., 4th ser., vol. 21, 1906, p. 459.—Knowlton, U. S. Geol. Surv., Prof. Paper 108, 1917, p. 85, pl. xxxiii, f. 6.
Frontier: 1 mile east of Cumberland, Wyo.

Equisetum sp. Knowlton, Washington Acad. Sci., Proc., vol. 11, 1909, p. 198.
Lance: Bluff east of Glendive, Mont.

Equisetum sp. Knowlton, Washingon Acad. Sci., Proc., vol. 11, 1909, p. 207.
Lance: Converse County, Wyo.

Equisetum sp. Knowlton, U. S. Geol. Surv., Prof. Paper 98, 1916, p. 88, pl. xviii, f. 2.
Fox Hills: Wildcat Mound, 2½ miles south of Milliken, Colo.

Equisetum sp: (Lesquereux) Knowlton, n. comb. ?
Physagenia sp. Lesquereux, Mus. Comp. Zool., Bull., vol. 16, 1888, p. 43.
Denver: Golden, Colo.

Equisetum, rootlets of, Lesquereux, Rept. on Clays in N. J., 1878, p. 28.
Raritan ?: Washington, N. J.

Equisetum sp. Lesquereux, Rept. U. S. Geol. Surv. Terr., vol. 8 (Cret. and Tert. Fl.), 1883, p. 239, pl. l, f. 7.
——: Contra Costa, Calif.

Equisetum sp. Lesquereux, Rept. U. S. Geol. Surv. Terr., vol. 8 (Cret. and Tert. Fl.), 1883, p. 239, pl. l, f. 8.
——: Corral Hollow, Calif.

Equisetum sp.? Newberry, U. S. Geol. Surv., Mon. 35, 1898, p. 16, pl. xxii, f. 3, 4.

Radicle tubers of *Equisetum* (not described) [Lesquereux], U. S. Geol. and Geog. Surv. Terr., Ill. Cret. and Tert. Pl., 1878, pl. vii, f. 4.

Roots of some ligneous plant (not described) [Lesquereux], idem, f. 3.

Formation and locality unknown.

Equisetum sp. Paisley, Can. Nat., 2d ser., vol. 7, 1875, p. 270.

Post-Pliocene [Pleistocene]: Bathurst, New Brunswick.

Equisetum sp. Penhallow, Roy. Soc. Canada, Trans., 2d ser., vol. 2, 1896, p. 71; Brit. Assoc. Adv. Sci., Bradford meeting, 1900, p. 335.

Pleistocene: Scarborough, Ontario, Canada.

EREMOPHYLLUM Lesquereux, Rept. U. S. Geol. Surv. Terr., vol. 6 (Cret. Fl.), 1874, p. 107. [Type, *Ficus fimbriatus*.]

Eremophyllum fimbriatum (Lesquereux) Lesquereux, Rept. U. S. Geol. Surv. Terr., vol. 6 (Cret. Fl.), 1874, p. 107, pl. viii, f. 1.

Ficus? fimbriatus Lesquereux, Am. Jour. Sci., vol. 46, 1868, p. 96.

Dakota: Decatur, Nebr.

ERIOCAULON Linné, Sp. pl., 1753, p. 87.

Eriocaulon? porosum Lesquereux, U. S. Geol. and Geog. Surv. Terr., Ann. Rept. 1873 [1874], p. 396; Rept. U. S. Geol. Surv. Terr., vol. 7 (Tert. Fl.), 1878, p. 106, pl. xvi, f. 2, 2a.

Denver: Sand Creek, Colo.

Eriocaulon sp. Penhallow, Brit. Assoc. Adv. Sci., Bradford meeting, 1900, p. 335.

Pleistocene: Don Valley, Canada.

EUCALYPTOPHYLLUM Fontaine, U. S. Geol. Surv., Mon. 15, 1889 [1890], p. 325. [Type, *E. oblongifolium*.]

Eucalyptophyllum oblongifolium Fontaine, U. S. Geol. Surv., Mon. 15, 1889 [1890], p. 325, pl. clxii, f. 4.

Patapsco: Near Brooke, Va.

EUCALYPTUS L'Héritier, Sert. Angl., 1788, p. 18.

Eucalyptus? americana Lesquereux, Rept. U. S. Geol. Surv. Terr., vol. 7 (Tert. Fl.), 1878, p. 296, pl. lix, f. 11, 12.

Eucalyptus americanus Lesquereux, U. S. Geol. and Geog. Surv. Terr., Ann. Rept. 1871 [1872], Suppl., p. 7.

Green River: Green River, Wyo.

Eucalyptus angusta Velenovsky, Fl. böhm. Kreidef., vol. 4, 1885, p. 3, pl. iii, f. 2–12.—Berry, Bull. Torr. Bot. Club, vol. 36, 1909, p. 260, pl. xviii, f. 5; idem, vol. 37, 1910, p. 504; U. S. Geol. Surv., Prof. Paper 84, 1914, pp. 55, 119, pl. xiv, f. 2; pl. xx, f. 2–4; Geol. Surv. N. J., Bull. 3, 1911, p. 193, pl. xxviii, f. 1–4.

Black Creek: Black Creek, S. C.
Raritan: South Amboy, N. J.
Eutaw: McBrides Ford, Columbus, Ga.
Ripley (Cusseta): Buena Vista, Ga.

Eucalyptus? angustifolia Newberry = *Eucalyptus geinitzi.*

Eucalyptus? attenuata Newberry, U. S. Geol. Surv., Mon. 26, 1895 [1896], p. 111, pl. xvi, f. 2, 3, 5.—Berry, Bull. Torr. Bot. Club vol. 33, 1906, p. 180; Geol. Surv. N. J., 1895 [1896], p. 138; Bull., Torr. Bot. Club, vol. 34, 1907, p. 203; Geol. Surv. N. J., Bull. 3, 1911, p. 195, pl. xxviii, f. 6; Md. Geol. Surv., Upper Cret., 1916, p. 869.

Raritan: South Amboy, N. J.

Magothy: Cliffwood, N. J.; Grove Point and Round Bay, Md.

Black Creek: Court House Bluff, Cape Fear River, N. C.

Eucalyptus dakotensis Lesquereux, U. S. Geol. Surv., Mon. 17, 1891 [1892], p. 137, pl. xxxvii, f. 14–19.

Dakota: Ellsworth County, Kans.

Eucalyptus? dubia Berry = *Eucalyptus wardiana.*

Eucalyptus geinitzi (Heer) Heer, Fl. foss. arct., vol. 6, Abt. 2, 1882, p. 93, pl. xix, f. 1c; pl. lxv, f. 4–9; pl. xlvi, f. 12c, d, 13.—Lesquereux, U. S. Geol. Surv., Mon. 17, 1891 [1892], p. 138, pl. xxxvii, f. 20.—Newberry, idem, Mon. 26, 1895 [1896], p. 110, pl. xxxii, f. 2, 12, 15, 16.—Berry, N. Y. Bot. Gard., Bull., vol. 3, 1903, p. 87, pl. liii, f. 3; Bull. Torr. Bot. Club, vol. 31, 1904, p. 78, pl. iv, f. 5; Geol. Surv. N. J., Ann. Rept. 1905 [1906], p. 138; Bull. Torr. Bot. Club, vol. 33, 1906, p. 180.—Hollick, U. S. Geol. Surv., Mon. 50, 1906, p. 96, pl. xxxv, f. 1–8, 10–12.—Berry, Bull. Torr. Bot. Club, vol. 34, 1907, p. 201, pl. xv, f. 4; idem, vol. 37, 1910, p. 26; idem, vol. 39, 1912, p. 402.—Hollick, N. Y. Bot. Gard., Bull., vol. 8, 1912, p. 166, pl. clxxx, f. 1, 2.—Berry, U. S. Geol. Surv., Prof. Paper 84, 1914, p. 56, pl. xiii, f. 8–12; pl. xiv, f. 1; Geol. Surv. N. J., Bull. 3, 1911, p. 189; Md. Geol. Surv., Upper Cret., 1916, p. 870, pl. lxxxi, f. 1–5; U. S. Geol. Surv., Prof. Paper 112, 1919, p. 126, pl. xxviii, f. 8.

Myrtophyllum geinitzi Heer, Fl. v. Moletein, 1872, p. 22, pl. xi, f. 3, 4.—Hollick, N. Y. Acad. Sci., Trans., vol. 12, 1893, p. 236, pl. vi, f. 2.

Myrtophyllum warderi Lesquereux, U. S. Geol. Surv., Mon. 17, 1891 [1892], p. 136, pl. liii, f. 10.—Hollick, idem, Mon. 50, 1906, p. 97, pl. xxxv, f. 13.

Eucalyptus? angustifolia Newberry, U. S. Geol. Surv., Mon. 26, 1895 [1896], p. 111, pl. xxxii, f. 1, 6, 7.—Hollick, N. Y. Bot. Gard., Bull., vol. 3, 1894, p. 408, pl. lxx, f. 8, 9; U. S. Geol. Surv., Mon. 50, 1906, p. 95, pl. xxxv, f. 9, 14, 15.

Dakota: Ellsworth County, Kans.

Tuscaloosa: Near Northport and Maplesville, Ala.

Raritan: Woodbridge, South Amboy, and Sayreville, N. J.; Tottenville, Roslyn, Glen Cove, and Northport, N. Y.

Magothy: Cliffwood, N. J.; Gay Head, Marthas Vineyard, Mass.; Grove Point and Round Bay, Md.; Deep Cut, Del.

Black Creek: Cape Fear River, N. C.; Langley, S. C.

Black Creek (Middendorf): Middendorf, S. C.

Woodbine: Arthurs Bluff, Tex.

Eucalyptus geinitzi Heer. White = *Dammara borealis.*

Eucalyptus geinitzi propinqua Hollick, N. Y. Bot. Gard., Bull., vol. 8, 1912, p. 166, pl. 170, f. 3.
Magothy: Roslyn, N. Y.

Eucalyptus gouldii Ward, Bull. Torr. Bot. Club, vol. 24, 1897, p. 566, text f. 1; Plant World, vol. 1, 1897, p. 75, f. 1.
Dakota: Clark County, Kans.

Eucalyptus haeringiana? Ettingshausen. Lesquereux, U. S. Geol. and Geog. Surv. Terr., Ann. Rept. 1872 [1873], p. 400; Rept. U. S. Geol. Surv. Terr., vol. 7 (Tert. Fl.), 1878, p. 296, pl. lix, f. 10.
Post-Laramie: Black Buttes, Wyo.

Eucalyptus latifolia Hollick, U. S. Geol. Surv., Mon. 50, 1906, p. 97, pl. xxxvi, f. 1–5.—Berry, Bull. Torr. Bot. Club, vol. 37, 1910, p. 26; Md. Geol. Surv., Upper Cret., 1916, p. 870, pl. lxxxi, f. 6, 7; U. S. Geol. Surv., Prof. Paper 112, 1919, p. 126.
Magothy: Glen Cove, Long Island, N. Y.; Gay Head, Marthas Vineyard, Mass.; Round Bay, Md.
Tuscaloosa: Snow place, Tuscaloosa County, Ala.

Eucalyptus linearifolia Berry, Bull. Torr. Bot. Club, vol. 34, 1907, p. 203; Geol. Surv. N. J., Bull. 3, 1911, p. 192, pl. xxviii, f. 8.
Eucalyptus? nervosa Newberry MS. Hollick, Bull. Torr. Bot. Club, vol. 21, 1894, p. 56, pl. clxxiv, f. 10.—Newberry, U. S. Geol. Surv., Mon. 26, 1895 [1896], p. 112, pl. xxxii, f. 3–5, 8.—Hollick, U. S. Geol. Surv., Mon. 50, 1906, p. 95, pl. viii, f. 63; pl. xxxv, f. 16.
Magothy: Sea Cliff, Long Island, N. Y.; ? Block Island, R. I.
Black Creek: Blackmans Bluff, Neuse River, N. C.
Raritan: South Amboy, N. J.

Eucalyptus? nervosa Newberry = *Eucalyptus linearifolia.*

Eucalyptus? parvifolia Newberry, U. S. Geol. Surv., Mon. 26, 1895 [1896], p. 112, pl. xxxii, f. 9, 10; Geol. Surv. N. J., Bull. 3, 1911, p. 193.
Raritan: South Amboy, N. J.

Eucalyptus rosieriana Ward = *Sapindopsis variabilis.*

Eucalyptus schübleri (Heer?) Hollick, U. S. Geol. Surv., Mon. 50, 1906, p. 96, pl. xxxvi, f. 6.
Myrtophyllum (Eucalyptus) schübleri Heer, Neue Denkschr. Schw. Gesellsch. Naturwissensch. vol. 23, 1869, p. 23, pl. xi, f. 2.
Magothy: Nashaquitsa, Marthas Vineyard, Mass.

Eucalyptus wardiana Berry, Bull. Torr. Bot. Club, vol. 32, 1905, p. 47; idem, vol. 33, 1906, p. 180; Geol. Surv. N. J., Ann. Rept. 1905 [1906], p. 138; U. S. Geol. Surv., Prof. Paper 84, 1914, p. 57, pl. xiv, f. 3, 4; Md. Geol. Surv., Upper Cret., 1916, p. 872.
Eucalyptus? dubia Berry, N. Y. Bot. Gard., Bull., vol. 3, 1903, p. 87, pl. lii, f. 1 [not *E. dubia* Ettingshausen, 1887].
Magothy: Cliffwood, N. J.; Deep Cut, Del.; Grove Point, Md.
Black Creek (Middendorf): Middendorf, S. C.

Eucalyptus sp. Knowlton, in Glenn, U. S. Geol. Surv., Water-Supply Paper 164, 1906, p. 38.
Lagrange: Wyckliff, Ballard County, Ky.

EUGEINITZIA Hollick and Jeffrey, N. Y. Bot. Gard., Mem., vol. 3, 1909, p. 43. [Type, *E. proxima*.]

Eugeinitzia proxima Hollick and Jeffrey, N. Y. Bot. Gard., Mem., vol. 3, 1909, p. 43, pl. x, f. 10; pl. xxv, f. 1–3.
Raritan: Kreischerville, Staten Island, N. Y.

EUGENIA (Micheli) Linné, Sp. pl., 1753, p. 470.

Eugenia? anceps Berry, Bull. Torr. Bot. Club, vol. 43, 1916, p. 301, pl. xvi, f. 2–4; U. S. Geol. Surv., Prof. Paper 112, 1919, p. 125, pl. xxxii, f. 3–5.
Ripley: Selmer, McNairy County, Tenn.

Eugenia densinervia (Lesquereux) Berry, U. S. Geol. Surv., Prof. Paper 91, 1916, p. 317.
Salix densinervis Lesquereux, Am. Jour. Sci., 2d ser., vol. 27, 1859, p. 364; in Safford, Geol. Tenn., 1869, p. 427, pl. K, f. 9.
Lagrange: Somerville, Fayette County, Tenn.

Eugenia grenadensis Berry, U. S. Geol. Surv., Prof. Paper 91, 1916, p. 319, pl. xci, f. 4, 5.
Wilcox (Grenada): Grenada, Miss.

Eugenia hilgardiana Berry, U. S. Geol. Surv., Prof. Paper 91, 1916, p. 318, pl. xc, f. 6.
Sapindus undulatus Al. Braun. Lesquereux, Am. Philos. Soc., Trans., vol. 13, 1869, p. 420, pl. xxii, f. 6.
Wilcox (Ackerman): Hurleys, Miss.

Eugenia primaeva Lesquereux, U. S. Geol. Surv., Mon. 17, 1891 [1892], p. 137, pl. liii, f. 5–9.
Dakota: Near Fort Harker, Kans.

Eugenia puryearensis Berry, U. S. Geol. Surv., Prof. Paper 91, 1916, p. 318, pl. xc, f. 11, 12.
Lagrange: Puryear, Tenn.

Eugenia tuscaloosensis Berry, U. S. Geol. Surv., Prof. Paper 112, 1919, p. 125, pl. xxviii, f. 6.
Tuscaloosa: Shirleys Mill and Cottondale, Ala.

EUONYMUS (Tournefort) Linné, Sp. pl., 1753, p. 197.

Euonymus flexifolius Lesquereux, Rept. U. S. Geol. Surv. Terr., vol. 8 (Cret. and Tert. Fl.), 1883, p. 183, pl. xxxviii, f. 13.
Green River: Uinta (not Randolph) County, Wyo.

Euonymus splendens Berry, U. S. Geol. Surv., Prof. Paper 91, 1916, p. 267, pl. lxi, f. 6; pl. lxii, f. 1–5.—Knowlton, U. S. Geol. Surv., Prof. Paper 101, 1917, p. 329, pl. lxix, f. 1.
Wilcox (Holly Springs): Early Grove, Marshall County, and Lamar, Benton County, Miss.

Euonymus splendens—Continued.

Lagrange: Grand Junction, Fayette County; Puryear, Henry County, and near Shandy, Hardeman County, Tenn.

Wilcox: Coushatta, La.; Pope Bend, Colorado River, Bastrop County, Tex.

Raton: South end Raton tunnel, Raton, N. Mex.

Euonymus xantholithensis Ward, U. S. Geol. Surv., Sixth Ann. Rept., 1884–85 [1886], p. 555, pl. liv, f. 1, 2; U. S. Geol. Surv., Bull. 37, 1887, p. 82, pl. xxxvii, f. 1, 2.

Fort Union: Burns's ranch, Glendive, Mont.

EUPHORBIOPHYLLUM Ettingshausen, Abh. K.-k. geol. Reichsanst., vol. 2, pt. 3, No. 2, 1853, p. 77.

Euphorbiophyllum fayettensis Berry, U. S. Geol. Surv., Prof. Paper 91, 1916, p. 259, pl. lvii, f. 1.

Lagrange: Near Grand Junction, Fayette County, Tenn.

EUPHORBOCARPUM Knowlton, U. S. Geol. Surv., Prof. Paper 101, 1917, p. 328. [Type, *E. richardsoni.*]

Euphorbocarpum richardsoni Knowlton, U. S. Geol. Surv., Prof. Paper 101, 1918, p. 328, pl. xcvi, f. 3, 4.

Raton: 5 miles south of Aguilar, Colo.

EXOSTEMA Richard, in Humboldt and Bonpland, Pl. Aequin., vol. 1, 1808, p. 131.

Exostema pseudocaribaeum Berry, U. S. Geol. Surv., Prof. Paper 91, 1916, p. 349, pl. cvi, f. 3.

Wilcox (Holly Springs): Early Grove, Miss.

Lagrange: Wickliffe, Ky.

Eysenhardtia nigrostipellata Cockerell = *Viborquia nigrostipellata.*

FAGARA Linné, Sp. pl., 1764, p. 172.

Fagara? delicatula Cockerell, Am. Mus. Nat. Hist., Bull., vol. 24, 1908, p. 98.

Miocene: Florissant, Colo.

Fagara diversifolia (Lesquereux) Cockerell, Am. Mus. Nat. Hist., Bull., vol. 24, 1908, p. 98.

Zanthoxylum diversifolium Lesquereux, Mus. Comp. Zool., Mem., vol. 6, No. 2, 1878, p. 33, pl. viii, f. 14, 15.

Miocene (auriferous gravels): Table Mountain, Calif.

Fagara eocenica Berry, U. S. Geol. Surv., Prof. Paper 91, 1916, p. 251, pl. lv, f. 6.

Lagrange: Puryear, Henry County, Tenn.

Fagara hurleyensis Berry, U. S. Geol. Surv., Prof. Paper 91, 1916, p. 252, pl. liv, f. 9.

Wilcox (Ackerman): Hurleys, Benton County, Miss.

Fagara puryearensis Berry, U. S. Geol. Surv., Prof. Paper 91, 1916, p. 251, pl. liv, f. 8.
Lagrange: Puryear, Henry County, Tenn.

Fagara spireaefolia (Lesquereux) Cockerell, Am. Mus. Nat. Hist., Bull., vol. 24, 1908, p. 98.
Zanthoxylum spireaefolium Lesquereux, Rept. U. S. Geol. Surv. Terr., vol. 8 (Cret. and Tert. Fl.), 1883, p. 196, pl. xl, f. 1-3.—Penhallow, Rept. Tert. Pl. Brit. Col., 1908, p. 97.
Miocene: Florissant, Colo.
Oligocene: Quilchena, British Columbia.

FAGOPHYLLUM Nathorst, Pal. Abhandl. v. Dames u. Kayser, vol. 4, Heft. 3, 1888, p. 5.

Fagophyllum nervosum Dawson, Roy. Soc. Canada, Trans., vol. 11, 1893 [1894], sec. 4, p. 58, pl. vii, f. 16.
Upper Cretaceous: Port McNeill, Vancouver Island, British Columbia.

Fagophyllum retosum Dawson, Roy. Soc. Canada, Trans., vol. 11, 1893 [1894], sec. 4, p. 57, pl. vii, f. 15.
Upper Cretaceous: Port McNeill, Vancouver Island, British Columbia.

FAGOPSIS Hollick, Torreya, vol. 9, 1909, p. 2. [Type, *Planera longifolia* Lesquereux.]

Fagopsis longifolia (Lesquereux) Hollick, Torreya, vol. 9, 1909, p. 2, text f. 1, 2.—Knowlton, U. S. Nat. Mus., Proc., vol. 51, 1916, p. 265, pl. xx, f. 5.
Planera longifolia Lesquereux, U. S. Geol. and Geog. Surv. Terr., Ann. Rept. 1872 [1873], p. 371; Rept. U. S. Geol. Surv. Terr., vol. 7 (Tert. Fl.), 1878, p. 189, pl. xxvii, f. 4-6.—Newberry, U. S. Geol. Surv., Mon. 35, 1898, p. 81, pl. lviii, f. 3.—Penhallow, Roy. Soc. Canada, Trans., 2d ser., vol. 8, 1902, p. 70; Rept. Tert. Pl. Brit. Col., 1908, p. 73.—Knowlton, U. S. Geol. Surv., Mon. 32, pt. 2, 1899, p. 712.
Quercus semi-elliptica Göppert. Lesquereux, U. S. Geol. and Geog. Surv. Terr., Ann. Rept. 1871 [1872], p. 286.
Fagus longifolia (Lesquereux) Hollick and Cockerell, Am. Mus. Nat. Hist., Bull., vol. 24, 1908, p. 88 (footnote).
Zelkowa longifolia (Lesquereux) Engler, in Engler and Prantl, Natür. Pflanzenfam., Teil 3, Abt. 1, 1888, p. 65.
Miocene: Elko station, Nev.; Tranquille and Horsefly rivers, British Columbia; Similkameen River; Fossil Forest Ridge, Yellowstone National Park; *Florissant, Colo.

FAGUS (Tournefort) Linné, Sp. pl., 1753, p. 997.

Fagus americana Sweet. Berry, Torreya, vol. 5, 1905, p. 88.—Hollick, Md. Geol. Surv., Pliocene and Pleistocene, 1906, p. 226.—Berry, Jour. Geol., vol. 15, 1907, p. 341; Am. Nat., vol. 41, 1907, p. 692, pl. ii, f. 7; idem, vol. 43, 1909, p. 435; Am. Jour. Sci.,

Fagus americana—Continued.
4th ser., vol. 29, 1910, p. 393; U. S. Nat. Mus., Proc., vol. 48, 1915, p. 299.—Torreya, vol. 5, 1915, p. 206.

Fagus ferruginea Aiton. Lesquereux, Am. Jour. Sci., 2d ser., vol. 27, 1859, p. 363; Geol. Tenn., 1869, p. 427, pl. K, f. 11.—Knowlton, Am. Geol., vol. 18, 1896, p. 371.—Meehan, in Mercer, Acad. Nat. Sci. Philadelphia, Jour., 2d ser., vol. 11, 1899, p. 281, text f. 8 (15).

Pleistocene: Somerville, Fayette County, Tenn.; Morgantown, W. Va.; Neuse River, N. C.; Abercrombies Landing, Chattahoochee River, Steels Bluff, Black Warrior River, and Alabama River near Montgomery, Ala.; Columbus, Ky.; Indian Head, Charles County, Md.; Port Kennedy, Pa.; Nomini Cliffs, Potomac River, Va.

Pleistocene (Talbot): Bodkin Point, Grace Point, and Grove Point, Md.; Tappahannock, Rappahannock River, Va.

Fagus antipofii Abich, Acad. sci. St.-Pétersbourg, Mém., 6th ser., vol. 7, 1858, p. 572 (32), pl. viii, f. 2.—Heer, Fl. foss. arct., vol. 2, Abt. 2, 1869, p. 30, pl. v, f. 4a; pl. vii, f. 4–8; pl. viii, f. 1.—Lesquereux, Mus. Comp. Zool., Mem., vol. 6, 1878, p. 3, pl. ii, f. 13.—Knowlton, U. S. Nat. Mus., Proc., vol. 17, 1894, p. 218; Geol. Soc. Am., Bull., vol. 5, 1893, p. 581; U. S. Geol. Surv., Mon. 32, pt. 2, 1899, p. 700.—Penhallow, Rept. Tert. Pl. Brit. Col., 1908, p. 54.—Knowlton, in Lindgren, U. S. Geol. Surv., Prof. Paper 73, 1911, p. 58.

Kenai: Port Graham, Alaska.

Fort Union: Elk Creek, Yellowstone National Park.

Oligocene: Quesnel River, British Columbia.

Fagus castaneaefolia Unger = *Castanea castaneaefolia*.

Fagus castaneaefolia Unger. Lesquereux, U. S. Nat. Mus., Proc., vol. 11, 1888, p. 18 = *Fagus?* sp. Knowlton.

Fagus cretacea Newberry, N. Y. Lyc. Nat. Hist., Ann., vol. 9, 1868, p. 23.—[Lesquereux], U. S. Geol. and Geog. Surv. Terr., Ill. Cret. and Tert. Pl., 1878, pl. ii, f. 3.—Newberry, U. S. Geol. Surv., Mon. 35, 1898, p. 68, pl. i, f. 3.

Dakota: Smoky Hill, Kans.

Fagus deucalionis Unger, Syn. pl. foss., 1845, p. 218.—Lesquereux, U. S. Nat. Mus., Proc., vol. 5, 1882 [1883], p. 447.—Knowlton, idem, vol. 17, 1894, p. 219; Geol. Soc. Am., Bull., vol. 5, 1893, p. 581.

Kenai: Kachemak Bay, Cook Inlet, Alaska.

Fagus deucalionis Unger. Lesquereux, U. S. Geol. and Geog. Surv. Terr., Ann. Rept. 1871 [1872], p. 292 = ? [Not afterward referred to by Lesquereux nor since found.]

Fagus feroniae Unger, Syn. pl. foss., 1845, p. 219.—Heer, Fl. foss. arct., vol. 2, Abt. 2, 1869, p. 31, pl. vi, f. 9.—Lesquereux, U. S. Geol. and Geog. Surv. Terr., Ann. Rept. 1873 [1874], p. 413; Rept. U. S. Geol. Surv. Terr., vol. 7 (Tert. Fl.), 1878, p. 146, pl. xix, f. 1-3.—Knowlton, U. S. Nat. Mus., Proc., vol. 17, 1894, p. 219; Geol. Soc. Am., Bull., vol. 5, 1893, p. 581.—Penhallow, Rept. Tert. Pl. Brit. Col., 1908, p. 54.
Miocene: Elko station, Nev.
Kenai: Port Graham, Alaska.
Oligocene: Quesnel and Blackwater rivers, British Columbia.

Fagus ferruginea Aiton = *Fagus americana*.

Fagus hitchcockii Lesquereux, Am. Jour. Sci., 2d ser., vol. 32, 1861, p. 357; Geol. Vt., vol. 2, 1861, p. 714, f. 126, 127.
Miocene: Brandon, Vt.

Fagus lambertensis Berry, U. S. Geol. Surv., Prof. Paper 98, 1916, p. 199, pl. xlvii, f. 5.
Citronelle: Lambert, Ala.

Fagus longifolia (Lesquereux) Hollick and Cockerell = *Fagopsis longifolia*.

Fagus macrophylla Unger, Denkschr. Wien. Akad., vol. 7, 1854, p. 175 [19 of reprint], pl. ii, f. 10.—Heer, Fl. foss. arct., vol. 2, Abt. 2 (Fl. foss. alask.), 1869, p. 31, pl. viii, f. 2.—Knowlton, U. S. Nat. Mus., Proc., vol. 17, 1894, p. 219; Geol. Soc. Am., Bull., vol. 5, 1893, p. 581.
Kenai: Port Graham, Alaska.

Fagus orbiculata(um) Lesquereux, U. S. Geol. Surv., Mon. 17, 1891 [1892], p. 51, pl. xlvii, f. 6.
Dakota: Ellsworth County, Kans.

Fagus papyracea Knowlton, U. S. Geol. Surv., Prof. Paper 101, 1918, p. 297, pl. lxviii, f. 1.
Raton: Near Aguilar, Colo.

Fagus polycladus Lesquereux, Am. Jour. Sci., 2d ser., vol. 46, 1868, p. 95.
Dakota: Decatur, Nebr.; Kansas.

Fagus proto-nucifera Dawson, Roy. Soc. Canada, Trans., vol. 1, 1882 [1883], sec. 4, p. 21, pl. ii, f. 6, 6a.
——: Peace River, British Columbia.

Fagus pseudo-ferruginea Lesquereux, Mus. Comp. Zool., Mem., vol. 6, No. 2, 1878, p. 3, pl. ii, f. 14.
Miocene (auriferous gravels): Chalk Bluffs, Nevada County, Calif.

Fagus undulata Knowlton, U. S. Geol. Surv., Mon. 32, pt. 2, 1899, p. 700, pl. lxxxv, f. 4, 5.
Fort Union: Elk Creek, Yellowstone National Park.

Fagus sp.? Hollick, Md. Geol. Surv., Pliocene and Pleistocene, 1906, p. 226, pl. lxx, f. 3.
Pleistocene (Sunderland): Point of Rocks, Calvert County, Md.

Fagus? sp. Knowlton, U. S. Geol. Surv., Bull. 204, 1902, p. 43.
Fagus castaneaefolia Unger. Lesquereux, U. S. Nat. Mus., Proc., vol. 11, 1888, p. 18.
Mascall: Van Horn's ranch, John Day Basin, Oreg.

FEILDENIA Heer, Fl. foss. arct., vol. 5, Abt. 1, 1878, p. 20.

Feildenia nordenskiöldi Nathorst, Kongl. Svenska Vetenskaps-Akad. Handl., vol. 30, 1897, p. 56, pl. iii, f. 16-27.—Knowlton, U. S. Geol. Surv., Prof. Paper 85, 1914, p. 54, pl. vii, f. 2.
Jurassic: Cape Lisburne, Alaska.

FEILDENIOPSIS Fontaine, U. S. Geol. Surv., Mon. 15, 1889 [1890], p. 204. [Type, *F. crassinervis.*]

Feildeniopsis crassinervis Fontaine, U. S. Geol. Surv., Mon. 15, 1889 [1890], p. 205, pl. lxxxvi, f. 5.
Patuxent: Kankeys, Va.

FEISTMANTELIA Ward, U. S. Geol. Surv., Nineteenth Ann. Rept., pt. 2, 1899, p. 693.

[*Feistmantelia* Ward = *Pinus*, bark of?.]

Feistmantelia oblonga Ward, U. S. Geol. Surv., Nineteenth Ann. Rept., pt. 2, 1899, p. 693, pl. clxix, f. 19.
Fuson: North side of Pine Creek, Crook County, Wyo.

Feistmantelia virginica Fontaine, in Ward, U. S. Geol. Surv., Mon. 48, 1905 [1906], p. 484, pl. cvii, f. 3.
Patuxent: Cockpit Point, Va.

FELIXIA Platen, Sitzungsb. Naturf. Ges., Leipzig, vol. 34, 1907 [1908], p. 66. [Type, *F. latiradiata.*]

Felixia latiradiata Platen, Sitzungsb. Naturf. Ges., Leipzig, vol. 34, 1907 [1908], p. 66, pl. ii, f. 3, 4.
——?: California.

Fern, undetermined, Emmons, American geology, pt. 6, 1857, p. 104, pl. iv, f. 3 = *Asplenites rösserti.*

Fern, undetermined, Emmons, American geology, pt. 6, 1857, pl. vi, f. 2 = *Laccopteris elegans.*

FESTUCA Linné, Sp. pl., 1753, p. 73.

Festuca ovina Linné. Penhallow, Brit. Assoc. Adv. Sci., Bradford meeting, 1900, p. 335.
Pleistocene: Don Valley, Toronto, Canada.

FICOPHYLLUM Fontaine, U. S. Geol. Surv., Mon. 15, 1889 [1890], p. 290. [Type, *F. crassinerve.*]

Ficophyllum crassinerve Fontaine, U. S. Geol. Surv., Mon. 15, 1889 [1890], p. 291, pl. cxliv, f. 3; pl. cxlv, f. 3; pl. cxlvi, f. 1; pl. cxlvii, f. 4; pl. cxlviii, f. 1, 2, 4; pl. clvii, f. 4; pl. clxxiii, f. 10.

Ficophyllum oblongifolium (Fontaine) Berry, Md. Geol. Surv., Lower Cret., 1912, p. 505.

Proteaephyllum oblongifolium Fontaine, U. S. Geol. Surv., Mon. 15, 1889 [1890], p. 284, pl. cxxxix, f. 5, pl. cxl, f. 1, 2, in Ward, idem, Mon. 48, 1905 [1906], p. 538.

Proteaephyllum sp., Fontaine, U. S. Geol. Surv., Mon. 15, 1889 [1890], p. 284, pl. cxxxix, f. 2.

Ficophyllum tenuinerve Fontaine, U. S. Geol. Surv., Mon. 15, 1889 [1890], p. 292, pl. cxl, f. 3; pl. cxli, f. 2; pl. cxlx, f. 1, 4; pl. cxlvii, f. 2; pl. cxlix, f. 1, 3, 5; pl. clvi, f. 1; in Ward, idem, Mon. 48, 1905 [1906], p. 520 (now pp. 504, 510).

Ficus virginiensis Fontaine, U. S. Geol. Surv., Mon. 15, 1889 [1890], p. 295, pl. cxliii, f. 1, 3; pl. cxliv, f. 1.

Ficus fredericksburgensis Fontaine, U. S. Geol. Surv., Mon. 15, 1889 [1890], p. 295, pl. cxlviii, f. 3, 5.

Sapindopsis cordata Fontaine, U. S. Geol. Surv., Mon. 15, 1889 [1890], p. 296, pl. cxlvii, f. 1.

Saliciphyllum ellipticum Fontaine, U. S. Geol. Surv., Mon. 15, 1889 [1890], p. 303 [part], pl. cxlvi, f. 2, 4; pl. cl, f. 3; pl. clxiii, f. 5; pl. clxvi, f. 2.

Patuxent: Fredericksburg, Dutch Gap, and Potomac Run, Va.
Arundel: Arlington, Md.; Langdon, D. C.

Ficophyllum eucalyptoides Fontaine = *Sapindopsis magnifolia.*

Ficophyllum? fontainei Knowlton, n. sp.

Ficophyllum serratum Fontaine. Fontaine, in Ward, U. S. Geol. Surv., Nineteenth Ann. Rept., pt. 2, 1899, p. 689, pl. clxix, f. 8.

Fuson: Oak Creek, Crook County, Wyo.

Ficophyllum oblongifolium (Fontaine) Berry = *Ficophyllum crassinerve.*

Ficophyllum serratum Fontaine, U. S. Geol. Surv., Mon. 15, 1889 [1890], p. 294, pl. cxlv, f. 2; pl. cxix, f. 9.—Berry, Md. Geol. Surv., Lower Cret., 1912, p. 504.

Quercophyllum tenuinerve Fontaine, U. S. Geol. Surv., Mon. 15, 1889 [1890], p. 308, pl. cxlix, f. 6, 7.—? Fontaine, in Ward, idem, Mon. 48, 1905 [1906], p. 521.

Patuxent: Fredericksburg, Va.
Arundel: Langdon?, D. C.

Ficophyllum serratum Fontaine. Fontaine, in Ward, Nineteenth Ann. Rept., pt. 2, 1899, p. 689, pl. clxix, f. 8 = *Ficophyllum? fontainei.*

Ficophyllum tenuinerve Fontaine = *Ficophyllum crassinerve.*

FICOXYLON Kaiser, Zeitschr. Gesammt. Naturw., vol. 53, 1880, p. 315.

Ficoxylon helictoxyloides Platen, Sitzungsb. Naturf. Ges., Leipzig, vol. 34, 1907 [1907], p. 51.

Upper Miocene: Plumas County, Calif.

FICUS (Tournefort) Linné, Sp. pl., 1753, p. 1059.

Ficus aguilar Knowlton, U. S. Geol. Surv., Prof. Paper 101, 1918, p. 300, pl. lxxi, f. 1.
Raton: Near Aguilar, Colo.

Ficus alabamensis Berry, U. S. Geol. Surv., Prof. Paper 112, 1919, p. 82, pl. xiv, f. 5.
Tuscaloosa: Shirleys Mill, Ala.

Ficus? alaskana Newberry, U. S. Nat. Mus., Proc., vol. 5, 1882 [1883], p. 512.—Knowlton, U. S. Nat. Mus., Proc., vol. 17, 1894, p. 222; Geol. Soc. Am., Bull., vol. 5, 1893, p. 584.—Newberry, U. S. Geol. Surv., Mon. 35, 1898, p. 84, pl. li, f. 1; pl. lii, f. 1; pl. lv, f. 1, 2.—?Knowlton, in Spurr, U. S. Geol. Surv., Eighteenth Ann. Rept., pt. 3, 1898, pp. 185, 192, 195.
Kenai: Cook Inlet and Admiralty Inlet; Millers Creek, Yukon River; 25 miles below Mission Creek; above Mynook Creek, Alaska.

Ficus aligera Lesquereux = *Citrophyllum aligerum.*

Ficus alkalina Lesquereux, Rept. U. S. Geol. Surv. Terr., vol. 8 (Cret. and Tert. Fl.), 1883, p. 164, pl. xliv, f. 7–9.
Green River: Alkali station, Wyo.

Ficus andraei Lesquereux, Mus. Comp. Zool., Bull., vol. 16, 1888, p. 50.
Denver: Golden, Colo.

Ficus? angustata Lesquereux = ***Ficus kansasensis.***

Ficus? apiculatus Knowlton, U. S. Geol. Surv., Prof. Paper 130, 19—, p. —, pl. xi, f. 6 [in preparation].
Laramie: Cowan station, near Denver, Colo.

Ficus appendiculata Heer, Fl. tert. Helv., vol. 2, 1856, p. 67, pl. lxxxv, f. 12, 13.—Lesquereux, U. S. Nat. Mus., Proc., vol. 11, 1888, p. 31.
Miocene?: Lassen County, Calif.

Ficus? applegatei Knowlton, Biol. Soc. Washington, Proc., vol. 15 1902, p. 86.
Ficus? hesperia Knowlton, U. S. Geol. Surv., Twentieth Ann. Rept., pt. 3, 1900, p. 45, pl. ii, f. 4.
Miocene: Ashland, Oreg.

Ficus arenacea Lesquereux, U. S. Geol. and Geog. Surv. Terr., Ann. Rept. 1871 [1872], p. 300; Rept. U. S. Geol. Surv. Terr., vol. 7 (Tert. Fl.), 1878, p. 195, pl. xxix, f. 1, 4, 5.—Knowlton, U. S. Geol. Surv., Prof. Paper 130, 19—, p. —, pl. x, f. 4, 5; pl. xxi, f. 6 [in preparation].
Ficus lanceolata Heer. Knowlton, U. S. Geol. Surv., Bull. 152, 1898, p. 102.
Laramie: Marshall, Colo.

Ficus arenacea brevipetiolata Lesquereux = *Ficus uncata.*

Ficus arenaceaeformis Cockerell, Am. Mus. Nat. Hist., Bull., vol. 24, 1908, p. 89, pl. viii, f. 24.
Miocene: Florissant, Colo.

Ficus artocarpoides Lesquereux, Rept. U. S. Geol. Surv. Terr., vol. 8 (Cret. and Tert. Fl.), 1883, p. 227, pl. xlvii, f. 1-5.—? Hollick, Geol. Surv. La., Special Rept. 5, 1899, p. 281, pl. xxxv, f. 4.—Knowlton, Washington Acad. Sci., Proc., vol. 11, 1907, p. 186.—Berry, U. S. Geol. Surv., Prof. Paper 91, 1916, p. 200, pl. xxxiv, f. 2.—Knowlton, U. S. Geol. Surv., Prof. Paper 101, 1918, p. 300, pl. lxxi, f. 3.
Fort Union: Badlands, N. Dak.
Lance: 60 miles south of Glasgow, Mont.
Wilcox: Coushatta, La.
Raton: Wootton, Colo.

Ficus asarifolia Ettingshausen, Denkschr. k. Akad. Wissensch., vol. 26, 1867, p. 156.—Lesquereux, U. S. Geol. and Geog. Surv. Terr., Bull., vol. 1, 1875 [1876], p. 366; U. S. Geol. and Geog. Surv. Terr., Ann. Rept. 1872 [1873], p. 378; Rept. U. S. Geol. Surv. Terr., vol. 7 (Tert. Fl.), 1878, p. 207, pl. xli, f. 18-21.—Newberry, U. S. Geol. Surv., Mon. 35, 1898, p. 85, pl. lxvii, f. 5, 6.—Knowlton, U. S. Geol. Surv., Bull. 163, 1900, p. 49, pl. xi, f. 4; pl. xiii, f. 2.—Penhallow, Rept. Tert. Pl. Brit. Col., 1908, p. 55.
Post-Laramie: ? Black Buttes, Wyo.
Mesaverde: Point of Rocks, Wyo.

Ficus asarifolia minor Lesquereux, U. S. Geol. and Geog. Surv. Terr., Bull., vol. 1, 1875 [1876], p. 367; Rept. U. S. Geol. Surv. Terr., vol. 7 (Tert. Fl.) 1878, p. 208.
Mesaverde: Point of Rocks, Wyo.

Ficus asiminaefolia Lesquereux, Rept. U. S. Geol. Surv. Terr., vol. 8 (Cret. and Tert. Fl.), 1883, p. 250, pl. lvi, f. 1-3.—Knowlton, in Lindgren, Jour. Geol., vol. 4, 1896, p. 889.—Penhallow, Rept. Tert. Pl. Brit. Col., 1908, p. 55.—Knowlton, U. S. Geol. Surv., Mon. 32, pt. 2, 1899, p. 716.
Miocene [auriferous gravels]: Independence Hill and Rock Corral, Placer County, Calif.
Miocene: Yellowstone National Park.
Oligocene: Tulameen River, British Columbia.

Ficus atavina Heer, Fl. foss. arct., vol. 6, Abt. 2, 1882, p. 69.—Lesquereux, U. S. Nat. Mus., Proc., vol. 10, 1887, p. 40.—Berry, Bull. Torr. Bot. Club., vol. 31, 1904, p. 75, pl. i, f. 8, 9; pl. iii, f. 6; Geol. Surv. N. J., Ann. Rept. 1905 [1906], p. 138.—Hollick, U. S. Geol. Surv., Mon. 50, 1906, p. 58, pl. x, f. 4-6.—Berry, U. S. Geol. Surv., Prof. Paper 84, 1914, p. 36, pl. x, f. 11.
Ficus protogaea Ettingshausen. Hollick, Bull. Torr. Bot. Club, vol. 21, 1894, p. 51, pl. 175, f. 4.
——: Upper Kanab Valley, Utah.
Magothy: Cliffwood, N. J.; Gay Head, Marthas Vineyard, Mass.; Glen Cove, Long Island, N. Y.
Black Creek (Middendorf): Middendorf, S. C.

Ficus atavina Heer? Hollick, N. Y. Acad. Sci., Trans., vol. 11, 1895, p. 103, pl. iv, f. 4, 6 = *Ficus krausiana*.

Ficus auriculata Lesquereux = *Ficus subtruncata*.

Ficus austiniana Lesquereux, in Winchell, Geol. and Nat. Hist. Surv. Minn., Thirteenth Ann. Rept., 1884 [1885], p. 76 [name].—Lesquereux, Geol. and Nat. Hist. Surv. Minn., Final Rept., vol. 3, pt. 1, 1893, p. 14, pl. A, f. 5.
  Dakota: North Star, Brown County, and Austin, Mower County, Minn.

Ficus baueri Knowlton, U. S. Geol. Surv., Prof. Paper 98, 1916, p. 338, pl. lxxxix, f. 2.
  Fruitland: 30 miles south of Farmington, San Juan County, N. Mex.

Ficus beckwithii Lesquereux = *Ficus krausiana*.

Ficus berryana Knowlton, U. S. Geol. Surv., Prof. Paper 130, 19—, p. —, pl. xi, f. 1 [in preparation].
  Laramie: Marshall, Colo.

Ficus berthoudi Lesquereux, Mus. Comp. Zool., Bull., vol. 16, 1888, p. 49.
  Denver: Golden, Colo.

Ficus berthoudi Lesquereux, U. S. Geol. Surv., Mon. 17, 1891 [1892], p. 78, pl. xii, f. 3 = *Ficus lesquereuxii*.

Ficus bruesi Cockerell, Torreya, vol. 10, 1910, p. 223, text f. 1.
  Miocene: Florissant, Colo.

Ficus cannoni Knowlton, U. S. Geol. Surv., Prof. Paper 130, 19—, p. —, pl. vi, f. 3; pl. x, f. 1 [in preparation].
  Laramie: Cowan station near Denver, and Popes Bluff, Pikeview, Colo.

Ficus cecilensis Berry, Md. Geol. Surv., Upper Cret., 1916, p. 821, pl. lviii, f. 4.
  Magothy: Grove Point, Md.

Ficus celtifolius Berry, U. S. Geol. Surv., Prof. Paper 84, 1914, p. 37, pl. xii, f. 4.
  Black Creek (Middendorf): Middendorf, S. C.

Ficus ceratops Knowlton, Bull. Torr. Bot. Club, vol. 38, 1911, p. 389, f. 1–4.
  Palmocarpon Knowlton, n. sp., in Stanton and Knowlton, Geol. Soc. Am., Bull., vol. 8, 1897, p. 136.
  Lance: Lance Creek, Converse County, Wyo.; Hell Creek, Mont.

Ficus cinnamomoides Lesquereux, Am. Philos. Soc., Trans., vol. 13, 1869, p. 417, pl. xvii, f. 8.—Berry, U. S. Geol. Surv., Prof. Paper 91, 1916, p. 198.
  Wilcox: Mississippi.

Ficus claibornensis Berry, U. S. Geol. Surv., Prof. Paper 84, 1914, p. 140, pl. xxiv, f. 6; pl. xxvii, f. 3, 4.
Barnwell (Twiggs): Grovetown and near Macon, Ga.

Ficus clintoni Lesquereux = *Ficus planicostata.*

Ficus cockerelli Knowlton, U. S. Geol. Surv., Prof. Paper 130, 19—, p. —, pl. xii, f. 2; pl. xxiii, f. 1, 2 [in preparation].
Ficus planicostata latifolia Lesquereux, U. S. Geol. and Geog. Surv. Terr., Ann. Rept. 1872 [1873], p. 393; Rept. U. S. Geol. Surv. Terr., vol. 7 (Tert. Fl.), 1878, p. 202, pl. xxxi, f. 9.—Berry, U. S. Geol. Surv., Prof. Paper 91, 1916, p. 199.
Ficus latifolia (Lesquereux) Knowlton, U. S. Geol. Surv., Bull. 152, 1898, p. 102; U. S. Geol. Surv., Prof. Paper 101, 1917, p. 304. [Homonym, Kunth, 1846.]
Post-Laramie: Black Buttes, Wyo.
Laramie: Near Golden and near Marshall, Colo.
Wilcox: Shreveport, La.
Raton?: Near Berwind, Colo.

Ficus coloradensis Cockerell, Torreya, vol. 10, 1910, p. 223.—Knowlton, U. S. Geol. Surv., Prof. Paper 130, 19—, p. —, pl. xxii, f. 1 [in preparation].
Ulmus? irregularis Lesquereux, Ann. Rept. U. S. Geol. and Geog. Surv. Terr., 1871 [1872], p. 378.
Ficus irregularis Lesquereux, U. S. Geol. and Geog. Surv. Terr., Bull., vol. 1, 1875 [1876], p. 368; Rept. U. S. Geol. Surv. Terr., vol. 7 (Tert. Fl.), 1878, p. 196, pl. xxxiv, f. 4–7; pl. lxiii, f. 9.—Ward, U. S. Geol. Surv., Sixth Ann. Rept., 1884–85 [1886], p. 552, pl. xliv, f. 4, 5; idem, Bull. 37, 1887, p. 38, pl. xx, f. 4, 5.—Knowlton, U. S. Geol. Surv., Bull. 162, 1900, p. 51. [Homonym, Miquel, Ann. Mus. Bot. Lugd. Bot., vol. 3, 1867, p. 224.]
Denver: *Golden, Colo.
Mesaverde: ?Point of Rocks, Wyo.
Laramie: Marshall and Golden, Colo.
Post-Laramie: Black Buttes, Wyo.
Hanna (?): Carbon, Wyo.

Ficus? condoni Newberry = *Platanus condoni.*

Ficus contorta Dawson, Roy. Soc. Canada, Trans., vol. 11, 1893 [1894], p. 60, pl. ix, f. 31.
Upper Cretaceous: Port McNeill, Vancouver Island.

Ficus? corylifolius Lesquereux, U. S. Geol. and Geog. Surv. Terr., Ann. Rept. 1872 [1873], p. 394. [Abandoned by author.]
Post-Laramie: Black Buttes, Wyo.

Ficus cowanensis Knowlton, U. S. Geol. Surv., Prof. Paper 130, 19—, p. —, pl. viii, f. 6; pl. ix, f. 3, 4 [in preparation].
Laramie: Cowan station, near Denver, Colo.

Ficus crassipes (Heer) Heer, Fl. foss. arct., vol. 6, Abt. 2, 1882, p. 70, pl. xvii, f. 9a; pl. xxiv, f. 1, 2.—Lesquereux, U. S. Geol. Surv., Mon. 17, 1891 [1892], p. 79, pl. xiii, f. 3.—Berry, Bull. Torr. Bot. Club, vol. 33, 1906, p. 172; idem, vol. 37, 1910, p. 505; U. S. Geol. Surv., Prof. Paper 84, 1914, pp. 37, 110, pl. x, f. 12; pl. xii, f. 8–10; Md. Geol. Surv., Upper Cret., 1916, p. 821, pl. lviii, f. 5; pl. lix, f. 2, 3; Bull. Torr. Bot. Club, vol. 43, 1916, p. 291; idem, vol. 44, 1917, p. 178; U. S. Geol. Surv., Prof. Paper 112, 1919, p. 79.

Proteoides crassipes Heer, Fl. foss. arct., vol. 3, Abt. 2, 1874, p. 110.

Dakota: Ellsworth County, Kans.

Magothy: Deep Cut, Del.; Grove Point and Little Round Bay, Md.

Eutaw: Chimney Bluff, below Columbus, Ga.; Coffee Bluff, Hardin County, Tenn.

Black Creek (Middendorf): Middendorf, Langley, Rocky Point, and Congaree River, S. C.

Tuscaloosa: Sanders Ferry Bluff and Snow place, Ala.

Bingen: Near Maxwell Spur and near Murfreesboro, Pike County, Ark.

Ficus crossii Ward, U. S. Geol Surv., Sixth Ann. Rept., 1884–85 [1886], p. 552, pl. xliv, f. 7; idem, Bull. 37, 1887, p. 39, pl. xxi, f. 2.—Knowlton, idem, Prof. Paper 130, 19—, p. —, pl. xi, f. 2 [in preparation].

Laramie: Golden, Colo.

Ficus? cuneatus Newberry = *Nyssa? cuneata.*

Ficus curta Knowlton, U. S. Geol. Surv., Prof. Paper 101, 1917, p. 266, pl. xlii, f. 5; idem, Prof. Paper 98, 1916, p. 338, pl. lxxxviii, f. 3 ?.

Vermejo: Near Rockvale, Colo.

Fruitland: 30 miles south of Farmington, San Juan County, N. Mex. [Species doubtful.]

Ficus dalli Cockerell, Torreya, vol. 10, 1910, p. 224.

Ficus membranacea Newberry, U. S. Nat. Mus., Proc., vol. 5, 1883, p. 512; U. S. Geol. Surv., Mon. 35, 1898, p. 87, pl. lix, f. 2.—Knowlton, U. S. Nat. Mus., Proc., vol. 17, 1894, p. 223; Geol. Soc. Am., Bull., vol. 5, 1893, p. 584. [Homonym, Wright, Sauvalle, Fl. Cub., 1873, p. 149.]

Kenai: Cook Inlet, Alaska.

Ficus dalmatica Ettingshausen, Denkschr. Wien. Akad., vol. 8, 1855, p. 13, pl. vii, f. 11.—Lesquereux, U. S. Geol. and Geog. Surv. Terr., Bull., vol. 1, 1875 [1876], p. 367; idem, Ann. Rept. 1874 [1876], p. 303; Rept. U. S. Geol. Surv. Terr., vol. 7 (Tert. Fl.), 1878, p. 199, pl. lxiii, f. 3–5.—Knowlton, U. S. Geol. Surv., Bull. 163, 1900, p. 51, pl. viii, f. 4; idem, Prof. Paper 101, 1918, p. 259; idem, Prof. Paper 130, 19—, p. —, pl. xxi, f. 9; pl. xxii, f. 5 [in preparation].

Mesaverde: Point of Rocks, Wyo.

Laramie: Coal Creek, Boulder County, Colo.

Vermejo: Rockvale, Colo.

Ficus daphnogenoides (Heer) Berry, Bull. Torr. Bot. Club, vol. 32, 1905, p. 329, pl. xxi; idem, vol. 33, 1906, p. 173, pl. i, f. 5; idem, vol. 34, 1907, p. 194, pl. xi, f. 10, 11; Geol. Surv. N. J., Ann. Rept. 1905 [1906], p. 138; Bull. Torr. Bot. Club, vol. 39, 1912, p. 394; Geol. Surv. N. J., Bull. 3, 1911, p. 122, pl. xii, f. 4; Md. Geol. Surv., Upper Cret., 1916, p. 818, pl. lviii, f. 3; Bull. Torr. Bot. Club, vol. 44, 1917, p. 177.

Proteoides daphnogenoides Heer, Phyll. crét. Nebr., 1866, p. 17, pl. iv, f. 9, 10.—Lesquereux, Rept. U. S. Geol. Surv. Terr., vol. 6 (Cret. Fl.), 1874, p. 85, pl. xv, f. 1, 2; U. S. Geol. Surv., Mon. 17, 1891 [1892], p. 90.—Hollick, N. Y. Acad. Sci. Trans., vol. 11, 1892, p. 98, pl. iii, f. 1, 2; idem, vol. 12, 1893, p. 36, pl. ii, f. 4, 9, 13; Bull. Torr. Bot. Club, vol. 21, 1894, p. 52, pl. 177, f. 1.—Smith, Geol. Coastal Plain Ala., 1894, p. 348.—Newberry, U. S. Geol. Surv., Mon. 26, 1895 [1896], p. 72, pl. xvii, f. 8, 9; pl. xxxii, f. 13, 14; pl. xxxiii, f. 3; pl. xli, f. 15.—Gould, Am. Jour. Sci., 4th ser., vol. 5, 1898, p. 175.—Berry, N. Y. Bot. Gard., Bull., vol. 3, 1903, p. 74, pl. li, f. 6-9.—Hollick, U. S. Geol. Surv., Mon. 50, 1906, p. 59, pl. xii, f. 1-5.—Berry, U. S. Geol. Surv., Prof. Paper 112, 1919, p. 80, pl. xiii, f. 6, 7.

Ficus proteoides Lesquereux, U. S. Geol. Surv., Mon. 17, 1891 [1892], p. 77, pl. xiii, f. 2.

Woodbine: Arthurs Bluff, Tex.
Dakota: *Decatur, Nebr.
Cheyenne: Chatman Creek, Kans.
Raritan: Sayreville, Woodbridge, Milltown, and South Amboy, N. J.; Tottenville, Staten Island, N. Y.
Magothy: Cliffwood, N. J.; Deep Cut, Del.; Grove Point, Md.; Gay Head, Mass.; Sea Cliff, Long Island, N. Y.
Black Creek: Court House Bluff, Cape Fear River, N. C.
Tuscaloosa: Shirleys Mill, Glen Allen, and Cottondale, Ala.
Mill Creek: Mill Creek, Northwest Territory.
Bingen: Near Maxwell Spur, Pike County, Ark.

Ficus dawsoni Knowlton and Cockerell, n. name.

Ficus maxima Dawson, Roy. Soc. Canada, Trans., vol. 1, sec. 4, 1882 [1883] p. 21, pl. ii, f. 5. [Homonym, Miller, 1768.]

——: Coal Brook, Northwest Territory.

Ficus decandolleana Heer. Penhallow, Rept. Tert. Pl. Brit. Col., 1908, p. 55.

Oligocene: Quilchena, British Columbia.

Ficus deflexa Lesquereux, U. S. Geol. Surv., Mon. 17, 1891 [1892], p. 80, pl. iii, f. 13; pl. xvi, f. 3.

Dakota: Ellsworth County, Kans.

Ficus deformata Knowlton, U. S. Geol. Surv., Mon. 32, pt. 2, 1899, p. 712, pl. xci, f. 2.

Fort Union: Elk Creek, Yellowstone National Park.

Ficus densifolia Knowlton, U. S. Geol. Surv., Mon. 32, pt. 2, 1899, p. 714, pl. lxxxix, f. 1; pl. xc, f. 1, 2; pl. xci, f. 1.

Fort Union: Elk Creek and Crescent Hill, Yellowstone National Park.
Miocene: Fossil Forest Ridge, Yellowstone National Park.

Ficus denveriana Cockerell, Torreya, vol. 10, 1910, p. 224.—Berry, U. S. Geol. Surv., Prof. Paper 91, 1916, pp. 11, 198.—Knowlton, idem, Prof. Paper 101, 1917, p. 302, pl. lxxv, f. 1, 2.

Ficus spectabilis Lesquereux, U. S. Geol. and Geog. Surv. Terr., Ann. Rept., 1872 [1873], p. 379; Rept. U. S. Geol. Surv. Terr., vol. 7 (Tert. Fl.), 1878, p. 199, pl. xxxiii, f. 4-6.—Ward, U. S. Geol. Surv., Sixth Ann. Rept., 1884-85 [1886], p. 552, pl. xliv, f. 6; idem, Bull. 37, 1887, p. 38, pl. xxi, f. 1.—Penhallow, Rept. Tert. Pl. Brit. Col., 1908, p. 56.—Knowlton, Washington Acad. Sci., Proc., vol. 11, 1909, p. 204. [Homonym, Kunth and Bouché, Ann. sci. nat., 3d ser., vol. 7, 1847, p. 235.]

Laurus utahensis Lesquereux. Lesquereux, U. S. Nat. Mus., Proc., vol. 11, 1888, p. 24.

Ficus goldiana Lesquereux. Lesquereux, U. S. Nat. Mus., Proc., vol. 11, 1888, p. 25.

Denver: *Golden, Colo.

Lagrange: Wickliffe, Ballard County, Ky.

Wilcox: Scarboroughs, Ark.; Campbells Quarry, Caddo Parish, McLees, and Coushatta, La.

Midway?: Earle, Bexar County, Tex.

Raton: Near Raton, Yankee, near Yankee, mouth of York Canyon, Vermejo Creek, Long Canyon, N. Mex.; Abeton, Wootton, near Trinidad, Berwind, and near Aguilar, Colo.

?Laramie: Popes Bluff, Pikeview, Colo.

Lance: Sevenmile Creek, 40 miles northwest of Edgemont, S. Dak.

Ficus distorta Lesquereux, U. S. Geol. and Geog. Surv. Terr., Bull. 1, 1875 [1876], p. 393; U. S. Geol. and Geog. Surv. Terr., Ann. Rept. 1874 [1876], p. 342, pl. v, f. 5.

Dakota: Fort Harker, Kans.

Ficus duplicata Knowlton, U. S. Geol. Surv., Prof. Paper 101, 1918, p. 302, pl. lxxiv, f. 1.

Raton: Near Aguilar, Colo.

Ficus eolignitica Berry, U. S. Geol. Surv., Prof. Paper 91, 1916, p. 203, pl. xxxi, f. 4.

Wilcox: Hardys Mill, Greene County, Ark.

Ficus eucalyptifolia Knowlton, U. S. Geol. Surv., Prof. Paper 101, 1918, p. 260, pl. xliv, f. 1, 2; idem, Prof. Paper 98, 1916, p. 340, pl. lxxxvii, f. 1, 2.

Vermejo: Rockvale, Colo.

Fruitland: San Juan County, N. Mex. [species doubtful].

Ficus evanstonensis Knowlton and Cockerell, n. name.

Ficus nervosa Newberry, U. S. Nat. Mus., Proc., vol. 5, 1882 [1883], p. 512. [Homonym, Heyne, 1821.]

Evanston: Evanston, Wyo.

Ficus? fimbriatus Lesquereux = *Eremophyllum fimbriatum*.

Ficus firma Knowlton, n. name.

Ficus montana Knowlton, U. S. Geol. Surv., Bull. 163, 1900, p. 48, pl. xi, f. 2, 3; pl. xii, f. 1. [Homonym, Burmeister, 1768.]

Mesaverde: Point of Rocks, Wyo.

Ficus florissantella Cockerell, Am. Mus. Nat. Hist., Bull., vol. 24, 1908, p. 88, pl. ix, f. 33.

Miocene: Florissant, Colo.

Ficus florissantia Knowlton, U. S. Nat. Mus., Proc., vol. 51, 1916, p. 267.

Ficus haydenii Lesquereux. Kirchner, St. Louis Acad. Sci., Trans., vol. 8, 1898, p. 179, pl. xii, f. 3.

Miocene: Florissant, Colo.

Ficus fontainii Berry, U. S. Geol. Surv., Prof. Paper 112, 1919, p. 82, pl. xi, f. 3.

Tuscaloosa: Cottondale and Shirleys Mill, Ala.

Ficus fracta Velenovsky, Fl. böhm. Kreideform., pt. 4, 1885, p. 10 (71), pl. viii (xxxi), f. 15.—Hollick, U. S. Geol. Surv., Mon. 50, 1906, p. 57, pl. xi, f. 7.

Aralia transversinerva Saporta and Marion. Hollick, Bull. Torr. Bot. Club, vol. 21, 1894, p. 54, pl. clxxvi, f. 1.

Magothy: Oak Neck, Long Island, N. Y.

Ficus fredericksburgensis Fontaine = *Ficophyllum crassinerve.*

Ficus fremonti Knowlton, U. S. Geol. Surv., Prof. Paper 108, 1917, p. 87, pl. xxxiv, f. 4–6; pl. xxxv, f. 4c, 5.

Glossopteris phillipsii? Brongniart. Hall, in Frémont, Rept. Exploring Expedition to the Rocky Mountains in 1842, etc., App. B, 1845, p. 305, pl. ii, f. 5, 5a, b, c.—Lesquereux, U. S. Nat. Mus., Proc., vol. 11, 1888, p. 37.

Frontier: About 1 mile east of Cumberland, Wyo.

Ficus gaudini Lesquereux = *Ficus uncata.*

Ficus georgiana Berry, U. S. Geol. Surv., Prof. Paper 84, 1914, p. 111, pl. xx, f. 1; idem, Prof. Paper 112, 1919, p. 79.

Ripley (Cusseta): Buena Vista, Ga.

Ficus gigantea Knowlton, U. S. Geol. Surv., Prof. Paper 101, 1918, p. 265, pl. xliii.

Vermejo?: Santa Clara Canyon, Colo.

Ficus glascoeana Lesquereux, Rept. U. S. Geol. Surv. Terr., vol. 8 (Cret. and Tert. Fl.), 1883, p. 48; U. S. Geol. Surv., Mon. 17, 91 [1892], p. 76, pl. xiii, f. 1, 2.

Dakota: Glascoe, Kans.

Ficus goldiana Lesquereux = *Ficus planicostata clintoni.*

Ficus haddeni Knowlton, U. S. Geol. Surv., Prof. Paper 101, 1918, p. 260, pl. xxxviii, f. 6, 7.

Vermejo: Rockvale, Colo.

Ficus haguei Knowlton, U. S. Geol. Surv., Mon. 32, pt. 2, 1899, p. 715, pl. xc, f. 3.
Miocene: Fossil Forest Ridge, Yellowstone National Park.

Ficus? halliana Lesquereux, Rept. U. S. Geol. Surv. Terr., vol. 6 (Cret. Fl.), 1874, p. 68, pl. xxviii, f. 3, 9.
Dakota: Minnesota.

Ficus harkeriana Knowlton, n. name.
Ficus sternbergii Lesquereux, U. S. Geol. Surv., Mon. 17, 1891 [1892], p. 82, pl. l, f. 1. [Homonym, Lesquereux, 1872.]
Dakota: Near Fort Harker, Kans.

Ficus harrisiana Hollick, Geol. Surv. La., Special Rept. 5, 1899, p. 281, pl. xlvi, f. 2.—Berry, U. S. Geol. Surv., Prof. Paper 91, 1916, p. 201, pl. xxxiv, f. 1.—Knowlton, U. S. Geol. Surv., Prof. Paper 101, 1918, p. 302.
Wilcox: Near Shreveport, Campbell's quarry, Cross Bayou, Caddo Parish, and De Soto Parish, La.
Raton?: Dillon Canyon, N. Mex.

Ficus haydenii Lesquereux, U. S. Geol. and Geog. Surv. Terr., Ann. Rept. 1872 [1873], p. 394; Rept. U. S. Geol. Surv. Terr., vol. 7 (Tert. Fl.), 1878, p. 197, pl. xxx, f. 1.
Post-Laramie: Black Buttes, Wyo.

Ficus haydenii Lesquereux. Kirchner, 1898 = *Ficus florissantia*.

Ficus hesperia Knowlton, U. S. Geol. Surv., Bull. 163, 1900, p. 45, pl. ix, f. 5.
Ficus sp. Knowlton, Geol. Soc. Am., Bull., vol. 8, 1897, p. 140.
Mesaverde: North Fork of Dutton Creek, Laramie Plains, Wyo.

Ficus? hesperia Knowlton, U. S. Geol. Surv., Twentieth Ann. Rept., pt. 3, 1900, p. 45, pl. ii, f. 4 = *Ficus? applegatei*.

Ficus impressa Knowlton, U. S. Geol. Surv., Prof. Paper 130, 19—, p. —, pl. vii, f. 1–3; pl. xvi, f. 3 [in preparation].
Laramie: Marshall, Colo.

Ficus inaequalis Lesquereux, U. S. Geol. Surv., Mon. 17, 1891 [1892], p. 82, pl. xlix, f. 6–8; pl. l, f. 3.—Berry, Bull. Torr. Bot. Club, vol. 34, 1907, p. 194, pl. xii, f. 2, 3; U. S. Geol. Surv., Prof. Paper 112, 1919, p. 80, pl. xii, f. 1.
Dakota: Near Fort Harker, Kans.
Black Creek: Court House Bluff, Cape Fear River, N. C.
Tuscaloosa: Shirleys Mill and Cottondale, Ala.

Ficus incompleta Knowlton, U. S. Geol. Surv., Bull. 163, 1900, p. 46, pl. ix, f. 2.
Mesaverde: Point of Rocks, Wyo.

Ficus interglacialis Hollick, N. Y. Bot. Gard., Jour., vol. 16, 1915, p. 44, pls. clii, cliii.
Pleistocene: Kootenay Valley, British Columbia.

Ficus irregularis Lesquereux = *Ficus coloradensis* and *Ficus regularis*.

Ficus irregularis? Lesquereux. Knowlton, Geol. Soc. Am., Bull., vol. 8, 1897, p. 150 = *Ficus multinervis*.

Ficus jynx Unger, Gen. et sp. pl. foss., 1850, p. 413.—Lesquereux, U. S. Geol. and Geog. Surv. Terr., Ann. Rept. 1873 [1874], p. 414; Rept. U. S. Geol. Surv. Terr., vol. 7 (Tert. Fl.), 1878, p. 193, pl. xxviii, f. 6.

Miocene: Elko station, Nev.

Ficus? kansasensis Knowlton and Cockerell, n. name.

Ficus? angustata Lesquereux, Rept. U. S. Geol. Surv. Terr., vol. 8 (Cret. and Tert. Fl.), 1883, p. 47; U. S. Geol. Surv., Mon. 17, 1892, p. 80. [Homonym, Miquel, 1848.]

Dakota: Bluff Creek, Kans.

Ficus krausiana Heer, Neue Denks. Schw. Ges., vol. 23, 1869, p. 15, pl. v, f. 3–6.—Lesquereux, U. S. Geol. Surv., Mon. 17, 1891 [1892], p. 81, pl. l, f. 5.—Hollick, Geol. Soc. Am., Bull., vol. 7, 1895, p. 13.—Berry, Bull. Torr. Bot. Club, vol. 33, 1906, p. 172.—Hollick, U. S. Geol. Surv., Mon. 50, 1906 [1907], p. 58, pl. ix, f. 9.—Berry, Bull. Torr. Bot. Club, vol. 37, 1910, p. 505; U. S. Geol. Surv., Prof. Paper 84, 1914, p. 38, pl. xi, f. 4–7; pl. xix, f. 4; Md. Geol. Surv., Upper Cret., 1916, p. 823, pl. lix, f. 1; Bull. Torr. Bot. Club, vol. 43, 1916, p. 291; U. S. Geol. Surv., Prof. Paper 112, 1919, p. 78.

Ficus beckwithii Lesquereux, Rept. U. S. Geol. Surv. Terr., vol. 8 (Cret. and Tert. Fl.), 1883, p. 46, pl. xvi, f. 5; pl. xvii, f. 3, 4.

Ficus atavina Heer? Hollick, N. Y. Acad. Sci., Trans., vol. 11, 1892, p. 103, pl. iv, f. 4, 6.

Black Creek (Middendorf): Middendorf, Rocky Point, and Congaree River, S. C.

Dakota: Near Fort Harker, Kans.

Eutaw: Chimney Bluff near Columbus, Ga.; Coffee Bluff, Hardin County, Tenn.; Havana, Ala.

Magothy: Grove Point, Md.; Gay Head, Marthas Vineyard, Mass.; Block Island, R. I.

Raritan: Tottenville, Staten Island, N. Y.

Tuscaloosa: Shirleys Mill, Glen Allen, Cottondale, and other localities, Ala.

Ficus krausiana subsimilis Hollick, N. Y. Bot. Gard., Bull., vol. 8, 1912, p. 158, pl. 164, figs. 1, 2.

Magothy: Roslyn, N. Y.

Ficus lacustris Knowlton, U. S. Geol. Surv., Twenty-first Ann. Rept., pt. 2, 1901, p. 215, pl. xxx, f. 26.

Esmeralda: 4.5 kilometers northeast of Emigrant Peak, Esmeralda County, Nev.

Ficus lanceolata Heer = *Ficus navicularis*.

Ficus lanceolata? Lesquereux. Knowlton, Geol. Soc. Am., Bull., vol. 8, 1897, p. 150 = *Ficus multinervis*.

Ficus lanceolato-acuminata Ettingshausen, Denkschr. Wien. Akad., vol. 32 (Foss. Fl. v. Sagor), 1872, p. 182, pl. vi, f. 3, 4.—Lesquereux, U. S. Geol. Surv., Mon. 17, 1891 [1892], p. 85, pl. xiii, f. 4.
Dakota: Pipe Creek, Cloud County, Kans.

Ficus latifolia (Lesquereux) Knowlton = *Ficus cockerelli*.

Ficus laurophylla Lesquereux.
Ficus laurophyllum Lesquereux, U. S. Geol. and Geog. Surv. Terr., Ann. Rept. 1874 [1876], p. 342, pl. v, f. 7; Rept. U. S. Geol. Surv. Terr., vol. 8 (Cret. and Tert. Fl.), 1883, p. 49, pl. i, f. 12, 13; U. S. Geol. Surv., Mon. 17, 1891 [1892], p. 85.
Laurophyllum reticulatum Lesquereux, U. S. Geol. and Geog. Surv. Terr., Ann. Rept. 1872 [1873], p. 425; Rept. U. S. Geol. Surv. Terr., vol. 6 (Cret. Fl.), 1874, p. 76, pl. xv, f. 4, 5.
Ficus reticulata (Lesquereux) Knowlton, U. S. Geol. Surv., Bull. 152, 1898, p. 104.—Newberry, idem, Mon. 35, 1898, p. 88, pl. xii, f. 2, 3.—Berry, N. Y. Bot. Gard., Bull., vol. 3, 1903, p. 73, pl. lii, f. 5; pl. liii, f. 1, 4; Geol. Surv. N. J., Rept. 1905 [1906], p. 138. [Homonym, Thunberg, 1786.]
Dakota: Fort Harker, Kans.; Blackbird Hill, Nebr.
Magothy: Cliffwood, N. J.

Ficus laurophyllidia Dawson, Roy. Soc. Canada, Trans., vol. 11, 1893 [1894], sec. 4, p. 60, pl. x, f. 37.
Upper Cretaceous: Wellington mine, Nanaimo, British Columbia.

Ficus laurophyllum Lesquereux = *Ficus laurophylla*.

Ficus leei Knowlton, U. S. Geol. Surv., Prof. Paper 101, 1918, p. 261, pl. xxxix, f. 1–6; pl. xl, f. 1, 2; idem, Prof. Paper 98, 1916, p. 338, pl. xc, f. 2.
Vermejo: Raton Mesa region, Colo. and N. Mex.
Fruitland: Amarillo Canyon, San Juan County, N. Mex.
Kirtland: Near Pina Veta China, San Juan County, N. Mex.

Ficus leonis Knowlton and Cockerell, n. name.
Ficus macrophylla Lesquereux, U. S. Geol. Surv., Mon. 17, 1891 [1892], p. 76, pl. xi, f. 1. [Homonym, Kunth and Bouché, 1846.]
Dakota: Near Fort Harker, Kans.

Ficus lesquereuxii Knowlton, U. S. Geol. Surv., Bull. 152, 1898, p. 102.
Ficus berthoudi Lesquereux, U. S. Geol. Surv., Mon. 17, 1891 [1892], p. 78, pl. xii, f. 3. [Name preoccupied by *F. berthoudi* Lesquereux, 1888.]
Dakota: *Ellsworth County, Kans.
Magothy: Marthas Vineyard, Mass.

Ficus? leyden Knowlton, U. S. Geol. Surv., Prof. Paper 130, 19—, p. —, pl. xiv, f. 1 [in preparation].
Laramie: Leyden Gulch, 6½ miles north of Golden, Colo.

Ficus limpida Ward, U. S. Geol. Surv., Sixth Ann. Rept., 1884–85 [1886], p. 552, pl. xlv, f. 4; U. S. Geol. Surv., Bull. 37, 1887, p. 42, pl. xxii, f. 3.
Fort Union: Clear Creek, Glendive, Mont.

Ficus macrophylla Lesquereux = *Ficus leonis.*

Ficus magnoliaefolia Lesquereux, Rept. U. S. Geol. Surv. Terr., vol. 8 (Cret. and Tert. Fl.), 1883, p. 47, pl. xvii, f. 5, 6; U. S. Geol. Surv., Mon. 17, 1891 [1892], p. 79, pl. xvi, f. 4.
Dakota: *Morrison, Colo.; Ellsworth County, Kans.
Upper Cretaceous: Vancouver Island, British Columbia.

Ficus mattawanensis Berry, Bull. Torr. Bot. Club, vol. 38, 1911, p. 399, pl. xix, f. 3.
Matawan: Woodbury clays, N. J.

Ficus maxima Dawson = *Ficus dawsoni.*

Ficus melanophylla Lesquereux, U. S. Geol. Surv., Mon. 17, 1891 [1892], p. 83, pl. l, f. 2.
Dakota: Kansas.

Ficus membranacea Newberry = *Ficus dalli.*

Ficus mensae Cockerell, Am. Nat., vol. 44, 1910, p. 45.
Ficus microphylla Lesquereux, Mus. Comp. Zool., Mem., vol. 6, No. 2, 1878, p. 18, pl. iv, f. 10, 11. [Homonym, Salzmann, Mart. Fl. Braz., vol. 4, 1853, p. 93.]
Miocene (auriferous gravels): Table Mountain and Shasta County, Calif.

Ficus microphylla Lesquereux = *Ficus mensae.*

Ficus minima Knowlton, U. S. Geol. Surv., Prof. Paper 101, 1918, p. 260, pl. xxxvi, f. 4; pl. xxxviii, f. 5.
Vermejo: Walsenburg, Colo.

Ficus minutidens Knowlton, U. S. Geol. Surv., Prof. Paper 101, 1918, p. 305, pl. lxxi, f. 2.
Raton: Wootton, Colo.

Ficus missouriensis Knowlton, U. S. Geol. Surv., Bull. 163, 1900, p. 12, pl. i, f. 5.
Eagle: Missouri River below Coal Banks, Mont.

Ficus monodon (Lesquereux) Berry, U. S. Geol. Surv., Prof. Paper 91, 1916, p. 201, pl. xxxii, f. 2; pl. xxxiii, f. 2.
Populus monodon Lesquereux, Am. Philos. Soc., Trans., vol. 13, 1869, p. 413, pl. xv, f. 1, 2.
Wilcox (Grenada): Grenada, Grenada County, Miss.
Wilcox (Ackerman): Hurleys, Benton County, Miss.
Lagrange: Puryear, Henry County, Tenn.

Ficus montana Knowlton = *Ficus firma.*

Ficus mudgei Lesquereux, U. S. Geol. Surv., Mon. 17, 1891 [1892], p. 83, pl. xii, f. 4.

Dakota: Ellsworth County, Kans.

Ficus multinervis Heer, Fl. tert. Helv., vol. 2, 1856, p. 63, pl. lxxxi, f. 6–10.—Lesquereux, U. S. Geol. and Geog. Surv. Terr., Ann. Rept. 1871 [1872], p. 300; Rept. U. S. Geol. Surv. Terr., vol. 7 (Tert. Fl.), 1878, p. 194, pl. xxviii, f. 7, 8.—Knowlton, U. S. Geol. Surv., Bull. 163, 1900, p. 43; idem, Prof. Paper 130, 19—, p. —, pl. xii, f. 3, 4 [in preparation].

Ficus irregularis? Lesquereux. Knowlton, Geol. Soc. Am., Bull., vol. 8, 1897, p. 150.

Ficus lanceolata? Lesquereux. Knowlton, op. cit., p. 150.

?Laramie: Coal Creek, Boulder County, Colo.

Montana: Coalville, Utah.

Ficus multinervis Heer. Lesquereux, U. S. Nat. Mus., Proc., vol. 11, 1888, p. 11 = *Ficus myrtifolia.*

Ficus myricoides Hollick, in Newberry, U. S. Geol. Surv., Mon. 26, 1895 [1896], p. 71, pl. xxxii, f. 18; pl. xli, f. 8, 9.—Fontaine, in Ward, U. S. Geol. Surv., Mon. 48, 1905 [1906], p. 531, pl. cxii, f. 12.—Hollick, idem, Mon. 50, 1906, p. 57, pl. xi, f. 8, 9; ?N. Y. Bot. Gard., Bull., vol. 8, 1912, p. 158, pl. clxvi, f. 2; Geol. Surv. N. J., Bull. 3, 1911, p. 121.

Raritan: Milltown, N. J.; Roslyn, N. Y.

Magothy: Gay Head, Marthas Vineyard, Mass.; Glen Cove, Long Island, N. Y.

Patapsco: Fort Foote, Md.

Ficus myricoides Hollick. Ward, Mon. 48, p. 531, pl. cxii, f. 12 = *Sapindopsis variabilis.*

Ficus myrtifolius Berry, U. S. Geol. Surv., Prof. Paper 91, 1916, p. 205, pl. xxx, f. 1–3.

Ficus multinervis Heer. Lesquereux, U. S. Nat. Mus., Proc., vol. 11, 1888, p. 11, pl. iv, f. 2, 3.

Wilcox (Holly Springs): Holly Springs and Early Grove, Marshall County, Miss.

Lagrange: Wickliffe, Ballard County, and Boaz, Graves County, Ky.

Ficus navicularis Cockerell, Am. Mus. Nat. Hist., Bull., vol. 24, 1908, p. 89; Torreya, vol. 9, 1909, p. 141, text f. A, C.—Knowlton, U. S. Geol. Surv., Prof. Paper 130, 19—, p. —, pl. vi, f. 4, 5; pl. xi, f. 3–5 [in preparation].

Ficus lanceolata Heer, Fl. tert. Helv., vol. 2, 1856, p. 62, pl. lxxxi, f. 2–5.—Lesquereux, U. S. Geol. and Geog. Surv. Terr., Ann. Rept. 1871 [1872], p. 300; Rept. U. S. Geol. Surv. Terr., vol. 7 (Tert. Fl.), 1878, p. 192, pl. xxviii, f. 1–5. [Homonym, Buch Ham, 1814.]

Laramie: Marshall's mine, Coal Creek, and near Mount Carbon, Colo.

Ficus neodalmatica Knowlton, U. S. Geol. Surv., Prof. Paper 130, 19—, p. —, pl. vii, f. 6 [in preparation].
Laramie: 6 miles south of Golden, Colo.

Ficus neoplanicostata Knowlton, U. S. Geol. Surv., Prof. Paper 101, 1918, p. 303, pl. lxxiii, f. 4; pl. lxxiv, f. 2, 3; pl. lxxvi, f. 4.—Berry, U. S. Geol. Surv., Prof. Paper 91, 1916, p. 198, pl. cxiv, f. 1.
Raton: Near Walsenburg, Colo.
Wilcox: De Soto Parish, La.

Ficus nervosa Newberry = *Ficus evanstonensis*.

Ficus neurocarpa Hollick, Bull. Torr. Bot. Club, vol. 30, 1903, p. 205, text f. B1.
Dakota: Ellsworth County, Kans.

Ficus newberryana Knowlton, U. S. Geol. Surv., Prof. Paper 101, 1918, p. 261, pl. xxxviii, f. 8.
Vermejo: Rockvale, Colo.

Ficus newberryana Heer, Fl. foss. arct., vol. 7, 1883, p. 28, pl. l, f. 2 = *Platanus asperaeformis*.

Ficus newtonensis Berry, in Cooke and Shearer, U. S. Geol. Surv., Prof. Paper 120, 1918, p. 55 [name].
Barnwell (Twiggs): Twiggs County, Ga.

Ficus oblanceolata Lesquereux, U. S. Geol. and Geog. Surv. Terr., Ann. Rept. 1872 [1873], p. 387; Rept. U. S. Geol. Surv. Terr., vol. 7 (Tert. Fl.), 1878, p. 194, pl. xxviii, f. 9–12.
Pistacia oblanceolata (Lesquereux) Knowlton, U. S. Geol. Surv., Bull. 152, 1898, p. 167.
——: Black Buttes, Wyo.

Ficus occidentalis (Lesquereux) Lesquereux, Rept. U. S. Geol. Surv. Terr., vol. 7 (Tert. Fl.), 1878, p. 200, pl. xxxii, f. 4; Mus. Comp. Zool., Bull., vol. 16, 1888, p. 50.—Penhallow, Rept. Tert. Pl. Brit. Col., 1908, p. 55.—Berry, U. S. Geol. Surv., Prof. Paper 91, 1916, pp. 121, 197, pl. xxviii, f. 3.—Knowlton, idem, Prof. Paper 101, 1918, p. 331, pl. lxxii, f. 1.
Dombeyopsis occidentalis Lesquereux, U. S. Geol. and Geog. Surv. Terr., Ann. Rept. 1872 [1873], p. 380.
Denver: Golden, Colo.
Eocene?: Stanley Park, Vancouver Island.
Raton: Dillon Canyon, near Blossburg, N. Mex.
Wilcox: Shreveport, La.; Hurley, Miss.
Midway?: Earle, Tex.

Ficus? oregoniana Lesquereux, U. S. Nat. Mus., Proc., vol. 11, 1888, p. 18, pl. ix, f. 3.—Knowlton, U. S. Geol. Surv., Bull. 204, 1902, p. 56, pl. x, f. 3.—Knowlton, in Smith, U. S. Geol. Surv. Folio 86 (Ellensburg, Wash.), 1903, p. 3.
Mascall: Van Horn's ranch, John Day Basin, Oreg.
Ellensburg: Kelly Hollow, Wenas Valley, Wash.

Ficus ovaliformis Cockerell, Torreya, vol. 10, 1910, p. 224.

Ficus ovalis Lesquereux, U. S. Geol. and Geog. Surv. Terr., Bull., vol. 1, 1875 [1876], p. 387. [Homonym, Miquel, Ann. Mus. Bot. Lugd. Bot., vol. 3, 1867, p. 298.]

——: Pleasant Park, Colo.

Ficus ovalis Lesquereux = *Ficus ovaliformis.*

Ficus ovata Newberry = *Ficus ovatifolia.*

Ficus ovatifolia Berry, Bull. Torr. Bot. Club, vol. 36, 1909, p. 253; idem, vol. 37, 1910, p. 504; idem, vol. 38, 1911, p. 410; U. S. Geol. Surv., Prof. Paper 84, 1914, p. 111, pl. xix, f. 5–7; Geol. Surv. N. J., Bull. 3, 1911, p. 123, pl. xii, f. 3; Md. Geol. Surv., Upper Cret., 1916, p. 820, pl. lix, f. 4; Bull. Torr. Bot. Club, vol. 43, 1916, p. 292; U. S. Geol. Surv., Prof. Paper 112, 1919, p. 79; ?Bull. Torr. Bot. Club, vol. 44, 1917, p. 179.

Ficus ovata Newberry, U. S. Geol. Surv., Mon. 26, 1896, p. 70, pl. xxiv, f. 1–3. [Homonym, Don, 1802–3.]

Black Creek: Cape Fear River, N. C.

Raritan: Milltown and Woodbridge, N. J.; East Washington Heights, D. C.

Eutaw: McBrides Ford, Columbus, Ga.; Coffee Bluff, Hardin County, Tenn.

Bingen: Mine Creek near Nashville, Ark.

Ficus pealei Knowlton, U. S. Geol. Surv., Prof. Paper 130, 19—, p. —, pl. xi, f. 6 [in preparation].

Laramie: Cowan station, near Denver, Colo.

Ficus planicostata Lesquereux, U. S. Geol. and Geog. Surv. Terr., Ann. Rept. 1872 [1873], p. 393; Rept. U. S. Geol. Surv. Terr., vol. 7 (Tert. Fl.), 1878, p. 201, pl. xxxi, f. 1–8, 10–12.—Newberry, U. S. Geol. Surv., Mon. 35, 1898, p. 88, pl. xlvi, f. 1.—Knowlton, U. S. Geol. Surv., Bull. 163, 1900, p. 52, pl. x, f. 4; pl. xii, f. 2–4; idem, Bull. 204, 1902, p. 56.—Cockerell, Univ. Colorado Studies, vol. 7, 1910, p. 151.—Hollick, Geol. Surv. La., Rept., 1899, p. 282, pl. xxxvi.—Knowlton, U. S. Geol. Surv., Prof. Paper 130 [in preparation].

Ficus clintoni Lesquereux, U. S. Geol. and Geog. Surv. Terr., Ann. Rept. 1872 [1873], p. 393.

Post-Laramie: *Black Buttes, Wyo.

Laramie: Coal Creek, Boulder County, and Golden, Colo.

Mesaverde?: Point of Rocks, Wyo.

Montana?: Near Meeker, Colo.

Wilcox: Shreveport, La.

Clarno (upper part): Bridge Creek, Oreg.

Ficus planicostata clintoni (Lesquereux) Knowlton = *Ficus planicostata goldiana.*

Ficus planicostata goldiana Lesquereux, U. S. Geol. and Geog. Surv. Terr., Ann. Rept. 1873 [1874], p. 399; Rept. U. S. Geol. Surv. Terr., vol. 7 (Tert. Fl.), 1878, p. 202, pl. xxxiii, f. 1–3.
Ficus planicostata clintoni (Lesquereux) Knowlton, U. S. Geol. Surv., Bull. 152, 1898, p. 103; idem, Prof. Paper 101, 1918, p. 303, pl. lxxvi, f. 3.
Denver: *Golden, Colo.
Raton: Near Raton, N. Mex.
Wilcox: Cross Lake, La.

Ficus planicostata latifolia Lesquereux = *Ficus cockerelli.*

Ficus planicostata magnifolia Knowlton, U. S. Geol. Surv., Prof. Paper 130, 19—, p. —, pl. x, f. 3 [in preparation].
Laramie: Marshall, Colo.

Ficus planicostata maxima Berry, U. S. Geol. Surv., Prof. Paper 91, 1916, p. 199, pl. xxxiv, f. 3.
Ficus planicostata Lesquereux. Hollick, Geol. Surv. La., Special Rept. 1888, p. 282, pl. xxxvi.
Wilcox: Slaughter Pen Bluff, Cross Bayou, Caddo Parish; De Soto Parish, La.; Old Port Caddo Landing, Harrison County, Tex.
Lagrange: Puryear, Henry County, Tenn.

Ficus populina Heer, Fl. tert. Helv., vol. 2, 1856, p. 66, pl. lxxxv, f. 1–7; pl. lxxxvi, f. 1–11.—Lesquereux, U. S. Geol. and Geog. Surv. Terr., Ann. Rept. 1871 [1872], Suppl., p. 6.—Penhallow, Rept. Tert. Pl. Brit. Col., 1908, p. 55.
Green River: Green River, Wyo.
Oligocene: Horsefly and Tulameen rivers, British Columbia.

Ficus populoides Knowlton, U. S. Geol. Surv., Bull. 163, 1900, p. 44, pl. viii, f. 3.
Mesaverde: Point of Rocks, Wyo.

Ficus post-trinervis Knowlton, U. S. Geol. Surv., Prof. Paper 130, p. —, pl. vi, f. 1, 2 [in preparation].
Laramie: Cowan station, near Denver, Colo.

Ficus praecursor Lesquereux, U. S. Geol. Surv., Mon. 17, 1891 [1892], p. 81, pl. xlix, f. 5.
Dakota: Ellsworth County, Kans.

Ficus praelatifolia Knowlton, U. S. Geol. Surv., Prof. Paper 98, 1916, p. 338, pl. lxxxvii, f. 4.
Fruitland: 30 miles south of Farmington, San Juan County, N. Mex.

Ficus praeplanicostata Knowlton, U. S. Geol. Surv., Prof. Paper 130, 19—, p. —, pl. xxii, f. 2 [in preparation].
Laramie: Coal Creek, Boulder County, Colo.

Ficus praesinuosa Knowlton and Cockerell, n. name.
Ficus sinuosa Ward, U. S. Geol. Surv., Sixth Ann. Rept., 1884–85 [1886], p. 552, pl. xlv, f. 3; idem, Bull. 37, 1887, p. 41, pl. xxii. f. 2. [Homonym, Miquel, 1848.]
Post-Laramie: Black Buttes, Wyo.

Ficus praetrinervis Knowlton, U. S. Geol. Surv., Prof. Paper 101, 1917, pp. 263, 304, pl. xli, f. 1–4; pl. xlii, f. 1; idem, Prof. Paper 98, 1916, p. 338.
Vermejo: Raton Mesa region, Colo. and N. Mex.
Raton: Cokedale, Wootton, and Dean, Colo.
Fruitland: 30 miles south of Farmington, San Juan County, N. Mex.

Ficus primordialis Heer, Phyll. crét. Nebr., 1866, p. 16, pl. iii, f. 1.
Dakota: Tekamah, Nebr.

Ficus problematica Knowlton, U. S. Geol. Surv., Bull. 163, 1900, p. 46, pl. ix, f. 3.
Mesaverde: Point of Rocks, Wyo.

Ficus proteoides Lesquereux = *Ficus daphnogenoides.*

Ficus protogaea Ettingshausen. Hollick, Bull. Torr. Bot. Club, vol. 21, 1894, p. 51, pl. 175, f. 4 = *Ficus atavina.*

Ficus pseudocuspidata Berry, U. S. Geol. Surv., Prof. Paper 91, 1916, p. 197, pl. xxviii, f. 1.
Lagrange: Puryear, Henry County, Tenn.

Ficus pseudolmediafolia Berry = *Laurus ratonensis.*

Ficus pesudo-populus Lesquereux, U. S. Geol. and Geog. Surv. Terr., Bull., vol. 1, 1875 [1876], p. 387; idem, Ann. Rept., 1874 [1876], p. 313; Rept. U. S. Geol. Surv. Terr., vol. 7 (Tert. Fl.), 1878, p. 204, pl. xxxiv, f. 1a, 2.—Berry, U. S. Geol. Surv., Prof. Paper 91, 1916, p. 200, pl. xxxvii, f. 3–5; pl. cxiii, f. 3.—Knowlton, U. S. Geol. Surv., Prof. Paper 101, 1918, p. 304, pl. lxxii, f. 2–4; pl. lxxiii, f. 1, 2; pl. cxii, f. 3.
Evanston: Evanston, Wyo.
Raton: Near Yankee, Vermejo Park, York Canyon, N. Mex.; near Trinidad, Berwind, and Wootton, Colo.
Wilcox: Near Mansfield and Naborton, De Soto Parish, La.
Lagrange: Puryear, Henry County, and Hatchie River near Sandy, Hardeman County, Tenn.

Ficus puryearensis Berry, U. S. Geol. Surv., Prof. Paper 91, 1916, p. 205, pl. xxvii, f. 4, 5; pl. xxviii, f. 5; pl. xxx, f. 4, 5.
Wilcox (Grenada): Grenada, Grenada County, Miss.
Wilcox (Ackerman): Hurleys, Benton County, Miss.
Lagrange: Puryear, Henry County, Tenn.

Ficus ratonensis Knowlton, U. S. Geol. Surv., Prof. Paper 101, 1918, p. 306, pl. lxxiv, f. 4.
Raton: Near Yankee, N. Mex.

Ficus regularis Knowlton, U. S. Geol. Surv., Prof. Paper 101, 1917, p. 264.

Ficus irregularis (Lesquereux) Lesquereux [part], U. S. Geol. and Geog. Surv. Terr., Bull., vol. 1, 1875, p. 368; U. S. Geol. Surv. Terr., Rept., vol. 7 (Tert. Fl.), 1878, p. 196, pl. lxiii, f. 9 [not pl. xxxiv, f. 4–7].

Ficus coloradensis Cockerell [part], Torreya, vol. 10, 1910, p. 223.

Mesaverde: Point of Rocks, Wyo.
Vermejo: Canon City field, Colo.

Ficus reticulata (Lesquereux) Knowlton = *Ficus laurophylla.*

Ficus rhamnoides Knowlton, U. S. Geol. Surv., Bull. 163, 1900, p. 47, pl. x, f. 1–3; pl. xi, f. 1; U. S. Geol. Surv., Prof. Paper 98, 1916, p. 339, pl. lxxxvii, f. 3; idem, Prof. Paper 101, 1918, p. 265.

Mesaverde: *Point of Rocks, Wyo.
Vermejo: Rockvale, Colo.
Fruitland: San Juan County, N. Mex.

Ficus? rhomboideus Lesquereux = *Phyllites rhomboideus.*

Ficus richardsoni Knowlton, U. S. Geol. Surv., Prof. Paper 101, 1918, p. 305, pl. lxxvi, f. 1.

Raton: Aguilar, Colo.

Ficus richmondensis Berry, Jour. Geol., vol. 17, 1909, p. 26, text f. 5.

Calvert: Richmond, Va.

Ficus rockvalensis Knowlton, U. S. Geol. Surv., Prof. Paper 101, 1918, p. 265, pl. xl, f. 3, 4.

Vermejo: Rockvale, Colo.

Ficus rotundata Dawson, Roy. Soc. Canada, Trans., vol. 11, 1893 [1894], sec. 4, p. 60, pl. ix, f. 32, 33a.

Upper Cretaceous: Vancouver Island, British Columbia.

Ficus russelli Knowlton, Bull. Torr. Bot. Club, vol. 38, 1911, p. 392.

Lance: Forsyth, Mont.

Ficus sapindifolia Hollick, N. Y. Bot. Gard., Bull., vol. 3, 1904, p. 411, pl. 78, f. 5; U. S. Geol. Surv., Mon. 50, 1906, p. 58, pl. xi, f. 1, 2.

Magothy: Mott Point, Long Island, N. Y.; Gay Head, Marthas Vineyard, Mass.; Manhasset Neck, Long Island, N. Y.

Ficus schimperi Lesquereux, Am. Philos. Soc., Trans., vol. 13, 1869, p. 417, pl. xviii, f. 1–3.—Berry, U. S. Geol. Surv., Prof. Paper 91, 1916, p. 204, pl. xxxi, f. 1–3.—Knowlton, U. S. Geol. Surv., Prof. Paper 101, 1918, p. 304, pl. lxxv, f. 3, 4.

Celtis brevifolia Lesquereux, Am. Philos. Soc., Trans., vol. 13, 1869, p. 416, pl. xx, f. 4, 5.—Berry, U. S. Geol. Surv., Prof. Paper 91, 1916, p. 204.

Raton: Near Yankee and Vermejo Valley, N. Mex.; Aguilar and near Trinidad, Colo.
Wilcox (Ackerman): Hurleys, Benton County, Miss.
Lagrange: Puryear, Henry County, and Baughs Bridge, Fayette County, Tenn.
Wilcox: Old Port Caddo Landing, Harrison County, Tex.; Coushatta, La.

Ficus shastensis Lesquereux, U. S. Nat. Mus., Proc., vol. 11, 1888, p. 28, pl. xi, f. 3.—Knowlton, in Turner, Am. Geol., vol. 15, 1895, p. 378.—Penhallow, Rept. Tert. Pl. Brit. Col., 1908, p. 56.—Knowlton, U. S. Geol. Surv., Mon. 32, pt. 2, 1899, p. 714.

——: Shasta County, Calif.

Miocene (auriferous gravels): ?Volcano Hill, Placer County, Calif.

?Eocene: Burrard Inlet, British Columbia.

Miocene: Lamar River, Yellowstone National Park. [Identification ?]

Ficus shirleyensis Berry, U. S. Geol. Surv., Prof. Paper 112, 1919, p. 81, pl. xiii, f. 8.

Tuscaloosa: Shirleys Mill, Ala.

Ficus sinuosa Ward = *Ficus praesinuosa.*

Ficus? smithsoniana (Lesquereux) Lesquereux, Rept. U. S. Geol. Surv. Terr., vol. 7 (Tert. Fl.), 1878, p. 200, pl. xxxii, f. 5.—Knowlton, U. S. Geol. Surv., Prof. Paper 101, 1918, p. 305, pl. —; idem, Prof. Paper 130, 19—, p. —, pl. xxi, f. 4 [in preparation].

Juglans smithsoniana Lesquereux, U. S. Geol. and Geog. Surv. Terr., Ann. Rept. 1871 [1872], Suppl., p. 16.

Laramie: Mount Carbon, near Morrison, Colo.

Raton: *Raton Mountains, near Trinidad, Colo.

Ficus sordida Lesquereux, Mus. Comp. Zool., Mem., vol. 6, No. 2, 1878, p. 17, pl. iv, f. 6, 7.—Knowlton, in Lindgren, Jour. Geol., vol. 4, 1896, p. 889; in Turner, Am. Geol., vol. 15, 1895, p. 378.—Knowlton, U. S. Geol. Surv., Mon. 32, pt. 2, 1899, p. 714.

Miocene (auriferous gravels): Chalk Bluffs, Nevada County, Volcano Hill and Independence Hill, Placer County, Calif.

Miocene: Specimen Ridge, Yellowstone National Park.

Ficus speciossima Ward, U. S. Geol. Surv., Sixth Ann. Rept., 1884–85 [1886], p. 552, pl. xlv, f. 1; U. S. Geol. Surv., Bull. 37, p. 39, pl. xxi, f. 3.—Knowlton, U. S. Geol. Surv., Bull. 163, 1900, p. 52.—Cockerell, Univ. Colo. Studies, vol. 7, 1910, p. 151.—Knowlton, U. S. Geol. Surv., Prof. Paper 98, 1916, p. 90, pl. xvi, f. 3; idem, Prof. Paper 101, 1918, p. 262.

Mesaverde: *Point of Rocks, Wyo.

Montana ?: 4 miles west of Meeker, Colo.

Vermejo: Raton Mesa region, Colo.–N. Mex.

Fox Hills: Wildcat Mound, 2½ miles south of Milliken, Colo.

Ficus spectabilis Lesquereux = *Ficus denveriana.*

Ficus squarrosa Knowlton, U. S. Geol. Surv., Bull. 163, 1900, p. 45, pl. viii, f. 2; U. S. Geol. Surv., Prof. Paper 98, 1916, p. 339, pl. lxxxvi, f. 10.

Mesaverde: *Point of Rocks, Wyo.

Fruitland: 30 miles south of Farmington, San Juan County, N. Mex. [Species doubtful.]

Ficus? starkvillensis Knowlton, U. S. Geol. Surv., Prof. Paper 101, 1918, p. 262, pl. xxxviii, f. 9.

Vermejo: Starkville, Forbes, and Gray Creek, Colo.

Ficus stephensoni Berry, Bull. Torr. Bot. Club, vol. 37, 1910, p. 194, pl. xxiii, f. 2, 3; U. S. Geol. Surv., Prof. Paper 84, 1914, p. 38, pl. xii, f. 1–3.

Black Creek: Court House Bluff, Cape Fear River, N. C.

Black Creek (Middendorf): Middendorf and Langley, S. C.

Ficus sternbergii Lesquereux, U. S. Geol. and Geog. Surv. Terr., Ann. Rept. 1872 [1873], p. 423 = *Persea sternbergii.*

Ficus sternbergii Lesquereux, U. S. Geol. Surv., Mon. 17, 1891 [1892], p. 82, pl. l, f. 1 = *Ficus harkeriana.*

Ficus subtruncata Lesquereux, Rept. U. S. Geol. Surv. Terr., vol. 7 (Tert. Fl.), 1878, p. 205, pl. xxx, f. 7–9.

Ficus auriculata Lesquereux, U. S. Geol. and Geog. Surv. Terr., Ann. Rept. 1872 [1873], p. 379; Rept. U. S. Geol. Surv. Terr., vol. 7 (Tert. Fl.), 1878, p. 206, pl. xxx, f. 4–8. [Homonym, Loureiro, 1834.]

Ficus truncata? Lesquereux, U. S. Geol. and Geog. Surv. Terr., Ann. Rept. 1873 [1874], p. 400.

Denver: Golden, Colo.

Ficus tenuinervis Lesquereux, Rept. U. S. Geol. Surv. Terr., vol. 8 (Cret. and Tert. Fl.), 1883, p. 164, pl. xliv, f. 4; U. S. Nat. Mus., Proc., vol. 11, 1888, p. 23.—Knowlton, U. S. Geol. Surv., Bull. 204, 1902, p. 55.

Green River: Alkali station, Wyo.

Clarno (lower part): Cherry Creek, Crook County, Oreg.

Ficus tessellata Lesquereux MS. Knowlton, U. S. Geol. Surv., Prof. Paper 101, 1918, p. 266, pl. xli, f. 5.

Vermejo: Rockvale, Colo.

Ficus tiliaefolia (Al. Braun) Heer, Fl. tert. Helv., vol. 2, 1856, p. 68, pl. lxxxiii, f. 3–12; pl. lxxxiv, f. 1–6; pl. lxxxv, f. 14.—Lesquereux, Rept. U. S. Geol. Surv. Terr., vol. 7 (Tert. Fl.), 1878, p. 203; Mus. Comp. Zool., Mem., vol. 6, No. 2, 1878, p. 18, pl. iv, f. 8, 9.—Ward, U. S. Geol. Surv., Sixth Ann. Rept., 1884–85 [1886], p. 552, pl. xlv, f. 2; idem, Bull. 37, 1887, p. 40, pl. xxii, f. 1.—Knowlton, U. S. Geol. Surv., Mon. 32, pt. 2, 1899, p. 716; idem, Bull. 163, 1900, p. 52, pl. xiii, f. 1; in Lindgren, Jour. Geol., vol. 4, 1896, p. 889.

Cordia? tiliaefolia Al. Braun. Leonhard and Brown, Neues Jahrb., 1845, p. 170.

Dombeyopsis aequifolia Göppert. Lesquereux, U. S. Geol. and Geog. Surv. Terr., Ann. Rept. 1871 [1872], p. 10.

Eocene: Southwestern Oregon.

Miocene (auriferous gravels): Chalk Bluffs, Nevada County. Independence Hill, Placer County, Calif.

Ficus tiliaefolia—Continued.

Fort Union: Burns's ranch, near Glendive, Mont.; Yanceys, Yellowstone National Park.

Denver: Golden, Colo.

Ione: Kosk Creek, Shasta County, Calif.

?Mesaverde: Point of Rocks, Wyo.

Ficus trinervis Knowlton, U. S. Geol. Surv., Bull. 163, 1900, p. 42.—Cockerell, Am. Mus. Nat. Hist., Bull., vol. 24, 1908, p. 89.—Knowlton, Washington Acad. Sci., Proc., vol. 11, 1909, p. 197.

Cinnamomum affine Lesquereux, in part, Rept. U. S. Geol. Surv. Terr., vol. 7 (Tert. Fl.), 1878, p. 219, pl. xxxvii, f. 5 [not f. 1–4, 7].

Cinnamomum affine Lesquereux, Ward, U. S. Geol. Surv., Sixth Ann. Rept., 1884–85 [1886], p. 553, pl. lxvii, f. 1–3; idem, Bull. 37, 1887, p. 50, pl. xxiv, f. 3–5.

?Mesaverde: Point of Rocks, Wyo.

Post-Laramie: *Black Buttes, Wyo.

Ficus truncata? Lesquereux = *Ficus subtruncata.*

Ficus ulmifolia Lesquereux = *Ficus uncata.*

Ficus uncata Lesquereux, Rept. U. S. Geol. Surv. Terr., vol. 7 (Tert. Fl.), 1878, p. 197, pl. xxxv, f. 1, 1a, 2.—Knowlton, U. S. Geol. Surv., Prof. Paper 101, 1918, p. 301, pl. lxxvi, f. 2.

Ficus ulmifolia Lesquereux, U. S. Geol. and Geog. Surv. Terr., Ann. Rept. 1871 [1872], Suppl., p. 14. [Homonym.]

Ficus gaudini Lesquereux [not Ettingshausen], U. S. Geol. and Geog. Surv. Terr., Ann. Rept. 1871 [1872], p. 300.—Cockerell, Torreya, vol. 9, 1909, p. 141.

Ficus arenacea brevipetiolata Lesquereux, Rept. U. S. Geol. Surv. Terr., vol. 7 (Tert. Fl.), 1878, p. 195, pl. xxix, f. 2, 3, 5.

Populus monodon Lesquereux. Lesquereux, Rept. U. S. Geol. Surv. Terr., vol. 7 (Tert. Fl.), 1878, p. 180, pl. xxiv, f. 1, 2.

Raton: *Fishers Peak, Raton Mountains, N. Mex.; near Abeton, Colo.

Ficus? undulata Lesquereux = *Ficus undulatiformis.*

Ficus undulatiformis Knowlton and Cockerell, n. name.

Ficus? undulata Lesquereux, U. S. Geol. Surv., Mon. 17, 1891 [1892], p. 84, pl. xii, f. 5. [Homonym, 1827.]

Dakota: Ellsworth County, Kans.

Ficus ungeri Lesquereux, U. S. Geol. and Geog. Surv. Terr., Ann. Rept. 1871 [1872], Suppl., p. 7; Rept. U. S. Geol. Surv. Terr., vol. 7 (Tert. Fl.), 1878, p. 195, pl. xxx, f. 3; idem, vol. 8 (Cret. and Tert. Fl.), 1883, p. 163, pl. xliv, f. 1–3.—Penhallow, Rept. Tert. Pl. Brit. Col., 1908, p. 56.—Knowlton, U. S. Geol. Surv., Mon. 32, pt. 2, 1899, p. 713, pl. xci, f. 3.

Green River: Green River, Wyo.?

Oligocene: Quilchena, British Columbia.

Fort Union?: Yellowstone National Park.

Ficus vaughani Berry, U. S. Geol. Surv., Prof. Paper 91, 1916, p. 203, pl. xxxii, f. 1; pl. xxxiii, f. 1; pl. xcii, f. 1.
Wilcox (Holly Springs): Oxford, Miss.
Lagrange: Puryear, Henry County, Tenn.
Wilcox: Near Gainesville, Ark.; Old Port Caddo Landing, Harrison County, Tex.; Calaveras Creek, Wilson County, Tex.

Ficus viburnifolia Ward, U. S. Geol. Surv., Sixth Ann. Rept., 1884–85 [1886], p. 552, pl. xlv, f. 5–9; U. S. Geol. Surv., Bull. 37, 1887, p. 42, pl. xxii, f. 4–8.
Fort Union: Clear Creek, Glendive, Mont.

Ficus virginiensis Fontaine = *Ficophyllum crassinerve*.

Ficus wardii Knowlton, U. S. Geol. Surv., Bull. 163, 1900, p. 48, pl. ix, f. 1; U. S. Geol. Surv., Prof. Paper 101, 1918, p. 266, pl. lxii, f. 2.
Mesaverde: *Point of Rocks, Wyo.
Vermejo: La Veta, Walsenburg, Bowen, and Rockvale, Colo.

Ficus wellingtoniae Dawson, Roy. Soc. Canada, Trans., vol. 11, 1893 [1894], sec. 4, p. 60, pl. ix, f. 33, 34.
Upper Cretaceous: Wellington mine and Port McNeill, Vancouver Island.

Ficus wilcoxensis Berry, U. S. Geol. Surv., Prof. Paper 91, 1916, p. 202, pl. xxvii, f. 6.
Sapindus falcifolius (A. Braun) A. Braun. Lesquereux, U. S. Nat. Mus., Proc., vol. 11, 1888, p. 12, pl. iv, f. 4.
Sapindus dubius Unger [part]. Lesquereux, U. S. Nat. Mus., Proc., vol. 11, 1888, p. 13.
Lagrange: Puryear, Henry County, Tenn., and Boaz, Graves County, Ky.

Ficus willisiana Hollick, Bull. Torr. Bot. Club, vol. 21, 1894, p. 52, pl. 176, f. 2, 5; U. S. Geol. Surv., Mon. 50, 1906, p. 59, pl. ix, f. 1, 2; N. Y. State Mus., Fifty-fifth Ann. Rept., 1901 [1903], p. r49.
Magothy: Sea Cliff and Glen Cove, Long Island, N. Y.

Ficus woolsoni Newberry, MS. Hollick, N. Y. Acad. Sci., Trans., vol. 12, 1892, p. 33[6], pl. ii, f. 1, 2c.—Newberry, U. S. Geol. Surv., Mon. 27, 1895 [1896], p. 70, pl. xx, f. 3; pl. xxiii, f. 1–6.—Hollick, N. Y. Acad. Sci., Ann., vol. 11, 1898, p. 419, pl. xxxvii, f. 9; U. S. Geol. Surv., Mon. 50, 1906, p. 59, pl. xi, f. 5, 6.—Berry, N. Y. Bot. Gard., Bull., vol. 3, 1903, p. 74, pl. xlvii, f. 7; Geol. Surv. N. J., Ann. Rept. 1905 [1906], p. 138, 139; Bull. Torr. Bot. Club, vol. 33, 1906, p. 172; idem, vol. 34, 1907, p. 194, pl. xii, f. 1.—Hollick, N. Y. Bot. Gard., Bull., vol. 8, 1912, p. 159, pl. clxv, f. 2.—Berry, Geol. Surv., N. J., Bull. 3, 1911, p. 124, pl. xii, f. 1, 2; U. S. Geol. Surv., Prof. Paper 112, 1919, p. 81, pl. xii, f. 2.

Ficus woolsoni—Continued.
Raritan: Woodbridge, Sayreville, N. J.; Tottenville and Kreischerville, Staten Island, and Glen Cove, Long Island, N. Y.
Magothy: Cliffwood, N. J.
Black Creek: Court House Bluff, Cape Fear River, N. C.
Tuscaloosa: Shirleys Mill, Glen Allen, and Cottondale, Ala.

Ficus wyomingiana Lesquereux, U. S. Geol. and Geog. Surv. Terr., Bull., vol. 1, 1875 [1876], p. 387; idem, Ann. Rept., 1874 [1876], p. 314; Rept. U. S. Geol. Surv. Terr., vol. 7 (Tert. Fl.), 1878, p. 205, pl. xxxiv, f. 3.
Green River: Green River station, Wyo.

Ficus zizyphoides Lesquereux, U. S. Geol. and Geog. Surv. Terr. Ann. Rept. 1873 (1874), p. 399.
Denver (?): Golden, Colo.

Ficus sp. Berry, Bull. Torr. Bot. Club, vol. 37, 1910, p. 505.
Ripley (Cusseta): Buena Vista, Ga.

Ficus sp. Berry, U. S. Geol. Surv., Prof. Paper 91, 1916, pp. 12, 206, pl. xxiv, f. 1.
Wilcox (Holly Spring): Holly Springs, Marshall County, Miss.
Midway (?): Earle, Bexar County, Tex.

Ficus n. sp. Cockerell, Univ. Colo. Studies, vol. 4, 1907, p. 152 = *Rhamnus williardi.*

Ficus sp. Dawson, Geol. Surv. Canada, 1887, p. —.
Eocene: Red Deer River, Alberta.

Ficus sp. Dawson, Roy. Soc. Canada, Trans., vol. 8, 1890, sec. 4, p. 89, text f. 27.
Oligocene: Similkameen River, British Columbia.

Ficus sp. Dawson, Geol. Surv. Canada, 1887; Royal Soc. Canada, Trans., vol. 8, 1890, pp. 75–91.—Penhallow, Rept. Tert. Pl. Brit. Col., 1908, p. 54.
Eocene: Red Deer River, Alberta.
Oligocene: Similkameen River, Coal Gully, and Horsefly River, British Columbia.

Ficus sp. Knowlton, in Lindgren, Jour. Geol., vol. 4, 1896, p. 889.
Miocene (auriferous gravels): Independence Hill, Placer County, Calif.

Ficus sp. Knowlton, in Washburne, U. S. Geol. Surv., Bull. 590, 1914, p. 36.
Eocene: Eugene, Oreg.

Ficus sp. Knowlton, U. S. Geol. Surv., Bull. 152, 1898, p. 68, under *Cinnamomum affine* Ward (not Lesquereux) = *Ficus trinervis.*

Ficus sp. Knowlton, Geol. Soc. Am., Bull., vol. 8, 1897, p. 140 = *Ficus hesperia.*

Ficus sp. Knowlton, U. S. Geol. Surv., Mon. 32, pt. 2, 1899, p. 713, pl. lxxxix, f. 3.
Fort Union: Elk Creek, Yellowstone National Park.

Ficus sp. Knowlton, Washington Acad. Sci., Proc., vol. 11, 1909, p. 207.
Lance: Converse County, Wyo.

Ficus sp. Knowlton, Washington Acad. Sci., Proc., vol. 11, 1909, p. 195.
Lance: 8 miles west of Bridger, Mont.

Ficus sp. Knowlton, Wash. Geol. Surv., Ann. Rept., vol. 1, 1902, p. 33.
Eocene: Skykomish, Snohomish County, Wash.

Ficus sp. Knowlton, Wash. Geol. Surv., Ann. Rept., vol. 1, 1902, p. 33.
Eocene: Skykomish, Snohomish County, Wash.

Ficus sp. Knowlton, Wash. Geol. Surv., Ann. Rept., vol. 1, 1902, p. 33.
Miocene: Black River Junction, King County, Wash.

Ficus sp. Knowlton, in Vaughan, Am. Geol., vol. 16, 1895, p. 308.
Wilcox: Old Port Caddo Landing, Harrison County, Tex.

Ficus? sp. Knowlton, U. S. Geol. Surv., Prof. Paper 108, 1917, p. 88, pl. xxxv, f. 1.
Frontier: About one-half mile east of Cumberland, Wyo.

Ficus sp. Knowlton, U. S. Geol. Surv., Prof. Paper 108, 1917, p. 88, pl. xxxiv, f. 2, 3.
Frontier: About 1 mile east of Cumberland, Wyo.

Ficus sp. Knowlton, U. S. Geol. Surv., Prof. Paper 98, 1916, p. 339, pl. lxxxviii, f. 1.
Fruitland: 30 miles south of Farmington, San Juan County, N. Mex.

Ficus sp. Knowlton, U. S. Geol. Surv., Prof. Paper 98, 1916, p. 339, pl. lxxxix, f. 1.
Fruitland: Amarillo Canyon, San Juan County, N. Mex.

Ficus sp. Knowlton, U. S. Geol. Surv., Prof. Paper 101, 1918, p. 266, pl. xlv, f. 1.
Vermejo: Near Rockvale, Colo.

Ficus sp. Lesquereux, Am. Jour. Sci., 2d ser., vol. 27, 1859, p. 361.
Cretaceous: Nanaimo, Vancouver Island, British Columbia.

Ficus, fruits of, Lesquereux, U. S. Geol. Surv., Mon. 17, 1891 [1892], p. 85, pl. x, f. 7, 8.
Dakota: Near Brookville, Kans.

Filicites nilsoniana Brongniart = *Sagenopteris nilsoniana.*

FLABELLARIA Sternberg, Fl. Vorw., vol. 1 (Versuch), 1822, p. 32.

Flabellaria communis Lesquereux = *Sabal? eocenica.*

Flabellaria eocenica Lesquereux = *Sabal? eocenica.*

Flabellaria florissanti Lesquereux, Rept. U. S. Geol. Surv. Terr., vol. 8 (Cret. and Tert. Fl.), 1883, p. 144, pl. xxiv, f. 1–2a.
Green River: Uinta County (not Randolph County), Wyo.

Flabellaria? fructifera Lesquereux = *Sabalites fructifer.*

Flabellaria latania Rossmässler, Verstein. Braunkohl. v. Altsattel, 1840, p. 39, pl. xi, f. 49.—Lesquereux, U. S. Geol. and Geog. Surv. Terr., Ann. Rept. 1872 [1873], p. 377. [Not afterward recognized in this country by Lesquereux.]
——: Golden, Colo.

Flabellaria longirachis Unger. Lesquereux, U. S. Geol. and Geog. Surv. Terr., Ann. Rept. 1873 [1874], p. 396 = *Geonomites schimperi* and *G. tenuirachis.*

Flabellaria magothiensis Berry = *Sabalites magothiensis.*

Flabellaria? minima Lesquereux, Rept. U. S. Geol. Surv. Terr., vol. 6 (Cret. Fl.), p. 56, pl. xxx, f. 12.
Dakota: Western Kansas.

Flabellaria zinckeni Heer. Lesquereux, U. S. Geol. and Geog. Surv. Terr., Ann. Rept. 1872 [1873], p. 377 = *Geonomites goldianus.*

FLORISSANTIA Knowlton, U. S. Nat. Mus., Proc., vol. 51, 1916, p. 270. [Type, *F. physalis.*]

Florissantia physalis Knowlton, U. S. Nat. Mus., Proc., vol. 51, 1916, p. 270.
Convolvulaceous (?) flower, Kirchner, St. Louis Acad. Sci., Trans., vol. 8, 1898, p. 187, pl. xv, f. 2.
Miocene: *Florissant, Colo.

FONTAINEA Newberry, U. S. Geol. Surv., Mon. 26, 1895 [1896], p. 94. [Type, *F. grandifolia.*]

Fontainea grandifolia Newberry, U. S. Geol. Surv., Mon. 26, 1895 [1896], p. 96, pl. xlv, f. 1–4; Geol. Surv. N. J., Bull. 3, 1911, p. 219; Md. Geol. Surv., Upper Cret., 1916, p. 899.
Raritan: Woodbridge, N. J.; Forked Creek, Md.

FONTINALIS Linné, Sp. pl., 1753, p. 1107.

Fontinalis pristina Lesquereux, Rept. U. S. Geol. Surv. Terr., vol. 8 (Cret. and Tert. Fl.), 1883, p. 135, pl. xxi, f. 9.—E. G. Britton and Hollick, Bull. Torr. Bot. Club, vol. 34, 1907, p. 139, pl. ix, f. 2, 2a.—Knowlton, U. S. Nat. Mus., Proc., vol. 51, 1916, p. 245 = feather.
Miocene: Florissant, Colo.

Fontinalis sp. Dawson, Canadian Nat., new ser., vol. 3, 1868, p. 73.
Pleistocene: Greens Creek, Ottawa River, Canada.

Fontinalis sp. Hinde, Canadian Jour. [Canada Inst.], vol. 15, 1878, p. 399.
Pleistocene: Scarborough Heights, Ontario.

Fontinalis sp. Penhallow, Roy. Soc. Canada, Trans., 2d ser., vol. 2, 1896, sec. 4, p. 71; Brit. Assoc. Adv. Sci., Bradford meeting, 1900, p. 335.
Pleistocene: Scarborough Heights and Greens Wharf, Ottawa River, Canada.

FRAXINUS (Tournefort) Linné, Sp. pl., 1753, p. 1057.

Fraxinus abbreviata Lesquereux, Rept. U. S. Geol. Surv. Terr., vol. 8 (Cret. and Tert. Fl.), 1883, p. 170, pl. xxviii, f. 5, 6.—Cockerell, Am. Mus. Nat. Hist., Bull., vol. 24, 1908, p. 107.
Miocene: Florissant, Colo.

Fraxinus affinis Newberry = *Quercus clarnensis*.

Fraxinus americana Linné. Penhallow, Brit. Assoc. Adv. Sci., Bradford meeting, 1900, p. 335.—Berry, U. S. Nat. Mus., Proc., vol. 48, 1915, p. 301; Torreya, vol. 15, 1915, p. 208.
Pleistocene: Don Valley, Toronto, Canada; Hickman, Ky.; Indian Head, Charles County, Md.

Fraxinus brownellii Lesquereux, Rept. U. S. Geol. Surv. Terr., vol. 7 (Tert. Fl.), 1878, p. 230.
Miocene: South Park, Colo.

Fraxinus denticulata Heer, Fl. foss. arct., vol. 1, 1868, p. 118, pl. xvi, f. 4.—Lesquereux, U. S. Geol. and Geog. Surv. Terr., Ann. Rept. 1872 [1873], p. 407; Rept. U. S. Geol. Surv. Terr., vol. 7 (Tert. Fl.), 1878, p. 228, pl. xl, f. 1, 2.—Knowlton, U. S. Geol. Surv., Bull. 105, 1893, p. 56; idem, Bull. 204, 1902, p. 84.—Newberry, U. S. Geol. Surv., Mon. 35, 1898, p. 128, pl. xlix, f. 6.—Knowlton, U. S. Geol. Surv., Mon. 32, pt. 2, 1899, p. 662, pl. lxxviii, f. 6.
Livingston: Bozeman coal field, Mont.
——: Mount Everts, Yellowstone National Park.
Clarno (upper part): Bridge Creek, Oreg.

Fraxinus eocenica Lesquereux, Rept. U. S. Geol. Surv. Terr., vol. 7 (Tert. Fl.), 1878, p. 229; idem, vol. 8 (Cret. and Tert. Fl.), 1883, p. 123, pl. xx, f. 1–3.
Denver: Golden, Colo.

Fraxinus heerii Lesquereux, Rept. U. S. Geol. Surv. Terr., vol. 8 (Cret. and Tert. Fl.), 1883, p. 169, pl. xxxiii, f. 5, 6.—Cockerell, Am. Mus. Nat. Hist., Bull., vol. 24, 1908, p. 102.
Miocene: Florissant, Colo.

Fraxinus herendeenensis Knowlton, U. S. Nat. Mus., Proc., vol. 17, 1894, p. 224, pl. ix, f. 7; Geol. Soc. Am., Bull., vol. 5, 1893, p. 584.
Kenai: Herendeen Bay, Alaska.

Fraxinus integrifolia Newberry = *Fraxinus oregonensis*.

Fraxinus johnstrupi Heer, Fl. foss. arct., vol. 7, 1878, p. 113, pl. lxxx, f. 1, 2.—?Hollick, Geol. Surv. La., Special Rept. 5, 1899, p. 287, pl. xliv, f. 1.—Berry, U. S. Geol. Surv., Prof. Paper 91, 1916, p. 341, pl. ci, f. 6.

Wilcox: Coushatta, La.
Wilcox (Grenada): Grenada, Miss.
Lagrange: Puryear, Tenn.

Fraxinus libbeyi Lesquereux, Rept. U. S. Geol. Surv. Terr., vol. 8 (Cret. and Tert. Fl.), 1883, p. 171, pl. xxvii, f. 5–7, 9.—Knowlton, U. S. Nat. Mus., Proc., vol. 51, 1916, p. 285, pl. xxiv, f. 5.

Ostrya betuloides Lesquereux, Rept. U. S. Geol. Surv. Terr., vol. 8 (Cret. and Tert. Fl.), 1883, p. 151.

Miocene: *Florissant, Colo.

Fraxinus libbeyana Knowlton, U. S. Geol. Surv., Prof. Paper 130, 19—, p. —, pl. xxii, f. 7 [in preparation].

Laramie?: Golden, Colo.

Fraxinus mespilifolia Lesquereux, Rept. U. S. Geol. Surv. Terr., vol. 8 (Cret. and Tert. Fl.), 1883, p. 169, pl. xxxiii, f. 7–12.—Cockerell, Am. Mus. Nat. Hist., Bull., vol. 24, 1908, p. 107.

Miocene: Florissant, Colo.

Fraxinus? myricaefolia Lesquereux, Rept. U. S. Geol. Surv. Terr., vol. 8 (Cret. and Tert. Fl.) 1883, p. 170, pl. xxxiii, f. 12, 14.—Cockerell, Am. Mus. Nat. Hist. Bull., vol. 24, 1908, p. 107.

Miocene: Florissant, Colo.

Fraxinus nigra Marshall.

Fraxinus sambucifolia Lamarck. Penhallow, Brit. Assoc. Adv. Sci., Bradford meeting, 1900, p. 335.

Pleistocene: Don Valley, Toronto, Canada.

Fraxinus oregonensis Knowlton and Cockerell, n. name.

Fraxinus integrifolia Newberry, U. S. Nat. Mus., Proc., vol. 5, 1882 [1883], p. 509; U. S. Geol. Surv., Mon. 35, 1898, p. 128, pl. xlix, f. 1–3.—Knowlton, U. S. Geol. Surv., Bull. 204, 1902, p. 84. [Homonym, Moench, 1785.]

Clarno (upper part): Bridge Creek, Oreg.

Fraxinus palaeophila Cockerell, Am. Mus. Nat. Hist., Bull., vol. 24, 1908, p. 106.

Miocene: Florissant, Colo.

Fraxinus praedicta Heer, Fl. tert. Helv., vol. 3, 1859, p. 22, pl. civ, f. 12, 13.—Lesquereux, U. S. Geol. and Geog. Surv. Terr., Ann. Rept. 1873 [1874], p. 414; Rept. U. S. Geol. Surv. Terr., vol. 7 (Tert. Fl.), 1878, p. 229, pl. xl, f. 3.—Cockerell, Am. Mus. Nat. Hist., Bull., vol. 24, 1908, p. 107.

Denver: Golden, Colo.
Miocene: ?Florissant, Colo. [Identification ?]

Fraxinus quadrangulata Michaux. Penhallow, in Coleman, Jour. Geol., vol. 3, 1895, p. 635.—Penhallow, Roy. Soc. Canada, Trans., 2d ser., vol. 2, 1896, p. 71; Brit. Assoc. Adv. Sci., Bradford meeting, 1900, p. 335.
Pleistocene: Don Valley, Toronto, Canada.

Fraxinus richmondensis Berry, Jour. Geol., vol. 17, 1909, p. 30, text f. 11.
Calvert: Richmond, Va.

Fraxinus sambucifolia Lamarck = *Fraxinus nigra.*

Fraxinus ungeri Lesquereux, Rept. U. S. Geol. Surv. Terr., vol 8 (Cret. and Tert. Fl.), 1883, p. 171.—Knowlton, U. S. Nat. Mus., Proc., vol. 51, 1916, p. 286, pl. xx, f. 3; pl. xxiii, f. 1, 2.
Miocene: Florissant, Colo.

Fraxinus wilcoxiana Berry, U. S. Geol. Surv., Prof. Paper 91, 1916, p. 340, pl. ci, f. 5.
Lagrange: Puryear, Tenn.

Fraxinus wrightii Knowlton, U. S. Geol. Surv., Mon. 32, pt. 2, 1899, p. 753, pl. xc, f. 4.
Fort Union: Elk Creek, Yellowstone National Park.

Fraxinus sp. Berry, U. S. Geol. Surv., Prof. Paper 98, 1916, p. 204, pl. xlvii, f. 12.
Citronelle: Lambert, Ala.

Fraxinus sp. Knowlton, Geol. Soc. Am., Bull., vol. 8, 1897, p. 142.
Medicine Bow: "Dutton Creek coal," Laramie Plains, Wyo.

Fraxinus? sp. Knowlton, U. S. Geol. Surv., Prof. Paper 101, 1918, p. 275, pl. xlix, f. 1.
Vermejo: Gray Creek, Colo.

Frenelites reichii Ettingshausen = *Widdringtonites reichii.*

FRENELOPSIS Schenk, Palaeontographica, vol. 19, 1869, p. 13.

Frenelopsis gracilis Newberry = *Raritania gracilis.*

Frenelopsis hoheneggeri (Ettingshausen) Schenk?, Palaeontog., vol. 9, Heft 1, 1869, p. 13, pl. iv, f. 5–7; pl. v, f. 1, 2; pl. vi, f. 1–6; pl. vii, f. 1.—Newberry, U. S. Geol. Surv., Mon. 26, 1895 [1896], p. 58, pl. xii, f. 4, 5.—Fontaine, U. S. Nat. Mus., Proc., vol. 16, 1893, p. 275, pl. xlii, f. 4, 4a.—Berry, Bull. Torr. Bot. Club, vol. 31, 1904, p. 71, pl. iv, f. 9, 10.—Hollick, N. Y. Bot. Gard., Bull., vol. 3, 1904, p. 410, pl. lxxii, f. 1.—Berry, Geol. Surv. N. J., Ann. Rept. 1905 [1906], p. 138.—Hollick, U. S. Geol. Surv., Mon. 50, 1906, p. 45, pl. iv, f. 9, 10; Geol. Surv. N. J., Bull. 3, 1911, p. 90.
Raritan: Woodbridge, N. J.
Magothy: Cliffwood, N. J.; Center Island, Oyster Bay, Long Island, N. Y.
Trinity: Glenrose, Tex.

Frenelopsis parceramosa Fontaine, U. S. Geol. Surv., Mon. 15, 1889 [1890], p. 218, pl. cxi, f. 1-5; pl. cxii, f. 1-5; pl. clxviii, f. 1; in Ward, idem, Mon. 48, 1905 [1906], pp. 544, 584.—Berry, Md. Geol. Surv., Lower Cret., 1912, p. 425, pl. lxx, f. 1-5.
 Patuxent: *Trents Reach, Va.
 Arundel: Hanover, Md.

Frenelopsis ramosissima Fontaine, U. S. Geol. Surv., Mon. 15, 1889 [1890], p. 215, pls. xcv-xcix; pl. c, f. 1-3; pl. ci, f. 1; in Ward, U. S. Geol. Surv., Mon. 48, 1905 [1906], p. 512, pl. cxi, f. 8.—Berry, Bot. Gaz., vol. 50, 1910, p. 305, text f. 1, 2; Md. Geol. Surv., Lower Cret., 1912, p. 422, pls. lxxi, lxxii.
 Patuxent: Fredericksburg, Va.
 Patapsco: Chinkapin Hollow and Hell Hole, Va.; Baltimore, Md.

Frenelopsis varians Fontaine, U. S. Nat. Mus., Proc., vol. 16, 1893, p. 273, pl. xl, f. 1, 2; pl. xli, f. 1-3a; in Ward, U. S. Geol. Surv., Mon. 48, 1905 [1906], p. 340.
 Trinity: Glen Rose, Tex.; Murfreesboro, Ark.

FUCUS (Tournefort) Linné, Sp. pl., 1753, p. 1158.

Fucus digitatus Penhallow, Roy. Soc. Canada, Trans., 2d ser., vol. 2, sec. 4, 1896, pp. 66, 71; Brit. Assoc. Adv. Sci., Bradford meeting, 1900, p. 335.
 Fucus sp. Penhallow, Geol. Soc. Am., Bull., vol. 1, 1890, p. 332.
 Pleistocene: Greens Creek and Besserer Wharf, Ottawa River, Canada.

Fucus lignitum Lesquereux, U. S. Geol. and Geog. Surv. Terr., Bull., vol. 1, 1875 [1876], p. 364; Rept. U. S. Geol. Surv. Terr., vol. 7 (Tert. Fl.), 1878, p. 42, pl. lxi, f. 24, 24a.—Ward, U. S. Geol. Surv., Sixth Ann. Rept., 1884-85 [1886], p. 549, pl. xxxi, f. 1, 2; U. S. Geol. Surv., Bull. 37, 1887, p. 13, pl. i, f. 1, 2.—Knowlton, U. S. Geol. Surv., Bull. 163, 1900, p. 17, pl. iii, f. 4.
 Fort Union?: Burns's ranch, Glendive, Mont.
 Mesaverde: Point of Rocks, Wyo.

Fucus sp. Penhallow = *Fucus digitatus*.

FUCOIDES Brongniart, Mém. Mus. hist. nat., vol. 8, 1822, p. 210.

Fucoides connecticutensis Hitchcock, Final Rept. Geol. Mass., vol. 2, 1841, p. 455, text, f. 95.—Ward, U. S. Geol. Surv., Twentieth Ann. Rept., pt. 2, 1900, p. 426.
 Triassic (Newark): Middletown and Enfield, Conn.; Suffield, Mass.

Fucoides erectus Bean = *Marchantites erectus*.

Fucoides shepardi Hitchcock, Final Rept. Geol. Mass., vol. 2, 1841, p. 453, text f. 92.—Ward, U. S. Geol. Surv., Twentieth Ann. Rept., pt. 2, 1900, p. 426.
 Triassic (Newark): Deerfield and Greenfield, Mass.

GALLA Ludwig, Palaeontographica, vol. 5, 1857, p. 108.

Galla quercina Lesquereux, U. S. Geol. Surv., Mon. 17, 1891 [1892], p. 58, pl. vii, f. 2.
Dakota: Ellsworth County, Kans.

GAYLUSSACIA Humboldt, Bonpland, and Kunth, Nov. gen. et sp., vol. 3, 1818, p. 275, pl. 257.

Gaylussacia dumosa (Andrae) Torrey and Gray. Berry, Torreya, vol. 17, 1917, p. 161, f. 1.
Pleistocene: Waterville, Maine.

Gaylussacia resinosa (Aiton) Torrey and Gray. Dawson, Canadian Nat., new ser., vol. 3, 1868, p. 71, text f. 1; Geol. Hist. Pl., 1888, p. 228, f. 77.—Penhallow, Roy. Soc. Canada, Trans., 2d ser., vol. 2, 1896, p. 71; Brit. Assoc. Adv. Sci., Bradford meeting, 1900, p. 335.
Pleistocene: Greens Wharf, Ottawa River, Canada.

GEASTER (Micheli) Scopoli, Carn., vol. 2, 1772, p. 489.

Geaster florissantensis Cockerell, Am. Jour. Sci., 4th ser., vol. 26, 1908, p. 538, f. 1.
Miocene: Florissant, Colo.

GEINITZIA Endlicher, Syn. conif., 1847, p. 280.

Geinitzia biforma (Lesquereux) Knowlton, U. S. Geol. Surv., Bull. 163, 1900, p. 28.
Sequoia biformis Lesquereux, U. S. Geol. and Geog. Surv. Terr., Bull., vol 1, 1875 [1876], p. 366; U. S. Geol. and Geog. Surv. Terr., Ann. Rept. 1874 [1876], p. 298; Rept. U. S. Geol. Surv. Terr., vol. 7 (Tert. Fl.), 1878, p. 80, pl. lxii, f. 15–18a.—Ward, U. S. Geol. Surv., Sixth Ann. Rept., 1884–85 [1886], p. 550, pl. xxxi, f. 7–12; idem, Bull. 37, 1887, p. 16, pl. i, f. 1-6.
Mesaverde: Point of Rocks, Wyo.

Geinitzia formosa Heer, Neue Denkschr. Schweiz. Ges., vol. 24, 1871, p. 6, pl. i, f. 9; pl. ii.—Newberry, U. S. Geol. Surv., Mon. 26, 1895 [1896], p. 51, pl. ix, f. 9.—Knowlton, idem, Bull. 163, 1900, p. 28, pl. v, f. 1, 2.—Berry, N. Y. Bot. Gard., Bull., vol. 3, 1903, p. 57; Bull. Torr. Bot. Club, vol. 31, 1904, p. 68, pl. iv, f. 2, 3; Geol. Surv. N. J., Ann. Rept. 1905 [1906], p. 138; Geol. Surv. N. J., Bull. 3, 1911, p. 97; Md. Geol. Surv., Upper Cret., 1916, p. 801, pl. liv, f. 6.—Knowlton, U. S. Geol. Surv., Prof. Paper 98, 1916, p. 333, pl. lxxxv, f. 3; idem, Prof. Paper 101, 1918, p. 251, pl. xxxi, f. 1–3.—Berry, U. S. Geol. Surv., Prof. Paper 112, 1919, p. 61.
Sequoia reichenbachi? (Geinitz) Heer. Knowlton, Geol. Soc. Am., Bull., vol. 8, 1897, p. 137.
Mesaverde: Harper station, Rock Springs, Mud Springs, and Big Horn Basin, Wyo.; Durango, Colo.
Montana: Bridger, Mont.
Magothy: Cliffwood, N. J.; Deep Cut, Del.

Geinitzia formosa—Continued.
Raritan: Woodbridge, N. J.
Fruitland: 17 miles south of San Juan River, San Juan County, N. Mex.
Vermejo: Near Walsenburg, Colo.
Tuscaloosa: Shirleys Mill and Cottondale, Ala.

Geinitzia jenneyi Fontaine, in Ward, U. S. Geol. Surv., Nineteenth Ann. Rept., pt. 2, 1899, p. 676, pl. clxvi, f. 5–11; pl. clxvii.
Fuson: Pine and Oak creeks, Crook County, Wyo.

Geinitzia longifolia (Lesquereux) Knowlton [part] = *Sequoia magnifolia.*

Geinitzia reichenbachi (Geinitz) Hollick and Jeffrey, N. Y. Bot. Gard., Mem., vol. 3, 1909, p. 38, pl. v, f. 7–10; pl. viii, f. 3, 4; pl. xvi, f. 2–4; pl. xvii, f. 1–4; pl. xviii, f. 1–4.—Cockerell, Univ. Colo. Studies, vol. 7, 1910, p. 150, f. 13.—Holden, Bot. Gaz., vol. 58, 1914, p. 171, pl. xii, figs. 9–14.
Araucarites reichenbachi Geinitz, Charakteristik, vol. 3, 1824, p. 98, pl. xxiv, f. 4.
Sequoia couttsiae Heer. Hollick, N. Y. Acad. Sci., Trans., vol. 12, 1892, p. 30, pl. i, f. 5.
Raritan: Kreischerville, Staten Island, N. Y.

Geinitzia sp.? Heer. Lesquereux, Rept. U. S. Geol. Surv. Terr., vol. 6 (Cret. Fl.), 1874, p. 54.
Dakota: Sioux City, Iowa.

Geinitzia sp. Newberry, Lyc. Nat. Hist. N. Y., Proc., 2d ser., Jan. 3 to Mar. 3, 1873, p. 10.
Raritan: Keyport, N. J.

GEONOMA Willdenow, Spec., iv, 1805, p. 593.

"Geonoma" gigantea Knowlton, U. S. Geol. Surv., Prof. Paper 101, 1918, p. 291, pl. lxi.
Raton: Rockland mine near Walsenburg, Colo.

GEONOMITES Visiani, Mem. Instituto Veneto, vol. 11 (Palmae pinnatae tertiariae agri veneti), 1864, p. 456.

Geonomites goldianus (Lesquereux) Lesquereux, Rept. U. S. Geol. Surv. Terr., vol. 7 (Tert. Fl.), 1878, p. 115, pl. ix, f. 9; U. S. Geol. and Geog. Surv. Terr., Ann. Rept. 1876 [1878], p. 502; Mus. Comp. Zool., Bull., vol. 16, 1888, p. 44.—Cockerell, Univ. Colo. Studies, vol. 7, 1910, p. 150.
Palmacites goldianus Lesquereux, U. S. Geol. and Geog. Surv. Terr., Ann. Rept. 1874 [1875], p. 311.
?*Flabellaria zinckeni* Heer. Lesquereux, U. S. Geol. and Geog. Surv. Terr., Ann. Rept. 1872 [1873], p. 377; idem, 1873 [1874], p. 380; idem, 1876 [1878], p. 502; Rept. U. S. Geol. Surv. Terr., vol. 7 (Tert. Fl.), 1878, p. 110, pl. ix, f. 6–8.
Denver: Golden, Colo.
Green River?: Barrell Springs, Wyo.

Geonomites graminifolius Lesquereux, Mus. Comp. Zool., Bull., vol. 16, 1888, p. 44.

Denver: Golden, Colo.

Geonomites schimperi Lesquereux, Rept. U. S. Geol. Surv. Terr., vol. 7 (Tert. Fl.), 1878, p. 116, pl. x, f. 1.—Knowlton, U. S. Geol. Surv., Mon. 32, pt. 2, 1899, p. 658.

Sabal major? Unger. Lesquereux, U. S. Geol. and Geog. Surv. Terr., Ann. Rept. 1871 [1872], p. 295.

Flabellaria longirachis? Unger. Lesquereux, idem, 1873 [1874], p. 396.

——: Yellowstone National Park.

Geonomites tenuirachis Lesquereux, Rept. U. S. Geol. Surv. Terr., vol. 7 (Tert. Fl.), 1878, p. 117, pl. xi, f. 1.—Knowlton, U.S. Geol., Surv., Prof. Paper 101, 1918, p. 291, pl. lxii.

Flabellaria longirachis? Unger. Lesquereux, U. S. Geol. and Geog. Surv. Terr., Ann. Rept. 1873 [1874], p. 396. [In part only.]

Raton: Raton Mountains, near Yankee, N. Mex.

Denver: Golden, Colo.

Geonomites ungeri Lesquereux = *Sabal ungeri*.

GILBERTINA Ulrich, Harriman Alaska Exped., vol. 4, 1904, p. 140. [Type, *G. spiralis*.]

Gilbertina spiralis Ulrich, Harriman Alaska Exped., vol. 4, 1904, p. 141, pl. xviii, f. 1, 2.

Yakutat: Pogibshi Island, opposite Kodiak, Alaska.

GINKGO (Kaempfer) Linné, Mantissa plantarum, vol. 2, 1771, p. 313.

Ginkgo? acetaria Ward, U. S. Geol. Surv., Mon. 48, 1905 [1906], p. 551, pl. cviii, f. 12.

Arundel: Vinegar Hill, Relay, Md.

Ginkgo adiantoides (Unger), Heer, Fl. foss. arct., vol. 5, Abt. 3, 1878, p. 21, pl. ii, f. 7–10.—Ward, U. S. Geol. Surv., Sixth Ann. Rept., 1884–85 [1886], p. 549, pl. xxxi, f. 5, 6; idem, Bull. 37, 1887. p. 15, pl. i, f. 5, 6.—Lesquereux, U. S. Nat. Mus., Proc., vol. 10, 1887, p. 35.—Knowlton, idem, vol. 17, 1894, p. 215; Geol. Soc. Am., Bull., vol. 5, 1893, p. 579; Washington Acad. Sci., Proc., vol. 11, 1909, pp. 197, 198, 204, 213; Jour. Geol., vol. 19, 1911, p. 370.—Penhallow, Rept. Tert. Pl. Brit. Col., 1908, p. 57, text f. 12.

Salisburia adiantoides Unger, Syn. pl. foss., 1845, p. 211.

Kenai: Sitka and Herendeen Bay, Alaska.

Fort Union: Sevenmile Creek near Glendive, Mont.

Lance: South Dakota [sec. 33, T. 20 N., R. 20 E.]; 60 miles south of Glasgow, Eagle Bluff, Sand Creek above Glendive, Mont.; Weston County, and Ilo, Big Horn Basin, Wyo.

Eocene: Porcupine Creek and Great Valley, Alberta.

Oligocene: Tulameen and Horsefly rivers, British Columbia.

Ginkgo baynesiana (Dawson) Knowlton, U. S. Geol. Surv., Bull. 152, 1898, p. 110.

Salisburia baynesiana Dawson, Roy. Soc. Canada, Trans., vol. 1, sec. 4, 1882 [1883], p. 25, pl. v, f. 21, 21a.

Upper Cretaceous: Baynes Sound and Beaver Harbor, Vancouver Island.

Ginkgo binervata (Lesquereux) Knowlton = *Lygodium binervatum.*

Ginkgo dawsoni Knowlton, n. name.

Salisburia pusilla Dawson, Roy. Soc. Canada, Trans., vol. 11, sec. 4, 1893 [1894], p. 56, pl. vi, f. 11–13.—Penhallow, Roy. Soc. Canada, Trans., 2d ser., vol. 8, sec. 4, 1902, p. 43, pl. xii, f. 13.

Ginkgo pusilla (Dawson) Knowlton, U. S. Geol. Surv., Bull. 152, 1898, p. 111.—Penhallow, Man. N. A. Gym., 1907, p. 209. [Homonym, Heer 1876.]

Upper Cretaceous: Port McNeill, Vancouver Island; Cumshaven Inlet, Queen Charlotte Islands.

Ginkgo digitata (Brongniart) Heer, in Regel's Gartenflora, Jahrg. 23, 1874, p. 261, pl. dcccvii, f. 1–4.—Fontaine, in Ward, U. S. Geol. Surv., Mon. 48, 1905 [1906], pp. 121, 170, pl. xxx, f. 1–7; pl. lxiv, f. 5, 6.

Cyclopteris digitata Brongniart, Hist. vég. foss., vol. 1, 1830, p. 219, pl. lxi, f. 2, 3.

Jurassic: Douglas County, Oreg.; Cape Lisburne, Alaska.

Ginkgo huttoni (Sternberg) Heer, in Regel's Gartenflora, Jahrg. 23, 1874, p. 261, pl. dcccvii, f. 4.—Fontaine, in Ward, U. S. Geol. Surv., Mon. 48, 1905 [1906], pp. 123, 170, pl. xxx, f. 8–12; pl. xxxi, f. 1–3.

Cyclopteris huttoni Sternberg, Fl. d. Vorwelt, vol. 2, 1833, p. 66.

Ginkgo multinervis Heer. Lesquereux, U. S. Nat. Mus., Proc., vol. 11, 1888, p. 31, pl. xvi, f. 6.

Jurassic: Douglas County, Oreg.; Cape Lisburne, Alaska.

Ginkgo huttoni magnifolia Fontaine, in Ward, U. S. Geol. Surv., Mon. 48, 1905 [1906], p. 124, pl. xxxi, f. 4–8; pl. xxxii, f. 1, 2; p. 170, pl. xliv, f. 7, 8.

Jurassic: Douglas County, Oreg.; Cape Lisburne, Alaska.

Ginkgo laramiensis Ward, Science, vol. 5, 1885, p. 496, f. 7; U. S. Geol. Surv., Sixth Ann. Rept., 1884–85 [1886], p. 549, pl. xxxi, f. 4; U. S. Geol. Surv., Bull. 37, 1887, p. 15, pl. i, f. 4.—Knowlton, U. S. Geol. Surv., Bull. 163, 1900, p. 31, pl. iv, f. 7–10; pl. v, f. 5.

Mesaverde: Point of Rocks, Wyo.

Ginkgo lepida (Heer) Heer, Fl. foss. arct., vol. 4, Abt. 2, 1876, p. 62, pl. vii, f. 7; pl. xii.—Fontaine, in Ward, U. S. Geol. Surv., Mon. 48, 1905 [1906], p. 125, pl. xxxii, f. 3–8.

Salisburia (Ginkgo) lepida Heer. Dawson, Roy. Soc. Canada, Trans., vol. 3, sec. 4, 1885 [1886], p. 8, pl. ii, f. 2.

Jurassic: Douglas County, Oreg.

Kootenai: Martin and Coal creeks, British Columbia.

Ginkgo multinervis Heer. Lesquereux, U. S. Nat. Mus., Proc., vol. 11, 1884, p. 31, pl. xvi, f. 6 = *Ginkgo huttoni.*

Ginkgo nana (Dawson) Knowlton, U. S. Geol. Surv., Bull. 152, 1898, p. 111.

Salisburia (Ginkgo) nana Dawson, Roy. Soc. Canada, Trans., vol. 3, sec. 4, 1885 [1886], p. 8, pl. ii, f. 3.

Kootenai: Coal Creek, British Columbia.

Ginkgo pusilla (Dawson) Knowlton = *Ginkgo dawsoni.*

Ginkgo sibirica (Heer) Heer, Fl. foss. arct., vol. 4, Abt. 2, 1876, p. 61, pl. vii, f. 6; pl. ix, f. 5f; pl. xi.—Fontaine, in Ward, U. S. Geol. Surv., Mon. 48, 1905 [1906], p. 125, pl. xxxiii.—Knowlton, Smithsonian Misc. Coll., vol. 50, 1907, p. 124, pl. xiii, f. 1–4; pl. xiv, f. 1, 2.

Salisburia (Ginkgo) sibirica Heer, Dawson, Roy. Soc. Canada, Trans., vol. 3, sec. 4, 1885 [1886], p. 8, pl. ii, f. 1.

Baiera brevifolia Newberry, Am. Jour. Sci., 3d ser., vol. 61, 1891, p. 199, pl. xiv, f. 3.

Kootenai: Martin Creek, British Columbia; near Geyser, Cascade County, Mont.

Jurassic: Douglas County, Oreg.

Ginkgo sp. (Fontaine).

Ginkgodium? alaskense Fontaine, in Ward, U. S. Geol. Surv., Mon. 48, 1905 [1906], p. 168, pl. xliv, f. 3, 4.

Jurassic: Cape Lisburne, Alaska.

Ginkgo sp. Knowlton, U. S. Geol. Surv., Bull. 152, 1898, p. 111.

Carpites myricarum Lesquereux, Rept. U. S. Geol. Surv. Terr., vol. 7 (Tert. Fl.), 1878, p. 303, pl. lx, f. 8–11.

Post-Laramie: Black Buttes, Wyo.

Ginkgo sp., nutlets of (Dawson).

Salisburia, nutlets of, Dawson, Roy. Soc. Canada, Trans., vol. 3, sec. 4, 1885 [1886], p. 17.

Edmonton: Canada.

Ginkgo sp. (Dawson) Knowlton, U. S. Geol. Surv., Bull. 152, 1898, p. 111.

Salisburia sp. Dawson, Roy. Soc. Canada, Trans., vol. 3, sec. 4, 1885 [1886], p. 18.

Paskapoo (Fort Union): Near Calgary, Alberta.

Ginkgo sp., nuts of (Dawson), Knowlton, U. S. Geol. Surv., Bull. 152, 1898, p. 111.

Salisburia sp., nuts of, Dawson, Roy. Soc. Canada, Trans., vol. 3, sec. 4, 1885 [1886], p. 9, pl. ii, f. 4.

Kootenai: British Columbia (no locality).

Ginkgo sp. [wood] Dawson, Roy. Soc. Canada, Trans., vol. 5, sec. 4, 1887, p. 33.

Belly River: South Saskatchewan; west of Medicine Hat, Alberta.

Ginkgo sp. [wood] Dawson, Roy. Soc. Canada, Trans., vol. 5, sec. 4, 1887, p. 34.
Paskapoo (Fort Union): Souris River; Mackenzie River, Northwest Territory.

Ginkgo sp. Fontaine, in Ward, U. S. Geol. Surv., Mon. 48, 1905 [1906], p. 127, pl. xxxiv, f. 1–12.
Jurassic: Douglas County, Oreg.

Ginkgo sp. Knowlton, U. S. Geol. Surv., Bull. 204, 1902, p. 23, pl. i, f. 5.
Mascall: Van Horn's ranch, John Day Basin, Oreg.

Ginkgodium? alaskense Fontaine = *Ginkgo* sp. (Fontaine).

GLEDITSIA Linné. Sp. pl., 1753, p. 1056.

Gleditschia donensis Penhallow, Am. Nat., vol. 41, 1907, p. 446.
Pleistocene: Don Valley, Toronto, Canada.

Gleditschia triacanthos Linné. Lesquereux, Am. Jour. Sci., 2d ser., vol. 27, 1859, p. 365.
Pleistocene: Banks of Mississippi near Columbus, Ky.

GLEDITSIOPHYLLUM Berry, Bull. Torr. Bot. Club, vol. 37, 1910, p. 197. [Type, *G. triacanthoides*.]

Gleditsiophyllum constrictum Berry, U. S. Geol. Surv., Prof. Paper 91, 1916, p. 239, pl. li, f. 4.
Lagrange: Puryear, Henry County, Tenn.

Gleditsiophyllum ellipticum Berry, U. S. Geol. Surv., Prof. Paper 91, 1916, p. 239, pl. li, f. 2, 3.
Lagrange: Puryear, Henry County, Tenn.

Gleditsiophyllum entadaformis Berry, U. S. Geol. Surv., Prof. Paper 91, 1916, p. 241, pl. liv, f. 5.
Wilcox (Holly Springs): Holly Springs, Marshall County, Miss.

Gleditsiophyllum eocenicum Berry, U. S. Geol. Surv., Prof. Paper 91, 1916, p. 238, pl. xlvi, f. 1–7.
Wilcox (Grenada): Grenada, Grenada County, Miss.
Lagrange: Puryear, Henry County, Tenn.
Wilcox: Calaveras Creek, Wilson County, Tex., and Mansfield, De Soto Parish, La.

Gleditsiophyllum fruticosum Berry, U. S. Geol. Surv., Prof. Paper 91, 1916, p. 240, pl. li, f. 7.
Wilcox (Holly Springs): Holly Springs, Marshall County, Miss.

Gleditsiophyllum hilgardianum Berry, U. S. Geol. Surv., Prof. Paper 91, 1916, p. 240, pl. li, f. 9.
Wilcox (Ackerman): Hurleys, Benton County, Miss.

Gleditsiophyllum minor Berry, U. S. Geol. Surv., Prof. Paper 91, 1916, p. 240, pl. li, f. 5, 6.
Lagrange: Puryear, Henry County, Tenn.

Gleditsiophyllum ovatum Berry, U. S. Geol. Surv., Prof. Paper 91, 1916, p. 239, pl. li, f. 1.
Lagrange: Puryear, Henry County, Tenn.

Gleditsiophyllum triacanthoides Berry, Bull. Torr. Bot. Club, vol. 37, 1910, p. 197.
Black Creek: Below Dunbars Bridge, Tar River, N. C.

GLEICHENIA Smith, Mém. Acad. Turin, vol. 5, 1793, p. 419.

Gleichenia delawarensis Berry, Johns Hopkins Univ. Circ., 2d ser., vol. 7, 1907, p. 82, f. 3, 3a; Bull. Torr. Bot. Club, vol. 37, 1910, p. 20; Md. Geol. Surv., Upper Cret., 1916, p. 762, pl. l, f. 5, 6.
Magothy: Deep Cut, Del.; Grove Point, Md.

Gleichenia delicatula? Heer, Fl. foss. arct., vol. 3, Abt. 2, 1874, p. 54, pl. ix, f. 11e; pl. x, f. 16, 17.—Hollick, Torreya, vol. 2, 1902, p. 147, pl. iii, f. 4.—Knowlton, U. S. Geol. Surv., Prof. Paper 101. 1918, p. 247.
Vermejo: "Florence, Colorado" = Rockvale, Colo.

Gleichenia giesekiana Heer, Fl. foss. arct., vol. 1, 1868, p. 78, pl. xliii, f. 1a, 2a, 3a.—Newberry, U. S. Geol. Surv., Mon. 26, 1895 [1896], p. 36, pl. iv, f. 12.—Berry, Geol. Surv. N. J., Bull. 3, 1911, p. 65.
Raritan: Woodbridge, N. J.

Gleichenia? gilbert-thompsoni Fontaine, in Ward, U. S. Geol. Surv., Mon. 48, 1905 [1906], p. 232, pl. lxvi, f. 11.—Penhallow, Roy. Soc. Canada, Trans., 3d ser., vol. 1, sec. 4, 1907, p. 302, pl. ix.—Knowlton, Am. Jour. Sci., 4th ser., vol. 30, 1910, p. 41.
Knoxville: 12 miles west of Red Bluff, Tehama County, Calif.
Lower Cretaceous: International boundary between Pasayten and Skagit rivers, British Columbia.

Gleichenia gracilis Heer, Fl. foss. arct., vol. 3, Abt. 3, 1874, p. 52, pl. x, f. 1-11.—Dawson, Roy. Soc. Canada, Trans., vol. 3, sec. 4, 1885 [1885], p. 11, pl. iii, f. 4.—?Hollick, U. S. Geol. Surv., Mon. 50, 1906, p. 31, pl. i, f. 9.
Mill Creek: Mill Creek, British Columbia.
Magothy: ?Black Rock Point, Block Island, R. I.

Gleichenia kurriana Heer, Neue Denkschr. Schw. Ges., vol. 23, 1869, p. 6, pl. ii, f. 1-4.—Lesquereux, U. S. Geol. and Geog. Surv. Terr., Ann. Rept. 1872 [1873], p. 421; U. S. Geol. Surv., Mon. 17, 1891 [1892], p. 25.
Dakota: Fort Harker, Kans.
Dakota(?): Drummond, Mont.
Mill Creek: Mill Creek, British Columbia.

Gleichenia micromera Heer? Fl. foss. arct., vol. 3, Abt. 2, 1874, p. 55, pl. x, f. 14, 15.—Newberry, U. S. Geol. Surv., Mon. 26, 1895 [1896], p. 36, pl. iii, f. 6.—Berry, Geol. Surv. N. J., Bull. 3, 1911, p. 66.

Raritan: Sayreville and Milltown, N. J.

Gleichenia nordenskiöldi Heer, Fl. foss. arct., vol. 3, Abt. 2, 1874, p. 50, pl. ix, f. 6–12.—Fontaine, U. S. Geol. Surv., Mon. 15, 1889 [1890], p. 119, pl. xxi, f. 11.—Lesquereux, U. S. Geol. Surv., Mon. 17, 1891 [1892], p. 25.—Fontaine, in Ward, U. S. Geol. Surv., Mon. 48, 1905 [1906], p. 231, pl. lxv, f. 24–29.—Knowlton, Am. Jour. Sci., 4th ser., vol. 30, 1910, p. 41.

Pecopteris strictinervis Fontaine? Fontaine, in Diller and Stanton, Geol. Soc. Am., Bull., vol. 5, 1895, p. 450; in Stanton, U. S. Geol. Surv., Bull. 133, 1895 [1896], p. 15.

Aspidium heterophyllum Fontaine. Fontaine, in Diller and Stanton, op. cit., p. 450; in Stanton, op. cit., p. 15.

Osmunda dicksonioides Fontaine? Fontaine, in Diller and Stanton, op. cit., p. 450; in Stanton, op. cit., p. 15.

Patuxent: Dutch Gap Canal, Va.
Knoxville: Tehama County, Calif.
Dakota: Kansas.

Gleichenia? obscura Knowlton, U. S. Geol. Surv., Twenty-first Ann. Rept., pt. 2, 1901, p. 210, pl. xxx, f. 1–4.

Esmeralda: 3.8 kilometers northeast of Emigrant Peak, Nev.

Gleichenia protogaea Debey and Ettingshausen? Hollick, U. S. Geol. Surv., Mon. 50, 1906, p. 31, pl. i, f. 8.

Magothy: Gay Head, Marthas Vineyard, Mass.

Gleichenia pulchella Knowlton, U. S. Nat. Mus., Proc., vol. 45, 1913, p. 557, pl. xliv, f. 1, 2.

Mesaverde: Half a mile north of Mine D, Superior, Wyo.

Gleichenia rhombifolia Hollick, Torreya, vol. 2, 1902, p. 147, pl. iii, f. 3.—Knowlton, U. S. Geol. Surv., Prof. Paper 101, 1918, p. 247.

Vermejo: "Florence, Colorado" = Rockvale, Colo.

Gleichenia saundersii Berry, Am. Nat., vol. 37, 1903, p. 679, f. 1–3; Geol. Surv. N. J., Ann. Rept. 1905 [1906], pp. 138, 139; Bull. Torr. Bot. Club, vol. 33, 1906, p. 163; idem, vol. 38, 1911, p. 402, pl. xviii, f. 2, 3, 3a; Md. Geol. Surv., Upper Cret., 1916, p. 762, pl. l, f. 7–9.

Magothy: Cliffwood and Kinkora, N. J.; Severn River, Md.

Gleichenia zippei (Corda) Heer, Fl. foss. arct., vol. 1, 1868, p. 79, pl. xlviii, f. 4.—Newberry, U. S. Geol. Surv., Mon. 26, 1895 [1896], p. 119, pl. iii, f. 5.—Fontaine, in Ward, U. S. Geol. Surv., Nineteenth Ann. Rept., pt. 2, 1899, p. 664, pl. clxii, f. 9.—Berry, Bull. Torr. Bot. Club, vol. 31, 1904, p. 67, pl. iv, f. 6; idem, vol. 33, 1906, p. 164; Geol. Surv. N. J., Ann. Rept. 1905 [1906], p.138;

Gleichenia zippei—Continued.
Geol. Surv. N. J., Bull. 3, 1911, p. 64; Md. Geol. Surv., Upper Cret., 1916, p. 760.
Pecopteris zippei Corda, in Reuss, Verst. böhm. Kreidef. Stuttgart, pt. 2, 1846, p. 95, pl. xlix, f. 2, 3.
Raritan: Woodbridge, N. J.
Magothy: Cliffwood, N. J.; Deep Cut, Del.
Lakota: Hay Creek coal field, Crook County, Wyo.

Gleichenia sp. Knowlton, in Veatch, Am. Jour. Sci., 4th ser., vol. 21, 1906, p. 459 = *Tapeinidium undulatum.*

Gleichenia sp. Penhallow, Roy. Soc. Canada, Trans., 3d ser., vol. 1, sec. 4, 1907, p. 306.
Lower Cretaceous: International boundary between Pasayten and Skagit rivers, British Columbia.

GLOSSOCARPELLITES Perkins, Geol. Soc. Am., Bull., vol. 16, 1905, p. 510. [Type, *Carpolithes parvus.*]

Glossocarpellites elongatus (Lesquereux) Perkins, Geol. Soc. Am., Bull., vol. 16, 1905, p. 511, pl. 87, f. 17; Rept. Vt. State Geol. 1905–6 [1906], pl. liii, f. 17.
Carpolithes brandoniana elongata Lesquereux, Am. Jour. Sci., 2d ser., vol. 32, 1861, p. 356.
Carpolithes elongatus (Lesquereux) Perkins, Rept. Vt. State Geol., 1903–4 [1904], p. 176, pl. lxxv, f. 1–3.
Miocene: Brandon, Vt.

Glossocarpellites grandis (Perkins) Perkins, Rept. Vt. State Geol., 1905–6 [1906], p. 206, pl. liv, f. 1–3.
Carpolithes grandis Perkins, Rept. Vt. State Geol., 1903–4 [1904], p. 178.
Miocene: Brandon, Vt.

Glossocarpellites obtusus (Lesquereux) Perkins, Geol. Soc. Am., Bull., vol. 16, 1905, p. 511, pl. 87, f. 16; Rept. Vt. State Geol., 1905–6 [1906], pl. liii, f. 16.
Carpolithes brandoniana obtusa Lesquereux, Am. Jour. Sci., 2d ser., vol. 32, p. 356, 1861; Hitchcock, Geol. Vt., vol. 2, p. 713, text f. 114–117, 1861.
Carpolithes obtusus (Lesquereux) Perkins, Rept. Vt. State Geol., 1903–4 [1904], p. 177, pl. lxxv, f. 5–8, 14.
Miocene: Brandon, Vt.

Glossocarpellites parvus (Perkins) Perkins, Geol. Soc. Am., Bull., vol. 16, 1905, p. 510, pl. 86, f. 15.—Perkins, Rept. Vt. State Geol., 1905–6 [1906], p. 207, pl. lii, f. 15.
Carpolithes parvus Perkins, Rept. Vt. State Geol., 1903–4 [1904], p. 179.
Miocene: Brandon, Vt.

Glossocarpellites? sp., Perkins, Rept. Vt. State Geol., 1905–6 [1906], p. 225, pl. lviii, f. 6.
Miocene: Brandon, Vt.

Glossopteris phillipsii? Brongniart. Hall, in Fremont = *Ficus fremonti.*

GLOSSOZAMITES Schimper, Pal. vég., vol. 2, 1870, p. 163.

Glossozamites distans Fontaine, U. S. Geol. Surv., Mon. 15, 1889 [1890], p. 176, pl. lxviii, f. 5.
Patuxent: Fredericksburg, Va.

Glossozamites fontaineanus Ward, U. S. Geol. Surv., Nineteenth Ann. Rept., pt. 2, 1899, p. 667, pl. clxii, f. 16–18.
Glossozamites? sp. Fontaine, in Ward, Jour. Geol., vol. 2, 1894, p. 260.
Lakota: Red Canyon, S. Dak.

Glossozamites klipsteinii (Dunker) Fontaine = *Sagenopteris mantelli.*

Glossozamites? sp. Fontaine = *Glossozamites fontaineanus.*

GLYPHOMITRIUM Bridel, Meth. Musc., 1819, p. 30.

Glyphomitrium cockerelleae E. G. Britton and Hollick, Bull. Torr. Bot. Club, vol. 34, 1907, p. 140, pl. 9, f. 6, 6a.
Miocene: Florissant, Colo.

GLYPTOSTROBUS Endlicher, Synop. Conif., 1847, p. 69.

Glyptostrobus brookensis (Fontaine) Ward [part] = *Widdringtonites ramosus.*

Glyptostrobus brookensis Fontaine [part] = *Sphenolepis kurriana.*

Glyptostrobus brookensis angustifolium (Fontaine) Knowlton = *Widdringtonites ramosus.*

Glyptostrobus europaeus (Brongniart) Heer, Fl. tert. Helv., vol. 1, 1855, p. 51, pl. xix; pl. xx.—Lesquereux, Rept. U. S. Geol. Surv. Terr., vol. 7 (Tert. Fl.), 1878, p. 74, pl. vii, f. 1, 2.—Newberry, U. S. Geol. Surv., Mon. 35, 1898, p. 24, pl. xxvi, f. 6–8a; pl. lv, f. 3, 4.—Knowlton, U. S. Geol. Surv., Seventeenth Ann. Rept., pt. 1, 1896, p. 899; Bull. Torr. Bot. Club, vol. 29, 1902, p. 705.—Penhallow, Roy. Soc. Canada, Trans., 2d ser., vol. 8, 1902, pp. 51, 68; Rept. Tert. Pl. Brit. Col., 1908, p. 58; Roy. Soc. Canada, Trans., 3d ser., vol. 1, 1907, p. 309.—Knowlton, Washington Acad. Sci., Proc., vol. 11, 1909, pp. 188, 197, 198, 211, 214; Jour. Geol., vol. 19, 1911, p. 369; in Calvert, U. S. Geol. Surv., Bull. 471, 1912, p. 16.—Berry, U. S. Geol. Surv., Prof. Paper 91, 1916, p. 169, pl. xv, f. 3.
Fort Union: Fort Union, N. Dak.; Porcupine Butte, Mont. Abundant and widespread in the Fort Union.
Lance: Cannonball River, S. Dak.; Miles City, 18 miles east of Miles City, 10 miles northeast of Glendive, Mont.; Dayton, Rairden, Wyo.
Kenai: Kenai Island near Sitka, Ninilchik, Herendeen Bay, De Groff tunnel, Admiralty Island, and other localities, Alaska.
—— Birch Bay, Wash.
Paskapoo?: Red Deer River, Alberta; Similkameen and Horsefly rivers, British Columbia.
Wilcox (Holly Springs): Oxford and Early Grove, Miss.

Glyptostrobus europaeus? Heer. Knowlton, in Stanton, U. S. Geol. Surv., Bull. 106, 1893, p. 42; Geol. Soc. Am., Bull., vol. 8, 1897, p. 150 = *Sequoia* sp.

Glyptostrobus europaeus ungeri (Heer) Heer, Fl. tert. Helv., vol. 3, 1859, p. 159.—Lesquereux, Rept. U. S. Geol. Surv. Terr., vol. 8 (Cret. and Tert. Fl.), 1883, p. 222, pl. xlvi, f. 1–1c.
Mascall: John Day Basin, Oreg.
Eocene?: Several Canadian localities.

Glyptostrobus expansus (Fontaine) Ward = *Sphenolepis kurriana.*

Glyptostrobus fastigiatus (Fontaine) Ward = *Sphenolepis kurriana.*

Glyptostrobus gracillimus Lesquereux = *Sequoia gracillima.*

Glyptostrobus groenlandicus Heer, Fl. foss. arct., vol. 3, Abt. 2, 1874, p. 76, pl. xvii, f. 9; pl. xx, f. 9, 10; pl. xxii, f. 12.—Dawson, Roy. Soc. Canada, Trans., vol. 3, sec. 4, 1885 [1886], p. 9, pl. iii, f. 8.
Kootenai: North branch of Old Man River, British Columbia.

Glyptostrobus ramosus (Fontaine) Ward = *Widdringtonites ramosus.*

Glyptostrobus ungeri Heer, Fl. tert. Helv., vol. 1, 1855, p. 52.—Dawson, Roy. Soc. Canada, Trans., vol. 7, 1889, p. 70.—Lesquereux, Rept. U. S. Geol. Surv. Terr., vol. 8 (Cret. and Tert. Fl.), 1883, p. 139, pl. xxii, f. 1–6a.—Knowlton, U. S. Geol. Surv., Bull. 204, 1900, p. 26.

[NOTE.—This is probably the same as that referred to *Glyptostrobus europaeus ungeri*. It is still left in this very unsatisfactory state pending a thorough revision.]

Mascall: Van Horn's ranch, John Day Basin, Oreg.
Eocene?: Mackenzie River valley, Northwest Territory.

Glyptostrobus sp. Dawson, Roy. Soc. Canada, Trans., vol. 1, sec. 4, 1882 [1883], p. 25.
Upper Cretaceous: Baynes Sound, Vancouver Island.

Glyptostrobus sp. Dawson, Roy. Soc. Canada, Trans., vol. 8, sec. 4, 1890, p. 79, text f. 7.
Miocene: Similkameen River and Stump Lake, British Columbia.

Glyptostrobus? sp. Knowlton, U. S. Geol. Surv., Bull. 163, 1900, p. 31, pl. v, f. 4.
Mesaverde: North fork of Dutton Creek, Laramie Plains, Wyo.

Glyptostrobus sp. Knowlton, Washington Geol. Surv., Ann. Rept., vol. 1, 1902, p. 33.
Eocene: Skykomish, Snohomish County, Wash.

GONIOPTERIS Presl, Tent. Pterid., 1836, p. 181.

Gonioptoris claiborniana Berry, Bull. Torr. Bot. Club, vol. 44, 1917, p. 331, pl. 22.
Yegua: Columbia, La.
Lisbon: Newton, Miss.

Goniopteris polypodioides Ettingshausen = *Dryopteris polypodioides.*

Goniopteris pulchella? Heer. Lesquereux, U. S. Geol. and Geog. Surv. Terr., Bull., vol. 1, 1875 [1876], p. 383. [Not afterward referred to by Lesquereux.]

——: Golden, Colo. Supposed to have come from Laramie but unknown.

GREWIA Linné, Sp. pl., 1753, p. 964.

Grewia auriculata Lesquereux, Rept. U. S. Geol. Surv. Terr., vol. 8 (Cret. and Tert. Fl.), 1883, p. 252, pl. lv, f. 1.—Knowlton, U. S. Geol. Surv., Bull. 204, 1902, p. 81.

Clarno (upper part): Bridge Creek, Oreg.

Grewia celastroides Ward, U. S. Geol. Surv., Sixth Ann. Rept., 1884–5 [1886], p. 555, pl. liv, f. 14; U. S. Geol. Surv., Bull. 37, 1887, p. 86, pl. xxxix, f. 2.

Fort Union: Iron Bluff, Glendive, Mont.

Grewia crenata (Unger) Heer, Fl. tert. Helv., vol. 3, 1859, p. 42, pl. cix, f. 12–21; cx, f. 1–11.—Ward, U. S. Geol. Surv., Sixth Ann. Rept., 1884–85 [1886], p. 555, pl. liv, f. 13; idem, Bull. 37, 1887, p. 85, pl. xxxix, f. 1.—Newberry, U. S. Geol. Surv., Mon. 35, 1898, p. 120, pl. xlvi, f. 2; pl. xlviii, f. 2, 3.—Knowlton, U. S. Geol. Surv., Bull. 204, 1902, p. 80.—Penhallow, Rept. Tert. Pl. Brit. Col., 1908, pp. 58, 66.

Paliurus colombi Heer. Lesquereux, U. S. Nat. Mus., Proc., vol. 11, 1888, p. 16.—Knowlton, U. S. Geol. Surv., Bull. 108, 1893, p. 104.

Mascall: Van Horn's ranch, John Day Basin, Oreg.
Clarno (upper part): Bridge Creek, Oreg.
Fort Union: Bull Mountains, Mont.
Oligocene and Miocene: Horsefly River, British Columbia.
Eocene: Porcupine Creek, Saskatchewan.

Grewia obovata Heer, Fl. foss. arct., vol. 4, Abt. 1, 1876, p. 86, pl. lix, f. 15, 15b.—Ward, U. S. Geol. Surv., Sixth Ann. Rept., 1884–85 [1886], p. 555, pl. lv, f. 4, 5; idem, Bull. 37, 1887, p. 88, pl. xxxix, f. 6, 7.

Fort Union: Sevenmile Creek, near Glendive, Mont.

Grewia pealei Ward, U. S. Geol. Surv., Sixth Ann. Rept., 1884–85 [1886], p. 555, pl. lv, f. 1–3; U. S. Geol. Surv., Bull. 37, 1887, p. 87, pl. xxxix, f. 3–5.

Fort Union: Bull Mountains, Mont.

GREWIOPSIS Saporta, Ann. sci. nat., Bot. (5), 1865, p. 49.

Grewiopsis acuminata Lesquereux, U. S. Nat. Mus., Proc., vol. 10, 1887, p. 44, pl. iii, f. 12, 13; pl. iv, f. 1, 2.

Locality and formation unknown.

Grewiopsis aequidentata Lesquereux, U. S. Geol. Surv., Mon. 17, 1891 [1892], p. 180, pl. lviii, f. 4.

Dakota: Ellsworth County, Kans.

Grewiopsis? aldersoni Knowlton, U. S. Geol. Surv., Mon. 32, pt. 2, 1899, p. 743.
Miocene (intermediate flora): Yellowstone National Park.

Gre(v)wiopsis cleburni Lesquereux, U. S. Geol. and Geog. Surv. Terr., Bull., vol. 1, 1875 [1876], p. 380; idem, Ann. Rept. 1874 [1876] p. 306; Rept. U. S. Geol. Surv. Terr., vol. 7 (Tert. Fl.), 1878, p. 259, pl. lxii, f. 12.—Knowlton, U. S. Geol. Surv., Bull. 163, 1900, p. 61.
Mesaverde: Point of Rocks, Wyo.
Adaville: Hodges Pass, Wyo.

Grewiopsis eocenica (Lesquereux) Knowlton, U. S. Geol. Surv., Bull. 152, 1898, p. 114; Washington Acad. Sci., Proc., vol. 11, 1909, p. 207.
Aleurites eocenica Lesquereux, U. S. Geol. and Geog. Surv. Terr., Ann. Rept. 1872 [1873], p. 397.
Grewiopsis saportanea Lesquereux, Rept. U. S. Geol. Surv. Terr., vol. 7 (Tert. Fl.), 1878, p. 257, pl. l, f. 10–12.
Grewiopsis ficifolia Ward, U. S. Geol. Surv., Sixth Ann. Rept., 1884–85 [1886], p. 556, pl. xlvi, f. 1, 2; idem, Bull. 37, 1887, p. 92, pl. xli, f. 1, 2.
Post-Laramie: Black Buttes, Wyo.
Lance: Buck Creek, Converse County, Wyo.

Grewiopsis ficifolia Ward = *Grewiopsis eocenica.*

Grewiopsis flabellata (Lesquereux) Knowlton, U. S. Geol. Surv., Bull. 152, 1898, p. 114.—Berry, Bull. Torr. Bot. Club, vol. 33, 1906, p. 177.
Populites flabellata Lesquereux, Am. Jour Sci., 2d ser., vol. 46, 1868, p. 94.
Populites fagifolia Lesquereux, U. S. Geol. and Geog. Surv. Terr., Ann. Rept 1872 [1873], p. 422.
Greviopsis haydenii Lesquereux, Rept. U. S. Geol. Surv. Terr., vol. 6 (Cret. Fl.), 1874, p. 97, pl. iii, f. 3, 4; pl. xxiv, p. 3.
Dakota: Fort Harker, Kans.
Magothy: Grove Point, Md.

Grewiopsis formosa Berry, U. S. Geol. Surv., Prof. Paper 112, 1919, p. 116, pl. xxv, f. 4, 5.
Tuscaloosa: Shirleys Mill, Ala.

Grewiopsis haydenii Lesquereux = *Grewiopsis flabellata.*

Grewiopsis mudgei Lesquereux, U. S. Geol. Surv., Mon. 17, 1891 [1892], p. 181, pl. lxvi, f. 3.
Dakota: Ellsworth County, Kans.

Grewiopsis paliurifolia Ward, U. S. Geol. Surv., Sixth Ann. Rept., 1884–85 [1886], p. 556, pl. lvi, f. 3; U. S. Geol. Surv., Bull. 37, 1887, p. 92, pl. xli, f. 3.
Post-Laramie: Black Buttes, Wyo.

Grewiopsis platanifolia Ward, U. S. Geol. Surv., Sixth Ann. Rept., 1884–85 [1886], p. 555, pl. lv, f. 6; U. S. Geol. Surv., Bull. 37, 1887, p. 89, pl. xl, f. 1.—Knowlton, Washington Acad. Sci., Proc., vol. 11, 1909, p. 189.
Fort Union: Sevenmile Creek, Glendive, Mont.
Lance: Near Miles City, Mont.

Grewiopsis populifolia Ward, U. S. Geol. Surv., Sixth Ann. Rept., 1884–85 [1886], p. 556, pl. lv, f. 8–10; U. S. Geol. Surv., Bull. 37, 1887, p. 90, pl. xl, f. 3–5.—Knowlton, Washington Acad. Sci., Proc., vol. 11, 1909, p. 189.
Fort Union: Burns's ranch, Glendive, Mont.
Lance: Near Miles City, Mont.

Grewiopsis saportanea Lesquereux = *Grewiopsis eocenica.*

Grewiopsis tennesseensis Berry, U. S. Geol. Surv., Prof. Paper 91, 1916, p. 285, pl. lxiv, f. 4, 5.
Wilcox: Hamilton, Sabine County, Tex.
Lagrange: Grand Junction, Tenn.

Grewiopsis tenuifolia Lesquereux, Rept. U. S. Geol. Surv. Terr., vol. 7 (Tert. Fl.), 1878, p. 258, pl. xl, f. 14.
Post-Laramie: Black Buttes, Wyo.

Grewiopsis tuscaloosensis Berry, U. S. Geol. Surv., Prof. Paper 112, 1919, p. 116, pl. xxv, f. 3.
Tuscaloosa: Shirleys Mill, Ala.

Grewiopsis viburnifolia Ward, U. S. Geol. Surv., Sixth Ann. Rept., 1884–85 [1886], p. 555, pl. lv, f. 7; U. S. Geol. Surv., Bull. 37, 1887, p. 89, pl. xl, f. 2.
Fort Union: Burns's ranch, Glendive, Mont.

Grewiopsis viburnifolia Ward. Hollick, Bull. Torr. Bot. Club, vol. 21, 1894, p. 59, pl. 174, f. 8 = *Viburnum hollickii.*

Grewiopsis walcotti Lesquereux, U. S. Nat. Mus., Proc., vol. 10, 1887, p. 45, pl. iv, f. 3, 4.
——: Upper Kanab Valley, Utah.

GROSSULARIA (Tournefort) Adanson, Fam., vol. 2, 1763, p. 243.

Grossularia menziesii (Pursh) Coville and Britton? Hannibal, Bull. Torr. Bot. Club, vol. 38, 1911, p. 337.
Pleistocene: Near Portola and in Calabasas Canyon, Santa Cruz Mountains, Calif.

GUATTERIA Ruiz and Pavon, Prodr., vol. 17, 1794, p. 85.

Guatteria cretacea Hollick, U. S. Geol. Surv., Mon. 50, 1906, p. 73, pl. xxi, f. 1–4.
Magothy: Gay Head, Marthas Vineyard, Mass.; Glen Cove, Long Island, N. Y.

GUETTARDA Linné, Sp. pl., vol. 2, 1753, p. 991.

Guettarda ellipticifolia Berry, U. S. Geol. Surv., Prof. Paper 91, 1916, p. 348, pl. cvi, f. 1, 2.
Wilcox (Holly Springs): Holly Springs, Miss.
Lagrange: Puryear, Tenn.

GYMINDA Sargent, Garden and Forest, vol. 4, 1891, p. 4.

Gyminda primordialis Hollick, U. S. Geol. Surv., Mon. 50, 1906, p. 88, pl. xxxiii, f. 5.
Magothy: Gay Head, Marthas Vineyard, Mass.

GYMNOCAULUS Emmons, Geol. Rept. Midland Counties N. C., 1856, p. 289.

Gymnocaulus alternatus Emmons, Geol. Rept. Midland Counties N. C., 1856, p. 289, pl. i, f. 4; American geology, pt. 6, 1857, p. 38, pl. v, f. 4.—Ward, U. S. Geol. Surv., Twentieth Ann. Rept., pt. ii, 1900, p. 426.
Triassic (Newark): Deep and Dan River coal fields, N. C.; Pennsylvania?.

GYMNOCLADUS Lamarck, Encycl., vol. 1, 1783, p. 733.

Gymnocladus casei Berry, U. S. Nat. Mus., Proc., vol. 54, 1918, p. 631, pl. 94, f. 2.
Miocene: Beaver Creek, 10 miles east of Beaver City, Okla.

Gymnogramma gardneri (Lesquereux) Lesquereux = *Saccoloma gardneri*.

Gymnogramma haydenii Lesquereux = *Anemia elongata*.

Gymnospermous cone, Berry, Bull. Torr. Bot. Club?, vol. 31, 1904, p. 72, pl. iv, f. 7.
Magothy: Cliffwood, N. J.

GYRODENDRON Ulrich, Harriman Alaska Exped., vol. 4, 1904, p. 140. [Type, *G. emersoni*.]

Gyrodendron emersoni Ulrich, Harriman Alaska Exped., vol. 4, 1904, p. 140, pl. xviii, f. 3; pl. xix, f. 1, 2.
Yakutat: Pogibshi Island, opposite Kodiak, Alaska.

HALYMENITES Sternberg, Versuch., vol. 2, fasc. 5 and 6, 1833, p. 29.

Halymenites major Lesquereux, U. S. Geol. and Geog. Surv. Terr., Ann. Rept. 1872 [1873], p. 373, 390; idem, 1873 [1874], pp. 379, 384; idem, 1876 [1878], p. 496; Rept. U. S. Geol. Surv. Terr., vol. 7 (Tert. Fl.), 1878, p. 38, pl. i, f. 7, 8.—Knowlton, U. S. Geol. Surv., Bull. 163, 1900, p. 17.—Henderson, Univ. Colo. Studies, vol. 5, 1908, p. 186.—Knowlton, U. S. Geol. Surv., Prof. Paper 98, 1916, p. 84.—Berry, Bull. Torr. Bot. Club, vol. 43, 1916, p. 303.—Knowlton, U. S. Geol. Surv., Prof. Paper 101, 1918, p. 243.
Halymenites minor? Fischer-Ooster. Lesquereux, U. S. Geol. and Geog. Surv. Terr., Ann. Rept. 1872 [1873], p. 373; Rept. U. S. Geol. Surv. Terr., vol. 7 (Tert. Fl.), 1878, p. 39, pl. i, f. 9.

Halymenites major—Continued.
Widely distributed from Colorado (Benton) to Fox Hills, being most abundant in or near the Fox Hills.
Trinidad, Colo.; Spring Canyon, N. Mex.
Fox Hills: Greeley quadrangle, Colo.
Eutaw: Coffee Bluff, Hardin County, Tenn.
Vermejo: Rockvale, Colo.
Ripley: McNairy County, Tenn.

Halymenites minor Fischer-Ooster. Lesquereux = *Halymenites major.*

Halymenites striatus Lesquereux, U. S. Geol. and Geog. Surv. Terr., Ann. Rept. 1872 [1873], p. 373; idem, 1873 [1874], p. 394; idem, 1876 [1878], p. 496; Rept. U. S. Geol. Surv. Terr., vol. 7 (Tert. Fl.). 1878, p. 37, pl. i, f. 6.—Knowlton, U. S. Geol. Surv., Prof. Paper 101, 1918, p. 343.
Trinidad: Raton Mountains, N. Mex.

HAMAMELITES Saporta, Ann. sci. nat., Bot., 5th ser., vol. 3, 1865, p. 47.

Hamamelites? cordatus Lesquereux, Rept. U. S. Geol. Surv. Terr., vol. 8 (Cret. and Tert. Fl.), 1883, p. 71, pl. iv, f. 3; U. S. Geol. Surv., Mon. 17, 1891 [1892], p. 139.—Berry, U. S. Geol. Surv., Prof. Paper 84, 1914, p. 45, pl. x, f. 2.
Dakota: Fort Harker, Kans.
Black Creek (Middendorf): Rocky Point, S. C.

Hamamelites fothergilloides Saporta, Ann. sci. nat., Bot., 5th ser., vol. 3, 1865, p. 47.—Ward, U. S. Geol. Surv., Sixth Ann. Rept., 1884–85 [1886], p. 554, pl. xlix, f. 6; idem, Bull. 37, 1887, p. 37, pl. xxix, f. 1.
Fort Union: Sevenmile Creek near Glendive, Mont.

Hamamelites quadrangularis (Lesquereux) Lesquereux, U. S. Geol. and Geog. Surv. Terr., Ann. Rept. 1874 [1876], p. 355; Rept. U. S. Geol. Surv. Terr., vol. 8 (Cret. and Tert. Fl.), 1883, p. 70; U. S. Geol. Surv., Mon. 17, 1891 [1892], p. 139.
Populites quadrangularis Lesquereux, Am. Jour. Sci., 2d ser., vol. 46, 1868, p. 94.
Alnites quadrangularis Lesquereux, Rept. U. S. Geol. Surv. Terr., vol. 6; (Cret. Fl.), 1874, p. 62, pl. iv, f. 1.
Dakota: Kansas?.

Hamamelites quercifolius Lesquereux, Rept. U. S. Geol. Surv. Terr., vol. 8 (Cret. and Tert. Fl.), 1883, p. 71; U. S. Geol. Surv., Mon. 17, 1891 [1892], p. 139.
Dakota: Kansas.

Hamamelites kansaseana Lesquereux = ***Quercus kanseana.***

Hamamelites tenuinervis Lesquereux, Rept. Geol. Surv. Terr., vol. 8 (Cret. and Tert. Fl.), 1883, p. 70; U. S. Geol. Surv., Mon. 17, 1891 [1892], p. 139.
Dakota: Kansas.

HAUSMANNIA Dunker, Mon. Norddeutsch. Wealdenb., 1846, p. 12.

Hausmannia? californica Fontaine, in Ward, U. S. Geol. Surv., Mon. 48, 1905 [1906], p. 238, pl. lxv, f. 47.—Knowlton, Am. Jour. Sci., 4th ser., vol. 30, 1911, p. 42.
Knoxville: Cold Fork of Cottonwood Creek, Tehama County, Calif.

Hausmannia rigida Newberry = *Newberryana rigida*.

HEDERA (Tournefort) Linné, Sp. pl., 1753, p. 202.

Hedera aquamara Ward = *Quercus aquamara*.

Hedera auriculata Heer, Fl. foss. arct., vol. 2, Abt. 2, 1869, p. 36, pl. ix, f. 6.—Knowlton, U. S. Nat. Mus., Proc., vol. 17, 1894, p. 226; Geol. Soc. Am., Bull., vol. 5, 1893, p. 585.
Kenai: Port Graham, Alaska.

Hedera bruneri Ward, U. S. Geol. Surv., Sixth Ann. Rept., 1884–85 [1886], p. 553, pl. xlviii, f. 6; U. S. Geol. Surv., Bull. 37, 1887, p. 58, pl. xxvi, f. 6.
Post-Laramie: Black Buttes, Wyo.

Hedera cecilensis Berry, Bull. Torr. Bot. Club, vol. 37, 1910, p. 28, pl. viii, f. 2; Md. Geol. Surv., Upper Cret., 1916, p. 874, pl. lxxviii, f. 1, 2.
Magothy: Grove Point, Md.

Hedera cretacea Lesquereux, U. S. Geol. Surv., Mon. 17, 1891 [1892], p. 127, pl. xviii, f. 1.—Berry, Bull. Torr. Bot. Club, vol. 33, 1906, p. 180; Md. Geol. Surv., Upper Cret., 1916, p. 873.
Dakota: Near Delphos, Kans.
Magothy: Deep Cut, Del.

Hedera decurrens Lesquereux, U. S. Geol. Surv., Mon. 17, 1891 [1892], p. 130, pl. xviii, f. 6.
Dakota: Near Delphos, Kans.

Hedera lucens Knowlton, U. S. Geol. Surv., Prof. Paper 130, 19—, p. —, pl. ix, f. 1 [in preparation].
Laramie: Erie, Colo.

Hedera macclurii Heer, Fl. foss. arct., vol. 1, 1868, p. 119, pl. xvii, f. 2, 3, 4, 5a; pl. xxi, f. 17a.—Penhallow, Rept. Tert. Pl. Brit. Col., 1908, p. 59.—Knowlton, in Atwood, U. S. Geol. Surv., Bull. 467, 1911, p. 54.
Eocene: Mackenzie River, Northwest Territory.
Kenai: Chignik Bay, Alaska.

Hedera marginata Lesquereux, Rept. U. S. Geol. Surv. Terr., vol. 8 (Cret. and Tert. Fl.), 1883, p. 177, pl. xl, f. 8.—Cockerell, Univ. Colo. Studies, vol. 3, 1906, p. 170.
Miocene: Florissant, Colo.

Hedera microphylla Lesquereux, U. S. Geol. Surv., Mon. 17, 1891 [1892], p. 127, pl. xviii, f. 2, 3.
Dakota: Near Delphos, Kans.

Hedera minima Ward, U. S. Geol. Surv., Sixth Ann. Rept., 1884–85 [1886], p. 553, pl. xlviii, f. 5; U. S. Geol. Surv., Bull. 37, 1887, p. 57, pl. xxvi, f. 5.
Fort Union: Clear Creek, Glendive, Mont.

Hedera obliqua Newberry, U. S. Geol. Surv., Mon. 26, 1891 [1892], p. 113, pl. xxxvii, f. 8; pl. xxxviii, f. 5.—Berry, N. J. Geol. Surv., Bull. 3, 1911, p. 183.
Raritan: Woodbridge, N. J.

Hedera orbiculata (Heer) Lesquereux, U. S. Geol. Surv., Mon. 17, 1891 [1892], p. 129, pl. xvii, f. 12–14.
?*Chondrophyllum orbiculatum* Heer, Fl. foss. arct., vol. 3, Abt. 2, 1874, p. 115, pl. xxi, f. 3c.
Dakota: Ellsworth County, Kans.

Hedera ovalis Lesquereux, Rept. U. S. Geol. Surv. Terr., vol. 6 (Cret. Fl.), 1874, p. 91, pl. xxv, f. 3; pl. xxvi, f. 4; U. S. Geol. Surv., Mon. 17, 1892, p. 129, pl. xvii, f. 15.
Chondrophyllum nordenskiöldi? Heer. Lesquereux, U. S. Geol. Surv., Mon. 17, 1892, p. 129.—Berry, in Stephenson, N. C. Geol. and Econ. Surv., vol. 3, 1912, p. 125.
Dakota: Ellsworth County, Kans.
Black Creek: Sykes Landing, Black River, N. C.

Hedera parvula Ward, U. S. Geol. Surv., Sixth Ann. Rept., 1884–85 [1886], p. 553, pl. xlviii, f. 4; U. S. Geol. Surv., Bull. 37, 1887, p. 57, pl. xxvi, f. 4.
Fort Union: Clear Creek, Glendive, Mont.

Hedera platanoidea Lesquereux, U. S. Geol. and Geog. Surv. Terr., Ann. Rept. 1874 [1876], p. 351, pl. iii, f. 3; U. S. Geol. Surv., Mon. 17, 1891 [1892], p. 128.
Dakota: Near Delphos, Kans.

Hedera primordialis Saporta, Le monde des plantes, 1879, p. 200, f. 29.—Newberry, U. S. Geol. Surv., Mon. 26, 1895 [1896], p. 113, pl. xix, f. 1, 9; pl. xxxvii, f. 1–7.—Berry, Bull. Torr. Bot. Club, vol. 34, 1907, p. 201, pl. xvi; N. J. Geol. Surv., Bull. 3, 1911, p. 184; U. S. Geol. Surv., Prof. Paper 84, 1914, p. 57.
Raritan: Woodbridge, N. J.
Black Creek: Court House Bluff, Cape Fear River, N. C.; Cheraw and Darlington, S. C.

Hedera rotundifolia Knowlton, U. S. Geol. Surv., Prof. Paper 101, 1918, p. 274, pl. xlvii, f. 8.
Vermejo: Near Raton, N. Mex.

Hedera schimperi Lesquereux, U. S. Geol. and Geog. Surv. Terr., Bull., vol. 1, 1875 [1876], p. 395; idem, Ann. Rept. 1874 [1876], p. 351, pl. vii, f. 5; Rept. U. S. Geol. Surv. Terr., vol. 8 (Cret. and Tert. Fl.), 1883, p. 65, pl. iv, f. 7.
Dakota: Fort Harker, Kans.

Hedera simplex Hollick, U. S. Geol. Surv., Mon. 50, 1906, p. 97, pl. xxxvii, f. 9.
Magothy: Nashaquitsa, Marthas Vineyard, Mass.

Hedera sp.? Hollick, N. Y. Acad. Sci., Ann., vol. 11, 1898, p. 421, pl. xxxviii, f. 5.
Raritan: Tottenville, Staten Island, N. Y.

Hedera sp.? Hollick, N. Y. Acad. Sci., Ann., vol. 11, p. 421, pl. 38, f. 5 = *Menispermites* sp.

HEDERAEPHYLLUM Fontaine, U. S. Geol. Surv., Mon. 15, 1889 [1890], p. 323. [Type, *H. crenulatum*.]

Hederaephyllum angulatum Fontaine = *Araliaephyllum magnifolium*.

Hederaephyllum crenulatum Fontaine, U. S. Geol. Surv., Mon. 15, 1889 [1890], p. 324, pl. clxii, f. 3.
Proteaephyllum dentatum Fontaine, U. S. Geol. Surv., Mon. 15, 1889 [1890], p. 286, pl. clvi, f. 7; pl. clxxii, f. 1, 4; pl. clxxiii, f. 12, 14; in Ward, idem, Mon. 48, 1905 [1906], p. 563, pl. cxviii, f. 3, 4.
Hederaephyllum dentatum (Fontaine) Berry, Md. Geol. Surv., Lower Cret., 1912, p. 493.
Patapsco: Federal Hill (Baltimore), Md.; near Brooke, Va.

Hederaephyllum dentatum (Fontaine) Berry = *Hederaephyllum crenulatum*.

HELMINTHOIDA Schafhäutl, Geognostische Untersuch. südbayer. Alpengebirges, 1851, p. 142.

Helminthoida abnormis Ulrich, Harriman Alaska Exped., vol. 4, 1904, p. 143, pl. xvi, f. 4.—Knowlton, in Johnson, U. S. Geol. Surv., Bull. 592, 1914, p. 209.
Yakutat: Pogibshi Island, opposite Kodiak, Alaska.
Greenstone tuff: Port Wells, Alaska.

Helminthoida exacta Ulrich, Harriman Alaska Exped., vol. 4, 1904, p. 142, pl. xvi, f. 5.
Yakutat: Pogibshi Island, opposite Kodiak, Alaska.

Helminthoida subcrassa Ulrich, Harriman Alaska Exped., vol. 4, 1904, p. 142, pl. xvi, f. 3.
Yakutat: Pogibshi Island, opposite Kodiak, Alaska.

Helminthoida vaga Ulrich, Harriman Alaska Exped., vol. 4, 1904, p. 143, pl. xvii.—Knowlton, in Johnson, U. S. Geol. Surv., Bull. 592, 1914, p. 209.
Yakutat: Pogibshi Island, opposite Kodiak, Alaska.
Greenstone tuff: Port Wells, Alaska.

HELMINTHOPSIS Heer, Fl. foss. Helv., 1877, p. 116.

Helminthopsis? labyrinthica Heer, Fl. foss. Helv., 1877, pl. xlvii, f. 3–5.—Ulrich, Harriman Alaska Exped., vol. 4, 1904, p. 144, pl. xx, f. 2, 3.
Yakutat: Pogibshi Island, opposite Kodiak, Alaska.

Helminthopsis magna Heer, Fl. foss. Helv., 1877, p. 116, pl. xlvii, f. 1, 2.—Ulrich, Harriman Alaska Exped., vol. 4, 1904, p. 144, pl. xxi, f. 1, 2.
Yakutat: Pogibshi Island, opposite Kodiak, Alaska.

HEMITELITES Goeppert, Syst. fil. foss., 1836, p. 329.

Hemitelites torelli Heer, Fl. foss. arct., vol. 2, Abt. 4, 1869, p. 462.—Lesquereux, U. S. Geol. and Geog. Surv. Terr., Ann. Rept. 1871, Suppl., 1871 [1872], p. 5. [Not afterward referred to by Lesquereux.]
Eocene?: Green River, Wyo.

Hemitelites torelli Heer = *Osmunda doroschkiana.*

HETERANTHERA Ruiz and Pavon, Prodr. Fl. Per., 1794, p. 9.

Heteranthera cretacea Knowlton, U. S. Geol. Surv., Prof. Paper 98, 1916, p. 333, pl. lxxxv, f. 5.
Fruitland: Coal Creek, 35 miles south of Farmington, San Juan County, N. Mex.

HETEROCALYX Saporta, Ann. sci. nat., 5th ser., Bot., vol. 18 (Études, Suppl., pt. 3), 1873, p. 110 (208).

Heterocalyx saportana Berry, U. S. Geol. Surv., Prof. Paper 91, 1916, p. 260, pl. lix, f. 1.
Wilcox (Holly Springs): Early Grove, Marshall County, Miss.

HETEROFILICITES Berry, Geol. Surv. N. J., Ann. Rept. 1905 [1906], p. 154. [Type, *H. anceps.*]

Heterofilicites anceps Berry, Geol. Surv. N. J., Ann. Rept. 1905 [1906], pp. 138, 139, 154, pl. xxvi, f. 1–16.
Magothy: Morgan and Kinkora, N. J.

HETEROLEPIS Berry, U. S. Geol. Surv., Prof. Paper 84, 1914, p. 27.

Heterolepis cretaceus Berry, U. S. Geol. Surv., Prof. Paper 84, 1914, p. 27, pl. iii, f. 3.
Black Creek (Middendorf): Rocky Point, S. C.

Heyderia coloradensis Cockerell = *Libocedrus coloradensis.*

HICORIA Rafinesque, Med. Rep., 2d ser., vol. 5, 1808, p. 352.

Hicoria alba (Linné) Britton.

Carya alba Nuttall. Penhallow, Roy. Soc. Canada, Trans., 2d ser., vol. 2, sec. 4, 1896, p. 70; Brit. Assoc. Adv. Sci., Bradford meeting, 1900, p. 335 [8]; Roy. Soc. Canada, Trans., 2d ser., vol. 10, sec. 4, 1904, p. 73; Am. Nat., vol. 41, 1907, p. 446.—Meehan, in Mercer, Acad. Nat. Sci. Philadelphia, Jour., 2d ser., vol. 11, 1899, pp. 278, 281, text f. 8 (8).

Pleistocene: Don Valley, Toronto, Canada; Port Kennedy, Pa.

Hicoria angulata Knowlton, U. S. Geol. Surv., Prof. Paper 130, 19—, p. —, pl. v, f. 4 [in preparation].

Laramie: Marshall and Mount Carbon, near Morrison, Colo.

Hicoria antiquora (Newberry) Knowlton, U. S. Geol. Surv., Bull. 152, 1898, p. 117.—Berry, U. S. Geol. Surv., Prof. Paper 91, 1916, p. 187.

Carya antiquorum Newberry, Lyc. Nat. Hist., Annals, vol. 9, 1868, p. 72.—[Lesquereux], Ill. Cret. and Tert. Pl., 1878, pl. xxiii, f. 1–4.—Newberry, U. S. Geol. Surv., Mon. 35, 1898, p. 35, pl. xxxi, f. 1–4.—Ward, U. S. Geol. Surv., Sixth Ann. Rept., 1884–85 [1886], p. 552, pl. lx, f. 7; idem, Bull. 37, 1887, p. 34, pl. xv, f. 2.—Knowlton, U. S. Geol. Surv., Mon. 32, pt. 2, 1899, p. 690.—Penhallow, Roy. Soc. Canada, Trans., 2d ser., vol. 8, 1902, p. 60; Rept. Tert. Pl. Brit. Col., 1908, p. 43.—Knowlton, Washington Acad. Sci., Proc., vol. 11, 1909, pp. 189, 190, 202, 211, 214.

Fort Union: *Mouth of Yellowstone, Mont.; near Jordan, and many localities in Montana; the Dakotas and Wyoming.

Paskapoo (Fort Union): Red Deer River, Porcupine Creek, Alberta; Horsefly River and Quilchena, British Columbia.

Lance: Near Miles City and Signal Butte, Mont.; York, N. Dak.; Ranchester and Big Horn Basin, Wyo.

Wilcox: Campbell's quarry, Cross Bayou, Caddo Parish, La.

Hicoria aquatica (Michaux f.) Britton. Berry, Torreya, vol. 9, 1909, p. 71; idem, vol. 14, 1914, p. 161.

Salix sp. Berry, Jour. Geol., vol. 15, 1907, p. 340.

Pleistocene: Neuse River, N. C.; Monroeville, Ala.

Hicoria biacuminata Perkins, Rept. Vt. State Geol., 1903–4 [1904], p. 193, pl. lxxviii, f. 77, 81; Geol. Soc. Am., Bull., vol. 16, 1905, p. 512, pl. 87, f. 22; Rept. Vt. State Geol., 1905–6 [1906], pl. liii, f. 22.

Miocene: Brandon, Vt.

Hicoria bruckmanni (Heer) Knowlton = *Hicoria princetonia*.

Hicoria crescentia Knowlton, U. S. Geol. Surv., Mon. 32, pt. 2, 1899, p. 690, pl. lxxxiv, f. 7.

Fort Union: Yellowstone National Park.

Hicoria culveri Knowlton, U. S. Geol. Surv., Mon. 32, pt. 2, 1899, p. 691, pl. lxxxiii, f. 7.

Fort Union: Yellowstone National Park.

Hicoria elaenoides (Unger) Knowlton, U. S. Geol. Surv., Bull. 152, 1898, p. 117; idem, Bull. 204, 1902, p. 38.

Carya elaenoides (Unger) Heer. Lesquereux, U. S. Nat. Mus., Proc., vol. 11, 1888, p. 18.

Mascall: Van Horn's ranch, John Day Basin, Oreg.

Hicoria glabra (Miller) Britton. Berry, Torreya, vol. 5, 1905, p. 89; Jour. Geol., vol. 15, 1907, p. 340; Torreya, vol. 9, 1909, p. 97, f. 1–5; idem, vol. 10, 1910, p. 264, f. 1; U. S. Nat. Mus., Proc., vol. 48, 1915, p. 297.

Carya porcina Nuttall. Meehan, in Mercer, Acad. Nat. Sci., Philadelphia Jour., vol. 11, 1699, pp. 277, 281, f. 8.

Pleistocene: Near Long Branch, N. J.; 1 mile south of Chesapeake Beach, Md.; Tappahannock, Rappahannock River, Va.; Neuse River, N. C.; Hickman and Columbus, Ky.; Port Kennedy, Pa.

Hicoria juglandiformis (Sternberg) Knowlton, U. S. Geol. Surv., Bull. 152, 1898, p. 117; Cockerell, Am. Mus. Nat. Hist., Bull., vol. 24, 1908, p. 80.

Phyllites juglandiformis Sternberg, Vers., vol. 4, 1825, index 40, pl. xxxv, f. 1.
Juglans bilinica Unger, Gen. et sp. pl. foss., 1850, p. 469.
Carya bilinica (Unger) Ettingshausen. Lesquereux, Rept. U. S. Geol. Surv. Terr., vol. 8 (Cret. and Tert. Fl.), 1883, p. 191, pl. xxxix, f. 1, 2, 13.

Miocene: Florissant, Colo.

Hicoria magnifica Knowlton, Harriman Alaska Exped., vol. 4, 1904, p. 151, pl. xxvi, f. 1; pl. xxvii, pl. xxix, f. 1.

Kenai: Kukak Bay, Alaska.

Hicoria minutula Knowlton, U. S. Geol. Surv., Prof Paper 130, 19—, p. —, pl. v, f. 5 [in preparation].

Laramie: Marshall, Colo.

Hicoria? oregoniana Knowlton, U. S. Geol. Surv., Bull. 204, 1902, p. 37, pl. v, f. 3, 4.

Hicoria n. sp. Knowlton, in Merriam, Univ. Calif. Bull. Dept. Geol., vol. 2, 1901, p. 289.

Clarno (lower part): Cherry Creek, Crook County, Oreg.

Hicoria ovata (Miller) Britton. Berry, Jour. Geol., vol. 15, 1907, p. 340.

Pleistocene: Neuse River, N. C.

Hicoria pecan (March) Britton. Hollick, Bull. Torr. Bot. Club, vol. 19, 1892, p. 332.—Berry, U. S. Nat. Mus., Proc., vol. 48, 1915, p. 297.

Carya oliraeformis Nuttall. Lesquereux, Am. Jour. Sci., 2d ser., vol. 27, 1859, p. 365.

Pleistocene: Near Columbus, Ky.; Bridgeton, N. J.

Hicoria pretexana Berry, U. S. Geol. Surv., Prof. Paper 98, 1916, p. 198, pl. xlv, f. 10–13.

Citronelle: Lambert and Red Bluff, Ala.

Hicoria princetonia Cockerell, Am. Mus. Nat. Hist., Bull., vol. 24, 1908, p. 80.

Carya bruckmanni Heer. Lesquereux, Rept. U. S. Geol. Surv. Terr., vol. 8 (Cret. and Tert. Fl.), 1883, p. 191, pl. xxxix, f. 6.

Miocene: Florissant, Colo.

Hicoria pseudo-glabra Hollick, Md. Geol. Surv., Pliocene and Pleistocene, 1906, p. 221, pl. lxxii, f. 1, 16, 17.

Pleistocene (Sunderland): Head of Island Creek, Calvert County, Md.

Hicoria rostrata (Göppert) Knowlton, U. S. Geol. Surv., Bull. 152, 1898, p. 118.

Juglans rostrata Göppert, Palaeontogr., vol. 8, 1861, p. 136, pl. lv.

Carya rostrata (Göppert) Schimper, Pal. vég., vol. 3, 1874, p. 257.—Lesquereux, Rept. U. S. Geol. Surv. Terr., vol. 8 (Cret. and Tert. Fl.), 1883, p. 191; pl. xxxix, f. 4.

Miocene: Florissant, Colo.

Hicoria vermontana (Lesquereux) Knowlton = *Bicarpellites vermontanus*.

Hicoria verrucosa (Lesquereux) Knowlton, U. S. Geol. Surv., Bull. 152, 1894, p. 118.

Carya verrucosa Lesquereux, Am. Jour. Sci., 2d ser., vol. 32, 1861, p. 357; in Hitchcock, Geol. Vt., vol. 2, 1861, p. 714, f. 129.

Miocene: Brandon, Vt.

Hicoria villosa (Sargent) Ashe. Berry, Am. Jour. Sci., 4th ser., vol. 29, 1910, p. 392.

Pleistocene: Black Warrior River, 311½ miles above Mobile, Ala.

Hicoria sp. cf. *H. microcarpa* (Nuttall) Britton. Berry, Jour. Geol., vol. 15, 1907, p. 340.

Pleistocene: Neuse River, N. C.

Hicoria sp. ? Hollick, Md. Geol. Surv., Pliocene and Pleistocene, 1906, p. 222.

Pleistocene (Talbot): Drum Point, Calvert County, Md.

Hicoria sp. ? Hollick, Md. Geol. Surv., Pliocene and Pleistocene, 1906, p. 222, pl. lxx, f. 4.

Pleistocene (Sunderland): Point of Rocks, Calvert County, Md.

Hicoria sp. (Dawson) Knowlton, U. S. Geol. Surv., Bull. 152, 1898, p. 118.

Carya sp. Dawson, Roy. Soc. Canada, Trans., vol. 5, sec. 4, 1887, p. 33.

Pierre ?: Head of Swift Current Creek, British Columbia.

Hicoria sp. Knowlton, Washington Acad. Sci., Proc., vol. 11, 1909, p. 207.

Lance: Converse County, Wyo.

Hicoria sp. (Dawson) Knowlton (?), n. comb.

Carya sp. Dawson, Geol. Surv. Canada, 1875–76.—Penhallow, Rept. Tert. Pl. Brit. Col., 1908, p. 43.

Eocene: Quesnel River, British Columbia.

Hicoria? sp. Knowlton, in Merriam, Univ. Calif., Bull. Dept. Geol., vol. 2, 1901, p. 289 = *Hicoria ? oregoniana*.

Hicoria sp.? Knowlton, U. S. Geol. Surv., Bull. 204, 1902, p. 38, pl. v, f. 2.

Clarno (upper part): Bridge Creek, Oreg.

Hicoria? sp. Knowlton, U. S. Geol. Surv., Bull. 316, 1907, p. 202.

Fort Union: Signal Butte, Miles City, Mont.

HICOROIDES Perkins, Rept. Vt. State Geol., 1903–4 [1904], p. 183. [Type, *H. angulata*.]

Hicoroides angulata Perkins, Rept. Vt. State Geol., 1903–4 [1904], p. 183, pl. lxxvi, f. 28, 32, 33; Geol. Soc. Am., Bull., vol. 16, 1905, p. 513, pl. lxxxvii, f. 27; Rept. Vt. State Geol., 1905–6 [1906], pl. liii, f. 27.

Miocene: Brandon, Vt.

Hicoroides ellipsoidea Perkins, Rept. Vt. State Geol., 1903–4 [1904], p. 184, pl. lxxvi, f. 41; Geol. Soc. Am., Bull., vol. 16, 1905, p. 513, pl. lxxxvii, f. 26; Rept. Vt. State Geol., 1905–6 [1906], pl. liii, f. 26.

Miocene: Brandon, Vt.

Hicoroides globulus Perkins, Rept. Vt. State Geol., 1903–4 [1904], p. 184, pl. lxxvi, f. 42, 43.

Miocene: Brandon, Vt.

Hicoroides levis Perkins, Rept. Vt. State Geol., 1905–6 [1906], p. 217, pl. lvi, f. 15.

Miocene: Brandon, Vt.

Hicoroides parva Perkins, Rept. Vt. State Geol., 1903–4 [1904], p. 184, pl. lxxxi, f. 172.

Miocene: Brandon, Vt.

Hicoroides triangularis Perkins, Rept. Vt. State Geol., 1903–4 [1904], p. 183, pl. lxxvi, f. 40.

Miocene: Brandon, Vt.

HIPPURIS Linné, Sp. pl., 1753, p. 4.

Hippuris vulgaris Linné. Penhallow, Roy. Soc. Canada, Trans., 2d ser., vol. 10, sec. 4, 1904, p. 73.

Pleistocene: Don River valley, Toronto, Canada.

HIRAEA Jacquin, Amer., 1763, p. 137.

Hiraea wilcoxiana Berry, U. S. Geol. Surv., Prof. Paper 91, 1916, p. 257, pl. lvii, f. 8; cix, f. 6.

Lagrange: Puryear, Henry County, Tenn.

HYDRANGEA Linné, Sp. pl., 1753, p. 397.

Hydrangea bendirei (Ward) Knowlton, in Merriam, Univ. Calif., Bull. Dept. Geol., vol. 2, 1901, p. 309.—Knowlton, U. S. Geol. Surv., Bull. 204, 1902, p. 60, pl. ix, f. 6, 7.

Marsilea bendirei Ward, U. S. Geol. Surv., Fifth Ann. Rept., 1885, p. 446.

Porana bendirei (Ward) Lesquereux, U. S. Nat. Mus., Proc., vol. 11, 1888, p. 16, pl. viii, f. 4.

Mascall: Van Horn's ranch, about 12 miles west of Mount Vernon, Grant County, Oreg.

Hydrangea florissantia Cockerell, Am. Jour. Sci., 4th ser., vol. 26, 1908, p. 67, f. 2; idem, p. 541.—Knowlton, U. S. Nat. Mus., Proc., vol. 51, 1916, p. 269.

Rhus rotundifolia Kirchner, St. Louis Acad. Sci., Trans., vol. 8, 1898, p. 184, pl. xii, f. 2.

Miocene: *Florissant, Colo.

Hydrangea? subincerta Cockerell, Am. Mus. Nat. Hist., Bull., vol. 24, 1908, p. 92, pl. ix, f. 32.—Knowlton, U. S. Nat. Mus., Proc., vol. 51, 1916, p. 269.

Miocene: *Florissant, Colo.

HYMENAEA Linné, Sp. pl., 1753, p. 1192.

Hymenaea dakotana Lesquereux, U. S. Geol. Surv., Mon. 17, 1891 [1892], p. 145, pl. lv, f. 3, 4; pl. lvi, f. 1, 2; pl. lxii, f. 2.—Berry, Geol. Surv. N. J., Ann. Rept. 1905 [1906], p. 138; Bull. Torr. Bot. Club, vol. 33, 1906, p. 176.—Hollick, U. S. Geol. Surv., Mon. 50, 1906, p. 83, pl. xxxii, f. 5–7.—Berry, N. J. Geol. Surv., Bull. 3, 1911, p. 165.

Dalbergia rinkiana Heer. Hollick, N. Y. Acad. Sci., Trans., vol. 12, 1893, p. 236.

Dakota: Kansas.

Raritan: Lloyd Neck, Long Island, N. Y.; New Jersey.

Magothy: Morgan and Cliffwood, N. J.; Sea Cliff, Long Island, N. Y.; Gay Head, Mass.

Hymenaea fayettensis Berry, Am. Jour. Sci., 4th ser., vol. 47, 1919, p. 66, text f.; U. S. Geol. Surv., Prof. Paper 112, 1919, p. 97, pl. xxiii, f. 2.

Tuscaloosa: Shirleys Mill, Fayette County, Ala.

Hymenaea primigenia Saporta, Le monde des plantes, 1879, p. 199, f. 2.—Hollick, U. S. Geol. Surv., Mon. 50, 1906, p. 84, pl. xxxii, f. 8, 9.

Magothy: Gay Head, Marthas Vineyard, Mass.

Hymenaea virginiensis Fontaine, U. S. Geol. Surv., Mon. 15, 1889 [1890], p. 320, pl. clxiii, f. 6.

——: Near Brooke, Va.

HYMENOPHYLLUM Smith, Mém. Acad. Turin, vol. 5, 1793, p. 418.

Hymenophyllum confusum Lesquereux, U. S. Geol. and Geog. Surv. Terr., Ann. Rept. 1873 [1874], p. 395; Rept. U. S. Geol. Surv. Terr., vol. 7 (Tert. Fl.), 1878, p. 51, pl. ii, f. 6–6a.
Denver: Golden, Colo.

Hymenophyllum cretaceum Lesquereux, U. S. Geol. and Geog. Surv. Terr., Ann. Rept. 1872 [1873], p. 421; Rept. U. S. Geol. Surv. Terr., vol. 6 (Cret. Fl.), 1874, p. 45, pl. xxix, f. 6 [not pl. i, f. 3, 4].—Knowlton, in Woodruff, U. S. Geol. Surv., Bull. 452, 1911, p. 20.
Dakota: Fort Harker, Kans.; 2 miles west of Lander, Wyo.

Hymenophyllum cretaceum Lesquereux, Rept. U. S. Geol. Surv. Terr., vol. 6 (Cret. Fl.), 1874, pl. xxix, f. 6 [not pl. i, f. 3, 4] = *Sphenopteris corrugata.*

Hymenopteris psilotoides Stokes and Webb = *Onychiopsis psilotoides.*

HYPNUM Linné, Sp. pl., 1753, p. 1122.

Hypnum brownii Kirchner, St. Louis Acad. Sci., Trans., vol. 8, 1898, p. 178, pl. xii, f. 4, 4a.—Cockerell, Am. Mus. Nat. Hist., Bull., vol. 24, 1908, p. 76.—Britton and Hollick, Bull. Torr. Bot. Club, vol. 34, 1907, p. 139, pl. ix, f. 3, 3a.—Knowlton, U. S. Nat. Mus., Proc., vol. 51, 1916, p. 245.
Miocene: *Florissant, Colo.

Hypnum columbianum Penhallow, in Dawson, Roy. Soc. Canada, Trans., vol. 8, sec. 4, 1890, p. 77, f. 3.—Penhallow, Rept. Tert. Pl. Brit. Col., 1908, p. 58.—Britton and Hollick, Bull. Torr. Bot. Club, vol. 34, 1907, p. 140, pl. ix, f. 4.
Miocene?: Similkameen River, British Columbia.

Hypnum commutatum Hedwig. Hinde, Canadian Jour. [Canadian Inst.], vol. 15, 1878, p. 399.
Pleistocene: Scarborough Heights, Ontario.

Hypnum fluitans Linné. Penhallow, Roy. Soc. Canada, Trans., 2d ser., vol. 2, sec. 4, 1896, p. 71; Brit. Assoc. Adv. Sci., Bradford meeting, 1900, p. 335.
Pleistocene: Besserer's wharf, Ottawa River, Canada.

Hypnum fluitans brachydictyon Holzinger, Bryologist, vol. 6, 1903, p. 94.
Pleistocene (under Kansan drift): Oelwein, Iowa.

Hypnum fluitans glaciale Holzinger, Bryologist, vol. 6, 1903, p. 94.
Pleistocene (under Kansan drift): Iowa City, Iowa.

Hypnum haydenii Lesquereux = *Juniperus? haydenii.*

Hypnum recurvans Schwegler. Penhallow, Roy. Soc. Canada, Trans., 2d ser., vol. 2, 1896, sec. 4, pp. 62, 71.
Pleistocene: Long Portage, Missinaibi River, Ontario.

Hypnum revalvens Swartz. Hinde, Canadian Jour. [Canadian Inst.], vol. 15, 1878, p. 399.—Penhallow, Roy. Soc. Canada, Trans., 2d ser., vol. 2, 1896, sec. 4, p. 71; Brit. Assoc. Adv. Sci., Bradford meeting, 1900, p. 336.—Holzinger, Bryologist, vol. 6, 1903, p. 94.
Pleistocene: Scarborough Heights, Ontario; Oelwein, Iowa (under Kansan drift).

Hypnum richardsoni Holzinger, Bryologist, vol. 6, 1903, p. 94.
Pleistocene (under Kansan drift): Oelwein, Iowa.

Hypnum sp. Penhallow, Brit. Assoc. Adv. Sci., Bradford meeting, 1900, p. 336.
Pleistocene: Don Valley, Ontario, Canada.

HYPOLEPIS Bernhardi, in Schraeder, Jour., vol. 1, pt. 2, 1806, p. 34.

Hypolepis coloradensis Cockerell, Torreya, vol. 9, 1909, p. 184, text f.
Miocene: Florissant, Colo.

ICACOREA Aublet, Guian., vol. 2, Suppl., 1775, p. 1.

Icacorea prepaniculata Berry, U. S. Geol. Surv., Prof. Paper 91, 1916, p. 332, pl. cvii, f. 5.
Lagrange: Puryear, Tenn.

ILEX (Tournefort), Linné, Sp. pl., 1753, p. 125.

Ilex affinis Lesquereux, U. S. Geol. and Geog. Surv. Terr., Ann. Rept. 1871 [1872], Suppl., p. 8; Rept. U. S. Geol. Surv. Terr., vol. 7 (Tert. Fl.), 1878, p. 270, pl. l, f. 2, 3.—?Hollick, Geol. Surv. La., Special Rept. 5, 1899, p. 285, pl. xliv, f. 2.—?Berry, U. S. Geol. Surv., Prof. Paper 91, 1916, p. 264.
Green River: *Green River station, Wyo.
Wilcox: Coushatta, La.

Ilex amboyensis Berry, Bull. Torr. Bot. Club, vol. 36, 1909, p. 259; N. J. Geol. Surv., Bull. 3, 1911, p. 171.
Ilex? ovata Newberry, U. S. Geol. Surv., Mon. 26, 1896, p. 98, pl. xviii, f. 2. [Homonym, Göppert, 1852.]
Raritan: Sayreville, N. J.

Ilex armata Lesquereux, U. S. Geol. Surv., Mon. 17, 1891 [1892], p. 176, pl. xxix, f. 8.
Dakota: Ellsworth County, Kans.

Ilex borealis Heer, Fl. foss. arct., vol. 7, 1883, p. 39, pl. lxiv, f. 3, 4.—Lesquereux, U. S. Geol. Surv., Mon. 17, 1891 [1892], p. 176, pl. xxxv, f. 8.
Dakota: Pipe Creek, Cloud County, Kans.

Ilex calvertensis Berry, U. S. Geol. Surv., Prof. Paper 98, 1916, p. 69, pl. xii, f. 8.
Calvert: Good Hope Hill, D. C.

Ilex cassine Linné. Hollick, Bull. Torr. Bot. Club, vol. 19, 1892, p. 331.—Berry, Am. Nat., vol. xliii, 1909, p. 436.
Pleistocene: Bridgeton, N. J.; Rappahannock River 1½ miles below Port Royal, Va.

Ilex dakotensis Lesquereux, U. S. Geol. Surv., Mon. 17, 1891 [1892], p. 178, pl. xxix, f. 11.
Dakota: Ellsworth County, Kans.

Ilex dissimilis Lesquereux, Rept. U. S. Geol. Surv. Terr., vol. 7 (Tert. Fl.), 1878, p. 271, pl. l, f. 7–9.
Quercus ilicoides? Heer. Lesquereux, U. S. Geol. and Geog. Surv. Terr., Ann. Rept. 1871 [1872], p. 291.
Green River (?): Sage Creek, Mont.

Ilex? elongata Newberry, U. S. Geol. Surv., Mon. 26, 1895 [1896], p. 98, pl. xviii, f. 5.—Berry, N. J. Geol. Surv., Bull. 3, 1911, p. 171.
Raritan: Sayreville, N. J.

Ilex eolignitica Berry, U. S. Geol. Surv., Prof. Paper 91, 1916, p. 263, pl. lix, f. 7.
Wilcox (Holly Springs): Holly Springs, Marshall County, Miss.

Ilex florissantensis Knowlton and Cockerell, n. name.
Ilex microphylla Lesquereux, Rept. U. S. Geol. Surv. Terr., vol. 8 (Cret. and Tert. Fl.), 1883, p. 186.—Cockerell, Am. Mus. Nat. Hist., Proc., vol. 24, 1908, p. 100.—Knowlton, U. S. Nat. Mus., Proc., vol. 51, 1916, p. 280, pl. xxi, f. 6. [Homonym, Hooker, 1862.]
Miocene: Florissant, Colo.

Ilex glabra (Linné) A. Gray. Berry, Jour. Geol., vol. 25, 1917, p. 662.
Pleistocene: Vero, Fla.

Ilex grandifolia Lesquereux, Rept. U. S. Geol. Surv. Terr., vol. 8 (Cret. and Tert. Fl.), 1883, p. 187, pl. xxxviii, f. 1.—Cockerell, Am. Mus. Nat. Hist., Bull., vol. 24, 1908, p. 100.
Miocene: Florissant, Colo.

Ilex insignis Heer, Fl. foss. arct., vol. 2, Abt. 2, 1869, p. 37, pl. x, f. 1.—Knowlton, U. S. Nat. Mus., Proc., vol. 17, 1894, p. 229; Geol. Soc. Am., Bull., vol. 5, 1893, p. 586.
Kenai: Port Graham, Alaska.

Ilex integrifolia (Elliott) Chapman.
Prinos integrifolia Elliott. Lesquereux, Am. Jour. Sci., 2d ser., vol. 27, 1859, p. 365.
Pleistocene: Near Columbus, Ky.

Ilex knightiaefolia Lesquereux, Rept. U. S. Geol. Surv. Terr., vol. 8 (Cret. and Tert. Fl.), 1883, p. 188, pl. xl, f. 4, 5.—Knowlton, U. S. Nat. Mus., Proc., vol. 51, 1916, p. 280, pl. xxvi, f. 3.
Ilex rigida Kirchner, St. Louis Acad. Sci., Trans., vol. 8, 1898, p. 182, pl. xiv, f. 2.
Miocene: Florissant, Colo.

Ilex laramiensis Knowlton, U. S. Geol. Surv., Prof. Paper 130, 1922, p. 152, pl. xxiv, f. 4–7 [in preparation].
Laramie: Erie, Colo.

Ilex leonis Cockerell, Bull. Torr. Bot. Club, vol. 33, 1906, p. 311; Univ. Colo. Studies, vol. 3, 1906, p. 170; Am. Mus. Nat. Hist., Bull., vol. 24, 1908, p. 99.

Ilex quercifolia Lesquereux, Rept. U. S. Geol. Surv. Terr., vol. 8 (Cret. and Tert. Fl.), 1883, p. 186, pl. xxxviii, f. 2–5. [Homonym, Meerburgh, 1798.]

Miocene: Florissant, Colo.

Ilex? longifolia Heer. Lesquereux, U. S. Nat. Mus., Proc., vol. 11, 1888, p. 21 = *Juglans? bendirei.*

Ilex maculata Lesquereux, Rept. U. S. Geol. Surv. Terr., vol. 8 (Cret. and Tert. Fl.), 1883, p. 186, pl. xliv, f. 5.

Green River: Alkali station, Wyo.

Ilex masoni Lesquereux, U. S. Geol. Surv., Mon. 17, 1891 [1892], p. 179, pl. vii, f. 6; pl. lxiii, f. 6.—Berry, U. S. Geol. Surv., Prof. Paper 112, 1919, p. 106.

Dakota: Pipe Creek, Cloud County, Kans.

Tuscaloosa: Glen Allen and Cottondale, Ala.

Ilex microphylla Lesquereux = *Ilex florissantensis.*

Ilex microphylla Newberry = *Ilex microphyllina.*

Ilex microphyllina Knowlton and Cockerell, n. name.

Ilex microphylla Newberry, U. S. Nat. Mus., Proc., vol. 5, 1882 [1883], p. 510. [Homonym, Hooker, 1862.]

Fort Union: Fort Union, N. Dak.

Ilex opaca Aiton. Hollick, Bull. Torr. Bot. Club, vol. 19, 1892, p. 331.—Berry, Jour. Geol., vol. 15, 1907, p. 345; Am. Nat., vol. 41, 1907, p. 696, pl. ii, f. 1.

Pleistocene: Neuse River, N. C.; Chattahoochee River, Ala. Bridgeton, N. J.

Ilex? ovata Newberry = *Ilex amboyensis.*

Ilex papillosa Lesquereux, U. S. Geol. Surv., Mon. 17, 1891 [1892], p. 177, pl. xxix, f. 9, 10; pl. lviii, f. 3.—Hollick, U. S. Geol. Surv., Mon. 50, 1906, p. 87, pl. xxxiii, f. 4.

Dakota: Ellsworth County, Kans.

Magothy: Gay Head, Marthas Vineyard, Mass.

Ilex prunifolia Lesquereux, Mus. Comp. Zool., Mem., vol. 6, 1878, p. 27, pl. ix, f. 7.

Miocene (auriferous gravels): Table Mountain, Tuolumne County, Calif.

Ilex pseudo-stenophylla Lesquereux, Rept. U. S. Geol. Surv. Terr., vol. 8 (Cret. and Tert. Fl.), 1883, p. 185.—Cockerell, Am. Mus. Nat. Hist., Proc., vol. 24, 1908, p. 100.—Knowlton, U. S. Nat. Mus., Proc., vol. 51, 1916, p. 280, pl. xxv, f. 3, 4.

Ilex stenophylla Unger. Lesquereux, U. S. Geol. and Geog. Surv. Terr. Ann. Rept. 1871 [1872], Suppl., p. 8.

Miocene: Florissant, Colo.

Ilex quercifolia Lesquereux = *Ilex leonis.*

Ilex rigida Kirchner = *Ilex knightiaefolia*.

Ilex scudderi Lesquereux, U. S. Geol. Surv., Mon. 17, 1891 [1892], p. 178, pl. lviii, f. 2.
Dakota: Ellsworth County, Kans.

Ilex severnensis Berry, Bull. Torr. Bot. Club, vol. 38, 1911, p. 407, pl. xix, f. 1, 1a; Md. Geol. Surv., Upper Cret., 1916, p. 849, pl. lxxvii, f. 1, 2.
Magothy: Little Round Bay, Severn River, Md.

Ilex sphenophylla? Heer. Lesquereux, U. S. Geol. and Geog. Surv. Terr., Ann. Rept. 1873 [1874], p. 415. [Not afterward mentioned.]
Miocene: Florissant, Colo.

Ilex stenophylla Unger. Lesquereux = *Ilex pseudo-stenophylla*.

Ilex strangulata Lesquereux, U. S. Geol. and Geog. Surv. Terr., Bull., vol. 1, 1875 [1876], p. 399; idem, Ann. Rept. 1874 [1876], p. 359, pl. viii, f. 3; idem, vol. 8 (Cret. and Tert. Fl.), 1883, p. 84, pl. iii, f. 7.—Berry, Bull. Torr. Bot. Club, vol. 33, 1906, p. 177; Geol. Surv. N. J., Ann. Rept. 1905 [1906], p. 139.
Dakota?: Southwestern Colorado.
Magothy: Cliffwood, N. J.

Ilex subdenticulata Lesquereux, U. S. Geol. and Geog. Surv. Terr., Ann. Rept. 1873 [1874], p. 416; Rept. U. S. Geol. Surv. Terr., vol. 7 (Tert. Fl.), 1878, p. 271, pl. l, f. 5, 6–6b.—Cockerell, Am. Mus. Nat. Hist., Bull., vol. 24, 1908, p. 100.
Miocene: Florissant, Colo.

Ilex undulata Lesquereux = *Morella bolanderi*.

Ilex verticillata (Linné) A. Gray. Berry, Torreya, vol. 17, 1917, p. 161, f. 3.
Pleistocene: Waterville, Maine.

Ilex vomitoriafolia Berry, U. S. Geol. Surv., Prof. Paper 91, 1916, p. 264, pl. lxiv, f. 6.
Wilcox (Holly Springs): Early Grove, Marshall County, Miss.

Ilex vyomingiana Lesquereux, Rept. U. S. Geol. Surv. Terr., vol. 7 (Tert. Fl.), 1878, p. 270, pl. l, f. 1.
Green River: Green River, Wyo.

Ilex? sp. Berry, U. S. Nat. Mus., Proc., vol. 48, 1915, p. 300, pl. xii, f. 2.
Pleistocene: Hickman, Ky.

Ilex sp. Hollick, Geol. Surv. La., Special Rept. 5, 1899, p. 285, pl. xliii, f. 2, 3.—Berry, U. S. Geol. Surv., Prof. Paper 91, 1916, p. 264.
Wilcox: Shreveport, La.

ILLICIUM Linné, Syst., 10th ed., 1759, p. 1050.

Illicium deletoides Berry, Md. Geol. Surv., Upper Cret., 1916, p. 838, pl. lxx, f. 6.
Magothy: Grove Point, Md.

Illicium lignitum Lesquereux, Am. Jour. Sci., 2d ser., vol. 33, 1861, p. 360.—Perkins, Rept. Vt. State Geol., 1903–4 [1904], p. 210, pl. lxxx, f. 146, 147.
Miocene: Brandon, Vt.

Illicium watereensis Berry, U. S. Geol. Surv., Prof. Paper 84, 1914, p. 44, pl. xiv, f. 8.
Black Creek (Middendorf): Rocky Point, S. C.

INGA Scopoli, Introd., 1777, p. 298.

Inga cretacea Lesquereux, U. S. Geol. Surv., Mon. 17, 1891 [1892], p. 153, pl. lv, f. 11.—Berry, U. S. Geol. Surv., Prof. Paper 112, 1919, p. 96, pl. xxii, f. 4, 5.
Dakota: Near Fort Harker, Kans.
Tuscaloosa: Shirleys Mill, Ala.

Inga heterophylla Knowlton, U. S. Geol. Surv., Prof. Paper 101, 1918, p. 327, pl. liv, f. 5.
Raton: Wootton, Colo.

Inga laurinafolia Berry, U. S. Geol. Surv., Prof. Paper 91, 1916, p. 224, pl. xlviii, f. 8.
Wilcox: Coushatta, Red River Parish, La.

Inga mississippiensis Berry, U. S. Geol. Surv., Prof. Paper 91, 1916, p. 222, pl. xlv, f. 1.
Prunus caroliniana Aiton. Knowlton, U. S. Geol. Surv., Bull. 152, 1898, p. 186.
Prunus caroliniana Michaux. Lesquereux, Am. Jour. Sci., 2d ser., vol. 27, 1859, p. 363; Geol. Tenn., 1869, p. 427, pl. K, f. 6.
Wilcox (Holly Spring): Early Grove, Marshall County, Miss.
Lagrange: Somerville, Fayette County, Tenn.

Inga puryearensis Berry, U. S. Geol. Surv., Prof. Paper 91, 1916, p. 223, pl. li, f. 12.
Lagrange: Puryear, Henry County, Tenn.

Inga wickliffensis Berry, U. S. Geol. Surv., Prof. Paper 91, 1916, p. 224, pl. l, f. 8.
Lagrange: Wickliffe, Ballard County, Ky.

INOLEPIS Heer, Fl. foss. arct., vol. 3, Abt. 2, 1874, p. 72.

Inolepis? sp. Lesquereux, U. S. Geol. and Geog. Surv. Terr., Bull., vol. 1, 1875 [1876], p. 392.
Dakota: Kansas.

IRIS (Tournefort) Linné, Sp. pl., 1753, p. 38.

Iris? sp. [Lesquereux], U. S. Geol. and Geog. Surv. Terr., Ill. Cret. and Tert. Pl., 1878, pl. viii, f. 6.—Newberry, U. S. Geol. Surv., Mon. 35, 1898, p. 33, pl. xxii, f. 6.
Formation and locality unknown.

Iritis alaskana Lesquereux = *Zamites megaphyllus*.

ISOETES Linné, Sp. pl., 1753, p. 1100.

Isoetes brevifolius Lesquereux, Rept. U. S. Geol. Surv. Terr., vol. 8 (Cret. and Tert. Fl.), 1883, p. 136.—Cockerell, Am. Mus. Nat. Hist., Bull., vol. 24, 1908, p. 77.
Miocene: Florissant, Colo.

JUGLANDIPHYLLUM Fontaine, U. S. Geol. Surv., Mon. 15, 1889 [1890], p. 315. [Type, *J. integrifolium.*]

Juglandiphyllum integrifolium Fontaine, U. S. Geol. Surv., Mon. 15, 1889 [1890], p. 315, pl. clvii, f. 3, 5, 6.
Patuxent: Deep Bottom, White House Bluff, Va.

JUGLANDITES Sternberg, Vers., vol. 1, 1825, p. xl.

Juglandites costatus Presl = ***Juglans? sepultus.***

Juglandites cretacea Dawson, Roy. Soc. Canada, Trans., vol. 3, 1885 [1886], p. 14.
Mill Creek: North Fork of Old Man River, Canada.

Juglandites ellsworthianus Lesquereux, U. S. Geol. Surv., Mon. 17, 1891 [1892], p. 70, pl. xxxvii, f. 1.
Dakota: Ellsworth County, Kans.

Juglandites fallax Dawson, Roy. Soc. Canada, Trans., vol. 11, sec. 4, 1893 [1894], p. 59, pl. xi, f. 48.
Upper Cretaceous: Vancouver Colliery, Vancouver Island, British Columbia.

Juglandites lacoei Lesquereux, U. S. Geol. Surv., Mon. 17, 1891 [1892], p. 71, pl. xlviii, f. 5.
Dakota: Kansas.

Juglandites primordialis Lesquereux, U. S. Geol. Surv., Mon. 17, 1891 [1892], p. 70, pl. xxxv, f. 15.
Dakota: Pipe Creek, Cloud County, Kans.

Juglandites sinuatus Lesquereux, U. S. Geol. Surv., Mon. 17, 1891 [1892], p. 71, pl. xxxv, f. 9–11.
Dakota: Pipe Creek, Cloud County, Kans.

Juglandites? sp. Dawson, Roy. Soc. Canada, Trans., vol. 11, 1893 [1894], sec. 4, p. 59, pl. x, f. 43; pl. xiii, f. 56.
Upper Cretaceous: Port McNeill, Vancouver Island, British Columbia.

JUGLANS Linné, Sp. pl., 1753, p. 997.

Juglans acuminata Al. Braun, Neues Jahrb., 1845, p. 170.—Heer, Fl. foss. arct., vol. 2, Abt. 2, 1869, p. 38, pl. iii, ix, f. 1.—Knowlton, U. S. Nat. Mus., Proc., vol. 17, 1894, p. 222; Geol. Soc. Am., Bull., vol. 5, 1893, p. 583; U. S. Geol. Surv., Bull. 204, 1900, p. 35.—Penhallow, Roy. Soc. Canada, Trans., 2d ser., vol. 8, 1902, p. 61.—Knowlton, Harriman Alaska Exped., vol. 4, 1904, p. 152, pl. xxxiii, f. 3.—?Hollick, Md. Geol. Surv., Pliocene and Pleistocene,

Juglans acuminata—Continued.
1906, p. 220, pl. lxxii, f. 15.—Penhallow, Rept. Tert. Pl. Brit. Col., 1908, p. 59.—Knowlton, U. S. Geol. Surv., Prof. Paper 101, 1918, p. 292, pl. lxv, f. 1.
Kenai: Port Graham and Kukak Bay, Alaska.
Clarno (upper part): Bridge Creek, Oreg.
Paskapoo (Fort Union): Red Deer River, Alberta.
?Pleistocene (Sunderland): Head of Island Creek, Calvert County, Md.
Raton: Primero, Colo.

Juglans acuminata? Heer. Lesquereux = *Juglans? sepultus.*

Juglans affinis Kirchner, St. Louis Acad. Sci., Trans., vol. 8, 1898, p. 184, pl. xiii, f. 2.—Knowlton, U. S. Nat. Mus., Proc., vol. 51, 1916, p. 254.
Miocene: *Florissant, Colo.

Juglans alkalina Lesquereux, U. S. Geol. and Geog. Surv. Terr., Bull., vol. 1, 1875 [1876], p. 382; U. S. Geol. and Geog. Surv. Terr., Ann. Rept. 1874 [1876], p. 308; Rept. U. S. Geol. Surv. Terr., vol. 7 (Tert. Fl.), 1878, p. 288, pl. lxii, f. 6–9.
Green River: Alkali station, Wyo.

Juglans appressa Lesquereux, Am. Philos. Soc., Trans., vol. 13, 1869, p. 420, pl. xx, f. 9; U. S. Geol. and Geog. Surv. Terr., Ann. Rept. 1871 [1872], p. 294.
Wilcox: Mississippi.

Juglans arctica Heer, Fl. foss. arct., vol. 6, Abt. 2, 1882, p. 71, pl. xi, f. 2.—Lesquereux, U. S. Geol. Surv., Mon. 17, 1892, p. 68, pl. xix, f. 3; pl. xxxix, f. 5.—Berry, Bull. Torr. Bot. Club, vol. 33, 1906, p. 170; Geol. Surv. N. J., Ann. Rept. 1905 [1906], p. 139.—Hollick, U. S. Geol. Surv., Mon. 50, 1906, p. 54, pl. ix, f. 6–8; N. Y. Bot. Gard., Bull., vol. 8, 1912, p. 157, pl. clxiv, f. 3, 4.—Berry, Bull. Torr. Bot. Club, vol. 37, 1910, pp. 192, 504; Geol. Surv. N. J., Bull. 3, 1911, p. 110; U. S. Geol. Surv., Prof. Paper 84, 1914, pp. 30, 109, pl. viii, f. 1, 2.
Dakota: Near Delphos, Kans.
Magothy: Cliffwood, N. J.; Nashaquitsa, Marthas Vineyard, Mass.; Black Rock Point, Block Island, R. I.
Raritan: Woodbridge, N. J.; Tottenville, Staten Island, Glen Cove, and Roslyn, N. Y.
Eutaw: McBrides Ford, Columbus, Ga.
Black Creek: Court House Bluff, Cape Fear River, N. C.
Black Creek (Middendorf): Middendorf and Rocky Point, S. C.
Tuscaloosa: Cottondale and Shirleys Mill, Ala.

Juglans baltica? Heer. Lesquereux, U. S. Geol. and Geog. Surv. Terr., Ann. Rept. 1872 [1873], p. 398. [Not afterward mentioned by Lesquereux.]
Post-Laramie: Black Buttes, Wyo.

Juglans? bendirei Knowlton, U. S. Geol. Surv., Bull. 204, 1902, p. 34, pl. iii, f. 3.

Ilex? longifolia Heer. Lesquereux, U. S. Nat. Mus., Proc., vol. 11, 1888, p. 21.

Clarno (lower part): Cherry Creek, Crook County, Oreg.

Juglans berryi Knowlton, in Berry, U. S. Geol. Surv., Prof. Paper 91, 1916, p. 183.—Knowlton, U. S. Geol. Surv., Prof. Paper 101, 1918, p. 293, pl. lxiii, f. 3; pl. lxiv, f. 3; pl. lxxiii, f. 3.

Raton: *Primero and Wootton, Colo.

Lagrange: Wickliffe, Ballard County, Ky.

Wilcox: Coushatta, La.

Juglans bilinica Unger = *Hicoria juglandiformis.*

Juglans brandonianus Perkins, Rept. Vt. State Geol., 1903–4 [1904], p. 182, pl. lxxvi, f. 36; Geol. Soc. Am., Bull., vol. 16, 1905, p. 511, pl. 87, f. 21; Rept. Vt. State Geol., 1905–6 [1906], pl. liii, f. 21.

Miocene: Brandon, Vt.

Juglans californica Lesquereux = *Juglans leonis.*

Juglans cinerea Linné ?. Dawson, Rept. Geol. and Res. near 49th Parallel, App. A, 1875, p. 330.—Penhallow, Rept. Tert. Pl. Brit. Col., 1908, p. 59.

Eocene: Porcupine Creek, British Columbia.

Juglans coloradensis Lesquereux (MS.). Knowlton, U. S. Geol. Surv., Prof. Paper 101, 1918, p. 255, pl. xxxvi, f. 1.

Vermejo: Coal Creek, Rockvale, Colo.

Juglans costatus (Presl) Brongniart = *Juglans? sepultus.*

Juglans crassifolia Knowlton, U. S. Geol. Surv., Bull. 204, 1902, p. 36, pl. iv, f. 3.

Juglans n. sp. Knowlton, in Merriam, Univ. Calif. Bull. Dept. Geol., vol. 2. 1901, p. 303.

Clarno (upper part): Near Lone Rock, Gilliam County, Oreg.

Juglans crassipes Heer, Neue Denkschr. Schw. Ges., vol. 22, Mem. 2, 1869, p. 23, pl. vi, f. 3.—Lesquereux, U. S. Geol. Surv., Mon. 17, 1892, p. 69, pl. xlix, f. 1–3.—Hollick, Bull. Torr. Bot. Club, vol. 21, 1894, p. 51, pl. clxxv, f. 3; U. S. Geol. Surv., Mon. 50, 1906, p. 55, pl. ix, f. 3–5; N. Y. Bot. Gard., Bull., vol. 8, 1912, p. 157, pl. clxiv, f. 5.

Dakota: Near Fort Harker, Kans.

Raritan: Brooklyn and Roslyn, N. Y.

Magothy: Gay Head, Marthas Vineyard, Mass.; Glen Cove, Long Island, N. Y.

Juglans crescentia Knowlton, U. S. Geol. Surv., Mon. 32, pt. 2, 1899, p. 689, pl. lxxxiv, f. 8.

Fort Union: Yellowstone National Park.

Juglans crossii Knowlton, U. S. Geol. Surv., Bull. 152, 1898, p. 122.—Penhallow, Rept. Tert. Pl. Brit. Col., 1908, p. 59.

Juglans denticulata Heer, Fl. foss. arct., vol. 2, Abt. 4, 1869, p. 483.—Lesquereux, U. S. Geol. and Geog. Surv. Terr., Ann. Rept. 1871 [1872], p. 298; Rept. U. S. Geol. Surv. Terr., vol. 7 (Tert. Fl.), 1878, p. 289, pl. lviii, f. 1. [Homonym, O. Weber, 1852.]

Green River: Green River, Wyo.

Hanna: Carbon, Wyo.

Eocene: Burrard Inlet, Vancouver Island, British Columbia.

Juglans crossii Knowlton. Kerchner, St. Louis Acad. Sci., Trans., vol. 8, 1898, p. 183, pl. xiv, f. 1 = *Juglans magnifica*.

Juglans cryptata Knowlton, U. S. Geol. Surv., Bull. 204 1902, p. 35, pl. vi, f. 4, 5.

Clarno (upper part): Bridge Creek, Oreg.

Juglans debeyana (Heer) Lesquereux, Am. Jour. Sci., 2d ser., vol. 46, 1868, p. 101.

Populus? debeyana Heer, Phyll. crét. Nebr., 1866, p. 14, pl. i, f. 1.—Newberry, Lyc. Nat. Hist., Annals, vol. 9, 1868, p. 17.—[Lesquereux], Ill. Cret. and Tert. Pl., 1878, pl. iv, f. 3; pl. v, f. 7.—Newberry, U. S. Geol. Surv., Mon. 35, 1898, p. 42, pl. iv, f. 3, pl. v, f. 7.

Dakota: Decatur, Nebr.; New Ulm, Minn.

Juglans dentata Newberry, U. S. Nat. Mus., Proc., vol. 5, 1882 [1883], p. 507.

Green River: Green River, Wyo.

Juglans denticulata Heer = *Juglans crossii*.

Juglans egregia Lesquereux, Mus. Comp. Zool., Mem., vol. 6, 1878, p. 36, pl. ix, f. 12; pl. x, f. 1.—Knowlton, in Lindgren, Jour. Geol., vol. 4, 1896, p. 889.

Miocene (auriferous gravels): Chalk Bluffs, Nevada County, and Independence Hill, Placer County, Calif.

Juglans elongata Hollick, U. S. Geol. Surv., Mon. 50, 1906, p. 55, pl. xi, f. 3, 4.

Magothy: Sea Cliff, Long Island, N. Y.; Gay Head, Marthas Vineyard, Mass.

Juglans falsifolia Al. Braun = *Sapindus falsifolius*.

Juglans florissanti Lesquereux, Rept. U. S. Geol. Surv. Terr., vol. 8 (Cret. and Tert. Fl.), 1883, p. 190.—Cockerell, Am. Mus. Nat. Hist., Bull., vol. 24, 1908, p. 80.—Knowlton, U. S. Nat. Mus., Proc., vol. 51, 1916, p. 254, pl. xvii, f. 2.

Miocene: *Florissant, Colo.

Juglans harwoodensis Dawson, Roy. Soc. Canada, Trans., vol. 1, sec. 4, 1882 [1883], p. 28, pl. viii, f. 31.

Upper Cretaceous: Nanaimo, Vancouver Island, British Columbia.

Juglans hesperia Knowlton = *Juglans oregonian*

Juglans laramiensis Knowlton, U. S. Geol. Surv., Prof. Paper 130, 19—, p. —, pl. xx, f. 12 [in preparation].
Laramie: Coal Creek, Boulder County, Colo.

Juglans laurifolia Knowlton, U. S. Geol. Surv., Mon. 32, pt. 2, 1899, p. 688, pl. lxxxiii, f. 2, 3.—Penhallow, Rept. Tert. Pl. Brit. Col., 1908, p. 60; Roy. Soc. Canada, Trans., 2d ser., vol. 8, 1902, sec. 4, p. 61.
Miocene (intermediate flora): Yellowstone National Park.
Paskapoo (Fort Union): Red Deer River, Alberta.

Juglans laurinea Lesquereux, Mus. Comp. Zool., Mem., vol. 6, No. 2, 1878, p. 35, pl. ix, f. 11.
Miocene (auriferous gravels): Chalk Bluffs, Nevada County, Calif.

Juglans leconteana Lesquereux, U. S. Geol. and Geog. Surv. Terr., Ann. Rept. 1870 [1871], p. 382 [name]; Rept. U. S. Geol. Surv. Terr., vol. 7 (Tert. Fl.), 1878, p. 285, pl. liv, f. 10–13.—Penhallow, Roy. Soc. Canada, Trans., 2d ser., vol. 8, 1902, sec. 4, p. 60; Rept. Tert. Pl. Brit. Col., 1908, p. 59.—Cockerell, Torreya, vol. 9, 1909, p. 142.—Knowlton, U. S. Geol. Surv., Prof. Paper 130, 19—, p. — pl. viii, f. 1–3 [in preparation].
Evanston: Evanston, Wyo.
Paskapoo (Fort Union): Red Deer River, Alberta.
Raton: Raton Pass, Colo.
Laramie: ?Marshall, Colo.

Juglans leonis Cockerell, Am. Jour. Sci., 4th ser., vol. 26, 1908, p. 543.
Juglans californica Lesquereux, Mus. Comp. Zool., Mem., vol. 6, No. 2, 1878, p. 34, pl. ix, f. 14; pl. x, f. 2, 3 [not *J. californica* S. Watson, 1875].—Knowlton, U. S. Geol. Surv., Mon. 32, pt. 2, 1899, p. 687.
?Eocene: Southwestern Oregon.
Miocene (auriferous gravels): Chalk Bluffs, Nevada County, Calif.
Miocene: Yellowstone National Park.

Juglans leydenianus Knowlton, U. S. Geol. Surv., Prof. Paper 130, 19—, p. —, pl. v, f. 1 [in preparation].
Laramie: Leyden Gulch, 6½ miles north of Golden, Colo.

Juglans magnifica Knowlton, U. S. Nat. Mus., Proc., vol. 51, 1916, p. 252, pl. xv.
Juglans crossii Knowlton. Kirchner, St. Louis Acad. Sci., Trans., vol. 8, 1898, p. 183, pl. xiv, f. 1.
Miocene: *Florissant, Colo.

Juglans minutidens Knowlton, U. S. Geol. Surv., Prof. Paper 101, 1918, p. 293, pl. lv, f. 3.
Raton: Turner mine, near Wootton, Colo.

Juglans? missouriensis Knowlton, U. S. Geol. Surv., Bull. 163, 1900, p. 12, pl. i, f. 7–9.

Eagle: Missouri River, below Coal Banks, Mont.

Juglans newberryi Knowlton, U. S. Geol. Surv., Prof. Paper 130, 19—, p. —, pl. xx, f. 8–10 [in preparation].

Laramie: Eric, Colo.

Juglans nigella Heer, Fl. foss. arct., vol. 2, Abt. 2, 1869, p. 38, pl. ix, f. 2–4.—Dawson, Geol. Surv. Canada, 1875–76, p. 57.—Lesquereux, Rept. U. S. Geol. Surv. Terr., vol. 8 (Cret. and Tert. Fl.), 1883, p. 235, pl. xlvi–A, f. 11.—Knowlton, U. S. Nat. Mus., Proc., vol. 17, 1894, p. 222; Geol. Soc. Am., Bull., vol. 5, 1893, p. 583.—Ward, U. S. Geol. Surv., Sixth Ann. Rept., 1884–85 [1886], p. 551, pl. xl, f. 6; idem, Bull. 37, 1887, p. 33, pl. xv, f. 1.—Newberry, U. S. Geol. Surv., Mon. 35, 1898, p. 33, pl. li, f. 2 (in part), 4.—Penhallow, Rept. Tert. Pl. Brit. Col., 1908, p. 60.—Knowlton, U. S. Geol. Surv., Prof. Paper 101, 1918, p. 292, pl. lv, f. 2; pl. lxiii, f. 2.

Kenai: *Port Graham and Admiralty Inlet, Alaska.

Fort Union: Near Glendive, Mont.; Badlands, N. Dak.

Paskapoo (Fort Union): Quilchena, British Columbia.

Raton: Near Weston, Abeton, and Sopris, Colo.

Juglans nigra Linné. Berry, Torreya, vol. 9, 1909, p. 98, text f. 6; Am. Jour. Sci., 4th ser., vol. 29, 1910, p. 392.

Pleistocene (Talbot): 1 mile south of Chesapeake Beach, Calvert County, Md.

Pleistocene: Black Warrior River, 311 miles above Mobile, Ala.

Juglans obtusifolia Heer, Fl. tert. Helv., vol. 3, 1859, p. 89, pl. cxxix, f. 9.—Lesquereux, U. S. Geol. and Geog. Surv. Terr., Ann. Rept., 1871 [1872], Suppl., p. 12. [Not afterward mentioned by Lesquereux.]

——: Evanston, Wyo.

Juglans occidentalis Newberry, U. S. Nat. Mus., Proc., vol. 5, 1882 [1883], p. 507; U. S. Geol. Surv., Mon. 35, 1898, p. 34, pl. lxv, f. 1; pl. lxvi, f. 1–4c.—Penhallow, Roy. Soc. Canada, Trans., 2d ser., vol. 8, 1902, p. 61; Rept. Tert. Pl. Brit. Col., 1908, p. 60.

Green River: Green River, Wyo.

Paskapoo (Fort Union): Red Deer River, Alberta.

Oligocene: Horsefly River, British Columbia.

Juglans oregoniana Lesquereux, Mus. Comp. Zool., Mem., vol. 6, 1878, p. 35, pl. ix, f. 10.—Knowlton, U. S. Geol. Surv., Bull. 204, 1902, p. 36.

Rhus bendirei Lesquereux, U. S. Nat. Mus., Proc., vol. 11, 1888, p. 15. [The small leaf described.]

Juglans hesperia Knowlton, U. S. Geol. Surv., Eighteenth Ann. Rept., pt. 3, 1898, p. 723, pl. xcix, f. 8.

Payette: Near Marsh, Boise County, Idaho.

Mascall: Van Horn's ranch, John Day Basin, Oreg.

Juglans picroides Heer, Fl. foss. arct., vol. 2, Abt. 2, 1869, p. 39, pl. ix, f. 5.—Knowlton, U. S. Nat. Mus., Proc., vol. 17, 1894, p. 222; Geol. Soc. Am., Bull., vol. 5, 1893, p. 583.

Kenai: Port Graham, Alaska.

Juglans praerugosa Knowlton, U. S. Geol. Surv., Prof. Paper 130, 19—, p. —, pl. v, f. 2; pl. xxi, f. 5 [in preparation].

Juglans rugosa Lesquereux, Am. Jour. Sci., 2d ser., vol. 45, 1868, p. 206; U. S. Geol. and Geog. Surv. Terr., Ann. Rept. 1869, p. 196; 2d ed., 1873.

Laramie: Marshall, Colo.

Juglans rhamnoides Lesquereux, U. S. Geol. and Geog. Surv. Terr., Ann. Rept. 1871 [1872], p. 294; idem, 1872 [1873], pp. 382, 400, 402; idem, 1874 [1876], p. 307; idem, 1876 [1878], p. 517; idem, Bull., vol. 1, 1875 [1876], p. 370; Rept. U. S. Geol. Surv. Terr., vol. 7 (Tert. Fl.), 1878, p. 284, pl. liv, f. 6–9; idem, vol. 8 (Cret. and Tert. Fl.), 1883, p. 235.—Dawson, Roy. Soc. Canada, Trans., vol. 3, 1886, p. 31.—Lesquereux, U. S. Nat. Mus., Proc., vol. 16, 1888, p. 56.—Knowlton, U. S. Geol. Surv., Bull. 105, 1893, p. 58.—Penhallow, Rept. Tert. Fl. Brit. Col., 1908, p. 60.—Knowlton, U. S. Geol. Surv., Prof. Paper 101, 1918, p. 294, pl. lxvi, f. 1.

Rhamnus rossmässleri Unger. Lesquereux, U. S. Geol. and Geog. Surv. Terr., Ann. Rept. 1874 [1876], p. 314; idem, Bull., vol. 1, 1875 [1876], p. 388; Rept. U. S. Geol. Surv. Terr., vol. 7 (Tert. Fl.), 1878, p. 283, pl. liv, f. 4.

Post-Laramie: Black Buttes, Wyo.

Livingston: Spring Canyon, 12 miles southeast of Bozeman, Mont.

Raton: Spring Canyon, north of Vermejo Park, N. Mex.

?Eocene: Porcupine Creek and Great Valley, British Columbia.

?Miocene: Tranquille River, British Columbia.

Juglans rostrata Göppert = *Hicoria rostrata.*

Juglans rugosa Lesquereux, Am. Jour. Sci., vol. 45, 1868, p. 206; Rept. U. S. Geol. Surv. Terr., vol. 7 (Tert. Fl.), 1878, p. 286, pl. liv, f. 4, 5; pl. lv, f. 1–9; pl. lvi, f. 1, 2.—Knowlton, U. S. Geol. Surv., Bull. 204, 1902, p. 34; idem, Mon. 32, pt. 2, 1899, p. 687; Washington Acad. Sci., Proc., vol. 11, 1909, p. 202; U. S. Geol. Surv., Prof. Paper 101, 1918, p. 293, pl. cxii, f. 4.

Denver: Golden, Colo.

Livingston: Bozeman coal field, Mont.

Evanston: Evanston, Wyo.

Wilcox: Coushatta, La.

Raton: Yankee, N. Mex.; Cucharas Canyon, near Abeton, Trinidad, Wootton, Primero, Weston, and Tercio, Colo.

Lagrange: Wickliffe, Ky.

Lance: Yule, N. Dak.

Fort Union: Yellowstone National Park.

Juglans rugosa Lesquereux [part] = *Juglans praerugosa.*

Juglans rugosa Lesquereux. Hollick, La. Geol. Surv., Special Rept. 5, 1899, p. 280, pl. xxxv, f. 1 = *Sapindus coushatta.*

Juglans saffordiana Lesquereux, Am. Philos. Soc., Trans., vol. 13, 1869, p. 421, pl. xx, f. 7.—Berry, U. S. Geol. Surv., Prof. Paper 91, 1916, p. 182.

Lagrange: Lagrange, Tenn.

Juglans sapindiformis Knowlton, U. S. Geol. Surv., Prof. Paper 101, 1918, p. 295, pl. lxv, f. 3.

Raton: Yankee, N. Mex.

Juglans sapindoides Knowlton, U. S. Geol. Surv., Prof. Paper 101, 1918, p. 295, pl. lxv, f. 4, 5.

Raton: Wootton, Colo.

Juglans schimperi Lesquereux, U. S. Geol. and Geog. Surv Terr., Ann. Rept. 1871 [1872], Suppl., p. 8; Rept. U. S. Geol. Surv. Terr., vol. 7 (Tert. Fl.), 1878, p. 287, pl. lvi, f. 5–10.—Knowlton, U. S. Geol. Surv., Bull. 204, 1902, p. 34.—Penhallow, Rept. Tert. Pl. Brit. Col., 1908, p. 60.—Knowlton, U. S. Geol. Surv., Mon. 32, pt. 2, 1899, p. 688.—Hollick, Geol. Surv. La., Special Rept. 5, 1899, p. 280, pl. xxxii, f. 5; pl. xxxiii, f. 1, 2; pl. xxxv, f. 3.—Berry, U. S. Geol. Surv., Prof. Paper 91, 1916, p. 182, pl. xviii, f. 3–5; pl. xix, f. 4.—Knowlton, U. S. Geol. Surv., Prof. Paper, 101, 1918, p. 296, pl. lxiv, f. 1.

Juglans rugosa Lesquereux. Lesquereux, U. S. Nat. Mus., Proc., vol. 11, 1888, p. 11.

Green River: Green River, Wyo.

Denver: Golden, Colo.

Raton: Dillon Canyon near Blossburg, Raton tunnel, and South Raton Canyon, N. Mex.; Aguilar and Wootton, Colo.

Miocene: Yellowstone National Park.

Clarno (upper part?): Bridge Creek, Oreg.

Eocene: Porcupine Creek, British Columbia.

Lagrange: Kentucky.

Wilcox: Coushatta, La.

Wilcox: Many localities in Arkansas, Louisiana, Texas, and Mississippi.

Juglans? sepultus Cockerell, Am. Mus. Nat. Hist., Bull., vol. 24, 1908, p. 80, pl. vi, f. 8.—Knowlton, U. S. Nat. Mus., Proc., vol. 51, 1916, p. 254, pl. xvii, f. 4.

Juglans costata (Presl) Brongniart. Lesquereux, Rept. U. S. Geol. Surv. Terr., vol. 8 (Cret. and Tert. Fl.), 1883, p. 190, pl. xxxix, f. 5.

Juglandites costatus Presl, in Sternburg, Vers., vol. 2, 1833, p. 207.

Juglans acuminata? Heer. Lesquereux, U. S. Geol. and Geog. Surv. Terr., Ann. Rept. 1871 [1872], Suppl., p. 8.

Miocene: *Florissant, Colo.

Juglans similis Knowlton, U. S. Geol. Surv., Prof. Paper 101, 1918, p. 255, pl. xxxvi, f. 2.
Vermejo: Coal Creek, Rockvale, Colo.

Juglans smithsoniana Lesquereux = *Ficus? smithsoniana.*

Juglans thermalis Lesquereux, U. S. Geol. and Geog. Surv. Terr., Ann. Rept. 1871 [1872], Suppl., p. 17; Rept. U. S. Geol. Surv. Terr., vol. 7 [Tert. Fl.], 1878, p. 287, pl. lvi, f. 3, 4.
Denver: Golden and Hot Springs, Middle Park, Colo.

Juglans townsendi Knowlton, U. S. Nat. Mus., Proc., vol. 17, 1894, p. 222, pl. ix, f. 5; Geol. Soc. Am., Bull., vol. 5, 1893, p. 584.
Kenai: Herendeen Bay, Alaska.

?Juglans ungeri Heer, Fl. tert. Helv., vol. 3, 1859, p. 199, pl. clv, f. 18.—Ward, U. S. Geol. Surv., Sixth Ann. Rept., 1884–85 [1886], p. 551, pl. xl, f. 5; idem, Bull. 37, 1887, p. 37, pl. xiv, f. 6.
Fort Union: Near Glendive, Mont.

Juglans vetusta Heer, Fl. tert. Helv., vol. 3, 1859, p. 90, pl. cxxvii, f. 40–44.—Lesquereux, U. S. Nat. Mus., Proc., vol. 11, 1888, p. 22.
Clarno (lower part): Cherry Creek, Oreg.

Juglans woodiana Heer, Neue Denkschr., vol. 21, 1865, p. 9, pl. ii, f. 4–7.—Lesquereux, Rept. U. S. Geol. Surv. Terr., vol. 8 (Cret. and Tert. Fl.), 1883, p. 236.—Knowlton, U. S. Nat. Mus., Proc., vol. 17, 1894, p. 222; Geol. Soc. Am., Bull., vol. 5, 1893, p. 584.
Kenai: Chignik Bay, Alaska.
Eocene?: Burrard Inlet, British Columbia.
Fort Union: Badlands, N. Dak.

Juglans sp. Knowlton, in Lindgren, Jour. Geol., vol. 4, 1896, p. 889.
Miocene (auriferous gravels): Independence Hill, Placer County, Calif.

Juglans sp. Knowlton, in Vaughan, Am. Geol., vol. 16, 1895, p. 308.
Wilcox: Old Port Caddo Landing, Harrison County, Tex.

Juglans, nut of, Knowlton, U. S. Geol. Surv., Bull. 204, 1902, p. 35.
Clarno (upper part): Bridge Creek, Oreg.

Juglans sp. Knowlton, Geol. Soc. Am., Bull., vol. 8, 1897, p. 145.
Fort Union: Black Buttes, Wyo.

Juglans sp. Knowlton, Washington Acad. Sci., Proc., vol. 11, 1909, p. 207.
Lance: Converse County, Wyo.

Juglans sp. Knowlton, in Washburne, U. S. Geol. Surv., Bull. 590, 1914, p. 36.
Eocene?: Eugene, Oreg.

JUNCUS (Tournefort) Linné, Sp. pl., 1753, p. 325.

Juncus crassulus Cockerell, Am. Mus. Nat. Hist., Bull., vol. 24, 1908, p. 79, pl. x, f. 44, 45.
Miocene: Florissant, Colo.

Juncus sp. Lesquereux, U. S. Geol. and Geog. Surv. Terr., Ann. Rept. 1871 [1872], Suppl., p. 5.
Green River: Green River, Wyo.

JUNGERMANNITES Göppert, in Berendt, Bernstein. Org. Vorw., vol. 1, 1845, p. 113.

Jungermannites cretaceus Berry, U. S. Geol. Surv., Prof. Paper 112, 1919, p. 49, pl. v, f. 2, 3.
Tuscaloosa: Shirleys Mill, Ala.

JUNIPERUS Linné, Sp. pl., 1753, p. 1038.

Juniperus californica Carrier. Penhallow, Roy. Soc. Canada, Trans., 2d ser., vol. 10, sec. 4, 1904, p. 70; Man. N. A. Gymnosp., 1907, p. 248.
Pleistocene: Near Orleans, Humboldt County, Calif.

Juniperus? haydenii (Lesquereux) Knowlton, U. S. Nat. Mus., Proc., vol. 51, 1916, p. 249.
Sequoia haydenii (Lesquereux) Cockerell, Science, vol. 26, 1907, p. 447; Pop. Sci. Mo., vol. 73, 1908, p. 122, text f.; Am. Nat., vol. 44, 1909, p. 36, f. 3; Am. Mus. Jour., vol. 16, 1916, p. 446.
Hypnum haydenii Lesquereux, U. S. Geol. and Geog. Surv., Terr. Bull., vol. 1, 1875 [1876], p. 383; idem, Ann. Rept. 1874 [1876], p. 309; U. S. Geol. Surv. Terr., Rept., vol. 7 (Tert. Fl.), 1878, p. 44, pl. v, f. 14–14b.—Britton and Hollick, Bull. Torr. Bot. Club, vol. 34, 1907, p. 139, pl. ix, f. 1, 1a.
Miocene: *South Park, Colo.

Juniperus hypnoides Heer, Fl. foss. arct., vol. 6, Abt. 2, 1882, p. 47, pl. xliv, f. 3; pl. xlvi, f. 18.—Hollick, N. Y. Acad. Sci., Trans., vol. 12, 1892, p. 29, pl. i, f. 1; N. Y. Bot. Gard., Bull. 2, 1902, p. 403, pl. xli, f. 7, 7a.—Berry, Bull. Torr. Bot. Club, vol. 33, 1906, p. 166; Geol. Surv. N. J., Ann. Rept. 1905 [1906], p. 139.—Hollick, U. S. Geol. Surv., Mon. 50, 1906, p. 46, pl. ii, f. 26, 27b, 28; pl. iii, f. 12–13a.—Hollick and Jeffrey, N. Y. Bot. Gard., Mem., vol. 3, 1909, p. 61, pl. v, f. 5, 6.—Berry, N. J. Geol. Surv., Bull. 3, 1911, p. 85; Md. Geol. Surv., Upper Cret., 1916, p. 792.
Magothy: Cliffwood, N. J.; Deep Cut, Del.; Chappaquiddick Island, Mass.
Raritan: Perth Amboy and Woodbridge, N. J.; Kreischerville, Staten Island, N. Y.

Juniperus virginiana Linné. Penhallow, Brit. Assoc. Adv. Sci., Bradford meeting, 1900, p. 336; Man. N. A. Gymnosp., 1907, p. 246.—Berry, Torreya, vol. 10, 1910, p. 264.
Pleistocene: Don Valley, Scarborough and Montreal, Canada; Moose and Missinaibi rivers, Manitoba; Long Branch, N. J.

Juniperus sp. Penhallow, Roy. Soc. Canada, Trans., 2d ser., vol. 10, sec. 4, 1904, p. 73.
Pleistocene: Don River valley, Toronto, Canada.

KALMIA Linné, Sp. pl., 1753, p. 391.

Kalmia brittoniana Hollick, N. Y. Acad. Sci., Trans., vol. 12, 1892, p. 34 [7], pl. ii, f. 6–8; U. S. Geol. Surv., Mon. 50, 1906, p. 100, pl. xxxix, f. 8, 9.—?Berry, Bull. Torr. Bot. Club, vol. 34, 1907, p. 204; U. S. Geol. Surv., Prof. Paper 112, 1919, p. 132.
Black Creek: Blackmans Bluff, Neuse River, N. C.
Raritan: Kreischerville, Staten Island, N. Y.
Tuscaloosa: Shirleys Mill, Ala.

KNIGHTIOPHYLLUM Berry, U. S. Geol. Surv., Prof. Paper 91, 1916, p. 208. [Type, *K. wilcoxianum*.]

Knightiophyllum wilcoxianum Berry, U. S. Geol. Surv., Prof. Paper 91, 1916, p. 208, pl. xxxv, f. 1–3.
Lagrange: Puryear, Henry County, Tenn.

KNOWLTONELLA Berry, Md. Geol. Surv., Lower Cret., 1911, p. 233. [Type, *K. maxoni*.]

Knowltonella maxoni Berry, Md. Geol. Surv., Lower Cret., 1911, p. 235, pls. xxv, xxvi, xxvii.
Patapsco: Stump Neck near Glymont, Md.; near Widewater, Va.

LACCOPTERIS Sternberg, Fl. Vorw., vol. 2 (Versuch), 1838, p. 115.

Laccopteris carolinensis (Emmons) Fontaine = ***Asterocarpus falcatus***.

Laccopteris elegans Presl? Fontaine, U. S. Geol. Surv., Mon. 6, 1883, p. 105.
Triassic: Virginia.

Laccopteris emmonsi Fontaine = ***Asterocarpus falcatus***.

Laccopteris lanceolata (Göppert) Ward, U. S. Geol. Surv., Twentieth Ann. Rept., pt. 2, 1900, p. 281, pl. xxxviii, f. 2–4.
Asterocarpus lanceolatus Göppert, Flora der Vorwelt, vol. 2, 1838, p. 115, pl. xxii, f. 8a, 8b, 8c.
Pecopteris sp.? Emmons, American geology, pt. 6, 1857, p. 104, pl. vi, f. 2.
Triassic (Newark): North Carolina?

LAGUNCULARIA Gärtner, Fruct., vol. 3, 1805, p. 217.

Laguncularia preracemosa Berry, U. S. Geol. Surv., Prof. Paper 91, 1916, p. 320, pl. xcv, f. 4–8.
Wilcox (Holly Springs): Holly Springs, Miss.
Lagrange: Puryear, Tenn.

LARICOPSIS Fontaine, U. S. Geol. Surv., Mon. 15, 1889 [1890], p. 232. [Type, *L. longifolia*.]

Laricopsis angustifolia Fontaine, U. S. Geol. Surv., Mon. 15, 1889 [1890], p. 233, pl. cii, f. 9, 10; pl. ciii, f. 1, 4; in Ward, idem, Mon. 48, 1905 [1906], p. 234, pl. cii, f. 5, 6.—Berry, Md. Geol. Surv., Lower Cret., 1912, p. 410.

Laricopsis brevifolia Fontaine, U. S. Geol. Surv., Mon. 15, 1889 [1890], p. 234, pl. cii, f. 5, 6.

Patuxent: Near Dutch Gap Canal and Trents Reach, Va.

Laricopsis brevifolia Fontaine = *Laricopsis angustifolia.*

Laricopsis longifolia Fontaine, U. S. Geol. Surv., Mon. 15, 1889 [1890], p. 233, pl. cii, f. 7, 8; pl. ciii, f, 2, 3; pl. clxv, f. 4; pl. clxviii, f. 5, 6; U. S. Nat. Mus., Proc., vol. 16, 1893, p. 268, pl. xxxvi, f. 9; in Ward, U. S. Geol. Surv., Mon. 48, 1905 [1906], p. 312, pl. lxxiii, f. 11, 14.

Patuxent: Dutch Gap Canal, Va.
Trinity: Glen Rose, Tex.

Laricopsis longifolia Fontaine? Fontaine, in Weed and Pirsson = *Laricopsis longifolia latifolia.*

Laricopsis longifolia latifolia Fontaine, in Ward, U. S. Geol. Surv., Mon. 48, 1905 [1906], p. 312, pl. lxxiii, f. 11–14.

Laricopsis longifolia Fontaine? Fontaine, in Weed and Pirsson, U. S. Geol. Surv., Eighteenth Ann. Rept., pt. 3, 1898, p. 451.

Kootenai: Near Geyser, Cascade County, and near Grafton, Mont.

LARIX (Tournefort) Adanson, Fam. Pl., vol. 2, 1763, p. 480.

Larix americana Michaux = *Larix laricina.*

Larix churchbridgensis Penhallow, Am. Geol., vol. 9, 1892, p. 370; Roy. Soc. Canada, Trans., 2d ser., vol. 2, sec. 4, 1896, p. 71; Brit. Assoc. Adv. Sci., Bradford meeting, 1900, p. 336.

Pleistocene: Churchbridge, Manitoba.

Larix laricina (Du Roy) Koch.

Larix americana Michaux. Penhallow, Geol. Soc. Am., Bull., vol. 1, 1890, p. 326; Roy. Soc. Canada, Trans., 2d ser., vol. 2, 1896, sec. 4, pp. 60, 71; Brit. Assoc. Adv. Sci., Bradford meeting, 1900, p. 336; Roy. Soc. Canada, Trans., 2d ser., vol. 10, 1904, sec. 4, pp. 66, 69, 75; Man. N. A. Gym., 1907, p. 278.

Pleistocene: Moose River, Manitoba; Don Valley and Montreal, Canada; Ithaca, N. Y.; Dahlonega, Ga.; Fort Madison, Iowa.

LASTREA Bory, Dict. class. hist. nat., vol. 6, 1824, p. 588.

Lastrea (Goniopteris) fischeri Heer, Fl. tert. Helv., vol. 1, 1855, p. 34, pl. ix, f. 3.—Lesquereux, Rept. U. S. Geol. Surv. Terr., vol. 8 (Cret. and Tert. Fl.), 1883, p. 239, pl. l, f. 1, 1a.—Newberry, U. S. Geol. Surv., Mon. 35, 1898, p. 10, pl. xlviii, f. 6.—Knowlton, U. S. Geol. Surv., Bull. 204, 1902, p. 22.—Penhallow, Roy. Soc. Canada,

Lastrea (Goniopteris) fischeri—Continued.
Trans., 2d ser., vol. 8, 1902, sec. 4, p. 48; Rept. Tert. Pl. Brit. Col., 1908, p. 61.—Knowlton, in Washburne, U. S. Geol. Surv., Bull. 590, 1914, p. 36.

Lastrea (Goniopteris) knightiana Newberry, U. S. Nat. Mus., Proc., vol. 5, 1882 [1883], p. 503.

Clarno (lower part): Currant Creek, Oreg.
Paskapoo (Fort Union): Red Deer River, Alberta.
Eocene: Burrard Inlet, British Columbia.
Eocene?: Eugene, Oreg.

Lastrea (Goniopteris) goldiana (Lesquereux) Lesquereux = *Dryopteris lesquereuxii.*

Lastrea (Goniopteris) intermedia Lesquereux = *Dryopteris laramiensis.*

Lastrea (Goniopteris) knightiana Newberry = *Lastrea (Goniopteris) fischeri?*

Lastrea (Goniopteris) polypodioides Ettingshausen = *Dryopteris polypodioides.*

Lastrea styriaca Heer, Lesquereux, U. S. Geol. and Geog. Surv. Terr., Ann. Rept. 1873 [1874], p. 418. [Not afterward referred to by Lesquereux.]
——: Blakes Fork, Uinta Mountains, Utah.

Lathraea arguta Lesquereux = *Dryopteris arguta.*

LAURINOXYLON Felix, Jahrb. K. geol. Anstalt (Holzopale Ungarns), vol. 7, 1884, p. 27.

Laurinoxylon aromaticum Felix = *Perseoxylon aromaticum.*

Laurinoxylon brandonianum Jeffrey and Chrysler, Rept. Vt. State Geol., 1905–6 [1906], p. 198, pl. xlix, f. 1–6.
Miocene: Brandon, Vt.

Laurinoxylon branneri Knowlton, Geol. Surv. Ark., Ann. Rept. 1889, vol. 2, p. 256, pl. ix, f. 8, 9; pl. x, f. 1, 2; pl. xi, f. 4.—Penhallow, Roy. Soc. Canada, Trans., 3d ser., vol. 1, 1907, sec. 4, p. 98, f. 6–8.—Berry, U. S. Geol. Surv., Prof. Paper 91, 1916, p. 314, pl. xvi, f. 6–10.
"Orange sand" [Wilcox?]: Bolivar Creek, Poinsett County, and Big Crow and Little Crow creeks, St. Francis County, Ark.
Yegua: Sommerville, Burleson County, Tex.

Laurinoxylon lesquereuxiana Knowlton, Geol. Surv. Ark., Ann. Rept. 1889, vol. 2, p. 258, pl. x, f. 3, 4; pl. xi, f. 3, 4.
Tertiary: Near Wittsburg, Cross County, Ark.

Laurinoxylon pulchrum Knowlton, U. S. Geol. Surv., Mon. 32, pt. 2, 1899, p. 765, pls. cxvi, cxix, f. 3–5, cxx, f. 1.—Platen, Sitzungsb. Naturf. Ges., Leipzig, vol. 34, 1907 [1908], p. 127.
Miocene (intermediate flora): Amethyst Mountain, Yellowstone National Park.

Laurinoxylon? sp. Knowlton, Geol. Surv. Ark., Ann. Rept. 1889, vol. 2, p. 259.
"Orange sand": Red Bluff, Jefferson County, Ark.

LAUROPHYLLUM Göppert, Neues Jahrb., 1853, p. 434.

Laurophyllum angustifolium Newberry, U. S. Geol. Surv., Mon. 27, 1895 [1896], p. 86, pl. xvii, f. 10, 11.—Berry, N. Y. Bot. Gard., Bull., vol. 3, 1903, p. 80, pl. lvii, f. 1, 5, 8; Bull. Torr. Bot. Club, vol. 33, 1906, p. 179; Geol. Surv. N. J., Ann. Rept. 1905 [1906], p. 138; idem, Bull. 3, 1911, p. 148; Md. Geol. Surv., Upper Cret., 1916, p. 865; U. S. Geol. Surv., Prof. Paper 112, 1919, p. 124.
Magothy: Cliffwood, N. J.; Grove Point, Md.
Raritan: Woodbridge, N. J.
Tuscaloosa: Snow place, Tuscaloosa County, Ala.

Laurophyllum debile Dawson, Roy. Soc. Canada, Trans., vol. 1, 1882 [1883], sec. 4, p. 22, pl. ii, f. 7, 7a.
——: Forks of Peace River and North Fork of Old Man River.

Laurophyllum elegans Hollick, U. S. Geol. Surv., Mon. 50, 1906, p. 81, pl. xxvii, f. 1–5.—Berry, Bull. Torr. Bot. Club, vol. 36, 1909, p. 255; idem, vol. 37, 1910, pp. 26, 198.—Hollick, N. Y. Bot. Gard., Bull., vol. 8, 1912, p. 162, pl. clxx, f. 5, 6.—Berry, U. S. Geol. Surv., Prof. Paper 84, 1914, p. 53, pl. xii, f. 6; Geol. Surv. N. J., Bull. 3, 1911, p. 147; Md. Geol. Surv. Upper Cret., 1916, p. 864, pl. lxxi, f. 1–3; Bull. Torr. Bot. Club, vol. 43, 1916, p. 297; U. S. Geol. Surv., Prof. Paper 112, 1919, p. 124.
Laurus plutonia Heer. Hollick, N. Y. Acad. Sci., Trans., vol. 11, 1892, p. 99, pl. iii, f. 3, 4; idem, vol. 12, 1893, p. 236, pl. vi, f. 1.
Proteoides daphnogenoides Heer. Hollick, N. Y. Acad. Sci., Annals, vol. 11, 1898, p. 420, pl. xxxvi, f. 2.
Raritan: Tottenville and Glen Cove, N. Y.; South Amboy, N. J.
Magothy: Grove Point and Round Bay, Md.
Black Creek: Cape Fear River, N. C.
Black Creek (Middendorf): Middendorf, S. C.
Eutaw: Coffee Bluff, Hardin County, Tenn.

Laurophyllum ellsworthianum Lesquereux = *Quercus? ellsworthianus.*

Laurophyllum florum Berry, U. S. Geol. Surv., Prof. Paper 91, 1916, p. 313, pl. lxxxvi, f. 6.
Lagrange: Grand Junction, Tenn.

Laurophyllum insigne Dawson, Roy. Soc. Canada, Trans., vol. 11, 1893 [1894], sec. 4, p. 61, pl. vii, f. 24, 25.—Penhallow, Roy. Soc. Canada, Trans., 2d ser., vol. 8, 1902, sec. 4, p. 46.
Upper Cretaceous: Port McNeill, Vancouver Island, British Columbia.

Laurophyllum juvenalis Berry, U. S. Geol. Surv., Prof. Paper 91, 1916, p. 313, pl. lxxxvi, f. 4.
Lagrange: Puryear, Tenn.

Laurophyllum lanceolatum Newberry, U. S. Geol. Surv., Mon. 26, 1895 [1896], p. 87, pl. xvii, f. 1, 12.—Berry, Geol. Surv. N. J., Bull. 3, 1911, p. 148.—Hollick, N. Y. Bot. Gard., Bull. 8, 1912, p. 163, pl. clxix, f. 2.

Raritan: Woodbridge and Milltown, N. J.

Magothy: Glen Cove, N. Y.

Laurophyllum minus Newberry, U. S. Geol. Surv., Mon. 26, 1895 [1896], p. 87, pl. xvii, f. 7–9.—Berry, Geol. Surv. N. J., Bull. 3, 1911, p. 149; Bull. Torr. Bot. Club, vol. 39, 1912, p. 402.

Raritan: South Amboy, N. J.

Woodbine: Arthurs Bluff, Tex.

Laurophyllum nervillosum Hollick, U. S. Geol. Surv., Mon. 50, 1906, p. 82, pl. xxvii, f. 6, 7.—Berry, Bull. Torr. Bot. Club, vol. 36, 1909, p. 255; N. J. Geol. Surv., Bull. 3, 1911, p. 146; U. S. Geol. Surv., Prof. Paper 84, 1914, p. 54, pl. xii, f. 7; idem, Prof. Paper 112, 1919, p. 124, pl. xxi, f. 1.

Proteoides daphnogenoides Heer. Hollick, N. Y. Acad. Sci., Annals, vol. 11, 1898, p. 420, pl. xxvi, f. 1, 3.

Tuscaloosa: Shirleys Mill, Ala.

Raritan: Tottenville, Staten Island, N. Y.; Milltown, N. J.

Black Creek (Middendorf): Middendorf and Darlington, S. C.

Laurophyllum ocoteaeoides Hollick, N. Y. Bot. Gard., Bull., vol. 8, 1912, p. 163, pl. clxix, f. 1, 6.

Magothy: Glen Cove, N. Y.

Laurophyllum preflorum Berry, U. S. Geol. Surv., Prof. Paper 91, 1916, p. 313, pl. lxxxvi, f. 5.

Lagrange: Grand Junction, Tenn.

Laurophyllum reticulatum Lesquereux = *Ficus laurophylla.*

LAURUS (Tournefort) Linné, Sp. pl., 1753, p. 369.

Laurus acuminata Newberry = *Laurus palaeophila.*

Laurus angusta Heer. Lesquereux = *Laurus atanensis.*

Laurus antecedens Lesquereux, U. S. Geol. Surv., Mon. 17, 1891 [1892], p. 92, pl. xi, f. 3.—Hollick, U. S. Geol. Surv., Mon. 50, 1906, p. 80, pl. xxviii, f. 9, 10.

Dakota: Kansas.

Magothy: Glen Cove, Long Island, N. Y.

Laurus atanensis Berry, Bull. Torr. Bot. Club, vol. 38, 1911, p. 420; U. S. Geol. Surv., Prof. Paper 84, 1914, p. 53, pl. xiii, f. 7.

Laurus angusta Heer, Fl. foss. arct., vol. 6, Abt. 2, 1882, p. 76, pl. xx, f. 1b, 7, pl. xliii, f. 1c.—Lesquereux, U. S. Geol. Surv., Mon. 17, 1891 [1892], p. 93; pl. xvi, f. 7.—Hollick, in Ries, School of Mines Quart., vol. 15, 1894, p. 355.—Hollick, N. Y. Bot. Gard., Bull., vol. 3, 1904, p. 408, pl. lxx, f. 10, 11; U. S. Geol. Surv., Mon. 50, p. 81, pl. xxvii, f. 11, 12.

Dakota: Ellsworth County, Kans.

Magothy: Northport, Long Island, N. Y.

Black Creek (Middendorf): Rocky Point, S. C.

Laurus baueri Knowlton, U. S. Geol. Surv., Prof. Paper 98, 1916, p. 340, pl. lxxxix, f. 5.
Fruitland: 2 miles east of Chaco River, San Juan Basin, N. Mex.

Laurus brossiana Lesquereux = *Persea brossiana.*

Laurus californica Lesquereux, Rept. U. S. Geol. Surv. Terr., vol. 8 (Cret. and Tert. Fl.), 1883, p. 252, pl. lvii, f. 3; pl. lviii, f. 6–8.—Knowlton, in Lindgren, Jour. Geol., vol. 4, 1896, p. 890.—Knowlton, U. S. Geol. Surv., Mon. 32, pt. 2, 1899, p. 725.
Miocene (auriferous gravels): Corral Hollow, Alameda County; Independence Hill, Placer County; and Little Cow Creek, Shasta County, Calif.
Miocene (intermediate flora): Yellowstone National Park.
?Eocene: Southwestern Oregon.

Laurus californica Lesquereux. Lesquereux, U. S. Nat. Mus., Proc., vol. 11, 1888, p. 12, pl. iv, f. 1 = *Mespilodaphne pseudoglauca.*

Laurus carolinensis Michaux = *Persea borbonia.*

Laurus cascadia Duror, Jour. Geol., vol. 24, 1916, p. 576, f. 6, 7A, B, B.[1]
Swauk: Skykomish Basin, Wash.

Laurus? caudata Knowlton, U. S. Geol. Surv., Prof. Paper 101, 1918, p. 316, pl. lxxxix, f. 1.
Raton: Wootton, Colo.

Laurus colombi Heer, Nouv. mém. Soc. helv. sci. nat., vol. 21, 1865, p. 7, pl. i, f. 6.
——: Nanaimo, Vancouver Island, British Columbia.

Laurus coloradensis Knowlton, U. S. Geol. Surv., Prof. Paper 101, 1917, p. 268, pl. xlv, f. 3; idem, Prof. Paper 98, 1916, p. 340, pl. lxxxviii, f. 4, 5.
Vermejo: Rockvale, Colo.
Fruitland: San Juan County, N. Mex.

Laurus crassinervis Dawson, Roy. Soc. Canada, Trans., vol. 1, 1882 [1883], sec. 4, p. 23.
——: Sisqua River.
Cretaceous: North Fork of Old Man River.
Mill Creek: Mill Creek.

Laurus fürstenbergii Al. Braun, in Unger, Gen. et sp. pl. foss., 1850, p. 423.—Lesquereux, U. S. Nat. Mus., Proc., vol. 11, 1888, p. 26.
Miocene (auriferous gravels): Corral Hollow, Alameda County, Calif.

Laurus grandis Lesquereux, Rept. U. S. Geol. Surv. Terr., vol. 8 (Cret. and Tert. Fl.), 1883, p. 251, pl. lviii, f. 1, 3.—Knowlton, U. S. Geol. Surv., Mon. 32, pt. 2, 1899, p. 725, pl. xciii, f. 3; pl. xcv, f. 1.
Miocene (auriferous gravels): Corral Hollow, Alameda County, Calif.
Miocene (intermediate flora): Yellowstone National Park.

Laurus hollae Heer, Fl. foss. arct., vol. 6, Abt. 2, 1882, p. 76, pl. xxxiii, f. 13; pl. xliv, f. 5b; pl. xlv, f. 3.—Lesquereux, U. S. Geol. Surv., Mon. 17, 1891 [1892], p. 92, pl. xii, f. 8.—Berry, N. Y. Bot. Gard., Bull., vol. 3, 1903, p. 78, pl. l, f. 7, 8; pl. lii, f. 7, 8; Geol. Surv. N. J., Ann. Rept. 1905 [1906], p. 138.—?Hollick, U. S. Geol. Surv., Mon. 50, 1906, p. 80, pl. xxviii, f. 11.
Dakota: Ellsworth County, Kans.
Magothy: Cliffwood, N. J.
Raritan: Kreischerville, Staten Island, N. Y.

Laurus hollickii Berry, N. Y. Bot. Gard., Bull., vol. 3, 1903, p. 79, pl. lii, f. 4; Bull. Torr. Bot. Club, vol. 31, 1904, p. 77, pl. iii, f. 2; idem, vol. 33, 1906, p. 178; Geol. Surv. N. J., Ann. Rept., 1905 [1906], pp. 138, 139; Bull. Torr. Bot. Club, vol. 37, 1910, p. 26; Md. Geol. Surv., Upper Cret., 1916, p. 863, pl. lxxi, f. 4.
Magothy: Cliffwood and Morgan, N. J.; Deep Cut, Del.; Grove Point and Round Bay, Md.

Laurus knowltoni Lesquereux, U. S. Geol. Surv., Mon. 17, 1891 [1892], p. 94, pl. l, f. 4.
Dakota: Near Fort Harker, Kans.

Laurus lakesii Knowlton, U. S. Geol. Surv., Prof. Paper 130, 19—, p. —, pl. xxii, f. 6 [in preparation].
Laramie: Coal Creek, Boulder County, Colo.

Laurus lanceolata Knowlton, U. S. Geol. Surv., Prof. Paper 130, 19—, p. —, pl. xxi, f. 7 [in preparation].
Laramie: Coal Creek, Boulder County, Colo.

Laurus macrocarpa Lesquereux, Am. Jour. Sci., 2d ser., vol. 46, 1868, p. 98; Rept. U. S. Geol. Surv. Terr., vol. 6 (Cret. Fl.), 1874, p. 74, pl. x, f. 2; U. S. Geol. Surv., Mon. 17, 1891 [1892], p. 95.
Dakota: Decatur, Nebr.

Laurus (Carpites) microcarpa Lesquereux, U. S. Geol. Surv., Mon. 17, 1891 [1892], p. 93, pl. xvi, f. 8.
Dakota: Ellsworth County, Kans.

Laurus modesta Lesquereux, Rept. U. S. Geol. Surv. Terr., vol. 8 (Cret. and Tert. Fl.), 1883, p. 53, pl. xvi, f. 4.
Dakota: Morrison, Colo.

Laurus montana Knowlton, U. S. Geol. Surv., Mon. 35, pt. 2, 1899, p. 724, pl. xcv, f. 2.
Fort Union: Yellowstone River below mouth of Elk Creek, Yellowstone National Park.

Laurus nebrascensis (Lesquereux) Lesquereux, Am. Jour. Sci., 2d ser., vol. 46, 1868, p. 98; Rept. U. S. Geol. Surv. Terr., vol. 6 (Cret. Fl.), 1874, p. 74, pl. x, f. 1; pl. xxviii, f. 14.—Hollick, U. S. Geol. Surv., Mon. 50, 1906, p. 79, pl. xxviii, f. 3-8.
Persea nebrascensis Lesquereux, Am. Philos. Soc., Trans., vol. 13, 1869, p. 431, pl. xxiii, f. 9, 10.

Laurus nebrascensis—Continued.
Dakota: Decatur, Nebr.; Kansas; New Ulm, Minn.
Raritan: Tottenville and Kreischerville, Staten Island, N. Y.
Magothy: Gay Head, Marthas Vineyard, and Chappaquiddick Island, Mass.

Laurus newberryana Hollick, Bull. Torr. Bot. Club, vol. 21, 1892; p. 52, pl. clxxix, f. 5; Geol. Soc. Am., Bull., vol. 7, 1895, p. 13; U. S. Geol. Surv., Mon. 50, 1906, p. 79, pl. xxxi, f. 2.
Magothy: Glen Cove, Long Island, N. Y.

Laurus obovata Weber. Lesquereux, U. S. Geol. and Geog. Surv. Terr., Ann. Rept. 1872 [1873], p. 399 = *Laurus primigenia*.

Laurus ocoteoides Lesquereux = *Laurus wardiana*.

Laurus omalii Saporta and Marion, Mém. cour. sav. étrang. Acad., vol. 37, No. 6, 1873, p. 49, pl. vi, f. 1.—Hollick, Bull. Torr. Bot. Club, vol. 21, 1894, p. 52, pl. clxxvi, f. 3, 6.
Magothy: Sea Cliff and Glen Cove, Long Island, N. Y.

Laurus oregoniana Knowlton, U. S. Geol. Surv., Bull. 204, 1902, p. 58, pl. ix, f. 2, 3.
urus n. sp. Knowlton, in Merriam, Univ. Calif., Bull. Dept. Geol., vol. 2, 1901, p. 309.
Mascall: Van Horn's ranch, John Day Basin, Oreg.

Laurus padata Lesquereux, U. S. Geol. and Geog. Surv. Terr., Ann. Rept. 1871 [1872], Suppl., p. 14 = misspelling for *Laurus pedatus*.

Laurus palaeophila Knowlton and Cockerell, n. name.
Laurus acuminata Newberry, U. S. Nat. Mus., Proc., vol. 5, 1882 [1883], p. 511. [Homonym, Robert Brown, 1864.]
Fort Union: Valley of Yellowstone River, Mont.

Laurus pedatus Lesquereux, Am. Philos. Soc., Trans., vol. 13, 1869, p. 418, pl. xix, f. 1 = *Osmanthus pedatus*.

Laurus perdita Knowlton, U. S. Geol. Surv., Mon. 32, pt. 2, 1899, p. 723, pl. xciv, f. 1–5.
Miocene (intermediate flora): Yanceys, Yellowstone National Park.

Laurus plutonia Heer, Fl. foss. arct., vol. 6, Abt. 2, 1882, p. 75, pl. xix, f. 1d, 2–4, etc.—Lesquereux, U. S. Geol. Surv., Mon. 17, 1891 [1892], p. 91, pl. xiii, f. 5, 6; pl. xxii, f. 5; Geol. and Nat. Hist. Surv. Minn., vol. 3, pt. 1, 1895, p. 14, pl. A, f. 6; pl. B, f. 5.—Newberry, U. S. Geol. Surv., Mon. 26, 1895 [1896], p. 85, pl. xvi, f. 10, 11.—Berry, N. Y. Bot. Gard., Bull., vol. 3, 1903, p. 79, pl. l, f. 9–11; Bull. Torr. Bot. Club, vol. 31, 1904, p. 77, pl. iii, f. 1; idem, vol. 33, 1906, p. 178; Geol. Surv. N. J., Ann. Rept. 1905 [1906], p. 138, 139.—Hollick, U. S. Geol. Surv., Mon. 50, 1906, p. 80, pl. xxvii, f. 9, 10; pl. xxviii, f. 1, 2.—Berry, U. S. Geol. Surv., Prof. Paper 84, 1914, p. 52, pl. xi, f. 2; pl. xii, f. 6; Bull. Torr. Bot. Club, vol. 39, 1912, p. 401.—Hollick, N. Y. Bot. Gard. Bull., vol. 8, 1912,

Laurus plutonia—Continued.
p. 162, pl. clxix, f. 3-5.—Berry, Md. Geol. Surv., Upper Cret., 1916, p. 861, pl. lxxi, f. 5; U. S. Geol. Surv., Prof. Paper 112, 1919, p. 123.
Dakota: Kansas; New Ulm, Minn.?
Raritan: Kriescherville, Staten Island, Roslyn, Long Island, N. Y.; New Jersey? [no locality given].
Magothy: Morgan and Cliffwood, N. J.; Sea Cliff and Glen Cove, Long Island, N. Y.; Black Rock Point, Block Island, R. I.; Grove Point and Round Bay, Md.
Woodbine: Arthurs Bluff, Tex.
Cheyenne: Chatman Creek, Kans.?
Black Creek (Middendorf): Middendorf, S. C.
Tuscaloosa: Shirleys Mill and Cottondale, Ala.
Eutaw: Near Havana, Ala.

Laurus plutonia Heer. Hollick = *Laurophyllum elegans*.

Laurus praestans Lesquereux, U. S. Geol. and Geog. Surv. Terr., Bull., vol. 1, 1875 [1876], p. 368; idem, Ann. Rept. 1874 [1876], p. 305; Rept. U. S. Geol. Surv. Terr., vol. 7 (Tert. Fl.), 1878, p. 215, pl. lxiii, f. 7.
Mesaverde: Point of Rocks, Wyo.

Laurus primigenia Unger, Gen. et sp. pl. foss., 1850, p. 423.—Lesquereux, U. S. Geol. and Geog. Surv. Terr., Ann. Rept. 1872 [1873], p. 406; Rept. U. S. Geol. Surv. Terr., vol. 7 (Tert. Fl.), 1878, p. 214, pl. xxxvi, f. 5, 6, 8.—Hollick, N. Y. Acad. Sci., Trans., vol. 12, 1892, p. 33, pl. ii, f. 20; pl. iii, f. 3.—Ward, U. S. Geol. Surv., Sixth Ann. Rept., 1884-85 [1886], p. 553, pl. xlvi, f. 8-10; idem, Bull. 37, 1887, p. 47, pl. xxiii, f. 8-10.—Knowlton, idem, Mon. 32, pt. 2, 1899, p. 722, pl. xci, f. 4, 5.—Hollick, Geol. Surv. La., 1899, p. 284, pl. xl, f. 1, 2.

Laurus obovata Weber. Lesquereux, U. S. Geol. and Geog. Surv. Terr., Ann. Rept. 1872 [1873], p. 399; Rept. U. S. Geol. Surv. Terr., vol. 7 (Tert. Fl.), 1878, p. 214.

Hanna: Carbon, Wyo.
Evanston: Evanston, Wyo.
Denver: Golden, Colo.
Wilcox: Coushatta, La.
Fort Union: Yellowstone National Park.

Laurus primigenia Unger. Hollick, La. Geol. Surv., Special Rept. 5, 1899, pl. xli, f. 1 [not fig. 2] = *Oreodaphne mississippiensis*.

Laurus princeps Heer, Fl. tert. Helv., vol. 2, 1856, p. 77, pl. lxxxix, f. 16, 17; pl. xc, f. 17, 20; xcvii, f. 1.—Lesquereux, Rept. U. S. Geol. Surv. Terr., vol. 8 (Cret. and Tert. Fl.), 1883, p. 250, pl. lviii, f. 2.—Knowlton, U. S. Geol. Surv., Mon. 32, pt. 2, 1899, p. 725, pl. xcv, f. 3.
Miocene (auriferous gravels): Corral Hollow, Alameda County, Calif.
Fort Union?: Yellowstone National Park.

Laurus proteaefolia Lesquereux, U. S. Geol. and Geog. Surv. Terr., Bull., vol. 1, 1875 [1876], p. 393; idem, Ann. Rept., 1874 [1876], p. 342, pl. v, f. 1, 2; Rept. U. S. Geol. Surv. Terr., vol. 8 (Cret. and Tert. Fl.), 1883, p. 52, pl. iii, f. 9, 10; pl. xvi, f. 6.—Berry, N. Y. Bot. Gard., Bull., vol. 3, 1903, p. 78, pl. xlvii, f. 9; pl. xlix, f. 6; Bull. Torr. Bot. Club, vol. 31, 1904, p. 78, pl. i, f. 10; idem, vol. 32, 1905, p. 46, pl. ii, f. 3; Geol. Surv. N. J., Ann. Rept. 1905 [1906], p. 138; Bull. Torr. Bot. Club, vol. 37, 1910, p. 26; Md. Geol. Surv., Upper Cret., 1916, p. 863, pl. lxxv, f. 1.

Dakota: Morrison, Colo.; Fort Harker, Kans.; near Lander, Wyo.

Magothy: Morgan and Cliffwood, N. J.; Grove Point and Round Bay, Md.

Laurus ratonensis Knowlton, U. S. Geol. Surv., Prof. Paper 101, 1918, p. 316, pl. xci, f. 1–4.

Ficus pseudolmediafolia Berry, U. S. Geol. Surv., Prof. Paper 91, 1916, p. 205, pl. xxvii, f. 2.

Raton: Near Yankee, N. Mex.

Wilcox: Near Harrisburg, Ark.

Lagrange: Puryear, Henry County, Tenn.

Laurus resurgens Saporta, Ann. sci. nat., Bot., 5th ser., vol. 4, 1866, p. 132, pl. vii, f. 9a, 9b.—Ward, U. S. Geol. Surv., Sixth Ann. Rept., 1884–85 [1886], p. 553, pl. xlvi, f. 7; idem, Bull. 37, 1887, p. 47, pl. xxiii, f. 7.

Fort Union: Bull Mountains, Mont.

Miocene (auriferous gravels): Corral Hollow, Alameda County, Calif.

Laurus salicifolia Lesquereux = *Laurus saliciformis.*

Laurus saliciformis Knowlton and Cockerell, n. name.

Laurus salicifolia Lesquereux, Rept. U. S. Geol. Surv. Terr., vol. 8 (Cret. and Tert. Fl.), 1883, p. 251, pl. lviii, f. 4, 5.—Knowlton, in Turner, Am. Geol., vol. 15, 1895, p. 377. [Homonym, Banks and Solander, 1838.]

Miocene (auriferous gravels): Corral Hollow, Alameda County, and Monte Cristo mine, Spanish Peak, Calif.

Laurus schmidtiana Heer, Fl. foss. arct., vol. 5, pt. 2, 1878, p. 50, pl. xv, f. 8.—Lesquereux, Mus. Comp. Zool., Bull., vol. 16, 1888, p. 50.

Denver: Golden, Colo.

Laurus sessiliflora Lesquereux = *Malapoenna sessiliflora.*

Laurus socialis Lesquereux, Rept. U. S. Geol. Surv. Terr., vol. 7 (Tert. Fl.), 1878, p. 213, pl. xxxvi, f. 1–4, 7; U. S. Geol. and Geog. Surv. Terr., Ann. Rept. 1876 [1878], p. 510; Mus. Comp. Zool., Bull., vol. 16, 1888, p. 50.—Knowlton, in Weed, U. S. Geol. Surv., Bull. 105, 1893, p. 56.—Knowlton, U. S. Geol. Surv., Prof. Paper 101, 1918, p. 317, pl. xci, f. 5.

Laurus socialis—Continued.
Evanston: Evanston, Wyo.
Denver: Golden, Colo.
——: Lassen County, Calif.
Livingston: Bozeman coal field, Mont.
Raton: Delagua, and near Aguilar, Colo.; York Canyon, N. Mex.

Laurus socialis Lesquereux. Lesquereux, U. S. Nat. Mus., Proc., vol. 11, 1888, p. 24 = *Mespilodaphne pseudoglauca*.

Laurus teliformis Lesquereux, U. S. Geol. Surv., Mon. 17, 1891 [1892], p. 94, pl. l, f. 9.—Hollick, U. S. Geol. Surv., Mon. 50, 1906, p. 80, pl. xxxi, f. 3.
Dakota: Kansas.
Magothy: Gay Head, Marthas Vineyard, Mass.

Laurus utahensis Lesquereux, Rept. U. S. Geol. Surv. Terr., vol. 7 (Tert. Fl.), 1878, p. 216, pl. xxvi, f. 11.—Knowlton, U. S. Geol. Surv., Prof. Paper 101, 1918, p. 318, pl. xc, f. 4.
Fort Union age?: *Bridgers Pass, Wyo.
Raton: Wet Canyon near Weston, Colo.

Laurus utahensis Lesquereux. Lesquereux, U. S. Nat. Mus., Proc., vol. 11, 1888, p. 24 = *Ficus denveriana*.

Laurus wardiana Knowlton, U. S. Geol. Surv., Bull. 152, 1898, p. 129.—Berry, U. S. Geol. Surv., Prof. Paper 91, 1916, p. 13.—Knowlton, idem, Prof. Paper 130, 19—, p. —, pl. xvi, f. 1 [in preparation].
Laurus ocoteoides Lesquereux, Rept. U. S. Geol. Surv. Terr., vol. 7 (Tert. Fl.), 1878, p. 215, pl. xxxvi, f. 10 [not *L. ocoteoides* Massalongo, 1858].
Laramie: *Golden, Colo.
Midway(?): Earle, Bexar County, Tex.
?Dawson: Near Mosley, Colo.

Laurus sp. Hollick, Bull. Torr. Bot. Club, vol. 19, 1892, p. 332.
Pleistocene: Bridgeton, N. J.

Laurus n. sp. Knowlton, in Merriam, Univ. Calif., Bull. Dept. Geol., vol. 2, 1901, p. 309 = *Laurus oregoniana*.

Laurus sp. Knowlton, Washington Acad. Sci., Proc., vol. 11, 1909, p. 195.
Lance: 8 miles west of Bridger, Mont.

Laurus sp. Knowlton, in Vaughan, Am. Geol., vol. 16, 1895, p. 308.
Wilcox: Old Port Caddo Landing, Harrison County, Tex.

Laurus? sp. Knowlton, U. S. Geol. Surv., Bull. 163, 1900, p. 13, pl. i, f. 4.
Eagle: Missouri River below Coal Banks, Mont.

Laurus sp. Knowlton, in Washburne, U. S. Geol. Surv., Bull. 590, 1914, p. 36.
Eocene?: Eugene, Oreg.

Laurus sp. Lesquereux, Rept. on Clays in New Jersey, 1878, p. 29.
Raritan: Sayreville, N. J.

Leaf, undetermined, Fontaine, in Ward, U. S. Geol. Surv., Mon. 48, 1905 [1906], p. 136, pl. xxxvii, f. 3, 4.
Jurassic: Douglas County, Oreg.

Leaf, undetermined, Fontaine, in Ward, U. S. Geol. Surv., Mon. 48, 1905 [1906], p. 136, pl. xxxvii, f. 5, 6.
Jurassic: Douglas County, Oreg.

LEBEPHYLLUM Wilson, Canada Geol. Surv., Victoria Mem. Mus., Bull. 1, 1913, p. 87. [Type, *L. reineckei*.]

Lebephyllum reineckei Wilson, Canada Geol. Surv., Victoria Mem. Mus., Bull. 1, 1913, p. 88, pl. ix, f. 1, 2.
Miocene: Kettle River, Beaverdell district, British Columbia.

LEGUMINOSITES Bowerbank, History Foss. Fr. and Seeds, London clay, 1840, p. 124.

Leguminosites alternans Lesquereux, U. S. Geol. and Geog. Surv. Terr., Bull., vol. 1, 1875 [1876], p. 388; idem, Ann. Rept., 1874 [1876], p. 315; Rept. U. S. Geol. Surv. Terr., vol. 8 (Cret. and Tert. Fl.), 1883, p. 202.
Green River: Mouth of White River, Wyo.

Leguminosites? arachioides (Lesquereux) Lesquereux, Rept. U. S. Geol. Surv. Terr., vol. 7 (Tert. Fl.), 1878, p. 301, pl. lix, f. 13, 14.—Dawson, Geol. Surv. Canada, Rept. of Progress, 1877–78 [1878], p. 186b.—Ward, U. S. Geol. Surv., Sixth Ann. Rept., 1884–85 [1886], p. 554, pl. xlix, f. 7; idem, Bull. 37, 1887, p. 65, pl. xxix, f. 2.—Penhallow, Rept. Tert. Pl. Brit. Col., 1908, p. 61, text f. 14.—Knowlton, Washington Acad. Sci., Proc., vol. 11, 1909, pp. 184, 212; Jour. Geol., vol. 19, 1911, p. 369.—Berry, U. S. Geol. Surv., Prof. Paper 91, 1916, p. 249, pl. xlviii, f. 9.—Knowlton, U. S. Geol. Surv., Prof. Paper 101, 1918, p. 326.

Carpolithes arachioides Lesquereux, U. S. Geol. and Geog. Surv. Terr., Ann. Rept. 1872 [1873], p. 403.

Raton: Near Berwynd, Colo.
Denver: Golden and Middle Park, Colo.
Evanston: *Evanston, Wyo.
Fort Union: Clear Creek, near Glendive, Mont.
Eocene: Finlay and Omineca rivers, British Columbia.
Eocene?: Similkameen and Horsefly rivers, British Columbia.
Lance: Melville, Mont.; Monarch, Wyo.; South Dakota [T. 23 N., R. 21 E.].
Wilcox: Hamilton, Sabine County, Tex., and Mansfield, De Soto Parish, La.

Leguminosites atanensis Heer, Fl. foss. arct., vol. 3, Abt. 2, 1874, p. 119, pl. xxxiv, f. 6.—Newberry, U. S. Geol. Surv., Mon. 26, 1895 [1896], p. 97, pl. xlii, f. 40.—Berry, Geol. Surv. N. J., Bull. 3, 1911, p. 154.

Raritan: New Jersey but no locality given.

Leguminosites? borealis Dawson, Roy. Soc. Canada, Trans., vol. 7, 1889, sec. 4, p. 72 pl. x, f. 7.—Penhallow, Rept. Tert. Pl. Brit. Col., 1908, p. 62.

Eocene: Mackenzie River basin, Northwest Territory.

Leguminosites canavalioides Berry, Md. Geol. Surv., Upper Cret., 1916, p. 842, pl. lxxvi, f. 6.

Magothy: Grove Point, Md.

Leguminosites cassioides Lesquereux = *Leguminosites lesquereuxiana.*

Leguminosites? coloradensis Knowlton, U. S. Geol. Surv., Prof. Paper 130, 19—, p. —, pl. xix, f. 9 [in preparation].

Laramie: Marshall, Colo.

Leguminosites columbianus Knowlton, U. S. Geol. Surv., Prof. Paper 130, 19—, p. —, pl. xix, f. 4, 5 [in preparation].

Laramie: Erie, Colo.

Leguminosites constrictus Lesquereux, U. S. Geol. Surv., Mon. 17, 1891 [1892], p. 151, pl. xliv, f. 3.—Hollick, U. S. Geol. Surv., Mon. 50, 1906, p. 86, pl. xxxii, f. 20.

Dakota: Ellsworth County, Kans.

Raritan: ?Oak Neck, Long Island, N. Y.

Leguminosites convolutus Lesquereux, U. S. Geol. Surv., Mon. 17, 1891 [1892], p. 151, pl. xliv, f. 4.—Hollick, U. S. Geol. Surv., Mon. 50, 1906, p. 86, pl. xxxii, f. 18, 19.

Dakota: Ellsworth County, Kans.

Magothy: Gay Head, Marthas Vineyard, Mass.; Glen Cove, Long Island, N. Y.

Leguminosites coronilloides Heer, Fl. foss. arct., vol. 3, 1874, Abt. 2, p. 119, pl. xxxiv, f. 14.—Lesquereux, U. S. Geol. Surv., Mon. 17, 1891 [1892], p. 149, pl. xiii, f. 10.—Newberry, idem, Mon. 27, 1896, p. 97, pl. xiii, f. 48.—Hollick, idem, Mon. 50, 1907, p. 86, pl. xxxii, f. 16, 17.—Berry, Bull. Torr. Bot. Club, vol. 37, 1910, p. 24; Geol. Surv. N. J., Bull. 3, 1911, p. 153; Md. Geol. Surv., Upper Cret., 1916, p. 841, pl. lxxvi, f. 4.

Leguminosites frigidus Heer. Hollick, N. Y. Acad. Sci., Trans., vol. 12, 1892, p. 34, pl. ii, f. 11.

Dakota: Near Delphos, Kans.

Magothy: Gay Head, Marthas Vineyard, Mass.; Grove Point, Md.

Raritan: Tottenville, Staten Island, N. Y.

Leguminosites cultriformis Lesquereux, Rept. U. S. Geol. Surv. Terr., vol. 8 (Cret. and Tert. Fl.), 1883, p. 86, pl. x, f. 4.
Dakota: Locality not given.

Leguminosites dakotensis Lesquereux, U. S. Geol. Surv., Mon. 17, 1891 [1892], p. 150, pl. xxxviii, f. 5.
Dakota: Ellsworth County, Kans.

Leguminosites frigidus Heer. Hollick, N. Y. Acad. Sci., Trans., vol. 12, 1892, p. 34, pl. ii, f. 11 = *Leguminosites coronilloides*.

Leguminosites hymenophyllus Lesquereux, U. S. Geol. Surv., Mon. 17, 1891 [1892], p. 152, pl. lv, f. 7–9; pl. lvi, f. 3.
Dakota: Kansas.

Leguminosites ingaefolia Berry, U. S. Geol. Surv., Prof. Paper 112, 1919, p. 103, pl. xxiii, f. 5.
Tuscaloosa: Shirleys Mill and Glen Allen, Ala.

Leguminosites insularis Heer, Fl. foss. arct., vol. 6, Abt. 2, 1882, p. 103, pl. xliv, f. 6.—Lesquereux, U. S. Geol. Surv., Mon. 17, 1891 [1892], p. 152, pl. liv, f. 4.
Dakota: Near Fort Harker, Kans.

Leguminosites lamarensis Knowlton, U. S. Geol. Surv., Mon. 32, pt. 2, 1899, p. 731, pl. lxxxix, f. 5, 6.
Miocene: Yellowstone National Park.

Leguminosites laramiensis Knowlton, U. S. Geol. Surv., Prof. Paper 130, 19—, p. —, pl. xvii, f. 4 [in preparation].
Laramie: 4 miles north of Colorado Springs, Colo.

Leguminosites lèsquereuxiana Knowlton, U. S. Geol. Surv., Bull. 152, 1898, p. 131; U. S. Geol. Surv., Mon. 32, pt. 2, 1899, p. 730, pl. lxxix, f. 4.

Leguminosites cassioides Lesquereux, Rept. U. S. Geol. Surv. Terr., vol. 7 (Tert. Fl.), 1878, p. 300, pl. lix, f. 1–4 [not *L. cassioides* Engelhardt, 1870].

Green River: Green River, Wyo.
Livingston: Bozeman coal field, Mont.
Fort Union: Yellowstone National Park.

Leguminosites marcouanus Heer = *Liriodendron marcouanum*.

Leguminosites middendorfensis Berry, U. S. Geol. Surv., Prof. Paper 84, 1914, p. 46, pl. viii, f. 13.
Black Creek (Middendorf): Middendorf, S. C.

Leguminosites? neomexicana Knowlton, U. S. Geol. Surv., Prof. Paper 98, 1916, p. 341, pl. xc, f. 3, 4.
Kirtland: Near Pina Veta China, San Juan County, N. Mex.

Leguminosites omphalobioides Lesquereux, U. S. Geol. Surv., Mon. 17, 1891 [1892], p. 149, pl. xxxviii, f. 4.—Newberry, U. S. Geol. Surv., Mon. 26, 1895 [1896], p. 97, pl. xlii, f. 39.—Berry, Bull. Torr. Bot. Club, vol. 37, 1910, p. 24; Geol. Surv. N. J., Bull. 3, 1911, p. 155; Md. Geol. Surv., Upper Cret., 1916, p. 843, pl. lxxvi, f. 5; U. S. Geol. Surv., Prof. Paper 112, 1919, p. 103; Bull. Torr. Bot. Club, vol. 44, 1917, p. 185.

Dakota: Ellsworth County, Kans.
Raritan: New Jersey? [no locality given].
Bingen: Near Maxwell Spur, Pike County, Ark.
Magothy: Grove Point, Md.
Tuscaloosa: Shirleys Mill, Ala.

Leguminosites perfoliatus Berry, U. S. Geol. Surv., Prof. Paper 91, 1916, p. 249, pl. xlviii, f. 1.

Lagrange: Puryear, Henry County, Tenn.

Leguminosites phaseolites? Heer, Fl. foss. arct., vol. 3, Abt. 2, 1874, p. 118, pl. xxxiv, f. 7–11.—Lesquereux, U. S. Geol. Surv., Mon. 17, 1891 [1892], p. 153, pl. lv, f. 10.

Dakota: Kansas.

Leguminosites pisiformis? Heer, Fl. tert. Helv., vol. 3, 1859, p. 129, pl. cxxxix, f. 37–39.—Lesquereux, Am. Jour. Sci., 2d ser., vol. 32, 1861, p. 361; Geol. Vt., vol. 2, 1861, p. 716, f. 152.—Perkins, Rept. Vt. State Geol., 1905–6 [1906], p. 224.

Miocene: Brandon, Vt.

Leguminosites podogonialis Lesquereux, U. S. Geol. Surv., Mon. 17, 1891 [1892], p. 148, pl. xiii, f. 11; pl. xxxviii, f. 16.

Dakota: Ellsworth County and near Delphos, Kans.

Leguminosites raritanensis Berry, Bull. Torr. Bot. Club, vol. 36, 1909, p. 257, pl. xviii, f. 4; Geol. Surv. N. J., Bull. 3, 1911, p. 155, pl. xx, f. 5.

Raritan: South Amboy, N. J.

Leguminosites reniformis Bowerbank, History of the fossil fruits and seeds of the London clay, 1840, p. 135, pl. xvii, f. 29, 30.—?Berry, U. S. Geol. Surv., Prof. Paper 91, 1916, p. 251.

Lagrange: Puryear, Henry County, Tenn.

Leguminosites robiniifolia Berry, Bull. Torr. Bot. Club, vol. 37, 1910, p. 196; U. S. Geol. Surv., Prof. Paper 84, 1914, p. 46, pl. ix, f. 11.

Black Creek: Court House Bluff, Cape Fear River, N. C.; Langley, S. C.

Leguminosites serrulatus Lesquereux = *Ptelea modesta.*

Leguminosites shirleyensis Berry, U. S. Geol. Surv., Prof. Paper 112, 1919, p. 104, pl. xxiii, f. 4.

Diospyros rotundifolia Lesquereux, U. S. Geol. Surv., Mon. 17, 1892, pl. xvii, f. 11 [not figs. 8–10].

Tuscaloosa: Shirleys Mill, Ala.

Leguminosites subovatus Bowerbank, History of the fossil fruits and seeds of the London clay, 1840, p. 125, pl. xvii, f. 1, 2.—?Berry, U. S. Geol. Surv., Prof. Paper 91, 1916, p. 251.
Lagrange: Puryear, Henry County, Tenn.

Leguminosites truncatus Knowlton, in Lesquereux, U. S. Geol. Surv., Mon. 17, 1891 [1892], p. 150, pl. xxi, f. 7.
Dakota: Near Delphos, Kans.

Leguminosites tuscaloosensis Berry, U. S. Geol. Surv., Prof. Paper 112, 1919, p. 104, pl. xxiii, f. 6.
Tuscaloosa: Shirleys Mill, Ala.

Leguminosites ungeri Lesquereux, U. S. Geol. Surv., Mon. 17, 1891 [1892], p. 153 [name].

Leguminosites wickliffensis Berry, U. S. Geol. Surv., Prof. Paper 91, 1916, p. 250, pl. li, f. 8.
Lagrange: Wickliffe, Ballard County, Ky.

Leguminosites sp. Knowlton, Jour. Geol., vol. 19, 1911, p. 371.
Lance: Near Thunder Butte, S. Dak.

LEITNERIA Chapman, Fl., 1860, p. 427.

Leitneria floridana Chapman ? Berry, Jour. Geol., vol. 25, 1917, p. 662.
Pleistocene: Vero, Fla.

LEMNA Linné, Sp. pl. 1753, p. 970.

Lemna? bullata Lesquereux, U. S. Geol. and Geog. Surv. Terr., Bull., vol. 1, 1875 [1876], p. 363.
Mesaverde: Point of Rocks, Wyo.

Lemna penicillata Lesquereux = *Spirodela penicillata.*

Lemna (Spirodela) scutata Dawson, Brit. N. A. Bound. Com. (Rept. Geol. and Res. Vic. 49th Parallel), 1875, App. A, p. 328, pl. xvi, f. 5, 6, 7a; Roy. Soc. Canada, Trans., vol. 4, sec. 4, 1886 [1887], p. 23, pl. i, f. 6.—Ward, U. S. Geol. Surv., Sixth Ann. Rept., 1884–85 [1886], p. 550, pl. xxxii, f. 4, 5; idem, Bull. 37, 1887, p. 17, pl. iii, f. 4, 5.
Eocene: *Badlands south of Woody Mountain, Alberta.
Fort Union: Near Glendive, Mont.

Lemna scutata Dawson. Lesquereux, Tert. Fl., p. 102, pl. lxi, f. 2 = *Nelumbo intermedia.*

LEPACYCLOTES Emmons, Geol. Rept. Midland Counties N. C., 1856, p. 332.

Lepacyclotes circularis Emmons, Geol. Rept. Midland Counties N. C., 1856, p. 332, pl. iii, f. 4; American geology, pt. 6, 1857, p. 130, pl. iii, f. 4.—Fontaine, in Ward, U. S. Geol. Surv., Twentieth Ann. Rept., pt. 2, 1900, p. 311, pl. xlvii, f. 4.
Araucarites carolinensis Fontaine, U. S. Geol. Surv., Mon. 6, 1883, p. 119, pl. xlix, f. 8.
Triassic (Newark): Lockville, N. C. ?

Lepacyclotes ellipticus Emmons, Geol. Rept. Midland Counties N. C., 1856, p. 332, pl. iii, f. 6; American geology, pt. 6, 1857, p. 129, pl. iii, f. 6.—Fontaine, in Ward, U. S. Geol. Surv., Twentieth Ann. Rept., pt. 2, 1900, p. 311, pl. xlvii, f. 5; pl. xlviii.

Araucarites carolinensis Fontaine, U. S. Geol. Surv., Mon. 6, 1883, p. 118, 119, pl. lii, f. 4, 4a.

Triassic (Newark): North Carolina.

Lepidodendron sp. Emmons = *Zamiostrobus virginiensis*.

Lepidodendron sp. Taylor = *Sphenolepidium sternbergianum*.

LEPTOSTROBUS Heer, Mém. Acad. imp. sci. St.-Pétersb. (Fl. foss. arct., vol. 4, Abt. 2), 7th ser., vol. 22, No. 12, 1876, p. 72.

Leptostrobus? alatus Ward, U. S. Geol. Surv., Nineteenth Ann. Rept., pt. 2, 1899, p. 673, pl. clxiii, f. 16, 17.

Fuson: Pine Creek, Crook County, Wyo.

Leptostrobus foliosus Fontaine = *Abietites foliosus*.

Leptostrobus longifolius Fontaine = *Abietites longifolius*.

Leptostrobus? mariposensis Fontaine, in Ward, U. S. Geol. Surv., Twentieth Ann. Rept., pt. 2, 1900, p. 340.

Mariposa: Near Princeton, Calif.

Leptostrobus? multiflorus Fontaine, U. S. Geol. Surv., Mon. 15, 1889, p. 230, pl. clxv, f. 6.

Patapsco: Near Brooke, Va.

Leptostrobus? ovalis Ward, U. S. Geol. Surv., Mon. 48, 1905 [1906], p. 514, pl. cviii, f. 9, 10.

Leptostrobus? (b) sp.? Fontaine, U. S. Geol. Surv., Mon. 15, 1889, p. 231, pl. cxxxvi, f. 10, 10a.

Patapsco: Chinkapin Hollow, Va.

Leptostrobus?, seeds of, Fontaine, U. S. Geol. Surv., Mon. 15, 1889 [1890], p. 332, pl. cxxxvi, f. 11.

——: Kankeys, Va.

Leptostrobus?, seeds of (*a*), Fontaine, U. S. Geol. Surv., Mon. 15, 1889, p. 331, pl. cxxxv, f. 6.

Patuxent: Near Dutch Gap Canal, Va.

Leptostrobus?, seeds of (*b*), Fontaine, U. S. Geol. Surv., Mon. 15, 1889, p. 231, pl. cxxxvi, f. 10.

Patuxent: Near Dutch Gap Canal, Va.

Leptostrobus? sp. Fontaine, in Ward, U. S. Geol. Surv., Twentieth Ann. Rept., pt. 2, 1900, p. 363, pl. xlvii, f. 1.

Undetermined cone, Fontaine, Am. Jour. Sci., 4th ser., vol. 2, 1896, p. 274.

Jurassic: Oroville, Calif.

LESCURIA Perkins, Rept. Vt. State Geol., 1905–6 [1906], p. 220. [Type, *L. attenuata*.]

Lescuria attenuata Perkins, Rept. Vt. State Geol., 1905–6 [1906], p. 220, pl. lvii, f. 7–10.
Miocene: Brandon, Vt.

LEUCAENA Bentham, in Hooker, Jour. Bot., vol. 4, 1842, p. 416.

Leucaena coloradensis Cockerell, Am. Mus. Nat. Hist., Bull., vol. 24, 1908, p. 96.
Miocene: Florissant, Colo.

LEUCOTHOE D. Don, Edinb. New Philos. Jour., vol. 17, 1834, p. 159.

Leucothoe racemosa (Linné) A. Gray. Hollick, Bull. Torr. Bot. Club, vol. 19, 1892, p. 332.
Pleistocene: Bridgeton, N. J.

LIBOCEDRUS Endlicher, Syn. Conif., 1847, p. 42.

Libocedrus coloradensis (Cockerell) Cockerell, Am. Nat., vol. 44, 1910, p. 32.
Heyderia coloradensis Cockerell, Science, new ser., vol. 26, 1907, p. 447; Am. Mus. Nat. Hist., Bull., vol. 24, 1908, p. 78; Pop. Sci. Mo., vol. 73, 1908, p. 122, text f.
Miocene: Florissant, Colo.

Libocedrus cretacea Heer = *Thuja cretacea.*

LIGUSTRUM (Tournefort) Linné, Sp. pl., 1753, p. 7.

Ligustrum subtile Hollick, N. Y. Bot. Gard., Bull., vol. 8, 1912, p. 167, pl. clxvii, f. 5.
Magothy: Roslyn, N. Y.

LIMNOBIUM L. C. Richard, Mém. Inst. Paris, vol. 32, 1811, p. 66, pl. viii.

Limnobium obliteratum Cockerell, Am. Jour. Sci., 4th ser., vol. 26, 1908, p. 65, f. 1a.
Miocene: Florissant, Colo.

Lindera masoni Lesquereux = *Benzoin masoni.*

Lindera venusta Lesquereux = *Benzoin venustum.*

Lindera vetusta Lesquereux. Ward, Jour. Geol., vol. 2, 1894, p. 261 = *Sassafras mudgii.*

LIQUIDAMBAR Linné, Sp. pl., 1753, p. 999.

Liquidambar californicum Lesquereux, Mus. Comp. Zool., Mem., vol. 6, No. 2, 1878, p. 14, pl. vi, f. 7c; pl. vii, f. 3, 6.
Miocene (auriferous gravels): Chalk Bluffs, Nevada County, Calif.

Liquidambar convexum Cockerell, Am. Mus. Nat. Hist., Bull., vol. 24, 1908, p. 94, pl. vii, f. 16.
Miocene: Florissant, Colo.

Liquidambar? cucharas Knowlton, U. S. Geol. Surv., Prof. Paper 101, 1918, p. 320, pl. xci, f. 6.
Raton: Cucharas Canyon, Walsenburg, Colo.

Liquidambar europaeum Al. Braun, in Buckland, Geol., vol. 1, 1836?, p. 115.—Heer, Fl. foss. arct., vol. 2, Abt. 2, 1869, p. 25, pl. ii, f. 7.—Lesquereux, Rept. U. S. Geol. Surv. Terr., vol. 8 (Cret. and Tert. Fl.), 1883, p. 159, pl. xxxii, f. 1.—Newberry, U. S. Geol. Surv., Mon. 35, 1898, p. 100, pl. xlvii, f. 1–3.—Knowlton, U. S. Geol. Surv., Bull. 204, 1902, p. 62; U. S. Nat. Mus., Proc., vol. 17, 1894, p. 226; Geol. Soc. Am., Bull., vol. 5, 1893, p. 585.

Clarno (upper part): Bridge Creek, Grant County, Oreg.

Kenai: Port Graham, Alaska.

Liquidambar europaeum Al. Braun. Lesquereux, U. S. Nat. Mus., Proc., vol. 11, 1888, p. 14 = *Liquidambar europaeum patulum*.

Liquidambar europaeum patulum Knowlton, U. S. Geol. Surv., Bull. 204, 1902, p. 62, pl. x, f. 5.

Liquidambar europaeum Al. Braun. Lesquereux, U. S. Nat. Mus., Proc., vol. 11, 1888, p. 14.

Mascall: Van Horn's ranch, John Day Basin, Oreg.

Liquidambar gracilis Lesquereux = *Aralia? gracilis*.

Liquidambar integrifolium Lesquereux = *Liquidambar obtusilobatus*.

Liquidambar obtusilobatus (Heer) Hollick, in Newberry, U. S. Geol. Surv., Mon. 35, 1898, p. 101, pl. v, f. 4; pl. xii, f. 4.

Phyllites obtusilobatus Heer, Acad. Nat. Sci. Philadelphia, Proc., vol. 10, 1858, p. 266.

Acerites pristinus Newberry, N. Y. Lyc. Nat. Hist., Ann., vol. 9, 1868, p. 15.—[Lesquereux] U. S. Geol. and Geog. Surv. Terr., Ill. Cret. and Tert. Pl., 1874, pl. v, f. 4.

Liquidambar integrifolium Lesquereux, Am. Jour. Sci., 2d ser., vol. 46, 1868, p. 93; U. S. Geol. Surv. Terr., Rept., vol. 6 (Cret. Fl.), 1874, p. 56, pl. ii, f. 1–3; pl. xxiv, f. 2; pl. xxix, f. 8; idem, vol. 8 (Cret. and Tert. Fl.), 1883, p. 45, pl. xiv, f. 3; U. S. Geol. Surv., Mon. 17, 1891 [1892], p. 75.

Dakota: Blackbird Hill, Nebr.; Fort Harker, Salina, Kans.

Liquidambar pachyphyllum Knowlton, U. S. Geol. Surv., Bull. 204, 1902, p. 63, pl. ix, f. 1.

Mascall: Van Horn's ranch, John Day Basin, Oreg.

Liquidambar protensum Unger? Lesquereux, U. S. Nat. Mus., Proc., vol. 11, 1888, p. 13, pl. viii, f. 3.—Knowlton, U. S. Geol. Surv., Bull. 204, 1902, p. 62.

Mascall: Van Horn's ranch, John Day Basin, Oreg.

Liquidambar styraciflua Linné. Hollick, Bull. Torr. Bot. Club, vol. 19, 1892, p. 331.—Knowlton, Am. Geol., vol. 18, 1896, p. 371.—Berry, Jour. Geol., vol. 15, 1907, p. 343; Am. Jour. Sci., 4th ser., vol. 29, 1910, p. 397.

Pleistocene: Morgantown, W. Va.; Neuse River, N. C.; Black Warrior River 311½ miles above Mobile, Ala.; Bridgeton, N. J.

Liquidambar sp. Knowlton, in Lindgren, Jour. Geol., vol. 4, 1896, p. 890.

Miocene (auriferous gravels): Independence Hill, Placer County, Calif.

Liquidambar sp. ? Knowlton, U. S. Geol. Surv., Bull. 204, 1902, p. 63, pl. xii, f. 4.
Mascall: Near Van Horn's ranch, John Day Basin, Oreg.

LIRIODENDRON Linné, Sp. pl., 1753, p. 535.

Liriodendron acuminatum Lesquereux, Mus. Comp. Zool., Bull., vol. 7, 1881, p. 227; U. S. Geol. Surv., Mon. 17, 1891 [1892], p. 207, pl. xxvii, f. 2, 3.
Dakota: Near Glencoe, Kans.

Liriodendron acuminatum bilobatum Lesquereux, U. S. Geol. Surv., Mon. 17, 1891 [1892], p. 207, pl. xxviii, f. 4.
Dakota: Near Fort Harker, Kans.

Liriodendron alatum Newberry. Hollick, Bull Torr. Bot. Club, vol. 21, 1894, p. 467, pl. ccxx, ccxxi.—Knowlton, U. S. Geol. Surv., Bull. 163, 1900, p. 13, pl. i, f. 6; U. S. Geol. Surv., Prof. Paper 101, 1918, p. 269, pl. xlviii, f. 3.
Vermejo: Walsenburg, Colo.
Eagle: Missouri River below coal banks, Mont.

Liriodendron attenuatum Hollick, U. S. Geol. Surv., Mon. 50, 1906, p. 68, pl. xxi, f. 9–11.
Liriodendron primaevum Newberry. Hollick, Bull. Torr. Bot. Club, vol. 21, 1894, p. 61, pl. 179, f. 4.
Magothy: Glen Cove, Long Island, N. Y.; Gay Head, Marthas Vineyard, Mass.

Liriodendron cruciforme Lesquereux = *Liriodendron giganteum cruciforme.*

Liriodendron dubium Berry, Bull. Torr. Bot. Club, vol. 34, 1907, p. 196, pl. 14, f. 3.
Black Creek: Court House Bluff, Cape Fear River, N. C.

Liriodendron giganteum Lesquereux, Am. Jour. Sci., 2d ser., vol. 46, 1868, p. 99; Rept. U. S. Geol. Surv. Terr., vol. 6 (Cret. Fl.), 1874, p. 93, pl. xxii, f. 2; U. S. Geol. Surv., Mon. 17, 1891 [1892], p. 206, pl. xxv, f. 1; pl. xxvi, f. 5; pl. xxvii, f. 1.
Dakota: Near Glencoe, Kans.

Liriodendron giganteum cruciforme (Lesquereux), Lesquereux, U. S. Geol. Surv., Mon. 17, 1891 [1892], p. 206, pl. xxviii, f. 1, 2.
Liriodendron cruciforme Lesquereux, Mus. Comp. Zool., Bull., vol. 7, 1881, p. 227.
Dakota: Elkhorn Creek, Kans.

Liriodendron intermedium Lesquereux, Am. Jour. Sci., 2d ser., vol. 46, 1868, p. 99; U. S. Geol. Surv., Mon. 17, 1891 [1892], p. 207, pl. xxv, f 5.
Dakota: Near Glencoe, Kans.

Liriodendron laramiense Ward, U. S. Geol. Surv., Sixth Ann. Rept., 1884–85 [1886], p. 556, pl. lix, f. 1–5; U. S. Geol. Surv., Bull. 37, 1887, p. 102, pl. xlviii, f. 2.—Knowlton, U. S. Geol. Surv., Bull. 163, 1900, p. 57.

Mesaverde: Point of Rocks, Wyo.

Liriodendron marcouanum (Heer) Knowlton, n. comb.

Leguminosites marcouanus Heer, Philadelphia Acad. Sci., Proc., vol. 10, 1858, p. 265.

Phyllites obcordatus Heer, Philadelphia Acad. Sci., Proc., vol. 10, 1858, p. 266.—Newberry, U. S. Geol. Surv., Mon. 35, 1898, p. 136, pl. v, f. 2.

Liriodendron primaevum Newberry, N. Y. Lyc. Nat. Hist., Ann., vol. 9, 1868, p. 12; Ill. Cret. and Tert. Pl., 1878, pl. vi, f. 7.—Lesquereux, U. S. Geol. Surv., Mon. 17, 1892, p. 203, pl. xxiv, f. 4; pl. xxvi, f. 1–4.—Newberry, idem, Mon. 35, 1898, p. 96, pl. vi, f. 7.

Bumelia marcouana (Heer) Lesquereux, Rept. U. S. Geol. Surv. Terr., vol. 6 (Cret. Fl.), 1874, p. 90, pl. xxviii, f. 2.—Hollick, U. S. Geol. Surv., Mon. 50, 1906, p. 68, pl. xxi, f. 7.

Dakota: Blackbird Hill, Nebr.; Minnesota; Pipe Creek, Cloud County, Kans.

Raritan: Tottenville, N. Y.

Liriodendron meekii Heer, Philadelphia Acad. Sci., Proc., vol. 10, 1858, p. 265.—Capellini and Heer, Phyll. crét. Nebr., 1866, p. 21, pl. iv, f. 3, 4.—Lesquereux, U. S. Geol. Surv., Mon. 17, 1891 [1892], p. 205, pl. xxviii, f. 5, 6.—Newberry, U. S. Geol. Surv., Mon. 35, 1898, p. 95, pl. vi, f. 5, 6.—Berry, U. S. Geol. Surv., Prof. Paper 112, 1919, p. 92, pl. xx, f. 2; Bull. Torr. Bot. Club, vol. 44, 1917, p. 180.

Tuscaloosa: Cottondale, Ala.

Dakota: Blackbird Hill, mouth of Big Sioux, Nebr.

Bingen: Near Maxwell Spur, Pike County, Ark.

Liriodendron morganensis Berry, Bull. Torr. Bot. Club, vol. 33, 1906, p. 174; Geol. Surv. N. J., Ann. Rept. 1905 [1906], pp. 138, 150, pl. xxi, f. 2–4; pl. xxiii, f. 1; pl. xxv, f. 1–4.

Magothy: Morgan, N. J.

Liriodendron oblongifolium Newberry, Bull. Torr. Bot. Club, vol. 14, 1887, p. 5, pl. lxi, f. 1.—Hollick, idem, vol. 21, 1894, p. 62, pl. 179, f. 3; U. S. Geol. Surv., Mon. 50, 1906, p. 68, pl. xxi, f. 8.—Berry, Geol. Surv. N. J., Bull. 3, 1911, p. 137.

Raritan: Woodbridge, N. J.

Magothy: Glen Cove, Long Island, N. Y.

Liriodendron pinnatifidum Lesquereux, Mus. Comp. Zool., Bull., vol. 7, 1881, p. 227; U. S. Geol. Surv., Mon. 17, 1891 [1892], p. 209, pl. xxvii, f. 4, 5.

Dakota: Near Glencoe, Kans.

Liriodendron praetulipiferum Dawson, Roy. Soc. Canada, Trans., vol. 11, 1893 [1894], sec. 4, p. 63, pl. viii, f. 27.

Cretaceous: Nanaimo, Vancouver Island, British Columbia.

Liriodendron primaevum Newberry = *Liriodendron marcouanum*.

Liriodendron primaevum Newberry. Hollick, Bull. Torr. Bot. Club, vol. 21, 1894, p. 61, pl. clxxix, f. 4 = *Liriodendron attenuatum*.

Liriodendron quercifolium Newberry, Bull. Torr. Bot. Club, vol. 14, 1887, p. 6, pl. lxii, f. 1; U. S. Geol. Surv., Mon. 26, 1895 [1896], p. 81, pl. li, f. 1–6.—Berry, Geol. Surv. N. J., Bull. 3, 1911, p. 138, pl. xvii, f. 1; Bull. Torr. Bot. Club, vol. 39, 1912, p. 395.
Raritan: Woodbridge, N. J.
Woodbine: Arthurs Bluff, Tex.

Liriodendron semialatum Lesquereux, Mus. Comp. Zool., Bull., vol. 7, 1881, p. 227; U. S. Geol. Surv., Mon. 17, 1891 [1892], p. 204, pl. xxv, f. 2–4; pl. xxix, f. 3.
Dakota: Near Glencoe, Kans.

Liriodendron simplex Newberry. Hollick, N. Y. Acad. Sci., Trans., vol. 12, 1893, p. 235, pl. v, f. 5 = *Liriodendropsis retusa*.

Liriodendron simplex Newberry [part] = *Liriodendropsis simplex*.

Liriodendron snowii Lesquereux, U. S. Geol. Surv., Mon. 17, 1891 [1892], p. 209, pl. xxix, f. 1, 2.
Dakota: Kansas.

Liriodendron succedens Dawson, Roy. Soc. Canada, Trans., vol. 11, 1893 [1894], sec. 4, p. 62, pl. viii, f. 26.
Upper Cretaceous: Port McNeill, Vancouver Island, British Columbia.

Liriodendron tulipifera Linné. Berry, Am. Nat., vol. 41, 1907, p. 695.—Torreya, vol. 9, 1909, p. 71, f. 1; Am. Jour. Sci., 4th ser., vol. 29, 1910, p. 396.—Torreya, vol. 15, 1915, p. 208, f. 1.
Pleistocene: Abercrombies Landing, Chattahoochee River, Ala.; Neuse River, N. C.; Black Warrior River, 356 and 328½ miles above Mobile, Ala.; Indian Head, Charles County, Md.

Liriodendron wellingtonii Lesquereux, U. S. Geol. Surv., Mon. 17, 1891 [1892], p. 208, pl. xxviii, f. 3.
Dakota: Near Fort Harker, Kans.

Liriodendron sp. (cf. *L. primaevum* Newberry) Berry, Bull. Torr. Bot. Club, vol. 34, 1907, p. 197.
Black Creek, Blackmans Bluff, Neuse River, N. C.

Liriodendron sp. Lesquereux, U. S. Geol. and Geog. Surv. Terr., Ann. Rept. 1871 [1872], p. 298. [Not afterward referred to by author.]
——: Spring Canyon, Mont.

LIRIODENDROPSIS Newberry MS. Hollick, N. Y. Acad. Sci., Trans., vol. 12, July 6, 1893, p. 235. [Type, *Liriodendron simplex* Newberry.] (See Newberry, U. S. Geol. Surv., Mon. 26, 1895, p. 82.)

Liriodendropsis angustifolia (Newberry) Newberry, U. S. Geol. Surv., Mon. 26, 1895 [1896], p. 84, pl. liii, f. 8.—Berry, Bull. Torr. Bot. Club, vol. 31, 1904, p. 77, pl. iv, f. 4; Geol. Surv. N. J., Ann. Rept. 1905 [1906], p. 138.—Hollick, U. S. Geol. Surv., Mon. 50, 1906, p. 71, pl. xxvi, f. 2–5.—Berry, U. S. Geol. Surv., Prof. Paper 112, 1919, p. 102.

Liriodendron simplex Newberry [part]. Newberry, Bull. Torr. Bot. Club, vol. 14, 1887, p. 6, pl. lxii, f. 4.

Liriodendropsis simplex angustifolia Newberry. Ward, U. S. Geol. Surv., Sixteenth Ann. Rept., pt. 1, 1895, p. 540, pl. cvii, f. 7.

Magothy: Cliffwood, N. J.; Gay Head, Mass.; Glen Cove, Long Island, N. Y.

Raritan: Woodbridge, N. J.

Tuscaloosa: Shirleys Mill and Glen Allen, Ala.

Liriodendropsis constricta Ward, U. S. Geol. Surv., Sixteenth Ann. Rept., pt. 1, 1894–95 [1896], p. 540, pl. cvii, f. 8 [on plate called *L. simplex*].—Hollick, U. S. Geol. Surv., Mon. 50, 1906, p. 71, pl. xxii, f. 7; pl. xxvi, f. 6–15; pl. xl, f. 15; N. Y. Bot. Gard., Bull., vol. 8, 1912, p. 164, pl. clxvi, f. 1; Md. Geol. Surv., Upper Cret., 1916, p. 844; U. S. Geol. Surv., Prof. Paper 112, 1919, p. 103.

Magothy: Glen Cove, Long Island, N. Y.; Gay Head, Marthas Vineyard, Mass.; Grove Point, Md.

Tuscaloosa: Shirleys Mill, Ala.

Liriodendropsis retusa (Heer) Hollick, U. S. Geol. Surv., Mon. 50, 1906, p. 72, pl. xxv, f. 8, 9.—Berry, Geol. Surv. N. J., Bull. 3, 1911, p. 157, pl. xix, f. 1.

Sapotacites retusus Heer, Fl. foss. arct., vol. 7, 1883, p. 32, pl. lxi, f. 10.—Newberry, U. S. Geol. Surv., Mon. 26, 1895 [1896], p. 123, pl. liii, f. 5, 6.

Liriodendron simplex Newberry. Hollick, N. Y. Acad. Sci., Trans., vol. 12, 1893, p. 235, pl. v, f. 5.

Magothy: Glen Cove, Long Island, N. Y.

Raritan: Woodbridge, N. J.

Liriodendropsis simplex (Newberry) Newberry, U. S. Geol. Surv., Mon. 26, 1895 [1896], p. 83, pl. xix, f. 2, 3; pl. liii, f. 1–4, 7.—Hollick, Geol. Soc. Am., Bull., vol. 7, 1895, p. 13.—Berry, Geol. Surv. N. J., Bull. 3, 1911, p. 158, pl. xix, f. 2.—Hollick, N. Y. Bot. Gard., Bull., vol. 8, 1912, p. 165, pl. clxviii, f. 2.—Berry, U. S. Geol. Surv., Prof. Paper 112, 1919, p. 101, pl. xxii, f. 6.

Liriodendron simplex Newberry [part]. Newberry, Bull. Torr. Bot. Club, vol. 14, 1887, p. 6, pl. lxii, f. 2, 3.

Magothy: Glen Cove, Long Island, N. Y.; Gay Head, Marthas Vineyard, Mass.

Raritan: Tottenville, Staten Island, N. Y.; Woodbridge and Milltown, N. J.; Roslyn, N. Y.

Tuscaloosa: Shirleys Mill and Glen Allen, Ala.

Liriodendropsis simplex angustifolia Newberry. Ward, U. S. Geol. Surv., Sixteenth Ann. Rept., pt. 1, 1895, p. 540, pl. cvii, f. 7 = *Liriodendropsis angustifolia.*

Liriodendropsis spectabilis Hollick, U. S. Geol. Surv., Mon. 50, 1906, p. 73, pl. xxii, f. 1–6.

Celastrophyllum decurrens Lesquereux? Hollick, Bull. Torr. Bot. Club, vol. 21, 1894, p. 59, pl. clxix, f. 1.

Magothy: Gay Head, Marthas Vineyard, Mass.; Glen Cove, Long Island, N. Y.

LIRIOPHYLLUM Lesquereux, U. S. Geol. and Geog. Surv. Terr., Ann. Rept. 1876 [1878], p. 482. [Type, *L. beckwithii.*]

Liriophyllum beckwithii Lesquereux, U. S. Geol. and Geog. Surv. Terr., Ann. Rept. 1876 [1878], p. 482; Rept. U. S. Geol. Surv. Terr., vol. 8 (Cret. and Tert. Fl.), 1883, p. 76, pl. x, f. 1; U. S. Geol. Surv., Mon. 17, 1891 [1892], p. 210.

Dakota: Morrison, Colo.

Liriophyllum obcordatum Lesquereux, Rept. U. S. Geol. Surv. Terr., vol. 8 (Cret. and Tert. Fl.), 1883, p. 77; U. S. Geol. Surv., Mon. 17, 1891 [1892], p. 210, pl. xxviii, f. 7.

Dakota: Morrison, Colo.; near Fort Harker, Kans.

Liriophyllum populoides Lesquereux, U. S. Geol. and Geog. Surv. Terr., Ann. Rept. 1876 [1878], p. 483; Rept. U. S. Geol. Surv. Terr., vol. 8 (Cret. and Tert. Fl.), 1883, p. 76, pl. xi, f. 1, 2.

Dakota: Colorado and Kansas.

Litsaea carbonensis Ward = *Malapoenna carbonensis.*

Litsea cretacea Lesquereux = *Malapoenna cretacea.*

Litsea cuneata Knowlton = *Malapoenna cuneata.*

Litsea falcifolia Lesquereux = *Malapoenna falcifolia.*

Litsea weediana Knowlton = *Malapoenna weediana.*

Litsea sp. Knowlton = *Malapoenna* sp.

LOMATIA Robert Brown, Linnaean Soc. London, Trans., vol. 10, 1809, p. 199.

Lomatia abbreviata Lesquereux, Rept. U. S. Geol. Surv. Terr., vol. 8 (Cret. and Tert. Fl.), 1883, p. 167, pl. xliii, f. 17.

Miocene: Florissant, Colo.

Lomatia acutiloba Lesquereux, Rept. U. S. Geol. Surv. Terr., vol. 8 (Cret. and Tert. Fl.), 1883, p. 167, pl. xliii, f. 11–16, 20.—Cockerell, Am. Nat., vol. 42, 1908, p. 579, text f. 10.

Miocene: Florissant, Colo.

Lomatia hakeaefolia Lesquereux = *Lomatites hakeaefolia.*

Lomatia interrupta Lesquereux, Rept. U. S. Geol. Surv. Terr., vol. 8 (Cret. and Tert. Fl.), 1883, p. 167, pl. xliii, f. 18, 19.—**Knowlton,** U. S. Nat. Mus., Proc., vol. 51, 1916, p. 267, pl. xxv, f. 5.
Miocene: Florissant, Colo.

Lomatia microphylla Lesquereux, U. S. Geol. and Geog. Surv. Terr., Bull., vol. 1, 1875 [1876], p. 389; idem, Ann. Rept. 1874 [1876], p. 315; Rept. U. S. Geol. Surv. Terr., vol. 7 (Tert. Fl.), 1878, p. 211, pl. lxv, f. 14, 15.
Green River (?): Mouth of White River, Wyo.

Lomatia? saportanea (Lesquereux) Lesquereux, U. S. Geol. and Geog. Surv. Terr., Ann. Rept. 1874 [1876], p. 346; Rept. U. S. Geol. Surv. Terr., vol. 8 (Cret. and Tert. Fl.), 1883, p. 51, pl. iii, f. 8; U. S. Geol. Surv., Mon. 17, 1891 [1892], p. 89.
Todea? saportanea Lesquereux, Rept. U. S. Geol. Surv. Terr., vol. 6 (Cret. Fl.), 1874, p. 48, pl. xxix, f. 1–4.
Dakota: Kansas.

Lomatia? saportanea longifolia Lesquereux, Rept. U. S. Geol. Surv. Terr., vol. 8 (Cret. and Tert. Fl.), 1883, p. 52.
Dakota: Morrison, Colo.

Lomatia spinosa Lesquereux = *Lomatites spinosa.*

Lomatia terminalis Lesquereux, Rept. U. S. Geol. Surv. Terr., vol. 8 (Cret. and Tert. Fl.), 1883, p. 166, pl. xliii, f. 2–7.
Miocene: Florissant, Colo.

Lomatia tripartita Lesquereux, Rept. U. S. Geol. Surv. Terr., vol. 8 (Cret. and Tert. Fl.), 1883, p. 166, pl. xliii, f. 8–10.—**Cockerell, Am.** Nat., vol. 42, 1908, p. 579, text f. 8, 9.
Miocene: Florissant, Colo.

LOMATITES Saporta, Ann. sci. nat., 4th ser., Bot., vol. 17 (Études, vol. 1, pt. 1), 1862, p. 252 (99).

Lomatites hakeaefolia (Lesquereux) Cockerell, Am. Mus. Nat. Hist., Bull., vol. 24, 1908, p. 89; Knowlton, U. S. Nat. Mus., Proc., vol. 51, 1916, p. 267, pl. xxvi, f. 1, 2.
Lomatia hakeaefolia Lesquereux, Rept. U. S. Geol. Surv. Terr., vol. 8 (Cret. and Tert. Fl.), 1883, p. 166, pl. xxxii, f. 19.
Carduus florissantensis Cockerell, Bull. Torr. Bot. Club, vol. 33, 1906, p. 311, text f. 6.
Miocene: *Florissant, Colo.

Lomatites spinosa (Lesquereux) Cockerell, Am. Mus. Nat. Hist., Bull., vol. 24, 1908, p. 89.
Lomatia spinosa Lesquereux, Rept. U. S. Geol. Surv. Terr., vol. 8 (Cret. and Tert. Fl.), 1883, p. 166, pl. xliii, f. 1.
Miocene: Florissant, Colo.

LONCHOCARPUS Humboldt, Bonpland and Kunth, Nov. gen. et sp., vol. 6, 1823, p. 383.

Lonchocarpus novae-caesareae Hollick, Bull. Torr. Bot. Club, vol. 23, 1896, p. 49, pl. cclix, f. 6–8.
Pleistocene: Bridgeton, N. J.

LONCHOPTERIS Brongniart, Dict. sci. nat., vol. 57 (Prodrome), 1828, p. 68 (59).

Lonchopteris oblonga (Emmons) Fontaine, U. S. Geol. Surv., Mon. 6, 1883, p. 103, pl. xlix, f. 1, 1a; in Ward, U. S. Geol. Surv., Twentieth Ann. Rept., pt. 2, 1900, p. 239, pl. xxv, f. 3–5; p. 285, pl. xxxviii, f. 8–10.

Acrostichites oblongus Emmons, Geol. Rept. Midland Counties N. C., 1856, p. 326, pl. iv, f. 6, 8; American geology, pt. 6, 1857, p. 101, pl. iv, f. 6, 8.—Fontaine, U. S. Geol. Surv., Mon. 6, 1883, p. 103.

Triassic (Newark): Ellingtons, N. C.; York County, Pa.

Lonchopteris virginiensis Fontaine, U. S. Geol. Surv., Mon. 6, 1883, p. 53, pl. xxviii, f. 1, 2; pl. xxix, f. 1–4.
Triassic (Newark): Manakin and Clover Hill, Va.

LOPERIA Newberry, N. Y. Acad. Sci., Trans., vol. 6, 1886–87 [1887], p. 126 [name]; U. S. Geol. Surv., Mon. 14, 1888, p. 93. [Type, *Bambusium carolinense*.]

Loperia carolinensis (Fontaine) Ward, U. S. Geol. Surv., Twentieth Ann. Rept., 1898–99 [1900], pt. 2, p. 427.

Bambusium carolinense Fontaine, U. S. Geol. Surv., Mon. 6, 1883, p. 120, pl. lii, f. 1.

Triassic (Newark): Connecticut; North Carolina.

Loperia simplex Newberry, N. Y. Acad. Sci., Trans., vol. 6, 1886–87 [1887], p. 126 [name]; U. S. Geol. Surv., Mon. 14, 1888, p. 93, pl. xxv, f. 1–3.

Bambusium? sp. Fontaine, U. S. Geol. Surv., Mon. 6, 1883, p. 90, pl. xlviii, f. 3, 3a.

Undetermined plant, Emmons, American geology, pt. 6, 1857, pp. 131, 132, f. 99, 100.

Triassic (Newark): Durham, Conn.; Cumberland coal field, Va.

LYCOPODITES Brongniart, Mém. Mus. hist. nat., Paris, vol. 8, 1822, p. 209.

Lycopodites? eolignitícus Berry, U. S. Geol. Surv., Prof. Paper 91, 1916, p. 163, pl. ix, f. 4, 5.
Wilcox (Holly Springs): Early Grove, Marshall County, Miss.

Lycopodites? montanensis Fontaine, in Ward, U. S. Geol. Surv., Mon. 48, 1905 [1906], p. 302, pl. lxxii, f. 15, 16.
Kootenai: Near Geyser, Cascade County, Mont.

Lycopodites tuscaloosensis Berry, U. S. Geol. Surv., Prof. Paper 112, 1919, p. 50, pl. v, f. 4–7.
Tuscaloosa: Shirleys Mill and Glen Allen, Ala.

LYCOPODIUM (Ruppius) Linné, Sp. pl., 1753, p. 1100.

Lycopodium cretaceum Berry, Am. Jour. Sci., 4th ser., vol. 30, 1910, pp. 275, 276, f. 1–6; U. S. Geol. Surv., Prof. Paper 84, 1914, p. 84, pl. ii, f. 1–6; Md. Geol. Surv., Upper Cret., 1916, p. 759, pl. l, f. 10, 11; U. S. Geol. Surv., Prof. Paper 112, 1919, p. 49.
Magothy: Little Round Bay, Anne Arundel County, Md.
Black Creek (Middendorf): Middendorf, S. C.
?Tuscaloosa: Shirleys Mill and Glen Allen, Ala.

Lycopodium lesquereuxiana Knowlton, Geol. Soc. Am., Bull., vol. 8, 1897, p. 154; U. S. Geol. Surv., Bull. 163, 1900, p. 24.
Selaginella falcata Lesquereux, ex. p., Rept. U. S. Geol. Surv. Terr., vol. 7 (Tert. Fl.), 1878, p. 46, pl. lxiv, f. 13, 13a [not pl. lxi, f. 12–15].
Lycopodium lesquereuxii Knowlton, U. S. Geol. Surv., Bull. 152, 1898, p. 136.
Mesaverde: Point of Rocks, Wyo.

Lycopodium prominens Lesquereux, U. S. Geol. and Geog. Surv. Terr., Ann. Rept. 1873 [1874], p. 409; Rept. U. S. Geol. Surv. Terr., vol. 7 (Tert. Fl.), 1878, p. 45, pl. v, f. 13–13b.
Miocene: Elko station, Nev.

Lycopodium sp. Hinde, Canadian Jour. (Canadian Inst.), vol. 15, 1878, p. 399.
Pleistocene: Scarborough Heights, Ontario.

Lycopodium sp.? (living) Penhallow, Roy. Soc., Canada, Trans., 2d ser., vol. 2, 1896, sec. 4, pp. 62, 72; Brit. Assoc. Adv. Sci., Bradford meeting, 1900, p. 335.
Pleistocene: Missinaibi River, Canada.

Lycopodolithes? sp. Taylor = *Frenelopsis ramosissima.*

LYGODIUM Swartz, Schrader, Jour. Bot., vol. 2, 1800, p. 106.

Lygodium binervatum (Lesquereux) Berry, U. S. Geol. Surv., Prof. Paper 91, 1916, p. 165, pl. x, f. 3–8.
Salisburia binervata Lesquereux, Am. Philos. Soc., Trans., vol. 13, 1869, p. 412, pl. xv, f. 3–6.
Ginkgo binervata (Lesquereux) Knowlton, U. S. Geol. Surv., Bull. 152, 1898, p. 110.
Wilcox (Ackerman): Hurleys, Benton County, Miss.
Wilcox: De Soto Parish, La.

Lygodium? compactum Lesquereux, Am. Jour. Sci., 2d ser., vol. 16, 1868, p. 206; Rept. U. S. Geol. Surv. Terr., vol. 7 (Tert. Fl.), 1878, p. 64, pl. v, f. 9.—Knowlton, U. S. Geol. Surv., Prof. Paper 130, 19—, p. —, pl. i, f. 1 [in preparation].
Laramie: Marshall, Colo.

Lygodium dentoni Lesquereux, U. S. Geol. and Geog. Surv. Terr., Bull., vol. 1, 1875 [1876], p. 383; idem, Ann. Rept. 1874 [1876], p. 309; Rept. U. S. Geol. Surv. Terr., vol. 7 (Tert. Fl.), 1878, p. 63, pl. lxv, f. 12, 13.
Green River (?): Mouth of White River, Wyo.

Lygodium kaulfussii Heer, Beiträge z. nähern Kennt. d. Sächs.-Thuring. Braunkohle, 1861, p. 400, pl. viii, f. 21.—Newberry, U. S. Geol. Surv., Mon. 35, 1898, p. 1, pl. lxii, f. 1–4.—Knowlton, U. S. Geol. Surv., Mon. 32, pt. 2, 1898, p. 672, pl. lxxx, f. 1–3; U. S. Geol. Surv., Bull. 204, 1900, p. 21.—Penhallow, Rept. Tert. Pl. Brit. Col., 1908, p. 62.

Lygodium neuropteroides Lesquereux, U. S. Geol. and Geog. Surv. Terr., Ann. Rept. 1870 [1871], p. 384; Rept. U. S. Geol. Surv. Terr., vol. 7 (Tert. Fl.), 1878, p. 61, pl. v, f. 4–7; pl. vi, f. 1.

Green River (?): Barrel Springs, Wyo. [Type locality for Lesquereux's *L. neuropteroides*.]

Green River: Green River, Wyo.

Fort Union: Elk Creek, Yellowstone National Park.

Clarno (lower part): Cherry Creek, Crook County, Oreg.

Eocene: Wilkinson coal field, Wash.; Burrard Inlet, British Columbia.

Lygodium marvinei Lesquereux, U. S. Geol. and Geog. Surv. Terr., Bull., vol. 1, 1875 [1876], p. 383; idem, Ann. Rept. 1874 [1876], p. 309.

——: Eagle River, Grand Junction, Colo. [In beds of probable Laramie age.]

Lygodium neuropteroides = *Lygodium kaulfussii*.

Lygodium trichomanoides Lesquereux, Rept. U. S. Geol. Surv. Terr., vol. 6 (Cret. Fl.), 1874, p. 45, pl. i, f. 2; U. S. Geol. Surv., Mon. 17, 1891 [1892], p. 25.

Lygodium? sp. Lesquereux, Am. Jour. Sci., vol. 46, 1868, p. 91.

Dakota: Fort Harker, Kans.

Lygodium? sp. Lesquereux = *Lygodium trichomanoides*.

LYSIMACHIA (Tournefort) Linné, Sp. pl., 1753, p. 146.

Lysimachia sp. Knowlton, Washington Acad. Sci., Proc., vol. 11, 1909, p. 207.

Lance: Converse County, Wyo.

MACCLINTOCKIA Heer, Fl. foss. arct., vol. 1, 1868, p. 114.

Macclintockia cretacea Heer, Fl. foss. arct., vol. 6, Abt. 2, 1882, p. 70, pl. xxxvi, f. 1a; pl. xxxvii, f. 2–4.—Dawson, Roy. Soc. Canada, Trans., vol. 3, 1885, sub. 4, p. 13, pl. iv, f. 3.—Lesquereux, U. S. Geol. Surv., Mon. 17, 1891 [1892], p. 197, pl. lix, f. 4.

Dakota: Ellsworth County, Kans.

Mill Creek: Mill Creek, British Columbia.

Macclintockia lyallii Heer, Fl. foss. arct., vol. 1, 1868, p. 115, pl. xv, f. 1a, 2; pl. xvi, f. 7a, b; pl. xvii, f. 2a, b; pl. xlvii, f. 13; pl. xlviii, f. 8.—Lesquereux, U. S. Geol. and Geog. Surv. Terr., Ann. Rept. 1872 [1873], p. 400. [Not afterward mentioned by Lesquereux.]

Post-Laramie: Black Buttes, Wyo.

Macclintockia trinervis Heer, Fl. foss. arct., vol. 1, 1868, p. 115, pl. xv, f. 7–9.—Dawson, Roy. Soc. Canada, Trans., vol. 11, 1893 [1894], p. 64, pl. x, f. 38.

Upper Cretaceous: Port McNeill, Vancouver Island, British Columbia.

Maclura auriantiaca Nuttall = *Toxylon pomiferum*.

Macreightia crassa Lesquereux = *Populus crassa*.

Macrospores Fontaine, U. S. Geol. Surv., Mon. 15, 1889, p. 274, pl. cxxxvi.

Patuxent: Dutch Gap Canal, Va.

MACROTAENIOPTERIS Schimper, Pal. vég., vol. 1, 1869, p. 610.

Macrotaeniopteris californica Fontaine, Am. Jour. Sci., 4th ser., vol. 2, 1896, p. 274 [name]; in Ward, U. S. Geol. Surv., Twentieth Ann. Rept., pt. 2, 1900, p. 349, pl. liii, f. 1; pl. liv, f. 1, 2; U. S. Geol. Surv., Mon. 48, 1905 [1906], p. 82, pl. xiv, f. 1–4.—Knowlton, in Martin and Katz, U. S. Geol. Surv., Bull. 485, 1912, p. 63.

Jurassic: Oroville, Calif.; Douglas County, Oreg.

Jurassic (Tuxedni): Tuxedni Bay, Alaska.

Macrotaeniopteris crassinervis Feistmantel, Pal. Indica, 2d ser., vol. 1 (Foss. Fl. Gond. Syst.), pt. 1, 1877, p. 102 (50), pl. xxxviii, f. 1–3.—Fontaine, U. S. Geol. Surv., Mon. 6, 1883, p. 22, pl. v, f. 5; pl. vi, f. 1, 2.

Triassic (Newark): Clover Hill, Va.

Macrotaeniopteris magnifolia (Rogers) Schimper, Pal. vég., vol. 1, 1869, p. 610; Fontaine, U. S. Geol. Surv., Mon. 6, 1883, pp. 18, 103, pl. ii, f. 1–3; pl. iii, f. 1–3; pl. iv, f. 1–4; pl. v, f. 1–4; in Ward, U. S. Geol. Surv., Twentieth Ann. Rept., pt. 2, 1900, pp. 238, 283, pl. xxii, f. 7–9; pl. xxiii, f. 1–3; pl. xxiv.—Russell, U. S. Geol. Surv., Bull. 85, 1892, pl. viii [restoration].

Taeniopteris magnifolia Rogers, Philadelphia Assoc. Am. Geol. and Nat., 1843, p. 306, pl. xiv, unnumbered figure on right.—Emmons, American geology, pt. 6, 1857, p. 102, f. 70.

Triassic (Newark): Virginia; North Carolina; York County, Pa.

Macrotaeniopteris nervosa Fontaine, Am. Jour. Sci., 4th ser., vol. 2, 1896, p. 274; in Ward, U. S. Geol. Surv., Twentieth Ann. Rept., pt. 2, 1900, p. 350, pl. liv, f. 3; pl. lv, f. 1.

Jurassic: Oroville, Calif.

Macrotaeniopteris vancouverensis Dawson, Roy. Soc. Canada, Trans., vol. 11, 1893 [1894], sec. 4, p. 55, pl. v, f. 1–3.

Upper Cretaceous: Nanaimo, Vancouver Island, British Columbia.

MAGNOLIA Linné, Sp. pl., 1753, p. 535.

Magnolia acuminata Linné, Hollick, Bull. Torr. Bot. Club, vol. 19, 1892, p. 331.

Pleistocene: Bridgeton, N. J.

Magnolia alternans Heer, Phyll. Crét. Nebr., 1866, p. 20, pl. iii, f. 2-4.—Lesquereux, Rept. U. S. Geol. Surv. Terr., vol. 6 (Cret. Fl.), 1874, p. 92, pl. xviii, f. 4; U. S. Geol. Surv., Mon. 17, 1891 [1892], p. 201, pl. xxxiv, f. 11.—Newberry, U. S. Geol. Surv., Mon. 35, 1895 [1896], p. 73, pl. lv, f. 1, 2, 4, 6; ?idem, Mon. 35, 1898, p. 94, pl. v, f. 6.—Hollick, N. Y. Bot. Gard., Bull., vol. 2, 1902, p. 405, pl. xli, f. 4, 5; idem, Mon. 50, 1906, p. 67.—Berry, Geol. Surv. N. J., Bull. 3, 1911, p. 130, pl. xv, f. 1.

Dakota: Kansas, Nebraska, and Minnesota.

Raritan: Woodbridge and Sayreville, N. J.

Magothy: Elm Point, Great Neck, Long Island, N. Y.; Chappaquiddick Island, Mass.

Tuscaloosa: Alabama.

Magnolia amplifolia Heer, Kreidefl. v. Moletein in Mähren: Neue Denkschr. Schw. Ges., vol. 23, 1869, p. 21, pl. viii, f. 2; pl. ix, f. 1.—Lesquereux, U. S. Geol. Surv., Mon. 17, 1891 [1892], p. 200, pl. xxiv, f. 3.—Hollick, U. S. Geol. Surv., Mon. 50, 1906, p. 65, pl. xviii, f. 1.

Dakota: Pipe Creek, Cloud County, Kans.

Magothy: Gay Head, Marthas Vineyard, Mass.

Magnolia angustifolia Newberry, U. S. Nat. Mus., Proc., vol. 5, 1882 [1883], p. 513.—Berry, U. S. Geol. Surv., Prof. Paper 91, 1916, p. 214.—Knowlton, U. S. Geol. Surv., Prof. Paper 101, 1917, p. 309, pl. lxxix, f. 1; pl. lxxx; pl. lxxxi, f. 1.

Magnolia attenuata Weber. Lesquereux, Rept. U. S. Geol. Surv. Terr., vol. 7 (Tert. Fl.), 1878, p. 250, pl. xlv, f. 6.

Terminalia radobojensis Heer (not Unger). Lesquereux, U. S. Geol. and Geog. Surv. Terr., Ann. Rept. 1871 [1872], Suppl., p. 15.

Magnolia lanceolata Lesquereux. Hollick, La. Geol. Surv., Special Rept. 5, 1899, p. 282, pl. xl.

Raton: Wootton, Weston, Starkville, Aguilar, Santa Clara, Trinidad, Tercio, and other localities, Colo.; Ute Park, Raton tunnel, Vermejo Creek, Yankee, and other localities, N. Mex.; *Fishers Peak, Raton Mountains, Colo.

Wilcox: Coushatta, Red River Parish, and Naborton, De Soto Parish, La.

Lagrange: Purycar, Henry County, Tenn.

Magnolia attenuata Weber. Lesquereux, 1878 = *Magnolia angustifolia.*

Magnolia auriculata Hollick = *Magnolia hollicki.*

Magnolia boulayana Lesquereux, U. S. Geol. Surv., Mon. 17, 1891 [1892], p. 202, pl. lx, f. 2.—Berry, Bull. Torr. Bot. Club, vol. 36, 1909, p. 254; idem, vol. 37, pp. 23, 504; U. S. Geol. Surv., Prof. Paper 84, 1914, p. 112, pl. xx, f. 5; Geol. Surv. N. J., Bull. 3, 1911, p. 131, pl. xiv, f. 2; Md. Geol. Surv., Upper Cret., 1916, p. 834, pl.

Magnolia boulayana—Continued.
lxix, f. 1; U. S. Geol. Surv., Prof. Paper 112, 1919, p. 90, pl. xviii f. 2.

Magnolia glaucoides Newberry MS. Hollick, Bull. Torr. Bot. Club, vol. 21, 1894, p. 60, pl. clxxv, f. 1, 7.—Newberry, U. S. Geol. Surv., Mon. 35, 1895 [1896], p. 74, pl. lvii, f. 1-4.—Hollick, idem, Mon. 50, 1906, p. 67, pl. xix, f. 6; pl. xx, f. 6.

Dakota: *Ellsworth County, Kans.
Woodbine: Texas.
Raritan: Woodbridge, N. J.
Magothy: Grove Point and Round Bay, Md.; ?Sea Cliff, Long Island, N. Y.
Eutaw: McBrides Ferry, Columbus, Ga.
Tuscaloosa: Shirleys Mill, Ala.

Magnolia californica Lesquereux, Mus. Comp. Zool., Mem., vol. 6, No. 2, 1878, p. 25, pl. vi, f. 6, 7 [not f. 5].—Knowlton, in Lindgren, Jour. Geol., vol. 4, 1896, p. 890; U. S. Geol. Surv., Mon. 32, pt. 2, 1899, p. 718; U. S. Geol. Surv., Prof. Paper 73, 1911, p. 59; in Diller, idem, Bull. 353, 1908, p. 77.

Miocene (auriferous gravels): Chalk Bluffs, Nevada County; Independence Hill, Placer County, and Contra Costa, Calif.
——: Lassen County, Calif.
Miocene: Fossil Forest Ridge, Yellowstone National Park.
?Ione: Kosk Creek, Shasta County, Calif.
?Eocene: Southwestern Oregon.

Magnolia californica Lesquereux, Mus. Comp. Zool., Mem., vol. 6, 1878, pl. vi, f. 5 = *Persea pseudo-carolinensis*.

Magnolia californica Lesquereux. Lesquereux, U. S. Nat. Mus., Proc., vol. 11, 1888, p. 29 = *Magnolia inglefieldi*.

Magnolia capellinii Heer, Phyll. crét. Nebr., 1866, p. 21, pl. iii, f. 5, 6.—Lesquereux, Rept. on Clays in N. J., 1878, p. 29; U. S. Geol. Surv., Mon. 17, 1891 [1892], p. 203, pl. lxvi, f. 1.—Berry, Bull. Torr. Bot. Club, vol. 31, 1904, p. 76, pl. iii, f. 3.—Hollick, N. Y. Bot. Gard., Bull., vol. 3, 1904, p. 413, pl. lxxviii, f. 3.—Berry, Geol. Surv. N. J., Ann. Rept. 1905 [1906], p. 138; Bull. Torr. Bot. Club, vol. 34, 1907, p. 195.—Hollick, U. S. Geol. Surv., Mon. 50, 1906, p. 63, pl. xvii, f. 3, 4.—Berry, Bull. Torr. Bot. Club, vol. 37, 1910, p. 504; idem, vol. 38, 1911, p. 406; U. S. Geol. Surv., Prof. Paper 84, 1914, pp. 43, 112, pl. x, f. 3; pl. xx, f. 6; Md. Geol. Surv., Upper Cret., 1916, p. 836, pl. lxix, f. 4; U. S. Geol. Surv., Prof. Paper 112, 1919, p. 89, pl. xviii, f. 1; pl. xxxii, f. 7.

Tuscaloosa: Cottondale, Ala.
Eutaw: McBrides Ferry, Ga.
Ripley: Near Selmer, Tenn.

Magnolia capellinii—Continued

Dakota: Ellsworth County, Kans.; *Tekamah, Nebr.

Magothy: Cliffwood and Morgan, N. J.; Glen Cove and Center Island, Long Island, N. Y.; Grove Point and along Severn River, Md.

Upper Cretaceous: Port McNeill, Vancouver Island, British Columbia.

Black Creek: Court House Bluff, Cape Fear River, N. C.

Black Creek (Middendorf): Middendorf and Rocky Point, S. C.

Magnolia? coalvillensis Knowlton, U. S. Geol. Surv., Prof. Paper 101, 1918, p. 312.

Magnolia tenuinervis Lesquereux. Knowlton, U. S. Geol. Surv., Bull. 163, 1900, p. 55, pl. xiv, f. 1.

Montana: *Coalville, Utah.

Magnolia cordifolia Lesquereux, Am. Philos. Soc., Trans., vol. 13, 1869, p. 422, pl. xxii, f. 1, 2.—Knowlton, U. S. Geol. Surv., Prof. Paper 101, 1918, p. 315, pl. lxxxvi; pl. lxxxviii, f. 1.

Wilcox: Mississippi.

Raton: Near Trinidad and Green Canyon mine, Aguilar, Colo.

Magnolia culveri Knowlton, U. S. Geol. Surv., Mon. 32, pt. 2, 1899, p. 720, pl. xcii, f. 5; U. S. Geol. Surv., Bull. 204, 1902, p. 58.

Populus monodon Lesquereux. Lesquereux, U. S. Nat. Mus., Proc., vol. 11, 1888, p. 21.

Miocene: Yellowstone National Park.

Clarno (lower part): Cherry Creek, Crook County, Oreg.

Magnolia dayana Cockerell, Am. Nat., vol. 44, 1910, p. 35.

Magnolia lanceolata Lesquereux, Mus. Comp. Zool., Mem., vol. 6, No. 2, 1878, p. 24, pl. vi, f. 4.—Knowlton, in Lindgren, Jour. Geol., vol. 4, 1896, p. 890.—Knowlton, U. S. Geol. Surv., Bull. 204, 1900, p. 58; in Smith, U. S. Geol. Surv., Geologic Folio 86 (Ellensburg, Wash.), 1903, p. 3; in Lindgren, U. S. Geol. Surv., Prof. Paper 73, 1911, p. 59; in Diller, idem, Bull. 353, 1908, p. 77.

Miocene (auriferous gravels): Chalk Bluffs, Nevada County; Independence Hill, Placer County, Calif.

Clarno (upper part): Cherry Creek, Oreg.

Ellensburg: Ellensburg, Wash.

Eocene: Southwestern Oregon.

Magnolia elliptica Newberry, U. S. Geol. Surv., Mon. 35, 1898, p. 94, pl. xii, f. 1.

Fort Union: Tongue River, Mont.

Magnolia ensifolia Lesquereux = *Celastrophyllum ensifolium.*

Magnolia florissanticola Cockerell, Torreya, vol. 10, 1910, p. 65, text f. 1.

Miocene: *Florissant, Colo.

Magnolia glauca Linné = *Magnolia virginiana.*

Magnolia glaucoides Newberry MS. = *Magnolia boulayana.*

Magnolia hilgardiana Lesquereux, Second Rept. Geol. Recon. Arkansas, 1860, p. 319, pl. vi, f. 1; Am. Philos. Soc., Trans., vol. 13, 1869, p. 421, pl. xxi, f. 1; Rept. U. S. Geol. Surv. Terr., vol. 7 (Tert. Fl.), 1878, p. 249, pl. xliv, f. 4.—Hollick, Geol. Surv., Third Ann. Rept., 1879, p. 282, pl. xxxix.—Knowlton, in Lindgren, U. S. Geol. Surv., Prof. Paper 73, 1911, pp. 60, 61.—Knowlton, U. S. Geol. Surv., Prof. Paper 101, 1918, p. 310, pl. lxxix, f. 2; pl. lxxxv, f. 1.

Quercus lyelli Heer. Lesquereux, Am. Philos. Soc., Trans., vol. 13, 1869, p. 415, pl. xvii, f. 3 [not f. 1, 2].

Terminalia hilgardiana (Lesquereux) Berry, U. S. Geol. Surv., Prof. Paper 91, 1916, p. 325, pl. xcii, f. 2.

Wilcox (Ackerman): Hurleys, Miss.

Wilcox: Campbell's quarry, Caddo Parish, Mansfield, and Coushatta, La.; Old Port Caddo Landing, Harrison County, Tex.

Lagrange: Puryear, Tenn.

Raton: "Fishers Peak, N. Mex."; near Yankee, N. Mex.; Cokedale, Trinidad, and Aguilar, Colo.

Miocene: Lassen County, Calif.

Midway?: Earle, Bexar County, Tex.

Magnolia hollicki Berry, Bull. Torr. Bot. Club, vol. 36, 1909, p. 253; N. J. Geol. Surv., Bull. 3, 1911, p. 136, pl. xv, f. 3; Md. Geol. Surv., Upper Cret., 1916, p. 831, pl. lxix, f. 3; U. S. Geol. Surv., Prof. Paper 112, 1919, p. 92.

Magnolia auriculata Newberry, MS. Hollick, Bull. Torr. Bot. Club, vol. 21, 1894, p. 61, pl. clxxix, f. 6, 7.—Newberry, U. S. Geol. Surv., Mon. 35, 1895 [1896], p. 75, pl. xli, f. 13; pl. lviii, f. 1-11.—Berry, Bull. Torr. Bot. Club, vol. 33, 1906, p. 174.—Hollick, U. S. Geol. Surv., Mon. 50, 1906, p. 67, pl. xix, f. 5; pl. xx, f. 5, 8.

Magothy: Glen Cove, Long Island, N. Y.; Gay Head, Marthas Vineyard, Mass.; Grove Point, Md.

Raritan: Woodbridge, N. J.

Tuscaloosa: Shirleys Mill, Ala.

Magnolia inglefieldi Heer, Fl. foss. arct., vol. 1, 1868, p. 120, pl. iii, f. 5c; pl. xvi, f. 5, 6, 8b; pl. xviii, f. 1-3.—Lesquereux, U. S. Nat. Mus., Proc., vol. 11, 1888, p. 13.—Knowlton, U. S. Geol. Surv., Bull. 204, 1902, p. 58; in Lindgren, U. S. Geol. Surv., Prof. Paper 73, 1911, p. 60.

Cornus hyperborea Heer. Lesquereux, U. S. Nat. Mus., Proc., vol. 11, 1888, p. 29, pl. xv, f. 3.

Magnolia californica Lesquereux. Lesquereux, U. S. Nat. Mus., Proc., vol. 11, 1888, p. 29.

Eocene: Near Susanville, Lassen County, Calif.

Mascall: ?Van Horn's ranch, John Day Basin, Oreg.

Magnolia? inquirenda Knowlton, U. S. Geol. Surv., Prof. Paper 101, 1918, p. 312.

Magnolia tenuinervis Lesquereux, Rept. U. S. Geol. Surv. Terr., vol. 7 (Tert. Fl.), 1878, p. 249, pl. xlv, f. 5.

——: Bridgers Pass, Wyo.

Magnolia isbergiana Heer, Fl. foss. arct., vol. 6, Abt. 2, 1882, p. 91, pl. xxxvi, f. 3.—Hollick, Bull. Torr. Bot. Club, vol. 21, 1894, p. 60; U. S. Geol. Surv., Mon. 50, 1906, p. 66, pl. xx, f. 4.—Berry, Geol. Surv. N. J., Bull. 3, 1911, p. 132.

Magothy: Glen Cove, Long Island, N. Y.
Raritan: Milltown, N. J.

Magnolia lacoeana Lesquereux, U. S. Geol. Surv., Mon. 17, 1891 [1892], p. 201, pl. lx, f. 1.—Newberry, U. S. Geol. Surv., Mon. 35, 1895 [1896], p. 73, pl. lv, f. 1, 2; Hollick, U. S. Geol. Surv., Mon. 50, 1906, p. 65, pl. xvii, f. 2.—Berry, Bull. Torr. Bot. Club, vol. 37, 1910, p. 23; Geol. Surv. N. J., Bull. 3, 1911, p. 134, pl. xvi, f. 2; Md. Geol. Surv., Upper Cret., 1916, p. 832, pl. lxx, f. 1, 2; U. S. Geol. Surv., Prof. Paper 112, 1919, p. 91, pl. xvii, f. 9.

Dakota: Ellsworth County, Kans.
Raritan: Woodbridge, N. J.
Magothy: Gay Head, Marthas Vineyard, Mass.; Grove Point, Md.
Tuscaloosa: Shirleys Mill and Glen Allen, Ala.

Magnolia lakesii Knowlton, U. S. Geol. Surv., Prof. Paper 130, 19—, p. —, pl. xiii, f. 2 [in preparation].

Laramie: Marshall, Colo.

Magnolia lanceolata Lesquereux = *Magnolia dayana*.

Magnolia lanceolata Lesquereux. Hollick, La. Geol. Surv., Special Rept. 5, 1899, p. 282, pl. xl = *Magnolia angustifolia*.

Magnolia laurifolia Lesquereux, Am. Philos. Soc., Trans., vol. 13, 1869, p. 421, pl. xx, f. 2, 3.—Lesquereux, U. S. Nat. Mus., Proc., vol. 11, 1888, p. 25.—Knowlton, U. S. Geol. Surv., Prof. Paper 101, 1918, p. 309, pl. lxxxv, f. 2; pl. cvi, f. 2.

Terminalia hilgardiana (Lesquereux) Berry, U. S. Geol. Surv., Prof. Paper 91, 1916, p. 325, pl. xcii, f. 2.

Wilcox: *Mississippi; Cross Lake and McLees, La.
Raton: Near Aguilar, Colo.

Magnolia leei Knowlton MS. Berry, U. S. Geol. Surv., Prof. Paper 91, 1916, p. 215, pl. xliii, f. 1, 2.—Knowlton, U. S. Geol. Surv., Prof. Paper 101, 1918, p. 313, pl. lxiv, f. 2; pl. lxv, f. 2; pl. lxxxi, f. 2.

Raton: Vermejo Park, N. Mex.; Primero, Colo.
Wilcox (Ackerman): Hurleys, Benton County, Miss.
Lagrange: Puryear, Henry County, Tenn.

Magnolia lesleyana Lesquereux, Am. Philos. Soc., Trans., vol. 13, 1869, p. 421, pl. xxi, f. 1, 2; U. S. Geol. and Geog. Surv. Terr., Ann. Rept. 1871 [1872], Suppl., p. 14; Rept. U. S. Geol. Surv. Terr., vol. 7 (Tert. Fl.), 1878, p. 248, pl. xliv, f. 1-3.

Terminalia lesleyana (Lesquereux) Berry, U. S. Geol. Surv., Prof. Paper 101, 1918, p. 323.

Wilcox: Mississippi.

Raton: "Raton Mountains, N. Mex."; Trinidad, Wootton, Weston, Berwynd, and other localities, Colo.

Magnolia longifolia Newberry = *Magnolia newberryi.*

Magnolia longipes Newberry, MS. Hollick, Bull. Torr. Bot. Club, vol. 21, 1894, p. 60, pl. clxxviii, f. 3.—Newberry, U. S. Geol. Surv., Mon. 26, 1895 [1896], p. 76, pl. liv, f. 1-3.—Hollick, U. S. Geol. Surv., Mon. 50, 1906, p. 64, pl. xxi, f. 5, 6; N. Y. Bot. Gard., Bull., vol. 8, 1912, p. 160, pl. clxvii, f. 1.—Berry, Bull. Torr. Bot. Club, vol. 37, 1910, p. 23; Geol. Surv. N. J., Bull. 3, 1911, p. 135, pl. xiv, f. 1; Md. Geol. Surv., Upper Cret., 1916, p. 833, pl. lxix, f. 2; U. S. Geol. Surv., Prof. Paper 112, 1919, p. 91.

Magothy: Glen Cove, Long Island, N. Y.; Grove Point, Md.

Raritan: Woodbridge, N. J.; Dosoris Island, Long Island, and Roslyn, N. Y.

Tuscaloosa: Shirleys Mill and Cottondale, Ala.

Magnolia magnifica Dawson, Roy. Soc. Canada, Trans., vol. 1, 1882 [1883], sec. 4, p. 22, pl. iii, f. 11.

——: Coal Brook.

Mill Creek: Mill Creek, British Columbia.

Magnolia magnifolia Knowlton, U. S. Geol. Surv., Prof. Paper 101, 1918, p. 311, pl. lxxxv.

Magnolia tenuinervis Lesquereux, Rept. U. S. Geol. Surv. Terr., vol. 7 (Tert. Fl.), 1878, p. 249, pl. xliv, f. 5, 6; pl. xlv, f. 1-3 [not pl. xlv, f. 4-6].

Denver: Golden, Colo.

Raton: Aguilar, Apishapa, Brilliant, Cokedale, Trinidad, Walsenburg, Weston, Wootton, Tercio, and other localities, Colo.; Blossburg, Raton, Raton tunnel, Saylors Creek, and other localities, N. Mex.

Post-Laramie: Black Buttes, Wyo.

Magnolia marshalli Knowlton, U. S. Geol. Surv., Prof. Paper 130, 19—, p. —, pl. xxi, f. 10 [in preparation].

Laramie: Marshall, Colo.

Magnolia microphylla Knowlton, U. S. Geol. Surv., Mon. 32, pt. 2, 1899, p. 720.

Miocene: Fossil Forest Ridge, Yellowstone National Park.

Magnolia newberryi Berry, Bull. Torr. Bot. Club, vol. 34, 1907, p. 195, pl. xiii, f. 6; U. S. Geol. Surv., Prof. Paper 84, 1914, p. 42; Geol. Soc. N. J., Bull. 3, 1911, p. 133, pl. xiii; U. S. Geol. Surv., Prof. Paper 112, 1919, p. 89, pl. xx, f. 4.

Magnolia longifolia Newberry, MS. Hollick, N. Y. Acad. Sci., Trans., vol. 12, 1892, p. 36[9], pl. iii, f. 9; N. Y. Acad. Sci., Annals, vol. 11, 1898, p. 422, pl. xxxvii, f. 3; U. S. Geol. Surv., Mon. 50, 1906, p. 66, pl. xx, f. 2, 3. [Homonym, Sweet, 1826.]

Raritan: Woodbridge, N. J.; Tottenville, N. Y.
Black Creek: Court House Bluff, Cape Fear River, N. C.
?Black Creek: Darlington, S. C.
Tuscaloosa: Glen Allen, Ala.

Magnolia nordenskiöldi Heer, Fl. foss. arct., vol. 4, Abt. 1, 1877, p. 82, pl. xxi, f. 3; pl. xxx, f. 1.—Dawson, Roy. Soc. Canada, Trans., vol. 1, 1882 [1883], p. 33.—Lesquereux, U. S. Nat. Mus., Proc., vol. 5, 1882 [1883], p. 448; Rept. U. S. Geol. Surv. Terr., vol. 8 (Cret. and Tert. Fl.), 1883, p. 262.—Penhallow, Rept. Tert. Pl. Brit. Col., 1908, p. 62.

Kenai: Chignik Bay, Alaska.

Magnolia obovata Newberry = *Nyssa vetusta*. [This is a complete duplication in all except name for *Nyssa vetusta*.]

Magnolia obtusata Heer, Fl. foss. arct., vol. 6, Abt. 2, 1882, p. 90, pl. xv, f. 12; pl. xxi, f. 3.—Lesquereux, U. S. Geol. Surv., Mon. 17, 1891 [1892], p. 201, pl. lx, f. 5, 6.—Berry, N. Y. Bot. Gard., Bull., vol. 3, 1903, p. 76, pl. xlvii, f. 4; Geol. Surv. N. J., Ann. Rept. 1905 [1906], p. 138; Bull. Torr. Bot. Club, vol. 37, 1910, p. 23; U. S. Geol. Surv., Prof. Paper 84, 1914, p. 44; Md. Geol. Surv., Upper Cret., 1916, p. 834, pl. lxviii, f. 2–4; U. S. Geol. Surv., Prof. Paper 112, 1919, p. 90, pl. xvii, f. 7, 8.

Dakota: *Near Fort Harker, Kans.
Magothy: Cliffwood, N. J.; Grove Point, Md.
Tuscaloosa: Shirleys Mill, Ala.
Black Creek (Middendorf): Middendorf, S. C.

Magnolia occidentalis Dawson, Roy. Soc. Canada, Trans., vol. 11, 1893 [1894], sec. 4, p. 63, pl. x, f. 36.

Upper Cretaceous: Nanaimo, Vancouver Island, British Columbia.

Magnolia ovalis Lesquereux, Am. Philos. Soc., Trans., vol. 13, 1869, p. 422, pl. xxi, f. 3, 4.

Wilcox: Mississippi.

Magnolia palaeopetala Hollick, Bull. Torr. Bot. Club, vol. 30, 1903, p. 202, text f. A.

Dakota: Ellsworth County, Kans.

Magnolia? pollardi Knowlton, U. S. Geol. Surv., Mon. 32, pt. 2, 1899, p. 721, pl. lxxxi, f. 9, 10.
Fort Union: Yellowstone River below mouth of Elk Creek, Yellowstone National Park.
Miocene (intermediate flora): Fossil Forest Ridge, Yellowstone National Park.

Magnolia pseudoacuminata Lesquereux, U. S. Geol. Surv., Mon. 17, 1891 [1892], p. 199, pl. xxiv, f. 2.—Hollick, U. S. Geol. Surv., Mon. 50, 1906, p. 65, pl. xviii, f. 2, 3.—Knowlton, in Woodruff, U. S. Geol. Surv., Bull. 452, 1911, p. 20.
Dakota: Near Delphos, Kans.; Lander, Wyo.
Magothy: Gay Head, Marthas Vineyard, Mass.

Magnolia pulchra Ward, U. S. Geol. Surv., Sixth Ann. Rept., 1884–85 [1886], p. 556, pl. lx, f. 2, 3; U. S. Geol. Surv., Bull. 37, 1887, p. 103, pl. xlviii, f. 3, 4.
Mesaverde: *Point of Rocks, Wyo.

Magnolia regalis? Heer, Fl. foss. arct., vol. 4, Abt. 1, p. 81, 1877, pl. 20.—Knowlton, U. S. Geol. Surv., Prof. Paper 101, 1918, p. 314, pl. lxxxvii.
Raton: Wet Canyon, near Weston, Colo.

Magnolia rotundifolia Newberry, U. S. Nat. Mus., Proc., vol. 5, 1882 [1883], p. 513; U. S. Geol. Surv., Mon. 35, 1898, p. 95, pl. lix, f. 1.—Knowlton, U. S. Geol. Surv., Prof. Paper 101, 1918, p. 314, pl. lxxxiii.
Raton: "*Fishers Peak, N. Mex."; Wootton and Aguilar, Colo.

Magnolia speciosa Heer, Neue Denkschr. Schw. Ges., vol. 23, 1869, p. 20, pl. vi, f. 1; pl. ix, f. 2; pl. x, f. 1.—Lesquereux, Rept. U. S. Geol. Surv. Terr., vol. 8 (Cret. and Tert. Fl.), 1883, p. 72; U. S. Geol. Surv., Mon. 17, 1891 [1892], p. 202, pl. lx, f. 3, 4.—Hollick, N. Y. Acad. Sci., Trans., vol. 12, 1893, p. 234, pl. vii, f. 4; Bull. Torr. Bot. Club, vol. 21, 1894, p. 60, pl. 178, f. 5; Biol. Soc. Am., Bull., vol. 7, 1895, p. 13.—Knowlton, in Hill, U. S. Geol. Surv., Twenty-first Ann. Rept., pt. 2, 1901, p. 318.—Berry, Bull. Torr. Bot. Club, vol. 31, 1904, p. 76, pl. iii, f. 10; idem, vol. 32, 1905, p. 46, pl. ii, f. 4, 5; Geol. Surv. N. J., Ann. Rept. 1905 [1906], p. 138.—Hollick, U. S. Geol. Surv., Mon. 50, 1906, p. 64, pl. xix, f. 1–4.—Berry, Bull. Torr. Bot. Club, vol. 39, 1912, p. 395; Geol. Surv. N. J., Bull. 3, 1911, p. 129, pl. xiv, f. 3; U. S. Geol. Surv., Prof. Paper 112, 1919, p. 88, pl. xviii, f. 3, 4.
Dakota: Morrison, Colo.; Ellsworth County, Kans.
Magothy: Cliffwood, N. J.; Glen Cove, Long Island, N. Y.; Gay Head, Marthas Vineyard, Mass.
Raritan: Woodbridge, N. J.
Tuscaloosa: Shirleys Mill, Glen Allen, and Cottondale, Ala.
Woodbine: Arthurs Bluff, Tex.

Magnolia spectabilis Knowlton, U. S. Geol. Surv., Mon. 32, pt. 2, 1899, p. 718, pl. xciii, f. 1, 2.
Miocene: Fossil Forest Ridge, Yellowstone National Park.

Magnolia tenuifolia Lesquereux, Am. Jour. Sci., 2d ser., vol. 46, 1868, p. 100; Rept. U. S. Geol. Surv. Terr., vol. 6 (Cret. Fl.), 1874, p. 92, pl. xxi, f. 1; U. S. Geol. Surv., Mon. 17, 1891 [1892], p. 198, pl. xxiv, f. 1.—Berry, N. Y. Bot. Gard., Bull., vol. 3, 1903, p. 77, pl. xlvii, f. 10.—Hollick, N. Y. Bot. Gard., Bull., vol. 3, 1903, p. 413, pl. 73, f. 2.—Berry, Bull. Torr. Bot. Club, vol. 31, 1904, p. 76, pl. i, f. 1; idem, vol. 33, 1906, p. 174, pl. vii, f. 1; Geol. Surv. N. J., Ann. Rept. 1905 [1906], p. 138.—Hollick, U. S. Geol. Surv., Mon. 50, 1906, p. 64, pl. xvii, f. 1; pl. xviii, f. 4, 5.—Berry, U. S. Geol. Surv., Prof. Paper 84, 1914, p. 44, pl. x, f. 2, 3; Md. Geol. Surv., Upper Cret., 1916, p. 835, pl. lxx, f. 2.
Dakota: Decatur, Nebr.; Kansas.
——: Peace River, Northwest Territory.
Magothy: ?Cliffwood, N. J.; Deep Cut, Del.; Sea Cliff and Glen Cove, Long Island, N. Y.; Gay Head, Marthas Vineyard, Mass.
Black Creek (Middendorf): Middendorf, S. C.

Magnolia tenuinervis Lesquereux, Am. Jour. Sci., 2d ser., vol. 45, 1868, p. 207; Rept. U. S. Geol. Surv. Terr., vol. 7 (Tert. Fl.), 1878, pl. xlv, f. 6 [not other figures].
Magnolia inglefieldi? Heer. Lesquereux, U. S. Geol. and Geog. Surv. Terr., Ann. Rept. 1872 [1873], p. 396.
——: "Golden City, Colo."

Magnolia tenuinervis Lesquereux, Rept. U. S. Geol. Surv. Terr., vol. 7 (Tert. Fl.), pl. xliv, f. 5, 6; pl. xlv, f. 1–3 [not pl. xlv, f. 4–6] = *Magnolia magnifolia.*

Magnolia tenuinervis Lesquereux, Rept. U. S. Geol. Surv. Terr., vol. 7 (Tert. Fl.), pl. xlv, f. 5 = *Magnolia inquirenda.*

Magnolia tenuinervis Lesquereux. Knowlton, U. S. Geol. Surv., Bull. 163, pl. xiv, f. 1 = *Magnolia? coalvillensis.*

Magnolia vaningenii Hollick, Bull. Torr. Bot. Club, vol. 21, 1894, p. 61, pl. clxxv, f. 6; U. S. Geol. Surv., Mon. 50, 1906, p. 67, pl. xx, f. 1; N. Y. Bot. Gard., Bull., vol. 8, 1912, p. 161, pl. clxviii, f. 1.
Magothy: Sea Cliff and Roslyn, Long Island, N. Y.

Magnolia virginiana Linné. Berry, Jour. Geol., vol. 25, 1917, p. 662.
Magnolia glauca Linné. Hollick, Bull. Torr. Bot. Club, vol. 19, 1892, p. 331.
Pleistocene: Bridgeton, N. J.; Vero, Fla.

Magnolia woodbridgensis Hollick, in Newberry, U. S. Geol. Surv., Mon. 26, 1895 [1896], p. 74, pl. xxxvi, f. 11; pl. lvii, f. 5–7.—Berry, N. Y. Bot. Gard., Bull., vol. 3, 1903, p. 77, pl. liii, f. 5; pl. lvii, f. 2; Geol. Surv. N. J., Ann. Rept. 1905 [1906], p. 138.—

Magnolia woodbridgensis—Continued.
Hollick, U. S. Geol. Surv., Mon. 50, 1906, p. 65, pl. xx, f. 7; N. Y. Bot. Gard., Bull., vol. 8, 1912, p. 161, pl. clxii [part], pl. clxvii, f. 2; Geol. Surv. N. J., Bull. 3, 1911, p. 136, pl. xv, f. 2.
Magothy: Cliffwood, N. J.; ? Balls Point, Block Island, R. I.
Raritan: Woodbridge, N. J.; Roslyn, N. Y.

Magnolia? sp. Dawson, Roy. Soc. Canada, Trans., vol. 8, 1890, sec. 4, p. 89, text f. 29.—Penhallow, Rept. Tert. Pl. Brit. Col., 1908, p. 62.
Oligocene: Similkameen River, British Columbia.

Magnolia sp. Knowlton, Geol. Soc. Am., Bull., vol. 8, 1897, p. 145.
Fort Union: Black Buttes, Wyo.

Magnolia sp. Lesquereux, U. S. Geol. and Geog. Surv. Terr., Ann. Rept. 1871 [1872], p. 287. [Not afterward referred to by Lesquereux.]
Green River: Washakie station, Wyo.

Magnolia sp. Lesquereux, Rept. U. S. Geol. Surv. Terr., vol. 8 (Cret. and Tert. Fl.), 1883, p. 73, pl. xi, f. 6.
Dakota: Morrison, Colo.

MAJANTHEMOPHYLLUM O. Weber, Palaeontographica [Tertiärfl. d. Niederrh. Braunkohlenf.], vol. 2, 1852, p. 156 [reprint, p. 42].

Majanthemophyllum grandifolium Penhallow, Roy. Soc. Canada, Trans., 2d ser., vol. 8, 1902, p. 54, f. 5; Rept. Tert. Pl. Brit. Col., 1908, p. 63.
Paskapoo (Fort Union): Red Deer River, Alberta.

Majanthemophyllum pusillum Heer, Fl. foss. arct., vol. 7, 1883, p. 18, pl. lv, f. 17.—Hollick, N. Y. Acad. Sci., Trans., vol. 12, 1892, p. 36 [9], pl. i, f. 7; U. S. Geol. Surv., Mon. 50, 1906, p. 48, pl. vi, f. 12.
Raritan: Kreischerville, Staten Island, N. Y.

MALAPOENNA Adanson, Fam. Pl., vol. 2, 1763, p. 447.

Malapoenna carbonensis (Ward) Knowlton, U. S. Geol. Surv., Bull. 152, 1898, p. 142.
Litsaea carbonensis Ward, U. S. Geol. Surv., Sixth Ann. Rept., 1884–85 [1886], p. 553, pl. xlvi, f. 11; idem, Bull. 37, 1887, p. 48, pl. xxiv, f. 1.
Hanna: Carbon, Wyo.

Malapoenna cottondalensis Berry, U. S. Geol. Surv., Prof. Paper 112, 1919, p. 121, pl. xxi, f. 4.
Tuscaloosa: Cottondale, Ala.

Malapoenna cretacea (Lesquereux) Knowlton, U. S. Geol. Surv., Bull. 152, 1898, p. 142.—Berry, U. S. Geol. Surv., Prof. Paper 112, 1919, p. 122, pl. xxi, f. 3.
Litsea cretacea Lesquereux, U. S. Geol. Surv., Mon. 17, 1891 [1892], p. 96, pl. xv, f. 2.
Dakota: *Near Delphos, Kans.
Tuscaloosa: Cottondale, Ala.

Malapoenna cuneata (Knowlton) Knowlton, n. comb.

Litsea cuneata Knowlton, U. S. Geol. Surv., Mon. 32, pt. 2, 1899, p. 726, pl. xcii, f. 2-4.

Fort Union: Yellowstone River below mouth of Elk Creek, Yellowstone National Park.

Malapoenna falcifolia (Lesquereux) Knowlton, U. S. Geol. Surv., Bull. 152, p. 142.—Berry, Bull. Torr. Bot. Club, vol. 33, 1906, p. 180; Geol. Surv. N. J., Ann. Rept. 1905 [1906], p. 139; Bull. Torr. Bot. Club, vol. 39, 1912, p. 399; U. S. Geol. Surv., Prof. Paper 112, 1919, p. 122, pl. xxi, f. 5.

Litsea falcifolia Lesquereux, U. S. Geol. Surv., Mon. 17, 1891 [1892], p. 97, pl. xi, f. 5.

Dakota: Near Delphos, Kans.
Magothy: Cliffwood, N. J.
Woodbine: Arthurs Bluff, Tex.
Tuscaloosa: Shirleys Mill, Ala.

Malapoenna harrellensis Berry, Bull. Torr. Bot. Club, vol. 37, 1910, pp. 198, 504, pl. xxiv, f. 19; U. S. Geol. Surv., Prof. Paper 84, 1914, p. 118; Bull. Torr. Bot. Club, vol. 43, 1916, p. 299; U. S. Geol. Surv., Prof. Paper 112, 1919, p. 122.

Ripley: Near Cypress and near Selmer, Tenn.
Black Creek: Harrell Landing, Corbits Bridge, and Parker Landing, Tar River, N. C.
Eutaw: Havana, Ala.; Chattahoochee River below Columbus, Ga.

Malapoenna lamarensis Knowlton, U. S. Geol. Surv., Mon. 32, pt. 2, 1899, p. 726, pl. xciii, f. 4, 5; pl. xcvi, f. 5.

Miocene: Yellowstone National Park.

Malapoenna louisvillensis Knowlton, U. S. Geol. Surv., Prof. Paper 130, 19—, p. —, pl. vii, f. 5 [in preparation].

Laramie: Rex mine, Louisville, Colo.

Malapoenna praecursoria (Lesquereux) Knowlton, n. comb.

Tetranthera praecursoria Lesquereux, Rept. U. S. Geol. Surv. Terr., vol. 8 (Cret. and Tert. Fl.), 1883, p. 228, pl. xlviii, f. 2.

Fort Union: Badlands, N. Dak.

Malapoenna sessiliflora (Lesquereux) Knowlton, U. S. Geol. Surv., Bull. 152, 1898, p. 142.

Laurus sessiliflora Lesquereux, U. S. Geol. and Geog. Surv. Terr., Ann. Rept. 1873 [1874], p. 407.

Tetranthera sessiliflora Lesquereux (part), Rept. U. S. Geol. Surv. Terr., vol. 7 (Tert. Fl.), 1878, p. 217, pl. xxxiv, f. 1c, 1d; pl. xxv, f. 8 [not f. 9].

Evanston: Evanston, Wyo.

Malapoenna weediana (Knowlton) Knowlton, U. S. Geol. Surv., Bull. 152, 1898, p. 142; idem, Mon. 32, pt. 2, 1899, p. 659.

Litsea weediana Knowlton, U. S. Geol. Surv., Bull. 105, 1893, p. 55.

Tetranthera sessiliflora Lesquereux (part), Rept. U. S. Geol. Surv. Terr., vol. 7 (Tert. Fl.), 1878, p. 217, pl. xxxv, f. 9.

Livingston: Bozeman coal field, Mont.
——: Mount Everts, Yellowstone National Park.

Malapoenna sp. Berry, U. S. Geol. Surv., Prof. Paper 84, 1914, p. 144, pl. xxvii, f. 10.
Barnwell (Twiggs): Near Macon, Ga.

Malapoenna sp. Hollick, U. S. Geol. Surv., Mon. 50, 1906, p. 78, pl. xxxi, f. 4.
Magothy: Gay Head, Marthas Vineyard, Mass.

Malapoenna sp. Knowlton, U. S. Geol. Surv., Bull. 152, 1898, p. 142.
Litsea sp. Knowlton, Geol. Soc. Am., Bull., vol. 8, 1897, p. 154.
Mesaverde: Point of Rocks, Wyo.

MALUS (Tournefort) Jussieu, Gen., 1789, p. 334.

Malus coronarifolia Berry, Jour. Geol., vol. 15, 1907, p. 344.
Pleistocene: Neuse River, N. C.

Malus pseudo-angustifolia Berry, in Stephenson, N. C. Geol. and Econ. Surv., vol. 3, 1912, p. 286 [name].
Pleistocene: Neuse River, N. C.

MALVASTRUM Asa Gray, Mem. Am. Acad., 2d ser., vol. 4, 1848, p. 21.

Malvastrum? exhumatum Cockerell, Bull. Torr. Bot. Club, vol. 33, 1906, p. 311, text f. 5.
Miocene: Florissant, Colo.

MANICARIA Gaertner, Fruct., vol. 2, 1791, p. 486.

Manicaria haydenii Newberry, U. S. Nat. Mus., Proc., vol. 5, 1882 [1883], p. 504; U. S. Geol. Surv., Mon. 35, 1898, p. 31, pl. lxiv, f. 3.
Green River: Green River, Wyo.

Manicaria sp. Dawson, Roy. Soc. Canada, Trans., 2d ser., vol. 1, 1895 [1896], sec. 4, p. 143.—Penhallow, Rept. Tert. Pl. Brit. Col., 1908, p. 63.
——: Burrard Inlet, Vancouver Island, British Columbia.

MANIHOTITES Berry, Bull. Torr. Bot. Club, vol. 37, 1910, p. 507. [Type, *M. georgiana.*]

Manihotites georgiana Berry, Bull. Torr. Bot. Club, vol. 37, 1910, p. 507, text f. 1, 2; U. S. Geol. Surv., Prof. Paper 84, 1914, p. 114, pl. xxii; pl. xxiii; pl. xxiv, f. 4, 5; text f. 2, 3; Bull. Torr. Bot. Club, vol. 43, 1916, p. 295; U. S. Geol. Surv., Prof. Paper 112, 1919, p. 106; Bull. Torr. Bot. Club, vol. 44, 1917, p. 185.
Eutaw: McBrides Ford, 10 miles southeast of Columbus, and near Buena Vista, Ga.; Coffee Bluff, Hardin County, Tenn.
Bingen: Near Maxwell Spur, Pike County, Ark.
Ripley: Selmer, McNairy County, Tenn.; Buena Vista, Ga.

MARATTIA Swartz, Prod. Fl. Ind. Occ., 1788, p. 128.

Marattia cretacea Velenovsky? Berry, U. S. Geol. Surv., Prof. Paper 112, 1919, p. 55.
Tuscaloosa: Glen Allen, Ala.

MARCHANTIA (Marchant) Linné, Sp. pl., 1753, p. 1137.

Marchantia pealei Knowlton, U. S. Nat. Mus., Proc., vol. 35, 1908, p. 157, pl. xxv; Washington Acad. Sci., Proc., vol. 11, 1909, p. 188.

Lance: Miles City, Custer County, Mont.

MARCHANTITES Brongniart, Dict. univ. hist. nat., vol. 13 (Tableau), 1849, p. 61 (12).

Marchantites erectus (Bean) Seward, Fossil plants for students of botany and geology, 1898, p. 233, text f. 49.—Fontaine, in Ward, U. S. Geol. Surv., Mon. 48, 1905 [1906], p. 53, pl. vi, f. 1, 2.

Fucoides erectus Bean, in Leckenby, Quart. Jour. Geol. Soc. London, vol. 20, 1863, p. 81.

Jurassic: Near Nichols station, Douglas County, Oreg.

MARSILEA Linné, Sp. pl., 1753, p. 1099.

Marsilea andersoni Hollick, N. Y. Bot. Gard., Bull., vol. 3, 1904, p. 409, pl. lxxi, f. 1–3; U. S. Geol. Surv., Mon. 50, 1906, p. 33, pl. i, f. 14–18.

Magothy: Manhasset Neck, Long Island, N. Y.; Gay Head, Marthas Vineyard, Mass.

Marsilea attenuata (Lesquereux) Hollick, Bull. Torr. Bot. Club, vol. 21, 1894, p. 256, pl. ccv, f. 10.—Knowlton, U. S. Geol. Surv., Bull. 163, 1900, p. 23.

Salvinia attenuata Lesquereux, U. S. Geol. and Geog. Surv. Terr., Bull., vol. 1, 1875 [1876], p. 377; idem, Ann. Rept. 1874 [1876], p. 296; Rept. U. S. Geol. Surv. Terr., vol. 7 (Tert. Fl.), 1878, p. 65, pl. lxiv, f. 14, 14a.

Mesaverde: Point of Rocks, Wyo.

Marsilea bendirei Ward = *Hydrangea bendirei*.

MATONIDIUM Schenk, Palaeontographica, vol. 19, 1871, p. 219.

Matonidium althausii (Dunker) Ward, U. S. Geol. Surv., Nineteenth Ann. Rept., pt. 2, 1899, p. 653, pl. clx, f. 5–8.—Fontaine, in Ward, U. S. Geol. Surv., Mon. 48, 1905, p. 230, pl. lxv, f. 22, 23.—Cockerell, Washington Acad. Sci., Jour., vol. 6, 1916, p. 111, f. 2.

Cycadites althausii Dunker, Programm d. höheren Gewerbschule in Cassel, 1843–44, p. 7.

Osmunda dicksonioides? Fontaine. Fontaine, in Diller and Stanton, Geol. Soc. Am., Bull., vol. 5, 1895, p. 450.

Fuson: Pine Creek, Crook County, Wyo.

Knoxville: Near Lowry, Tehama County, Calif.

Chico: Elder Creek, near Lowry, Tehama County, Calif.

Lower Cretaceous: Near Hovenweep Canyon, latitude 37° 53′ N., longitude 108° 57′ W., southwestern Colorado.

MAYTENUS Molina, Saggio, 1782, p. 177.

Maytenus puryearensis Berry, U. S. Geol. Surv., Prof. Paper 91, 1916, p. 264, pl. lxi, f. 5.

Lagrange: Puryear, Henry County, Tenn.

MELASTOMITES Unger, Gen. et sp. pl. foss., 1850, p. 480.

Melastomites americanus Berry, U. S. Geol. Surv., Prof. Paper 91, 1916, p. 327, pl. xcvii, f. 1–3.
Wilcox (Grenada): Grenada, Miss.
Wilcox: Mansfield, La.
Lagrange: Puryear and Grand Junction, Tenn.

MELIA Linné, Sp. pl., 1753, p. 384.

Melia? expulsa Cockerell, Am. Mus. Nat. Hist., Bull., vol. 24, 1908, p. 98; Am. Nat., vol. 44, 1910, p. 46, f. 11.
Miocene: Florissant, Colo.

MELOTHRIA Linné, Sp. pl., 1753, p. 35.

Melothria? coloradensis Cockerell, Am. Mus. Nat. Hist., Bull., vol. 24, 1908, p. 108, pl. x, f. 40.
Miocene: *Florissant, Colo.

MENIPHYLLOIDES Berry, U. S. Geol. Surv., Prof. Paper 91, 1916, p. 166. [Type, *M. ettingshauseni.*]

Meniphylloides ettingshauseni Berry, U. S. Geol. Surv., Prof. Paper 91, 1916, p. 166, pl. xi, f. 4–7.
Wilcox: Old Port Caddo Landing, Harrison County, Tex.; Naborton, De Soto Parish, La.
Wilcox (Grenada): Grenada, Grenada County, Miss.

MENISPERMITES Lesquereux, Rept. U. S. Geol. Surv. Terr., vol. 6 (Cret. Fl.), 1874, p. 94. [Type, *M. obtusiloba.*]

Menispermites acerifolia Lesquereux = *Menispermites menispermifolius.*

Menispermites acutilobus Lesquereux, Rept. U. S. Geol. Surv. Terr., vol. 8 (Cret. and Tert. Fl.), 1883, p. 78, pl. xiv, f. 2.—Hollick, U. S. Geol. Surv., Mon. 50, 1906, p. 62, pl. xii, f. 8.
Dakota: Kansas.
Magothy: ?Nashaquitsa, Marthas Vineyard, Mass.

Menispermites borealis Heer?, Fl. foss. arct., vol. 6, Abt. 2, 1882, p. 91, pl. xxxix, f. 2.—Newberry, U. S. Geol. Surv., Mon. 35, 1895 [1896], p. 84, pl. l, f. 1–6.—Berry, Geol. Surv. N. J., Bull. 3, 1911, p. 151, pl. xviii, f. 4.
Raritan: Woodbridge, N. J.

Menispermites brysoniana Hollick, Bull. Torr. Bot. Club, vol. 21, 1894, p. 59, pl. clxx, f. 10.
Magothy: Glen Cove, Long Island, N. Y.

Menispermites californicus Fontaine, in Ward, U. S. Geol. Surv., Mon. 48, 1905 [1906], p. 268, pl. lxix, f. 12–14.
Horsetown: Elder Creek, Tehama County, Calif.

Menispermites cyclophyllus Lesquereux, U. S. Geol. and Geog. Surv. Terr., Ann. Rept. 1874 [1876], p. 358, pl. vi, f. 4.
Dakota: Kansas.

Menispermites grandis Lesquereux, Rept. U. S. Geol. Surv. Terr., vol. 8 (Cret. and Tert. Fl.), 1883, p. 80, pl. xv, f. 1, 2; U. S. Geol. Surv., Mon. 17, 1891 [1892], p. 196.
Dakota: Kansas.

Menispermites integrifolia Berry, Bull. Torr. Bot. Club, vol. 44, 1917, p. 183; U. S. Geol. Surv., Prof. Paper 112, 1919, p. 94, pl. xx, f. 1.
Bingen: Near Maxwell Spur, Pike County, Ark.
Tuscaloosa: Cottondale, Ala.

Menispermites menispermifolius (Lesquereux) Knowlton, U. S. Geol. Surv., Bull. 152, 1898, p. 143.
Acorites menispermifolius Lesquereux, Am. Jour. Sci., 2d ser., vol. 46, 1868, p. 101.
Menispermites acerifolia Lesquereux, Rept. U. S. Geol. Surv. Terr., vol. 6 (Cret. Fl.), 1874, p. 96, pl. xx, f. 2, 3; U. S. Geol. Surv., Mon. 17, 1891 [1892], p. 196.
Dakota: Decatur, Nebr.

Menispermites obtusiloba Lesquereux, Rept. U. S. Geol. Surv. Terr., vol. 6 (Cret. Fl.), 1874, p. 94, pl. xxv, f. 1, 2; pl. xxvi, f. 3; idem, vol. 8 (Cret. and Tert. Fl.), 1883, p. 78.
Dombeyopsis obtusiloba Lesquereux, Am. Jour. Sci., 2d ser., vol. 46, 1868, p. 100.
Menispermites obtusiloba var.? Lesquereux, Rept. U. S. Geol. Surv. Terr., vol. 6 (Cret. Fl.), 1874, p. 95, pl. xxii, f. 1.
Dakota: Salina River, Kans.

Menispermites ovalis Lesquereux, U. S. Geol. and Geog. Surv. Terr., Bull., vol. 1, 1875 [1876], p. 398; idem, Ann. Rept. 1874 [1876], p. 357, pl. v, f. 4.
Dakota: Kansas.

Menispermites populifolius Lesquereux, U. S. Geol. and Geog. Surv. Terr., Ann. Rept. 1874 [1876], p. 357, pl. v, f. 3; Rept. U. S. Geol. Surv. Terr., vol. 8 (Cret. and Tert. Fl.), 1883, p. 79, pl. iv, f. 4; U. S. Geol. Surv., Mon. 17, 1891 [1892], p. 196.
Dakota: Kansas.

Menispermites potomacensis Berry, Md. Geol. Surv., Lower Cret., 1912, p. 466, pl. xciii, f. 3, 4.
Patapsco: Stump Neck, Md.; Widewater, Va.

Menispermites reniformis Dawson, Roy. Soc. Canada, Trans., vol. 1, 1882 [1883], sec. 4, p. 23, pl. iii, f. 12.
——: Coal Brook, British Columbia.

Menispermites rugosus Lesquereux, U. S. Geol. Surv., Mon. 17, 1891 [1892], p. 196, pl. xxix, f. 7.
Dakota: Ellsworth County, Kans.

Menispermites salinae (Lesquereux) Knowlton, U. S. Geol. Surv., Bull. 152, 1898, p. 143.
Populites salinae Lesquereux, U. S. Geol. and Geog. Surv. Terr., Ann. Rept. 1872 [1873], p. 423.

Menispermites salinae—Continued.

Acer obtusilobium? Unger. Lesquereux, Am. Jour. Sci., 2d ser., vol. 46, 1868, p. 100.

Menispermites salinensis Lesquereux, Rept. U. S. Geol. Surv. Terr., vol. 6 (Cret. Fl.), 1874, p. 95, pl. xx, f. 1, 4; U. S. Geol. Surv., Mon. 17, 1891 [1892], p. 196.

Dakota: Salina River, Kans.

Menispermites salinensis Lesquereux = *Menispermites salinae.*

Menispermites tenuinervis Fontaine = *Nelumbites tenuinervis.*

Menispermites trilobatus Berry, U. S. Geol. Surv., Prof. Paper 112, 1919, p. 95, pl. xvii, f. 5, 6.

Tuscaloosa: Shirleys Mill, Ala.

Menispermites variabilis Berry, U. S. Geol. Surv., Prof. Paper 84, 1914, p. 113, pl. xxi, f. 1–4; idem, Prof. Paper 112, 1919, p. 95.

Eutaw: McBrides Ford, Chattahoochee County, Ga.

Menispermites virginiensis Fontaine [part] = *Nelumbites virginiensis* [part] = *Nelumbites tenuinervis.*

Menispermites wardianus Hollick, in Newberry, U. S. Geol. Surv., Mon. 35, 1895 [1896], p. 85, pl. xxix, f. 9, 11.—Berry, Geol. Surv. N. J., Bull. 3, 1911, p. 152, pl. xviii, f. 3.

Raritan: No locality given.

Menispermites wilcoxensis Berry, U. S. Geol. Surv., Prof. Paper 91, 1916, p. 218, pl. cxv, f. 1, 2; pl. cxvi, f. 2, 3.

Wilcox: Near Naborton, De Soto Parish, La.

Menispermites sp. Berry, Bull. Torr. Bot. Club, vol. 37, 1910, p. 504.

Tuscaloosa: McBrides Ford, near Columbus, Ga.

Menispermites sp. Dawson, Roy. Soc. Canada, Trans., vol. 11, 1893 [1894], sec. 4, p. 62, pl. xi, f. 50.

Upper Cretaceous: Port McNeill, Vancouver Island, British Columbia.

Menispermites sp. Hollick, U. S. Geol. Surv., Mon. 50, 1906, p. 62, pl. xii, f. 7.

Hedera sp.? Hollick, N. Y. Acad. Sci., Ann., vol. 11, 1898, p. 421, pl. xxxviii, f. 5.

Raritan: Tottenville, Staten Island, N. Y.

Menispermites sp. Knowlton, Geol. Soc. Am., Bull., vol. 8, 1897, p. 131.

Lance: Lance Creek, Converse County, Wyo.

Menispermum canadense Linné. Knowlton = *Cebotha carolina.*

MENYANTHES Linné, Sp. pl., 1753, p. 145.

Menyanthes coloradensis Cockerell, Am. Jour. Sci., 4th ser., vol. 26, 1908, p. 543, f. 9.—Knowlton, U. S. Nat. Mus., Proc., vol. 51, 1916, p. 288.

Miocene: *Florissant, Colo.

Menyanthes trifoliata Linné. Penhallow, Geol. Soc. Am., Bull., vol. 1, 1890, p. 327; Roy. Soc. Canada, Trans., 2d ser., vol. 2, 1896, p. 72; Brit. Assoc. Adv. Sci., Bradford meeting, 1900, p. 336.

Pleistocene: Montreal, Canada.

MERTENSIDES Fontaine, U. S. Geol. Surv., Mon. 6, 1883, p. 35. [Type, *M. bullatus*, the *Pecopteris bullata* of Bunbury.]

Mertensides bullatus (Bunbury) Fontaine, U. S. Geol. Surv., Mon. 6, 1883, pp. 35, 102, pl. xv, f. 2–5; pl. xvi, f. 1–3; pl. xvii, f. 1, 2; pl. xviii, f. 1, 2; pl. xix, f. 1.

Pecopteris bullatus Bunbury, Geol. Soc. London, Quart. Jour., vol. 3, 1847, p. 282.—Emmons, Geol. Rept. Midland Counties N. C., 1856, p. 328, pl. ii, f. 8; American geology, pt. 6, 1857, p. 101, pl. vi, f. 8.

Triassic (Newark): Carbon Hill, Clover Hill, near Midlothian and Deep Run, Va.; North Carolina ?; Pennsylvania ?

Mertensides distans Fontaine, U. S. Geol. Surv., Mon. 6, 1883, p. 39, pl. xv, f. 1, 1a.

Triassic (Newark): Clover Hill, Va.

MESPILODAPHNE Nees, Linnaea, vol. 8, 1833, p. 45.

Mespilodaphne coushatta Berry, U. S. Geol. Surv., Prof. Paper 91, 1916, p. 307, pl. lxxx, f. 6; pl. lxxxvii, f. 3.

Andromeda delicatula Lesquereux. Hollick, La. Geol. Surv., Special Rept. 5, 1899, p. 487, pl. xlv, f. 1.

Lagrange: Puryear, Tenn.

Wilcox: Hardys Mill, near Gainesville, Ark.; Mansfield and Coushatta, La.

Mespilodaphne eolignitica (Hollick) Berry, U. S. Geol. Surv., Prof. Paper 91, 1916, p. 307, pl. lxxxi, f. 2, 3.

Andromeda eolignitica Hollick, La. Geol. Surv., Special Rept. 5, 1899, p. 287, pl. xlvii, f. 2.

Wilcox (Grenada): Grenada, Miss.

Wilcox (Ackerman): Hurleys, Miss.

Wilcox: *Coushatta, Naborton, La., and Calaveras Creek, Wilson County, Tex.

Lagrange: Near Shandy, Baughs Bridge, and Puryear, Tenn.

Mespilodaphne pseudoglauca Berry, U. S. Geol. Surv., Prof. Paper 91, 1916, p. 306, pl. lxxx, f. 4.

Laurus socialis Lesquereux. Lesquereux, U. S. Nat. Mus., Proc., vol. 11, 1888, p. 24.

Laurus californica Lesquereux. Lesquereux, idem, p. 12, pl. iv, f. 1.

Wilcox: Campbell's quarry, Caddo Parish, and Naborton, La.

Wilcox (Holly Springs): Vaughans, Miss.

Lagrange: Puryear, Tenn.; Wickliffe, Ky.

Mespilodaphne puryearensis Berry, U. S. Geol. Surv., Prof. Paper 91, 1916, p. 305, pl. lxxxvii, f. 1.

Lagrange: Puryear, Tenn.

METOPIUM P. Browne, Jamaic., 1756, p. 177, pl. xiii, f. 3.

Metopium wilcoxianum Berry, U. S. Geol. Surv., Prof. Paper 91, 1916, p. 260, pl. lvii, f. 2, 3; pl. cxi, f. 5.
Wilcox (Grenada): Grenada, Grenada County, Miss.
Wilcox: Old Port Caddo Landing, Harrison County, Tex.
Lagrange: Puryear, Henry County, Tenn.

MEZONEURUM Desfontaines, Mém. Mus. Paris, vol. 4, 1818, p. 245.

Mezoneurum bridgetonense Hollick, Bull. Torr. Bot. Club, vol. 23, 1896, p. 49, pl. cclviii, f. 1–7.
Pleistocene: Bridgeton, N. J.

MICROTAENIA Knowlton, U. S. Geol. Surv., Prof. Paper 108, 1917, p. 81. [Type, *M. variabilis*.]

Microtaenia paucifolia (Hall) Knowlton, U. S. Geol. Surv., Prof. Paper 108, 1917, p. 82, pl. xxx, f. 1, 2.
Sphenopteris? paucifolia Hall, in Frémont, Rept. Exploring Expedition to the Rocky Mountains in 1842, etc., App. B, 1845, p. 304, pl. ii, f. 1a, b, c, d.
Sphenopteris? trifoliata Hall, idem, pl. ii, f. 2.
Frontier: About 1½ miles east of Cumberland, Wyo.

Microtaenia variabilis Knowlton, U. S. Geol. Surv., Prof. Paper 108, 1917, p. 81, pl. xxix, f. 1–4a.
Frontier: About 1½ miles east of Cumberland, Wyo.

MICROZAMIA Corda, in Reuss, Verst. böhm. Kreidefl., Abt. 2, 1846, p. 85.

Microzamia? dubia Berry, Bull. Torr. Bot. Club, vol. 32, 1905, p. 43, pl. i, f. 2; Geol. Surv. N. J., Ann. Rept. 1905 [1906], p. 138.
Magothy: Cliffwood Bluff, N. J.

Microzamia gibba (Reuss) Corda, Verst. böhm. Kreidefl., Abt. 2. 1846, p. 85.—Newberry, U. S. Geol. Surv., Mon. 26, 1895 [1896], p. 45, pl. xii, f. 6, 7; Geol. Surv. N. J., Bull. 3, 1911, p. 78.
Conites gibbus Reuss, Geogn. Skizz. Böhm., vol. 2, 1844, p. 169.
Raritan: Woodbridge, N. J.

MIMOSA (Tournefort) Linné, Sp. pl., 1753, p. 516.

Mimosa americana Berry, U. S. Geol. Surv., Prof. Paper 84, 1914, p. 139, pl. xxvii, f. 13.
Barnwell (Twiggs): Phinizy Gully, Columbia County, Ga.

Mimosa carolinensis Berry, U. S. Geol. Surv., Prof. Paper 84, 1914, p. 36, pl. xii, f. 5.
Black Creek (Middendorf): Middendorf, S. C.

MIMOSITES Bowerbank, History of the fossil fruits and seeds of the London clay, 1840, p. 140.

Mimosites acaciaefolius Berry, U. S. Geol. Surv., Prof. Paper 91, 1916, p. 226, pl. xlv, f. 14.
Lagrange: Puryear, Henry County, Tenn.

Mimosites georgianus Berry, U. S. Geol. Surv., Prof. Paper 84, 1914, p. 142, pl. xxvii, f. 5–9.

Barnwell (Twiggs): Phinizy Gully, Columbia County, Grovetown, and near Macon, Ga.

Mimosites inaequilateralis Berry, U. S. Geol. Surv., Prof. Paper 91, 1916, p. 226, pl. xlv, f. 12.

Lagrange: Near Grand Junction, Fayette County, Tenn.

Mimosites lanceolatus Berry, U. S. Geol. Surv., Prof. Paper 91, 1916, p. 226, pl. xlv, f. 13.

Lagrange: Near Grand Junction, Fayette County, Tenn.

Mimosites linearifolius Lesquereux = *Mimosites linearis.*

Mimosites linearis (Lesquereux), Knowlton, U. S. Geol. Surv., Bull. 152, 1898, p. 144.

Caesalpinia? linearis Lesquereux, U. S. Geol. and Geog. Surv. Terr., Ann. Rept. 1873 [1874], p. 417.

Mimosites linearifolius Lesquereux, Rept. U. S. Geol. Surv. Terr., vol. 7 (Tert. Fl.), 1878, p. 300, pl. lix, f. 7.

Miocene: *Florissant, Colo.

Mimosites marshallanus Knowlton, U. S. Geol. Surv., Prof. Paper 130, 19—, p. —, pl. xvi, f. 4 [in preparation].

Laramie: Marshall, Colo.

Mimosites variabilis Berry, U. S. Geol. Surv., Prof. Paper 91, 1916, p. 227, pl. xlv, f. 6–11.

Wilcox (Grenada): Grenada, Grenada County; Holly Springs and Early Grove, Marshall County, Miss.

Lagrange: Puryear, Henry County; Grand Junction, Fayette County, Tenn., and Wickliffe, Ballard County, Ky.

MIMULUS Linné, Sp. pl., 1753, p. 634.

Mimulus saxorum Cockerell, Am. Mus. Nat. Hist., Bull., vol. 24, 1908, p. 107, pl. x, f. 47.

Miocene: *Florissant, Colo.

MIMUSOPS Linné, Syst., ed. 1759, p. 1000.

Mimusops eolignitica Berry = *Mimusops truncatus.*

Mimusops mississippiensis Berry, U. S. Geol. Surv., Prof. Paper 91, 1916, p. 340, pl. cviii, f. 1.

Wilcox (Grenada): Grenada, Miss.

Mimusops siberifolia Berry, U. S. Geol. Surv., Prof. Paper 91, 1916, p. 339, pl. xcix, f. 2; pl. c, f. 3.

Lagrange: Puryear, Tenn.

Mimusops truncata (Lesquereux) Knowlton, n. comb.

?Phyllites truncatus Lesquereux, Am. Philos. Soc., Trans., vol. 13, 1869, p. 423, pl. xvii, f. 9.

Quercus chlorophylla Unger. Lesquereux, Am. Philos. Soc., Trans., vol. 13, 1869, p. 416, pl. xvii, f. 6 [not f. 5, 7].

Mimusops eolignitica Berry, U. S. Geol. Surv., Prof. Paper 91, 1916, p. 339, pl. xcix, f. 3.

Wilcox (Ackerman): Hurleys, Miss.

Lagrange: Puryear, Tenn.

MONIMIOPSIS Saporta, Mém. Soc. géol. France (Proc. fl. foss. d. Sézanne), 2d ser., vol. 7, 1868, p. 361.

Monimiopsis amboraefolia Saporta, Fl. foss. de Sézanne, 1868, p. (73) 361, pl. (vii) xxix, f. 13.—Ward, U. S. Geol. Surv., Sixth Ann. Rept., 1884–85 [1886], p. 553, pl. xlvii, f. 5; U. S. Geol. Surv., Bull. 37, 1887, p. 51, pl. xxv, f. 2.
Fort Union: Sevenmile Creek, near Glendive, Mont.

Monimiopsis fraterna Saporta, Fl. foss. de Sézanne, 1868, p. (74) 362, pl. (viii) xxix, f. 14.—Ward, U. S. Geol. Surv., Sixth Ann. Rept., 1884–85 [1886], p. 553, pl. xlvii, f. 6; U. S. Geol. Surv., Bull. 37, 1887, p. 52, pl. xxv, f. 3.
Fort Union: Sevenmile Creek, near Glendive, Mont.

MONOCARPELLITES Perkins, Rept. Vt. State Geol. 1903–4 [1904], p. 180. [Type, *M. whitfieldii.*]

Monocarpellites amygdaloidus Perkins, Rept. Vt. State Geol. 1905–6 [1906], p. 208, pl. liv, f. 4.
Miocene: Brandon, Vt.

Monocarpellites elegans Perkins, Rept. Vt. State Geol. 1903–4 [1904], p. 181, pl. lxxvi, f. 25.
Miocene: Brandon, Vt.

Monocarpellites gibbosus Perkins, Rept. Vt. State Geol. 1903–4 [1904], p. 181, pl. lxxvi, f. 26; Geol. Soc. Am., Bull., vol. 16, 1905, p. 512, pl. lxxxvii, f. 18; Rept. Vt. State Geol. 1905–6 [1906], pl. liii, f. 18.
Miocene: Brandon, Vt.

Monocarpellites hitchcockii Perkins, Rept. Vt. State Geol. 1903–4 [1904], p. 182, pl. lxxvi, f. 29.
Miocene: Brandon, Vt.

Monocarpellites irregularis Perkins, Rept. Vt. State Geol. 1903–4 [1904], p. 181, pl. lxxvi, f. 27.
Miocene: Brandon, Vt.

Monocarpellites medius Perkins, Rept. Vt. State Geol. 1903–4 [1904], p. 182, pl. lxxvi, f. 30, 31.
Miocene: Brandon, Vt.

Monocarpellites multicostatus Perkins, Rept. Vt. State Geol. 1905–6 [1906], p. 208, pl. liv, f. 5–7.
Miocene: Brandon, Vt.

Monocarpellites orbicularis Perkins, Rept. Vt. State Geol. 1903–4 [1904], p. 181, pl. lxxvi, f. 24.
Miocene: Brandon, Vt.

Monocarpellites ovalis Perkins, Rept. Vt. State Geol. 1903–4 [1904], p. 182, pl. lxxvi, f. 35.
Miocene: Brandon, Vt.

Monocarpellites pruniformis Perkins, Rept. Vt. State Geol. 1905–6 [1906], p. 208, pl. liv, f. 8.
Miocene: Brandon, Vt.

Monocarpellites pyramidalis Perkins, Rept. Vt. State Geol. 1903–4 [1904], p. 180, pl. lxxvi, f. 22.
Miocene: Brandon, Vt.

Monocarpellites sulcatus Perkins, Rept. Vt. State Geol. 1903–4 [1904], p. 180, pl. lxxvi, f. 23; Geol. Soc. Am., Bull., vol. 16, 1905, p. 512, pl. lxxxvii, f. 20; Rept. Vt. State Geol. 1905–6 [1906], pl. liii, f. 20.
Miocene: Brandon, Vt.

Monocarpellites vermontanus Perkins, Rept. Vt. State Geol. 1903–4 [1904], p. 182, pl. lxxvi, f. 34.
Miocene: Brandon, Vt.

Monocarpellites whitfieldii Perkins, Rept. Vt. State Geol., 1903–4 [1904], p. 180, pl. lxxvi, f. 21.
Miocene: Brandon, Vt.

Monocotyledon, gen. and sp. ? Hollick, in Newberry, U. S. Geol. Surv., Mon. 35, 1898, p. 33, pl. xlvi, f. 9.
Formation and locality unknown.

Monocotyledon, gen. and sp. ? Hollick, Md. Geol. Surv., Pliocene and Pleistocene, 1906, p. 220, pl. lxx, f. 8–11.
Pleistocene (Sunderland): Head of Island Creek, Calvert County, Md.

Monocotyledon, n. gen. and sp., Berry, Bull. Torr. Bot. Club, vol. 37, 1910, p. 505.
Ripley (Cusseta): Buena Vista, Ga.

Monocotyledonous plant, Knowlton, U. S. Geol. Surv., Bull. 204, 1902, p. 28, pl. i, f. 7, 8.
Clarno (upper part): Bridge Creek, Oreg.

MORELLA Loureiro, Fl. Cochinch., vol. 2, 1790, p. 548.

Morella bolanderi (Lesquereux) Cockerell, Univ. Colo. Studies, vol. 3, 1906, p. 173.

Myrica bolanderi Lesquereux, Rept. U. S. Geol. Surv. Terr., vol. 7 (Tert. Fl.), 1878, p. 133, pl. xvii, f. 17.—Knowlton, U. S. Geol. Surv., Mon. 32, pt. 2, 1899, p. 658, pl. lxxviii, f. 4.

Ilex undulata Lesquereux, U. S. Geol. and Geog. Surv. Terr., Ann. Rept., 1873 [1874], p. 416 [not *Ilex undulata* Heer, 1859].

——: Original locality unknown.
——: Mount Everts, Yellowstone National Park.

Morella hendersoni (Cockerell) Cockerell = ***Myrica hendersoni***.

MORICONIA Debey and Ettingshausen, Denkschr. Wien. Akad. (Urweltlichen Acrobryen v. Aachen), vol. 17, 1859, p. (59) 339.

Moriconia americana Berry, Bull. Torr. Bot. Club, vol. 37, 1910, p. 20; idem, p. 186, pl. xx, f. 5; U. S. Geol. Surv., Prof. Paper 84, 1914, p. 26, pl. vii, f. 1–4; Md. Geol. Surv., Upper Cret., 1916, p. 802, pl. lvi, f. 1.

Moriconia cyclotoxon Debey and Ettingshausen. Berry, N. Y. Bot. Gard., Bull., vol. 3, 1903, p. 65, pl. xliii, f. 4; pl. xlviii, f. 1–4; Bull. Torr. Bot. Club, vol. 31, 1904, p. 70; idem, vol. 33, 1906, p. 165–167.

Magothy: Deep Cut, Del.; Grove Point and Round Bay, Severn River, Md.

Black Creek: Elizabethtown, Cape Fear River, N. C.

Black Creek (Middendorf): Middendorf, S. C.

Moriconia cyclotoxon Debey and Ettingshausen, Denkschr. Wien. Akad. (Urweltlichen Acrobryen v. Aachen), vol. 17, 1859, p. (59) 239, pl. (vii); pl. xxvii, f. 23–27.—Newberry, U. S. Geol. Surv., Mon. 26, 1895 [1896], p. 55, pl. x, f. 11–21.—Hollick, N. Y. Acad. Sci., Ann., vol. 11, 1898, p. 418, pl. xxxvii, f. 8.—Berry, Bull. Torr. Bot. Club, vol. 31, 1894, p. 70; N. Y. Bot. Gard, Bull., vol. 3, 1903, p. 65, pl xliii, f. 4; pl. xlviii, f. 1–4; Bull. Torr. Bot. Club, vol. 33, 1906, p. 165; Geol. Surv. N. J., Ann. Rept. 1905 [1906], p. 138.—Hollick, U. S. Geol. Surv., Mon. 50, 1906, p. 46, pl. iii, f. 16, 17.—Berry, Geol. Surv., N. J., Bull. 3, 1911, p. 86, pl. viii, f. 3–6.

Magothy: Cliffwood Bluff, N. J.; Deep Cut, Del.; Grove Point, Md.; ?Black Rock Point, Block Island, R. I.

Raritan: South Amboy, N. J.; Princess Bay, Staten Island, N. Y.

Moriconia cyclotoxon Debey and Ettingshausen [part] = *Moriconia americana.*

MORUS Linné, Sp. pl., 1753, p. 986.

Morus affinis Lesquereux, U. S. Geol. and Geog. Surv. Terr., Ann. Rept. 1871 [1872], Suppl., p. 11. [Not afterward mentioned by author.]

Evanston: *Evanston, Wyo.

Morus symmetrica Cockerell, Am. Mus. Nat. Hist., Bull., vol. 24, 1908, p. 88, pl. vii, f. 19; Am. Mus. Jour., vol. 16, 1916, p. 442.

Miocene: *Florissant, Colo.

Morus sp. Hollick, Bull. Torr. Bot. Club, vol. 19, 1892, p. 332.

Pleistocene: Bridgeton, N. J.

MUHLENBERGIA Schreber, Gram., vol. 50, 1779, p. 51.

Muhlenbergia florissanti Knowlton, U. S. Nat. Mus., Proc., vol. 51, 1916, p. 250, pl. xiii, f. 1–3.

Miocene: *Florissant, Colo.

MUSOPHYLLUM Göppert, Tertiärfl. d. Insel Java, 1854, p. 39.

Musophyllum complicatum Lesquereux, U. S. Geol. and Geog. Surv. Terr., Ann. Rept. 1873 [1874], p. 418; Rept. U. S. Geol. Surv. Terr., vol. 7 (Tert. Fl.), 1878, p. 96, pl. xv, f. 1–6.—Penhallow, Rept. Tert. Pl. Brit. Col., 1908, p. 63.—Knowlton, U. S. Geol. Surv., Mon. 32, pt. 2, 1899, p. 686, pl. lxxxiii, f. 1.
Green River: *Green River station, Wyo.
Miocene: Yellowstone National Park.

Musophyllum sp. Knowlton, Washington Acad. Sci., Proc., vol. 11, 1909, p. 204.
Lance: Weston County, Wyo.

MYELOPHYCUS Ulrich, Harriman Alaska Exped., vol. 4, 1904, p. 145. [Type, *M. curvatum*.]

Myelophycus curvatum Ulrich, Harriman Alaska Exped., vol. 4, 1904, p. 145, pl. xiii, f. 2.
Yakutat: Woody Island, Kodiak, Alaska.

MYRCIA De Candolle, Dict. class., vol. 11, 1826, p. 224.

Myrcia bentonensis Berry, U. S. Geol. Surv., Prof. Paper 91, 1916, p. 317, pl. xc, f. 7–9.
Quercus retracta Lesquereux, Am. Philos. Soc., Trans., vol. 13, 1869, p. 416, pl. xvi, f. 4 [not f. 5].
Wilcox (Grenada): Grenada, Miss.
Wilcox (Ackerman): Hurleys, Miss.
Wilcox (Holly Springs): Oxford, Miss.
Wilcox: Benton, Ark.
Lagrange: Puryear and Grand Junction, Tenn.

Myrcia granadensis Berry, U. S. Geol. Surv., Prof. Paper 91, 1916, p. 316, pl. xci, f. 3.
Wilcox (Grenada): Grenada, Miss.

Myrcia parvifolia Berry, U. S. Geol. Surv., Prof. Paper 91, 1916, p. 315, pl. xc, f. 1.
Lagrange: Puryear, Tenn.

Myrcia puryearensis Berry, U. S. Geol. Surv., Prof. Paper 91, 1916, p. 316, pl. xci, f. 1, 2.
Lagrange: Puryear, Tenn.

Myrcia vera Berry, U. S. Geol. Surv., Prof. Paper 91, 1916, p. 314, pl. xc, f. 3.
Wilcox (Holly Springs): Oxford, Miss.
Lagrange: Puryear, Tenn.

Myrcia worthenii (Lesquereux) Berry, U. S. Geol. Surv., Prof. Paper 91, 1916, p. 315, pl. xc, f. 2, 10.
Salix worthenii Lesquereux, Am. Philos. Soc., Trans., vol. 13, 1869, p. 414, pl. xv, f. 7.
Lagrange: Mound City, Ill.; Lagrange and Puryear, Tenn.

MYRICA Linné, Sp. pl., 1753, p. 1024.

Myrica acuminata Unger. Lesquereux = *Myrica drymeja* and *Myrica scottii.*

Myrica acuta Hollick, in Newberry, U. S. Geol. Surv., Mon. 26, 1895 [1896], p. 65, pl. xlii, f. 35.—Berry, Geol. Surv. N. J., Bull. 3, 1911, p. 107, pl. x, f. 1.
Raritan: No locality given.

Myrica alkalina Lesquereux = *Comptonia insignis.*

Myrica ambigua Lesquereux = *Celastrinites ambiguus.*

Myrica amygdalina Saporta. Lesquereux, 1883 = *Myrica coloradensis.*

Myrica aspera Lesquereux, U. S. Geol. Surv., Mon. 17, 1891 [1892], p. 66, pl. ii, f. 11.—Knowlton, in Woodruff, U. S. Geol. Surv., Bull. 452, 1911, p. 20.
Dakota: *Pipe Creek, Cloud County, Kans.; near Lander, Wyo.

Myrica banksiaefolia Unger, Syn. pl. foss., 1845, p. 214.—Heer, Fl. foss. arct., vol. 2, Abt. 2, 1869, p. 28, pl. ii, f. 11.—Knowlton, U. S. Nat. Mus., Proc., vol. 17, 1894, p. 221; Geol. Soc. Am., Bull., vol. 5, 1893, p. 583.
Kenai: Port Graham, Alaska.

Myrica bolanderi Lesquereux = *Morella bolanderi.*

Myrica brittoniana Berry, Bull. Torr. Bot. Club, vol. 32, 1905, p. 46; Geol. Surv. N. J., Ann. Rept. 1905. [1906], p. 138; U. S. Geol. Surv., Prof. Paper 84, 1914, p. 31, pl. vii, f. 17, 18.
Myrica heerii Berry, Am. Nat., vol. 37, 1903, p. 682, f. 7, 8. [Homonym, Boulay, 1887.]
Magothy: Cliffwood, N. J.
Black Creek: Black Creek, S. C.

Myrica brongniarti? Ettingshausen. Lesquereux, Rept. U. S. Geol. Surv. Terr., vol. 7 (Tert. Fl.), 1878, p. 135, pl. xvii, f. 15.
Miocene: Elko station, Nev.

Myrica brookensis Fontaine = *Celastrophyllum acutidens.*

Myrica calicomaefolia Lesquereux = *Myrica drymeja.*

Myrica cerifera Linné. Berry, Jour. Geol., vol. 25, 1917, p. 662.
Pleistocene: Vero, Fla.

Myrica cinnamomifolia Newberry, U. S. Geol. Surv., Mon. 26, 1895 [1896], p. 64, pl. xxii, f. 9–14.—Berry, Geol. Surv. N. J., Bull. 3, 1911, p. 106, pl. x, f. 10; Bull. Torr. Bot. Club, vol. 44, 1917, p. 175.
Raritan: South Amboy, N. J.
Bingen: Mine Creek near Nashville, Ark.

Myrica cliffwoodensis Berry, Bull. Torr. Bot. Club, vol. 31, 1904, p. 73, pl. 4, f. 1; Geol. Surv. N. J., Ann. Rept. 1905 [1906], pp, 138, 139; Bull. Torr. Bot. Club, vol. 37, 1910, p. 191; idem, vol. 44, 1917, p. 174.
Magothy: Cliffwood, N. J.
Black Creek: Parker Landing, Tar River, N. C.
Bingen: Near Maxwell Spur, Pike County, Ark.

Myrica coloradensis Knowlton, U. S. Nat. Mus., Proc., vol. 51, 1916, p. 257, pl. xxi, f. 1.
Myrica amygdalina Saporta. Lesquereux, Rept. U. S. Geol. Surv. Terr., vol. 8 (Cret. and Tert. Fl.), 1883, p. 147, pl. xxvi, f. 1–4.
Miocene: Florissant, Colo.

Myrica copeana Lesquereux, U. S. Geol. and Geog. Surv. Terr., Ann. Rept. 1873 [1874], p. 411; Rept. U. S. Geol. Surv. Terr., vol. 7 (Tert. Fl.), 1878, p. 131, pl. xvii, f. 5.—Cockerell, Am. Mus. Nat. Hist., Bull., vol. 24, 1908, p. 81.—Knowlton, U. S. Nat. Mus., Proc., vol. 51, 1916, p. 259.
Miocene: *Florissant, Colo.

Myrica copeana Lesquereux. Lesquereux, U. S. Nat. Mus., Proc., vol. 11, 1888, p. 12 = *Cupanites loughridgii.*

Myrica coriacea Knowlton, U. S. Geol. Surv., Prof. Paper 101, 1918, p. 256, pl. xxxvii, f. 7.
Vermejo: Coal Creek, Rockvale, Colo.

Myrica cretacea Lesquereux = *Myrica dakotensis.*

Myrica (Comptonia) cuspidata Lesquereux. Dawson, Roy. Soc. Canada, Trans., vol. 8, 1890, sec. 4, p. 80, f. 9 = *Comptonia dryandroides.*

Myrica (Comptonia) cuspidata (Lesquereux) Dawson = *Comptonia cuspidata.*

Myrica dakotensis Lesquereux, Rept. U. S. Geol. Surv. Terr., vol. 8 (Cret. and Tert. Fl.), 1883, p. 35, pl. iv, f. 9.
Myrica cretacea Lesquereux, U. S. Geol. and Geog. Surv. Terr., Bull., vol. 1, 1875 [1876], p. 392; idem, Ann. Rept. 1874 [1876], p. 339, pl. iii, f. 4 [not *M. cretacea* Heer, 1871].
Dakota: Near Fort Harker, Kans.

Myrica dakotensis minima Berry, U. S. Geol. Surv., Prof. Paper 112, 1919, p. 74, pl. xiii, f. 5.
Tuscaloosa: Glen Allen, Ala.

Myrica davisii Hollick, N. Y. Acad. Sci., Trans., vol. 12, 1892, p. 32, pl. ii, f. 3; U. S. Geol. Surv., Mon. 50, 1906, p. 53, pl. vii, f. 25.
Raritan: Kreischerville, Staten Island, N. Y.

Myrica diversifolia Lesquereux, Rept. U. S. Geol. Surv. Terr., vol. 8 (Cret. and Tert. Fl.), pl. xxv, f. 6, 10–12 = *Sorbus diversifolia;* f. 7, 8, 14 = *Sorbus nupta;* f. 13 = *Ribes? florissanti.*

Myrica drymeja (Lesquereux) Knowlton, U. S. Geol. Surv., Bull. 152, 1898, p. 146.—Cockerell, Am. Nat., vol. 44, 1910, p. 42, f. 6.—Knowlton, U. S. Nat. Mus., Proc., vol. 51, 1916, p. 256, pl. xx, f. 1, 2.

Rhus? drymeja Lesquereux, U. S. Geol. and Geog. Surv. Terr., Ann. Rept. 1873 [1874], p. 416.

Callicoma microphylla Ettingshausen. Lesquereux, Rept. U. S. Geol. Surv. Terr., vol. 7 (Tert. Fl.), 1878, p. 246, pl. xliii, f. 2–4.

Myrica callicomaefolia Lesquereux, Rept. U. S. Geol. Surv. Terr., vol. 8 (Cret. and Tert. Fl.), 1883, p. 146, pl. xxvi, f. 5–14.

Myrica fallax Lesquereux, idem, p. 147, pl. xxxii, f. 11–16.

Myrica zachariensis Saporta. Lesquereux, idem, p. 146, pl. xxv, f. 5 [not pl. xlvA, f. 6–9].

Myrica acuminata Unger. Lesquereux, U. S. Geol. and Geog. Surv. Terr., Ann. Rept. 1873 [1874], p. 411; Rept. U. S. Geol. Surv. Terr., vol. 7 (Tert. Fl.), 1878, p. 130, pl. xvii. f. 2, 3 [not f. 1, 4, which = *M. scottii*].

Ceratopetalum americanum Ettingshausen, Denkschr. K. Akad. Wiss., Matt.-naturw. Classe, vol. 47, 1883, p. 137 [37]; Geol. Surv. N. S. W., Mem. Palaeont. No. 2 (Tert. Fl. Australia), 1888, p. 58.

Miocene: Florissant, Colo.

Miocene: Elko, Nev.

Myrica dubia Knowlton, U. S. Geol. Surv., Prof. Paper 130, 19—, p. —, pl. v, f. 3 [in preparation].

Laramie: Popes Bluff, Pikeview, Colo.

Myrica elaeanoides Lesquereux, U. S. Nat. Mus., Proc., vol. 11, 1888, p. 12, pl. iv, f. 5.

Wilcox (Ackerman): Hurleys, Benton County, Miss.

Lagrange: Boaz, Graves County, Ky., and Grand Junction, Tenn.

Myrica elegans Berry, Bull. Torr. Bot. Club, vol. 34, 1907, p. 191, pl. xi, f. 1–4, 6; U. S. Geol. Surv., Prof. Paper 84, 1914, p. 31, pl. ix, f. 4.

Black Creek: Court House Landing, Cape Fear River, N. C.; Darlington, S. C.

Myrica emarginata Heer, Fl. foss. arct., vol. 6, Abt. 2, 1882, p. 66, pl. xli, f. 2; Lesquereux, U. S. Geol. Surv., Mon. 17, 1891 [1892], p. 67, pl. xii, f. 1.—Newberry, U. S. Geol. Surv., Mon. 26, 1895 [1896], p. 62, pl. xli, f. 10, 11.—Berry, Bull. Torr. Bot. Club, vol. 39, 1912, p. 393; Geol. Surv. N. J., Bull. 3, 1911, p. 104, pl. x, f. 5; U. S. Geol. Surv., Prof. Paper 112, 1919, p. 73, pl. xiii, f. 4.

Dakota: Ellsworth County, Kans.

Raritan: New Jersey [no definite locality].

Woodbine: Arthurs Bluff, Tex.

Tuscaloosa: Cottondale and Shirleys Mill, Ala.

Myrica fallax Lesquereux = *Myrica drymeja* (according to Cockerell).

Myrica fenestrata Newberry, U. S. Geol. Surv., Mon. 26, 1895 [1896], p. 63, pl. xliii, f. 32.—Berry, Geol. Surv. N. J., Bull. 3, 1911, p. 105, pl. x, f. 3.

Raritan: Sayreville and Milltown, N. J.

Myrica grandifolia Hollick = *Myrica hollicki.*

Myrica havanensis Berry, Bull. Torr. Bot. Club, vol. 43, 1916, p. 300, U. S. Geol. Surv., Prof. Paper 112, 1919, p. 125, pl. xi, f. 4; pl. xxviii, f. 7.

Eutaw: Havana, Hale County, Ala.

Ripley: Near Camden, near Cypress, and near Selmer, Tenn.

Myrica heerii Berry = *Myrica brittoniana.*

Myrica hendersoni Cockerell, Bull. Torr. Bot. Club, vol. 33, 1906, p. 308, text f. 1.—Knowlton, U. S. Nat. Mus., Proc., vol. 51, 1916, p. 258.

Morella hendersoni (Cockerell) Cockerell, Univ. Colo., Studies, vol. 3, 1906, p. 173.

Miocene: *Florissant, Colo.

Myrica hollicki Ward, Am. Jour. Sci., 3d ser., vol. 45, 1893, p. 437.—Hollick, U. S. Geol. Surv., Mon. 50, 1907, p. 53, pl. vii, f. 24.—Berry, Geol. Surv. N. J., Bull. 3, 1901, p. 103, pl. x, f. 6.

Myrica grandifolia Hollick, N. Y. Acad. Sci., Trans., vol. 12, 1892, p. 32, pl. iii, f. 1.—Berry, Bull. Torr. Bot. Club, vol. 36, 1909, p. 249, pl. xviii, f. 2. [Homonym, *M. grandifolia* (Unger) Schimper, Pal. vég., vol. 2, 1872, p. 559.]

Raritan: Tottenville, Staten Island, N. Y.; Milltown, N J.

Myrica? idahoensis Knowlton, U. S. Geol. Surv., Eighteenth Ann. Rept., pt. 3, 1898, p. 724, pl. xcix, f. 7.

Payette: Marsh, Idaho.

Myrica insignis Lesquereux = *Comptonia insignis.*

Myrica lamarensis Knowlton, U. S. Geol. Surv., Mon. 32, pt. 2, 1899, p. 693, pl. lxxxiv, f. 5.

Miocene: Yellowstone National Park.

Myrica lanceolata Knowlton, U. S. Geol. Surv., Eighteenth Ann. Rept., pt. 3, 1898, p. 724, pl. xcix, f. 5, 6.

Payette: Shafer Creek, Boise County, Idaho.

Myrica latiloba acutiloba Lesquereux = *Comptonia acutiloba.*

Myrica? lessigiana Lesquereux = *Artocarpus lessigiana.*

Myrica? lessigii Lesquereux = *Artocarpus lessigiana.*

Myrica (Aralia) lessigii? Lesquereux, U. S. Nat. Mus., Proc., vol. 11, 1888, p. 16 = *Artocarpus californica.*

Myrica lignitum (Unger) Saporta, Ann. Sci. Nat., Bot., 5th ser., vol. 4, 1866, p. 102, pl. v, f. 10.

Dryandroides lignitum (Unger) Ettingshausen. Lesquereux, U. S. Nat. Mus., Proc., vol. 10, 1887, p. 41.

——: Deer Creek coal field, Ariz.

Myrica longa (Heer) Heer, Fl. foss. arct., vol. 6, 1883, Abt. 2, p. 65, pl. xviii, f. 9*b*; pl. xxix, f. 15–17; pl. xxxiii, f. 10; pl. xli, f. 4*d*; idem, vol. 7, 1883, p. 21.—Lesquereux, U. S. Geol. Surv., Mon. 17, 1891 [1892], p. 67, pl. iii, f. 1–6.—Knowlton, U. S. Geol. Surv., Twenty-first Ann. Rept., pt. 7, 1901, p. 314, pl. xxxix, f. 7.—Berry, Bull. Torr. Bot. Club, vol. 33, 1906, p. 170; Md. Geol. Surv., Upper Cret., 1916, p. 812, pl. lvii, f. 1–3; Bull. Torr. Bot, Club, vol. 44, 1917, p. 175; U. S. Geol. Surv., Prof. Paper 112, 1919, p. 74.

Proteoides longus Heer, Fl. foss. arct., vol. 3, Abt. 2, 1874, p. 110, pl. xxix, f. 8*b*; pl. xxxi, f. 4, 5.—Dawson, Roy. Soc. Canada, Trans., vol. 1, 1882 [1883], sec. 4, p. 22, pl. ii, f. 8.

Bingen: Mine Creek, near Nashville, Ark.
Dakota: Ellsworth County, Kans.; near Lander, Wyo.
——: Fork of Peace River, Northwest Territory.
Magothy: Grove Point, Round Bay, and Little Round Bay, Md.
Tuscaloosa: Snow place, Tuscaloosa County, Ala.

Myrica ludwigii Schimper, Pal. vég., vol. 2, 1872, p. 545.—Lesquereux, U. S. Geol. and Geog. Surv. Terr., Bull., vol. 1 [1875], 1876, p. 385; idem, Ann. Rept. 1874 [1876], p. 311; Rept. U. S. Geol. Surv. Terr., vol. 7 (Tert. Fl.), 1878, p. 133, pl. lxv, f. 9.
Green River: Mouth of White River, Wyo.

Myrica? neomexicana Knowlton, U. S. Geol. Surv., Prof. Paper 98, 1916, p. 336, pl. lxxxvi, f. 2–4.
Kirtland: Near Pina Veta China, San Juan County, N. Mex.

Myrica newberryana Hollick, in Newberry, U. S. Geol. Surv., Mon. 26, 1895 [1896], p. 63, pl. xliii, f. 5.—Berry, Geol. Surv. N. J., Bull. 3, 1911, p. 105, pl. x, f. 2.
Raritan: South Amboy and Milltown, N. J.

Myrica nervosa Knowlton, U. S. Geol. Surv., Prof. Paper 108, 1917, p. 85, pl. xxxiv, f. 1.
Frontier: About 1 mile east of Cumberland, Wyo.

Myrica nigricans Lesquereux, U. S. Geol. and Geog. Surv. Terr., Ann. Rept. 1871 [1872], Suppl., p. 6; Rept. U. S. Geol. Surv. Terr., vol. 7 (Tert. Fl.), 1883, p. 132, pl. xvii, f. 9–12.
Green River: Green River, Wyo.

Myrica obliqua Knowlton, in Lesquereux, U. S. Geol. Surv., Mon. 17, 1891 [1892], p. 68, pl. xliv, f. 16.
Dakota: Ellsworth County, Kans.

Myrica oblongifolia Knowlton, U. S. Geol. Surv., Prof. Paper 130, 19—, p. —, pl. xxi, f. 1 [in preparation].
Laramie: Marshall, Colo.

Myrica obscura Lesquereux, Rept. U. S. Geol. Surv. Terr., vol. 8 (Cret. and Tert. Fl.), 1883, p. 145, pl. xxxii, f. 8–10; Cockerell, Am. Mus. Nat. Hist., Bull., vol. 24, 1908, p. 108; Knowlton, U. S. Nat. Mus., Proc., vol. 51, 1916, p. 258.
Miocene: *Florissant, Colo.

Myrica obtusa Lesquereux, Rept. U. S. Geol. Surv. Terr., vol. 6 (Cret. Fl.), 1874, p. 63, pl. xxix, f. 10.
Dakota: Kansas.

Myrica oregoniana Knowlton, U. S. Geol. Surv., Bull. 204, 1902, p. 33, pl. iii, f. 4.
Mascall: Near Van Horn's ranch, John Day Basin, Oreg.

Myrica partita Lesquereux = *Comptonia partita.*

Myrica (Comptonia) parvula Heer = *Comptonia microphylla.*

Myrica? personata Knowlton, U. S. Geol. Surv., Bull. 204, 1902, p. 33, pl. iii, f. 2.—Penhallow, Rept. Tert. Pl. Brit. Col., 1908, p. 63.
Clarno (upper part): Near Fossil, Gilliam County, Oreg.
Oligocene or Miocene: Horsefly River, British Columbia.

Myrica polymorpha Schimper. Lesquereux, Rept. U. S. Geol. Surv. Terr., vol. 8 (Cret. and Tert. Fl.), 1883, p. 146, pl. xxv, f. 1, 2 = *Myrica scottii.*

Myrica (Comptonia) praemissa (Lesquereux) Knowlton = *Comptonia praemissa.*

Myrica raritanensis Hollick, in Newberry, U. S. Geol. Surv., Mon. 26, 1895 [1896], p. 65, pl. xlii, f. 34.—Berry, Geol. Surv. N. J., Bull. 3, 1911, p. 108, pl. x, f. 4.
Raritan: New Jersey [no definite locality given].

Myrica rigida Lesquereux = *Myrica drymeja* [according to Cockerell].

Myrica ripleyensis Berry, Bull. Torr. Bot. Club, vol. 43, 1916, p. 288; U. S. Geol. Surv., Prof. Paper 112, 1919, p. 74, pl. xi, f. 2.
Ripley: Near Camden, Benton County, and Selmer, McNairy County, Tenn.

Myrica salicina Unger, Gen. et sp. pl. foss., 1850, p. 396.—Lesquereux, U. S. Geol. and Geog. Surv. Terr., Ann. Rept. 1871 [1872], Suppl., p. 6.
Green River: Green River, Wyo.

Myrica saportana Lesquereux, U. S. Geol. and Geog. Surv. Terr., Bull., vol. 1, 1875 [1876], p. 385. [Not afterward mentioned.]
——: Middle Park, Colo.

Myrica schimperi Lesquereux, U. S. Geol. Surv., Mon. 17, 1891 [1892], p. 66, pl. ii, f. 12.
Dakota: Near Brookville, Kans.

Myrica scottii Lesquereux, Rept. U. S. Geol. Surv. Terr., vol. 8 (Cret. and Tert. Fl.), 1883, p. 147, pl. xxxii, f. 17, 18.—Knowlton, U. S. Geol. Surv., Mon. 32, pt. 2, 1899, p. 692, pl. lxxxiv, f. 6.—Cockerell, Am. Mus. Nat. Hist., Bull., vol. 24, 1908, p. 81.—Knowlton, U. S. Nat. Mus., Proc., vol. 51, 1916, p. 259, pl. xx, f. 3, 4.

Myrica acuminata Unger. Lesquereux, U. S. Geol. and Geog. Surv. Terr., Ann. Rept. 1873 [1874], p. 411; Rept. U. S. Geol. Surv. Terr., vol. 7 (Tert. Fl.), 1878, p. 130, pl. xvii, f. 1, 4 [not f. 2, 3, which = *Myrica drymeja*].

Myrica polymorpha Schimper. Lesquereux, Rept. U. S. Geol. Surv. Terr., vol. 8 (Cret. and Tert. Fl.), 1883, p. 146, pl. xxv, f. 1, 2.

Miocene: *Florissant, Colo.; ?Elk Creek, Yellowstone National Park.

Myrica serrata Penhallow, Roy. Soc. Canada, Trans., 3d ser., vol. 1, sec. 4, 1907, p. 309, text f. 4.

Lower Cretaceous(?): International boundary between Pasayten and Skagit rivers, British Columbia.

Myrica sternbergii Lesquereux, Rept. U. S. Geol. Surv. Terr., vol. 8 (Cret. and Tert. Fl.), 1883, p. 35; U. S. Geol. Surv., Mon. 17, 1891 [1892], p. 68.

Dakota: Kansas.

Myrica studeri? Heer. Lesquereux, U. S. Nat. Mus., Proc., vol. 10, 1887, p. 38.

Green River?: White River, Wyo.?

Myrica torreyi Lesquereux, U. S. Geol. and Geog. Surv. Terr., Ann. Rept. 1872 [1873], p. 392; Rept. U. S. Geol. Surv. Terr., vol. 7 (Tert. Fl.), 1878, p. 129, pl. xvi, f. 3–10.—Ward, U. S. Geol. Surv., Sixth Ann. Rept., 1884–85 [1886], p. 551, pl. xl, f. 4; idem, Bull, 37, 1887, p. 32, pl. xiv, f. 5.—Knowlton, U. S. Geol. Surv., Bull. 163, 1900, p. 34, pl. vi, f. 1–3.—Cockerell, Univ. Colo. Studies, vol. 7, 1910, p. 150.—Knowlton, U. S. Geol. Surv., Prof. Paper 98, 1916, p. 90, pl. xvii, f. 7; p. 336, pl. lxxxvi, f. 1; idem, Prof. Paper 101, 1918, p. 256, pl. xxxvii, f. 2–4.

Post-Laramie: *Black Buttes, Wyo.
?Lance: Converse County, Wyo.
Laramie: Crow Creek, near Greeley, Colo.
Mesaverde: Near Meeker, Colo.; Point of Rocks, Wyo.
Fruitland: 30 miles south of Farmington, N. Mex.
Fox Hills: 2½ miles south of Milliken, Colo.
Vermejo: Rockvale, Canon City field, La Veta, near Trinidad, and Walsenburg, Colo.

Myrica torreyi minor Lesquereux, U. S. Geol. and Geog. Surv. Terr., Ann. Rept. 1873 [1874], p. 397.

Denver: Sand Creek, Colo.

Myrica? trifoliata Newberry, U. S. Geol. Surv., Mon. 35, 1898, p. 37, pl. xiv, f. 2.
Dakota: Whetstone Creek, N. Mex.

Myrica undulata (Heer) Schimper, Pal. vég., vol. 2, 1872, p. 556.—Lesquereux, U. S. Geol. and Geog. Surv. Terr., Ann. Rept. 1873 [1874], p. 412; Rept. U. S. Geol. Surv. Terr., vol. 7 (Tert. Fl.), 1878, p. 131, pl. xvii, f. 6-8.
Dryandroides undulata Heer, Fl. tert. Helv., vol. 3, 1859, p. 188.
Miocene: Elko station, Nev.

Myrica ungeri Heer, Fl. tert. Helv., vol. 2, 1856, p. 35, pl. lxx, f. 7, 8.—Lesquereux, U. S. Nat. Mus., Proc., vol. 11, 1888, p. 27.
Miocene (auriferous gravels): Spanish Peak, Plumas County, Calif.

Myrica vindobonensis (Ettingshausen) Heer, Fl. foss. arct., vol. 2, Abt. 2, 1869, p. 27, pl. iii, f. 4, 5.—Knowlton, U. S. Nat. Mus., Proc., vol. 17, 1894, p. 222; Geol. Soc. Am., Bull., vol. 5, 1893, p. 583.
Dryandra vindobonensis Ettingshausen, Tert. Fl. Wien, 1851, p. 18, pl. iii, f. 6.
Kenai?: Ninilchik, Alaska.

Myrica wardii Knowlton, U. S. Geol. Surv., Mon. 32, pt. 2, 1899, p. 692, pl. lxxxiv, f. 4.
Miocene: Yellowstone National Park.

Myrica wilcoxensis Berry, U. S. Geol. Surv., Prof. Paper 91, 1916, p. 188, pl. xviii, f. 1.
Wilcox (Grenada): Grenada, Grenada County, Miss.

Myrica zachariensis Saporta. Lesquereux = *Myrica drymeja*.

Myrica zenkeri (Ettingshausen) Heer, Fl. foss. arct., vol. 3, Abt. 2, 1874, p. 108, pl. xxxi, f. 2.—?Hollick, U. S. Geol. Surv., Mon. 50, 1906, p. 54, pl. vii, f. 23.
Magothy: ?Glen Cove, Long Island, N. Y.

Myrica?, seeds of, Lesquereux, Rept. U. S. Geol. Surv. Terr., vol. 6 (Cret. Fl.), 1874, p. 63, pl. xxvii, f. 4, 4a; U. S. Geol. Surv., Mon. 17, 1891 [1892], p. 68.
Dakota: Decatur, Nebr.

Myrica sp. Dawson, Geol. Surv. Canada, Rept. Prog. 1877-78, p. 64.—Penhallow, Rept. Tert. Pl. Brit. Col., 1908, p. 63.
Oligocene?: Ninemile Creek, Similkameen River, British Columbia.

Myrica sp., ament of, Hollick, U. S. Geol. Surv., Mon. 50, 1906, p. 54, pl. vii, f. 22.
Magothy: Gay Head, Marthas Vineyard, Mass.

Myrica? sp. Knowlton, U. S. Geol. Surv., Bull. 163, 1900, p. 34, pl. vi, f. 4.
Mesaverde: North Fork of Dutton Creek, Laramie Plains, Wyo.

Myrica sp. Lesquereux, Rept. on Clays in N. J., 1878, p. 29.
Raritan: South Amboy, N. J.

Myrica? sp. Penhallow, Rept. Tert. Pl. Brit. Col., 1908, p. 63.
Oligocene or lower Miocene: Horsefly River, British Columbia.

MYRICAEPHYLLUM Fontaine, U. S. Geol. Surv., Mon. 15, 1889 [1890], p. 315. [Type, *M. dentatum*.]

Myricaephyllum dentatum Fontaine, U. S. Geol. Surv., Mon. 15, 1889 [1890], p. 316, pl. clvi, f. 6.
Patapsco: Near Brooke, Va.

MYRSINE Linné, Sp. pl., 1753, p. 196.

Myrsine borealis Heer, Fl. foss. arct., vol. 6, Abt. 2, 1882, p. 81, pl. xxiv, f. 7b, 8; pl. xxvii, f. 1b; pl. xliv, f. 5a; pl. xlvi, f. 19, 20.—White, Am. Jour. Sci., 3d ser., vol. 39, 1890, p. 98, pl. ii, f. 5.—Smith, Geol. Coastal Plain, Ala., 1894, p. 348.—Newberry, U. S. Geol. Surv., Mon. 26, 1895 [1896], p. 122, pl. xxiv, f. 4–6.—Hollick, Geol. Soc. Am., Bull., vol. 7, 1895, p. 13; U. S. Geol. Surv., Mon. 50, 1906, p. 102, pl. xxxix, f. 10, 11.—Berry, Bull. Torr. Bot. Club, vol. 37, 1910, pp. 29, 199; N. J. Geol. Surv., Bull. 3, 1911, p. 208; Md. Geol. Surv., Upper Cret., 1916, p. 890, pl. lxxxix, f. 5; U. S. Geol. Surv., Prof. Paper 112, 1919, p. 133.

Diospyros rotundifolia Lesquereux. Hollick, Bull. Torr. Bot. Club, vol. 21, 1894, p. 53, pl. 179, f. 2.

Black Creek: Cape Fear River, N. C.
Magothy: Glen Cove, Long Island, N. Y.; Gay Head, Marthas Vineyard, Mass.; Grove Point, Md.
Raritan: South Amboy, Milltown, N. J.
Tuscaloosa: Shirleys Mill and Glen Allen, Ala.

Myrsine crassa Lesquereux = *Myrsine crassula*.

Myrsine crassula Knowlton and Cockerell, n. name.

Myrsine crassa Lesquereux, U. S. Geol. Surv., Mon. 17, 1891 [1892], p. 114, pl. lii, f. 2, 3.—Berry, N. Y. Bot. Gard., Bull., vol. 3, 1903, p. 98, pl. lii, f. 6; Geol. Surv. N. J., Ann. Rept. 1905 [1906], p. 138. [Homonym, Dietrich, 1839.]

Dakota: Near Fort Harker, Kans.
Magothy: Cliffwood, N. J.

Myrsine elongata Newberry. Hollick, N. Y. Bot. Gard., Bull., vol. 2, 1902, p. 405, pl. xli, f. 2 = *Diospyros provecta*.

Myrsine elongata Hollick = *Myrsine gaudini*.

Myrsine gaudini (Lesquereux) Berry, Bull. Torr. Bot. Club, vol. 36, 1909, p. 262; idem, vol. 37, 1910, p. 200; idem, vol. 38, 1911, p. 408; Geol. Surv. N. J., Bull. 3, 1911, p. 210, pl. xxiv, f. 3, 4; U. S. Geol. Surv., Prof. Paper 84, 1914, p. 61, pl. xiv, f. 9; Md. Geol. Surv., Upper Cret., 1916, p. 891, pl. lxxxix, f. 6, 7; U. S. Geol. Surv., Prof. Paper 112, 1919, p. 133.

Myrsinites? gaudini Lesquereux, U. S. Geol. Surv., Mon. 17, 1891 [1892], p. 115, pl. lii, f. 4.—Hollick, U. S. Geol. Surv., Mon. 50, 1906, p. 103, pl. xxxix, f. 12.

Myrsine gaudini—Continued.

Rhamnus rossmässleri Unger. Hollick, N. Y. Acad. Sci., Trans., vol. 12, 1892, p. 35, pl. iii, f. 5.

Myrsine elongata Hollick, Bull. Torr. Bot. Club, vol. 21, 1894, p. 54, pl. clxxvii, f. 2.—Newberry, U. S. Geol. Surv., Mon. 26, 1896, p. 122, pl. xxii, f. 1-3.—Hollick, N. Y. Acad. Sci., Annals, vol. 11, 1898, p. 420, pl. xxxviii, f. 3, 4b, 4c; N. Y. Bot. Gard., Bull., vol. 2, 1902, p. 405, pl. xli, f. 2; U. S. Geol. Surv., Mon. 50, 1906, p. 102, pl. viii, f. 1b; pl. xxxix, f. 13, 14.

Raritan: South Amboy and Milltown, N. J.; Tottenville, Staten Island, N. Y.

Magothy: Grove Point, Md.

Black Creek: Court House Bluff, Cape Fear River, N. C.; Langley, S. C.

Tuscaloosa: Near Northport, Glen Allen, and Cottondale, Ala.

Myrsine laminarum Cockerell, Am. Mus. Nat. Hist., Bull., vol. 24, 1908, p. 105.

Myrsine latifolia Lesquereux, Rept. U. S. Geol. Surv. Terr., vol. 8 (Cret. and Tert. Fl.), 1883, p. 173, pl. xxxviii, f. 16. [Homonym, Springe, 1825.]

Miocene: Florissant, Colo.

Myrsine latifolia Lesquereux = *Myrsine laminarum.*

Myrsine oblongata Hollick, in Newberry, U. S. Geol. Surv., Mon. 26, 1895 [1896], p. 122, pl. xlii, f. 15; Geol. Surv. N. J., Bull. 3, 1911, p. 209, pl. xxiv, f. 1.

Raritan ?: New Jersey [no definite locality].

Myrsine sp. Hollick, in Ries, School of Mines Quarterly, vol. 15, 1894, p. 355.

Magothy: Northport, Long Island, N. Y.

Myrsinites? gaudini Lesquereux = *Myrsine gaudini.*

MYRTOPHYLLUM Heer, Neue Denkschr. Schw. Ges., vol. 23, Mém. 2, 1869, p. 22.

Myrtophyllum (Eucalyptus?) geinitzi Heer = *Eucalyptus geinitzi.*

Myrtophyllum sapindoides Hollick, N. Y. Bot. Gard., Bull., vol. 8, 1912, p. 167, pl. clxvii, f. 4.

Raritan: Glen Cove, N. Y.

Myrtophyllum (Eucalyptus) schübleri Heer = *Eucalyptus schübleri.*

Myrtophyllum warderi Lesquereux = *Eucalyptus geinitzi.*

MYRTUS (Tournefort) Linné, Sp. pl., 1753, p. 471.

Myrtus oregonensis Lesquereux, Rept. U. S. Geol. Surv. Terr., vol. 8 (Cret. and Tert. Fl.), 1883, p. 254, pl. lviii, f. 10.

Miocene (auriferous gravels): Corral Hollow, Alameda County, Calif.

NAGEIOPSIS Fontaine, U. S. Geol. Surv., Mon. 15, 1889 [1890], p. 194. [Type, *N. longifolia.*]

Nageiopsis acuminata Fontaine = *Podozamites acutifolius.*

Nageiopsis angustifolia Fontaine, U. S. Geol. Surv., Mon. 15, 1889 [1890], p. 202, pl. lxxxvi, f. 8, 9.; pl. lxxxvii, f. 2–6; pl. lxxxviii, f. 1, 3, 4, 6–8; pl. lxxxix, f. 2; in Ward, U. S. Geol. Surv., Nineteenth Ann. Rept., pt. 2, 1899, p. 684, pl. clxviii, f. 7; U. S. Geol. Surv., Mon. 48, 1905 [1906], pp. 219, 491, 516, 528, 560, pl. cxvii, f. 4, 5.—Berry, U. S. Nat. Mus., Proc., vol. 38, 1910, p. 190; Md. Geol. Surv., Lower Cret., 1912, p. 389, pl. lxiii, f. 3, 4.

Cephalotaxopsis ramosa Fontaine? Fontaine, in Ward, U. S. Geol. Surv., Mon. 48, 1905 [1906], p. 547.

Patuxent: Fredericksburg, near Dutch Gap, near Potomac Run, Va.; Sixteenth Street, D. C.

Arundel: Bay View, Md.

Patapsco: Near Brooke, Mount Vernon, Va.; Federal Hill, Fort Foote, Fort Washington ?, Md.

?Lakota: Barrett, Cook County, Wyo.

Knoxville and Horsetown: California.

Nageiopsis crassicaulis Fontaine = *Nageiopsis longifolia.*

Nageiopsis decrescens Fontaine = *Nageiopsis zamioides.*

Nageiopsis heterophylla Fontaine = *Nageiopsis zamioides.*

Nageiopsis inaequilateralis Fontaine = *Podozamites inaequilateralis.*

Nageiopsis latifolia Fontaine = *Phyllites latifolius.*

Nageiopsis longifolia Fontaine, U. S. Geol. Surv., Mon. 15, 1889 [1890], p. 195, pl. lxxv, f. 1; pl. lxxvi, f. 2–6; pl. lxxvii, f. 1, 2; pl. lxxviii, f. 1–5; pl. lxxix, f. 7; pl. lxxxv, f. 1, 2, 8, 9.—?Fontaine, in Diller and Stanton, Geol. Soc. Am., Bull., vol. 5, 1894, p. 450.—?Fontaine, in Stanton, U. S. Geol. Surv., Bull. 133, 1896, p. 15.—Fontaine, in Ward, U. S. Geol. Surv., Mon. 48, 1905 [1906], pp. 259, 311, 484, 491, 510, 528, 548, 557, pl. lxviii, f. 9–12; pl. lxxiii, f. 9 [not pl. xlv, f. 1–5].—?Knowlton, in Diller, Geol. Soc. Am., Bull., vol. —, 1908, p. 386.—Knowlton, Am. Jour. Sci., 4th ser., vol. 30, 1910, p. 42.—Berry, U. S. Nat. Mus., Proc., vol. 38, 1910, p. 189; Md. Geol. Surv., Lower Cret., 1912, p. 384, pl. lxi.

Nageiopsis crassicaulis Fontaine, U. S. Geol. Surv., Mon. 15, 1889 [1890], p. 198, pl. lxxix, f. 2, 6; pl. lxxxii, f. 1; pl. lxxxiv, f. 3, 9, 11.—Berry, U. S. Nat. Mus., Proc., vol. 38, 1910, p. 189.

Angiopteridium strictinerve Fontaine. Fontaine?, in Diller and Stanton, Geol. Soc. Am., Bull., vol. 5, 1894, p. 450.—Fontaine, in Stanton, U. S. Geol. Surv., Bull. 133, 1896, p. 15.

Patuxent: Dutch Gap, Kankeys, Cockpit Point, telegraph station near Lorton, Va.

Arundel: Langdon, D. C.

Patapsco: Near Brooke, Mount Vernon, Deep Bottom ?, Chinkapin Hollow ?, Va.; Fort Foote, Federal Hill, Vinegar Hill ?, Md.

?Lakota: Barrett and South Fork of Hay Creek, Crook County, Wyo.

?Kootenai: Near Geyser, Cascade County, Mont.

Knoxville and Horsetown: Tehama County. Calif

Nageiopsis longifolia Fontaine. Fontaine, in Ward, U. S. Geol. Surv., Mon. 48, 1905 [1906], p. 171, pl. xlv, f. 1–5 = *Zamites megaphyllus.*

Nageiopsis microphylla Fontaine = *Nageiopsis zamioides.*

Nageiopsis montanensis Fontaine = *Podozamites inaequilateralis.*

Nageiopsis obtusifolia Fontaine, U. S. Geol. Surv., Mon. 15, 1889, p. 200, pl. lxxxv, f. 7.

Patuxent: Potomac River, Va.

Nageiopsis ovata Fontaine = *Nageiopsis zamioides.*

Nageiopsis recurvata Fontaine = *Nageiopsis zamioides.*

Nageiopsis recurvata Fontaine. Fontaine, in Ward = *Podozamites knowltoni.*

Nageiopsis subfalcata Fontaine = *Sequoia?* sp.

Nageiopsis zamioides Fontaine, U. S. Geol. Surv., Mon. 15, 1889 [1890], p. 196, pl. lxxix, f. 1, 3; pl. lxxx, f. 1, 2, 4; pl. lxxxi, f. 1–6; in Ward, U. S. Geol. Surv., Mon. 48, 1905 [1906], pp. 510, 521, 528, 545.—Berry, U. S. Nat. Mus., Proc., vol. 38, 1910, p. 191; Md. Geol. Surv., Lower Cret., 1912, p. 386, pl. lxii, f. 1, 2; pl. lxiii.

Nageiopsis recurvata Fontaine, U. S. Geol. Surv., Mon. 15, 1889 [1890], p. 197, pl. lxxv, f. 2; pl. lxxix, f. 4; pl. lxxx, f. 3.

Nageiopsis decrescens Fontaine, U. S. Geol. Surv., Mon. 15, 1889 [1890], p. 199, pl. lxxvii, f. 3.

Nageiopsis ovata Fontaine, U. S. Geol. Surv., Mon. 15, 1889 [1890], p. 199, pl. lxxvii, f. 4; pl. lxxx, f. 5.

Nageiopsis heterophylla Fontaine, U. S. Geol. Surv., Mon. 15, 1889 [1890], p. 201, pl. lxxxiv, f. 4; pl. lxxxvi, f. 6, 7; pl. lxxxviii, f. 2, 5; in Ward, U. S. Geol. Surv., Mon. 48, 1905 [1906], pp. 219?, 483, 520, 548, 561, pl. cxvii, f. 6.

Nageiopsis microphylla Fontaine, U. S. Geol. Surv., Mon. 15, 1889 [1890], p. 201, pl. lxxxiv, f. 6; pl. lxxxv, f. 14; pl. lxxxvi, f. 1–3, 5; in Ward, U. S. Geol. Surv., Mon. 48, 1905 [1906], p. 484.

Patuxent: Fredericksburg, Dutch Gap, Cockpit Point, near Potomac Run, Va.

Arundel: Langdon, D. C.; Bay View, Hobb's iron mine(?), Md.

Patapsco: Vinegar Hill, Federal Hill, Fort Foote, and Overlook Inn Road, Md.

?Shasta: Tehama County, Calif.

NAJADOPSIS Heer, Fl. tert. Helv., vol. 1, 1855, p. 104.

Najadopsis rugulosa Lesquereux, Rept. U. S. Geol. Surv. Terr., vol. 8 (Cret. and Tert. Fl.), 1883, p. 142, pl. xxiii, f. 7.—Knowlton, U. S. Nat. Mus., Proc., vol. 51, 1916, p. 252.

Miocene: *Florissant, Colo.

NECTANDRA Bergius, Fl. Cap., 1767, p. 131.

Nectandra glenni Berry, U. S. Geol. Surv., Prof. Paper 91, 1916, p. 309, pl. lxxxv, f. 1.

Lagrange: Near Grand Junction, Tenn.

Nectandra imperfecta Hollick, U. S. Geol. Surv., Mon. 50, 1906, p. 76, pl. xxvii, f. 13, 14.

Magothy: Gay Head, Marthas Vineyard, Mass.

Nectandra lancifolia (Lesquereux) Berry, U. S. Geol. Surv., Prof. Paper 91, 1916, p. 308, pl. lxxxv, f. 2.—Knowlton, U. S. Geol. Surv., Prof. Paper 101, 1918, p. 318, pl. xc, f. 1, 2.

Persea lancifolia Lesquereux, Am. Philos. Soc., Trans., vol. 13, 1869, p. 419, pl. xix, f. 3, 4.

Quercus leyelli Heer. Lesquereux, idem, p. 415, pl. xvii, f. 1, 2 [not f. 3].

Laurus carolinensis? Michaux. Lesquereux, in Safford, Geol. Tenn., 1869, p. 426, pl. K, f. 10.

Rhamnus eridani Unger. Lesquereux, U. S. Nat. Mus., Proc., vol. 11, 1888, p. 25.

Wilcox (Ackerman): Hurleys and Colemans Mill, Miss.

Wilcox (Grenada): Grenada, Miss.

Wilcox: Campbell's quarry, Caddo Parish, and Naborton, De Soto Parish, La.; Old Port Caddo Landing, Harrison County, Tex.

Lagrange: Puryear and Baughs Bridge, Tenn.

Raton: Salyers Creek, Vermejo Valley, N. Mex.; near Weston and Aguilar, Colo.

Nectandra lowii Berry, U. S. Geol. Surv., Prof. Paper 91, 1916, p. 310, pl. lxxxviii, f. 4, 5.

Wilcox (Holly Springs): Oxford, Miss.

Wilcox: Boydsville, Ark., Mansfield, La.

Lagrange: Near La Grange, Tenn.

Nectandra pseudocoriacea Berry, U. S. Geol. Surv., Prof. Paper 91, 1916, p. 311, pl. lxxxvii, f. 2; pl. lxxxviii, f. 1–3.

Laurus primigenia Unger. Hollick, La. Geol. Surv., Special Rept. 5, 1899, p. 284, pl. xli, f. 2 [not f. 1].

Wilcox (Ackerman): Hurleys, Miss.

Wilcox (Grenada): Grenada, Miss.

Wilcox (Holly Springs): Oxford, Holly Springs, and Vaughans, Miss.

Wilcox: Coushatta, Naborton, La.; Boydsville and Benton, Ark.

Lagrange: Puryear, Grand Junction, and Baughs Bridge, Tenn.

Nectandra puryearensis Berry, U. S. Geol. Surv., Prof. Paper 91. 1916, p. 310, pl. cv, f. 2.

Lagrange: Puryear, Tenn.

Wilcox: Near Naborton, La.

Nectandra sp. Berry, U. S. Geol. Surv., Prof. Paper 91, 1916, p. 312, pl. cx, f. 3.

Lagrange: Puryear, Tenn.

Wilcox: De Soto Parish, La.; Old Port Caddo Landing, Harrison County, Tex.

NEGUNDO Mönch, Meth., 1794, p. 334.

Negundo acutifolia (Lesquereux) Pax, Engler's Bot. Jahrb., vol. 6, 1885, pt. iv, p. 346.

Negundoides acutifolius Lesquereux, Am. Jour. Sci., 2d ser., vol. 46, 1868, p. 101; Rept. U. S. Geol. Surv. Terr., vol. 6 (Cret. Fl.), 1874, p. 97, pl. xxi, f. 5; U. S. Geol. Surv., Mon. 17, 1892, p. 156.

Dakota: Lancaster, Nebr.

Negundo brittoni Knowlton, U. S. Geol. Surv., Prof. Paper 130, 19—, p. —, pl. xxv, f. 8–10 [in preparation].

Laramie: Marshall, Colo.

Negundo decurrens Lesquereux, Mus. Comp. Zool., Bull., vol. 16, 1888, p. 54.

Denver: Golden, Colo.

Negundo triloba Newberry, N. Y. Lyc. Nat. Hist., Ann., vol. 9, 1868, p. 57.—[Lesquereux], U. S. Geol. and Geog. Surv. Terr., Ill. Cret. and Tert. Pl., 1878, pl. xxiii, f. 5; U. S. Geol. Surv., Mon. 35, 1898, p. 115, pl. xxxi, f. 5.

Fort Union: Fort Union, N. Dak.

Negundoides Lesquereux = *Negundo* Mönch.

Negundoides acutifolius Lesquereux = *Negundo acutifolia.*

NELUMBITES Berry, Md. Geol. Surv., Lower Cret., 1912, p. 462. [Type, *Menispermites virginiensis.*]

Nelumbites primaeva (Berry) Berry, Md. Geol. Surv., Upper Cret., 1916, p. 840, pl. lxxv, f. 4.

Nelumbo primaeva Berry, N. Y. Bot. Gard., Bull., vol. 3, 1903, p. 75, pl. xliii, f. 1; Geol. Surv. N. J., Ann. Rept. 1905 [1906], p. 138; Bull. Torr. Bot. Club, vol. 37, 1910, p. 23.

Magothy: Cliffwood, N. J.; Round Bay, Md.

Nelumbites tenuinervis (Fontaine) Berry, Md. Geol. Surv., Lower Cret., 1912, p. 464, pl. lxxxii, f. 1, 2.

Menispermites tenuinervis Fontaine, U. S. Geol. Surv., Mon. 15, 1889 [1890], p. 322, pl. clxii. f. 8; in Ward, idem, Mon. 48, 1905 [1906], pp. 496, 557, pl. cix, f. 2, 3.

Menispermites virginiensis Fontaine, U. S. Geol. Surv., Fifteenth Ann. Rept., 1895, p. 360, pl. iv, f. 7 [not f. 8].

Patapsco: Federal Hill (Baltimore), Md.; Mount Vernon, Aquia Creek, and White House Bluff, Va.

Nelumbites virginiensis (Fontaine) Berry, Md. Geol. Surv., Lower Cret., 1912, p. 463, pl. lxxxii, f. 3–5.

Menispermites virginiensis Fontaine, U. S. Geol. Surv., Mon. 15, 1889 [1890], p. 321, pl. clxi, f. 1, 2.—Ward, U. S. Geol. Surv., Fifteenth Ann. Rept., 1895, p. 360, pl. iv, f. 8 [not f. 7].—Fontaine, in Ward, U. S. Geol. Surv., Mon. 48, 1905 [1906], pp. 504, 528, 534, 557.

Patapsco: Fort Foote, Federal Hill (Baltimore), and Overlook Inn Road, Md.; Mount Vernon, Hell Hole, Brooke, and 72d milepost, Va.

Nelumbium Jussieu = *Nelumbo* Adanson.

Nelumbium lakesianum (Lesquereux) Knowlton = *Nelumbo lakesiana.*

Nelumbium lakesii Lesquereux = *Nelumbo lakesiana.*

Nelumbium [mirabile] James = *Cycadeoidea mirabilis.*

Nelumbium pygmaeum Dawson = *Nelumbo pygmaea.*

Nelumbium saskatchuense Dawson = *Nelumbo saskatchuense.*

Nelumbium tenuifolium Lesquereux = *Nelumbo tenuifolia.*

NELUMBO (Tournefort) Adanson, Fam. Pl., vol. 2, 1763, p. 76.

Nelumbo dawsoni Hollick, Bull. Torr. Bot. Club, vol. 21, 1894, p. 309.

Brasenia antiqua Dawson, Roy. Soc. Canada, Trans., vol. 3, 1885 [1886], sec. 4, p. 15, text f. [Homonym, Newberry, 1883.]

Belly River: Medicine Hat, Alberta.

Nelumbo intermedia Knowlton, U. S. Geol. Surv., Bull. 163, 1900, p. 53, pl. xiii, f. 3-5.

Lemna scutata Dawson. Lesquereux, Rept. U. S. Geol. Surv. Terr., vol. 7 (Tert. Fl.), 1878, p. 102, pl. lxi, f. 2.

Mesaverde: Point of Rocks, Wyo.

Nelumbo kempii (Hollick) Hollick, N. Y. Bot. Gard., Bull., vol. 3, 1904, p. 412, pl. lxxiv, f. 1, 2; pl. lxxv; pl. lxxvi; pl. lxxvii, f. 1.—Berry, Bull. Torr. Bot. Club, vol. 33, 1906, p. 173; Geol. Surv. N. J., Ann. Rept. 1905 [1906], p. 138.—Hollick, U. S. Geol. Surv., Mon. 50, 1906, p. 61, pl. xiii, f. 1-4; pl. xiv, f. 1, 2; pl. xv; pl. xvi, f. 1-6, N. Y. Bot. Gard., Bull., vol. 8, 1912, p. 160, pl. clxvi, f. 3, 4.

Serenopsis kempii Hollick, Bull. Torr. Bot. Club, vol. 20, 1893, p. 168, pl. cxlix; p. 334, pl. clxvi.

Magothy: Glen Cove, Roslyn, and Manhasset Neck, Long Island, N. Y.; Gay Head, Mass.; Morgan, N. J.

Nelumbo lakesiana (Lesquereux) Knowlton, U. S. Geol. Surv., Prof. Paper 101, 1918, p. 308.

Nelumbium lakesianum Lesquereux, U. S. Geol. and Geog. Surv. Terr., Ann. Rept. 1873 [1874], p. 403.

Nelumbium lakesii Lesquereux, Rept. U. S. Geol. Surv. Terr., vol. 7 (Tert. Fl.), 1878, p. 252, pl. xlvi, f. 1, 2.

Nelumbium lakesianum (Lesquereux) Knowlton, U. S. Geol. Surv., Bull. 152, 1898, p. 151.

Denver: *Golden, Colo.

Dawson: Sedalia, Colo.

Raton: Near Trinidad, Colo.

Nelumbo laramiensis Hollick, Bull. Torr. Bot. Club, vol. 21, 1894, p. 307, text f.

Vermejo: Florence, Colo.

Nelumbo primaeva Berry = *Nelumbites primaeva.*

Nelumbo protolutea Berry, U. S. Geol. Surv., Prof. Paper 108, 1918, p. 64, pls. xxiv-xxvi.

Wilcox (Grenada): Meridian, Miss.

Nelumbo pygmaea (Dawson) Knowlton.

Nelumbium pygmaeum Dawson, Roy. Soc. Canada, Trans., vol. 8, 1890, sec. 4, p. 87, text f. 22.—Penhallow, Roy. Soc. Canada, Trans., 2d ser., vol. 8, 1902, sec. 4, p. 70; Rept. Tert. Pl. Brit. Col., 1908, p. 64.

Miocene: Mill Creek, Similkameen, Tulameen, and Horsefly rivers, British Columbia.

Nelumbo saskatchuense (Dawson) Knowlton.

Nelumbium saskatchuense Dawson, Roy. Soc. Canada, Trans., vol. 5, 1887 [1888], sec. 4, p. 35.—Penhallow, Rept. Tert. Pl. Brit. Col., 1908, p. 64.

Paskapoo(?): Red Deer River, Alberta.

Nelumbo tenuifolia (Lesquereux) Knowlton, U. S. Geol. Surv., Prof. Paper 130, 19—, p. —, pl. xxvi, f. 7 [in preparation].

Nelumbium tenuifolium Lesquereux, U. S. Geol. and Geog. Surv. Terr., Ann. Rept. 1873 [1874], p. 402; Rept. U. S. Geol. Surv. Terr., vol. 7 (Tert. Fl.), 1878, p. 253, pl. xlvi, f. 3.

Denver: Sand Creek, Colo.

Laramie: Erie, Colo.

Nelumbo sp. (Dawson).

Nelumbium sp. Dawson, Geol. Surv. Canada, 1877–78, p. 168.—Penhallow, Rept. Tert. Pl. Brit. Col., 1908, p. 64.

Miocene?: Ninemile Creek, Similkameen River, British Columbia.

Nelumbo? sp. Knowlton, U. S. Geol. Surv., Bull. 163, 1900, p. 54, pl. xiii, f. 6.

Castalia sp. Knowlton, in Stanton and Knowlton, Geol. Soc. Am., Bull., vol. 8, 1897, p. 140.

Mesaverde: Point of Rocks, Wyo.

Nelumbo sp. Knowlton, U. S. Geol. Surv., Bull. 316, 1907, p. 202.

Fort Union: Miles City, Mont.

Nelumbo sp. Knowlton, Washington Acad. Sci., Proc., vol. 11, 1909, p. 211.

Lance: Bluffs of Yellowstone River opposite Miles City, Mont., and Ranchester, Wyo.

Nelumbo sp. Knowlton, U. S. Geol. Surv., Prof. Paper 98, 1916, p. 341, pl. lxxxvi, f. 11.

Fruitland: 30 miles south of Farmington, San Juan County, N. Mex.

NEOCALAMITES Halle, Kgl. Sv. Vet. Akad., Stockholm, Handl., vol. 43, No. 1, 1908, p. 6.

Neocalamites knowltoni Berry, Bot. Gaz., vol. 53, 1912, p. 177, pl. xvii, text f. 1.

Triassic (Newark): Clover Hill mine, 1 mile south of Gayton, on Tuckahoe Creek, Richmond coal field, Va.

Neocalamites virginiensis (Fontaine) Berry, Bot. Gaz., vol. 53, 1912, p. 176.

Schizonema virginiensis Fontaine, U. S. Geol. Surv., Mon. 6, 1883, p. 17, pl. i, f. 4–6.

Triassic (Newark): Clover Hill, Va.

NEUROPTERIS (Brongniart) Sternberg, Fl. Vorw., vol. 1 (Tentamen), 1825, p. xvii.

"Neuropteris acutifolia" Grewingk, Verhandl. d. Russ.-k. mineralog. Gesell. St. Petersb., 1848–49 [1850], p. 170.
Kenai: Unga Island, Alaska.

Neuropteris albertsii Dunker = *Cladophlebis albertsii.*

Neuropteris angulata Newberry = *Trapa microphylla?*

"Neuropteris castor" Dawson, Roy. Soc. Canada, Trans., vol. 1, 1882 [1883], p. 24, f. 14, 14a. [Not a *Neuropteris.*]
Upper Cretaceous: Beaver Harbor, Vancouver Island, British Columbia.

Neuropteris civica Dawson = *Dryopteris? civica.*

Neuropteris heterophylla cretacea Penhallow, Roy. Soc. Canada, Trans., 2d ser., vol. 8, 1902, sec. 4, p. 36. [Not *Neuropteris.*]
——: Alliford Bay, Queen Charlotte Islands.

Neuropteris huttoni Dunker = *Cladophlebis huttoni.*

Neuropteris linnaeaefolius Bunbury = *Acrostichites linnaeaefolius.*

Neuropteris sp. Emmons, American geology, pt. 6, 1857, p. 102, f. 71 = *Asterocarpus platyrachis.*

NEWBERRYANA Berry, Jour. Geol., vol. 18, 1910, p. 254 [name]; Geol. Surv. N. J., Bull. 3, 1911, p. 220. [Type, *Hausmannia rigida.*]

Newberryana rigida (Newberry) Berry, Jour. Geol., vol. 18, 1910, p. 254 [name], Geol. Surv. N. J., Bull. 3, 1911, p. 220.
Hausmannia rigida Newberry, U. S. Geol. Surv., Mon. 26, 1896, p. 35, pl. i, f. 2, 3, 5.
Raritan: Woodbridge and South Amboy, N. J.

NILSONIA Brongniart, Ann. sci. nat., vol. 4, 1825, p. 218.

Nilsonia californica (Fontaine) Fontaine, in Ward, U. S. Geol. Surv., Mon. 48, 1905 [1906], p. 252, pl. lxvii, f. 7.
Pterophyllum californicum Fontaine, in Diller and Stanton, Geol. Soc. Am., Bull., vol. 5, 1894, p. 450 [name]; in Stanton, U. S. Geol. Surv., Bull. 133, 1895 [1896]. p. 17 [name].
Knoxville: McCarty Creek, Paskenta, Tehama County, Calif.

Nilsonia compta (Phillips) Göppert. Fontaine, in Ward, U. S. Geol. Surv., Mon. 48, 1905 [1906], p. 94, pl. xvii, f. 11–14.
Jurassic: Douglas County, Oreg.

Nilsonia gibbsii (Newberry) Hollick, in Newberry, U. S Geol. Surv., Mon. 35, 1898, p. 16, pl. xv, f. 2, 2a.
Taeniopteris gibbsii Newberry, Boston Jour. Nat. Hist., vol. 7, 1863, p. 512.
Nilsonia johnstrupi Heer, Fl. foss. arct., vol. 6, Abt. 2, 1882, p. 44, pl. vi, f. 1–6.
Cretaceous: Point Doughty, Orcas Island, Wash.

Nilsonia johnstrupi Heer = *Nilsonia gibbsii.*

Nilsonia lata Dawson, Roy. Soc. Canada, Trans., vol. 1, 1882 [1883], sec. 4, p. 24, pl. iv, f. 15 bis.

Upper Cretaceous: Baynes Sound, Vancouver Island, British Columbia.

Nilsonia nigracollensis Wieland, in Ward, U. S. Geol. Surv., Mon. 48, 1905 [1906], p. 319, pl. lxxiii, f. 15a–d.—Knowlton, Washington Acad. Sci., Jour., vol. 6, 1916, p. 181.

Lakota: 5 miles north of Sturgis, S. Dak.

Morrison: Cottonwood Creek, 5 miles west of Tensleep, Big Horn Basin, Wyo.

Nilsonia nipponensis Yokoyama, Jour. Coll. Sci. Tokio, vol. 3, 1889. pt. 1, p. 42, pl. vi, f. 8d; pl. vii, f. 2–7, 8a; pl. xii, f. 6; pl. xiii, f. 1.—Fontaine, in Ward, U. S. Geol. Surv., Mon. 48, 1905, p. 94, pl. xvii, f. 8–10.

Jurassic: Douglas County, Oreg.

Nilsonia oregonensis (Fontaine) Berry, U. S. Nat. Mus., Proc., vol. 38, 1910, p. 637.

Angiopteridium strictinerve Fontaine. Fontaine, in Ward, U. S. Geol. Surv., Mon. 48, 1905 [1906], pp. 240, 511, pl. lxvi, f. 5–7; pl. cx, f. 12.

Sapindopsis oregonensis Fontaine, in Ward, U. S. Geol. Surv., Mon. 48, 1905 [1906], p. 268, pl. lxix, f. 15–17.

Patuxent?: Chinkapin Hollow, Va.

Knoxville: Tehama County, Calif.

Horsetown: Tehama County, Calif.; Riddles, Oreg.

Nilsonia orientalis Heer, Acad. imp. sci. St.-Pétersb., Mém. (Fl. foss. arct., vol. 5, Abt. 2), 1878, p. 18, pl. iv, f. 4–9.—Fontaine, in Ward, U. S. Geol. Surv., Mon. 48, 1905 [1906], p. 90, pl. xvi, f. 3–9.

Jurassic: Douglas County, Oreg.

Nilsonia orientalis minor Fontaine, in Ward, U. S. Geol. Surv., Mon. 48, 1905 [1906], p. 92, pl. xvi, f. 10–13.

Jurassic: Douglas County, Oreg.

Nilsonia parvula (Heer) Fontaine, in Ward, U. S. Geol. Surv., Mon. 48, 1905 [1906], p. 92, pl. xvii, f. 1–7.

Taeniopteris parvula Heer, Fl. foss. arct., vol. 4, Abt. 2, 1876, p. 98, pl. xxi, f. 5.

Jurassic: Douglas County, Oreg.

Nilsonia pasaytensis Penhallow, Roy. Soc. Canada, Trans., 3d ser., vol. 1, sec. 4, 1907, p. 307, text f. 3.

Lower Cretaceous: International boundary between Pasayten and Skagit rivers, British Columbia.

Nilsonia polymorpha Schenk, Foss. Flora Grenzschichtens d. Keupers u. Lias Frankens, 1867, p. 127, pl. xxix; pl. xxx, f. 1–5.—Knowlton, U. S. Nat. Mus., Proc., vol. 51, 1916, p. 459, pl. lxxxi, f. 4.

Lower Jurassic: Upper Matanuska Valley, Alaska.

Nilsonia polymorpha cretacea Penhallow, Roy. Soc. Canada, Trans., 2d ser., vol. 8, 1902, sec. 4, p. 42.

——: Maud Island, Queen Charlotte Islands, British Columbia.

Nilsonia pterophylloides Nathorst [not Yokoyama], Abh. Sv. Geol. Unders., ser. C, No. 27 (Fl. Bjuf, pt. 1), 1878, p. 11; idem, No. 33, 1879, p. 72, pl. xvi, f. 1; pl. xvii, f. 2, 3.—Fontaine, in Ward, U. S. Geol. Surv., Mon. 48, 1905 [1906], p. 96, pl. xviii.

Jurassic: Douglas County, Oreg.

Nilsonia? sambucensis Ward, U. S. Geol. Surv., Mon. 48, 1905 [1906], p. 254, pl. lxvii, f. 8.

Knoxville: Tehama County, Calif.

Nilsonia schaumbergensis (Dunker) Nathorst, Anzeiger K. Akad. Wiss. Wien, Jahrg. 26, 1889, p. 237.—Fontaine, in Ward, U. S. Geol. Surv., Mon. 48, 1905 [1906], p. 303, pl. lxxii, f. 17–21.—Knowlton, Smithsonian Misc. Coll., vol. 50, 1907, p. 123.

Angiopteridium strictinerve Fontaine. Fontaine. in Weed and Pirsson, U. S. Geol. Surv., Eighteenth Ann. Rept., pt. 3, 1898, p. 481.

Kootenai: Near Geyser, Cascade County, and near Giltedge, Fergus County, Mont.

Knoxville and Horsetown: California.

Nilsonia stantoni Ward, U. S. Geol. Surv., Mon. 48, 1905 [1906], p. 251, pl. lxvii, f. 5, 6.

Knoxville: Elder Creek, Tehama County, Calif.

NIPADITES Bowerbank, History of the fossil fruits and seeds of the London clay, 1840, p. 1.

Nipadites burtini umbonatus Bowerbank, History of the fossil fruits and seeds of the London clay, 1840, p. 9, pl. i.—Berry, U. S. Geol. Surv., Prof. Paper 91, 1916, p. 176, pl. cxii, f. 11, 14.

Wilcox (Grenada): Grenada, Grenada County, Miss.

Noeggerathia stricta Emmons = *Baiera multifida?*

NOEGGERATHIOPSIS Feistmantel, Mem. Geol. Surv. India (Foss. Fl. Gondwana Syst.), 12th ser., vol. 3, pt. 1, 1879, p. 23.

Noeggerathiopsis robinsi Dawson, Roy. Soc. Canada, Trans., vol. 11, 1893 [1894], sec. 4, p. 56, pl. vi, f. 7.

Upper Cretaceous: Nanaimo, Vancouver Island, British Columbia.

NORDENSKIÖLDIA Heer, Kongl. Svenska Vetensk.-Acad., Handl. (Fl. foss. arct., vol. 2, Abt. 3), vol. 8, No. 7, 1870, p. 65.

Nordenskiöldia borealis Heer, Fl. foss. arct., vol. 2, Abt. 3, 1870, p. 65, pl. vii, f. 1–13.—Dawson, Geol. Surv. Canada, 1875–76, p. 54; idem, 1888–89, p. 97D.—Newberry, U. S. Geol. Surv., Mon. 35, 1898, p. 137, pl. lxviii, f. 4–6.—Penhallow, Rept. Tert. Pl. Brit. Col., 1908, p. 64.

Green River: Green River, Wyo.

Eocene: Quesnel River, British Columbia; Mackenzie River, Northwest Territory.

Nordenskiöldia borealis Heer. Lesquereux, U. S. Geol. Surv., Mon. 17, 1892, p. 219, pl. xliv, f. 6 = *Carpites dakotensis*.

Nymphaea Linné = *Castalia*.

NYSSA Linné, Sp. pl., 1753, p. 1058.

Nyssa acuticostata Perkins, Rept. Vt. State Geol. 1905–6 [1906], p. 218, pl. lvi, f. 16, 17.
Miocene: Brandon, Vt.

Nyssa aquatica Marshall.
Nyssa caroliniana Poir. Hollick, Bull. Torr. Bot. Club, vol. 19, 1892, p. 331.
Pleistocene: Bridgeton, N. J.

Nyssa aquaticaformis Berry, U. S. Geol. Surv., Prof. Paper 98–L, 1916, p. 203, pl. xlvii, f. 8.
Citronelle: Lambert and Red Bluff, Ala.

Nyssa arctica Heer, Fl. foss. arct., vol. 2, Abt. 4, 1869, p. 477, pl. xliii, f. 12c; pl. l, f. 5–7.—Lesquereux, U. S. Nat. Mus., Proc., vol. 5, 1882 [1883], p. 477; Rept. U. S. Geol. Surv. Terr., vol. 8 (Cret. and Tert. Fl.), 1883, p. 261.
Kenai: Unga Island, Alaska.

Nyssa ascoidea Perkins, Rept. Vt. State Geol. 1903–4 [1904], p. 196, pl. lxxix, f. 96.
Miocene: Brandon, Vt.

Nyssa biflora Walter. Berry, Torreya, vol. 5, 1905, p. 90.—Hollick, Md. Geol. Surv., Pliocene and Pleistocene, 1906, p. 235, pl. lxix, f. 5.—Berry, Jour. Geol., vol. 15, 1907, p. 345; Torreya, vol. 10, 1910, p. 266; Am. Jour. Sci., 4th ser., vol. 29, 1910, p. 398.
Pleistocene: Tappahannock River, Va.; Neuse and Roanoke rivers, N. C.; Drum Point and Bodkin Point, Md.; Long Branch, N. J.; near Mobile, Ala.

Nyssa buddiana Ward, U. S. Geol. Surv., Sixth Ann. Rept., 1884–85 [1886], p. 553, pl. xlvii, f. 7; U. S. Geol. Surv., Bull. 37, 1887, p. 53, pl. xxv, f. 4.
Adaville: Hodges Pass, Wyo.

Nyssa caroliniana Poir = *Nyssa aquatica*.

Nyssa clarkii Perkins, Rept. Vt. State Geol. 1903–4 [1904], p. 199, pl. lxxxi, f. 167.
Miocene: Brandon, Vt.

Nyssa complanata Lesquereux, Am. Jour. Sci., 2d ser., vol. 32, 1861, p. 361; Geol. Vt., vol. 2, 1861, p. 717, f. 153.—Perkins, Rept. Vt. State Geol. 1903–4 [1904], p. 198, pl. lxxix, f. 112; idem, 1905–6 [1906], p. 219, pl. lvii, f. 3, 5.
Miocene: Brandon, Vt.

Nyssa crassicostata Perkins, Rept. Vt. State Geol. 1903–4 [1904], p. 196, pl. lxxix, f. 97; Geol. Soc. Am., Bull., vol. 16, 1905, p. 509, pl. 86, f. 11; Rept. Vt. State Geol. 1905–6 [1906], pl. lii, f. 11
Miocene: Brandon, Vt.

Nyssa? cuneata (Newberry) Newberry, U. S. Geol. Surv., Mon. 35, 1898, p. 125, pl. xvii, f. 4–6.

Ficus? cuneatus Newberry, Boston Jour. Nat. Hist., vol. 7, 1863, p. 524.

——: Orcas Island, Wash.

Nyssa curta Perkins, Rept. Vt. State Geol. 1903–4 [1904], p. 199, pl. lxxix, f. 111; idem, 1905–6 [1906], p. 219, pl. lvii, f. 4, 6.
Miocene: Brandon, Vt.

Nyssa cylindrica Perkins, Rept. Vt. State Geol. 1903–4 [1904], p. 195, pl. lxxix, f. 91.
Miocene: Brandon, Vt.

Nyssa denveriana Knowlton, n. sp.

Nyssa lanceolata Lesquereux. Lesquereux, Rept. U. S. Geol. Surv. Terr., vol. 7 (Tert. Fl.), 1878, p. 245, pl. xxxv, f. 6 [not f. 5].

Denver: *Golden, Colo.
Tertiary: White Oaks, N. Mex.

Nyssa elongata Perkins, Rept. Vt. State Geol. 1903–4 [1904], p. 197, pl. lxxix, f. 102.
Miocene: Brandon, Vt.

Nyssa eolignitica Berry, U. S. Geol. Surv., Prof. Paper 91, 1916, p. 332, pl. xcix, f. 8.
Lagrange: Puryear, Tenn.

Nyssa equicostata Perkins, Rept. Vt. State Geol. 1903–4 [1904], p. 198, pl. lxxix, f. 110.
Miocene: Brandon, Vt.

Nyssa europaea Unger, Syn. pl. foss., 1845, p. 228.—Lesquereux, Mus. Comp. Zool., Bull., vol. 16, 1888, p. 53.
Denver: Golden, Colo.

Nyssa excavata Perkins, Rept. Vt. State Geol. 1903–4 [1904], p. 199, pl. lxxxi, f. 166.
Miocene: Brandon, Vt.

Nyssa gracilis Berry, Jour. Geol., vol. 17, 1909, p. 29, text f. 10.
Calvert: Richmond, Va.

Nyssa jonesii Perkins, Rept. Vt. State Geol. 1903–4 [1904], p. 197, pl. lxxix, f. 101; Geol. Soc. Am., Bull., vol. 16, 1905, p. 509, pl. lxxxvi, f. 8; Rept. Vt. State Geol. 1905–6 [1906], pl. lii, f. 8.
Miocene: Brandon, Vt.

Nyssa laevigata Lesquereux, Am. Jour. Sci., 2d ser., vol. 32, 1861, p. 361; Geol. Vt., vol. 1, 1861, p. 717, f. 151.—Perkins, Rept. Vt. State Geol. 1903–4 [1904], p. 198, pl. lxxix, f. 99, 109, 113, 114.
Miocene: Brandon, Vt.

Nyssa lamellosa Perkins, Rept. Vt. State Geol. 1903–4 [1904], p. 195, pl. lxxix, f. 93, 94; Geol. Soc. Am., Bull., vol. 16, 1905, p. 509, pl. lxxxvi, f. 10; Rept. Vt. State Geol. 1905–6 [1906], pl. lii, f. 10.
Miocene: Brandon, Vt.

Nyssa lanceolata Lesquereux, U. S. Geol. and Geog. Surv. Terr., Ann. Rept. 1872 [1873], p. 407; idem, Rept. 1873 [1874], p. 385, idem, Rept. 1876 [1878], p. 513; Rept. U. S. Geol. Surv. Terr.; vol. 7 (Tert. Fl.), 1878, p. 245, pl. xxxv, f. 5 [not f. 6].—Knowlton, U. S. Geol. Surv., Prof. Paper 101, 1918, p. 343, pl. cviii, f. 1; pl. cxiii, f. 2.

Denver: *Golden, Colo.

Livingston?: *Meadow Creek, 12 miles east of Bozeman, Mont.

Dawson: Mosby, Colo.

Raton: Mayne and Riley Canyon, Colo.

Nyssa lanceolata Lesquereux, Rept. U. S. Geol. Surv. Terr., vol. 7 (Tert. Fl.), 1878, p. 245, pl. xxxv, f. 6 = *Nyssa denveriana*.

Nyssa lescurii (Hitchcock) Perkins, Rept. Vt. State Geol. 1903–4 [1904], p. 197, pl. lxxix, f. 100; Geol. Soc. Am., Bull., vol. 16, 1905, p. 509, pl. lxxxvi, f. 9; Rept. Vt. State Geol. 1905–6 [1906], p. 218, pl. lvii, f. 2; pl. lii, f. 9.

Carpolithes lescurii Hitchcock, Portland Soc. Nat. Hist., vol. 1, 1862, p. 95, pl. i, f. 5.

Miocene: Brandon, Vt.

Nyssa microcarpa Lesquereux, Am. Jour. Sci., 2d ser., vol. 32, 1861, p. 361; Geol. Vt., vol. 2, 1861, p. 717, f. 154.—Perkins, Rept. Vt. State Geol. 1903–4 [1904], p. 194, pl. lxxix, f. 90.

Miocene: Brandon, Vt.

Nyssa multicostata Perkins, Rept. Vt. State Geol. 1903–4 [1904], p. 197, pl. lxxix, f. 103.

Miocene: Brandon, Vt.

Nyssa ovalis Perkins, Rept. Vt. State Geol. 1905–6 [1906], p. 218, pl. lvii, f. 1.

Miocene: Brandon, Vt.

Nyssa ovata Perkins, Rept. Vt. State Geol. 1903–4 [1904], p. 196, pl. lxxix, f. 98.

Miocene: Brandon, Vt.

Nyssa? racemosa Knowlton, U. S. Geol. Surv., Bull. 152, 1898, p. 153; U. S. Geol. Surv., Prof. Paper 101, 1918, p. 343.

Sabalites? fructifer Lesquereux [part], Rept. U. S. Geol. Surv. Terr., vol. 7 (Tert. Fl.), 1878, p. 114, pl. xi, f. 3a [not f. 3].

Denver: *Golden, Colo.

Raton: Yankee mine near Raton, N. Mex.

Post-Laramie: Black Buttes, Wyo.

Nyssa snowiana Lesquereux, U. S. Geol. Surv., Mon. 17, 1891 [1892], p. 126, pl. lii, f. 11.—Berry, U. S. Geol. Surv., Prof. Paper 112, 1919, p. 129, pl. xxvi, f. 6.

Dakota: Ellsworth County, Kans.

Tuscaloosa: Shirleys Mill, Ala.

Nyssa solea Perkins, Rept. Vt. State Geol. 1903–4 [1904], p. 194, pl. lxxviii, f. 78.
Miocene: Brandon, Vt.

Nyssa sylvatica Marshall. Berry, U. S. Nat. Mus., Proc., vol. 48, 1915, p. 301.
Pleistocene: Hickman, Ky.

Nyssa uniflora Walter. Hollick, Bull. Torr. Bot. Club, vol. 19, 1892, p. 331.
Pleistocene: Bridgeton, N. J.

Nyssa vetusta Newberry, N. Y. Lyc. Nat. Hist., Ann., vol. 9, 1868, p. 11; U. S. Geol. Surv., Mon. 35, 1898, p. 125, pl. l, f. 2; pl. iv, f. 4.

Magnolia obovata Newberry, N. Y. Lyc. Nat. Hist., Ann., vol. 9, 1868, p. 15.—[Lesquereux], U. S. Geol. and Geog. Surv. Terr., Ill. Cret. and Tert. Pl., 1878, pl. ii, f. 2; pl. iv, f. 4; U. S. Geol. Surv., Mon. 17, 1891 [1892], p. 203.—Newberry, U. S. Geol. Surv., Mon. 35, 1898, p. 94.

Dakota: Blackbird Hill, Nebr.

· **Nyssa wilcoxiana** Berry, U. S. Geol. Surv., Prof. Paper 91, 1916, p. 331, pl. xcix, f. 5–7.
Lagrange: Puryear, Tenn.
Wilcox: Naborton, La.

Nyssa sp.? Hollick, Md. Geol. Surv., Pliocene and Pleistocene, 1906, p. 236.
Pleistocene (Talbot): Bodkin, Grace, and Drum points, Md.

NYSSIDIUM Heer, Kgl. Sv. Vet. Akad., Handl., vol. 8, No. 7 (Fl. foss. arct., vol. 2, Abt. 3), 1870, p. 61.

Nyssidium? sp. Dawson, Geol. Surv. Canada, 1875–76, p. 55.—Penhallow, Rept. Tert. Pl. Brit. Col., 1908, p. 64.
Eocene: Quesnel River, British Columbia.

OCOTEA Aublet, Guian., vol. 2, 1775, p. 780.

Ocotea nassauensis Hollick, U. S. Geol. Surv., Mon. 50, 1906, p. 76, pl. xxvii, f. 8.
Magothy: Glen Cove, Long Island, N. Y.

Odontopteris tenuifolius Emmons = *Acrostichites tenuifolius.*

ODOSTEMON Rafinesque, Am. Monthly Mag., 1817, p. 192.

Odostemon florissantensis Cockerell, Am. Mus. Nat. Hist., Bull., vol. 24, 1908, p. 91.
Miocene: Florissant, Colo.

Odostemon simplex (Newberry) Cockerell, Am. Mus. Nat. Hist., Bull., vol. 24, 1908, p. 91.

Berberis simplex Newberry, U. S. Nat. Mus., Proc., vol. 5, 1882 [1883], p. 514; U. S. Geol. Surv., Mon. 35, 1898, p. 97, pl. lvi, f. 2.—Knowlton, U. S. Geol. Surv., Bull. 204, 1902, p. 56.

Clarno (upper part): Bridge Creek, Grant County, Oreg.

Olea americana Hilgard = *Osmanthus pedatus.*

Olea praemissa Lesquereux = *Osmanthus praemissa.*

OLEANDRA Cavanilles, Prael., 1801, p. 623.

Oleandra arctica Heer, Fl. foss. arct., vol. 3, 1874, p. 38, pl. xii, f. 3–11.—Newberry, Am. Jour. Sci., 3d ser., vol. 41, 1891, p. 201, pl. xiv, f. 9.

Kootenai: Great Falls, Mont.

Oleandra graminaefolia Knowlton, Smithsonian Misc. Coll., vol. 50, 1907, p. 113, pl. xi, f. 5, 5a, 6, 6a.

Cephalotaxopsis ramosa Fontaine? Fontaine, in Ward, U. S. Geol. Surv., Mon. 48, 1905 [1906], p. 311, pl. lxxiii, f. 8.

?*Pinus* (*Cyclopitus*) *nordenskiöldi* Heer. Dawson, Roy. Soc. Canada, Trans., vol. 10, 1892, sec. 4, p. 88, text f. 9.

Kootenai: 6 miles southwest of Geyser, Cascade County, Mont.

ONOCLEA Linné, Sp. pl., 1753, p. 1062.

Onoclea? fecunda (Lesquereux) Knowlton, U. S. Geol. Surv., Bull. 152, 1898, p. 153; idem, Prof. Paper 130, 19—, p. —, pl. i, f. 2, 3 [in preparation].

Caulinites fecunda Lesquereux, U. S. Geol. and Geog. Surv. Terr., Ann. Rept. 1872 [1873], p. 384; Rept. U. S. Geol. Surv. Terr., vol. 7 (Tert. Fl.), 1878, p. 101, pl. xiv, f. 1–3.

Laramie: Erie, Colo.

Onoclea inquirenda (Hollick) Hollick, U. S. Geol. Surv., Mon. 50, 1906, p. 32, pl. i, f. 1–7.—Berry, U. S. Geol. Surv., Prof. Paper 84, 1914, p. 14, pl. ii, f. 7, 8; Md. Geol. Surv., Upper Cret., 1916, p. 764, pl. li, f. 1, 2.

Caulinites inquirendus Hollick, N. Y. Bot. Gard., Bull., vol. 3, 1904, p. 406, pl. lxx, f. 3.

Magothy: Glen Cove and Little Neck, Northport Harbor, Long Island, N. Y.; Gay Head and Nashaquitsa, Marthas Vineyard, Mass.; Round Bay Md.

Black Creek (Middendorf): Middendorf, S. C.

Onoclea minima Knowlton, U. S. Geol. Surv., Mon. 32, pt. 2, 1899, p. 656, pl. lxxvii, f. 11–15.

"Laramie": Pinyon Peak, Yellowstone National Park.

Onoclea neomexicana Knowlton, U. S. Geol. Surv., Prof. Paper 98, 1916, p. 332, pl. lxxxiv, f. 1, 2.

Kirtland: Near Pina Veta China, San Juan County, N. Mex.

Onoclea reducta Cockerell = *Sorbus diversifolia.*

Onoclea sensibilis fossilis Newberry, N. Y. Lyc. Nat. Hist., Ann., vol. 9, 1868, p. 39.—[Lesquereux], U. S. Geol. and Geog. Surv. Terr., Ill. Cret. and Tert. Pl., 1878, pl. lviii, f. 1; pl. ix, f. 1–3.—Newberry, U. S. Geol. Surv., Mon. 35, 1898, p. 8, pl. xxiii, f. 3; pl. xxiv, f. 1–5.—Knowlton, Bull. Torr. Bot. Club, vol. 29, 1902, p. 705, pl. xxvi, f. 1–4.—Penhallow, Rept. Tert. Pl. Brit. Col.,

Onoclea sensibilis fossilis—Continued.
1908, p. 47.—Knowlton, Washington Acad. Sci., Proc., vol. 11, 1909, pp. 188, 189, 198, 213.
Fort Union: *Fort Union, N. Dak. Abundant and widespread in the Dakotas, Montana, and Wyoming.
Lance: Bluffs of Yellowstone opposite Miles City; 9 miles west of Miles City; 9 miles above Glendive, Mont.
Paskapoo (Fort Union): Porcupine Creek and Red Deer River, Alberta.

ONYCHIOPSIS Yokoyama, Jour. Coll. Sci. Japan, vol. 3, 1890, p. 26.

Onychiopsis brevifolia (Fontaine) Berry, U. S. Nat. Mus., Proc., vol. 41, 1911, p. 329; Md. Geol. Surv., Lower Cret., 1911, p. 278, pl. xxxiv, f. 1, 2.
Thyrsopteris brevifolia Fontaine, U. S. Geol. Surv., Mon. 15, 1889 [1890], p. 121, pl. xxiv, f. 5, 10; in Ward, U. S. Geol. Surv., Nineteenth Ann. Rept., pt. 2, 1899, p. 660, pl. clxi, f. 10–15.
Thyrsopteris dentata Fontaine, U. S. Geol. Surv., Mon. 15, 1889 [1890], p. 121, pl. xxiv, f. 4, 6, 7, 9; pl. xxv, f. 1, 2.
Thyrsopteris pachyphylla Fontaine, idem, p. 135, pl. l, f. 3.
Thyrsopteris nana Fontaine, idem, p. 141, pl. lvi, f. 8.
Thyrsopteris heterophylla Fontaine, idem, p. 142, pl. lviii, f. 3.
Thyrsopteris sphenopteroides Fontaine, idem, p. 143, pl. lviii, f. 6.
Thyrsopteris squarrosa Fontaine, idem, p. 143, pl. lix, f. 3.
Thyrsopteris retusa Fontaine, idem, p. 144, pl. lix, f. 10.
Patuxent: Fredericksburg, Dutch Gap, Potomac Run, telegraph station near Lorton, Va.
Lakota: Near Barrett, Crook County, Wyo.

Onychiopsis göpperti (Schenk) Berry, U. S. Nat. Mus., Proc., vol. 41, 1911, p. 325; Md. Geol. Surv., Lower Cret., 1911, p. 281, pl. xxxiv, f. 3, 4.
Sphenopteris göpperti Schenk (part), Palaeont., vol. 19, 1871, p. 209 (7), pl. xxx (iv), f. 2, 2a.
Thyrsopteris rarinervis Fontaine, U. S. Geol. Surv., Mon. 15, 1889 [1890], p. 123, pl. xxvi, f. 6, 7; pl. xliii, f. 4–6; pl. xliv, f. 1, 2, 5; pl. xlix, f. 2; pl. clxix, f. 6, 7; in Ward, U. S. Geol. Surv., Mon. 48, 1905 [1906], pp. 225, 484, 491, 514, 517, 518, 521, 528, 548, pl. lxv, f. 2–4; pl. cxiii, f. 2, 3.
Thyrsopteris alata Fontaine, op. cit. (Mon. 15), p. 124, pl. xxxvi, f. 3.
Thyrsopteris meekiana angustiloba Fontaine, op. cit. (Mon. 15), p. 126, pl. xxxviii, f. 5–7, 9; pl. xliii, f. 8; pl. xliv, f. 3; pl. xlvii, f. 4; pl. xlviii, f. 1; pl. liv, f. 2, 11; pl. lv, f. 1; pl. lvi, f. 1–3; in Ward, op. cit. (Mon. 48), p. 557.
Thyrsopteris angustiloba Fontaine, op. cit. (Mon. 15), p. 134, pl. xlviii, f. 3–5 pl. lv, f. 3.
Thyrsopteris densifolia Fontaine, op. cit. (Mon. 15), p. 129, pl. xxxix, f. 3; pl. xl, f. 2–5; pl. li, f. 5; in Ward, op. cit. (Mon. 48), pp. 484, 511, 517.
Thyrsopteris decurrens Fontaine, op. cit. (Mon. 15), p. 130, pl. xliii, f. 7; pl. xlvi, f. 2, 4; pl. xlix, f. 5–7; in Ward, op. cit. (Mon. 48), pp. 484, 491, 511, 525, pl. cxi, f. 11.
Thyrsopteris virginica Fontaine, op. cit. (Mon. 15), p. 120, pl. xxiv, f. 1.

Onychiopsis göpperti—Continued.

Thyrsopteris pachyrachis Fontaine, op. cit. (Mon. 15), p. 132, pl. xlvi, f. 3, 5; pl. xlvii, f. 1, 2; pl. xlix, f. 1; in Ward, op. cit. (Mon. 48) pp. 487, 538, 557.

Thyrsopteris elliptica Fontaine, op. cit. (Mon. 15), p. 133, pl. xxiv, f. 3; pl. xlvi, f. 1; pl. l, f. 6, 9; pl. li, f. 4, 6, 7; pl. liv, f. 6; pl. lv, f. 4; pl. lvi, f. 6, 7; pl. lvii, f. 6; pl. lviii, f. 2; in Ward, U. S. Geol. Surv., Nineteenth Ann. Rept., pt. 3, 1898, p. 482; U. S. Geol. Surv., Mon. 48, 1905 [1906], pp. 290, 484, 514, 517, 528, 557, pl. lxxi, f. 12, 13.—Knowlton, Smithsonian Misc. Coll., vol. 50, 1907, p. 110.

Thyrsopteris distans Fontaine, op. cit. (Mon. 15), p. 134, pl. xlvii, f. 3; pl. liv, f. 8.

Thyrsopteris pinnatifida Fontaine, op. cit. (Mon. 15), p. 136, pl. li, f. 2; pl. liv, f. 4, 5, 7; pl. lvii, f. 7; in Ward, U. S. Geol. Surv., Nineteenth Ann. Rept., pt. 2, 1898, p. 658, pl. clxi, f. 1, 2; U. S. Geol. Surv., Mon. 48, 1905 [1906], p. 511.

Thyrsopteris varians Fontaine, op. cit. (Mon. 15), p. 137, pl. lii, f. 2–4; pl. liii, f. 1–3; pl. liv, f. 10; pl. lvii, f. 2.

Thyrsopteris rhombifolia Fontaine, op. cit. (Mon. 15), p. 138, pl. lii, f. 5; pl. liv, f. 1.

Thyrsopteris bella Fontaine, op. cit. (Mon. 15), p. 139, pl. liii, f. 5; pl. lv, f. 6, 7; pl. lvi, f. 2, 5; pl. lvii, f. 1, 5; pl. lviii, f. 4; in Ward, op. cit. (Mon. 48), pp. 491, 511.

Thyrsopteris microloba Fontaine, op. cit. (Mon. 15), p. 140, pl. lvii, f. 4.

Thyrsopteris microloba alata Fontaine, op. cit. (Mon. 15), p. 140, pl. lv, f. 5; pl. lviii, f. 1; U. S. Nat. Mus., Proc., vol. 15, 1892, p. 493; in Ward, op. cit. (Mon. 48), p. 281.

Thyrsopteris inaequipinnata Fontaine, U. S. Geol. Surv., Mon. 15, 1889 [1890], p. 142, pl. lvii, f. 3, 8.

Thyrsopteris rhombiloba Fontaine, op. cit. (Mon. 15), p. 144, pl. lix, f. 7; pl. lx, f. 8.

Patuxent: Fredericksburg, Trents Reach, Cockpit Point, Dutch Gap, Potomac Run, Colchester Road, Va.; New Reservoir, Ivy City, D. C.

Arundel: Langdon, D. C.; Arlington? Md.

Patapsco: Federal Hill (Baltimore), Fort Foote, Vinegar Hill, Md.; near Brooke, White House Bluff, Mount Vernon, Chinkapin Hollow, Va.

Kootenai: Great Falls, near Geyser, Mont.

Lakota: Crook County, Wyo.

Onychiopsis latiloba (Fontaine) Berry, U. S. Nat. Mus., Proc., vol. 41, 1911, p. 332; Md. Geol. Surv., Lower Cret., 1911, p. 273, pl. xxxiii, f. 1, 2.

Sphenopteris latiloba Fontaine, U. S. Geol. Surv., Mon. 15, 1889 [1890], p. 90, pl. xxxv, f. 3–5; pl. xxxvi, f. 4–9; pl. xxxvii, f. 1; in Ward, U. S. Geol. Surv., Mon. 48, 1905 [1906], pp. 281, 479, 491, 511, 534, 557.

Thyrsopteris brevipennis Fontaine, op. cit. (Mon. 15), p. 124, pl. xxxiv, f. 3; pl. xxxvi, f. 2; pl. xxxvii, f. 3, 9; pl. xxxviii, f. 1; pl. xli, f. 4; in Ward, U. S. Geol. Surv., Nineteenth Ann. Rept., pt. 2, 1899, p. 662, pl. clxii, f. 1a.

Onychiopsis latiloba—Continued.

Thyrsopteris divaricata Fontaine, op. cit. (Mon. 15), p. 125, pl. xxxvii, f. 5-8; pl. lxx, f. 1; in Ward, U. S. Geol. Surv., Mon. 48, 1905 [1906], pp. 504, 511, 517, 521.

Thyrsopteris crenata Fontaine, op. cit. [1890], p. 127, pl. xxxix, f. 1, 2.

Lakota: Crook County, Wyo.

Patuxent: Fredericksburg, Dutch Gap, telegraph station near Lorton, Va.; New Reservoir ?, D. C.

Arundel: Langdon ?, D. C.; Bewley estate ?, Bay View, Md.

Patapsco: Deep Bottom, Mount Vernon, Hell Hole, Chinkapin Hollow, Va.; Federal Hill (Baltimore), Md.

Onychiopsis nervosa (Fontaine) Berry, U. S. Nat. Mus., Proc., vol. 41, 1911, p. 327; Md. Geol. Surv., Lower Cret., 1911, p. 279, pl. xxxvi, f. 1–6.

Thyrsopteris nervosa Fontaine, U. S. Geol. Surv., Mon. 15, 1889 [1890], p. 122. pl. xxv, f. 4, 5, 16; pl. xxxvii, f. 2, 4; pl. xxxix, f. 5; pl. xl, f. 6; in Ward, U. S. Geol. Surv., Mon. 48, 1905 [1906], pp. 511, 517, 519, 521, 528, 548, 571.

Thyrsopteris meekiana Fontaine, op. cit. (Mon. 15), p. 125, pl. xxxviii, f. 2–4, 8; pl. l, f. 7, 8; pl. li, f. 3; in Ward, op. cit. (Mon. 48), pp. 519, 565, pl. cxix, f. 1.

Thyrsopteris crassinervis Fontaine, op. cit. (Mon. 15), p. 130, pl. xli, f. 1–3; in Ward, U. S. Geol. Surv., Nineteenth Ann. Rept., pt. 2, 1899, p. 658, pl. clxi, f. 3, 4; U. S. Geol. Surv., Mon. 48, 1905 [1906], pp. 513, 528, pl. cxii, f. 5, 6.

Thyrsopteris pecopteroides Fontaine, op. cit. (Mon. 15), p. 135, pl. li, f. 1; in Ward, op. cit. (Nineteenth Ann. Rept., pt. 2), p. 661, pl. clxi, f. 16–19.

Adiantites parvifolius Fontaine, in Ward, op. cit. (Mon. 48), p. 558, pl. lxvii, f. 1.

Thyrsopteris heteroloba Fontaine, op. cit. (Mon. 15), p. 139, pl. liii, f. 4.

Thyrsopteris obtusiloba Fontaine, op. cit. (Mon. 15), p. 143, pl. lviii, f. 7, 10.

Thyrsopteris maakiana Heer, Fl. foss. arct., vol. 4, Abt. 2, 1876, pp. 23, 31, 118, pl. i, f. 1a, 1b; pl. ii, f. 5, 5b, 6.—Fontaine, in Ward, U. S. Geol. Surv., Twentieth Ann. Rept., pt. 2, 1900, p. 343, pl. xlix, f. 1.

Patuxent: Fredericksburg, Dutch Gap, Potomac Run, Va.; New Reservoir, Ivy City, D. C.; Springfield, Md.

Arundel: Langdon, D. C.

Patapsco: Chinkapin Hollow, Va.; Fort Foote, Glymont, Vinegar Hill ?, Federal Hill (Baltimore), Md.

Lakota: South Fork of Hay Creek, Crook County, Wyo.

Onychiopsis psilotoides (Stokes and Webb) Ward, U. S. Geol. Surv., Mon. 48, 1905 [1906], p. 155, pl. xxxix, f. 3–6.—Fontaine, in Ward, idem, pp. 506, 518, 528, pl. cxi, f. 4; pl. cxiii, f. 1.—Knowlton, in Diller, Geol. Soc. Am., Bull., vol. 19, 1908, p. 380.—Berry, U. S. Nat. Mus., Proc., vol. 41, 1911, p. 330; Md. Geol. Surv., Lower Cret., 1911, p. 274, pl. xxxv; pl. xxxvi, f. 7–9.

Hymenopteris psilotoides Stokes and Webb, Geol. Soc. London, Trans., 2d ser., vol. 1, 1824, p. 424, pl. xlvi, f. 7; pl. xlvii, f. 2.

Aspidium oerstedi Heer. Lesquereux, U. S. Nat. Mus., Proc., vol. 11, 1888, p. 32 (in part).

Onychiopsis psilotoides—Continued.

Sphenopteris mantelli Brongniart. Fontaine, U. S. Geol. Surv., Mon. 15, 1889 [1890], p. 91, pl. l, f. 1, 2.

Thyrsopteris insignis Fontaine, op. cit. (Mon. 15), p. 127, pl. xxxix, f. 4; pl. xl, f. 6; pl. xliii, f. 1, 2, 4; pl. liii, f. 1, 3; in Ward, op. cit. (Mon. 48), p. 521.

Thyrsopteris insignis angustipennis Fontaine, op. cit. (Mon. 15), p. 128, pl. xliii, f. 2.

Thyrsopteris angustifolia Fontaine, op. cit. (Mon. 15), p. 131, pl. xliv, f. 4; pl. xlv, f. 3; pl. xlviii, f. 2; pl. xlix, f. 3, 4; pl. lv, f. 2; pl. lviii, f. 8; in Ward, op. cit. (Mon. 48), p. 516.

Thyrsopteris microphylla Fontaine, op. cit. (Mon. 15), p. 151, pl. xlv, f, 1, 2, 4, 5.

Thyrsopteris rarinervis Fontaine, op. cit. (Mon. 15), p. 123, pl. xxvi, f. 6, 7; pl. xliii, f. 4–6; pl. xliv, f. 1, 2, 5; pl. xlix, f. 2; pl. clxix, f. 6, 7; in Ward, op. cit. (Mon. 48), pp. 225, 484, 491, 514, 517, 518, 519, 521, 528, 548, 558, pl. lxv, f. 2–4; pl. cxiii, f. 2, 3.

Thyrsopteris dentifolia Fontaine, in Ward, U. S. Geol. Surv., Nineteenth Ann. Rept., pt. 2, 1899, p. 660, pl. clxi, f. 6–9.

Kootenai: Great Falls, Mont.

Lakota: Crook County, Wyo.

Jurassic: Cape Lisburne, Alaska.

Patuxent: Fredericksburg, Dutch Gap, Trents Reach, and near Potomac Run, Va.; New Reservoir and Sixteenth Street, D. C.

Arundel: Langdon, D. C.; Bay View, Md.

Patapsco: Federal Hill (Baltimore), Stump Neck, Wellhams, Md.; near Brooke, Widewater, and Hell Hole, Va.

Knoxville: Elder Creek and South Fork of Cottonwood Creek, Tehama County, Calif.

OPEGRAPHA Persoon, in Usteri, Ann., vol. 1, 1794, p. 23.

Opegrapha antiqua Lesquereux, U. S. Geol. and Geog. Surv. Terr., Ann. Rept. 1872 [1873], p. 390; Rept. U. S. Geol. Surv. Terr., vol. 7 (Tert. Fl.), 1878, p. 36, pl. i, f. 1–1c.

Post-Laramie: Black Buttes, Wyo.

Ophioglossum alleni Lesquereux = *Tmesipteris alleni.*

Ophioglossum granulatum Heer = *Pinus granulatum.*

OREODAPHNE Nees, Linnaea, vol. 8, 1833, p. 39.

Oreodaphne alabamensis Berry, Bull. Torr. Bot. Club, vol. 39, 1912, p. 400, pl. xxxii; U. S. Geol. Surv., Prof. Paper 112, 1919, p. 119, pl. xix, f. 3–5.

Woodbine: Arthurs Bluff, Lamar County, Tex.

Tuscaloosa: Cottondale, Ala.

Oreodaphne coushatta Berry, U. S. Geol. Surv., Prof. Paper 91, 1916, p. 302, pl. lxxxi, f. 1, 2.

Tetranthera praecursoria Lesquereux. Hollick, La. Geol. Surv., Special Rept. 5, 1899, p. 284, pl. xliv, f. 3, 4.

Wilcox: Coushatta and near Naborton, La.

Oreodaphne cretacea Lesquereux, Rept. U. S. Geol. Surv. Terr., vol. 6 (Cret. Fl.), 1874, p. 84, pl. xxx, f. 5; idem, vol. 8 (Cret. and Tert. Fl.), 1883, p. 55; U. S. Geol. Surv., Mon. 17, 1891 [1892], p. 108.

Dakota: Kansas.

Oreodaphne heerii Gaudin. Lesquereux, U. S. Nat. Mus., Proc., vol. 11, 1888, p. 30 = *Oreodaphne litsaeformis*.

Oreodaphne litsaeformis [printed *lithaeformis*] Lesquereux, U. S. Nat. Mus., Proc., vol. 11, 1888, p. 30, pl. xiv, f. 4.

Oreodaphne heerii Gaudin. Lesquereux, U. S. Nat. Mus., Proc., vol. 11, 1888, p. 30.

Eocene: Near Susanville, Lassen County, Calif.

Oreodaphne mississippiensis Berry, U. S. Geol. Surv., Prof. Paper 91, 1916, p. 303, pl. lxxxii, f. 3-5.

Laurus primigenia Unger. Hollick, La. Geol. Surv., Special Rept. 5, 1899, p. 284, pl. xli, f. 1 [not f. 2].

Wilcox (Holly Springs): Holly Springs and Lockhart, Miss.

Wilcox: Coushatta, Shreveport, Naborton, and near Texarkana, Ark.

Oreodaphne obtusifolia Berry, U. S. Geol. Surv., Prof. Paper 91, 1916, p. 301, pl. lxxx, f. 1; pl. lxxxiii, f. 2-5; pl. lxxxiv, f. 1, 2.

Cinnamomum sezannense Watelet. Hollick, La. Geol. Surv., Special Rept. 5, 1899, p. 283, pl. xlii, f. 2.

Wilcox (Holly Springs): Holly Springs, Miss.

Wilcox (Ackerman): Hurleys, Miss.

Wilcox (Grenada): Grenada, Miss.

Wilcox: Vineyard Bluff, Caddo Parish, La.; Old Port Caddo Landing, Harrison County, Tex.

Lagrange: Puryear, Tenn.

Oreodaphne pseudoguianensis Berry, U. S. Geol. Surv., Prof. Paper 91, 1916, p. 305, pl. lxxxi, f. 3, 4.

Wilcox: Coushatta, La.

Lagrange: Puryear, Tenn.

Oreodaphne puryearensis Berry, U. S. Geol. Surv., Prof. Paper 91, 1916, p. 301, pl. lxxxiii, f. 1.

Wilcox (Ackerman): Hurleys, Benton County, Miss.

Lagrange: Puryear, Tenn.

Oreodaphne ?ratonensis Knowlton, U. S. Geol. Surv., Prof. Paper 101, 1918, p. 318, pl. lxxxviii, f. 2.

Raton: Near Trujillo Plaza, Colo.

Oreodaphne salinensis Berry, U. S. Geol. Surv., Prof. Paper 91, 1916, p. 313, pl. lxxxii, f. 1, 2.

Wilcox: Benton and Malvern, Ark.

Oreodaphne shirleyensis Berry, U. S. Geol. Surv., Prof. Paper 112, 1919, p. 119, pl. xix, f. 1, 2.
Tuscaloosa: Shirleys Mill, Ala.

Oreodaphne wilcoxensis Berry, U. S. Geol. Surv., Prof. Paper 91, 1916, p. 305, pl. lxxxvii, f. 6.
Lagrange: Puryear, Tenn.

OREODOXITES Göppert, Foss. Fl. Perm. Form., 1864, p. 146.

Oreodoxites plicatus Lesquereux, Rept. U. S. Geol. Surv. Terr., vol. 8 (Cret. and Tert. Fl.), 1883, p. 122, pl. xviii, f. 1–4.—Knowlton, U. S. Geol. Surv., Prof. Paper 101, 1918, p. 287, pl. lxiii, f. 1.
Denver: Golden, Colo.
Raton: Tercio, Colo.

OREOPANAX Decaisne and Planchon, Revue hort., 4th ser., vol. 3, 1854, p. 127.

Oreopanax minor Berry, U. S. Geol. Surv., Prof. Paper 91, 1916, p. 331, pl. xcix, f. 1.
Lagrange: Puryear, Tenn.

Oreopanax oxfordensis Berry, U. S. Geol. Surv., Prof. Paper 91, 1916, p. 329, pl. xcviii, f. 1–5.
Wilcox (Holly Springs): Oxford, Miss.
Wilcox: Benton, Ark.

ORYZOPSIS Michaux, Fl. bor. Am., vol. 1, 1803, p. 51.

Oryzopsis asperifolia Michaux. Penhallow, Roy. Soc. Canada, Trans., 2d ser., vol. 2, 1896, p. 72; Brit. Assoc. Adv. Sci., Bradford meeting, 1900, p. 336.
Pleistocene: Greens Wharf, Ottawa River, Canada.

OSMANTHUS Loureiro, Fl. Cochinchina, vol. 1, 1790, p. 28.

Osmanthus pedatus (Lesquereux) Berry, U. S. Geol. Surv., Prof. Paper 91, 1916, p. 341, pl. civ, f. 1.
Laurus pedatus Lesquereux, Am. Philos. Soc. Trans., vol. 13, 1869, p. 418, pl. xix, f. 1.—?Knowlton, U. S. Geol. Surv., Prof. Paper 101, 1918, p. 317.
Olea americana Hilgard, Rept. Geol and Agr. Miss., 1860, p. 108, 113.
Wilcox (Ackerman): Hurleys, Miss.
Lagrange: Puryear, Tenn.
Raton: Near Aguilar, Colo. [Identification?]

Osmanthus praemissa (Lesquereux) Cockerell, Univ. Colo. Studies, vol. 3, 1906, p. 169.
Olea praemissa Lesquereux, Rept. U. S. Geol. Surv. Terr., vol. 8 (Cret. and Tert. Fl.), 1883, p. 168, pl. xxxiii, f. 1.
Miocene: Florissant, Colo.

OSMUNDA (Tournefort) Linné, Sp. pl., 1753, p. 1063.

Osmunda affinis (Lesquereux) = *Salpichlaena anceps*.

Osmunda arctica Heer, Fl. foss. arct., vol. 7, 1883, p. 7, pl. xlix, f. 4–7; pl. l, f. 6–8.—Knowlton, in Atwood, U. S. Geol. Surv., Bull. 467, 1911, p. 45.

Chignik (Upper Cretaceous): Chignik Lagoon, Alaska.

Osmunda delawarensis Berry, Bull. Torr. Bot. Club, vol. 33, 1906, p. 164, pl. viii, f. 2–4; Md. Geol. Surv., vol. 6, pl. xx, f. 17: idem, Upper Cretaceous, 1916, p. 763, pl. l, f. 2–4.

Magothy: Deep Cut, Del.

Osmunda dicksonioides Fontaine, U. S. Geol. Surv., Mon. 15, 1889 [1890], p. 146, pl. xli, f. 5; pl. lviii, f. 9; pl. lix, f. 1, 4, 8, 9, 11; pl. lx, f. 2, 4, 5, 9; pl. lxi, f. 1, 2; U. S. Nat. Mus., Proc., vol. 15, 1892, p. 493.

Patuxent: Near Potomac Run, Va.
Kootenai: Great Falls, Mont.
Knoxville: Tehama County, Calif.

Osmunda dicksonioides Fontaine. Fontaine, in Diller and Stanton = *Gleichenia nordenskiöldi*.

Osmunda dicksonioides latipennis Fontaine, U. S. Geol. Surv., Mon. 15, 1889 [1890], p. 147, pl. lx, f. 1, 3; pl. lxi, f. 3.

Patuxent: Near Potomac Run, Va.

Osmunda doroschkiana Göppert, Abh. Schles. Ges. vaterl. Cult., pt. 2, 1861, p. 203.

Osmunda torelli Heer, Fl. foss. arct., vol. 5, Abt. 3, 1878, p. 19, pl. i, f. 4.—Lesquereux, U. S. Nat. Mus., Proc., vol. 5, 1882 [1883], p. 444, pl. vi, f. 3–6.

Hemitelites torelli Heer, op. cit., vol. 2, Abt. 4, 1869, p. 462.

Pecopteris torellii Heer, op. cit., vol. 1, 1868, p. 88, pl. i, f. 15.

Kenai: Coal Harbor, Unga Island, Alaska.

Osmunda heerii Gaudin. Penhallow, Rept. Tert. Pl. Brit. Col., 1908, p. 65.

Oligocene: Tulameen River, British Columbia.

Osmunda hollicki Knowlton, U. S. Geol. Surv., Prof. Paper 101, 1918, p. 246, pl. xxx, f. 6.

Pecopteris (*Cheilanthes*) *sepulta* Newberry? Hollick, Torreya, vol. 2, 1902, p. 147, pl. iii, f. 5, 5a.

Vermejo: Coal Creek mine, near Rockvale, Colo.

Osmunda macrophylla Penhallow, Rept. Tert. Pl. Brit. Col., 1908, p. 65, text f. 15, 16.

Eocene: Red Deer River, Alberta.

Osmunda major Lesquereux = *Allantodiopsis erosa*.

Osmunda montanensis Knowlton, U. S. Geol. Surv., Bull. 257, 1905, p. 129, pl. xiv, f. 6.

Judith River: Willow Creek, Fergus County, Mont.

Osmunda novae-caesarae Berry, Johns Hopkins Univ., Circ., new ser., No. 7, 1907, p. 83, f. 2.
Magothy: Cliffwood, N. J.

Osmunda obergiana Heer. Ward, U. S. Geol. Surv., Fifteenth Ann. Rept., 1895, p. 390.
Potomac: No locality.

Osmunda spectabilis Willdenow. Berry, Am. Jour. Sci., 4th ser., vol. 29, 1910, p. 391.
Pleistocene: Steels Bluff, Black Warrior River, Ala.

Osmunda sphenopteroides Fontaine, U. S. Geol. Surv., Mon. 15, 1889 [1890], p. 145, pl. xxv, f. 13.
Patuxent: Dutch Gap Canal, Va.

Osmunda torellii Heer = *Osmunda doroschkiana.*

Osmunda sp.? Hollick, Md. Geol. Surv., Pliocene and Pleistocene, 1906, p. 217, pl. xlvii, f. 3.
Pleistocene (Talbot): Tolly Point (Bay Ridge), Md.

OSMUNDITES Unger, Denkschr. K. Akad. Wiss., vol. 6, 1854, p. 143.

Osmundites skidegatensis Penhallow, Roy. Soc. Canada, Trans., 2d ser., vol. 8, 1902, sec. 4, pp. 3–29, pls. [not numbered] vi, with f. 1–12.
Lower Cretaceous: Alliford Bay, Skidegate Inlet, Queen Charlotte Islands, British Columbia.

OSTRYA (Michaux) Scopoli, Carn., vol. 2, 1772, p. 243.

Ostrya betuloides Lesquereux = *Fraxinus libbeyi.*

Ostrya virginiana (Miller) Willdenow. Berry, Am. Jour. Sci., 4th ser., vol. 29, 1910, p. 395.
Ostrya virginica (Miller) Britton, Sterns, and Poggenberg. Hollick, Bull. Torr. Bot. Club, vol. 19, 1892, p. 332.—Penhallow, Am. Nat., vol. 41, 1907, p. 447.
Pleistocene: Bridgeton, N. J.; Black Warrior River, 311 miles above Mobile, Ala.

Ostrya walkeri? Heer. Lesquereux, U. S. Nat. Mus., Proc., vol. 10, 1887, p. 38.
Pleistocene?: Wytheville, Va.

OTOZAMITES Braun, in Münster, Beitr. Petr.-Kunde, pt. 6, 1843, p. 29.

Otozamites bornholmensis? Möller, Bidrag till Bornholms fossila flora, Gymnospermer: Kongl. Svenska Vetenskaps-Ak., Handl., vol. 36, 1902, p. 12, pl. ii, f. 1–7.—Knowlton, U. S. Nat. Mus., Proc., vol. 51, 1916, p. 456, pl. lxxxi, f. 5, 6.
Lower Jurassic: Upper Matanuska Valley, Alaska.

Otozamites brevifolius (Braun) Braun, in Münster, Beitr. Petref.-Kunde, pt. 6, 1843, p. 36.—Newberry, N. Y. Acad. Sci., Trans., vol. 6, 1886–87 [1887], p. 126; U. S. Geol. Surv., Mon. 14, 1888, p. 91, pl. xxiv, f. 3.

Zamites brevifolius Braun, in Münster, Beitr. Petref.-Kunde, pt. 6, 1843, p. 23.

Triassic (Newark): Durham, Conn.

Otozamites carolinensis Fontaine, U. S. Geol. Surv., Mon. 6, 1883, pp. 117, 118, pl. lii, f. 6; in Ward, U. S. Geol. Surv., Twentieth Ann. Rept., pt. 2, 1900, p. 298, pl. xlii, f. 5, 6.

Albertia latifolia? Emmons, American geology, pt. 6, 1857, p. 126, f. 95. [Homonym, Schimper, 1837.]

Triassic (Newark): Lockville, N. C.

Otozamites giganteus Thomas, Jurassic flora of Kamenka: Com. géol., Mém., new ser., pt. 71, 1911, p. 84, pl. vi, f. 1, 2.—Knowlton, U. S. Geol. Surv., Prof. Paper 85, 1914, p. 50, pl. vi, f. 2.

Jurassic: Cape Lisburne, Alaska.

Otozamites latior Saporta, Pal. franç. (Pl. jurassiques), vol. 2, 1873, p. 130, pl. xcvii, f. 1; pl. xcviii, f. 1–3.—Newberry, N. Y. Acad. Sci., Trans., vol. 6, 1886–87 [1887], p. 126; U. S. Geol. Surv., Mon. 14, 1888, p. 90, pl. xxiv, f. 1, 2, 2a.—Rice and Gregory, Mon. Geol. Conn. (State Geol. and Nat. Hist. Surv., Bull. 6), 1906, p. 171, pl. xvi.

Triassic (Newark): Durham, Conn.

Otozamites macombii Newberry, in Macomb, Rept. Expl. Exped. Santa Fe, 1859 [1876], p. 141, pl. iv, f. 1, 2.

Triassic: Abiquiu, N. Mex.

Otozamites oregonensis Fontaine, in Ward, U. S. Geol. Surv., Mon. 48, 1905 [1906], p. 150, pl. xxxviii, f. 13, 14.

Jurassic: Curry County, Oreg.

Otozamites pterophylloides Brongniart, MS. Schimper, Pal. vég., vol. 2, 1870, p. 173.—Knowlton, U. S. Nat. Mus., Proc., vol. 51, 1916, p. 456, pl. 79.

Lower Jurassic: Upper Matanuska Valley, Alaska.

OTTELIA Persoon, Encl., vol. 1, 1805, p. 400.

Ottelia americana Lesquereux, U. S. Geol. and Geog. Surv. Terr., Ann. Rept. 1874 [1876], p. 300; Rept. U. S. Geol. Surv. Terr., vol. 7 (Tert. Fl.), 1878, p. 98, pl. lxi, f. 8.—Knowlton, U. S. Geol. Surv., Bull. 163, 1900, p. 32.

Mesaverde: Point of Rocks, Wyo.

Ottelia? n. sp. Knowlton, Washington Acad. Sci., Proc., vol. 11, 1909, p. 207.

Lance: Converse County, Wyo.

OVULARITES Whitford, Nebr. Geol. Surv., vol. 7, 1916, p. 85, text f. 1–5. [Type, *O. barbouri.*]

Ovularites barbouri Whitford, Nebr. Geol. Surv., vol. 7, 1916, p. 85, f. 1–5.
Graneros: Rose Creek, 1¼ miles west of Reynolds, Jefferson County, Nebr.

OXYCOCCUS (Tournefort) Adanson, Fam., vol. 2, 1763, p. 164.

Oxycoccus palustris Persoon. Penhallow, Roy. Soc. Canada, Trans., 2d ser., vol. 2, 1896, sec. 4, p. 72; Brit. Assoc. Adv. Sci., Bradford meeting, 1900, p. 336.
Pleistocene: Scarborough Heights, Ontario.

OXYPOLIS Rafinesque, in Seringe, Bull. Bot., vol. 1, 1830, p. 217.

Oxypolis destructus Cockerell, Am. Mus. Nat. Hist., Bull., vol. 24, 1908, p. 105.
Miocene: Florissant, Colo.

Pachyphyllum brevifolium Newberry = *Pagiophyllum brevifolium.*

Pachyphyllum peregrinum (Lindley and Hutton) Fontaine = *Pagiophyllum peregrinum.*

Pachyphyllum simile Newberry = *Pagiophyllum simile.*

Pachyphyllum sp. Newberry = *Pagiophyllum newberryi.*

Pachypteris sp.? Emmons = *Abietites carolinensis.*

PACHYSTIMA Rafinesque, Am. Month. Mag., vol. 2, 1818, p. 176.

Pachystima? cretacea Berry, U. S. Geol. Surv., Prof. Paper 84, 1914, p. 49, pl. x, f. 6.
Black Creek (Middendorf): Middendorf, S. C.

Pachystima? integra Cockerell, Am. Mus. Nat. Hist., Bull., vol. 24, 1908, p. 100, pl. viii, f. 27.
Miocene: Florissant, Colo.

PADUS Mönch, Meth., 1794, p. 671.

Padus demissa (Nuttall) Roemer. Hannibal, Bull. Torr. Bot. Club, vol. 38, 1911, p. 337.
Pliocene: Near Portola and in Calabasas Canyon, Santa Cruz Mountains, and Bear Valley, Gavilan Range, Calif.

PAGIOPHYLLUM Heer, Contr. fl. foss. Portugal, 1881, p. 11.

Pagiophyllum brevifolium (Newberry) Ward, U. S. Geol. Surv., Twentieth Ann. Rept., pt. 2, 1900, p. 427.
Pachyphyllum brevifolium Newberry, U. S. Geol. Surv., Mon. 14, 1888, p. 89, pl. xxii, f. 3–3c.
Triassic (Newark): Turners Falls, Mass.; Durham, Conn.

Pagiophyllum dubium Fontaine, U. S. Nat. Mus., Proc., vol. 16, 1893, p. 271, pl. xxxix, f. 2–11.
Trinity: Glen Rose, Tex.

Pagiophyllum falcatum Bartolin, Nogle i den bornholmske Juraformation forekommende Planteforsteninger.—Knowlton, U. S. Nat. Mus., Proc., vol. 51, 1916, p. 459, pl. lxxxi, f. 1.
Lower Jurassic: Upper Matanuska Valley, Alaska.

Pagiophyllum kurrii (Pomel) Schimper, Pal. vég., vol. 2, 1872, p. 250.—Knowlton, U. S. Geol. Surv., Prof. Paper 85, 1914, p. 53, pl. vi, f. 1.
Jurassic: Cape Lisburne, Alaska.

Pagiophyllum newberryi Ward, U. S. Geol. Surv., Twentieth Ann. Rept., pt. 2, 1900, pp. 318, 427.
Pachyphyllum sp. Newberry, Rept. Expl. Exped. from Santa Fe. N. Mex., in 1859, under Capt. J. N. Macomb. 1876, p. 69, pl. v, f. 4, 5; pl. vi, f. 9.
Triassic: New Mexico.

Pagiophyllum peregrinum (Lindley and Hutton) Schenk, in Zittel, Handb. Paleontologie, Abt. 2, 1884, p. 276, f. 192a.—Fontaine, in Ward, U. S. Geol. Surv., Twentieth Ann. Rept., pt. 2, 1900, pp. 308, 340, pl. xlvi.
Araucaria peregrina Lindley and Hutton, Foss. Fl. Great Britain, vol. 2, 1833, p. 19, pl. lxxxviii.
Walchia variabilis Emmons, American geology, pt. 6, 1857, p. 108, f. 76.
Pachyphyllum peregrinum (Lindley and Hutton). Fontaine, U. S. Geol. Surv., Mon. 6, 1883, p. 108, pl. l, f. 4.
Triassic (Newark): North Carolina.
Jurassic: Near Princeton, Calif.

Pagiophyllum simile (Newberry) Ward, U. S. Geol. Surv., Twentieth Ann. Rept., pt. 2, 1900, p. 427.
Pachyphyllum simile Newberry, U. S. Geol. Surv., Mon. 14, 1888, p. 88, pl. xxii, f. 2.
Triassic (Newark): New Jersey; Connecticut; Massachusetts.

Pagiophyllum steenstrupi Bartolin? Knowlton, U. S. Geol. Surv., Prof. Paper 85, 1914, p. 54.
Jurassic: Cape Lisburne, Alaska.

Pagiophyllum williamsoni (Schimper) Fontaine = *Pagiophyllum williamsonis*.

Pagiophyllum williamsonis (Brongniart) Fontaine, in Ward, U. S. Geol. Surv., Twentieth Ann. Rept., pt. 2, 1900, p. 362, 427, pl. lxvi, f. 1, 2.
Palissya? williamsonis Brongniart, Tableau, 1849, pp. 68, 106.
Pagiophyllum williamsoni (Schimper) Fontaine, Am. Jour. Sci., 4th ser., vol. 2, 1896, p. 274.
Jurassic: Oroville, Calif.; Virginia?; North Carolina?

Pagiophyllum sp. Dawson, Roy. Soc. Canada, Trans., vol. 10, 1892, sec. 4, p. 90, text f. 14.
Kootenai: Anthracite?, British Columbia.

PALAEANTHUS Newberry, Bull. Torr. Bot. Club, vol. 13, 1886, p. 37. [Type, *P. problematicus*.]

Palaeanthus (Williamsonia) problematicus Newberry, U. S. Geol. Surv., Mon. 26, 1895 [1896], p. 125, pl. xxxv, f. 1-9.

Williamsonia problematica (Newberry) Ward, U. S. Geol. Surv., Fifteenth Ann. Rept., 1895, p. 382.—Hollick, U. S. Geol. Surv., Mon. 50, 1906, p. 107, pl. v, f. 27-32; N. Y. Bot. Gard., Bull., vol. 8, 1912, p. 156, pl. clxiii, f. 4.—Berry, Geol. Surv. N. J., Bull. 3, 1911, p. 225.

Raritan: New Jersey; Marthas Vineyard, Mass.; Glen Cove, Long Island, N. Y.

PALAEOASTER Knowlton, U. S. Geol. Surv., Prof. Paper 101, 1918, p. 278. [Type, *P. inquirenda.*]

Palaeoaster inquirenda Knowlton, U. S. Geol. Surv., Prof. Paper 101, 1918, pp. 278, 348, pl. xlix, f. 5, 6.

Vermejo: Alkali Gap, Canon City field, and Walsenburg, Colo.; Koehler, N. Mex.

Raton: Bowen mine near Trinidad, Colo.

Palaeoaster? similis Knowlton, U. S. Geol. Surv., Prof. Paper 130, 19—, p. —, pl. xxiv, f. 10, 11 [in preparation].

Laramie: Near Golden, Colo.

PALAEOCASSIA Ettingshausen, Sitzb. Wien. Akad., vol. 55, Abt. 1, 1867, p. 261.

Palaeocassia laurinea Lesquereux, U. S. Geol. Surv., Mon. 17, 1891 [1892], p. 147, pl. lxiv, f. 12.—Berry, Bull. Torr. Bot. Club, vol. 39, 1912, p. 396; U. S. Geol. Surv., Prof. Paper 112, 1919, p. 100, pl. xxiii, f. 1.

Dakota: Ellsworth County, Kans.

Woodbine: Arthurs Bluff, Tex.

Tuscaloosa: Shirleys Mill, Ala.

PALAEODENDRON Saporta, Ann. sci. nat., 4th ser., Bot., vol. 17 (Études, vol. 1, pt. 2), 1862, p. 250 (97).

Palaeodendron americanum Berry, U. S. Geol. Surv., Prof. Paper 91, 1916, p. 207, pl. xxxviii, f. 1.

Wilcox (Holly Springs): Oxford, Miss.

Wilcox: Bolivar Creek near Harrisburg, Ark.

PALAEODICTYON Heer, Urwelt der Schweiz, 1865, p. 245.

Palaeodictyon magnum laxum Ulrich, Harriman Alaska Exped., vol. 4, 1904, p. 137, pl. xv, f. 1.

Yakutat: Woody Island near Kodiak, Alaska.

Palaeodictyon singulare Heer, Urwelt der Schweiz, 1865, p. 245, pl. x, f. 10.—Ulrich, Harriman Alaska Exped., vol. 4, 1904, p. 137, pl. xv, f. 2.

Yakutat: Woody Island near Kodiak, Alaska.

PALAEOPHYCUS Hall, Palaeont. N. Y., vol. 1, 1847, p. 7.

Palaeophycus limaciformis Lewis, Philadelphia Acad. Sci., Proc., vol. 32, 1880, p. 293, text f.—Ward, U. S. Geol. Surv., Twentieth Ann. Rept., 1898–99, pt. 2, 1900, p. 427.
Triassic (Newark): Near Milford, N. J.

PALAEOPOTAMOGETON Knowlton, U. S. Nat. Mus., Proc., vol. 51, 1916, p. 251. [Type, *P. florissanti*.]

Palaeopotamogeton florissanti Knowlton, U. S. Nat. Mus., Proc., vol. 51, 1916, p. 251, pl. xvi, f. 1; pl. xvii, f. 3.
Miocene: *Florissant, Colo.

Palaeozamia megaphylla Phillips = *Zamites megaphyllus.*

PALEOHILLIA Knowlton, Bull. Torr. Bot. Club, vol. 22, 1895, p. 387. [Type, *P. arkansana*.]

Paleohillia arkansana Knowlton, Bull. Torr. Bot. Club, vol. 22, 1895, p. 387, text f. 1–3.
Trinity: 6 miles northwest of Center Point, Howard County, Ark.

PALISSYA Endlicher, Syn. Conif., 1847, p. 306.

Palissya braunii Endlicher. Fontaine, U. S. Geol. Surv., Mon. 6, 1883, p. 107 = *Palissya sphenolepis.*

Palissya brevifolia (Emmons) Fontaine, in Ward, U. S. Geol. Surv., Twentieth Ann. Rept., pt. 2, 1900, p. 507, pl. xlv, f. 4.
Walchia brevifolia Emmons, American geology, pt. 6, 1857, p. 107, f. 74.
Cheirolepis muensteri Fontaine [not (Schenk) Schimper], U. S. Geol. Surv., Mon. 6, 1883, p. 108, pl. liii, f. 3.
Triassic (Newark): Lockville, N. C.

Palissya carolinensis Fontaine = *Abietites carolinensis.*

Palissya diffusa (Emmons) Fontaine, U. S. Geol. Surv., Mon. 6, 1883, p. 107, pl. li, f. 4; in Ward, U. S. Geol. Surv., Twentieth Ann. Rept., pt. 2, 1900, pp. 250, 306, pl. xxxi, f. 3–5; pl. xlv, f. 2, 3.—Brown, Acad. Nat. Sci. Philadelphia, Proc., vol. 63, 1911, p. 19, pl. ii.
Walchia diffusus Emmons, Geol. Rept. Midland Counties N. C., 1856, p. 333, pl. iii, f. 2; American geology, pt. 6, 1857, p. 105, pl. iii, f. 2.
Walchia gracile Emmons, Amerian geology, pt. 6, 1857, p. 106, f. 75.
Cheirolepis muensteri (Schenk) Schimper. Fontaine, U. S. Geol. Surv., Mon. 6, 1883, p. 108, pl. l, f. 3; U. S. Nat. Mus., Proc., vol. 13, 1890, p. 284.—Newberry, N. Y. Acad. Sci., Trans., vol. 6, 1886–87 [1887], p. 126.
Triassic (Newark): Ellingtons, N. C.; York County and Bucks County, Pa.; Massachusetts?; Connecticut?; New Jersey
Triassic: Abiquiu, N. Mex.

Palissya longifolia Wherry, U. S. Nat. Mus., Proc., vol. 51, 1916, p. 327, pl. xxix.
Triassic (Newark, Lockatong): Carversville, Bucks County, Pa.

Palissya sphenolepis (F. Braun) Brongniart, Tableau, 1848, p. 68; Fontaine, in Ward, U. S. Geol. Surv., Twentieth Ann. Rept., pt. 2, 1900, pp. 249, 305, pl. xxxii; pl. xliv; pl. xlv, f. 1.

Cunninghamites sphenolepis F. Braun, Programm. z. Jahresber. d. Kön. Kreis.-Landw. u. Gewerbsschule z. Bayreuth, 1843, pp. 17, 18, pl. ii, f. 16–20.

Walchia longifolius Emmons, Geol. Rept. Midland Counties N. C., 1856, p. 333; American geology, pt. 6, 1857, p. 105, pl. iv-a.

Palissya braunii Endlicher. Fontaine, U. S. Geol. Surv., Mon. 6, 1883, p. 107, pl. l, f. 1, 2; pl. li, f. 1; U. S. Nat. Mus., Proc., vol. 13, 1890, p. 284.

Triassic (Newark): York County, Pa.; North Carolina; Newark, N. J.

Triassic: Abiquiu, N. Mex.?

Palissya? williamsonis Brongniart = *Pagiophyllum williamsonis.*

Palissya sp. (cone?) Fontaine, U. S. Nat. Mus., Proc., vol. 13, 1890, p. 284.

Triassic: Abiquiu, N. Mex.

Palissya? sp. Newberry = *Cycadeomyelon yorkense.*

PALIURUS (Tournefort) Jussieu, in Ht. Trian., 1759, p. — [not seen].

Paliurus affinis? Heer. Hollick, N. Y. Acad. Sci., Trans., vol. 12, 1892, p. 35, pl. ii, f. 12, 14, 18, 19; pl. iii, f. 7; U. S. Geol. Surv., Mon. 50, 1906, p. 92, pl. xxxiv, f. 6, 7.

Raritan: Tottenville and Kreischerville, Staten Island, N. Y.

Paliurus anceps Lesquereux, U. S. Geol. Surv., Mon. 17, 1891 [1892], p. 166, pl. xxxv, f. 4.

Dakota: Near Delphos, Kans.

Paliurus angustus Berry, U. S. Geol. Surv., Prof. Paper 91, 1916, p. 280, pl. lxxi, f. 5, 6.

Wilcox (Holly Springs): Holly Springs, Miss.

Paliurus colombi Heer, Fl. foss. arct., vol. 1, 1868, p. 122, pl. xvii, f. 2d; pl. xix, f. 2–4.—Lesquereux, U. S. Geol. and Geog. Surv. Terr., Ann. Rept. 1871 [1872], p. 288; Rept. U. S. Geol. Surv. Terr., vol. 7 (Tert. Fl.), 1878, p. 273, pl. l, f. 13–17.—Ward, U. S. Geol. Surv., Sixth Ann. Rept., 1884-85 [1886], p. 555, pl. lii, f. 4–6; idem, Bull. 37, 1887, p. 75, pl. xxxiii, f. 8–10.—Knowlton, idem, Mon. 32, pt. 2, 1899, p. 740, pl. ci, f. 7; Jour. Geol., vol. 19, 1911, p. 269.

[NOTE.—This species is in need of critical revision.]

Fort Union: Yellowstone National Park.

Hanna: Creston and Carbon, Wyo.

Lance: Cannonball River, S. Dak.

Kenai: Alaska.

Paliurus colombi Heer. Lesquereux, U. S. Nat. Mus., Proc., vol. 11, 1888, p. 16 = *Grewia crenata.*

Paliurus coloradensis Lesquereux, Mus. Comp. Zool., Bull., vol. 16, 1888, p. 55.
Denver: Golden, Colo.

Paliurus cretaceus Lesquereux, U. S. Geol. Surv., Mon. 17, 1891 [1892], p. 165, pl. xxxv, f. 3.
Dakota: Near Delphos, Kans.

Paliurus florissanti Lesquereux, U. S. Geol. and Geog. Surv. Terr., Ann. Rept. 1873 [1874], p. 416; Rept. U. S. Geol. Surv. Terr., vol. 7 (Tert. Fl.), 1878, p. 274, pl. l, f. 18.
Miocene: Florissant, Colo.

Paliurus haydeni Cockerell, Am. Mus. Nat. Hist., Bull., vol. 24, 1908, p. 102.
Paliurus orbiculatus Saporta. Lesquereux, Rept. U. S. Geol. Surv. Terr., vol. 8 (Cret. and Tert. Fl.), 1883, p. 188, pl. xxxviii, f. 12.
Miocene: Florissant, Colo.

Paliurus integrifolius Hollick, Bull. Torr. Bot. Club, vol. 21, 1894, p. 57, pl. clxxvii, f. 5, 8, 12.—Hollick, in Ries, School of Mines Quart., vol. 15, 1894, p. 355.—Hollick, N. Y. Bot. Gard., Bull., vol. 3, 1904, p. 408, pl. lxx, f. 7.—Berry, N. Y. Bot. Gard., Bull., vol. 3, 1903, p. 86; Geol. Surv. N. J., Ann. Rept. 1905 [1906], p. 138.—Hollick, U. S. Geol. Surv., Mon. 50, 1906, p. 91, pl. xxxiv, f. 2–5.
Raritan: Oak Neck and Lloyds Neck, Long Island, N. Y.
Magothy: Cliffwood, N. J.; Glen Cove, Little Neck, and Northport, Long Island, N. Y.

Paliurus membranaceus Lesquereux = *Cinnamomum membranaceum.*

Paliurus minimus Knowlton, U. S. Geol. Surv., Mon. 32, pt. 2, 1899, p. 759, pl. lxxvii, f. 7–9.
"Laramie": Wolverine Creek, Yellowstone National Park.

Paliurus mississippiensis Berry, U. S. Geol. Surv., Prof. Paper 91, 1916, p. 279, pl. lxxi, f. 4.
Wilcox (Holly Springs): Early Grove and Holly Springs, Miss.

Paliurus montanus Dawson, Roy. Soc. Canada, Trans., vol. 3, 1885 [1886], sec. 4, p. 14.
——: North Fork of Old Man River, Alberta.

Paliurus neillii Dawson, Roy. Soc. Canada, Trans., vol. 11, 1893 [1894], sec. 4, p. 62, pl. xi, f. 44, 45.
Upper Cretaceous: Port McNeill, Vancouver Island, British Columbia.

Paliurus obovatus Lesquereux, U. S. Geol. Surv., Mon. 17, 1891 [1892], p. 165, pl. xxxv, f. 6.
Dakota: Near Delphos, Kans.

Paliurus orbiculatus Saporta. Lesquereux = *Paliurus haydeni.*

Paliurus ovalis Dawson, Roy. Soc. Canada, Trans., vol. 3, 1885 [1886], sec. 4, p. 14, pl. iv, f. 4, 8.—Newberry, U. S. Geol. Surv., Mon. 26, 1895 [1896], p. 107, pl. xxiii, f. 8.—Lesquereux, U. S. Geol. Surv., Mon. 17, 1892, p. 166, pl. xxxv, f. 7.—Hollick, U. S. Geol. Surv., Mon. 50, 1906, p. 91, pl. xxxiv, f. 14.
Dakota: Near Delphos, Kans.
Raritan: New Jersey.
Magothy: Gay Head, Marthas Vineyard, Mass.
Mill Creek: Mill Creek, British Columbia.

Paliurus ovalis Newberry = *Smilax raritanensis.*

Paliurus pealei Ward, U. S. Geol. Surv., Sixth Ann. Rept., 1884–85 [1886], p. 555, pl. lii, f. 8–10; U. S. Geol. Surv., Bull. 37, 1887, p. 76, pl. xxxiii, f. 12–14.
Fort Union: Little Missouri River, N. Dak.

Paliurus pinsonensis Berry, U. S. Geol. Surv., Prof. Paper 91, 1916, p. 280, pl. lxxi, f. 7.
Lagrange: Pinson, Tenn.

Paliurus populiferus Berry, Bull. Torr. Bot. Club, vol. 33, 1906, p. 177; Geol. Surv. N. J., Ann. Rept. 1905 [1906], pp. 139, 153, pl. xx, f. 1.
Magothy: Pit of Cliffwood Brick Co., Whale River, N. J.

Paliurus pulcherrimus Ward, U. S. Geol. Surv., Sixth Ann. Rept., 1884–85 [1886], p. 555, pl. lii, f. 7; U. S. Geol. Surv., Bull. 37, 1887, p. 75, pl. xxxiii, f. 11.—Knowlton, Washington Acad. Sci., Proc., vol. 11, 1909, p. 213.
Hanna: Carbon, Wyo.
Lance: Near Ilo, Big Horn Basin, Wyo.

Paliurus upatoiensis Berry, U. S. Geol. Surv., Prof. Paper 84, 1914, p. 116, pl. xxi, f. 5, 6; idem, Prof. Paper 112, 1919, p. 113.
Eutaw: McBrides Ford, Ga.

Paliurus zizyphoides Lesquereux, U. S. Geol. and Geog. Surv. Terr. Ann. Rept. 1872 [1873], p. 397; Rept. U. S. Geol. Surv. Terr., vol. 7 (Tert. Fl.), 1878, p. 274, pl. li, f. 1–6.—?Knowlton, U. S. Geol. Surv., Mon. 32, pt. 2, 1899, p. 660, pl. lxxviii, f. 3; idem, Prof. Paper 101, 1918, p. 334, pl. civ, f. 2.
Denver: Golden, Colo.
?Laramie: Erie, Colo.
Post-Laramie: *Black Buttes, Wyo.
"Laramie": Wolverine Creek, Yellowstone National Park.
Raton: Wootton, Colo.

Paliurus sp. Berry, Bull. Torr. Bot. Club, vol. 37, 1910, p. 504.
Tuscaloosa: McBrides Ford, near Columbus, Ga.

Paliurus? sp. Dawson, Geol. Surv. Canada, 1877–78, p. 186B.—Penhallow, Rept. Tert. Pl. Brit. Col., 1908, p. 66.
Oligocene: Tulameen River, British Columbia.

Paliurus sp. Hollick, in Ries, School of Mines Quart., vol. 15, 1894, p. 355.

Magothy?: Northport, Long Island, N. Y.

Palmacites goldianus Lesquereux = *Geonomites goldianus.*

PALMOCARPON Lesquereux, Rept. U. S. Geol. Surv. Terr., vol. 7 (Tert. Fl.), 1878, p. 119. [Type, *Carpolithes compositus.*]

Palmocarpon commune Lesquereux = *Palmocarpon palmarum.*

Palmocarpon compositum (Lesquereux) Lesquereux, Rept. U. S. Geol. Surv. Terr., vol. 7 (Tert. Fl.), 1878, p. 119, pl. xi, f. 4.

Carpolithes compositus Lesquereux, U. S. Geol. and Geog. Surv. Terr., Ann. Rept. 1871 [1872], Suppl., p. 16.

Raton?: Placer Mountain, N. Mex.

Palmocarpon corrugatum Lesquereux = *Carpolithes corrugatus.*

Palmocarpon? globosum Lesquereux, Rept. U. S. Geol. Surv. Terr., vol. 8 (Cret. and Tert. Fl.), 1883, p. 144, pl. xxiv, f. 3.—Knowlton, U. S. Nat. Mus., Proc., vol. 51, 1916, p. 252.

Miocene: *Florissant, Colo.

Palmocarpon lineatum Lesquereux, Mus. Comp. Zool., Bull., vol. 16, 1888, p. 44.

Denver: Golden, Colo.

Palmocarpon mexicanum (Lesquereux) Lesquereux, Rept. U. S. Geol. Surv. Terr., vol. 7 (Tert. Fl.), 1878, p. 119, pl. xi, f. 5.

Carpolithes mexicanus Lesquereux, U. S. Geol. and Geog. Surv. Terr., Ann. Rept. 1871 [1872], Suppl., p. 17.

Raton?: Placer Mountain, N. Mex.

Palmocarpon palmarum (Lesquereux) Knowlton, U. S. Geol. Surv., Bull. 152, 1898, p. 158; Washington Acad. Sci., Proc., vol. 11, 1909, p. 204; U. S. Geol. Surv., Prof. Paper 101, 1918, p. 292.

Carpolithes palmarum Lesquereux, U. S. Geol. and Geog. Surv. Terr., Ann. Rept. 1871 [1872], Suppl., p. 13; idem, 1872 [1873], pp. 382, 398.

Palmocarpon commune Lesquereux, Rept. U. S. Geol. Surv. Terr., vol. 7 (Tert. Fl.), 1878, p. 119, pl. xiii, f. 4–7.

Raton: *Fishers Peak, Raton Mountains, Colo.; near Blossburg, Spring Canyon, Vermejo Park, and Riley Canyon, N. Mex.; Cokedale, Trinidad, Dean, etc., Colo.

Lance: Western and Converse counties, Wyo.

Denver: South Table Mountain, Golden, Colo.

Palmocarpon subcylindricum Lesquereux, Rept. U. S. Geol. Surv. Terr., vol. 7 (Tert. Fl.), 1878, p. 121, pl. xi, f. 12.

Denver: Golden, Colo.

Palmocarpon truncatum Lesquereux, Rept. U. S. Geol. Surv. Terr., vol. 7 (Tert. Fl.), 1878, p. 120, pl. xi, f. 6–9.

Denver (?): Golden, Colo.

Palmocarpon n. sp. Knowlton = *Ficus ceratops.*

PALMOXYLON Schenk, Engler's bot. Jahrb., vol. 3, 1882, p. 486.

Palmoxylon anchorus Stevens, Am. Jour. Sci., 4th ser., vol. 34, 1912, p. 435, text f. 1–24.
Monmouth: Seabright, N. J.

Palmoxylon cellulosum Knowlton, U. S. Nat. Mus., Proc., vol. 11, 1888, p. 90, pl. xxx, f. 2.—Berry, Mycologia, vol. 8, 1916, p. 74.
Tertiary: Rapides Parish, La.
Oligocene: Mississippi.

Palmoxylon cliffwoodensis Berry, Am. Jour. Sci., 4th ser., vol. 41, 1916, p. 196, f. 1–4.
Magothy: Cliffwood, N. J.

Palmoxylon quenstedti Felix, Foss. Hölz. Westindiens, 1883, p. 25, pl. iv, f. 4.—Knowlton, U. S. Nat. Mus., Proc., vol. 11, 1888, p. 90, pl. iv, f. 4.
Oligocene? Rapides Parish, La.

Palmoxylon sp. Hollick, N. Y. Acad. Sci., Trans., vol. 15, 1905, p. 9.
——: Nantucket, Mass.

PANAX Linné, Sp. pl., 1753, p. 1058.

Panax andrewsii Cockerell, Am. Nat., vol. 42, 1908, p. 581, text f. 12.
Miocene: Florissant, Colo.

Panax cretacea Heer. Hollick, U. S. Geol. Surv., Mon. 50, 1906, p. 100, pl. xxxviii, f. 7.—Berry, U. S. Geol. Surv., Prof. Paper 112, 1919, p. 128, pl. xxvi, f. 4, 5.
Magothy: Gay Head, Marthas Vineyard, Mass.
Tuscaloosa: Shirleys Mill, Ala.

PARACEDROXYLON Sinnott, Rhodora, vol. 11, 1909, p. 171. [Type, *P. scituatense.*]

Paracedroxylon scituatense Sinnott, Rhodora, vol. 11, 1909, p. 171, pl. lxxx, f. 1–6; pl. lxxxi, f. 7–12.
Matawan(?): Second Cliff, Scituate, Mass.

PARACUPRESSINOXYLON Holden, Ann. Bot., London, vol. 27, 1913, p. 538.

Paracupressinoxylon cupressoides Holden, Bot. Gaz., vol. 58, 1914, p. 175, pl. xvi, f. 20–24.
Magothy: Cliffwood, N. J.

Paracupressinoxylon potomacense Sinnott and Bartlett, Am. Jour. Sci., 4th ser., vol. 41, 1916, p. 287, text f. 11–18.
Patuxent: Central High School and Meridian Hill Park, Washington, D. C.

Paracupressinoxylon sp. Holden, Bot. Gaz., vol. 58, 1914, p. 175, pl. xv, f. 25–32.
Magothy: Cliffwood, N. J.

PARROTIA C. A. Meyer, Verz. Pfl. Canc., 1831, p. 46.

Parrotia canfieldi Lesquereux, U. S. Geol. Surv., Mon. 17, 1891 [1892], p. 141, pl. xxx, f. 6.
Dakota: Kansas.

Parrotia cuneata (Newberry) Berry, U. S. Geol. Surv., Prof. Paper 91, 1916, p. 219.
Viburnum cuneatum Newberry, U. S. Nat. Mus., Proc., vol. 5, 1882 [1883], p. 511; U. S. Geol. Surv., Mon. 35, 1898, p. 130, pl. lvii, f. 2.
Fort Union: Tongue River, Mont.
Lagrange: Near Shandy, Hardeman County, Tenn.

Parrotia grandidentata Lesquereux, U. S. Geol. Surv., Mon. 17, 1891 [1892], p. 140, pl. xxix, f. 2–4.
Dakota: Pipe Creek, Cloud County, Kans.

Parrotia? winchelli Lesquereux, U. S. Geol. Surv., Mon. 17, 1891 [1892], p. 140, pl. xxix, f. 5, 6.
Dakota: Minnesota; Ellsworth County, Kans.

PARTHENOCISSUS Planchon, in De Candolle, Mon. Phan., vol. 5, pt. 2, 1887, p. 447.

Parthenocissus osborni Cockerell, Am. Mus. Nat. Hist., Bull., vol. 24, 1908, p. 103.
Miocene: Florissant, Colo.

PASANIA Miquel, Fl. Ind. Bat., vol. 1, 1855, p. 848.

Pasania densiflora (Hooker and Arnott) Oersted. Hannibal, Bull. Torr. Bot. Club, vol. 38, 1911, p. 336, pl. xv, f. 8.
Pliocene: Near Portola and in Calabasas Canyon, Santa Cruz Mountains, and near Cook, Bear Valley, Gavilan Range, Calif.

PASSIFLORA Linné, Sp. pl., 1753, p. 955.

Passiflora antiqua Newberry, U. S. Geol. Surv., Mon. 26, 1895 [1896], p. 109. pl. xxiii, f. 7.—Berry, Geol. Surv. N. J., Bull. 3, 1911, p. 188, pl. xxiii, f. 5.
Raritan: Woodbridge, N. J.

PECOPTERIS (Brongniart) Sternberg, Fl. Vorw., vol. 1 (Tentamen), 1825, p. 17.

Pecopteris angustipennis Fontaine = *Cladophlebis browniana.*

Pecopteris arctica? Heer. Knowlton, in Spurr, U. S. Geol. Surv., Eighteenth Ann. Rept., pt. 3, 1898, p. 195. [Not a *Pecopteris.*]
Eocene?: Below Melozokakat, Yukon River, Alaska.

Pecopteris borealis Brongniart, Hist. vég. foss., vol. 1, 1828, p. 351, pl. cxix, f. 1, 2.—Fontaine, in Ward, U. S. Geol. Surv., Nineteenth Ann. Rept., pt. 2, 1899, p. 655, pl. clx, f. 14, 15. [Not *Pecopteris.*]
Fuson: Oak Creek, Crook County, Wyo.

Pecopteris brevipennis Fontaine = *Cladophlebis ungeri.*

Pecopteris browniana Dunker = *Cladophlebis browniana.*

Pecopteris bullatus Bunbury = *Mertensides bullatus.*

Pecopteris carolinensis Emmons = *Asterocarpus falcatus.*

Pecopteris constricta Fontaine = *Cladophlebis browniana.*

Pecopteris cycloloba Newberry, in Ives, Rept. on Colorado River of the West, explored in 1857, 1858, pt. 3, 1861, p. 129, pl. iii, f. 3, 4.—Fontaine, in Ward, U. S. Geol. Surv., Twentieth Ann. Rept., pt. 2, 1900, p. 337. [Probably not a *Pecopteris.*]
Triassic (?): Moqui Villages, N. Mex.

?*Pecopteris denticulata* Heer. Lesquereux, U. S. Nat. Mus., Proc., vol. 11, 1888, p. 32 = *Cladophlebis vaccensis.*

Pecopteris denticulata Brongniart = *Cladophlebis denticulata.*

Pecopteris falcatus Emmons = *Asterocarpus falcatus.*

Pecopteris falcatus variabilis Emmons = *Asterocarpus falcatus.*

Pecopteris geyleriana Nathorst, Denkschr. Wien. Akad. Wiss., vol. 57, 1890, p. 48, pl. iv, f. 1; pl. vi, f. 1.—Fontaine, in Ward, U. S. Geol. Surv., Nineteenth Ann. Rept., pt. 2, 1899, p. 654, pl. clx, f. 9–13. [Not *Pecopteris.*]
Fuson: Oak Creek, Pine Creek, Rollins Tunnel, etc., Crook County, Wyo.

Pecopteris haiburnensis Lindley and Hutton = *Cladophlebis haiburnensis.*

Pecopteris (*Alethopteris*) *indica* Oldham and Morris = *Cladophlebis indica.*

Pecopteris microdonta Fontaine = *Cladophlebis browniana.*

Pecopteris montanensis Fontaine, U. S. Geol. Surv., Proc., vol. 15, 1892, p. 492, pl. lxxxiii, f. 1, 1a. [Not a true *Pecopteris.*]
Kootenai: Great Falls, Mont.

Pecopteris murrayana Brongniart = *Thyrsopteris murrayana.*

Pecopteris nebraskana? Heer = *Cyatheites nebraskana.*

Pecopteris? odontopteroides Hall = *Tapeinidium? undulatum.*

Pecopteris ovodentata Fontaine = *Cladophlebis browniana.*

Pecopteris pachyphylla Fontaine = *Cladophlebis ungeri.*

Pecopteris paucifolia Phillips = *Sagenopteris paucifolia.*

Pecopteris rarinervis Fontaine, U. S. Geol. Surv., Mon. 6, 1883, p. 48, pl. xxvi, f. 3–4a.
Triassic (Newark): Manakin and Carbon Hill, Va.

Pecopteris reticulata Stokes and Webb = *Weichselia reticulata.*

Pecopteris (Cheilanthes) sepulta Newberry, U. S. Nat. Mus., Proc., vol. 5, 1882 [1883], p. 503. [Not *Pecopteris.*]
Green River: Green River, Wyo.

Pecopteris (*Cheilanthes*) *sepulta* Newberry. Hollick = *Osmunda hollicki*.

Pecopteris socialis Fontaine = *Cladophlebis browniana*.

Pecopteris socialis Heer = *Cladophlebis socialis*.

Pecopteris strictinervis Fontaine = *Cladophlebis browniana*.

Pecopteris strictinervis Fontaine? Fontaine, in Diller and Stanton = *Gleichenia nordenskiöldi*.

Pecopteris torelli Heer = *Osmunda doroschkiana*.

Pecopteris undulata Hall = *Tapeinidium? undulatum*.

Pecopteris ungeri Dunker = *Cladophlebis ungeri*.

Pecopteris virginiensis Fontaine = *Cladophlebis browniana*.

Pecopteris zippei Corda = *Gleichenia zippei*.

Pecopteris sp. Dawson, Geol. Surv. Canada, Rept. Progress, 1872–73 [1873], App. 1, 1873, p. 70.
Upper Cretaceous: Hornby Island, British Columbia.

Pecopteris sp. Dawson, Roy. Soc. Canada, Trans., vol. 1, 1882 [1883], sec. 4, p. 25.
Upper Cretaceous: Baynes Sound, Vancouver Island, British Columbia.

Pecopteris sp.? Emmons, American geology, pt. 6, 1857, p. 104, pl. vi, f. 2 = *Laccopteris lanceolata*.

Pecopteris? sp. Taylor = *Cladophlebis constricta*.

PELLAEA Link, Fil. Hort. Berol., 1841, p. 59.

Pellaea antiquella Cockerell, Am. Mus. Nat. Hist., Bull., vol. 24, 1908, p. 77, pl. viii, f. 20.
Miocene: Florissant, Colo.

PERIPLOCA (Tournefort) Linné, Sp. pl., 1753, p. 211.

Periploca cretacea Hollick, U. S. Geol. Surv., Mon. 50, 1906, p. 105, pl. xl, f. 16.
Magothy: Gay Head, Marthas Vineyard, Mass.

PERONOSPOROIDES Berry, Mycologia, vol. 8, 1916, p. 74. [Type, *P. palmi*.]

Peronosporoides palmi Berry, Mycologia, vol. 8, 1916, p. 74, pl. clxxx, f. 1–3; pl. clxxxi, f. 9–11.
Lower Oligocene: Bayou Pierre, Miss.

PERSEA Gaertner f., Fr. and Sem., vol. 3, 1805, p. 222.

Persea borbonia (Linné) Spreng. Hollick, Bull. Torr. Bot. Club, vol. 19, 1892, p. 332.
Laurus carolinensis Michaux. Lesquereux, Am. Jour. Sci., 2d ser., vol. 27, 1859, p. 363; Geol. Tenn., 1869, p. 426, pl. K, f. 10.
Pleistocene: Somerville, Fayette County, Tenn.; Bridgeton, N. J.

Persea brossiana (Lesquereux) Lesquereux, U. S. Geol. and Geog. Surv. Terr., Ann. Rept. 1873 [1874], p, 407.

Laurus brossiana Lesquereux, Rept. U. S. Geol. Surv. Terr., vol. 7 (Tert. Fl.), 1878, p. 216, pl. xxxvi, f. 9.

Denver: Mount Bross, Middle Park, Colo.

Persea coloradica Cockerell, Am. Mus. Nat. Hist., Bull., vol. 24, 1908, p. 91, pl. vii, f. 12, 13.

Miocene: Florissant, Colo.

Persea dilleri Lesquereux, U. S. Nat. Mus., Proc., vol. 11, 1888, p. 27, pl. xiii, f. 2–4.—Knowlton, in Turner, Am. Geol., vol. 15, 1895, p. 378.

Ione ?: Volcano Hill, Placer County, Calif.

Persea hayana Lesquereux, U. S. Geol. Surv., Mon. 17, 1891 [1892], p. 103, pl. xvi, f. 6.

Dakota: Ellsworth County, Kans.

Persea lancifolia Lesquereux = *Nectandra lancifolia.*

Persea leconteana (Lesquereux) Lesquereux, Rept. U. S. Geol. Surv. Terr., vol. 6 (Cret. Fl.), 1874, p. 75, pl. xxviii, f. 1; U. S. Geol. Surv., Mon. 17, 1891 [1892], p. 104, pl. xi, f. 2.—Hollick, U. S. Geol. Surv., Mon. 50, 1906, p. 76, pl. xxxi, f. 1.

Sassafras leconteanum Lesquereux, Am. Philos. Soc., Trans., vol. 13, 1869, p. 431, pl. xxiii, f. 1.

Quercus benzoin? Lesquereux. Lesquereux, Am. Jour. Sci., 2d ser., vol. 27, 1859, p. 360.

Dakota: Ellsworth County, Kans.

Magothy: Gay Head, Marthas Vineyard, Mass.

Upper Cretaceous: Nanaimo, Vancouver Island, British Columbia.

Persea longipetiolat[um]a (Hollick) Berry, U. S. Geol. Surv., Prof. Paper 91, 1916, p. 300, pl. lxxxvi, f. 1, 2.

Toxylon longipetiolatum Hollick, La. Geol. Surv., Special Rept. 5, 1899, p. 282, pl. xxxviii.

Persea speciosa Heer. Hollick, idem, p. 284, pl. xli, f. 5.

Wilcox: Coushatta, La.; Old Port Caddo Landing, Harrison County, Tex.

Lagrange: Puryear, Tenn.

Persea nebrascensis Lesquereux = *Laurus nebrascensis.*

Persea pseudo-carolinensis Lesquereux, Mus. Comp. Zool., Mem., vol. 6, No. 2, 1878, p. 19, pl. vii, f. 1, 2.—Knowlton, in Lindgren, Jour. Geol., vol. 4, 1896, p. 890.—Knowlton, U. S. Geol. Surv., Mon. 32, p. 2, 1899, p. 725, pl. xcv, f. 4.

Magnolia californica Lesquereux, Mus. Comp. Zool., Mem., vol. 6, 1878, p, 25, pl. vi, f. 5 [not f. 6, 7].

Miocene (auriferous gravels): Table Mountain, Tuolumne County; Corral Hollow, Alameda County; and Independence Hill, Placer County, Calif.

Miocene: Yellowstone National Park.

Persea pubescens (Pursh) Sargent. Berry, Jour. Geol., vol. 15, 1907, p. 345.—Torreya, vol. 14, 1914, p. 161.
Pleistocene: Neuse River, N. C.; Monroeville, Mound County, Ala.

Persea punctulata Lesquereux, U. S. Nat. Mus., Proc., vol. 11, 1888, p. 26, pl. xiv, f. 1.
Miocene (auriferous gravels): Corral Hollow, Alameda County, Calif.

Persea schimperi Lesquereux, U. S. Geol. Surv., Mon. 17, 1891 [1892] p. 103, pl. xvi, f. 5.
Dakota: Ellsworth County, Kans.

Persea speciosa? Heer. Lesquereux, U. S. Nat. Mus., Proc., vol. 11, 1888, p. 36.
——: Selma, Tex.

Persea speciosa Heer. Hollick, 1899 = *Persea longipetiolatum.*

Persea sternbergii (Lesquereux) Lesquereux, Rept. U. S. Geol. Surv. Terr., vol. 6 (Cret. Fl.), 1874, p. 76, pl. vii, f. 1; U. S. Geol. Surv., Mon. 17, 1891 [1892], p. 104.
Ficus sternbergii Lesquereux, U. S. Geol. and Geog. Surv. Terr., Ann. Rept. 1872 [1873], p. 423.
Dakota: Near Fort Harker, Kans.

Persea valida Hollick, U. S. Geol. Surv., Mon. 50, 1906, p. 76, pl. xxix, f. 8, 9.—Berry, U. S. Geol. Surv., Prof. Paper 112, 1919, p. 119, pl. xxi, f. 2.
Magothy: Glen Cove, Long Island, N. Y.
Tuscaloosa: Cottondale, Ala.

Persea wilcoxiana Berry, U. S. Geol. Surv., Prof. Paper 91, 1916, p. 300, pl. lxxxvi, f. 3.
Wilcox: Frierson Mill, De Soto Parish, La.

PERSEOXYLON Felix, Jahrb. d. Kgl. ungar. geol. Anstalt, vol. 8, 1887, p. 153[11].

Perseoxylon aromaticum (Felix) Felix, Jahrb. Kgl. ungar. geol. Anstalt, vol. 8, 1887, p. 157; Zeitschr. Deutschen geol. Gesell., Jahr. 1896, p. 254.—Knowlton, U. S. Geol. Surv., Mon. 32, pt. 2, 1899, p. 767.
Laurinoxylon aromaticum Felix, Jahrb. Kgl. ungar. geol. Anstalt, vol. 7, 1884, p. 27, pl. i, f. 7; pl. ii, f. 7, 9.
Miocene: Yanceys, Yellowstone National Park.

Perseoxylon californicum Platen, Sitzbungsb. Naturf. Ges., Leipzig, vol. 34, 1907 [1908], p. 52.
Miocene (auriferous gravels): Nevada County, Calif.

Perseoxylon eberi Platen, Sitzungsb. Naturf. Ges., Leipzig, vol. 34, 1907 [1908], p. 135.
"Monument Creek": Bijou Basin, near Colorado Springs, Colo.

PERSICARIA (Tournefort) Linné, Sp. pl., 1753, p. 360.

Persicaria tertiaria (Small) Cockerell, Am. Mus. Nat. Hist., Bull., vol. 24, 1908, p. 90.

Polygonum tertiarium Small, Dept. Botany, Columbia College, Mem., vol. 1, 1895, p. 16, pl. A.

Miocene: Florissant, Colo.

PERSOONIA Smith, Linnaean Soc., vol. 4, 1798, p. 215.

Persoonia lesquereuxii Knowlton, in Lesquereux, U. S. Geol. Surv., Mon. 17, 1891 [1892], p. 89, pl. xx, f. 10–12.—Berry, Bull. Torr. Bot. Club, vol. 33, 1906, p. 173; Geol. Surv. N J., Bull. 3, 1911, p. 126, pl. xx, f. 6; U. S. Geol. Surv., Prof. Paper 112, 1919, p. 85, pl. xiv, f. 2.

Andromeda latifolia Newberry, U. S. Geol. Surv., Mon. 26, 1892, p. 120 [part], pl. xxxiii, f. 9.

Magothy: Cliffwood, N. J.
Raritan: South Amboy, N. J.
Dakota: Ellsworth County, Kans.
Tuscaloosa: Shirleys Mill and Glen Allen, Ala.

Persoonia lesquereuxii minor Berry, U. S. Geol. Surv., Prof. Paper 112, 1919, p. 86, pl. xiv, f. 3.

Tuscaloosa: Shirleys Mill, Ala.

Persoonia oviformis Lesquereux, Am. Jour. Sci., 2d ser., vol. 27, 1859, p. 361.

Puget: Bellingham Bay, Wash.

Persoonia spatulata Hollick, in Newberry, U. S. Geol. Surv., Mon. 26, 1895 [1896], p. 71, pl. xlii, f. 14.—Berry, Geol. Surv. N. J., Bull. 3, 1911, p. 125.

Raritan: South Amboy, N. J.

Pestalozzites minor Berry, in Cooke and Shearer, U. S. Geol. Surv., Prof. Paper 120, 1918, p. 55 [name].

Barnwell (Twiggs): Twiggs County, Ga.

Peuce pannonica Unger = *Pinites pannonicus*.

PHACA Linné, Sp. pl., 1753, p. 755.

Phaca wilamattae Cockerell, Am. Jour. Sci., 4th ser., vol. 26, 1908, p. 67, f. 1d.

Miocene: Florissant, Colo.

PHALARIS Linné, Sp. pl., 1753, p. 54.

Phalaris? geometrorum Cockerell, Torreya, vol. 13, 1913, p. 76, f. 2.

Miocene: *Florissant, Colo.

PHANEROPHLEBITES Knowlton, U. S. Geol. Surv., Prof. Paper 130, 19—, p. — [in preparation]. [Type, *P. pealei*.]

Phanerophlebites pealei Knowlton, U. S. Geol. Surv., Prof. Paper 130, 19—, p. —, pl. iii, f. 3 [in preparation].

Laramie: Lafayette, Colo.

PHASEOLITES Unger, Syn. pl. foss., 1845, p. 244.

Phaseolites crassus Knowlton, U. S. Geol. Surv., Prof. Paper 101, 1918, p. 269, pl. xlv, f. 5.
Vermejo: Rockvale, Colo.

Phaseolites elegans Hollick, U. S. Geol. Surv., Mon. 50, 1906, p. 85, pl. xxxii, f. 4.
Raritan: Brooklyn, Long Island, N. Y.

Phaseolites formus Lesquereux, U. S. Geol. Surv., Mon. 17, 1891 [1892], p. 147, pl. lv, f. 5, 6, 12.—Berry, Bull. Torr. Bot. Club, vol. 37, 1910, p. 198; U. S. Geol. Surv., Prof. Paper 112, 1919, p. 101, pl. xxii, f. 7.
Tuscaloosa: Shirleys Mill and Cottondale, Ala.
Dakota: Ellsworth County, Kans.
Black Creek: Cape Fear River, N. C.

Phaseolites juglandinus? Heer. Lesquereux, U. S. Geol. and Geog. Surv. Terr., Bull., vol. 1, 1875 [1876], p. 388; idem, Ann. Rept. 1874 [1876], p. 314. [Probably error for *Phyllites juglandinus* Heer.]
Green River?: Green River near mouth of White River, Wyo.

Phaseolites leei Knowlton, U. S. Geol. Surv., Prof. Paper 101, 1918, p. 270, pl. xlv, f. 6.
Vermejo: Rockvale, Colo.

Phaseolites manhassettensis Hollick, N. Y. Bot. Gard., Bull., vol. 3, 1904, p. 414, pl. lxxviii, f. 1, 2; U. S. Geol. Surv., Mon. 50, 1906, p. 86, pl. xxxii, f. 2, 3.—Berry, Bull. Torr. Bot. Club, vol. 36, 1909, p. 256, pl. xviii, f. 3; Geol. Surv. N. J., Bull. 3, 1911, p. 167, pl. xxii, f. 2.
Magothy: Manhasset Neck, Long Island, N. Y.
Raritan: Milltown, N. J.

Phaseolites minutus Knowlton, U. S. Geol. Surv., Prof. Paper 101, 1918, p. 270, pl. liii, f. 4.
Trinidad: Raton, N. Mex.

PHEGOPTERIS Fée, Gen. Fil., 1850–1852, p. 242.

Phegopteris grothiana Heer, Fl. foss. arct., vol. 7, 1883, p. 3, pl. xlviii, f. 12, 13.—Newberry, U. S. Geol. Surv., Mon. 26, 1895 [1896], p. 42, pl. iii, f. 4.—Berry, Geol. Surv. N. J., Bull. 3, 1911, p. 67.
Raritan: Woodbridge, N. J.

Phegopteris guyottii (Lesquereux) Cockerell, Am. Mus. Nat. Hist., Bull., vol. 24, 1908, p. 76; Pop. Sci. Mo., vol. 73, 1908, p. 120, text f.—Knowlton, U. S. Nat. Mus., Proc., vol. 51, 1916, p. 246, pl. xii, f. 1.

Sphenopteris guyottii Lesquereux, Rept. U. S. Geol. Surv. Terr., vol. 8 (Cret. and Tert. Fl.), 1883, p. 137, pl. xxi, f. 1–7.—Penhallow, Roy. Soc. Canada, Trans., 2d ser., vol. 8, 1902, p. 48; Rept. Tert. Pl. Brit. Col., 1908, p. 90.

Miocene: *Florissant, Colo.
Paskapoo (Fort Union): Red Deer River, Canada.

PHENANTHERA Hollick, Torreya, vol. 7, 1907, p. 182. [Type, *P. petalifera.*]

Phenanthera petalifera Hollick, Torreya, vol. 7, 1907, p. 182, text f. 1, 2.—Cockerell, Am. Nat., vol. 42, 1908, p. 576, text f. 7; Am. Mus. Nat. Hist., Bull., vol. 24, 1908, p. 108.
Miocene: Florissant, Colo.

PHILADELPHUS Linné, Sp. pl., 1753, p. 470.

Philadelphus palaeophilus Cockerell, Am. Mus. Nat. Hist., Bull., vol. 24, 1908, p. 92, pl. x, f. 37
Miocene: Florissant, Colo.

PHILOTRIA Rafinesque, Am. Monthly Mag., vol. 2, 1818, p. 175.

Philotria canadensis (Michaux) Britton.
Elodia canadensis Michaux. Penhallow, Geol. Soc. Am., Bull., vol. 1, 1890, p. 325; Roy. Soc. Canada, Trans., 2d ser., vol. 2, 1896, p. 70; Brit. Assoc. Adv. Sci., Bradford meeting, 1900, p. 335.
Pleistocene: Rolling River, Manitoba; Greens Creek, Ontario.

PHOENICITES Brongniart, Dict. sci. nat., vol. 57 (Prodrome), 1828, p. 123 (121).

Phoenicites occidentalis Berry, Am. Jour. Sci., 4th ser., vol. 37, 1914, p. 403, text f. 1–3.
Catahoula: Cut on International & Great Northern Railroad, Trinity County, Tex.

PHOENICOPSIS Heer, Mém. Acad. imp. sci. St.-Pétersbourg, vol. 22, No. 12 (Fl. foss. arct., vol. 4, Abt. 2), 1876, p. 49.

Phoenicopsis angustifolia Heer, Fl. foss. arct., vol. 4, Abt. 2, 1876, pp. 51, 113, pl. i, f. 1d; pl. ii, f. 3b.—Knowlton, U. S. Geol. Surv., Prof. Paper 85, 1914, pl. liii.
Jurassic: Cape Lisburne, Alaska.

Phoenicopsis speciosa Heer, Fl. foss. arct., vol. 4, Abt. 2, 1876, p. 112, pl. xix, f. 1, 2; pl. xxx, f. 1–6.—Knowlton, U. S. Geol. Surv., Prof. Paper 85, 1914, p. 52, pl. viii, f. 2–4.
Jurassic: Cape Lisburne, Alaska.

Phoenicopsis? sp. Fontaine, in Ward, U. S. Geol. Surv., Mon. 48, 1905 [1906], p. 128, pl. xxxiv, f. 13, 14.
Jurassic: Douglas County, Oreg.

PHORADENDRON Nuttall, Jour. Acad. Philadelphia, 2d ser., vol. 1, 1847–1850, p. 185.

Phoradendron flavescens (Pursh) Nuttall. Berry, Am. Jour. Sci., 4th ser., vol. 29, 1910, p. 396.
Pleistocene: Black Warrior River, 356 miles above Mobile, and Abercrombie Landing, Chattahoochee River, Ala.

PHRAGMITES Trinius, Fund. Agrost., 1820, p. 134.

Phragmites alaskana Heer, Fl. foss. alask., 1869, p. 24, pl. i, f. 12.—Lesquereux, U. S. Geol. and Geog. Surv. Terr., Ann. Rept. 1871 [1872], p. 296; Rept. U. S. Geol. Surv. Terr., vol. 7 (Tert. Fl.), 1878, p. 90, pl. viii, f. 10–12.—Ward, U. S. Geol. Surv., Sixth Ann. Rept., 1884–85 [1886], p. 550, pl. xxxii, f. 1–3; U. S. Geol. Surv., Bull. 37, 1887, p. 17, pl. iii, f. 1–3.—?Knowlton, in Spurr, U. S. Geol. Surv., Eighteenth Ann. Rept., pt. 3, 1898, p. 192; U. S. Nat. Mus., Proc., vol. 17, 1894, p. 216; Geol. Soc. Am., Bull., vol. 5, 1893, p. 579.

Phragmites oeningensis Braun. Göppert, Abh. schles. Gesell. vaterländ. Kultur, pt. 2, 1861, p. 201.

Kenai: Port Graham, Yukon River, Alaska.

Fort Union: Burns's ranch, below Glendive, Mont.

Phragmites? cliffwoodensis Berry, N. Y. Bot. Gard., Bull., vol. 3, 1903, p. 99, pl. 46, f. 5; Geol. Surv. N. J., Ann. Rept. 1905 [1906], p. 138.

Magothy: Cliffwood, N. J.

Phragmites cordaiformis Dawson, Roy. Soc. Canada, Trans., vol. 1, 1882 [1883], sec. 4, p. 26, pl. v, f. 22.

Upper Cretaceous: North Saanich and Nanaimo; Baynes Sound. Vancouver Island, British Columbia.

Phragmites cretaceus (Lesquereux) Lesquereux, Rept. U. S. Geol. Surv. Terr., vol. 6 (Cret. Fl.), 1874, p. 55, pl. i, f. 13, 14; pl. xxix, f. 7; U. S. Geol. Surv., Mon. 17, 1891 [1892], p. 37, pl. ii, f. 8.

Arundo cretaceus Lesquereux, Am. Jour. Sci., 2d ser., vol. 46, 1868, p. 92.

Equisetum nodosum Lesquereux, Rept. U. S. Geol. Surv. Terr., vol. 8 (Cret. and Tert. Fl.), 1883, p. 25.

Dakota: Near Glencoe, Kans.

Phragmites falcata Knowlton, U. S. Geol. Surv., Mon. 32, pt. 2, 1899, p. 658, pl. lxxviii, f. 5.

"Laramie": Mount Everts, Yellowstone National Park.

Phragmites laramianus Cockerell, Torreya, vol. 9, 1909, p. 141.—Knowlton, U. S. Geol. Surv., Prof. Paper 130, 19—, p. — [in preparation].

Phragmites oeningensis Al. Braun. Lesquereux, Rept. U. S. Geol. Surv. Terr., vol. 7 (Tert. Fl.), 1878, p. 88, pl. viii, f. 1.

Laramie: *Golden, Hoyt's mine, 1 mile north of Golden, and Marshall, Colo.

Phragmites? latissima Knowlton, U. S. Geol. Surv., Mon. 32, pt. 2, 1899, p. 683, pl. lxxxiii, f. 5.

Fort Union: Yellowstone National Park.

Phragmites oeningensis Al. Braun = *Phragmites laramianus.*

Phragmites prattii Berry, Bull. Torr. Bot. Club, vol. 37, 1910, pp. 191, 504; U. S. Geol. Surv., Prof. Paper 84, 1914, pp. 28, 109; Bull. Torr. Bot. Club, vol. 43, 1916, p. 287; U. S. Geol. Surv., Prof. Paper 112, 1919, p. 71.

Phragmites sp. Berry, Bull. Torr. Bot. Club, vol. 34, 1907, p. 190, pl. xi, f. 5.

Black Creek: Court House Bluff, Prospect Hall, mouth of Harrisons Creek, Cape Fear River, N. C.

Black Creek (Middendorf): Rocky Point, S. C.

Eutaw: Chattahoochee River below Columbus, Ga.; Coffee Bluff, Tenn.

Phragmites sp. Berry = *Phragmites prattii.*

Phragmites sp. Berry, Bull. Torr. Bot. Club, vol. 34, 1907, p. 190, pl. 11, f. 5.

Black Creek: Court House Landing, Cape Fear River, N. C.

Phragmites? sp. Dawson, Brit. N. A. Bound. Com. (Rept. Geol. and Res. Vic., 49th Parallel), 1875, App. A, p. 329.

——: Porcupine Creek, Alberta.

Phragmites sp.? Newberry, N. Y. Lyc. Nat. Hist., Ann., vol. 9, 1868, p. 38.—[Lesquereux], U. S. Geol. and Geog. Surv. Terr., Ill. Cret. and Tert. Pl., 1878, pl. vii, f. 5, 5a.—Newberry, U. S. Geol. Surv., Mon. 35, 1898, p. 27, pl. xxii, f. 5, 5a.

Fort Union: Fort Union, N. Dak.

Phragmites sp. Penhallow, Roy. Soc. Canada, Trans., 3d ser., vol. 1, sec. 4, 1907, p. 300.

Oligocene: Kettle River near Midway, British Columbia.

PHYLLITES Brongniart, Mém. Mus. hist. nat., Paris, vol. 8, 1822, p. 237.

Phyllites amissus Lesquereux, U. S. Geol. Surv., Mon. 17, 1891 [1892], p. 217, pl. lxii, f. 1.

Dakota: Near Fort Harker, Kans.

Phyllites aceroides Heer, Fl. foss. arct., vol. 1, 1868, p. 139, pl. xxiii, f. 5.—Penhallow, Rept. Tert. Pl. Brit. Col., 1908, p. 66.

Eocene: Mackenzie River, Northwest Territory.

Phyllites amorphus Lesquereux, Am. Jour. Sci., vol. 46, 1868, p. 102; Rept. U. S. Geol. Surv. Terr., vol. 6 (Cret. Fl.), 1874, p. 113, pl. xxii, f. 3, 4; U. S. Geol. Surv., Mon. 17, 1891 [1892], p. 219.

Dakota: Decatur, Nebr.

Phyllites arctica Knowlton, U. S. Nat. Mus., Proc., vol. 17, 1894, p. 230, pl. ix, f. 10, 11; Geol. Soc. Am., Bull., vol. 5, 1893, p. 587.

Kenai: Herendeen Bay, Alaska.

Phyllites aristolochiaeformis Lesquereux, U. S. Geol. Surv., Mon. 17, 1891 [1892], p. 217, pl. lix, f. 8.

Dakota: Ellsworth County, Kans.

Phyllites asplenioides Berry, U. S. Geol. Surv., Prof. Paper 112, 1919, p. 140, pl. xxxiii, f. 1-3.
Eutaw: Coffee Bluff, Tenn.

Phyllites aurantiacus Knowlton, U. S. Geol. Surv., Prof. Paper 101, 1918, p. 280, pl. l, f. 5.
Vermejo: Walsenburg, Colo.

Phyllites betulaefolius Lesquereux, Am. Philos. Soc., Trans., vol. 13, 1869, p. 430, pl. xxiii, f. 2-4; Rept. U. S. Geol. Surv. Terr., vol. 6 (Cret. Fl.), 1874, p. 112, pl. xxviii, f. 4-7; U. S. Geol. Surv., Mon. 17, 1891 [1892], p. 66.
Dakota: Kansas and Nebraska.

Phyllites bifurcies Knowlton U. S. Geol. Surv., Bull. 204, 1902, p. 85, pl. xvi, f. 2.
Mascall: Near Van Horn s ranch, John Day Basin, Oreg.

Phyllites carneosus Newberry, N. Y. Lyc. Nat. Hist., Ann., vol. 9, 1868, p. 75.—[Lesquereux], U. S. Geol. and Geog. Surv. Terr., Ill. Cret. and Tert. Pl., 1878, pl. xxvi, f. 1, 2.—Newberry, U. S. Geol. Surv., Mon. 35, 1898, p. 134, pl. xli, f. 1, 2.—Penhallow, Roy. Soc. Canada, Trans., 2d ser., vol. 8, 1902, p. 63; Rept. Tert. Pl. Brit. Col., 1908, p. 66.
Fort Union: Fort Union, N. Dak.
Paskapoo (Fort Union): Red Deer River, Alberta.

Phyllites castalioides Knowlton, U. S. Geol. Surv., Prof. Paper 101, 1918, p. 282.
Vermejo: Rockvale, Colo.

Phyllites celatus Lesquereux, U. S. Geol. Surv., Mon. 17, 1891 [1892], p. 215, pl. lxi, f. 1.
Dakota: Near Fort Harker, Kans.

Phyllites cercocarpifolia Berry, U. S. Geol. Surv., Prof. Paper 98, 1916, p. 68, pl. xii, f. 3, 4.
Calvert: Benning Road, D. C.

Phyllites cliffwoodensis Berry, Bull. Torr. Bot. Club, vol. 32, 1905, p. 47, pl. ii, f. 8; Geol. Surv. N. J., Ann. Rept. 1905 [1906], p. 138.
Magothy: Cliffwood, N. J.

Phyllites cockerelli Knowlton, U. S. Geol. Surv., Prof. Paper 98, 1916, p. 93, pl. xvii, f. 5, 6.
Fox Hills: Wildcat Mound, 2½ miles south of Milliken, Colo.

Phyllites coriaceus Newberry, in Ives, Rept. Colorado River of the West, pt. 3, 1861, p. 132, pl. iii, f. 7, 7a.
Triassic (?): Moqui Villages, Ariz.

Phyllites crassifolia Knowlton, U. S. Geol. Surv., Mon. 32, pt. 2, 1899, p. 753, pl. cii, f. 5; pl. ciii, f. 1.
Miocene: Yellowstone National Park.

Phyllites cupanioides Newberry = *Pterospermites cupanioides.*

Phyllites cuspidatus Rossmässler = *Quercus* cf. *Q. cuspidatus.*

Phyllites cyclophylla[us] (Lesquereux) Hollick, Bull. Torr. Bot. Club, vol. 21, 1894, p. 256, pl. ccv, f. 11.

Salvinia cyclophylla Lesquereux, U. S. Geol. and Geog. Surv. Terr., Ann. Rept. 1873 [1874], p. 408; Rept. U. S. Geol. Surv. Terr., vol. 7 (Tert. Fl.), 1878, p. 64, pl. v, f. 10, 10a.

Denver: Middle Park, Colo.

Phyllites dentata Knowlton, U. S. Geol. Surv., Prof. Paper 108, 1917, p. 94, pl. xxxiv, f. 8.

Frontier: 1 mile east of Cumberland, Wyo.

Phyllites denticulatus Knowlton, U. S. Geol. Surv., Bull. 257, p. 148, pl. xix, f. 2.

Judith River: Willow Creek, Fergus County, Mont.

Phyllites dombeyopsoides Knowlton, U. S. Geol. Surv., Prof. Paper 130, 19—, p. —, pl. xviii, f. 2 [in preparation].

Laramie: Cowan station, near Denver, Colo.

Phyllites dubius Sternberg = *Taxodium dubium.*

Phyllites durescens Lesquereux, U. S. Geol. Surv., Mon. 17, 1891 [1892], p. 218, pl. lxi, f. 5; pl. lxii, f. 3.

Dakota: Ellsworth County, Kans.

Phyllites ellipticus Newberry = *Diospyros amboyensis.*

Phyllites erosus Lesquereux, U. S. Geol. Surv., Mon. 17, 1891 [1892], p. 216, pl. lxi, f. 4.

Dakota: Near Fort Harker, Kans.

Phyllites ficifolia Knowlton, U. S. Geol. Surv., Prof. Paper 108, 1917, p. 94, pl. xxxv, f. 4b.

Frontier: 1 mile east of Cumberland, Wyo.

Phyllites flexuosus Knowlton, U. S. Geol. Surv., Eighteenth Ann. Rept., pt. 3, 1898, p. 735, pl. cii, f. 8.

Payette: Near Marsh, Boise County, Idaho.

Phyllites fraxineus Lesquereux, U. S. Nat. Mus., Proc., vol. 10, 1887, p. 46 [name].

Pleistocene: Bridgeton, N. J.

Phyllites fremonti Unger, Gen. et sp. pl. foss., 1850, p. 503.—Knowlton, U. S. Geol. Surv., Prof. Paper 108, 1917, p. 94.

Leaf of a dicotyledonous plant (?) Hall, in Frémont, Rept. Exploring Expedition to the Rocky Mountains in 1842, etc., App. B, 1845, p. 307, pl. ii, f. 4.

Green River ?: Black Fork of Green River, Wyo.

Phyllites ilicifolius Lesquereux, U. S. Geol. Surv., Mon. 17, 1891 [1892], p. 213, pl. x, f. 9.

Dakota: Ellsworth County, Kans.

Phyllites improbatus Lesquereux, Rept. U. S. Geol. Surv. Terr., vol. 7 (Tert. Fl.), 1878, p. 107, pl. xiv, f. 18.

Rhizocaulon gracile Lesquereux, U. S. Geol. and Geog. Surv. Terr., Ann. Rept. 1873 [1874], p. 396. [Homonym, Saporta, 1861.]

Post-Laramie: Black Buttes, Wyo.

Phyllites inexpectans Knowlton, U. S. Geol. Surv., Bull. 204, 1902, p. 86, pl. xvi, f. 6.

Mascall: Van Horn's ranch, John Day Basin, Oreg.

Phyllites innectens Lesquereux, U. S. Geol. Surv., Mon. 17, 1891 [1892], p. 219, pl. lxv, f. 6.

Dakota: Ellsworth County, Kans.

Phyllites intricata Knowlton, U. S. Geol. Surv., Bull. 257, p. 148, pl. xviii, f. 6.

Judith River: Willow Creek, Fergus County, Mont.

Phyllites juglandiformis Sternberg [1825] = *Hicoria juglandiformis*.

Phyllites lacoei Lesquereux, U. S. Geol. Surv., Mon. 17, 1891 [1892], p. 213, pl. xlv, f. 6.

Dakota: Kansas.

Phyllites latifolius (Fontaine) Berry, U. S. Nat. Mus., Proc., vol. 38, 1910, p. 194.

Nageiopsis latifolia Fontaine, U. S. Geol. Surv., Mon. 15, 1889 [1890], p. 198, pl. lxxxii, f. 3; in Ward, U. S. Geol. Surv., Mon. 48, 1905 [1906], p. 260, pl. lxviii, f. 13.

Patuxent: Near Dutch Gap and near Potomac Run, Va.

?Shasta: Near Riddles, Oreg.

Phyllites laurencianus Lesquereux, U. S. Geol. Surv., Mon. 17, 1891 [1892], p. 215, pl. xliv, f. 5.

Dakota: Ellsworth County, Kans.

Phyllites leydenianus Knowlton, U. S. Geol. Surv., Prof. Paper 130, 19—, p. —, pl. xviii, f. 5 [in preparation].

Laramie: Leyden Gulch, 6½ miles north of Golden, Colo.

Phyllites mahoniaeformis Heer, Neue Denkschr., vol. 21, 1865, p. 10, pl. ii, f. 9.

——: Burrard Inlet, British Columbia.

Phyllites marshallensis Knowlton, U. S. Geol. Surv., Prof. Paper 130, 19—, p. —, pl. xxviii, f. 7 [in preparation].

Laramie: Marshall, Colo.

Phyllites mimusopsoides Lesquereux, U. S. Nat. Mus., Proc., vol. 10, 1887, p. 46 [name].

Pleistocene: Bridgeton, N. J.

Phyllites nanus Knowlton, U. S. Geol. Surv., Prof. Paper 101, 1918, p. 280, pl. l, f. 3, 4.

Vermejo: Walsenburg, Colo.

Phyllites neomexicanus Knowlton, U. S. Geol. Surv., Prof. Paper 98, 1916, p. 343, pl. xci, f. 2.

Fruitland: 30 miles south of Farmington, San Juan County, N. Mex.

Phyllites obcordatus Heer = *Liriodendron marcouanum.*

Phyllites obscurus Hollick, in Newberry, U. S. Geol. Surv., Mon. 26, 1895 [1896], p. 131, pl. xlii, f. 33.

Raritan: New Jersey [no definite locality].

Phyllites obscurus Knowlton = *Phyllites payettensis.*

Phyllites obtusilobatus Heer = *Liquidambar obtusilobatus.*

Phyllites orbicularis Newberry = *Populus orbicularis.*

Phyllites oregonianus Knowlton, U. S. Geol. Surv., Bull. 204, 1902, p. 85, pl. xvi, f. 1.

Phyllites n. sp. Knowlton, in Merriam, Univ. Calif., Bull. Dept. Geol., vol. 2, 1901, p. 303.

Clarno (upper part): Near Lone Rock, Gilliam County, Oreg.

Phyllites pachyphyllus Fontaine = *Proteaephyllum reniforme.*

Phyllites payettensis Knowlton, U. S. Geol. Surv., Bull. 152, 1898, p. 163.

Phyllites obscurus Knowlton, U. S. Geol. Surv., Eighteenth Ann. Rept., pt. 3, 1898, p. 735, pl. xcix, f. 10, 11. [Homonym, Hollick, 1896.]

Payette: Near Marsh, Boise County, Idaho.

Phyllites perplexus Lesquereux, U. S. Geol. Surv., Mon. 17, 1891 [1892], p. 215, pl. xxxviii, f. 15.

Dakota: Ellsworth County, Kans.

Phyllites personatus Knowlton, U. S. Geol. Surv., Bull. 204, 1902, p. 87, pl. xvi, f. 4.

Clarno (lower part): Cherry Creek, Crook County, Oreg.

Phyllites petiolatus Knowlton, U. S. Geol. Surv., Prof. Paper 98, 1916, p. 342, pl. xci, f. 3.

Fruitland: 30 miles south of Farmington, San Juan County, N. Mex.

Phyllites pistiaeformis Berry, U. S. Geol. Surv., Prof. Paper 112, 1919, p. 140, pl. xxix, f. 2.

Tuscaloosa: Glen Allen, Ala.; Iuka, Miss.

Phyllites poinsettioides Hollick, N. Y. Acad. Sci., Trans., vol. 12, 1892, p. 37 [10], pl. i, f. 10; U. S. Geol. Surv., Mon. 50. 1906, p. 106, pl. xxxiii, f. 1.

Raritan: Kreischerville, Staten Island, N. Y.

Phyllites populoides Knowlton, U. S. Geol. Surv., Prof. Paper 101, 1918, p. 280, pl. l, f. 1, 2.

Vermejo: Tercio, Colo.

Phyllites protophylloides Knowlton, U. S. Geol. Surv., Prof. Paper 101, 1918, p. 281, pl. l, f. 6.
Vermejo: Rockvale, Colo.

Phyllites ratonensis Knowlton, U. S. Geol. Surv., Prof. Paper 101, 1918, p. 283, pl. liii, f. 1, 2.
Trinidad: Trinidad, Colo.

Phyllites retusoides Knowlton, U. S. Geol. Surv., Prof. Paper 101, 1918, p. 349, pl. cii, f. 5.
Raton: Near Wootton, Colo.

Phyllites rhoifolius Lesquereux, Am. Jour. Sci., vol. 46, 1868, p. 101; Rept. U. S. Geol. Surv. Terr., vol. 6 (Cret. Fl.), 1874, p. 111, pl. xxii, f. 5, 6; U. S. Geol. Surv., Mon. 17, 1891 [1892], p. 219.
Dakota: Lancaster County, Nebr.

Phyllites rhomboideus (Lesquereux) Lesquereux, Rept. U. S. Geol. Surv. Terr., vol. 6 (Cret. Fl.), 1874, p. 112, pl. vi, f. 7; U. S. Geol. Surv., Mon. 17, 1892, p. 119.
Ficus? rhomboideus Lesquereux, Am. Jour. Sci., 2d ser., vol. 46, 1868, p. 96.
Dakota: Decatur, Nebr.

Phyllites rosaefolius Knowlton, U. S. Geol. Surv., Prof. Paper 101, 1918, p. 281, pl. xlix, f. 7.
Vermejo: Vermejo Park, N. Mex.

Phyllites sapindiformis Lesquereux, Rept. U. S. Geol. Surv. Terr., vol. 7 (Tert. Fl.), 1878, p. 301, pl. xxix, f. 6, 7.
Green River?: Green River?, Wyo.

Phyllites sapindus Knowlton, U. S. Geol. Surv., Prof. Paper 101, 1918, p. 281, pl. l, f. 7.
Vermejo: Vermejo Park, N. Mex.

Phyllites saundersi Knowlton, Harriman Alaska Exped., vol. 4, 1904, p. 157, pl. xxv, f. 2; pl. xxxiii, f. 2.
Kenai: Kukak Bay, Alaska.

Phyllites shirleyensis Berry, U. S. Geol. Surv., Prof. Paper 112, 1919, p. 139, pl. xxix, f. 3.
Tuscaloosa: Shirleys Mill, Ala.

Phyllites snowii Lesquereux, U. S. Geol. Surv., Mon. 17, 1891 [1892], p. 214, pl. xxxviii, f. 2.
Dakota: Ellsworth County, Kans.

Phyllites stipulaeformis Lesquereux, U. S. Geol. Surv., Mon. 17, 1891 [1892], p. 216, pl. lxi, f. 2.
Dakota: Kansas.

Phyllites sulcatus Lesquereux, Am. Jour. Sci., 2d ser., vol. 45, 1868, p. 206. [Not afterward employed by Lesquereux.]
Laramie?: Marshall, Colo.

Phyllites trapaformis Berry, Geol. Surv. N. J., Bull. 3, 1911, p. 227, pl. xx, f. 1, 2.
Raritan: South Amboy, N. J.

Phyllites trinervis Knowlton, U. S. Geol. Surv., Prof. Paper 130, 19—, p. —, pl. xxviii, f. 12 [in preparation].
Laramie: Coal Creek, Boulder County, Colo.

Phyllites truncatus Lesquereux = *Mimusops truncatus.*

Phyllites umbonatus Lesquereux, Am. Jour. Sci., 2d ser., vol. 46, 1868, p. 102; Rept. U. S. Geol. Surv. Terr., vol. 6 (Cret. Fl.), 1874, p. 113, pl. xix, f. 4; U. S. Geol. Surv., Mon. 17, 1891 [1892], p. 219.
Dakota: Beatrice, Nebr.

Phyllites undulatus Newberry, U. S. Geol. Surv., Mon. 26, 1895 [1896], p. 131, pl. xxiv, f. 10.—Berry, Geol. Surv. N. J., Bull. 3, 1911, p. 228.
Raritan: Woodbridge, N. J.

Phyllites unitus Knowlton and Cockerell, n. name.
Phyllites venosus Newberry, N. Y. Lyc. Nat. Hist., Ann., vol. 9, 1868, p. 75.—[Lesquereux], U. S. Geol. and Geog. Surv. Terr., Ill. Cret. and Tert. Pl., 1878, pl. xxiv, f. 4.—Newberry, U. S. Geol. Surv., Mon. 35, 1898, p. 136, pl. xxx, f. 4.—Penhallow, Rept. Tert. Pl. Brit. Col., 1908, p. 67. [Homonym, Roemässler, 1840.]
Fort Union: Fort Union, N. Dak.
Paskapoo (Fort Union): Porcupine Creek, Saskatchewan.

Phyllites vanonae Heer, Phyll. crét. Nebr., 1866, p. 22, pl. i, f. 8.—Lesquereux, Rept. U. S. Geol. Surv. Terr., vol. 6 (Cret. Fl.), 1874, p. 113, pl. xx, f. 7; pl. xxviii, f. 8; U. S. Geol. Surv., Mon. 17, 1891 [1892], p. 214, pl. xx, f. 9; pl. xlii, f. 5.—Newberry, U. S. Geol. Surv., Mon. 35, 1898, p. 136, pl. iii, f. 8.
Diospyros primaeva Newberry. [Lesquereux], Ill. Cret. and Tert. Pl., 1878, pl. iii, f. 8.
Dakota: Nebraska and Kansas.

Phyllites venosissimus Newberry, in Ives, Rept. on Colorado River of the West, explored in 1857 and 1858, pt. 3, 1861, p. 131, pl. iii, f. 6.—Fontaine, in Ward, U. S. Geol. Surv., Twentieth Ann. Rept., pt. 2, 1900, p. 337.
Triassic(?): Moqui Villages, Ariz.

Phyllites venosus Newberry = *Phyllites unitus.*

Phyllites venosus Newberry [Lesquereux], Ill. Cret. and Tert. Pl., 1878, pl. xxvi, f. 3, 4 = *Pterospermites cupanioides.*

Phyllites vermejoensis Knowlton, U. S. Geol. Surv., Prof. Paper 101, 1918, p. 282, pl. lii, f. 1.
Vermejo: Vermejo Park, N. Mex.

Phyllites walsenburgensis Knowlton, U. S. Geol. Surv., Prof. Paper 101, 1918, p. 282, pl. li.
Vermejo: Walsenburg, Colo.

Phyllites wascoensis Lesquereux, U. S. Nat. Mus., Proc., vol. 11, 1888, p. 22, pl. xv, f. 3.—Knowlton, U. S. Geol. Surv., Bull. 204, 1902, p. 84.
Clarno (lower part): Cherry Creek, Crook County, Oreg.

Phyllites zamiaeformis Lesquereux, U. S. Geol. Surv., Mon. 17, 1891 [1892], p. 28, pl. ii, f. 7.
Dakota: Near Delphos, Kans.

Phyllites sp. Berry, U. S. Geol. Surv., Prof. Paper 91, 1916, p. 353, pl. civ, f. 2.
Lagrange: Puryear, Tenn.

Phyllites sp. Dawson, Roy. Soc. Canada, Trans., vol. 4, 1886 [1887], sec. 4, p. 32, pl. ii, f. 20.—Penhallow, Rept. Tert. Pl. Brit. Col., 1908, p. 66.
Paskapoo (Fort Union): Great Valley and Porcupine Creek, Saskatchewan.

Phyllites sp. Dawson, Roy. Soc. Canada, Trans., vol. 11, 1893 [1894], sec. 4, p. 64.
Upper Cretaceous: Vancouver Island, British Columbia.

Phyllites sp. Hollick, Md. Geol. Surv., Miocene, 1904, p. 486.
Calvert: Good Hope Hill, D. C.

Phyllites sp. Knowlton, Wash. Geol. Surv., Ann. Rept., vol. 1, 1902, p. 33.
Eocene: Coal Creek near Keese, Whatcom County, Wash.

Phyllites n. sp. Knowlton, in Merriam = *Phyllites oregonianus*.

Phyllites n. sp. Knowlton, in Calvert, U. S. Geol. Surv., Bull. 471, 1912, p. 16.
Lance: Near Glendive, Mont.

Phyllites 3 sp. Knowlton, Washington Acad. Sci., Proc., vol. 11, 1909, p. 207.
Lance: Converse County, Wyo.

Phyllites sp. Knowlton, U. S. Geol. Surv., Bull. 204, 1902, p. 85, pl. xvii.
Clarno (lower part): Cherry Creek, Crook County, Oreg.

Phyllites sp. Knowlton, U. S. Geol. Surv., Prof. Paper 130, 19—, p. —, pl. xix, f. 1 [in preparation].
Laramie: Moffat Railroad, 8 miles north of Golden, Colo.

Phyllites sp. Knowlton, U. S. Geol. Surv., Prof. Paper 130, 19—, p. —, pl. xvi, f. 5 [in preparation].
Laramie: Marshall, Colo.

Phyllites sp. Knowlton, U. S. Geol. Surv., Prof. Paper 130, 19—, p. —, pl. viii, f. 5 [in preparation].
Laramie: Cowan station, near Denver, Colo.

Phyllites sp. Knowlton, U. S. Geol. Surv., Prof. Paper 108, 1917, p. 94, pl. xxxiv, f. 4b.
Frontier: About 1½ miles south of Cumberland, Wyo.

Phyllites sp. Lesquereux, U. S. Geol. Surv., Mon. 17, 1891 [1892], p. 216, pl. lix, f. 7.
Dakota: Kansas.

PHYLLOCLADOPSIS Fontaine, U. S. Geol. Surv., Mon. 15, 1889 [1890], p. 204. [Type, *P. heterophylla*.]

Phyllocladopsis heterophylla Fontaine, U. S. Geol. Surv., Mon. 15, 1889 [1890], p. 204, pl. lxxxiv, f. 5; pl. clxvii, f. 4.
Patuxent: Near Dutch Gap Canal, Va.

Phyllocladus lesquereuxiana Heer = *Protophyllocladus subintegrifolius.*

Phyllocladus subintegrifolius Lesquereux = *Protophyllocladus subintegrifolius.*

Physagenia sp. Lesquereux = *Equisetum* sp.

Physagenia parlatorii Heer = *Equisetum parlatorii.*

PHYSEMATOPITYS Göppert, Natuurk. Verh. Holl. Maatsch. Wetensch. Leiden, vol. 6 (Mon. Foss. Conif.), 1850, p. 242.

Physematopitys goepperti Platen, Sitzungsb. Naturf. Ges., Leipzig, vol. 34, 1907 [1908], p. 143.
Miocene: Milam County, Tex.

Phytolithus striaticulmis Martin = *Equisetum rogersii.*

PICEA Link, Abhand. Akad. Wiss. Berlin, 1827–1830 [1827], p. 179.

Picea alba Link = *Picea canadensis.*

Picea albertensis Penhallow, Ottawa Nat., vol. 22, 1908, p. 82.
Edmonton: Red Deer River, 100 miles west of Gleichen, Alberta.

Picea canadensis (Miller) Britton, Sterns, and Poggenberg. Knowlton, U. S. Geol. Surv., Bull. 152, 1898, p. 164.—McBride, Davenport Acad. Sci., Proc., vol. 10, 1907, p. 161, pls. x, xi.

Picea alba Link. Penhallow, Geol. Soc. Am., Bull., vol. 1, 1890, p. 333; Roy. Soc. Canada, Trans., 2d ser., vol. 2, 1896, p. 72; Brit. Assoc. Adv. Sci., Bradford meeting, 1900, p. 336; Man. N. A. Gym., 1907, p. 285.

Pleistocene: Bloomington, Ill.; Scarborough, Ontario; What Cheer, Keokuk County, Iowa.

Picea cliffwoodensis Berry, Bull. Torr. Bot. Club, vol. 33, 1906, p. 169; Geol. Surv. N. J., Ann. Rept. 1905 [1906], pp. 138, 139, 143, pl. xix, f. 1.
Magothy: Cliffwood Brick Co.'s pit, Whale River, N. J.

Picea columbiensis Penhallow, Roy. Soc. Canada, Trans., 3d ser., vol. 1, sec. 4, 1907, p. 290, pl. 1; Rept. Tert. Pl. Brit. Col., 1908, p. 67.

Oligocene: Kettle River, 6 miles above Midway, British Columbia.

Picea harrimani Knowlton, Harriman Alaska Exped., vol. 4, 1904, p. 150, pl. xxii, f. 3, 4.

Kenai: Kukak Bay, Alaska.

Picea mariana (Miller) Britton, Sterns, and Poggenberg. Knowlton, U. S. Geol. Surv., Bull. 152, 1898, p. 164.—McBride, Davenport Acad. Sci., Proc., vol. 10, 1907, p. 161, pl. x.

Picea nigra Link or Aiton. Penhallow, in Dawson, Canadian Rev. Sci., vol. 5, 1895, p. 353.—Penhallow, Roy. Soc. Canada, Trans., 2d ser., vol. 2, 1896, p. 72; idem, vol. 10, 1904, pp. 74, 75; Brit. Assoc. Adv. Sci., Bradford meeting, 1900, p. 336; Am. Nat., vol. 41, 1907, p. 447; Man. N. A. Gym., 1907, p. 289.

Pleistocene: Don Valley, Toronto, Canada; Hamilton, Ontario; Leda clays, Montreal, Canada; Moose and Missinaibi rivers, Manitoba; beneath Kansas drift, Washington County, Iowa.

Picea nigra Link or Aiton = *Picea mariana*.

Picea quilchensis Penhallow, Rept. Tert. Pl. Brit. Col., 1908, p. 67, text f. 18.

Miocene: Quilchena, British Columbia.

Picea sitchensis (Bongard) Trautvetter and Mayer. Knowlton, in Herrick, Nat. Geog. Mag., vol. 4, 1894, pp. 75–78, f. 4, 5.—Knowlton, Jour. Geol., vol. 3, 1895, p. 529, text f. 1; U. S. Nat. Mus., Proc., vol. 17, 1894, p. 215; Geol. Soc. Am., Bull., vol. 5, 1893–94, p. 579.—Penhallow, in Coleman, Geol. Soc. Am., Bull., vol. 3, 1895, p. 635.

Recent: Muir Glacier, Alaska.

Picea tranquillensis Penhallow, Rept. Tert. Pl. Brit. Col., 1908, p. 67, text f. 17.

Miocene: Tranquille River, British Columbia.

Picea sp. (branches) Knowlton, Harriman Alaska Exped., vol. 4, 1904, p. 151, pl. xxiv, f. 3; pl. xxv, f. 3, 4.

Kenai: Kukak Bay, Alaska.

Picea? sp. (seed) Knowlton, Harriman Alaska Exped., vol. 4, 1904, p. 151, pl. xxxiii, f. 1.

Kenai: Kukak Bay, Alaska.

Picea sp.? (living) Penhallow, Brit. Assoc. Adv. Sci., Bradford meeting, 1900, p. 336.

Pleistocene: Don Valley, Toronto, Canada.

PIERIS D. Don, Edinburgh New Philos. Jour., vol. 17, 1834, p. 159.

Pieris scorbiculata Hollick, Md. Geol. Surv., Miocene, 1904, p. 486, f. 1g.—Berry, U. S. Geol. Surv., Prof. Paper 98, 1916, p. 70, pl. xii, f. 13.
Calvert: Good Hope Hill, D. C.

PIMELIA Gaertner, Fruct., vol. 1, 1788, p. 186.

Pimelia delicatula Lesquereux, Rept. U. S. Geol. Surv. Terr., vol. 8, (Cret. and Tert. Fl.), 1883, p. 168, pl. xxxiii, f. 15, 16.
Miocene: Florissant, Colo.

PINITES Wilham, Int. Struct. Foss. Veg., 1833, p. 72.

Pinites leei Fontaine = *Cedrus leei.*

Pinites palaeostrobus Ettingshausen = *Pinus palaeostrobus.*

Pinites pannonicus (Unger) Göppert, Abhandl. Schles. Gesell., 1861, p. 203.—Heer, Fl. foss. arct., vol. 2, Abt. 2, 1869, p. 23.—Knowlton, U. S. Nat. Mus., Proc., vol. 17, 1894, p. 215; Geol. Soc. Am., Bull., vol. 5, 1893, p. 579.
Peuce pannonica Unger, Syn. pl. foss., 1845, p. 208.
Kenai ?: Mega Island, Alaska.

Pinoxylon Knowlton, in Ward, U. S. Geol. Surv., Twentieth Ann. Rept., pt. 2, 1900, p. 420 [type, *P. dacotense*] = *Pityoxylon.*

Pinoxylon dacotense Knowlton = *Pityoxylon dacotense.*

PINUS (Tournefort) Linné, Sp. pl., 1753, p. 1000.

Pinus andraei Coemans, Mém. Acad. roy. Belgique, vol. 36, 1867, p. 12, pl. iv, f. 4; pl. v, f. 1.—Berry, Bull. Torr. Bot. Club, vol. 32, 1905, p. 45, pl. i, f. 4–6; Geol. Surv. N. J., Ann. Rept. 1905 [1906], p. 138.
Magothy: Cliffwood, N. J. ?

Pinus anthraciticus Dawson, Roy. Soc. Canada, Trans., vol. 10, 1892 [1893], sec. 4, p. 89, text f. 10.
Kootenai: Anthracite, British Columbia.

Pinus armstrongi Heer, Fl. foss. arct., vol. 1, 1868, p. 134, pl. xx, f. 19.
——: Banks Land.

Pinus attenuata Lemmon. Knowlton, Washington Acad. Sci., Jour., vol. 6, 1916, p. 85.
Pleistocene: Rancho La Brea, Los Angeles, Calif.

Pinus bathursti Heer, Fl. foss. arct., vol. 1, 1868, p. 134, pl. xx, f. 14.
——: Bathurst Land.

Pinus caribaea Morelet. Berry, Jour. Geol., vol. 25, 1917, p. 662.
Pleistocene: Vero, Fla.

Pinus columbiana Penhallow, Roy. Soc. Canada, Trans., 3d ser., vol. 1, sec. 4, 1907, p. 294, pl. iii, f. 1, 2; pl. iv, f. 1; Rept. Tert. Pl. Brit. Col., 1908, p. 68.
Oligocene: Kettle River, 6 miles above Midway, British Columbia.

Pinus conoides Perkins, Rept. Vt. State Geol., 1903–4 [1904], p. 209, pl. lxxx, f. 150.
Miocene: Brandon, Vt.

Pinus cuneatus Perkins, Rept. Vt. State Geol., 1903–4 [1904], p. 209, pl. lxxx, f. 142, 143.
Miocene: Brandon, Vt.

Pinus delicatulus Berry, Bull. Torr. Bot. Club, vol. 31, 1904, p. 68, pl. i, f. 12; Geol. Surv. N. J., Ann. Rept. 1905 [1906], p. 138.
Magothy: Cliffwood, N. J.

Pinus echinata Miller. Hollick, Md. Geol. Surv., Pliocene and Pleistocene, 1906, p. 217, pl. lxvii, f. 1.—Berry, Am. Jour. Sci., 4th ser., vol. 29, 1910, p. 392.
Pleistocene (Talbot): Bodkin Point, Md.
Pleistocene: Black Warrior River, 342 miles above Mobile, Ala.

Pinus florissanti Lesquereux, Rept. U. S. Geol. Surv. Terr., vol. 8 (Cret. and Tert. Fl.), 1883, p. 138, pl. xxi, f. 13.—Cockerell, Univ. Colo. Studies, vol. 3, 1906, p. 175; Am. Jour. Sci., 4th ser., vol. 26, 1908, p. 538.—Knowlton, U. S. Nat. Mus., Proc., vol. 51, 1916, p. 247.
Miocene: Florissant, Colo.

Pinus gracilistrobus Knowlton, U. S. Geol. Surv., Mon. 32, pt. 2. 1899, p. 676, pl. lxxx, f. 12.
Miocene: Yellowstone National Park.

Pinus granulatum (Heer) Stopes, Ann. Bot., vol. 25, 1911, p. 905, f. 1, 2.
Ophioglossum granulatum Heer, Fl. foss. arct., vol. 7, 1883, p. 8, pl. lvii, f. 8, 9.—Newberry, U. S. Geol. Surv., Mon. 26, 1895 [1896], p. 43, pl. ix, f. 11–13.—Berry, Geol. Surv. N. J., Bull. 3, 1911, p. 72.
Raritan: New Jersey [no exact locality given].

Pinus hambachi Kirchner, St. Louis Acad. Sci., Trans., vol. 8, 1898; p. 179, pl. xiii, f. 3.—Cockerell, Am. Jour. Sci., 4th ser., vol. 26, 1908, p. 538.
Miocene: Florissant, Colo.

Pinus iddingsi Knowlton, U. S. Geol. Surv., Mon. 32, pt. 2, 1899, p. 680, pl. lxxxii, f. 8, 9.
Fort Union: Yellowstone National Park.

Pinus insignis Loudon = ***Pinus radiata***.

Pinus lardyana Heer, Fl. tert. Helv., vol. 1, 1855, p. 58, pl. xx, f. 5.—Penhallow, Rept. Tert. Pl. Brit. Col., 1908, p. 68, text f. 19.
Oligocene?: Quilchena River, British Columbia.

Pinus lindgrenii Knowlton, Torreya, vol. 1, 1901, p. 114, f. 1–3.
Pliocene: Bernards Ferry, Idaho.

Pinus macclurii Heer, Fl. foss. arct., vol. 1, 1868, p 134, pl. xx, f. 16–18; pl. xxxv, f. 1; pl. xxxvi, f. 1–5.—Cramer, in Heer, Fl. foss. arct., vol. 1, 1868, p. 170, pl. xxxv, f. 1; pl. xxxvi, f. 1–5.—?Knowlton, in Spurr, U. S. Geol. Surv., Eighteenth Ann. Rept., pt. 3, 1898, p. 195.

Pliocene?: Banks Land; 35 miles below? Tanana, Yukon River, Alaska.

Pinus macrolepis Knowlton, U. S. Geol. Surv., Mon. 32, pt. 2, 1899, p. 679, pl. lxxx, f. 11.

Miocene: Yellowstone National Park.

Pinus mattewanensis Berry, Am. Nat., vol. 37, 1903, p. 680, f. 4, Geol. Surv. N. J., Ann. Rept. 1905 [1906], p. 138.

Magothy: Cliffwood, N. J.

Pinus nordenskiöldi Heer, Fl. foss. arct., vol. 4, Abt. 1, 1876, p. 45, pl. ix, f. 1–6.—Fontaine, in Ward, U. S. Geol. Surv., Twentieth Ann. Rept., pt. 2, 1900, p. 362, pl. lxv, f. 3; U. S. Geol. Surv., Mon. 48, 1906, p. 131, pl. xxxv, f. 10–17.

Jurassic: Oroville, Calif.; Douglas County, Oreg.

Pinus (Cyclopitus) nordenskiöldi Heer. Dawson, Roy. Soc. Canada, Trans., vol. 10, p. 88, text f. 9 = *Oleandra graminaefolia?*

Pinus palaeostrobus? (Ettingshausen) Heer. Lesquereux, Rept. U. S. Geol. Surv. Terr., vol. 7 (Tert. Fl.), 1878, p. 83, pl. vii, f. 25, 31 = *Pinus wheeleri*.

Pinus polaris Heer. Lesquereux = *Pinus wheeleri*.

Pinus premurrayana Knowlton, U. S. Geol. Surv., Mon. 32, pt. 2, 1899, p. 677, pl. lxxxii, f. 5.

Miocene: Yellowstone National Park.

Pinus protoscleropitys Holden, Am. Acad. Arts and Sci., Proc., vol. 48, 1913, p. 609, pl. i, f. a–f; pl. ii, f. a–f; pl. iv, f. c, d.

Magothy: Cliffwood, N. J.

Pinus quenstedti Heer, Neue Denkschr. Schw. Ges., vol. 23, 1869, p. 13, pl. ii, f. 5, 9; pl. iii, f. 1, 2.—Lesquereux, U. S. Geol. and Geog. Surv. Terr., Bull., vol. 1, 1875 [1876], p. 392; Rept. U. S. Geol. Surv. Terr., vol. 8 (Cret. and Tert. Fl.), 1883, p. 33, pl. i, f. 3, 4; U. S. Geol. Surv., Mon. 17, 1891 [1892], p. 32.

Dakota: Near Fort Harker and Clay Center, Kans.

?Mesaverde: Near Harper station, Wyo.

Pinus quinquefolia Hollick and Jeffrey, N. Y. Bot. Gard., Mem., vol. 3, 1909, p. 16, pl. xxii, f. 2.

Raritan: Kreischerville, Staten Island, N. Y.

Pinus radiata Don.

Pinus insignis Loudon. Green, in Lawson, Univ. Calif., Bull. Dept. Geol., vol. 1, 1893, p. 146.

Pliocene (Merced): Mussel Rock, south of San Francisco, Calif.

Pinus raritanensis Berry, Bull. Torr. Bot. Club, vol. 37, 1910, p. 189, U. S. Geol. Surv., Prof. Paper 84, 1914, p. 22; Geol. Surv. N. J., Bull. 3, 1911, p. 92; U. S. Geol. Surv., Prof. Paper 112, 1919, p. 70.

Pinus sp. Newberry, U. S. Geol. Surv., Mon. 26, 1896, p. 47, pl. ix, f. 5–8.

Raritan: South Amboy, N. J.
Tuscaloosa: Cottondale, Ala.
Black Creek: Court House Bluff, Cape Fear River, N. C.
Black Creek (Middendorf): Middendorf, S. C.

Pinus rigida Miller. Penhallow, Roy. Soc. Canada, Trans., 2d ser., vol. 10, 1904, sec. 4, p. 69.—Berry, Jour. Geol., vol. 15, 1907, p. 339.—Meehan, in Mercer, Acad. Nat. Sci. Philadelphia, Jour., 2d ser., vol. 11, 1899, pp. 277, 281.

Pleistocene: Ithaca, N. Y.; Neuse River, N. C.; Port Kennedy, Pa.

Pinus schista Ward = *Pinus vernonensis.*

Pinus scituatensiformis Bailey, Ann. Bot., London, vol. 25, 1911, p. 323, pl. xxvi, f. 1–4, 6, 8–9.

Matawan (?): Scituate, Mass.

Pinus shastensis Fontaine, in Ward, U. S. Geol. Surv., Mon. 48, 1905 [1906], p. 262, pl. lxix, f. 1–3.

Horsetown: Near Horsetown, Shasta County, Calif.

?*Pinus staratschini* Heer. Lesquereux, 1888 = *Cladophlebis alata.*

Pinus steenstrupiana Heer. Penhallow, Rept. Tert. Pl. Brit. Col., 1908, p. 68, f. 20–23.

Oligocene or Upper Eocene: Quilchena, British Columbia.

Pinus strobus Linné. Penhallow, Brit. Assoc. Adv. Sci., Bradford meeting, 1900, p. 336.—Hollick, Md. Geol. Surv., Pliocene and Pleistocene, 1906, p. 218, pl. lxvii, f. 2.—Penhallow, Man. N. A. Gym., 1907, p. 315.

Pleistocene: Don Valley, Toronto, Canada.
Pleistocene (Talbot): Bodkin Point, Md.

Pinus sturgisi Cockerell, Am. Jour. Sci., 4th ser., vol. 26, 1908, p. 538, f. 2.

Miocene: Florissant, Colo.

Pinus susquaensis Dawson. Roy. Soc. Canada, Trans., vol. 1, 1882 [1883], sec. 4, p. 23, pl. iii, f. 36.—Fontaine, in Ward, U. S. Geol. Surv., Nineteenth Ann. Rept., pt. 2, 1899, p. 670, pl. clxiii, f. 11a, 12, 13.

Kootenai: Susqua River, Martin Creek, Coal Creek, and Crows Nest Pass, British Columbia.
Lakota: Crook County, Wyo.
Fuson: Pine and Oak creeks, Crook County, Wyo.

Pinus taeda Linné. Berry, Am. Jour. Sci., 4th ser., vol. 29, 1910, p. 391; Torreya, vol. 10, 1910, p. 263; Jour. Geol., vol. 25, 1917, p. 662.

Pleistocene: Steels Bluff, Black Warrior River, and Abercrombies Landing, Chattahoochee River, Ala.; near Long Branch, N. J.; Vero, Fla.

Pinus tetraphylla Jeffrey, Ann. Bot., vol. 22, 1908, p. 220, pl. xiv, f. 17.—Hollick and Jeffrey, N. Y. Bot. Gard., Mem., vol. 3, 1909, p. 15, pl. xxii, f. 4.

Raritan: Kreischerville, Staten Island, N. Y.

Pinus triphylla Hollick and Jeffrey, N. Y. Bot. Gard., Mem., vol. 3, 1909, p. 14, pl. iii, f. 6, 7[?]; pl. xxii, f. 1.

Pinus sp.? Hollick, N. Y. Acad. Sci., Trans., vol. 12, 1892, p. 31, pl. i, f. 20, 22.

Leaf of *Pinus* sp. Jeffrey, Ann. Bot., vol. 22, 1908, pp. 215, 220, pl. xiv, f. 22, 24.

Raritan: Kreischerville, Staten Island, N. Y.

Pinus trunculus Dawson, Roy. Soc. Canada, Trans., vol. 8, sec. 4, 1888, p. 78, text f. 5.—Penhallow, Rept. Tert. Pl. Brit. Col., 1908, p. 69, text f. 24–28.

Oligocene: Stump Lake; Quilchena; Tulameen and Horsefly rivers, British Columbia.

Miocene: Tranquille River, British Columbia.

Pinus tulameenensis Penhallow, Rept. Tert. Pl. Brit. Col., 1908, p. 73, text f. 34.

Oligocene: Tulameen River, British Columbia.

Pinus vernonensis Ward, in Fontaine, U. S. Geol. Surv., Mon. 48, 1905 [1906], p. 497, pl. cix, f. 4–6.—Berry, U. S. Nat. Mus., Proc., vol. 40, 1911, p. 316; Md. Geol. Surv., Lower Cret., 1911, p. 401, pl. lxvi.

Pinus schista Ward, U. S. Geol. Surv., Mon. 48, 1905 [1906], p. 521, pl. cxii, f. 13–15.

Araucarites virginicus Fontaine. Fontaine, in Ward, U. S. Geol. Surv., Mon. 48, 1905 [1906], p. 572, pl. cxix, f. 8.

Seed of *Pinus?* sp. Fontaine, U. S. Geol. Surv., Mon. 15, 1889 [1890], p. 272, pl. clxx, f. 4.

Patapsco: Mount Vernon and Widewater, Va.; Federal Hill (Baltimore), Fort Foote, Muddy Creek, near Wellhams, Md.

Pinus wardii Knowlton, U. S. Geol. Surv., Mon. 32, pt. 2, 1899, p. 679.

Miocene: Yellowstone National Park.

Pinus wheeleri Cockerell, Am. Mus. Nat. Hist., Bull., vol. 24, 1908, p. 78, pl. vi, f. 5, 11; Am. Jour. Sci., 4th ser., vol. 26, 1908, p. 538.—Knowlton, U. S. Nat. Mus., Proc., vol. 51, 1916, p. 248, pl. xii, f. 3.

Pinus palaeostrobus (Ettingshausen) Heer. Lesquereux, Rept. U. S. Geol. Surv. Terr., vol. 7 (Tert. Fl.), 1878, p. 83, pl. vii, f. 25, 31.

Pinus polaris Heer. Lesquereux, U. S. Geol. and Geog. Surv. Terr., Ann. Rept. 1873 [1874], p. 410.

Miocene: Florissant, Colo.

Pinus sp. Berry, Am. Jour. Sci., 4th ser., vol. 34, 1912, p. 218.
Pleistocene: **Buena Vista, Va.**

Pinus sp. Berry, Torreya, vol. 14, 1914, p. 160.
Pleistocene: **Milton, Santa Rosa County, Fla.**

Pinus sp. Berry, U. S. Geol. Surv., Prof. Paper 98, 1916, p. 66, pl. xii, f. 1.
Calvert: Good Hope Hill, Anacostia Heights, D. C.

Pinus sp. Berry, U. S. Geol. Surv., Prof. Paper 98, 1916, p. 197, pl. xlv, f. 8, 9.
Citronelle: Lambert, Ala.

Pinus sp. Berry, Jour. Geol., vol. 25, 1917, p. 662.
Pleistocene: Vero, Fla.

Pinus sp. Dawson, Geol. Surv. Canada, 1875–76, p. 87.—Penhallow, Rept. Tert. Pl. Brit. Col., 1908, p. 67.
——: Blackwater River, British Columbia.

Pinus? sp. Dawson, Roy. Soc. Canada, Trans., vol. 5, 1887, sec. 4, p. 34 [wood].
Eocene: Mackenzie River, Northwest Territory.

Pinus, seeds of, Fontaine = *Pinus vernonensis*.

Pinus sp. Fontaine, U. S. Nat. Mus., Proc., vol. 16, 1893, p. 269, pl. xxxvi, f. 11.
Trinity: Glen Rose, Tex.

Pinus sp. Heer, Fl. foss. arct., vol. 1, 1868, p. 136, pl. xxi, f. 9.
Eocene: Mackenzie River, Northwest Territory.

Pinus sp. Heer, Fl. foss. alask., 1869, p. 23, pl. i, f. 11.
Kenai: Port Graham, Alaska.

Pinus, bark of ?, Hollick and Jeffrey, N. Y. Bot. Gard., Mem., vol. 3, 1909, p. 17, pl. iii, f. 8; pl. xxii, f. 5.
Feistmantelia Ward, U. S. Geol. Surv., Nineteenth Ann. Rept., 1897–98 [1899], pt. 2, p. 693.
Raritan: Kreischerville, Staten Island, N. Y.

"*Pinus* sp." Hollick, N. Y. Acad. Sci., Trans., vol. 12, 1892, p. 31, pl. i, f. 19 = *Tricalycites major*.

Pinus sp. ? Hollick, N. Y. Acad. Sci., Trans., vol. 12, 1892, p. 31, pl. i, f. 20, 22 = *Pinus tryphylla*.

Pinus sp. Hollick, Md. Geol. Surv., Pliocene and Pleistocene, 1906, p. 218.
Pleistocene (Talbot): Bodkin Point and Grace Point, Md.

Pinus sp. Hollick, U. S. Geol. Surv., Mon. 50, 1906, p. 40, pl. ii, f. 39, 47, 48.
Magothy: Gay Head, Marthas Vineyard, Mass.
Raritan: Kreischerville, Staten Island, N. Y.

Pinus, cone scales of, Hollick and Jeffrey, N. Y. Bot. Gard., Mem., vol. 3, 1909, p. 16, pl. ix, f. 11, 12; pl. xxiii, f. 6.
Raritan: Kreischerville, Staten Island, N. Y.

Pinus sp., leaf of, Jeffrey, Ann. Bot., vol. 22, 1908, pp. 215, 220, pl. xiv, f. 22, 24 = *Pinus triphylla*.

Pinus sp. Knowlton, U. S. Geol. Surv., Eighteenth Ann. Rept., pt. 3, 1898, p. 722, pl. cxix, f. 3.
Payette: Idaho City, Idaho.

Pinus sp. Knowlton, U. S. Geol. Surv., Mon. 32, pt. 2, 1899, p. 678.
Miocene (intermediate flora): Yellowstone National Park.

Pinus? sp. (leaves) Knowlton, Harriman Alaska Exped., vol. 4, 1904, p. 151, pl. xxiii, f. 2.
Kenai: Kukak Bay, Alaska.

Pinus? sp. (scales) Knowlton, Harriman Alaska Exped., vol. 4, 1904, p. 151, pl. xxiv, f. 1.
Kenai: Kukak Bay, Alaska.

Pinus sp. Lesquereux, Geol. and Nat. Hist. Surv. Minn., vol. 3, 1893, p. 10.
Dakota: New Ulm, Minn.

Pinus sp.? Newberry, U. S. Geol. Surv., Mon. 26, 1895 [1896], p. 47, pl. ix, f. 5–8 = *Pinus raritanensis*.

Pinus sp. Penhallow, Roy. Soc. Canada, Trans., 3d ser., vol. 1, sec. 4, 1907, p. 296, pl. ii.
Oligocene: Kettle River near Midway, British Columbia.

Pinus sp. Penhallow, Roy. Soc. Canada, Trans., 3d ser., vol. 1, sec. 4, 1907, p. 302, text f. 1.
Lower Cretaceous: International boundary between Pasayten and Skagit rivers, British Columbia.

PIPER Linné, Sp. pl., 1753, p. 28.

Piper heerii Lesquereux, Mus. Comp. Zool., Bull., vol. 16, 1888, p. 44.
Denver: Golden, Colo.

Piper sp. Knowlton, Wash. Geol. Surv., Ann. Rept., vol. 1, 1902, p. 33.
Eocene: Coal Creek near Keese, Whatcom County, Wash.

PIPERITES Göppert, Tertiärflora Insel Java, 1854, p. 40.

Piperites tuscaloosensis Berry, U. S. Geol. Surv., Prof. Paper 112, 1919, p. 72, pl. xii, f. 3.
Tuscaloosa: State University and Shirleys Mill, Ala.

PISONIA (Plumier) Linné, Sp. pl., 1753, p. 1026.

Pisonia chlorophylloides Berry, U. S. Geol. Surv., Prof. Paper 91, 1916, p. 214, pl. xxxvii, f. 1; pl. xlii, f. 1.
Quercus chlorophylla Unger. Lesquereux, Am. Philos. Soc., Trans., vol. 13, 1869, p. 416, pl. xvii, f. 7 [not f. 5, 6].
Raton: Fishers Peak, N. Mex.
Wilcox (Ackerman): Hurleys, Benton County, Miss.
Denver: [Locality not given].

Pisonia clairbornensis Berry, U. S. Geol. Surv., Prof. Paper 84, 1914, p. 140, pl. xxviii, f. 3.
Barnwell (Twiggs): Phinizy Gully, Columbia County, and near Macon, Ga.

Pisonia cretacea Berry, Bull. Torr. Bot. Club, vol. 37, 1910, p. 191.
Black Creek: Near Dunbars Bridge, Tar River, N. C.

Pisonia eolignitica Berry, U. S. Geol. Surv., Prof. Paper 91, 1916, p. 213, pl. xxxviii, f. 5, 6.
Lagrange: Puryear, Henry County, Tenn.

Pisonia puryearensis Berry, U. S. Geol. Surv., Prof. Paper 91, 1916, p. 214, pl. xxxviii, f. 7.
Lagrange: Puryear, Henry County, Tenn.

Pisonia racemosa Lesquereux, U. S. Geol. and Geog. Surv. Terr., Ann. Rept. 1873 [1874], p. 400; Rept. U. S. Geol. Surv. Terr., vol. 7 (Tert. Fl.), 1878, p. 209, pl. xxxv, f. 4.
Post-Laramie: Black Buttes, Wyo.

PISTACIA Linné, Sp. pl., 1753, p. 1025.

Pistacia aquehongensis Hollick, N. Y. Acad. Sci., Ann., vol. 11, 1898, p. 421, pl. xxxvi, f. 5; U. S. Geol. Surv., Mon. 50, 1906, p. 87, pl. xxxiii, f. 3.
Raritan: Tottenville, Staten Island, N. Y.

Pistacia eriensis Knowlton, U. S. Geol. Surv., Prof. Paper 130, 19—, pl. xxviii, f. 1–4 [in preparation].
Laramie: Erie, Colo.

Pistacia hollicki Knowlton, U. S. Geol. Surv., Prof. Paper 130, 19—, p. —, pl. xxviii, f. 5–6 [in preparation].
Laramie: Erie, Colo.

Pistacia oblanceolata (Lesquereux) Knowlton = ***Ficus oblanceolata.***

PISTIA Linné, Sp. pl., 1753, p. 963.

Pistia claibornensis Berry, U. S. Geol. Surv., Prof. Paper 84, 1914, p. 137, pl. xxvi, f. 1, 2.
Barnwell (Twiggs): Grovetown, Ga.

Pistia corrugata Lesquereux, U. S. Geol. and Geog. Surv. Terr., Ann. Rept. 1874 [1876], p. 299; Rept. U. S. Geol. Surv. Terr., vol. 7 (Tert. Fl.), 1878, p. 103, pl. lxi, f. 1, 3, 4, 6, 7, 9–11.—Knowlton, U. S. Geol. Surv., Bull. 163, 1900, p. 31; idem, Prof. Paper 98, 1916, p. 334, pl. lxxxv, f. 4.
Montana: Superior, Wyo.
Mesaverde: Mud River Basin and Point of Rocks, Wyo.
Judith River: Milk River, Mont.
Kirtland: San Juan Basin, N. Mex.

Pistia nordenskiöldi (Heer) Berry, Bull. Torr. Bot. Club, vol. 37, 1910, p. 189, pl. xxi, f. 1–15; idem, vol. 38, 1911, p. 405; Md. Geol. Surv., Upper Cret., 1916, p. 809, pl. lvi, f. 3.

Chondrophyllum nordenskiöldi Heer, Fl. foss. arct., vol. 3, Abt. 2, 1874, p. 114, pl. xxx, f. 43; pl. xxii, f. 11, 12.—Berry, Bull. Torr. Bot. Club, vol. 34, 1907, p. 198, pl. xiii, f. 1.

Black Creek: Blackmans Bluff, Neuse River; Parker Landing, Tar River; and several points on Black River, all in North Carolina.

Magothy: Grove Point, Md.

Pistia spathulata Michaux. Berry, Jour. Geol., vol. 25, 1917, p. 662.
Pleistocene: Vero, Fla.

Pistia wilcoxensis Berry, U. S. Geol. Surv., Prof. Paper 91, 1916, p. 175, pl. cxiii, f. 4.
Wilcox: Naborton, De Soto Parish, La.

PITHECOLOBIUM Martius, Catal. h. Monac., 1829, p. 188.

Pithecolobium eocenicum Berry, U. S. Geol. Surv., Prof. Paper 91, 1916, p. 225, pl. xlv, f. 2.
Lagrange: Puryear, Henry County, Tenn.

Pithecolobium oxfordensis Berry, U. S. Geol. Surv., Prof. Paper 91, 1916, p. 225, pl. xlv, f. 3.
Wilcox (Holly Springs): Oxford, Fayette County, Miss.

PITYOIDOLEPIS Hollick and Jeffrey, N. Y. Bot. Gard., Mem., vol. 3, 1909, p. 53. [Type, *P. statenensis*.]

Pityoidolepis statenensis Hollick and Jeffrey, N. Y. Bot. Gard., Mem., vol. 3, 1909, p. 53, pl. ix, f. 13, 14; pl. xxvii, f. 1–3.
Raritan: Kreischerville, Staten Island, N. Y.

PITYOPHYLLUM Nathorst, Norwegian N. Pol. Exp., 1893–1896 [1899], p. 19.

Pityophyllum nordenskiöldi (Heer) Seward, Jurassic plants from Amurland: Com. géol. Mém., new ser., pt. 81, 1912, p. 30, pl. ii, f. 17; pl. iii, f. 22.—Knowlton, U. S. Geol. Surv., Prof. Paper 85, 1914, p. 54.
Jurassic: Cape Lisburne, Alaska.

PITYOXYLON Kraus, in Schimper, Pal. vég., vol. 2, 1870, p. 377.

Pityoxylon aldersoni Knowlton, U. S. Geol. Surv., Mon. 32, pt. 2, 1898, p. 763, pls. cvi, cix, cxii, cxiii, cxviii, f. 3, 4; pl. cxix, f. 2.—Penhallow, Man. N. A. Gym., 1907, p. 346.
Miocene: Yellowstone National Park.

Pityoxylon amethystinum Knowlton, U. S. Geol. Surv., Mon. 32, pt. 2, 1899, p. 764, pls. cvii, cviii, cxiv, cxv, cxviii, f. 1, 2.—Penhallow, Man. N. A. Gym., 1907, p. 347.
Miocene (intermediate flora): Yellowstone National Park.

Pityoxylon annulatum Platen, Sitzungsb. Naturf. Ges., Leipzig, vol. 34, 1907 [1908], p. 20.
——: Calistoga, Calif.

Pityoxylon annulatum Platen, Sitzungsb. Naturf. Ges., Leipzig, vol. 34, 1907 [1908], p. 109.
——?: Arizona [no locality].

Pityoxylon anomalum Holden, Am. Acad. Arts and Sci., Proc., vol. 48, 1913, p. 619, pl. iii, f. f; pl. iv, f. a, b, e–g.
Magothy: Cliffwood, N. J.

Pityoxylon columbina Penhallow, Man. N. A. Gym., 1907, p. 348.
Tertiary: Midway, British Columbia.

Pityoxylon dakotense (Knowlton) Penhallow, Man. N. A. Gym., 1907, p. 346.
Pinoxylon dakotense Knowlton, in Ward, U. S. Geol. Surv., Twentieth Ann. Rept., pt. 2, 1900, p. 420, pl. clxxix.
Lakota: Near Sturgis, S. Dak.

Pityoxylon fallax Felix, Zeitschr. d. Deutsch. geol. Ges., Jahrg. 1896, p. 254.—Platen, Sitzb. Naturf. Ges., Leipzig, vol. 34, 1907 [1908], p. 118.
Miocene: Yanceys, and Amethyst Mountain, Yellowstone National Park.

Pityoxylon foliosum Holden, Am. Acad. Arts and Sci., Proc., vol. 48, 1913, p. 615, pl. iii, f. a–e.
Magothy: Cliffwood, N. J.

Pityoxylon hollicki Knowlton, in Hollick, N. Y. Acad. Sci., Trans., vol. 16, 1897, p. 134, text f. 1, 2.—Berry, N. Y. Bot. Gard., Bull., vol. 3, 1903, p. 101; Geol. Surv. N. J., Ann. Rept. 1905 [1906], p. 139.
Magothy: Cliffwood, N. J.

Pityoxylon inaequale Felix, Zeitschr. Deutsch. geol. Ges., Jahrg. 1886, p. 483, pl. xii, f. 3.—Knowlton, U. S. Nat. Mus., Proc., vol. 17, 1894, p. 215; Geol. Soc. Am., Bull., vol. 5, 1893, p. 579.
Kenai?: Danaákua, Alaska.

Pityoxylon krausei Felix, Zeitschr. Deutsch. geol. Ges., Jahrg. 1886, p. 486, pl. xii, f. 1, 2.
Fort Union: Little Missouri River, N. Dak.

Pityoxylon macclurii (Cramer) Kraus, in Schimper, Pal. vég., vol. 2, 1870, p. 379.—Platen, Sitzb. Naturf. Ges., Leipzig, vol. 34, 1907 [1908], p. 146.
——: Alaska.

Pityoxylon microporosum brandonianum Knowlton, Bull. Torr. Bot. Club, vol. 29, 1902, p. 639, pl. xxv, f. 9, 10.—Perkins, Rept. Vt. State Geol., 1903–4 [1904], p. 159, text f. 9, 10.
Miocene: Brandon, Vt.

Pityoxylon pealei Knowlton, Bull. Torr. Bot. Club, vol. 23, 1896, p. 251, pl. cclxxi.—Penhallow, Man. N. A. Gym., 1907, p. 349.
Miocene: Upper Gallatin Basin, Mont.

Pityoxylon scituatense Jeffrey and Chrysler, Bot. Gaz., vol. 42, 1906, p. 11, pl. ii, f. 8–12.—Penhallow, Man. N. A. Gym., 1907, p. 350.
Matawan (?): Scituate, Mass.

Pityoxylon statenense Jeffrey and Chrysler, Bot. Gaz., vol. 42, 1906, p. 8, pl. i, f. 1–6; pl. vii, f. 7.—Hollick and Jeffrey, N. Y. Bot. Gard., Mem., vol. 3, 1909, p. 20, pl. xxiii, f. 1–4.—Penhallow, Man. N. A. Gym., 1907, p. 349.
Raritan: Kreischerville, Staten Island, N. Y.

Pityoxylon vateri Platen, Sitzungsb. Naturf. Ges., Leipzig, 1907 [1908], p. 22.
——: Calistoga, Calif.

Pityoxylon sp. Dawson, Brit. N. A. Bound Com. (Rept. Geol. and Res. 49th Parallel), 1875, App. A, p. 331, pl. xv, f. 2.
Fort Union (?): Vicinity of 49th Parallel.

Pityoxylon sp. Dawson, Roy. Soc. Canada, Trans., vol. 5, 1887, sec. 4, p. 33 [wood].
Belly River: West of Medicine Hat, Alberta.

Pityoxylon sp. Jeffrey and Chrysler, Rept. Vt. State Geol. 1905–6 [1906], p. 201, pl. li, f. 16–18.
Miocene: Brandon, Vt.

PLAGIOPODOPSIS Britton and Hollick, Bull. Torr. Bot. Club, vol. 42, 1915, p. 10. [Type, *P. scudderi*.]

Plagiopodopsis scudderi Britton and Hollick, Bull. Torr. Bot. Club, vol. 42, 1915, p. 10, text f. 1–2, p. 9.—Knowlton, U. S. Nat. Mus., Proc., vol. 51, 1916, p. 246, pl. xii, f. 2.
Miocene: *Florissant, Colo.

PLANERA Gmelin, Syst., vol. 2, pt. 1, 1719.

Planera aquatica (Walter) T. F. Gmelin. Hollick, Bull. Torr. Bot. Club, vol. 19, 1892, p. 332.—Berry, Jour. Geol., vol. 15, 1907, p. 343; U. S. Nat. Mus., Proc., vol. 48, 1915, p. 300; U. S. Geol. Surv., Prof. Paper 98, 1916, p. 201, pl. xlvii, f. 1–4.
Planera gmelini Michaux. Lesquereux, Am. Jour. Sci., 2d ser., vol. 27, 1859, p. 365.
Pleistocene: Mississippi River near Columbus, Ky.; Neuse River, N. C.; Bridgeton, N. J.
Citronelle: Lambert, Ala.

Planera betuloides Hollick, U. S. Geol. Surv., Mon. 50, 1906, p. 57, pl. viii, f. 22.
Magothy: Gay Head, Marthas Vineyard, Mass.

Planera crenata Newberry = *Planera lingualis*.

Planera cretacea Berry, Bull. Torr. Bot. Club, vol. 34, 1907, p. 193, pl. xi, f. 7, 8.
Black Creek: Blackmans Bluff, Neuse River, N. C.

Planera dubia Lesquereux = *Planera ungeri.*

Planera gmelini Michaux = *Planera aquatica.*

Planera knowltoniana Hollick, in Newberry, U. S. Geol. Surv., Mon. 26, 1895 [1896], p. 69, pl. xlii, f. 1–4.—Berry, Geol. Surv. N. J., Bull. 3, 1911, p. 120.
Raritan: Woodbridge, N. J.

Planera lingualis Knowlton and Cockerell, n. name.
Planera crenata Newberry, U. S. Nat. Mus., Proc., vol. 5, 1882 [1883], p. 508; U. S. Geol. Surv., Mon. 35, 1898, p. 81, pl. lvii, f. 3.—Penhallow, Rept. Tert. Pl. Brit. Col., 1908, p. 73.—Berry, U. S. Geol. Surv., Prof. Paper 91, 1916, p. 193. [Homonym, Desfontaines, 1815.]
Fort Union: Tongue River, Wyo.
Paskapoo (Fort Union): Burrard Inlet, Vancouver Island; Horsefly River, British Columbia.
Wilcox (Grenada): Grenada, Grenada County, Miss.

Planera longifolia Lesquereux = *Fagopsis longifolia.*

Planera longifolia myricaefolia Lesquereux = *Planera myricaefolia.*

Planera microphylla Newberry, N. Y. Lyc. Nat. Hist., Ann., vol. 9, 1868, p. 55.—[Lesquereux], U. S. Geol. and Geog. Surv. Terr., Ill. Cret. and Tert. Pl., 1878, pl. xvi, f. 3, 4.—Newberry, U. S. Geol. Surv., Mon. 35, 1898, p. 81, pl. xxxiii, f. 3, 4.
Fort Union: Fort Union, N. Dak.

Planera myricaefolia (Lesquereux) Cockerell, Am. Mus. Nat. Hist., Bull., vol. 24, 1908, p. 88.—Knowlton, U. S. Nat. Mus., Proc., vol. 51, 1916, p. 266, pl. xxi, f. 2.
Planera longifolia myricaefolia Lesquereux, Rept. U. S. Geol. Surv. Terr., vol. 8 (Cret. and Tert. Fl.), 1883, p. 161, pl. xxix, f. 15–27.
Podocarpus eocenica? Unger. Lesquereux, Rept. U. S. Geol. Surv. Terr., vol. 8 (Cret. and Tert. Fl.), 1883, p. 140.
Miocene: Florissant, Colo.

Planera nervosa Newberry, U. S. Nat. Mus., Proc., vol. 5, 1882 [1883], p. 508; U. S. Geol. Surv., Mon. 35, 1898, p. 82, pl. xlvii, f. 2, 3.
Green River: Green River, Wyo.

Planera ungeri Ettingshausen, Foss. Fl. Wien, 1851, p. 14, pl. ii, f. 5–18.—Heer, Fl. foss. arct., vol. 2, Abt. 2, 1869, p. 34, pl. v, f. 2.—Lesquereux, U. S. Geol. and Geog. Surv. Terr., Bull., vol. 1, 1875 [1876], p. 387; Rept. U. S. Geol. Surv. Terr., vol. 7 (Tert. Fl.), 1878, p. 190, pl. xxvii, f. 7; U. S. Nat. Mus., Proc., vol. 11, 1888, p. 19.—Knowlton, U. S. Nat. Mus., Proc., vol. 17, 1894, p. 224; Geol. Soc. Am., Bull., vol. 5, 1893, p. 584; U. S. Geol. Surv., Bull. 204, 1902, p. 55.—Hollick, Md. Geol. Surv., Pliocene and

Planera ungeri—Continued.
Pleistocene, 1906, p. 229, pl. lxxi, f. 14, 15.—Berry, Jour. Geol., vol. 17, 1909, p. 25, f. 4.
Planera dubia Lesquereux, Am. Jour. Sci., 2d ser., vol. 27, 1859, p. 361.
Planera longifolia? Lesquereux. Lesquereux, U. S. Geol. and Geog. Surv. Terr., Ann. Rept. 1872 [1873], p. 371.
Pleistocene (Sunderland): Head of Island Creek, Calvert County, Md.
Mascall: Van Horn's ranch, John Day Basin, Oreg.
Kenai ?: Port Graham, Alaska.
——: ?Bellingham Bay, Wash.
Miocene: Elko, Nev.; Florissant, Colo.

Planera variabilis Newberry, U. S. Nat. Mus., Proc., vol. 5, 1882 [1883], p. 508; U. S. Geol. Surv., Mon. 35, 1898, p. 83, pl. lxvi, f. 5–7.
Green River: Green River, Wyo.

PLANTAGINOPSIS Fontaine, in Ward, U. S. Geol. Surv., Mon. 48, 1905 [1906], p. 561. [Type, *P. marylandica*.]

Plantaginopsis marylandica Fontaine, in Ward, U. S. Geol. Surv., Mon. 48, 1905 [1906], p. 561, pl. cxvii, f. 7; pl. cxviii, f. 1, 2.—Berry, Md. Geol. Surv., Lower Cret., 1912, p. 456, pl. lxxix, f. 1–4; pl. lxxx.
Celastrophyllum marylandicum Fontaine, in Ward, U. S. Geol. Surv., Mon. 48, 1905 [1906], p. 559, pl. cxvi, f. 7.
Patapsco: Federal Hill (Baltimore), Md.

"*Plataeanthus* (*Williamsonia*) *problematicus* Newberry." Knowlton, U. S. Geol. Surv., Bull. 152, 1898, p. 168. Error for *Palaeanthus* (*Williamsonia*) *problematicus*.

PLATANINIUM Unger, in Endlicher, Gen. Pl., Suppl. 2, 1842, p. 101.

Plataninium crystallophilum Platen, Sitzungsb. Naturf. Ges., Leipzig, vol. 34, 1907 [1908], p. 111.
——: Arizona [no locality].

Plataninium haydeni Felix, Zeitschr. Deutsch. geol. Ges., vol. 48, 1896, p. 251.—Knowlton, U. S. Geol. Surv., Mon. 32, pt. 2, 1899, p. 767, pl. cxx, f. 3–5.—Platen, Sitzungsb. Naturf. Ges., Leipzig, vol. 34, 1907 [1908], p. 129.
Miocene: Amethyst Mountain, Yellowstone National Park.

Plataninium knowltoni Platen, Sitzungsb. Naturf. Ges., Leipzig, vol. 34, 1907 [1908], p. 130.
Miocene: Amethyst Mountain, Yellowstone National Park.

Plataninium pacificum Platen, Sitzungsb. Naturf. Ges., Leipzig, vol. 34, 1907 [1908], p. 65.
Miocene (auriferous gravels): Nevada County, Calif.

Platanophyllum crassinerve Fontaine = *Araliaephyllum crassinerve*.

PLATANUS (Tournefort) Linné, Sp. pl., 1753, p. 999.

Platanus aceroides Göppert, Zeitschr. Deutsch. geol. Ges., vol. 5, 1852, p. 492.—Lesquereux, Am. Jour. Sci., 2d ser., vol. 45, 1869, p. 206; Rept. U. S. Geol. Surv. Terr., vol. 7 (Tert. Fl.), 1878, p. 184, pl. xxv, f. 4-6.—?Knowlton, in Smith, U. S. Geol. Surv., Geol. Folio 86 (Ellensburg, Wash.), 1903, p. 3?; idem, Bull. 204, 1902, p. 65.—Penhallow, Rept. Tert. Pl. Brit. Col., 1908, p. 74.—Berry, Jour. Geol., vol. 17, 1909, p. 27.—Knowlton, U. S. Geol. Surv., Prof. Paper 101, 1918, p. 321, pl. lxiii, f. 4; pl. xcvii, f. 2, 3.—Berry, U. S. Nat. Mus., Proc., vol. 54, 1918, p. 630, pl. xciv, f. 3; pl. xcv, f. 5.

Fort Union?: Badlands, N. Dak.
Denver: Golden, Colo.
Hanna: Carbon, Wyo.
Post-Laramie: Black Buttes, Wyo.
Mascall: Van Horn's ranch, John Day Basin, Oreg.
Ellensburg: Ellensburg, Wash.
Calvert: Richmond, Va.
Miocene: Beaver Creek, 10 miles east of Beaver City, Okla.
Eocene?: Mackenzie River, Northwest Territory.
Raton: Cokedale, Wootton, and Morely, Colo.; Raton, N. Mex.

Platanus aceroides Göppert. Hollick = *Platanus occidentalis.*

Platanus aceroides cuneata Knowlton, U. S. Geol. Surv., Prof. Paper 101, 1918, p. 321, pl. cxiii, f. 1.

Raton: South Raton Canyon, N. Mex.

Platanus aceroides latifolia Knowlton, U. S. Geol. Surv., Prof. Paper 101, 1918, p. 321, pl. xcii; pl. xciii, f. 3; pl. xciv.—Berry, U. S. Geol. Surv., Prof. Paper 91, 1916, p. 13.

Midway?: Earle, Bexar County, Tex.
Raton: Raton Mesa region, Colo. and N. Mex.

Platanus aceroides latior Lesquereux = *Platanus latior.*

Platanus affinis Lesquereux = *Cissites affinis.*

Platanus affinis ampla Dawson = *Cissites affinis ampla.*

Platanus appendiculata Lesquereux, Mus. Comp. Zool., Mem., vol. 6, No. 2, 1878, p. 12, pl. iii, f. 1-6; pl. vi, f. 7b.—?Knowlton, in Turner, Am. Geol., vol. 15, 1895, p. 378; in Lindgren, Jour. Geol., vol. 4, 1896, p. 889.

Miocene (auriferous gravels): Chalk Bluffs, Nevada County, Independence Hill, Placer County, Calif.
Ione?: Volcanic Hill, Placer County, Calif.

Platanus aquehongensis Hollick, N. Y. Acad. Sci., Trans., vol. 12, 1892, p. 32, pl. iv; U. S. Geol. Surv., Mon. 50, 1906, p. 82, pl. xxxi, f. 6.

Raritan: Richmond Valley, Staten Island, N. Y.

Platanus aspera Newberry, U. S. Nat. Mus., Proc., vol. 5, 1882 [1883], p. 509; U. S. Geol. Surv., Mon. 35, 1898, p. 102, pl. xlii, f. 1–3; pl. xliv, f. 5; pl. lix, f. 3.—Knowlton, U. S. Geol. Surv., Bull. 204, 1902, p. 64.

Clarno (upper part): Bridge Creek, Grant County, Oreg.

Platanus asperaeformis Berry, U. S. Geol. Surv., Prof. Paper 112, 1919, p. 83, pl. xvi, f. 1.

Ficus newberryana Heer, Fl. foss. arct., vol. 7, 1883, p. 28, pl. l, f. 2 [not other figures].

Tuscaloosa: Shirleys Mill, Ala.

Platanus basilobata Ward, U. S. Geol. Surv., Sixth Ann. Rept., 1884–85 [1886], p. 552, pl. xlii, f. 1–4; pl. xliii, f. 1; idem, Bull. 37, 1887, p. 35, pl. xvii, f. 1; pl. xviii, f. 1–3; pl. xix, f. 1.

Fort Union: Sevenmile and Clear creeks, Glendive, Mont.

Platanus cissoides Lesquereux, U. S. Geol. Surv., Mon. 17, 1891 [1892], p. 75, pl. lxi, f. 3.—?Ward, U. S. Geol. Surv., Nineteenth Ann. Rept., pt. 2, 1899, p. 706, pl. clxxi, f. 2.

Dakota: Near Fort Harker, Kans.; Evans Quarry, 5 miles east of Hot Springs, S. Dak.

Platanus condoni (Newberry) Knowlton, in Merriam, Univ. Calif., Bull. Dept. Geol., vol. 2, 1901, p. 289.—Knowlton, U. S. Geol. Surv., Bull. 204, 1902, p. 64.

Ficus? condoni Newberry, U. S. Nat. Mus., Proc., vol. 5, 1883, p. 512; U. S. Geol. Surv., Mon. 1898, p. 85, pl. lvi, f. 1; pl. lviii, f. 1.

Clarno (upper part): Bridge Creek, Oreg.

Platanus cordata Knowlton, U. S. Geol. Surv., Bull. 152, 1898, p. 169.

Quercus platania Heer. Lesquereux, U. S. Geol. and Geog. Surv. Terr., Ann. Rept. 1872 [1873], p. 386; Rept. U. S. Geol. Surv. Terr., vol. 7 (Tert. Fl.), 1878, p. 160, pl. xxi, f. 1.

Hanna: Carbon, Wyo.

Platanus diminutiva Lesquereux, Rept. U. S. Geol. Surv. Terr., vol. 6 [Cret. Fl.], 1874, p. 73, pl. viii, f. 5.

Platanus diminutivus Lesquereux, Am. Jour. Sci., 2d ser., vol. 46, 1868, p. 98.

Dakota: Nebraska?.

Platanus dissecta Lesquereux, Mus. Comp. Zool., Mem., vol. 6, No. 2, 1878, p. 13, pl. vii, f. 12; pl. x, f. 4, 5; Rept. U. S. Geol. Surv. Terr., vol. 8 (Cret. and Tert. Fl.), 1883, p. 249, pl. lvi, f. 4; pl. lvii, f. 1, 2.—Knowlton, in Smith, U. S. Geol. Surv., Geol. Folio 86 (Ellensburg, Wash.), 1903, p. 3.

Acer trilobatum productum (Al. Braun) Heer. Lesquereux, Cret. and Tert. Fl., 1883, p. 253, pl. lix, f. 3.

Miocene (auriferous gravels): Chalk Bluffs, Table Mountain, Corral Hollow, and Spanish Peak, Calif.

Ellensburg: Ellensburg, Wash.

Platanus dubia Lesquereux = *Aralia nonata*.

Platanus guillelmae Göppert, Zeitschr. Deutsch. geol. Ges., vol. 4, 1852, p. 492.—Lesquereux, U. S. Geol. and Geog. Surv. Terr., Ann. Rept. 1871 [1872], p. 289; Rept. U. S. Geol. Surv. Terr., vol. 7 (Tert. Fl.), 1878, p. 183, pl. xxv, f. 1-3.—?Ward, U. S. Geol. Surv., Sixth Ann. Rept., 1884-85 [1886], p. 552, pl. xliv, f. 1; idem, Bull. 37, 1887, p. 37, pl. xx, f. 1.—Knowlton, idem, Mon. 32, pt. 2, 1898, p. 727, pl. xcvi, f. 1; pl. xcvii, f. 5; Washington Acad. Sci., Proc., vol. 11, 1909, pp. 198, 211, 215; U. S. Geol. Surv., Prof. Paper 101, 1918, p. 322, pl. xciii, f. 1.

Acer indivisum Weber. Ward, U. S. Geol. Surv., Sixth Ann. Rept., 1884-85 [1886], p. 554, pl. l, f. 1; idem, Bull. 37, 1887, p. 66, pl. xxix, f. 5.

Hanna: Carbon, Wyo.

Denver: Golden, Colo.

Fort Union: Near Glendive, Mont.; Yellowstone National Park.

Lance: Ten miles northeast of Glendive, Mont.; Dayton and Kirbey, Wyo.

Raton: Raton Mesa region, Colorado and New Mexico.

Platanus guillelmae heerii Knowlton, U. S. Geol. Surv., Prof. Paper 101, 1918, p. 323, pl. xcvi, f. 5; pl. xcvii, f 1; pl. xcviii, f. 2.

Raton: Tercio, Colo.

Platanus haydenii Newberry, N. Y. Lyc. Nat. Hist., Ann., vol. 9, 1868, p. 70.—[Lesquereux], U. S. Geol. and Geog. Surv. Terr., Ill. Cret. and Tert. Pl., 1878, pl. xix, pl. xxi.—Lesquereux, U. S. Geol. Surv. Terr., Rept., vol. 7 (Tert. Fl.), 1878, p. 182.—Newberry, U. S. Geol. Surv., Mon. 35, 1898, p. 103, pl. xxxvi; pl. xxxviii; pl. lvi, f. 3.—Penhallow, Rept. Tert. Pl. Brit. Col., 1908, p. 74.—Knowlton, Washington Acad. Sci., Proc., vol. 11, 1909, pp. 195, 202, 213, 214; in Calvert, U. S. Geol. Surv., Bull. 471, 1912, p. 16.

Fort Union: *Yellowstone River, Mont.; abundant and widely distributed in the Fort Union.

Denver: Golden, Colo.

Eocene: Porcupine Creek, Great Valley, and Omineca River, British Columbia.

Lance: Near Bridger, Ilo, and mouth of Sage Creek, Big Horn Basin, Wyo.; near Glendive, Mont.

Platanus heerii Lesquereux, U. S. Geol. and Geog. Surv. Terr., Ann. Rept. 1871 [1872], p. 303; Rept. U. S. Geol. Surv. Terr., vol. 6 (Cret. Fl.), 1874, p. 70, pl. viii, f. 4; pl. ix, f. 1, 2; Rept. on Clays in N. J., 1878, p. 29.—Berry, Bull. Torr. Bot. Club, vol. 37, 1910, p. 23; idem, vol. 38, 1911, p. 411; Md. Geol. Surv., Upper Cret., 1916, p. 824, pls. lxv, lxvi, lxvii.

Sassafras recurvatus Lesquereux, U. S. Geol. and Geog. Surv. Terr., Ann. Rept. 1872 [1873], p. 424; Rept. U. S. Geol. Surv. Terr., vol. 6 (Cret. Fl.), 1874, p. 71, pl. x, f. 4, 5 [not f. 3].

Platanus heerii—Continued.

Sassafras (Araliopsis) recurvatum Lesquereux, Rept. U. S. Geol. Surv. Terr., vol. 8 (Cret. and Tert. Fl.), 1883, p. 57 (part).

Sassafras cretaceum recurvatum Lesquereux. Berry, Bot. Gaz., vol. 34, 1902, p. 438.

Dakota: Salina River, Kans.

Raritan: Drum Point, Md.; East Washington Heights, D. C.

Platanus heerii Lesquereux. Ward = *Platanus platanoides.*

Platanus heterophyllus Newberry. Dawson, Brit. N. W. Bound. Com. (Geol. and Res. near 49th Parallel), App. A, 1875, p. 330; Roy. Soc. Canada, Trans., vol. 1, 1882–83, p. 32.—Penhallow, Rept. Tert. Pl. Brit. Col., 1908, p. 74.

——: Wood End, Alberta.

Platanus integrifolia Lesquereux = *Platanus raynoldsii integrifolia.*

Platanus kümmellii Berry, Bull. Torr. Bot. Club, vol. 33, 1906, p. 174; Geol. Surv. N. J., Ann. Rept. 1905 [1906], pp. 139, 146, pl. xxiii, f. 2, 3; pl. xxiv.

Magothy: Cliffwood Brick Co. pit, Whale River, N. J.

Platanus latiloba Newberry = *Sassafras latilobum.*

Platanus latior (Lesquereux) Knowlton, U. S. Geol. Surv., Bull. 152, 1898, p. 170.—Berry, U. S. Geol. Surv., Prof. Paper 112, 1919, p. 84.

Platanus aceroides? Göppert var. *latior* Lesquereux, Am. Jour. Sci., 2d ser., vol. 46, 1868, p. 97.

Platanus primaeva Lesquereux, Rept. U. S. Geol. Surv. Terr., vol. 6 (Cret. Fl.), 1874, p. 69, pl. vii, f. 2; pl. xxvi, f. 2; U. S. Geol. Surv., Mon. 17, 1892, p. 72, pl. viii, f. 7, 8b; pl. x, f. 1.

Dakota: Kansas, Nebraska, and Minnesota.

Tuscaloosa: Cottondale, Ala.

Platanus latior grandidentata (Lesquereux) Knowlton, U. S. Geol. Surv., Bull. 152, 1898, p. 170.

Platanus primaeva grandidentata Lesquereux, U. S. Geol. Surv., Mon. 17, 1891 [1892], p. 73, pl. ix, f. 1, 2.

Dakota: Ellsworth County, Kans.

Platanus latior integrifolia (Lesquereux) Knowlton, U. S. Geol. Surv., Bull. 152, 1898, p. 170.

Platanus primaeva integrifolia Lesquereux, U. S. Geol. Surv., Mon. 17, 1891 [1892], p. 74, pl. xlix, f. 4.

Dakota: Near Fort Harker, Kans.

Platanus latior subintegrifolia (Lesquereux) Knowlton, U. S. Geol. Surv., Bull. 152, 1898, p. 170.

Platanus primaeva subintegrifolia Lesquereux, U. S. Geol. Surv., Mon. 17, 1891 [1892], p. 73, pl. ix, f. 3, 4.

Dakota: Ellsworth County, Kans.

Platanus marginata (Lesquereux) Heer, Fl. foss. arct., vol. 7, 1883, p. 97, pl. xcviii, f. 3–5; pl. xcix, f. 2, 3; pl. ci, f. 5.—Jankó, Engler's bot. Jahrb., vol. 11, 1889, p. 454.—Cross, Colo. Sci. Soc., Proc., vol. 4, 1892, p. 211 [25 of reprint].

Viburnum marginatum Lesquereux, U. S. Geol. Surv. Terr., Ann. Rept., 1872 [1873], p. 395; idem, 1873 [1874], pp. 382, 401; idem, 1874 [1876], p. 306; idem, 1876 [1878], p. 510; idem, Bull., vol. 1, 1875, p. 380; Rept. U. S. Geol. Surv. Terr., vol. 7 (Tert. Fl.), 1878, p. 223, pl. xxxviii, f. 1, 4 [not f. 2, 3, 5]; Mus. Comp. Zool., Bull., vol. 16, 1888, p. 51.—Knowlton, Geol. Soc. Am., Bull., vol. 8, 1897, p. 145; Jour. Geol., vol. 19, 1911, p. 361.

Post-Laramie: *Black Buttes, Wyo.

Medicine Bow: Carbon County, Wyo.

Lance: Cheyenne Indian Reservation, S. Dak.

Dawson: Mosby, Colo.

Platanus montana Knowlton, U. S. Geol. Surv., Mon. 32, pt. 2, 1899, p. 726, pl. xcvi, f. 2, 3.

Miocene (intermediate flora): Yellowstone National Park.

Platanus newberrii Lesquereux = *Platanus newberryana.*

Platanus? newberryana Heer, Phyl. crét. Nebr., 1866, p. 16, pl. i, f. 4.—Hollick, U. S. Geol. Surv., Mon. 50, 1906, p. 82.

Platanus newberrii Lesquereux, Am. Jour. Sci., 2d ser., vol. 46, 1868, p. 97.

Dakota: Beatrice, Nebr., and other localities.

Raritan: Elm Point, Great Neck, Long Island, N. Y.

Platanus newberryana Heer. Hollick, N. Y. Acad. Sci., Trans., vol. 11, 1892, p. 103, pl. iv, f. 5 = *Platanus* sp.

Platanus nobilis Newberry, N. Y. Lyc. Nat. Hist., Ann., vol. 9, 1868, p. 67.—[Lesquereux], U. S. Geol. and Geog. Surv. Terr., Ill. Cret. and Tert. Pl., 1878, pl. xvii; pl. xx, f. 1, under *Platanus haydenii.*—Dawson, Roy. Soc. Canada, Trans., vol. 4, sec. 4, 1886 [1887], p. 24, pl. i, f. 7.—Ward, U. S. Geol. Surv., Sixth Ann. Rept., 1884–85 [1886], p. 552, pl. xli, f. 1; idem, Bull. 37, 1887, p. 35, pl. xvi, f. 1.—?Knowlton, idem, Bull. 204, 1902, p. 65; Washington Acad. Sci., Proc., vol. 11, 1909, pp. 192, 195, 211, 212, 213, 214.

Fort Union: *Fort Clarke, N. Dak. Abundant and widely distributed in Fort Union.

Lance: Near Big Horn, Mont.; near Bridger; between Buffalo and Klondike, Ranchester; Monarch, near Ilo, Big Horn Basin, and other localities, Wyo.

?Mascall: John Day Basin, Oreg.

Eocene: Souris River, Porcupine Creek, and Great Valley, Saskatchewan; Calgary, Alberta.

Platanus obtusiloba Lesquereux, Am. Jour. Sci., 2d ser., vol. 46, 1868, p. 97; U. S. Geol. Surv., Mon. 17, 1892, p. 74, pl. x, f. 2.

Dakota: Beatrice, Nebr.; near Carnerio, Ellsworth County, Kans.

Platanus occidentalis Linné. Knowlton, Am. Geol., vol. 18, 1896, p. 371.—Penhallow, Roy. Soc. Canada, Trans., 2d ser., vol. 2, sec. 4, 1897, pp. 68, 72.—Mercer, Jour. Philadelphia Acad., 2d ser., vol. 2, 1899, p. 277.—Penhallow, Brit. Assoc. Adv. Sci., Bradford meeting, 1900, p. 336; Am. Nat., vol. 41, 1907, p. 448.—Berry, Jour. Geol., vol. 15, 1907, p. 344; Am. Nat., vol. 41, 1907, p. 695, pl. ii, f. 5; Am. Jour. Sci., 4th ser., vol. 29, 1910, p. 397.—Torreya, vol. 15, 1915, p. 207.

Platanus aceroides Göppert. Hollick, Md. Geol. Surv., Pliocene and Pleistocene, 1906, p. 231, pl. lxxiii; pl. lxxiv.

Pleistocene: Morgantown, W. Va.; Don Valley, Toronto, Canada; Neuse River, N. C.; Black Warrior River, 356 and 328½ miles above Mobile, Abercrombie Landing, Chattahoochee River, and Alabama River, 12 miles above Montgomery, Ala.; head of Island Creek, Calvert County, and Indian Head, Charles County, Md.

Platanus platanoides (Lesquereux) Knowlton, U. S. Geol. Surv., Bull. 152, 1898, p. 171; Jour. Geol., vol. 19, 1911, p. 370; U. S. Geol. Surv., Prof. Paper 101, 1918, p. 323, pl. xcv, f. 4.

Viburnum platanoides Lesquereux, U. S. Geol. and Geog. Surv. Terr., Ann. Rept. 1874 [1876], p. 314; Rept. U. S. Geol. Surv. Terr., vol. 7 (Tert. Fl.), 1878, p. 224, pl. xxxviii, f. 8, 9.

Platanus heerii Lesquereux. Ward, U. S. Geol. Surv., Sixth Ann. Rept., 1884–85 [1886], p. 552, pl. xl, f. 8, 9; idem, Bull. 37, 1887, p. 34, pl. xv, f. 3, 4.

Post-Laramie: Black Buttes, Wyo.
?Lance: Sec. 20, T. 14 N., R. 19 E., S. Dak.
Medicine Bow: Carbon County, Wyo.
?Laramie: Erie, Colo.
Raton: Rouse, Colo.

Platanus primaeva Lesquereux = *Platanus latior.*

Platanus primaeva grandidentata Lesquereux = *P. latior grandidentata.*

Platanus primaeva integrifolia Lesquereux = *P. latior integrifolia.*

Platanus primaeva subintegrifolia Lesquereux = *P. latior subintegrifolia.*

Platanus raynoldsii Newberry, N. Y. Lyc. Nat. Hist., Ann., vol. 9, 1868, p. 69.—[Lesquereux], U. S. Geol. and Geog. Surv. Terr., Ill. Cret. and Tert. Pl., 1878, pl. xviii.—Lesquereux, U. S. Geol. Surv. Terr., Rept., vol. 7 (Tert. Fl.), 1878, p. 185, pl. xxvi, f. 4, 5; pl. xxvii, f. 1–3.—Ward, U. S. Geol. Surv., Sixth Ann. Rept., 1884–85 [1886], p. 552, pl. xliv, f. 2, 3; U. S. Geol. Surv., Bull. 37, 1887, p. 37, pl. xx, f. 2, 3.—Newberry, U. S. Geol. Surv., Mon. 35, 1898, p. 109, pl. xxxv.—Penhallow, Rept. Tert. Pl. Brit. Col., 1908, p. 75.—Knowlton, Washington Acad. Sci., Proc., vol. 11, 1909, pp. 189, 196, 198, 204, 207, 214; Jour. Geol., vol. 19, 1911, p. 370; U. S. Geol. Surv., Prof. Paper 101, 1918, p. 324, pl. xcv, f. 1.

Platanus raynoldsii—Continued.

Fort Union: *Banks of Yellowstone River, Mont.; near Jordan, Mont.; Sevenmile Creek, Weston County, Wyo.

Eocene: Porcupine Creek, Saskatchewan; Calgary, Alberta, Canada.

Lance: South Dakota (sec. 33, T. 20 N., R. 20 E.); 10 miles northeast of Glendive, Signal Butte, Miles City, 25 miles northwest of Red Lodge, 9 miles above Glendive, Mont.; Converse and Winston counties, Big Horn Basin, and other localities, Wyo. (This is one of the most abundant and widespread forms in the Fort Union and Lance formations.)

Raton: Morley, Cokedale, and Wootton, Colo.; York Canyon, N. Mex.

Platanus raynoldsii Newberry. Lesquereux, U. S. Nat. Mus., Proc., vol. 11, 1888, p. 19 = *Platanus* sp.

Platanus raynoldsii integrifolia Lesquereux, Rept. U. S. Geol. Surv. Terr., vol. 7 (Tert. Fl.), 1878, p. 185.—Knowlton, Washington Acad. Sci., Proc., vol. 11, 1909, p. 207.

Platanus integrifolia Lesquereux MS. [Not published.]

Denver: Golden, Colo.

Post-Laramie: Black Buttes, Wyo.

Lance: Converse County, Wyo.

Platanus recurvata Lesquereux = *Sassafras recurvatum.*

Platanus? regularis Knowlton, U. S. Geol. Surv., Prof. Paper 101, 1918, p. 325, pl. cxiii, f. 4.

Raton?: Locality unknown.

Platanus rhomboidea Lesquereux, U. S. Geol. and Geog. Surv. Terr., Ann. Rept. 1873 [1874], p. 400; Rept. U. S. Geol. Surv. Terr., vol. 7 [Tert. Fl.], 1878, p. 186, pl. xxvi, f. 6, 7.—Knowlton, Washington Acad. Sci., Proc., vol. 11, 1909, p. 204; U. S. Geol. Surv., Prof. Paper 101, 1918, p. 324.

Viburnum anceps Lesquereux, Rept. U. S. Geol. Surv. Terr., vol. 7 (Tert. Fl.), 1878, p. 227, pl. xxxviii, f. 11.

Denver: Golden, Colo.

Lance: Weston County, Wyo.

Raton: Raton and York Canyon, N. Mex.; Wootton, Colo.

Platanus ripleyensis Berry, U. S. Geol. Surv., Prof. Paper 112, 1919, p. 84, pl. xxxii, f. 6.

Ripley: Near Eufaula, Ala.

Platanus shirleyensis Berry, Plant World, vol. 17, 1914, p. 5, f. 4; U. S. Geol. Surv., Prof. Paper 112, 1919, p. 83, pl. xv, f. 1–5.

Tuscaloosa: Shirleys Mill and Cottondale, Ala.

Platanus? wardii Knowlton, U. S. Geol. Surv., Bull. 163, 1900, p. 14, pl. ii, f. 1–4.

Eagle: Missouri River below Coal Banks, Mont.

Platanus sp. Berry, U. S. Geol. Surv., Prof. Paper 112, 1919, p. 85, pl. xxxi, f. 7, 8.
Ripley: Mouth of Cowikee Creek, Barbour County, Ala.

Platanus sp. Cragin, Washburn Coll. Lab., Bull., vol. 2, 1889, p. 66.
Dakota ?: Cimarron Valley, south of Meade County, Kans.

Platanus sp. Dawson, Geol. Surv. Canada, 1871–72 [1872], p. 59; idem, 1877–78, p. 187B; Roy. Soc. Canada, Trans., vol. 1, 1885, p. 17.—Penhallow, Rept. Tert. Fl. Brit. Col., 1908, p. 73.
Eocene: Quesnel River, British Columbia.

Platanus sp. Hollick, U. S. Geol. Surv., Mon. 50, 1906, p. 83, pl. xxxi, f. 5.
Platanus newberryana Heer. Hollick, N. Y. Acad. Sci., Trans., vol. 11, 1892, p. 103, pl. iv, f. 9.
Raritan: Princess Bay, Staten Island, N. Y.

Platanus sp. ? Hollick, Md. Geol. Surv., Pliocene and Pleistocene, 1906, p. 232, pl. lxxv.
Pleistocene (Sunderland): Point of Rocks, Calvert County, Md.

Platanus sp. Knowlton, U. S. Geol. Surv., Bull. 204, 1902, p. 66.
Platanus raynoldsii Newberry. Lesquereux, U. S. Nat. Mus., Proc., vol. 11, 1888, p. 19.
Mascall: Van Horn's ranch, John Day Basin, Oreg.

Platanus sp. ? Knowlton, in Spurr, U. S. Geol. Surv., Eighteenth Ann. Rept., pt. 3, 1898, p. 192.
Kenai: Miller's mine, Yukon River, Alaska.

Platanus sp. Knowlton, U. S. Geol. Surv., Prof. Paper 101, 1918, p. 269, pl. xlii, f. 3.
Vermejo: Rockvale, Colo.

Platanus sp., probably *P. haydenii* Newberry. Knowlton, in Calvert, U. S. Geol. Surv., Bull. 471, 1912, p. 13.
Lance (Colgate): Iron Bluff, near Glendive, Mont.

PLATYPTERYGIUM (Schimper) Feistmantel, Mem. Geol. Surv. India, Pal. Indica, vol. 4 (Foss. Fl. Gond. Syst., pt. 2), 1886, p. 37.

Platypterygium densinerve Fontaine, U. S. Geol. Surv., Mon. 15, 1889 [1890], p. 169, pl. xxx, f. 8; pl. xxxi, f. 1, 4; pl. xxxii, f. 1, 2; pl. xxxiii, f. 1; pl. xxxiv, f. 1; pl. xxxv, f. 1, 2.—?Fontaine, in Ward, U. S. Geol. Surv., Mon. 48, 1905 [1906], p. 521, pl. cxii, f. 8.
Patuxent: Fredericksburg, Va.

Platypterygium rogersianum Fontaine, U. S. Geol. Surv., Mon. 15, 1889 [1890], p. 171, pl. xxxi, f. 2; pl. xxxiii, f. 2; pl. xxxiv, f. 2.
Patuxent: Fredericksburg, Va.

POACITES Brongniart, Prod. hist. vég. foss., 1828, p. 137.

Poacites laevis Al. Braun. Lesquereux, U. S. Geol. and Geog. Surv. Terr., Ann. Rept. 1871 [1872], p. 285.
Green River ?: Barrell Spring, Wyo.

Poacites mengeanus Heer, Mioc. Balt. Fl., 1859, p. 59, pl. xv, f. 2–11.—Lesquereux, U. S. Nat. Mus.. Proc., vol. 10, 1887, p. 35.

Pliocene?: Clear Lake, Calif.

Poacites tenue-striatus Heer, Fl. foss. arct., vol. 2, Abt. 2 (Fl. foss. alask.), 1869, p. 24, pl. i, f. 14.—Eichwald, Geognost-Paleontolog. Bemerk. ü. Halbinsel Mangischlak u. Aleutischen Inseln, 1871, p. 114, pl. iv, f. 7.—Knowlton, U. S. Nat. Mus., Proc., vol. 17, 1894, p. 216; Geol. Soc. Am., Bull., vol. 5. 1893, p. 580.

Kenai: Port Graham, Alaska.

Poacites sp. Hollick, N. Y. Bot. Gard., Bull., vol. 3, 1904, p. 411, pl. 73, f. 1.

Magothy: Glen Cove, Long Island, N. Y.

Poacites sp. Hollick, U. S. Geol. Surv., Mon. 50, 1906, p. 48, pl. ii, f. 11; pl. vi, f. 9–11.

Magothy: Gay Head, Marthas Vineyard, Mass.; Glen Cove, Long Island, N. Y.

Poacites sp. Hollick, Geol. Surv. La., Special Rept. 5, 1899, p. 279, pl. xxxii, f. 2.—Berry, U. S. Geol. Surv., Prof. Paper 91, 1916, p. 174.

Wilcox: Shreveport, La.

PODOCARPITES Andrae, Abh. K. k. geol. Reichsanst., vol. 2, Abt. 3, 1853, p. 45.

Podocarpites tyrrellii Dawson, Roy. Soc. Canada, Trans., vol. 5, 1887, sec. 4, p. 35.

Belly River: Vermilion River, Alberta.

PODOCARPOXYLON Gothan, Abh. K. preuss. geol. Landesanst., Neue Folge, Heft 44, 1905, p. 103.

Podocarpoxylon mcgeei (Knowlton) Sinnott and Bartlett, Am. Jour. Sci., 4th ser., vol. 41, 1916, p. 280, text f. 1–10.

Cupressinoxylon mcgeei Knowlton, Am. Geol., vol. 3, 1889, p. 106 [name]; U. S. Geol. Surv., Bull. 56, 1889, p. 46, pl. ii, f. 5; pl. iii, f. 1–5.—Penhallow, Man. N. A. Gym., 1907, p. 243.—Berry, Md. Geol. Surv., Lower Cret., 1912, p. 417, pl. lxix, f. 1–6.

Patuxent: New Reservoir,* New Central High School, and Meridian Hill Park, Washington, D. C.

PODOCARPUS L'Héritier, MS., 1788.

Podocarpus eocenica? Unger. Lesquereux, Rept. U. S. Geol. Surv. Terr., vol. 8 (Cret. and Tert. Fl.), 1883, p. 140 = *Planera myricaefolia*.

Podocarpus? stantoni Knowlton, U. S. Geol. Surv., Prof. Paper 98, 1916, p. 89, pl. xv, f. 5.

Fox Hills: About 1 mile east of Milliken, Colo.

PODOGONIUM Heer, Fl. tert. Helv., vol. 3, 1859, p. 201.

Podogonium acuminatum Lesquereux, Rept. U. S. Geol. Surv. Terr., vol. 8 (Cret. and Tert. Fl.), 1883, p. 201.

Miocene: Florissant, Colo.

Podogonium americanum Lesquereux, Rept. U. S. Geol. Surv. Terr., vol. 7 (Tert. Fl.), 1878, p. 298, pl. lix, f. 5; pl. lxiii, f. 5; pl. lxv, f. 6.

Podogonium sp.? Lesquereux, U. S. Geol. and Geog. Surv. Terr., Ann. Rept. 1873 [1874], p. 417.

Cassia podogonioides Ettingshausen, Denks. K. Akad. Wiss., Matt.-naturw. Classe, vol. 47, 1883, p. 143 [43]; Geol. Surv. N. S. W., Memoir Palaeontology 2 (Tert. Fl. Australia), 1888, p. 69.

Post-Laramie: Black Buttes, Wyo.

Podogonium? virginianum Berry, Jour. Geol., vol. 17, 1909, p. 27, text f. 6.

Calvert: Richmond, Va.

Podogonium sp.? Lesquereux = *Podogonium americanum.*

PODOZAMITES F. Braun, in Münster, Beitr. Petref., Heft 6, 1843, p. 36.

Podozamites acuminatus Hollick, in Newberry, U. S. Geol. Surv., Mon. 26, 1895 [1896], p. 45, pl. xiii, f. 7.—Berry, Geol. Surv. N. J., Bull. 3, 1911, p. 77.

Raritan: Woodbridge, N. J.

Podozamites acutifolius Fontaine, U. S. Geol. Surv., Mon. 15, 1889 [1890], p. 181, pl. lxxx, f. 6; pl. lxxxv, f. 10, 15; pl. lxxxvii, f. 1; pl. clxx, f. 2; ?U. S. Nat. Mus., Proc., vol. 16, 1893, p. 266, pl. xxxvi, f. 7.—Berry, Md. Geol. Surv., Lower Cret., 1911, p. 338, pl. liii f. 4.

Nageiopsis acuminata Fontaine, U. S. Geol. Surv., Mon. 15, 1889 [1890], p. 201, pl. lxxxv, f. 11.

Patapsco: Deep Bottom, Va.

Patuxent: Near Dutch Gap Canal; telegraph station near Lorton; near Brooke, Va.

Trinity: Glen Rose, Tex.

Podozamites angustifolius (Eichwald) Schimper. American authors = *Podozamites knowltoni.*

Podozamites angustifolius (Eichwald) Schimper. Newberry, U. S. Geol. Surv., Mon. 26, 1895 [1896], p. 44, pl. xiii, f. 1–4 = *Podozamites proximans.*

Podozamites? carolinensis Fontaine, in Ward, U. S. Geol. Surv., Twentieth Ann. Rept., pt. 2, 1900, p. 298, pl. xlii, f. 4.

Triassic (Newark): North Carolina.

Podozamites caudatus Lesquereux = *Dammarites caudatus.*

Podozamites distans? (Presl) F. Braun, in Münster, Beiträge zur Petrefactenkunde, vol. 2, pt. 6, 1843, p. 28.—Fontaine, in Ward, U. S. Geol. Surv., Twentieth Ann. Rept., pt. 2, 1900, p. 246, pl. xxix, f. 5–7.

Zamites distans Presl, in Sternberg, Flora der Vorwelt, vol. 2, 1833, p. 196, pl. xli, f. 1.

Triassic (Newark): York County, Pa.?

Podozamites distantinervis Fontaine, U. S. Geol. Surv., Mon. 15, 1889 [1890], p. 179, pl. lxxix, f. 9; pl. lxxxii, f. 4; pl. lxxxiii, f. 1, 2, 6, 7; pl. lxxxiv, f. 1, 2, 8, 10, 14, 15; pl. lxxxv, f. 12, 16; in Ward, U. S. Geol. Surv., Mon. 48, 1905 [1906], pp. 479, 516 [not pp. 165, 281, 573].—Berry, Md. Geol. Surv., Lower Cret., 1911, p. 340, pl. liii, f. 8, 9.

Podozamites pedicellatus Fontaine, U. S. Geol. Surv., Mon. 15, 1889 [1890], p. 180, pl. lxxvi, f. 1; pl. lxxviii, f. 7; pl. lxxxii, f. 5; in Ward, U. S. Geol. Surv., Mon. 48, 1905 [1906], p. 532, pl. cxiv, f. 1.

Patuxent: Fredericksburg, Potomac Run, Dutch Gap, Va.; Sixteenth Street, D. C.; Fort Foote, Md.

Podozamites distantinervis Fontaine. Fontaine, in Ward, U. S. Geol. Surv., Mon. 48, 1905 [1906], pp. 165, 282 = *Podozamites lanceolatus.*

Podozamites eichwaldi Schimper = *Podozamites lanceolatus eichwaldi.*

Podozamites emarginatus Lesquereux = *Dammarites emarginatus.*

Podozamites emmonsii Newberry, in Pumpelly, Smithsonian Cont. Knowl., No. 202, 1866, p. 121, pl. ix, f. 2.—Fontaine, in Ward, U. S. Geol. Surv., Twentieth Ann. Rept., pt. 2, 1900, p. 297, pl. xlii, f. 1, 2.

Podozamites lanceolatus Emmons [not Lindley and Hutton], Geol. Rept. Midland Counties N. C., 1856, p. 331, pl. iii, f. 7; American geology, pt. 6, 1857, p. 116, pl. iii, f. 7.

Triassic (Newark): North Carolina.

Podozamites emmonsi Newberry. Fontaine, in Turner, Jour. Geol., vol. 3, 1895, p. 395 = *Podozamites lanceolatus?*

Podozamites emmonsi Fontaine [not Newberry] = *Podozamites longifolius.*

Podozamites formosus Brown, Acad. Nat. Sci., Philadelphia, Proc., vol. 63, 1911, p. 17, pl. i.

Triassic (Newark): Bucks County, Pa.

Podozamites grandifolius Fontaine, U. S. Geol. Surv., Mon. 15, 1889 [1890], p. 180, pl. lxxxii, f. 2; pl. lxxxiii, f. 5.

Patuxent: Fredericksburg and near Potomac Run, Va.

Podozamites grandifolius Fontaine [part] = *Ctenopsis latifolia.*

Podozamites grandifolius Fontaine? Fontaine, in Ward, U. S. Geol. Surv., Mon. 48, 1905 [1906], p. 167, pl. xliv, f. 1 = *Zamites megaphyllus.*

Podozamites haydenii (Lesquereux) Lesquereux, Rept. U. S. Geol. Surv. Terr., vol. 8 (Cret. and Tert. Fl.), 1883, p. 27; U. S. Geol. Surv., Mon. 17, 1891 [1892], p. 26.

Pterophyllum haydenii Lesquereux, Am. Jour. Sci., 2d ser., vol. 46, 1868, p. 91; Rept. U. S. Geol. Surv. Terr., vol. 6 (Cret. Fl.), 1874, p. 49, pl. i, f. 6, 6b.

Dakota: Decatur, Nebr.

Podozamites inaequilateralis (Fontaine) Berry, U. S. Nat. Mus., Proc., vol. 38, 1910, p. 194; Md. Geol. Surv., Lower Cret., 1911, p. 336, pl. liii, f. 1.

Nageiopsis inaequilateralis Fontaine, U. S. Geol. Surv., Mon. 15, 1889 [1890], p. 200, pl. lxxxv, f. 6.

Nageiopsis obtusifolia Fontaine, U. S. Geol. Surv., Mon. 15, 1889 [1890], p. 200, pl. lxxxv, f. 7; in Ward, U. S. Geol. Surv., Mon. 48, 1905 [1906], p. 484.

Nageiopsis montanensis Fontaine, in Ward, op. cit. (Mon. 48), p. 312, pl. lxxiii, f. 7.

Patuxent: Kankeys, Cockpit Point, near Potomac Run, Va.

Kootenai: Geyser, Cascade County, Mont.

Podozamites knowltoni Berry, Bull. Torr. Bot. Club, vol. 36, 1906, p. 247; idem, vol. 37, 1910, p. 182; Md. Geol. Surv., Lower Cret., 1911, p. 339, pl. liii, f. 7; U. S. Geol. Surv., Prof. Paper 87, 1914, p. 16, pl. iv, f. 5.

Podozamites angustifolius (Eichwald) Schimper, Pal. vég., vol. 2, 1872, p. 160 [not Schenk, 1868].—Lesquereux, Rept. U. S. Geol. Surv. Terr., vol. 8 (Cret. and Tert. Fl.), 1884, p. 28; U. S. Geol. Surv., Mon. 17, 1892, p. 17, pl. i, f. 4.—Newberry, U. S. Geol. Surv., Mon. 26, 1895, p. 44, pl. xiii, f. 1, 3, 4 [not f. 2].—Berry, Md. Geol. Surv., Upper Cret., 1916, p. 774.

Zamites angustifolius Eichwald, Lethaea rossica, vol. 2, 1860, p. 39, pl. ii, f. 7.

Nageiopsis recurvata Fontaine. Fontaine, in Ward, U. S. Geol. Surv., Mon. 48, 1905 [1906], p. 552, pl. cxvi, f. 2.

Zamites tenuinervis Fontaine, in Ward, idem, p. 528.

Patuxent: Fredericksburg, Potomac Run, and Dutch Gap, Va.; Sixteenth Street, D. C.

Dakota: Near Fort Harker, Kans.

Raritan: Woodbridge, N. J.

Black Creek: Rockfish Creek near Hope Mills, N. C.

Magothy: Round Bay, Md.

Podozamites lanceolatus Emmons [not Lindley and Hutton] = *Podozamites emmonsii* Newberry [not Fontaine].

Podozamites lanceolatus (Lindley and Hutton) Fr. Braun, in Münster, Beitr. Petrefactenkunde, vol. 2, 1843, pt. 6, p. 33.—Dawson, Roy. Soc. Canada, Trans., vol. 3, 1886, sec. 4, p. 6, pl. i, f. 3.—Lesquereux, U. S. Geol. Surv., Mon. 17, 1892, p. 28, pl. i, f. 5, 6.—Newberry, U. S. Geol. Surv., Mon. 26, 1896, p. 44, pl. xiii, f. 2 [not f. 1, 3, 4].—Penhallow, Geol. Surv. Canada, Summary, 1904 [1905], p. 9.—Fontaine, in Ward, U. S. Geol. Surv., Mon. 48, 1905 [1905], p. 110, pl. xxiv, f. 17–20; U. S. Geol. Surv., Twentieth Ann. Rept., pt. 2, 1900, p. 360, pl. lxiii, f. 4; pl. lxiv, f. 1; pl. lxvi, f. 4; pl. lxvii, f. 3, 4.—Knowlton, Smithsonian Misc. Coll., vol. 50, 1907, p. 120.—Hollick, U. S. Geol. Surv., Mon. 50, 1907, p. 35, pl. ii, f. 1.—Berry, Bull. Torr. Bot. Club, vol. 39, 1912, p. 391; Md. Geol. Surv., Lower Cret., 1911, p. 341, pl. liii, f. 5, 6.—Hollick, N. Y.

Podozamites lanceolatus—Continued.
Bot. Gard., Bull., vol. 8, 1912, p. 155, pl. clxii; pl. clxiii, f. 2, 3.—Berry, Bull. Torr. Bot. Club, vol. 38, 1911, p. 410; Md. Geol. Surv., Upper Cret., 1916, p. 772.—Knowlton, U. S. Geol. Surv., Prof. Paper 85, 1914, p. 52.

Podozamites pedicellatus Fontaine. Fontaine, in Ward, U. S. Geol. Surv., Mon. 48, 1905 [1906], p. 532, pl. cxiv, f. 1.
Podozamites distantinervis Fontaine. Fontaine, in Ward, idem, pp. 165, 282.
Zamia washingtoniana Ward. Fontaine, in Ward, idem, p. 503, pl. cxi, f. 2 [not f. 1].

Jurassic: Oroville, Calif.; Cape Lisburne, Alaska; Douglas County, Oreg.
Kootenai: Geyser, Cascade County, Mont.
Dakota: Near Fort Harker, Kans.
Woodbine: Arthurs Bluff, Tex.
Raritan: Woodbridge, N. J.; Roslyn, N. Y.; Shannon Hill, Cecil County, Md.
Magothy: Glen Cove, Long Island, N. Y.

Podozamites lanceolatus eichwaldi (Schimper) Heer, Fl. foss. arct., vol. 4, Abt. 2, 1876, p. 109, pl. xxiii, f. 4; pl. xxvi, f. 2, 3, 9; pl. xxvii, f. 1, 5c.—Knowlton, U. S. Geol. Surv., Prof. Paper 85, 1914, p. 53, pl. v, f. 6; pl. vi, f. 5 (part).

Podozamites eichwaldi Schimper, Pal. veg., vol. 2, 1870, p. 160.
Podozamites lanceolatus latifolius (Schenk) Heer. Knowlton, in Collier, U. S. Geol. Surv., Bull. 278, 1906, p. 29.

Jurassic: Cape Lisburne, Alaska.

Podozamites lanceolatus latifolius (Fr. Braun) Heer, Fl. foss. arct., vol. 4, Abt. 2, 1876, p. 109, pl. xxvi, f. 5, 6, 8b, 8c.—Fontaine, in Ward, U. S. Geol. Surv., Twentieth Ann. Rept., pt. 2, 1900, p. 361, pl. lxiv, f. 2; U. S. Geol. Surv., Mon. 48, 1905 [1906], p. 112, pl. xxv, f. 5–7.—Knowlton, Am. Jour. Sci., 4th ser., vol. 30, 1910, p. 45.

Jurassic: Oroville, Calif.; Douglas County, Oreg.

Podozamites lanceolatus latifolius (Schenk) Heer. Knowlton, in Collier, 1906 = *Podozamites lanceolatus eichwaldi.*

Podozamites lanceolatus minor (Schenk) Heer, Fl. foss. arct., vol. 4, Abt. 2, 1876, p. 110, pl. xxvi, f. 5a, b, 6–8.—Fontaine, in Ward, U. S. Geol. Surv., Mon. 48, 1905 [1906], p. 111, pl. xxv, f. 1–4, p. 150; pl. xxxviii, f. 11, 12.—Knowlton, Am. Jour. Sci., 4th ser., vol. 30, 1910, p. 45.

Jurassic: Douglas and Curry counties, Oreg.

Podozamites latipennis Heer, Fl. foss. arct., vol. 6, Abt. 2, 1882, p. 42, pl. xiv, f. 1–9; pl. xv, f. 2a, 3b.—Weed, Geol. Soc. Am., Bull., vol. 3, 1892, p. 323 [doubtful].

Kootenai: Sun River, Mont.

Podozamites latipennis Heer. Lesquereux, U. S. Nat. Mus., Proc., vol. 11, 1888, p. 31, pl. xvi, f. 2, 3 =*Podozamites* sp.?

Podozamites longifolius Emmons, Geol. Rept. Midland Counties N. C., 1856, p. 331; American geology, pt. 6, 1857, p. 116, f. 83.—Fontaine, in Ward, U. S. Geol. Surv., Twentieth Ann. Rept., pt. 2, 1900, p. 294, pl. xl; pl. xli.

Cycadites longifolius Emmons, Geol. Rept. Midland Counties N. C., 1856, p. 330; American geology, pt. 6, 1857, p. 115, f. 82.

Dioonites longifolius (Emmons) Fontaine, U. S. Geol. Surv., Mon. 6, 1883, p. 111, pl. liii, f. 5.—Newberry, idem, Mon. 14, 1888, p. 92, pl. xxv, f. 4.

Podozamites emmonsi Fontaine, op. cit., p. 77, pl. xxxiii, f. 2.

Triassic (Newark): Ellingtons and Lockwood, N. C.; Clover Hill, Va.; Newark, N. J.

Podozamites marginatus Heer, Fl. foss. arct., vol. 6, Abt. 2, 1882, p. 43, pl. xvi, f. 10.—Berry, N. Y. Bot. Gard., Bull., vol. 3, 1903, p. 99, pl. xlvi, f. 1–3; Geol. Surv. N. J., Ann. Rept. 1905 [1906], p. 139; Bull. Torr. Bot. Club, vol. 38, 1911, p. 410; Geol. Surv. N. J., Bull. 3, 1911, p. 74; Md. Geol. Surv., Upper Cret., 1916, p. 775, pl. li, f. 8; U. S. Geol. Surv., Prof. Paper 112, 1919, p. 55, pl. vi, f. 1; Bull. Torr. Bot. Club, vol. 44, 1917, p. 171.

Magothy: Cliffwood, N. J.; Drum Point Railroad, Md.

Raritan: Woodbridge, N. J.

Bingen: Near Murfreesboro, Ark.

Tuscaloosa: Soap Hill, Bibb County, and Snow place, Tuscaloosa County, Ala.

Podozamites nervosa Newberry =*Podozamites newberryi.*

Podozamites newberryi, Knowlton, n. name.

Podozamites nervosa Newberry, Am. Jour. Sci., 3d ser., vol. 41, 1891, p. 200, pl. xiv, f. 6.—Knowlton, Smithsonian Misc. Coll., vol. 50, 1907, p. 120. [Homonym, (Schenck) Schimper, Pal. vég., vol. 2, 1875, p. 161.]

Kootenai: Great Falls, Mont.

Podozamites oblongus Lesquereux, Rept. U. S. Geol. Surv. Terr., vol. 8 (Cret. and Tert. Fl.), 1883, p. 28, pl. i, f. 10, 11; U. S. Geol. Surv., Mon. 17, 1892, p. 26.

Dakota: Kansas.

Podozamites? pachynervis Fontaine, in Ward, U. S. Geol. Surv., Mon. 48, 1905 [1906], p. 112, pl. xxv, f. 8.

Jurassic: Douglas County, Oreg.

Podozamites pachyphyllus Fontaine, in Ward, U. S. Geol. Surv., Mon. 48, 1905 [1906], p. 109, pl. xxiv, f. 11–16.

Jurassic: Douglas County, Oreg.

Podozamites pedicellatus Fontaine =*Podozamites distantinervis.*

Podozamites pedicellatus Fontaine. Fontaine, in Ward, U. S. Geol. Surv., Mon. 48, 1905 [1906], pl. cxiv, f. 1 =*Podozamites lanceolatus.*

Podozamites praelongus Lesquereux = *Dammarites caudatus*.

Podozamites proximans Conrad, Am. Jour. Sci., 2d ser., vol. 47, 1869, p. 361, text f.

Podozamites angustifolius (Eichwald) Schimper. Newberry, U. S. Geol. Surv., Mon. 26, 1895 [1896], p. 44 pl. xiii, f. 1–4; Hollick, N. Y. Bot. Gard., Bull. 3, 1904, p. 410, pl. lxxi, f. 8.

Raritan: Near Washington, on South River, N. J.
Magothy: Glen Cove, Long Island, N. Y.

Podozamites pulchellus Heer, Fl. foss. arct., vol. 4, pt. 1, 1876, p. 38, pl. ix, f. 10–14.—Fontaine, in Ward, U. S. Geol. Surv., Mon. 48, 1905 [1906], p. 108, pl. xxiv, f. 1–10.

Jurassic: Douglas County, Oreg.

Podozamites stenopus Lesquereux, U. S. Geol. Surv., Mon. 17, 1891 [1892], p. 27, pl. i, f. 7.

Dakota: Ellsworth County, Kans.

Podozamites subfalcatus Fontaine, U. S. Geol. Surv., Mon. 15, 1889 [1890], p. 179, pl. lxviii, f. 6; pl. clxx, f. 9.—Berry, Md. Geol. Surv., Lower Cret., 1911, p. 338, pl. liii, f. 2, 3.

Zamites ovalis Fontaine, U. S. Geol. Surv., Mon. 15, 1889 [1890], p. 173 [part].

Patapsco: Near Brooke, Va.
Patuxent: Telegraph station near Lorton, Va.

Podozamites taylorsvillensis Ward = *Pterophyllum taylorsvillensis*.

Podozamites tenuinervis Heer, Fl. foss. arct., vol. 6, Abt. 2, 1882, p. 44, pl. xvi, f. 9.—Newberry, Bull. Torr. Bot. Club, vol. 13, 1886, p. 35.

Raritan: New Jersey.

Podozamites tenuistriatus (Rogers) Fontaine, U. S. Geol. Surv., Mon. 6, 1883, p. 78, pl. xlii, f. 2–5; pl. xliv, f. 3; in Ward, U. S. Geol. Surv., Twentieth Ann. Rept., pt. 2, 1900, p. 297, pl. xlii, f. 3.

Zamites tenuistriatus Rogers, Trans. Assoc. Am. Nat. and Geol., Philadelphia, 1843, p. 314.

Triassic (Newark): Richmond coal field and Cumberland area, Va.; North Carolina.

Podozamites sp. Berry, Bull. Torr. Bot. Club, vol. 44, 1917, p. 172.

Bingen: Mine Creek near Nashville, Ark.

Podozamites sp. Fontaine, U. S. Nat. Mus., Proc., vol. 16, 1893, p. 267, pl. xxxvi, f. 8.

Trinity: Glen Rose, Tex.

Podozamites sp.? Hollick, N. Y. Bot. Gard., Bull., vol. 2, 1902, p. 401, pl. 41, f. 8, 9.

Magothy: Chappaquiddick Island, Marthas Vineyard, Mass.

Podozamites sp. Hollick, U. S. Geol. Surv., Mon. 50, 1906, p. 35, pl. vi, f. 1–3.
Magothy: Chappaquiddick, Marthas Vineyard, Mass.; Glen Cove, Long Island, N. Y.

POLYGONUM (Tournefort) Linné, Sp. pl., 1753, p. 359.

Polygonum tertiarium Small = *Persicaria tertiaria*.

Polygonum sp. Berry, Torreya, vol. 14, 1914, p. 160.
Pleistocene: Seven Springs, Neuse River, Wayne County, N. C.

Polygonum sp. Berry, Jour. Geol., vol. 25, 1917, p. 662.
Pleistocene: Vero, Fla.

Polygonum sp.? Hollick, Md. Geol. Surv., Pliocene and Pleistocene, 1906, p. 231.
Pleistocene (Talbot): Grove Point, Cecil County; Grace Point, Baltimore County, Md.

Polygonum sp. Knowlton, Washington Acad. Sci., Proc., vol. 11, 1909, p. 199; Jour. Geol., vol. 19, 1911, p. 370.
Lance: Yellowstone River, 10 miles northeast of Glendive, Mont.; South Dakota (sec. 7, T. 17 N., R. 24 E.).

POLYPODIUM (Tournefort) Linné, Sp. pl., 1753, p. 1082.

Polypodium dentatum Fontaine, U. S. Geol. Surv., Mon. 15, 1889 [1890], p. 105, pl. xxii, f. 4, 5.
Patuxent: Fredericksburg, Va.

Polypodium fadyenioide Fontaine, U. S. Geol. Surv., Mon. 15, 1889 [1890], p. 104, pl. xvi, f. 4, 5.
Patuxent: Near Potomac Run, Va.

Polypodium oregonense Fontaine, in Ward, U. S. Geol. Surv., Mon. 48, 1905 [1906], p. 63, pl. viii, f. 8–15; pl. ix; pl. x, f. 1–7.
Jurassic: Douglas County, Oreg.

POLYSTICHUM Roth, Tent. Fl. Germ., vol. 3, 1797, p. 69.

Polystichum hillsianum Hollick, Torreya, vol. 2, 1902, p. 146, pl. 4, f. 7.—Knowlton, U. S. Geol. Surv., Prof. Paper 101, 1918, p. 244.
Vermejo: *Florence and Rockvale, Colo.

POLYTRICHUM Linné, Sp. pl., vol. 2, 1753, p. 1109.

Polytrichum? florissanti Knowlton, U. S. Nat. Mus., Proc., vol. 51, 1916, p. 245, pl. xii, f. 4.
Miocene: *Florissant, Colo.

POPULITES Viviani, Mém. Soc. géol. France, vol. 1, 1833, p. 133.

Populites affinis Lesquereux = *Cissites affinis*.

Populites amplus Knowlton, U. S. Geol. Surv., Bull. 257, p. 140, pl. xviii, f. 1.
Judith River: Willow Creek, Fergus County, Mont.

Populites cyclophylla (Heer) Lesquereux, Am. Jour. Sci., 2d ser., vol. 46, 1868, p. 93; Rept. U. S. Geol. Surv. Terr., vol. 6 (Cret. Fl.), 1874, p. 59; U. S. Geol. Surv., Mon. 17, 1891, p. 48.

Populus cyclophylla Heer, Philadelphia Acad. Nat. Sci., Proc., vol. 10, 1858, p. 266.—Newberry, U. S. Geol. Surv., Mon. 35, 1898, p. 41, pl. iii, f. 3, 4; pl. iv, f. 1.—[Lesquereux], U. S. Geol. and Geog. Surv. Terr., Ill. Cret. and Tert. Pl., 1898, pl. iii, f. 3, 4.—Penhallow, Roy. Soc. Canada, Trans., 3d ser., vol. 1, 1907, p. 309.

Populus litigiosa Heer. [Lesquereux], U. S. Geol. and Geog. Surv. Terr., Ill. Cret. and Tert. Pl., 1878, pl. iv, f. 1.

Cissites cyclophylla Lesquereux, U. S. Geol. and Geog. Surv. Terr., Ann. Rept. 1874 [1876], p. 353.

Dakota: Blackbird Hill, Lancaster, and Decatur, Nebr.; Fort Harker, Kans.; New Ulm, Minn.

Lower Cretaceous(?): International boundary between Pasayten and Skagit rivers, British Columbia.

Populites elegans Lesquereux, Am. Jour. Sci., 2d ser., vol. 46, 1868, p. 94; Geol. and Nat. Hist. Surv. Minn., Final Rept., vol. 3, 1893, pt. 1, p. 10, pl. A, f. 2; pl. B, f. 1.—?Newberry, U. S. Geol. Surv., Mon. 35, 1898, p. 54, pl. viii, f. 3.—Lesquereux, U. S. Geol. Surv., Mon. 17, 1891 [1892], p. 47, pl. xlvi, f. 5; pl. xlvii, f. 2, 3.

Dakota: Decatur, Nebr.; Fort Harker, Kans.; New Ulm, Minn.

Populites fagifolia Lesquereux = ***Grewiopsis flabellata.***

Populites flabellata Lesquereux = *Grewiopsis flabellata.*

Populites lancastriensis (Lesquereux) Lesquereux, Rept. U. S. Geol. Surv. Terr., vol. 6 (Cret. Fl.), 1874, p. 58, pl. vii, f. 1.

Populus lancastriensis Lesquereux, Am. Jour. Sci., 2d ser., vol. 46, 1868, p. 93.

Dakota: Lancaster, Nebr.; New Ulm, Minn.

Populites litigiosus (Heer) Lesquereux, U. S. Geol. Surv., Mon. 17, 1891 [1892], p. 46, pl. vii, f. 7; pl. viii, f. 5; pl. xlvi, f. 6; pl. xlvii, f. 1.

Populus litigiosa Heer, Phyll. crét. Nebr., 1866, p. 13, pl. i, f. 2.

Dakota: Nebraska and Kansas; New Ulm, Minn.

Populites litigiosus denticulata (Lesquereux) Knowlton, n. comb.

Populus litigiosus denticulata Lesquereux, U. S. Geol. Surv., Mon. 17, 1891, [1892], p. 47.

Dakota: Kansas?

Populites microphyllus Lesquereux, Am. Philos. Soc., Trans., vol. 13, 1869, p. 430, pl. xxiii, f. 2, 3.

Dakota: Kansas.

Populites ovata Lesquereux = ***Ampelophyllites ovatus.***

Populites probalsamifera Dawson, Roy. Soc. Canada, Trans., vol. 11, 1893 [1894], p. 57, pl. vii, f. 23.

Upper Cretaceous: Port McNeill, Vancouver Island, British Columbia.

Populites quadrangularis Lesquereux = *Hamamelites quadrangularis.*

Populites salinae Lesquereux = *Menispermites salinae.*

Populites salisburiaefolius Lesquereux = *Araliopsis cretacea salisburaefolia.*

Populites sternbergii Lesquereux, U. S. Geol. Surv., Mon. 17, 1891 [1892], p. 45, pl. vii, f. 8, 9.
Dakota: Near Glascoe, Kans.

Populites tenuifolius Berry, N. Y. Bot. Gard., Bull., vol. 3, 1903, p. 69, pl. xlix, f. 7; Bull. Torr. Bot. Club, vol. 31, 1904, p. 74; Geol. Surv. N. J., Ann. Rept. 1905 [1906], p. 139.
Magothy: Cliffwood Bluff, N. J.

Populites tuscaloosensis Berry, U. S. Geol. Surv., Prof. Paper 112, 1919, p. 77, pl. xiii, f. 2.
Tuscaloosa: Glen Allen and Tuscaloosa, Ala.

Populites winchelli Lesquereux, Geol. and Nat. Hist. Surv. Minn., Final Rept., vol. 3, pt. 1, 1893, p. 12, pl. B, f. 2.
Dakota: New Ulm, Minn.

POPULOPHYLLUM Fontaine, U. S. Geol. Surv., Mon. 15, 1889 [1900], p. 311. [Type, *P. reniforme.*]

Populophyllum crassinerve Fontaine, U. S. Geol. Surv., Mon. 15, 1889 [1890], p. 312, pl. clviii, f. 4.
Patapsco: Deep Bottom, Va.

Populophyllum hederaeforme Fontaine = *Populophyllum reniforme.*

Populophyllum menispermoides Ward, Mon. 48, 1905 [1906], pl. cx, f. 2 [not f. 3, 4] = *Populus potomacensis.*

Populophyllum menispermoides Ward, Mon. 48, 1905 [1906], pl. cx, f. 3, 4 = *Populophyllum reniforme.*

Populophyllum minutum Ward, U. S. Geol. Surv., Mon. 48, 1905 [1906], p. 499, pl. cvii, f. 9; p. 532, pl. cviii, f. 11.—Berry, Md. Geol. Surv., Lower Cret., 1912, p. 460, pl. lxxxi, f. 2.
Patapsco: Mount Vernon, Va.; Fort Foote and Wellhams, Md.

Populophyllum reniforme Fontaine, U. S. Geol. Surv., Mon. 15, 1889 [1890], p. 311, pl. clv, f. 9; pl. clvi, f. 3.—Ward, U. S. Geol. Surv., Sixteenth Ann. Rept., 1894–95, pt. 1, 1896, p. 539, pl. cvi, f. 6, 7; idem, Fifteenth Ann. Rept., 1895, p. 360, pl. iv, f. 5, 6.—Berry, Md. Geol. Surv., Lower Cret., 1912, p. 461, pl. lxxxi, f. 3–6.

Populophyllum hederaeforme Fontaine, U. S. Geol. Surv., Mon. 15, 1889 [1890], p. 311, pl. clxvi, f. 3.

Populophyllum menispermoides Ward, U. S. Geol. Surv., Mon. 48, 1905 [1906], p. 498, pl. cx, f. 3, 4 [not f. 2, which = *Populus potomacensis*].

Patapsco: Mount Vernon, near Brooke, White House Bluff, Widewater, and near Aquia Creek, Va.; Wellhams, Md.

POPULUS (Tournefort) Linné, Sp. pl., 1753, p. 1034.

Populus acerifolia Newberry = *Populus newberryi*.

Populus aequalis Lesquereux, U. S. Geol. and Geog. Surv. Terr., Ann. Rept. 1870 [1871], p. 383 [name].—Schimper, Pal. vég., vol. 2, 1872, p. 693.—Cockerell, Bull. Torr. Bot. Club, vol. 33, 1906, p. 312.

Populus laevigata Lesquereux, Am. Jour. Sci., 2d ser., vol. 45, 1868, p. 205; Rept. U. S. Geol. Surv. Terr., vol. 7 (Tert. Fl.), 1898, p. 175, pl. xxii, f. 9. [Homonym, Aiton, 1789.]

Hanna: Rock Creek, Laramie Plains, Wyo.

Populus amblyrhyncha Ward, U. S. Geol. Surv., Sixth Ann. Rept., 1884–85 [1886], p. 550, pl. xxxiv, f. 5–9; pl. xxxv, f. 1–6; idem, Bull. 37, 1887, p. 20, pl. vi, f. 1–8; pl. vii, f. 1–3.—Knowlton, Washington Acad. Sci., Proc., vol. 11, 1909, pp. 188, 189, 194, 195, 198, 201, 202; Jour. Geol., vol. 19, 1911, p. 361; in Calvert, U. S. Geol. Surv., Bull. 471, 1912, p. 16.

Fort Union: *Sevenmile Creek, near Glendive, Mont. (Abundant and widely distributed in the Fort Union.)

Lance: Cannonball River, sec. 4, T. 19 N., R. 18 E., S. Dak.; 60 miles east of Glasgow, 6 miles east of Miles City; 3 miles east of Miles City; 18 miles east of Miles City; near Melville; 8 miles west of Bridger; near Glendive, Mont.; Monarch, between Buffalo and Klondike, near Ilo, near Kirby, Converse County, Wyo.

Lance (Colgate): Iron Bluff near Glendive, Mont.

Populus anomala Ward, U. S. Geol. Surv., Sixth Ann. Rept., 1884–85 [1886], p. 550, pl. xxxvi, f. 5; U. S. Geol. Surv., Bull. 37, 1887, p. 23, pl. viii, f. 7.

Fort Union: Burns's ranch, Glendive, Mont.

Populus apiculata Hollick = *Cordia apiculata*.

Populus arctica Heer, Fl. foss. arct., vol. 1, 1868, pp. 100, 137, pl. iv, f. 6a, 7; pl. v; pl. vi, f. 5, 6; pl. viii, f. 5, 6; pl. xvii, f. 5f, c; pl. xxi, f. 14, 15a.—Lesquereux, U. S. Nat. Mus., vol. 5, 1882 [1883], p. 447, pl. ix, f. 2; U. S. Geol. and Geog. Surv., Ann. Rept. 1871 [1872], pp. 289, 300; Rept. U. S. Geol. Surv. Terr., vol. 7 (Tert. Fl.), 1878, p. 178, pl. xxiii, f. 1–6.—Knowlton, U. S. Nat. Mus., Proc., vol. 17, 1894, p. 217; Geol. Soc. Am., Bull., vol. 5, 1893, p. 581; U. S. Geol. Surv., Seventeenth Ann. Rept., pt. 1, 1896, p. 881.—Penhallow, Rept. Tert. Pl. Brit. Col., 1908, p. 76.—Knowlton, Washington, Acad. Sci., Proc., vol. 11, 1909, pp. 189, 211, 213.

[Note.—This species is greatly in need of critical revision, pending which it may stand as above.]

Kenai: Chignik Bay and very many localities in Alaska.

Eocene: Coal Brook, Red Deer River, Omineca River, Porcupine Creek, Great Valley, Mackenzie Basin, Bear River, Quesnel, etc., Canada.

Oligocene: Similkameen River, British Columbia.

Lance: 9 miles west of Miles City, Mont.; between Buffalo and Klondike, Dayton, and near Ilo, Wyo.

Populus arctica decipiens (Lesquereux) Cockerell, Am. Mus. Nat. Hist., Bull., vol. 24, 1908, p. 83.
Miocene: Florissant, Colo.

"**Populus arctica latior** Heer." Dawson, in McConnell, Geol. Surv. Canada, new ser., vol. 4, 1891, p. 96D.—Penhallow, Rept. Tert. Pl. Brit. Col., 1908, p. 76.
Eocene: Bear River, Mackenzie Basin, Northwest Territory.

Populus aristolochioides Lesquereux, U. S. Geol. and Geog. Surv. Terr., Bull., vol. 1, 1875, p. 393.
Dakota: Kansas.

Populus attenuata Al. Braun. Lesquereux = *Populus subrotunda.*

Populus auriculata Ward, U. S. Geol. Surv., Fifteenth Ann. Rept., 1905, p. 356, pl. iv, f. 4.—Fontaine, in Ward, U. S. Geol. Surv., Mon. 48, 1905 [1906], p. 499, pl. cx, f. 5.
Patapsco: Mount Vernon, Va.

Populus balsamifera Linné. Dawson, Canadian Nat., vol. 2, 1857, p. 422; idem, new ser., vol. 3, 1868, p. 72, f. 2–4; Geol. Hist. Pl., 1888, p. 228, f. 78.—Penhallow, Roy. Soc. Canada, Trans., 2d ser., vol. 2, 1896, pp. 64, 72; idem, vol. 10, 1904, p. 74; Brit. Assoc. Adv. Sci., Bradford meeting, 1900, p. 336.—Berry, Torreya, vol. 17, 1917, p. 161.
Pleistocene: Don Valley, Besserer's wharf, Ottawa River; Greens River, Canada; Waterville, Maine.

Populus balsamoides Göppert. Heer, Fl. foss. arct., vol. 2, Abt. 2, 1869, p. 26, pl. ii, f. 3.—Knowlton, U. S. Geol. Surv., Proc., vol. 17, 1894, p. 217; Geol. Soc. Am., Bull., vol. 5, 1893, p. 580; in Spurr, U. S. Geol. Surv., Eighteenth Ann. Rept., pt. 3, 1898, p. 192.—Knowlton, U. S. Geol. Surv., Mon. 32, pt. 2, 1899, p. 696, pl. lxxxvi, f. 1.—Penhallow, Rept. Tert. Pl. Brit. Col., 1908, p. 76.
Kenai: Port Graham, Kootznahoo near Sitka, and Yukon River, Alaska.
Miocene: Yellowstone National Park.
Eocene?: Burrard Inlet, British Columbia.

Populus balsamoides eximia (Göppert) Lesquereux, Rept. U. S. Geol. Surv. Terr., vol. 8 (Cret. and Tert. Fl.), 1883, p. 226, pl. xlviA, f. 10.
Populus eximia Göppert, Tert. Fl. v. Schossnitz, 1855, p. 23; Abhandl. Schles. Gesell. vaterländ. Kultur, 1861, p. 204.
Fort Union: Badlands, N. Dak.
Kenai: Kootznahoo, Alaska.

Populus balsamoides? Göppert var. *latifolia* Lesquereux, Rept. U. S. Geol. Surv. Terr., vol. 8 (Cret. and Tert. Fl.), 1883, p. 158, pl. xxxi, f. 4 = *Populus scudderi.*

Populus berggreni Heer, Öfversigt. K. Vetenskaps-Akad. Förhandl., 1871, p. 1183.—Lesquereux, U. S. Geol. Surv., Mon. 17, 1891 [1892], p. 42, pl. viii, f. 2–4.

Dakota: Ellsworth County, Kans.; New Ulm, Minn.

Populus clarkiana Hollick, Md. Geol. Surv., Pliocene and Pleistocene, 1906, p. 223, pl. lxx, f. 6.

Pleistocene (Sunderland): Head of Island Creek, Calvert County, Md.

Populus cordata Newberry, Lyc. Nat. Hist. N. Y., Ann., vol. 9, 1868, p. 60; U. S. Geol. Surv., Mon. 35, 1898, p. 38, pl. xxix, f. 6.—Penhallow, Rept. Tert. Pl. Brit. Col., 1908, p. 76.

Fort Union: Yellowstone River, Mont.

Paskapoo (Fort Union): Quilchena and Tulameen rivers, British Columbia.

Populus cordifolia Newberry = *Populus dakotana.*

Populus craspedodroma Ward, U. S. Geol. Surv., Sixth Ann. Rept., 1884–85 [1886], p. 550, pl. xxxvi, f. 1; U. S. Geol. Surv., Bull. 37, 1887, p. 21, pl. viii, f. 3.

Fort Union: Burns's ranch, Glendive, Mont.

Populus crassa (Lesquereux) Cockerell, Am. Mus. Nat. Hist., Bull., vol. 24, 1908, p. 83, pl. viii, f. 22; pl. x, f. 42; Am. Nat., vol. 44, 1910, p. 44, f. 7.—Knowlton, U. S. Nat. Mus., Proc., vol. 51, 1916, p. 262, pl. xviii.

Macreightia crassa Lesquereux, Rept. U. S. Geol. Surv. Terr., vol. 8 (Cret. and Tert. Fl.), 1883, p. 175, pl. xxxiv, f. 16, 17.

Diospyros cuspidata Kirchner, Acad. Sci. St. Louis, Trans., vol. 8, 1898, p. 185, pl. xii, f. 1.

Miocene: Florissant, Colo.

Populus cretacea Knowlton, U. S. Geol. Surv., Bull. 257, 1905, p. 138, pl. xvii, f. 1–5.

Judith River: Willow Creek, Fergus County, Mont.

Populus cuneata Newberry, N. Y. Lyc. Nat. Hist., Ann., vol. 9, 1868, p. 64.—Ward, U. S. Geol. Surv., Mon. 35, 1898, p. 41, pl. xxviii, f. 2–4; pl. xxix, f. 7; U. S. Geol. Surv., Sixth Ann. Rept., 1884–85 [1886], p. 550, pl. xxxiii, f. 5–11; U. S. Geol. Surv., Bull. 37, 1887, p. 19, pl. iv, f. 5–8; pl. v, f. 1–3.—Penhallow, Rept. Tert. Pl. Brit. Col., 1908, p. 77.—Knowlton, Washington Acad. Sci., Proc., vol. 11, 1909, pp. 185–215.

"*Populus nervosa* var. *β. elongata* Ny." [Lesquereux], U. S. Geol. and Geog. Surv. Terr., Ill. Cret. and Tert. Pl., 1878, pl. xiii, f. 2–4.

"*Populus nebrascensis* Ny." [Lesquereux], op. cit., pl. xiv, f. 7.

Fort Union: Yellowstone River, Sevenmile and Clear creeks below Glendive, Mont.

Lance: Near Thunder Butte post office, S. Dak.; 60 miles south of Glasgow, 6 miles east of Miles City, 9 miles east of Miles

Populus cuneata—Continued.
City, 18 miles east of Miles City, 10 miles northeast of Glendive, near Melville, near Bridger, Mont.; Ilo, Nowater mine, Big Horn County, near Kirby, Wyo.
Lance (Colgate): Iron Bluff near Glendive, Mont.
Eocene: Red Deer River, Alberta.
?Oligocene: Tulameen River, British Columbia.
?Miocene: Tranquille River, British Columbia.

Populus cuneata Newberry [Lesquereux], U. S. Geol. and Geog. Surv. Terr., Ill. Cret. and Tert. Pl., 1878, pl. xiv, f. 1–4 = *Populus rotundifolia*.

Populus cyclomorpha Knowlton and Cockerell, n. name.
Populus rotundifolia Newberry, U. S. Nat. Mus., Proc., vol. 5, 1882 [1883], p. 506; U. S. Geol. Surv., Mon. 35, 1898, p. 51, pl. xxix, f. 1–4.—Penhallow, Rept. Tert. Pl. Brit. Col., 1908, p. 79.—Knowlton, Washington Acad. Sci., Proc., vol. 11, 1909, p. 189. [Homonym, Griffith, 1847.]
Populus cuneata Newberry. [Lesquereux], U. S. Geol. and Geog. Surv. Terr., Ill. Cret. and Tert. Pl., 1878, pl. xiv, f. 1–4.
Hanna: Carbon, Wyo.
Fort Union: Yellowstone River, Mont.; Fort Union, N. Dak.
Lance: Near Miles City, Mont.

Populus cyclophylla Heer = *Populites cyclophylla*.

Populus dakotana Cockerell, Bull. Torr. Bot. Club, vol. 33, 1906, p. 312.
Populus? cordifolia Newberry, N. Y. Lyc. Nat. Hist. Ann., vol. 9, 1868, p. 18.—[Lesquereux], U. S. Geol. and Geog. Surv. Terr., Ill. Cret. and Tert. Pl., 1878, pl. v, f. 5.—Newberry, U. S. Geol. Surv., Mon. 35, 1898, p. 40, pl. iii, f. 7; pl. v, f. 5.—Dawson, Roy. Soc. Canada, Trans., vol. 3, 1885 [1886], p. 17.—Penhallow, Rept. Tert. Pl., Brit. Col., 1908, p. 76.—Knowlton, in Woodruff, U. S. Geol. Surv., Bull. 452, 1911, p. 20. [Homonym, Burgsdorf, 1787.]
Dakota: Blackbird Hill, Nebr.; near Lander, Wyo.
Dakota (?): Calgary, Alberta.

Populus daphnogenoides Ward, U. S. Geol. Surv., Sixth Ann. Rept., 1884–85 [1886], p. 550, pl. xxxv, f. 7–9; U. S. Geol. Surv., Bull. 37, 1887, p. 20, pl. vii, f. 4–6.—Penhallow, Roy. Soc. Canada, Trans., 2d ser., vol. 8, 1902, p. 57; ?Rept. Tert. Pl. Brit. Col., 1908, p. 77.—Knowlton, U. S. Geol. Surv., Mon. 32, pt. 2, 1899, p. 696, pl. lxxxiv, f. 2; Washington Acad. Sci., Proc., vol. 11, 1909, pp. 189, 194, 213.
Fort Union: Sevenmile Creek, Glendive, Mont.; Elk Creek, Yellowstone National Park.
Paskapoo (Fort Union): Red Deer River, Saskatchewan; Similkameen River, British Columbia.
Lance: Near Miles City, and near Melville, Mont.; near Ilo, Big Horn Basin, Wyo.

Populus? daturaefolia (Ward) Cockerell = *Credneria daturaefolia*.

Populus? debeyana Heer = *Juglans debeyana*.

Populus decipiens Lesquereux, U. S. Geol. and Geog. Surv. Terr., Ann. Rept. 1872 [1873], p. 385; Rept. U. S. Geol. Surv. Terr., vol. 7 (Tert. Fl.), 1878, p. 179, pl. xxiii, f. 7–11.
Hanna: Creston and Carbon, Wyo.

Populus decipiens Lesquereux. See ***Populus arctica decipiens*** (Lesquereux) Cockerell.

Populus deltoides Marshall. Berry, Am. Jour. Sci., 4th ser., vol. 29, 1910, p. 393.
Pleistocene: Warrior River 328½ and 311½ miles above Mobile, and Abercrombie Landing, Chattahoochee River, Ala.

Populus cf. **P. deltoides** Marshall. Berry, Torreya, vol. 14, 1914, p. 160.
Pleistocene: Seven Springs, Neuse River, Wayne County, N. C.

Populus denticulata Heer, Fl. foss. arct., vol. 7, 1883, p. 20, pl. lv, f. 5.—Lesquereux, U. S. Nat. Mus., Proc., vol. 10, 1887, p. 40.
——: Upper Kanab Valley, Utah.

Populus? distorta Knowlton, U. S. Geol. Surv., Prof. Paper 130, 19—, p. —, pl. iv, f. 6 [in preparation].
Laramie: Cowan station, near Denver, Colo.

Populus elliptica Newberry, N. Y. Lyc. Nat. Hist., Ann., vol. 9, 1868, p. 16.—[Lesquereux], U. S. Geol. and Geog. Surv. Terr., Ill. Cret. and Tert. Pl., 1878, pl. iii, f. 1, 2.—Newberry, U. S. Geol. Surv., Mon. 35, 1898, p. 43, pl. iii, f. 1, 2.—Lesquereux, U. S. Geol. Surv., Mon. 17, 1891 [1892], p. 45.
Dakota: Blackbird Hill, Nebr.

Populus eotremuloides Knowlton, U. S. Geol. Surv., Eighteenth Ann. Rept., pt. 3, 1898, p. 725, pl. c, f. 1, 2; pl. ci, f. 1, 2.
Payette: Marsh, Idaho.

Populus eximia Göppert = ***Populus balsamoides eximia.***

Populus flabellum Newberry, Boston Jour. Nat. Hist., vol. 7, 1863, p. 524; U. S. Geol. Surv., Mon. 35, 1898, p. 44, pl. xx, f. 4.
Puget: Chuckanutz, near Bellingham Bay, Wash.

Populus genetrix Newberry, N. Y. Lyc. Nat. Hist., Ann., vol. 9, 1868, p. 64.—[Lesquereux], U. S. Geol. and Geog. Surv. Terr., Ill. Cret. and Tert. Pl., 1878, pl. xii, f. 1.—Newberry, U. S. Geol. Surv., Mon. 35, 1898, p. 44, pl. xxvii, f. 1.—Dawson, Roy. Soc. Canada, Trans., vol. 3, 1885 [1886], p. 17.—Penhallow, Rept. Tert. Pl. Brit. Col., 1908, p. 77.—Knowlton, Washington Acad. Sci., Proc., vol. 11, 1909, pp. 190, 194, 198, 211.
Fort Union: Banks of Yellowstone, Mont.
Lance: Signal Butte near Miles City, near Melville and mouth of Cedar Creek, near Glendive, Mont.; Monarch and near Kirby, Wyo.
Eocene: Coal Brook, British Columbia; Porcupine Creek and Great Valley; Souris River and Quilchena, British Columbia.

Populus glandulifera Heer, Fl. tert. Helv., vol. 2, 1856, p. 17, pl. lviii, f. 5–11; pl. lxviii, f. 7; Fl. foss. arct., vol. 2, Abt. 2, 1869, p. 25, pl. ii, f. 1, 2.—Lesquereux, Rept. U. S. Geol. Surv. Terr., vol. 8 (Cret. and Tert. Fl.), 1883, p. 226, pl. xlvi–A, f. 3, 4.—Ward, U. S. Geol. Surv., Sixth Ann. Rept., 1884–85 [1886], p. 550, pl. xxxiii, f. 1–4; U. S. Geol. Surv., Bull. 37, 1887, p. 19, pl. iv, f. 1–4.—Knowlton, U. S. Nat. Mus., Proc., vol. 17, 1894, p. 216; Geol. Soc. Am., Bull., vol. 5, 1893, p. 580; in Russell, U. S. Geol. Surv., Bull. 108, 1893, p. 103.—Knowlton, U. S. Geol. Surv., Bull. 204, 1902, p. 109; in Smith, U. S. Geol. Surv., Geol. Folio 86 (Ellensburg, Wash.), 1903, p. 3.—Knowlton, U. S. Geol. Surv., Mon. 32, pt. 2, 1899, p. 694, pl. lxxxiv, f. 1.

Fort Union: Near Glendive, Mont.; Badlands, N. Dak.; Elk Creek, Yellowstone National Park.

Kenai: Port Graham, Alaska.

Ellensburg: Ellensburg, Wash.

Populus grandidentata Michaux. Penhallow, Geol. Soc. Am., Bull., vol. 1, 1890, p. 326; Roy. Soc. Canada, Trans., 2d ser., vol. 2, 1896, p. 72; Brit. Assoc. Adv. Sci., Bradford meeting, 1900, p. 336; Am. Nat., vol. 41, 1907, p. 448.

Pleistocene: Don Valley, Toronto; Greens Creek and Montreal, Canada.

Populus grewiopsis Ward, U. S. Geol. Surv., Sixth Ann. Rept., 1884–85 [1886], p. 550, pl. xxxvi, f. 6; U. S. Geol. Surv., Bull. 37, 1887, p. 23, pl. ix, f. 1.

Fort Union: Sevenmile Creek, Glendive, Mont.

Populus harkeriana Lesquereux, U. S. Geol. Surv., Mon. 17, 1891 [1892], p. 44, pl. xlvi, f. 4.—Hollick, N. Y. Acad. Sci., Ann., vol. 11, 1898, p. 419, pl. xxxvi, f. 8; U. S. Geol. Surv., Mon. 50, 1906, p. 49, pl. vii, f. 31.—Berry, Bull. Torr. Bot. Club, vol. 39, 1912, p. 394.

Dakota: Near Fort Harker, Kans.

Raritan: Tottenville, Staten Island, N. Y.

Woodbine: Arthurs Bluff, Tex.

Populus hederoides Ward, U. S. Geol. Surv., Sixth Ann. Rept., 1884–85 [1886], p. 550, pl. xxxvi, f. 3; U. S. Geol. Surv., Bull. 37, 1887, p. 22, pl. viii, f. 5.

Fort Union: Sevenmile Creek, Glendive, Mont.

Populus heerii Saporta. Lesquereux, Rept. U. S. Geol. Surv. Terr., vol. 8 (Cret. and Tert. Fl.), 1883, p. 157 = *Populus lesquereuxi*.

Populus heliadum Unger. Lesquereux, 1873 = *Populus ungeri*.

Populus hookeri Heer, Fl. foss. arct., vol. 1, 1868, p. 137, pl. xxi, f. 16.—Dawson, Roy. Soc. Canada, Trans., vol. 7, 1889, p. 71, pl. x, f. 5.—Penhallow, Rept. Tert. Pl. Brit. Col., 1908, p. 77.

Eocene: Bear River, Mackenzie Basin, Northwest Territory.

Populus hyperborea Heer, Fl. foss. arct., vol. 3, pt. 2, 1873, p. 106.—Lesquereux, U. S. Geol. Surv., Mon. 17, 1891 [1892], p. 43, pl. iii, f. 9–11; pl. viii, f. 1; pl. xlvii, f. 5; U. S. Geol. Surv., Prof. Paper 112, 1919, p. 77.

Dakota: Near Delphos, and Ellsworth County, Kans.

Tuscaloosa: Cottondale, Ala.

Populus inaequalis Ward, U. S. Geol. Surv., Sixth Ann. Rept., 1884–85 [1886], p. 550, pl. xxxvi, f. 7; U. S. Geol. Surv., Bull. 37, 1887, p. 24, pl. ix, f. 2.—Knowlton, Washington Acad. Sci., Proc., vol. 11, 1909, p. 196.

Fort Union: Burns's ranch, Glendive, Mont.

Lance: Near Bridger, Mont.

Populus kansaseana Lesquereux, U. S. Geol. Surv., Mon. 17, 1891 [1892], p. 42, pl. xvii, f. 1–7.

Dakota: Ellsworth County, Kans.

Populus knightii Knowlton, Geol. Soc. Am., Bull., vol. 8, 1897, p. 142.

Mesaverde: Dutton Creek coal, Laramie Plains, Wyo.

Populus laevigata Lesquereux = *Populus aequalis.*

Populus lancastriensis Lesquereux = *Populites lancastriensis.*

Populus latidentata Dawson, Roy. Soc. Canada, Trans., vol. 3, sec. 4, 1885 [1886], p. 16.

Belly River: Medicine Hat, Alberta.

Populus latior Al. Braun, in Buckland's Geol., 1835 ?, p. 512.—Heer, Fl. foss. alask., 1869, p. 25, pl. ii, f. 4.— ?Hollick, Md. Geol. Surv., Pliocene and Pleistocene, 1906, p. 224, pl. lxx, f. 7.—Penhallow, Rept. Tert. Pl. Brit. Col., 1908, p. 77.—Knowlton, U. S. Nat. Mus., Proc., vol. 17, 1894, p. 216; Geol. Soc. Am., Bull., vol. 5, 1893, p. 580.

Kenai: Port Graham, Alaska.

Oligocene: Similkameen River, Horsefly River, Tulameen River, British Columbia.

?Pleistocene (Sunderland): Head of Island Creek, Calvert County, Md.

Populus latior cordifolia Heer, Fl. tert. Helv., vol. 2, 1856, p. 12, pl. lv.—Lesquereux, U. S. Geol. and Geog. Surv. Terr., Ann. Rept. 1871 [1872], p. 289.—Penhallow, Rept. Tert. Pl., Brit. Col., 1908, p. 78.

Populus latior transversa Heer. Lesquereux, U. S. Geol. and Geog. Surv. Terr., Ann. Rept. 1871 [1872], p. 287.

Post-Laramie ?: Medicine Bow, Wyo.

Oligocene: Tulameen River, British Columbia.

Populus latior transversa Heer. Lesquereux = *Populus latior cordifolia.*

Populus latior truncata Al. Braun, in Stizenberger, Uebers., 1851, p. 79.—Lesquereux, Rept. U. S. Geol. Surv. Terr., vol. 8 (Cret. and Tert. Fl.), 1883, p. 226, pl. xlvi, f. 14.
Fort Union: Badlands, N. Dak.

Populus lesquereuxi Cockerell, Bull. Torr. Bot. Club, vol. 33, 1906, p. 307; Univ. Colo. Studies, vol. 3, 1906, p. 172; Am. Nat., vol. 44, 1910, p. 44, f. 8.—Knowlton, U. S. Nat. Mus., Proc., vol. 51, 1916, p. 261.
Populus heerii Saporta. Lesquereux, Rept. U. S. Geol. Surv. Terr., vol. 8 (Cret. and Tert. Fl.), 1883, p. 157, pl. xxx, f. 1-8; pl. xxxi, f. 11.
Miocene: *Florissant, Colo.

Populus leuce (Rossmässler) Unger, Gen. et sp. pl. foss., 1850, p. 417.—Heer, Acad. Nat. Sci., Philadelphia, Proc., vol. 10, 1858, p. 265.
Dakota: Kansas?

Populus leucophylla Unger, Gen. et sp. pl. foss., 1850, p. 417.—Heer, Fl. foss. arct., vol. 2, Abt. 2, 1869, p. 26, pl. iii, f. 4, 5.—Lesquereux, U. S. Geol. Surv. Terr., Ann. Rept. 1871 [1872], p. 296; idem, 1872 [1873], p. 392.—Knowlton, U. S. Nat. Mus., Proc., vol. 17, 1894, p. 217; Geol. Soc. Am., Bull., vol. 5, 1893, p. 580.
Kenai: Alaska.
Post-Laramie: Black Buttes, Wyo.

Populus leucophylla Unger. Lesquereux, U. S. Geol. and Geog. Surv. Terr., Ann. Rept. 1871 [1872], p. 296. [Not afterward recognized by Lesquereux.]
——: Spring Canyon, Mont.

Populus lindgreni Knowlton, U. S. Geol. Surv., Eighteenth Ann. Rept., pt. 3, 1898, p. 725, pl. c, f. 3; U. S. Geol. Surv., Bull. 204, 1902, p. 29, pl. ii, f. 1.
Payette: Marsh, Idaho.
Mascall: Near Van Horn's ranch, Grant County, Oreg.

Populus litigiosa Heer [Lesquereux], U. S. Geol. and Geog. Surv. Terr., Ill. Cret. and Tert. Pl., 1878, pl. iv, f. 1 = *Populites cyclophylla.*

Populus litigiosa Lesquereux = *Populites litigiosus.*

Populus litigiosus denticulata Lesquereux = *Populites litigiosus denticulata.*

Populus longior Dawson, Roy. Soc. Canada, Trans., vol. 1, 1882 [1883], p. 27.
Upper Cretaceous: Nanaimo, Baynes Sound, Vancouver Island, British Columbia.

Populus meedsii Knowlton, U. S. Nat. Mus., Proc., vol. 16, 1893, p. 34, pl. i, f. 1, 2.
Fort Union: Near Glendive?, Mont.

Populus melanaria Heer. Lesquereux = *Populus melanarioides.*

Populus melanarioides Lesquereux, U. S. Geol. and Geog. Surv. Terr., Bull., vol. 1, 1875, p. 379; idem, Ann. Rept. 1874 [1876], p. 302; Rept. U. S. Geol. Surv. Terr., vol. 7 (Tert. Fl.), 1878, p. 174, pl. lxii, f. 5.—Knowlton, U. S. Geol. Surv., Bull. 163, 1900, p. 35.

Populus melanaria Heer. Lesquereux, U. S. Geol. and Geog. Surv. Terr., Ann. Rept. 1874 [1876], p. 302; Rept. U. S. Geol. Surv. Terr., vol. 7 (Tert. Fl.), 1878, p. 173, pl. lxiv, f. 5.

Mesaverde: Point of Rocks, Wyo.

Populus microphylla Newberry, N. Y. Lyc. Nat. Hist., Ann., vol. 9, 1868, p. 17.—[Lesquereux], U. S. Geol. and Geog. Surv. Terr., Ill. Cret. and Tert. Pl., 1878, pl. iii, f. 5.—Newberry, U. S. Geol. Surv., Mon. 35, 1898, p. 46, pl. iii, f. 5.—Lesquereux, U. S. Geol. Surv., Mon. 17, 1891 [1892], p. 45.

Dakota: Blackbird Hill, Nebr.

Populus micro-tremuloides Knowlton, U. S. Nat. Mus., Proc., vol. 51, 1916, p. 261, pl. xix, f. 2.

Miocene: *Florissant, Colo.

Populus monodon Lesquereux, Am. Philos. Soc., Trans., vol. 13, 1869, p. 413, pl. xv, f. 1, 2 = *Ficus monodon.*

Populus monodon Lesquereux. Lesquereux, Rept. U. S. Geol. Surv. Terr., vol. 7 (Tert. Fl.), 1878, p. 180, pl. xxiv, f. 1, 2 = *Ficus uncata.*

Populus monodon Lesquereux. Lesquereux, U. S. Nat. Mus., Proc., vol. 11, 1888, p. 21 = *Magnolia culveri.*

Populus mutabilis crenata Heer, Fl. tert. Helv., vol. 2, 1856, p. 21, pl. lx, f. 5, 8, 11, 12a.—Lesquereux, U. S. Geol. and Geog. Surv. Terr., Ann. Rept. 1872 [1873], p. 386. [Not afterward referred to by Lesquereux.]

Hanna: Carbon, Wyo.

Populus mutabilis lancifolia Heer, Fl. tert. Helv., vol. 2, 1856, p. 23, pl. lxi, f. 7, 8, 10.—Lesquereux, U. S. Geol. and Geog. Surv. Terr., Ann. Rept. 1871 [1872], p. 296. [Not afterward referred to by Lesquereux.]

——: Spring Canyon, Mont.

Populus mutabilis oblonga (Braun) Heer, Fl. tert. Helv., vol. 2, 1856, p. 21, pl. ix, f. 6, 7, 9, 10, 13–16.—Penhallow, Rept. Tert. Pl. Brit. Col., 1908, p. 78.

Oligocene: Quilchena River, British Columbia.

Miocene: Tranquille River, British Columbia.

Populus nebrascensis Newberry, N. Y. Lyc. Nat. Hist., Ann., vol. 9, 1868, p. 62.—[Lesquereux], U. S. Geol. and Geog. Surv. Terr., Ill. Cret. and Tert. Pl., 1878, pl. xii, f. 4, 5.—Newberry, U. S. Geol. Surv., Mon. 35, 1898, p. 47, pl. xxvii, f. 4, 5.—Penhallow, Rept.

Populus nebrascensis—Continued.
Tert. Pl. Brit. Col., 1908, p. 78.—Knowlton, Washington Acad. Sci., Proc., vol. 11, 1909, p. 189.
Fort Union: Banks of Yellowstone River, Mont.
Paskapoo (Fort Union): Quilchena, British Columbia.
Lance: Near Miles City, Mont.
Denver: Golden, Colo.

"*Populus nebrascensis* Ny." [Lesquereux], U. S. Geol. and Geog. Surv. Terr., Ill. Cret. and Tert. Pl., 1878, pl. xiv, f. 7 = *Populus cuneata*.

Populus nebrascensis acute-dentata Lesquereux, Mus. Comp. Zool., Bull., vol. 16, 1888, p. 47.
Denver: Golden, Colo.

Populus nebrascensis grandidentata Lesquereux, Mus. Comp. Zool., Bull., vol. 16, 1888, p. 47.
Denver: Golden, Colo.

Populus nebrascensis longifolia Lesquereux, Mus. Comp. Zool., Bull., vol. 16, 1888, p. 48.
Denver: Golden, Colo.

Populus nebrascensis rotunda Lesquereux, Mus. Comp. Zool., Bull., vol. 16, 1888, p. 47.
Denver: Golden, Colo.

Populus? neomexicana Knowlton, U. S. Geol. Surv., Prof. Paper 101, 1918, p. 258, pl. liii, f. 3.
Trinidad: Raton, N. Mex.

Populus neotremuloides Knowlton, U. S. Geol. Surv., Prof. Paper 101, 1918, p. 296, pl. lxvi, f. 2.
Raton: Primero, Colo.

Populus nervosa Newberry, N. Y. Lyc. Nat. Hist., Ann., vol. 9, 1868, p. 61.—[Lesquereux], U. S. Geol. and Geog. Surv. Terr., Ill. Cret. and Tert. Pl., 1878, pl. xii, f. 2, 3.—Newberry, U. S. Geol. Surv., Mon. 35, 1898, p. 48, pl. xxvii, f. 2, 3.—Penhallow, Rept. Tert. Pl. Brit. Col., 1908, p. 78.
Fort Union: Banks of Yellowstone River, Mont.
Eocene: Rocky Mountain House, northern Saskatchewan.

Populus nervosa elongata Newberry, N. Y. Lyc. Nat. Hist., Ann., vol. 9, 1868, p. 62.—[Lesquereux], U. S. Geol. and Geog. Surv. Terr., Ill. Cret. and Tert. Pl., 1878, pl. xiii, f. 1.—Newberry, U. S. Geol. Surv., Mon. 35, 1898, p. 49, pl. xxviii, f. 1.—Knowlton, Washington Acad. Sci., Proc., vol. 11, 1909, p. 189.
Fort Union: *Yellowstone River, Mont.
Lance: Near Miles City, Mont.

"*Populus nervosa* var. *β. elongata* Ny." [Lesquereux], U. S. Geol. and Geog. Surv. Terr., Ill. Cret. and Tert. Pl., 1878, pl. xiii, f. 2–4 = *Populus cuneata*.

Populus newberryi Cockerell, Bull. Torr. Bot. Club, vol. 33, 1906, p. 312.

Populus acerifolia Newberry, N. Y. Lyc. Nat. Hist., Ann., vol. 9, 1868, p. 65.—Newberry, U. S. Geol. Surv., Mon. 35, 1898, p. 37, pl. xxviii, f. 5–8.—Penhallow, Rept. Tert. Pl., Brit. Col., 1908, p. 75.—Knowlton, Washington Acad. Sci., Proc., vol. 11, 1909, pp. 190, 211. [Homonym, Loddiges, 1838.]

Fort Union: Fort Union, N. Dak.

Lance: Near Miles City, Mont.

Eocene: Calgary, Alberta; Porcupine Creek and Great Valley, Saskatchewan.

Oligocene: Quilchena, Coal Gully, and Horsefly River, British Columbia.

Miocene: Tranquille River, British Columbia.

Populus? apiculata Hollick = *Cordia apiculata*.

Populus obovata Knowlton, U. S. Geol. Surv., Bull. 163, 1900, p. 34, pl. vii, f. 4.

Mesaverde: Near Harper station, Wyo.

Populus obtrita Dawson, Roy. Soc. Canada, Trans., vol. 8, 1890, p. 82, text f. 12.—Penhallow, Roy. Soc. Canada, Trans., 2d ser., vol 8, 1902, p. 57; Rept. Tert. Pl. Brit. Col., 1908, p. 78.

Eocene: Similkameen Valley, Red Deer River, Quilchena, British Columbia.

Populus occidentalis Knowlton, U. S. Geol. Surv., Eighteenth Ann. Rept., pt. 3, 1898, p. 727, pl. xcix, f. 14.

Payette: Marsh, Idaho.

Populus orbicularis (Newberry) Berry, Bull. Torr. Bot. Club, vol. 36, 1909, p. 250; Geol. Surv. N. J., Bull. 3, 1911, p. 112, pl. xi, f. 5, 6.

Phyllites orbicularis Newberry, U. S. Geol. Surv., Mon. 26, 1896, p. 130, pl. xxiv, f. 7, 8.

Raritan: Sayreville, N. J.

Populus ovalis? Göppert. Lesquereux [1871] = *Populus mutabilis ovalis*.

Populus ovata Lesquereux = *Ampelophyllum ovatum*.

Populus oxyphylla Saporta. Lesquereux, Rept. U. S. Geol. Surv. Terr., vol. 8 (Cret. and Tert. Fl.), 1883, p. 159, pl. xxxviii, f. 9–11 = *Populus pyrifolia*.

Populus oxyrhyncha Ward, U. S. Geol. Surv., Sixth Ann. Rept., 1884–85 [1886], p. 550, pl. xxxv, f. 10, 11; U. S. Geol. Surv., Bull. 37, 1887, p. 21, pl. viii, f. 1, 2.

Fort Union: Sevenmile Creek, Glendive, Mont.

Populus polymorpha Newberry, U. S. Nat. Mus., Proc., vol. 5, 1882 [1883], p. 506; U. S. Geol. Surv., Mon. 35, 1898, p. 50, pl. xlvi, f. 3; pl. xlvii, f. 4; pl. xlix, f. 4, 8, 9 [misprinted 1]; pl. lviii, f. 4.

Clarno (upper part): Bridge Creek, Oreg.

Populus polymorpha Newberry [part] = *Betula heteromorpha.*

Populus polymorpha Newberry [part], U. S. Geol. Surv., Mon. 35, 1898 [1899], p. 50, pl. xlvi, f. 4 = *Quercus oregoniana.*

Populus polymorpha Newberry, U. S. Geol. Surv., Mon. 35, 1898 [1899], p. 50, pl. xlix, f. 7 = *Quercus pseudo-alnus.*

Populus potomacensis Ward, U. S. Geol. Surv., Fifteenth Ann. Rept., 1895, p. 356, pl. iv, f. 1–3.—Fontaine, in Ward, U. S. Geol. Surv., Mon. 48, 1905 [1906], p. 500.—Berry, Md. Geol. Surv., Lower Cret., 1912, p. 458, pl. lxxxi, f. 1–1e.

Populus auriculata Ward, U. S. Geol. Surv., Fifteenth Ann. Rept., 1895, p. 356, pl. iv, f. 4.—Fontaine, in Ward, U. S. Geol. Surv., Mon. 48, 1905 [1906], p. 499, pl. cx, f. 5.

Populophyllum menispermoides Ward U. S. Geol. Surv., Mon. 48, 1905 [1906], p. 498, pl. cx, f. 2 [not f. 3, 4].

Patapsco: Mount Vernon and White House Bluff, Va.

Populus? problematica Knowlton, U. S. Geol. Surv., Bull. 105, 1893, p. 51, pl. vi, f. 5, 6.

———: Head of Fir Canyon, near Bozeman, Mont.

Populus protozaddachi Dawson, Roy. Soc. Canada, Trans., vol. 1, 1882 [1883], p. 26, pl. vii, f. 25.

Upper Cretaceous: Newcastle Island, British Columbia.

Populus pseudocredneria Cockerell, Am. Mus. Nat. Hist., Bull., vol. 24, 1908, p. 83, pl. viii, f. 28.

Miocene: Florissant, Colo.

Populus pseudo-tremuloides Hollick, Md. Geol. Surv., Pliocene and Pleistocene, 1906, p. 224, pl. lxx, f. 5.

Pleistocene (Sunderland): Head of Island Creek, Calvert County, Md.

Populus pyrifolia Kirchner, St. Louis Acad. Sci., Trans., vol. 8, 1898, p. 185, pl. xv, f. 4.—Cockerell, Univ. Colo. Studies, vol. 3, 1906, p. 173.—Knowlton, U. S. Nat. Mus., Proc., vol. 51, 1916, p. 261.

Populus oxyphylla Saporta. Lesquereux, Rept. U. S. Geol. Surv. Terr., vol. 8 (Cret. and Tert. Fl.), 1883, p. 159, pl. xxxviii, f. 9–11.

Miocene: Florissant, Colo.

Populus rectinervata Dawson, Roy. Soc. Canada, Trans., vol. 1, 1882 [1883], p. 27, pl. vii, f. 26.

Upper Cretaceous: Baynes Sound, Vancouver Island, British Columbia.

Populus rhomboidea Newberry, Am. Jour. Sci., 2d ser., vol. 27, 1859, p. 360.—Newberry, U. S. Geol. Surv., Mon. 35, 1898, p. 51, pl. xx, f. 1, 2.

Upper Cretaceous: Nanaimo, Vancouver Island, British Columbia.

Populus? ricei Fontaine, in Ward, U. S. Geol. Surv., Mon. 48, 1905 [1906], p. 266, pl. lxix, f. 10.

Horsetown: Wilson Creek, 25 miles southeast of Buck Mountain, Oreg.

Populus richardsoni Heer, Fl. foss. arct., vol. 1, 1868, p. 98, pl. iv, f. 1–5; pl. vi, f. 7, 8; pl. xv, f. 1c.—Lesquereux, U. S. Geol. and Geog. Surv. Terr., Ann. Rept. 1873 [1874], p. 411; Rept. U. S. Geol. Surv. Terr., vol. 7 (Tert. Fl.), 1878, p. 177, pl. xxii, f. 10–12; U. S. Nat. Mus., Proc., vol. 5, 1882 [1883], p. 441, pl. ix, f. 1.—Ward, U. S. Geol. Surv., Sixth Ann. Rept., 1884–85 [1886], p. 550, pl. xxxvi, f. 4; U. S. Geol. Surv., Bull. 37, 1887, p. 23, pl. viii, f. 6.—Penhallow, Roy. Soc. Canada, Trans., 2d ser., vol. 8, 1902, p. 58; Rept. Tert. Pl. Brit. Col., 1908, p. 79.—Knowlton, U. S. Nat. Mus., Proc., vol. 17, 1894, p. 217.

Miocene: Elko station, Nev.

Fort Union: Near Glendive, Mont.

Paskapoo (Fort Union): Red Deer River, Porcupine Creek, Calgary, and Mackenzie River, Canada.

Kenai: Chignik Bay, Alaska.

Populus rotundifolia Newberry = *Populus cyclomorpha.*

Populus russelli Knowlton, U. S. Geol. Surv., Bull. 108, 1893, p. 104; Smith, U. S. Geol. Surv., Folio 86 (Ellensburg, Wash.), 1903, p. 3.

Ellensburg: Near Ellensburg, Wash.

Populus scudderi Cockerell, Bull. Torr. Bot. Club, vol. 33, 1906, p. 307; Univ. Colo. Studies, vol. 3, 1906, p. 172.

Populus balsamoides? Göppert var. *latifolia* Lesquereux, Rept. U. S. Geol. Surv. Terr., vol. 8 (Cret. and Tert. Fl.), 1883, p. 158, pl. xxxi, f. 4. [Homonym, *P. latifolia* Moench, 1794.]

Miocene: Florissant, Colo.

Populus smilacifolia Newberry, N. Y. Lyc. Nat. Hist., Ann., vol. 9, 1868, p. 66.—[Lesquereux], U. S. Geol. and Geog. Surv. Terr., Ill. Cret. and Tert. Pl., 1878, pl. xiv, f. 5.—Newberry, U. S. Geol. Surv., Mon. 35, 1898, p. 53, pl. xxix, f. 5.

Fort Union: Fort Union, N. Dak.

Populus speciosa Ward, U. S. Geol. Surv., Sixth Ann. Rept., 1884–85 [1886], p. 550, pl. xxxiv, f. 1–4; U. S. Geol. Surv., Bull. 37, 1887, p. 20, pl. v, f. 4–7.—Penhallow, Rept. Tert. Pl. Brit. Col., 1908, p. 79.—Knowlton, U. S. Geol. Surv., Mon. 32, pt. 2, 1899, p. 694, pl. lxxxiv, f. 3; Washington Acad. Sci., Proc., vol. 11, 1909, p. 189; Jour. Geol., vol. 19, 1911, p. 369.

Fort Union: Clear Creek, Glendive, Mont.; Elk Creek, Yellowstone National Park.

Paskapoo (Fort Union): Omineca River, British Columbia.

Lance: Cannonball River, S. Dak.; near Miles City, Mont.

Eocene: Omineca River, British Columbia.

Populus stygia Heer, Fl. foss. arct., vol. 3, Abt. 2, 1873, p. 107.—Lesquereux, U. S. Geol. Surv., Mon. 17, 1891 [1892], p. 44, pl. iii, f. 12.—?Hollick, U. S. Geol. Surv., Mon. 50, 1906, p. 49, pl. vii, f. 30.—Berry, Md. Geol. Surv., Upper Cret., 1916, p. 816, pl. lviii, f. 1.

Dakota: Ellsworth County, Kans.

Magothy: Gay Head, Marthas Vineyard, Mass.; Bodkin Point, Anne Arundel County, Md.

Populus subrotunda Lesquereux, Am. Jour. Sci., 2d ser., vol. 45, 1868, p. 205.—Knowlton, Washington Acad. Sci., Proc., vol. 11, 1909, p. 207.

Populus attenuata Braun. Lesquereux, Am. Jour. Sci., 2d ser., vol. 45, 1868, p. 205.

Populus subrotundata Lesquereux, Rept. U. S. Geol. Surv. Terr., vol. 7 (Tert. Fl.), 1878, p. 173, pl. xxiv, f. 6–8.

Hanna: Carbon, Wyo.

Evanston: Evanston, Wyo.

Denver: ?Golden, Colo.

Eocene: Coal Brook, British Columbia.

Lance: Converse County, Wyo.

Populus tenuinervata Lesquereux, Mus. Comp. Zool., Bull., vol. 16, 1888, p. 48.

Denver: Golden, Colo.

Populus trichocarpa Torrey and Gray. Hannibal, Bull. Torr. Bot. Club, vol. 38, 1911, p. 336.

Pliocene: Near Portola, Santa Cruz Mountains, Calif.

Populus trinervis Dawson, Roy. Soc. Canada, Trans., vol. 1, 1882 [1883], p. 26.

Upper Cretaceous: Nanaimo, Baynes Sound, Vancouver Island, British Columbia.

Populus ungeri Lesquereux, Rept. U. S. Geol. Surv. Terr., vol. 7 (Tert. Fl.), 1878, p. 175, pl. xxiv, f. 5.—Penhallow, Roy. Soc. Canada, Trans., 2d ser., vol. 8, 1902, pp. 46, 57; Rept. Tert. Pl., Brit. Col., 1908, p. 79.

Populus helidanum Unger. Lesquereux, U. S. Geol. and Geog. Surv. Terr., Ann. Rept. 1873 [1874], p. 397.

Denver: Golden, Colo.

Paskapoo (Fort Union): Red Deer River, Alberta.

Populus? vivaria Knowlton, U. S. Geol. Surv., Mon. 32, pt. 2, 1899, p. 696, pl. lxxxvi, f. 2.

Miocene: Yellowstone National Park.

Populus wardii Knowlton, U. S. Geol. Surv., Bull. 163, 1900, p. 36, pl. vi, f. 7.

Mesaverde: Point of Rocks, Wyo.

116234°—19—Bull. 696——32

Populus whitei Ward, U. S. Geol. Surv., Sixth Ann. Rept., 1884–85 [1886], p. 550, pl. xxxvi, f. 2; U. S. Geol. Surv., Bull. 37, 1887, p. 22, pl. viii, f. 4.
Fort Union: Burns's ranch, Glendive, Mont.

Populus xantholithensis Knowlton, U. S. Geol. Surv., Mon. 32, pt. 2, 1899, p. 695, pl. lxxxv, f. 1, 2.
Fort Union: Yellowstone National Park.

Populus zaddachi Heer, Fl. tert. Helv., vol. 3, 1859, p. 307; Fl. foss. arct., vol. 2, Abt. 2, 1869, p. 26, pl. ii, f. 5a.—Lesquereux, U. S. Geol. and Geog. Surv. Terr., Ann. Rept. 1871 [1872], p. 292; Rept. U. S. Geol. Surv. Terr., vol. 8 (Cret. and Tert. Fl.), 1883, p. 158, pl. xxxi, f. 8.—?Knowlton, in Turner, Am. Geol., vol. 15, 1895, p. 378; in Lindgren, Jour. Geol., vol. 4, 1896, p. 889.—Penhallow, Rept. Tert. Pl., Brit. Col., 1908, p. 79.—Knowlton, U. S. Geol. Surv., Mon. 32, pt. 2, 1899, p. 694; U. S. Nat. Mus., Proc., vol. 17, 1894, p. 217; Geol. Soc. Am., Bull., vol. 5, 1893, p. 580; in Washburne, U. S. Geol. Surv., Bull. 590, 1914, p. 36.
Kenai: Port Graham, Alaska.
Miocene (auriferous gravels): Chalk Bluffs, Nevada County; Independence Hill, Placer County; Little Cow Creek, and Moonlight, Calif.; Yellowstone National Park.
Ione ?: Volcano Hill, Placer County, Calif.
Miocene: Tranquille River, British Columbia.
Oligocene: Tulameen River, British Columbia.
Eocene ?: Eugene and southwestern Oregon.

Populus sp. Berry, U. S. Nat. Mus., Proc., vol. 48, 1915, p. 298.
Pleistocene: Hickman, Ky.

Populus sp. Dawson, Brit. N. A. Bound. Com., Rept. Geol. and Res. Vic. 49th Parallel, 1875, App. A, p. 331 = *Rhamnacinium triseriatim* and *R. porcupinianum*.

Populus sp. Dawson, Geol. Surv. Canada, 1871–72, p. 59.—Penhallow, Rept. Tert. Pl. Brit. Col., 1908, p. 75.
Eocene: Quesnel, British Columbia.

Populus sp. Dawson, Geol. Surv. Canada, Rept. Progress 1872–73, App. A, p. 68.
Upper Cretaceous: Vancouver Island, British Columbia.

Populus sp. Dawson, Roy. Soc. Canada, Trans., vol. 1, 1882 [1883], p. 27.
Upper Cretaceous: Baynes Sound, Vancouver Island, British Columbia.

Populus sp. Dawson, Roy. Soc. Canada, Trans., vol. 5, 1887, p. 33.
Pierre: Head of Swift Current Creek.

Populus sp. (aments of) Hollick, U. S. Geol. Surv., Mon. 50, 1906, p. 50, pl. vii, f. 16–18.
Magothy: Glen Cove, Long Island, N. Y.; Gay Head, Marthas Vineyard, Mass.

Populus sp. Knowlton, U. S. Geol. Surv., Bull. 105, 1893, p. 50, pl. vi, f. 7–9.
——: Head of Bear Canyon, near Bozeman, Mont.

Populus sp. Knowlton, U. S. Geol. Surv., Bull. 163, 1900, p. 37, pl. vii, f. 5.
Mesaverde: North Fork of Dutton Creek, Laramie Plains, Wyo.

Populus sp. Knowlton, Wash. Geol. Surv., Ann. Rept., vol. 1, 1902, p. 33.
Eocene: Coal Creek, near Keese, Whatcom County, Wash.

Populus n. sp. Knowlton, Washington Acad. Sci., Proc., vol. 11, 1909, p. 213.
Lance: Near Ilo post office, Bighorn Basin, Wyo.

Populus sp. Knowlton, U. S. Geol. Surv., Bull. 257, 1905, p. 140, pl. xix, f. 1.
Judith River: Willow Creek, Fergus County, Mont.

Populus sp. Knowlton, U. S. Nat. Mus., Proc., vol. 51, 1916, p. 268, pl. xvi, f. 4.
Miocene: *Florissant, Colo.

Populus, female ament of, Knowlton, U. S. Geol. Surv., Prof. Paper 101, 1918, p. 297, pl. lxvi, f. 3.
Raton: Near Aguilar, Colo.

PORANA Burmann, Fl. Ind., 1768, p. 51.

Porana bendirei (Ward) Lesquereux = *Hydrangea bendirei.*

Porana cockerelli Knowlton, U. S. Nat. Mus., Proc., vol. 51, 1916, p. 287, pl. xxvii, f. 3.
Porana tenuis Lesquereux. Cockerell, Am. Mus. Nat. Hist., Bull., vol. 24, 1906, p. 107, pl. vi, f. 10.
Miocene: *Florissant, Colo.

Porana similis Knowlton, U. S. Nat. Mus., Proc., vol. 51, 1916, p. 288, pl. xxvii, f. 1, 2.
Miocene: *Florissant, Colo.

Porana speirii Lesquereux, Rept. U. S. Geol. Surv. Terr., vol. 8 (Cret. and Tert. Fl.), 1883, p. 172, pl. xxviii, f. 15.
Miocene: Florissant, Colo.

Porana tenuis Lesquereux, Rept. U. S. Geol. Surv. Terr., vol. 8 (Cret. and Tert. Fl.), 1883, p. 173.—Knowlton, U. S. Nat. Mus., Proc., vol. 51, 1916, p. 286, pl. xxvii, f. 4–6.—Cockerell, Am. Mus. Jour., vol. 16, 1916, p. 448.
Miocene: *Florissant, Colo.

Porana tenuis Lesquereux. Cockerell, Am. Mus. Nat. Hist., Bull., vol. 24, 1906, p. 107, pl. vi, f. 10 = *Porana cockerelli.*

Porana sp. Lesquereux, U. S. Nat. Mus., Proc., vol. 11, 1888, p. 13.
Lagrange: Wickliffe, Ky.

POTAMOGETON (Tournefort) Linné, Sp. pl., 1753, p. 126.

Potamogeton geniculatus Al. Braun, Neues Jahrb., 1845, p. 168.—Lesquereux, Rept. U. S. Geol. Surv. Terr., vol. 8 (Cret. and Tert. Fl.), 1883, p. 142.—Cockerell, Univ. Colo. Studies, vol. 3, 1906, p. 175.
Miocene: Florissant, Colo.

Potamogeton megaphyllus Berry, U. S. Geol. Surv., Prof. Paper 84, 1914, p. 135, pl. xxvii, f. 2.
Barnwell (Twiggs): Phinizy Gully, Columbia County, Ga.

Potamogeton middendorfensis Berry, U. S. Geol. Surv., Prof. Paper 84, 1914, p. 27, pl. iv, f. 6.
Black Creek (Middendorf): Middendorf, S. C.

Potamogeton natans Linné. Penhallow, Brit. Assoc. Adv. Sci., Bradford meeting, 1900, p. 336.
Pleistocene: Don Valley, Toronto, Canada.

Potamogeton pectinatus Linné. Penhallow, Roy. Soc. Canada, Trans., 2d ser., vol. 2, 1896, sec. 4, pp. 66, 72.
Pleistocene: Besserer's wharf, Ottawa River, Canada.

Potamogeton perfoliatus Linné. Dawson, Canadian Nat., new ser., vol. 3, 1868, p. 72; Geol. Hist. Plants, 1888, p. 229.—Penhallow, Roy. Soc. Canada, Trans., 2d ser., vol. 2, 1896, sec. 4, pp. 65, 72.
Pleistocene: Besserer's wharf and Greens Creek, Ottawa River, Canada.

Potamogeton pusillus Linné. Dawson, Canadian Nat., new ser., vol. 3, 1868, p. 73; Geol. Hist. Plants, 1888, p. 229.—Penhallow, Roy. Soc. Canada, Trans., 2d ser., vol. 2, 1896, sec. 4, pp. 65, 72.
Pleistocene: Besserer's wharf and Greens Creek, Ottawa River, Canada.

Potamogeton robbinsii Oakes. Knowlton, Am. Geol., vol. 18, 1896, p. 371.
Pleistocene (Carmichaels): Along Monongahela River, Morgantown, W. Va.

Potamogeton rutilans Wolfgang = *Potamogeton rutilus.*

Potamogeton rutilus Wolfgang. Knowlton, U. S. Geol. Surv., Bull. 152, 1898, p. 182.
Potamogeton rutilans? Wolfg—. Penhallow, Geol. Soc. Am., Bull., vol. 1, 1890, p. 327; Roy. Soc. Canada, Trans., 2d ser., vol. 2, 1896, sec. 4, pp. 65, 72.
Pleistocene: Besserer's wharf and Greens Creek, Ottawa River, Canada.

Potamogeton? verticillatus Lesquereux, Rept. U. S. Geol. Surv. Terr., vol. 8 (Cret. and Tert. Fl.), 1883, p. 142, pl. xxiii, f. 5, 6.—Penhallow, Rept. Tert. Pl. Brit. Col., 1908, p. 80.—Cockerell, Univ. Colo. Studies, vol. 3, 1906, p. 175.
Miocene: Florissant, Colo.
Oligocene: Tulameen River, British Columbia.

Potamogeton sp. Penhallow, Roy. Soc. Canada, Trans., 3d ser., vol. 1, sec. 4, 1907, p. 300; Rept. Tert. Pl. Brit. Col., 1908, p. 80.
Oligocene: Kettle River near Midway, British Columbia.

POTAMOGETOPHYLLUM Fontaine, in Ward, U. S. Geol. Surv., Mon. 48, 1905 [1906], p. 500.

Potamogetophyllum vernonensis Fontaine, in Ward, U. S. Geol. Surv., Mon. 48, 1905 [1906], p. 500, pl. cix, f. 7.
Patapsco: Mount Vernon, Va.

POTENTILLA Linné, Sp. pl., 1753, p. 495.

Potentilla anserina Linné. Penhallow, Geol. Soc. Am., Bull., vol. 1, 1890, p. 330; Roy. Soc. Canada, Trans., 2d ser., vol. 2, 1896, pp. 66, 73; Brit. Assoc. Adv. Sci., Bradford meeting, 1900, p. 336.
Potentilla norvegica Linné. Dawson, Canadian Nat., vol. 2, 1857, p. 422.
Potentilla canadensis Linné. Dawson, idem, new ser., vol. 3, 1878, p. 71.
Potentilla simplex? Michaux. Dawson, Geol. Hist. Pl., 1888, p. 228.
Pleistocene: Greens Creek and Besserer's wharf, Ottawa River, Canada.

Potentilla canadensis Linné = *Potentilla anserina*.

Potentilla norvegica Linné = *Potentilla anserina*.

Potentilla simplex? Michaux = *Potentilla anserina*.

POUROUMA Aublet, Gnian., vol. 2, 1775, p. 891.

Pourouma texana Berry, U. S. Geol. Surv., Prof. Paper 91, 1916, p. 11, pls. i, ii.
Midway(?): Earle, Bexar County, Tex.

PRAEENGELHARDTIA Berry, U. S. Geol. Surv., Prof. Paper 91, 1916, p. 186. [Type, *P. eocenica*.]

Praeengelhardtia eocenica Berry, U. S. Geol. Surv., Prof. Paper 91, 1916, p. 186, pl. xvii, f. 2–5.
Lagrange: Puryear, Henry County, Tenn.

PREISSITES Knowlton, Bull. Torr. Bot. Club, vol. 21, 1894, p. 458. [Type, *P. wardii*.]

Preissites wardii Knowlton, Bull. Torr. Bot. Club, vol. 21, 1894, p. 458, pl. ccxix.
Fort Union: Burns's ranch, 30 miles below Glendive, Mont.

PREMNOPHYLLUM Velenovsky, Beitr. Pal. Österr.-Ungarns u. d. Orients, vol. 4, No. 1 (Böhm. Kreidef., pt. 3), 1884, p. 4 (51).

Premnophyllum trigonum Velenovsky, Beitr. Pal. Österr.-Ungarns u. d. Orients, vol. 4, No. 1 (Böhm. Kreidef.), 1884, p. 4 (51), pl. iii (xviii), f. 2.—Hollick, N. Y. Bot. Gard., Bull. 3, 1904, p. 416, pl. lxxix, f. 1; U. S. Geol. Surv., Mon. 50, 1906, p. 106, pl. xl, f. 13, 14.
Magothy: Glen Cove, Long Island, N. Y.

PREPINUS Jeffrey, Annals Bot., vol. 22, 1908, p. 209. [Type, *P. statenensis*.]

Prepinus statenensis Jeffrey, Annals Bot., vol. 22, 1908, p. 209, pl. xiii, f. 1–15.—Hollick and Jeffrey, N. Y. Bot. Gard., Mem., vol. 3, 1909, p. 19, pl. ix, f. 9, 10; pl. xxii, f. 3; pl. xxiii, f. 5; pl. xxiv, f. 1.
Raritan: Kreischerville, Staten Island, N. Y.

Prepinus viticetensis Jeffrey, Boston Soc. Nat. Hist., Proc., vol. 34, 1910, p. 336, pl. xxxiii, f. 1–6.
Magothy: Gay Head, Marthas Vineyard, Mass.

Prinos integrifolia Ellis. Lesquereux = *Ilex integrifolia*.

PROTEAEPHYLLUM Fontaine, U. S. Geol. Surv., Mon. 15, 1889 [1890], p. 281. [Type, *P. reniforme?*.]

Proteaephyllum californicum Fontaine, in Ward, U. S. Geol. Surv., Mon. 48, 1905 [1906], p. 267, pl. lxix, f. 11.
Knoxville: Elder Creek, 3½ miles above Lowry, Tehama County, Calif.

Proteaephyllum dentatum Fontaine = *Hederaephyllum crenulatum*.

Proteaephyllum ellipticum Fontaine = *Proteaephyllum ovatum*.

Proteaephyllum oblongifolium Fontaine = *Ficophyllum crassinerve*.

Proteaephyllum orbiculare Fontaine = *Proteaephyllum reniforme*.

Proteaephyllum ovatum Fontaine, U. S. Geol. Surv., Mon. 15, 1889 [1890], p. 285, pl. cxli, f. 1.—Berry, Md. Geol. Surv., Lower Cret., 1912, p. 499.

Proteaephyllum ellipticum Fontaine, U. S. Geol. Surv., Mon. 15, 1889 [1890], p. 285, pl. cxlii, f. 1, 2.

Proteaephyllum tenuinerve Fontaine, U. S. Geol. Surv., Mon. 15, 1889 [1890], p. 286, pl. cl, f. 13; pl. clvi, f. 2.

Patuxent: Fredericksburg, Dutch Gap, and Potomac Run, Va.

Proteaephyllum reniforme Fontaine, U. S. Geol. Surv., Mon. 15, 1889 [1890], p. 282, pl. cxxxix, f. 3; pl. clvi, f. 4; pl. clx, f. 1, 2.—Berry, Md. Geol. Surv., Lower Cret., 1912, p. 498.—Ward, U. S. Geol. Surv., Sixteenth Ann. Rept., 1894–95, pt. 1, 1896, p. 539, pl. cvi, f. 8, 9.

Proteaephyllum orbiculare Fontaine, U. S. Geol. Surv., Mon. 15, 1889 [1890], p. 283, pl. cxxxix, f. 4.

Phyllites pachyphyllus Fontaine, U. S. Geol. Surv., Mon. 15, 1889 [1890], p. 325, pl. cxlix, f. 2.

Patuxent: Fredericksburg, Mount Vernon, and Dutch Gap, Va.

Proteaephyllum tenuinerve Fontaine = *Proteaephyllum ovatum*.

Proteaephyllum uhleri Fontaine = *Celastrophyllum latifolium*.

Proteaephyllum sp. Fontaine = *Ficophyllum crassinerve*.

PROTEOIDES Heer, Nouv. mém. Soc. helv. sci. nat., vol. 22, No. 1. 1866, p. 17.

Proteoides acuta Heer, Phyll. crét. Nebr., 1866, p. 17, pl. iv, f. 7, 8; Am. Philos. Soc., Trans., vol. 13, 1869, p. 431, pl. xxiii, f. 5–7.
Dakota: Kansas and Nebraska.

Proteoides conospermaefolia Berry, U. S. Geol. Surv., Prof. Paper 112, 1919, p. 85, pl. xiv, f. 4.
Tuscaloosa: Shirleys Mill, Ala.

Proteoides crassipes Heer = *Ficus crassipes.*

Proteoides daphnogenoides Heer = *Ficus daphnogenoides.*

Proteoides daphnogenoides Heer. Newberry, U. S. Geol. Surv., Mon. 26, 1895, pl. xxxii, f. 11 [not other figures] = *Salix lesquereuxii.*

Proteoides daphnogenoides Heer. Hollick, N. Y. Acad. Sci., Annals, vol. 11, 1898, p. 420, pl. xxxvi, f. 2 = *Laurophyllum elegans.*

Proteoides daphnogenoides Heer. Hollick, N. Y. Acad. Sci., Annals, vol. 11, 1898, p. 420, pl. xxxvi, f. 1, 3 = *Laurophyllum nervillosum.*

Proteoides grevilleaeformis Heer, Phyl. crét. Nebr., 1866, p. 17, pl. iv, f. 11.—Lesquereux, Am. Philos. Soc., Trans., vol. 13, 1869, p. 432, pl. xxiii, f. 8; Rept. U. S. Geol. Surv. Terr., vol. 6 (Cret. Fl.), 1874, p. 86, pl. xxviii, f. 12; U. S. Geol. Surv., Mon. 17, 1891 [1892], p. 90.
Dakota: Decatur, Nebr.; Kansas.

Proteoides lancifolius Heer, Neue Denkschr. Schweiz. Ges., vol. 24, 1871, p. 12, pl. iii, f. 5, 6.—Lesquereux, Rept. U. S. Geol. Surv. Terr., vol. 8 (Cret. and Tert. Fl.), 1883, p. 50; U. S. Geol. Surv., Mon. 17, 1891 [1892], p. 90, pl. xv, f. 5; pl. l, f. 8.—Berry, U. S. Geol. Surv., Prof. Paper 84, 1914, p. 40, pl. x, f. 1.
Dakota: Near Fort Harker, Kans.
Black Creek (Middendorf): Middendorf, Rocky Point, and Darlington, S. C.

Proteoides longus Heer = *Myrica longa.*

Proteoides major Dawson, Roy. Soc. Canada, Trans., vol. 11, 1893 [1894], sec. 4, p. 61, pl. xii, f. 54.
Upper Cretaceous: Wellington mine, Vancouver Island, British Columbia.

Proteoides neillii Dawson, Roy. Soc. Canada, Trans., vol. 11, 1893 [1894], sec. 4, p. 61, pl. xii, f. 53.
Upper Cretaceous: Port McNeill, Vancouver Island, British Columbia.

Proteoides parvula Berry, U. S. Geol. Surv., Prof. Paper 84, 1914, p. 40, pl. x, f. 5.
Black Creek (Middendorf): Middendorf, S. C.

Proteoides wilcoxensis Berry, U. S. Geol. Surv., Prof. Paper 91, 1916, p. 207, pl. xxxv, f. 4–6.
Wilcox (Grenada): Grenada, Grenada County, Miss.
Wilcox: Near Naborton, De Soto Parish, La.
Lagrange: Puryear, Henry County, Tenn.

Proteoides sp. Dawson, Roy. Soc. Canada, Trans., vol. 11, 1893 [1894], sec. 4, p. 61, pl. xiii, f. 55.

Upper Cretaceous: Wellington mine, Vancouver Island, British Columbia.

Proteoides sp. Lesquereux, Rept. on Clays in N. J., 1878, p. 29.

Raritan: Sayreville, N. J.

PROTODAMMARA Hollick and Jeffrey, Am. Nat., vol. 40, 1906, p. 199. [Type, *P. speciosa.*]

Protodammara speciosa Hollick and Jeffrey, Am. Nat., vol. 40, 1906, p. 199, pl. i, f. 5–13; pl. ii, f. 1–5; N. Y. Bot. Gard., Mem., vol. 3, 1909, p. 46, pl. iv, f. 1–11; pl. x, f. 1–3; pl. xiv, f. 1, 4, 5; pl. xv, f. 1–6; pl. xvi, f. 1.—Berry, U. S. Geol. Surv., Prof. Paper 112, 1919, p. 60.

Dammara microlepis Heer (?). Hollick, N. Y. Acad. Sci., Ann., vol. 11, 1898, p. 57, pl. iii, f. 9 a, b.

Dammara minor Hollick, U. S. Geol. Surv., Mon. 50, 1906, p. 40, pl. ii, f. 35–37.

Raritan: Kreischerville, Staten Island, N. Y.; Balls Point, Block Island, R. I.

Tuskaloosa: Shirleys Mill, Ala.

PROTOFICUS Saporta, Soc. géol. France, Mém., vol. 7 (Fl. foss. Sézanne), 1868, p. 355 (67).

Protoficus fossi Duror, Jour. Geol., vol. 24, 1916, p. 578, f. 9, 10.

Swauk: Skykomish Basin, Wash.

Protoficus inaequalis Newberry, U. S. Nat. Mus., Proc., vol. 5, 1882 [1883], p. 512; U. S. Geol. Surv., Mon. 35, 1898, p. 89, pl. lviii, f. 2; pl. lx, f. 1.

Fort Union: Tongue River, Mont.

Protoficus zeilleri Lesquereux, Mus. Comp. Zool., Bull., vol. 16, 1888, p. 50.

Denver: Golden, Colo.

PROTOPHYLLOCLADUS Berry, Bull. Torr. Bot. Club, vol. 30, 1903, p. 440. [Type, *Salisburia polymorpha* Lesquereux.]

Protophyllocladus lanceolatus (Knowlton) Berry, Bull. Torr. Bot. Club, vol. 30, 1903, p. 441.

Thinnfeldia lanceolata Knowlton, U. S. Geol. Surv., Bull. 105, 1893, p. 49, pl. v, f. 5.

Livingston: Bozeman coal field, Mont.

Protophyllocladus lobatus Berry, Bull. Torr. Bot. Club, vol. 38, 1911, p. 403; U. S. Geol. Surv., Prof. Paper 84, 1914, p. 17, pl. ii, f. 9–13; Md. Geol. Surv., Upper Cret., 1916, p. 798.

Thinnfeldia n. sp. Berry, Johns Hopkins Univ. Circ., new ser., No. 7, 1907, p. 81.

Magothy: Sullivans Cove, Severn River, Md.

Black Creek (Middendorf): Rocky Point, Sumter County, S. C.

Protophyllocladus polymorphus (Lesquereux) Berry, Bull. Torr. Bot. Club, vol. 30, 1903, p. 442.

Salisburia polymorpha Lesquereux, Am. Jour. Sci., 2d ser., vol. 27, 1859, p. 362 [name]; U. S. Geol. and Geog. Surv. Terr., Ann. Rept. 1872 [1873], p. 404; Rept. U. S. Geol. Surv. Terr., vol. 7 (Tert. Fl.), 1878, p. 84, pl. vi, f. 40, 41.

Thinnfeldia polymorpha (Lesquereux) Knowlton, Biol. Soc. Washington, Proc., vol. 7, 1892, p. 153; U. S. Geol. Surv., Bull. 105, 1893, p. 47, pl. v[i], f. 1–4 [not *T. polymorpha* Ettingshausen, 1860].

Thinnfeldia montana Knowlton, U. S. Geol. Surv., Bull. 152, 1898, p. 227; idem, Bull. 163, 1900, p. 11, pl. i, f. 1–3.

Livingston: Bozeman coal field, Mont.

Eagle: Missouri River below Coal Banks, Mont.

Protophyllocladus subintegrifolius (Lesquereux) Berry, Bull. Torr. Bot. Club, vol. 30, 1903, p. 440; idem, vol. 31, 1904, p. 69, pl. i, f. 5; Geol. Surv. N. J., Ann. Rept. 1905 [1906], p. 139.—Hollick, U. S. Geol. Surv., Mon. 50, 1906, p. 36, pl. v, f. 1–6.—Berry, Johns Hopkins Univ. Cir., new ser., No. 7, 1907, p. 89, f. 6; Geol. Surv. N. J., Bull. 3, 1911, p. 98, pl. ix; Md. Geol. Surv., Upper Cret., 1916, p. 796, pl. lvi, f. 2; U. S. Geol. Surv., Prof. Paper 112, 1919, p. 57.

Phyllocladus subintegrifolius Lesquereux, Am. Jour. Sci., 2d ser., vol. 46, 1868, p. 92; Rept. U. S. Geol. Surv. Terr., vol. 6 (Cret. Fl.), 1874, p. 54, pl. i, f. 12; U. S. Geol. Surv., Mon. 17, 1891 [1892], p. 34, pl. ii, f. 1–3.

Thinnfeldia lesquereuxiana Heer, Fl. foss. arct., vol. 6, Abt. 2, 1882, p. 37, pl. xliv, f. 9, 10; pl. xlvi, f. 1–12a, 12b.—Newberry, U. S. Geol. Surv., Mon. 26, 1896, p. 59, pl. xi, f. 1–17.—Hollick, N. Y. Acad. Sci., Trans., vol. 11, 1892, p. 99, pl. iii, f. 6; Geol. Soc. Am., Bull., vol. 7, 1895, p. 12–14; N. Y. Acad. Sci., Ann., vol. 11, 1898, p. 58, pl. iii, f. 4, 5; p. 419, pl. xxxvi, f. 6.

Thinnfeldia subintegrifolia (Lesquereux) Knowlton, U. S. Geol. Surv., Bull. 152, 1898, p. 228.—Hollick, N. Y. Bot. Gard., Bull., vol. 2, 1898, p. 403, pl. xli, f. 13, 14.

Magothy: Cliffwood and Florida Grove, N. J.; Chappaquiddick Island, Marthas Vineyard, Mass.; Grove Point, Md.

Magothy ?: Block Island, R. I.

Raritan: Tottenville and Princess Bay, Staten Island, N. Y.

Tuscaloosa: Cottondale, Ala.

Dakota: Kansas; Decatur, Nebr.

Atane: Greenland.

Protophyllocladus? n. sp. Knowlton, Washington Acad. Sci., Proc., vol. 11, 1909, p. 213.

Lance: Near Ilo, Big Horn Basin, Wyo.

PROTOPHYLLUM Lesquereux, U. S. Geol. Surv. Terr., Rept., vol. 6 (Cret. Fl.), 1874, p. 100. [Type, *Pterospermites sternbergii*.]

Protophyllum boreale Dawson, Roy. Soc. Canada, Trans., vol. 1, 1882 [1883], sec. 4, p. 23, pl. iv, f. 13.

Upper Cretaceous: Peace River, Athabasca (now Saskatchewan).

Protophyllum crassum Lesquereux, U. S. Geol. Surv., Mon. 17, 1891 [1892], p. 193, pl. lxv, f. 4.
Dakota: Ellsworth County, Kans.

Protophyllum crednerioides Lesquereux, U. S. Geol. and Geog. Surv. Terr., Ann. Rept. 1874 [1876], p. 363, pl. iii, f. 1; pl. viii, f. 4; Rept. U. S. Geol. Surv. Terr., vol. 8 (Cret. and Tert. Fl.), 1883, p. 90, pl. ii, f. 1–3; U. S. Geol. Surv., Mon. 17, 1891 [1892], p. 194; pl. xxxvi, f. 11; pl. xliii, f. 4, 5.
Dakota: Ellsworth County, Kans.; New Ulm, Minn.

Protophyllum crenatum Knowlton, in Lesquereux, U. S. Geol. Surv., Mon. 17, 1891 [1892], p. 190, pl. lxv, f. 7.
Dakota: Ellsworth County, Kans.

Protophyllum denticulatum Lesquereux, U. S. Geol. Surv., Mon. 17, 1891 [1892], p. 193, pl. xxxvi, f. 9.
Dakota: Near Delphos, Kans.

Protophyllum dimorphum Lesquereux, U. S. Geol. Surv., Mon. 17, 1891 [1892], p. 190, pl. xli, f. 1.
Dakota: Ellsworth County, Kans.

Protophyllum haydenii (Lesquereux) Lesquereux, Rept. U. S. Geol. Surv. Terr., vol. 6 (Cret. Fl.), 1874, p. 106, pl. xvii, f. 3; U. S. Geol. Surv., Mon. 17, 1891 [1892], p. 192, pl. xliii, f. 3.
Pterospermites haydenii Lesquereux, U. S. Geol. and Geog. Surv. Terr., Ann. Rept. 1871 [1872], p. 302.
Dakota: Ellsworth County, Kans.

Protophyllum integerrimum Lesquereux, U. S. Geol. Surv., Mon. 17, 1891 [1892], p. 192, pl. xliii, f. 3.
Dakota: Mankato, Minn.

Protophyllum leconteanum (Lesquereux) Lesquereux, Rept. U. S. Geol. Surv. Terr., vol. 6 (Cret. Fl.), 1874, p. 103, pl. xvii, f. 4; pl. xxvi, f. 1; U. S. Geol. Surv., Mon. 17, 1891 [1892], p. 187, pl. xl, f. 1.
Credneria leconteana Lesquereux, Am. Jour. Sci., vol. 46, 1868, p. 98.
Dakota: Ellsworth County, Kans.
Upper Cretaceous: Peace River, Athabasca (now Saskatchewan).

Protophyllum minus Lesquereux, Rept. U. S. Geol. Surv. Terr., vol. 6 (Cret. Fl.), 1874, p. 104, pl. xix, f. 2; pl. xxvii, f. 1; idem, vol. 8 (Cret. and Tert. Fl.), 1883, p. 89, pl. iv, f. 6.—Newberry, U. S. Geol. Surv., Mon. 35, 1898, p. 132, pl. ix, f. 3.
Dakota: Fort Harker, Kans.

Protophyllum? mudgii (Lesquereux) Knowlton, U. S. Geol. Surv., Bull. 152, 1898, p. 185.
Protophyllum mudgii (Lesquereux) Lesquereux, Rept. U. S. Geol. Surv. Terr., vol. 6 (Cret. Fl.), 1874, p. 106, pl. xviii, f. 3.
Quercus mudgii Lesquereux, U. S. Geol. and Geog. Surv. Terr., Ann. Rept. 1871 [1872], p. 302.
Dakota: Kansas.

Protophyllum mudgii Lesquereux = *Protophyllum? mudgii.*

Protophyllum multinerve (Lesquereux) Lesquereux, Rept. U. S. Geol. Surv. Terr., vol. 6 (Cret. Fl.), 1874, p. 105, pl. xviii, f. 1; U. S. Geol. Surv., Mon. 17, 1891 [1892], p. 191, pl. xliii, f. 2; pl. xlv, f. 1.—Newberry, U. S. Geol. Surv., Mon. 35, 1898, p. 132, pl. vii, f. 4.—Berry, Bull. Torr. Bot. Club, vol. 38, 1911, p. 411; Md. Geol. Surv., Upper Cret., 1916, p. 829, pl. lxiii, f. 3; pl. lxiv, f. 1, 2.

Pterospermites multinervis Lesquereux, U. S. Geol. and Geog. Surv. Terr., Ann. Rept. 1871 [1872].

Raritan: Cedar Point, Baltimore County, Md.; East Washington Heights, D. C.

Dakota: Fort Harker, Kans.

Protophyllum nanaimo Dawson, Roy. Soc. Canada, Trans., vol. 1, 1882 [1883], p. 28, pl. viii, f. 35.

Upper Cretaceous: Nanaimo; Baynes Sound, Vancouver Island, British Columbia.

Protophyllum nebrascense Lesquereux, Rept. U. S. Geol. Surv. Terr., vol. 6 (Cret. Fl.), 1874, p. 103, pl. xxvii, f. 3; U. S. Geol. Surv., Mon. 17, 1891 [1892], p. 195.

Dakota: Decatur, Nebr.

Protophyllum obovatum Newberry, U. S. Geol. Surv., Mon. 26, 1895 [1896], p. 128, pl. xxxviii, f. 4.

Raritan: Woodbridge, N. J.

Protophyllum praestans Lesquereux, U. S. Geol. Surv., Mon. 17, 1891 [1892], p. 188, pl. xli, f. 2, 3; pl. xlii, f. 3, 4.

Dakota: Ellsworth County, Kans.

Protophyllum pseudospermoides Lesquereux, U. S. Geol. Surv., Mon. 17, 1891 [1892], p. 194, pl. lix, f. 2.

Dakota: Kansas.

Protophyllum pterospermifolium Lesquereux, U. S. Geol. Surv., Mon. 17, 1891 [1892], p. 195, pl. lix, f. 1.

Dakota: Kansas.

Protophyllum quadratum (Lesquereux) Lesquereux, Rept. U. S. Geol. Surv. Terr., vol. 6 (Cret. Fl.), 1874, p. 104, pl. xix, f. 1; U. S. Geol. Surv., Mon. 17, 1891 [1892], p. 195.

Pterospermites quadratus Lesquereux, U. S. Geol. and Geog. Surv. Terr., Ann. Rept. 1871 [1872], p. 301.

Dakota: South of Fort Harker, Kans.

Protophyllum querciforme Hollick, Bull. Torr. Bot. Club, vol. 22, 1895, p. 227, pl. ccxxxvii, f. 1.

Dakota: Fort Harker, Kans.

Protophyllum rugosum (Lesquereux) Lesquereux, Rept. U. S. Geol. Surv. Terr., vol. 6 (Cret. Fl.), 1874, p. 105, pl. xvii, f. 1, 2; pl. xix, f. 3; U. S. Geol. Surv., Mon. 17, 1891 [1892], p. 195.

Pterospermites rugosus Lesquereux, U. S. Geol. and Geog. Surv. Terr., Ann. Rept., 1872 [1873], p. 426.

Dakota: South of Fort Harker, Kans.

Upper Cretaceous: Coal Brook, British Columbia.

Protophyllum sternbergii (Lesquereux) Lesquereux, Rept. U. S. Geol. Surv. Terr., vol. 6 (Cret. Fl.), 1874, p. 101, pl. xvi; pl. xviii, f. 2; U. S. Geol. Surv., Mon. 17, 1891 [1892], p. 189, pl. xlii, f. 1.—Newberry, U. S. Geol. Surv., Mon. 35, 1898, p. 133, pl. x; pl. xi.—Berry, Bull. Torr. Bot. Club, vol. 34, 1911, p. 411; Md. Geol. Surv., Upper Cret., 1916, p. 828, pl. lxii, f. 1–3; pl. lxiii, f. 1, 2; pl. lxiv, f. 3.

Pterospermites sternbergii Lesquereux, U. S. Geol. and Geog. Surv. Terr., Ann. Rept., 1872 [1873], p. 425.

Dakota: Ellsworth County, Kans.

Raritan: Shannon Hill and Bull Mountain, Cecil County, Md.; East Washington Heights, D. C.

Protophyllum undulatum Lesquereux, U. S. Geol. Surv., Mon. 17, 1891 [1892], p. 189, pl. xlii, f. 2.

Dakota: Ellsworth County, Kans.

Protophyllum sp. Dawson, Roy. Soc. Canada, Trans., vol. 11, 1893 [1894], p. 63, pl. xi, f. 46, 47.

Upper Cretaceous: Port McNeill; Nanaimo.

PROTORHIPIS Andrae, Abh. K. k. geol. Reichsanst., vol. 2, pt. 3, No. 4, 1855, p. 35.

Protorhipis fisheri Knowlton, Smithsonian Misc. Coll., vol. 50, 1907, p. 114, pl. xii, f. 3, 4.

Kootenai: Skull Butte, near Stanford, Fergus County, Mont.

PRUINIUM Platen, Sitzungsb. Naturf. Ges., Leipzig, vol. 34, 1907 [1908], p. 122. [Type, *P. gummosum*.]

Pruinium gummosum Platen, Sitzungsb. Naturf. Ges., Leipzig, vol. 34, 1907 [1908], p. 122, pl. iii, f. 2–6.

Miocene: Amethyst Mountain, Yellowstone National Park.

PRUNOIDES Perkins, Rept. Vt. State Geol. 1903–4 [1904], p. 208. [Type, *Carpolithes bursaeformis* Lesquereux.]

Prunoides bursaeformis (Lesquereux) Perkins, Rept. Vt. State Geol. 1903–4 [1904], p. 208, pl. lxxx, f. 133.

Carpolithes bursaeformis Lesquereux, Am. Jour. Sci., 2d ser., vol. 32, 1861, p. 359; in Hitchcock, Geol. Vt., vol. 1, 1861, p. 231, f. 146, 147; vol. 2, 1861, p. 715.

Miocene: Brandon, Vt.

Prunoides inequalis Perkins, Rept. Vt. State Geol. 1905–6 [1906], p. 221, pl. lvii, f. 13.

Miocene: Brandon, Vt.

Prunoides seelyi Perkins, Rept. Vt. State Geol. 1903–4 [1904], p. 209, pl. lxxx, f. 141; Geol. Soc. Am., Bull., vol. 16, 1905, p. 509, pl. 86, f. 12; Rept. Vt. State Geol. 1905–6 [1906], pl. lii, f. 12.
Miocene: Brandon, Vt.

PRUNUS (Tournefort) Linné, Sp. pl., 1753, p. 473.

Prunus? acutifolia Newberry, U. S. Geol. Surv., Mon. 26, 1895 [1896], p. 90, pl. xiv, f. 1.—Berry, Geol. Surv. N. J., Bull. 3, 1911, p. 168, pl. xxii, f. 1.
Raritan: Woodbridge and South Amboy, N. J.

Prunus (Amygdalus)? antecedens Lesquereux, U. S. Geol. Surv., Mon. 17, 1891 [1892], p. 144, pl. lv, f. 4.
Dakota: Kansas.

Prunus caroliniana Michaux. Lesquereux = *Inga mississippiensis.*

Prunus coloradensis Knowlton, U. S. Geol. Surv., Prof. Paper 101, 1918, p. 326, pl. xcvi, f. 2.
Raton: Near Wootton, Colo.

Prunus cretacea Lesquereux, Am. Jour. Sci., 2d ser., vol. 46, 1868, p. 102; U. S. Geol. Surv., Mon. 17, 1891 [1892], p. 254.
Dakota: Lancaster County, Nebr.

Prunus dakotensis Lesquereux, Rept. U. S. Geol. Surv. Terr., vol. 8 (Cret. and Tert. Fl.), 1883, p. 237, pl. xlvi–A, f. 8.
Fort Union: Badlands, N. Dak.

Prunus? merriami Knowlton, U. S. Geol. Surv., Bull. 204, 1902, p. 67, pl. xi, f. 2, 3, 6, 7.—Hollick, Md. Geol. Surv., Pliocene and Pleistocene, 1906, p. 233, pl. lxxii, f. 2.—Penhallow, Rept. Tert. Pl. Brit. Col., 1908, p. 80.
Prunus n. sp. Knowlton, in Merriam, Univ. Calif., Bull. Dept. Geol., vol. 2, 1901, p. 309.
Pleistocene (Sunderland): Calvert County, Md.
Mascall: Van Horn's ranch, John Day Basin, Oreg.
Miocene: Quilchena, British Columbia.

Prunus nabortensis Berry, U. S. Geol. Surv., Prof. Paper 91, 1916, p. 221, pl. cxvi, f. 1.
Wilcox: Near Naborton, De Soto Parish, La.

Prunus parlatorii (Heer) Lesquereux = *Andromeda parlatorii.*

Prunus? tufacea Knowlton, U. S. Geol. Surv., Bull. 204, 1902, p. 68, pl. xi, f. 4.
Prunus n. sp. Knowlton, in Merriam, Univ. Calif., Bull. Dept. Geol., vol. 2, 1901, p. 309.
Mascall: Van Horn's ranch, John Day Basin, Oreg.

Prunus variabilis Newberry, U. S. Nat. Mus., Proc., vol. 5, 1882 [1883], p. 509; U. S. Geol. Surv., Mon. 35, 1898, p. 112, pl. lii, f. 3, 4 (in part), 5.—Knowlton, U. S. Nat. Mus., Proc., vol. 17, 1894, p. 226; Geol. Soc. Am., Bull., vol. 5, 1893, p. 585.
Kenai: Cook Inlet, Alaska.

Prunus sp. Berry, U. S. Geol. Surv., Prof. Paper 98, 1916, p. 202, pl. xlvii, f. 7.

Citronelle: Lambert, Ala.

Prunus sp. Dawson, Brit. N. A. Bound. Com. (Rept. Geol. and Res. Vic. 49th Parallel), 1875, App. A, p. 330.

——: Porcupine Creek, Alberta ?.

Prunus n. sp. Knowlton, in Merriam, Univ. Calif., Bull. Dept. Geol., vol. 2, 1901, p. 309 = *Prunus ? merriami.*

Prunus n. sp. Knowlton, in Merriam, Univ. Calif., Bull. Dept. Geol., vol. 2, 1901, p. 309 = *Prunus ? tufacea.*

Prunus sp. Penhallow, Brit. Assoc. Adv. Sci., Bradford meeting, 1900, p. 336; Am. Nat., vol. 41, 1907, p. 448.

Pleistocene: Don Valley, Toronto, Canada.

PSEUDODANAEOPSIS Fontaine, U. S. Geol. Surv., Mon. 6, 1883, p. 58. [Type, *Strangerites obliquus* Emmons.]

Pseudodanaeopsis nervosa Fontaine = *Pseudodanaeopsis obliqua.*

Pseudodanaeopsis obliqua (Emmons) Fontaine, in Ward, U. S. Geol. Surv., Twentieth Ann. Rept., pt. 2, 1900, p. 285.

Strangerites obliquus Emmons, Geol. Rept. Midland Counties N. C., 1856, p. 325; American geology, pt. 6, 1857, p. 121, f. 89.

Pseudodanaeopsis nervosa Fontaine, U. S. Geol. Surv., Mon. 6, 1883, pp. 61, 116, pl. xxxi, f. 1, 2; pl. liv, f. 3.

Triassic (Newark): Clover Hill, Va.; North Carolina.

Pseudodanaeopsis plana (Emmons) Fontaine, in Ward, U. S. Geol. Surv., Twentieth Ann. Rept., pt. 2, 1900, p. 238, pl. xxv, f. 1, 2; p. 284.

Strangerites planus Emmons, American geology, pt. 6, 1857, p. 122, f. 90.

Pseudodanaeopsis reticulata Fontaine, U. S. Geol. Surv., Mon. 6, 1883, pp. 59, 116, pl. xxx, f. 1-4a; pl. liv, f. 3.

Triassic (Newark): Clover Hill, Midlothian, and Carbon Hill, Va.; Ellingtons, N. C.; York County, Pa.

Pseudodanaeopsis reticulata Fontaine = *Pseudodanaeopsis plana.*

PSEUDOGEINITZIA Hollick and Jeffrey, N. Y. Bot. Gard., Mem., vol. 3, 1909, p. 45. [Type, *P. sequoiiformis.*]

Pseudogeinitzia sequoiiformis Hollick and Jeffrey, N. Y. Bot. Gard., Mem., vol. 3, 1909, p. 45, pl. x, f. 11; pl. xxv, f. 4.

Raritan: Kreischerville, Staten Island, N. Y.

PSEUDOLMEDIA Trécul, Ann. sci. nat., 3d ser., vol. 8, 1847, p. 129.

Pseudolmedia eocenica Berry, U. S. Geol. Surv., Prof. Paper 91, 1916, p. 196, pl. xxvii, f. 3; pl. xxviii, f. 2.

Lagrange: Puryear, Henry County, Tenn.

PSEUDOTSUGA Carrière, Traité conif., 2d ed., 1867, p. 256.

Pseudotsuga douglasii Carrière = *Pseudotsuga taxifolia.*

Pseudotsuga macrocarpa Mayr. Penhallow, Roy. Soc. Canada, Trans., 2d ser., vol. 10, 1904, sec. 4, p. 70.

Pleistocene: Orleans, Humboldt County, Calif.

Pseudotsuga miocena Penhallow, Roy. Soc. Canada, Trans., 2d ser., vol. 8, 1903, sec. 4, p. 68; op. cit., vol. 9, 1904, sec. 4, p. 47, f. 12, 13; Mon. N. A. Gym., 1907, p. 276; Rept. Tert. Pl. Brit. Col., 1908, p. 80.

Cupressoxylon sp. (f.) Dawson, Brit. N. A. Bound. Com. (Rept. Geol. and Res. Vicinity 49th Parallel), 1875, App. A, p. 331.

Miocene: Horsefly River, British Columbia.

Miocene ?: Porcupine Creek, Alberta ?

Pseudotsuga taxifolia (Poiret) Britton. Hannibal, Bull. Torr. Bot. Club, vol. 38, 1911, p. 335.

Pseudotsuga douglasii Carrière. Penhallow, Roy. Soc. Canada, Trans. 2d ser., vol. 10, 1904, sec. 4, p. 62; Mon. N. A. Gym., 1907, p. 272.

Pleistocene: Mystic Lake, near Bozeman, Mont.

Pliocene ?: Calabasas Canyon, Santa Cruz Mountains, Calif.

Psilotites inermis (Newberry) Schimper = *Cabomba inermis.*

Psilotum inerme Newberry = *Cabomba inermis.*

PSORALEA Linné, Sp. pl., 1753, p. 762.

Psoralea physodes Douglas. Hannibal, Bull. Torr. Bot. Club, vol. 38, 1911, p. 338.

Pliocene: Calabasas Canyon, Santa Cruz Mountains, Calif.

PSYCHOTRIA Linné, Syst., 10th ed., 1759, p. 929.

Psychotria grandifolia Engelhardt, Senckenb. naturf. Gesell., Abh., vol. 16, 1891, p. 656, pl. xi, f. 4.—Berry, U. S. Geol. Surv., Prof. Paper 91, 1916, p. 349, pl. cv, f. 1.

Lagrange: Puryear, Tenn.

PTELEA Linné, Sp. pl., 1753, p. 118.

Ptelea modesta (Lesquereux) Cockerell, Am. Mus. Nat. Hist., Bull., vol. 24, 1908, p. 98.—Knowlton, U. S. Nat. Mus., Proc., vol. 51, 1916, p. 275.

Cytisus modestus Lesquereux, Rept. U. S. Geol. Surv. Terr., vol. 8 (Cret. and Tert. Fl.), 1883, p. 200, pl. xxxix, f. 9–11.

Leguminosites serrulatus Lesquereux, op. cit., p. 202, pl. xxxix, f. 7, 8.

Miocene: *Florissant, Colo.

PTENOSTROBUS Lesquereux, Rept. U. S. Geol. Surv. Terr., vol. 6 (Cret. Fl.), 1874, p. 114. [Type, *P. nebrascensis.*]

Ptenostrobus nebrascensis Lesquereux, Rept. U. S. Geol. Surv. Terr., vol. 6 (Cret. Fl.), 1874, p. 114, pl. xxiv, f. 1; U. S. Geol. Surv., Mon. 17, 1891 [1892], p. 36.

?Dakota: Winnebago Village, Missouri River, N. Dak. ?

PTERIS Linné, Sp. pl., 1753, p. 1073.

Pteris affinis Lesquereux = *Salpichlaena anceps.*

Pteris anceps Lesquereux = *Salpichlaena anceps.*

Pteris dakotensis Lesquereux, U. S. Geol. Surv., Mon. 17, 1891 [1892], p. 24, pl. i, f. 2, 3.
Dakota: Near Delphos, Kans.

Pteris elegans Newberry, U. S. Nat. Mus., Proc., vol. 5, 1882 [1883], p. 503. [Not afterward recognized.]
Clarno (lower part): Current Creek, Oreg.

Pteris erosa Lesquereux = *Allantodiopsis erosa.*

Pteris gardneri Lesquereux = *Saccoloma gardneri.*

Pteris (Oleandra) glossopteroides Dawson, Roy. Soc. Canada, Trans., vol. 1, 1882 [1883], sec. 4, p. 24, pl. iv, f. 16.
Upper Cretaceous: Protection Island, British Columbia.

Pteris goldmani Knowlton, U. S. Geol. Surv., Prof. Paper 130, 19—, p. —, pl. ii, f. 3 [in preparation].
Laramie: Popes Bluff, Pikeview, Colo.

Pteris inquirenda Berry, in Cooke and Shearer, U. S. Geol. Surv., Prof. Paper 120, 1918, p. 55 [name].
Barnwell (Twiggs): Twiggs County, Ga.

Pteris linearis Knowlton, U. S. Geol. Surv., Prof. Paper 101, 1918, p. 284, pl. liv, f. 3.
Raton: 3 miles southeast of Trinidad, Colo.

Pteris pinnaeformis Heer. Lesquereux = *Pteris pseudopinnaeformis.*

Pteris pseudopinnaeformis Lesquereux, Rept. U. S. Geol. Surv. Terr., vol. 7 (Tert. Fl.), 1878, p. 52, pl. iv, f. 3, 4.—?Newberry, U. S. Geol. Surv., Mon. 35, 1898, p. 7, pl. xlviii, f. 5.—Knowlton, U. S. Geol. Surv., Bull. 204, 1902, p. 22.—Hollick, Geol. Surv. La., Special Rept., 1899, p. 279, pl. xxxii, f. 1.—Berry, U. S. Geol. Surv., Prof. Paper 91, 1916, p. 168, pl. ix, f. 6.
Pteris pinnaeformis Heer. Lesquereux, U. S. Geol. and Geog. Surv. Terr., Ann. Rept. 1870 [1871], p. 384; idem, 1871 [1872], p. 283.
Denver: Golden, Colo.
Clarno (lower part): Current Creek, Oreg.
Wilcox: Vineyard Bluff, Caddo Parish, and Naborton and Mansfield, De Soto Parish, La.
Wilcox (Grenada): Grenada County, Miss.

Pteris russellii Newberry, U. S. Nat. Mus., Proc., vol. 5, 1882 [1883], p. 503; U. S. Geol. Surv., Mon. 35, 1898, p. 7, pl. lxi, f. 1, 1a.—Knowlton, U. S. Geol. Surv., Prof. Paper 101, 1918, pp. 245, 284.
Vermejo: Vermejo Canyon, N. Mex.; 4 miles south of Palisade, Colo.
Raton ?: Vermejo Canyon, N. Mex.

Pteris sitkensis Heer, Fl. foss. alask., 1869, p. 21, pl. i, f. 7a.—Knowlton, Geol. Soc. Am., Bull., vol. 5, 1893, p. 577; U. S. Nat. Mus., Proc., vol. 17, 1894, p. 212.
Kenai?: Near Sitka, Alaska.
Eocene: Mackenzie River, Northwest Territory.

Pteris subsimplex Lesquereux = *Allantodiopsis erosa.*

Pteris undulata Lesquereux = *Allantodiopsis erosa.*

Pteris sp. Knowlton, U. S. Geol. Surv., Prof. Paper 130, 19—, p. —, pl. ii, f. 5 [in preparation].
Laramie: Marshall, Colo.

PTEROCARYA Kunth, Ann. sci. nat., vol. 2, 1824, p. 345.

Pterocarya americana Lesquereux, U. S. Geol. and Geog. Surv. Terr., Ann. Rept. 1873 [1874], p. 417; Rept. U. S. Geol. Surv. Terr., vol. 7 (Tert. Fl.), 1878, p. 290, pl. lviii, f. 3.
Denver?: Middle Park, Colo.

Pterocarya denticulata (Weber) Heer, Fl. tert. Helv., vol. 3, 1859, p. 94, pl. cxxxi, f. 5-7.—Hollick, Md. Geol. Surv. Pliocene and Pleistocene, 1906, p. 222, pl. lxxii, f. 6-10.
Pleistocene (Sunderland): Head of Island Creek, Calvert County, Md.

Pterocarya retusa Lesquereux, Mus. Comp. Zool., Bull., vol. 16, 1888, p. 56.
Denver: Golden, Colo.

PTEROPHYLLUM Brongniart, Prodr. hist. vég. foss., 1828, p. 95.

Pterophyllum abietinum Göppert = *Dioonites buchianus abietinus.*

Pterophyllum aequale (Brongniart) Nathorst, Afh Sv. Geol. Unders., ser. C, No. 27 (Fl. Bjuf, pt. 1), 1878, p. 11.—Fontaine, in Ward, U. S. Geol. Surv., Mon. 48, 1905 [1906], p. 100, pl. xx.—Knowlton, U. S. Nat. Mus., Proc., vol. 51, 1916, p. 457, pl. lxxx, f. 3.
Jurassic: Douglas County, Oreg.
Lower Jurassic: Upper Matanuska Valley, Alaska.

Pterophyllum affine (Nathorst), Nathorst, Afh. Sv. Geol. Unders., ser. C, No. 33 (Fl. Bjuf, pt. 2), 1879, p. 68, pl. xv, f. 12-14.—Fontaine, U. S. Geol. Surv., Mon. 6, 1886, p. 66, pl. xxxii, f. 2-4.
Triassic (Newark): Midlothian, Va.

Pterophyllum alaskense Fontaine, in Ward, U. S. Geol. Surv., Mon. 48, 1905 [1906], p. 152, pl. xxxviii, f. 19, 20.
Jurassic: Herendeen Bay, Alaska.

Pterophyllum buchianum Ettingshausen = *Dioonites buchianus.*

Pterophyllum californicum Fontaine = *Nilsonia californica.*

Pterophyllum carnallianum Göppert = *Dioonites carnallianus.*

Pterophyllum contiguum Schenk, in Richthofen, China, vol. 4, No. 10 (Pflänzl. Versteinerungen), 1883, p. 262, pl. liii, f. 6.—Fontaine, in Ward, U. S. Geol. Surv., Mon. 48, 1905 [1906], p. 99, pl. xix, f. 7-11.
Jurassic: Douglas County, Oreg.

Pterophyllum daleanum Ward, U. S. Geol. Surv., Twentieth Ann. Rept., pt. 2, 1900, p. 290.
Pterozamites pectinatus Emmons, American geology, pt. 6, 1857, p. 117, f. 84. [Homonym of *Zamites* (=*Pterozamites*) *pectinatus* Brongniart, Prodrome, 1828, p. 94.]
Pterophyllum pectinatum Fontaine, U. S. Geol. Surv., Mon. 6, 1883, p. 112, pl. liii, f. 4.
Triassic (Newark): Lockville, N. C.

Pterophyllum decussatum (Emmons) Fontaine = *Ctenophyllum braunianum angustum.*

Pterophyllum dunkerianum Göppert = *Dioonites dunkerianus.*

Pterophyllum haydenii Lesquereux [part] = *Abietites ernestinae.*

Pterophyllum? haydenii Lesquereux [part] = *Podozamites haydenii.*

Pterophyllum inaequale Fontaine, U. S. Geol. Surv., Mon. 6, 1883, p. 64, pl. xxxvi, f. 1; in Ward, U. S. Geol. Surv., Twentieth Ann. Rept., pt. 2, 1900, p. 242, pl. xxvi, f. 2, 3.
Triassic (Newark): Clover Hill, Va.; York County, Pa.

Pterophyllum lepidum? Heer. Knowlton, in Atwood, U. S. Geol. Surv., Bull. 467, 1911, p. 44.
Chignik (Upper Cretaceous): Chignik Bay, Alaska.

Pterophyllum? lowryanum Ward, U. S. Geol. Surv., Mon. 48, 1905 [1906], p. 254, pl. lxvii, f. 9.
Knoxville: Near Lowry, Tehama County, Calif.

Pterophyllum minus Brongniart, Ann. sci. nat., vol. 4, 1825, p. 219, pl. xii, f. 8.—Fontaine, in Ward, U. S. Geol. Surv., Mon. 48, 1905 [1906], p. 104, pl. xxi, f. 8, 9.
Jurassic: Douglas County, Oreg.

Pterophyllum montanense (Fontaine) Knowlton, Smithsonian Misc. Coll., vol. 50, 1907, p. 122, pl. xiv, f. 3.
Zamites montanensis Fontaine, U. S. Nat. Mus., Proc., vol. 15, 1892, p. 494, pl. lxxxiv, f. 4.
Kootenai: Great Falls and Spanish Coulée, east of Cascade, Mont.

Pterophyllum nathorsti Schenk, Pflanliche Verst. a. Richthofen's China, vol. 4, 1883, p. 261, pl. liii, f. 5, 7.—Fontaine, in Ward, U. S. Geol. Surv., Mon. 48, 1905 [1906], p. 97, pl. xix, f. 1-6.
Jurassic: Douglas County, Oreg.

Pterophyllum obovatum Newberry = *Pterospermites obovatus.*

Pterophyllum pectinatum Fontaine = *Pterophyllum daleanum.*

Pterophyllum princeps Oldham and Morris = *Anomozamites princeps.*

Pterophyllum rajmahalense Morris. Fontaine, in Ward, U. S. Geol. Surv., Twentieth Ann. Rept., pt. 2, 1900, p. 354, pl. lvi, f. 4, 5; U. S. Geol. Surv., Mon. 48, 1905 [1906], p. 102, pl. xxi, f. 1–7.—Knowlton, in Stanton and Martin, Geol. Soc. Am., Bull., vol. 161, 1905, p. 402.—Knowlton, U. S. Nat. Mus., Proc., vol. 51, 1916, p. 457, pl. 80, f. 1; in Martin and Katz, U. S. Geol. Surv., Bull. 485, 1912, p. 63.

Jurassic: Oroville, Calif.; Douglas County, Oreg.
Jurassic (Tuxedni): Cook Inlet, Alaska.
Lower Jurassic: Upper Matanuska Valley, Alaska.

Pterophyllum robustum Emmons = *Ctenophyllum robustum.*

Pterophyllum robustum var. ? Emmons = *Ctenophyllum robustum.*

Pterophyllum spatulatum (Emmons) Fontaine = *Ctenophyllum braunianum angustum.*

Pterophyllum taylorsvillensis (Ward) Knowlton, n. comb.

Podozamites taylorsvillensis Ward, in Fontaine, in Ward, U. S. Geol. Surv., Twentieth Ann. Rept., pt. 2, 1900, p. 333.—Knowlton, in Diller, U. S. Geol. Surv., Bull. 353, 1908, p. 55.

Foreman: Taylorsville region, Calif.

Pterophyllum? trilobatum Lesquereux, U. S. Geol. and Geog. Surv. Terr., Bull., vol. 1, 1875 [1876], p. 425. [Apparently not afterward recognized by author.]

Dakota: Fort Harker, Kans.

PTEROSPERMITES Heer, Fl. tert. Helv., vol. 3, 1859, p. 36.

Pterospermites alaskana Knowlton, Harriman Alaska Exped., vol. 4, 1904, p. 156, pl. xxvi, f. 2; pl. xxxii.

Kenai: Kukak Bay, Alaska.

Pterospermites carolinensis Berry, Bull. Torr. Bot. Club, vol. 34, 1907, p. 198, pl. 14, f. 2; idem, vol. 43, 1916, p. 297; U. S. Geol. Surv., Prof. Paper 112, 1919, p. 117, pl. xxv, f. 6–8.

Black Creek: Court House Bluff, Cape Fear River, N. C.
Eutaw: Coffee Bluff, Hardin County, Tenn.
Tuscaloosa: Shirleys Mill, Glen Allen, and Cottondale, Ala.

Pterospermites cordatus Ward, U. S. Geol. Surv., Sixth Ann. Rept., 1884–85 [1886], p. 556, pl. lvi, f. 4; U. S. Geol. Surv., Bull. 37, 1887, p. 93, pl. xli, f. 4.

Fort Union: Sevenmile Creek, Glendive, Mont.

Pterospermites credneriifolius Berry, Bull. Torr. Bot. Club, vol. 34, 1907, p. 199, pl. 13, f. 4.

Black Creek: Court House Bluff, Cape Fear River, N. C.

Pterospermites cupanioides (Newberry) Knowlton, U. S. Nat. Mus., Proc., vol. 16, 1893, p. 35, pl. ii, f. 1.—Penhallow, Rept. Tert. Pl., Brit. Col., 1908, p. 81.

Phyllites cupanioides Newberry, N. Y. Lyc. Nat. Hist., Ann., vol. 9, 1868, p. 74; U. S. Geol. Surv., Mon. 35, 1898, p. 135, pl. xli, f. 3, 4.—Knowlton, Washington Acad. Sci., Proc., vol. 11, 1909, p. 194.

Phyllites venosus Newberry [Lesquereux], Ill. Cret. and Tert. Pl., 1878, pl. xxvi, f. 3, 4.

Fort Union: Yellowstone River, Mont.
Lance: Near Melville, Mont.

Peterospermites dentatus Heer, Fl. foss. arct., vol. 1, 1868, p. 138, pl. xxi, f. 15b; pl. xxiii, f. 6, 7.—Lesquereux, U. S. Nat. Mus., Proc., vol. 10, 1887, p. 45.—Knowlton, in Spurr, U. S. Geol. Surv., Eighteenth Ann. Rept., pt. 3, 1898, p. 185.—Newberry, U. S. Geol. Surv., Mon. 35, 1898, p. 133, pl. liii, f. 1, 2; pl. liv, f. 4.—Penhallow, Rept. Tert. Pl., Brit. Col., 1908, p. 81.

Eocene ?: Mackenzie River, Northwest Territory.
Kenai: Twenty-five miles below Mission Creek and Yukon River, Alaska.
——: ?Upper Kanab Valley, Utah.

Pterospermites grandidentatus Lesquereux, Mus. Comp. Zool., Bull., vol. 16, 1888, p. 53.

Denver: Golden, Colo.

Pterospermites haydenii Lesquereux = *Protophyllum haydenii.*

Pterospermites haguei Knowlton, U. S. Geol. Surv., Mon. 32, pt. 2, 1899, p. 742, pl. xcix, f. 4.

Miocene: Yellowstone National Park.

Pterospermites longeacuminatus Lesquereux, U. S. Geol. Surv., Mon. 17, 1891 [1892], p. 186, pl. lix, f. 3.

Dakota: Near Fort Harker, Kans.

Pterospermites magnifolia Knowlton, Harriman Alaska Exped., vol. 4, 1904, p. 156, pl. xxxi.

Kenai: Kukak Bay, Alaska.

Pterospermites minor Ward, U. S. Geol. Surv., Sixth Ann. Rept., 1884–85 [1886], p. 556, pl. lvi, f. 7–9; U. S. Geol. Surv., Bull. 37, 1887, p. 95, pl. xlii, f. 1–3.—Knowlton, Washington Acad. Sci., Proc., vol. 11, 1909, p. 198.

Fort Union: Burns's ranch, Glendive, Mont.
Lance: Near Glendive, Mont.

Pterospermites modestus Lesquereux, U. S. Geol. Surv., Mon. 17, 1891 [1892], p. 186, pl. lviii, f. 5.—Hollick, N. Y. Acad. Sci., Ann., vol. 11, 1898, p. 422, pl. xxxvii, f. 6; U. S. Geol. Surv., Mon. 50, 1906, p. 95, pl. xxxviii, f. 8.

Dakota: Near Fort Harker, Kans.
Raritan: Tottenville, Staten Island, N. Y.

Pterospermites multinervis Lesquereux = *Protophyllum multinerve*.

Pterospermites neomexicanus Knowlton, U. S. Geol. Surv., Prof. Paper 98, 1916, p. 341, pl. xc, f. 6.
Fruitland: Thirty miles south of Farmington, San Juan County, N. Mex.

Pterospermites nervosus Knowlton, U. S. Geol. Surv., Prof. Paper 101, 1918, p. 273, pl. xlviii, f. 4.
Vermejo: Walsenburg, Colo.

Pterospermites obovatus (Newberry) Berry, Bull. Torr. Bot. Club, vol. 36, 1909, p. 259; Geol. Surv. N. J., Bull. 3, 1911, p. 187.
Pterophyllum obovatum Newberry, U. S. Geol. Surv., Mon. 26, 1896, p. 128, pl. xxxviii, f. 4.
Raritan: Woodbridge, N. J.

Pterospermites quadratus Lesquereux = *Protophyllum quadratum*.

Pterospermites rugosus Lesquereux = *Protophyllum rugosum*.

Pterospermites spectabilis Heer, Fl. foss. arct., vol. 2, Abt. 4, 1869, p. 480, pl. xliii, f. 15b; pl. liii, f. 1–4.—Dawson, Roy. Soc. Canada, Trans., vol. 1, 1882 [1883], p. 33; idem, vol. 7, 1889, p. 69.—Penhallow, Rept. Tert. Pl. Brit. Col., 1908, p. 81.
Eocene: Mackenzie River, Northwest Territory.
——: Spanish Peak, Calif.

Pterospermites sternbergii Lesquereux = *Pterophyllum sternbergii*.

Pterospermites undulatus Knowlton, U. S. Geol. Surv., Bull. 163, 1900, p. 67, pl. xvi, f. 3; pl. xvii, f. 2; pl. xviii, f. 4; U. S. Geol. Surv., Prof. Paper 98, 1916, p. 341, pl. xc, f. 5; idem, Prof. Paper 101, 1918, p. 273.
Mesaverde: *Point of Rocks, Wyo.
Vermejo: Rockvale, Colo., Vermejo Park, N. Mex.
Fruitland: Thirty miles south of Farmington, San Juan County, N. Mex.

Pterospermites wardii Knowlton, U. S. Geol. Surv., Bull. 163, 1900, p. 66, pl. xvi, f. 1; idem, Prof. Paper 101, 1918, p. 273.
Mesaverde: *Point of Rocks, Wyo.
Vermejo: Walsenburg, Colo.

Pterospermites whitei Ward, U. S. Geol. Surv., Sixth Ann. Rept., 1884–85 [1886], p. 556, pl. lvi, f. 5, 6; U. S. Geol. Surv., Bull. 37, 1887, p. 94, pl. xli, f. 5, 6.—Knowlton, Washington Acad. Sci., Proc., vol. 11, 1909, p. 198.
Fort Union: Burns's ranch, Glendive, Mont.
Lance: Cedar Creek near Glendive, Mont.

Pterospermites sp. Dawson, Geol. Surv. Canada, 1871–72, p. 87.—Penhallow, Rept. Tert. Pl. Brit. Col., 1908, p. 81.
Eocene: Quesnel River, British Columbia.

Pterospermites sp. Knowlton, Geol. Soc. Am., Bull., vol. 8, 1897, p. 154.

Pterospermites sp. Knowlton, Geol. Soc. Am., Bull., vol. 8, 1897, p. 154.

Pterospermites sp. Knowlton, U. S. Geol. Surv., Prof. Paper 98, 1916, p. 341, pl. lxxxix, f. 3; pl. xc, f. 1.

Kirtland: Near Pina Veta China, San Juan County, N. Mex.

Pterospermites sp. Lesquereux, Mus. Comp. Zool., Bull., vol. 16, 1888, p. 53.

Denver: Golden, Colo.

Pterozamites abbreviatus F. Braun = *Ctenophyllum braunianum abbreviatum.*

Pterozamites decussatus Emmons = *Ctenophyllum braunianum angustum.*

Pterozamites gracilis Emmons = *Ctenophyllum braunianum abbreviatum.*

Pterozamites linearis Emmons = *Ctenophyllum lineare.*

Pterozamites obtusifolius (Rogers) Emmons = *Ctenophyllum braunianum abbreviatum.*

Pterozamites obtusus Emmons = *Ctenophyllum robustum.*

Pterozamites pectinatus Emmons = *Pterophyllum daleanum.*

Pterozamites spatulatus Emmons = *Ctenophyllum braunianum angustum.*

Pterozamites sp. Emmons = *Ctenophyllum braunianum angustum.*

PTILOZAMITES Morris, in Grant, Trans. Geol. Soc. London, 2d ser., vol. 5, 1840, on unnumbered page.

Ptilozamites leckenbyi (Bean) Nathorst, Öfv. K. Svensk. Vet.-Akad. Förh., vol. 37, 1880, pp. 65, 83.—Fontaine, in Ward, U. S. Geol. Surv., Mon. 48, 1905 [1906], p. 89, pl. xvi, f. 1, 2.

Ctenis leckenbyi Bean, in Leckenby, Quart. Jour. Geol. Soc. London, vol. 20, 1863, p. 78, pl. x, f. 1a, 1b.

Jurassic: Douglas County, Oreg.

PUCCINITES Ettingshausen, Die Tertiär Flora v. Häring in Tirol: Abh. K. k. geol. Reichenstalt, vol. 2, pt. 3, No. 2, 1853, p. 26.

Puccinites cretaceum Whitford = *Puccinites whitfordi.*

Puccinites whitfordi Knowlton, n. name.

Puccinites cretaceum Whitford, Nebraska Geol. Surv., vol. 7, 1916, p. 88, f. 6–13. [Homonym, *Puccinites cretaceus* Velenovsky, Abh. K. böhm. Ges. Wiss., 7th ser., vol. 3, 1889, p. 56.]

Graneros: Rose Creek, 1½ miles west of Reynolds, Jefferson County, Nebr.

PYRUS Linné, Sp. pl., 1753, p. 479.

Pyrus cretacea Newberry, N. Y. Lyc. Nat. Hist., Ann., vol. 9, 1868, p. 12.—[Lesquereux], U. S. Geol. and Geog. Surv. Terr., Ill. Cret. and Tert. Pl., 1878, pl. ii, f. 7.—Newberry, U. S. Geol. Surv., Mon. 35, 1898, p. 110, pl. i, f. 7.
Dakota: Smoky Hill, Kans.

Pyrus sp. Dawson, Geol. Surv. Canada, 1877–78, p. 186B.—Penhallow, Rept. Tert. Pl. Brit. Col., 1908, p. 81.
Oligocene: Coal Gully, British Columbia.

QUERCINIUM Unger, Neues Jahrb., 1842, p. 173.

Quercinium abromeiti Platen, Sitzungsb. Naturf. Ges., Leipzig, vol. 34, 1907 [1908], p. 23.
Miocene: Calistoga, Calif.

Quercinium anomalum Platen, Sitzungsb. Naturf. Ges., Leipzig, vol. 34, 1907 [1908], p. 47.
Miocene (auriferous gravels): Nevada County, Calif.

Quercinium knowltoni Felix, Zeitschr. Deutschen geol. Ges., 1896, p. 250, pl. v, f. 2.—Knowlton, U. S. Geol. Surv., Mon. 32, pt. 2, 1899, p. 773.
Miocene: Amethyst Mountain, Yellowstone National Park.

Quercinium lamarense Knowlton, U. S. Geol. Surv., Mon. 32, pt. 2, 1899, p. 771, pl. cxviii, f. 5; pl. cxx, f. 2; pl. cxxi, f. 1, 2.
Miocene (intermediate flora): Yellowstone National Park.

Quercinium lesquereuxi Platen, Sitzungsb. Naturf. Ges., Leipzig, vol. 34, 1907 [1908].
Miocene ?: Calistoga, Calif.
Miocene: Nevada County, Calif.

Quercinium solerederi Platen, Sitzungsb. Naturf. Ges., Leipzig, vol. 34, 1907 [1908], p. 41.
——?: California.

Quercinium wardi Platen, Sitzungsb. Naturf. Ges., Leipzig, vol. 34, 1907 [1908], p. 49.
——?: California.

QUERCOPHYLLUM Fontaine, U. S. Geol. Surv., Mon. 15, 1889 [1890], p. 307. [Type, *Q. grossedentatum.*]

Quercophyllum chinkapinense Ward, U. S. Geol. Surv., Mon. 48, 1905 [1906], p. 513, pl. cxii, f. 3, 4.
Patapsco: Chinkapin Hollow, Va.

Quercophyllum grossedentatum Fontaine, U. S. Geol. Surv., Mon. 15, 1889 [1890], p. 307, pl. clvi, f. 9.
Patapsco: Near Brooke, Va.

Quercophyllum tenuinerve Fontaine = *Ficophyllum serratum.*

Quercophyllum wyomingense Fontaine, in Ward, U. S. Geol. Surv., Nineteenth Ann. Rept., pt. 2, 1899, p. 688, pl. clxix, f. 6.
Fuson: North side Pine Creek, Crook County, Wyo.

QUERCUS Linné, Sp. pl., 1753, p. 994.

Quercus abnormalis Berry, Jour. Geol., vol. 15, 1907, p. 342.
Pleistocene: Neuse River, N. C.

Quercus acrodon Lesquereux = *Quercus lesquereuxiana.*

Quercus aemulans Lesquereux, U. S. Geol. and Geog. Surv. Terr., Ann. Rept. 1871 [1872], p. 288.
Fort Union (?): Washakie station, Wyo.

Quercus acuminata (Michaux) Sargent. Penhallow, Brit. Assoc. Adv. Sci., Bradford meeting, 1900, p. 336.
Pleistocene: Don Valley, near Toronto, Canada.

Quercus affinis (Newberry) Knowlton = *Quercus clarnensis.*

Quercus agrifolia Née. Hannibal, Bull. Torr. Bot. Club, vol. 38, 1911, p. 337.
Pliocene: Calabasas Canyon, Santa Cruz Mountains, Calif.

Quercus alaskana Trelease, Brooklyn Bot. Gard., Mem., vol. 1, 1918, p. 499.
Quercus pandurata Heer, Öfversigt K. Vetensk.-Akad., Forhandl., 1868, p. 64; Fl. foss. arct. [Fl. foss. alask.], 1869, p. 33, pl. vi, f. 6. [Homonym, Humboldt and Bonpland, 1813.]
Kenai: Port Graham, Alaska.

Quercus alba Linné. Penhallow, Brit. Assoc. Adv. Sci., Bradford meeting, 1900, p. 336; Roy. Soc. Canada, Trans., 2d ser., vol. 10, 1904, p. 74; Am. Nat., vol. 41, 1907, p. 448.—Berry, Jour. Geol., vol. 15, 1907, p. 342; Am. Jour. Sci., 4th ser., vol. 34, 1912, p. 221.
Pleistocene: Don Valley, Toronto, Canada; Neuse River, N. C.; Buena Vista, Va.

Quercus alnoides Lesquereux, U. S. Geol. Surv., Mon. 17, 1891 [1892], p. 54, pl. vii, f. 3.
Dakota: Kansas.

Quercus anceps Lesquereux = *Diospyros nebraskana.*

Quercus angustiloba Al. Braun. Lesquereux = *Quercus praeangustiloba.*

Quercus antiqua Newberry, N. Y. Lyc. Nat. Hist., Ann., vol. 9, 1868, p. 26; U. S. Geol. Surv., Mon. 35, 1898, p. 69, pl. xiii, f. 2.
Dakota ?: Banks of Rio Dolores, Utah.

Quercus aquamara (Ward) Knowlton, U. S. Geol. Surv., Bull. 152, 1898, p. 189.
Hedera aquamara Ward, U. S. Geol. Surv., Sixth Ann. Rept., 1884–85 [1886], p. 553, pl. xlviii, f. 7; U. S. Geol. Surv., Bull. 37, 1887, p. 59, pl. xxvi, f. 7.
Post-Laramie: Black Buttes, Wyo.

Quercus argentum Knowlton, U. S. Geol. Surv., Twenty-first Ann. Rept., pt. 2, 1901, p. 215, pl. xxx, f. 12.
Esmeralda: 4.5 kilometers northeast of Emigrant Peak, Esmeralda County, Nev.

Quercus attenuata? Göppert. Lesquereux = *Quercus viburnifolia.*

Quercus balaninorum Cockerell, Am. Mus. Nat. Hist., Bull., vol. 24, 1908, p. 86, pl. viii, f. 21.
Miocene: Florissant, Colo.

Quercus banksiaefolia Newberry, Boston Jour. Nat. Hist., vol. 7, 1863, p. 522; U. S. Geol. Surv., Mon. 35, 1898, p. 69, pl. xviii, f. 2–5.
Puget: Chuckanutz, near Bellingham Bay, Discovery mine, Glacier coal field, Whatcom County, Wash.

Quercus baueri Knowlton, U. S. Geol. Surv., Prof. Paper 98, 1916, p. 337, pl. lxxxvi, f. 5, 6.
Fruitland: 30 miles south of Farmington, San Juan County, N. Mex.

Quercus benzoin? Lesquereux = *Persea leconteana.*

Quercus berryi Trelease, Brooklyn Bot. Gard., Mem., vol. 1, 1918, p. 499.
Quercus breweri Lesquereux, Rept. U. S. Geol. Surv. Terr. (Cret. and Tert. Fl.), 1883, p. 246, pl. liv, f. 5–8 [not f. 9].—Knowlton, Washington Acad. Sci., Proc., vol. 11, 1911, p. 197. [Homonym, Engelmann, 1880.]
Clarno (upper part): Bridge Creek, Oreg.
Lance: Sand Creek near Glendive, Mont.

Quercus bicornis Ward, U. S. Geol. Surv., Sixth Ann. Rept., 1884–85 [1886], p. 551, pl. xxxvi, f. 8; U. S. Geol. Surv., Bull. 37, 1887, p. 24, pl. ix, f. 3.
Fort Union: Sevenmile Creek, Glendive, Mont.

Quercus boweniana Lesquereux, Mus. Comp. Zool., Mem., vol. 6, No. 2, 1878, p. 6, pl. ii, f. 5, 6.—Knowlton, in Lindgren, Jour. Geol., vol. 4, 1896, p. 890.
Miocene (auriferous gravels): Independence Hill, Placer County, Calif.

Quercus brevifolia (Lamarck) Sargent. Berry, Jour. Geol., vol. 25, 1917, p. 662.
Pleistocene: Vero, Fla.

Quercus breweri Lesquereux [part] = *Quercus berryi.*

Quercus breweri Lesquereux, Rept. U. S. Geol. Surv. Terr. (Cret. and Tert. Fl.), 1883, pl. liv, f. 9 [not other figures] = *Quercus clarnensis.*

Quercus carbonensis Ward, U. S. Geol. Surv., Sixth Ann. Rept., 1884–85 [1886], p. 551, pl. xxxvii, f. 1; U. S. Geol. Surv., Bull. 37, 1887, p. 25, pl. ix, f. 6.
Dryophyllum basidentatum Ward, U. S. Geol. Surv., Sixth Ann. Rept., 1884–85 [1886], p. 551, pl. xxxvii, f. 11; idem, Bull. 37, 1887, p. 27, pl. xi, f. 2.
Hanna: Carbon, Wyo.

Quercus castaneopsis Lesquereux, Rept. U. S. Geol. Surv. Terr., vol. 8 (Cret. and Tert. Fl.), 1883, p. 155, pl. xxviii, f. 10.
Green River: Uinta County, Wyo.

Quercus castanoides Newberry, U. S. Nat. Mus., Proc., vol. 5, 1882 [1883], p. 506; U. S. Geol. Surv., Mon. 35, 1898, p. 70, pl. lxv, f. 6.
Green River: Green River, Wyo.

Quercus castanopsis Newberry, U. S. Nat. Mus., Proc., vol. 5, 1882 [1883], p. 505; U. S. Geol. Surv., Mon. 35, 1898, p. 71, pl. lvi. f. 4.—?Knowlton, in Lindgren, Jour. Geol., vol. 4, 1896, p. 890.—Penhallow, Rept. Tert. Pl. Brit. Col., 1908, p. 82.
Fort Union: Yellowstone River, Mont.
?Miocene (auriferous gravels): Independence Hill, Placer County, Calif.
Paskapoo ?: Quilchena, British Columbia.

Quercus catesbaeifolia Berry, U. S. Geol. Surv., Prof. Paper 98, 1916, p. 200, pl. xlvi, f. 12, 13.
Citronelle: Lambert, Ala.

Quercus celastrifolia Lesquereux, Mus. Comp. Zool., Bull., vol. 16, 1888, p. 46.
Denver: Golden, Colo.

Quercus chamissonis Heer, Öfversigt K. Vetensk.-Akad. Förhandl., 1868, p. 64; Fl. foss. arct., vol. 2, Abt. 2, 1869, p. 33, pl. vi, f. 7, 8.—Knowlton, U. S. Nat. Mus., Proc., vol. 17, 1894, p. 219; Geol. Soc. Am., Bull., vol. 5, 1893, p. 581.
Kenai: Port Graham, Alaska.

Quercus chapmani Sargent ? Berry, Jour. Geol., vol. 25, 1917, p. 662.
Pleistocene: Vero, Fla.

Quercus chapmanifolia Berry, U. S. Geol. Surv., Prof. Paper 98, 1916, p. 66, pl. xi, f. 1, 2.
Calvert: Good Hope Road, Anacostia Heights, D. C.

Quercus chlorophylla Unger = ***Quercus eucalyptifolia***.

Quercus chlorophylla Unger. Lesquereux, Am. Philos. Soc., Trans., vol. 13, 1869, p. 416, pl. xvii, f. 7 = ***Pisonia chlorophylloides;*** pl. xvii, f. 6 = ***Mimusops truncata***.

Quercus chlorophylloides Knowlton = ***Quercus eucalyptifolia***.

Quercus chrysolepis Liebmann. Hannibal, Bull. Torr. Bot. Club, vol. 38, 1911, p. 337, pl. xv, f. 2, 9.
Pliocene: Near Portola, Santa Cruz Mountains, and Calabasas Canyon, Calif.

Quercus chrysolepis montana Lesquereux, U. S. Nat. Mus., Proc., vol. 10, 1887, p. 38.
Miocene (auriferous gravels): Placer County, Calif.

Quercus cinereoides Lesquereux, Rept. U. S. Geol. Surv. Terr., vol. 7 (Tert. Fl.), 1878, p. 152, pl. xxi, f. 6.

——: Original locality and formation unknown.

Lance: Converse County, Wyo.

Quercus clarnensis Trelease, Brooklyn Bot. Gard., Mem., vol. 1, 1918, p. 499.

Fraxinus affinis Newberry, U. S. Nat. Mus., Proc., vol. 5, 1883, p. 510; U. S. Geol. Surv., Mon., vol. 35, 1898 [1899], p. 127, pl. xlix, f. 5.

Quercus affinis (Newberry) Knowlton, U. S. Geol. Surv., Bull. 204, 1902, p. 45. [Homonym, Scheider, 1837.]

Quercus furcinervis Rossmässler. Lesquereux, Rept. U. S. Geol. Surv. Terr. (Cret. and Tert. Fl.), 1883, p. 244, pl. liii, f. 10–12.

Quercus breweri Lesquereux, Rept. U. S. Geol. Surv. Terr. (Cret. and Tert. Fl.), 1883, pl. liv, f. 9 [not other figures].

Clarno (upper part): Bridge Creek, Oreg.

Quercus cleburni Lesquereux = *Dryandroides cleburni*.

Quercus cockerelli Trelease = *Quercus hatcheri*.

Quercus coloradensis Lesquereux, Mus. Comp. Zool., Bull., vol. 16, 1888, p. 46.

Denver: Golden, Colo.

Quercus competens Lesquereux, U. S. Geol. and Geog. Surv. Terr., Bull., vol. 1, 1875 [1876], p. 370. [Apparently not afterward referred to by author.]

Mesaverde: Point of Rocks, Wyo.

Quercus consimilis Newberry, U. S. Nat. Mus., Proc., vol. 5, 1882 [1883], p. 505; U. S. Geol. Surv., Mon. 35, 1898, p. 71, pl. xliii, f. 2–5, 7–10.—Knowlton, U. S. Geol. Surv., Bull. 204, 1902, p. 45.—Penhallow, Rept. Tert. Pl. Brit. Col., 1908, p. 82.—Knowlton, U. S. Geol. Surv., Mon. 32, pt. 2, 1899, p. 704, pl. lxxxvii, f. 6.

Fort Union ?: Yellowstone National Park.

Clarno (upper part): Bridge Creek, and Officer's ranch, Butler Basin, Oreg.

Paskapoo ?: Quilchena, British Columbia.

Quercus convexa Lesquereux, Mus. Comp. Zool., Mem., vol. 6, No. 2, 1878, p. 4, pl. i, f. 13–17; Rept. U. S. Geol. Surv. Terr., vol. 8 (Cret. and Tert. Fl.), 1883, p. 265, pl. xlv–B, f. 5, 6.—Knowlton, in Lindgren, Jour. Geol., vol. 4, 1896, p. 889.

Miocene (auriferous gravels): Table Mountain, Tuolumne County, and Independence Hill, Placer County, Calif.

Quercus coriacea Newberry, Boston Jour. Nat. Hist., vol. 7, 1863, p. 521; U. S. Geol. Surv., Mon. 35, 1898, p. 73, pl. xix, f. 1–3; pl. xx, f. 5.

Puget: Chuckanutz, near Bellingham Bay; Discovery mine, Glacier coal field, Whatcom County, Wash.

Quercus! crassinervis? Göppert. Lesquereux, Am. Jour. Sci., 2d ser., vol. 27, 1859, p. 364 = *Dryophyllum tennesseensis.*

Quercus crossii Lesquereux, U. S. Nat. Mus., Proc., vol. 10, 1887, p. 39, pl. ii, f. 5, 6.
Denver: Silver Cliff, Colo.

Quercus culveri Knowlton, U. S. Geol. Surv., Mon. 32, pt. 2, 1899, p. 708, pl. lxxxvii, f. 5.
Fort Union: Elk Creek, Yellowstone National Park.

Quercus cuneata Newberry = *Quercus newberryi.*

Quercus (Dryophyllum) dakotensis Lesquereux, Rept. U. S. Geol. Surv. Terr., vol. 8 (Cret. and Tert. Fl.), 1883, p. 39; U. S. Geol. Surv., Mon. 17, 1891 [1892], p. 56, pl. vii, f. 4.
Dakota: Kansas.

Quercus dallii Lesquereux, U. S. Nat. Mus., Proc., vol. 5, 1882 [1883], p. 446, pl. viii, f. 2–5; Rept. U. S. Geol. Surv. Terr., vol. 8 (Cret. and Tert. Fl.), 1883, p. 259.—Knowlton, U. S. Nat. Mus., Proc., vol. 17, 1894, p. 219; Geol. Soc. Am., Bull., vol. 5, 1893, p. 582.—Penhallow, Rept. Tert. Pl. Brit. Col., 1908, p. 82.
Kenai: Cook Inlet, Alaska.
——: North fork of Similkameen River, British Columbia.

Quercus dawsoni Knowlton, U. S. Geol. Surv., Bull. 152, 1898, p. 191,
Quercus platania Heer. Dawson, Roy. Soc. Canada, Trans., vol. 7, 1889, p. 72, pl. xi. [Not *Q. platania* Heer.]
Paskapoo ?: Near Calgary and Bow River, Canada.

Quercus dayana Knowlton, U. S. Geol. Surv., Bull. 204, 1902, p. 51, pl. vi, f. 1; in Smith, U. S. Geol. Surv., Folio 86 (Ellensburg, Wash.) 1903, p. 3.
Mascall: Near Van Horn's ranch, John Day Basin, Oreg.
Ellensburg: Kelly Hollow, Wenas Valley, Wash.

Quercus dentoni Lesquereux, Rept. U. S. Geol. Surv. Terr., vol. 8 (Cret. and Tert. Fl.), 1883, p. 224, pl. xlviii, f. 1, 11.—Penhallow. Rept. Tert. Pl. Brit. Col., 1908, p. 82.
Fort Union: Badlands, N. Dak.

Quercus dentoni Lesquereux. Ward, U. S. Geol. Surv., Sixth Ann. Rept., 1884–85 [1886], p. 551, pl. xxxvii, f. 2 = *Quercus dentonoides,*

Quercus dentonoides Knowlton, U. S. Geol. Surv., Bull. 163, 1900, p. 40, pl. vii, f. 7.
Quercus dentoni Lesquereux. Ward, U. S. Geol. Surv., Sixth Ann. Rept., 1884–85 [1886], p. 551, pl. xxxvii, f. 2; idem, Bull. 37, 1887, p. 26, pl. x, f. 1.
Mesaverde: Point of Rocks, Wyo.
Adaville: Hodges Pass, Wyo.

Quercus digitata (Marshall) Sudworth.
Quercus falcata Michaux. Knowlton, Am. Geol., vol. 18, 1896, p. 371.
Pleistocene: Morgantown, W. Va.

Quercus distincta Lesquereux, Mus. Comp. Zool., Mem., vol. 6, No. 2, 1878, p. 6, pl. ii, f. 7–10.—Knowlton, in Lindgren, Jour. Geol., vol. 4, 1896, p. 890.

Quercus voyana Lesquereux, Mus. Comp. Zool., Mem., vol. 6, No. 2, 1878, p. 8, pl. ii, f. 12.

Miocene (auriferous gravels): Chalk Bluffs, Tuolumne County, and Independence Hill, Placer County, Calif.

Quercus doljensis Pilar, Fl. foss. susedana, 1883, p. 37, pl. vii, f. 14.—Ward, U. S. Geol. Surv., Sixth Ann. Rept., 1884–85 [1886], p. 551, pl. xxxvi, f. 9, 10; U. S. Geol. Surv., Bull. 37, 1887, p. 25; pl. ix, f. 4, 5.

Rhus pseudo-meriani Lesquereux, Rept. U. S. Geol. Surv. Terr., vol. 7 (Tert. Fl.), 1878, p. 293, pl. lviii, f. 11.

Post-Laramie: Black Buttes, Wyo.

Adaville: Hodges Pass, Wyo.

Quercus dryophyllopsis Trelease, Brooklyn Bot. Gard., Mem., vol. 1, 1918, p. 499.

Dryophyllum (Quercus) latifolium Lesquereux, U. S. Geol. and Geog. Surv. Terr., Bull., vol. 1, 1875, p. 393; idem, Ann. Rept. 1874 [1876], p. 340, pl. vi, f. 1; Rept. U. S. Geol. Surv. Terr., vol. 8 (Cret. and Tert. Fl.), 1883, p. 37, pl. iv, f. 1, 2.

Quercus latifolia (Lesquereux) Lesquereux, U. S. Geol. Surv., Mon. 17, 1891 [1892], p. 58. [Homonym, Steudel, 1841.]

Dakota: Kansas.

Quercus drymeja Unger, Chlor. Prot., 1847, p. 113, pl. xxxii, f. 1–4.—Lesquereux, Rept. U. S. Geol. Surv. Terr., vol. 7 (Tert. Fl.), 1878, p. 157, pl. xix, f. 14; idem, vol. 8 (Cret. and Tert. Fl.), 1883, p. 154, pl. xxviii, f. 12 [not p. 245, pl. liv, f. 4].—Knowlton, U. S. Geol. Surv., Bull. 204, 1902, p. 44; ?U. S. Nat. Mus., Proc., vol. 51, 1916, p. 265.

Clarno (upper part): Bridge Creek, Oreg.

Miocene: Florissant, Colo.

Quercus drymeja Unger, var.? Lesquereux, U. S. Geol. and Geog. Surv. Terr., Ann. Rept. 1871 [1872], p. 292.

Evanston: Evanston, Wyo.

Quercus dubia Newberry, U. S. Nat. Mus., Proc., vol. 5, 1882 [1883], p. 506; U. S. Geol. Surv., Mon. 35, 1898, p. 73, pl. xxxvii, f. 5.

Phyllites cupanioides Newberry. [Lesquereux], U. S. Geol. and Geog. Surv. Terr., Ill. Cret. and Tert. Pl., 1878, pl. xx, f. 5.

Fort Union: Tongue River, Mont.

Quercus duriuscula Knowlton, U. S. Geol. Surv., Bull. 204, 1902, p. 50, pl. viii, f. 2.

Mascall: Near Van Horn's ranch, John Day Basin, Oreg.

Quercus eamesi Trelease, Brooklyn Bot. Gard., Mem., vol. 1, 1918, p. 499.

Quercus salicifolia Newberry, N. Y. Lyc. Nat. Hist., Ann., vol. 9, 1868, p. 24.—[Lesquereux], U. S. Geol. and Geog. Surv. Terr., Ill. Cret. and Tert. Pl., 1878, pl. ii, f. 1.—Newberry, U. S. Geol. Surv., Mon. 35, 1898, p. 77, pl. l, f. 1. [Homonym, Née, 1801.]

Dakota: Blackbird Hill, Nebr.

Quercus elaena Unger, Chlor. Prot., 1847, p. 112, pl. xxxii, f. 4.—Lesquereux, Rept. U. S. Geol. Surv. Terr., vol. 8 (Cret. and Tert. Fl.), 1883, p. 155, pl. xxviii, f. 11, 13.

Miocene: Florissant, Colo.

Quercus elaenoides Lesquereux, Mus. Comp. Zool., Mem., vol. 6, No. 2, 1878, p. 4, pl. i, f. 9–12.—Knowlton, in Lindgren, Jour. Geol., vol. 4, 1896, p. 890.

Miocene (auriforous gravels): Table Mountain, Tuolumne County, Independence Hill, Placer County, Calif.

Quercus elkoana Lesquereux = *Carpinus elkoana*.

Quercus eucalyptifolia Ettingshausen, Denks. K. Akad. W., Math.-naturw. Classe, vol. 47, 1883, p. 116 [16]; Geol. Surv. N. S. W., Mem. Palaeont., No. 2 (Tert. Fl. Australia), 1888, p. 26.

Quercus chlorophylla Unger. Lesquereux, Am. Jour. Sci., 2d ser., vol. 45, 1868, p. 206; Am. Philos. Soc., Trans., vol. 13, 1869, p. 416, pl. xvii, f. 5, 6 [not f. 7]; Rept. U. S. Geol. Surv. Terr., vol. 7 (Tert. Fl.), 1878, p. 151, pl. xxi, f. 3.

Quercus? chlorophylloides Knowlton, U. S. Geol. Surv., Bull. 152, 1898, p. 190.

Denver: [No locality given.]

——: "Erie and Marshall, Colo." [Reported by Lesquereux from the Laramie, but these specimens are not in the National Museum and none others have been found.]

Quercus elliptica Newberry = *Quercus washingtonensis*.

Quercus ellisiana Lesquereux, U. S. Geol. and Geog. Surv. Terr., Ann. Rept. 1871 [1872], p. 297; Rept. U. S. Geol. Surv. Terr., vol. 7 (Tert. Fl.), 1878, p. 155, pl. xx, f. 4–8.—Knowlton, U. S. Geol. Surv., Mon. 32, pt. 2, 1899, p. 659, pl. lxxvii, f. 6.—Penhallow, Roy. Soc. Canada, Trans., 2d ser., vol. 8, 1902, p. 58; Rept. Tert. Pl. Brit. Col., 1908, p. 82.

Quercus pealei Lesquereux, U. S. Geol. and Geog. Surv. Terr., Ann. Rept. 1871 [1872], p. 297.—Knowlton, U. S. Geol. Surv., Bull. 105, 1893, p. 53.

Livingston: Hodson's coal mine and head of Fir Canyon, near Bozeman, Mont.

——: Mount Everts, Yellowstone National Park.

Paskapoo (Fort Union): Red Deer River, Alberta.

Quercus? ellsworthianus Lesquereux, Rept. U. S. Geol. Surv. Terr. vol. 6 (Cret. Fl.), 1874, p. 65, pl. vi, f. 7.

Quercus ellsworthianus Lesquereux, Am. Jour. Sci., 2d ser., vol. 46, 1868, p. 96.

Laurophyllum ellsworthianum Lesquereux, U. S. Geol. Surv., Mon. 17, 1891 [1892], p. 95, pl. xiii, f. 7.

Dakota: Kansas and Nebraska.

Quercus eoprinoides Berry, Bull. Torr. Bot. Club, vol. 31, 1904, p. 74, pl. iv, f. 11; Geol. Surv. N. J., Ann. Rept. 1905 [1906], p. 139.
Magothy: Cliffwood, N. J.

Quercus evansii Lesquereux, Am. Jour. Sci., 2d ser., vol. 27, 1859, p. 360.
Puget: Bellingham Bay, Wash.

Quercus eximia Knowlton, U. S. Geol. Surv., Prof. Paper 130, 19—, p. —, pl. xxi, f. 2 [in preparation].
Laramie: Coal Creek, Boulder County, Colo.

Quercus falcata Michaux = *Quercus digitata*.

Quercus fisheriana Knowlton, U. S. Geol. Surv., Prof. Paper 101, 1918, p. 297, pl. lxviii, f. 3, 4.
Raton: Near Trinidad, Colo.

Quercus flexuosa Newberry, Boston Jour. Nat. Hist., vol. 7, 1863, p. 521; U. S. Geol. Surv., Mon. 35, 1898, p. 74, pl. xix, f. 4–6.—?Penhallow, Roy. Soc. Canada, Trans., 3d ser., vol. 1, 1907, p. 309.
Puget: Chuckanutz, near Bellingham Bay, Wash.
Lower Cretaceous(?): International Boundary between Pasayten and Skagit rivers, British Columbia.

Quercus? florissantensis Cockerell, Am. Mus. Nat. Hist., Bull., vol. 24, 1908, p. 85.
Quercus pyrifolia Lesquereux, Rept. U. S. Geol. Surv. Terr., vol. 8 (Cret. and Tert. Fl.), 1883, p. 154, pl. xxviii, f. 14.
Miocene: Florissant, Colo.

Quercus? fraxinifolia Lesquereux, Rept. U. S. Geol. Surv. Terr., vol. 7 (Tert. Fl.), 1878, p. 154, pl. xx, f. 3.
Quercus laharpi Gaudin. Lesquereux, U. S. Geol. and Geog. Surv. Terr., Ann. Rept. 1871 [1872], p. 297.
Livingston?: Bozeman coal field, Mont.

Quercus furcinervis (Rossmässler) Unger. Lesquereux, U. S. Nat. Mus., Proc., vol. 11, 1888, p. 22 = *Quercus furcinervis americana*.

Quercus furcinervis Rossmässler. Lesquereux, Rept. U. S. Geol. Surv. Terr. (Cret. and Tert. Fl.), 1883, pl. liii, f. 10–12 = *Quercus clarnensis*.

Quercus furcinervis americana Knowlton, U. S. Geol. Surv., Bull. 152, 1898, p. 192; idem, Bull. 204, 1902, p. 43; idem, Mon. 32, pt. 2, 1899, p. 705, pl. lxxxviii, f. 5.
Quercus furcinervis (Rossmässler) Unger. Lesquereux, U. S. Nat. Mus., Proc., vol. 11, 1888, p. 22.
Clarno (lower part): Cherry Creek, Crook County, Oreg.
Miocene: Lamar Valley, Yellowstone National Park.

Quercus furuhjelmi Heer, Öfversigt K. Vetensk.-Akad. Förhandl., 1868, p. 64; Fl. foss. arct., vol. 2, Abt. 2, 1869, p. 32, pl. v, f. 10; pl. vi, f. 1, 2.—Knowlton, U. S. Nat. Mus., Proc., vol. 17, 1894, p. 219; Geol. Soc. Am., Bull., vol. 5, 1893, p. 581.
Kenai: Port Graham, Alaska.

Quercus gardneri Knowlton, U. S. Geol. Surv., Prof. Paper 101, 1918, p. 259, pl. xxxviii, f. 3.
Vermejo: Walsenburg, Colo.

Quercus gaudini Lesquereux, Am. Jour. Sci., 2d ser., vol. 27, 1859, p. 360; U. S. Geol. and Geog. Surv. Terr., Ann. Rept. 1871 [1872], p. 296; U. S. Nat. Mus., Proc., vol. 10, 1887, p. 39, pl. ii, f. 7, 8.
Puget: Bellingham Bay, Wash.

Quercus glascoena Lesquereux, U. S. Geol. Surv., Mon. 17, 1891 [1892], p. 55, pl. vi, f. 6.
Dakota: Near Glascoe, Kans.

Quercus glennii Hollick, Md. Geol. Surv., Pliocene and Pleistocene, 1906, p. 226, pl. lxxii, f. 3–5.
Pleistocene (Sunderland): Head of Island Creek, Calvert County, Md.

Quercus godeti? Heer, Fl. tert. Helv., vol. 2, 1856, p. 50, pl. lxxviii, f. 10, 11.—Lesquereux, U. S. Geol. and Geog. Surv. Terr., Ann. Rept. 1871 [1872], p. 297; Rept. U. S. Geol. Surv. Terr., vol. 7 (Tert. Fl.), 1878, p. 153, pl. xx, f. 1.
Livingston: Bozeman coal field, Mont.

Quercus goldianus Lesquereux, U. S. Geol. and Geog. Surv. Terr., Ann. Rept. 1873 [1874], p. 398. [Not recognized by Lesquereux after first description.]
——: Golden, Colo.

Quercus göpperti Lesquereux, Mus. Comp. Zool., Mem., vol. 6, No. 2, 1878, p. 7, pl. ii, f. 11.
Miocene (auriferous gravels): Chalk Bluffs, Tuolumne County, Calif.

Quercus gracilis Newberry = *Dryophyllum subfalcatum.*

Quercus grönlandica Heer, Fl. foss. arct., vol. 1, 1868, p. 108, pl. viii, f. 8; pl. x, f. 3, 4; pl. xi, f. 4; pl. xlvii, f. 1.—Newberry, U. S. Geol. Surv., Mon. 35, 1898, p. 75, pl. li, f. 3 [in part]; pl. liv, f. 1, 2.
Kenai: Cook Inlet, Alaska.

Quercus grossidentata Knowlton, U. S. Geol. Surv., Mon. 32, pt. 2, 1899, p. 704, pl. lxxxvii, f. 7.
Miocene: Yellowstone National Park.

Quercus haidingeri Ettingshausen, Foss. Fl. v. Wien, 1851, p. 12, pl. ii, f. 1.—Lesquereux, U. S. Geol. and Geog. Surv. Terr., Bull., vol. 1, 1875 [1876], p. 387; idem, Ann. Rept. 1874 [1876], p. 313, Rept. U. S. Geol. Surv. Terr., vol. 7 (Tert. Fl.), 1878, p. 156, pl. xx, f. 9, 10.
Denver ?: Golden, Colo.
Locality of American types unknown.

Quercus hatcheri Knowlton, Biol. Soc. Washington, Proc., vol. 19, 1906, p. 95.

Quercus montana Knowlton, U. S. Geol. Surv., Bull. 257, 1905, p. 143, pl. xvii, f. 6. [Homonym, Willdenow, 1801.]

Quercus cockerelli Trelease, Brooklyn Bot. Gard., Mem., vol. 1, 1918, p. 499.

Judith River: Willow Creek, Fergus County, Mont.

Quercus haydeni Lesquereux, Am. Jour. Sci., 2d ser., vol. 45, 1869, p. 205; Rept. U. S. Geol. Surv. Terr., vol. 7 (Tert. Fl.), 1878, p. 157, pl. xix, f. 10.

Hanna: Rock Creek, Laramie Plains, Wyo.

Quercus hesperia Knowlton, U. S. Geol. Surv., Mon. 32, pt. 2, 1899, p. 709.

Fort Union: Below Elk Creek, Yellowstone National Park.

Quercus hexagona Lesquereux, Am. Jour. Sci., 2d ser., vol. 46, 1868, p. 95; Rept. U. S. Geol. Surv. Terr., vol. 6 (Cret. Fl.), 1874, p. 64, pl. v, f. 8; U. S. Geol. Surv., Mon. 17, 1891 [1892], p. 56, pl. vii, f. 5.

Dakota: Cass County, Nebr.; Delphos, Kans.

Quercus (Dryophyllum) hieracifolia (Debey), Hosius and Von der Marck, Fl. d. Westf. Kreid. Form., 1880, p. 166, pl. xxxi, f. 85–88.—Lesquereux, U. S. Geol. Surv., Mon. 17, 1891 [1892], p. 58, pl. iii, f. 15.

Dakota: Ellsworth County, Kans.

Quercus hollickii Berry, N. Y. Bot. Gard., Bull., vol. 3, 1903, p. 71, pl. li, f. 1, 2; Bull. Torr. Bot. Club, vol. 31, 1904, p. 74, pl. iii, f. 4, 5; Geol. Surv. N. J., Ann. Rept. 1905 [1906], p. 139.

Magothy: Cliffwood, N. J.

Quercus holmesii (Lesquereux) Lesquereux, U. S. Geol. Surv., Mon. 17, 1891 [1892], p. 58.—Penhallow, Roy. Soc. Canada, Trans., 2d ser., vol. 8, 1902, p. 46.—Berry, N. Y. Bot. Gard., Bull., vol. 3, 1903, p. 72, pl. xlviii, f. 13; Geol. Surv. N. J., Ann. Rept. 1905 [1906], p. 139.

Dryophyllum (Quercus) salicifolium Lesquereux, U. S. Geol. and Geog. Surv. Terr., Ann. Rept. 1874 [1876], p. 340, pl. viii, f. 2. [Name preoccupied by *Quercus salicifolia* Newberry.]

Dryophyllum (Quercus) holmesii Lesquereux, Rept. U. S. Geol. Surv. Terr., vol. 8 (Cret. and Tert. Fl.), 1883, p. 38, pl. iv, f. 8.

Dakota ?: San Juan River, southwestern Colorado.

Magothy: Cliffwood, N. J.

Upper Cretaceous: Port McNeill, Vancouver Island, British Columbia.

Quercus horniana Lesquereux, U. S. Nat. Mus., Proc., vol. 11, 1888, p. 17, pl. v, f. 6, which purports to show this species, is a figure of a specimen of *Ulmus californica*.—Knowlton, U. S. Geol. Surv., Bull. 204, 1902, p. 52, pl. viii, f. 1 [this is a figure of the type].

Castanea atavia Unger. Lesquereux, Rept. U. S. Geol. Surv. Terr., vol. 8 (Cret. and Tert. Fl.), 1883, p. 247, pl. lii, f. 2.

Mascall: Van Horn's ranch, John Day Valley, Oreg.

Quercus (Dryophyllum) hosiana Lesquereux, U. S. Geol. Surv., Mon. 17, 1891 [1892], p. 57, pl. iii, f. 14.
Dakota: Near Delphos, Kans.

Quercus idahoensis Knowlton, U. S. Geol. Surv., Eighteenth Ann. Rept., pt. 3, 1898, p. 729, pl. cii, f. 4.
Payette: Marsh, Idaho.

Quercus ilicoides? Heer. Lesquereux = *Ilex dissimilis.*

Quercus imbricaria fossilis Lesquereux, U. S. Nat. Mus., Proc., vol. 10, 1887, p. 39.
Pleistocene: Bridgeton, N. J.

Quercus johnstrupi Heer, Fl. foss. arct., vol. 7, 1883, p. 24, pl. lvi, f. 7–12a.—Knowlton, in Atwood, U. S. Geol. Surv., Bull. 467, 1911, p. 45.
Chignik (Upper Cretaceous): Chignik Lagoon, Alaska.

Quercus johnstrupi Heer. Newberry, U. S. Geol. Surv., Mon. 26, 1896, p. 69, pl. xix, f. 7 = *Quercus raritanensis.*

Quercus judithae Knowlton, U. S. Geol. Surv., Bull. 257, 1905, p. 143, pl. xviii, f. 2.
Judith River: Willow Creek, Fergus County, Mont.

Quercus kanseana (Lesquereux) Knowlton, U. S. Geol. Surv., Bull. 152, 1898, p. 194.
Hamamelites kansaseana Lesquereux, U. S. Geol. and Geog. Surv. Terr., Ann. Rept. 1874 [1876], p. 355, pl. vii, f. 4.
Hamamelites kansaseanus Lesquereux, Rept. U. S. Geol. Surv. Terr., vol. 8 (Cret. and Tert. Fl.), 1883, p. 70, pl. iv, f. 5.
Alnus kanseana Lesquereux, Rept. U. S. Geol. Surv. Terr., vol. 6 (Cret. Fl.), 1874, p. 62, pl. xxx, f. 8.
Dakota: Kansas.

Quercus knowltoniana Cockerell, Am. Jour. Sci., 4th ser., vol. 26, 1908, p. 540, f. 4.
Miocene: Florissant, Colo.

Quercus laharpi Gaudin. Lesquereux = *Quercus? fraxinifolia.*

Quercus lambertensis Berry, U. S. Geol. Surv., Prof. Paper 98, 1916, p. 201, pl. xlvi, f. 9, 10.
Citronelle: Lambert, Ala.

Quercus latifolia (Lesquereux) Lesquereux = *Quercus dryophyllopsis.*

Quercus laurifolia Michaux. Berry, Jour. Geol., vol. 25, 1917, p. 662.
Pleistocene: Vero, Fla.

Quercus laurifolia Newberry = *Quercus penhallowi.*

Quercus lehmani Hollick, Md. Geol. Surv., Miocene, 1904, p. 483, f. 1a, 1b.—Berry, U. S. Geol. Surv., Prof. Paper 98, 1916, p. 67, pl. xi, f. 9–11.
Calvert: Good Hope Road, Anacostia Heights, D. C.

Quercus lesquereuxiana Knowlton, U. S. Geol. Surv., Bull. 152, 1898, p. 194; idem, Bull. 163, 1900, p. 39, pl. vii, f. 1.

Quercus acrodon Lesquereux, Am. Jour. Sci., 2d ser., vol. 45, 1868, p. 205; Rept. U. S. Geol. Surv. Terr., vol. 7 (Tert. Fl.), 1878, p. 158, pl. xix, f. 11–13. [Homonym, Massalongo, 1853.]

Mesaverde: "Rock Creek, Laramie Plains"; near Harper station, Wyo.

Quercus lonchitis Unger, Gen. et sp. pl. foss., 1850, p. 403.—Lesquereux, U. S. Geol. and Geog. Surv. Terr., Ann. Rept. 1871 [1872], Suppl., p. 6.

Green River: Green River, Wyo.

Quercus lyellii Heer. Lesquereux, Am. Philos. Soc., Trans., vol. 13, 1869, p. 415, pl. xvii, f. 1, 2 = *Nectandra lancifolia;* f. 3 = *Magnolia hilgardiana.*

Quercus lyrata Walter. Berry, Jour. Geol., vol. 15, 1907, p. 343.

Pleistocene: Neuse River, N. C.

Quercus lyratiformis Cockerell, Am. Mus. Nat. Hist., Bull., vol. 24, 1908, p. 85, pl. vii, f. 17.

Miocene: Florissant, Colo.

Quercus macrocarpa Michaux. Penhallow, Brit. Assoc. Adv. Sci., Bradford meeting, 1900, p. 336; Roy. Soc. Canada, Trans., 2d ser., vol. 10, 1904, p. 74.—Meehan, in Mercer, Acad. Nat. Sci. Philadelphia, Jour., 2d ser., vol. 11, 1899, pp. 277, 281, text f. 8 (2).

Pleistocene: Don Valley, Toronto, Canada; Port Kennedy, Pa.

Quercus? magnifolia Knowlton, U. S. Geol. Surv., Mon. 32, pt. 2, 1899, p. 704, pl. lxxxviii, f. 1.

Fort Union: Elk Creek, Yellowstone National Park.

Quercus marylandica Muench. Berry, Jour. Geol., vol. 15, 1907, p. 342.

Pleistocene: Neuse River, N. C.

Quercus mediterranea Unger. Lesquereux, Cret. and Tert. Fl., pl. xxviii, f. 9 = *Quercus peritula.*

Quercus merriami Knowlton, U. S. Geol. Surv., Bull. 204, 1902, p. 49, pl. vi, f. 6, 7; pl. vii, f. 4, 5.

Quercus n. sp. Knowlton, in Merriam, Univ. Calif., Bull. Dept Geol., vol. 2, 1902, p. 308.

Quercus pseudo-lyrata angustiloba Lesquereux, U. S. Nat. Mus., Proc., vol 11, 1888, p. 17, pl. xi, f. 2. [Varietal name homonym of *Quercus angustiloba* Al. Braun, 1861.]

Mascall: Van Horn's ranch, John Day Basin, Oreg.

Quercus michauxii Nuttall. Berry, Torreya, vol. 9, 1909, p. 71; idem, vol. 15, 1915, p. 207.

Pleistocene (Chowan): Neuse River, N. C.

Pleistocene (Talbot): Indian Head, Charles County, Md.

Quercus microdentata Hollick = *Dillenites microdentatus*.

Quercus milleri Berry, Jour. Geol., vol. 17, 1909, p. 24, text f. 3.
Calvert: Richmond, Va.

Quercus minor (Marshall) Sargent. Knowlton, U. S. Geol. Surv., Bull. 152, 1898, p. 194.

Quercus obtusiloba Michaux. Penhallow, in Coleman, Jour. Geol., vol. 3, 1895, p. 635.—Penhallow, Am. Geol., vol. 13, 1894, p. 95; Brit. Assoc. Adv. Sci., 1900, p. 336; Roy. Soc. Canada, Trans., 2d ser., vol. 2, 1896, p. 73.

Pleistocene: Don Valley, Toronto, Canada.

Quercus montana Knowlton = *Quercus hatcheri*.

Quercus? montanensis Knowlton, U. S. Geol. Surv., Bull. 163, 1900, p. 11, pl. i, f. 10.
Eagle: Missouri River, below Coal Banks, Mont.

Quercus moorii Lesquereux = *Dryophyllum moorii*.

Quercus morrisoniana Lesquereux, Rept. U. S. Geol. Surv. Terr., vol. 8 (Cret. and Tert. Fl.), 1883, p. 40, pl. xvii, f. 1, 2.—Hollick, N. Y. Acad. Sci., Trans., vol. 16, 1897, p. 131, pl. xiii, f. 11, 12; N. Y. Bot. Gard., Bull., vol. 3, 1903, p. 72; idem, vol. 4, 1904, p. 411, pl. lxxiii, f. 5; U. S. Geol. Surv., Mon. 50, 1906, p. 56, pl. viii, f. 14.—Berry, Geol. Surv. N. J., Ann. Rept. 1905 [1906], p. 139; Bull. Torr. Bot. Club, vol. 37, 1910, p. 21; Md. Geol. Surv. Upper Cret., 1916, p. 816, pl. lviii, f. 2.
Dakota: Morrison, Colo.
Magothy: Center Island, Long Island, N. Y.; Cliffwood, N. J.; Round Bay, Md.

Quercus mudgii Lesquereux = *Protophyllum mudgii*.

Quercus multinervis. Lesquereux, Am. Jour. Sci., 2d ser., vol. 27, 1859, p. 360. Upper Cretaceous: Nanaimo, Vancouver Island, British Columbia.

Quercus myrtifolia Willdenow. Lesquereux, Am. Jour. Sci., 2d ser., vol. 27, 1859, p. 363 = *Sophora lesquereuxi*.

Quercus negundoides Lesquereux, U. S. Geol. and Geog. Surv. Terr., Ann. Rept. 1871 [1872], p. 292; Rept. U. S. Geol. Surv. Terr., vol. 7 (Tert. Fl.), 1878, p. 161, pl. xxi, f. 2.
Evanston: Evanston, Wyo.

Quercus? neomexicana Knowlton, U. S. Geol. Surv., Prof. Paper 101, 1918, p. 298, pl. lxx, f. 5.
Raton: Near Yankee, N. Mex.

Quercus neriifolia Al. Braun. Lesquereux, Tert. Fl., pl. xix, f. 4 = *Apocynophyllum pealii*.

Quercus neriifolia Al. Braun. Lesquereux, Tert. Fl., pl. xix, f. 5 = *Apocynophyllum lesquereuxii*.

Quercus nevadensis Lesquereux, Mus. Comp. Zool., Mem., vol. 6, No. 2, 1878, p. 5, pl. ii, f. 3, 4.

Quercus olafseni Heer. Lesquereux, Rept. U. S. Geol. Surv. Terr., vol. 8 (Cret. and Tert. Fl.), 1883, p. 245, pl. liv, f. 3.

Miocene (auriferous gravels): Chalk Bluffs, Nevada County, Calif.

Quercus newberryi Trelease, Brooklyn Bot. Gard., Mem., vol. 1, 1918, p. 499.

Quercus cuneata Newberry, N. Y. Lyc. Nat. Hist., Ann., vol. 9, 1868, p. 25. [Homonym, Wangenheim, 1820.]

Dakota: Blackbird Hill, Nebr.

Quercus nigra Linné. Berry, Jour. Geol., vol. 15, 1907, p. 342; Am. Nat., vol. 41, 1907, p. 693, pl. i, f. 3, 4; Am. Jour. Sci., 4th ser., vol. 29, 1910, p. 394; U. S. Geol. Surv., Prof. Paper 98, 1916, p. 200, pl. xlvi, f. 11.

Pleistocene: Neuse River, N. C.; Abercrombies Landing, Chattahoochee River, Black Warrior and Alabama rivers, Ala.

Citronelle: Lambert, Ala.

Quercus? novae-caesareae Hollick, N. Y. Acad. Sci., Trans., vol. 16, 1897, p. 131, pl. xiii, f. 9, 10; U. S. Geol. Surv., Mon. 50, 1906, p. 56, pl. viii, f. 15, 16.—Berry, N. Y. Bot. Gard., Bull., vol. 3, 1903, p. 72, pl. li, f. 4; Geol. Surv. N. J., Ann. Rept. 1905 [1906], p. 139.

Magothy: Cliffwood, N. J.

Raritan: Tottenville, Staten Island, N. Y.

Quercus obtusiloba Michaux = *Quercus minor.*

Quercus? (Dryophyllum) occidentalis (Dawson) Knowlton = *Quercus vancouveriana.*

Quercus olafseni Heer, Fl. foss. arct., vol. 1, 1868, p. 109, pl. x, f. 5; pl. xi, f. 7; pl. xlvi, f. 10; pl. xxii, f. 17.—Lesquereux, Rept. U. S. Geol. Surv., Terr., vol. 8 (Cret. and Tert. Fl.), 1883, p. 224, pl. xlviii, f. 4.—Knowlton, U. S. Geol. Surv., Mon. 32, pt. 2, 1899, p. 707.

Fort Union: Badlands, N. Dak.; Yellowstone National Park.

Quercus olafseni Heer. Lesquereux, Rept. U. S. Geol. and Geog. Surv. Terr., vol. 8 (Cret. and Tert. Fl.), 1878, p. 245, pl. liv, f. 3 = *Quercus nevadensis.*

Quercus oregoniana Knowlton, U. S. Geol. Surv., Bull. 204, 1902, p. 47, pl. vi, f. 2, 3; pl. vii, f. 1.

Quercus n. sp. Knowlton, in Merriam, Univ. Calif., Bull. Dept. Geol., vol. 2, 1901, p. 288.

Populus polymorpha Newberry [part], U. S. Geol. Surv., Mon. 35, 1898 [1899], p. 50, pl. xlvi, f. 4.

Clarno (upper part): Bridge Creek, Oreg.

Quercus osbornii Lesquereux, Rept. U. S. Geol. Surv. Terr., vol. 8 (Cret. and Tert. Fl.), 1883, p. 154, pl. xxxviii, f. 17.

Miocene: Florissant, Colo.

Quercus palustris Du Roi. Meehan, in Mercer, Acad. Nat. Sci., Philadelphia, Jour., 2d ser., vol. 11, 1899, pp. 278, 281, text f. 6 (7).—Berry, Jour. Geol., vol. 15, 1907, p. 342; idem, vol. 15, 1915, p. 207.
Pleistocene: Neuse River, N. C.; Port Kennedy, Pa.; Indian Head, Charles County, Md.

Quercus pandurata Heer = *Quercus alaskana.*

Quercus paucidentata Newberry, U. S. Nat. Mus., Proc., vol. 5, 1882 [1883], p. 505; U. S. Geol. Surv., Mon. 35, 1898, p. 76, pl. xliii, f. 1.—Knowlton, U. S. Geol. Surv., Bull. 204, 1902, p. 44.
Clarno (upper part): Bridge Creek, Grant County, Oreg.

Quercus payettensis Knowlton, U. S. Geol. Surv., Eighteenth Ann. Rept., pt. 3, 1898, p. 730, pl. cii, f. 9.
Payette: Jackass Creek, Boise County, Idaho.

Quercus pealei Lesquereux = *Quercus ellisiana.*

Quercus penhallowi Trelease, Brooklyn Bot. Gard., Mem., vol. 1, 1918, p. 499.
Quercus laurifolia Newberry, U. S. Nat. Mus., Proc., vol. 5, 1883, p. 505; U. S. Geol. Surv., Mon. 35, 1898, p. 76, pl. lix, f. 4; pl. lx, f. 3.—Penhallow, Rept. Tert. Pl. Brit. Col., 1908, p. 83. [Homonym, Michaux, 1803.]
Fort Union: Fort Berthold, N. Dak.
——: ?Tulameen River, British Columbia.

Quercus peritula Cockerell, Am. Mus. Nat. Hist., Bull., vol. 24, 1908, p. 85.
Quercus mediterranea Unger. Lesquereux, Rept. U. S. Geol. Surv. Terr., vol. 8 (Cret. and Tert. Fl.), 1883, p. 153, pl. xxviii, f. 9.
Miocene: Florissant, Colo.

Quercus phellos Linné. Berry, Jour. Geol., vol. 15, 1907, p. 342; Am. Nat., vol. 41, 1907, p. 694, pl. i, f. 1; Am. Jour. Sci., 4th ser., vol. 29, 1910, p. 394; Torreya, vol. 14, 1914, pp. 161, 162.
Pleistocene: Neuse River, N. C.; Abercrombies Landing, Chattahoochee River, and Black Warrior River, 342 and 311½ miles above Mobile, and Monroeville, Monroe County, Ala.; Waynesboro, Miss.

Quercus cf. **Q. phellos** Linné. Berry, Torreya, vol. 10, 1910, p. 265.
Pleistocene: Near Long Branch, N. J.

Quercus platania Heer. Dawson = *Quercus dawsoni.*

Quercus platania Heer. Lesquereux = *Platanus cordata.*

Quercus platania var. *rotundifolia* Lesquereux = *Dombeyopsis platanoides.*

Quercus platanoides (Lamarck) Sudworth. Berry, Jour. Geol., vol. 15, 1907, p. 343.
Pleistocene: Neuse River, N. C.

Quercus platinervis Lesquereux, Am. Jour. Sci., 2d ser., vol. 32, 1859, p. 361.

Upper Cretaceous: Nanaimo, Vancouver Island, British Columbia.

Quercus poranoides Lesquereux, Rept. U. S. Geol. Surv. Terr., vol. 6 (Cret. Fl.), 1874, p. 66, pl. xxx, f. 9.

Dakota: Kansas.

Quercus praeangustiloba Knowlton, U. S. Geol. Surv., Prof. Paper 130, 19—, p. —, pl. v, f. 6, 7 [in preparation].

Quercus angustiloba Al. Braun, in Ludwig, Palaeontogr., vol. 8, 1861, p. 103, pl. xxxvi, f. 3.—Lesquereux, U. S. Geol. and Geog. Surv. Terr., Ann. Rept. 1872 [1873], p. 378; Rept. U. S. Geol. Surv. Terr., vol. 7 (Tert. Fl.), 1878, p. 161, pl. xxi, f. 4, 5.

Laramie: Golden, Colo.

Quercus praedumosa Trelease, Brooklyn Bot. Gard., Mem., vol. 1, 1918, p. 499.

Quercus turneri Knowlton, U. S. Geol. Surv., Twenty-first Ann. Rept., pt. 2, 1901, p. 214, pl. xxx, f. 1. [Homonym, Willdenow, 1809.]

Esmeralda: 4.5 kilometers northeast of Emigrant Peak, Esmeralda County, Nev.

Quercus praeundulata Trelease, Brooklyn Bot. Gard., Mem., vol. 1, 1918, p. 499.

Quercus sinuata Newberry, N. Y. Lyc. Nat. Hist., Ann., vol. 9, 1868, p. 27; U. S. Geol. Surv., Mon. 35, 1898, p. 78, pl. xiii, f. 1. [Homonym, 1864.]

Dakota ?: Banks of Rio Dolores, Utah.

Quercus pratti Berry, Bull. Torr. Bot. Club, vol. 34, 1907, p. 192, pl. 11, f. 9.

Black Creek: Court House Bluff, Cape Fear River, N. C.

Quercus predigitata Berry, Jour. Geol., vol. 15, 1907, p. 342; Am. Jour. Sci., 4th ser., vol. 34, 1912, p. 221, f. 4, 5.

Pleistocene: Neuse River, N. C.; Buena Vista, Va.

Quercus previrginiana Berry, U. S. Geol. Surv., Prof. Paper 98, 1916, p. 199, pl. xlvi, f. 1–8.

Citronelle: Lambert and Red Bluff, Ala.

Quercus primordialis Lesquereux, Am. Jour. Sci., 2d ser., vol. 46, 1868, p. 95; Rept. U. S. Geol. Surv. Terr., vol. 6 (Cret. Fl.), 1874, p. 64, pl. v, f. 7.

Dryophyllum (*Quercus*) *primordiale* (Lesquereux) Lesquereux, U. S. Geol. and Geog. Surv. Terr., Ann. Rept. 1874 [1876], p. 340; idem, 1876 [1878], p. 490.

Dakota: Fort Harker, Kans.

Quercus prinoides Willdenow. Berry, Jour. Geol., vol. 15, 1907, p. 343.

Pleistocene: Neuse River, N. C.

Quercus prinus Linné. Berry, Jour. Geol., vol. 15, 1907, p. 342; Am. Nat., vol. 41, 1907, p. 693, pl. i, f. 5; Am. Jour. Sci., 4th ser., vol. 29, 1910, p. 394.

Pleistocene: Neuse River, N. C.; Abercrombies Landing, Chattahoochee River, and Black Warrior River 311½ miles above Mobile, Ala.

Quercus pseudo-alba Hollick, Md. Geol. Surv., Pliocene and Pleistocene, 1906, p. 227, pl. lxx, f. 2; pl. lxxi, f. 1–6.

Pleistocene (Sunderland): Point of Rocks and Head of Island Creek, Calvert County, Md.

Quercus pseudo-alnus Ettingshausen, Abhandl. K. k. geol. Reichsanstalt, vol. 1, Abt. 3, 1852, p. 5, pl. i, f. 7.—Lesquereux, Rept. U. S. Geol. Surv. Terr., vol. 8 (Cret. and Tert. Fl.), 1883, p. 244, pl. liii, f. 1–7.—Knowlton, U. S. Geol. Surv., Bull. 204, p. 46.

Populus polymorpha Newberry [part], U. S. Geol. Surv., Mon. 35, 1898, p. 50, pl. xlix, f. 7.

Clarno (upper part): Bridge Creek and Officer's ranch, Butler Basin, Oreg.

Quercus pseudo-castanea Göppert, in Dunker and Meyer, Palaeontogr., vol. 2, 1853, pl. xxxv, f. 1.—Heer, Öfversigt K. Vetensk., Akad. Förhandl., 1868, p. 64 [name]; Fl. foss. arct., vol. 2, Abt. 2, 1869, p. 32, pl. vi, f. 3–5.—Dawson, Geol. Surv. Canada, 1875–76, pp. 259–260.—Knowlton, U. S. Nat. Mus., Proc., vol. 17, 1894, p. 219; Geol. Soc. Am., Bull., vol. 5, 1893, p. 581.—Penhallow, Rept. Tert. Pl. Brit. Col., 1908, p. 83.

Kenai: Port Graham, Alaska.

Eocene: Quesnel River, British Columbia.

Quercus pseudo-chrysophylla Lesquereux, Mus. Comp. Zool., Mem., vol. 6, No. 2, 1878, p. 60, appendix.

Miocene (auriferous gravels): Oregon Creek, 20 miles north of Chalk Bluff, Calif.

Quercus pseudo-lyrata Lesquereux, Mus. Comp. Zool., Mem., vol. 6, No. 2, 1878, p. 8, pl. ii, f. 1, 2; U. S. Nat. Mus., Proc., vol. 11, 1888, p. 17, pl. x, f. 1.—Knowlton, Univ. Calif., Bull. Dept. Geol., vol. 2, 1901, p. 308; U. S. Geol. Surv., Bull. 204, 1902, p. 48; in Smith, U. S. Geol. Surv., Folio 86 (Ellensburg, Wash.), 1903, p. 3.

Quercus pseudo-lyrata acutiloba Lesquereux, U. S. Nat. Mus., Proc., vol. 11, 1888, p. 17, pl. xi, f. 1.
Quercus pseudo-lyrata brevifolia Lesquereux, idem, p. 18, pl. x, f. 2.
Quercus pseudo-lyrata latifolia Lesquereux, idem, p. 18, pl. xii, f. 1.
Quercus pseudo-lyrata obtusiloba Lesquereux, idem, p. 18, pl. x, f. 3.

Mascall: Van Horn's ranch, John Day Basin, Oreg. [erroneously stated by Lesquereux to be from auriferous gravels of California].

Ellensburg: Kelly Hollow, Wenas Valley, Wash.

Quercus pseudo-lyrata acutiloba Lesquereux = *Quercus pseudo-lyrata*.

Quercus pseudo-lyrata angustiloba Lesquereux = *Quercus merriami.*

Quercus pseudo-lyrata brevifolia Lesquereux = *Quercus pseudo-lyrata.*

Quercus pseudo-lyrata latifolia Lesquereux = *Quercus pseudo-lyrata.*

Quercus pseudo-lyrata obtusiloba Lesquereux = *Quercus pseudo-lyrata.*

Quercus pseudowestfalica Berry, U. S. Geol. Surv., Prof. Paper 84, 1914, p. 35, pl. ix, f. 5.
Black Creek (Middendorf): Middendorf, Rocky Point and Miles Mill, S. C.

Quercus pyrifolia Lesquereux = *Quercus? florissantensis.*

Quercus ramaleyi Cockerell, Bull. Torr. Bot. Club, vol. 33, 1906, p. 309, text f. 2.
Miocene: Florissant, Colo.

Quercus raritanensis Berry, Bull. Torr. Bot. Club, vol. 36, 1909, p. 249; Geol. Surv. N. J., Bull. 3, 1911, p. 119.
Quercus johnstrupi Heer. Newberry, U. S. Geol. Surv., Mon. 26, 1896, p. 69, pl. xix, f. 7.
Raritan: Sayreville, N. J.

Quercus? ratonensis Knowlton, U. S. Geol. Surv., Prof. Paper 101, 1918, p. 298, pl. lxix, f. 6, 7.
Raton: Near Yankee, N. Mex.
Wilcox: Coushatta, La.

Quercus retracta Lesquereux, Am. Philos. Soc., Trans., vol. 12, 1869, p. 416, pl. xvi, f. 4 = *Myrcia bentonensis.*

Quercus (Dryophyllum) rhamnoides Lesquereux, U. S. Geol. Surv., Mon. 17, 1891 [1892], p. 57, pl. xlviii, f. 4.
Dakota: Ellsworth County, Kans.

Quercus rockvalensis Knowlton, U. S. Geol. Surv., Prof. Paper 101, 1918, p. 259, pl. xxxviii, f. 4.
Vermejo: Rockvale, Colo.

Quercus rubra Linné. Penhallow, Brit. Assoc. Adv. Sci., Bradford meeting, 1900, p. 336.
Pleistocene: Don Valley, Toronto, Canada.

Quercus saffordi Lesquereux = *Banksia saffordi.*

Quercus salicifolia Newberry = *Quercus eamesi.*

Quercus scudderi Knowlton, U. S. Nat. Mus., Proc., vol. 51, 1916, p. 265, pl. xxi, f. 3.
Miocene: *Florissant, Colo.

Quercus semialatus Lesquereux = *Anisophyllum semialatum.*

Quercus semi-elliptica Göppert. Lesquereux = *Fagopsis longifolia.*

Quercus serra Unger, Chlor. Prot., 1847, p. 109, pl. xxx, f. 5-7.—Lesquereux, Rept. U. S. Geol. Surv. Terr., vol. 8 (Cret. and Tert. Fl.), 1883, p. 153.
Miocene: Florissant, Colo.

Quercus severnensis Berry, Bull. Torr. Bot. Club, vol. 37, 1910, p. 22, pl. viii, f. 3; Md. Geol. Surv., Upper Cret., 1916, p. 817, pl. lvii, f. 9.
Magothy: Round Bay on Severn River, Md.

Quercus simplex Newberry, U. S. Nat. Mus., Proc., vol. 5, 1882 [1883], p. 505; U. S. Geol. Surv., Mon. 35, 1898, p. 78, pl. xliii, f. 6.—Knowlton, U. S. Geol. Surv., Bull. 204, 1902, p. 44; in Washburne, U. S. Geol. Surv., Bull. 590, 1914, p. 36.—Knowlton, U. S. Geol. Surv., Prof. Paper 101, 1918, p. 298, pl. lxx, f. 3.
Clarno (upper part): Bridge Creek, Grant County, and Officer's ranch, Butler Basin, Oreg.
Eocene ?: Eugene, Oreg.
Raton: Salyers Creek, N. Mex.; Rouse, Colo.

Quercus simulata Knowlton, U. S. Geol. Surv., Eighteenth Ann. Rept., pt. 3, 1898, p. 728, pl. cii, f. 1, 2.
Payette: Marsh, Idaho.

Quercus sinuata Newberry = *Quercus praeundulata.*

Quercus spurio-ilex Knowlton, in Lesquereux, U. S. Geol. Surv., Mon. 17, 1891 [1892], p. 53, pl. xlviii, f. 3.
Dakota: Near Delphos, Kans.

Quercus stantoni Knowlton, U. S. Geol. Surv., Prof. Paper 108, 1917, pp. 86–87, pl. xxiii, f. 2–4.
Frontier: About 1½ miles south and also 1 mile east of Cumberland, Wyo.

Quercus steenstrupiana? Heer, Fl. foss. arct., vol. 1, 1868, p. 109, pl. xi, f. 5; pl. xlvi, f. 8, 9.—Lesquereux, Mus. Comp. Zool., Mem., vol. 6, No. 2, 1878, p. 59.
Miocene (auriferous gravels): Oregon Creek, 20 miles north of Chalk Bluffs, Calif.

Quercus stramineus Lesquereux, U. S. Geol. and Geog. Surv. Terr., Ann. Rept. 1872 [1873], p. 378; Rept. U. S. Geol. Surv. Terr., vol. 7 (Tert. Fl.), 1878, p. 151, pl. xix, f. 6, 7.—Knowlton, U. S. Geol. Surv., Prof. Paper 130, 19—, p. — [in preparation].
Laramie: Golden, Colo.

Quercus sullyi Newberry, U. S. Nat. Mus., Proc., vol. 5, 1882 [1883], p. 506; U. S. Geol. Surv., Mon. 35, 1898, p. 79, pl. lx, f. 2.
Fort Union: Fort Berthold, N. Dak.

Quercus sumpterensis Berry, U. S. Geol. Surv., Prof. Paper 84, 1914, p. 35, pl. x, f. 9, 10.
Black Creek (Middendorf): Rocky Point, S. C.

Quercus suspecta Lesquereux, U. S. Geol. Surv., Mon. 17, 1891 [1892], p. 52, pl. xlvii, f. 7; pl. xlviii, f. 1, 2.
Dakota: Near Fort Harker and Ellsworth County, Kans.

Quercus tinctoria Bartram. Penhallow, Roy. Soc. Canada, Trans., 2d ser., vol. 10, 1904, sec. 4, p. 74 = *Quercus velutina.*

Quercus transgressus Lesquereux, Mus. Comp. Zool., Mem., vol. 6, No. 2, 1878, p. 59.

Miocene (auriferous gravels): Oregon Creek, 20 miles north of Chalk Bluffs, Calif.

Quercus triangularis Göppert. Lesquereux = *Quercus viburnifolia.*

Quercus turneri Knowlton = *Quercus prae-dumosa.*

Quercus ursina Knowlton, U. S. Geol. Surv., Bull. 204, 1902, p. 51, pl. vii, f. 2, 3.

Quercus n. sp. Knowlton, in Merriam, Univ. Calif. Bull. Dept. Geol., vol. 2, 1901, p. 308.

Mascall: Van Horn's ranch, John Day Basin, Oreg.

Quercus valdensis Heer, Fl. tert. Helv., vol. 2, 1856, p. 49, pl. lxxviii, f. 15.—Lesquereux, Rept. U. S. Geol. Surv. Terr., vol. 7 (Tert. Fl.), 1878, p. 153, pl. xix, f. 8.

Mesaverde: Rock Creek, Laramie Plains, Wyo.

Quercus vancouveriana Trelease, Brooklyn Bot. Gard., Mem., vol. 1, 1918, p. 499.

Quercus? (Dryophyllum) occidentalis (Dawson) Knowlton, U. S. Geol. Surv., Bull. 152, 1898, p. 195.

Dryophyllum occidentale Dawson, Roy. Soc. Canada, Trans., vol. 11, 1893 [1894], sec. 4, p. 58, pl. vii, f. 17, 18.

Dryophyllum neillianum Dawson, idem, p. 58, pl. vii, f. 19.

Dryophyllum elongatum Dawson, idem, p. 58, pl. vii, f. 20.

Dryophyllum sp. Dawson, idem, p. 59, pl. xii, f. 59.

Upper Cretaceous: Port McNeill, Vancouver Island, British Columbia.

Quercus velutina Lamarck. Penhallow, Brit. Assoc. Adv. Sci., Bradford meeting, 1900, p. 336.

Quercus tinctoria Bartram. Penhallow, Roy. Soc. Canada, Trans., 2d ser., vol. 10, 1904, p. 74.

Pleistocene: Don Valley, Toronto, Canada.

Quercus viburnifolia Lesquereux, Rept. U. S. Geol. Surv. Terr., vol. 7 (Tert. Fl.), 1878, p. 159, pl. xx, f. 11, 12.—Knowlton, Washington Acad. Sci., Proc., 1909, p. 207, U. S. Geol. Surv., Prof. Paper 130, 19—, p. — [in preparation].

Quercus triangularis Göppert [ex. p.]. Lesquereux, U. S. Geol. and Geog. Surv. Terr., Ann. Rept. 1872 [1873], p. 377.

Quercus attenuata? Göppert. Lesquereux, idem, 1873 [1874], p. 398.

Laramie: Crow Creek, Colo.

Post-Laramie: Black Buttes, Wyo.

Denver: *Sand Creek 10 miles east of Denver, and Golden, Colo.

Lance: Converse County, Wyo.; Forsyth, Mont.

Quercus victoriae Dawson, Roy. Soc. Canada, Trans., vol. 1, 1882 [1883], sec. 4, p. 27, pl. vii, f. 27.
Upper Cretaceous: Newcastle Island, Nanaimo, Protection Island, Port McNeill, Vancouver Island, British Columbia.

Quercus virens Michaux = *Quercus virginiana.*

Quercus virginiana Miller. Berry, Am. Nat., vol. 41, 1907, p. 693, pl. i, f. 2; Torreya, vol. 14, 1914, p. 161; Jour. Geol., vol. 25, 1917, p. 662.
Quercus virens Michaux. Lesquereux, Am. Jour. Sci., 2d ser., vol. 27, 1859, p. 364.
Pleistocene: Abercrombies Landing, Chattahoochee River, and Monroeville, Ala.; banks of Mississippi River near Columbus, Ky.; Vero, Fla.

Quercus voyana Lesquereux = *Quercus distincta.*

Quercus wardiana Lesquereux, U. S. Geol. Surv., Mon. 17, 1891 [1892], p. 53, pl. vii, f. 1.—?Ward, U. S. Geol. Surv., Nineteenth Ann. Rept., pt. 2, 1899, p. 704, pl. clxx, f. 2, 3.
Dakota: Ellsworth County, Kans.; ?Evans quarry, 5 miles east of Hot Springs, S. Dak.

Quercus washingtonensis Trelease, Brooklyn Bot. Gard., Mem., vol. 1, 1918, p. 499.
Quercus elliptica Newberry, Boston Jour. Nat. Hist., vol. 7, 1863, p. 523; U. S. Geol. Surv. Mon. 35, 1898 [1899], p. 74, pl. xviii, f. 1; pl. xx, f. 3. [Homonym, Née, 1801.]
Puget: Chuckanutz, near Bellingham Bay, Wash.

Quercus weedii Knowlton, U. S. Geol. Surv., Mon. 32, pt. 2, 1899, p. 705, pl. lxxxvii, f. 4.
Miocene: Yellowstone National Park.

Quercus whitei Lesquereux, Mus. Comp. Zool., Bull., vol. 16, p. 46, 1888.
Denver: Golden, Colo.

Quercus wyomingiana Lesquereux, U. S. Geol. and Geog. Surv. Terr., Ann. Rept. 1872 [1873], p. 400. [Not afterward recorded by Lesquereux.]
Post-Laramie: Black Buttes, Wyo.

Quercus yanceyi Knowlton, U. S. Geol. Surv., Mon. 32, pt. 2, 1899, p. 707, pl. lxxxix, f. 2.
Fort Union: Yellowstone National Park.

Quercus sp. Berry, Am. Nat., vol. 43, 1909, p. 435.
Pleistocene: Rappahannock River, 1½ miles below Port Royal, and Nomini Cliffs, Potomac River, Va.

Quercus sp. Berry, Geol. Surv. N. J., Ann. Rept. 1905 [1906], p. 139.
Magothy: Cliffwood, N. J.

Quercus sp. Berry, U. S. Nat. Mus., Proc., vol. 48, 1915, p. 299.
Pleistocene: Columbus, Ky.

Quercus sp. Dawson, Geol. Surv. Canada, Rept. Prog. 1872–73, App. I, 1873, p. 68.
——: Hornby Island, British Columbia.
Upper Cretaceous: Vancouver Island, British Columbia.

Quercus sp. Dawson, Geol. Surv. Canada, Rept. Prog. 1879–80 [1881], p. 54A.
——: Roche Percée, Souris River, British Columbia.

Quercus sp. Dawson, Roy. Soc. Canada, Trans., vol. 4, 1886 [1887], sec. 4, p. 27.
Paskapoo (Fort Union): Great Valley, Saskatchewan.

Quercus? sp. Dawson, Roy. Soc. Canada, Trans., vol. 8, 1890, sec. 4, p. 89, text f. 28.
Miocene: Similkameen Valley, British Columbia.

Quercus sp. Eames, Bot. Gaz., vol. 49, 1910, p. 165, pl. viii, f. 5; pl. ix, f. 10.
Miocene (auriferous gravels): California.

Quercus sp. Hollick, Bull. Torr. Bot. Club, vol. 19, 1892, p. 332.
Pleistocene: Bridgeton, N. J.

Quercus sp. Hollick, N. Y. Acad. Sci., Trans., vol. 16, 1897, p. 131, pl. xiv, f. 9.—Berry, N. Y. Bot. Gard., Bull., vol. 3, 1903, p. 72, pl. xlvii, f. 6; pl. lii, f. 9.
Magothy: Cliffwood, N. J.

Quercus sp. Hollick, U. S. Geol. Surv., Mon. 50, 1906, p. 56, pl. viii, f. 17.
Magothy: Gay Head, Marthas Vineyard, Mass.

Quercus sp. Knowlton, in Turner, Am. Geol., vol. 15, 1895, p. 378.
Miocene (auriferous gravels): Volcano Hill, Placer County, Calif.

Quercus sp. Knowlton, Washington Acad. Sci., Proc., vol. 11, 1909, p. 197.
Lance: Eagle Bluff near Glendive, Mont.

Quercus sp. Knowlton, Washington Acad. Sci., Proc., vol. 9, 1909, p. 191.
Lance: Forsyth, Mont.

Quercus sp. Knowlton, in Spurr, U. S. Geol. Surv., Eighteenth Ann. Rept., pt. 3, 1898, p. 192.
Kenai: Miller's mine, Yukon River, Alaska.

Quercus sp. Knowlton, Geol. Soc. Am., Bull., vol. 8, 1897, p. 157.
Lance: Lance Creek, Converse County, Wyo.

Quercus? sp. Knowlton, U. S. Geol. Surv., Eighteenth Ann. Rept., pt. 3, 1898, p. 730.
Payette: Marsh, Idaho.

Quercus n. sp. Knowlton, in Merriam, Univ. Calif., Bull. Dept. Geol., vol. 2, 1901, p. 288 = *Quercus oregoniana*.

Quercus n. sp. Knowlton, in Merriam, Univ. Calif., Bull. Dept. Geol., vol. 2, 1901, p. 303 = *Quercus ursina*.

Quercus? sp. Knowlton, U. S. Geol. Surv., Bull. 204, 1902, p. 43, pl. viii, f. 4.

Clarno (lower part): Cherry Creek, Crook County, Oreg.

Quercus n. sp. Knowlton, in Merriam, Univ. Calif., Bull. Dept. Geol., vol. 2, 1901, p. 308 = *Quercus merriami*.

Quercus sp. Knowlton, U. S. Geol. Surv., Mon. 32, pt. 2, 1899, p. 707, pl. lxxxix, f. 7.

Fort Union: Elk Creek, Yellowstone National Park.

Quercus sp. Knowlton, in Merriam, Univ. Calif., Bull. Dept. Geol., vol. 2, 1901, p. 308 = *Quercus?* sp. Knowlton.

Quercus? sp. Knowlton, U. S. Geol. Surv., Bull. 204, 1902, p. 53, pl. viii, f. 3.

Quercus n. sp. Knowlton, in Merriam, Univ. Calif., Bull. Dept. Geol., vol. 2, 1901, p. 308.

Mascall: Van Horn's ranch, John Day Valley, Oreg.

Quercus sp. Knowlton, in Glenn, U. S. Geol. Surv., Water-Supply Paper 164, 1906, p. 38.

Lagrange: Wickliffe, Ballard County, Ky.

Quercus sp. Lesquereux, Rept. on Clays in N. J., 1878, p. 29.

Raritan: South Amboy, N. J.

Quercus sp. cf. **Q. cuspidata** (Rossmässler) Unger. Lesquereux, U. S. Nat. Mus., Proc., vol. 11, 1888, p. 12.

Lagrange: Boaz, Graves County, Ky.

RAMULUS Wanner, Pa. Geol. Surv., Ann. Rept. 1887 [1889], p. 27. [Type, *R. rugosus*.]

Ramulus rugosus Wanner, Pa. Geol. Surv., Ann. Rept. 1887 [1889], pp. 27, 35, pl. xiii, f. 1–3.—Ward, U. S. Geol. Surv., Twentieth Ann. Rept., 1898–99 [1900], pt. 2, p. 428.

Triassic (Newark): York County, Pa.

RARITANIA Hollick and Jeffrey, N. Y. Bot. Gard., Mem., vol. 3, 1909, p. 26. [Type, *Frenelopsis gracilis* Newberry.]

Raritania gracilis (Newberry) Hollick and Jeffrey, N. Y. Bot. Gard., Mem., vol. 3, 1909, p. 26, pl. vi, f. 4–7; pl. ix, f. 1–4; pl. x, f. 14–17; pl. xix, f. 3–6; pl. xx, f. 1.—Berry, Geol. Surv. N. J., Bull. 3, 1911, p. 92; Md. Geol. Surv., Upper Cret., 1916, p. 800, pl. lv, f. 2, 3.

Frenelopsis gracilis Newberry, U. S. Geol. Surv., Mon. 26, 1895, p. 59, pl. xii, f. 1–3a.

Raritan: Kreischerville, Staten Island, N. Y.; South Amboy, N. J.

Magothy: Deep Cut, Del.; Grove Point, Md.

RETIPHYCUS Ulrich, Harriman Alaska Exped., vol. 4, 1904, p. 139. [Type, *R. hexagonale*.]

Retiphycus hexagonale Ulrich, Harriman Alaska Exped., vol. 4, 1904, p. 139, pl. xviii, f. 5.
Yakutat: Pogibshi Island opposite Kodiak, Alaska.

REYNOSIA Grisebach, Cat. Pl. Cub., 1866, p. 33.

Reynosia praenuntia Berry, U. S. Geol. Surv., Prof. Paper 91, 1916, p. 281, pl. lxviii, f. 4; pl. lxix, f. 2, 3.
Wilcox (Holly Spring): Holly Spring, Miss.
Lagrange: Puryear, Tenn.

Reynosia texana Penhallow, Roy. Soc. Canada, Trans., 3d ser., vol. 1, sec. 4, 1907, p. 97, f. 4, 5.
Yegua: Sommerville, Burleson County, Tex.

Reynosia wilcoxiana Berry, U. S. Geol. Surv., Prof. Paper 91, 1916, p. 282, pl. lxv, f. 6, 7.
Lagrange: Puryear, Tenn.

RHAMNACINIUM Felix, Zeitschr. Deutsch. geol. Gesell., vol. 46, 1894, p. 89.

Rhamnacinium porcupinianum Penhallow, Roy. Soc. Canada, Trans., 2d ser., vol. 9, 1903, sec. 4, p. 48, f. 14–16, 21, 22; Rept. Tert. Pl. Brit. Col., 1908, p. 83.
Populus sp. Dawson, Brit. N. A. Bound. Com. (Rept. Geol. and Res. Vic. 49th Parallel), 1875, App. A, p. 331.
Eocene: Porcupine Creek and Great Valley, Saskatchewan.

Rhamnacinium radiatum Felix, Zeitschr. Deutsch. geol. Gesell., Jahr. 1896, p. 252, pl. vi, f. 3.—Knowlton, U. S. Geol. Surv., Mon. 32, pt. 2, 1899, p. 769, pl. cxviii, f. 6, 7; pl. cxix, f. 1.
Miocene: Amethyst Mountain, Yellowstone National Park.

Rhamnacinium texanum Penhallow, Roy. Soc. Canada, Trans., 3d ser., vol. 1, sec. 4, 1907, p. 96, f. 1–3.
Yegua: Sommerville, Burleson County, Tex.

Rhamnacinium triseriatum Penhallow, Roy. Soc. Canada, Trans., 2d ser., vol. 9, sec. 4, 1903, p. 54, f. 17–20; Rept. Tert. Pl. Brit. Col., 1908, p. 83.
Populus sp. Dawson, Brit. N. A. Bound. Com. (Rept. Geol. and Res. Vic. 49th Parallel), 1875, App. A, p. 331.
Eocene: Porcupine Creek and Great Valley, Saskatchewan.

RHAMNITES Forbes, Quart. Jour. Geol. Soc., London, vol. 7, 1851, p. 103.

Rhamnites apiculatus Lesquereux, U. S. Geol. Surv., Mon. 17, 1891 [1892], p. 171, pl. xxxvii, f. 8–13.—Berry, Bull. Torr. Bot. Club, vol. 37, 1910, p. 25; Md. Geol. Surv., Upper Cret., 1916, p. 854, pl. lxxviii, f. 3.
Dakota: Ellsworth County, Kans.
Magothy: Round Bay, Md.

Rhamnites berchemiaformis Berry, U. S. Geol. Surv., Prof. Paper 91, 1916, p. 285, pl. lxxi, f. 3.
Wilcox: Calaveras Creek, Wilson County, Tex.

Rhamnites concinnus Newberry, N. Y. Lyc. Nat. Hist., Ann., vol. 9, 1868, p. 50.—[Lesquereux], U. S. Geol. and Geog. Surv. Terr., Ill. Cret. and Tert. Pl., 1878, pl. xvi, f. 7, 9 [f. 9 under *Viburnum asperum*].—Newberry, U. S. Geol. Surv., Mon. 35, 1898, p. 118, pl. xxxiii, f. 7 (8?).—Dawson, Brit. N. A. Boundary Com. 1875, App. A, p. 330; Roy. Soc. Canada, Trans., vol. 1, 1882 [1883], p. 32.—Penhallow, Rept. Tert. Pl. Brit. Col., 1908, p. 83.
Fort Union: Fort Union, N. Dak.
Paskapoo (Fort Union): Porcupine Creek, Saskatchewan.

Rhamnites minor Hollick, in Newberry, U. S. Geol. Surv., Mon. 26, 1895 [1896], p. 106, pl. xlii, f. 36.—Berry, Geol. Surv. N. J., Bull. 3, 1911, p. 182.
Raritan: New Jersey.

RHAMNUS Linné, Sp. pl., 1753, p. 193.

Rhamnus acuminatifolius Weber. Lesquereux, U. S. Geol. and Geog. Surv. Terr., Ann. Rept. 1872 [1873], p. 407. [Not afterward mentioned by Lesquereux.]
——: 6 miles above Spring Canyon, Mont.; Golden, Colo.

Rhamnus? acuta Heer, Fl. foss. arct., vol. 6, Abt. 2, 1882, p. 98, pl. xli, f. 6; pl. xlv, f. 13c.—Hollick, Bull. Torr. Bot. Club, vol. 21, 1894, p. 58, pl. clxxvii, f. 6; U. S. Geol. Surv., Mon. 50, 1906, p. 93, pl. xxxiv, f. 1.
Raritan: Lloyds Neck, Long Island, N. Y.

Rhamnus alaternoides Heer, Fl. tert. Helv., vol. 3, 1859, p. 78, pl. cxxiii, f. 21–23.—Lesquereux, Rept. U. S. Geol. Surv. Terr., vol. 7 (Tert. Fl.), 1878, p. 278, pl. lii, f. 11, 11a.
Denver: Golden, Colo.

Rhamnus belmontensis Knowlton and Cockerell, n. name.
Rhamnus elegans Newberry, N. Y. Lyc. Nat. Hist., Ann., vol. 9, 1868, p. 49; U. S. Geol. Surv., Mon. 35, 1898, p. 117, pl. l, f. 2.—Penhallow, Rept. Tert. Pl. Brit. Col., 1908, p. 84. [Homonym, Humboldt, Bonpland, and Kunth, 1825.]
Laramie: Belmont [=Marshall], Colo.
Eocene?: Quilchena, British Columbia.

Rhamnus brittoni Knowlton, U. S. Geol. Surv., Prof. Paper 130, 19—, p. —, pl. xv, f. 6; pl. xxiv, f. 8 [in preparation].
Laramie: Erie, Colo.

Rhamnus californica Eschscholtz. Hannibal, Bull. Torr. Bot. Club, vol. 38, 1911, p. 338, pl. xv, f. 7.
Pliocene: Calabasas Canyon, Santa Cruz Mountain, Calif.

Rhamnus cleburni Lesquereux, U. S. Geol. and Geog. Surv. Terr., Ann. Rept. 1872 [1873], p. 381; Rept. U. S. Geol. Surv. Terr., vol. 7 (Tert. Fl.), 1878, p. 280, pl. liii, f. 1-3.—?Knowlton, U. S. Geol. Surv., Bull. 204, 1902, p. 80.—Berry, U. S. Geol. Surv., Prof. Paper 91, 1916, p. 283.—Knowlton, idem, Prof. Paper 101, 1918, p. 332, pl. cxiii, f. 3.

Denver: *Golden, Colo.

Raton: Wootton, Berwynd, Aguilar, etc., Colo.; Vermejo Park, Raton Tunnel, N. Mex.

Wilcox (Grenada): Grenada, Miss.

Wilcox: Campbell's quarry, Caddo Parish, La.

?Clarno (lower part): Cherry Creek, Crook County, Oreg.

Rhamnus cleburni Lesquereux. Hollick, La. Geol. Surv., Special Rept. 5, 1899, pl. xlvii, f. 1 = *Rhamnus coushatta*.

Rhamnus concinnus Newberry. Dawson = *Rhamnites concinnus*.

Rhamnus coushatta Berry, U. S. Geol. Surv., Prof. Paper 91, 1916, p. 284, pl. lxviii, f. 1.

Rhamnus cleburni Lesquereux. Hollick, La. Geol. Surv., Special Rept. 5, 1899, p. 286, pl. xlvii, f. 1.

Andromeda eolignitica Hollick. Veatch, U. S. Geol. Surv., Prof. Paper 46, 1906, pl. xvii, f. 2.

Wilcox (Grenada): Grenada, Miss.

Wilcox: Coushatta and Naborton, La.

Rhamnus crenatulus Knowlton and Cockerell, n. name.

Rhamnus crenatus Lesquereux, Mus. Comp. Zool., Bull., vol. 16, 1888, p. 55. [Homonym, Sieboldt and Zuccarini, 1843.]

Denver: Golden, Colo.

Rhamnus dechenii Weber, Palaeontogr., vol. 2, 1852, p. 204, pl. xxiii, f. 2.—Lesquereux, U. S. Geol. and Geog. Surv. Terr., Ann. Rept. 1872 [1873], p. 397; U. S. Nat. Mus., Proc., vol. 11, 1888, p. 20.

Post-Laramie: Black Buttes, Wyo.

Rhamnus deformatus Lesquereux, Rept. U. S. Geol. Surv. Terr., vol. 8 (Cret. and Tert. Fl.), 1883, p. 126, pl. xx, f. 6.

Denver: Golden, Colo.

Rhamnus deletus? Heer. Lesquereux, U. S. Geol. and Geog. Surv. Terr., Ann. Rept. 1871 [1872], Suppl., p. 15. [Not afterward recognized by Lesquereux.]

Raton ?: Raton Mountains, N. Mex.

Rhamnus? discolor Lesquereux, U. S. Geol. and Geog. Surv. Terr., Ann. Rept. 1872 [1873], p. 398; Rept. U. S. Geol. Surv. Terr., vol. 7 (Tert. Fl.), 1878, p. 280, pl. lii, f. 17. [See *Apeibopsis discolor*.]

Post-Laramie: *Black Buttes, Wyo.

Rhamnus discolor Lesquereux [part] = *Apeibopsis discolor*.

Rhamnus elegans Newberry = *Rhamnus belmontensis*.

Rhamnus ellipticus Kirchner = *Rhamnus kirchneri*.

Rhamnus eolignitieus Berry, U. S. Geol. Surv., Prof. Paper 91, 1916, p. 284, pl. lxix, f. 4; pl. lxxi, f. 2.
Lagrange: Puryear, Tenn.
Wilcox: De Soto Parish, La.

Rhamnus eorectinervis Knowlton, n. sp.
Rhamnus rectinervis Heer. Knowlton, U. S. Geol. Surv., Mon. 32, pt. 2, 1899, p. 740.
Miocene: Lamar River, Yellowstone National Park.

Rhamnus eridani Unger, Gen. et sp. pl. foss., 1850, p. 465.—Newberry, U. S. Geol. Surv., Mon. 35, 1898, p. 118, pl. xlviii, f. 7.—Knowlton, U. S. Geol. Surv., Bull. 204, 1902, p. 80.—Penhallow, Rept. Tert. Pl. Brit. Col., 1908, p. 84.
Clarno (upper part): Bridge Creek, Oreg.
Eocene: Tranquille River, British Columbia.

Rhamnus eridani Unger. Lesquereux, U. S. Nat. Mus., Proc., vol. 11, 1888, p. 25 = *Nectandra lancifolia*.

Rhamnus fischeri Lesquereux, U. S. Geol. and Geog. Surv. Terr., Ann. Rept. 1871 [1872], Suppl., p. 15 [not afterward referred to by Lesquereux].—Knowlton, U. S. Geol. Surv., Prof. Paper 101, 1918, p. 333.
Raton?: "Fishers Peak, Raton Mountains, Colo."

Rhamnus florissantensis Cockerell, Am. Mus. Nat. Hist., Bull., vol. 24, 1908, p. 102.
Rhamnus oleaefolius Lesquereux, Rept. U. S. Geol. Surv. Terr., vol. 8 (Cret. and Tert. Fl.), 1883, p. 189, pl. xxxviii, f. 14. [Homonym of *R. oleifolia* Hooker, 1829.]
Miocene: Florissant, Colo.

Rhamnus gaudini Heer, Fl. tert. Helv., vol. 3, 1859, p. 79, pl. cxxiv, f. 4–15; pl. cxxv, f. 1, 7, 13.—Newberry, Boston Jour. Nat. Hist., vol. 7, 1863, p. 520.—Penhallow, Rept. Tert. Pl. Brit. Col., 1908, p. 84.
Puget?: Birch Bay, Wash.
Eocene: Quilchena, British Columbia.

Rhamnus goldianus Lesquereux, U. S. Geol. and Geog. Surv. Terr., Ann. Rept. 1872 [1873], p. 381; Rept. U. S. Geol. Surv. Terr., vol. 7 (Tert. Fl.), 1878, p. 281, pl. liii, f. 4–8.—Knowlton, U. S. Geol. Surv., Prof. Paper 101, 1918, p. 332, pl. ci, f. 4; pl. cxii, f. 5; idem, Prof. Paper 130, 19—, p. —, pl. xviii, f. 3 [in preparation].
Berchemia multinervis (Al. Braun) Heer. Ward, U. S. Geol. Surv., Sixth Ann. Rept., 1884–85 [1886], p. 554, pl. li, f. 13; idem, Bull. 37, 1887, p. 73, pl. xxxiii, f. 2.
Raton: Wootton and Aguilar, Colo.
Denver: *Golden and Silver Cliff, Colo.
?Laramie: Marshall, Colo.

Rhamnus goldianus latior Lesquereux, U. S. Geol. and Geog. Surv. Terr., Ann. Rept. 1872 [1873], p. 381. [Not afterward referred to by Lesquereux.]
——: Golden, Colo.

Rhamnus inaequalis Lesquereux = *Diospyros brachysepala.*

Rhamnus inaequilateralis Lesquereux, U. S. Geol. Surv., Mon. 17, 1892, p. 170, pl. xxxvii, f. 4–7.—Berry, N. Y. Bot. Gard., Bull., vol. 3, 1903, p. 86; Geol. Surv. N. J., Ann. Rept. 1905 [1906], p. 139.
Dakota: Kansas.
Magothy: Cliffwood, N. J.

Rhamnus intermedius Lesquereux = *Rhamnus washakiensis.*

Rhamnus kansensis Knowlton and Cockerell, n. name.
Rhamnus prunifolius Lesquereux, Rept. U. S. Geol. Surv. Terr., vol. 8 (Cret. and Tert. Fl.), 1883, p. 85; U. S. Geol. Surv., Mon. 17, 1891 [1892], p. 169, pl. xxxv, f. 14. [Homonym, Sieber, 1841.]
Dakota: Kansas.

Rhamnus kirchneri Cockerell, Bull. Torr. Bot. Club, vol. 33, 1906, p. 311.—Univ. Colo. Studies, vol. 3, 1906, p. 170.—Knowlton, U. S. Nat. Mus., Proc., vol. 51, 1916, p. 283.
Rhamnus ellipticus Kirchner, Acad. Sci. St. Louis, Trans., vol. 8, 1898, p. 183, pl. xv, f. 3. [Homonym, Swartz, 1788.]
Miocene: *Florissant, Colo.

Rhamnus lesquereuxi Berry, U. S. Nat. Mus., Proc., vol. 54, 1918, p. 633, pl. 95, f. 4.
Rhamnus notatus? Saporta. Lesquereux, Rept. U. S. Geol. Surv. Terr., vol. 8 (Cret. and Tert. Fl.), 1883, p. 189, pl. xxxviii, f. 15.
Miocene: Florissant, Colo.; Beaver Creek 10 miles east of Beaver City, Okla.

Rhamnus marginatus Lesquereux, Geol. Surv. Ark., vol. 2, 1860, p. 319, pl. vi, f. 1; Am. Philos. Soc., Trans., vol. 13, 1869, p. 420, pl. xxii, f. 3–5.—Berry, U. S. Geol. Surv., Prof. Paper 91, 1916, p. 282, pl. lxxi, f. 1; pl. lxxii, f. 1.
Wilcox: *Mississippi.
Wilcox (Ackerman): Hurleys, Miss.
Lagrange: Puryear, Tenn.

Rhamnus marginatus apiculatus Berry, U. S. Geol. Surv., Prof. Paper 91, 1916, p. 283, pl. lxix, f. 1.
Wilcox (Ackerman): Hurleys, Miss.

Rhamnus marshallensis Knowlton, U. S. Geol. Surv., Prof. Paper 130, 19—, p. —, pl. xv, f. 3 [in preparation].
Laramie: Marshall, Colo.

Rhamnus meriani? Heer. Lesquereux, U. S. Geol. and Geog. Surv. Terr., Ann. Rept. 1873 [1874], p. 405. [Not afterward recognized.]
——: Black Buttes, Wyo.

Rhamnus minutus Knowlton, U. S. Geol. Surv., Prof. Paper 130, 19—, p. —, pl. xvii, f. 2 [in preparation].
  Laramie: Erie, Colo.

Rhamnus mudgei Lesquereux, U. S. Geol. Surv., Mon. 17, 1891 [1892], p. 169, pl. xxxvii, f. 2, 3.
  Dakota: Ellsworth County, Kans.

Rhamnus newberryi Knowlton and Cockerell, n. name.
  Fort Union: Valley of Yellowstone River, Mont.
    Rhamnus parvifolius Newberry, U. S. Nat. Mus., Proc., vol. 5, 1882 [1883], p. 511. [Homonym, Bunge, 1830.]

Rhamnus notatus? Saporta. Lesquereux, Rept. U. S. Geol. Surv. Terr., vol. 8 (Cret. and Tert. Fl.), 1883, p. 189, pl. xxxviii, f. 15 = *Rhamnus lesquereuxi*.

Rhamnus novae-caesareae Berry, N. Y. Bot. Gard., Bull., vol. 3, 1903, p. 85, pl. 50, f. 5, 6; Geol. Surv. N. J., Ann. Rept. 1905 [1906], p. 139.
  Magothy: Cliffwood, N. J.

Rhamnus obovatus Lesquereux, Am. Jour. Sci., 2d ser., vol. 45, 1868, p. 207; Rept. U. S. Geol. Surv. Terr., vol. 7 (Tert. Fl.), 1878, p. 281, pl. liv, f. 1, 2.—Knowlton, U. S. Geol. Surv., Prof. Paper 101, 1918, p. 333.
  Raton?: *Purgatory Canyon, N. Mex.
  Denver?: Golden, Colo.

Rhamnus oleaefolius Lesquereux = *Rhamnus florissantensis*.

Rhamnus parvifolius Newberry = *Rhamnus newberryi*.

Rhamnus? pealei Knowlton, U. S. Geol. Surv., Prof. Paper 130, 19—, p. —, pl. xv, f. 7 [in preparation].
  Laramie: Moffat Railroad cut, about 8 miles north of Golden, Colo.

Rhamnus pfaffiana Heer. Hollick, N. Y. Acad. Sci., Trans., vol. 11, 1892, p. 103, pl. iv, f. 3 = *Diospyros apiculata?*

Rhamnus pfaffiana Heer. Hollick, N. Y. Acad. Sci., Trans., vol. 11, 1892, p. 103, pl. iv, f. 1 = *Diospyros provecta*.

Rhamnus pfaffiana Heer. Hollick, N. Y. Acad. Sci., Trans., vol. 11, 1892, p. 103, pl. iv, f. 2 = *Andromeda tenuinervis*.

Rhamnus prunifolius Lesquereux = *Rhamnus kansensis*.

Rhamnus purshiana De Candolle. Hannibal, Bull. Torr. Bot. Club, vol. 38, 1911, p. 338, pl. xv, f. 10.
  Pliocene: Near Portola, Santa Cruz Mountains, Calif.

Rhamnus puryearensis Berry, U. S. Geol. Surv., Prof. Paper 91, 1916, p. 284, pl. lxiv, f. 7.
  Lagrange: Puryear, Tenn.

Rhamnus quilchensis Penhallow, Rept. Tert. Pl. Brit. Col., 1908, p. 83, text f. 29.
Oligocene: Quilchena and Horsefly rivers, British Columbia.

Rhamnus rectinervis Heer, Fl. tert. Helv., vol. 3, 1859, p. 80, pl. cxxv, f. 2–6.—Lesquereux, U. S. Geol. and Geog. Surv. Terr., Ann. Rept. 1871 [1872], p. 295; Rept. U. S. Geol. Surv. Terr., vol. 7 (Tert. Fl.), 1878, p. 279, pl. lii, f. 12–15.
Laramie: Golden, Colo.
Denver: Golden, Colo.
——: Black Buttes, Wyo.

Rhamnus rectinervis Heer. Knowlton, U. S. Geol. Surv., Mon. 32, pt. 2, 1899, p. 740 = *Rhamnus eorectinervis.*

Rhamnus revoluta Lesquereux, U. S. Geol. Surv., Mon. 17, 1891 [1892], p. 171, pl. lxv, f. 5.
Dakota: Near Delphos, Kans.

Rhamnus rossmaessleri Unger. Lesquereux = *Juglans rhamnoides.*

Rhamnus rossmässleri Unger. Hollick, N. Y. Acad. Sci., Trans., vol. 12, 1892, p. 35, pl. iii, f. 5 = *Myrsinites? gaudini.*

Rhamnus salicifolius Lesquereux, Am. Jour. Sci., 2d ser., vol. 45, 1868, p. 206; U. S. Geol. and Geog. Surv. Terr., Ann. Rept. 1869 (reprint, 1873), p. 196; Rept. U. S. Geol. Surv. Terr., vol. 7 (Tert. Fl.), 1878, p. 282, pl. liii, f. 9, 10.—Knowlton, U. S. Geol. Surv., Prof. Paper 101, 1918, p. 271; idem, Prof. Paper 130, 19—, p. —, pl. xv, f. 4; pl. xix, f. 2b [in preparation].
Laramie: *Marshall, Coal Creek, Boulder County, and Colorado Springs, Colo.
Vermejo: Coal Creek, Rockvale, and near Trinidad, Colo.
Dawson: Templeton Gap, 4 miles northeast of Colorado Springs, Colo.
Mesaverde: Harper station, Wyo.

Rhamnus similis Lesquereux, U. S. Geol. Surv., Mon. 17, 1891 [1892], p. 168, pl. xxxv, f. 12, 13.
Dakota: Ellsworth County, Kans.

Rhamnus tenax Lesquereux, Am. Jour. Sci., 2d ser., vol. 46, 1868, p. 101; Rept. U. S. Geol. Surv. Terr., vol. 6 (Cret. Fl.), 1874, p. 109, pl. xxi, f. 4; U. S. Geol. Surv., Mon. 17, 1891 [1892], p. 170, pl. xxxviii, f. 6.—Berry, Bull. Torr. Bot. Club, vol. 39, 1912, p. 398; U. S. Geol. Surv., Prof. Paper 112, 1919, p. 114, pl. xxv, f. 1, 2.
Dakota: Ellsworth County, Kans.
Woodbine: Arthurs Bluff, Tex.
Tuscaloosa: Shirleys Mill, Ala.

Rhamnus washakiensis Cockerell, Am. Mus. Nat. Hist., Bull., vol. 24, 1908, p. 74.

Rhamnus intermedius Lesquereux, U. S. Geol. and Geog. Surv. Terr., Ann. Rept. 1871 [1872], p. 286; Rept. U. S. Geol. Surv. Terr., vol. 7 (Tert. Fl.), 1878, p. 282, pl. liv, f. 3. [Homonym, Steudel and Hochstetter, 1827.]

Green River: *Bridgers Pass, Wyo.

Rhamnus? williardi Knowlton, U. S. Geol. Surv., Prof. Paper 98, 1916, p. 91, pl. xvi, f. 1, 2; pl. xvii, f. 1–4.

Ficus sp. Cockerell, Colo. Univ. Studies, vol. 4, 1907, p. 152.

Fox Hills: Wildcat Mound and Thompson Creek, near Milliken, Colo.

Rhamnus? woottonensis Knowlton, U. S. Geol. Surv., Prof. Paper 101, 1918, p. 332, pl. ci. f. 6.

Raton: Wootton, Colo.

Rhamnus sp. Dawson, Brit. N. A. Bound. Com. (Rept. Geol. and Res. Vic. 49th Parallel), 1875, App. A, p. 330.

Paskapoo?: Great Valley, Northwest Territory.

Rhamnus sp. Knowlton, U. S. Geol. Surv., Prof. Paper 130, 19—, p. —, pl. xvii, f. 1 [in preparation].

Laramie: Popes Bluff, Pikeview, Colo.

Rhizocaulon gracile Lesquereux = *Phyllites improbatus.*

Rhizomorphs, Hollick, N. Y. Acad. Sci., Ann., vol. 11, 1898, p. 423, pl. xxxviii, f. 1; U. S. Geol. Surv., Mon. 50, 1906, p. 112, pl. vi, f. 13.

Raritan: Tottenville, Staten Island, N. Y.

RHIZOPHORA Linné, Sp. pl., vol. 1, 1753, p. 443.

Rhizophora eocenica Berry, U. S. Geol. Surv., Prof. Paper 84, 1914, p. 144, pl. xxix, f. 1, 2.

Barnwell (Twiggs): Phinizy Gully, Columbia County, near Sandersville and near Macon, Ga.

RHUS Linné, Sp. pl., 1753, p. 265.

Rhus acuminata Lesquereux = *Rhus lesquereuxii.*

Rhus bella? Heer. Lesquereux, U. S. Geol. and Geog. Surv. Terr., Ann. Rept. 1872 [1873], p. 407. [Not afterward mentioned by Lesquereux.]

——: 6 miles above Spring Canyon, Mont.

Rhus bendirei Lesquereux, U. S. Nat. Mus., Proc., vol. 11, 1888, p. 15, pl. ix, f. 2.—Knowlton, U. S. Geol. Surv., Bull. 204, 1902, p. 70.

Mascall: Van Horn's ranch, John Day Basin, Oreg.

Rhus bendirei Lesquereux, U. S. Nat. Mus., Proc., vol. 11, 1888, p. 15 [the small leaf described] = *Juglans oregoniana.*

Rhus boweniana Lesquereux, Mus. Comp. Zool., Mem., vol. 6, No. 2, 1878, p. 29, pl. ix, f. 9, 10.—Knowlton, in Lindgren, Jour. Geol., vol. 4, 1896, p. 890; U. S. Geol. Surv., Prof. Paper 73, 1911, p. 59.

Miocene (auriferous gravels): Chalk Bluffs, Independence Hill, Placer County, Calif.

Rhus cassioides Lesquereux, Rept. U. S. Geol. Surv. Terr., vol. 8 (Cret. and Tert. Fl.), 1883, p. 193, pl. xli, f. 11.
Miocene: Florissant, Colo.

Rhus coriarioides Lesquereux, Rept. U. S. Geol. Surv. Terr., vol. 8 (Cret. and Tert. Fl.), 1883, p. 193, pl. xli, f. 3.—Cockerell, Am. Nat., vol. 44, 1910, p. 38, f. 4.
Miocene: Florissant, Colo.

Rhus cretacea Heer? Hollick, U. S. Geol. Surv., Mon. 50, 1906, p. 87, pl. xxxiii, f. 2.
Magothy: Glen Cove, Long Island, N. Y.

Rhus darlingtonensis Berry, U. S. Geol. Surv., Prof. Paper 84, 1914, p. 51, pl. ix, f. 7, 8.
Black Creek: Darlington, S. C.

Rhus deleta Heer, Fl. tert. Helv., vol. 3, 1859, p. 83, pl. cxxvii, f. 8; pl. cliv, f. 26.—Lesquereux, U. S. Geol. and Geog. Surv. Terr., Ann. Rept. 1871 [1872], Suppl., p. 12. [Not afterward referred to by Lesquereux.]
——: Evanston, Wyo.

Rhus dispersa Lesquereux, Mus. Comp. Zool., Mem., vol. 6, No. 2, 1878, p. 32, pl. i, f. 23.—Knowlton, in Lindgren, Jour. Geol., vol. 4, 1896, p. 890; U. S. Geol. Surv., Prof. Paper 73, 1911, p. 60.
Miocene (auriferous gravels): Table Mountain, Tuolumne County; Independence Hill, Placer County, Calif.

Rhus? drymeja Lesquereux = *Myrica drymeja.*

Rhus evansii Lesquereux, U. S. Geol. and Geog. Surv. Terr., Ann. Rept. 1871 [1872], p. 293; Rept. U. S. Geol. Surv. Terr., vol. 7 (Tert. Fl.), 1878, p. 291, pl. l, f. 4; pl. lviii, f. 5–9.
Evanston: Evanston, Wyo.

Rhus fraterna Lesquereux = *Cotinus fraterna.*

Rhus frigida Knowlton, U. S. Nat. Mus., Proc., vol. 17, 1894, p. 227, pl. ix, f. 6; Geol. Soc. Am., Bull., vol. 5, 1893, p. 586.
Kenai: Herendeen Bay, Alaska.

Rhus haydenii Lesquereux = *Weinmannia haydenii.*

Rhus heüfleri Heer, Fl. tert. Helv., vol. 3, 1859, p. 85, pl. cxxvii, f. 3–6.—Lesquereux, U. S. Nat. Mus., Proc., vol. 11, 1888, p. 26.
Miocene: Corral Hollow, Calif.

Rhus hilliae Lesquereux, Rept. U. S. Geol. Surv. Terr., vol. 8 (Cret. and Tert. Fl.), 1883, p. 194, pl. xli, f. 12–15.—Cockerell, Am. Mus. Nat. Hist., Bull., vol. 24, 1908, p. 99.—Knowlton, U. S. Nat. Mus., Proc., vol. 51, 1916, p. 278.
Miocene: Florissant, Colo.

Rhus lesquereuxii Knowlton and Cockerell, n. name.

Rhus acuminata Lesquereux, U. S. Geol. and Geog. Surv. Terr., Ann. Rept. 1871 [1872], Suppl., p. 8; Rept. U. S. Geol. Surv. Terr., vol. 8 (Cret. and Tert. Fl.), 1883, p. 194, pl. xlii, f. 14-17.—Cockerell, Am. Mus. Nat. Hist., Bull., vol. 24, 1908, p. 94. [Homonym, De Candolle, 1865.]

Miocene: Florissant, Colo.

Green River: Green River, Wyo.

Rhus membranacea Lesquereux, U. S. Geol. and Geog. Surv. Terr., Bull., vol. 1, 1875 [1876], p. 369; idem, Ann. Rept. 1874 [1876], p. 306; Rept. U. S. Geol. Surv. Terr., vol. 7 (Tert. Fl.), 1878, p. 292, pl. lxiv, f. 6, 7.—Knowlton, U. S. Geol. Surv., Bull. 163, 1900, p. 61.

Mesaverde: Point of Rocks, Wyo.

Rhus mensae Cockerell, Am. Jour. Sci., 4th ser., vol. 26, 1908, p. 543.

Rhus metopioides Lesquereux, Mus. Comp. Zool., Mem., vol. 6, No. 2, 1878, p. 31, pl. viii, f. 12, 13.—Knowlton, in Lindgren, U. S. Geol. Surv., Prof. Paper 73, 1901, p. 60. [Homonym, Turczaninow, 1858.]

Miocene (auriferous gravels): Table Mountain, Tuolumne County, Calif.

Rhus metopioides Lesquereux = *Rhus mensae*.

Rhus microphylla Heer = *Comptonia microphylla*.

Rhus milleri Hollick, Md. Geol. Surv., Miocene, 1904, p. 485, f. c, d; Jour. Geol., vol. 17, 1909, p. 29, text f. 9.

Calvert: Good Hope Hill Road, Prince Georges County, Md.; Richmond, Va.

Rhus mixta Lesquereux, Mus. Comp. Zool., Mem., vol. 6, No. 2, 1878, p. 31, pl. i, f. 5-8.—?Knowlton, U. S. Geol. Surv., Mon. 32, pt. 2, 1899, p. 731; in Lindgren, U. S. Geol. Surv., Prof. Paper 73, 1911, p. 59.

Miocene (auriferous gravels): Chalk Bluffs, Nevada County, Calif.

Miocene: ?Yellowstone National Park. [Identification?]

Rhus myricaefolia Lesquereux, Mus. Comp. Zool., Mem., vol. 6, No. 2, 1878, p. 30, pl. ix, f. 13.—Knowlton, in Lindgren, Jour. Geol., vol. 4, 1896, p. 890; U. S. Geol. Surv., Prof. Paper 73, 1911, p. 59.

Miocene (auriferous gravels): Chalk Bluffs, Nevada County, and Independence Hill, Placer County, Calif.

Rhus nervosa Newberry = *Rhus unitus*.

Rhus? nevadensis Knowlton, U. S. Geol. Surv., Twenty-first Ann. Rept., pt. 2, 1901, p. 218, pl. xxx, f. 15.

Esmeralda: 4.5 kilometers northwest of Emigrant Peak, Esmeralda County, Nev.

Rhus payettensis Knowlton, U. S. Geol. Surv., Eighteenth Ann. Rept., pt. 3, 1898, p. 733, pl. ci, f. 6, 7.

Payette: Marsh, Idaho.

Rhus powelliana Lesquereux, U. S. Geol. Surv., Mon. 17, 1891 [1892], p. 155, pl. lvi, f. 4, 5.—Knowlton, in Woodruff, U. S. Geol. Surv., Bull. 452, 1911, p. 20.

Dakota: Near *Fort Harker, Kans.; near Lander, Wyo.

Rhus pseudo-meriani Lesquereux = *Quercus doljensis*.

Rhus redditeformis Berry, Bull. Torr. Bot. Club, vol. 39, 1912, p. 397, pl. xxxi, f. 2.

Woodbine: Arthurs Bluff, Lamar County, Tex.

Rhus rosaefolia Lesquereux = *Weinmannia? dubiosa*.

Rhus rotundifolia Kirchner = *Hydrangea florissantia*.

Rhus subrhomboidalis Lesquereux, Rept. U. S. Geol. Surv. Terr., vol. 8 (Cret. and Tert. Fl.), 1883, p. 195, pl. xli, f. 16–19.

Miocene: Florissant, Colo.

Rhus? trifolioides Lesquereux, Rept. U. S. Geol. Surv. Terr., vol. 8 (Cret. and Tert. Fl.), 1883, p. 196.—Cockerell, Am. Mus. Nat. Hist., Bull., vol. 24, 1908, p. 99.—Knowlton, U. S. Nat. Mus., Proc., vol. 51, 1916, p. 279.

Miocene: *Florissant, Colo.

Rhus typhinoides Lesquereux, Mus. Comp. Zool., Mem., vol. 6, No. 2 1878, p. 29, pl. ix, f. 1–6.—Knowlton, in Lindgren, U. S. Geol. Surv., Prof. Paper 73, 1911, p. 59.

Miocene (auriferous gravels): Table Mountain, Tuolumne County, Calif.

Rhus uddeni Lesquereux, U. S. Geol. Surv., Mon. 17, 1891 [1892], p. 154, pl. lvii, f. 2.—Knowlton, in Hill, Am. Jour. Sci., 3d ser., vol. 50, 1895, p. 212.

Dakota: Smoky Hill Buttes, near Salemsburg, Salina County, Kans.

Cheyenne: Black Hills, near Belvidere, Kans.

Rhus unitus Knowlton and Cockerell, n. name.

Rhus? nervosa Newberry, N. Y. Lyc. Nat. Hist., Ann., vol. 9, 1868, p. 53.—[Lesquereux], U. S. Geol. and Geog. Surv. Terr., Ill. Cret. and Tert. Pl., 1878, pl. xvi, f. 5, 6.—Newberry, U. S. Geol. Surv., Mon. 35, 1898, p. 114, pl. xxxiii, f. 5, 6. [Homonym, Ecklon and Zeyher, 1834.].

Fort Union: Fort Union, N. Dak.

Rhus vexans Lesquereux = *Schmaltzia vexans*.

Rhus? viburnoides Knowlton, U. S. Geol. Surv., Prof. Paper 101, 1918, p. 328, pl. xcviii, f. 5.

Raton: Near Wootton, Colo.

Rhus? westii Knowlton, in Lesquereux, U. S. Geol. Surv., Mon. 17, 1891 [1892], p. 154, pl. xxxviii, f. 9, 10.

Dakota: *Ellsworth County, Kans.

Rhus winchellii Lesquereux, Rept. U. S. Geol. Surv. Terr., vol. 8 (Cret. and Tert. Fl.), 1883, p. 236.

Fort Union: *Yellowstone Valley, Mont.

Rhus sp. Knowlton, in Lindgren, Jour. Geol., vol. 4, 1896, p. 890.
Miocene (auriferous gravels): Independence Hill, Placer County, Calif.

Rhus sp. Knowlton, Geol. Soc. Am., Bull., vol. 8, 1897, p. 145.
Fort Union: Black Buttes, Wyo.

Rhus? sp. Knowlton, Geol. Soc. Am., Bull., vol. 8, 1897, p. 133.
Lance: Lance Creek, Converse County, Wyo.

Rhus sp. Knowlton, in Washburne, U. S. Geol. Surv., Bull. 590, 1914, p. 36.
Eocene?: Eugene, Oreg.

Rhus? sp. Lesquereux, U. S. Nat. Mus., Proc., vol. 11, 1888, p. 15.—Knowlton, U. S. Geol. Surv., Bull. 204, 1902, p. 70, pl. xiv, f. 6.
Mascall: Van Horn's ranch, John Day Basin, Oreg.

RHYNCHOSTEGIUM Schimper, Bryol. Eur., vol. 5, 1852.

Rhynchostegium knowltoni E. G. Britton, Bull. Torr. Bot. Club, vol. 26, 1899, p. 79, text f.; E. G. Britton and Hollick, idem, vol. 34, 1907, p. 140, pl. ix, f. 5.
Puget: Cle Elum, Wash.

RIBES Linné, Sp. pl., 1753, p. 200.

Ribes? florissanti Knowlton, U. S. Nat. Mus., Proc., vol. 51, 1916, p. 272.
Myrica diversifolia Lesquereux, Rept. U. S. Geol. Surv. Terr., vol. 8 (Cret. and Tert. Fl.), 1883, p. 148, pl. xxv, f. 13 [not other figures].
Miocene: *Florissant, Colo.

Ribes neomexicana Knowlton, U. S. Geol. Surv., Prof. Paper 98, 1916, p. 342, pl. lxxxix, f. 4.
Fruitland: 10 miles south of San Juan River and 4 miles east of Chaco River, San Juan County, N. Mex.

Ribes protomelaenum Cockerell, Am. Mus. Nat. Hist., Bull., vol. 24, 1908, p. 93, pl. vii, f. 15.
Miocene: *Florissant, Colo.

ROBINIA Linné, Sp. pl., 1753, p. 722.

Robinia brittoni Cockerell, Am. Jour. Sci., 4th ser., vol. 26, 1908, p. 543, f. 8.—Knowlton, U. S. Nat. Mus., Proc., vol. 51, 1916, p. 277, pl. xxiv, f. 2.
Miocene: *Florissant, Colo.

Robinia mesozoica Cockerell, Torreya, vol. 12, 1912, p. 32, text f. 1.
Laramie?: *25 miles north of Kremmling, Colo.

Robinia pseudacacia Linné. Penhallow, Brit. Assoc. Adv. Sci., Bradford meeting, 1900, p. 336; Am. Nat., vol. 41, 1907, p. 448.—Hollick, Md. Geol. Surv., Pliocene and Pleistocene, 1906, p. 234, pl. lxix, f. 4.
Pleistocene: Don Valley, Toronto, Canada.
Pleistocene (Talbot): Bodkin Point, Anne Arundel County, Md.

ROGERSIA Fontaine, U. S. Geol. Surv., Mon. 15, 1889 [1890], p. 287. [Type *R. longifolia*.]

Rogersia angustifolia Fontaine, U. S. Geol. Surv., Mon. 15, 1889 [1890], p. 288, pl. cxliii, f. 2; pl. cxlix, f. 4, 8; pl. cl, f. 2, 7; in Ward, idem, Mon. 48, 1905 [1906], p. 521 [not pp. 491, 510].—Berry, Md. Geol. Surv., Lower Cret., 1912, p. 501.

Saliciphyllum longifolium Fontaine, U. S. Geol. Surv., Mon. 15, 1889 [1890], p. 302, pl. cl, f. 12.

Patuxent: Fredericksburg and Potomac Run, Va.
Arundel: Langdon ?, D. C.

Rogersia angustifolia parva Fontaine, in Ward, U. S. Geol. Surv., Mon. 48, 1905 [1906], p. 523, pl. cxi, f. 9.—Berry, Md. Geol. Surv., Lower Cret., 1912, p. 501.

Arundel: Langdon, D. C.

Rogersia angustifolia Fontaine. Fontaine, in Ward, U. S. Geol. Surv., Mon. 48, pp. 491, 510 = *Sapindopsis variabilis*.

Rogersia longifolia Fontaine, U. S. Geol. Surv., Mon. 15, 1889 [1890], p. 287, pl. cxxxix, f. 6; pl. cxliv, f. 2; pl. cl, f. 1; pl. clix, f. 1, 2; in Ward, U. S. Geol. Surv., Mon. 48, 1905 [1906], p. 533 [not pp. 511, 523, pl. cxii, f. 9].—Berry, Md. Geol. Surv., Lower Cret., 1912, p. 500.

Sapindopsis elliptica Fontaine, U. S. Geol. Surv., Mon. 15, 1889 [1890], p. 297, pl. cxlvii, f. 3.

Patuxent: Kankeys, near Dutch Gap Canal, Va.
Arundel: Arlington ?, Md.

ROSA (Tournefort) Linné, Sp. pl., 1753, p. 491.

Rosa hilliae Lesquereux, Rept. U. S. Geol. Surv. Terr., vol. 8 (Cret. and Tert. Fl.), 1883, p. 199, pl. xl, f. 16, 17.

Miocene: Florissant, Colo.

Rosa? inquirenda Knowlton, U. S. Nat. Mus., Proc., vol. 51, 1916, p. 273, pl. xvii, f. 1.

Miocene: *Florissant, Colo.

Rosa ruskiniana Cockerell, Am. Jour. Sci., 4th ser., vol. 26, 1908, p. 540, f. 5.

Miocene: *Florissant, Colo.

Rosa scudderi Knowlton, U. S. Nat. Mus., Proc., vol. 51, 1916, p. 272, pl. xxii, f. 4.

Miocene: *Florissant, Colo.

Rosa wilmattae Cockerell, Am. Mus. Nat. Hist., Bull., vol. 24, 1908, p. 94, pl. vii, f. 14; Am. Mus. Jour., vol. 16, 1916, p. 450.

Miocene: *Florissant, Colo.

ROSELLINITES (De Notaris) Meschinelli, Sylloge Fung. Foss., 1892, p. 14.

Rosellinites lapideus (Lesquereux) Knowlton, U. S. Geol. Surv., Bull. 152, 1898, p. 204; U. S. Geol. Surv., Prof. Paper 101, 1918, p. 242.

Spheria lapidea Lesquereux, U. S. Geol. and Geog. Surv. Terr., Ann. Rept. 1872 [1873], p. 373; Rept. U. S. Geol. Surv. Terr., vol. 7 (Tert. Fl.), 1878, p. 34, pl. i, f. 3.

Sphaerites lapideus (Lesquereux) Meschinelli, Sylloge Fung. Foss., 1892, p. 30.

Trinidad: Near Trinidad, Colo.

RUBIOIDES Perkins, Rept. Vt. State Geol. 1903–4 [1904], p. 193. [Type, *R. lignita.*]

Rubioides lignita Perkins, Rept. Vt. State Geol. 1903–4 [1904], p. 193, pl. lxxviii, f. 80, 84.

Miocene: Brandon, Vt.

RUBUS (Tournefort) Linné, Sp. pl., 1753, p. 492.

Rubus sp. Berry, Jour. Geol., vol. 15, 1907, p. 344.

Pleistocene: Neuse River, N. C.

RUFFORDIA Seward, Cat. Mesozoic Pl. Brit. Mus., Wealden Fl., pt. 1, 1894, p. 75.

Ruffordia acrodenta (Fontaine) Berry, Md. Geol. Surv., Lower Cret., 1911, p. 230, pl. xxiii, f. 5, 6.

Sphenopteris acrodenta Fontaine, U. S. Geol. Surv., Mon. 15, 1890, p. 90, pl. xxxiv, f. 4.

Ruffordia göpperti latifolia Seward, Wealden Fl., pt. 1, 1894, p. 85, pl. vi, f. 1, 1a.

Patuxent: Dutch Gap, Va.

Patapsco: Federal Hill (Baltimore), Md.

Ruffordia göpperti latifolia Seward = *Ruffordia acrodenta.*

Ruffordia göpperti (Dunker) Seward, Cat. Mesoz. Pl. Brit. Mus., Wealden Fl., pt. 1, 1894, p. 76, pl. iv; pl. v; pl. x, f. 1, 2.—Fontaine, in Ward, U. S. Geol. Surv., Mon. 48, 1906, p. 75, pl. xii, f. 4–8.—Berry, Md. Geol. Surv., Lower Cret., 1911, p. 231, pl. xxiii, f. 3, 4.

Cheilanthites göpperti Dunker, Norddeutsch. Wälder thon., 1844, p. 6.

Cladophlebis sphenopteroides Fontaine, U. S. Geol. Surv., Mon. 15, 1889 [1890], p. 79, pl. xxi, f. 4; in Ward, U. S. Geol. Surv., Mon. 48, 1905 [1906], p. 519.

Thyrsopteris heteromorpha Fontaine, op. cit. (Mon. 15), p. 136, pl. lii, f. 1.

Sphenopteris thysopteroides Fontaine, op. cit. (Mon. 15), p. 89, pl. xxv, f. 3; pl. lviii, f. 5.

Sphenopteris spatulata Fontaine, op. cit. (Mon. 15), p. 93, pl. l, f. 4.

Sphenopteris pachyphylla Fontaine, op. cit. (Mon. 15), p. 93, pl. l, f. 5.

Sphenopteris valdensis? Heer. Fontaine, U. S. Nat. Mus., Proc., vol. 16, 1893, p. 263, pl. xxxvi, f. 2?.

Patuxent: Fredericksburg, Dutch Gap, and Trents Reach, Va.; Ivy City ?, D. C.

Trinity: Glen Rose, Tex. [Identification ?]

Jurassic: Douglas County, Oreg.

RULAC Adanson, Fam. Pl., vol. 2, 1763, p. 383.

Rulac crataegifolium Knowlton, U. S. Geol. Surv., Bull. 204, 1902, p. 77, pl. xvi, f. 7.

Mascall: Van Horn's ranch, John Day Basin, Oreg.

SABAL Adanson, Fam. Pl., vol. 2, 1763, p. 495.

Sabal campbelli Newberry [part] = *Sabalites campbelli.*

Sabal campbelli Newberry [part] = *Sabal ungeri.*

Sabal communis Lesquereux = *Sabal? eocenica.*

Sabal? eocenica (Lesquereux) Knowlton, n. comb.

Flabellaria eocenica Lesquereux, U. S. Geol. and Geog. Surv. Terr., Ann. Rept. 1872 [1873], p. 391; idem, 1873 [1874], p. 380; idem, 1876 [1878], p. 502; Rept. U. S. Geol. Surv. Terr., vol. 7 (Tert. Fl.), 1878, p. 111, pl. xiii, f. 1-3.—Knowlton, Washington Acad. Sci., Proc., vol. 11, 1909, pp. 204, 207.

Flabellaria communis Lesquereux, U. S. Geol. and Geog. Surv. Terr., Bull., vol. 1, 1875 [1876], p. 385.

Sabal communis Lesquereux, U. S. Geol. and Geog. Surv. Terr., Ann. Rept. 1874 [1876], p. 311.

Sabal inquirenda Knowlton, U. S. Geol. Surv., Prof. Paper 101, 1918, p. 288, pl. lvi.

Post-Laramie: *Black Buttes, Wyo.

Denver: Golden, Colo.

Lance: Lance and Lightning creeks, Converse County, and Weston County, Wyo.

Raton: Near Blossburg and Raton tunnel near Raton, N. Mex.

Dawson: Jimmy Camp Creek, 9 miles east of Colorado Springs, Colo.

Sabal goldiana Lesquereux, U. S. Geol. and Geog. Surv. Terr., Ann. Rept. 1872 [1873], p. 377. [Not afterward referred to by Lesquereux. Same as *Sabalites campbelli?*]

——: Golden, Colo.

Sabal grandifolia Newberry, U. S. Geol. Surv., Mon. 35, 1898, p. 28, pl. xxv; pl. lxiii, f. 5 [not pl. lxiv, f. 2, 2a].—Knowlton, in Calvert, U. S. Geol. Surv., Bull. 471, 1912, p. 16.

Sabal campbelli Newberry [part], Boston Jour. Nat. Hist., vol. 7, 1863, p. 515.

Sabal campbelli Newberry [Lesquereux], U. S. Geol. and Geog. Surv. Terr., Ill. Cret. and Tert. Pl., 1878, pl. x.

Fort Union: Yellowstone River, Mont.

Lance: Near Glendive, Mont.; Carbon County, Wyo.

Sabal grandifolia Newberry [part] = *Sabal ungeri.*

Sabal grayana Lesquereux = *Sabalites grayanus.*

Sabal imperialis Dawson, Roy. Soc. Canada, Trans., vol. 1, 1882 [1883], sec. 4, p. 26, pl. vi, f. 23-23b; idem, vol. 11, 1893 [1894], p. 57, pl. xiv, f. 61.—Newberry, U. S. Geol. Surv., Mon. 35, 1898, p. 30, pl. xvi, f. 6, 6a.

Sabal sp. Newberry, Boston Jour. Nat. Hist., vol. 7, 1863, p. 515.

Upper Cretaceous: Nanaimo and Port McNeill, Vancouver Island, British Columbia.

Sabal inquirenda Knowlton = *Sabal? eocenica.*

Sabal? leei Knowlton, U. S. Geol. Surv., Prof. Paper 101, 1918, p. 289, pl. lx.

Raton: Near Walsenburg, Colo.; Jones Canyon, and near Trinidad, N. Mex.

Sabal major? Unger. Lesquereux, U. S. Geol. Surv. Terr., Ann. Rept. 1871 [1872], p. 295 = *Geonomites schimperi.*

Sabal? montana Knowlton, U. S. Geol. Surv., Prof. Paper 101, 1918, p. 253, pl. xxxii, f. 3; idem, Prof. Paper 98, 1916, p. 335, pl. lxxxv, f. 2; idem, Prof. Paper 130, 19—, p. —, pl. iii, f. 4 [in preparation].

Sabalites grayanus (Lesquereux) Lesquereux, Rept. U. S. Geol. Surv. Terr., vol. 7 (Tert. Fl.), 1878, p. 112, pl. xii, f. 1 [not pl. xii, f. 2].—Cockerell, Univ. Colo. Studies, vol. 7, 1910, p. 150, f. 14.—Knowlton, U. S. Geol. Surv., Bull. 163, 1900, p. 32, pl. vi, f. 5.

Mesaverde: Point of Rocks, Wyo.; near Meeker, Colo.

Vermejo: Rockvale and Florence, Canon City field, and Forbes, Colo.; Vermejo Park, N. Mex.

Fruitland: 18 miles south of San Juan River, San Juan County, N. Mex.

Laramie: Reliance mine, Erie, Colo.

Sabal palmetto (Walter) Roemer and Schultes. Berry, Jour. Geol., vol. 25, 1917, p. 662.

Pleistocene: Vero, Fla.

Sabal powellii Newberry, U. S. Nat. Mus., Proc., vol. 5, 1882 [1883], p. 504; U. S. Geol. Surv., Mon. 35, 1898, p. 30, pl. lxiii, f. 6; pl. lxiv, f. 1, 1a.

Green River: Green River, Wyo.

Sabal rigida Hatcher, Carnegie Mus., Ann., vol. 1, 1901, p. 263, text f. 1.

Lance: Converse County, Wyo.

Sabal? rugosa Knowlton, U. S. Geol. Surv., Prof. Paper 101, 1918, p. 288, pl. lviii.

Raton: Near Yankee, N. Mex.

Sabal? ungeri (Lesquereux) Knowlton, U. S. Geol. Surv., Prof. Paper 101, 1918, pp. 254, 289, pls. lvii, lix.

Geonomites ungeri Lesquereux, Rept. U. S. Geol. Surv. Terr., vol. 7 (Tert. Fl.), 1878, p. 118, pl. xi, f. 2.

Sabal campbelli Newberry [part], Boston Soc. Nat. Hist., Jour., vol. 7, 1863, p. 515.

"*Sabal campbelli* Newberry." Lesquereux, Ill. Cret. and Tert. Pl., 1878, pl. x.

Sabal grandifolia Newberry [part], U. S. Geol. Surv., Mon. 35, 1898, p. 28, pl. lxiii, f. 5 [not pl. xxv; pl. lxiv, f. 2, 2a].

Raton: *Raton Mountains, N. Mex. Very abundant in the Raton formation of Colorado and New Mexico.

Vermejo: Ponil Canyon, N. Mex.

Sabal sp. Knowlton, U. S. Geol. Surv., Prof. Paper 98, 1916, p. 336, pl. lxxxv, f. 1.
Fruitland: San Juan River near Fruitland, San Juan County, N. Mex.

Sabal sp. Newberry = ***Sabal imperialis.***

SABALITES Saporta, Ann. sci. nat., 5th ser., vol. 3 [Études, vol. 2], 1865, p. 77.

Sabalites californicus Lesquereux, Mus. Comp. Zool., Mem., vol. 6, No. 2, 1878, p. 1, pl. i, f. 1.
Miocene (auriferous gravels): Chalk Bluffs, Nevada County, Calif.
Ione: Kosk Creek, Shasta County, Calif.
?Eocene: Southwestern Oregon.

Sabalites campbelli (Newberry) Lesquereux, Rept. U. S. Geol. Surv. Terr., vol. 7 (Tert. Fl.), 1878, p. 113.—Penhallow, Rept. Tert. Pl. Brit. Col., 1908, p. 85.
Sabal campbelli Newberry [part], Boston Jour. Nat. Hist., vol. 7, 1863, p. 515; U. S. Geol. Surv., Mon. 35, 1898, p. 27, pl. xxi, f. 1, 2.
Puget: Bellingham Bay, and Glacier coal field, Whatcom County, Wash.
——: Coal Brook, British Columbia.

Sabalites carolinensis Berry, U. S. Geol. Surv., Prof. Paper 84, 1914, p. 29, pls. v, vi.
Black Creek (Middendorf): Langley, S. C.

Sabalites fructifer (Lesquereux) Lesquereux [part], Rept. U. S. Geol. Surv. Terr., vol. 7 (Tert. Fl.), 1878, p. 114, pl. xi, f. 3 [not f. 3a].
Flabellaria? fructifera Lesquereux, U. S. Geol. and Geog. Surv. Terr., Ann. Rept. 1873 [1874], p. 396.
Denver: *Golden, Colo.

Sabalites? fructifer Lesquereux, Rept. U. S. Geol. Surv. Terr., vol. 7 (Tert. Fl.), 1878, pl. xi, f. 3a = ***Nyssa? racemosa.***

Sabalites grayanus (Lesquereux) Lesquereux, Rept. U. S. Geol. Surv. Terr., vol. 7 (Tert. Fl.), 1878, p. 112, pl. xii, f. 2.—Berry, U. S. Geol. Surv., Prof. Paper 91, 1916, p. 177, pl. xii, f. 1–3; pl. xiv, f. 1.—Knowlton, U. S. Geol. Surv., Prof. Paper 101, 1918, p. 288.
[illegible] grayana Lesquereux, Am. Philos. Soc., Trans., vol. 13, 1869, p. 412, pl. xiv, f. 4–6.
Wilcox: *Mississippi, Arkansas, and Texas
Lagrange: Tennessee.
Laramie: Golden, Colo.
Raton: Near Blossburg and Raton, N. Mex.
Lance: Weston County, Wyo.

Sabalites grayanus (Lesquereux) Lesquereux, Tert. Fl., pl. xii, f. 1 [not f. 2] = ***Sabal? montana.***

Sabalites magothiensis (Berry) Berry, Bull. Torr. Bot. Club, vol. 38, 1911, p. 405; Md. Geol. Surv., Upper Cret., 1916, p. 811, pl. lvi, f. 4, 5.

Flabellaria magothiensis Berry, Torreya, vol. 5, 1905, p. 32, f. 1, 2; Geol. Surv. N. J., Ann. Rept. 1905 [1906], p. 139; Bull. Torr. Bot. Club, vol. 33, 1906, p. 170; idem, vol. 37, 1907, p. 21.

Magothy: Cliffwood, N. J.; Grove Point and Round Bay, Md.; Deep Cut, Del.

Sabalites sp. Berry, Bull. Torr. Bot. Club, vol. 43, 1916, p. 283; U. S. Geol. Surv., Prof. Paper 112, 1919, p. 72, pl. xi, f. 1.

Ripley: Camden, Benton County, and Selmer, McNairy County, Tenn.

SABINA (Hall) Spach, Ann. sci. nat., 2d ser., vol. 16, 1841, p. 291.

Sabina linguaefolia (Lesquereux) Cockerell, Univ. Colo. Studies, vol. 3, 1906, p. 175; Am. Mus. Nat. Hist., Bull., vol. 24, 1908, p. 79.—Knowlton, U. S. Nat. Mus., Proc., vol. 51, 1916, p. 249, pl. xiv.

Widdringtonia linguaefolia Lesquereux, Rept. U. S. Geol. Surv. Terr., vol. 8. (Cret. and Tert. Fl.), 1883, p. 139, pl. xxi, f. 14, 14a.

Miocene: *Florissant, Colo.

Sabina linguaefolia gracilis (Lesquereux) Knowlton, n. comb.

Widdringtonia linguaefolia gracilis Lesquereux, Rept. U. S. Geol. Surv. Terr., vol. 8 (Cret. and Tert. Fl.), 1883, p. 139.

Thuites callitrina Unger. Lesquereux, U. S. Geol. and Geog. Surv. Terr., Ann. Rept. 1872 [1873], p. 371.

Miocene: *Florissant, Colo.

SACCOLOMA Kaulfuss, Berlin Jahrb. Pharmacie, 1820, p. 51.

Saccoloma gardneri (Lesquereux) Knowlton, n. comb.

Pteris gardneri Lesquereux, U. S. Geol. and Geog. Surv. Terr., Ann. Rept. 1873 [1873], p. 393.

Gymnogramma gardneri (Lesquereux) Lesquereux, Rept. U. S. Geol. Surv. Terr., vol. 7 (Tert. Fl.), 1878, p. 58, pl. iv, f. 2.

Denver: *Sand Creek, Colo.

Dawson: Near Ramah, Jimmy Camp Creek, and Mosby, Colo.

SAGENOPTERIS Presl. Sternberg, Flora der Vorwelt, vol. 7, 1838, p. 164.

Sagenopteris alaskensis Fontaine, in Ward, U. S. Geol. Surv., Mon. 48, 1905 [1906], p. 152, pl. xxxviii, f. 21.

Jurassic: Nikolai Creek, Copper River region, Alaska.

Sagenopteris elliptica Fontaine, U. S. Geol. Surv., Mon. 15, 1889 [1890], p. 149, pl. xxviii, f. 9, 11–16a; in Ward, U. S. Geol. Surv., Mon. 48, 1905 [1906], p. 236, pl. lxv, f. 39, 40.—Berry, Md. Geol. Surv., Lower Cret., 1911, p. 287, text f. 4.—Penhallow, Roy. Soc. Canada, Trans., 11th ser., vol. 8, 1902, sec. 4, p. 41.

Sagenopteris sp. Fontaine, in Diller and Stanton, Geol. Soc. Am., Bull., vol. 5, 1894, p. 450; in Stanton, U. S. Geol. Surv., Bull. 133, 1896, p. 15.

Sagenopteris elliptica—Continued.

Sagenopteris sp. Fontaine, in Ward, U. S. Geol. Surv., Mon. 48, 1905 [1906], p. 238, pl. lxv, f. 46.

Cheiropteris spatulata Newberry, Am. Jour. Sci., 3d ser., vol. 41, 1891, p. 199, pl. xiv, f. 1, 2.—Knowlton, Smithsonian Misc. Coll., vol. 50, pt. 1, 1907, p. 114.

Patuxent: Potomac Run and Kankeys, Va.
Patapsco: Federal Hill, Baltimore, Md.
Horsetown: Near Horsetown, Shasta County, Calif.
Knoxville: Tehama County, Calif.
Kootenai: Great Falls, Mont.
Lower Cretaceous: Alliford Bay, Queen Charlotte Islands, British Columbia.

Sagenopteris emmonsi Fontaine, in Ward, U. S. Geol. Surv., Twentieth Ann. Rept., pt. 2, 1900, p. 286, pl. xxxix, f. 1–3. [See Ward, op. cit., p. 286, footnote, on reasons for renaming this plant.]

?*Cyclopteris* sp. Emmons, Geol. Rept. Midland Counties N. C., 1856, p. 329, pl. iv, f. 10.

?*Cyclopteris obscurus* Emmons, American geology, pt. 6, 1857, p. 104, pl. iv, f. 10.

?*Sagenopteris rhoifolia* Presl. Fontaine, U. S. Geol. Surv., Mon. 6, 1883, p. 104, pl. xlix, f. 5.

Triassic (Newark): North Carolina.

Sagenopteris goeppertiana Zigno. Fontaine, in Ward, U. S. Geol. Surv., Mon. 48, 1905 [1906], p. 83, pl. xiv, f. 5–11.—Knowlton, in Martin and Katz, U. S. Geol. Surv., Bull. 485, 1912, p. 63.

Jurassic: Douglas County, Oreg.
Jurassic (Tuxedni): Tuxedni Bay, Alaska.

Sagenopteris grandifolia Fontaine, in Ward, U. S. Geol. Surv., Mon. 48, 1905 [1906], p. 87, pl. xv, f. 4, 5.

Jurassic: Douglas County, Oreg.

Sagenopteris latifolia Fontaine, U. S. Geol. Surv., Mon. 15, 1889 [1890], p. 148, pl. xxvii, f. 10.—Berry, Md. Geol. Surv., Lower Cret., 1911, p. 286.

Patuxent: Telegraph station near Lorton, Va.

Sagenopteris latifolia Fontaine. Fontaine, in Diller and Stanton, Geol. Soc. Am., Bull., vol. 5, 1894, p. 450 = *Sagenopteris oregonensis*.

Sagenopteris? magnifolia Ward, in Fontaine, in Ward, U. S. Geol. Surv., Twentieth Ann. Rept., pt. 2, 1900, p. 334.

Jurassic (Foreman): Taylorsville, Calif.

Sagenopteris mantelli (Dunker) Schenk, Palaeontogr., vol. 19, 1871, p. 222.—Fontaine, in Diller and Stanton, Geol. Soc. Am., Bull., vol. 5, 1894, p. 450; in Stanton, U. S. Geol. Surv., Bull. 133, 1895 [1896], p. 15; in Ward, U. S. Geol. Surv., Mon. 48, 1905, p. 233, pl. lxv, f. 30–35.

Sagenopteris mantelli—Continued.

Cyclopteris mantelli Dunker, Mon. Nordd. Wealdenb., 1846, p. 10, pl. ix, f. 4, 5.

Glossozamites klipstinii (Dunker) Fontaine, in Diller and Stanton, Geol. Soc. Am., Bull., vol. 5, 1895, p. 450; in Stanton, U. S. Geol. Surv., Bull. 133, 1895 [1896], p. 15.

Knoxville: Wilcox's, 4 miles south of Lowry, Tehama County, Calif.

Horsetown: Tehama and Shasta counties, Calif.

Sagenopteris nervosa Fontaine, in Ward, U. S. Geol. Surv., Mon. 48, 1905 [1906], p. 237, pl. lxv, f. 41–45.

Horsetown: Riddles, Oreg.; North Fork Cottonwood Creek, Shasta County, Calif.

Knoxville: Cold Fork Cottonwood Creek, Tehama County, Calif.

Sagenopteris nilsoniana (Brongniart) Ward, U. S. Geol. Surv., Twentieth Ann. Rept., pt. 2, 1900, p. 352, pl. lvi, f. 1; pl. lxvii, f. 2.

Filicites nilsoniana Brongniart, Ann. sci. nat. de Paris, vol. 4, 1825; p. 218, pl. xii, f. 1.

Sagenopteris rhoifolia, Presl, in Sternberg, Fl. Vorwelt, vol. 2, 1838, pp. 165, 210, pl. xxxv, f. 1.

Jurassic: Oroville, Calif.

Sagenopteris nilsoniana (Brongniart) Ward. Penhallow, Roy. Soc. Canada, Trans., 2d ser., vol. 8, 1902, p. 39.

"Lower Cretaceous: Maud Island and Alliford Bay, Queen Charlotte Islands, British Columbia."

Sagenopteris oblongifolia Penhallow, Roy. Soc. Canada, Trans., 2d ser., vol. 8, 1902, sec. 4, p. 40, f. 2.

"Lower Cretaceous": Alliford Bay, Queen Charlotte Islands, British Columbia.

Sagenopteris oregonensis (Fontaine) Fontaine, in Ward, U. S. Geol. Surv., Mon. 48, 1905 [1906], p. 235, pl. lxv, f. 36–38.

Sagenopteris latifolia Fontaine. Fontaine, in Diller and Stanton, Geol. Soc. Am., Bull., vol. 5, 1894, p. 450.

Angiopteridium oregonense Fontaine, in Stanton, U. S. Geol. Surv., Bull. 133, 1895 [1896], p. 22 [name].

Horsetown: Riddles, Oreg.; Eagle Creek near Ono, Shasta County, Calif.

Sagenopteris paucifolia (Phillips) Ward, U. S. Geol. Surv., Mon. 48, 1905 [1906], p. 85, pl. xv, f. 1–3.

Pecopteris paucifolia Phillips, Geol. Yorkshire, 1829, p. 148, pl. viii, f. 8.

Jurassic: Douglas County, Oreg.

Sagenopteris rhoifolia Presl = *Sagenopteris nilsoniana*.

Sagenopteris rhoifolia Presl. Fontaine, U. S. Geol. Surv., Mon. 6, 1883, p. 104, pl. xlix, f. 5 = *Sagenopteris emmonsi*.

Sagenopteris variabilis (Velenovsky) Velenovsky? Abh. K. böhm. Gesell. Wiss., vol. 3, 1889, p. 40.—Hollick, U. S. Geol. Surv., Mon. 50, 1906, p. 34, pl. i, f. 22.

Thinnfeldia variabilis Velenovsky, Gymnosp. böhm. Kreidef., 1885, p. 6, pl. ii, f. 1-5.—Hollick, N. Y. Bot. Gard., Bull., vol. 2, 1902, p. 403, pl. 41, f. 12.

Magothy?: Chappaquiddick, Marthas Vineyard, Mass.

Sagenopteris virginiensis Fontaine, U. S. Geol. Surv., Mon. 15, 1889 [1890], p. 150, pl. cxxxviii, f. 13; pl. cxxxix, f. 1.—Berry, Md. Geol. Surv., Lower Cret., 1911, p. 289.

Patuxent: Fredericksburg, Va.

Sagenopteris? sp. Fontaine, in Diller and Stanton, Geol. Soc. Am., Bull., vol. 5, 1894, p. 450 [name]; in Stanton, U. S. Geol. Surv., Bull. 133, 1895 [1896], p. 15 [name]; in Ward, U. S. Geol. Surv., Mon. 48, 1905 [1906], p. 238, pl. xv, f. 46.

Knoxville: Elder Creek, 3 miles above Lowry, Tehama County, Calif.

Sagenopteris sp. Fontaine, in Ward, U. S. Geol. Surv., Twentieth Ann. Rept., pt. 2, 1900, p. 239, pl. xxv, f. 6.

Triassic (Newark): York County, Pa.

Sagenopteris? sp. Fontaine, in Ward, U. S. Geol. Surv., Twentieth Ann. Rept., pt. 2, 1900, p. 334.—Knowlton, in Diller, U. S. Geol. Surv., Bull. 353, 1908, p. 55.

Jurassic (Foreman): Taylorsville region, Calif.

Sagenopteris sp. Fontaine = *Sagenopteris elliptica*.

Sagenopteris sp. Knowlton, U. S. Nat. Mus., Proc., vol. 51, 1916, p. 455, pl. 81, f. 2.

Lower Jurassic: Upper Matanuska Valley, Alaska.

SAGITTARIA Linné, Sp. pl., 1753, p. 993.

Sagittaria pulchella Heer, Fl. foss. alask., 1869, p. 25, pl. i, f. 15.—Knowlton, U. S. Nat. Mus., Proc., vol. 17, 1894, p. 216; Geol. Soc. Am., Bull., vol. 5, 1893, p. 580.

Kenai: Alaska.

Sagittaria victor-masoni Ward = *Alismaphyllum victor-masoni*.

Sagittaria n. sp.? Lesquereux, U. S. Nat. Mus., Proc., vol. 10, 1887, p. 37.—Knowlton, idem, vol. 17, 1894, p. 216; Geol. Soc. Am., Bull., vol. 5, 1893, p. 590.

Kenai?: Sitka, Alaska.

SALICIPHYLLUM Fontaine, U. S. Geol. Surv., Mon. 15, 1889 [1890], p. 302. [Type *S. longifolium*.]

Saliciphyllum californicum Fontaine, in Ward, U. S. Geol. Surv., Mon. 48, 1905 [1906], p. 266, pl. lxix, f. 9.

Horsetown: Elder Creek, Tehama County, Calif.

Saliciphyllum ellipticum Fontaine [part] = *Celastrophyllum parvifolium.*

Saliciphyllum ellipticum Fontaine [part] = ***Ficophyllum crassinerve.***

Saliciphyllum longifolium Fontaine = ***Rogersia angustifolia.***

Saliciphyllum pachyphyllum Fontaine, in Ward, U. S. Geol. Surv., Mon. 48, 1905 [1906], p. 265, pl. lxix, f. 8.
  Horsetown: Elder Creek, Tehama County, Calif.

Saliciphyllum parvifolium Fontaine = *Celastrophyllum parvifolium.*

Salisburia Smith = *Ginkgo* Linné, Linn. Trans., vol. 3, 1797, p. 340.

Salisburia adiantoides Unger = *Ginkgo adiantoides.*

Salisburia baynesiana Dawson = *Ginkgo baynesiana.*

Salisburia binervata Lesquereux = *Lygodium binervatum.*

Salisburia (Ginkgo) lepida Heer. Dawson = *Ginkgo lepida.*

Salisburia nana Dawson = *Ginkgo nana.*

Salisburia polymorpha Lesquereux = *Protophyllocladus polymorphus.*

Salisburia pusilla Dawson = *Ginkgo dawsoni.*

Salisburia (Ginkgo) siberica Heer. Dawson = *Ginkgo siberica.*

Salisburia sp. Dawson = *Ginkgo* sp.

Salisburia, nuts of, Dawson = *Ginkgo* sp.

Salisburia, nutlets of, Dawson = *Ginkgo* sp.

SALIX (Tournefort) Linné, Sp. pl., 1753, p. 1015.

Salix amygdalaefolia Lesquereux = *Salix florissanti.*

Salix angusta Al. Braun, in Buckland, Geol., also in Stizenberger, Verst., 1851, p. 77.—Lesquereux, U. S. Geol. and Geog. Surv. Terr., Ann. Rept. 1871 [1872], Suppl., p. 6; Rept. U. S. Geol. Surv. Terr., vol. 7 (Tert. Fl.), 1878, p. 168, pl. xxii, f. 4, 5.—?Newberry, U. S. Geol. Surv., Mon. 35, 1898, p. 54, pl. lxv, f. 2.—?Knowlton, U. S. Geol. Surv., Bull. 163, 1900, p. 38, pl. vii, f. 6; U. S. Geol. Surv., Twenty-first Ann. Rept., pt. 2, 1901, p. 212, pl. xxx, f. 22; U. S. Geol. Surv., Bull. 204, 1902, p. 30; Washington Acad. Sci., Proc., vol. 11, 1909, p. 207; U. S. Geol. Surv., Mon. 32, pt. 2, 1899, p. 697.
  Green River: Green River station, Wyo.
  Esmeralda: Near Emigrant Peak, Nev.
  Mascall: Van Horn's ranch, John Day Basin, Oreg.
  Tertiary: Montana, Wyoming, California, Kentucky, etc.
  Mesaverde: Point of Rocks, Wyo.
  Montana: Coalville, Utah.
  Lance: Converse County, Wyo.; Sand Creek, Glendive, Mont.
  Miocene: Yellowstone National Park.

Salix angusta Al. Braun. Lesquereux, U. S. Nat. Mus., Proc., vol. 11, 1888, p. 13 = *Sapindus linearifolius.*

Salix baueri Knowlton, U. S. Geol. Surv., Prof. Paper 98, 1916, p. 337, pl. lxxxvi, f. 7, 8.
Fruitland: 10 miles south of San Juan River and 4 miles east of Chaco River, San Juan River, N. Mex.

Salix brittoniana Knowlton, U. S. Geol. Surv., Prof. Paper 130, 19—, p. —, pl. xxi, f. 8 [in preparation].
Laramie: Coal Creek, Boulder County, Colo.

Salix californica Lesquereux, Mus. Comp. Zool., Mem., vol. 6, No. 2, 1878, p. 10, pl. i, f. 18–20.—Knowlton, in Lindgren, Jour. Geol., vol. 4, 1896, p. 889.
Miocene (auriferous gravels): Table Mountain, Tuolumne County, and Independence Hill, Placer County, Calif.

Salix cumberlandensis Knowlton, U. S. Geol. Surv., Prof. Paper 108, 1917, p. 86, pl. xxxviii, f. 3.
Frontier: About 1½ miles south and also 1 mile east of Cumberland, Wyo.

Salix cuneata Newberry = *Salix siouxiana.*

Salix dayana Knowlton, U. S. Geol. Surv., Bull. 204, 1902, p. 31, pl. ii, f. 9, 10.
Mascall: Van Horn's ranch, John Day Basin, Oreg.

Salix deleta Lesquereux, U. S. Geol. Surv., Mon. 17, 1891 [1892], p. 49, pl. iii, f. 8.
Dakota: Pipe Creek, Cloud County, Kans.

Salix? densinervis Lesquereux = *Eugenia densinervia.*

Salix elliptica Lesquereux = *Salix merriami.*

Salix elongata O. Weber, Palaeontogr., vol. 2, 1853, p. 177, pl. xix, f. 10.—Lesquereux, U. S. Geol. and Geog. Surv. Terr., Ann. Rept. 1872 [1873], p. 372; Rept. U. S. Geol. Surv. Terr., vol. 7 (Tert. Fl.), 1878, p. 169, pl. xxii, f. 6, 7.—Knowlton, U. S. Geol. Surv., Mon. 32, pt. 2, 1899, p. 698.
Miocene: Elko, Nev.; Yellowstone National Park.

Salix elongata O. Weber. Knowlton = *Salix* sp. Knowlton.

Salix engelhardti Lesquereux, U. S. Nat. Mus., Proc., vol. 11, 1888, p. 17, pl. viii, f. 2.—Knowlton, U. S. Geol. Surv., Bull. 204, 1902, p. 29; idem, Folio 86, Ellensburg, Wash., 1903, p. 3.
Cassia phaseolites? Unger. Lesquereux, U. S. Nat. Mus., Proc., vol. 11, 1888, p. 16.
Mascall: *Van Horn's ranch, John Day Basin, Oreg.
Ellensburg: Kelly Hollow, Wenas Valley, Wash.

Salix eutawensis Berry, Bull. Torr. Bot. Club, vol. 37, 1910, p. 193, pl. xxii, f. 1–11; idem, p. 504; U. S. Geol. Surv., Prof. Paper 84, 1914, p. 109, pl. xix, f. 3; Bull. Torr. Bot. Club, vol. 43, 1916, p. 289; U. S. Geol. Surv., Prof. Paper 112, 1919, p. 76.
Black Creek: Below Dunbars Bridge, Tar River, N. C.
Eutaw: Chattahoochee River below Columbus, Ga.; near Parsons, Tenn.

Salix evanstoniana Lesquereux, U. S. Geol. and Geog. Surv. Terr., Ann. Rept. 1871 [1872], Suppl., p. 10. [Not afterward mentioned by Lesquereux.]

——: Evanston, Wyo.

Salix flexuosa Newberry, N. Y. Lyc. Nat. Hist., Ann., vol. 9, 1868, p. 21.—[Lesquereux], U. S. Geol. and Geog. Surv. Terr., Ill. Cret. and Tert. Pl., 1878, pl. 1, f. 4.—Lesquereux, U. S. Geol. Surv., Mon. 17, 1891 [1892], p. 50.—Newberry, U. S. Geol. Surv., Mon. 35, 1898, p. 56, pl. ii, f. 4; pl. xiii, f. 3, 4; pl. xiv, f. 1.—Berry, Bull. Torr. Bot. Club, vol. 33, 1906, p. 171; Geol. Surv. N. J., Ann. Rept. 1905 [1906], pp. 138, 139; Bull. Torr. Bot. Club, vol. 37, 1910, pp. 193, 504; idem, vol. 36, 1909, p. 252; U. S. Geol. Surv., Prof. Paper 84, 1914, pp. 32, 109, pl. vii, f. 14–16; pl. xi, f. 1; Md. Geol. Surv., Upper Cret., 1916, p. 813, pl. lvii, f. 4; U. S. Geol. Surv., Prof. Paper 112, 1919, p. 75, pl. xiii, f. 3.

Salix proteaefolia flexuosa (Newberry) Lesquereux, U. S. Geol. Surv., Mon. 17, 1891 [1892], p. 50, pl. xliv, f. 4, 5.—Hollick, idem, Mon. 50, 1906, p. 51, pl. viii, f. 5, 6a; pl. xxxvii, f. 8b.—Berry, N. Y. Bot. Gard., Bull., vol. 3, 1903, p. 67, pl. xlviii, f. 12; pl. lii, f. 2.

Salix proteaefolia linearifolia Lesquereux, U. S. Geol. Surv., Mon. 17, 1892, p. 49, pl. lxiv, f. 1–3.—Hollick, idem, Mon. 50, 1906, p. 52, pl. viii, f. 12.

Dakota: Big Sioux River, Blackbird Hill, Cedar Spring, Nebr.; Whetstone Creek, N. Mex.; Ellsworth County, Kans.

Raritan: South Amboy, N. J.; Kreischerville, Staten Island, N. Y.

Eutaw: Chattahoochee River below Columbus, Ga.

Black Creek (Middendorf): Middendorf, S. C.

Black Creek: Cape Fear River, Black River, N. C.; Langley and Columbia, S. C.

Magothy: Cliffwood and Morgan, N. J.; Deep Cut, Del.; Grove Point, Md.; Gay Head, Marthas Vineyard, Mass.; Sea Cliff, Long Island, N. Y.; Black Rock Point, Block Island, R. I.

Tuscaloosa: Shirleys Mill and Glen Allen, Fayette County; Snow place and Sanders Bluff, Tuscaloosa County; Whites Bluff, Greene County, Ala.

Bingen: Near Maxwell Spur; near Murfreesboro and Mine Creek; near Nashville, Ark.

Salix florissanti Knowlton and Cockerell, n. name.

Salix amygdalaefolia Lesquereux, Rept. U. S. Geol. Surv. Terr., vol. 8 (Cret. and Tert. Fl.), 1883, p. 156, pl. xxxi, f. 1, 2.—Knowlton, U. S. Geol. Surv., Bull. 204, 1902, p. 30.—Lesquereux, U. S. Nat. Mus., Proc., vol. 11, 1888, p. 17.—Cockerell, Am. Mus. Nat. Hist., Bull., vol. 24, 1908, p. 82. [Homonym, Gilib, 1792.]

Miocene: *Florissant, Colo.

Mascall: Van Horn's ranch, John Day Basin, Oreg.

Salix fluviatilis Nuttall? Hannibal, Bull. Torr. Bot. Club, vol. 38, 1911, p. 336.

Pliocene: Calabasas Canyon, Santa Cruz Mountains, Calif.

Salix foliosa Newberry, U. S. Geol. Surv., Mon. 35, 1898, p. 57, pl. xiii, f. 5, 6.
Dakota: Whetstone Creek, N. Mex.

Salix frontierensis Knowlton, U. S. Geol. Surv., Prof. Paper 108, 1917, p. 86, pl. xxxv, f. 4a.
Frontier: About 1½ miles south and 1 mile east of Cumberland, Wyo.

Salix gardneri Knowlton, U. S. Geol. Surv., Prof. Paper 101, 1918, p. 257, pl. xxxvii, f. 1.
Vermejo: Near Walsenburg, Colo.

Salix grönlandica Heer. Lesquereux, U. S. Geol. and Geog. Surv. Terr., Ann. Rept. 1871 [1872], p. 296. [Not afterward recognized by Lesquereux.]
——: 6 miles above Spring Canyon, Mont.

Salix hayei Lesquereux, U. S. Geol. Surv., Mon. 17, 1891 [1892], p. 48, pl. iii, f. 7.
Dakota: Ellsworth County, Kans.

Salix inaequalis Newberry, U. S. Geol. Surv., Mon. 26, 1895 [1896], p. 67, pl. xvi, f. 1, 4, 6; pl. xvii, f. 2-7.—Hollick, N. Y. Acad. Sci., Annals, vol. 11, 1898, p. 419, pl. xxxviii, f. 4a.—Berry, Geol. Surv. N. J., Bull. 3, 1911, p. 117, pl. xi, f. 3.
Magothy: ?Arrochar, Staten Island, N. Y.
Raritan: Woodbridge, N. J.

Salix integra Göppert = *Salix wyomingensis*.

Salix islandicus Lesquereux, Am. Jour. Sci., 2d ser., vol. 27, 1859, p. 360.
Puget: Bellingham Bay, Wash.

Salix kamloopsiana Dawson, Roy. Soc. Canada, Trans., vol. 8, 1890, sec. 4, p. 90, text f. 32.—Penhallow, Rept. Tert. Pl. Brit. Col., 1908, p. 85.
Eocene?: Kamloops, British Columbia.

Salix laevigata Bebb. Hannibal, Bull. Torr. Bot. Club, vol. 38, 1911, p. 336.
Pliocene: Near Portola, Calabasas Canyon, and Stevens Canyon, Santa Cruz Mountains; and near Hollister, Gavilan Range, Calif.

Salix laramina Dawson, Roy. Soc. Canada, Trans., vol. 4, 1886 [1887], sec. 4, p. 28, pl. i, f. 10.—Penhallow, Rept. Tert. Pl. Brit. Col., 1908, p. 85.
Paskapoo?: Porcupine Creek and Red Deer River, British Columbia.

Salix lavateri Heer, Fl. tert. Helv., vol. 2, 1856, p. 28, pl. lxvi, f. 1–12; Fl. foss. arct., vol. 2, Abt. 2, 1869, p. 27, pl. ii, f. 10.—Knowlton, U. S. Nat. Mus., Proc., vol. 17, 1894, p. 218; Geol. Soc. Am., Bull., vol. 5, 1893, p. 581; U. S. Geol. Surv., Mon. 32, pt. 2, 1899, p. 697.
Kenai: Port Graham, Alaska.
Fort Union: Yellowstone National Park.

Salix lesquereuxii Berry, Bull. Torr. Bot. Club, vol. 36, 1909, p. 252; idem, vol. 37, 1910, pp. 21, 194, 505; U. S. Geol. Surv., Prof. Paper 84, 1914, pp. 33, 109, pl. vii, f. 11–13; Md. Geol. Surv., Upper Cret., 1916, p. 814, pl. lviii, f. 5–8: Bull. Torr. Bot. Club, vol. 44, 1917, p. 176; U. S. Geol. Surv., Prof. Paper 112, 1919, p. 76.
Salix proteaefolia Lesquereux, Am. Jour. Sci., 2d ser., vol. 46, 1868, p. 94; Rept. on Clay in N. J., 1878, p. 29; Rept. U. S. Geol. Surv. Terr., vol. 6 (Cret. Fl.), 1874, p. 60, pl. v, f. 1–4; idem, vol. 8 (Cret. and Tert. Fl.), 1883, p. 42, pl. i, f. 14–16; U. S. Geol. Surv., Mon. 17, 1892, p. 49. [Homonym, Forbes, 1829.]
Salix proteaefolia longifolia Lesquereux, U. S. Geol. Surv., Mon. 17, 1891 [1892], p. 50, pl. lxiv, f. 9.
Proteoides daphnogenoides Heer. Newberry [part], U. S. Geol. Surv., Mon. 26, 1895, p. 72, pl. xxxii, f. 11.
Dewalquea grönlandica Heer. Newberry [part], U. S. Geol. Surv., Mon. 26, 1895, p. 129, pl. xli, f. 12 [not f. 2, 3].
Raritan: Sayreville, Woodbridge, South Amboy, N. J.
Magothy: Pennsylvania Avenue extended, D. C.; Deep Cut, Del.
Black Creek (Middendorf): Middendorf, Rocky Point, Langley, S. C.
Black Creek: Big Bend of Black River, N. C.
Eutaw: Chimney Bluff below Columbus, Ga.
Tuscaloosa: Shirleys Mill, Glen Allen, Cottondale, and Sanders Ferry Bluff, Ala.

Salix libbeyi Lesquereux, Rept. U. S. Geol. Surv. Terr., vol. 8 (Cret. and Tert. Fl.), 1883, p. 156, pl. xxxi, f. 3.
Miocene: *Florissant, Colo.

Salix macrophylla Heer, Fl. tert. Helv., vol. 2, 1856, p. 29, pl. lxvii; Fl. foss. arct., vol. 2, Abt. 2, 1869, p. 27, pl. ii, f. 9.—Eichwald, Geognost.-Palaeontolog. Bemerk. ü. Halbinsel Mangischlak u. Aleutischen Inseln, 1871, p. 113, pl. iv, f. 5.—Knowlton, U. S. Nat. Mus., Proc., vol. 17, 1894, p. 217; Geol. Soc. Am., Bull., vol. 5, 1893, p. 581.
Kenai: *Port Graham, Alaska.

Salix mattewanensis Berry, N. Y. Bot. Gard., Bull., vol. 3, 1903, p. 68, pl. 51, f. 5; Geol. Surv. N. J., Ann. Rept. 1905 [1906], p. 139.
Magothy: Cliffwood Bluff, N. J.

Salix media Heer, Fl. tert. Helv., vol. 2, 1856, p. 32, pl. i, f. 9; pl. lxviii, f. 14–19.—Lesquereux, U. S. Geol. and Geog. Surv. Terr., Ann. Rept. 1871 [1872], Suppl., p. 6; Rept. U. S. Geol. Surv. Terr., vol. 7 (Tert. Fl.), 1878, p. 168, pl. xxii, f. 3.

Green River: Green River, Wyo.

Miocene: Elko station, Nev.

Lagrange: Wickliffe, Ky.

Salix meekii Newberry, N. Y. Lyc. Nat. Hist., Ann., vol. 9, 1868, p. 19; U. S. Geol. Surv., Mon. 35, 1898, p. 58, pl. ii, f. 3.—Hollick, N. Y. Bot. Gard., Bull., vol. 2, 1902, p. 404, pl. xli, f. 1.—Berry, op. cit., p. 68.—Berry, Geol. Surv. N. J., Ann. Rept. 1905 [1906], p. 139.—Hollick, U. S. Geol. Surv., Mon. 50, 1906, p. 51, pl. viii, f. 1c, 8, 9.—Berry, U. S. Geol. Surv., Prof. Paper 112, 1919, p. 77.

Salix cuneata Newberry. [Lesquereux], U. S. Geol. and Geog. Surv. Terr., Ill. Cret. and Tert. Pl., 1878, pl. i, f. 3.

Salix proteaefolia lanceolata Lesquereux, U. S. Geol. Surv., Mon. 17, 1891 [1892], p. 50, pl. lxiv, f. 6–8.

Dakota: Blackbird Hill, Nebr.; Ellsworth County, Kans.

Magothy: Cliffwood, N. J.; Chappaquiddick Island, Marthas Vineyard, Mass.

Raritan: Arrochar, Staten Island, N. Y.

Tuscaloosa: Shirleys Mill, Ala.

Salix meekii Newberry. [Lesquereux], U. S. Geol. and Geog. Surv. Terr., Ill. Cret. and Tert. Pl., 1878, pl. i, f. 1 = *Salix siouxiana*.

Salix membranacea Newberry = *Salix raritanensis*.

Salix merriami Cockerell, Am. Jour. Sci., 4th ser., vol. 26, 1908, p. 543.

Salix elliptica Lesquereux, Mus. Comp. Zool., Mem., vol. 6, No. 2, 1878, p. 10, pl. i, f. 22. [Homonym, Sleich, 1815.]

Miocene (auriferous gravels): Chalk Bluffs, Nevada County, Calif.

Salix minuta Knowlton, U. S. Nat. Mus., Proc., vol. 17, 1894, p. 218, pl. ix, f. 1; Geol. Soc. Am., Bull., vol. 5, 1893, p. 581.

Kenai: Herendeen Bay, Alaska.

Salix mixta Knowlton, U. S. Geol. Surv., Bull. 204, 1902, p. 32, pl. ii, f. 11, 12.

Mascall: Van Horn's ranch, John Day Basin, Oreg.

Salix myricoides Knowlton, U. S. Geol. Surv., Prof. Paper 130, 19—, p. —, pl. iv, f. 7 [in preparation].

Laramie: Moffat Railroad, 6 miles north of Golden, Colo.

Salix nervillosa Heer, Phyll. crét. Nebr., 1866, p. 15, pl. i, f. 3.—Lesquereux, U. S. Geol. Surv., Mon. 17, 1891 [1892], p. 49.

Dakota: Tekamah, Nebr.

Salix newberryana Hollick, in Newberry, U. S. Geol. Surv., Mon. 26, 1895 [1896], p. 68, pl. xiv, f. 2–7.—Berry, Bull. Torr. Bot. Club, vol. 37, 1910, p. 193; Geol. Surv. N. J., Bull. 3, 1911, p. 113, pl. xi, f. 2.
Raritan: Sayreville, Woodbridge, South Amboy, and Milltown, N. J.
Black River: Court House Bluff, Cape Fear River, N. C.

Salix orbicularis Penhallow, Rept. Tert. Pl. Brit. Col., 1908, p. 85, text f. 30.
Oligocene: Quilchena, British Columbia.

Salix pacifica Dawson, Roy. Soc. Canada, Trans., vol. 1, 1882 [1883], p. 26, pl. vii, f. 24.
Upper Cretaceous: Baynes Sound, Vancouver Island, British Columbia.

Salix perplexa Knowlton, U. S. Geol. Surv., Bull. 204, 1902, p. 31, pl. ii, f. 5–8.—Penhallow, Rept. Tert. Pl. Brit. Col., 1908, p. 86.
Mascall: Van Horn's ranch, John Day Basin, Oreg.
Oligocene: Quilchena, British Columbia.

Salix pilosula Göppert, Acad. imp. sci. St.-Pétersbourg, Bull., vol. 3, 1861, p. 456.
Kenai?: Alaska.

Salix plicata Knowlton, U. S. Geol. Surv., Prof. Paper 101, 1918, p. 257, pl. xxxvi, f. 6–8.
Vermejo: Rockvale and Walsenburg, Colo.

Salix proteaefolia Lesquereux = *Salix lesquereuxii.*

Salix proteaefolia flexuosa (Newberry) Lesquereux = *Salix flexuosa.*

Salix proteaefolia lanceolata Lesquereux = *Salix meekii.*

Salix proteaefolia linearifolia Lesquereux = *Salix flexuosa.*

Salix proteaefolia longifolia Lesquereux = *Salix lesquereuxii.*

Salix pseudo-argentea Knowlton, U. S. Geol. Surv., Bull. 204, 1902, p. 31, pl. ii, f. 2–4; in Smith, U. S. Geol. Surv., Folio 86 (Ellensburg, Wash.), 1902, p. 3.
Mascall: Van Horn's ranch, John Day Basin, Oreg.
Ellensburg: Ellensburg and Kelly Hollow, Wenas Valley, Wash.

Salix pseudo-hayei Berry, Bull. Torr. Bot. Club, vol. 36, 1909, p. 251; U. S. Geol. Surv., Prof. Paper 84, 1914, p. 34, pl. x, f. 8; Geol. Surv. N. J., Bull. 3, 1911, p. 118, pl. xi, f. 1.
Salix sp. Newberry, U. S. Geol. Surv., Mon. 26, 1896, p. 68, pl. xlii, f. 6–8.
Raritan: Milltown, N. J.
Black Creek (Middendorf): Middendorf, S. C.

Salix purpuroides Hollick, Bull. Torr. Bot. Club, vol. 21, 1894, p. 50, pl. clxxiv, f. 9; U. S. Geol. Surv., Mon. 50, 1906, p. 50, pl. viii, f. 11.
Magothy: Sea Cliff, Long Island, N. Y.

Salix raeana Heer, Fl. foss. arct., vol. 1, 1868, pp. 102, 139, pl. iv, f. 11–13; pl. xxi, f. 13; pl. xlvii, f. 11.—Lesquereux, U. S. Nat. Mus., Proc., vol. 5, 1882 [1883], p. 447, pl. viii, f. 6.—Knowlton, idem, vol. 17, 1894, p. 218; Geol. Soc. Am., Bull., vol. 5, 1893, p. 581; ?U. S. Geol. Surv., Bull. 204, 1902, p. 30.—Penhallow, Rept. Tert. Pl. Brit. Col., 1908, p. 86.—Berry, Jour. Geol., vol. 17, 1909, p. 23, text f. 2.

Kenai: Cook Inlet, Alaska.

Eocene: Porcupine Creek and Great Valley, Saskatchewan; Mackenzie River, Northwest Territory.

Calvert: Richmond, Va.

Mascall: Van Horn's ranch, John Day Basin, Oreg.

Salix ramaleyi Cockerell, Bull. Torr. Bot. Club, vol. 33, 1906, p. 307; Am. Nat., vol. 44, 1910, p. 44, f. 9.

Miocene: *Florissant, Colo.

Salix ramaleyi rohweri Cockerell, Am. Mus. Nat. Hist., Bull., vol. 24, 1908, p. 82, pl. ix, f. 34.

Miocene: *Florissant, Colo.

Salix raritanensis Berry, Bull. Torr. Bot. Club, vol. 36, 1909, p. 250; Geol. Surv. N. J., Bull. 3, 1911, p. 116.

Salix membranacea Newberry, N. Y. Lyc. Nat. Hist., Ann., vol. 9, 1868, p. 19.—[Lesquereux], U. S. Geol. and Geog. Surv. Terr., Ill. Cret. and Tert. Pl., 1878, pl. i, f. 5–8a.—Newberry, U. S. Geol. Surv., Mon. 26, 1895 [1896], p. 66, pl. xxix, f. 12; idem, Mon. 35, 1898, p. 59, pl. ii, f. 5–8a.—Hollick, U. S. Geol. Surv., Mon. 50, 1906, p. 50, pl. viii, f 10, 23. [Homonym, Thuill, 1799.]

Raritan: Sayreville, Woodbridge, Milltown, N. J.; Kreischerville, Staten Island, N. Y.

Magothy: Gay Head, Marthas Vineyard, Mass.

Salix schimperi Lesquereux, U. S. Nat. Mus., Proc., vol. 11, 1888, p. 21, pl. xiii, f. 5.—Knowlton, U. S. Geol. Surv., Bull. 204, 1902, p. 29.

Clarno (lower part): *Cherry Creek, Oreg.

Salix siouxiana Knowlton and Cockerell, n. name.

Salix cuneata Newberry, N. Y. Lyc. Nat. Hist., Ann., vol. 9, 1868, p. 21; U. S. Geol. Surv., Mon. 35, 1898, p. 55, pl. ii, f. 1, 2.—Hollick, idem, Mon. 50, 1906, p. 50, pl. vii, f. 26, 27; pl. viii, f. 7. [Homonym, Nuttall, 1865.]

Salix meekii Newberry [Lesquereux], U. S. Geol. and Geog. Surv. Terr., Ill. Cret. and Tert. Pl., 1878, pl. i, f. 1 [f. 2=*Populus cuneata*.]

Magothy: Glen Cove, Long Island, N. Y.

Dakota: *Mouth of Big Sioux River, Nebr.

Raritan: Arrochar and Kreischerville, Staten Island, N. Y.

Salix sloani Berry, U. S. Geol. Surv., Prof. Paper 84, 1914, p. 34, pl. viii, f. 10–12.

Black Creek (Middendorf): Near Langley and Miles Hall, S. C.

Salix stantoni Knowlton, U. S. Geol. Surv., Bull. 163, 1900, p. 38, pl. vi, f. 6.
 Montana: *Coalville, Utah.

Salix tabellaris Lesquereux = *Apocynophyllum tabellarum.*

Salix tulameenensis Penhallow, Rept. Tert. Pl. Brit. Col., 1908, p. 86, text f. 31.
 Oligocene: Tulameen River, British Columbia.

Salix vaccinifolia Knowlton, U. S. Geol. Surv., Twenty-first Ann. Rept., pt. 2, 1901, p. 212, pl. xxx, f. 8, 20.
 Esmeralda: 3.8 and 4.5 kilometers northeast of Emigrant Peak, Esmeralda County, Nev.

Salix varians Göppert, Zeitschr. Deutsch. geol. Gesell., vol. 4, 1852, p. 493.—Heer, Fl. foss. arct., vol. 2, Abt. 2, 1869, p. 27, pl. ii, f. 8; pl. iii, f. 1–3.—Knowlton, U. S. Nat. Mus., Proc., vol. 17, 1894, p. 217; Geol. Soc. Am., Bull., vol. 5, 1893, p. 581; U. S. Geol. Surv., Bull. 204, 1902, p. 30; U. S. Geol. Surv., Mon. 32, pt. 2, 1899, p. 697, pl. lxxxv, f. 3; in Smith, U. S. Geol. Surv., Folio 86 (Ellensburg, Wash.), 1903, p. 3.—Penhallow, Rept. Tert. Pl. Brit. Col., 1908, p. 86.
 Kenai: English Bay and Ninilchik, Alaska.
 Mascall: John Day Basin, Oreg.
 Miocene: Yellowstone National Park.
 Ellensburg: Ellensburg, Wash.
 Miocene?: Tranquille River; Horsefly River, British Columbia.

Salix viminalifolia Berry, U. S. Nat. Mus., Proc., vol. 48, 1915, p. 297.
 Pleistocene: Hickman, Ky.

?Salix wimmeriana Göppert. Göppert, Abhandl. Schles. Gesell. vaterländ. Kultur, 1861, p. 201.
 Kenai: Ninilchik, Alaska.

Salix worthenii Lesquereux = *Myrica worthenii.*

Salix wyomingensis Knowlton and Cockerell, n. name.
 Salix integra Goppert, Zeitschr. Deutsch. geol. Gesell., vol. 4, 1852, p. 493; Abhandl. Schles. Gesell., 1861, p. 202; idem, 1867, p. 50.—Lesquereux, U. S. Geol. and Geog. Surv. Terr., Ann. Rept. 1873 [1874], p. 397; Rept. U. S. Geol. Surv. Terr., vol. 7 (Tert. Fl.), 1878, p. 167, pl. xxii, f. 1, 2.—Penhallow, Rept. Tert. Pl. Brit. Col., 1908, p. 85. [Homonym, Thunberg, 1784.]
 Kenai?: Ninilchik, Alaska.
 Eocene?: Burrard Inlet, British Columbia.
 Post-Laramie: Black Buttes, Wyo.
 Laramie: *Golden, Coal Creek, and Marshall, Colo.

Salix sp. Berry, Jour. Geol., vol. 15, 1907, p. 340 = *Hicoria aquatica.*

Salix sp. Berry, U. S. Nat. Mus., Proc., vol. 54, 1919, p. 629.
 Miocene: Beaver Creek, 10 miles east of Beaver City, Okla.

Salix sp. Dawson, Roy. Soc. Canada, Trans., vol. 11, 1893 [1894], sec. 4, p. 58, pl. vii, f. 22.
Upper Cretaceous: Port McNeill, Vancouver Island, British Columbia.

Salix sp. Hollick, N. Y. Acad. Sci., Trans., vol. 12, 1892, p. 32, pl. ii, f. 15; U. S. Geol. Surv., Mon. 50, 1906, p. 53, pl. viii, f. 13.
Magothy: Kreischerville, Staten Island, N. Y.

Salix sp. Knowlton, Geol. Soc. Am., Bull., vol. 8, 1897, p. 133.
Lance: Lightning Creek, Converse County, Wyo.

Salix sp. Knowlton, Geol. Soc. Am., Bull., vol. 8, 1897, p. 148.
Mesaverde: Rock Springs, Wyo.

Salix sp. Knowlton, U. S. Geol. Surv., Bull. 163, 1900, p. 39, pl. vii, f. 2.
Salix elongata O. Weber. Knowlton, U. S. Geol. Surv., Bull. 106, 1893, p. 42; Geol. Soc. Am., Bull., vol. 8, 1897, p. 150.
Montana: Coalville, Utah.

Salix sp. Knowlton, U. S. Geol. Surv., Twenty-first Ann. Rept., pt. 2, 1901, p. 213, pl. xxx, f. 13.
Esmeralda: 4.5 kilometers northeast of Emigrant Peak, Esmeralda County, Nev.

Salix? sp. Knowlton, U. S. Geol. Surv., Twenty-first Ann. Rept., pt. 2, 1901, p. 213, pl. xxx, f. 14.
Esmeralda: 3.8 kilometers northeast of Emigrant Peak, Esmeralda County, Nev.

Salix sp. Knowlton, U. S. Geol. Surv., Bull. 204, 1902, p. 32, pl. iii, f. 1.
Clarno (upper part): Near Lone Rock, John Day Basin, Oreg.

Salix sp. *a* Knowlton, U. S. Geol. Surv., Prof. Paper 101, 1918, p. 258, pl. xxxviii, f. 1; idem, Prof. Paper 98, 1916, p. 337, pl. lxxxvi, f. 9.
Vermejo: Vermejo Park, N. Mex.
Fruitland: San Juan County, N. Mex.

Salix sp. *b* Knowlton, U. S. Geol. Surv., Prof. Paper 101, 1918, p. 258, pl. xxxviii, f. 2.
Vermejo: Ponil Canyon, N. Mex.

Salix sp. *c* Knowlton, U. S. Geol. Surv., Prof. Paper 101, 1918, p. 258.
Vermejo: Rockvale, Colo.

Salix sp., Knowlton, U. S. Nat. Mus., Proc., vol. 51, 1916, p. 260, pl. xiii, f. 4, 5.
Miocene: *Florissant, Colo.

Salix, fruiting catkin of, Lesquereux, U. S. Geol. Surv., Mon. 17, 1891 [1892], p. 51, pl. viii, f. 6.
Dakota: Near Delphos, Kans.

Salix sp. Newberry, U. S. Geol. Surv., Mon. 26, 1896, p. 68, pl. xlii, f. 6–8 = *Salix pseudo-hayei.*

Salix sp. Penhallow, Jour. Geol., vol. 3, 1895, pp. 626, 635; Roy. Soc. Canada, Trans., 2d ser., vol. 2, 1896, sec. 4, p. 73; British Assoc. Adv. Sci., Bradford meeting, 1900, p. 336.
Pleistocene: Don Valley, Toronto, Canada.

SALPICHLAENA Smith, in Hooker, Gen. fil., 1842, t. 93.

Salpichlaena anceps (Lesquereux) Knowlton, n. comb.
Pteris anceps Lesquereux, U. S. Geol. and Geog. Surv. Terr., Ann. Rept. 1872 [1873], p. 376.
Pteris affinis Lesquereux, idem, Ann. Rept. 1873 [1874], p. 392.
Osmunda affinis (Lesquereux) Lesquereux, Rept. U. S. Geol. Surv. Terr., vol. 7 (Tert. Fl.), 1878, p. 60, pl. iv, f. 1.—Knowlton, U. S. Geol. Surv., Mon. 32, pt. 2, 1898, p. 673, pl. lxxx, f. 4, 5.
Denver: Golden, Colo.
Fort Union: Yellowstone National Park.

SALVINIA Adanson, Fam. Pl., vol. 2, 1763, p. 15.

Salvinia alleni Lesquereux = *Tmesipteris alleni.*

Salvinia attenuata Lesquereux = *Marsilea attenuata.*

Salvinia cyclophylla Lesquereux = *Phyllites cyclophylla.*

Salvinia elliptica Newberry, MS. Hollick, Bull. Torr. Bot. Club, vol. 21, 1894, p. 255, pl. ccv, f. 14, 14a, 15.
Puget: Carbonado, Wash.

Salvinia formosa Heer ?, Fl. tert. Helv., vol. 3, 1859, p. 156, pl. cxlv, f. 13–15.—Berry, Jour. Geol., vol. 17, 1909, p. 21.
Calvert: Richmond, Va.

Salvinia sp. Knowlton, Geol. Soc. Am., Bull., vol. 8, 1897, p. 133; Washington Acad. Sci., Proc., vol. 11, 1909, p. 207.
Lance: Lance Creek, Converse County, Wyo.

SAMAROPSIS Göppert, Palaeontographica, vol. 12 (Perm. Fl.), 1864, p. 177.

Samaropsis? oregonensis Fontaine, in Ward, U. S. Geol. Surv., Mon. 48, 1905 [1906], p. 134, pl. xxxvi, f. 9–12.
Jurassic: *Douglas County, Oreg.

SAMBUCUS (Tournefort) Linné, Sp. pl., 1753, p. 269.

Sambucus amabilis Cockerell, Torreya, vol. 10, 1910, p. 126, f. 1 (on p. 127).
Miocene: *Florissant, Colo.

Sambucus ellisiae Cockerell, Torreya, vol. 13, 1913, p. 75, f. 1.
Miocene: *Florissant, Colo.

Sambucus newtoni Cockerell, Am. Jour. Sci., 4th ser., vol. 26, 1908, p. 541, f. 6.
Miocene: *Florissant, Colo.

SANTALUM Linné, Sp. pl., 1753, p. 349.

Santalum americanum Lesquereux, Rept. U. S. Geol. Surv. Terr., vol. 8 (Cret. and Tert. Fl.), 1883, p. 164, pl. xxxii, f. 7.
Miocene: *Florissant, Colo.

Santalum novae caesareae Berry, Bull. Torr. Bot. Club, vol. 33, 1906, p. 182; Geol. Surv. N. J., Ann. Rept. 1905 [1906], p. 153, pl. xx, f. 7; pl. xxii, f. 3.
Magothy: Morgan, and pits of Cliffwood Brick Co. on Whale River, N. J.

SAPINDOIDES Perkins, Rept. Vt. State Geol. 1903–4 [1904], p. 206. [Type, *S. varius*.]

Sapindoides americanus (Lesquereux) Perkins, Rept. Vt. State Geol. 1903–4 [1904], p. 207, pl. lxxx, f. 119, 140; Geol. Soc. Am., Bull., vol. 16, 1905, p. 508, pl. 86, f. 3; Rept. Vt. State Geol. 1905–6 [1906], p. 221, pl. lii, f. 3; pl. lvii, f. 14, 15.
Sapindus americanus Lesquereux, Am. Jour. Sci , 2d ser., vol. 32, 1861, p. 359; Geol. Vt., vol. 2, 1861, p. 715, f. 142–145.
Miocene: Brandon, Vt.

Sapindoides cylindricus Perkins, Rept. Vt. State Geol. 1903–4 [1904], p. 208, pl. lxxx, f. 131, 139.
Miocene: Brandon, Vt.

Sapindoides medius Perkins, Rept. Vt. State Geol. 1903–4 [1904], p. 207, pl. lxxx, f. 130; Geol. Soc. Am., Bull., vol. 16, 1905, p. 507, pl. lxxxvi, f. 1; Rept. Vt. State Geol. 1905–6 [1906], pl. lii, f. 1.
Miocene: Brandon, Vt.

Sapindoides minimus Perkins, Rept. Vt. State Geol. 1903–4 [1904], p. 208, pl. lxxx, f. 127.
Miocene: Brandon, Vt.

Sapindoides parva Perkins, Rept. Vt. State Geol. 1903–4 [1904], p. 207, pl. lxxx, f. 126.
Miocene: Brandon, Vt.

Sapindoides urceolatus Perkins, Rept. Vt. State Geol. 1905–6 [1906], p. 220, pl. lvii, f. 11, 12.
Miocene: Brandon, Vt.

Sapindoides varius Perkins, Rept. Vt. State Geol. 1903–4 [1904], p. 206, pl. lxxx, f. 116, 117, 122; Geol. Soc. Am., Bull., vol. 16, 1905, p. 508, pl. lxxxvi, f. 2; Rept. Vt. State Geol. 1905–6 [1906], pl. lii, f. 2.
Miocene: Brandon, Vt.

Sapindoides vermontanus Perkins, Rept. Vt. State Geol. 1903–4 [1904], p. 207, pl. lxxx, f. 132, 135.
Miocene: Brandon, Vt.

SAPINDOPSIS Fontaine, U. S. Geol. Surv., Mon. 15, 1889 [1890], p. 296.

Sapindopsis brevifolia Fontaine, U. S. Geol. Surv., Mon. 15, 1889 [1890], p. 300, pl. cliii, f. 4; pl. cliv, f. 1, 7; pl. clxiii, f. 3; in Ward, U. S. Geol. Surv., Mon. 48, 1905 [1906], pp. 481, 482, 528.—Berry, U. S. Nat. Mus., Proc., vol. 38, 1910, p. 644; Md. Geol. Surv., Lower Cret., 1912, p. 473, pl. lxxxvii, f. 2–5.

Patapsco: Near Brooke, 72d milepost, Aquia Creek, Va.; Fort Foote, Md.

Sapindopsis cordata Fontaine = *Ficophyllum crassinerve.*

Sapindopsis elliptica Fontaine = *Rogersia longifolia.*

Sapindopsis magnifolia Fontaine emend., Berry, U. S. Nat. Mus., Proc., vol. 38, 1910, p. 642; Md. Geol. Surv., Lower Cret., 1912, p. 471, pls. lxxxvi, lxxxvii, lxxxviii.

Sapindopsis magnifolia Fontaine, U. S. Geol. Surv., Mon. 15, 1889 [1890], p. 297, pl. cli, f. 2, 3; pl. clii, f. 2, 3; pl. cliii, f. 2; pl. cliv, f. 1, 5; pl. clv, f. 6; in Ward, U. S. Geol. Surv., Mon. 48, 1905 [1906], pp. 481, 482, 548.

?*Aralia dubia* Fontaine, op. cit. (Mon. 15), p. 314, pl. clvii, f. 1, 7 [not Schimper, 1874].

Ficophyllum eucalyptoides Fontaine, op. cit. (Mon. 15), p. 294, pl. clxiv, f. 1, 2; in Ward, op. cit. (Mon. 48), p. 489.

Sapindopsis tenuinervis Fontaine, op. cit. (Mon. 15), p. 301, pl. cliii, f. 1; in Ward, op. cit. (Mon. 48), pp. 489, 528.

?*Aralia fontainei* Knowlton, U. S. Geol. Surv., Bull. 152, 1898, p. 37.

Sapindopsis obtusifolia Fontaine, op. cit. (Mon. 15), p. 301, pl. clvi, f. 13; pl. clix, f. 3–6.

Patapsco: Near Brooke, 72d milepost, Deep Bottom?, near Widewater and Aquia Creek, White House Bluff, Va.; Stump Neck and Fort Foote, Md.

Fuson: Oak Creek and Robbin's ranch, Crook County, Wyo.

Sapindopsis obtusifolia Fontaine = *Sapindopsis magnifolia.*

Sapindopsis oregonensis Fontaine = *Nilsonia oregonensis.*

Sapindopsis parvifolia Fontaine = *Sapindopsis variabilis.*

Sapindopsis tenuinervis Fontaine = *Sapindopsis magnifolia.*

Sapindopsis variabilis Fontaine emend., Berry, U. S. Nat. Mus., Proc., vol. 38, 1910, p. 641; Md. Geol. Surv., Lower Cret., 1912, p. 469, pls. lxxxiii, lxxxiv, lxxxv.

Sapindopsis variabilis Fontaine, U. S. Geol. Surv., Mon. 15, 1889 [1890], p. 298, pl. cli, f. 1; pl. clii, f. 1, 4; pl. cliii, f. 3; pl. cliv, f. 2–4; pl. clv, f. 2–5; in Ward, U. S. Geol. Surv., Mon. 48, 1905 [1906], pp. 481, 482, 489, 532, pl. cxiv, f. 2; U. S. Geol. Surv., Nineteenth Ann. Rept., pt. 2, 1899, p. 690, pl. clxix, f. 9.—?Cockerell, Washington Acad. Sci., Jour., vol. 6, 1916, p. 110.

Sapindopsis parvifolia Fontaine, op. cit. (Mon. 15), p. 300, pl. cliv, f. 6.

Eucalyptus rosieriana Ward, U. S. Geol. Surv., Mon. 48, 1905 [1906], p. 530, pl. cxiii, f. 9, 10.

Ficus myricoides Hollick. Ward, U. S. Geol. Surv., Mon. 48, 1905 [1906], p. 531, pl. cxii, f. 12.

Rogersia angustifolia Fontaine. Fontaine, in Ward, U. S. Geol. Surv., Mon. 48, 1905 [1906], pp. 491, 510 [not p. 521].

Sapindopsis variabilis—Continued.

Patapsco: Fort Foote, Md.; near Bróoke, White House Bluff, Mount Vernon, 72d milepost, Aquia Creek, and near Widewater, Va.

Tuson: Oak Creek, Crook County, Wyo.

Lower Cretaceous: Hovenweep Canyon, latitude 37° 53′ N., longitude 108° 57′ W., Colorado.

SAPINDUS (Tournefort) Linné, Sp. pl., 1753, p. 367.

Sapindus affinis Newberry, N. Y. Lyc. Nat. Hist., Ann., vol. 9, 1868, p. 51.—[Lesquereux] U. S. Geol. and Geog. Surv. Terr., Ill. Cret. and Tert. Pl., 1878, pl. xxiv, f. 1; pl. xxv, f. 2.—Ward, U. S. Geol. Surv., Sixth Ann. Rept., 1884–85 [1886], p. 554, pl. l, f. 2, 3; idem, Bull. 37, 1887, p. 67, pl. xxxi, f. 1, 2.—Knowlton, idem, Mon. 32, pt. 2, 1899, p. 736, pl. cii, f. 1–3.—Newberry, idem, Mon. 35, 1898, p. 116, pl. xxx, f. 1; pl. xl, f. 2.—Knowlton, Washington Acad. Sci., Proc., vol. 11, 1909, pp. 185, 189, 191, 201, 202, 211, 214; in Calvert, U. S. Geol. Surv., Bull. 471, 1912, p. 16.—Knowlton, U. S. Geol. Surv., Prof. Paper 101, 1918, p. 331, pl. xcix, f. 3.

Raton: Primero, Colo.

Fort Union: *Mouth of Yellowstone River, Mont.; Elk Creek, Yellowstone National Park. [Common everywhere in the Fort Union.]

Lance: 60 miles south of Glasgow, west of Miles City, Forsyth, Glendive, Mont.; Bismarck, Yule, N. Dak.; Bighorn Basin, Ranchester, Wyo.

Eocene: Porcupine Creek and Great Valley, Saskatchewan.

Sapindus alatus Ward = *Sapindus glendivensis*.

Sapindus americanus Lesquereux = *Sapindoides americanus*.

"Sapindus angustifolius" Lesquereux, U. S. Geol. and Geog. Surv. Terr., Ann. Rept. 1873 [1874], p. 415; Rept. U. S. Geol. Surv. Terr., vol. 7 (Tert. Fl.), 1878, p. 265, pl. xlix, f. 2–7.—Ward, U. S. Geol. Surv., Sixth Ann. Rept., 1884–85 [1886], p. 554, pl. li, f. 1–3; U. S. Geol. Surv., Bull. 37, 1887, p. 68, pl. xxxi, f. 5–7.—Lesquereux, U. S. Nat. Mus., Proc., vol. 11, 1888, p. 15.—?Knowlton, U. S. Geol. Surv., Bull. 204, 1902, p. 79.

[NOTE.—This is a homonym of *S. angustifolius* Blume. The leaves that have long been known as *Sapindus angustifolius* Lesquereux are obviously in confusion. Some have been removed under the names *S. coloradense* Cockerell and *S. leonis* Cockerell; the remainder are unassigned. Pending critical studies the species is permitted to stand as above.]

Miocene: *Florissant and *Middle Park, Colo.

Mascall: John Day Basin, Oreg.

Fort Union: Near Glendive, Mont.

Sapindus angustifolius Lesquereux. Hollick, Lesquereux, and Veatch = *Sapindus mississippiensis*.

Sapindus angustifolius Lesquereux. Lesquereux, U. S. Nat. Mus., Proc., vol. 11, 1888, p. 24 = *Sapindus formosus.*

Sapindus apiculatus Velenovsky, Beitr. Pal. Österr.-Ungarns u. d. Orients (Böhm. Kreidef., pt. 3), 1884, p. 53 (6), pl. xxii (vii), f. 1-8.—Hollick, N. Y. Acad. Sci., Trans., vol. 16, 1897, p. 133, pl. xiii, f. 1-2.—Berry, N. Y. Bot. Gard., Bull., vol. 3, 1903, p. 84; Geol. Surv. N. J., Ann. Rept. 1905 [1906], p. 139.—Hollick, U. S. Geol. Surv., Mon. 50, 1906, p. 91, pl. xxxiii, f. 21.

Magothy: Cliffwood, N. J.; Glen Cove, Long Island, N. Y.

Sapindus bentonensis Berry, U. S. Geol. Surv., Prof. Paper 91, 1916, p. 273, pl. lxvii, f. 4.

Wilcox: Benton, Ark., and Calaveras Creek, Wilson County, Tex.

Sapindus caudatus Lesquereux, U. S. Geol. and Geog. Surv. Terr., Ann. Rept. 1872 [1873], p. 380; Rept. U. S. Geol. Surv. Terr., vol. 7 (Tert. Fl.), 1878, p. 264, pl. xlviii, f. 6.—Knowlton, U. S. Geol. Surv., Prof. Paper 101, 1918, p. 330, pl. C, f. 2.

Denver: *Golden, Colo.
Wilcox: Cross Lakes, La.
Raton: Yankee, N. Mex.

Sapindus coloradensis Cockerell, Am. Mus. Nat. Hist., Bull., vol. 24, 1908, p. 101, pl. ix, f. 31.—Knowlton, U. S. Nat. Mus., Proc., vol. 51, 1916, p. 283.

Sapindus angustifolius Lesquereux [part], Rept. U. S. Geol. Surv. Terr., vol. 7 (Tert. Fl.), 1878, pl. xlix, f. 2; idem, vol. 8 (Cret. and Tert. Fl.), 1883, pl. xxxvii, f. 3-5.

Miocene: *Florissant, Colo.

Sapindus coriaceus Lesquereux = *Sapotacites coriaceus.*

Sapindus coushatta Berry, U. S. Geol. Surv., Prof. Paper 91, 1916, p. 273, pl. lxv, f. 5.

Juglans rugosa Lesquereux. Hollick, La. Geol. Surv., Special Rept. 5, 1899, p. 280, pl. xxxv, f. 1 [not f. 2].

Wilcox: Coushatta, La.

Sapindus dentoni Lesquereux, U. S. Geol. and Geog. Surv. Terr., Bull., vol. 1, 1875 [1876], p. 388; idem, Ann. Rept. 1874 [1876], p. 315; Rept. U. S. Geol. Surv. Terr., vol. 7 (Tert. Fl.), 1878, p. 265, pl. lxiv, f. 2-4.

Green River?: *Mouth of White River, Utah.

Sapindus diversifolius Lesquereux, U. S. Geol. Surv., Mon. 17, 1891 [1892], p. 158, pl. lxiv, f. 18.

Dakota: *Ellsworth County, Kans.

Sapindus dubius Unger. Lesquereux, U. S. Nat. Mus., Proc., vol. 11, 1888, p. 13 [part] = *Ficus wilcoxensis.*

Sapindus eoligniticus Berry, U. S. Geol. Surv., Prof. Paper 91, 1916, p. 276, pl. lxvii, f. 1-3; pl. cix, f. 3.

Lagrange: Puryear, Tenn., and Wickliffe, Ky.

Sapindus falcifolius Al. Braun. Lesquereux, U. S. Nat. Mus., Proc., vol. 11, 1888, p. 12 = *Ficus wilcoxensis*.

Sapindus formosus Berry, U. S. Geol. Surv., Prof. Paper 91, 1916, p. 276, pl. lxvi, f. 3–7.

Sapindus angustifolius Lesquereux. Lesquereux, U. S. Nat. Mus., Proc., vol. 11, 1888, p. 24.

Wilcox (Grenada): Grenada, Miss.
Wilcox (Holly Springs): Holly Springs, Miss.
Wilcox: Campbell's quarry, Caddo Parish, La.
Lagrange: Puryear, Tenn.; Wickliffe, Ky.

Sapindus georgiana Berry, U. S. Geol. Surv., Prof. Paper 84, 1914, p. 143, pl. xxvii, f. 11, 12.

Barnwell (Twiggs): Phinizy Gully, Columbia County, Ga.

Sapindus glendivensis Knowlton, n. name.

Sapindus alatus Ward, U. S. Geol. Surv., Sixth Ann. Rept., 1884–85 [1886], p. 554, pl. l, f. 9, 10; idem, Bull. 37, 1887, p. 68, pl. xxxi, f. 3, 4.—Knowlton, idem, Mon. 32, pt. 2, 1899, p. 737. [Homonym, Salisbury, 1776.]

Fort Union: Sevenmile Creek near Glendive, Mont.; Elk Creek, Yellowstone National Park.

Sapindus grandifolioloides Knowlton, U. S. Geol. Surv., Mon. 32, pt 2, 1899, p. 738, pl. C, f. 2.

Miocene: *Yellowstone National Park.

Sapindus grandifoliolus Ward, U. S. Geol. Surv., Sixth Ann. Rept., 1884–85 [1886], p. 554, pl. l, f. 4–8; U. S. Geol. Surv., Bull. 37, 1887, p. 67, pl. xxx, f. 3–5.—Knowlton, U. S. Geol. Surv., Mon. 32, pt. 2, 1899, p. 737, pl. xcix, f. 1, 2; pl. cii, f. 4; Washington Acad. Sci., Proc., vol. 11, 1909, pp. 190, 195, 198, 211, 213, 214; Jour. Geol., vol. 19, 1911, p. 369.

Fort Union: *Sevenmile Creek, Glendive, Mont. Very common throughout the Fort Union.

Lance: Cannonball River, N. Dak.; Signal Butte, Miles City, 8 miles west of Bridger, Yellowstone opposite Glendive, 10 miles northeast of Glendive, Mont.; Bighorn Basin, between Buffalo and Klondike, and Ranchester, Wyo.

Sapindus imperfectus Hollick, N. Y. Bot. Gard., Bull., vol. 3, 1904, p. 415, pl. 78, f. 4; U. S. Geol. Surv., Mon. 50, 1906, p. 90, pl. xxxiii, f. 15.

Magothy: *Manhasset Neck, Long Island, N. Y.

Sapindus inexpectans Knowlton, U. S. Geol. Surv., Bull. 257, p. 144, pl. xvii, f. 7.

Judith River: *Willow Creek, Fergus County, Mont.

Sapindus inflexus Lesquereux, Rept. U. S. Geol. Surv. Terr., vol. 8 (Cret. and Tert. Fl.), 1883, p. 182, pl. xxxii, f. 2.

Miocene: *Florissant, Colo.

Sapindus knowltoni Berry, U. S. Geol. Surv., Prof. Paper 91, 1916, p. 274, pl. lxiii, f. 6.
Wilcox: Benton, Ark.
Lagrange: Puryear, Henry County, Tenn.

Sapindus lancifolius Lesquereux, Rept. U. S. Geol. Surv. Terr., vol. 8 (Cret. and Tert. Fl.), 1883, p. 182, pl. xxxii, f. 3–6; pl. xxxvii, f. 9.—Knowlton, U. S. Nat. Mus., Proc., vol. 51, 1916, p. 283.
Miocene: *Florissant, Colo.

Sapindus leonis Cockerell, Am. Mus. Nat. Hist., Bull., vol. 24, 1908, p. 102.
Sapindus angustifolius Lesquereux [part], Rept. U. S. Geol. Surv. Terr., vol. 8 (Cret. and Tert. Fl.), 1883, pl. xxxvii, f. 7.
Miocene: *Florissant, Colo.

Sapindus linearifolius Berry, U. S. Geol. Surv., Prof. Paper 91, 1916, p. 275, pl. lxiii, f. 2–5; pl. cix, f. 4.
Salix angusta A. Braun. Lesquereux, U. S. Nat. Mus., Proc., vol. 11, 1888, p. 13.
Wilcox (Holly Springs): Early Grove, Miss.
Wilcox: Calaveras Creek, Wilson County, Tex.; Boydsville, Ark.
Lagrange: Puryear, Tenn.; Wickliffe, Ky.

Sapindus marylandicus Hollick, Md. Geol. Surv., Pliocene and Pleistocene, 1906, p. 234, pl. lxxii, f. 11–14.
Pleistocene (Sunderland): *Head of Island Creek, Calvert County, Md.

Sapindus? membranaceus Newberry, N. Y. Lyc. Nat. Hist., Ann., vol. 9, 1868, p. 52.—[Lesquereux], U. S. Geol. and Geog. Surv. Terr., Ill. Cret. and Tert. Pl., 1878, pl. xxiv, f. 2, 3.—Newberry, U. S. Geol. Surv., Mon. 35, 1898, p. 117, pl. xxx, f. 2, 3.
Fort Union: *Fort Union, N. Dak.

Sapindus merriami Knowlton, U. S. Geol. Surv., Bull. 204, 1902, p. 78, pl. ix, f. 5.
Clarno (upper part): *Bridge Creek, Oreg.

Sapindus mississippiensis Berry, U. S. Geol. Surv., Prof. Paper 91, 1916, p. 274, pl. lxiii, f. 1; pl. lxiv, f. 10; pl. lxvi, f. 1, 2; pl. cix, f. 1.
Sapindus angustifolius Lesquereux. Hollick, La. Geol. Surv., Special Rept. No. 5, 1899, p. 286, pl. xxxv, f. 5.—Lesquereux, U. S. Nat. Mus., Proc., vol. 11, 1888, p. 12.—Veatch, U. S. Geol. Surv., Prof. Paper 46, 1906, pl. xvii, f. 6.
Wilcox (Holly Springs): Early Grove, Miss.
Wilcox (Grenada): Grenada, Grenada County, Miss.
Wilcox: Shreveport, La.
Lagrange: Wickliffe, Ky., and Puryear, Henry County, Tenn.

Sapindus morrisoni Lesquereux, Rept. U. S. Geol. Surv. Terr., vol. 8 (Cret. and Tert. Fl.), 1883, p. 83, pl. xvi, f. 1, 2; U. S. Geol. Surv., Mon. 17, 1891 [1892], p. 158, pl. xxxv, f. 1, 2.—Hollick, N. Y. Acad. Sci., Ann., vol. 11, 1898, p. 422, pl. xxxvi, f. 4.—Berry, N. Y.

Sapindus morrisoni—Continued.
Bot. Gard., Bull., vol. 3, 1903, p. 83, pl. xlvii, f. 2, 3; Bull. Torr. Bot. Club, vol. 31, 1904, p. 78; Geol. Surv. N. J., Ann. Rept. 1905 [1906], p. 138.—Hollick, U. S. Geol. Surv., Mon. 50, 1906, p. 90, pl. xxxiii, f. 16–20.—Berry, Bull. Torr. Bot. Club, vol. 39, 1912, p. 396; U. S. Geol. Surv., Prof. Paper 84, 1914, p. 49, pl. ix, f. 6; idem, Prof. Paper 112, 1919, p. 112; Bull. Torr. Bot. Club, vol. 44, 1917, p. 186.
Dakota: *Morrison, Colo.; Pipe Creek, Kans.; New Ulm, Minn.
Raritan: Princes Bay and Tottenville, Staten Island, N. Y.
Magothy: Cliffwood and Morgan, N. J.; Glen Cove, Long Island, N. Y.
Woodbine: Arthurs Bluff, Tex.
Black Creek (Middendorf): Langley, S. C.
Tuscaloosa: Northport, Ala.
Bingen: Near Maxwell Spur, Pike County, and Mine Creek, near Nashville, Howard County, Ark.

Sapindus obtusifolius Lesquereux, U. S. Geol. and Geog. Surv. Terr., Ann. Rept. 1873 [1874], p. 419; Rept. U. S. Geol. Surv. Terr., vol. 7 (Tert. Fl.), 1878, p. 266, pl. xlix, f. 8–11.—Knowlton, U. S. Geol. Surv., Bull. 204, 1902, p. 79.
Fort Union: *Badlands, S. Dak.
Mascall: Van Horn's ranch, John Day Basin, Oreg.

Sapindus oklahomensis Berry, U. S. Nat. Mus., Proc., vol. 54, 1918, p. 632, pl. 95, f. 1, 2.
Miocene: Beaver Creek, 10 miles east of Beaver City, Okla.

Sapindus oregonianus Knowlton, U. S. Geol. Surv., Bull. 204, 1902, p. 79, pl. xv, f. 3.
Mascall: *Van Horn's ranch, John Day Valley, Oreg.

Sapindus oxfordensis Berry, U. S. Geol. Surv., Prof. Paper 91, 1916, p. 273, pl. lxvii, f. 5.
Wilcox (Holly Springs): Oxford, Miss.
Wilcox (Grenada): Grenada, Miss.

Sapindus pseudaffinis Berry, U. S. Geol. Surv., Prof. Paper 91, 1916, p. 272, pl. lxvii, f. 6.
Lagrange: Puryear, Henry County, Tenn.

Sapindus rocklandensis Knowlton, U. S. Geol. Surv., Prof. Paper 101, 1918, p. 331, pl. xcviii, f. 4.
Raton: Cucharas Canyon, Walsenburg, Colo.

Sapindus stellariaefolius Lesquereux, Rept. U. S. Geol. Surv. Terr., vol. 7 (Tert. Fl.), 1878, p. 264, pl. xlix, f. 1.—Cockerell, Am. Mus. Nat. Hist., Bull., vol. 24, 1908, p. 101; Pop. Sci. Mo., vol. 73, 1908, p. 121, text f.
Miocene: *Florissant, Colo.

Sapindus undulatus Al. Braun. Lesquereux = *Eugenia hilgardiana.*

Sapindus variabilis Berry, U. S. Geol. Surv., Prof. Paper 112, 1919, p. 111, pl. xxvii, f. 1–3.
Tuscaloosa: Shirleys Mill and Glen Allen, Ala.

Sapindus wardii Knowlton, U. S. Geol. Surv., Mon. 32, pt. 2, 1899, p. 738, pl. xcviii, f. 1, 2; pl. xcix, f. 5.
Fort Union: Yellowstone National Park.

Sapindus sp. Dawson, Geol. Surv. Canada, 1877–78 [1887]; Roy. Soc. Canada, Trans., vol. 5, 1887 [1888], sec. 4, p. 35.—Penhallow, Rept. Tert. Pl. Brit. Col., 1908, p. 87.
Oligocene: Ninemile Creek, Similkameen River, British Columbia.

Sapindus sp. Knowlton, Geol. Soc. Am., Bull., vol. 8, 1897, p. 142.
Hanna: "Dutton Creek coal," Laramie Plains, Wyo.

SAPOTACITES Ettingshausen, Abhandl. K. k. geol. Reichsanstalt [Tert. Fl. Häring], vol. 2, Abt. 3, 1853, p. 61.

Sapotacites americanus Lesquereux = *Bumelia americana.*

Sapotacites copeanus Ettingshausen = *Sapotacites coriaceus.*

Sapotacites coriaceus (Lesquereux) Knowlton, n. comb.
Sapotacites copeanus Ettingshausen, Denks. K. Akad. Wiss., Math.-naturw. Classe, vol. 47, 1883, p. 136 [36]; Geol. Surv. N. S. W., Mem. Palaeont. No. 2 (Tert. Fl. Australia), 1888, p. 57.
Sapindus coriaceus Lesquereux, U. S. Geol. and Geog. Surv. Terr., Ann. Rept. 1873 [1874], p. 415; Rept. U. S. Geol. Surv. Terr., vol. 7 (Tert. Fl.), 1878, p. 265, pl. xlix, f. 12–14.
Miocene: *Elko station, Nev.

Sapotacites ettingshauseni Berry, U. S. Geol. Surv., Prof. Paper 112, 1919, p. 135, pl. xxix, f. 7.
Tuscaloosa: Shirleys Mill, Ala.

Sapotacites formosus Berry, U. S. Geol. Surv., Prof. Paper 112, 1919, p. 136, pl. xxx, f. 6.
Tuscaloosa: Shirleys Mill, Ala.

Sapotacites haydenii Heer, Philadelphia Acad. Sci., Proc., vol. 10, 1858, p. 265.—Newberry, U. S. Geol. Surv., Mon. 35, 1898, p. 126, pl. v, f. 1.
Dakota: *Blackbird Hill, Nebr.

Sapotacites knowltoni Berry, Bull. Torr. Bot. Club, vol. 33, 1906, p. 181, pl. 8, f. 1; Md. Geol. Surv., Upper Cret., 1916, p. 892, pl. xc, f. 2.
Sapotacites sp.? Lesquereux, U. S. Geol. Surv., Mon. 17, 1891 [1892], p. 114, pl. lxv, f. 3.
Magothy: Deep Cut, Del.
Dakota: *Ellsworth County, Kans.

Sapotacites retusus Heer = *Liriodendropsis retusa.*

Sapotacites shirleyensis Berry, U. S. Geol. Surv., Prof. Paper 112, 1919, p. 135, pl. xxviii, f. 11; pl. xxix, f. 4–6.
Tuscaloosa: Shirleys Mill, Ala.

Sapotacites sp.? Lesquereux = *Sapotacites knowltoni.*

SASSAFRAS Nees, Handb. Med. Pharm. Bot., vol. 2, 1831, p. 418.

Sassafras acutilobum Lesquereux, Rept. U. S. Geol. Surv. Terr., vol. 6 (Cret. Fl.), 1874, p. 79, pl. xiv, f. 1, 2.—Hollick, N. Y. Acad. Sci., Trans., vol. 12, 1893, p. 236, pl. vii, f. 1.—Berry, N. Y. Bot. Gard., Bull., vol. 3, 1903, p. 81, pl. xlv, f. 1, 2; Geol. Surv. N. J., Ann. Rept. 1905 [1906], p. 139.—Hollick, U. S. Geol. Surv., Mon. 50, 1906, p. 77, pl. xxx, f. 8, 9; N. Y. Bot. Gard., Bull., vol. 8, 1912, p. 161, pl. clxviii, f. 3.—Berry, Bull. Torr. Bot. Club, vol. 37, 1910, p. 22; Geol. Surv. N. J., Bull. 3, 1911, p. 140, pl. xviii, f. 2.—Knowlton, in Woodruff, U. S. Geol. Surv., Bull. 452, 1911, p. 20.—Berry, Md. Geol. Surv., Upper Cret., 1916, p. 866, pls. lxxii, lxxiii, lxxiv, f. 1, 2; lxxv, f. 2; U. S. Geol. Surv., Prof. Paper 112, 1919, p. 120.
Dakota: *Salina, Kans.; 2 miles west of Lander, Wyo.
Magothy: Cliffwood, N. J., Gay Head, Marthas Vineyard, Mass.; Glen Cove and Roslyn, N. Y.; Grove Point, Md.
Raritan: Woodbridge, N. J.; Brightseat, Md.; East Washington Heights, D. C.
Black Creek: Court House Bluff, Cape Fear River, N. C.
Tuscaloosa: Cottondale, Ala.

Sassafras angustilobum Hollick, U. S. Geol. Surv., Mon. 50, 1906, p. 77, pl. xxix, f. 1–3.
Magothy: Gay Head, Marthas Vineyard, Mass.

Sassafras? bilobatum Fontaine, U. S. Geol. Surv., Mon. 15, 1889 [1890], p. 290, pl. clvi, f. 12; pl. clxiv, f. 4; in Ward, U. S. Geol. Surv., Mon. 48, 1905 [1906], p. 506, pl. cxi, f. 5.—Berry, Bot. Gaz., vol. 34, 1902, p. 435; Md. Geol. Surv., Lower Cret., 1912, p. 484, pl. xciii, f. 1.
Patapsco: Near Brooke, Va.

Sassafras burpeana Dawson, Roy. Soc. Canada, Trans., vol. 3, 1885 [1886], sec. 4, p. 17; idem, vol. 4, 1886 [1887], sec. 4, p. 28, pl. ii, f. 12.—Penhallow, Rept. Tert. Pl. Brit. Col., 1908, p. 87.
Paskapoo (Fort Union): *Calgary, Alberta; Porcupine Creek, Saskatchewan.

Sassafras cretaceum Newberry.

[Note.—In his last treatment of this species Dr. Newberry referred the following forms to it: *Sassafras mudgii*, *S. subintegrifolius*, *S. harkeriana*, *S. obtusus*, *S. acutilobum*, *S. cretaceum dentatum*, and *S. cretaceum obtusum*. I have not made these transfers in this book, on the ground that a thorough revision based on full modern material has not been undertaken.]

Sassafras (Araliopsis) cretaceum Newberry = *Araliopsis cretacea.*

Sassafras cretaceum Penhallow, Geol. Surv. Canada, Ann. Rept., new ser., vol. 14, 1904 [1906], p. 390A [name] = *Sassafras* sp. Penhallow, Rept. Tert. Pl. Brit. Col., 1908, p. 87.

Sassafras (Araliopsis) cretaceum dentatum Lesquereux = *Araliopsis cretacea dentata.*

Sassafras (Araliopsis) cretaceum grossedentatum Lesquereux, U. S. Geol. Surv., Mon. 17, 1891 [1892], p. 101, pl. li, f. 5.
  Dakota: Kansas.

Sassafras cretaceum heterolobum Fontaine = *Sassafras parvifolium.*

Sassafras (Araliopsis) cretaceum obtusum Lesquereux = *Araliopsis cretacea salisburiaefolia.*

Sassafras cretaceum recurvatum (Lesquereux) Newberry, U. S. Geol. Surv., Mon. 35, 1898, p. 99, pl. ix, f. 2 = *Sassafras recurvatum.*

Sassafras cretaceum recurvatum Lesquereux. Berry = *Platanus heerii.*

Sassafras (Araliopsis) dissectum Lesquereux, Rept. U. S. Geol. Surv. Terr., vol. 8 (Cret. and Tert. Fl.), 1883, p. 57; U. S. Geol. Surv., Mon. 17, 1891 [1892], p. 101.
  Dakota: *Near Fort Harker, Kans.

Sassafras (Araliopsis) dissectum symmetricum Hollick, Bull. Torr. Bot. Club, vol. 22, 1895, p. 226, pl. ccxxxvi.
  Dakota: *Fort Harker, Kans.

Sassafras giganteum Gould, Kans. Acad. Sci., Trans., vol. 17, 1899–1900 [1901], p. 160 [39], pl. xii.
  Dakota: Kansas ?.

Sassafras (Araliopsis) harkerianum Lesquereux = *Cissites harkeriana.*

Sassafras hastatum Newberry, U. S. Geol. Surv., Mon. 26, 1895 [1896], p. 88, pl. xxvii, f. 4–6; pl. xxviii, f. 1, 2; pl. xl, f. 4.—?Hollick, N. Y. Bot. Gard., Bull., vol. 3, 1904, p. 414, pl. lxxix, f. 4; ?U. S. Geol. Surv., Mon. 50, 1906, p. 78, pl. xxix, f. 4; pl. xxx, f. 12.—Berry, Geol. Surv. N. J., Bull. 3, 1911, p. 143, pl. xvii, f. 2.
  Magothy: ?Gay Head, Marthas Vineyard, Mass.; Glen Cove, Long Island, N. Y.
  Raritan: *Woodbridge, N. J.

Sassafras latilobum (Newberry) Knowlton, n. comb.
  Platanus latiloba Newberry, N. Y. Lyc. Nat. Hist., Ann., vol. 9, 1868, p. 23.—[Lesquereux], U. S. Geol. and Geog. Surv. Terr., Ill. Cret. and Tert. Pl., 1878, pl. ii, f. 4; U. S. Geol. Surv., Mon. 35, 1898, p. 105, pl. i, f. 4.
  Sassafras mirabilis Lesquereux, U. S. Geol. and Geog. Surv. Terr., Ann. Rept. 1872 [1873], p. 424.
  Sassafras (Araliopsis) mirabile Lesquereux, U. S. Geol. Surv. Terr., Rept., vol. 6 (Cret. Fl.), 1874, p. 80, pl. xii, f. 1; idem, vol. 8 (Cret. and Tert. Fl.), 1883, p. 56; U. S. Geol. Surv., Mon. 17, 1891 [1892], p. 102.—Knowlton, in Woodruff, U. S. Geol. Surv., Bull. 452, 1911, p. 20.
  Dakota: *Blackbird Hill, Nebr.; Fort Harker, Kans.; 2 miles west of Lander, Wyo.

Sassafras leconteanum Lesquereux = *Persea leconteanum.*

Sassafras mirabile Lesquereux = *Sassafras latilobum.*

Sassafras mirabilis Lesquereux = *Sassafras latilobum.*

Sassafras mudgii Lesquereux, Am. Jour. Sci., 2d ser., vol. 46, 1868, p. 99; U. S. Geol. Surv. Terr., Rept., vol. 6 (Cret. Fl.), 1874, p. 78, pl. xiv, f. 3, 4; pl. xxx, f. 7.—Ward, U. S. Geol. Surv., Nineteenth Ann. Rept., pt. 2, 1899, p. 705, pl. clxx, f. 4, 5; pl. clxxi, f. 1.

Lindera vetusta Lesquereux. Ward, Jour. Geol., vol. 2, 1894, p. 262.

Aralia towneri Lesquereux. Ward, op. cit., p. 262.

Dakota: *Salina River, Kans.; Evans's quarry, 5 miles east of Hot Springs, S. Dak.

Sassafras (Araliopsis) obtusum Lesquereux = *Araliopsis cretacea salisburiaefolia.*

Sassafras (Araliopsis) papillosum Lesquereux, U. S. Geol. Surv., Mon. 17, 1891 [1892], p. 102, pl. vi, f. 7.

Dakota: *Ellsworth County, Kans.

Sassafras parvifolium Fontaine, U. S. Geol. Surv., Mon. 15, 1889 [1890], p. 289, pl. cxxxix, f. 7.—Berry, Bot. Gaz., vol. 34, 1902, p. 434; Md. Geol. Surv., Lower Cret., 1912, p. 486, pl. xciii, f. 2; pl. xciv, f. 2.

Sassafras cretaceum heterolobum Fontaine, U. S. Geol. Surv., Mon. 15, 1889 [1890], p. 289, pl. clii, f. 5; pl. clix, f. 8; pl. clxiv, f. 5.—Berry, Bot. Gaz., vol. 34, 1902, p. 435.

Patapsco: Baltimore, Md.; near Brooke, Va.

Sassafras (Araliopsis) platanoides Lesquereux, Rept. U. S. Geol. Surv. Terr., vol. 8 (Cret. and Tert. Fl.), 1883, p. 58, pl. vii, f. 1.

Dakota: *Kansas.

Sassafras potomacensis Berry, Md. Geol. Surv., Lower Cret., 1912, p. 487, pl. xciv, f. 1.

Patapsco: *Widewater, Dumfries Landing, Va.

Sassafras? primordiale Lesquereux, U. S. Geol. Surv., Mon. 17, 1891 [1892], p. 100, pl. xvi, f. 10.

Dakota: *Ellsworth County, Kans.

Sassafras progenitor Newberry MS. Hollick, Bull. Torr. Bot. Club, vol. 21, 1894, p. 53, pl. clxxiv, f. 1.—Newberry, U. S. Geol. Surv., Mon. 26, 1895 [1896], p. 88, pl. xxvii, f. 1–3.—Hollick, Geol. Soc. Am., Bull., vol. 7, 1895, p. 13.—Berry, Bull. Torr. Bot. Club, vol. 31, 1904, p. 78, pl. i, f. 3; Geol. Surv. N. J., Ann. Rept. 1905 [1906], p. 139.—Hollick, U. S. Geol. Surv., Mon. 50, 1906, p. 78, pl. xxx, f. 11.—Berry, Geol. Surv. N. J., Bull. 3, 1911, p. 142, pl. xviii, f. 1.

Magothy: Cliffwood, N. J.

Magothy or Raritan: Oak Neck, Long Island, N. Y.

Raritan: Woodbridge, N. J.

Sassafras (Araliopsis) recurvatum Lesquereux, Rept. U. S. Geol. Surv. Terr., vol. 8 (Cret. and Tert. Fl.), 1883, p. 57.

Sassafras recurvata Lesquereux, Rept. U. S. Geol. Surv. Terr., vol. 6 (Cret. Fl.), 1874, p. 71, pl. x, f. 3–5.

Sassafras recurvatus Lesquereux, U. S. Geol. and Geog. Surv. Terr., Ann. Rept. 1872 [1873], p. 424.

Sassafras cretaceum recurvatum (Lesquereux), U. S. Geol. Surv., Mon. 35, 1898, p. 99, pl. ix, f. 2.

Dakota: *Near Fort Harker, Kans.

Sassafras recurvatus Lesquereux = *Platanus heerii.*

Sassafras selwynii Dawson, Geol. Surv. Canada, Rept. Prog. 1879–80, p. 53A; Roy. Soc. Canada, Trans., vol. 4, 1886 [1887], sec. 4, p. 28, pl. ii, f. 13.—Penhallow, Rept. Tert. Pl. Brit. Col., 1908, p. 87.

Paskapoo (Fort Union): Souris River and Porcupine Creek, Saskatchewan.

Sassafras subintegrifolium Lesquereux, Am. Jour. Sci., vol. 46, 1868, p. 99; U. S. Geol. Surv., Mon. 17, 1891 [1892], p. 99, pl. xiv, f. 2.

Dakota: *Ellsworth County, Kans.

Sassafras sp. Dawson, Roy. Soc. Canada, Trans., vol. 1, 1882 [1883], sec. 4, p. 27, pl. vii, f. 30, 30a.

Upper Cretaceous: Baynes Sound, Vancouver Island, British Columbia.

Sassafras sp. Knowlton, in Hill, Am. Jour. Sci., 3d ser., vol. 50, 1895, p. 212.

Dakota: Black Hills near Belvidere, Kans.

Sassafras sp. Knowlton, Washington Acad. Sci., Proc., vol. 11, 1909, p. 201.

Lance: 20 miles below Mandan, N. Dak.; Converse County, Wyo.; Hell Creek, Mont.

Sassafras sp. Lesquereux, U. S. Geol. and Geog. Surv. Terr., Ann. Rept. 1871 [1872], p. 298. [Not afterward mentioned by Lesquereux.]

——: 6 miles above Spring Canyon, Mont.

Sassafras (Araliopsis) sp. Lesquereux, Rept. on Clays in N. J., 1878, p. 29.

Raritan ?: Burts Neck, N. J.

Sassafras sp. Penhallow, Rept. Tert. Pl. Brit. Col., 1908, p. 87.

Oligocene: Quilchena, British Columbia.

SAXIFRAGA (Tournefort) Linné, Sp. pl., 1753, p. 398.

Saxifraga? peritula Cockerell, Am. Mus. Nat. Hist., Bull., vol. 24, 1908, p. 92, pl. x, f. 41.

Miocene: *Florissant, Colo.

SCHIZAEOPSIS Berry, Annals Bot., vol. 25, 1911, p. 194. [Type, *Baieropsis expansa*, part.]

Schizaeopsis americana Berry, Md. Geol. Surv., Lower Cret., 1917, p. 216, pl. xxii, f. 1–9, text f. 2.

Baieropsis expansa Fontaine, U. S. Geol. Surv., Mon. 15, 1889 [1890], p. 207, pl. lxxxix, f. 1 [not other figures which=*Acrostichopteris expansa*].

Baieropsis macrophylla Fontaine, U. S. Geol. Surv., Mon. 15, 1889 [1890], p. 212, pl. xc, f. 6.

Schizaeopsis expansa (Fontaine) Berry, Annals Bot., vol. 25, 1911, p. 194, pl. 6, f. 1–6, text f. 1.

Patuxent: Fredericksburg, Va.

Schizaeopsis expansa (Fontaine) Berry = *Schizaeopsis americana.*

SCHIZOLEPIS F. Braun, Flora, vol. 1, 1847, p. 186.

Schizolepis liaso-keuperiana F. Braun, Flora, Neue Reihe, vol. 5, 1847, p. 86.—Fontaine, in Ward, U. S. Geol. Surv., Twentieth Ann. Rept., pt. 2, 1900, p. 252, pl. xxxiii, f. 3–5.

Triassic (Newark): York County, Pa.

SCHIZONEURA Schimper and Mougeot, Mon. pl. foss. grès bigarré Vosges, 1844, p. 48.

Schizoneura planicostata (Rogers) Fontaine, U. S. Geol. Surv., Mon. 6, 1883, p. 14, pl. i, f. 1.—Newberry, idem, Mon. 14, 1888, p. 87.—Fontaine, in Ward, U. S. Geol. Surv., Twentieth Ann. Rept., pt. 2, 1900, p. 289.

Calamites planicostatus Rogers, Trans. Assoc. Am. Geol. and Nat., Philadelphia, 1843, p. 305.

Triassic (Newark): Carbon Hill and Clover Hill, Va.; North Carolina; New Jersey; Connecticut; Massachusetts.

Schizoneura virginiensis Fontaine = *Neocalamites virginiensis.*

Schizoneura sp.? Fontaine, U. S. Geol. Surv., Mon. 6, 1883, p. 16, pl. i, f. 3.

Triassic (Newark): Clover Hill, Va.

Schizopteris gracilis Bean = *Baiera gracilis.*

SCHMALTZIA Desveaux, Jour. bot., vol. 2, 1809, p. 170.

Schmaltzia vexans (Lesquereux) Cockerell, Torreya, vol. 6, 1906, p. 12; Univ. Colo. Studies, vol. 3, 1906, p. 170.

Rhus vexans Lesquereux, U. S. Geol. Surv. Terr., Rept., vol. 8 (Cret. and Tert. Fl.), 1883, p. 195, pl. xli, f. 20.

Miocene: *Florissant, Colo.

SCIRPUS (Tournefort) Linné, Sp. pl., 1753, p. 47.

Scirpus sp. Dawson, Brit. N. A. Bound. Com. (Rept. Geol. and Res. Vic. 49th Parallel), 1875, App. A, p. 329.

Paskapoo (Fort Union): Badlands, near Woody Mountain; Alberta.

Paskapoo (Fort Union): Porcupine Creek, Saskatchewan.

Schlerophyllina dichotoma Heer = *Czekanowskia dichotoma.*

SCLEROPTERIS Saporta, Paléont. française, pl. jurass., vol. 1, 1873, p. 364.

Scleropteris dentata Fontaine = *Zamiopsis dentata.*

Scleropteris distantifolia Fontaine, in Ward, U. S. Geol. Surv. Nineteenth Ann. Rept., pt. 2, 1899, p. 662, pl. clxii, f. 2, 3.
Lakota: *South Fork of Hay Creek, Crook County, Wyo.

Scleropteris elliptica Fontaine, U. S. Geol. Surv., Mon. 15, 1889 [1890], p. 151, pl. xxviii, f. 2, 4, 6; pl. xxix, f. 1; in Ward, U. S. Geol. Surv., Mon. 48, 1905 [1906], p. 511.—Berry, Md. Geol. Surv., Lower Cret., 1911, p. 300, pl. xxxix, f. 1, 2.

Scleropteris elliptica longifolia Fontaine, op. cit. (Mon. 15), p. 152, pl. xxviii, f. 7.

Scleropteris virginica Fontaine, op. cit. (Mon. 15), p. 152, pl. xxviii, f. 3, 5; in Ward, op. cit. (Mon. 48), p. 484.

Ctenopteris integrifolia Fontaine, op. cit. (Mon. 15), p. 158, pl. lxii, f. 2; pl. lxv, f. 3; in Ward, op. cit. (Mon. 48), p. 242, pl. lxvi, f. 12, 13.

Patuxent: *Fredericksburg, Potomac Run, Dutch Gap, Kankeys, Chinkapin Hollow, and telegraph station near Lorton, Va.; Cockpit Point ?, Md.

?Horsetown: Near Horsetown and near Ono, Shasta County, Calif.

Scleropteris elliptica longifolia Fontaine = *Scleropteris elliptica.*

Scleropteris oregonensis Fontaine, in Ward, U. S. Geol. Surv., Mon. 48, 1905 [1906], p. 74, pl. xii, f. 1–3.
Jurassic: *Douglas County, Oreg.

Scleropteris rotundifolia Fontaine, in Ward, U. S. Geol. Surv., Nineteenth Ann. Rept., pt. 2, 1899, p. 663, pl. clxii, f. 4, 5.
Lakota: *South Fork of Hay Creek, Crook County, Wyo.

Scleropteris vernonensis Ward = *Dicksoniopteris vernonensis.*

Scleropteris virginica Fontaine = *Scleropteris elliptica.*

SCLEROTITES Meschinelli, Syllog. Fung. Foss., 1892, p. 67 [803].

Sclerotites brandonianus Jeffrey and Chrysler, Rept. Vt. State Geol. 1905–6 [1906], p. 200, pl. l, f. 12; pl. li, f. 13–15.
Miocene: Brandon, Vt.

Sclerotites rubellus (Lesquereux) Meschinelli, Syllog. Fung. Foss., 1892, p. 69 [805] ?.

Sclerotium rubellum Lesquereux, U. S. Geol. and Geog. Surv. Terr., Ann, Rept. 1872 [1873], p. 375; U. S. Geol. Surv. Terr., Rept., vol. 7 (Tert. Fl.), 1878, p. 35, pl. i, f. 2, 2f.

Denver: *Golden, Colo.

Sclerotites sp. (Lesquereux) Knowlton, U. S. Geol. Surv., Bull. 152, 1898, p. 215.

Sclerotium? sp. Lesquereux, U. S. Geol. Surv., Mon. 17, 1891 [1892], p. 23, pl. lix, f. 4, 4a.

Dakota: *Ellsworth County, Kans.

Sclerotium rubellum Lesquereux = *Sclerotites rubellus*.

Sclerotium? sp. Lesquereux = *Sclerotites* sp.

Seedling plant? Knowlton, U. S. Geol. Surv., Prof. Paper 101, 1918, p. 280, pl. xlix, f. 4.
Vermejo: Tercio Park at Cuatro, Colo.

SELAGINELLA Beauvois, Prodr. Aetheog., 1805, p. 101.

Selaginella berthoudi Lesquereux, U. S. Geol. and Geog. Surv. Terr., Ann. Rept. 1873 [1874], p. 395; U. S. Geol. Surv. Terr., Rept., vol. 7 (Tert. Fl.), 1878, p. 46, pl. v, f. 12, 12a.
Denver: *Golden, Colo.

Selaginella collieri Knowlton, Torreya, vol. 16, 1916, p. 201, pl. i, f. 1–6.
Fort Union: Northeast Montana near international boundary.

Selaginella falcata Lesquereux emend. Knowlton, U. S. Geol. Surv., Bull. 152, 1898, p. 215; idem, Bull. 163, 1900, p. 25.
Selaginella? falcata Lesquereux, U. S. Geol. and Geog. Surv. Terr., Bull., vol. 1, 1875 [1876], p. 365; idem, Ann. Rept. 1874 [1876], p. 297; Rept. U. S. Geol. Surv. Terr., vol. 7 (Tert. Fl.), 1878, p. 46, pl. lxi, f. 12–15 [not pl. lxiv, f. 13, 13a].
Mesaverde: *Point of Rocks, Wyo.

Selaginella falcata Lesquereux [part] = *Lycopodium lesquereuxiana*.

Selaginella laciniata Lesquereux, U. S. Geol. and Geog. Surv. Terr., Bull., vol. 1, 1875 [1876], p. 378; idem, Ann. Rept. 1874 [1876], p. 297; Rept. U. S. Geol. Surv. Terr., vol. 7 (Tert. Fl.), 1878, p. 47, pl. lxiv, f. 12, 12a.—Knowlton, U. S. Geol. Surv., Bull. 163, 1900, p. 24, pl. iii, f. 5–8.
Mesaverde: *Point of Rocks, Wyo.

Selaginella marylandica Fontaine, in Ward, U. S. Geol. Surv., Mon. 48, 1905 [1906], p. 553, pl. cxv, f. 9, 10.—Berry, Md. Geol. Surv., Lower Cret., 1911, p. 307, pl. xli, f. 1, 2.
Arundel: *Vinegar Hill, Relay, Md.

SEQUOIA Endlicher, Synop. Conif., 1847, p. 197.

Sequoia acuminata Lesquereux, U. S. Geol. and Geog. Surv. Terr., Bull., vol. 1, 1875 [1876], p. 384; U. S. Geol. Surv. Terr., Rept., vol. 7 (Tert. Fl.), 1878, p. 80, pl. vii, f. 15–16a.—Knowlton, Jour. Geol., vol. 19, 1911, p. 371; U. S. Geol. Surv., Prof. Paper 130, 19—, p. —, pl. ii, f. 7, 8 [in preparation].
Post-Laramie: *Black Buttes, Wyo.
Dawson: Colorado Springs, Colo.
Lance: ?Thunder Butte, S. Dak.
?Laramie: Cowan station near Denver, Colo.

Sequoia acutifolia Newberry, Am. Jour. Sci., 3d ser., vol. 41, 1891, p. 200, pl. xiv, f. 7, 7a.
Kootenai: Great Falls, Mont.

Sequoia affinis Lesquereux, U. S. Geol. and Geog. Surv. Terr., Bull., vol. 1, 1875 [1876], p. 384; idem, Ann. Rept. 1874 [1876], p. 310; Rept. U. S. Geol. Surv. Terr., Rept., vol. 7 (Tert. Fl.), 1878, p. 75, pl. vii, f. 3–5; pl. lxv, f. 1–4; idem, vol. 8 (Cret. and Tert. Fl.), 1883, p. 138.—Knowlton, U. S. Nat. Mus., Proc., vol. 51, 1916, p. 148.

Sequoia langsdorfii Brongniart. Lesquereux, U. S. Geol. and Geog. Surv. Terr., Ann. Rept. 1874 [1876], p. 400; U. S. Geol. Surv. Terr., Rept., vol. 7 (Tert. Fl.), 1878, p. 76.

Glyptostrobus ungeri? Heer. Lesquereux, Rept. U. S. Geol. Surv. Terr., vol. 8 (Cret. and Tert. Fl.), 1883, p. 139, pl. xxii, f. 1–6a.

Sequoia haydenii (Lesquereux) Cockerell, Science, vol. 26, 1907, p. 447; Pop. Sci. Mo., vol. 73, 1908, p. 122, text f.; Am. Mus. Nat. Hist., Bull., vol. 24, 1908, p. 78.

Miocene: Florissant, Colo.; Elko station, Nev.

Sequoia albertensis Penhallow, Ottawa Nat., vol. 22, 1908, p. 83, f. 1–6.

Belly River: Red Deer River, Alberta.

Sequoia ambigua Heer, Fl. foss. arct., vol. 3, 1878, p. 78, pl. xxi, f. 1–11.—Fontaine, U. S. Geol. Surv., Mon. 15, 1889 [1890], p. 245, pl. cxviii, f. 2; pl. cxx, f. 1–6; pl. cxxvii, f. 5; pl. cxxxii, f. 3; U. S. Nat. Mus., Proc., vol. 15, 1892, p. 494; in Ward, U. S. Geol. Surv., Mon. 48, 1905 [1906], pp. 264, 555, pl. lxix, f. 6; pl. cx, f. 13.—Knowlton, Smithsonian Misc. Coll., vol. 50, 1907, p. 126.—Hollick, U. S. Geol. Surv., Mon. 50, 1906, p. 41, pl. iii, f. 7, 8.—Berry, Bull. Torr. Bot. Club, vol. 37, 1910, p. 20; U. S. Nat. Mus., Proc., vol. 40, 1911, p. 310; Md. Geol. Surv., Lower Cret., 1911, p. 449, pl. lxxviii, f. 1–7; idem, Upper Cret., 1916, p. 786; U. S. Geol. Surv., Prof. Paper 112, 1919, p. 66, pl. vi, f. 3, 4.

Sphenolepidium recurvifolium Fontaine, U. S. Geol. Surv., Mon. 15, 1889 [1890], p. 258, pl. cxxvii, f. 2; pl. cxxx, f. 2, 7.

Sphenolepidium dentifolium Fontaine, op. cit. (Mon. 15), p. 258, pl. cxxvii, f. 3, 4; pl. xxviii, f. 2–6; pl. cxxix, f. 5; pl. cxxx, f. 4–6, 10; in Ward, U. S. Geol. Surv., Mon. 48, 1905 [1906], pp. 484, 528, 538, 546, 555.

Sequoia gracilis Heer. Fontaine, in Ward, U. S. Geol. Surv., Nineteenth Ann. Rept., pt. 2, 1899, p. 675, pl. clxii, f. 2.

Arthrotaxopsis expansa Fontaine, in Ward, U. S. Geol. Surv., Mon. 48, 1906, pp. 533, 535, 538, 555, pl. cix, f. 12, 13 [not pp. 504, 520, 546, 547, 571].

Patuxent: Fredericksburg, Dutch Gap, Cockpit Point, Potomac Run, and telegraph station near Lorton, Va.

Arundel: Soper Hall, Riverdale, Arlington, Muirkirk, Hanover, Md.

Patapsco: Federal Hill (Baltimore), Locust Point, Fort Foote, Md.

Kootenai: Great Falls, Mont.

Horsetown: Shasta County, Calif.

Magothy: Round Bay, Severn River, N. J.; Gay Head, Marthas Vineyard, Mass.

Eutaw: Near Havana, Hale County, Ala.

Tuscaloosa: Snow place, Tuscaloosa County, Ala.

Sequoia angustifolia Lesquereux, U. S. Geol. and Geog. Surv. Terr., Ann. Rept. 1872 [1873], p. 372; U. S. Geol. Surv. Terr., Rept., vol. 7 (Tert. Fl.), 1878, p. 77, pl. vii, f. 6–10; idem, Rept., vol. 8 (Cret. and Tert. Fl.), 1883, p. 240, pl. l, f. 5.—Knowlton, U. S. Geol. Surv., Bull. 204, 1902, p. 24.—Hollick, Md. Geol. Surv., Pliocene and Pleistocene, 1906, p. 219, pl. lxxi, f. 16, 17.—Penhallow, Rept. Tert. Pl. Brit. Col., 1908, p. 88.

Miocene: *Elko station, Nev.; Corral Hollow, Calif.

Pleistocene (Sunderland): Head of Island Creek, Calvert County, Md.

Mascall: Van Horn's ranch, John Day Basin, Oreg.

Oligocene: Quilchena, Similkameen, Horsefly, and Tulameen rivers, British Columbia.

Sequoia biformis Lesquereux = *Geinitzia biforma.*

Sequoia brevifolia Heer, Fl. foss. arct., vol. 1, 1868, p. 93, pl. ii, f. 23.—Penhallow, Rept. Tert. Pl. Brit. Col., 1908, p. 88.

Oligocene ?: Ninemile Creek and Tulameen River; Tranquille River, British Columbia.

Sequoia burgessii Penhallow, Roy. Soc. Canada, Trans., 2d ser., vol. 9, 1903, sec. 4, p. 42, f. 5–8; Mon. N. A. Gym., 1907, p. 226; Rept. Tert. Pl. Brit. Col., 1908, p. 88.

Cupressoxylon sp. (a), Dawson, Brit. N. A. Bound. Com. (Rept. Geol. and Res. Vic. 49th Parallel), 1875, App. A, p. 331, pl. xv, f. 3.

Paskapoo (Fort Union): Porcupine Creek, Saskatchewan.

Sequoia charlottensis Knowlton, n. name.

Sequoia langsdorfii (Brongniart) Heer. Penhallow, Roy. Soc. Canada, Trans., 2d ser., vol. 8, 1902, sec. 4, p. 44, pl. xiii, f. 14; pl. xiv.

Upper Cretaceous: Queen Charlotte Islands, British Columbia.

Sequoia concinna Heer (cone). Hollick, U. S. Geol. Surv., Mon. 50, 1906, p. 43, pl. ii, f. 41.—Berry, Geol. Surv. N. J., Bull. 3, 1911, p. 96; Bull. Torr. Bot. Club, vol. 44, 1917, p. 172, pl. vii, f. 1–5.

Magothy: Gay Head, Marthas Vineyard, Mass.

Raritan ?: No locality given.

Bingen: Near Maxwell Spur, Pike County, Ark.

Sequoia condita Lesquereux, U. S. Geol. and Geog. Surv. Terr., Bull., vol. 1, 1875 [1876], p. 391; idem, Ann. Rept. 1874 [1876], p. 355, pl. iv, f. 5–7; U. S. Geol. Surv. Terr., Rept., vol. 8 (Cret. and Tert. Fl.), 1883, p. 32, pl. i, f. 5–7; Rept. on Clays in N. J., 1878, p. 29.

Dakota: Kansas.

Raritan: Sayreville, N. J. ?

Sequoia couttsiae Heer, Philos. Trans., London, vol. 152, 1862, pt. 2, p. 1051, pls. lix, lx, lxi.—Knowlton, U. S. Geol. Surv., Mon. 32, pt. 2, 1899, p. 681.—Penhallow, Roy. Soc. Canada, Trans., vol. 8, 1902, p. 50; Rept. Tert. Pl. Brit. Col., 1908, p. 88.

Fort Union: Yellowstone National Park.

Paskapoo (Fort Union): Red Deer River, Alberta

Sequoia couttsiae Heer. Hollick = *Geinitzia reichenbachi.*

Sequoia cuneata (Newberry) Newberry, U. S. Geol. Surv., Mon. 35, 1898, p. 18, pl. xiv, f. 3–4a.

Taxodium cuneatum Newberry, Boston Jour. Nat. Hist., vol. 7, 1863, p. 517.—Dawson, Roy. Soc. Canada, Trans., vol. 1, 1883, p. 25.

Upper Cretaceous: Nanaimo, Vancouver Island, British Columbia.

Kootenai: British Columbia.

Sequoia cycadopsis Fontaine = *Dichotozamites cycadopsis.*

Sequoia delicatula Fontaine, U. S. Geol. Surv., Mon. 15, 1889 [1890], p. 247, pl. cxxi, f. 3.—Berry, U. S. Nat. Mus., Proc., vol. 40, 1911, p. 310; Md. Geol. Surv., Lower Cret., 1911, p. 448.

Patuxent: *Near Dutch Gap Canal, Va.

Sequoia densifolia Fontaine = *Sequoia reichenbachi.*

Sequoia fairbanksi Fontaine, in Ward, U. S. Geol. Surv., Mon. 48, 1905 [1906], p. 178, pl. xlv, f. 9–11.

Jurassic: *Slate Springs, Calif.

Sequoia fastigiata (Sternberg) Heer, Neue Deutschr. Schw. Ges., vol. 23, Mém. 2, 1869, p. 11, pl. i, f. 10–13.—Lesquereux, U. S. Geol. and Geog. Surv. Terr., Bull., vol. 1, 1875 [1876], p. 391; U. S. Geol. Surv. Terr., Rept., vol. 8 (Cret. and Tert. Fl.), 1883, p. 31; U. S. Geol. and Geog. Surv. Terr., Ann. Rept. 1874 [1876], p. 335, pl. iii, f. 2, 8.—Hollick, U. S. Geol. Surv., Mon. 50, 1906, p. 43, pl. iii, f. 15.—Lesquereux, idem, Mon. 17, 1891 [1892], p. 36.—Berry, U. S. Geol. Surv., Prof. Paper 112, 1919, p. 67.

Caulerpites fastigiatus Sternberg, Vers., vol. 2, 1833, p. 23.

Dakota: *Kansas.

Magothy: Gay Head, Marthas Vineyard, Mass.

Tuscaloosa: Snow place, Tuscaloosa County, Ala.

Sequoia formosa Lesquereux, Am. Jour. Sci., 2d ser., vol. 46, 1868, p. 92; U. S. Geol. Surv. Terr., Rept., vol. 6 (Cret. Fl.), 1874, p. 50, pl. i, f. 9, 9b; U. S. Geol. Surv., Mon. 17, 1891 [1892], p. 36.

Dakota: *Decatur, Nebr.

Sequoia gracillima (Lesquereux) Newberry, U. S. Geol. Surv., Mon. 26, 1895 [1896], p. 50, pl. ix, f. 1–3; idem, Mon. 35, 1898, p. 19, pl. xiv, f. 6; pl. xxvi, f. 9?.—Berry, N. Y. Bot. Gard., Bull., vol. 3, 1903, p. 57, pl. xlviii, f. 21, 22; Bull. Torr. Bot. Club, vol. 31, 1904, p. 69, pl. ii, f. 1–7; idem, vol. 32, 1905, p. 44; Geol. Surv. N. J., Ann. Rept. 1905 [1906], p. 139; Bull. Torr. Bot. Club, vol. 33, 1906, p. 165.

Glyptostrobus gracillimus Lesquereux, Am. Jour. Sci., 2d ser., vol. 46, 1868, p. 92; U. S. Geol. Surv. Terr., Rept., vol. 6 (Cret. Fl.), 1874, p. 52, pl. i, f. 8, 11–11f; idem, vol. 8 (Cret. and Tert. Fl.), 1883, p. 32, pl. i, f. 6–6b.

Cone of *Sequoia* (not described), [Lesquereux], U. S. Geol. and Geog. Surv. Terr., Ill. Cret. and Tert. Pl., 1878, pl. xi, f. 9.

Sequoia gracillima—Continued.
Dakota: *Near Sioux City, Iowa; Whetstone Creek, N. Mex.
Cheyenne: Belvidere, Kans.
Raritan ?: Keyport, N. J.
Kootenai: Forks of Pine River, Northwest Territory.
Magothy: Cliffwood and Kinkora, N. J.; Deep Cut, Del.

Sequoia gracilis Heer, Fl. foss. arct., vol. 3, Abt. 2, 1873, p. 80.—Fontaine, U. S. Geol. Surv., Mon. 15, 1889 [1890], p. 247, pl. cxxvi, f. 3, 4.—Newberry, Am. Jour. Sci., 3d ser., vol. 61, 1891, p. 201.—Fontaine, in Ward, U. S. Geol. Surv., Nineteenth Ann. Rept., pt. 2, 1899, p. 675, pl. clxvi, f. 2.—Knowlton, Smithsonian Misc. Coll., vol. 50, 1907, p. 126.—?Hollick, U. S. Geol. Surv., Mon. 50, 1906, p. 43, pl. iii, f. 14.
Patuxent: Near Brooke, Va.
Kootenai: Great Falls and near Cascade, Mont.
Fuson: Pine Creek, Crook County, Wyo.
Magothy: Gay Head, Marthas Vineyard, Mass.

Sequoia gracilis Heer. Fontaine = *Sequoia ambigua.*

Sequoia haydenii (Lesquereux) Cockerell = *Juniperus haydenii.*

Sequoia heerii Lesquereux, U. S. Geol. and Geog. Surv. Terr., Ann. Rept. 1871 [1872], p. 290; U. S. Geol. Surv. Terr., Rept., vol. 7 (Tert. Fl.), 1878, p. 77, pl. vii, f. 11–13.—Newberry, U. S. Geol. Surv., Mon. 35, 1898, p. 20, pl. xlvii, f. 7.—Knowlton, U. S. Geol. Surv., Bull. 204, 1902, p. 23; Harriman Alaska Exped., vol. 4, 1904, p. 152.—Penhallow, Rept. Tert. Pl. Brit. Col., 1908, p. 89, text f. 32.—Knowlton, Washington Acad. Sci., Proc., vol. 11, 1909, pp. 185, 204.
Green River ?: *Sage Creek, Mont.
Clarno (upper part): Bridge Creek, Oreg.
Kenai: Kukak Bay, Alaska.
Oligocene: Quilchena and Tulameen River, British Columbia.
Lance: Big Muddy near Lisner, Milk River, Hell Creek, Mont.; Weston County, Table Mountain, Wyo.

Sequoia heterophylla Velenovsky, Gymnosp. böhm. Kreidef., 1885, p. 22, pl. xiii, f. 2–4; pl. xii, f. 12.—Newberry, U. S. Geol. Surv., Mon. 26, 1895 [1896], p. 49, pl. vi, f. 1–13.—?Knowlton, U. S. Geol. Surv., Bull. 257, 1905, p. 132, pl. xvi, f. 5.—Berry, Geol. Surv. N. J., Ann. Rept. 1905 [1906], p. 139; Bull. Torr. Bot. Club, vol. 33, 1906, p. 165; idem, vol. 34, 1907, p. 189.—Hollick, U. S. Geol. Surv., Mon. 50, 1906, p. 41, pl. iii, f. 23.—Hollick and Jeffrey, N. Y. Bot. Gard., Mem., vol. 3, 1909, p. 61, pl. iii, f. 11–13.—Berry, Geol. Surv. N. J., Bull. 3, 1911, p. 95, pl. vi; Md. Geol. Surv., Upper Cret., 1916, p. 785, pl. liii, f. 2; pl. liv, f. 7; U. S.

Sequoia heterophylla—Continued.
Geol. Surv., Prof. Paper 112, 1919, p. 65; Bull. Torr. Bot. Club, vol. 44, 1917, p. 172.
Bingen: Near Maxwell Spur, Pike County, Ark.
Raritan: Kreischerville, Staten Island, N. Y.; South Amboy, N. J.
Judith River: Willow Creek, Fergus County, Chinook ?, Mont.
Magothy: Cliffwood, N. J.; Deep Cut, Del.; Grove Point and Little Round Bay, Md.
Black Creek: Big Sugar Loaf Landing, Cape Fear River, N. C.
Mesaverde: Rock Springs area, Wyo.
Tuscaloosa: Shirleys Mill and Whites Bluff, Ala.

Sequoia? inferna Ward = *Sequoia reichenbachi.*

Sequoia langsdorfii (Brongniart) Heer, Fl. tert. Helv., vol. 1, 1855, p. 54, pl. xx, f. 2; pl. xxi, f. 4.—Lesquereux, U. S. Geol. Surv. Terr., Rept., vol. 7 (Tert. Fl.), 1878, p. 76.—?Knowlton, U. S. Nat. Mus., Proc., vol. 17, 1894, p. 213; Geol. Soc. Am., Bull., vol. 5, 1893, p. 578; U. S. Geol. Surv., Mon. 32, pt. 2, 1898, p. 657, pl. lxxvii, f. 5; in Spurr, U. S. Geol. Surv., Eighteenth Ann. Rept., pt. 3, 1898, p. 185; idem, Bull. 204, 1902, p. 25.—Penhallow, Roy. Soc. Canada, Trans., 2d ser., vol. 8, 1902, p. 68; idem, vol. 9, 1903, sec. 4, p. 41, f. 2–4; Man. N. A. Gym., 1907, p. 226; Roy. Soc. Canada, Trans., 3d ser., vol. 1, 1907, sec. 4, p. 94; Rept. Tert. Pl. Brit. Col., 1908, p. 89.—Knowlton, Jour. Geol., vol. 19, 1911, p. 370.
Taxites langsdorfii Brongniart, Prod. hist. veg. foss., 1828, p. 108.
Cedroxylon sp. Dawson, Brit. N. A. Bound. Com. (Rept. Geol. and Res. Vic. 49th Parallel), 1875, App. A, p. 331, pl. xvi, f. 1.
Kenai: Mega Island, Herendeen Bay, and 25 miles below Mission Creek, Alaska.
Mascall: Van Horn's ranch, John Day Basin, Oreg.
Clarno (upper part): Bridge Creek, Clarnos Ferry, and near Fossil, Oreg.
Miocene: Florissant, Colo.
Paskapoo (Fort Union): Blackwater River, Red Deer River, Finlay and Omineca rivers, Alberta; Bear River, Mackenzie Valley, Northwest Territory; Porcupine Creek, Saskatchewan.
Fort Union: Montana; Yellowstone National Park.
Yegua: Sommerville, Tex.
Livingston ?: Wolverine Creek, Yellowstone National Park.
Lance: Sec. 20, T. 14 N., R. 19 E., South Dakota.

Sequoia langsdorfii var. Heer, Fl. foss. arct., vol. 2, 1869 (Fl. foss. alask.), p. 23, pl. i, f. 10b.
Kenai ?: Near Sitka, Alaska.

"*Sequoia langsdorfii?* Br." Newberry, N. Y. Lyc. Nat. Hist., Ann., vol. 9, 1868, p. 46.—[Lesquereux], U. S. Geol. and Geog. Surv. Terr., Ill Cret. and Tert. Pl., 1878, pl. xi, f. 4=*Sequoia nordenskioldii.*

Sequoia longifolia Lesquereux, U. S. Geol. and Geog. Surv. Terr., Bull., vol. 1, 1875 [1876], p. 365; idem, Ann. Rept. 1874 [1876], p. 298; U. S. Geol. Surv. Terr., Rept., vol. 7 (Tert. Fl.), 1878, p. 79, pl. lxi, f. 28, 29 [not pl. vii, f. 14, 14a=*S. magnifolia*].—Cockerell, Torreya, vol. 9, 1909, p. 142.—Knowlton, U. S. Geol. Surv., Prof. Paper ——, 19—, p. —, pl. iii, f. 3; pl. iv, f. 2 [in preparation].

Geinitzia longifolia (Lesquereux) Knowlton, U. S. Geol. Surv., Bull. 163, 1900, p. 28 [in part].

Mesaverde: *Point of Rocks, Wyo.
Montana: Coalville, Utah.
Laramie: Marshall, Colo.

Sequoia longifolia Lesquereux [part]=*Sequoia magnifolia.*

Sequoia longifolia Lesquereux, or n. sp. Knowlton, in Calvert, U. S. Geol. Surv., Bull. 471, 1912, p. 16.

Lance: Near Glendive, Mont.

Sequoia magnifica Knowlton, U. S. Geol. Surv., Mon. 32, pt. 2, 1898, p. 761, pls. civ, cv, cx, cxi, cxvii, f. 1–6.—Penhallow, Mon. N. A. Gym., 1907, p. 227.

Miocene: Yellowstone National Park.

Sequoia magnifolia Knowlton, U. S. Geol. Surv., Prof. Paper 98, 1916, p. 88, pl. xv, f. 1–3.

Sequoia longifolia Lesquereux, Rept. U. S. Geol. Surv. Terr., vol. 7 (Tert. Fl.), 1878, p. 79, pl. vii, f. 14, 14a [not pl. lxiii, f. 28, 29].

Geinitzia longifolia (Lesquereux) Knowlton, U. S. Geol. Surv., Bull. 152, 1898, p. 28 [part].

Fox Hills: Near Milliken, Colo.

Sequoia minor Velenovsky, Sitzb. K. böhm. Ges. Wiss., 1886 [1887], p. 638, f. 11, 12.—Berry, Bull. Torr. Bot. Club, vol. 37, 1910, p. 185.

Black Creek: Big Bend of Black River and Parker Landing, Tar River, N. C.

Sequoia nordenskiöldii Heer, Fl. foss. arct., vol. 2, Abt. 2, 1870, p. 36, pl. ii, f. 13b; pl. iv, f. 1a, b, 4–38.—Newberry, U. S. Geol. Surv., Mon. 35, 1898, p. 20, pl. xxvi, f. 4.—Penhallow, Roy. Soc. Canada, Trans., 2d ser., vol. 8, 1902, sec. 4, p. 50; Rept. Tert. Fl. Brit. Col., 1908, p. 90.—Knowlton, Washington Acad. Sci., Proc., vol. 11, 1909, pp. 185, 189, 190, 198, 203, 211, 213, 214; Jour. Geol., vol. 19, 1911, pp. 369, 370, 371.

"*Sequoia langsdorfii?* Br." Newberry, N. Y. Lyc. Nat. Hist., Ann., vol. 9, 1868, p. 46.—[Lesquereux], U. S. Geol. and Geog. Surv. Terr., Ill. Cret. and Tert. Pl., 1878, pl. xi, f. 4.

Sequoia nordenskiöldii—Continued.

Fort Union: *Yellowstone River, Mont. [Common and widely distributed in Fort Union.]

Paskapoo (Fort Union): Porcupine Creek and Great Valley, Saskatchewan; Mackenzie River; Red Deer River, Alberta.

Lance: Glasgow, near Miles City, Glendive, etc., Mont.; Cannonball River, between Buffalo and Klondike, etc., N. Dak.; Converse and Weston counties, Bighorn Basin, etc., Wyo.

Sequoia obovata Knowlton, U. S. Geol. Surv., Prof. Paper 98, 1916, p. 333; idem, Prof. Paper 101, 1918, p. 250, pl. xxx, f. 7.

Sequoia brevifolia Heer. Lesquereux, U. S. Geol. and Geog. Surv. Terr., Bull., vol. 1, 1875 [1876], p. 365; idem, Ann. Rept. 1874 [1876], p. 298; idem, Ann. Rept. 1876 [1878], p. 500; Rept. U. S. Geol. Surv. Terr., vol. 7 (Tert. Fl.), 1878, p. 78, pl. lxi, f. 25–27.—Knowlton, U. S. Geol. Surv., Bull. 163, 1900, p. 27, pl. iv, f. 1–4.

Mesaverde: *Point of Rocks, Rock Springs area, Cody, Wyo.; Grand Mesa, 3½ miles southeast of Newcastle, San Juan River, Colo.

Vermejo: La Veta, Walsenburg, Forbes, Morley, Cuatro, Rockvale, Colo.; Vermejo Park, Dawson, Ponil Canyon, N. Mex.

Dawson: Colorado Springs, Colo.

Fruitland: 30 miles south of Farmington, San Juan County, N. Mex.

Sequoia pagiophylloides Fontaine, U. S. Geol. Surv., Proc., vol. 16, 1893, p. 276, pl. xlii, f. 1–3a.

Trinity: Glen Rose, Tex.

Sequoia penhallowi Jeffrey, Bot. Gaz., vol. 38, 1904, p. 328, pl. xviii, pl. xix.—Penhallow, Man. N. A. Gym., 1907, p. 228.

Miocene: Blue Gap, Sierra Nevada, Calif.

Sequoia reichenbachi (Geinitz) Heer, Fl. foss. arct., vol. 1, 1868, p. 83, pl. xliii, f. 1d, 2b, 5a.—Lesquereux, U. S. Geol. Surv. Terr., Rept., vol. 6 (Cret. Fl.), 1874, p. 51, pl. i, f. 10–10b; U. S. Geol. Surv., Mon. 17, 1891 [1892], p. 35, pl. ii, f. 4.—Hollick, N. Y. Acad. Sci., Trans., vol. 12, 1892, p. 30, pl. 1, f. 18.—Fontaine, U. S. Geol. Surv., Mon. 15, 1889 [1890], p. 243, pl. cxviii, f. 1, 4; pl. cxix, f. 1–5; pl. cxx, f. 7, 8; pl. cxxii, f. 2; pl. clxvii, f. 5.—Dawson, Roy. Soc. Canada, Trans., vol. 1, 1883, p. 21.—Newberry, U. S. Geol. Surv., Mon. 26, 1895 [1896], p. 49, pl. ix, f. 19.—Knowlton, idem, Mon. 32, pt. 2, 1898, p. 657.—Fontaine, in Ward, U. S. Geol. Surv., Nineteenth Ann. Rept., pt. 2, 1899, p. 674, pl. clxv, f. 1, 2; pl. clxvi, f. 1.—Berry, N. Y. Bot. Gard., Bull., vol. 3, 1903, p. 59, pl. xlviii, f. 15–17, 20; Bull. Torr. Bot. Club, vol. 31, 1904, p. 69, pl. iv, f. 8.—Knowlton, U. S. Geol. Surv., Bull. 257, 1905, p. 131, pl. xiv, f. 3–5.—Berry, Bull. Torr. Bot. Club, vol. 32, 1905, p. 44, pl. i, f. 3; Geol. Surv. N. J., Ann. Rept. 1905 [1906], p. 139; Bull. Torr.

Sequoia reichenbachi—Continued.

Bot. Club, vol. 33, 1906, p. 165.—Knowlton, Smithsonian Misc. Coll., vol. 50, 1907, p. 126, pl. xii, f. 7, 8.—Fontaine, in Ward, U. S. Geol. Surv., Mon. 48, 1905 [1906], p. 177, pl. xlv, f. 7, 8; p. 263, pl. lxix, f. 4, 5.—Hollick, idem, Mon. 50, 1906, p. 42, pl. ii, f. 40; pl. iii, f. 4, 5.—Berry, Geol. Surv. N. J., Rept. 1905 [1906], p. 139; Bull. Torr. Bot. Club, vol. 33, 1896, p. 165; idem, vol. 37, 1910, p. 20; idem, p. 185; idem, p. 504; Geol. Surv. N. J., Bull. 3, 1911, p. 93; U. S. Nat. Mus., Proc., vol. 40, 1911, p. 308; Md. Geol. Surv., Lower Cret., 1912, p. 444, pl. lxxvii, f. 7; U. S. Geol. Surv., Prof. Paper 84, 1914, pp. 23, 107, pl. iv, f. 1–4; Md. Geol. Surv., Upper Cret., 1916, p. 788.—Knowlton, U. S. Geol. Surv., Prof. Paper 98, 1916, pp. 88, 333; idem, Prof. Paper 101, 1918, p. 250.—Berry, U. S. Geol. Surv., Prof. Paper 112, 1919, p. 64, pl. vi, f. 2.

Araucarites reichenbachi Geinitz, Charak. Schichten u. Petrefact. säch.-böhm. Kreidegebirges, pt. 3, 1842, p. 98, pl. xxiv, f. 4.

Sequoia reichenbachi longifolia Fontaine, U. S. Geol. Surv., Mon. 15, 1889 [1890], p. 244, pl. cxvii, f. 8.

Sequoia densifolia Fontaine, op. cit. (Mon. 15), p. 246, pl. cxxi, f. 4.

Sequoia? sp. Fontaine, op. cit. (Mon. 15), p. 248, pl. cxvi, f. 7, pl. cxxxii, f. 2, 5, 6.

Sequoia sp. Fontaine, op. cit. (Mon. 15), p. 248, pl. cxxxii, f. 10.

Sequoia? inferna Ward, in Fontaine, in Ward, U. S. Geol. Surv., Mon. 48, 1905 [1906], p. 507.

Sequoia couttsiae Heer. Hollick, N. Y. Acad. Sci., Trans., vol. 12, 1892, p. 30, pl. i, f. 5.

Jurassic: Bridger Creek near Bozeman, Mont.

Horsetown: Shasta County, Calif.

Kootenai: Skull Butte, Fergus County, Mont.

Fuson: Pine and Oak creeks, Crook County, Wyo.

Raritan: Woodbridge, N. J.; Kreischerville, Staten Island, N. Y.

Magothy: Cliffwood, N. J.; Round Bay and Little Round Bay, Severn River, Md.; Deep Cut, Del.

Patuxent: Fredericksburg, Lorton, and Dutch Gap, Va.; Springfield, Md.

Arundel: Reynolds ore pit, Maryland.

Patapsco: Near Brooke, Va.

Eutaw: McBrides Ford and Chimney Bluff, Ga.; Havana, Hale County, Ala.

Black Creek: Rockfish Creek near Hope Mills, Parker Landing on Tar River, 92d milepost, Neuse River, N. C.

Black Creek (Middendorf): Middendorf, S. C.

Dakota: Near Fort Harker, Kans.

Belly River: Belly River, Canada.

Judith River: Willow Creek, Fergus County, 6½ miles east of Billings, Bridger, Scribner, Milk River, Chinook, Crazy Moun-

Sequoia reichenbachi—Continued.

tain field, Bull Mountain field, Livingston field, Summit, 3 miles east of Dorsey, Mont.

Mesaverde: Point of Rocks, Wyo.; near Newcastle, near Lay post office, Grand Mesa region, Crested Butte, Rifle Gap, near Meeker, Colo.; Rock Springs region, Big Horn County, Wind River Basin, Table Mountain, southwest of Rawlins, Cody, Wyo.

Laramie: Coal Creek, Boulder County, Crow Creek, and Morrison, Colo.

Fruitland: 30 miles south of Farmington, San Juan County, N. Mex.

Fox Hills: Near Milliken, Colo.

Vermejo: Raton Mesa, Colo., and N. Mex.

Tuscaloosa: Snow place and Cottondale, Tuscaloosa County; Whites Bluff, Green County, Ala.; Iuka, Miss.

Sequoia reichenbachi longifolia Fontaine = *Sequoia reichenbachi.*

Sequoia rigida Heer, Fl. foss. arct., vol. 3, Abt. 2, 1873, pp. 80, 91, 102, 128, pl. xxii, f. 5g, 11a; pl. xxv, f. 6; pl. xxvii, f. 8–14; pl. xxxviii, f. 9a, 10.—Fontaine, U. S. Geol. Surv., Mon. 15, 1889 [1890], p. 246, pl. cxviii, f. 3; pl. cxxi, f. 2; pl. cxxvi, f. 2; pl. cxxx, f. 3; U. S. Nat. Mus., Proc., vol. 15, 1892, p. 494.—Knowlton, in Stanton and Martin, Geol. Soc. Am., Bull., vol. 16, 1905, p. 408.—Fontaine, in Ward, U. S. Geol. Surv., Mon. 48, 1905 [1906], p. 219.—Berry, U. S. Nat. Mus., Proc., vol. 40, 1911, p. 309; Md. Geol. Surv., Lower Cret., 1911, p. 447, pl. lxxviii, f. 8.—?Knowlton, in Atwood, U. S. Geol. Surv., Bull. 467, 1911, p. 44.

Sequoia subulata Heer. Fontaine, U. S. Geol. Surv., Mon. 15, 1889 [1890], p. 245, pl. cxvii, f. 7; pl. cxviii, f. 5, 6; in Ward, U. S. Geol. Surv., Mon. 48, 1905 [1906], pp. 486, 571.

Patuxent: Near Potomac Run, telegraph station near Lorton, Va.; Springfield, Md.

Kootenai: Great Falls, Mont.

Chignik (Upper Cretaceous): Chignik Bay, Alaska.

Sequoia sempervirens (Lamb) Endlicher. Hannibal, Bull. Torr. Bot. Club, vol. 38, 1911, p. 336, pl. xv, f. 3.

Pliocene: Near Portola, Santa Cruz Mountains, Calif.

Sequoia smittiana Heer, Zeitschr. Deutsch. geol. Gesell., 1872, p. 161.—Dawson, Roy. Soc. Canada, Trans., vol. 3, 1885 [1886], sec. 4, p. 9, pl. ii, f. 7, 7a.

Raritan ?: Sayreville, N. J. ?

Kootenai: Coal Creek, Crows Nest Pass, and Middle Fork of North Branch of Old Man River, British Columbia; Great Falls, Mont.

Sequoia spinosa Newberry, U. S. Nat. Mus., Proc., vol. 5, 1882 [1883], p. 504.—Knowlton, idem, vol. 17, 1894, p. 214; Geol. Soc. Am., Bull., vol. 5, 1893, p. 578.—Newberry, U. S. Geol. Surv., Mon. 35, 1898, p. 21, pl. liii, f. 4, 5.
Kenai?: Cook Inlet, Alaska.

Sequoia subulata Heer. Fontaine = *Sequoia rigida*.

Sequoia winchellii Lesquereux, in Winchell, Geol. and Nat. Hist. Surv. Minn., Thirteenth Ann. Rept., 1884 [1885], p. 76 [name].—Lesquereux, Geol. and Nat. Hist. Surv. Minn., Final Rept., vol. 3, pt. 1, 1893, p. 10, pl. A, f. 1.
Dakota: Austin, Minn.

Sequoia sp. Dawson, Roy. Soc. Canada, Trans., vol. 5, 1887, p. 32 (wood).
Belly River: Ribstone Creek and Saskatchewan River, British Columbia.

Sequoia sp. Dawson, Roy. Soc. Canada, Trans., vol. 5, 1887, p. 32 (wood).
Pierre: Bow River, Canada.
Belly River: Ribstone Creek and west of Medicine Hat, British Columbia.

Sequoia sp. Dawson, Roy. Soc. Canada, Trans., vol. 5, 1887, p. 34 (wood).
Paskapoo (Fort Union): Wood End Depot, Mackenzie River, Alberta.

Sequoia sp. Dawson, Roy. Soc. Canada, Trans., vol. 5, 1887, sec. 4, p. 34 (wood).
Paskapoo (Fort Union): Turtle Mountain, Souris Valley, 245-Mile Valley, Wood End Depot, Mackenzie River, Edmonton, British Columbia.

Sequoia sp. Dawson, Roy. Soc. Canada, Trans., vol. 8, 1890 [1891], p. 80, text f. 8.
——: Stump Lake, British Columbia.

Sequoia sp. Fontaine, U. S. Geol. Surv., Mon. 15, 1889 [1890], p. 247, pl. cxx, f. 9.
Patuxent: Fredericksburg, Va.

Sequoia sp.? Fontaine, U. S. Geol. Surv., Mon. 15, 1889 [1890], p. 248, pl. cxvi, f. 7; pl. cxxxii, f. 2, 5, 6 = *Sequoia reichenbachi*.

Sequoia sp. Fontaine, U. S. Geol. Surv., Mon. 15, 1889 [1890], p. 248, pl. cxxxii, f. 10.
Patuxent: Fredericksburg, Va.

Sequoia sp. (cone) Fontaine, in Ward, U. S. Geol. Surv., Nineteenth Ann. Rept., pt. 2, 1899, p. 676, pl. clxvi, f. 3, 4.
Lakota: South Fork of Hay Creek, Crook County, Wyo.

Sequoia sp. (cone) Hollick, U. S. Geol. Surv., Mon. 50, 1906, p. 44, pl. ii, f. 42.
Magothy: Gay Head, Marthas Vineyard, Mass.

Sequoia sp. Hollick, N. Y. Bot. Gard., Bull., vol. 3, 1904, p. 410, pl. 72, f. 2; U. S. Geol. Surv., Mon. 50, 1906, p. 43, pl. iii, f. 6.
Raritan: Glen Cove, Long Island, N. Y.

Sequoia (cones) Knowlton, U. S. Geol. Surv., Mon. 32, pt. 2, 1899, p. 683, pl. lxxxi, f. 8; pl. lxxxii, f. 6, 7.
Miocene: Yellowstone National Park.

Sequoia sp. Knowlton, in Hill, Am. Jour. Sci., 3d ser., vol. 50, 1895, p. 212.
Cheyenne: Belvidere, Kans.

Sequoia sp.? Knowlton, Geol. Soc. Am., Bull., vol. 8, 1897, p. 134.
Fort Union: Lightning Creek, Converse County, Wyo.

Sequoia sp. Knowlton, U. S. Geol. Surv., Eighteenth Ann. Rept., pt. 3, 1898, p. 723.
Payette: Marsh, Idaho.

Sequoia sp. (cone) Knowlton, U. S. Geol. Surv., Bull. 204, 1902, p. 26, pl. i, f. 2.
Mascall: Van Horn's ranch, John Day Basin, Oreg.

Sequoia sp.? (cone) Knowlton, U. S. Geol. Surv., Bull. 257, p. 131, pl. xiv, f. 2.
Belly River: Near Wild Horse Lake, Assiniboia.

Sequoia sp. Knowlton, U. S. Geol. Surv., Bull. 163, p. 27.
Glyptostrobus europaeus? Heer. Knowlton, in Stanton, U. S. Geol. Surv., Bull. 106, 1893, p. 42.—Knowlton, Geol. Soc. Am., Bull., vol. 8, 1897, p. 150.
Montana: Coalville, Utah.

Sequoia sp. Knowlton, in Turner, Am. Geol., vol. 15, 1895, p. 373.
Ione: Near Ione, Calif.

Sequoia sp. (*S. sempervirens* type) Knowlton, in Turner, Jour. Geol., vol. 3, 1895, p. 408.
Pliocene?: Canyon of Fall River, south of Cammel Peak, Calif.

Sequoia sp. (cone) Knowlton, Harriman Alaska Exped., vol. 4, 1904, p. 152, pl. xxii, f. 1.
Kenai: Kukak Bay, Alaska.

Sequoia sp. Lesquereux, U. S. Geol. and Geog. Surv. Terr., Bull., vol. 1, 1875 [1876], p. 384.
Post-Laramie, Black Buttes, Wyo.

Sequoia? sp. Lesquereux, U. S. Geol. and Geog. Surv. Terr., Ann. Rept. 1874 [1876], p. 310.
Denver: Middle Park, Colo.

Sequoia sp. Lesquereux, Rept. on Clays in N. J., 1878, p. 29.
Magothy?: Burts Creek, N. J.

Sequoia? sp. Knowlton.
Nageiopsis subfalcata Fontaine, U. S. Geol. Surv., Mon. 15, 1889 [1890], p. 203, pl. clxviii, f. 4.
Patuxent: Near Dutch Gap Canal, Va.

SERENOA Hooker f., Hooker's Gen. Pl., vol. 3, 1883, p. 1228.

Serenoa serrulata (Michaux) Hooker. Berry, Jour. Geol., vol. 25, 1917, p. 662.
Pleistocene: Vero, Fla.

Serenopsis Hollick = *Nelumbo.*

Serenopsis kempii Hollick = *Nelumbo kempii.*

SICYOS Linné, Sp. pl., 1753, p. 1013.

Sicyos? florissantia Cockerell, Am. Mus. Nat. Hist., Bull., vol. 24, 1908, p. 107, pl. x, f. 43.
Miocene: Florissant, Colo.

SIDEROXYLON Linné, Sp. pl., vol. 1, 1753, p. 192.

Sideroxylon ellipticus Berry, U. S. Geol. Surv., Prof. Paper 91, 1916, p. 334, pl. c, f. 8.
Lagrange: Puryear, Tenn.

Sideroxylon premastichodendron Berry, U. S. Geol. Surv., Prof. Paper 91, 1916, p. 335, pl. xcix, f. 4.
Lagrange: Puryear, Tenn.

SIMARUBA Aublet, Guian., vol. 2, 1775, p. 859.

Simaruba eocenica Berry, U. S. Geol. Surv., Prof. Paper 91, 1916, p. 252, pl. liv, f. 7.
Lagrange: Puryear, Henry County, Tenn.

SIMARUBINIUM Platen, Sitzungsb. Naturf. Ges., Leipzig, vol. 34, 1907 [1908], p. 54. [Type, *S. crystallophorum.*]

Simarubinium crystallophorum Platen, Sitzungsb. Naturf. Ges., Leipzig, vol. 34, 1907 [1908], p. 54.
Miocene (auriferous gravels): Nevada County, Calif.

Simarubinium engelhardti Platen, Sitzungsb. Naturf. Ges., Leipzig, vol. 34, 1907 [1908], p. 56.
Miocene (auriferous gravels): Nevada County, Calif.

Smilacites grandifolia Unger = *Smilax grandifolia.*

SMILAX (Tournefort) Linné, Sp. pl., 1753, p. 1028.

Smilax carbonensis Cockerell, Torreya, vol. 14, 1914, p. 135.
Smilax grandifolia Unger. Lesquereux, Rept. U. S. Geol. Surv. Terr., vol. 7 (Tert. Fl.), 1878, p. 94, pl. ix, f. 5.
Hanna: Carbon, Wyo.

Smilax? coloradensis Knowlton, U. S. Geol. Surv., Prof. Paper 108, 1917, p. 85, pl. xxxiii, f. 1.
Frontier: Dump of mine No. 1, Cumberland, Wyo.

Smilax cyclophylla Newberry, Boston Jour. Nat. Hist., vol. 7, 1863, p. 520; U. S. Geol. Surv., Mon. 35, 1898, p. 32, pl. liv, f. 3 [part].
——: Birch Bay, Wash.

Smilax franklini Heer, Fl. foss. arct., vol. 1, 1868, p. 136, pl. xxi, f. 18.
Miocene?: Mackenzie River, Northwest Territory.

Smilax grandifolia Unger. Lesquereux = *Smilax carbonensis.*

Smilax grandifolia-cretacea Lesquereux, U. S. Geol. Surv., Mon. 17, 1891 [1892], p. 40, pl. xlvi, f. 3.
Dakota: Kansas.

Smilax? inquirenda Knowlton, U. S. Geol. Surv., Prof. Paper 130, 19—, p. —, pl. iv, f. 5 [in preparation].
Laramie: Popes Bluff, Pikeview, Colo.

Smilax kansana Cockerell, Torreya, vol. 24, 1914, p. 137.
Smilax undulata Lesquereux, U. S. Geol. Surv., Mon. 17, 1892, p. 39, pl. xlvi, f. 2. [Homonym, *S. undulata* Pohl, A. De Candolle, in De Candolle, Mon. phan., vol. 1, 1878, p. 135.]
Dakota: Near Fort Harker, Kans.

Smilax labidurommae Cockerell, Torreya, vol. 14, 1914, p. 135, text f. 1.
Miocene: Florissant, Colo.

Smilax lamarensis Knowlton, U. S. Geol. Surv., Mon. 32, pt. 2, 1899, p. 685, pl. cxxi, f. 3, 4.
Miocene: Yellowstone National Park.

Smilax obtusangula? Heer, Fl. tert. Helv., vol. 3, 1859, p. 166, pl. cxlvii, f. 25.—Lesquereux, U. S. Geol. and Geog. Surv. Terr., 1872 [1873], p. 391. [Not afterward mentioned by Lesquereux.]
——: Black Buttes, Wyo.

Smilax raritanensis Berry, Bull. Torr. Bot. Club, vol. 36, 1909, p. 248; Geol. Surv. N. J., Bull. 3, 1911, p. 102, pl. xxiii, f. 3.
Paliurus ovalis Newberry, U. S. Geol. Surv., Mon. 26, 1896, p. 107, pl. xxiii, f. 8, 9. [Homonym, Dawson, 1886.]
Magothy?: Locality unknown.

Smilax undulata Lesquereux = *Smilax kansana.*

Smilax wardii Lesquereux, U. S. Nat. Mus., Proc., vol. 11, 1888, p. 19, pl. xiii, f. 1.—Knowlton, U. S. Geol. Surv., Bull. 204, 1902, p. 28.
Mascall: Van Horn's ranch, John Day Basin, Oreg.

SOLANITES Saporta, Ann. sci. nat., 4th ser., Bot., vol. 17 (Études, vol. 1, pt. 2), 1862, p. 262 (109).

Solanites saportana Berry, U. S. Geol. Surv., Prof. Paper 91, 1916, p. 348, pl. cvi, f. 4, 5.
Wilcox (Holly Springs): Holly Springs, Miss.

SOPHORA Linné, Sp. pl., vol. 1, 1753, p. 373.

Sophora claiborneana Berry, in Cooke and Shearer, U. S. Geol. Surv., Prof. Paper 120, 1918, p. 55 [name].
Barnwell (Twiggs): Twiggs County, Ga.

Sophora henryensis Berry, U. S. Geol. Surv., Prof. Paper 91, 1916, p. 243, pl. lii, f. 2.
Lagrange: Puryear, Henry County, Tenn.

Sophora lesquereuxi Berry, U. S. Geol. Surv., Prof. Paper 91, 1916, p. 244.
Quercus myrtifolia Willdenow. Lesquereux, Am. Jour. Sci., 2d ser., vol. 27, 1859, p. 363; in Safford, Geol. Tenn., 1869, p. 427, pl. K, f. 3.
Lagrange: Somerville, Fayette County, Tenn.

Sophora mucronata Berry, U. S. Geol. Surv., Prof. Paper 91, 1916, p. 244, pl. lii, f. 4.
Lagrange: Puryear, Henry County, Tenn.

Sophora nervosa Knowlton, U. S. Geol. Surv., Prof. Paper 101, 1918, p. 326, pl. xcv, f. 3.
Raton: Near Strong, Colo.

Sophora palaelobifolia Berry, U. S. Geol. Surv., Prof. Paper 91, 1916, p. 243, pl. lii, f. 1.
Lagrange: Near Grand Junction, Fayette County, Tenn.

Sophora puryearensis Berry, U. S. Geol. Surv., Prof. Paper 91, 1916, p. 242, pl. lii, f. 3; pl. cix, f. 3.
Lagrange: Puryear, Henry County, Tenn.

Sophora repandifolia Berry, U. S. Geol. Surv., Prof. Paper 91, 1916, p. 244, pl. xlviii, f. 6, 7.
Lagrange: Puryear, Henry County, Tenn.

Sophora wilcoxiana Berry, U. S. Geol. Surv., Prof. Paper 91, 1916, p. 241, pl. xlvii, f. 1–13.
Wilcox (Holly Springs): Holly Springs, Marshall County, Miss.
Wilcox (Grenada): Grenada, Grenada County, Miss.
Lagrange: Puryear, Henry County, and Grand Junction, Fayette County, Tenn.
Wilcox: Malvern, Ark.; Bolivar Creek, Harrisburg, Ark.
Barnwell (Twiggs): Georgia.

SORBUS (Tournefort) Linné, Sp. pl., 1753, p. 447.

Sorbus diversifolia (Lesquereux) Cockerell, Am. Jour. Sci., 4th ser., vol. 29, 1910, p. 76, f. 1.—Knowlton, U. S. Nat. Mus., Proc., vol. 51, 1916, p. 273.
Myrica diversifolia Lesquereux, Rept. U. S. Geol. Surv. Terr., vol. 8 (Cret. and Tert. Fl.), 1883, p. 148, pl. xxv, f. 6, 10–12.
Crataegus acerifolia Lesquereux, Rept. U. S. Geol. Surv. Terr., vol. 8 (Cret. and Tert. Fl.), 1883, p. 198, pl. xxxvi, f. 10.
Crataegus lesquereuxi Cockerell, Bull. Torr. Bot. Club, vol. 33, 1906, p. 311; Univ. Colo. Studies, vol. 3, 1906, p. 171.
Onoclea reducta Cockerell, Am. Mus. Nat. Hist., Bull., vol. 24, 1908, p. 108, pl. vi, f. 4.
Miocene: *Florissant, Colo.

Sorbus megaphylla Cockerell, Am. Mus. Nat. Hist., Bull., vol. 24, 1908, p. 95, pl. ix, f. 29.
Miocene: *Florissant, Colo.

Sorbus nupta Cockerell, Am. Jour. Sci., 4th ser., vol. 29, 1910, p. 78, text f. 2.—Knowlton, U. S. Nat. Mus., Proc., vol. 51, 1916, p. 274.
Myrica diversifolia Lesquereux, Rept. U. S. Geol. Surv. Terr., vol. 8 (Cret. and Tert. Fl.), 1883, p. 148, pl. xxv, f. 7, 8, 14.
Miocene: *Florissant, Colo.

SPARGANIUM Linné, Sp. pl., 1753, p. 971.

Sparganium stygium Heer, Fl. tert. Helv., vol. 1, 1855, p. 101, pl. xlv, f. 1–4.—Ward, U. S. Geol. Surv., Sixth Ann. Rept., 1884–85 [1886], p. 550, pl. xxxii, f. 6, 7; idem, Bull. 37, 1887, p. 18, pl. iii, f. 6, 7.—Knowlton, U. S. Geol. Surv., Mon. 32, pt. 2, 1899, p. 683; Washington Acad. Sci., Proc., vol. 11, 1909, p. 211.
Fort Union: Sevenmile Creek, Mont.; Elk Creek, Yellowstone National Park.
Lance: Dayton, Wyo.

Sparganium sp. Berry, Torreya, vol. 14, 1914, p. 160.
Pleistocene: Seven Springs, Neuse River, Wayne County, N. C.

Sparganium? sp. Knowlton, U. S. Geol. Surv., Prof. Paper 101, 1918, p. 253, pl. xxxii, f. 6.
Vermejo: Near La Veta, Colo.

Sparganium? sp. Lesquereux, U. S. Geol. and Geog. Surv. Terr., Ann. Rept. 1871 [1872], p. 285. [Not afterward mentioned by Lesquereux.]
Livingston: 3 miles above Spring Canyon, Mont.

SPATHITES Knowlton, Geol. Soc. Am., Bull., vol. 8, 1897, p. 140.

Spathites sp. Knowlton, Geol. Soc. Am., Bull., vol. 8, 1897, p. 140.
Mesaverde: Near Harpers, Wyo.

SPATHYEMA Rafinesque, in Desvaux, Jour. Bot., vol. 2, 1809, p. 171.

Spathyema? nevadensis Knowlton, U. S. Geol. Surv., Twenty-first Ann. Rept., pt. 2, 1901, p. 210, pl. xxx, f. 17, 18.
Esmeralda: 3.8 kilometers northeast of Emigrant Peak, Esmeralda County, Nev.

SPEGAZZINITES Felix, Zeitschr. Deutsch. geol. Gesell., 1894, p. 279.

Spegazzinites cruciformis Felix, Zeitschr. Deutsch. geol. Gesell., 1894, p. 279, pl. xix, f. 18; op. cit., 1896, p. 255.
Miocene: Yanceys, Yellowstone National Park.

Sphaeria problematica Knowlton = *Sphaerites problematicus.*

SPHAERITES Unger, Gen. et sp. pl. foss., 1850, p. 37.

Sphaerites alabamensis Berry, U. S. Geol. Surv., Prof. Paper 112, 1919, p. 43, pl. v, f. 1.
Tuscaloosa: Shirleys Mill, Glen Allen, and Tuscaloosa, Ala.

Sphaerites claibornensis Berry, U. S. Geol. Surv., Prof. Paper 84, 1914, p. 132, pl. xxiv, f. 6.
Barnwell (Twiggs): Grovetown, Ga.

Sphaerites lapideus (Lesquereux) Meschinelli = *Rostellinites lapideus.*

Sphaerites lesquereuxi Meschinelli, Syllog. Fung. Foss., 1892, p. 31.
Spheria rhytismoides Lesquereux, U. S. Geol. and Geog. Surv. Terr., Bull., vol. 1, 1875 [1876], p. 382; idem, Ann. Rept. 1874 [1876], p. 308; U. S. Geol. Surv. Terr., Rept., vol. 7 (Tert. Fl.), 1878, p. 35, pl. i, f. 5, 5a. [Homonym, Ettingshausen, 1869.]
Post-Laramie: *Black Buttes, Wyo.

Sphaerites myricae (Lesquereux) Meschinelli, Syllog. Fung. Foss., 1892, p. 23.
Spheria myricae Lesquereux, U. S. Geol. and Geog. Surv. Terr., Ann. Rept. 1872 [1873], p. 390; U. S. Geol. Surv. Terr., Rept., vol. 7 (Tert. Fl.), 1878, p. 34, pl. i, f. 4.
Green River: Green River station, Wyo.
Post-Laramie: *Black Buttes, Wyo.

Sphaerites problematicus (Knowlton) Meschinelli.
Sphaeria problematica Knowlton, in Lesquereux, U. S. Geol. Surv., Mon. 17, 1891 [1892], p. 23, pl. xxxi, f. 2, 2a.
Dakota: *Ellsworth County, Kans.

Sphaerites raritanensis Berry, Jour. Geol., vol. 18, 1910, p. 254 [name]; Geol. Surv. N. J., Bull. 3, 1911, p. 62; Md. Geol. Surv., Upper Cret. 1916, p. 757, pl. lxxxi, f. 3.
Raritan: *Hylton Pits on Pensauken Creek, N. J.
Magothy: Sullivans Cove, Anne Arundel County, Md.

Sphaerococcites münsterianus Presl = *Baiera münsteriana.*

SPHENASPIS Hollick and Jeffrey, N. Y. Bot. Gard., Mem., vol. 3, 1909, p. 51. [Type, *S. statenensis.*]

Sphenaspis statenensis Hollick and Jeffrey, N. Y. Bot. Gard., Mem., vol. 3, 1909, p. 51, pl. x, f. 22, 23; pl. xxvi, f. 2–4.
Raritan: *Kreischerville, Staten Island, N. Y.

Sphenoglossum quadrifolium Emmons = *Actinopteris quadrifolia.*

SPHENOLEPIDIUM Heer, Fl. foss. du Portugal, 1881, p. 19.

Sphenolepidium dentifolium Fontaine = *Sequoia ambigua.*

Sphenolepidium kurrianum (Dunker) Heer = *Sphenolepis kurriana.*

Sphenolepidium oregonense Fontaine, in Ward, U. S. Geol. Surv., Mon. 48, 1905 [1906], p. 133, pl. xxxvi, f. 3–8.
Jurassic: Douglas County, Oreg.

Sphenolepidium pachyphyllum Fontaine, U. S. Geol. Surv., Mon. 15, 1889 [1890], p. 259, pl. cxxxi, f. 6, 7.
Patuxent: Near telegraph station, near Lorton, Va.
Kootenai: Anthracite, British Columbia.

Sphenolepidium parceramosum Fontaine = *Sphenolepis kurriana.*

Sphenolepidium recurvifolium Fontaine = *Sequoia ambigua*.

Sphenolepidium sternbergianum (Dunker) Heer, Fl. foss. du Portugal, 1881, p. 19, pl. xiii, f. 1a, 2–8; pl. xiv.—Fontaine, U. S. Geol. Surv., Mon. 15, 1889 [1890], p. 261, pl. cxxi, f. 8, 10, 11; pl. cxxx, f. 9; in Ward, U. S. Geol. Surv., Mon. 48, 1905 [1906], pp. 264, 507, 515, 524; pl. lxix, f. 7; pl. cxii, f. 1; pl. cix, f. 8, 9; pl. cxii, f. 10, 11.

Patuxent: Potomac Run, Fredericksburg, Hell Hole, and Chinkapin Hollow, Va.

Horsetown: Ono, Shasta County, Calif.

Trinity: Glen Rose, Tex.

Sphenolepidium sternbergianum densifolium Fontaine = *Athrotaxopsis expansa*.

Sphenolepidium virginicum Fontaine = *Sphenolepis kurriana*.

Sphenolepidium sp. Dawson, Roy. Soc. Canada, Trans., vol. 10, 1892, p. 90, text f. 13.

Kootenai: Anthracite, British Columbia.

SPHENOLEPIS Schenk, Palaeontographica, vol. 19, 1871, p. 243.

Sphenolepis kurriana (Dunker) Schenk, Palaeontogr., vol. 19, 1871, p. 243, pl. xxxvii, f. 5–8; pl. xxxviii, f. 1.—Berry, U. S. Nat. Mus., Proc., vol. 40, 1911, p. 291.

Thuites kurrianus Dunker, Monogr. Norddeutsch. Weald-bild., 1846, p. 20, pl. vii, f. 8.

Sphenolepidium kurrianum (Dunker) Heer, Fl. foss. du Portugal, 1881, p. 19, pl. xii, f. 1b; pl. xiii, f. 1b, 8b; pl. xviii, f. 1–8.—Fontaine, U. S. Geol. Surv., Mon. 15, 1889 [1890], p. 260, pl. cxxvi, f. 1, 5, 6; pl. cxxviii, f. 1, 7; pl. cxxix, f. 1, 4, 6, 8; pl. cxxx, f. 11; pl. cxxxi, f. 4; pl. clxvii, f. 2; in Ward, U. S. Geol. Surv., Nineteenth Ann. Rept., pt. 2, 1890, p. 681, pl. clxvi, f. 12, 13; U. S. Geol. Surv., Mon. 48, 1905 [1906], pp. 484, 489, 515, 538, 543.

Sphenolepidium parceramosum Fontaine, U. S. Geol. Surv., Mon. 15, 1889 [1890], p. 257, pl. cxxix, f. 7; pl. cxxx, f. 8; pl. cxxxi, f. 2; in Ward, U. S. Geol. Surv., Nineteenth Ann. Rept., pt. 2, 1899, p. 682, pl. clxiii, f. 11b; pl. clxvii, f. 1–3; U. S. Geol. Surv., Mon. 48, 1905 [1906], pp. 484, 538, 548.

Sphenolepidium virginicum Fontaine, U. S. Geol. Surv., Mon. 15, 1889 [1890], p. 259, pl. cxxv, f. 4; pl. clxvi, f. 6; U. S. Nat. Mus., Proc., vol. 15, 1892, p. 494; in Ward, U. S. Geol. Surv., Mon. 48, 1905 [1906], pp. 481, 484, 517.

Arthrotaxopsis grandis Fontaine, U. S. Geol. Surv., Mon. 15, 1889 [1890], p. 240 [part].

Arthrotaxopsis expansa Fontaine, U. S. Geol. Surv., Mon. 15, 1889 [1890], p. 241, pl. clxxxv, f. 15, 18, 22.

Taxodium expansum Fontaine, U. S. Geol. Surv., Mon. 15, 1889 [1890], p. 252, pl. cxxiii, f. 1.

Glyptostrobus expansus (Fontaine) Ward. Fontaine, in Ward, U. S. Geol. Surv., Mon. 48, 1905 [1906], p. 543.

Taxodium denticulatum Fontaine, U. S. Geol. Surv., Mon. 15, 1889 [1890], p. 253, pl. cxxiv, f. 1.

Sphenolepis kurriana—Continued.

Taxodium fastigiatum Fontaine, U. S. Geol. Surv., Mon. 15, 1889 [1890], p. 253, pl. cxxv, f. 1, 3.

Glyptostrobus fastigiatus (Fontaine) Ward, U. S. Geol. Surv., Fifteenth Ann. Rept., 1895, p. 380.

Glyptostrobus brookensis Fontaine, in Ward, U. S. Geol. Surv., Nineteenth Ann. Rept., pt. 2, 1899, p. 682, pl. clxv, f. 4; pl. clxviii, f. 4.

Patuxent: New Reservoir and Ivy City, D. C.; Fredericksburg, Potomac Run, Trents Reach, Dutch Gap, telegraph station near Lorton, Cockpit Point, Kankeys, Va.

Arundel: Bay View and Arlington, Md.

Patapsco: Federal Hill (Baltimore), Vinegar Hill, Md.; Brooke, 72d milepost, Mount Vernon, and Hell Hole, Va.

Fuson: Pine Creek and Oak Creek, Crook County, Wyo.

Kootenai: Great Falls, Mont.

SPHENOPTERIS (Brongniart) Sternberg, Fl. Vorw., vol. 1 (Tentamen), 1825, p. 15.

Sphenopteris acrodenta Fontaine = *Ruffordia acrodenta.*

Sphenopteris blomstrandi Heer = *Dennstaedtia americana.*

Sphenopteris corrugata Newberry, N. Y. Lyc. Nat. Hist., Ann., vol. 9, 1868, p. 10.—[Lesquereux], U. S. Geol. and Geog. Surv. Terr., Ill. Cret. and Tert. Pl., 1878, pl. ii, f. 6.—Newberry, U. S. Geol. Surv., Mon. 35, 1898, p. 14, pl. i, f. 6.

Hymenophyllum cretaceum Lesquereux, Rept. U. S. Geol. Surv. Terr., vol. 6 (Cret. Fl.), 1874, p. 45, pl. xxix, f. 6 [not pl. i, f. 3, 4].

Dakota: Blackbird Hill, Nebr.; Fort Harker, Kans.

Sphenopteris egyptiaca Emmons, American geology, pt. 6, 1857, p. 36, f. 8, 9 (on p. 37).—Fontaine, in Ward, U. S. Geol. Surv., Twentieth Ann. Rept., pt. 2, 1900, p. 280, pl. xxxviii, f. 1.

Acrostichides egyptiacus (Emmons) Fontaine, U. S. Geol. Surv., Mon. 6, 1883, p. 99, pl. xlviii, f. 8, 8a.

Triassic (Newark): Egypt, N. C.

Sphenopteris (Asplenium) elongatum Newberry = *Anemia elongata.*

Sphenopteris eocenica Ettingshausen = *Dryopteris arguta.*

Sphenopteris fremonti Hall = *Dennstaedtia? fremonti.*

Sphenopteris goepperti Schenk = *Onychiopsis goepperti.*

Sphenopteris grevillioides Heer = *Thyrsopteris grevillioides.*

Sphenopteris guyottii Lesquereux = *Phegopteris guyottii.*

Sphenopteris hymenophylloides Brongniart = *Coniopteris hymenophylloides.*

Sphenopteris lakesii Lesquereux = *Dryopteris arguta.*

Sphenopteris latiloba Fontaine = *Onychiopsis latiloba.*

Sphenopteris mantelli Brongniart = *Onychiopsis psilotoides.*

Sphenopteris membranacea Lesquereux = *Dryopteris arguta.*

Sphenopteris nigricans Lesquereux = *Dryopteris nigricans.*

Sphenopteris obtusifolia Andrae = *Cladophlebis obtusifolia.*

Sphenopteris pachyphylla Fontaine = *Ruffordia göpperti.*

Sphenopteris? paucifolia Hall = *Microtaenia paucifolia.*

Sphenopteris plurinervia Heer?, Contr. fl. foss. du Portugal, 1881, p. 13, pl. xi, f. 6, 6b; pl. xv, f. 8.—Fontaine, in Ward, U. S. Geol. Surv., Nineteenth Ann. Rept., pt. 2, 1899, p. 657, pl. cxx, f. 19, 20.
Lakota: Hay Creek, Crook County, Wyo.

Sphenopteris spatulata Fontaine = *Ruffordia göpperti.*

Sphenopteris thyrsopteroides Fontaine = *Ruffordia göpperti.*

Sphenopteris? trifoliata Hall = *Microtaenia paucifolia.*

Sphenopteris valdensis? Heer. Fontaine, U. S. Nat. Mus., Proc., vol. 16, 1893, p. 263, pl. xxxvi, f. 2 = *Ruffordia göpperti.*

Sphenopteris sp. Taylor = *Scleropteris elliptica.*

SPHENOZAMITES Brongniart, Tableau vég. foss., 1849, p. 61 [110].

Sphenozamites? oblanceolatus Penhallow, Roy. Soc. Canada, Trans., 2d ser., vol. 8, 1902, sec. 4, p. 63, f. 11; Summ. Rept. Geol. Surv. Canada, 1904 [1905], p. 391; Rept. Tert. Pl. Brit. Col., 1908, p. 90 [generic reference very questionable].
Paskapoo (Fort Union): Red Deer Run, Alberta.

Sphenozamites rogersianus Fontaine, U. S. Geol. Surv., Mon. 6, 1883, pp. 80, 98, pl. xliii, f. 1; pl. xliv, f. 1, 2; pl. xlv, f. 1, 2; pl. xlix, f. 4; in Ward, U. S. Geol. Surv., Twentieth Ann. Rept., pt. 2, 1900, p. 247, pl. xxix, f. 8, 9.
Calamites punctatus Emmons, American geology, pt. 6, 1857, p. 35, pl. vi, f. 5.
Triassic (Newark): Clover Hill, Va.; North Carolina; York County, Pa.

Sphenozamites sp. Dawson, Roy. Soc. Canada, Trans., vol. 3, 1885 [1886], sec. 4, p. 7.
Kootenai: Martin Creek, British Columbia.

Spheria lapidea Lesquereux = *Rossellinites lapideus.*

Spheria myricae Lesquereux = *Sphaerites myricae.*

Spheria rhytismoides Lesquereux = *Sphaerites lesquereuxi.*

SPIRAEA Linné, Sp. pl., 1753, p. 195.

Spiraea andersoni Heer, Fl. foss. arct., vol. 2, Abt. 2, 1869, p. 39, pl. viii, f. 3.—Knowlton, U. S. Nat. Mus., Proc., vol. 17, 1894, p. 226; Geol. Soc. Am., Bull., vol. 5, 1893, p. 585.
Kenai: Port Graham, Alaska.

SPIRAXIS Newberry, N. Y. Acad. Sci., Annals, vol. 3, 1885, p. 219. [Type, *S. major.*]

Spiraxis bivalvis Ward, U. S. Geol. Surv., Sixth Ann. Rept., 1884–85 [1886], p. 549, pl. xxxi, f. 3; U. S. Geol. Surv., Bull. 37, 1887, p. 14, pl. i, f. 3.
Fort Union: Clear Creek, Mont.

SPIRODELA Schleiden, Linnaea, vol. 13, 1839, p. 391.

Spirodela penicillata (Lesquereux) Cockerell, Univ. Colo. Studies, vol. 3, 1906, p. 174; Am. Mus. Nat. Hist., Bull., vol. 24, 1908, p. 79.—Knowlton, U. S. Nat. Mus., Proc., vol. 51, 1916, p. 252.
Lemna penicillata Lesquereux, U. S. Geol. Surv. Terr., Rept., vol. 8 (Cret. and Tert. Fl.), 1883, p. 143, pl. xxiii, f. 8.
Miocene: *Florissant, Colo.

STAPHIDOIDES Perkins, Rept. Vt. State Geol. 1905–6 [1906], p. 222. [Type, *S. venosus*.]

Staphidoides ovalis Perkins, Rept. Vt. State Geol. 1905–6 [1906], p. 223, pl. lviii, f. 2–5.
Miocene: Brandon, Vt.

Staphidoides perkinsi Knowlton, n. name.
Staphidoides venosus (Lesquereux) Perkins, Rept. Vt. State Geol. 1905–6 [1906], p. 223, pl. lviii, f. 1.
Carpolithes venosus? Sternberg. Lesquereux, Am. Jour. Sci., 2d ser., vol. 32, 1861, p. 361; Geol. Vt., vol. 2, 1861, p. 717, f. 159, 160. [Not *C. venosus* of Sternberg.]
Miocene: Brandon, Vt.

Staphidoides venosus (Lesquereux) Perkins = *Staphidoides perkinsi*.

STAPHYLEA Linné, Sp. pl., 1753, p. 270.

Staphylea acuminata Lesquereux, U. S. Geol. and Geog. Surv. Terr., Ann. Rept. 1873 [1874], p. 415; U. S. Geol. Surv. Terr., Rept., vol. 7 (Tert. Fl.), 1878, p. 267, pl. xlviii, f. 4, 5.—Knowlton, U. S. Nat. Mus., Proc., vol. 51, 1916, p. 282.
Miocene: Middle Park and Costello's ranch, *Florissant, Colo.

Staphylea? fremonti Knowlton, U. S. Geol. Surv., Prof. Paper 108, 1917, p. 93, pl. xxxii, f. 4, 5; pl. xxxiii, f. 5.
Frontier: About 1 mile east of Cumberland, Wyo.

STENOPTERIS Saporta, Paléont. française, Pl. jur., vol. 1, 1870, p. 290.

Stenopteris? cretacea Hollick, Torreya, vol. 2, 1902, p. 148.—Knowlton, U. S. Geol. Surv., Prof. Paper 101, 1918, p. 246.
Vermejo: Florence, Colo.

Stenopteris virginica Fontaine, U. S. Geol. Surv., Mon. 15, 1889, p. 112, pl. xxi, f. 8.
Patapsco: Near Brooke, Va.

STERCULIA Linné, Sp. pl., 1753, p. 1007.

Sterculia aperta Lesquereux, U. S. Geol. Surv. Terr., Rept., vol. 8 (Cret. and Tert. Fl.), 1883, p. 82, pl. x, f. 2, 3; U. S. Geol. Surv., Mon. 17, 1891 [1892], p. 185, pl. xxii, f. 4.
Dakota: *Morrison, Colo.; Ellsworth County, Kans.

Sterculia berryana Knowlton, U. S. Geol. Surv., Prof. Paper 101, 1918, p. 337, pl. cii, f. 3, 4.
Raton: Near Yankee, N. Mex.

Sterculia cliffwoodensis Berry, N. Y. Bot. Gard., Bull., vol. 3, 1903, p. 88, pl. xliii, f. 5; Geol. Surv. N. J., Ann. Rept. 1905 [1906], p. 139; Bull. Torr. Bot. Club, vol. 33, 1906, p. 178; Md. Geol. Surv., Upper Cret., 1916, p. 858, pl. lxxx, f. 4.
Magothy: *Cliffwood, N. J.; Deep Cut, Del.

Sterculia coriacea Knowlton, U. S. Geol. Surv., Prof. Paper 101, 1918, p. 272, pl. xlviii, f. 1.
Vermejo: Vermejo Park, N. Mex.

Sterculia drakei Cummins = *Sterculia snowii.*

Sterculia elegans Fontaine, U. S. Geol. Surv., Mon. 15, 1889 [1890], p. 314, pl. clvii, f. 2; pl. clviii, f. 2, 3; in Ward, U. S. Geol. Surv., Mon. 48, 1905 [1906], p. 502, pl. cx, f. 6.
Patapsco: *Deep Bottom and Mount Vernon, Va.

Sterculia engleri Kirchner, St. Louis Acad. Sci., Trans., vol. 8, 1898, p. 180, pl. xiv, f. 3.—Knowlton, U. S. Nat. Mus., Proc., vol. 51, 1916, p. 285.
Miocene: *Florissant, Colo.

Sterculia krejcii Velenovsky. Hollick, Geol. Soc. Am., Bull., vol. 7, 1895, p. 13 = *Aralia grönlandica.*

Sterculia labrusca Unger. Hollick, Geol. Soc. Am., Bull., vol. 7, 1895, p. 13 = *Sterculia praelabrusca.*

Sterculialeariloba Lesquereux, U. S. Geol. and Geog. Surv. Terr., Ann. Rept. 1874 [1876], p. 358. [Not afterward referred to by Lesquereux.]
Dakota: Clay Center, Kans.

Sterculia lugubris Lesquereux, U. S. Geol. Surv. Terr., Rept., vol. 8 (Cret. and Tert. Fl.), 1883, p. 81, pl. vi, f. 1–3.—?Berry, Bull. Torr. Bot. Club, vol. 39, 1912, p. 399, pl. xxxi, f. 3.
Dakota: Near *Golden, Colo., and Smith ranch, Apishapa quadrangle, Colo.
Woodbine: ?Arthurs Bluff, Tex.

Sterculia minima Berry, Bull. Torr. Bot. Club, vol. 33, 1906, p. 177; Geol. Surv. N. J., Ann. Rept. 1905 [1906], p. 139; Md. Geol. Surv., Upper Cret., 1916, p. 857, pl. lxxx, f. 1–3.
Sterculia mucronata Lesquereux. Berry, N. Y. Bot. Gard., Bull., vol. 3, 1903, p. 90, pl. 43, f. 3.
Magothy: Cliffwood Bluff, pits of Cliffwood Brick Co., N. J.; Deep Cut, Del.; Grove Point, Md.

Sterculia modesta Saporta. Lesquereux = *Sterculia saportanea*.

Sterculia mucronata Lesquereux, U. S. Geol. Surv., Mon. 17, 1891 [1892], p. 182, pl. xxx, f. 1–4.

Dakota: Ellsworth County, Kans.

Sterculia mucronata Lesquereux. Berry, N. Y. Bot. Gard., Bull., vol. 3, 1903, p. 90, pl. xliii, f. 3 = *Sterculia minima*.

Sterculia obtusiloba Lesquereux = *Sterculia tripartita*.

Sterculia pre-labrusca Hollick, U. S. Geol. Surv., Mon. 50, 1906, p. 94, pl. xxxiv, f. 21, 22.

Sterculia labrusca Unger. Hollick, Geol. Soc. Am., Bull., vol. 7, 1895, p. 13.

Magothy: Gay Head, Marthas Vineyard, Mass.

Sterculia puryearensis Berry, U. S. Geol. Surv., Prof. Paper 91, 1916, p. 286, pl. lxxi, f. 2, 3; pl. lxxiii, f. 1; pl. lxxiv, f. 4.

Wilcox (Grenada): Grenada, Miss.
Wilcox: Naborton, Miss.
Lagrange: Puryear, Tenn.

Sterculia reticulata Lesquereux, U. S. Geol. Surv., Mon. 17, 1891 [1892], p. 185, pl. xxxiv, f. 10.

Dakota: Near Delphos, Kans.

Sterculia rigida Lesquereux, U. S. Geol. Surv. Terr., Rept., vol. 8 (Cret. and Tert. Fl.), 1883, p. 179, pl. xxxiv, f. 12.—Cockerell, Am. Mus. Nat. Hist., Bull., vol. 24, 1908, p. 104.

Miocene: *Florissant, Colo.

Sterculia saportanea Knowlton, U. S. Geol. Surv., Bull. 152, 1898, p. 224.

Sterculia modesta Saporta. Lesquereux, U. S. Geol. Surv. Terr., Rept., vol. 8 (Cret. and Tert. Fl.), 1883, p. 125, pl. xx, f. 5. [Homonym, Heer, 1853–1855.]

Denver: *Golden, Colo.

Sterculia snowii Lesquereux, U. S. Geol. Surv., Mon. 17, 1891 [1892], p. 183, pl. xxx, f. 5; pl. xxxi, f. 2, 3; pl. xxxii; pl. xxxiii, f. 1–4.—?Hollick, N. Y. Acad. Sci., Ann., vol. 11, 1898, p. 422, pl. xxxvii, f. 4; ?U. S. Geol. Surv., Mon. 50, 1906, p. 94, pl. xxxiv, f. 20.

Sterculia drakei Cummins, Geol. Surv. Tex., Third Ann. Rept., 1891, p. 210, f. 8.—Knowlton, in Hill, Am. Jour. Sci., 3d ser., vol. 50, 1895, p. 213.

Dakota: *Ellsworth County, Kans.; Big Tucumcari Mountain, N. Mex.
Cheyenne: Belvidere, Kans.
?Raritan: Tottenville, Staten Island, N. Y.
?Magothy: Gay Head, Marthas Vineyard, Mass.

Sterculia snowii? Lesquereux. Hollick, 1898 = *Aralia ravniana*.

Sterculia snowii bilobata Berry, N. Y. Bot. Gard., Bull., vol. 3, 1903, p. 89, pl. 43, f. 7; Geol. Surv. N. J., Ann. Rept. 1905 [1906], p. 139.

Magothy: Cliffwood, N. J.

Sterculia snowii disjuncta Lesquereux, U. S. Geol. Surv., Mon. 17, 1891 [1892], p. 184, pl. lviii, f. 6.
Dakota: Near Fort Harker, Kans.

Sterculia snowii tennesseensis Berry, Bull. Torr. Bot. Club, vol. 43, 1916, p. 297, pl. xvi, f. 5; U. S. Geol. Surv., Prof. Paper 112, 1919, p. 117, pl. xxxii, f. 1.
Ripley: Near Selmer, McNairy County, Tenn.

Sterculia tripartita (Lesquereux) Knowlton, U. S. Geol. Surv., Bull. 152, 1898, p. 224.
Aralia tripartita Lesquereux, U. S. Geol. and Geog. Surv. Terr., Ann. Rept. 1874 [1876], p. 248, pl. i, f. 1.
Sterculia obtusiloba Lesquereux, U. S. Geol. Surv. Terr., Rept., vol. 8 (Cret. and Tert. Fl.), 1883, p. 82, pl. viii, f. 3.
Dakota: *Near Fort Harker, Kans.

Sterculia vetustula Dawson, Roy. Soc. Canada, Trans., vol. 3, 1885 [1886], sec. 4, p. 10, pl. iii, f. 2.
——: North Fork of Old Man River, British Columbia.

Sterculia sp. Hollick, N. Y. Acad. Sci., Trans., vol. 16, 1897, p. 133, pl. xiv, f. 4–7.
Magothy: Cliffwood, N. J.

Sterculia sp. Hollick, N. Y. Acad. Sci., Ann., vol. 11, 1898, p. 422, pl. xxxvii, f. 5; U. S. Geol. Surv., Mon. 50, 1906, p. 95, pl. xxxiv, f. 18, 19.—Berry, Geol. Surv. N. J., Ann. Rept. 1905 [1906], p. 139.
Magothy: Gay Head, Marthas Vineyard, Mass.; Cliffwood, N. J.
Raritan: Tottenville, Staten Island, N. Y.

Sterculia sp. Knowlton, in Washburne, U. S. Geol. Surv., Bull. 590, 1914, p. 36.
Eocene?: Eugene, Oreg.

Sterculia sp. Lesquereux, Rept. on Clays in N. J., 1878, p. 28.
Raritan?: Washington, N. J.

STERCULIOCARPUS Berry, U. S. Geol. Surv., Prof. Paper 91, 1916, p. 287. [Type, *S. eocenicus*.]

Sterculiocarpus eocenicus Berry, U. S. Geol. Surv., Prof. Paper 91, 1916, p. 288, pl. lxxiv, f. 1–3.
Wilcox: Frierson Mill and near Naborton, La.

Sterculiocarpus sezannelloides Berry, U. S. Geol. Surv., Prof. Paper 91, 1916, p. 288, pl. lxxii, f. 4–6.
Lagrange: Puryear, Tenn.

STIPA Linné, Sp. pl., 1753, p. 78.

Stipa laminarium Cockerell, Am. Mus. Nat. Hist., Bull., vol. 24, 1908, p. 79, pl. vi, f. 1, 3.
Miocene: *Florissant, Colo.

Strangerites obliquus Emmons = *Pseudo-danaeopsis obliqua.*

Strangerites planus Emmons = *Pseudo-danaeopsis plana*.

STROBILITES Lindley and Hutton, Foss. Fl. Great Brit., vol. 2, 1833–1835, p. 23, pl. lxxxix.

Strobilites anceps Berry, U. S. Geol. Surv., Prof. Paper 84, 1914, p. 27, pl. iii, f. 5.
Black Creek: Black Creek near Darlington, S. C.

Strobilites davisii Hollick and Jeffrey, N. Y. Bot. Gard., Mem., vol. 3, 1909, p. 68, pl. iii, f. 10.
Raritan: Kreischerville, Staten Island, N. Y.

Strobilites inquirendus Hollick, N. Y. Acad. Sci., Trans., vol. 16, 1897, p. 130, pl. xi, f. 1.—Berry, N. Y. Bot. Gard., Bull., vol. 3, 1903, p. 101; Geol. Surv. N. J., Ann. Rept. 1905 [1906], p. 139.
Magothy: Cliffwood, N. J.

Strobilites microsporophorus Hollick and Jeffrey, N. Y. Bot. Gard., Mem., vol. 3, 1909, p. 66, pl. x, f. 18–21; pl. xxiv, f. 2–6.
Raritan: Kreischerville, Staten Island, N. Y.

Strobilites perplexus Hollick, U. S. Geol. Surv., Mon. 50, 1906, p. 107, pl. ii, f. 43.
Magothy: Gay Head, Marthas Vineyard, Mass.

Strobilites sp. Hollick and Jeffrey, N. Y. Bot. Gard., Mem., vol. 3, 1909, p. 69, pl. iii, f. 9.
Raritan: Kreischerville, Staten Island, N. Y.

STYRAX (Tournefort) Linné, Sp. pl., 1753, p. 444.

Styrax ambra Unger, Syll. Pl. Foss., pt. 3, 1865, p. 34, pl. xxiv, f. 19, 20.—Lesquereux, Mus. Comp. Zool., Bull., vol. 16, 1888, p. 51.
Denver: *Golden, Colo.

Styrax laramiense Lesquereux, Mus. Comp. Zool., Bull., vol. 16, 1888, p. 51.
Denver: *Golden, Colo.

SYMPHOROCARPOPHYLLUM Dawson, Roy. Soc. Canada, Trans., vol. 4, 1886 [1887], sec. 4, p. 29. [Type, *S. albertum.*]

Symphorocarpophyllum albertum Dawson, Roy. Soc. Canada, Trans., vol. 4, 1886 [1887], sec. 4, p. 30, pl. ii, f. 17.—Penhallow, Rept. Tert. Pl. Brit. Col., 1908, p. 91.
Paskapoo (Fort Union): Porcupine Creek and Great Valley, Alberta.

Symphorocarpophyllum linnaeiforme Dawson, Roy. Soc. Canada, Trans., vol. 4, 1886 [1887], sec. 4, p. 30, pl. ii, f. 18.—Penhallow, Rept. Tert. Pl. Brit. Col., 1908, p. 91.
Paskapoo (Fort Union): Porcupine Creek, Alberta.

Symphorocarpophyllum sp. Dawson, Roy. Soc. Canada, Trans., vol. 4, 1886 [1887] ?.—Penhallow, Rept. Tert. Pl. Brit. Col., 1908, p. 91.
Paskapoo (Fort Union): Porcupine Creek, Alberta.

TAENIOPTERIS Brongniart, Prodrome, 1828, p. 61.

Taeniopteris auriculata (Fontaine) Berry, U. S. Nat. Mus., Proc., vol. 38, 1910, p. 634; Md. Geol. Surv., Lower Cret., 1911, p. 293.

Angiopteridium auriculatum Fontaine, U. S. Geol. Surv., Mon. 15, 1889, p. 113, pl. vii, f. 8–11; pl. xxviii, f. 1.

Patuxent: Fredericksburg and Potomac Run, Va.

Taeniopteris gibbsii Newberry = *Nilsonia gibbsii.*

Taeniopteris magnifolia Rogers = *Macrotaeniopteris magnifolia.*

Taeniopteris major Lindley and Hutton. Fontaine, in Ward, U. S. Geol. Surv., Mon. 48, 1905 [1906], p. 79, pl. xiii, f. 1–3.

Jurassic: Douglas County, Oreg.

Taeniopteris nervosa (Fontaine) Berry, U. S. Nat. Mus., Proc., vol. 38, 1910, p. 634; Md. Geol. Surv., Lower Cret., 1911, p. 293, pl. lxxvii, f. 1.

Angiopteridium nervosum Fontaine, U. S. Geol. Surv., Mon. 15, 1889 [1890], p. 114, pl. xxix, f. 2.

Angiopteridium densinerve Fontaine, op. cit. (Mon. 15), p. 115, pl. xxix, f. 4.

Angiopteridium pachyphyllum Fontaine, op. cit. (Mon. 15), p. 115, pl. xxix, f. 5.

Angiopteridium strictinerve Fontaine, op. cit. (Mon. 15), p. 116, pl. xxix, f. 8, 9. [Not Fontaine, in Ward, 1906.]

Angiopteridium strictinerve latifolium Fontaine, op. cit. (Mon. 15), p. 116, pl. xxx, f. 1, 5; in Ward, U. S. Geol. Surv., Mon. 48, 1905 [1906], p. 241, pl. lxvi, f. 8–10.

Anomozamites angustifolius Fontaine, op. cit. (Mon. 15), p. 167, pl. xxx, f. 3 [not f. 2].

Anomozamites virginicus Fontaine, op. cit. (Mon. 15), p. 168, pl. xxx, f. 4; pl. xxxi, f. 3.

Patuxent: Fredericksburg, telegraph station near Lorton, and near Potomac Run, Va.

Taeniopteris? oregonensis Fontaine, in Ward, U. S. Geol. Surv., Mon. 48, 1905 [1906], p. 82, pl. xiii, f. 9, 10.

Jurassic: Douglas County, Oreg.

Taeniopteris orovillensis Fontaine, Am. Jour. Sci., 4th ser., vol. 2, 1896, p. 274; in Ward, U. S. Geol. Surv., Twentieth Ann. Rept., pt. 2, 1900, p. 348, pl. lii, f. 2–4; U. S. Geol. Surv., Mon. 48, 1905 [1906], p. 78, pl. xii, f. 12–17.

Taeniopteris stenoneura Schenk? Fontaine, in Turner, Jour. Geol., vol. 3 1895, p. 395.

Jurassic: Oroville, Calif.; Douglas County, Oreg.

Taeniopteris parvula Heer = *Nilsonia parvula.*

Taeniopteris plumosa Dawson, Roy. Soc. Canada, Trans., vol. 1, 1882 [1883], sec. 4, p. 24, pl. iv, f. 15.—Penhallow, idem, 2d ser., vol. 8, 1902, p. 37.

Upper Cretaceous: Baynes Sound, Vancouver Island, Alliford Bay, Queen Charlotte Island, British Columbia.

Taeniopteris shastensis Knowlton, n. sp.

Angiopteridium strictinerve Fontaine. Fontaine, in Ward, U. S. Geol. Surv., Mon. 48, 1905 [1906], p. 240, pl. lxvi, f. 5–7.

Chico: Elder Creek, Tehama County, Calif.
Horsetown: Tehama County, Calif.
Knoxville: Tehama County, Calif.

Taeniopteris stenoneura Schenk? Fontaine, in Turner, Jour. Geol., vol. 3, 1895, p. 395 = *Taeniopteris orovillensis*.

Taeniopteris vittata Brongniart, Dict. sci. nat., vol. 57 (Prodrome), 1828, p. 70 [name].—Fontaine, in Ward, U. S. Geol. Surv., Mon. 48, 1905 [1906], p. 80, pl. xiii, f. 4–8.

Jurassic: Douglas County, Oreg.

Taeniopteris? yorkensis Fontaine, in Ward, U. S. Geol. Surv., Twentieth Ann. Rept., pt. 2, 1900, p. 237, pl. xxii, f. 4–6.

Triassic (Newark): *York County, Pa.

TAONURUS Fischer-Ooster, Foss. Fuc. Schweizer-Alpen, 1858, p. 41.

Taonurus incertus Dawson, Roy. Soc. Canada, Trans., vol. 1, 1882 [1883], p. 24, pl. iv, f. 15.

Upper Cretaceous: Baynes Sound, Vancouver Island, British Columbia.

TAPEINIDIUM (Presl) Christensen, Index, 1906, p. 631.

Tapeinidium? undulatum (Hall) Knowlton, U. S. Geol. Surv., Prof. Paper 108, 1917, p. 80, pl. xxviii, f. 1–4.

Pecopteris undulata Hall, in Frémont, Rept. Exploring Exped. to the Rocky Mountains in 1842, etc., App. B, 1845, p. 306, pl. i, f. 1a, 1b.
Pecopteris undulata Hall, var., idem, p. 306, pl. i, f. 2, 2a, 2b.
Pecopteris? odontopteroides Hall, idem, p. 306, pl. i, f. 3, 4.

Frontier: About 1 mile east of Cumberland, Wyo.

TAXITES Brongniart, Dict. sci. nat., vol. 57 (Prodrome), 1828, p. 111 (108).

Taxites langsdorfii Brongniart = *Sequoia langsdorfii*.

Taxites microphyllus Heer, Fl. foss. alask., 1869, p. 24, pl. i, f. 9, 9b.

Kenai: Port Graham, Alaska.

Taxites olriki Heer, Fl. foss. arct., vol. 1, 1868, p. 95, pl. i, f. 21–24; pl. xlv, f. a, b, c; idem, vol. 2, Abt. 2, 1869 (Fl. foss. alask.), p. 23, pl. i, f. 8; pl. ii, f. 5b.—Knowlton, U. S. Nat. Mus., Proc., vol. 17, 1894, p. 214; Geol. Soc. Am., Bull., vol. 5, 1893, p. 578; U. S. Geol. Surv., Mon. 32, pt. 2, 1899, p. 680, pl. lxxxii, f. 1, 3, 4.—Penhallow, Rept. Tert. Pl. Brit. Col., 1908, p. 91.

Kenai: Port Graham, Alaska.
Fort Union: Yellowstone National Park.
Eocene: Mackenzie River, Northwest Territory; Souris River and Porcupine Creek, Saskatchewan.

Taxites zamioides (Leckenby) Seward, Brit. Mus. Nat. Hist. (Cat. Mus. Pl. Jurassic Flora, pt. 1), 1900, p. 300, pl. x, f. 5.—Fontaine, in Ward, U. S. Geol. Surv., Mon. 48, 1905 [1906], pp. 129, 151, pl. xxxiv, f. 15–17; pl. xxxv, f. 1–3; pl. xxxviii, f. 15–18.
Jurassic: Curry and Douglas counties, Oreg.

Taxites sp. Dawson, Roy. Soc. Canada, Trans., vol. 5, 1887, sec. 4, p. 33 [wood].
Belly River: South Saskatchewan.

Taxites sp. Dawson, Roy. Soc. Canada, Trans., vol. 5, 1887, sec. 4, p. 34 [wood].
Paskapoo (Fort Union): Red Deer River and Wood End Depot, Alberta.

TAXODIOXYLON Hartig, Bot. Zeitung, vol. 6, 1848, p. 169.

Taxodioxylon credneri Platen, Sitzungsb. Naturf. Ges., Leipzig, vol. 34, 1907 [1908], p. 97, pl. iii, f. 1.
Miocene?: Big Smoky Valley, near Austin, Nev.

Taxodites europaeus (Brongniart) Endlicher = *Glyptostrobus europaeus.*

TAXODIUM L. C. Richard, Ann. Mus., Paris, vol. 16, 1810, p. 298.

Taxodium (Glyptostrobus) brookensis Fontaine = *Widdringtonites ramosus.*

Taxodium (Glyptostrobus) brookense angustifolium Fontaine = *Widdringtonites ramosus.*

Taxodium cuneatum Newberry = *Sequoia cuneata.*

Taxodium denticulatum Fontaine = *Sphenolepis kurriana.*

Taxodium distichum (Linné) Richard. Penhallow, Roy. Soc. Canada, Trans., 2d ser., vol. 9, 1903, sec. 4, p. 36, f. 1.—Berry, Torreya, vol. 5, 1905, p. 89.—Hollick, Md. Geol. Surv., Pliocene and Pleistocene, 1906, p. 218, pl. lxviii, f. 1, 2.—Penhallow, Roy. Soc. Canada, Trans., 3d ser., vol. 1, sec. 4, 1907, p. 301.—Berry, Jour. Geol., vol. 15, 1907, p. 339; Am. Jour. Sci., 4th ser., vol. 29, 1910, p. 391; Torreya, vol. 10, 1910, p. 263; Am. Jour. Sci., 4th ser., vol. 34, 1912, p. 219, f. 1, 2; Torreya, vol. 14, 1914, p. 160; idem, vol. 15, 1915, p. 206; U. S. Nat. Mus., Proc., vol. 48, 1915, p. 296; U. S. Geol. Surv., Prof. Paper 98, 1916, p. 195, pl. xlv, f. 1–6; Jour. Geol., vol. 25, 1917, p. 662.
Cupressoxylon sp. (C) Dawson, Brit. N. A. Bound. Com. (Rept. Geol. and Res. Vic. 49th Parallel), 1875, App. A, p. 331, pl. xv, f. 4.
Pleistocene: Bodkin Point, Grace Point, Pond Neck, and Indian Head, Md.; near Long Branch, N. J.; Tappahannock and Buena Vista, Va.; Neuse and Roanoke rivers, N. C.; Hickman, Ky.; Mitten and Vero, Fla.; Chicoria, Miss.
Citronelle: Lambert and Red Bluff, Ala.
Oligocene?: Kettle River near Midway, British Columbia.

Taxodium distichum dubium (Sternberg) Cockerell = *Taxodium dubium.*

Taxodium distichum miocenum Heer. Newberry, U. S. Geol. Surv., Mon. 35, 1898, p. 22, pl. xlvii, f. 6 = *Sequoia langsdorfii.*

Taxodium dubium (Sternberg) Heer, Fl. tert. Helv., vol. 1, 1855, p. 49, pl. xvii, f. 3, 15.—Lesquereux, U. S. Geol. and Geog. Surv. Terr., Ann. Rept. 1873 [1874], p. 409.—Prindle, U. S. Geol. Surv., Bull. 375, 1909, p. 26.—Berry, U. S. Geol. Surv., Prof. Paper 91, 1916, p. 171, pl. xv, f. 4–6.

Phyllites dubius Sternberg, Flora der Vorwelt, vol. 1, 1824, p. 37, pl. xxiv, f. 2; pl. xxxvi, f. 3, 4.

Taxodium distichum miocenum Heer, Miocene baltische Flora, 1869, p. 18, pl. ii; pl. iii, f. 6, 7.—Dawson, Roy. Soc. Canada, Trans., vol. 8, 1882, p. 79.—Knowlton, Geol. Soc. Am., Bull., vol. 5, 1893, p. 578; U. S. Nat. Mus., Proc., vol. 17, 1894, p. 214.—Newberry, U. S. Geol. Surv., Mon. 35, 1898, p. 22, pl. xlvii, f. 6; pl. li, f. 3; pl. lii, f. 2, 3; pl. lv, f. 5.—Knowlton, U. S. Geol. Surv., Bull. 204, 1902, p. 27.—Penhallow, Roy. Soc. Canada, Trans., 2d ser., vol. 8, 1902, pp. 51, 68.—Knowlton, Harriman Alaska Exped., vol. 4, 1904, p. 152.—Penhallow, Man. N. A. Gym., 1907, p. 217; Roy. Soc. Canada, Trans., 3d ser., vol. 8, 1908, pp. 301, 312, 314, 315; Rept. Tert. Pl. Brit. Col., 1908, p. 91.—Berry, Jour. Geol., vol. 17, 1909, p. 22, f. 1.—Knowlton, Washington Acad. Sci., Proc., vol. 11, 1909, pp. 204, 207, 215.

Taxodium distichum dubium (Sternberg) Cockerell, Am. Nat., vol. 44, 1910, p. 35.

Taxodium nevadensis Lesquereux, U. S. Geol. and Geog. Surv. Terr., Ann. Rept. 1873 [1874], p. 372.

Lance: Kirby, Wyo.; Weston and Converse counties, Wyo.
Kenai: Port Graham, Sitka, Herendeen and Kukak bays, Alaska.
Clarno (lower part): Bridge Creek, Oreg.
Puget: Birch Bay, Wash.
Paskapoo (Fort Union): Red Deer River, Alberta.
Mascall: Van Horn's ranch, John Day Basin, Oreg.
Miocene: Horsefly River, British Columbia.
Calvert: Richmond, Va.
Lagrange: Pinson, Madison County, Tenn.

Taxodium europaeum Brongniart = *Glyptostrobus europaeus.*

Taxodium expansum Fontaine = *Sphenolepis kurriana.*

Taxodium fastigiatum Fontaine = *Sphenolepis kurriana.*

Taxodium laramianum Penhallow, Roy. Soc. Canada, Trans., 2d ser., vol. 10, 1904, sec. 4, p. 57; Man. N. A. Gym., 1907, p. 218; Rept. Tert. Pl. Brit. Col., 1908, p. 91.

Paskapoo (Fort Union): Cochran, Alberta.

Taxodium nevadensis Lesquereux = *Taxodium dubium.*

Taxodium occidentale Newberry, Boston Jour. Nat. Hist., vol. 7, 1863, p. 517; N. Y. Lyc. Nat. Hist., Ann., vol. 9, 1868, p. 45.—[Lesquereux], U. S. Geol. and Geog. Surv. Terr., Ill. Cret. and Tert. Pl., 1878, pl. xi, f. 1-3.—Newberry, U. S. Geol. Surv., Mon. 35, 1898, p. 23, pl. xxvi, f. 1-3; pl. lv, f. 5 [part].—Knowlton, Washington Acad. Sci., Proc., vol. 11, 1909, pp. 189, 195, 202, 203, 211, 214; Jour. Geol., vol. 19, 1911, p. 369.—Penhallow, Rept. Tert. Pl. Brit. Col., 1908, p. 91.

Fort Union: Yellowstone River, Mont. Are abundant Fort Union species.

Lance: Rattlesnake Butte, Cheyenne Indian Reservation, S. Dak.; 60 miles south of Glasgow, west of Miles City, 5 miles east of Miles City, 8 miles west of Bridger, 9 miles west of Bridger, Mont.; Yule, N. Dak.; Weston County, Dayton, Kirby, Rairden, Bighorn Basin, Wyo.

Eocene: Mackenzie, Souris, Red Deer, and Blackwater rivers, and Porcupine Creek, British Columbia.

Oligocene: Quilchena, Similkameen River, British Columbia.

Miocene: Tranquille and Horsefly rivers, British Columbia.

Taxodium (Glyptostrobus) ramosum Fontaine = *Widdringtonites ramosus.*

Taxodium tinajorum Heer var. Eichwald, Geognost.-palaeont. Bemerk. ü. Halbinsel Mangischlak u. Aleutischen Inseln, 1871, p. 116, pl. iv, f. 4.—Knowlton, U. S. Nat. Mus., Proc., vol. 17, 1894, p. 214; Geol. Soc. Am., Bull., vol. 5, 1893, p. 578; in Atwood, U. S. Geol. Surv., Bull. 467, 1911, p. 54.

Kenai: Port Graham, Chignik Bay, and Ninilchik, Alaska.

Taxodium tinajorum? Heer. Lesquereux, U. S. Geol. and Geog. Surv. Terr., Ann. Rept. 1871 [1872], p. 285 = (?). [Not afterward referred to by Lesquereux.]

Taxodium (Glyptostrobus) virginicum Fontaine, U. S. Geol. Surv., Mon. 15, 1889 [1890], p. 252, pl. cxxi, f. 6.

Patuxent: Near Potomac Run, Va.

Taxodium sp. Berry, U. S. Geol. Surv., Prof. Paper 91, 1916, p. 173, pl. xv, f. 9.

Lagrange: Near Grand Junction, Fayette County, Tenn.

Taxodium sp. Dawson, Roy. Soc. Canada, Trans., vol. 11, 1893 [1894], sec. 4, p. 56, pl. vi, f. 10.

Upper Cretaceous: Nanaimo, Vancouver Island, British Columbia.

Taxodium, male aments of, Knowlton, U. S. Geol. Surv., Bull. 204, 1902, p. 27, pl. i, f. 4, 6.

Mascall: Van Horn's ranch, John Day Basin, Oreg.

Taxodium? sp. Knowlton, U. S. Geol. Surv., Prof. Paper 101, 1918, p. 252, pl. xxxii, f. 1, 2.

Vermejo: Spring Canyon, Vermejo Park, N. Mex.

TAXOXYLUM Unger, Chlor. Prot., 1842, p. 33.

Taxoxylon sp. Dawson, Geol. Surv. Canada, Rept. Prog. 1872–73, App. I, p. 67.

Lower Cretaceous: British Columbia.

Taxoxylon sp. Dawson, Brit. N. A. Bound. Com. (Rept. Geol. and Res. Vic. 49th Parallel), 1875, App. A, p. 331.

Paskapoo (Fort Union): Vicinity of 49th Parallel, Alberta.

TAXUS (Tournefort) Linné, Sp. pl., 1753, p. 1040.

Taxus baccata Linné. Penhallow, 1890 = *Taxus minor*.

Taxus canadensis Willdenow. Penhallow, 1896 = *Taxus minor*.

Taxus minor (Michaux) Britton.

Taxus baccata Linné. Penhallow, Geol. Soc. Am., Bull., vol. 1, 1890, pp. 321, 333.

Taxus canadensis Willdenow. Penhallow, Roy. Soc. Canada, Trans., 2d ser., vol. 2, 1896, sec. 4, p. 73; Brit. Assoc. Adv. Sci., Bradford meeting, 1900, p. 336; Man. N. A. Gym., 1907, p. 213.

Pleistocene: Don River and Don Valley, Toronto; Salsgirth and Rolling River, Manitoba; Cape Breton; Fort Madison, Iowa; Bloomington, Ill.

Taxus sp. Dawson, Geol. Surv. Canada, Rept. Prog. 1873–74, p. 51.—Penhallow, Rept. Tert. Pl. Brit. Col., 1908, p. 92.

Eocene: Edmonton, Alberta.

TECOMA Jussieu, Gen., 1789, p. 139.

Tecoma preradicans Berry, U. S. Nat. Mus., Proc., vol. 48, 1915, p. 302, pl. xiii, f. 1–5.

Tecoma radicans Linné. Knowlton, in Glenn, U. S. Geol. Surv., Water-Supply Paper 164, 1906, p. 38.

Pleistocene: Hickman and Columbus, Ky.

Tecoma radicans Linné. Knowlton = *Tecoma preradicans*.

TEMPSKYA Corda, Beitr. Fl. der Vorwelt, 1845, p. 81.

Tempskya whitei Berry, Md. Geol. Surv., Lower Cret., 1911, p. 298, pls. xxxvii, xxxviii.

Patapsco: Valleys of Stony Run and Deep Run near Severn, Anne Arundel County, and Patuxent Neck, Prince Georges County, Md.

TERMINALIA Linné, Mantissa plantarum, 1767, p. 21.

Terminalia hilgardiana (Lesquereux) Berry = *Magnolia hilgardiana*.

Terminalia phaeocarpoides Berry, U. S. Geol. Surv., Prof. Paper 84, 1914, p. 146, pl. xxix, f. 3.

Barnwell (Twiggs): Grovetown, Ga.

Terminalia radobojensis Heer. Lesquereux = *Magnolia angustifolia.*

Terminalia wilcoxiana Berry, U. S. Geol. Surv., Prof. Paper 91, 1916, pls. 325, xxxix, f. 3.
Wilcox (Grenada): Grenada, Miss.

TERNSTROEMITES Berry, U. S. Geol. Surv., Prof. Paper 91, 1916, p. 294. [Type, *T. eolігniticus.*]

Ternstroemites eolignіticus Berry, U. S. Geol. Surv., Prof. Paper 91, 1916, p. 294, pl. lxxvi, f. 1, 2; pl. lxxviii, f. 5.
Lagrange: Puryear, Tenn.

Ternstroemites lanceolatus Berry, U. S. Geol. Surv., Prof. Paper 91, 1916, p. 296, pl. lxxvii, f. 5.
Wilcox (Grenada): Grenada, Miss.
Lagrange: Puryear, Tenn.

Ternstroemites ovatus Berry, U. S. Geol. Surv., Prof. Paper 91, 1916, p. 295, pl. lxxvii, f. 2–4.
Wilcox (Grenada): Grenada, Miss.
Lagrange: Puryear, Tenn.
Wilcox: De Soto Parish, La.

Ternstroemites preclaibornensis Berry, U. S. Geol. Surv., Prof. Paper 91, 1916, p. 295, pl. lxxviii, f. 1–4.
Lagrange: Puryear, Tenn.
Wilcox: Near Mansfield, La.

Tetranthera Jacques = *Malapoenna.*

Tetranthera praecursoria Lesquereux, 1883 = *Malapoenna praecursoria.*

Tetranthera praecursoria Lesquereux. Hollick, 1899 = *Oreodaphne coushatta.*

Tetranthera sessiliflora Lesquereux = *Litsea sessiliflora* and *Litsea weediana.*

THALICTRUM (Tournefort) Linné, Sp. pl., 1753, p. 545.

Thalictrum sp. Dawson, Brit. N. A. Bound. Com. (Rept. Geol. and Res. Vic. 49th Parallel), 1875, App. A, p. 330.
Paskapoo (?): Porcupine Creek, Alberta.

THINNFELDIA Ettingshausen, Abhandl. K. k. geol. Reichsanstalt, vol. 1, 1852, Abt. 3, No. 3, p. 2.

Thinnfeldia fontainei Berry, Bull. Torr. Bot. Club, vol. 30, 1903, p. 443; Md. Geol. Surv., Lower Cret., 1911, p. 302, pl. xl, f. 4–7.
Thinnfeldia variabilis Fontaine, U. S. Geol. Surv., Mon. 15, 1889 [1890], p. 110, pl. xvii, f. 3–7; pl. xviii, f. 1–6; in Ward, U. S. Geol. Surv., Mon. 48, 1905 [1906], p. 502, pl. cx, f. 7, 8. [Homonym, Velenovsky, 1885.]
Patapsco: Mount Vernon, White House Bluff, near Brooke and near Potomac River, Va.; Fort Foote, Md.

Thinnfeldia granulata Fontaine, U. S. Geol. Surv., Mon. 15, 1889 [1890], p. 111, pl. xxvi, f. 10–12; pl. xxvii, f. 1–5, 8; pl. clxix, f. 1.—Berry, Bull. Torr. Bot. Club, vol. 30, 1903, p. 443; Md. Geol. Surv., Lower Cret., 1911, p. 303, pl. xl, f. 1, 2.

Acaciaephyllum longifolium Fontaine, U. S. Geol. Surv., Mon. 15, 1889 [1890], p. 279, pl. cxxxvii, f. 6, pl. cxxxviii, f. 1–3.

Acaciaephyllum spatulatum Fontaine, op. cit. (Mon. 15), p. 280, pl. cxxxviii, f. 4, 6–9.

Acaciaephyllum microphyllum Fontaine, op. cit. (Mon. 15), p. 280, pl. cxxxviii, f. 5; in Ward, U. S. Geol. Surv., Mon. 48, 1905 [1906], p. 486.

?*Acaciaephyllum variabile* Fontaine, op. cit. (Mon. 15), p. 281, pl. clxx, f. 7.

Celastrophyllum proteoides Fontaine, op. cit. (Mon. 15), p. 304, pl. cxlvi, f. 5.

Patuxent: Dutch Gap, Potomac Run, and telegraph station near Lorton, Va.

Patapsco: Federal Hill (Baltimore), Md.

Thinnfeldia lanceolata Knowlton = *Protophyllocladus lanceolatus.*

Thinnfeldia lesquereuxiana Heer = *Protophyllocladus subintegrifolius.*

Thinnfeldia marylandica Fontaine, in Ward, U. S. Geol. Surv., Mon. 48, 1905 [1906], p. 541, pl. cxiv, f. 8. 9.—Berry, Md. Geol. Surv., Lower Cret., 1911, p. 305, pl. xl, f. 8, 9.

Arundel: Arlington, Md.

Thinnfeldia montana Knowlton = *Protophyllocladus polymorphus.*

Thinnfeldia montanensis Fontaine = *Cladophlebis virginiensis.*

Thinnfeldia polymorpha (Lesquereux) Knowlton = *Protophyllocladus polymorphus.*

Thinnfeldia? reticulata Fontaine, in Ward, U. S. Geol. Surv., Twentieth Ann. Rept., pt. 2, 1900, p. 235, pl. xxii, f. 1, 2.—Berry, Bull. Torr. Bot. Club, vol. 30, 1903, p. 444.

Triassic (Newark): York County, Pa.

Thinnfeldia rotundiloba Fontaine, U. S. Geol. Surv., Mon. 15, 1889 [1890], p. 111, pl. xxvii, f. 6, 7.—Berry, Bull. Torr. Bot. Club, vol. 30, 1903, p. 443; Md. Geol. Surv., Lower Cret., 1911, p. 305, pl. xl, . 3.

Patuxent: Fredericksburg and near Potomac Run, Va.

Thinnfeldia subintegrifolia (Lesquereux) Knowlton = *Protophyllocladus subintegrifolius.*

Thinnfeldia variabilis Fontaine? Fontaine, in Diller and Stanton, Geol. Soc. Am., Bull., vol. 5, 1894, p. 450; in Stanton, U. S. Geol. Surv., Bull. 133, 1895 [1896], p. 15 = *Cladophlebis virginiensis.*

Thinnfeldia variabilis Fontaine = *Thinnfeldia fontainei.*

Thinnfeldia variabilis Fontaine = *Cladophlebis virginiensis.*

Thinnfeldia variabilis Velenovsky = *Sagenopteris variabilis.*

Thinnfeldia n. sp. Berry, 1907 = *Protophyllocladus lobatus.*

THRINAX Linné, Sw. Prod. Veg. Ind. Occ., p. 57, 1788.

Thrinax eocenica Berry, U. S. Geol. Surv., Prof. Paper 84, 1914, p. 136, pl. xxv; pl. xxvi, f. 3; text f. 10.
 Barnwell (Twiggs): Grovetown and Phinizy Gully, Columbia County, Ga.

THUITES Sternberg, Fl. Vorw., vol. 1 (Versuch), 1823, p. 39.

Thuites alaskensis Lesquereux, U. S. Nat. Mus., Proc., vol. 5, 1882 [1883], p. 445, pl. vi, f. 7–9; Rept. U. S. Geol. Surv. Terr., vol. 8 (Cret. and Tert. Fl.), 1883, p. 257.—Knowlton, U. S. Nat. Mus., Proc., vol. 17, 1894, p. 215; Geol. Soc. Am., Bull., vol. 5, 1893, p. 579.
 Kenai: Unga Island, Alaska.

Thuites callitrina Unger. Lesquereux, U. S. Geol. and Geog. Surv. Terr., Ann. Rept. 1872 [1873], p. 371 = *Widdringtonia linguaefolia gracilis*.

Thuites crassus Lesquereux = *Brachyphyllum macrocarpum*.

Thuites kurrianus Dunker = *Sphenolepidium kurrianum*.

Thuites meriani Heer, Fl. foss. arct., vol. 3, Abt. 2, 1873, p. 73, pl. xvi, f. 17, 18.—Newberry, U. S. Geol. Surv., Mon. 26, 1895 [1896], p. 54, pl. x, f. 5.—Berry, Geol. Surv. N. J., Bull. 3, 1911, p. 84.
 Raritan: New Jersey.

Thuites sp.? Hollick and Jeffrey, N. Y. Bot. Gard., Mem., vol. 3, 1909, p. 31, pl. viii, f. 12–18; pl. xxvii, f. 4–6; pl. xxviii, f. 1–4.
 Raritan: Kreischerville, Staten Island, N. Y.

Thuites sp. Knowlton, U. S. Geol. Surv., Bull. 204, 1902, p. 26, pl. i, f. 3.
 Mascall: Van Horn's ranch, John Day Basin, Oreg.

Thuites? sp. Taylor = *Sphenolepidium dentifolium*.

THUJA Linné, Sp. pl., 1753, p. 1002.

Thuja cretacea? (Heer) Newberry, U. S. Geol. Surv., Mon. 26, 1895 [1896], p. 53, pl. x, f. 1, 1a.—Knowlton, U. S. Geol. Surv., Bull. 257, 1905, p. 133, pl. xvi, f. 3, 3a.—Berry, Bull. Torr. Bot. Club, vol. 33, 1906, p. 169; Geol. Surv. N. J., Bull. 3, 1911, p. 83; Md. Geol. Surv., Upper Cret., 1916, p. 791.
 Libocedrus cretacea Heer, Fl. foss. arct., vol. 6, Abt. 2, 1882, p. 49, pl. xxix, f. 1, 2; pl. xliii, f. 1d.
 Magothy: Grove Point, Md.; Deep Cut, Del.
 Raritan: South Amboy, N. J.
 ?Judith River: Willow Creek, Fergus County, Mont.

Thuja garmani Lesquereux, U. S. Geol. and Geog. Surv. Terr., Ann. Rept. 1872 [1873], p. 372; Rept. U. S. Geol. Surv. Terr., vol. 8 (Cret. and Tert. Fl.), 1883, p. 139.
 Miocene: Elko, Nev.

Thuja gracilis Newberry, N. Y. Lyc. Nat. Hist., Ann., vol. 9, 1868, p. 30 [name].
 Miocene: Locality not given.

Thuja interrupta Newberry, N. Y. Lyc. Nat. Hist., Ann., vol. 9, 1868, p. 42.—[Lesquereux], Ill. Cret. and Tert. Pl., 1878, pl. xi, f. 5, 5a.—Newberry, U. S. Geol. Surv., Mon. 35, 1898, p. 25, pl. xxvi, f. 5–5d.—Dawson, Roy. Soc. Canada, Trans., vol. 4, 1886 [1887], sec. 4, p. 22, pl. i, f. 3, 4.—Penhallow, Rept. Tert. Pl. Brit. Col., 1908, p. 92.—Knowlton, Washington Acad. Sci., Proc., vol. 11, 1909, pp. 198, 202; Jour. Geol., vol. 19, 1911, pp. 369, 370, 371.

Fort Union: *Fort Union, N. Dak. Abundant everywhere in the Fort Union.

Paskapoo (Fort Union): Porcupine Creek, Saskatchewan; Nine-mile Creek, Similkameen River, British Columbia.

Lance: Cannonball River and Yule, N. Dak.; Glendive and vicinity, Mont.

Puget: Discovery mine, Glacier coal field, Whatcom County, Wash.

Thuja occidentalis Linné. Dawson, Canadian Nat., vol. 2, 1857, p. 422; idem, new ser., vol. 3, 1885, p. 72; idem, vol. 6, 1888, p. 404; Geol. Hist. Pl., 1888, p. 229.—Britton, N. Y. Acad. Sci., Trans., vol. 5, 1886, p. 218.—Penhallow, Geol. Soc. Am., Bull., vol. 1, 1890, pp. 324, 334; Roy. Soc. Canada, Trans., 2d ser., vol. 2, 1896, p. 73; Brit. Assoc. Adv. Sci., Bradford meeting, 1900, p. 336; Man. N. A. Gym., 1907, p. 221.

Pleistocene: Don Valley, Toronto; "Leda clays," Montreal; Leda River, Manitoba; Marietta, Ohio; Cedar Creek and Natural Bridge, Va.

Thuja sp. Dawson, Roy. Soc. Canada, Trans., vol. 5, 1887, sec. 4, p. 33.

Belly River: Old Man River, Ribstone Creek, Twentymile Creek, British Columbia.

Thuja sp. Dawson, Roy. Soc. Canada, Trans., vol. 5, 1887, p. 34.

Paskapoo (Fort Union): 400-mile point, 49th Parallel, Middle Fork of Old Man River, British Columbia.

Thuja sp. Knowlton, Geol. Soc. Am., Bull., vol. 8, 1897, p. 134.

Lance: Lightning Creek, Converse County, Wyo.

Thuja sp. Penhallow, Roy. Soc. Canada, Trans., 2d ser., vol. 10, 1904, sec. 4, p. 61.

"Lignite Tertiary" (Paskapoo ?): Assiniboia.

Thuyites sp. Knowlton, in Merriam, Univ. Calif., Bull. Dept. Geol., vol. 2, 1901, p. 308 = *Thuites* sp. Knowlton.

THUYOXYLON Unger, Chloris Protogaea, 1842, p. 31.

Thuyoxylon americanum Unger, in Roemer, Kreideb. v. Texas, 1852, p. 95.

Trinity ?: New Braunfels, Tex.

THYRSOPTERIS Kunze, Linnaea, vol. 9, 1834, p. 507.

Thyrsopteris alata Fontaine = *Onychiopsis goepperti.*

Thyrsopteris angustifolia Fontaine = *Onychiopsis psilotoides.*

Thyrsopteris angustiloba Fontaine = *Onychiopsis goepperti.*

Thyrsopteris bella Fontaine = *Onychiopsis goepperti.*

Thyrsopteris brevifolia Fontaine = *Onychiopsis brevifolia.*

Thyrsopteris brevipennis Fontaine = *Onychiopsis latiloba.*

Thyrsopteris crassinervis Fontaine = *Onychiopsis nervosa.*

Thyrsopteris crenata Fontaine = *Onychiopsis latiloba.*

Thyrsopteris decurrens Fontaine = *Onychiopsis goepperti.*

Thyrsopteris densifolia Fontaine = *Onychiopsis goepperti.*

Thyrsopteris dentata Fontaine = *Onychiopsis brevifolia.*

Thyrsopteris dentifolia Fontaine = *Onychiopsis psilotoides.*

Thyrsopteris dentifolia Fontaine, in Ward, U. S. Geol. Surv., Nineteenth Ann. Rept., pt. 2, 1899, p. 660, pl. clxi, f. 6–9 = *Onychiopsis psilotoides.*

Thyrsopteris distans Fontaine = *Onychiopsis goepperti.*

Thyrsopteris divaricata Fontaine = *Onychiopsis latiloba.*

Thyrsopteris elliptica Fontaine = *Onychiopsis goepperti.*

Thyrsopteris grevillioides (Heer) Hollick, U. S. Geol. Surv., Mon. 50, 1906, p. 31, pl. i, f. 10–13.

Sphenopteris grevillioides Heer, Handl. Kgl. Sv. Vet. Akad., vol. 12, No. 6 (Fl. foss. arct., vol. 3, Abt. 2), 1874, p. 34, pl. xi, f. 10-11.

Magothy: Gay Head, Marthas Vineyard, Mass.

Thyrsopteris heteroloba Fontaine = *Onychiopsis nervosa.*

Thyrsopteris heteromorpha Fontaine = *Ruffordia goepperti.*

Thyrsopteris heterophylla Fontaine = *Onychiopsis brevifolia.*

Thyrsopteris inaequipinnata Fontaine = *Onychiopsis goepperti.*

Thyrsopteris insignis Fontaine = *Onychiopsis psilotoides.*

Thyrsopteris insignis angustipennis Fontaine = *Onychiopsis psilotoides.*

Thyrsopteris meekiana Fontaine = *Onychiopsis nervosa.*

Thyrsopteris meekiana angustiloba Fontaine = *Onychiopsis goepperti.*

Thyrsopteris microloba Fontaine = *Onychiopsis goepperti.*

Thyrsopteris microloba alata Fontaine = *Onychiopsis goepperti.*

Thyrsopteris microphylla Fontaine = *Onychiopsis psilotoides.*

Thyrsopteris murrayana (Brongniart) Heer, Fl. foss. arct., vol. 4, 1876, Abt. 2, p. 30, pl. i, f. 4, 4b, 4c; pl. ii, f. 1, 2a, 3, 4, 4b; pl. viii, f. 11b.—Fontaine, in Ward, U. S. Geol. Surv., Mon. 48, 1905 [1906], p. 61, pl. viii, f. 4–11; p. 148, pl. xxxviii, f. 3, 4.

Pecopteris murrayana Brongniart, Hist. vég. foss., 1836, p. 358, pl. cxxvi, f. 1, 1A, 2–4, 4A, 5, 5A.

Jurassic: Douglas and Curry counties, Oreg.

Thyrsopteris nana Fontaine = *Onychiopsis brevifolia.*

Thyrsopteris nervosa Fontaine = *Onychiopsis nervosa.*

Thyrsopteris obtusiloba Fontaine = *Onychiopsis nervosa.*

Thyrsopteris pachyphylla Fontaine = *Onychiopsis brevifolia.*

Thyrsopteris pachyrachis Fontaine = *Onychiopsis goepperti.*

Thyrsopteris pecopteroides Fontaine = *Onychiopsis nervosa.*

Thyrsopteris pinnatifida Fontaine = *Onychiopsis goepperti.*

Thyrsopteris rarinervis Fontaine = *Onychiopsis goepperti.*

Thyrsopteris rarinervis Fontaine = *Onychiopsis psilotoides.*

Thyrsopteris retusa Fontaine = *Onychiopsis brevifolia.*

Thyrsopteris rhombifolia Fontaine = *Onychiopsis goepperti.*

Thyrsopteris rhombiloba Fontaine = *Onychiopsis goepperti.*

Thyrsopteris sphenopteroides Fontaine = *Onychiopsis brevifolia.*

Thyrsopteris squarrosa Fontaine = *Onychiopsis brevifolia.*

Thyrsopteris varians Fontaine = *Onychiopsis goepperti.*

Thyrsopteris virginica Fontaine = *Onychiopsis goepperti.*

TILIA (Tournefort) Linné, Sp. pl., 1753, p. 514.

Tilia alaskana Heer, Fl. foss. arct., vol. 2, Abt. 2, 1869, p. 36, pl. x, f. 2, 3.—Knowlton, U. S. Nat. Mus., Proc., vol. 17, 1894, p. 230; Geol. Soc. Am., Bull., vol. 5, 1893, p. 586.
Kenai: Port Graham, Alaska.

Tilia americana Linné. Penhallow, Brit. Assoc. Adv. Sci., Bradford meeting, 1900, p. 336; Roy. Soc. Canada, Trans., 2d ser., vol. 10, sec. 4, p. 75; Am. Nat., vol. 41, 1907, p. 449.
Pleistocene: Don Valley, Toronto, Canada.

Tilia antiqua Newberry = *Viburnum antiquum.*

Tilia dubia (Newberry) Berry, Torreya, vol. 7, 1907, p. 81.
Tiliaephyllum dubium Newberry, U. S. Geol. Surv., Mon. 26, 1895 [1896], p. 109, pl. xv, f. 5.—Woolman, Ann. Rept. State Geol. N. J., 1896 [1897], p. 212.
Pleistocene: Fish House, N. J.

Tilia populifolia Lesquereux, Rept. U. S. Geol. Surv. Terr., vol. 8 (Cret. and Tert. Fl.), 1883, p. 179, pl. xxxiv, f. 8, 9.—Knowlton, U. S. Geol. Surv., Mon. 32, pt. 2, 1899, p. 743; U. S. Nat. Mus., Proc., vol. 51, 1916, p. 289.
Miocene: Florissant, Colo.
Fort Union: Elk Creek, Yellowstone National Park.

Tilia speciosissima Knowlton, U. S. Geol. Surv., Prof. Paper 101, 1918, p. 336, pl. lxvii.
Raton: Cucharas Canyon, Walsenburg, Colo.

Tilia weedii Knowlton, Bull. Torr. Bot. Club, vol. 29, 1902, p 706, text f. 1.
Fort Union: Porcupine Butte, near Melville, Mont.

Tiliaephyllum dubium Newberry = *Tilia dubia.*

TITHYMALUS (Tournefort) Adanson, Fam. Pl., vol. 2, 1763, 355.

Tithymalus phenacodorum Cockerell, Torreya, vol. 14, 1914, p. 137, text f. 2A, B.
Wasatch?: Five miles southeast of mouth of Pat O'Hara Creek, Clark Fork basin, Wyo.

Tithymalus willistoni Cockerell, Torreya, vol. 9, 1909, p. 119; idem, vol. 14, 1914, p. 137, text f. 2c.
"Loup Fork": Long Island, Kans.

TMESIPTERIS Bernardi, Schrader's Jour., vol. 2, 1800, p. 131.

Tmesipteris alleni (Lesquereux) Hollick, Bull. Torr. Bot. Club, vol. 21, 1894, p. 256, pl. ccv, f. 12.—Cockerell, Am. Mus. Nat. Hist., Bull., vol. 24, 1908, p. 77.—Knowlton, U. S. Nat. Mus., Proc., vol. 51, 1916, p. 247.
Ophioglossum alleni Lesquereux, U. S. Geol. and Geog. Surv. Terr., Ann. Rept. 1872 [1873], p. 371.
Salvinia alleni Lesquereux, Rept. U. S. Geol. Surv. Terr., vol. 7 (Tert. Fl.), 1878, p. 65, pl. v, f. 11.—Knowlton, in Lindgren, Jour. Geol., vol. 4, 1896, p. 889.
Miocene: *Florissant, Colo.
Miocene (auriferous gravels): Independence Hill, Placer County, Calif.

Todia? saportanea Lesquereux = *Lomatia? saportanea.*

Torreya densifolia Dawson = *Tumion densifolium.*

Torreya dicksonioides Dawson = *Tumion dicksonioides.*

Torreya falcata Fontaine = *Tumion falcatum.*

Torreya oblanceolata Lesquereux = *Tumion oblanceolatum.*

Torreya virginica Fontaine = *Tumion virginicum.*

TOXYLON Rafinesque, Am. Month. Mag., vol. 2, 1817, p. 118.

Toxylon longipetiolatum Hollick = *Persea longipetiolatum.*

Toxylon pomiferum Rafinesque. Knowlton, U. S. Geol. Surv., Bull. 152, 1898, p. 232.
Maclura auriantiaca Nuttall. Penhallow, in Coleman, Jour. Geol., vol. 3, 1895, p. 635.—Penhallow, Roy. Soc. Canada, Trans., 2d ser., vol. 2, 1896, pp. 69, 72; Brit. Assoc. Adv. Sci., Bradford meeting, 1900, p. 336; Am. Nat., vol. 41, 1907, p. 446.
Pleistocene: Don Valley, Toronto, Canada; Bridgeton, N. J.

TRAPA Linné, Sp. pl., 1753, p. 120.

Trapa alabamensis Berry, Torreya, vol. 14, 1914, p. 107, text f. 4, 5; U. S. Geol. Surv., Prof. Paper 98, 1916, p. 203, pl. xlvii, f. 9, 10.
Pliocene: Red Bluff, Perdido Bay, Baldwin County, Ala.
Pliocene (Citronelle): Lambert, Mobile County, Ala.

Trapa americana Knowlton, U. S. Geol. Surv., Eighteenth Ann. Rept., pt. 3, 1898, p. 733, pl. cii, f. 7a.

Payette: Idaho City, Idaho.

Trapa borealis Heer, Fl. foss. arct., vol. 2, Abt. 2, 1869, p. 38, pl. viii, f. 9–14.—?Dawson, Brit. N. A. Bound. Com. (Rept. Geol. and Res. Vic. 49th Parallel), 1875, App. A, p. 330, pl. xvi, f. 10.—Penhallow, Rept. Tert. Pl. Brit. Col., 1908, p. 92.—Knowlton, U. S. Nat. Mus., Proc., vol. 17, 1894, p. 226; Geol. Soc. Am., Bull., vol. 5, 1893, p. 585.

Kenai: Port Graham, Alaska.

Eocene: ?Badlands, west of Woody Mountain; Great Valley, Porcupine Creek, British Columbia.

Trapa? cuneata Knowlton, U. S. Geol. Surv., Bull. 163, 1900, p. 64, pl. v, f. 6; idem, Bull. 257, 1905, p. 145.

Mesaverde: Point of Rocks, Wyo.

Judith River: Willow Creek, Fergus County, Mont.

Trapa? microphylla Lesquereux, U. S. Geol. and Geog. Surv. Terr., Bull., vol. 1, 1875 [1876], p. 369; idem, Ann. Rept. 1874 [1876], p. 304; Rept. U. S. Geol. Surv. Terr., vol. 7 (Tert. Fl.), 1878, p. 295, pl. lxi, f. 16–17a.—Ward, U. S. Geol. Surv., Sixth Ann. Rept., 1884–85 [1886], p. 554, pl. xlix, f. 2–5; idem, Bull. 37, 1887, p. 64, pl. xxviii, f. 2–5.—Knowlton, idem, Mon. 32, pt. 2, 1898, p. 761, pl. lxxvii, f. 3, 4; idem, Bull. 163, 1900, p. 62, pl. v, f. 7; idem, Bull. 257, 1905, p. 144; Washington Acad. Sci., Proc., vol. 11, 1909, pp. 189, 202, 207; in Atwood, U. S. Geol. Surv., Bull. 467, 1911, p. 44.

?*Neuropteris angulata* Newberry, in Ives, Rept. Colorado River of the West, explored in 1857, 1858, pt. 3, 1861, p. 131, pl. iii, f. 5.—Fontaine, in Ward, U. S. Geol. Surv., Twentieth Ann. Rept., pt. 2, 1900, p. 337.

Mesaverde: *Point of Rocks, Wyo.

Belly River: Red Deer River, Alberta.

Judith River: Willow Creek, Fergus County, Mont.; Bighorn Basin, Wyo.

——: Wolverine Creek, Yellowstone National Park.

Fort Union: Near Glendive, Mont. Abundant in Fort Union.

Lance: Near Miles City, Mont.; Yule, N. Dak.; Converse County, Wyo.

Chignik (Upper Cretaceous): Chignik River, Alaska.

Trapa? occidentalis Knowlton, U. S. Geol. Surv., Eighteenth Ann. Rept., pt. 3, 1898, p. 734, pl. cii, f. 7b.

Payette: Idaho City, Idaho.

Trapa wilcoxensis Berry, Torreya, vol. 14, 1914, p. 106, text f. 1–3; U. S. Geol. Surv., Prof. Paper 91, 1916, p. 326, pl. ci, f. 7–9.

Lagrange: Puryear, Henry County, Tenn.

Trapa sp. Knowlton, Geol. Soc. Am., Bull., vol. 8, 1897, p. 142.
Hanna: "Dutton Creek coal," Laramie Plains, Wyo.

TRICALYCITES Newberry, U. S. Geol. Surv., Mon. 26, 1895 [1896], p. 132. [Type, *T. papyraceus*.]

Tricalycites major Hollick, N. Y. Bot. Gard., Bull., vol. 3, 1904, p. 416, pl. 72, f. 3–7; U. S. Geol. Surv., Mon. 50, 1906, p. 108, pl. v, f. 13–22; N. Y. Bot. Gard., Bull., vol. 8, 1912, p. 168, pl. clxiii, f. 1.
"*Pinus*" sp. Hollick, N. Y. Acad. Sci., Trans., vol. 12, 1892, p. 31, pl. i, f. 19.
Winged seed, Hollick, Bull. Torr. Bot. Club, vol. 21, 1894, p. 62, pl. clxxx, f. 1.
Magothy: Gay Head, Marthas Vineyard, Mass.; Glen Cove, Long Island, N. Y.; Nashaquitsa, R. I.
Raritan: Tottenville, Staten Island; Roslyn, Long Island, N. Y.

Tricalycites papyraceus Newberry, in Hollick, Bull. Torr. Bot. Club, vol. 21, 1894, p. 63, pl. clxxx, f.(?) 8.—Hollick, N. Y. Acad. Sci., Ann., vol. 11, 1898, p. 423, pl. xxxvii, f. 1, 2; N. Y. Bot. Gard., Bull., vol. 2, 1902, p. 405, pl. xli, f. 3.—Berry, Bull. Torr. Bot. Club, vol. 31, 1904, p. 81, pl. i, f. 4; Geol. Surv. N. J., Ann. Rept. 1905 [1906], p. 139.—Hollick, U. S. Geol. Surv., Mon. 50, 1906, p. 109, pl. v, f. 8–12.—Berry, Bull. Torr. Bot. Club, vol. 39, 1912, p. 405; Geol. Surv. N. J., Bull. 3, 1911, p. 221; U. S. Geol. Surv., Prof. Paper 112, 1919, p. 137, pl. xxviii, f. 1–5.
Tuscaloosa: Shirleys Mill and Glen Allen, Ala.
Magothy: Chappaquiddick Island, Marthas Vineyard, Mass.
Magothy?: Lloyd Neck, Long Island, N. Y.; Balls Point, Block Island, R. I.; Cliffwood, N. J.
Raritan: Tottenville, Staten Island, N. Y.; Woodbridge and South Amboy, N. J.

Tricalycites papyraceus Newberry. Hollick, N. Y. Acad. Sci., Trans., vol. 15, 1895, p. 6 = *Calycites alatus*.

TRICARPELLITES Bowerbank, History of the fossil fruits and seeds of the London clay, 1840, p. 76.

Tricarpellites acuminatus Perkins, Rept. Vt. State Geol. 1903–4 [1904], p. 190, pl. lxxviii, f. 83.
Miocene: Brandon, Vt.

Tricarpellites alatus Perkins, Rept. Vt. State Geol. 1905–6 [1906], p. 216, pl. lvi, f. 11, 12.
Miocene: Brandon, Vt.

Tricarpellites amygdaloideus Perkins, Rept. Vt. State Geol. 1903–4 [1904], p. 188, pl. lxxvii, f. 58.
Miocene: Brandon, Vt.

Tricarpellites angularis Perkins, Rept. Vt. State Geol. 1903–4 [1904], p. 187, pl. lxxvii, f. 57.
Miocene: Brandon, Vt.

Tricarpellites brandonianus Perkins, Rept. Vt. State Geol. 1905–6 [1906], p. 217, pl. lvi, f. 13, 14.
Miocene: Brandon, Vt.

Tricarpellites carinatus Perkins, Rept. Vt. State Geol. 1903–4 [1904], p. 186, pl. lxxvii, f. 47.
Miocene: Brandon, Vt.

Tricarpellites castanoides Perkins, Rept. Vt. State Geol. 1903–4 [1904], p. 187, pl. lxxvii, f. 54.
Miocene: Brandon, Vt.

Tricarpellites contractus Perkins, Rept. Vt. State Geol. 1903–4 [1904], p. 189, pl. lxxvii, f. 67.
Miocene: Brandon, Vt.

Tricarpellites curtus Perkins, Rept. Vt. State Geol. 1905–6 [1906], p. 216, pl. lvi, f. 7, 8.
Miocene: Brandon, Vt.

Tricarpellites daleii Perkins, Rept. Vt. State Geol. 1903–4 [1904], p. 186, pl. lxxvii, f. 48.
Miocene: Brandon, Vt.

Tricarpellites elongatus Perkins, Rept. Vt. State Geol. 1903–4 [1904], p. 186, pl. lxxvii, f. 45.
Miocene: Brandon, Vt.

Tricarpellites fagoides Perkins, Rept. Vt. State Geol. 1903–4 [1904], p. 188, pl. lxxvii, f. 59.
Miocene: Brandon, Vt.

Tricarpellites fissilis (Lesquereux) Perkins, Rept. Vt. State Geol. 1903–4 [1904], p. 188, pl. lxxvii, f. 61–64; text f. 7, 8 (p. 185); Geol. Soc. Am., Bull., vol. 16, 1905, p. 512, pl. lxxxvii, f. 19; Rept. Vt. State Geol. 1905–6 [1906], pl. liii, f. 19.—Knowlton, Bull. Torr. Bot. Club, vol. 29, 1902, pl. xxv, f. 7, 8.

Carpolithes fissilis Lesquereux, Am. Jour. Sci., 2d ser., vol. 32, 1861, p. 365; Geol. Vt., vol. 2, 1861, p. 713, f. 118, 119, 124.

Miocene: Brandon, Vt.

Tricarpellites hemiovalis Perkins, Rept. Vt. State Geol. 1903–4 [1904], p. 190, pl. lxxxi, f. 171.
Miocene: Brandon, Vt.

Tricarpellites inequalis Perkins, Rept. Vt. State Geol. 1903–4 [1904], p. 186, pl. lxxvii, f. 44.
Miocene: Brandon, Vt.

Tricarpellites lignitus Perkins, Rept. Vt. State Geol. 1903–4 [1904], p. 186, pl. lxxvii, f. 46.
Miocene: Brandon, Vt.

Tricarpellites major Perkins, Rept. Vt. State Geol. 1903–4 [1904], p. 189, pl. lxxvii, f. 66.
Miocene: Brandon, Vt.

Tricarpellites obesus Perkins, Rept. Vt. State Geol. 1903–4 [1904], p. 188, pl. lxxvii, f. 60.
Miocene: Brandon, Vt.

Tricarpellites ovalis Perkins, Rept. Vt. State Geol. 1903–4 [1904], p. 187, pl. lxxvii, f. 53.
Miocene: Brandon, Vt.

Tricarpellites pringlei Perkins, Rept. Vt. State Geol. 1903–4 [1904, p. 189, pl. lxxvii, f. 68.
Miocene: Brandon, Vt.

Tricarpellites rostratus Perkins, Rept. Vt. State Geol. 1903–4 [1904], p. 187, pl. lxxvii, f. 52.
Miocene: Brandon, Vt.

Tricarpellites rugosus Perkins, Rept. Vt. State Geol. 1903–4 [1904], p. 187, pl. lxxvii, f. 55.
Miocene: Brandon, Vt.

Tricarpellites seelyi Perkins, Rept. Vt. State Geol. 1903–4 [1904], p. 189, pl. lxxvii, f. 65.
Miocene: Brandon, Vt.

Tricarpellites striatus Newberry, U. S. Geol. Surv., Mon. 26, 1895 [1896], p. 132, pl. xlvi, f. 9–13.—Hollick, idem, Mon. 50, 1906, p. 108, pl. vii, f. 1.—Berry, Geol. Surv. N. J., Bull. 3, 1911, p. 223.
Raritan: Woodbridge, N. J.
Magothy: Gay Head, Marthas Vineyard, Mass.

Tricarpellites triangularis Perkins, Rept. Vt. State Geol. 1905–6 [1906], p. 216, pl. lvi, f. 9, 10.
Miocene: Brandon, Vt.

Trichopteris filamentosa Hall, in Frémont = rootlets; discarded.

Trichopteris gracilis Hall, in Frémont = rootlets; discarded.

Trunk No. 1, Fontaine, U. S. Geol. Surv., Mon. 15, 1889 [1890], p. 189, pls. clxxiv–clxxviii.

Trunk No. 2, Fontaine, U. S. Geol. Surv., Mon. 15, 1889 [1890], p. 191, pls. clxxix, clxxx.
——: Maryland.

TSUGA Endlicher, Syn. Conif., 1847, p. 83.

Tsuga mertensiana (Bongard) Sargent. Knowlton, Jour. Geol., vol. 3, 1895, p. 531; U. S. Nat. Mus., Proc., vol. 17, 1894, p. 215; Geol. Soc. Am., Bull., vol. 5, 1893, p. 579.
Recent: Muir Glacier, Alaska.

TUMION Rafinesque, Amen. Nat., 1840, p. 63.

Tumion carolinianum Berry, Am. Jour. Sci., 4th ser., vol. 25, 1908, p. 383, text f. 1–3; ?Bull. Torr. Bot. Club, vol. 37, 1910, p. 504; U. S. Geol. Surv., Prof. Paper 84, 1914, p. 107; idem, Prof. Paper 112, 1919, p. 70.
Black Creek: Rockfish Creek near Hope Mills, Cumberland County, N. C.
Eutaw: ?McBrides Ford, near Columbus, Ga.

Tumion densifolium (Dawson) Knowlton, U. S. Geol. Surv., Bull. 152, 1898, p. 234.

Torreia densifolia Dawson, Roy. Soc. Canada, Trans., vol. 1, 1882 [1883], p. 25, sec. 4, pl. v, f. 20, 20a.

Upper Cretaceous: Protection Island.

Tumion dicksonioides (Dawson) Knowlton, U. S. Geol. Surv., Bull. 152, 1898, p. 234.

Torreia dicksonioides Dawson, Roy. Soc. Canada, Trans., vol. 1, 1882 [1883], sec. 4, p. 21, pl. ii, f. 4.

Colorado ?: Pine River, Alberta.

Tumion falcatum (Fontaine) Knowlton, U. S. Geol. Surv., Bull. 152, 1898, p. 234.

Torreya falcata Fontaine, U. S. Geol. Surv., Mon. 15, 1889 [1890], p. 235, pl. cxiii, f. 4.

Patuxent: Near Potomac Run, Va.

Tumion oblanceolatum (Lesquereux) Knowlton, U. S. Geol. Surv., Bull. 152, 1898, p. 234.

Torreya oblanceolata Lesquereux, Rept. U. S. Geol. Surv. Terr., vol. 8 (Cret. and Tert. Fl.), 1883, p. 30, pl. i, f. 2.

Dakota: Near Golden, Colo.

Tumion virginicum (Fontaine) Knowlton, U. S. Geol. Surv., Bull. 152, 1898, p. 234.

Torreya virginica Fontaine, U. S. Geol. Surv., Mon. 15, 1889 [1890], p. 234, pl. cix, f. 8.

Patuxent: Near Brooke, Va.

TYPHA (Tournefort) Linné, Sp. pl., 1753, p. 971.

Typha latifolia? Linné. Penhallow, Roy. Soc. Canada, Trans., 2d ser., vol. 2, 1896, sec. 4, pp. 64, 73; Brit. Assoc. Adv. Sci., Bradford meeting, 1900, p. 336.

Pleistocene: Besserer's Wharf, Ottawa River, Canada.

Typha latissima "Al. Braun." Lesquereux, Rept. U. S. Geol. Surv. Terr., vol. 8 (Cret. and Tert. Fl.), 1883, p. 141, pl. xxiii, f. 4, 4a = *Typha lesquereuxi*.

Typha lesquereuxi Cockerell, Bull. Torr. Bot. Club, vol. 33, 1906, p. 307; Univ. Colo. Studies, vol. 3, 1906, p. 175.—Knowlton, U. S. Nat. Mus., Proc., vol. 51, 1916, p. 251.

Typha latissima "Al. Braun." Lesquereux, Rept. U. S. Geol. Surv. Terr., vol. 8 (Cret. and Tert. Fl.), 1883, p. 141, pl. xxiii, f. 4, 4a [not *T. latissima* Al. Braun].—Penhallow, Rept. Tert. Pl. Brit. Col., 1908, p. 93.

Miocene: Florissant, Colo.; Coal Gully, Horsefly, Tulameen, and Tranquille rivers, British Columbia.

Typha sp. Hollick, U. S. Geol. Surv., Mon. 50, 1906, p. 47, pl. vi, f. 4–6.

Magothy: Gay Head, Marthas Vineyard, Mass.

Raritan or Magothy: Lloyd Neck, Long Island, N. Y.

Typha sp. Penhallow, Roy. Soc. Canada, Trans., 2d ser., vol. 8, 1902, p. 51, f. 3, 4; Rept. Tert. Pl. Brit. Col., 1908, p. 93.
Paskapoo (Fort Union): Red Deer River, Alberta.

Tysonia Fontaine = *Cycadeoidea.*

Tysonia marylandica Fontaine = *Cycadeoidea marylandica.*

Ulmania Newberry = *Ullmannia.*

ULLMANNIA Göppert, Mon. Foss. Conif., 1850, p. 185.

Ullmannia sp. (Newberry) Knowlton, U. S. Geol. Surv., Bull. 152, 1898, p. 234.
Ulmania sp. Newberry, N. Y. Lyc. Nat. Hist., Proc., 2d ser., 1873, p. 10.
Raritan ?: Keyport, N. J.

Ulmiphyllum Fontaine = *Ulmophyllum.*

Ulmiphyllum brookense Fontaine = *Ulmophyllum brookense.*

Ulmiphyllum crassinerve Fontaine = *Ulmophyllum crassinerve.*

Ulmiphyllum tenuinerve Fontaine = *Ulmophyllum tenuinerve.*

ULMITES Dawson, Roy. Soc. Canada, Trans., vol. 8, 1890, sec. 4, p. 88. [Type, *U. pusillus.*]

Ulmites pusillus Dawson, Roy. Soc. Canada, Trans., vol. 8, 1890, sec. 4, p. 88, text f. 24.
Miocene: North Similkameen River, British Columbia.

ULMOPHYLLUM Ettingshausen, Denkschr. K. Akad. Wiss., vol. 53, 1886, p. 103 (23).

Ulmophyllum brookense (Fontaine) Knowlton, U. S. Geol. Surv., Bull. 152, 1898, p. 235.
Ulmiphyllum brookense Fontaine, U. S. Geol. Surv., Mon. 15, 1889 [1890], p. 312, pl. clv, f. 8; pl. clxiii, f. 7.
Patapsco: Near Brooke, Va.

Ulmophyllum crassinerve (Fontaine) Knowlton, U. S. Geol. Surv., Bull. 152, 1898, p. 235.
Ulmiphyllum crassinerve Fontaine, U. S. Geol. Surv., Mon. 15, 1889 [1890], p. 313, pl. clviii, f. 6, 7.
Patapsco: Deep Bottom, Va.

Ulmophyllum densinerve Fontaine, in Ward, U. S. Geol. Surv., Nineteenth Ann. Rept., pt. 2, 1899, p. 689, pl. clxix, f. 7.
Fuson: Rollin's tunnel, Crook County, Wyo.

Ulmophyllum priscum Dawson, Roy. Soc. Canada, Trans., vol. 11, 1893 [1894], sec. 4, p. 59, pl. viii, f. 28.
Upper Cretaceous: Port McNeill, Vancouver Island, British Columbia.

Ulmophyllum tenuinerve (Fontaine) Knowlton, U. S. Geol. Surv., Bull. 152, 1898, p. 235.
Ulmiphyllum tenuinerve Fontaine, U. S. Geol. Surv., Mon. 15, 1889 [1890], p. 313, pl. clviii, f. 1.
Patapsco: Near Brooke, Va.

ULMOXYLON Kaiser, Zeitschr. gesammt. Naturw., Berlin, vol. 52, 1879, p. 100.

Ulmoxylon simrothi Platen, Sitzungsb. Naturf. Ges., Leipzig, vol. 34, 1907 [1908], p. 26, pl. i, f. 5–6.
Miocene?: Calistoga, Calif.

ULMUS (Tournefort) Linné, Sp. pl., 1753, p. 225.

Ulmus affinis Lesquereux, Mus. Comp. Zool., Mem., vol. 6, No. 2, 1878, p. 16, pl. iv, f. 4, 5.—Knowlton, in Lindgren, Jour. Geol., vol. 4, 1896, p. 889.
Miocene (auriferous gravels): Table Mountain, Independence Hill, Placer County, Calif.

Ulmus alata Michaux. Lesquereux, Am. Jour. Sci., 2d ser., vol. 27, 1859, p. 365.—Berry, Jour. Geol., vol. 15, 1907, p. 343; Am. Nat., vol. 41, 1907, p. 694, pl. i, f. 6, 7; Am. Jour. Sci., 4th ser., 1910, p. 396; U. S. Nat. Mus., Proc., vol. 48, 1915, p. 299, pl. xii, f. 6.
Pleistocene: Mississippi River near Columbus, Ky.; Neuse River, N. C.; Abercrombies Landing, Chattahoochee River, Black Warrior River above Mobile, Ala.

Ulmus americana Linné. Hollick, Bull. Torr. Bot. Club, vol. 19, 1892, p. 332.—Penhallow, in Coleman, Jour. Geol., vol. 3, 1895, p. 635.—Penhallow, Roy. Soc. Canada, Trans., 2d ser., vol. 2, 1896, sec. 4, pp. 68, 73; idem, vol. 10, 1904, sec. 4, p. 75; Am. Nat., vol. 41, 1907, p. 449.—Berry, Torreya, vol. 15, 1915, p. 207.
Pleistocene: Don Valley, Toronto, Canada; Bridgeton, N. J.
Pleistocene (Talbot): Indian Head, Charles County, Md.

Ulmus antecedens Lesquereux, Mus. Comp. Zool., Bull., vol. 16, 1888, p. 49.
Denver: Golden, Colo.

Ulmus basicordata Hollick, Md. Geol. Surv., Miocene, 1904, p. 484, f. 1f.—Berry, U. S. Geol. Surv., Prof. Paper 98, 1916, p. 68, pl. xii, f. 2.
Calvert: Good Hope Hill, D. C.

Ulmus betuloides Hollick, Md. Geol. Surv., Pliocene and Pleistocene, 1906, p. 228, pl. lxx, f. 1.
Pleistocene (Sunderland): Point of Rocks, Calvert County, Md.

Ulmus braunii Heer, Fl. tert. Helv., vol. 2, 1856, p. 59, pl. lxxix, f. 14–21.—Lesquereux, Rept. U. S. Geol. Surv. Terr., vol. 8 (Cret. and Tert. Fl.), 1883, p. 161, pl. xxvii, f. 1–4, 8.—Knowlton, Harriman Alaska Exped., vol. 4, 1904, p. 154.
Miocene: Florissant, Colo.
Kenai: Kukak Bay, Alaska.
— —: Kamloops, British Columbia.

Ulmus brownellii Lesquereux, Rept. U. S. Geol. Surv. Terr., vol. 8 (Cret. and Tert. Fl.), 1883, p. 160, pl. xxviii, f. 2, 4.
Miocene: *Florissant, Colo.; White River, Wyo.

Ulmus californica Lesquereux, Mus. Comp. Zool., Mem., vol. 6, No. 2, 1878, p. 15, pl. iv, f. 1, 2; pl. vi, f. 7a; Rept. U. S. Geol. Surv. Terr., vol. 8 (Cret. and Tert. Fl.), 1883, p. 265, pl. xlvB, f. 3, 4, 7; U. S. Nat. Mus., Proc., vol. 11, 1888, p. 18, pl. v, f. 6 [there wrongly stated to be a figure of *Quercus horniana*].—Knowlton, in Lindgren, Jour. Geol., vol. 4, 1896, p. 889; in Smith, U. S. Geol. Surv., Folio 86 (Ellensburg, Wash.), 1903, p. 3.
Miocene (auriferous gravels): Chalk Bluffs, Table Mountain, and Independence Hill, Calif.
Mascall: Van Horn's ranch, John Day Basin, Oreg.
Ellensburg: Ellensburg, Wash.
Eocene: Southwestern Oregon.

Ulmus columbiana Penhallow, Roy. Soc. Canada, Trans., 3d ser., vol. 1, sec. 4, 1907, p. 299, pl. viii, f. 1, 2; Rept. Tert. Pl. Brit. Col., 1908, p. 93.
Oligocene: Kettle River, 6 miles above Midway, British Columbia.

Ulmus dubia Dawson, Roy. Soc. Canada, Trans., vol. 1, 1882 [1883], sec. 4, p. 27, pl. vii, f. 29.
Upper Cretaceous: Baynes Sound, Vancouver Island, British Columbia.

Ulmus grandifolia Newberry, U. S. Nat. Mus., Proc., vol. 5, 1882 [1883], p. 508.
Fort Union: Tongue River, Wyo.

Ulmus hilliae Lesquereux, Rept. U. S. Geol. Surv. Terr., vol. 8 (Cret. and Tert. Fl.), 1883, p. 160, pl. xxviii, f. 1, 3.—Cockerell, Am. Nat., vol. 44, 1910, p. 40, f. 5.
Miocene: *Florissant, Colo.

Ulmus irregularis Lesquereux = *Ficus coloradensis*.

Ulmus minima Ward, U. S. Geol. Surv., Sixth Ann. Rept., 1884–85 [1886], p. 552, pl. xlvi, f. 3, 4; U. S. Geol. Surv., Bull. 37, 1887, p. 45, pl. xxiii, f. 3, 4.—Penhallow, Rept. Tert. Pl. Brit. Col., 1908, p. 93.—?Knowlton, U. S. Geol. Surv., Mon. 32, pt. 2, 1899, p. 711.
Fort Union: Clear Creek, Glendive, Mont.; Yanceys, Yellowstone National Park.

Ulmus newberryi Knowlton, U. S. Geol. Surv., Bull. 204, 1902, p. 54, pl. ix, f. 4.
Ulmus speciosa Newberry, U. S. Geol. Surv., Mon. 26, 1898 [1899], p. 80, pl. xlv, f. 5, 8 [not f. 2–4, 7].
Ulmus n. sp. Knowlton, in Merriam, Univ. Calif., Bull. Dept. Geol., vol. 2, 1901, p. 288.
Clarno (upper part): Bridge Creek, Oreg.

Ulmus "officinis" Lesquereux. Knowlton, in Lindgren, Jour. Geol., vol. 4, 1896, p. 889 = misspelling for *Ulmus affinis*.

Ulmus orbicularis Ward, U. S. Geol. Surv., Sixth Ann. Rept., 1884–85 [1886], p. 553, pl. xlvi, f. 6; U. S. Geol. Surv., Bull. 37, 1887, p. 46, pl. xxiii, f. 6.

Fort Union: Clear Creek, Glendive, Mont.

Ulmus planeroides Ward = *Ulmus wardii*.

Ulmus plurinervia Unger, Chlor. Prot., 1847, p. 95, pl. xxv, f. 1–4.—Heer, Fl. foss. arct., vol. 2, Abt. 2, 1869, p. 34, pl. v, f. 1.—Lesquereux, U. S. Nat. Mus., Proc., vol. 11, 1888, p. 18.—Knowlton, idem, vol. 17, 1894, p. 224; Geol. Soc. Am., Bull., vol. 5, 1893, p. 584; U. S. Geol. Surv., Bull. 204, 1902, p. 55.

Mascall: Van Horn's ranch, John Day Basin, Oreg.

Kenai: Port Graham, Alaska.

Ulmus praecursor Dawson, Roy. Soc. Canada, Trans., vol. 4, 1886 [1887], sec. 4, p. 28, pl. ii, f. 11.

Paskapoo (Fort Union): Canada [exact locality not given].

Ulmus protoamericana Penhallow, Roy. Soc. Canada, Trans., 3d ser., vol. 1, sec. 4, 1907, p. 298, pl. vii, f. 1, 2; Rept. Tert. Pl. Brit. Col., 1908, p. 94.

Oligocene: Kettle River, 6 miles above Midway, British Columbia.

Ulmus protoracemosa Penhallow, Roy. Soc. Canada, Trans., 3d ser., vol. 1, sec. 4, 1907, p. 297, pl. iv, f. 2; pl. v, f. 1, 2; pl. vi, f. 1, 2; Am. Nat., vol. 41, 1907, p. 450; Rept. Tert. Pl. Brit. Col., 1908, p. 94.

Oligocene: Kettle River, 6 miles above Midway, British Columbia.

Pleistocene: Elmira, N. Y.

Ulmus pseudo-americana Lesquereux = *Ulmus speciosa*.

Ulmus pseudo-fulva Lesquereux, Mus. Comp. Zool., Mem., vol. 6, No. 2, 1878, p. 16, pl. iv, f. 3.—Knowlton, in Lindgren, Jour. Geol., vol. 4, 1896, p. 889.—?Knowlton, U. S. Geol. Surv., Mon. 32, pt. 2, 1899, p. 711, pl. lxxxviii, f. 2; in Smith, U. S. Geol. Surv., Folio 86 (Ellensburg, Wash.), 1903, p. 3.

Miocene (auriferous gravels): Chalk Bluffs, Nevada County; Independence Hill, Placer County, Calif.

Miocene: Lamar River, Yellowstone National Park.

Ulmus pseudo-racemosa Hollick, Md. Geol. Surv., Pliocene and Pleistocene, 1906, p. 228, pl. lxxi, f. 11–13.

Pleistocene (Sunderland): Head of Island Creek, Calvert County, Md.

Ulmus quercifolia Unger, Chlor. Prot., 1847, p. 96, pl. xxv, f. 5.—Lesquereux, Mus. Comp. Zool., Bull., vol. 16, 1888, p. 49.

Denver: Golden, Colo.

Ulmus racemosa Thomas. Penhallow, Roy. Soc. Canada, Trans., 2d ser., vol. 2, 1896, p. 73; Brit. Assoc. Adv. Sci., Bradford meeting, 1900, p. 336.

Pleistocene: Don Valley near Toronto, Canada.

Ulmus rhamnifolia Ward, U. S. Geol. Surv., Sixth Ann. Rept., 1884–85 [1886], p. 552, pl. xlvi, f. 5; U. S. Geol. Surv., Bull. 37, 1887, p. 45, pl. xxiii, f. 5.—?Knowlton, U. S. Geol. Surv., Mon. 32, pt. 2, 1899, p. 712.

Fort Union: Clear Creek, Glendive, Mont.; below Elk Creek, Yellowstone National Park.

Ulmus sorbifolia Göppert, Zeitschr. Deutsch. geol. Gesell., vol. 4, 1852, p. 492.—Lesquereux, U. S. Nat. Mus., Proc., vol. 5, 1882, p. 447; Rept. U. S. Geog. Surv. Terr., vol. 8 (Cret. and Tert. Fl.), 1883, p. 260.—Knowlton, U. S. Nat. Mus., Proc., vol. 17, 1894, p. 224; Geol. Soc. Am., Bull., vol. 5, 1893, p. 584.

Kenai: Cook Inlet, Alaska.

Ulmus speciosa Newberry, U. S. Nat. Mus., Proc., vol. 5, 1883, p. 507; U. S. Geol. Surv., Mon. 26, 1898 [1899], p. 80, pl. xlv, f. 2–4, 7 [not f. 5, 8].—Knowlton, U. S. Geol. Surv., Bull. 204, 1902, p. 53.—Penhallow, Rept. Tert. Pl. Brit. Col., 1908, p. 94.

Ulmus pseudo-americana Lesquereux, Rept. U. S. Geol. Surv. Terr., vol. 8 (Cret. and Tert. Fl.), 1883, p. 249, pl. liv, f. 10.

Clarno (upper part): Bridge Creek, Oreg.

——: Coal Gully, Quilchena; Horsefly River, British Columbia.

Ulmus speciosa Newberry, U. S. Geol. Surv., Mon. 35, 1898, p. 80, pl. xlv, f. 5, 8 = *Ulmus newberryi*.

Ulmus tenuinervis Lesquereux, U. S. Geol. and Geog. Surv. Terr., Ann. Rept. 1873 [1874], p. 412; Rept. U. S. Geol. Surv. Terr., vol. 7 (Tert. Fl.), 1878, p. 188, pl. xxvi, f. 1–3.—Penhallow, Rept. Tert. Pl. Brit. Col., 1908, p. 94.—Knowlton, U. S. Nat. Mus., Proc., vol. 51, 1916, p. 266.

Miocene: *Florissant, Colo.; Tranquille River, British Columbia.

Ulmus tenuinervis Lesquereux. Hollick, Louisiana = *Dillenites ovatus*.

Ulmus wardii Knowlton and Cockerell, n. name.

Ulmus planeroides Ward, U. S. Geol. Surv., Sixth Ann. Rept., 1884–85 [1886], p. 552, pl. xlvi, f. 1, 2; U. S. Geol. Surv., Bull. 37, 1887, p. 44, pl. xxiii, f. 1, 2. [Homonym, Carrière, 1875.]

Fort Union: Clear Creek, Glendive, Mont.

Ulmus? sp. Dawson, Roy. Soc. Canada, Trans., vol. 5, 1887, sec. 4, p. 33 [wood].

Pierre: Head of Swift Current Creek, British Columbia.

Ulmus sp. ? Hollick, Md. Geol. Surv., Pliocene and Pleistocene, 1906, p. 229, pl. lxix, f. 10.

Pleistocene (Talbot): Grove Point, Cecil County, Md.

Ulmus sp. Knowlton, in Lindgren, Jour. Geol., vol. 4, 1896, p. 889.
Miocene (auriferous gravels): Independence Hill, Placer County, Calif.

Ulmus n. sp. Knowlton, in Merriam, Univ. Calif., Bull. Dept. Geol., vol. 2, 1901, p. 288 = *Ulmus newberryi.*

Ulmus, fruits of, Knowlton, U. S. Geol. Surv., Mon. 32, pt. 2, 1899, p. 712, pl. lxxxviii, f. 3, 4.
Fort Union: Below Elk Creek, Yellowstone National Park.

Ulmus sp. Knowlton, Washington Acad. Sci., Proc., vol. 11, 1909, p. 207.
Lance: Converse County, Wyo.

Ulmus sp. Knowlton, U. S. Geol. Surv., Prof. Paper 101, 1918, p. 300, pl. lxx, f. 4.
Raton: Near Bowen, Colo.

Ulmus sp. Penhallow, Roy. Soc. Canada, Trans., 3d ser., vol. 1, sec. 4, 1907, p. 301; Rept. Tert. Pl. Brit. Col., 1908, p. 93.
Oligocene: Coal Gully, Tulameen River, and Kettle River near Midway, British Columbia.

Undetermined fern, Fontaine, U. S. Geol. Surv., Mon. 15, 1889 [1890], p. 119, pl. xxi, f. 12.
Patuxent: Near Dutch Gap Canal, Va.

Undetermined fern, Fontaine, U. S. Geol. Surv., Mon. 15, 1889 [1890], p. 145, pl. lix, f. 5.
Patuxent: Near Potomac Run, Va.

Undetermined plant (a), Fontaine, U. S. Geol. Surv., Mon. 15, 1889 [1890], p. 274, pl. cxxxvi, f. 13.
Patuxent: Fredericksburg, Va.

Undetermined plant (b), Fontaine, U. S. Geol. Surv., Mon. 15, 1889 [1890], p. 274, pl. cxxxvi, f. 14.
Patuxent: Fredericksburg, Va.

Undetermined plant (c), Fontaine, U. S. Geol. Surv., Mon. 15, 1889 [1890], p. 275, pl. cxxxvii, f. 1.
Patuxent: Fredericksburg, Va.

Undetermined plant (d), Fontaine, U. S. Geol. Surv., Mon. 15, 1889 [1890], p. 275, pl. cxxxvii, f. 2.
Patuxent: Fredericksburg, Va.

Undetermined plant (e), Fontaine, U. S. Geol. Surv., Mon. 15, 1889 [1890], p. 275, pl. cxxxvii, f. 4.
Patuxent: Fredericksburg, Va.

Undetermined plant (f), Fontaine, U. S. Geol. Surv., Mon. 15, 1889 [1890], p. 275, pl. cxxxvii, f. 5.
Patuxent: Fredericksburg, Va.

Undetermined plant (g), Fontaine, U. S. Geol. Surv., Mon. 15, 1889 [1890], p. 275, pl. clxix, f. 4, 5.
Patapsco: Baltimore, Md.

Undetermined plant (h), Fontaine, U. S. Geol. Surv., Mon. 15, 1889 [1890], p. 275, pl. clxix, f. 9.
Patapsco: Baltimore, Md.

Undetermined plant (i), Fontaine, U. S. Geol. Surv., Mon. 15, 1889 [1890], p. 276, pl. clxxiii, f. 11.
Patapsco: Baltimore, Md.

Undetermined plant (*Araucarites?*) Hill, Biol. Soc. Washington, Proc., vol. 8, 1893, p. 39, pl. i, f. 1–1d.
Trinity: Glen Rose, Tex.

Unknown plant, Knowlton, U. S. Geol. Surv., Twenty-first Ann. Rept., pt. 2, 1901, p. 212, pl. xxx, f. 16, 24, 25.
Esmeralda: 3.8 kilometers northeast of Emigrant Peak, Esmeralda County, Nev.

VACCINIUM Linné, Sp. pl., 1753, p. 349.

Vaccinium alaskanum Knowlton, Harriman Alaska Exped., vol. 4, 1904, p. 157, pl. xxv, f. 1.
Kenai: Kukak Bay, Alaska.

Vaccinium arboreum Marshall. Berry, Torreya, vol. 9, 1909, p. 73; Am. Jour. Sci., 4th ser., vol. 29, 1910, p. 398; idem, vol. 34, 1912, p. 222.
Berberis sp. Berry, Jour. Geol., vol. 15, 1907, p. 343.
Pleistocene: Neuse River, N. C.; Buena Vista, Va.; Alabama River above Montgomery, Ala.

Vaccinium coloradense Lesquereux, U. S. Nat. Mus., Proc., vol. 10, 1887, p. 42, pl. iii, f. 4, 5.
Denver: Silver Cliff, Colo.

Vaccinium corymbosum Linné. Hollick, Md. Geol. Surv., Pliocene and Pleistocene, 1906, p. 236, pl. lxix, f. 7–9.—Berry, Jour. Geol., vol. 15, 1907, p. 346; Am. Jour. Sci., 4th ser., vol. 29, 1910, p. 398; Torreya, vol. 17, 1917, p. 161, f. 2.
Pleistocene (Talbot): Drum Point, Calvert County, Md.
Pleistocene: Neuse and Roanoke rivers, N. C.; Black Warrior River, 311 miles above Mobile, Ala.; Waterville, Maine.

Vaccinium friesii Heer, Fl. foss. arct., vol. 2, Abt. 2, 1869, p. 35, pl. viii, f. 4.—Knowlton, U. S. Nat. Mus., Proc., vol. 17, 1894, p. 225; Geol. Soc. Am., Bull., vol. 5, 1893, p. 585.
Kenai: Port Graham, Alaska.

Vaccinium hollicki Knowlton, n. name.
Vaccinium reticulatum Al. Braun. Lesquereux, U. S. Nat. Mus., Proc., vol. 5, 1882 [1883], p. 448, pl. x, f. 3–5; Rept. U. S. Geol. Surv. Terr., vol. 8 (Cret. and Tert. Fl.), 1883, p. 261.—Knowlton, U. S. Nat. Mus., Proc., vol. 17, 1894, p. 225; Geol. Soc. Am., Bull., vol. 5, 1893, p. 585.
Kenai: Cook Inlet, Alaska.

Vaccinium reticulatum? Al. Braun. Lesquereux (Tert. Fl., pl. lix, f. 6) = *Dalbergia lesquereuxii.*

Vaccinium reticulatum Al. Braun. Lesquereux, U. S. Nat. Mus., Proc., vol. 5, 1882 [1883], p. 448, pl. x, f. 3–5 = *Vaccinium hollicki.*

Vaccinium retigerum Cockerell = *Dalbergia lesquereuxii.*

Vaccinium spathulata Berry, Jour. Geol., vol. 15, 1907, p. 346.
Pleistocene: Neuse River, N. C.

Vaccinium uliginosum Linné. Macoun, in Coleman, Jour. Geol., vol. 3, 1895, p. 626.—Penhallow, Roy. Soc. Canada, Trans., 2d ser., vol. 2, 1896, p. 73; Brit. Assoc. Adv. Sci., Proc., Bradford meeting, 1900, p. 336.
Pleistocene: Scarborough Heights, Ontario.

Vaccinium cf. **V. textum** Heer. Berry, U. S. Geol. Surv., Prof. Paper '98' 1916, p. 70, pl. xii, f. 14, 15.
Calvert: Good Hope Hill, D. C.

VACCINOPHYLLUM Dawson, Roy. Soc. Canada, Trans., vol. 8, 1890, sec. 4, p. 88. [Type, *V. quaestum.*]

Vaccinophyllum quaestum Dawson, Roy. Soc. Canada, Trans., vol. 8, 1890, sec. 4, p. 88, text f. 23.—Penhallow, Rept. Tert. Pl. Brit. Col., 1908, p. 95.
Miocene: Similkameen River, British Columbia.

VALLISNERIA Linné, Sp. pl., 1753, p. 1015.

Vallisneria spiralis Linné. Penhallow, Brit. Assoc. Adv. Sci., Bradford meeting, 1900, p. 336.
Vallisneria sp. Penhallow, Geol. Soc. Am., Bull., vol. 1, 1890, p. 325; Roy. Soc. Canada, Trans., 2d ser., vol. 2, 1896, sec. 4, pp. 64, 73.
Pleistocene: Rolling River, Manitoba; Greens Creek and Besserer's wharf, Ottawa River, Canada.

Vallisneria sp. Penhallow = *Vallisneria spiralis.*

VANTANEA Aublet, Guian., vol. 2, 1775, p. 572.

Vantanea wilcoxiana Berry, U. S. Geol. Surv., Prof. Paper 91, 1916, p. 255, pl. liv, f. 6
Lagrange: Puryear, Henry County, Tenn.
Wilcox: Naborton, De Soto Parish, La.

VIBORQUIA Ortega, Hort. Matr., Dec. 66, 1798.

Viborquia nigrostipellata (Cockerell) Cockerell, Am. Mus. Nat. Hist., Bull., vol. 24, 1908, p. 97.
Eysenhardtia nigrostipellata Cockerell, Am. Jour. Sci., 4th ser., vol. 25, 1908, p. 232 (footnote).
Miocene: Florissant, Colo.

VIBURNITES Lesquereux, U. S. Geol. Surv., Mon. 17, 1891 [1892], p. 124. [Type, *V. crassus.*]

Viburnites crassus Lesquereux, U. S. Geol. Surv., Mon. 17, 1891 [1892], p. 124, pl. xlv, f. 1–4.
Dakota: Near Delphos, Kans.

Viburnites evansanus Ward, Jour. Geol., vol. 2, 1894, pp. 261, 262; U. S. Geol. Surv., Nineteenth Ann. Rept., pt. 2, 1899, p. 709, pl. clxxii, f. 3, 4.
Dakota: Evans quarry, 5 miles east of Hot Springs, S. Dak.

Viburnites masoni Lesquereux, U. S. Geol. Surv., Mon. 17, 1891 [1892], p. 125, pl. xlv, f. 5.
Dakota: Ellsworth County, Kans.

VIBURNUM (Tournefort) Linné, Sp. pl., 1753, p. 383.

Viburnum anceps Lesquereux = *Platanus rhomboidea.*

Viburnum anomalinervum Knowlton, U. S. Geol. Surv., Prof. Paper. 101, 1918, p. 275, pl. xlv, f. 4.
Viburnum whymperi Heer. Lesquereux, Rept. U. S. Geol. Surv. Terr., vol. 7 (Tert. Fl.), 1878, p. 225, pl. lxi, f. 23.
Mesaverde: Point of Rocks, Wyo.
Vermejo: Gray Creek, Colo.

Viburnum antiquum (Newberry) Hollick, in Newberry, U. S. Geol. Surv., Mon. 35, 1898, p. 128, pl. xxxiii, f. 1, 2.—Knowlton, Washington Acad. Sci., Proc., vol. 11, 1909, pp. 186, 189, 213; in Calvert, U. S. Geol. Surv., Bull. 47, 1912, p. 16.
Tilia antiqua Newberry, N. Y. Lyc. Nat. Hist., Ann., vol. 9, 1868, p. 52.—[Lesquereux], U. S. Geol. and Geog. Surv. Terr., Ill. Cret. and Tert. Pl., 1878, pl. xvi, f. 1, 2.
Viburnum tilioides Ward, U. S. Geol. Surv., Sixth Ann. Rept., 1884–85 [1886], p. 556, pl. lxi, f. 1–7; pl. lxii, f. 1–6; U. S. Geol. Surv., Bull. 37, 1887, p. 107, pl. l, f. 1–3; pl. li, f. 1–8; pl. lii, f. 1, 2.
Fort Union: Fort Clark, N. Dak.; Clear Creek near Glendive, Mont.
Lance: Big Muddy Creek, 28 miles south of Lisner, 9 miles west of Miles City, near Glendive, Mont.; Ilo, Big Horn Basin, Wyo.

Viburnum asperum Newberry, N. Y. Lyc. Nat. Hist., Ann., vol. 9, 1868, p. 54.—[Lesquereux], U. S. Geol. and Geog. Surv. Terr., Ill. Cret. and Tert. Pl., 1878, pl. xvi, f. 8.—Ward, U. S. Geol. Surv., Sixth Ann. Rept., 1884–85 [1886], p. 557, pl. lxiv, f. 4–9; U. S. Geol. Surv., Bull. 37, 1887, p. 113, pl. lv, f. 4–9.—Newberry, U. S. Geol. Surv., Mon. 37, 1898, p. 129, pl. xxxiii, f. 9.—Penhallow, Rept. Tert. Pl. Brit. Col., 1908, p. 95.
Fort Union: Fort Union, N. Dak.; Sevenmile Creek and Crackerbox Creek, near Glendive, Mont.
Paskapoo (Fort Union): Porcupine Creek, Red Deer River, Finlay River, Omineca River, British Columbia.

Viburnum betulaefolium Ward, U. S. Geol. Surv., Sixth Ann. Rept., 1884–85 [1886], p. 557, pl. lxv, f. 7; U. S. Geol. Surv., Bull. 37, 1887, p. 114, pl. lvii, f. 4.
Fort Union: Burns's ranch near Glendive, Mont.

Viburnum bridgetonense Britton, in Hollick, Bull. Torr. Bot. Club, vol. 19, 1892, p. 331.
Pleistocene: Bridgeton, N. J.

Viburnum calgarianum Dawson, Roy. Soc. Canada, Trans., vol. 3, 1885 [1886], sec. 4, p. 18; idem, vol. 4, 1886 [1887], sec. 4, p. 29, pl. ii, f. 14.—Penhallow, Rept. Tert. Pl. Brit. Col., 1908, p. 95.
Paskapoo (Fort Union): Calgary, Alberta; Porcupine Creek, Saskatchewan.

Viburnum castrae Knowlton and Cockerell, n. name.
Viburnum lanceolatum Newberry, N. Y. Lyc. Nat. Hist., Ann., vol. 9, 1868, p. 54.—[Lesquereux], U. S. Geol. and Geog. Surv. Terr., Ill. Cret. and Tert. Pl., 1878, pl. xvi, f. 10.—Newberry, U. S. Geol. Surv., Mon. 35, 1898, p. 131, pl. xxxiii, f. 10.—Penhallow, Rept. Tert. Pl. Brit. Col., 1908, p. 96. [Homonym, Hill, 1868.]
Fort Union: Fort Union, N. Dak.
Paskapoo (Fort Union): Porcupine Creek, Saskatchewan.

Viburnum contortum Lesquereux, U. S. Geol. and Geog. Surv. Terr., Ann. Rept. 1872 [1873], p. 396.—Knowlton, U. S. Geol. Surv., Prof. Paper 101, 1918, p. 346, pl. cviii, f. 3.
Viburnum marginatum Lesquereux, Rept. U. S. Geol. Surv. Terr., vol. 7 (Tert. Fl.), 1878, p. 223, pl. xxxviii, f. 2 [not f. 3].
Post-Laramie: *Black Buttes, Wyo.
Raton: Rockland mine, Cucharas Canyon, Colo.

Viburnum crassum Knowlton, U. S. Geol. Surv., Prof. Paper 101, 1918, p. 277, pl. lii, f. 3, 4.
Trinidad: Raton, N. Mex.

Viburnum cuneatum Newberry = *Parrotia cuneata.*

Viburnum dakotense Lesquereux, Rept. U. S. Geol. Surv. Terr., vol. 8 (Cret. and Tert. Fl.), 1883, p. 231, pl. xlvi–A, f. 9.
Fort Union: Badlands, N. Dak.

Viburnum cf. **V. dentatum** Linné. Berry, Jour. Geol., vol. 25, 1917, p. 662.
Pleistocene: Vero, Fla.

Viburnum dentoni Lesquereux, Rept. U. S. Geol. Surv. Terr., vol. 8 (Cret. and Tert. Fl.), 1883, p. 231, pl. xlix, f. 2, 3.—Penhallow, Rept. Tert. Pl. Brit. Col., 1908, p. 95.
Fort Union: Badlands, N. Dak.
Paskapoo (Fort Union): Tranquille River, British Columbia.

Viburnum dichotomum Lesquereux = *Viburnum melaenum.*

Viburnum ellsworthianum Lesquereux, U. S. Geol. Surv., Mon. 17, p. 121, pl. xxi, f. 6.
Dakota: Ellsworth County, Kans.

Viburnum elongatum Ward, U. S. Geol. Surv., Sixth Ann. Rept., 1884–85 [1886], p. 557, pl. lxiii, f. 8, 9; U. S. Geol. Surv., Bull. 37, 1887, p. 112, pl. liv, f. 4, 5.—?Knowlton, in Lindgren, Jour. Geol., vol. 4, 1896, p. 890.—Knowlton, Washington Acad. Sci., Proc., vol. 11, 1909, p. 370; in Calvert, U. S. Geol. Surv., Bull. 471, 1912, p. 16.
Fort Union: Clear Creek, Mont.
?Miocene (auriferous gravels): Independence Hill, Placer County, Calif.
Lance: Near Glendive, Mont.; South Dakota.

Viburnum erectum Ward, U. S. Geol. Surv., Sixth Ann. Rept., 1884–85 [1886], p. 557, pl. lxiv, f. 3; U. S. Geol. Surv., Bull. 37, 1887, p. 112, pl. lv, f. 3.
Fort Union: Clear Creek near Glendive, Mont.

Viburnum finale Ward, U. S. Geol. Surv., Sixth Ann. Rept., 1884–85 [1886], p. 557, pl. lxv, f. 8; U. S. Geol. Surv., Bull. 37, 1887, p. 115, pl. lvii, f. 5.
Fort Union: Iron Bluff, Glendive, Mont.

Viburnum goldianum Lesquereux, Rept. U. S. Geol. Surv. Terr., vol. 7 (Tert. Fl.), 1878, p. 227, pl. lx, f. 2–2c.
Denver: Golden, Colo.

Viburnum grandedentatum Newberry, U. S. Nat. Mus., Proc., vol. 5, 1882 [1883], p. 511.
Fort Union: Tongue River, Wyo.

Viburnum grewiopsideum Lesquereux, U. S. Geol. Surv., Mon. 17, 1891 [1892], p. 120, pl. xxi, f. 4.
Dakota: Ellsworth County, Kans.

Viburnum? hesperium Knowlton, U. S. Geol. Surv., Prof. Paper 101, 1918, p. 276, pl. xlix, f. 2.
Vermejo: Near Rockvale, Colo.

Viburnum hollickii Berry, Am. Nat., vol. 37, 1903, p. 683, f. 5, 6; Geol. Surv. N. J., Ann. Rept. 1905 [1906], p. 139.—Hollick, U. S. Geol. Surv., Mon. 50, 1906, p. 105, pl. xl, f. 17.

Grewiopsis viburnifolia Ward. Hollick, Bull. Torr. Bot. Club, vol. 21, 1894, p. 59, pl. 174, f. 8.

Viburnum whymperi Heer. Knowlton, U. S. Geol. Surv., Bull. 163, 1900, p. 72, pl. xix, f. 3 [not pl. xvii, f. 1; pl. xviii, f. 1].

Magothy: Cliffwood, N. J.
Raritan or Magothy: Lloyd Neck, Long Island, N. Y.
Mesaverde: Point of Rocks, Wyo.

Viburnum inaequilaterale Lesquereux, U. S. Geol. Surv., Mon. 17, 1891 [1892], p. 119, pl. xxi, f. 2, 3.
Dakota: Near Delphos, Kans.

Viburnum integrifolium Newberry, in Hollick, Bull. Torr. Bot. Club, vol. 21, 1894, p. 54, pl. clxxvii, f. 7.—Newberry, U. S. Geol. Surv., Mon. 26, 1895 [1896], p. 125, pl. xli, f. 1.—Hollick, U. S. Geol. Surv., Mon. 50, 1906, p. 105, pl. xl, f. 1.—Berry, Geol. Surv. N. J., Bull. 3, 1911, p. 224.
Magothy: Glen Cove, Long Island, N. Y.
Raritan: Woodbridge, N. J.

Viburnum lakesii Lesquereux, U. S. Geol. and Geog. Surv. Terr., Ann. Rept. 1873 [1874], p. 401; Rept. U. S. Geol. Surv. Terr., vol. 7 (Tert. Fl.), 1878, p. 226, pl. xxxvii, f. 13.—Dawson, Roy. Soc. Canada, Trans., vol. 1, 1882 [1883], sec. 4, p. 32.—Penhallow, Rept. Tert. Pl. Brit. Col., 1908, p. 95.
Denver: Golden, Colo.

Viburnum lanceolatum Newberry = *Viburnum castrae.*

Viburnum lesquereuxii Ward, in Lesquereux, U. S. Geol. Surv., Mon. 17, 1891 [1892], p. 121.
Dakota: Kansas.

Viburnum lesquereuxii commune Lesquereux, U. S. Geol. Surv., Mon. 17, 1891 [1892], p. 122, pl. liii, f. 2.
Dakota: Kansas.

Viburnum lesquereuxii cordifolium Lesquereux, U. S. Geol. Surv., Mon. 17, 1891 [1892], p. 122, pl. lii, f. 9.
Dakota: Kansas.

Viburnum lesquereuxii lanceolatum Lesquereux, U. S. Geol. Surv., Mon. 17, 1891 [1892], p. 123, pl. liii, f. 3.
Dakota: Kansas.

Viburnum lesquereuxii latius Lesquereux, U. S. Geol. Surv., Mon. 17, 1891 [1892], p. 123, pl. lii, f. 10.
Dakota: Kansas.

Viburnum lesquereuxii longifolium Lesquereux, U. S. Geol. Surv., Mon. 17, 1891 [1892], p. 122, pl. liii, f. 1.
Dakota: Near Fort Harker, Kans.

Viburnum lesquereuxii rotundifolium Lesquereux, U. S. Geol. Surv., Mon. 17, 1891 [1892], p. 122, pl. lii, f. 8.
Dakota: Kansas.

Viburnum? lesquereuxii tenuifolium Lesquereux, U. S. Geol. Surv., Mon. 17, 1891 [1892], p. 123, pl. lxiv, f. 13.
Dakota: Ellsworth County, Kans.

Viburnum limpidum Ward, U. S. Geol. Surv., Sixth Ann. Rept., 1884–85 [1886], p. 556, pl. lxiii, f. 1–4; U. S. Geol. Surv., Bull. 37, 1887, p. 110, pl. liii, f. 3–6.
Fort Union: Clear Creek near Clendive, Mont.

Viburnum macrodontum Ward, U. S. Geol. Surv., Sixth Ann. Rept., 1884–85 [1886], p. 556, pl. lxii, f. 10; U. S. Geol. Surv., Bull. 37, 1887, p. 110, pl. liii, f. 2.
Fort Union: Clear Creek near Glendive, Mont.

Viburnum magnum Knowlton, U. S. Geol. Surv., Prof. Paper 101, 1918, p. 347, pl. cx, f. 2.
Raton: Near Wootton and Walsenburg, Colo.

Viburnum marginatum Lesquereux = *Platanus marginata.*

Viburnum mattawanense Berry, Bull. Torr. Bot. Club, vol. 31, 1904, p. 80, pl. iv, f. 13.
Magothy: Cliffwood, N. J.

Viburnum melaenum Knowlton and Cockerell, n. name.
Viburnum dichotomum Lesquereux, U. S. Geol. and Geog. Surv. Terr., Ann. Rept. 1872 [1873], p. 399; Rept. U. S. Geol. Surv. Terr., vol. 7 (Tert. Fl.), 1878, p. 225, pl. xxxviii, f. 6. [Homonym, Buch-Ham, 1825.]
Post-Laramie: Black Buttes, Wyo.

Viburnum cf. **V. molle** Michaux. Berry, Torreya, vol. 14, 1914, p. 160.
Pleistocene: Seven Springs, Neuse River, Wayne County, N. C.

Viburnum montanum Knowlton, U. S. Geol. Surv., Bull. 163, 1900, p. 73, pl. xix, f. 1, 2; idem, Prof. Paper 101, 1918, p. 276, pl. lii, f. 2.
Vermejo: Rockvale, Colo.; Vermejo Park, N. Mex.
Mesaverde: Point of Rocks, Wyo.

Viburnum newberryanum Ward, U. S. Geol. Surv., Sixth Ann. Rept., 1884–85 [1886], p. 557, pl. lxiv, f. 10–12; pl. lxv, f. 1–3; U. S. Geol. Surv., Bull. 37, 1887, p. 113, pl. lvi, f. 1–6.—Knowlton, Washington Acad. Sci., Proc., vol. 11, 1909, p. 313.
Fort Union: Crackerbox Creek near Glendive, Mont.
Lance (Colgate): Iron Bluff near Glendive, Mont.
Lance: Near Ilo, Big Horn Basin, Wyo.

Viburnum nordenskiöldi Heer, Fl. foss. arct., vol. 2, Abt. 2, 1869, p. 36, pl. iii, f. 13.—Ward, U. S. Geol. Surv., Sixth Ann. Rept., 1884–85 [1886], p. 557, pl. lxv, f. 4–6; U. S. Geol. Surv., Bull. 37, 1887, p. 114, pl. lvii, f. 1–3.—Knowlton, U. S. Nat. Mus., Proc., vol. 17, 1894, p. 225; Geol. Soc. Am., Bull., vol. 5, 1893, p. 545.
Kenai ?: Ninilchik, Alaska.
Fort Union: Clear Creek near Glendive, Mont.; Little Missouri River and Gladstone, N. Dak.

Viburnum nudum Linné. Berry, Jour. Geol., vol. 25, 1917, p. 662.
Pleistocene: Vero, Fla.

Viburnum oppositinerve Ward, U. S. Geol. Surv., Sixth Ann. Rept., 1884–85 [1886], p. 557, pl. lxiv, f. 1, 2; U. S. Geol. Surv., Bull. 37, 1887, p. 112, pl. lv, f. 1, 2.
Fort Union: Clear Creek near Glendive, Mont.

Viburnum ovatum Penhallow, Roy. Soc. Canada, Trans., 2d ser., vol. 8, 1902, p. 62, f. 7; Rept. Tert. Pl. Brit. Col., 1908, p. 96.
Paskapoo (Fort Union): Red Deer River, Alberta.

Viburnum oxycoccoides Dawson, Roy. Soc. Canada, Trans., vol. 3, 1885 [1886], sec. 4, p. 17; idem, vol. 4, 1886 [1887], sec. 4, p. 29, pl. ii, f. 15.—Penhallow, Rept. Tert. Pl. Brit. Col., 1908, p. 96.
Paskapoo (?): Calgary, Alberta; Porcupine Creek, British Columbia.

Viburnum paucidentatum Newberry, U. S. Nat. Mus., Proc., vol. 5, 1882 [1883], p. 511.
Fort Union: Tongue River, Wyo.

Viburnum perfectum Ward, U. S. Geol. Surv., Sixth Ann. Rept., 1884–85 [1886], p. 556, pl. lxii, f. 7–9; U. S. Geol. Surv., Bull. 37, 1887, p. 109, pl. lii, f. 3, 4; pl. liii, f. 1.
Fort Union: Clear Creek near Glendive, Mont.

Viburnum perplexum Ward, U. S. Geol. Surv., Sixth Ann. Rept., 1884–85 [1886], p. 557, pl. lxiii, f. 6, 7; U. S. Geol. Surv., Bull. 37, 1887, p. 111, pl. liv, f. 2, 3.—Knowlton, Washington Acad. Sci., Proc., vol. 11, 1919, p. 215.
Fort Union: Burns's ranch, Glendive, Mont.
Lance: Near Kirby, Wyo.

Viburnum platanoides Lesquereux = *Platanus platanoides*.

Viburnum? problematicum Knowlton, U. S. Geol. Surv., Bull. 163, 1900, p. 71, pl. xix, f. 4; idem, Prof. Paper 101, 1918, p. 276, pl. xlix, f. 9.
Montana: Coalville, Utah.
Vermejo: Rockvale, Colo.

Viburnum pubescens Pursh. Dawson, Brit. N. A. Bound. Com. (Rept. Geol. and Res. Vic. 49th Parallel), 1875, App. A, p. 330.—Penhallow, Rept. Tert. Pl. Brit. Col., 1908, p. 96.
Miocene?: Porcupine Creek, Saskatchewan.

Viburnum rhamnifolium Knowlton, U. S. Geol. Surv., Prof. Paper 101, 1918, p. 277, pl. xlvii, f. 5.
Vermejo: Vermejo Park, N. Mex.

Viburnum robustum Lesquereux, U. S. Geol. Surv., Mon. 17, 1891 [1892], p. 120, pl. xx, f. 4–6.—Berry, Bull. Torr. Bot. Club, vol. 39, 1912, p. 405.
Dakota: Ellsworth County, Kans.
Woodbine: Arthurs Bluff, Tex.

Viburnum rotundifolium Lesquereux, U. S. Geol. and Geog. Surv. Terr., Ann. Rept. 1874 [1875], p. 305; Rept. U. S. Geol. Surv. Terr., vol. 7 (Tert. Fl.), 1878, p. 225, pl. xxxvii, f. 12; pl. xxxviii, f. 10; pl. lxi, f. 22.

Post-Laramie: Black Buttes, Wyo.

Viburnum saskatchuense Dawson, Roy. Soc. Canada, Trans., vol. 5, 1887 [1888], sec. 4, p. 35.—Penhallow, Rept. Tert. Pl. Brit. Col., 1908, p. 96.

Paskapoo (Fort Union): Red Deer and North Saskatchewan rivers.

Viburnum simile Knowlton, U. S. Geol. Surv., Prof. Paper 101, 1918, p. 277, pl. xlix, f. 3.

Vermejo: Rockvale, Colo.

Viburnum solitarium Lesquereux, Rept. U. S. Geol. Surv. Terr., vol. 7 (Tert. Fl.), 1878, p. 227, pl. lx, f. 3.

Denver: Golden, Colo.

Viburnum speciosum Knowlton, U. S. Geol. Surv., Prof. Paper 101, 1918, p. 347, pl. cxi, f. 1–5.

Raton: Riley Canyon near Cokedale, Colo.

Viburnum sphenophyllum Knowlton, in Lesquereux, U. S. Geol. Surv., Mon. 17, 1891 [1892], p. 123, pl. liii, f. 4.

Dakota: Kansas.

Viburnum tilioides Ward = *Viburnum antiquum.*

Viburnum vulpinum Knowlton, U. S. Geol. Surv., Prof. Paper 98, 1916, p. 92, pl. xviii, f. 1.

Fox Hills: Wildcat Mound, 2½ miles south of Milliken, Colo.

Viburnum whymperi Heer, Fl. foss. arct., vol. 2, Abt. 1b, 1869, p. 475, pl. xlvi, f. 1b.—Lesquereux, U. S. Geol. Surv. Terr., vol. 7 (Tert. Fl.), 1878, p. 225, pl. xxxviii, f. 7.—Ward, U. S. Geol. Surv., Sixth Ann. Rept., 1884–85 [1886], p. 557, pl. lxiii, f. 5; U. S. Geol. Surv., Bull. 37, 1887, p. 111, pl. liv, f. 1.—Knowlton, in Lindgren, Jour. Geol., vol. 4, 1896, p. 890.—Knowlton, ?U. S. Geol. Surv., Bull. 163, 1900, p. 72, pl. xvii, f. 1; pl. xviii, f. 1; Washington Acad. Sci., Proc., vol. 11, 1909, pp. 197, 202, 212, 214.

Fort Union: Clear Creek, Mont.

?Miocene (auriferous gravels): Independence Hill, Placer County, Calif.

Lance: Eagle Bluff near Glendive, Mont.; Yule, N. Dak.; Monarch, Big Horn County, and Converse County, Wyo.

Viburnum whymperi Heer. Knowlton, U. S. Geol. Surv., Bull. 163, 1900, p. 72, pl. xix, f. 3 = *Viburnum hollickii.*

Viburnum whymperi Heer. Lesquereux, Tert. Fl., pl. lxi, f. 23 = *Viburnum anomalinervum.*

Viburnum woottonianum Knowlton, U. S. Geol. Surv., Prof. Paper 101, 1918, p. 347, pl. cxi, f. 6.
Raton: Turner mine, Wootton, Colo.

Viburnum sp. Knowlton, in Lindgren, Jour. Geol., vol. 4, 1896, p. 890.
Miocene (auriferous gravels): Independence Hill, Placer County, Calif.

Viburnum sp. Knowlton, Geol. Soc. Am., Bull., vol. 8, 1897, p. 134.
Fort Union: Lightning Creek, Converse County, Wyo.

Viburnum sp. Knowlton, Washington Acad. Sci., Proc., vol. 11, 1909, p. 207.
Lance: Converse County, Wyo.

Viburnum n. sp. Knowlton, Washington Acad. Sci., Proc., vol. 11, 1909, p. 197.
Lance: Eagle Bluff near Glendive, Mont.

Viburnum sp. Knowlton, Jour. Geol., vol. 19, 1911, p. 370.
Lance: Sec. 20, T. 14 N., R. 19 E.; sec. 33, T. 20 N., R. 20 E., S. Dak.

Viburnum sp. Knowlton, U. S. Geol. Surv., Prof. Paper 101, 1918, p. 278, pl. xliv, f. 3.
Vermejo: Coal Creek, Rockvale, Colo.

VICIA Linné, sp. pl., 1753, p. 734.

Vicia sp. Knowlton, U. S. Nat. Mus., Proc., vol. 51, 1916, p. 277, pl. xxiii, f. 4.
Miocene: Florissant, Colo.

VITIPHYLLUM Fontaine, U. S. Geol. Surv., Mon. 15, 1889 [1890], p. 308. [Type, *V. crassifolium.*]

Vitiphyllum crassifolium Fontaine, U. S. Geol. Surv., Mon. 15, 1889 [1890], p. 308, pl. cl, f. 9, 10.
Patuxent: Potomac Run, Va.

Vitiphyllum multifidum Fontaine = *Cissites parvifolius.*

Vitiphyllum parvifolium Fontaine = *Cissites parvifolius.*

VITIS (Tournefort) Linné, Sp. pl., 1753, p. 202.

Vitis cf. **V. aestivalis Michaux.** Berry, Torreya, vol. 10, 1910, p. 266.
Pleistocene: Near Long Branch, N. J.

Vitis alaskana Cockerell, Am. Mus. Nat. Hist., Bull., vol. 24, 1908, p. 103.
Vitis rotundifolia Newberry, U. S. Nat. Mus., Proc., vol. 5, 1882 [1883], p. 513.—Penhallow, Rept. Tert. Pl. Brit. Col., 1908, p. 96. [Homonym, Michaux, 1803.]
Kenai?: Admiralty Inlet, Alaska.
——: Quilchena, British Columbia.

Vitis bruneri Ward = *Ampelopsis bruneri.*

Vitis carbonensis Ward = *Ampelopsis bruneri carbonensis.*

Vitis crenata Heer = *Vitis heeriana.*

Vitis cuspidata Ward = *Ampelopsis montanensis.*

Vitis florissantella Cockerell, Am. Mus. Nat. Hist., Bull., vol. 24, 1908, p. 102, pl. vii, f. 18.
Miocene: *Florissant, Colo.

Vitis fragmenta Knowlton, U. S. Geol. Surv., Prof. Paper 101, 1918, p. 274, pl. xlvii, f. 7.
Trinidad: Near Raton, N. Mex.

Vitis heeriana Knowlton and Cockerell, n. name.
Vitis crenata Heer, Fl. foss. arct., vol. 2, Abt. 2, 1869, p. 36, pl. viii, f. 6.—Knowlton, in Spurr, U. S. Geol. Surv., Eighteenth Ann. Rept., pt. 3, 1898, p. 192; U. S. Nat. Mus., Proc., vol. 17, 1894, p. 228; Geol. Soc. Am., Bull., vol. 5, 1893, p. 586. [Homonym, Thunberg, 1825.]
Kenai: Port Graham, Miller's mine, Yukon River, Alaska.

Vitis hesperia Knowlton, U. S. Nat. Mus., Proc., vol. 51, 1916, p. 284, pl. xxvi, f. 4.
Miocene: *Florissant, Colo.

Vitis inominata Knowlton, U. S. Geol. Surv., Prof. Paper 101, 1918, p. 339, pl. cvii, f. 1.
Raton: Cucharas Canyon, Walsenburg, Colo.

Vitis islandica? Heer, Fl. foss. arct., vol. 1, 1868, p. 150, pl. xxvi, f. 1c, 7a.—Lesquereux, U. S. Geol. and Geog. Surv. Terr., Ann. Rept. 1871 [1872], Suppl., p. 10. [Not afterward referred to by Lesquereux.]
?Mesaverde: Point of Rocks, Wyo.

Vitis leei Knowlton, U. S. Geol. Surv., Prof. Paper 101, 1918, p. 338, pl. lxvi, f. 4.
Raton: Green Canyon mine, Aguilar, Colo.

Vitis olriki Heer, Fl. foss. arct., vol. 1, 1868, p. 120, pl. xlviii, f. 1.—Lesquereux, U. S. Geol. and Geog. Surv. Terr., Ann. Rept. 1871 [1872], Suppl., p. 12; Rept. U. S. Geol. Surv. Terr., vol. 7 (Tert. Fl.), 1878, p. 241, pl. xli, f. 8.—?Penhallow, Rept. Tert. Pl. Brit. Col., 1908, p. 96.—Knowlton, U. S. Geol. Surv., Prof. Paper 101, 1918, p. 338.
Raton: Mesa north of Raton, N. Mex.
Oligocene: Horsefly River, British Columbia.
Evanston: Evanston, Wyo.

Vitis? platanifolia Knowlton, U. S. Geol. Surv., Prof. Paper 101, 1918, p. 339, pl. ciii, f. 2.
Raton: Riley Canyon, Cokedale, Colo.

Vitis pseudo-rotundifolia Berry, Torreya, vol. 10, 1910, p. 265, text f. 2.
Pleistocene: Near Long Branch, N. J.

Vitis rotundifolia Newberry = *Vitis alaskana*.

Vitis cf. **V. rotundifolia** Michaux. Berry, Jour. Geol., vol. 25, 1917, p. 662.
Pleistocene: Vero, Fla.

Vitis sparsa Lesquereux, Rept. U. S. Geol. Surv. Terr., vol. 7 (Tert. Fl.), 1878, p. 241, pl. lx, f. 24.
Post-Laramie: Black Buttes, Wyo.

Vitis tricuspidata Heer. Lesquereux = *Cissus lesquereuxii*.

Vitis xantholithensis Ward = *Ampelopsis xantholithensis*.

Vitis sp. Berry, Torreya, vol. 5, 1905, p. 89.
Pleistocene (Talbot): Tappahannock, Rappahannock River, Va.

Vitis sp. Berry, Jour. Geol., vol. 15, 1907, p. 345.
Pleistocene: Neuse River, N. C.

Vitis sp. Berry, U. S. Geol. Surv., Prof. Paper 98, 1916, p. 202.
Citronelle: Red Bluff, Ala.

Vitis sp. Berry, Jour. Geol., vol. 25, 1917, p. 662.
Pleistocene: Vero, Fla.

Vitis sp.? Hollick, Md. Geol. Surv., Pliocene and Pleistocene, 1906, p. 235.
Pleistocene (Talbot): Bodkin, Grace, and Grove points, Md.

Vitis sp.? Lesquereux, U. S. Nat. Mus., Proc., vol. 11, 1888, p. 35.
Miocene?: Contra Costa, Calif.

VOLTZIA Brongniart, Dict. sci. nat., vol. 57 (Prodrome), 1828, p. 112 (108).

Voltzia coburgensis Schauroth, Zeitschr. Deutsch. geol. Ges., vol. 4, 1852, p. 540, text f. p. 539.—Holden, Ann. Bot., vol. 27, 1913, p. 250, pl. xxii, f. 11; pl. xxiii, f. 20–24.
Triassic (Newark): Martins Head, New Brunswick.

Walchia angustifolia Emmons = *Cheirolepis muensteri*.

Walchia brevifolia Emmons = *Palissya brevifolia*.

Walchia diffusus Emmons = *Palissya diffusa*.

Walchia gracile Emmons = *Palissya diffusa*.

Walchia longifolius Emmons = *Palissya sphenolepis*.

Walchia variabilis Emmons = *Pagiophyllum peregrinum*.

WEICHSELIA Stiehler, Zeitschr. gesammt. Naturw., Berlin, vol. 9, 1857, p. 453.

Weichselia reticulata (Stokes and Webb) Ward, U. S. Geol. Surv., Nineteenth Ann. Rept., pt. 2, 1899, p. 651, pl. clx, f. 2–4.
Pecopteris reticulata Stokes and Webb, Geol. Soc. London, Trans., 2d ser., vol. 1, 1824, p. 424, pl. xlvi, f. 5; pl. xlvii, f. 3.
Fuson: Hay Creek coal field, Crook County, Wyo.

WEINMANNIA Linné, Syst., 10th ed., 1759, p. 1005.

Weinmannia? dubiosa Cockerell, Am. Mus. Nat. Hist., Bull., vol. 24, 1908, p. 94.

Weinmannia rosaefolia Lesquereux, U. S. Geol. and Geog. Surv. Terr., Ann. Rept. 1873 [1874], p. 415. [Homonym, A. Cunningham, 1839.]

Rhus rosaefolia Lesquereux, Rept. U. S. Geol. Surv. Terr., vol. 7 (Tert. Fl.), 1878, p. 293, pl. xlii, f. 7–9.

Miocene: *Florissant, Colo.

Weinmannia haydenii (Lesquereux) Lesquereux, Rept. U. S. Geol. Surv. Terr., vol. 8 (Cret. and Tert. Fl.), 1883, p. 178, pl. xlii, f. 1–7.—Cockerell, Am. Mus. Nat. Hist., Bull., vol. 24, 1908, p. 93.—Knowlton, U. S. Nat. Mus., Proc., vol. 51, 1916, p. 270.

Rhus haydenii Lesquereux, U. S. Geol. and Geog. Surv. Terr., Ann. Rept. 1873 [1874], p. 417; Rept. U. S. Geol. Surv. Terr., vol. 7 (Tert. Fl.), 1878, p. 294, pl. lviii, f. 12.

Weinmannia phenacophylla Cockerell, Am. Mus. Nat. Hist., Bull., vol. 24, 1908, p. 93; Am. Nat., vol. 44, 1910, p. 34, f. 1.

Miocene: *Florissant, Colo.

Weinmannia integrifolia Lesquereux, Rept. U. S. Geol. Surv. Terr., vol. 8 (Cret. and Tert. Fl.), 1883, p. 178, pl. xlii, f. 8–13.—Knowlton, U. S. Nat. Mus., Proc., vol. 51, 1916, p. 270.

Miocene: *Florissant, Colo.

Weinmannia lesquereuxi Cockerell, Am. Mus. Nat. Hist., Bull., vol. 24, 1908, p. 93; Am. Nat., vol. 44, 1910, p. 34, f. 2.

Rhus acuminata Lesquereux. Cockerell. Am. Mus. Nat. Hist., Bull., vol. 24, 1908, p. 94.

Miocene: *Florissant, Colo.

Weinmannia obtusifolia Lesquereux, Rept. U. S. Geol. Surv. Terr., vol. 8 (Cret. and Tert. Fl.), 1883, p. 178, pl. xli, f. 4–10.—Cockerell, Am. Mus. Nat. Hist., Bull., vol. 24, 1908, p. 93.—Knowlton, U. S. Nat. Mus., Proc., vol. 51, 1916, p. 270.

Miocene: *Florissant, Colo.

Weinmannia phenacophylla Cockerell = ***Weinmannia haydenii.***

Weinmannia rosaefolia Lesquereux = ***Weinmannia? dubiosa*** (according to Cockerell).

WIDDRINGTONIA Endlicher, Gen. Suppl., vol. 2, 1842, p. 25.

Widdringtonia? complanata Lesquereux, U. S. Geol. and Geog. Surv. Terr., Bull., vol. 1, 1875 [1876], p. 366; idem, Ann. Rept. 1874 [1876], p. 299; Rept. U. S. Geol. Surv. Terr., vol. 7 (Tert. Fl.), 1878, p. 72, pl. lxii, f. 13, 14.—Knowlton, U. S. Geol. Surv., Bull. 163, 1900, p. 30; idem, Prof. Paper 101, 1918, p. 252, pl. xxxii, f. 4, 5.

Mesaverde: Point of Rocks, Wyo.

Vermejo: Near Walsenburg, Colo.

Widdringtonia linguaefolia Lesquereux = ***Sabina linguaefolia.***

Widdringtonia linguaefolia var. *gracilis* Lesquereux = ***Sabina linguaefolia gracilis.***

WIDDRINGTONITES Endlicher, Synop. Conif., 1847, p. 271.

Widdringtonites fasciculatus Hollick, U. S. Geol. Surv., Mon. 50, 1906, p. 45, pl. iv, f. 1.

Magothy: Gay Head, Marthas Vineyard, Mass.

Widdringtonites gracilis Heer = *Cyparissidium gracile.*

Widdringtonites ramosus (Fontaine) Berry, U. S. Nat. Mus., Proc., vol. 40, 1911, p. 302; Md. Geol. Surv., Lower Cret., 1912, p. 428, pl. lxxiii, f. 1–6.

Taxodium (Glyptostrobus) ramosum Fontaine, U. S. Geol. Surv., Mon. 15, 1889 [1890], p. 251, pl. cxxiii, f. 2, 3; pl. cxxiv, f. 2; pl. cxxvii, f. 1; pl. cxxxii, f. 1; pl. clxvi, f. 1; U. S. Nat. Mus., Proc., vol. 15, 1892, p. 494.

Taxodium (Glyptostrobus) brookense Fontaine, U. S. Geol. Surv., Mon. 15, 1889 [1890], p. 254, pl. cxxii, f. 1; pl. cxxiv, f. 3–9; pl. cxxxi, f. 5; pl. clxv, f. 1–3; pl. clxvi, f. 4, 7; pl. clxvii, f. 3.

Taxodium (Glyptostrobus) brookense angustifolium Fontaine, U. S. Geol. Surv., Mon. 15, 1889 [1890], p. 256, pl. clxvii, f. 1.

Glyptostrobus brookensis (Fontaine) Ward, U. S. Geol. Surv., Fifteenth Ann. Rept., 1895, p. 359.—Fontaine, in Ward, U. S. Geol. Surv., Mon. 48, 1905 [1906], pp. 481, 489, 495, pl. cx, f. 1 [not pp. 483, 486, 520, which = *Arthrotaxopsis expansa*].

Glyptostrobus brookensis angustifolium (Fontaine) Knowlton, U. S. Geol. Surv., Bull. 152, 1898, p. 112.—Fontaine, in Ward, idem, Mon. 48, 1905 [1906], p. 489, pl. cviii, f. 4.

Glyptostrobus ramosus (Fontaine) Ward, in Fontaine, in Ward, U. S. Geol. Surv., Mon. 48, 1905 [1906], pp. 281?, 489, 544.

Athrotaxopsis expansa Fontaine. Fontaine, in Ward, U. S. Geol. Surv., Mon. 48, 1905 [1906], p. 547 [not other citations].

Patapsco: Fort Foote, Overlook Inn Road, Stump Neck, Williams, and Vinegar Hill, Md.; Mount Vernon, Hell Hole, White House Bluff, near Brooke and 72d milepost, Va.

Kootenai: Great Falls, Mont.

Trinity: ?Glen Rose, Tex.

Widdringtonites reichii (Ettingshausen) Heer, Fl. foss. arct., vol. 6, Abt. 2, 1882, p. 52, pl. xxviii, f. 5.—Hollick, U. S. Geol. Surv., Mon. 50, 1906, p. 44, pl. iv, f. 6, 7.—Hollick and Jeffrey, N. Y. Bot. Gard., Mem., vol. 3, 1909, p. 29, pl. v, f. 1–4; pl. viii, f. 7–11; pl. xx, f. 3–5.—Berry, Bull. Torr. Bot. Club, vol. 33, 1906, p. 169; idem, vol. 37, 1910, p. 21; Geol. Surv. N. J., Bull. 3, 1911, p. 87, pl. viii, f. 1, 2; Md. Geol. Surv., Upper Cret., 1916, p. 793, pl. lv, f. 1; U. S. Geol. Surv., Prof. Paper 112, 1919, p. 69.

Frenelites reichii Ettingshausen, Kreidefl. Niederschöna, 1867, p. 246, pl. i, f. 10a, 10c.—Hollick, N. Y. Acad. Sci., Trans., vol. 12, 1892, p. 29, pl. i, f. 23.

Raritan: Kreischerville, Staten Island, N. Y.; Sayreville, Woodbridge, Milltown, and South Amboy, N. J.

Magothy: Morgan, N. J.; Deep Cut, Del.; Grove Point, Md.; Good Hope Hill, D. C.; Gay Head, Marthas Vineyard, Mass.

Tuscaloosa: Shirleys Mill, Glen Allen, and Whites Bluff, Ala.

Widdringtonites subtilis Heer, Fl. foss. arct., vol. 3, Abt. 2, 1874, p. 101, pl. xxviii, f. 1, 1b.—Newberry, U. S. Geol. Surv., Mon. 26, 1895 [1896], p. 57, pl. x, f. 2–4.—Hollick, U. S. Geol. Surv., Mon. 50, 1906, p. 45, pl. iv, f. 2–5.—Berry, Bull. Torr. Bot. Club, vol. 39, 1911, p. 344, pl. xxiv, f. 1–2a; pl. xxv, f. 1–8; U. S. Geol. Surv., Prof. Paper 84, 1914, p. 25, pl. ii, f. 14–17; Geol. Surv. N. J., Bull. 3, 1916, p. 90; U. S. Geol. Surv., Prof. Paper 112, 1919, p. 67, pl. viii, f. 1–12.

Magothy: Gay Head, Marthas Vineyard, Mass.
Magothy?: Black Rock Point, Block Island, R. I.
Raritan: Woodbridge and South Amboy, N. J.
Tuscaloosa: Shirleys Mill, Fayette County, Snow place, Tuscaloosa County, Ala.
Black Creek (Middendorf): Middendorf and Rocky Point, S. C.
Bingen: Near Maxwell Spur, Pike County, Ark.

WILLIAMSONIA Carruthers, Linn. Soc. London, Trans., vol. 26, 1868, p. 680.

Williamsonia? bibbinsi Ward = ***Abietites macrocarpus.***

Williamsonia delawarensis Berry, Johns Hopkins Univ. Cir., new ser., 1907, No. 7, p. 84, f. 4 (on p. 85); Md. Geol. Surv., Upper Cret., 1916, p. 771, pl. li, f. 7.

Magothy: Deep Cut, Del.; Grove Point, Md.

Williamsonia elocata Lesquereux, U. S. Geol. Surv., Mon. 17, 1891 [1892], p. 87, pl. ii, f. 9, 9a.

Dakota: Ellsworth County, Kans.

Williamsonia? gallinacea Ward, U. S. Geol. Surv., Mon. 48, 1905 [1906], p. 485, pl. cvii, f. 4.

——: Cockpit Point, Va.

Williamsonia marylandica Berry, Bull. Torr. Bot. Club, vol. 41, 1914, p. 297 [name]; Md. Geol. Surv., Upper Cret., 1916, p. 769, pl. li, f. 5, 6.

Magothy: Round Bay, Md.

Williamsonia oregonensis Fontaine, in Ward, U. S. Geol Surv., Mon. 48, 1905 [1906], p. 118, pl. xxix, f. 6.

Jurassic: Douglas County, Oreg.

Williamsonia? phoenicopsoides Ward, U. S. Geol. Surv., Nineteenth Ann. Rept., pt. 2, 1899, p. 668, pl. clxii, f. 20.

Lakota: South Fork of Hay Creek, Crook County, Wyo.

Williamsonia problematica (Newberry) Ward = ***Palaeanthus problematicus.***

Williamsonia recentior Dawson, Roy. Soc. Canada, Trans., vol. 3, 1885 [1886], sec. 4, p. 12, pl. iv, f. 1.

Mill Creek: Mill Creek, British Columbia.

Williamsonia riesii Hollick, N. Y. Acad. Sci., Trans., vol. 12, 1892, p. 37 [10], pl. i, f. 2, 3; U. S. Geol. Surv., Mon. 50, 1906, p. 107, pl. ii, f. 43.
Raritan: Kreischerville, Staten Island, N. Y.
Magothy?: Gay Head, Marthas Vineyard, Mass.

Williamsonia smockii Newberry, U. S. Geol. Surv., Mon. 26, 1895 [1896], p. 127, pl. xxxvi, f. 1-8.—Berry, Geol. Surv. N. J., Bull. 3, 1911, p. 224.
Raritan: New Jersey.

Williamsonia texana Fontaine, U. S. Nat. Mus., Proc., vol. 16, 1893, p. 278, pl. xliii, f. 1, 2.
Trinity: Glen Rose, Tex.

Williamsonia virginiensis Fontaine, U. S. Geol. Surv., Mon. 15, 1889, p. 273, pl. cxxxiii, f. 5-7; pl. clxv, f. 5.
Patuxent: Trents Reach and near Dutch Gap Canal, Va.

Williamsonia? sp. Dawson, Roy. Soc. Canada, Trans., vol. 10, 1892 [1893], sec. 4, p. 87.
Kootenai: Anthracite, British Columbia.

Williamsonia? sp. Fontaine, in Ward, U. S. Geol. Surv., Mon. 48, 1905 [1906], p. 119, pl. xxix, f. 7.
Jurassic: Douglas County, Oreg.

Williamsonia? sp. Fontaine, in Ward, U. S. Geol. Surv., Mon. 48, 1905 [1906], p. 119, pl. xxix, f. 8-12.
Jurassic: Douglas County, Oreg.

Williamsonia sp. Hollick, in Ries, School of Mines Quart., vol. 15, 1894, p. 355.
Magothy: Northport, Long Island, N. Y.

WINCHELLIA Lesquereux, Am. Geol., vol. 12, 1893, p. 209. [Type, *W. triphylla*.]

Winchellia triphylla Lesquereux, Am. Geol., vol. 12, 1893, p. 209, pl. viii.
Fort Union: Yellowstone River near mouth of Powder River, Mont.

"Winged seed," Hollick, Bull. Torr. Bot. Club, vol. 21, 1894, p. 62, pl. 180, f. 1 = *Tricalycites major*.

WOODWARDIA J. E. Smith, Mem. Acad. Turin, vol. 5, 1793, p. 411.

Woodwardia columbiana Knowlton, Smithsonian Misc. Coll. (Quart. Issue), vol. 53, 1910, p. 491, pl. lxiii, f. 1, 2.
Pliocene?: Cascades of Columbia River, Oreg.

Woodwardia crenata Knowlton, U. S. Geol. Surv., Bull. 163, 1900, p. 22, pl. iii, f. 3; idem, Prof. Paper 101, 1918, p. 246.
Woodwardia sp. Knowlton, Geol. Soc. Am., Bull., vol. 8, 1897, p. 154.
Mesaverde: Point of Rocks, Wyo.
Vermejo: La Veta, Colo.

Woodwardia florissantia Cockerell, Am. Mus. Nat. Hist., Bull., vol. 24, 1908, p. 77, pl. vi, f. 2.
Miocene: Florissant, Colo.

Woodwardia latiloba Lesquereux, U. S. Geol. and Geog. Surv. Terr., Ann. Rept. 1873 [1874], p. 391; Rept. U. S. Geol. Surv. Terr., vol. 7 (Tert. Fl.), 1878, p. 54, pl. iii, f. 1, 1a.
Denver: Golden, Colo.

Woodwardia maxoni Knowlton, Smithsonian Misc. Coll. (Quart. Issue), vol. 53, 1910, p. 489, pl. lxiii, f. 3; pl. lxiv, f. 1, 2.
Fort Union: 35 and 47 miles southeast of Rock Springs, Wyo.

Woodwardia preareolata Knowlton, U. S. Geol. Surv., Mon. 32, pt. 2, 1899, p. 665, pl. lxxix, f. 1.
Fort Union: Yellowstone National Park.

Woodwardia sp. Knowlton, Geol. Soc. Am., Bull., vol. 8, 1897, p. 140; U. S. Geol. Surv., Bull. 163, 1900, p. 23, pl. iii, f. 9.
Mesaverde: North Fork of Dutton Creek, Laramie Plains, Wyo.

Woodwardia sp. Knowlton, in Washburne, U. S. Geol. Surv., Bull. 590, 1914, p. 36.
Eocene?: Eugene, Oreg.

Woodwardia sp. Knowlton = ***Woodwardia crenata.***

WOODWORTHIA Jeffrey, Boston Soc. Nat. Hist., Proc., vol. 34, 1910, p. 330. [Type, *W. arizonica.*]

Woodworthia arizonica Jeffrey, Boston Soc. Nat. Hist., Proc., vol. 34, 1910, p. 330, pl. xxxi; pl. xxxii.—Knowlton, Am. Forestry, vol. 19, 1913, p. 214.
Triassic: Fossil forest near Adamana, Ariz.

XANTHIUM Linné, sp. pl., 1753, p. 987.

Xanthium sp. Berry, Jour. Geol., vol. 25, 1917, p. 662.
Pleistocene: Vero, Fla.

XOLISMA Rafinesque, Jour. Phys., vol. 89, 1819, p. 259.

Xolisma ligustrina (Linné) Britton. Hollick, Md. Geol. Surv., Pliocene and Pleistocene, 1906, p. 236, pl. lxix, f. 6.—Berry, Jour. Geol., vol. 15, 1907, p. 346; Am. Nat., vol. 41, 1907, p. 696, pl. ii, f. 6; Am. Jour. Sci., 4th ser., vol. 29, 1910, p. 398; U. S. Nat. Mus., Proc., vol. 48, 1915, p. 301, pl. xii, f. 1.
Pleistocene: Bodkin Point, Md.; Neuse River, N. C.; Abercrombies Landing, Chattahoochee River, and Black Warrior River 311 miles above Mobile, Ala.; Hickman, Ky.

XYLOMITES Unger, Chloris Protogaea, 1841, pt. 1, p. 3.

Xylomites borealis Heer, Fl. foss. arct., vol. 6 (North Canada), 1880, p. 12, pl. ii, f. 1.
Eocene: North Canada (Mackenzie River?).

YORKIA Wanner, in Ward, U. S. Geol. Surv., Twentieth Ann. Rept., pt. 2, 1900, p. 254. [Type, *Y. gramineoides.*]

Yorkia gramineoides Ward, U. S. Geol. Surv., Twentieth Ann. Rept., pt. 2, 1900, p. 254, pl. xxxiv, f. 4–6.
Triassic (Newark): York County, Pa.

YUCCA Linné, Sp. pl., vol. 1, 1753, p. 319.

Yucca sp. Berry, U. S. Geol. Surv., Prof. Paper 98, 1916, p. 197, pl. xlv, f. 7.
Citronelle: Lambert, Ala.

YUCCITES Martius, Denkschr. K. bay. bot. Ges., vol. 2, 1822, p. 136.

Yuccites hettangensis Saporta MS., in Schimper, Pal. vég., vol. 2, 1870, p. 427.—Fontaine, in Ward, U. S. Geol. Surv., Mon. 48, 1905 [1906], p. 135, pl. xxxvii, f. 1, 2.
Jurassic: Douglas County, Oreg.

ZAMIA Linné, Sp. pl., 2d ed., 1763, p. 1659.

Zamia mississippiensis Berry, Torreya, vol. 16, 1916, p. 177, f. 1–3; U. S. Geol. Surv., Prof. Paper 108, 1917, p. 63, f. 17a, b, c.
Wilcox (Grenada): Meridian, Miss.

Zamia taxina Lindley and Hutton = *Ctenophyllum taxinum.*

Zamia washingtoniana Ward [part] = *Podozamites lanceolatus.*

Zamia washingtoniana Ward [part] = *Zamites tenuinervis.*

Zamia? wilcoxensis Berry, U. S. Geol. Surv., Prof. Paper 91, 1916, p. 169, pl. cxiv, f. 2.
Wilcox: Near Naborton, De Soto Parish, La.

ZAMIOPSIS Fontaine, U. S. Geol. Surv., Mon. 15, 1890, p. 160. [Type, *Z. pinnatifida.*]

Zamiopsis dentata (Fontaine) Berry, Md. Geol. Surv., Lower Cret., 1911, p. 355, pl. lvi, f. 1, 2.

Scleropteris dentata Fontaine, U. S. Geol. Surv., Mon. 15, 1889 [1890], p. 153, pl. lxiii, f. 3, 4.

Zamiopsis pinnatifida Fontaine, U. S. Geol. Surv., Mon. 15, 1889 [1890], p. 161, pl. lxi, f. 7; pl. lxii, f. 5; pl. lxiv, f. 2; pl. lxvii, f. 2.

Zamiopsis insignis Fontaine, U. S. Geol. Surv., Mon. 15, 1889 [1890], p. 162, pl. lxii, f. 3; pl. lxiv, f. 1, 3; pl. lxv, f. 4–6; pl. lxvi, f. 2; pl. lxvii, f. 7; in Ward, U. S. Geol. Surv., Mon. 48, 1905 [1906], pp. 511?, 517.

Zamiopsis longipennis Fontaine, U. S. Geol. Surv., Mon. 15, 1889 [1890], p. 164, pl. lxi, f. 8.

Patuxent: Fredericksburg, Chinkapin Hollow ?, Va.; New Reservoir, D. C.

Zamiopsis insignis Fontaine [part] = *Ctenopteris insignis.*

Zamiopsis insignis Fontaine [part] = *Zamiopsis dentata.*

Zamiopsis laciniata Fontaine, U. S. Geol. Surv., Mon. 15, 1889 [1890], p. 164, pl. lxvi, f. 1, 5–8.—Berry, Md. Geol. Surv., Lower Cret., 1912, p. 358.
Patuxent: *Fredericksburg, Va.

Zamiopsis longipennis Fontaine = *Zamiopsis dentata.*

Zamiopsis petiolata Fontaine, U. S. Geol. Surv., Mon. 15, 1889 [1890], p. 166, pl. lxvi, f. 3.—Berry, Md. Geol. Surv., Lower Cret., 1912, p. 357.

Patuxent: *Fredericksburg, Va.

Zamiopsis pinnatifida Fontaine = *Zamiopsis dentata.*

ZAMIOSTROBUS Endlicher, Gen. Pl., 1836, p. 72.

Zamiostrobus emmonsi Fontaine = *Cycadeoidea emmonsi.*

Zamiostrobus? mirabilis Lesquereux = *Cycadeoidea mirabilis.*

Zamiostrobus virginiensis Fontaine, U. S. Geol. Surv., Mon. 6, 1883, p. 85, pl. xlvii, f. 4, 5; in Ward, U. S. Geol. Surv., Twentieth Ann. Rept., pt. 2, 1900, p. 301, pl. xliii, f. 2.

Lepidodendron sp. Emmons, American geology, pt. 6, 1857, p. 124, f. 93, 94.

Zamiostrobus sp. Fontaine, U. S. Geol. Surv., Mon. 6, 1883, p. 117, pl. liv, f. 10.

Triassic (Newark): Near Midlothian, Va.; Haw River and Lockville, N. C.

Zamiostrobus sp. Fontaine = *Zamiostrobus virginiensis.*

ZAMITES Brongniart, Prod. hist. vég. foss., 1828, p. 94.

Zamites acutipennis Heer, Zeitschr. Deutsch. geol. Gesell., vol. 24, 1872, p. 161.—Dawson, Roy. Soc. Canada, Trans., vol. 3, 1885 [1886], sec. 4, p. 7, pl. i, f. 5.—Newberry, Am. Jour. Sci., 3d ser., vol. 41, 1891, p. 197.

Kootenai: Martin Brook, British Columbia; Great Falls, Mont.

Zamites alaskana Fontaine, U. S. Nat. Mus., Proc., vol. 11, 1888, p. 32, pl. x, f. 10.—Knowlton, idem, vol. 17, 1894, p. 215; Geol. Soc. Am., Bull., vol. 5, 1893, p. 579.

Jurassic: Cape Lisburne, Alaska.

Zamites angustifolius Eichwald = *Podozamites knowltoni.*

Zamites apertus Newberry, Am. Jour. Sci., 3d ser., vol. 61, 1891, p. 199, pl. xiv, f. 4, 5.—Knowlton, Smithsonian Misc. Coll., vol. 50, 1907, p. 121, pl. xiii, f. 5.

Kootenai: Great Falls and Spanish Coulee east of Cascade, Cascade County, Mont.

Zamites arcticus Göppert, 41st Jahresb. Schles. Gesell. vaterl. Kultur, 1863 [1864], p. 84.—Knowlton, Smithsonian Misc. Coll., vol. 50, 1907, p. 121.—Fontaine, in Ward, U. S. Geol. Surv., Mon. 48, 1905, pp. 256, 306, pl. lxviii, f. 1; pl. lxxiii, f. 1–6.—Knowlton, Washington Acad. Sci., Jour., vol. 6, 1916, p. 181.

Zamites weedii Fontaine, in Weed and Pirsson, U. S. Geol. Surv., Nineteenth Ann. Rept., pt. 3, 1898, p. 481 [name].

Zamites sp. Dawson, Roy. Soc. Canada, Trans., vol. 3, 1885, p. 7, pl. i, f. 4.

Kootenai: Giltedge, Belt, Geyser, and Hazlet, Mont.

Knoxville: Near Lowry, Tehama County, Calif.

Morrison: Cottonwood Creek, 5 miles west of Tensleep, Big Horn Basin, Wyo.

Zamites borealis Heer, Fl. foss. arct., vol. 3, Abt. 2, 1873, p. 66, pl. xiv, f. 13, 14; pl. xv, f. 1, 2.—Newberry, Am. Jour. Sci., vol. 41, 1891, p. 197.—Fontaine, in Ward, U. S. Geol. Surv., Nineteenth Ann. Rept., pt. 2, 1899, p. 666, pl. clxii, f. 14.

Kootenai: Great Falls, Mont.

Lakota: Barr's tunnel, Webster's ranch, and Larsbee's shaft, Crook County, Wyo.

Zamites brevipennis Heer. Fontaine, in Ward, U. S. Geol. Surv., Nineteenth Ann. Rept., pt. 2, 1899, p. 665, pl. clxii, f. 10–13.

Lakota: Barr's tunnel, Crook County, Wyo.

Zamites brevifolius Braun = *Otozamites brevifolius.*

Zamites crassinervis Fontaine, U. S. Geol. Surv., Mon. 15, 1889 [1890], p. 172, pl. lxix, f. 4; pl. lxxxiii, f. 3.—Penhallow, Roy. Soc. Canada, 2d ser., vol. 8, 1902, p. 41.—Berry, Md. Geol. Surv., Lower Cret., 1911, p. 347, pl. liv, f. 6.

Zamites ovalis Fontaine [part], U. S. Geol. Surv., Mon. 15, 1889 [1890], p. 173, pl. lxxxv, f. 4; pl. clxx, f. 3.

Zamites sp. Fontaine, idem, p. 173, pl. lxxxiv, f. 12.

Patuxent: Fredericksburg, Dutch Gap, Kankeys, Potomac Run, Va.

Lower Cretaceous: Alliford Bay, Queen Charlotte Islands, British Columbia.

Zamites distans Presl = *Podozamites distans.*

Zamites distantinervis Fontaine, U. S. Geol. Surv., Mon. 15, 1889. p. 172, pl. lxxxiii, f. 4.

Patuxent: *Fredericksburg, Va.

Zamites distantinervis Fontaine [part] = *Zamites tenuinervis.*

Zamites graminioides Emmons = *Ctenophyllum braunianum angustum,*

Zamites megaphyllus (Phillips) Seward, Jurassic Fl., 1904, pt. 2, pl. x, f. 4, 5; pl. xii, f. 1, 3–5.—Knowlton, U. S. Geol. Surv., Prof. Paper 85, 1914, p. 51, pl. vii, f. 1.

Palaeozamia megaphylla Phillips, Geol. Oxford, 1871, p. 169, diag. xxx, f. 6.

Irites alaskana Lesquereux, U. S. Nat. Mus., Proc., vol. 10, 1887, p. 36.—Knowlton, idem, vol. 17, 1894, p. 216; Geol. Soc. Am., Bull., vol. 5, 1893, p. 580.

Baiera palmata Heer. Lesquereux, U. S. Nat. Mus., Proc., vol. 11, 1888, p. 31, pl. xvi, f. 4, 5.

Nageiopsis longifolia Fontaine. Fontaine, in Ward, U. S. Geol. Surv., Mon. 48, 1905 [1906], p. 171, pl. xlv, f. 1–5.

Podozamites grandifolius Fontaine? Fontaine, in Ward, U. S. Geol. Surv., Mon. 48, 1905 [1906], p. 167, pl. xliv, f. 1.

Jurassic (Corwin): Cape Lisburne, Alaska.

Zamites montana Dawson, Roy. Soc. Canada, Trans., vol. 3, 1885 [1886], sec. 4, p. 7, pl. i, f. 6, 6a.

Kootenai: Martin Creek, Kootanie Pass, and Anthracite, British Columbia.

Zamites montanensis Fontaine = *Pterophyllum montanense.*

Zamites obtusifolius Rogers = *Ctenophyllum braunianum abbreviatum.*

Zamites occidentalis Newberry, in Macomb, Rept. Expl. Exped. Santa Fe, 1859 [1876], p. 142, pl. v, f. 1–2.—Fontaine, U. S. Nat. Mus., Proc., vol. 13, 1890, p. 384.

Triassic: Abiquiu, N. Mex.

Zamites ovalis Fontaine [part] = *Podozamites subfalcatus.*

Zamites ovalis Fontaine [part] = *Zamites crassinervis.*

Zamites pennsylvanicus Fontaine, in Ward, U. S. Geol. Surv., Twentieth Ann. Rept., pt. 2, 1900, p. 245, pl. xxviii, f. 3, 4.

Triassic (Newark): York County, Pa.

Zamites powelli Fontaine, U. S. Nat. Mus., Proc., vol. 13, 1890, p. 284, pl. xxv, f. 5; pl. xxvi, f. 6, 7.

Triassic: Abiquiu, N. Mex.

Zamites subfalcatus Fontaine = *Zamites tenuinervis.*

Zamites tenuinervis Fontaine, U. S. Geol. Surv., Mon. 15, 1889 [1890], p. 171, pl. lxvii, f. 1; pl. lxix, f. 2; pl. lxx, f. 1; pl. lxxv, f. 3; pl. lxxvi, f. 7; pl. lxxviii, f. 6; pl. lxxxiv, f. 7; U. S. Nat. Mus., Proc., vol. 16, 1893, p. 267, pl. xxxvii, f. 3, 4; pl. xxxviii, f. 1, 2.—Penhallow, Roy. Soc. Canada, Trans., 2d ser., vol. 8, sec. 4, 1902, p. 42.—Fontaine, in Ward, U. S. Geol. Surv., Mon. 48, 1906, pp. 251, 548?, pl. lxviii, f. 2, 3 [not f. 528].—Berry, Md. Geol. Surv., Lower Cret., 1911, p. 345, pl. liv, f. 1–5.

Zamites distantinervis Fontaine, U. S. Geol. Surv., Mon. 15, 1889 [1890], p. 172, pl. lxxxiv, f. 13; pl. lxxxv, f. 3; in Ward, U. S. Geol. Surv., Mon. 48, 1905 [1906], p. 573 [not other citations].

Zamites subfalcatus Fontaine, U. S. Geol. Surv., Mon. 15, 1889 [1890], p. 173, pl. lxxxiv, f. 13; pl. lxxxv, f. 3.

Zamia washingtoniana Ward, U. S. Geol. Surv., Fifteenth Ann. Rept., 1895, p. 350, pl. ii, f. 6.—Fontaine, in Ward, U. S. Geol. Surv., Mon. 48, 1905 [1906], p. 563, pl. cxi, f. 1 [not f. 2].

Patuxent: Fredericksburg, Kankeys, Dutch Gap, Va.

Patapsco: Mount Vernon, Dumfries Landing, Widewater, Va.; Grays Hill, Vinegar Hill, Stump Neck, Md.

Trinity: Glen Rose, Tex.

Horsetown and Knoxville: Shasta and Tehama counties, Calif.

Lower Cretaceous: Alliford Bay, Queen Charlotte Islands, British Columbia.

Zamites tenuinervis Fontaine. Fontaine, in Ward = *Podozamites knowltoni.*

Zamites tenuistriatus Rogers = *Podozamites tenuistriatus.*

Zamites velderi Brown, Acad. Nat. Sci. Philadelphia, Proc., vol. 63, 1911, p. 18, pl. ii.

Zamites weedii Fontaine = *Zamites arcticus.*

Zamites yorkensis Fontaine, in Ward, U. S. Geol. Surv., Twentieth Ann. Rept., pt. 2, 1900, p. 245, pl. xxix, f. 1–4.

Triassic (Newark): York County, Pa.

Zamites sp. Dawson, Roy. Soc. Canada, Trans., vol. 3, 1885, sec. 4, p. 7, pl. i, f. 4 = *Zamites arcticus*.

Zamites sp. Fontaine, U. S. Geol. Surv., Mon. 15, 1889 [1890], p. 173, pl. lxxxiv, f. 12 = *Zamites crassinervis*.

Zamites? sp. Fontaine, in Ward, U. S. Geol. Surv., Nineteenth Ann. Rept., pt. 2, 1899, p. 666, pl. clxii, f. 15.
Fuson: Oak Creek, Crook County, Wyo.

Zamites sp. Lesquereux, U. S. Geol. Surv., Mon. 17, 1891 [1892], p. 26, pl. i, f. 8.
Dakota: Near Delphos, Kans.

ZANTHOXYLUM Linné, Sp. pl., 1753, p. 270.

Zanthoxylum diversifolium Lesquereux = *Fagara diversifolia*.

Zanthoxylum dubium Lesquereux, U. S. Geol. and Geog. Surv. Terr., Ann. Rept. 1871 [1872], Suppl., p. 15 [not afterward mentioned by Lesquereux].—Knowlton, U. S. Geol. Surv., Prof. Paper 101, 1918, p. 328.
Raton?: "Fishers Peak, Raton Mountains, N. Mex."

Zanthoxylon juglandinum? Al. Braun. Lesquereux, Rept. U. S. Geol. Surv. Terr., vol. 7 (Tert. Fl.), 1878, p. 294, pl. lviii, f. 10.
Green River?: Washakie, Wyo.

Zanthoxylum spireaefolium Lesquereux = *Fagara spireaefolia*.

Zelkowa longifolia (Lesquereux) Engler = *Fagopsis longifolia*.

ZINGIBERITES Heer, Fl. tert. Helv., vol. 3, 1859, p. 172.

Zingiberites dubius Lesquereux, Rept. U. S. Geol. Surv. Terr., vol. 7 (Tert. Fl.), 1878, p. 95, pl. xvi, f. 1.
Zingiberites? undulatus Lesquereux, U. S. Geol. and Geog. Surv. Terr., Ann. Rept. 1873 [1874], p. 396. [Homonym, Heer, 1856.]
Denver: Golden, Colo.

Zingiberites? undulatus Lesquereux = *Zingiberites dubius*.

ZIZYPHUS Adanson, Fam. Pl., vol. 2, 1763, p. 304.

Zizyphus beckwithii Lesquereux, Rept. U. S. Geol. Surv. Terr., vol. 8 (Cret. and Tert. Fl.), 1883, p. 125, pl. xix, f. 5.
Denver: Golden, Colo.

Zizyphus californicus Knowlton and Cockerell, n. name.
Zizyphus microphyllus Lesquereux, Mus. Comp. Zool., Mem., vol. 6, No. 2, 1878, p. 28, pl. viii, f. 9. [Homonym, Roxburgh, 1814.]
Miocene (auriferous gravels): Chalk Bluffs, Nevada County, Calif.

Zizyphus cinnamomoides. (Lesquereux) Lesquereux, Rept. U. S. Geol. Surv. Terr., vol. 7 (Tert. Fl.), 1878, p. 277, pl. lii, f. 7, 8.—Ward, U. S. Geol. Surv., Sixth Ann. Rept., 1884–85 [1886], p. 554, pl. lii, f. 3; idem, Bull. 37, 1887, p. 74, pl. xxxiii, f. 7.
Ceanothus cinnamomoides Lesquereux, U. S. Geol. and Geog. Surv. Terr., Ann. Rept. 1871 [1872], p. 289.
Green River: Green River, Wyo.
Fort Union: Sevenmile Creek near Glendive, Mont.

Zizyphus cliffwoodensis Berry, Johns Hopkins Univ. Circ., new ser., 1907, No. 7, p. 88, f. 5 (on p. 85).
Magothy: Cliffwood Beach, N. J.

Zizyphus coloradensis Knowlton, U. S. Geol. Surv., Prof. Paper 130, 19—, p. —, pl. xv, f. 5 [in preparation].
Laramie: Popes Bluff, Pikeview, Colo.

Zizyphus corrugatus Knowlton, U. S. Geol. Surv., Prof. Paper 130, 19—, p. —, pl. xvii, f. 3 [in preparation].
Laramie: Cowan station near Denver, Colo.

Zizyphus dakotensis Lesquereux, U. S. Geol. Surv., Mon. 17, 1891 [1892], p. 167, pl. xxxvi, f. 4–7.
Dakota: Ellsworth County, Kans.

Zizyphus distortus Lesquereux, U. S. Geol. and Geog. Surv. Terr., Ann. Rept. 1873 [1874], p. 404; Rept. U. S. Geol. Surv. Terr., vol. 7 (Tert. Fl.), 1878, p. 275, pl. li, f. 7–9.
Denver: Golden, Colo.

Zizyphus elegans Hollick, Bull. Torr. Bot. Club, vol. 21, 1894, p. 58, pl. clxxvii, f. 9 [not f. 10]; U. S. Geol. Surv., Mon. 50, 1906, pl. xxxiv, f. 8.
Magothy: Glen Cove, Long Island, N. Y.

Zizyphus elegans Hollick [part], Bull. Torr. Bot. Club, vol. 21, 1894, p. 58, pl. 177, f. 10 = *Zizyphus oblongus*.

Zizyphus falcatus Berry, U. S. Geol. Surv., Prof. Paper 91, 1916, p. 277, pl. lxix, f. 5; pl. lxx, f. 1, 2.
Lagrange: Puryear, Tenn.

Zizyphus fibrillosus (Lesquereux) Lesquereux, Rept. U. S. Geol. Surv. Terr., vol. 7 (Tert. Fl.), 1898, p. 276, pl. lii, f. 1–6.—Knowlton, U. S. Geol. Surv., Prof. Paper 101, 1918, p. 335, pl. cii, f. 1.
Ceanothus fibrillosus Lesquereux, U. S. Geol. and Geog. Surv. Terr., Ann. Rept. 1872 [1873], p. 381.
Denver: Golden, Colo.
Raton: Raton Mesa region, Colorado and New Mexico.

Zizyphus grönlandicus Heer, Fl. foss. arct., vol. 7, 1883, p. 42, pl. lxii, f. 20.—Hollick, U. S. Geol. Surv., Mon. 50, 1906, p. 93, pl. xxxiv, f. 11, 12.
Magothy: Gay Head and Nashaquitsa, Marthas Vineyard, Mass.

Zizyphus hendersoni Knowlton, U. S. Geol. Surv., Prof. Paper 130, 19—, p. —, pl. xv, f. 1, 2 [in preparation].
Laramie: Cowan station near Denver and 1½ miles south of Golden, Colo.

Zizyphus hyperboreus? Heer. Lesquereux = *Zizyphus lesquereuxii*.

Zizyphus lamarensis Berry, Bull. Torr. Bot. Club, vol. 39, 1912, p. 398, pl. xxxi, f. 1; U. S. Geol. Surv., Prof. Paper 112, 1919, p. 112.
Woodbine: Arthurs Bluff, Lamar County, Tex.
Tuscaloosa: Glen Allen, Ala.

Zizyphus laurifolius Berry, U. S. Geol. Surv., Prof. Paper 84, 1914, p. 116, pl. xxi, f. 7; idem, Prof. Paper 112, 1919, p. 113.
Eutaw: McBrides Ford, Ga.

Zizyphus lesquereuxii Knowlton, U. S. Geol. Surv., Bull. 152, 1898, p. 246.
Zizyphus hyperboreus? Heer. Lesquereux, U. S. Geol. and Geog. Surv. Terr., Ann. Rept. 1872 [1873], p. 389; Rept. U. S. Geol. Surv. Terr., vol. 7 (Tert. Fl.), 1878, p. 276, pl. li, f. 15. [Not *Z. hyperboreus* of Heer.]
Hanna: Carbon, Wyo.
Denver: Golden, Colo.

Zizyphus lewisiana Hollick, Bull. Torr. Bot. Club, vol. 21, 1894, p. 58, pl. clxxx, f. 13; U. S. Geol. Surv., Mon. 50, 1906, p. 93, pl. xxxiv, f. 13.
Magothy: Oak Neck, Long Island, N. Y.

Zizyphus longifolia Newberry, U. S. Nat. Mus., Proc., vol. 5, 1882 [1883], p. 513; U. S. Geol. Surv., Mon. 35, 1898, p. 119, pl. lxv, f. 3–5.
Green River: Green River, Wyo.

Zizyphus meekii Lesquereux, U. S. Geol. and Geog. Surv. Terr., Ann. Rept. 1872 [1873], p. 388; Rept. U. S. Geol. Surv. Terr., vol. 7 (Tert. Fl.), 1878, p. 275, pl. li, f. 10–14.—Ward, U. S. Geol. Surv., 6th Ann. Rept., 1884–85 [1886], p. 554, pl. lii, f. 1, 2; idem, Bull. 37, 1887, p. 74, pl. xxxiii, f. 5, 6.
Denver: Mount Bross, Middle Park, Colo.
Post-Laramie: Black Buttes and Carbon, Wyo.
Livingston: ?Bozeman coal field, Mont.

Zizyphus meigsii (Lesquereux) Berry, U. S. Geol. Surv., Prof. Paper 91, 1916, p. 278, pl. lxx, f. 3–5.—Knowlton, U. S. Geol. Surv., Prof. Paper 101, 1918, p. 336, pl. xcix, f. 12; pl. c, f. 3.
Ceanothus meigsii Lesquereux, Am. Philos. Soc., Trans., vol. 13, 1869, p. 419, pl. xix, f. 5–7.
Wilcox (Ackerman): Colemans Mill, Miss.
Wilcox (Holly Springs): Oxford, Miss.
Lagrange: Grand Junction, La Grange, and Puryear, Tenn.
Raton: Salyers Creek and near Yankee, N. Mex.

Zizyphus microphyllus Lesquereux = *Zizyphus californicus*.

Zizyphus minutus Knowlton, U. S. Geol. Surv., Prof. Paper 130, 19—, p. —, pl. xviii, f. 1 [in preparation].
Laramie: Marshall, Colo.

Zizyphus oblongus Hollick, U. S. Geol. Surv., Mon. 50, 1906, p. 92, pl. xxxiv, f. 9, 10.
Zizyphus elegans Hollick [part], Bull. Torr. Bot. Club, vol. 21, 1894, p. 53, pl. clxxvii, f. 10; N. Y. Bot. Gard., Bull., vol. 3, 1904, p. 415, pl. lxxiii, f. 4.
Magothy: Glen Cove, Long Island, N. Y.

Zizyphus obtusa Kirchner, St. Louis Acad. Sci., Trans., vol. 8, 1898, p. 182, pl. xiii, f. 1.—Knowlton, U. S. Nat. Mus., Proc., vol. 51, 1916, p. 284, pl. xvi, f. 3.
Miocene: *Florissant, Colo.

Zizyphus paliurifolius Knowlton, U. S. Geol. Surv., Prof. Paper 101, 1918, p. 272, pl. xlvii, f. 1–4.
Vermejo: Vermejo Park, N. Mex.

Zizyphus piperoides Lesquereux, Mus. Comp. Zool., Mem., vol. 6, No. 2, 1878, p. 28, pl. viii, f. 10, 11 —Knowlton, in Lindgren, Jour. Geol., vol. 4, 1896, p. 890.
Miocene (auriferous gravels): Chalk Bluffs, Nevada County, and Independence Hill, Placer County, Calif.

Zizyphus serrulatus Ward, U. S. Geol. Surv., Sixth Ann. Rept., 1884–85 [1886], p. 554, pl. li, f. 14, 15; U. S. Geol. Surv., Bull. 37, 1887, p. 73, pl. xxxiii, f. 3, 4.—Knowlton, U. S. Geol. Surv., Mon. 32, pt. 2, 1899, p. 740, pl. ci, f. 4, 5.
Fort Union: Burns's ranch, Glendive, Mont.; Elk Creek, Yellowstone National Park.

Zizyphus townsendi Knowlton, U. S. Nat. Mus., Proc., vol. 17, 1894, p. 229, pl. ix, f. 8, 9; Geol. Soc. Am., Bull., vol. 5, 1893, p. 586.
Kenai: Herendeen Bay, Alaska.

Zizyphus sp. Berry, Torreya, vol. 10, 1910, p. 266.
Pleistocene: Near Long Branch, N. J.

Zizyphus sp. Berry, Bull. Torr. Bot. Club, vol. 37, 1910, p. 504.
——: McBrides Ford, below Columbus, Ga.

Zizyphus sp. Berry, Jour. Geol., vol. 25, 1917, p. 662.
Pleistocene: Vero, Fla.

Zizyphus sp. Knowlton, in Lindgren, Jour. Geol., vol. 4, 1896, p. 890.
Miocene (auriferous gravels): Independence Hill, Placer County, Calif.

Zizyphus sp. Knowlton, Geol. Soc. Am., Bull., vol. 8, 1897, p. 134.
Fort Union: Lightning Creek, Converse County, Wyo.

Zizyphus sp. Knowlton, Geol. Soc. Am., Bull., vol. 8, 1897, p. 148.
Mesaverde: Rock Springs, Wyo.

ZONARITES Sternberg, Flora d. Vorwelt, vol. 2, 1833, p. 34.

Zonarites digitatus (Brongniart) Sternberg, Flora d. Vorwelt, vol. 2, pt. 5, 1833, p. 34.—Lesquereux, U. S. Geol. and Geog. Surv. Terr., Ann. Rept. 1872 [1873], p. 421; idem, 1874 [1876], p. 333; Rept. U. S. Geol. Surv. Terr., vol. 6, 1874, p. 44, pl. i, f. 1.
Dakota: Near Fort Harker, Kans.

ZOSTERA Linné, Sp. pl., 1753, p. 968.

Zostera marina Linné. Penhallow, Brit. Assoc. Adv. Sci., Bradford meeting, 1900, p. 336.
Pleistocene: Montreal, Canada.

BIOLOGIC CLASSIFICATION OF GENERA.

The following biologic classification is introduced for convenience in quickly classifying any of the genera enumerated in the catalogue. The arrangement adopted is in the main that promulgated by Engler and Prantl in their "Natürlichen Pflanzenfamilien." Certain genera as to whose family affinity there is more or less doubt are indicated by a question mark placed in front of the generic name in the following list. An index of generic and family names is appended to the classification.

PLANT KINGDOM.

Phylum THALLOPHYTA.

Class FUNGI.

Series PHYCOMYCETES.

Order VAUCHERIALES.

Family PERONOSPORACEAE.

Peronosporoides Berry.

Series ASCOMYCETES.

Order LABOULBENIALES.

Family LABOULBENIACEAE.

Caenomyces Berry.

Order SPHAERIALES.

Family SPHAERIACEAE.

Didymosphaeria Rostrock. | Sphaerites Unger.

Order HYSTERIALES.

Family GRAPHIDIACEAE.

Opegrapha Persoon.

Order UREDINALES.

Family UREDINACEAE.

Puccinites Ettingshausen.

Series BASIDIOMYCETES.

Order AGARICALES.

Family AGARICACEAE.

Agaricites Meschinelli.

Order LYCOPERDIALES.

Family LYCOPERDIACEAE.

Geaster Scopoli.

FUNGI IMPERFECTI.

Order MONILIALES.

Family MONILIACEAE.

Ovularites Whitford.

Order MELANCONIALES.

Family DEMATIACEAE.

Cladosporites Felix.

Family TUBERCULARIACEAE.

Spegazzinites Felix.

STERILE MYCELIA.

Sclerotites Meschinelli.

Class ALGAE.

Order CHARACEALES.

Family CHARACEAE.

Chara Linné.

Algae of uncertain position.

Algites Seward.
Arthrodendron Ulrich.
Cancellophycus Ulrich.
Caulerpites Sternberg.
Chondrites Sternberg.
Confervites Brongniart.
Delesseria Lamarck.
Dendrophycus Lesquereux.
Fucoides Brongniart.
Fucus Linné.
Gilbertina Ulrich.
Gyrodendron Ulrich.
Halymenites Sternberg.
Helminthoida Schafhäutl.
Helminthopsis Heer.
Loperia Newberry.
Myelophycus Ulrich.
Palaeodictyon Heer.
Palaeophycus Hall.
Retiphycus Ulrich.

Phylum BRYOPHYTA.

Class HEPATICAE.

Order MARCHANTIALES.

Family MARCHANTIACEAE.

Marchantia (Linné) Raddi.
Marchantites Brongniart.
Preissites Knowlton.

Class MUSCI.

Order BRYALES.

Family BRYACEAE.

Bryum Linné.

Family DICRANACEAE.

Distichium Bruchman, Schimper, and Gümbel.

Family GRIMMIACEAE.

Plagiopodopsis Britton and Hollick.

Family POLYTRICHACEAE.

?Polytrichum Linné.

Family FONTINALACEAE.

Fontinalis Linné.

Family HYPNACEAE.

Hypnum Linné.

Family BRACHYTHECIACEAE.

Rhynchostegium Schimper.

Phylum PTERIDOPHYTA.

Class FILICES.

Order FILICALES.

Family HYMENOPHYLLACEAE.

Hymenophyllum Smith.

Family CYATHEACEAE.

Coniopteris Brongniart.
Cyatheites Göppert.
Dicksonia L'Héritier.
Dicksoniopsis Berry.
Hemitelites Göppert.
Thyrsopteris Kunze.

Family POLYPODIACEAE.

Acrostichum Linné.
Adiantites Göppert.
Adiantum Linné.
Allantodiopsis Knowlton and Maxon.
Anomalofilicites Hollick.
Aspidium Swartz.
Aspleniopteris Fontaine.
Asplenites Göppert.
Asplenium Linné.
Blechnum Linné.
Cladophlebis Brongniart.
Davallia Smith.
Davallites Dawson.
Dennstaedtia Bernhardi.
Dryopteris Adanson.
Dryopterites Berry.
Gymnogramma Desvaux.
Hypolepis Bernhardi.
Lastrea Bory.
Meniphylloides Berry.
Microtaenia Knowlton.
Oleandra Cavanilles.
Onoclea Linné.
Onychiopsis Yokoyama.
Pellaea Link.
Phanerophlebites Knowlton.
Phegopteris Fée.
Polypodium Linné.
Polystichum Roth.
Pteris Linné.
Salpichlaena Smith.
Tapeinidium (Presl) Christensen.
Woodwardia Smith.

Subfamily DIPTERIDINAE.

Clathropteris Brongniart.
Dictyophyllum Lindley and Hutton.
Hausmannia Dunker.
Protorhipis Andrae.

Family MATONIACEAE.

?Knowltonella Berry.
Laccopteris Presl.
Matonidium Schenk.

Family GLEICHENIACEAE.

Dicranopteris Bernhardi.
Gleichenia Smith.

Family SCHIZAEACEAE.

Acrostichopteris Fontaine.
Anemia Swartz.
Lygodium Swartz.
Ruffordia Seward.
Schizaeopsis Berry.

Family OSMUNDACEAE.

Acrostichites Göppert.
Osmunda Linné.
Osmundites Unger.

Family SALVINIACEAE.

Azollophyllum Penhallow.
Salvinia Adanson.

Family MARSILEACEAE.

Marsilea Linné.

Order MARATTIALES.

Family MARATTIACEAE.

Angiopteridium Schimper.
Danaeopsis Heer.
?Marattia Swartz.
?Mertensides Fontaine.
?Taeniopteris Brongniart.

Family HYDROPTERACEAE.

Sagenopteris Presl.

Ferns of uncertain affinity.

Asterocarpus Göppert.
Didymosarus Debey.
Heterofilicites Berry.
Laccopteris Presl.
Lonchopteris Brongniart.
Macrotaeniopteris Fontaine.
?Newberrya Berry.
Pecopteris Sternberg.
Pseudodanaeopsis Fontaine.
Scleropteris Saporta.
Sphenopteris Sternberg.
Stenopteris Saporta.
Tempskya Corda.
Thinnfeldia Ettingshausen.
Weichselia Stiehler.

Order OPHIOGLOSSALES.

Family OPHIOGLOSSACEAE.

?Cheiropteris Kurr.
Ophioglossum Linné.

Order EQUISETALES.

Family EQUISETACEAE.

Equisetum Linné.
Neocalamites Halle.
Schizoneura Schimper and Mougeot.

Order LYCOPODIALES.

Family LYCOPODIACEAE.

Lycopodites Brongniart.
Lycopodium Linné.

Family PSILOTACEAE.

?Tmesipteris Bernhardi.

Family SELAGINELLACEAE.

Selaginella Beauvois.

Order ISOETALES.

Family ISOETACEAE.

Isoetes Linné.

Phylum SPERMATOPHYTA.

Class GYMNOSPERMAE.

Order CYCADALES.

Family CYCADACEAE.

Anomozamites Schimper.
Ctenis Lindley and Hutton.
Ctenophyllum Schimper.
Ctenopsis Berry.
Ctenopteris Saporta.
?Cycadeomyelon Saporta.
Cycadeospermum Saporta.
Cycadinocarpus Schimper.
Cycadites Buckland.
Cycadospadix Schimper.
Cycas Linné.
Dichotozamites Berry.
Dioonites Miquel.
Encephalartopsis Fontaine.
Encephalartos Lehman.
Glossozamites Schimper.
Microzamia Corda.
Nilsonia Brongniart.
Otozamites F. Braun.
Platypterygium (Schimper) Feistmantel.
Podozamites F. Braun.
Pterophyllum Brongniart.
Ptilozamites Morris.
Sphenozamites Brongniart.
Williamsonia Carruthers.
Zamia Linné.
Zamiopsis Fontaine.
Zamiostrobus Endlicher.
Zamites Brongniart.

Order BENNETTITALES.

Family BENNETTITACEAE.

Bennettites Carruthers.
Cycadella Ward.
Cycadeoidea Buckland.

Order GINKGOALES.

Family GINKGOACEAE.

Baiera Braun.
Ginkgo Linné.

Order CONIFERALES.

Family TAXACEAE.

Subfamily PODOCARPEAE.

Nageiopsis Fontaine.
Palissya Endlicher.
Podocarpites Andrae.
Podocarpoxylon Gothan.
Podocarpus L'Héritier.

Subfamily TAXEAE.

Cephalotaxopsis Fontaine.
Cephalotaxospermum Berry.
Cephalotaxus Linné.
Protophyllocladus Berry.
Taxites Brongniart.
Tumion Rafinesque.

Family PINACEAE.

Subfamily ABIETINEAE.

Abies Adanson.
Abietites Hisinger.
Cedroxylon Kraus.
Cedrus Miller.
Elatides Heer.
Laricopsis Fontaine.
Larix Adanson.
Paracedroxylon Sennott.
Picea Link.
Pinites Witham.
Pinus Linné.
Pityophyllum Nathorst.
Pityoxylon Kraus.
Prepinus Jeffrey.
Pseudotsuga Carrière.
Tsuga Endlicher.

Subfamily TAXODIEAE.

Athrotaxis Don.
Athrotaxopsis Fontaine.
Cunninghamites Presl.
Cyparissidium Heer.
Geinitzia Endlicher.
Glyptostrobus Endlicher.
Inolepis Heer.
Pseudogeinitzia Hollick and Jeffrey.
Sequoia Endlicher.
Sphenolepidium Heer.
Taxodioxylon Hartig.
Taxodium Richard.

Subfamily CUPRESSINEAE.

Callitris Ventenat.
Cupressinoxylon Göppert.
Cupressites Brongniart.
Cupressus Linné.
Frenelopsis Schenk.
Juniperus Linné.
Libocedrus Endlicher.
Moriconia Debey and Ettingshausen.
Paracupressinoxylon Sennott.
Sabina Spach.
Thuites Sternberg.
Thuya Linné.
Thuyoxylon Unger.
Voltzia Brongniart.
Widdringtonia Endlicher.
Widdringtonites Endlicher.

Family ARAUCARIACEAE.

Araucaria Jussieu.
Araucariopitys Jeffrey.
Araucarioxylon Kraus.
Araucarites Presl.
Dammara Gaertner.
Dammarites Presl.
Protodammara Hollick and Jeffrey.
Woodworthia Jeffrey.

Family BRACHYPHYLLACEAE.

Brachyoxylon Hollick and Jeffrey.
Brachyphyllum Brongniart.

Coniferae of uncertain position.

Androvettia Hollick and Jeffrey.
Anomaspis Hollick and Jeffrey.
Brunswickia Wherry.
Cheirolepis Schimper.
Cyclopitys Schmalhausen.
Czekanowskia Heer.
Dactyolepis Hollick and Jeffrey.
Feildenia Heer.
Feildeniopsis Fontaine.
Heterolepis Berry.
Lepidostrobus Heer.
Noeggerathiopsis Feistmantel.
Pagiophyllum Heer.
Phoenicopsis Heer.
?Phyllocladopsis Fontaine.
Physematopitys Göppert.
Pityoidolepis Hollick and Jeffrey.
Ptenostrobus Lesquereux.
Raritania Hollick and Jeffrey.
Schizolepis Braun.
Sphenaspis Hollick and Jeffrey.
Strobilites Lindley and Hutton.

Order GNETALES.

Family GNETACEAE.

Ephedrites Göppert and Berger.

Class ANGIOSPERMAE.

Subclass MONOCOTYLEDONES.

Order PANDANALES.

Family TYPHACEAE.

Typha Linné.

Family SPARGANIACEAE.

Sparganium Linné.

Order NAIADALES.

Family NAIADACEAE.

Najadopsis Heer.
Zostera Linné.

Family POTAMOGETONACEAE.

?Palaeopotamogeton Knowlton.
Potamogeton Linné.
?Potamogetophyllum Fontaine.

Family ALISMACEAE.

Alismacites Saporta.
Alismaphyllites Knowlton.
Alismaphyllum Berry.
Sagittaria Linné.

Family VALLISNERIACEAE.

Limnobium Richard.
Philotria Rafinesque.
Vallisneria Linné.

Family HYDROCHARITACEAE.

Ottelia Persoon.

Order GRAMINALES.

Family GRAMINEAE.

Arundinaria Richard.
Arundo Linné.
Bambusium Unger.
Bromus Linné.
Chaetochloa Scribner.
Festuca Linné.
Muhlenbergia Schreber.
Oryzopsis Michaux.
?Phalaris Linné.
Phragmites Trinius.
Poacites Brongniart.
Stipa Linné.

Family CYPERACEAE.

Carex Linné.
Cyperacites Schimper.
Cyperus Linné.
Scirpus Linné.

Order ARECALES.

Family ARECACEAE.

Calamopsis Heer.
Chamaedorea Willdenow.
Flabellaria Sternberg.
"Geonoma" of authors.
Geonomites Visiani.
Manicaria Gaertner.
Nipadites Bowerbank.
Oreodoxites Göppert.
?Palmocarpon Lesquereux.
Palmoxylon Schenk.
Phoenicites Brongniart.
Sabal Adanson.
Sabalites Saporta.

Order ARALES.

Family ARACEAE.

Acorus Linné.
Araceaeites Fritel.
Arisaema Martius.
Arisaemites Knowlton.
Pistia Linné.
Thrinax Linné.

Family LEMNACEAE.

Lemna Linné. | Spirodela Schleiden.

Order XYRIDALES.

Family XYRIDACEAE?

Plantaginopsis Fontaine.

Family ERIOCAULACEAE.

Eriocaulon Linné.

Family BROMELIACEAE.

Bromelia Linné.

Family PONTEDERIACEAE.

Heteranthera Ruiz and Pavon.

Order LILIALES.

Family JUNCACEAE.

Juncus Linné.

Family LILIACEAE.

Majanthemophyllum Weber.

Family CONVALLARIACEAE.

Clintonia Rafinesque.

Family SMILACEAE.

Smilax Linné.

Family DIOSCOREACEAE.

Dioscorea Linné.

Family IRIDACEAE.

Iris Linné.

Family DRACAENACEAE.

Yucca Linné.

Order SCITAMINALES.

Family CANNACEAE.

Canna Linné.

Order ZINGIBERALES.

Family ZINGIBERACEAE.

Zingiberites Heer.

Monocotyledonae of uncertain position.

Anomophyllites Watelet.
Doryanthes Berry.
Musophyllum Göppert.
?Yorkia Wanner.

Subclass DICOTYLEDONES.

Series CHORIPETALAE.

Order CASUARINALES.

Family CASUARINACEAE.

Casuarina Adanson.

Order PIPERALES.

Family PIPERACEAE.

Piper Linné.

Piperites Göppert.

Order SALICALES.

Family SALICACEAE.

Populites Viviani.
?Populophyllum Fontaine.
Populus Linné.

Saliciphyllum Fontaine.
Salix Linné.

Order MYRICALES.

Family MYRICACEAE.

Comptonia Banks.
Morella Louereiro.

Myrica Linné.

Order JUGLANDALES.

Family JUGLANDACEAE.

Carya Linné.
Cryptocarya R. Brown.
Engelhardtia Leschen.
Hicoria Rafinesque.
Hicorioides Perkins.

Juglandiphyllum Fontaine.
Juglandites Sternberg.
Juglans Linné.
Praeengelhardtia Berry.
Pterocarya Kunth.

Order FAGALES.

Family BETULACEAE.

Alnites Hisinger.
Alnus Gaertner.
Betula Linné.
Betulites Göppert.
Betuloxylon Kaiser.

Carpinoxylon Vater.
Carpinus Linné.
Corylus Linné.
Ostrya Scopoli.

Family FAGACEAE.

Castanea Adanson.
Castanopsis Spach.
Dryophyllum Debey.
Fagophyllum Nathorst.
Fagopsis Hollick.

Fagus Linné.
Pasania Miquel.
Quercinium Unger.
Quercophyllum Fontaine.
Quercus Linné.

Order URTICALES.

Family ULMACEAE.

Celtis Linné.
Planera Gmelin.
Ulmites Dawson.

Ulmophyllum Ettingshausen.
Ulmoxylon Kaiser.
Ulmus Linné.

Family MORACEAE.

Artocarpidium Unger.
Artocarpoides Saporta.
Artocarpophyllum Dawson.
Artocarpus Forster.
Dorstenia Linné.
Eremophyllum Lesquereux.
Ficophyllum Fontaine.

Ficoxylon Kaiser.
Ficus Linné.
Morus Linné.
Pourouma Aublet.
Protoficus Saporta.
Pseudolmedia Trécul.
Toxylon Rafinesque.

Family URTICACEAE?

Lebephyllum Wilson.

Order PLATANALES.

Family PLATANACEAE.

Aspidiophyllum Lesquereux.
Credneria Zenker.
Plataninium Unger.
Platanus Linné.
Protophyllum Lesquereux

Order PROTEALES.

Family PROTEACEAE.

Banksia Forster.
Banksites Saporta.
Dryandroides Unger.
Embothriopsis Hollick.
Embothrites Unger.
Knightiophyllum Berry.
Lomatia R. Brown.
Lomatites Saporta.
?Macclintockia Heer.
Palaeodendron Saporta.
Persoonia Smith.
Proteoides Heer.

Order SANTALALES.

Family SANTALACEAE.

Santalum Linné.

Family LORANTHACEAE.

Phoradendron Nuttall.

Order ARISTOLOCHIALES.

Family ARISTOLOCHIACEAE.

Aristolochia Linné.
Aristolochiaephyllum Fontaine.
Aristolochites Heer.

Order POLYGONALES.

Family POLYGONACEAE.

Coccoloba Linné.
Coccolobis P. Brown.
Coccolobites Berry.
Persicaria Linné.
Polygonum Linné.

Order CHENOPODIALES.

Family NYCTAGINACEAE.

Pisonia Linné.

Order PAPAVERALES.

Family CAPPARIDACEAE.

Capparis Linné.
Capparites Berry.

Order THYMELEALES.

Family LAURACEAE.

Benzoin Fabricius.
Cinnamomum R. Brown.
Daphnogene Unger.
Daphnophyllum Heer.
Laurinoxylon Felix.
Laurophyllum Göppert.
Laurus Linné.
Malapoenna Adanson.
Mespilodaphne Nees.
Nectandra Bergius.
Ocotea Aublet.
Oreodaphne Nees.
Persea Gaertner.
Perseoxylon Felix.
Sassafras Nees.

Order RANALES.

Family NYMPHAEACEAE.

Brasenia Schreber.
Cabomba Aublet.
Castalia Salisbury.
Nelumbites Berry.
Nelumbo Adanson.

Family RANUNCULACEAE.

Thalictrum Linné.

Family BERBERIDACEAE.

Berberis Linné.
Odostemon Rafinesque.
Winchellia Lesquereux.

Family MENISPERMACEAE.

Cebatha Forskål.
Cocculites Heer.
Cocculus Linné.
Menispermites Lesquereux.

Family MAGNOLIACEAE.

Illicium Linné.
Liriodendron Linné.
Liriodendropsis Newberry.
Magnolia Linné.

Family ANONACEAE.

Anona Linné.
Asimina Adanson.
Guatteria Ruiz and Pavon.

Family MONIMIACEAE.

Monimiopsis Saporta.

Order SARRACENIALES.

Family DROSERACEAE.

Drosera Linné.

Order ROSALES.

Family SAXIFRAGACEAE.

Hydrangea Linné.
Philadelphus Linné.
Saxifraga Linné.

Family GROSSULARIACEAE.

Grossularia Adanson.
Ribes Linné.

Family CUNONIACEAE.

Weinmannia Linné.

Family HAMAMELIDACEAE.

Hamamelites Saporta.
Liquidambar Linné.
Parrotia Meyer.

Family ROSACEAE.

Cercocarpus Humboldt, Bonpland, and Kunth.
Chrysobalanus Linné.
Potentilla Linné.
Rosa Linné.
Rubus Linné.
Spiraea Linné.

Family POMACEAE.

Amelanchier Medicus.
Crataegus Linné.
Malus Jussieu.
Pyrus Linné.
Sorbus Linné.

Family DRUPACEAE.

Amygdalus Linné.
Padus Mönch.
Pruinium Platen.
?Prunoides Perkins.
Prunus Linné.

Family LEGUMINOSAE.

Bauhinia Linné.
Colutea Linné.
Cytisus Linné.
Phaca Linné.
Podogonium Heer.

Family PAPILIONACEAE.

Dalbergia Linné.
Dalbergites Berry.
Dolichites Unger.
Leguminosites Bowerbank.
Lonchocarpus Humboldt, Bonpland, and Kunth.
Phaseolites Unger.
Psoralea Linné.
Robinia Linné.
Sophora Linné.

Family CAESALPINIACEAE.

Caesalpinia Linné.
Caesalpinites Saporta.
Canavalia Adanson.
Cassia Linné.
Cercidoxylon Platen.
Cercis Linné.
Gleditsia Linné.
Gleditsiophyllum Berry.
Hymenaea Linné.
Palaeocassia Ettingshausen.

Family MIMOSACEAE.

Acacia Adanson.
Acaciaephyllum Fontaine.
Acaciaphyllites Berry.
Inga Scopoli.
Mimosa Linné.
Mimosites Bowerbank.
Pithecolobium Martius.
Viborquia Ortega.

Order GERANIALES.

Family RUTACEAE.

Citrophyllum Berry.
Fagara Linné.
Ptelea Linné.
Zanthoxylum Linné.

Family SIMARUBACEAE.

Ailanthophyllum Dawson.
Ailanthus Desfontaines.
Simaruba Aublet.
?Simarubinium Platen.

Family MELIACEAE.

Carapa Aublet.
Cedrela Linné.
Melia Linné.

Family MALPIGHIACEAE.

Banisteria Linné.
Hiraea Jacquin.

Family EUPHORBIACEAE.

Acalypha Linné.
Croton Linné.
Crotonophyllum Velenovsky.
Drypetes Vahl.
Euphorbiophyllum Ettingshausen.
Euphorbocarpum Knowlton.
Tithymalus Adanson.

Family HUMIRIACEAE.

Vantanea Aublet.

Order SAPINDALES.

Family ANACARDIACEAE.

Anacardioxylon Felix.
Anacardites Saporta.
Cotinus Adanson.
Heterocalyx Saporta.

Metopium P. Browne.
Pistacea Linné.
Rhus Linné.
Schmaltzia Desveaux.

Family CELASTRACEAE.

Celastrinites Saporta.
Celastrophyllum Göppert.
Celastrus Linné.
Elaeodendron Jacques.
Elaeodendroxylon Platen.

Euonymus Linné.
Gyminda Sargent.
Maytenus Molina.
Pachystima Rafinesque.

Family HIPPOCASTANACEAE.

Aesculophyllum Dawson.

Aesculus Linné.

Family STAPHYLEACEAE.

Staphylea Linné.

Family ACERACEAE.

Acer Linné.
Acerites Viviani.

Negundo Mönch.
Rulac Adanson.

Family SAPINDACEAE.

Cupanites Schimper.
Dodonaea Linné.
Sapindoides Perkins.

Sapindopsis Fontaine.
Sapindus Linné.

Family ILICACEAE.

Ilex Linné.

Order RHAMNALES.

Family RHAMNACEAE.

Berchemia Necker.
Ceanothus Linné.
Eorhamnidium Berry.
Paliurus Miller.
Reynosia Grisebach.

Rhamnacinium Felix.
Rhamnites Forbes.
Rhamnus Linné.
Zizyphus Adanson.

Family VITACEAE.

Ampelopsis Michaux.
Cissites Heer.
Cissus Linné.
Hedera Linné.

?Hederaephyllum Fontaine.
Parthenocissus Planchon.
?Vitiphyllum Fontaine.
Vitis Linné.

Order MALVALES.

Family TILIACEAE.

Apeibopsis Heer.
Grewia Linné.

Grewiopsis Saporta.
Tilia Linné.

Family MALVACEAE.

Malvastrum A. Gray.

Manihotites Berry.

Family BOMBACEAE.

Bombacites Berry.

Bombax Linné.

Family STERCULIACEAE.

Pterospermites Heer.
Sterculia Linné.

Sterculinocarpus Berry.

Family BUETTNERIACEAE.

Buettneria Löfling.

Order PARIETALES.

Family DILLENIACEAE.

Dillenites Berry.

Family PASSIFLORACEAE.

Passiflora Linné.

Family TERNSTROEMIACEAE.

Ternstroemites Berry.

Order MYRTALES.

Family RHIZOPHORACEAE.

Rhizophora Linné.

Family MYRTACEAE.

Callistemophyllum Ettingshausen.
Calyptranthes Swartz.
Eucalyptophyllum Fontaine.
Eucalyptus L'Héritier.
Eugenia Linné.
Myrcia De Candolle.
Myrtus Linné.

Family COMBRETACEAE.

Combretanthites Berry.
Combretum Linné.
Conocarpites Berry.
Conocarpus Linné.
Laguncularia Gaertner.
Terminalia Linné.

Family HALORAGIDACEAE.

Hippuris Linné.

Family HYDROCARYACEAE.

Trapa Linné.

Family MELASTOMACEAE.

Melastomites Unger

Order UMBELLALES.

Family ARALIACEAE.

Aralia Linné.
Araliaephyllum Fontaine.
Aralinium Platen.
Araliopsis (Lesquereux) Berry.
Araliopsoides Berry.
Chondrophyllum Necker.
?Dewalquea Saporta and Marion.
Oreopanax Decaisne and Planchon.
Panax Linné.

Family UMBELLIFERAE.

Oxypolis Rafinesque.

Family CORNACEAE.

Cornophyllum Newberry.
Cornus Linné.
Nyssa Linné.
Nyssidium Heer.

Series GAMOPETALAE.

Order ERICALES.

Family CLETHRACEAE.

Clethra Linné.

Family PIROLACEAE.

Osmanthus Loureiro.

Family ERICACEAE.

Andromeda Linné.
Arbutus Linné.
Arctostaphylos Adanson.
Dendrium Desvaux.
Gaylussacia Humboldt, Bonpland, and Kunth.
Kalmia Linné.
Leucothoe Don.
Xolisma Rafinesque.

Family VACCINIACEAE.

Oxycoccus Adanson.
Vaccinium Linné
Vaccinophyllum Dawson.

Order PRIMULALES.

Family MYRSINACEAE.

Icacorea Aublet.
Myrsine Linné.

Family PRIMULACEAE.

Lysimachia Linné.

Order EBENALES.

Family SAPOTACEAE.

Bumelia Swartz.
Chrysophyllum Linné.
Eoachras Berry.
Mimusops Linné.
Sapotacites Ettingshausen.
Sideroxylon Linné.

Family EBENACEAE.

Diospyros Linné.
Ebenoxylon Felix.

Family STYRACACEAE.

Styrax Linné.

Order OLEALES.

Family OLEACEAE.

Chionanthus Linné.
Fraxinus Linné.
Ligustrum Linné

Order GENTIANALES.

Family APOCYNACEAE.

Apocynophyllum Unger.
Echitonium Unger.

Family ASCLEPIADACEAE.

Acerates Elliott.
Periploca Linné.

Family MENYANTHACEAE.

Menyanthes Linné.

Order POLEMONIALES.

Family BORAGINACEAE.

Cordia Linné.

Order TUBIFLORAE.

Family CONVOLVULACEAE.

Porana Burmann.

Family VERBENACEAE.

Avicennia Linné.
Citharexylon Linné.

Family SOLANACEAE.

?Florissantia Knowlton.
Solanites Saporta.

Family SCROPHULARIACEAE.

Mimulus Linné.

Family BIGNONIACEAE.

Catalpa Scopoli.
Dombeyopsis Unger.
Tecoma Jussieu.

Order RUBIALES.

Family RUBIACEAE.

Cephalanthus Linné.
Exostema Richard.
Guettarda Linné.
Psychotria Linné.

Family CAPRIFOLIACEAE.

Sambucus Linné.
Viburnites Lesquereux.
Viburnum Linné.

Order CAMPANULALES.

Family CUCURBITACEAE.

Melothria Linné.
Sicyos Linné.

Family COMPOSITAE.

Aster Linné.
Carduus Linné.

Dicotyledonae of uncertain position

Anisophyllum Lesquereux.
Anomalophyllum Massalongo.
Fontainea Newberry.
Liriophyllum Lesquereux.
Palaeanthus Newberry.
Phyllites Brongniart.
Premnophyllum Velenovsky.
Proteaephyllum Fontaine.
Rogersia Fontaine.
Symphorocarpophyllum Dawson.

Genera based on fruits of doubtful or unrecognized affinity.

Bicarpellites Perkins.
Brandonea Perkins.
Carpites Schimper.
Carpolithes Schlotheim.
Carpolithus of authors.
Cucumites Bowerbank.
Drupa Lesquereux.
Glossocarpellites Perkins.
Lescuria Perkins.
Monocarpellites Perkins.
Palaeoaster Knowlton.
Palmocarpon Lesquereux.
Rubioides Perkins.
Samaropsis Göppert.
Staphidoides Perkins.
Tricarpellites Perkins.

Plants of uncertain position.

Actinopteris Schenk.
Anabacaulus Emmons.
Antholithes Brongniart.
Calycites of authors.
Caudex Lesquereux.
Caulinites Brongniart.
Cornephyllum Emmons.
Dictyocaulus Emmons.
Galla Ludwig.
Gymnocaulus Emmons.
Lepacyclotes Emmons.
Ramulus Wanner.
Spiraxis Newberry.
Tricalycites Newberry
Yuccites Martius.

INDEX OF GENERA AND FAMILIES.

		Page.
Abies Adanson	Abietineae	668
Abietites Hisinger	Abietineae	668
Acacia Adanson	Mimosaceae	674
Acaciaephyllum Fontaine	Mimosaceae	674
Acaciaphyllites Berry	Mimosaceae	674
Acalypha Linné	Euphorbiaceae	674
Acer Linné	Aceraceae	675
Aceraceae		675
Acerates Elliott	Asclepiadaceae	677
Acerites Viviani	Aceraceae	675
Acorus Linné	Araceae	669
Acrostichites Göppert	Osmundaceae	666
Acrostichopteris Fontaine	Schizaeaceae	666
Acrostichum Linné	Polypodiaceae	665
Actinopteris Schenk	Plant	678
Adiantites Göppert	Polypodiaceae	665
Adiantum Linné	Polypodiaceae	665
Aesculophyllum Dawson	Hippocastanaceae	675
Aesculus Linné	Hippocastanaceae	675
Agaricaceae		663
Agaricites Meschinelli	Agaricaceae	663
Ailanthophyllum Dawson	Simarubaceae	674
Ailanthus Desfontaines	Simarubaceae	674
Algites Seward	Alga	664
Alismaceae		669
Alismacites Saporta	Alismaceae	669
Alismaphyllites Knowlton	Alismaceae	669
Alismaphyllum Berry	Alismaceae	669
Allantodiopsis Knowlton and Maxon	Polypodiaceae	665
Alnites Hisinger	Betulaceae	671
Alnus Gaertner	Betulaceae	671
Amelanchier Medicus	Pomaceae	673
Ampelopsis Michaux	Vitaceae	675
Amygdalus Linné	Drupaceae	674
Anabacaulus Emmons	Plant	678
Anacardiaceae		675
Anacardioxylon Felix	Anacardiaceae	675
Anacardites Saporta	Anacardiaceae	675
Andromeda Linné	Ericaceae	677
Androvettia Hollick and Jeffrey	Coniferae	668
Anemia Swartz	Schizaeaceae	666
Angiopteridium Schimper	Marattiaceae	666
Anisophyllum Lesquereux	Dicotyledon	678
Anomalofilicites Hollick	Polypodiaceae	665
Anomalophyllum Massalongo	Dicotyledon	678

		Page
Anomaspis Hollick and Jeffrey	Coniferae	668
Anomophyllites Watelet	Monocotyledon	670
Anomozamites Schimper	Cycadaceae	667
Anona Linné	Anonaceae	673
Anonaceae		673
Antholithes Brongniart	Plant	678
Apeibopsis Heer	Tiliaceae	675
Apocynaceae		677
Apocynophyllum Unger	Apocynaceae	677
Araceae		669
Araceaeites Fritel	Araceae	669
Aralia Linné	Araliaceae	676
Araliaceae		676
Araliaephyllum Fontaine	Araliaceae	676
Aralinium Platen	Araliaceae	676
Araliopsis (Lesquereux) Berry	Araliaceae	676
Araliopsoides Berry	Araliaceae	676
Araucaria Jussieu	Araucariaceae	668
Araucariaceae		668
Araucariopitys Jeffrey	Araucariaceae	668
Araucarioxylòn Kraus	Araucariaceae	668
Araucarites Presl	Araucariaceae	668
Arbutus Linné	Ericaceae	677
Arctostaphylos Adanson	Ericaceae	677
Arecaceae		669
Arisaema Martius	Araceae	669
Arisaemites Knowlton	Araceae	669
Aristolochia Linné	Aristolochiaceae	672
Aristolochiaceae		672
Aristolochiaephyllum Fontaine	Aristolochiaceae	672
Aristolochites Heer	Aristolochiaceae	672
Arthrodendron Ulrich	Alga	664
Artocarpidium Unger	Moraceae	671
Artocarpoides Saporta	Moraceae	671
Artocarpophyllum Dawson	Moraceae	671
Artocarpus Forster	Moraceae	671
Arundinaria Richard	Gramineae	669
Arundo Linné	Gramineae	669
Asclepiadaceae		677
Asimina Adanson	Anonaceae	673
Aspidiophyllum Lesquereux	Platanaceae	672
Aspidium Swartz	Polypodiaceae	665
Aspleniopteris Fontaine	Polypodiaceae	665
Asplenites Göppert	Polypodiaceae	665
Asplenium Linné	Polypodiaceae	665
Aster Linné	Compositae	678
Asterocarpus Göppert	Fern	666
Athrotaxis Don	Taxodieae	668
Athrotaxopsis Fontaine	Taxodieae	668
Avicennia Linné	Verbenaceae	678
Azollophyllum Penhallow	Salviniaceae	666
Baiera Braun	Ginkgoaceae	667
Bambusium Unger	Gramineae	669

		Page.
Banisteria Linné	Malpighiaceae	674
Banksia Forster	Proteaceae	672
Banksites Saporta	Proteaceae	672
Bauhinia Linné	Leguminosae	674
Bennettitaceae		667
Bennettites Carruthers	Bennettitaceae	667
Benzoin Fabricius	Lauraceae	672
Berberidaceae		673
Berberis Linné	Berberidaceae	673
Berchemia Necker	Rhamnaceae	675
Betula Linné	Betulaceae	671
Betulaceae		671
Betulites Göppert	Betulaceae	671
Betuloxylon Kaiser	Betulaceae	671
Bicarpellites Perkins	Fruit	678
Bignoniaceae		678
Blechnum Linné	Polypodiaceae	665
Bombaceae		675
Bombacites Berry	Bombaceae	675
Bombax Linné	Bombaceae	675
Boraginaceae		677
Brachyoxylon Hollick and Jeffrey	Brachyphyllaceae	668
Brachyphyllaceae		668
Brachyphyllum Brongniart	Brachyphyllaceae	668
Brachytheiaceae		665
Brandonea Perkins	Fruit	678
Brasenia Schreber	Nymphaeaceae	673
Bromelia Linné	Bromeliaceae	670
Bromeliaceae		670
Bromus Linné	Gramineae	669
Brunswickia Wherry	Coniferae	668
Bryaceae		664
Bryum Linné	Bryaceae	664
Buettneria Löfling	Buettneriaceae	675
Buettneriaceae		675
Bumelia Swartz	Sapotaceae	677
Cabomba Aublet	Nymphaeaceae	673
Caenomyces Berry	Laboulbeniaceae	663
Caesalpinia Linné	Caesalpiniaceae	674
Caesalpiniaceae		674
Caesalpinites Saporta	Caesalpiniaceae	674
Calamopsis Heer	Arecaceae	669
Callistemophyllum Ettingshausen	Myrtaceae	676
Callitris Ventenat	Cupressineae	668
Calycites of authors	Plant	678
Calyptranthes Swartz	Myrtaceae	676
Canavalia Adanson	Caesalpiniaceae	674
Cancellophycus Ulrich	Alga	664
Canna Linné	Cannaceae	670
Cannaceae		670
Capparidaceae		672
Capparis Linné	Capparidaceae	672
Capparites Berry	Capparidaceae	672

		Page.
Caprifoliaceae		678
Carapa Aublet	Meliaceae	674
Carduus Linné	Compositae	678
Carex Linné	Cyperaceae	669
Carpinoxylon Vater	Betulaceae	671
Carpinus Linné	Betulaceae	671
Carpites Schimper	Fruit	678
Carpolithes Schlotheim	Fruit	678
Carpolithus of authors	Fruit	678
Carya Linné	Juglandaceae	671
Cassia Linné	Caesalpiniaceae	674
Castalia Salisbury	Nymphaeaceae	673
Castanea Adanson	Fagaceae	671
Castanopsis Spach	Fagaceae	671
Casuarina Adanson	Casuarinaceae	670
Casuarinaceae		670
Catalpa Scopoli	Bignoniaceae	678
Caudex Lesquereux	Plant	678
Caulerpites Sternberg	Alga	664
Caulinites Brongniart	Plant	678
Ceanothus Linné	Rhamnaceae	675
Cebatha Forskål	Menispermaceae	673
Cedrela Linné	Meliaceae	674
Cedroxylon Kraus	Abietineae	668
Cedrus Miller	Abietineae	668
Celastraceae		675
Celastrinites Saporta	Celastraceae	675
Celastrophyllum Göppert	Celastraceae	675
Celastrus Linné	Celastraceae	675
Celtis Linné	Ulmaceae	671
Cephalanthus Linné	Rubiaceae	678
Cephalotaxopsis Fontaine	Taxeae	667
Cephalotaxospermum Berry	Taxeae	667
Cephalotaxus Linné	Taxeae	667
Cercidoxylon Platen	Caesalpiniaceae	674
Cercis Linné	Caesalpiniaceae	674
Cercocarpus Humboldt, Bonpland, and Kunth	Rosaceae	673
Chaetochloa Scribner	Gramineae	669
Chamaedorea Willdenow	Arecaceae	669
Chara Linné	Characeae	664
Characeae		664
Cheirolepis Schimper	Coniferae	668
?Cheiropteris Kurr	Ophioglossaceae	666
Chionanthus Linné	Oleaceae	677
Chondrites Sternberg	Alga	664
Chondrophyllum Necker	Araliaceae	676
Chrysobalanus Linné	Rosaceae	673
Chrysophyllum Linné	Sapotaceae	677
Cinnamomum R. Brown	Lauraceae	672
Cissites Heer	Vitaceae	675
Cissus Linné	Vitaceae	675
Citharexylon Linné	Verbenaceae	678
Citrophyllum Berry	Rutaceae	674

		Page.
Cladophlebis Brongniart	Polypodiaceae	665
Cladosporites Felix	Dematiaceae	664
Clathropteris Brongniart	Dipteridinae	665
Clethra Linné	Clethraceae	676
Clethraceae		676
Clintonia Rafinesque	Convallariaceae	670
Coccoloba Linné	Polygonaceae	672
Coccolobis P. Brown	Polygonaceae	622
Coccolobites Berry	Polygonaceae	672
Cocculites Heer	Menispermaceae	673
Cocculus Linné	Menispérmaceae	673
Colutea Linné	Leguminosae	674
Combretaceae		676
Combretanthites Berry	Combretaceae	676
Combretum Linné	Combretaceae	676
Compositae		678
Comptonia Banks	Myricaceae	671
Confervites Brongniart	Alga	664
Coniopteris Brongniart	Cyatheaceae	665
Conocarpites Berry	Combretaceae	676
Conocarpus Linné	Combretaceae	676
Convallariaceae		670
Convolvulaceae		677
Cordia Linné	Boraginaceae	677
Cornaceae		676
Cornephyllum Emmons	Plant	678
Cornophyllum Newberry	Cornaceae	676
Cornus Linné	Cornaceae	676
Corylus Linné	Betulaceae	671
Cotinus Adanson	Anacardiaceae	675
Crataegus Linné	Pomaceae	673
Credneria Zenker	Platanaceae	672
Croton Linné	Euphorbiaceae	674
Crotonophyllum Velenovsky	Euphorbiaceae	674
Cryptocarya R. Brown	Juglandaceae	671
Ctenis Lindley and Hutton	Cycadaceae	667
Ctenophyllum Schimper	Cycadaceae	667
Ctenopsis Berry	Cycadaceae	667
Ctenopteris Saporta	Cycadaceae	667
Cucumites Bowerbank	Fruit	678
Cucurbitaceae		678
Cunninghamites Presl	Taxodieae	668
Cunoniaceae		673
Cupanites Schimper	Sapindaceae	675
Cupressinoxylon Göppert	Cupressineae	668
Cupressites Brongniart	Cupressineae	668
Cupressus Linné	Cupressineae	668
Cyatheaceae		665
Cyatheites Göppert	Cyatheaceae	665
Cycadaceae		667
Cycadella Ward	Bennettitaceae	667
Cycadeoidea Buckland	Bennettitaceae	667
?Cycadeomyelon Saporta	Cycadaceae	667

		Page.
Cycadeospermum Saporta	Cycadaceae	667
Cycadinocarpus Schimper	Cycadaceae	667
Cycadites Buckland	Cycadaceae	667
Cycadospadix Schimper	Cycadaceae	667
Cycas Linné	Cycadaceae	667
Cyclopitys Schmalhausen	Coniferae	668
Cyparissidium Heer	Taxodieae	668
Cyperaceae		669
Cyperacites Schimper	Cyperaceae	669
Cyperus Linné	Cyperaceae	669
Cytisus Linné	Leguminosae	674
Czekanowskia Heer	Coniferae	668
Dactyolepis Hollick and Jeffrey	Coniferae	668
Dalbergia Linné	Papilionaceae	674
Dalbergites Berry	Papilionaceae	674
Dammara Gaertner	Araucariaceae	668
Dammarites Presl	Araucariaceae	668
Danaeopsis Heer	Marattiaceae	666
Daphnogene Unger	Lauraceae	672
Daphnophyllum Heer	Lauraceae	672
Davallia Smith	Polypodiaceae	665
Davallites Dawson	Polypodiaceae	665
Delesseria Lamarck	Alga	664
Dematiaceae		664
Dendrium Desvaux	Ericaceae	677
Dendrophycus Lesquereux	Alga	664
Dennstaedtia Bernhardi	Polypodiaceae	665
Dewalquea Saporta and Marion	Araliaceae	676
Dichotozamites Berry	Cycadaceae	667
Dicksonia L'Héritier	Cyatheaceae	665
Dicksoniopsis Berry	Cyatheaceae	665
Dicranaceae		665
Dicranopteris Bernhardi	Gleicheniaceae	666
Dictyocaulus Emmons	Plant	678
Dictyophyllum Lindley and Hutton	Dipteridinae	665
Didymosarus Debey	Fern	666
Didymosphaeria Rostrock	Sphaeriaceae	663
Dilleniaceae		676
Dillenites Berry	Dilleniaceae	676
Dioonites Miquel	Cycadaceae	667
Dioscorea Linné	Dioscoreaceae	670
Dioscoreaceae		670
Diospyros Linné	Ebenaceae	677
Distichium Bruchman, Schimper, and Gümbel	Dicranaceae	665
Dodonaea Linné	Sapindaceae	675
Dolichites Unger	Papilionaceae	674
Dombeyopsis Unger	Bignoniaceae	678
Dorstenia Linné	Moraceae	671
Doryanthes Berry	Monocotyledon	670
Dracaenaceae		670
Drosera Linné	Droseraceae	673
Droseraceae		673
Drupa Lesquereux	Fruit	678

		Page.
Drupaceae		674
Dryandroides Unger	Proteaceae	672
Dryophyllum Debey	Fagaceae	671
Dryopteris Adanson	Polypodiaceae	665
Dryopterites Berry	Polypodiaceae	665
Drypetes Vahl	Euphorbiaceae	674
Ebenaceae		677
Ebenoxylon Felix	Ebenaceae	677
Echitonium Unger	Apocynaceae	677
Elaeodendron Jacques	Celastraceae	675
Elaeodendroxylon Platen	Celastraceae	675
Elatides Heer	Abietineae	668
Embothriopsis Hollick	Proteaceae	672
Embothrites Unger	Proteaceae	672
Encephalartopsis Fontaine	Cycadaceae	667
Encephalartos Lehman	Cycadaceae	667
Engelhardtia Leschen	Juglandaceae	671
Eoachras Berry	Sapotaceae	677
Eorhamnidium Berry	Rhamnaceae	675
Ephedrites Göppert and Berger	Gnetaceae	668
Equisetaceae		666
Equisetum Linné	Equisetaceae	666
Eremophyllum Lesquereux	Moraceae	671
Ericaceae		677
Eriocaulaceae		670
Eriocaulon Linné	Eriocaulaceae	670
Eucalyptophyllum Fontaine	Myrtaceae	676
Eucalyptus L'Héritier	Myrtaceae	676
Eugenia Linné	Myrtaceae	676
Euonymus Linné	Celastraceae	675
Euphorbiaceae		674
Euphorbiophyllum Ettingshausen	Euphorbiaceae	674
Euphorbocarpum Knowlton	Euphorbiaceae	674
Exostema Richard	Rubiaceae	678
Fagaceae		671
Fagara Linné	Rutaceae	674
Fagophyllum Nathorst	Fagaceae	671
Fagopsis Hollick	Fagaceae	671
Fagus Linné	Fagaceae	671
Feildenia Heer	Coniferae	668
Feildeniopsis Fontaine	Coniferae	668
Festuca Linné	Gramineae	669
Ficophyllum Fontaine	Moraceae	671
Ficoxylon Kaiser	Moraceae	671
Ficus Linné	Moraceae	671
Flabellaria Sternberg	Arecaceae	669
?Florissantia Knowlton	Solanaceae	678
Fontainea Newberry	Dicotyledon	678
Fontinalaceae		665
Fontinalis Linné	Fontinalaceae	665
Fraxinus Linné	Oleaceae	677
Frenelopsis Schenk	Cupressineae	668
Fucoides Brongniart	Alga	664

		Page.
Fucus Linné	Alga	664
Galla Ludwig	Plant	678
Gaylussacia Humboldt, Bonpland, and Kunth	Ericaceae	677
Geaster Scopoli	Lycoperdiaceae	664
Geinitzia Endlicher	Taxodieae	668
"Geonoma" of authors	Arecaceae	669
Geonomites Visiani	Arecaceae	669
Gilbertina Ulrich	Alga	664
Ginkgo Linné	Ginkgoaceae	667
Ginkgoaceae		667
Gleditsia Linné	Caesalpiniaceae	674
Gleditsiophyllum Berry	Caesalpiniaceae	674
Gleichenia Smith	Gleicheniaceae	666
Gleicheniaceae		666
Glossocarpellites Perkins	Fruit	678
Glossozamites Schimper	Cycadaceae	667
Glyptostrobus Endlicher	Taxodieae	668
Gnetaceae		668
Gramineae		669
Graphidiaceae		663
Grewia Linné	Tiliaceae	675
Grewiopsis Saporta	Tiliaceae	675
Grimmiaceae		665
Grossularia Adanson	Grossulariaceae	673
Grossulariaceae		673
Guatteria Ruiz and Pavon	Anonaceae	673
Guettarda Linné	Rubiaceae	678
Gyminda Sargent	Celastraceae	675
Gymnocaulus Emmons	Plant	678
Gymnogramma Desvaux	Polypodiaceae	665
Gyrodendron Ulrich	Alga	664
Haloragidaceae		676
Halymenites Sternberg	Alga	664
Hamamelites Saporta	Hamamelidaceae	673
Hamamelidaceae		673
Hausmannia Dunker	Dipteridinae	665
Hedera Linné	Vitaceae	675
?Hederaephyllum Fontaine	Vitaceae	675
Helminthoida Schafhäutl	Alga	664
Helminthopsis Heer	Alga	664
Hemitelites Göppert	Cyatheaceae	665
Heteranthera Ruiz and Pavon	Pontederiaceae	670
Heterocalyx Saporta	Anacardiaceae	675
Heterofilicites Berry	Fern	666
Heterolepis Berry	Coniferae	668
Hicoria Rafinesque	Juglandaceae	671
Hicorioides Perkins	Juglandaceae	671
Hippocastanaceae		675
Hippuris Linné	Haloragidaceae	676
Hiraea Jacquin	Malpighiaceae	674
Humiriaceae		674
Hydrangea Linné	Saxifragaceae	673
Hydrocaryaceae		676

		Page.
Hydrocharitaceae		669
Hydropteraceae		666
Hymenaea Linné	Caesalpiniaceae	674
Hymenophyllaceae		665
Hymenophyllum Smith	Hymenophyllaceae	665
Hypnaceae		665
Hypnum Linné	Hypnaceae	665
Hypolepis Bernhardi	Polypodiaceae	665
Icacorea Aublet	Myrsinaceae	677
Ilex Linné	Ilicaceae	675
Ilicaceae		675
Illicium Linné	Magnoliaceae	673
Inga Scopoli	Mimosaceae	674
Inolepis Heer	Taxodieae	668
Iridaceae		670
Iris Linné	Iridaceae	670
Isoetaceae		667
Isoetes Linné	Isoetaceae	667
Juglandaceae		671
Juglandiphyllum Fontaine	Juglandaceae	671
Juglandites Sternberg	Juglandaceae	671
Juglans Linné	Juglandaceae	671
Juncaceae		670
Juncus Linné	Juncaceae	670
Juniperus Linné	Cupressineae	668
Kalmia Linné	Ericaceae	677
Knightiophyllum Berry	Proteaceae	672
?Knowltonella Berry	Matoniaceae	666
Laboulbiliaceae		663
Laccopteris Presl	Matoniaceae	666
Laguncularia Gaertner	Combretaceae	676
Laricopsis Fontaine	Abietineae	668
Larix Adanson	Abietineae	668
Lastrea Bory	Polypodiaceae	665
Lauraceae		672
Laurinoxylon Felix	Lauraceae	672
Laurophyllum Göppert	Lauraceae	672
Laurus Linné	Lauraceae	672
Lebephyllum Wilson	Urticaceae	671
Leguminosae		674
Leguminosites Bowerbank	Papilionaceae	674
Lemna Linné	Lemnaceae	670
Lemnaceae		670
Lepacyclotes Emmons	Plant	678
Lepidostrobus Heer	Coniferae	668
Lescuria Perkins	Fruit	678
Leucothoe Don	Ericaceae	677
Libocedrus Endlicher	Cupressineae	668
Ligustrum Linné	Oleaceae	677
Liliaceae		670
Limnobium Richard	Vallisneriaceae	669
Liquidambar Linné	Hamamelidaceae	673
Liriodendron Linné	Magnoliaceae	673

		Page.
Liriodendropsis Newberry	Magnoliaceae	673
Liriophyllum Lesquereux	Dicotyledon	678
Lomatia R. Brown	Proteaceae	672
Lomatites Saporta	Proteaceae	672
Lonchocarpus Humboldt, Bonpland, and Kunth	Papilionaceae	674
Lonchopteris Brongniart	Fern	666
Loperia Newberry	Alga	664
Loranthaceae		672
Lycoperdiaceae		664
Lycopodiaceae		666
Lycopodites Brongniart	Lycopodiaceae	666
Lycopodium Linné	Lycopodiaceae	666
Lygodium Swartz	Schizaeaceae	666
Lysimachia Linné	Primulaceae	677
?Macclintockia Heer	Proteaceae	672
Macrotaeniopteris Fontaine	Fern	666
Magnolia Linné	Magnoliaceae	673
Magnoliaceae		673
Majanthemophyllum Weber	Liliaceae	670
Malapoenna Adanson	Lauraceae	672
Malpighiaceae		674
Malus Jussieu	Pomaceae	673
Malvaceae		675
Malvastrum A. Gray	Malvaceae	675
Manicaria Gaertner	Arecaceae	669
Manihotites Berry	Malvaceae	675
Marattia Swartz	Marattiaceae	666
Marattiaceae		666
Marchantia (Linné) Raddi	Marchantiaceae	664
Marchantiaceae		664
Marchantites Brongniart	Marchantiaceae	664
Marsileaceae		666
Marsilea Linné	Marsileaceae	666
Matoniaceae		666
Matonidium Schenk	Matoniaceae	666
Maytenus Molina	Celastraceae	675
Melastomaceae		676
Melastomites Unger	Melastomaceae	676
Melia Linné	Meliaceae	674
Meliaceae		674
Melothria Linné	Cucurbitaceae	678
Meniphylloides Berry	Polypodiaceae	665
Menispermaceae		673
Menispermites Lesquereux	Menispermaceae	673
Menyanthaceae		677
Menyanthes Linné	Menyanthaceae	677
?Mertensides Fontaine	Marattiaceae	666
Mespilodaphne Nees	Lauraceae	672
Metopium P. Browne	Anacardiaceae	675
Microtaenia Knowlton	Polypodiaceae	665
Microzamia Corda	Cycadaceae	667
Mimosa Linné	Mimosaceae	674
Mimosaceae		674
Mimosites Bowerbank	Mimosaceae	674

		Page.
Mimulus Linné	Scrophulariaceae	678
Mimusops Linné	Sapotaceae	677
Moniliaceae		664
Monimiaceae		673
Monimiopsis Saporta	Monimiaceae	673
Monocarpellites Perkins	Fruit	678
Moraceae		671
Morella Loureiro	Myricaceae	670
Moriconia Debey and Ettingshausen	Cupressineae	668
Morus Linné	Moraceae	671
Muhlenbergia Schreber	Gramineae	669
Musophyllum Göppert	Monocotyledon	670
Myelophycus Ulrich	Alga	664
Myrcia De Candolle	Myrtaceae	676
Myrica Linné	Myricaceae	671
Myricaceae		671
Myrsinaceae		677
Myrsine Linné	Myrsinaceae	677
Myrtaceae		676
Myrtus Linné	Myrtaceae	676
Nageiopsis Fontaine	Podocarpeae	667
Naiadaceae		669
Najadopsis Heer	Naiadaceae	669
Nectandra Bergius	Lauraceae	672
Negundo Mönch	Aceraceae	675
Nelumbites Berry	Nymphaeaceae	673
Nelumbo Adanson	Nymphaeaceae	673
Neocalamites Halle	Equisetaceae	666
?Newberrya Berry	Fern	666
Nilsonia Brongniart	Cycadaceae	667
Nipadites Bowerbank	Arecaceae	669
Noeggerathiopsis Feistmantel	Coniferae	668
Nyctaginaceae		672
Nymphaeaceae		673
Nyssa Linné	Cornaceae	676
Nyssidium Heer	Cornaceae	676
Ocotea Aublet	Lauraceae	672
Odostemon Rafinesque	Berberidaceae	673
Oleaceae		677
Oleandra Cavanilles	Polypodiaceae	665
Onoclea Linné	Polypodiaceae	665
Onychiopsis Yokoyama	Polypodiaceae	665
Opegrapha Persoon	Graphidiaceae	663
Ophioglossaceae		666
Ophioglossum Linné	Ophioglossaceae	666
Oreodaphne Nees	Lauraceae	672
Oreodoxites Göppert	Arecaceae	669
Oreopanax Decaisne and Planchon	Araliaceae	676
Oryzopsis Michaux	Gramineae	669
Osmanthus Loureiro	Pirolaceae	676
Osmunda Linné	Osmundaceae	666
Osmundaceae		666
Osmundites Unger	Osmundaceae	666

		Page.
Ostrya Scopoli	Betulaceae	671
Otozamites F. Braun	Cycadaceae	667
Ottelia Persoon	Hydrocharitaceae	669
Ovularites Whitford	Moniliaceae	664
Oxycoccus Adanson	Vacciniaceae	677
Oxypolis Rafinesque	Umbelliferae	676
Pachystima Rafinesque	Celastraceae	675
Padus Mönch	Drupaceae	674
Pagiophyllum Heer	Coniferae	668
Palaeanthus Newberry	Dicotyledon	678
Palaeoaster Knowlton	Fruit	678
Palaeocassia Ettingshausen	Caesalpiniaceae	674
Palaeodendron Saporta	Proteaceae	672
Palaeodictyon Heer	Alga	664
Palaeophycus Hall	Alga	664
?Palaeopotamogeton Knowlton	Potamogetonaceae	669
Palissya Endlicher	Podocarpeae	667
Paliurus Miller	Rhamnaceae	665
?Palmocarpon Lesquereux	Arecaceae	669, 678
Palmoxylon Schenk	Arecaceae	669
Panax Linné	Araliaceae	676
Papilionaceae		674
Paracedroxylon Sinnott	Abietineae	668
Paracupressinoxylon Sinnott	Cupressineae	668
Parrotia Meyer	Hamamelidaceae	673
Parthenocissus Planchon	Vitaceae	675
Pasania Miquel	Fagaceae	671
Passiflora Linné	Passifloraceae	676
Passifloraceae		676
Pecopteris Sternberg	Fern	666
Pellaea Link	Polypodiaceae	665
Periploca Linné	Asclepiadaceae	677
Peronosporaceae		663
Peronosporoides Berry	Peronosporaceae	663
Persea Gaertner	Lauraceae	672
Perseoxylon Felix	Lauraceae	672
Persicaria Linné	Polygonaceae	672
Persoonia Smith	Proteaceae	672
Phaca Linné	Leguminosae	674
?Phalaris Linné	Gramineae	669
Phanerophlebites Knowlton	Polypodiaceae	665
Phaseolites Unger	Papilionaceae	674
Phegopteris Fée	Polypodiaceae	665
Philadelphus Linné	Saxifragaceae	673
Philotria Rafinesque	Vallisneriaceae	669
Phoenicites Brongniart	Arecaceae	669
Phoenicopsis Heer	Coniferae	668
Phoradendron Nuttall	Loranthaceae	672
Phragmites Trinius	Gramineae	669
Phyllites Brongniart	Plant	678
?Phyllocladopsis Fontaine	Coniferae	668
Physematopitys Göppert	Coniferae	668
Picea Link	Abietineae	668

		Page.
Pinaceae		668
Pinites Witham	Abietineae	668
Pinus Linné	Abietineae	668
Piper Linné	Piperaceae	671
Piperaceae		671
Piperites Göppert	Piperaceae	671
Pirolaceae		676
Pisonia Linné	Nyctaginaceae	672
Pistacea Linné	Anacardiaceae	675
Pistia Linné	Araceae	669
Pithecolobium Martius	Mimosaceae	674
Pityoidolepis Hollick and Jeffrey	Coniferae	668
Pityophyllum Nathorst	Abietineae	668
Pityoxylon Kraus	Abietineae	668
Plagiopodopsis Britton and Hollick	Grimmiaceae	665
Planera Gmelin	Ulmaceae	671
Plantaginopsis Fontaine	Xyridaceae?	670
Plataninium Unger	Platanaceae	672
Platanus Linné	Platanaceae	672
Platanaceae		672
Platypterygium (Schimper) Feistmantel	Cycadaceae	667
Poacites Brongniart	Gramineae	669
Podocarpites Andrae	Podocarpeae	667
Podocarpoxylon Gothan	Podocarpeae	667
Podocarpus L'Héritier	Podocarpeae	667
Podogonium Heer	Leguminosae	674
Podozamites F. Braun	Cycadaceae	667
Polygonaceae		672
Polygonum Linné	Polygonaceae	672
Polypodiaceae		665
Polypodium Linné	Polypodiaceae	665
Polystichum Roth	Polypodiaceae	665
Polytrichaceae		665
?Polytrichum Linné	Polytrichaceae	665
Pomaceae		673
Pontederiaceae		670
Populites Viviani	Salicaceae	671
?Populophyllum Fontaine	Salicaceae	671
Populus Linné	Salicaceae	671
Porana Burmann	Convolvulaceae	677
Potamogeton Linné	Potamogetonaceae	669
Potamogetonaceae		669
?Potamogetophyllum Fontaine	Potamogetonaceae	669
Potentilla Linné	Rosaceae	673
Pourouma Aublet	Moraceae	671
Praeengelhardtia Berry	Juglandaceae	671
Preissites Knowlton	Marchantiaceae	664
Premnophyllum Velenovsky	Dicotyledon	678
Primulaceae		677
Prepinus Jeffrey	Abietineae	668
Proteaceae		672
Proteaephyllum Fontaine	Dicotyledon	678
Proteoides Heer	Proteaceae	672
Protodammara Hollick and Jeffrey	Araucariaceae	668

		Page.
Protoficus Saporta	Moraceae	671
Protophyllocladus Berry	Taxeae	667
Protophyllum Lesquereux	Platanaceae	672
Protorhipis Andrae	Dipteridinae	665
Pruinium Platen	Drupaceae	674
?Prunoides Perkins	Drupaceae	674
Prunus Linné	Drupaceae	674
Pseudodanaeopsis Fontaine	Fern	666
Pseudogeinitzia Hollick and Jeffrey	Taxodieae	668
Pseudolmedia Trécul	Moraceae	671
Pseudotsuga Carrière	Abietineae	668
Psilotaceae		667
Psoralea Linné	Papilionaceae	674
Psychotria Linné	Rubiaceae	678
Ptelea Linné	Rutaceae	674
Ptenostrobus Lesquereux	Coniferae	668
Pteris Linné	Polypodiaceae	665
Pterocarya Kunth	Juglandaceae	671
Pterophyllum Brongniart	Cycadaceae	667
Pterospermites Heer	Sterculiaceae	675
Ptilozamites Morris	Cycadaceae	667
Puccinites Ettingshausen	Uredinaceae	663
Pyrus Linné	Pomaceae	673
Quercinium Unger	Fagaceae	671
Quercophyllum Fontaine	Fagaceae	671
Quercus Linné	Fagaceae	671
Ramulus Wanner	Plant	678
Ranunculaceae		673
Raritania Hollick and Jeffrey	Coniferae	668
Retiphycus Ulrich	Alga	664
Reynosia Grisebach	Rhamnaceae	675
Rhamnaceae		665
Rhamnacinium Felix	Rhamnaceae	675
Rhamnites Forbes	Rhamnaceae	675
Rhamnus Linné	Rhamnaceae	675
Rhizophora Linné	Rhizophoraceae	676
Rhizophoraceae		676
Rhus Linné	Anacardiaceae	675
Rhynchostegium Schimper	Brachytheiaceae	675
Ribes Linné	Grossulariaceae	673
Robinia Linné	Papilionaceae	674
Rogersia Fontaine	Dicotyledon?	678
Rosa Linné	Rosaceae	673
Rosaceae		673
Rubiaceae		678
Rubioides Perkins	Fruit	678
Rubus Linné	Rosaceae	673
Ruffordia Seward	Schizaeaceae	666
Rulac Adanson	Aceraceae	675
Rutaceae		674
Sabal Adanson	Arecaceae	669
Sabalites Saporta	Arecaceae	669
Sabina Spach	Cupressineae	668
Sagenopteris Presl	Hydropteraceae	666

		Page.
Sagittaria Linné	Alismaceae	669
Salicaceae		671
Saliciphyllum Fontaine	Salicaceae	671
Salix Linné	Salicaceae	671
Salpichlaena Smith	Polypodiaceae	665
Salvinia Adanson	Salviniaceae	666
Salviniaceae		666
Samaropsis Göppert	Fruit	678
Sambucus Linné	Caprifoliaceae	678
Santalaceae		672
Santalum Linné	Santalaceae	672
Sapindaceae		675
Sapindoides Perkins	Sapindaceae	675
Sapindopsis Fontaine	Sapindaceae	675
Sapindus Linné	Sapindaceae	675
Sapotaceae		677
Sapotacites Ettingshausen	Sapotaceae	677
Sassafras Nees	Lauraceae	672
Saxifraga Linné	Saxifragaceae	673
Saxifragaceae		673
Schizaeaceae		666
Schizaeopsis Berry	Schizaeaceae	666
Schizolepis Braun	Coniferae	668
Schizoneura Schimper and Mougeot	Equisetaceae	666
Schmaltzia Desvaux	Anacardiaceae	675
Scirpus Linné	Cyperaceae	669
Scleropteris Saporta	Fern	666
Sclerotites Meschinelli	Sterile mycelia	664
Scrophulariaceae		678
Selaginella Beauvois	Selaginellaceae	667
Selaginellaceae		667
Sequoia Endlicher	Taxodieae	668
Sicyos Linné	Cucurbitaceae	678
Sideroxylon Linné	Sapotaceae	677
Simaruba Aublet	Simarubaceae	674
Simarubaceae		674
?Simarubinium Platen	Simarubaceae	674
Smilaceae		670
Smilax Linné	Smilaceae	670
Solanaceae		678
Solanites Saporta	Solanaceae	678
Sophora Linné	Papilionaceae	674
Sorbus Linné	Pomaceae	673
Sparganiaceae		669
Sparganium Linné	Sparganiaceae	669
Spegazzinites Felix	Tuberculariaceae	664
Sphaeriaceae		663
Sphaerites Unger	Sphaeriaceae	663
Sphenaspis Hollick and Jeffrey	Coniferae	668
Sphenolepidium Heer	Taxodieae	668
Sphenopteris Sternberg	Fern	666
Sphenozamites Brongniart	Cycadaceae	667
Spiraea Linné	Rosaceae	673

		Page.
Spiraxis Newberry	Plant	678
Spirodela Schleiden	Lemnaceae	670
Staphidoides Perkins	Fruit	678
Staphylea Linné	Staphyleaceae	675
Staphyleaceae		675
Stenopteris Saporta	Fern	666
Sterculia Linné	Sterculiaceae	675
Sterculiaceae		675
Sterculinocarpus Berry	Sterculiaceae	655
Stipa Linné	Gramineae	669
Strobilites Lindley and Hutton	Coniferae	668
Styracaceae		677
Styrax Linné	Styracaceae	677
Symphorocarpophyllum Dawson	Dicotyledon	678
?Taeniopteris Brongniart	Marattiaceae	666
Tapeinidium (Presl) Christensen	Polypodiaceae	665
Taxaceae		667
Taxites Brongniart	Taxeae	667
Taxodioxylon Hartig	Taxodieae	668
Taxodium Richard	Taxodieae	668
Tecoma Jussieu	Bignoniaceae	678
Tempskya Corda	Fern	666
Terminalia Linné	Combretaceae	676
Ternstroemiaceae		676
Ternstroemites Berry	Ternstroemiaceae	676
Thalictrum Linné	Ranunculaceae	673
Thinnfeldia Ettingshausen	Fern	666
Thrinax Linné	Araceae	669
Thuites Sternberg	Cupressineae	668
Thuya Linné	Cupressineae	668
Thuyoxylon Unger	Cupressineae	668
Thyrsopteris Kunze	Cyatheaceae	665
Tilia Linné	Tiliaceae	675
Tiliaceae		675
Tithymalus Adanson	Euphorbiaceae	674
?Tmesipteris Bernhardi	Psilotaceae	667
Toxylon Rafinesque	Moraceae	671
Trapa Linné	Hydrocaryaceae	676
Tricalycites Newberry	Plant	678
Tricarpellites Perkins	Fruit	678
Tsuga Endlicher	Abietineae	668
Tuberculariaceae		664
Tumion Rafinesque	Taxeae	667
Typha Linné	Typhaceae	669
Typhaceae		669
Ulmaceae		671
Ulmites Dawson	Ulmaceae	671
Ulmophyllum Ettingshausen	Ulmaceae	671
Ulmoxylon Kaiser	Ulmaceae	671
Ulmus Linné	Ulmaceae	671
Umbelliferae		676
Uredinaceae		663
Urticaceae?		671

		Page.
Vacciniaceae		677
Vaccinium Linné	Vacciniaceae	677
Vaccinophyllum Dawson	Vacciniaceae	677
Vallisneria Linné	Vallisneriaceae	669
Vallisneriaceae		669
Vantanea Aublet	Humiriaceae	674
Verbenaceae		678
Viborquia Ortega	Mimosaceae	674
Viburnites Lesquereux	Caprifoliaceae	678
Viburnum Linné	Caprifoliaceae	678
Vitaceae		675
?Vitiphyllum Fontaine	Vitaceae	675
Vitis Linné	Vitaceae	675
Voltzia Brongniart	Cupressineae	668
Weichselia Stiehler	Fern	666
Weinmannia Linné	Cunoniaceae	673
Widdringtonia Endlicher	Cupressineae	668
Widdringtonites Endlicher	Cupressineae	668
Williamsonia Carruthers	Cycadaceae	667
Winchellia Lesquereux	Berberidaceae	673
Woodwardia Smith	Polypodiaceae	665
Woodworthia Jeffrey	Araucariaceae	668
Xolisma Rafinesque	Ericaceae	677
Xyridaceae?		670
?Yorkia Wanner	Monocotyledon	670
Yucca Linné	Dracaenaceae	670
Yuccites Martius	Plant	678
Zamia Linné	Cycadaceae	667
Zamiopsis Fontaine	Cycadaceae	667
Zamiostrobus Endlicher	Cycadaceae	667
Zamites Brongniart	Cycadaceae	667
Zanthoxylum Linné	Rutaceae	674
Zingiberaceae		670
Zingiberites Heer	Zingiberaceae	670
Zizyphus Adanson	Rhamnaceae	675
Zostera Linné	Naiadaceae	669

FLORAL LISTS OF NORTH AMERICAN MESOZOIC AND CENOZOIC PLANT-BEARING FORMATIONS.

TRIASSIC FLORAS.[1]

Abietites carolinensis (Fontaine) Fontaine. (N. C.)
Acrostichites densifolius Fontaine. (Va.)
Acrostichites linnaeaefolius (Bunbury) Fontaine. (N. C., Va., Pa.)
Acrostichites microphyllus Fontaine. (Va., Pa.)
Acrostichites tenuifolius (Emmons) Fontaine. (N. C., Va.)
Acrostichites tenuifolius rarinervis (Fontaine) Ward. (Va.)
Actinopteris quadrifolia (Emmons) Fontaine. (N. C.)
Anabacaulus duplicatus Emmons. (N. C.)
Anabacaulus sulcatus Emmons. (N. C.)
Anomozamites? egyptiacus Fontaine. (N. C.)
Anomozamites princeps (Oldham and Morris) Schimper. (Pa.)
Araucarioxylon arizonicum Knowlton. (N. Mex., Ariz., Utah.)
Araucarioxylon virginianum Knowlton. (N. C., Va., Pa.)
Araucarioxylon woodworthi Knowlton. (N. C.)
Araucarioxylon sp. Platen. (Ariz.)
Araucarites chiquito Ward. (Ariz.)
Araucarites? pennsylvanicus Fontaine. (Pa.)
Araucarites yorkensis Fontaine. (Pa.)
Asplenites rösserti (Presl) Schenk. (N. C.)
Asterocarpus falcatus (Emmons) Fontaine. (N. C., Va., Pa.)
Asterocarpus falcatus obtusilobus (Fontaine) Knowlton. (Va.)
Asterocarpus penticarpus Fontaine. (Va.)
Asterocarpus platyrachis Fontaine. (N. C., Va.)
Baiera münsteriana (Presl) Heer. (N. C., Pa., Conn.)
Baiera multifida Fontaine. (N. C., Va.)
Bambusium carolinense Fontaine. (Va.)
Brachyoxylon pennsylvanicum Wherry. (Pa.)
Brachyphyllum yorkense Fontaine. (Pa.)
Brachyphyllum sp. Newberry. (N. Mex.)
Brunswickia dubia Wherry. (Pa.)
Cephalotaxopsis carolinensis Fontaine. (N. C.)
Cheirolepis latus Brown. (Pa.)
Cheirolepis muensteri (Schenk) Schimper. (Va., N. C., Pa.)
Chondrites gracilis Emmons. (N. C.)
Chondrites heerii Eichwald. (Alaska. Age doubtful.)
Chondrites interruptus Emmons. (N. C.)
Chondrites ramosus Emmons. (N. C.)
Cladophlebis auriculata Fontaine. (Va.)
Cladophlebis microphylla Fontaine. (Va.)

---

[1] So little work has been done on the stratigraphic position and correlation of the plant-bearing beds of the American Triassic that no attempt has been made in this catalogue to allocate them within definite formational limits. The known areal distribution of each species is shown within the parentheses following the species name.

Cladophlebis obtusifolia (Andrae) Ward. (N. C.)
Cladophlebis ovata Fontaine. (Va.)
Cladophlebis pseudowhitbiensis Fontaine. (Va.)
Cladophlebis reticulata Fontaine. (Pa.)
Cladophlebis rotundiloba Fontaine. (Va.)
Cladophlebis subfalcata Fontaine. (Va.)
Clathropteris platyphylla (Göppert) Brongniart. (Conn., Mass., N. J.)
Clathropteris platyphylla expansa Saporta. (Va.)
Clathropteris rectiusculus Hitchcock. (Mass.)
Comephyllum cristatum Emmons. (N. C.)
Ctenophyllum braunianum abbreviatum (Braun) Schimper. (N. C.)
Ctenophyllum braunianum angustifolium (Braun) Schimper. (N. C., Va.)
Ctenophyllum braunianum var. a Göppert. (Va.)
Ctenophyllum giganteum Fontaine. (Va.)
Ctenophyllum grandifolium Fontaine. (N. C., Va., Pa.)
Ctenophyllum lineare (Emmons) Fontaine. (N. C.)
Ctenophyllum robustum (Emmons) Fontaine. (N. C.)
Ctenophyllum taxinum (Lindley and Hutton) Fontaine. (Va.)
Ctenophyllum truncatum Fontaine. (Va.)
Ctenophyllum wannerianum Fontaine. (Pa.)
Ctenophyllum? sp. Fontaine. (N. Mex.)
Cycadeoidea emmonsi (Fontaine) Ward. (N. C.)
Cycadeomyelon yorkense Fontaine. (N. J., Pa.)
Cycadeospermum wanneri Fontaine. (Pa.)
Cycadinocarpus chapini Newberry. (Conn.)
Cycadites acutus Emmons. (N. C.)
Cycadites tenuinervis Fontaine. (N. C., Va.)
Cycadites? sp. Fontaine. (N. Mex.)
Danaeopsis? sp. Fontaine. (N. C.)
Dendrophycus schoemakeri Ward. (Md.)
Dendrophycus triassicus Newberry. (Conn.)
Dicranopteris sp. Fontaine. (Va.)
Dioonites carnallianus (Göppert) Bornemann. (Pa.)
Dyctuocaulus striatus Emmons. (N. C., Pa.)
Equisetum abiquiensis Fontaine. (N. Mex.)
Equisetum knowltoni Fontaine. (N. Mex.)
Equisetum meriani? Brongniart. (N. J.)
Equisetum rogersii (Bunbury) Schimper. (Pa., Va., N. C., N. J., N. B.)
Fucoides connecticutensis Hitchcock. (Conn., Mass.)
Fucoides shepardi Hitchcock. (Mass.)
Gymnocaulus alternatus Emmons. (N. C., Pa.?)
Laccopteris lanceolata (Göppert) Ward. (N. C.?)
Lepacyclotes circularis Emmons. (N. C.)
Lepacyclotes ellipticus Emmons. (N. C.)
Lonchopteris oblonga (Emmons) Fontaine. (N. C., Pa.)
Lonchopteris virginiensis Fontaine. (Va.)
Loperia carolinensis (Fontaine) Ward. (Conn., N. C.)
Loperia simplex Newberry. (Conn., Va.)
Macrotaeniopteris crassinervis Feistmantel. (Va.)
Macrotaeniopteris magnifolia (Rogers) Schimper. (Va., N. C., Pa.)
Mertensides bullatus (Bunbury) Fontaine. (Va., N. C.?, Pa.?)
Mertensides distans Fontaine. (Va.)
Neocalamites knowltoni Berry. (Va.)

Neocalamites virginiensis (Fontaine) Berry. (Va.)
Otozamites brevifolius (Braun) Braun. (Conn.)
Otozamites carolinensis Fontaine. (N. C.)
Otozamites latior Saporta. (Conn.)
Otozamites macombii Newberry. (N. Mex.)
Pagiophyllum brevifolium (Newberry) Ward. (Mass., Conn.)
Pagiophyllum newberryi Ward. (N. Mex.)
Pagiophyllum peregrinum (Lindley and Hutton) Schenk. (N. C.)
Pagiophyllum simile (Newberry) Ward. (N. J., Conn., Mass.)
Palaeophycus limaciformis Lewis. (N. J.)
Palissya brevifolia (Emmons) Fontaine. (N. C.)
Palissya diffusa (Emmons) Fontaine. (N. C., N. J., Mass., N. Mex., Pa.)
Palissya longifolia Wherry. (Pa.)
Palissya sphenolepis (Braun) Brongniart. (Pa., N. C., N. J., N. Mex.?)
Palissya sp. Fontaine. (N. Mex.)
Pecopteris rarinervis Fontaine. (Va.)
Podozamites? carolinensis Fontaine. (N. C.)
Podozamites distans? (Presl) F. Braun. (Pa.)
Podozamites emmonsii Newberry. (N. C.)
Podozamites formosus Brown. (Pa.)
Podozamites longifolius Emmons. (N. C., Va., N. J.)
Podozamites tenuistriatus (Rogers) Fontaine. (Va., N. C.)
Pesudodanaeopsis obliqua (Emmons) Fontaine. (Va., N. C.)
Pseudodanaeopsis plana (Emmons) Fontaine. (Va., N. C., Pa.)
Pterophyllum affine (Nathorst) Nathorst. (Va.)
Pterophyllum daleanum Ward. (N. C.)
Pterophyllum inaequale Fontaine. (Va., Pa.)
Ramulus rugosus Wanner. (Pa.)
Sagenopteris emmonsi Fontaine. (N. C.)
Sagenopteris sp. Fontaine. (Pa.)
Schizolepis liaso-keuperiana Braun. (Pa.)
Schizoneura planicostata (Rogers) Fontaine. (Mass., N. J., Va., N. C., Conn.)
Schizoneura sp.? Fontaine. (Va.)
Sphenopteris egyptiaca Emmons. (N. C.)
Sphenozamites rogersianus Fontaine. (Va., N. C., Pa.)
Taeniopteris? yorkensis Fontaine. (Pa.)
Thinnfeldia? reticulata Fontaine. (Pa.)
Voltzia coburgensis Schaurolt. (New Brunswick.)
Woodworthia arizonica Jeffrey. (Ariz.)
Yorkia gramineoides Ward. (Pa.)
Zamiostrobus virginiensis Fontaine. (Va., N. C.)
Zamites occidentalis Newberry. (N. Mex.)
Zamites pennsylvanicus Fontaine. (Pa.)
Zamites powelli Fontaine. (N. Mex.)
Zamites velderi Brown. (Pa.)
Zamites yorkensis Fontaine. (Pa.)

TRIASSIC (?) FLORA (MOQUI VILLAGES, ARIZ.)

Cyclopteris moquensis Newberry.
Pecopteris cycloloba Newberry.
Phyllites coriaceus Newberry.
Phyllites venosissimus Newberry.

JURASSIC FLORAS.

CALIFORNIA AND OREGON, FORMATIONS NOT SETTLED.

Adiantites nympharum Heer. (Oreg.)
Adiantites orovillensis Fontaine. (Calif.)
Angiopteridium californicum Fontaine. (Calif.)
Araucarites? sp. Fontaine. (Oreg.)
Baiera multifida Fontaine. (Calif.?)
Brachyphyllum mamillaria Brongniart. (Oreg.)
Brachyphyllum? storrsii Ward. (Calif. Age doubtful.)
Carpolithes bucklandii? Williamson. (Oreg.)
Carpolithus douglasensis Fontaine. (Oreg.)
Carpolithus elongatus Fontaine. (Oreg.)
Carpolithus olallensis Ward. (Oreg.)
Carpolithus oregonensis Fontaine. (Oreg.)
Carpolithus storrsii Fontaine. (Calif.)
Cladophlebis acutiloba (Heer) Fontaine. (Oreg.)
Cladophlebis argutula (Heer) Fontaine. (Calif.)
Cladophlebis densifolia Fontaine. (Calif.)
Cladophlebis denticulata (Brongniart) Nathorst. (Oreg.)
Cladophlebis haiburnensis (Lindley and Hutton) Brongniart. (Oreg.)
Cladophlebis indica (Oldham and Morris) Fontaine. (Calif.)
Cladophlebis pecopteroides Fontaine. (Oreg.)
Cladophlebis spectabilis (Heer) Fontaine. (Calif.)
Cladophlebis vaccensis Ward. (Calif., Oreg.)
Coniopteris hymenophylloides (Brongniart) Seward. (Oreg.)
Ctenis auriculata Fontaine. (Calif., Oreg.)
Ctenis grandifolia Fontaine. (Calif., Oreg.)
Ctenis orovillensis Fontaine. (Calif., Oreg.)
Ctenis sulcicaulis (Phillips) Ward. (Oreg.)
Ctenis wardii Fontaine. (Calif.)
Ctenophyllum angustifolium Fontaine. (Calif., Oreg.)
Ctenophyllum densifolium Fontaine. (Calif.)
Ctenophyllum grandifolium storrsii Fontaine. (Calif.)
Ctenophyllum pachynerve Fontaine. (Oreg.)
Ctenophyllum wardii Fontaine. (Calif., Oreg.)
Ctenophyllum? sp. Fontaine. (Oreg.)
Cycadeospermum oregonense Fontaine. (Oreg.)
Cycadeospermum ovatum Fontaine. (Oreg.)
Cyclopitys oregonensis Fontaine. (Oreg.)
Danaeopsis storrsii Fontaine. (Oreg.)
Dicksonia oregonensis Fontaine. (Oreg.)
Didymosorus? bindrabundensis acutifolius Fontaine. (Calif.)
Encephalartopsis? oregonensis Fontaine. (Oreg.)
Equisetum? sp. Fontaine. (Oreg.)
Ginkgo digitata (Brongniart) Heer. (Oreg.)
Ginkgo huttoni (Sternberg) Heer. (Oreg.)
Ginkgo huttoni magnifolia Fontaine. (Oreg.)
Ginkgo lepida (Heer) Heer. (Oreg.)
Ginkgo siberica (Heer) Heer. (Oreg.)
Ginkgo sp. Fontaine. (Oreg.)
Leptostrobus? sp. Fontaine. (Calif.)
Macrotaeniopteris californica Fontaine. (Calif., Oreg.)

Macrotaeniopteris nervosa Fontaine. (Calif.)
Marchantites erectus (Bean) Seward. (Oreg.)
Nilsonia compta (Phillips) Göppert. (Oreg.)
Nilsonia nipponensis Yokoyama. (Oreg.)
Nilsonia orientalis Heer. (Oreg.)
Nilsonia orientalis minor Fontaine. (Oreg.)
Nilsonia parvula (Heer) Fontaine. (Oreg.)
Nilsonia pterophylloides Nathorst. (Oreg.)
Otozamites oregonensis Fontaine. (Oreg.)
Pagiophyllum peregrinum (Lindley and Hutton) Schenk. (Calif.)
Pagiophyllum williamsonis (Brongniart) Fontaine. (Calif., Va.?, N. C.?)
Phoenicopsis? sp. Fontaine. (Oreg.)
Pinus nordenskiöldi Heer. (Calif., Oreg.)
Podozamites lanceolatus (Lindley and Hutton) F. Braun. (Calif., Oreg.)
Podozamites lanceolatus latifolius (Braun) Heer. (Calif., Oreg.)
Podozamites lanceolatus minor (Schenk) Heer. (Oreg.)
Podozamites? pachynervis Fontaine. (Oreg.)
Podozamites pachyphyllus Fontaine. (Oreg.)
Podozamites pulchellus Heer. (Oreg.)
Polypodium oregonense Fontaine. (Oreg.)
Pterophyllum aequale (Brongniart) Nathorst. (Oreg.)
Pterophyllum contiguum Schenk. (Oreg.)
Pterophyllum minus Brongniart. (Oreg.)
Pterophyllum nathorsti Schenk. (Oreg.)
Pterophyllum rajmahalense Morris? (Calif., Oreg.)
Ptilozamites leckenbyi (Bean) Nathorst. (Oreg.)
Ruffordia göpperti (Dunker) Seward. (Oreg.)
Sagenopteris göppertiana Zigno. (Oreg.)
Sagenopteris grandifolia Fontaine. (Oreg.)
Sagenopteris nilsoniana (Brongniart) Ward. (Calif.)
Sagenopteris paucifolia (Phillips) Ward. (Oreg.)
Samaropsis? oregonensis Fontaine. (Oreg.)
Scleropteris oregonensis Fontaine. (Oreg.)
Sphenolepidium oregonense Fontaine. (Oreg.)
Taeniopteris major Lindley and Hutton. (Oreg.)
Taeniopteris? oregonensis Fontaine. (Oreg.)
Taeniopteris orovillensis Fontaine. (Calif., Oreg.)
Taeniopteris vittata Brongniart. (Oreg.)
Taxites zamioides (Leckenby) Seward. (Oreg.)
Thyrsopteris murrayana (Brongniart) Heer. (Oreg.)
Williamsonia oregonensis Fontaine. (Oreg.)
Williamsonia? sp. Fontaine. (Oreg.)
Yuccites hettangensis Saporta. (Oreg.)

FOREMAN FORMATION, CALIFORNIA.

Acrostichites? brevipennis Ward.
Acrostichites? coniopteroides Ward.
Acrostichites? fructifera Ward.
Acrostichites? princeps (Presl) Schenk.
Equisetum muensteri (Sternberg) Brongniart.
Pterophyllum taylorsvillensis (Ward) Knowlton.
Sagenopteris? magnifolia Ward.
Sagenopteris? sp. Fontaine.

MARIPOSA SLATE, CALIFORNIA.

Dicksonia saportana Heer. (Calif. Formation doubtful.)
Leptostrobus? mariposensis Fontaine.

SLATE SPRINGS, CALIF.

Sequoia fairbanksi Fontaine. (Age doubtful.)

BOULDER, COLO.

Cycadeoidea nigra Ward. (Age doubtful.)

BRIDGER CREEK, NEAR BOZEMAN, MONT.

Sequoia reichenbachii (Geinitz) Heer.

YAKUTAT GROUP (LIAS?), KODIAK, ALASKA.

Arthrodendron diffusum Ulrich.
Cancellophycus rhombicum Ulrich.
Chondrites alpestris Heer.
Chondrites divaricatus Fischer-Ooster.
Gilbertina spiralis Ulrich.
Gyrodendron emersoni Ulrich.
Helminthoida abnormis Ulrich.
Helminthoida exacta Ulrich.
Helminthoida subcrassa Ulrich.
Helminthoida vaga Ulrich.
Helminthopsis? labyrinthica Heer.
Helminthopsis magna Heer.
Melophycus curvatum Ulrich.
Palaeodictyon magnum laxum Ulrich.
Palaeodictyon singulare Heer.
Retiphycus hexagonale Ulrich.

LOWER JURASSIC (LIAS), MATANUSKA VALLEY, ALASKA.

Cladophlebis hirta? Möller.
Ctenophyllum angustifolium? Fontaine.
Dictyophyllum nilssoni (Brongniart) Göppert.
Nilsonia polymorpha Schenk.
Otozamites bornholmiensis? Möller.
Otozamites pterophylloides Brongniart.
Pagiophyllum falcatum Bartolin.
Pterophyllum aequale (Brongniart) Nathorst.
Pterophyllum rajmahalense Morris.
Sagenopteris? sp. Knowlton.

CAPE LISBURNE, ALASKA.

Baiera gracilis (Bean) Bunbury.
Chondrites filiciformis Lesquereux.
Cladophlebis alata Fontaine.
Cladophlebis huttoni (Dunker) Fontaine.
Cladophlebis vaccensis Ward.
Coniopteris burejensis (Zalessky) Seward.
Coniopteris hymenophylloides (Brongniart) Seward.
Dicksonia saportanea Heer.
Elatides curvifolia (Dunker) Nathorst.
Equisetum collieri Knowlton.
Fieldenia nordenskiöldi Nathorst.

Ginkgo digitata (Brongniart) Heer.
Ginkgo huttoni (Sternberg) Heer.
Ginkgo huttoni magnifolia Fontaine.
Ginkgo sp. (Fontaine).
?Onychiopsis psilotoides (Stokes and Webb) Ward.
Otozamites giganteus Thomas.
Pagiophyllum kurrii (Pomel) Schimper.
Pagiophyllum steenstrupi Bartolin?
Phoenicopsis angustifolia Heer.
Phoenicopsis speciosa Heer.
Pityophyllum nordenskiöldi (Heer) Seward.
Podozamites lanceolatus (Lindley and Hutton) Fr. Braun.
Podozamites lanceolatus eichwaldi (Schimper) Heer.
Zamites alaskana Fontaine.
Zamites megaphyllus (Phillips) Seward.

COPPER RIVER REGION, ALASKA.

Sagenopteris alaskensis Fontaine.

HERENDEEN BAY, ALASKA.

Pterophyllum alaskense Fontaine.

TUXEDNI SANDSTONE (MIDDLE JURASSIC), TUXEDNI BAY, ALASKA.

Macrotaeniopteris californica Fontaine.
Pterophyllum rajmahalense Morris.
Sagenopteris goeppertiana Zigno.

CRETACEOUS FLORAS.

NORTH AND MIDDLE ATLANTIC REGIONS.

POTOMAC GROUP (LOWER CRETACEOUS).

PATUXENT FORMATION.

Abietites foliosus (Fontaine) Berry. (Va.)
Abietites longifolius (Fontaine) Berry. (Va.)
Abietites macrocarpus Fontaine. (Va., Md.)
Acrostichopteris adiantifolia (Fontaine) Berry. (Va.)
Acrostichopteris cyclopteroides (Fontaine) Berry. (Va.)
Acrostichopteris expansa (Fontaine) Berry. (Va.)
Acrostichopteris parvifolia (Fontaine) Berry. (Va., Md.)
Acrostichopteris pluripartita (Fontaine) Berry. (Va., D. C.)
Acrostichum crassifolium Fontaine. (Va.)
Angiopteridium dentatum Fontaine. (Va.)
Angiopteridium ellipticum Fontaine. (Va.)
Angiopteridium ovatum Fontaine. (Va.)
Angiosperm, ament of? (Va.)
Araucaria obtusifolia Fontaine. (Va.)
Araucaria zamioides Fontaine. (Va.)
Araucarites virginicus Fontaine. (Va.)
Aspleniopteris adiantifolia Fontaine. (Va.)
Aspleniopteris pinnatifida Fontaine. (Va.)
Asplenium dubium Fontaine. (Va.)
Athrotaxopsis expansa Fontaine. (Va., Md.)
Athrotaxopsis grandis Fontaine. (Va., Md., D. C.)
Baiera foliosa Fontaine. (Va.)

Brachyphyllum crassicaule Fontaine. (Va.)
Brachyphyllum parceramosum Fontaine. (Va., D. C.)
Brachyphyllum sp. Fontaine. (Va.)
Callitris? sp. Fontaine. (Va.)
Carpolithus agglomeratus Fontaine. (Va.)
Carpolithus brookensis Fontaine. (Va.)
Carpolithus conjugatus Fontaine. (Va.)
Carpolithus curvatus Fontaine. (Va.)
Carpolithus fasciculatus Fontaine. (Va.)
Carpolithus geminatus Fontaine. (Va.)
Carpolithus latus Fontaine. (Va.)
Carpolithus sessilis Fontaine. (Va.)
Carpolithus ternatus Fontaine. (Va.)
Carpolithus virginiensis Fontaine. (Va.)
Celastrophyllum arcinerve Fontaine. (Va.)
Cephalotaxopsis brevifolia Fontaine. (Va.)
Cephalotaxopsis magnifolia Fontaine. (Va.)
Cephalotaxopsis ramosa Fontaine. (Va.)
Cladophlebis albertsii (Dunker) Brongniart. (Va.)
Cladophlebis browniana (Dunker) Seward. (Va., D. C., Md.)
Cladophlebis constricta Fontaine. (Va.)
Cladophlebis distans Fontaine. (Va., Md.)
Cladophlebis inaequiloba Fontaine. (Va.)
Cladophlebis parva Fontaine. (Va.)
Cladophlebis rotundata Fontaine. (Va.)
Cladophlebis ungeri (Dunker) Ward. (Va.)
Cladophlebis virginiensis Fontaine. (Va.)
Cladophlebis? sp. Fontaine. (Va.)
Coniospermites ellipticus Fontaine. (Va.)
Ctenis imbricata Fontaine. (Va.)
Ctenopsis latifolia (Fontaine) Berry. (Va.)
Ctenopteris angustifolia Fontaine. (Va.)
Ctenopteris insignis Fontaine. (Va.)
Ctenopteris longifolia Fontaine. (Va.)
Cupressinoxylon columbianum Knowlton. (Va.)
Cupressinoxylon pulchellum Knowlton. (Va.)
Cupressinoxylon wardii Knowlton. (Va., D. C., Md.)
Cycadeoidea bibbinsi Ward. (Md.)
Cycadeoidea clarkiana Ward. (Md. Formation doubtful.)
Cycadeoidea fisherae Ward. (Md.)
Cycadeoidea fontaineana Ward. (Md.)
Cycadeoidea goucheriana Ward. (Md.)
Cycadeoidea marylandica (Fontaine) Capillini and Solms-Laubach. (Md.)
Cycadeoidea mcgeeana Ward. (Md.)
Cycadeoidea tysoniana Ward. (Md.)
Cycadeoidea uhleri Ward. (Md. Formation doubtful.)
Cycadeospermum acutum Fontaine. (Va.)
Cycadeospermum angustum Fontaine. (Va.)
Cycadeospermum rotundatum Fontaine. (Va.)
Cycadeospermum spatulatum Fontaine. (Va.)
Dichotozamites cycadopsis (Fontaine) Berry. (Va., Md.)
Dioonites buchianus (Ettingshausen) Bornemann. (Va., D. C.)
Dioonites buchianus abietinus (Göppert) Ward. (Va.)
Dioonites buchianus obtusifolius Fontaine. (Va.)

Dryopterites cystopteroides (Fontaine) Berry. (Va.)
Dryopterites dentata (Fontaine) Berry. (Va.)
Dryopterites elliptica (Fontaine) Berry. (Va.)
Dryopterites macrocarpa (Fontaine) Berry. (Va.)
Dryopterites virginica (Fontaine) Berry. (Va.)
Encephalartopsis nervosa Fontaine. (Va.)
Equisetum burchardti (Dunker) Brongniart. (D. C., Va.)
Equisetum lyelli Mantell. (Md., Va.)
Equisetum sp. Fontaine. (Va.)
Fieldenopsis crassinervis Fontaine. (Va.)
Feistmantelia virginica Fontaine. (Va.)
Ficophyllum crassinerve Fontaine. (Va.)
Ficophyllum serratum Fontaine. (Va.)
Frenelopsis parceramosa Fontaine. (Va.)
Frenelopsis ramosissima Fontaine. (Va.)
Gleichenia nordenskiöldi Heer. (Va.)
Glossozamites distans Fontaine. (Va.)
Juglandiphyllum integrifolium Fontaine. (Va.)
Laricopsis angustifolia Fontaine. (Va.)
Laricopsis longifolia Fontaine. (Va.)
Leptostrobus? sp. Fontaine. (Va.)
Macrospores Fontaine. (Va.)
Nageiopsis angustifolia Fontaine. (Va., D. C.)
Nageiopsis longifolia Fontaine. (Va.)
Nageiopsis obtusifolia Fontaine. (Va.)
Nageiopsis zamioides Fontaine. (Va.)
Nilsonia oregonensis (Fontaine) Berry. (Va. Formation doubtful.)
Onychiopsis brevifolia (Fontaine) Berry. (Va.)
Onychiopsis göpperti (Schenck) Berry. (Va., D. C.)
Onychiopsis latiloba (Fontaine) Berry. (Va., D. C.)
Onychiopsis nervosa (Fontaine) Berry. (Va., D. C., Md.)
Onychiopsis psilotoides (Stokes and Webb) Ward. (Va., D. C.)
Osmunda dicksonioides Fontaine. (Va.)
Osmunda dicksonioides latipennis Fontaine. (Va.)
Osmunda sphenopteroides Fontaine. (Va.)
Paracupressinoxylon potomacense Sinnott and Bartlett. (D. C.)
Phyllites latifolius (Fontaine) Berry. (Va.)
Phyllocladopsis heterophylla Fontaine. (Va.)
Platypterigium densinerve Fontaine. (Va.)
Platypterigium rogersianum Fontaine. (Va.)
Podocarpoxylon mcgeei (Knowlton) Sinnott and Bartlett. (D. C.)
Podozamites acutifolius Fontaine. (Va.)
Podozamites distantinervis Fontaine. (Va., D. C., Md.)
Podozamites grandifolius Fontaine. (Va.)
Podozamites inaequilateralis (Fontaine) Berry. (Va.)
Podozamites knowltoni Berry. (Va., D. C.)
Podozamites subfalcatus Fontaine. (Va.)
Polypodium dentatum Fontaine. (Va.)
Polypodium fadyenioide Fontaine. (Va.)
Proteaephyllum ovatum Fontaine. (Va.)
Proteaephyllum reniforme Fontaine. (Va.)
Rogersia angustifolia Fontaine. (Va.)
Rogersia longifolia Fontaine. (Va.)
Ruffordia acrodenta (Fontaine) Berry. (Va.)

Ruffordia göpperti (Dunker) Seward. (Va., D. C.)
Sagenopteris elliptica Fontaine. (Va.)
Sagenopteris latifolia Fontaine. (Va.)
Sagenopteris virginiensis Fontaine. (Va.)
Schizaeopsis americana Berry. (Va.)
Scleropteris elliptica Fontaine. (Va., Md.)
Sequoia ambigua Heer. (Va.)
Sequoia delicatula Fontaine. (Va.)
Sequoia gracilis Heer. (Va.)
Sequoia reichenbachi (Geinitz) Heer. (Va., Md.)
Sequoia rigida Heer. (Va., Md.)
Sequoia? sp. Fontaine. (Va.)
Sphenolepidium pachyphyllum Fontaine. (Va.)
Sphenolepidium sternbergianum (Dunker) Heer. (Va.)
Sphenolepis kurriana (Dunker) Schenk. (D. C., Va.)
Taeniopteris auriculata (Fontaine) Berry. (Va.)
Taeniopteris nervosa (Fontaine) Berry. (Va.)
Taxodium virginicum Fontaine. (Va.)
Thinnfeldia granulata Fontaine. (Va.)
Thinnfeldia rotundiloba Fontaine. (Va.)
Tumion falcatum (Fontaine) Knowlton. (Va.)
Tumion virginicum (Fontaine) Knowlton. (Va.)
Undetermined ferns. Several species. (Va.)
Vitiphyllum crassifolium Fontaine. (Va.)
Williamsonia virginiensis Fontaine. (Va.)
Zamiopsis dentata (Fontaine) Berry. (Va., D. C.)
Zamiopsis laciniata Fontaine. (Va.)
Zamiopsis petiolata Fontaine. (Va.)
Zamites crassinervis Fontaine. (Va.)
Zamites distantinervis Fontaine. (Va.)
Zamites tenuinervis Fontaine. (Va.)

ARUNDEL FORMATION.

Abietites macrocarpus Fontaine. (Md.)
Acrostichopteris adiantifolia (Fontaine) Berry. (Md.)
Athrotaxopsis expansa Fontaine. (D. C., Md.)
Athrotaxopsis grandis Fontaine. (D. C., Md.)
Brachyphyllum parceramosum Fontaine. (Md.)
Cedrus leei (Fontaine) Berry. (Md.)
Cladophlebis albertsii (Dunker) Brongniart. (Md.)
Cladophlebis browniana (Dunker) Seward. (Md.)
Cladophlebis distans Fontaine. (Md.)
Cladophlebis parva Fontaine. (Md.)
Cladophlebis virginiensis Fontaine. (Md.)
Ctenopteris insignis Fontaine. (D. C.)
Cycadeospermum acutum Fontaine. (Md.)
Cycadeospermum marylandicum Berry. (Md.)
Cycadeospermum obovatum Fontaine. (D. C., Md.)
Cycadeospermum rotundatum Fontaine. (Md.)
Cycadeospermum spatulatum Fontaine. (D. C.)
Dicksoniopsis vernonensis (Ward) Berry. (Md.)
Dioonites buchianus (Ettingshausen) Bornemann. (Md.)
Ficophyllum crassinerve Fontaine. (Md., D. C.)
Ficophyllum serratum Fontaine. (D. C.)

Frenelopsis parceramosa Fontaine. (Md.)
Ginkgo? acetaria Ward. (Md.)
Nageiopsis angustifolia Fontaine. (Md.)
Nageiopsis longifolia Fontaine. (D. C.)
Nageiopsis zamioides Fontaine. (D. C., Md.)
Onychiopsis göpperti (Schenk) Berry. (D. C., Md.)
Onychiopsis latiloba (Fontaine) Berry. (D. C., Md.)
Onychiopsis nervosa (Fontaine) Berry. (D. C.)
Onychiopsis psilotoides (Stokes and Webb) Ward. (D. C., Md.)
Rogersia angustifolia Fontaine. (D. C.)
Rogersia angustifolia parva Fontaine. (D. C.)
Rogersia longifolia Fontaine. (Md.)
Selaginella marylandica Fontaine. (Md.)
Sequoia ambigua Heer. (Md.)
Sequoia reichenbachi (Geinitz) Heer. (Md.)
Sphenolepis kurriana (Dunker) Schenk. (Md.)
Thinnfeldia marylandica Fontaine. (Md.)

PATAPSCO FORMATION.

Abietites folioaus (Fontaine) Berry. (Va.)
Abietites longifolius (Fontaine) Berry. (Va., Md.)
Abietites macrocarpus Fontaine. (Md.)
Abietites marylandicus Fontaine. (Md.)
Acrostichopteris adiantifolia (Fontaine) Berry. (Md., Va.)
Acrostichopteris longipennis Fontaine. (Md., Va.)
Acrostichopteris pluripartita (Fontaine) Berry. (Md., Va.)
Alismaphyllum victor-masoni (Ward) Berry. (Va.)
Antholithes gaudium-rosae Ward. (Va.)
Aralia? vernonensis Fontaine. (Va.)
Araliaephyllum crassinerve (Fontaine) Berry. (Va., Md.)
Araliaephyllum magnifolium Fontaine. (Va., Md.)
Araucaria podocarpoides Fontaine. (Va.)
Araucarites aquiensis Fontaine. (Va., Md.)
Araucarites patapscoensis Berry. (Va.)
Aristolochiaephyllum? cellulare Ward. (Va.)
Aristolochiaephyllum crassinerve Fontaine. (Va., Md.)
Asplenium dicksonianum Heer. (Md., D. C.)
Bombax virginiensis Fontaine. (Va.)
Brachyphyllum crassicaule Fontaine. (Va., Md.)
Carpolithus brookensis Fontaine. (Va.)
Carpolithus mucronatus Fontaine. (Va.)
Casuarina covillei Ward. (Va.)
Celastrophyllum acutidens Fontaine. (Md., Va.)
Celastrophyllum albaedomus Ward. (Va.)
Celastrophyllum brittonianum Hollick. (Va.)
Celastrophyllum denticulatum Fontaine. (Md.)
Celastrophyllum hunteri Ward. (Va.)
Celastrophyllum latifolium Fontaine. (Md.)
Celastrophyllum parvifolium (Fontaine) Berry. (Md., Va.)
Cissites parvifolius (Fontaine) Berry. (Md.)
Cladophlebis browniana (Dunker) Seward. (Va., Md.)
Cladophlebis constricta Fontaine. (Va., Md.)
Cladophlebis distans Fontaine. (Va.)
Cladophlebis parva Fontaine. (Md.)

Cladophlebis rotundata Fontaine. (Va., Md.)
Cladophlebis ungeri (Dunker) Ward. (Va.)
Cladophlebis virginiensis Fontaine. (Va.)
Cladophlebis sp. Fontaine. (Va.)
Cycadeospermum obovatum Fontaine. (Md.)
Dicksoniopsis vernonensis (Ward) Berry. (Va.)
Dryopterites elliptica (Fontaine) Berry. (Va.)
Dryopterites pinnatifida (Fontaine) Berry. (Va.)
Dryopterites virginica (Fontaine) Berry. (Va.)
Ephedrites? vernonensis Fontaine. (Va.)
Equisetum burchardti (Dunker) Brongniart. (Md.)
Eucalyptophyllum oblongifolium Fontaine. (Va.)
Ficus myricoides Hollick. (Md.)
Frenelopsis ramosissima Fontaine. (Va., Md.)
Hederaephyllum crenulatum Fontaine. (Md., Va.)
Knowltonella maxoni Berry. (Va.)
Leptostrobus? multiformis Fontaine. (Va.)
Leptostrobus? ovalis Ward. (Va.)
Menispermites potomacensis Berry. (Md., Va.)
Myricaephyllum dentatum Fontaine. (Va.)
Nageiopsis angustifolia Fontaine. (Va., Md.)
Nageiopsis longifolia Fontaine. (Va., Md.)
Nageiopsis zamioides Fontaine. (Md.)
Nelumbites tenuinervis (Fontaine) Berry. (Va., Md.)
Nelumbites virginiensis (Fontaine) Berry. (Va., Md.)
Onychiopsis göpperti (Schenk) Berry. (Md., Va.)
Onychiopsis latiloba (Fontaine) Berry. (Md., Va.)
Onychiopsis nervosa (Fontaine) Berry. (Va., Md.)
Onychiopsis psilotoides (Stokes and Webb) Ward. (Md., Va.)
Pinus vernonensis Ward. (Va., Md.)
Plantaginopsis marylandica Fontaine. (Md.)
Podozamites acutifolius Fontaine. (Va.)
Podozamites subfalcatus Fontaine. (Va.)
Populophyllum crassinerve Fontaine. (Va.)
Populophyllum minutum Ward. (Va., Md.)
Populophyllum reniforme Fontaine. (Va., Md.)
Populus auriculata Ward. (Va.)
Populus potomacensis Ward. (Va.)
Potamogetophyllum vernonensis Fontaine. (Va.)
Quercophyllum chinkapinense Ward. (Va.)
Quercophyllum grossedentatum Fontaine. (Va.)
Ruffordia acrodenta (Fontaine) Berry.
Sagenopteris elliptica Fontaine. (Md.)
Sapindopsis brevifolia Fontaine. (Va., Md.)
Sapindopsis magnifolia Fontaine. (Va., Md.)
Sapindopsis variabilis Fontaine emend Berry. (Md., Va.)
Sassafras? bilobatum Fontaine. (Va.)
Sassafras parvifolium Fontaine. (Md., Va.)
Sassafras potomacensis Berry. (Va.)
Sequoia ambigua Heer. (Md.)
Sequoia reichenbachi (Geinitz) Heer. (Va.)
Sphenolepis kurriana (Dunker) Schenk. (Md., Va.)
Stenopteris virginica Fontaine. (Va.)
Sterculia elegans Fontaine. (Va.)

Tempskya whitei Berry. (Md.)
Thinnfeldia fontainei Berry. (Va., Md.)
Thinnfeldia granulata Fontaine. (Md.)
Ulmophyllum brookense (Fontaine) Knowlton. (Va.)
Ulmophyllum crassinerve (Fontaine) Knowlton. (Va.)
Ulmophyllum tenuinerve (Fontaine) Knowlton. (Va.)
Undetermined plants. (Md.)
Widdringtonites ramosus (Fontaine) Berry. (Md., Va.)
Zamites tenuinervis Fontaine. (Va., Md.)

POTOMAC GROUP, FORMATION NOT IDENTIFIED.

Osmunda obergiana Heer. (Locality?)

RARITAN FORMATION.

Acer amboyense Newberry. (N. J.)
Acer minutum Hollick. (N. Y.)
Acerates amboyensis Berry. (N. J.)
Andromeda cookii Berry. (N. J., N. Y.)
Andromeda grandifolia Berry. (N. Y., N. J.)
Andromeda novaecaesareae Hollick. (N. J.)
Andromeda parlatorii Heer. (N. Y., N. J.)
Andromeda tenuinervis Lesquereux. (N. Y.)
Andromeda sp. Lesquereux. (N. J. Formation doubtful.)
Androvettia statenensis Hollick and Jeffrey. (N. Y.)
Anomaspis hispida Hollick and Jeffrey. (N. Y.)
Anomaspis tuberculata Hollick and Jeffrey. (N. Y.)
Aralia formosa Heer. (N. J.)
Aralia groenlandica Heer. (N. J.)
Aralia newberryi Berry. (N. Y., N. J.)
Aralia patens Newberry. (N. J.)
Aralia polymorpha Newberry. (N. J.)
Aralia quinquepartita Lesquereux. (N. J.)
Aralia ravniana Heer. (N. Y.)
Aralia rotundiloba Newberry. (Newberry. (N. J.)
Aralia washingtoniana Berry. (D. C.)
Aralia wellingtoniana Lesquereux. (N. J.)
Araliopsis sp. Lesquereux. (Formation doubtful.) (N. J.)
Araliopsoides breviloba (Berry) Berry. (Md.)
Araliopsoides cretacea (Newberry) Berry. (Md., D. C.)
Araliopsoides cretacea dentata (Lesquereux) Berry. (Md.)
Araliopsoides cretacea salisburiaefolia (Lesquereux) Berry. (Md.)
Araucariopitys americana Jeffrey. (N. Y.)
Araucarioxylon noveboracense Hollick and Jeffrey. (N. Y.)
Araucarioxylon sp. Hollick and Jeffrey. (N. Y.)
Aspidiophyllum trilobatum Lesquereux. (Md., D. C.)
Asplenium dicksonianum Heer. (N. J.)
Asplenium foersteri Debey and Ettingshausen. (N. J., Md.)
Asplenium jerseyensis Berry. (N. J.)
Asplenium raritanense Berry. (N. J.)
Baiera incurvata Heer. (N. J.)
Bauhinia cretacea Newberry. (N. J.)
Bauhinia? gigantea Newberry. (N. J.)
Brachyoxylon notabile Hollick and Jeffrey. (N. Y.)
Brachyphyllum macrocarpum Newberry. (N. J., N. Y.)

Brachyphyllum sp. Hollick and Jeffrey. (N. Y.)
Caesalpinia cookiana Hollick. (N. J.)
Caesalpinia raritanensis Berry. (N. J.)
Calycites diospyriformis Newberry. (N. J.)
Calycites parvus Newberry. (N. J.)
Carpolithus euonymoides Hollick. (N. Y.)
Carpolithus floribundus Newberry. (N. J.)
Carpolithus hirsutus Newberry. (N. J.)
Carpolithus ovaeformis Newberry. (N. J.)
Carpolithus pruniformis Newberry. (N. J.)
Carpolithus vaccinioides Hollick. (N. Y.)
Carpolithus woodbridgensis Newberry. (N. J.)
Carpolithus sp. Hollick. Hollick. (N. Y.)
Celastrophyllum angustifolium Newberry. (N. J.)
Celastrophyllum brittonianum Hollick. (N. J.)
Celastrophyllum crenatum Heer. (N. J.)
Celastrophyllum cretaceum Lesquereux. (N. J.)
Celastrophyllum decurrens Lesquereux. (N. J.)
Celastrophyllum grandifolium Newberry. (N. J.)
Celastrophyllum minus Hollick. (N. J.)
Celastrophyllum newberryanum Hollick. (N. J.)
Celastrophyllum robustum Newberry. (Formation doubtful.) (N. J.)
Celastrophyllum spatulatum Newberry. (N. J.)
Celastrophyllum undulatum Newberry. (N. J.)
Celastrophyllum sp. Hollick. (N. Y.)
Celastrus arctica Heer. (N. J.)
Celastrus? sp. Hollick. (N. Y.)
Chondrites flexuosus Newberry. (N. J.)
Chondrophyllum obovatum Newberry. (N. J.)
Chondrophyllum orbiculatum Heer. (N. Y.)
Chondrophyllum reticulatum Hollick. (N. J.)
Cinnamomum heerii (Lesquereux) Lesquereux. (N. J.)
Cinnamomum newberryi Berry. (N. J., D. C.)
Cissites formosus Heer. (N. J., N. Y.)
Cissites newberryi Berry. (N. J.)
Citrophyllum aligerum (Lesquereux) Berry. (N. J.)
Cladophlebis socialis (Heer) Berry. (Md.)
Colutea primordialis Heer. (N. Y., N. J.)
Comptonia microphylla (Heer) Berry. (N. J.)
Cordia apiculata (Hollick) Berry. (N. J., N. Y.)
Cornophyllum vetustum Newberry. (N. J.)
Cunninghamites elegans (Corda) Endlicher. (N. J.)
Cupressinoxylon sp. Hollick. (N. Y.)
Cupressites cookii Newberry. (N. J. Formation doubtful.)
Cycadinocarpus circularis Newberry. (N. J.)
Cyparissidium gracile? (Heer) Heer. (N. Y.)
Czekanowskia capillaris Newberry. (N. J., N. Y., Md.)
Czekanowskia dichotoma? (Heer) Heer. (N. J.)
Dactyolepis cryptomerioides Hollick and Jeffrey. (N. Y.)
Dalbergia apiculata Newberry. (N. J.)
Dalbergia hyperborea Heer. (N. Y.)
Dammara borealis Heer. (N. J., N. Y.)
Daphnophyllum sp. Lesquereux. (N. J.)
Dewalquea grönlandica Heer. (N. J., N. Y.)

Dewalquea insignis Hosius and Von der Marck. (N. Y.)
Dewalquea trifoliata Newberry. (N. J.)
Dicksonia groenlandica Heer. (N. J.)
Diospyros amboyensis Berry. (N. J.)
Diospyros apiculata Lesquereux. (N. Y.)
Diospyros primaeva Heer. (N. J., Md.)
Diospyros provecta Velenovsky. (N. Y.)
Diospyros raritanensis Berry. (N. J.)
Diospyros vera Berry. (D. C.)
Equisetum, rootlets of, Lesquereux. (N. J. Formation doubtful.)
Eucalyptus angusta Velenovsky. (N. J.)
Eucalyptus? attenuata Newberry. (N. J.)
Eucalyptus geinitzi (Heer) Heer. (N. J., N. Y.)
Eucalyptus linearifolia Berry. (N. J.)
Eucalyptus? parvifolia Newberry. (N. J.)
Eugeinitzia proxima Hollick and Jeffrey. (N. Y.)
Ficus daphnogenoides (Heer) Berry. (N. J., N. Y.)
Ficus krausiana Heer. (N. Y.)
Ficus myricoides Hollick. (N. J., N. Y.)
Ficus ovatifolia Berry. (D. C., N. J.)
Ficus woolsoni Newberry. (N. J., N. Y.)
Fontainea grandifolia Newberry. (Md., N. J.)
Frenelopsis hoheneggeri (Ettingshausen) Schenk. (N. J.)
Geinitzia formosa Heer. (N. J.)
Geinitzia reichenbachi (Geinitz) Hollick and Jeffrey. (N. Y.)
Geinitzia sp. Newberry. (N. J.)
Gleichenia giesekiana Heer. (N. J.)
Gleichenia micromera Heer. (N. J.)
Gleichenia zippei (Corda) Heer. (N. J.)
Hedera obliqua Newberry. (N. J.)
Hedera primordialis Saporta. (N. J.)
Hedera sp.? Hollick. (N. Y.)
Hymenaea dakotana Lesquereux. (N. Y., N. J.)
Ilex amboyensis Berry. (N. J.)
Ilex? elongata Newberry. (N. J.)
Juglans arctica Heer. (N. J., N. Y.)
Juglans crassipes Heer. (N. Y.)
Juniperus hypnoides Heer. (N. Y., N. J.)
Kalmia brittoniana Hollick. (N. Y.)
Laurophyllum angustifolium Newberry. (N. J.)
Laurophyllum elegans Hollick. (N. Y., N. J.)
Laurophyllum lanceolatum Newberry. (N. J.)
Laurophyllum minus Newberry. (N. J.)
Laurophyllum nervillosum Hollick. (N. Y., N. J.)
Laurus hollae Heer? (N. Y.)
Laurus nebrascensis (Lesquereux) Lesquereux. (N. Y.)
Laurus plutonia Heer. (N. Y., N. J.?)
Laurus sp. Lesquereux. (N. J.)
Leguminosites atanensis Heer. (N. J.)
Leguminosites constrictus Lesquereux? (N. Y.)
Leguminosites coronilloides Heer. (N. Y.)
Leguminosites omphaloboides Lesquereux. (N. J.)
Leguminosites raritanensis Berry. (N. J.)
Liriodendron marcouanum (Heer) Knowlton. (N. Y.)

Liriodendron oblongifolium Newberry. (N. J.)
Liriodendron quercifolium Newberry. (N. J.)
Liriodendropsis angustifolia (Newberry) Newberry. (N. J.)
Liriodendropsis retusa (Heer) Hollick. (N. J.)
Liriodendropsis simplex (Newberry) Newberry. (N. Y., N. J.)
Magnolia alternans Heer. (N. J.)
Magnolia boulayana Lesquereux. (N. J.)
Magnolia hollicki Berry. (N. J.)
Magnolia isbergiana Heer. (N. J.)
Magnolia lacoeana Lesquereux. (N. J.)
Magnolia longipes Newberry. (N. Y., N. J.)
Magnolia newberryi Berry. (N. Y., N. J.)
Magnolia speciosa Heer. (N. J.)
Magnolia woodbridgensis Hollick. (N. Y., N. J.)
Majanthemophyllum pusillum Heer. (N. Y.)
Menispermites borealis? Heer. (N. J.)
Menispermites wardianus Hollick. (Locality?)
Menispermites sp. Hollick. (N. Y.)
Microzamia gibba (Reuss) Corda. (N. J.)
Moriconia cyclotoxon Debey and Ettingshausen. (N. J., N. Y.)
Myrica acuta Hollick. (Locality?)
Myrica cinnamomifolia Newberry. (N. J.)
Myrica davisii Hollick. (N. Y.)
Myrica emarginata Heer. (N. J.)
Myrica fenestrata Newberry. (N. J.)
Myrica hollicki Ward. (N. Y., N. J.)
Myrica newberryana Hollick. (N. J.)
Myrica raritanensis Hollick. (N. J.)
Myrica sp. Lesquereux. (N. J.)
Myrsine borealis Heer. (N. J.)
Myrsine gaudini (Lesquereux) Berry. (N. Y., N. J.)
Myrsine oblongata Hollick. (N. J. Formation doubtful.)
Myrtophyllum sapindoides Hollick. (N. Y.)
Newberryana rigida (Newberry) Berry. (N. J.)
Palaeanthus problematicus Newberry. (N. Y., N. J., Mass.)
Paliurus affinis? Heer. (N. Y.)
Paliurus integrifolius Hollick. (N. Y. Formation doubtful.)
Paliurus ovalis Dawson. (N. J.)
Passiflora antiqua Newberry. (N. J.)
Persoonia lesquereuxii Knowlton. (N. J.)
Persoonia spatulata Hollick. (N. J.)
Phaseolites elegans Hollick. (N. Y.)
Phaseolites manhassettensis Hollick. (N. J.)
Phegopteris grothiana Heer. (N. J.)
Phyllites obscurus Hollick. (N. J.)
Phyllites poinsettioides Hollick. (N. Y.)
Phyllites trapaformis Berry. (N. J.)
Phyllites undulatus Newberry. (N. J.)
Pinus granulatum (Heer) Stopes. (N. J.)
Pinus quinquefolia Hollick and Jeffrey. (N. Y.)
Pinus raritanensis Berry. (N. J.)
Pinus tetraphylla Jeffrey. (N. Y.)
Pinus triphylla Hollick and Jeffrey. (N. Y.)
Pinus sp. Hollick and Jeffrey. (N. Y.)

Pinus sp. Hollick. (N. Y.)
Pistacia aquehongensis Hollick. (N. Y.)
Pityoidolepis statenensis Hollick and Jeffrey. (N. Y.)
Pityoxylon statenense Jeffrey and Chrysler. (N. Y.)
Planera knowltoniana Hollick. (N. J.)
Platanus aquehongensis Hollick. (N. Y.)
Platanus heerii Lesquereux. (Md., D. C.)
Platanus? newberryana Heer. (N. Y.)
Platanus sp. Hollick. (N. Y.)
Podozamites acuminatus Hollick. (N. J.)
Podozamites knowltoni Berry. (N. J.)
Podozamites lanceolatus (Lindley and Hutton) F. Braun. (N. J., Md.)
Podozamites marginatus Heer. (N. J.)
Podozamites proximans Conrad. (N. J.)
Podozamites tenuinervis Heer. (N. J.)
Populus harkeriana Lesquereux. (N. Y.)
Populus orbicularis (Newberry) Berry. (N. J.)
Prepinus statenensis Jeffrey. (N. Y.)
Proteoides sp. Lesquereux. (N. J.)
Protodammara speciosa Hollick and Jeffrey. (R. I., N. Y.)
Protophyllocladus subintegrifolius (Lesquereux) Berry. (N. Y.)
Protophyllum multinerve (Lesquereux) Lesquereux. (D. C., Md.)
Protophyllum obovatum Newberry. (N. J.)
Protophyllum sternbergii (Lesquereux) Lesquereux. (Md., D. C.)
Prunus? acutifolia Newberry. (N. J.)
Pseudogeinitzia sequoiiformis Hollick and Jeffrey. (N. Y.)
Pterospermites modestus Lesquereux. (N. Y.)
Pterospermites obovatus (Newberry) Berry. (N. J.)
Quercus? novae-caesareae Hollick. (N. Y.)
Quercus raritanensis Berry. (N. J.)
Quercus sp. Lesquereux. (N. J.)
Raritania gracilis (Newberry) Hollick and Jeffrey. (N. Y., N. J.)
Rhamnites minor Hollick. (N. J.)
Rhamnus? acuta Heer. (N. Y.)
Salix flexuosa Newberry. (N. Y., N. J.)
Salix inaequalis Newberry. (N. J.)
Salix lesquereuxii Berry. (N. J.)
Salix meekei Newberry. (N. Y.)
Salix newberryana Hollick. (N. J.)
Salix pseudo-hayei Berry. (N. J.)
Salix raritanensis Berry. (N. Y., N. J.)
Salix siouxiana Knowlton and Cockerell. (N. Y.)
Sapindus morrisoni Lesquereux. (N. Y.)
Sassafras acutilobum Lesquereux. (N. J., Md., D. C.)
Sassafras hastatum Newberry. (N. J.)
Sassafras progenitor Newberry. (N. J., N. Y.?)
Sassafras sp. Lesquereux. (N. J.)
Sequoia concinna Heer. (Locality?)
Sequoia condita Lesquereux. (N. J.?)
Sequoia gracillima (Lesquereux) Newberry. (N. J. Formation doubtful.)
Sequoia heterophylla Velenovsky. (N. Y., N. J.)
Sequoia reichenbachi (Geinitz) Heer. (N. Y., N. J.)
Sequoia smittiana Heer. (N. J.? Formation doubtful.)
Sequoia sp. Hollick. (N. Y.)

Sphaerites raritanensis Berry. (N. J.)
Sphenaspis statenensis Hollick and Jeffrey. (N. Y.)
?Sterculia snowii Lesquereux. (N. Y.)
Sterculia sp. Hollick. (N. Y.)
Sterculia sp. Lesquereux. (N. J. Formation doubtful.)
Strobilites davisii Hollick and Jeffrey. (N. Y.)
Strobilites microsporophorus Hollick and Jeffrey. (N. Y.)
Strobilites sp. Hollick and Jeffrey. (N. Y.)
Thuites meriani Heer. (N. J.)
Thuites sp.? Hollick and Jeffrey. (N. Y.)
Thuja cretacea? (Heer) Newberry. (N. J.)
Tricalycites major Hollick. (N. Y.)
Tricalycites papyraceus Newberry. (N. Y., N. J.)
Tricarpellites striatus Newberry. (N. J.)
Typha sp. Hollick. (N. Y. Formation doubtful.)
Ulmannia sp. (Newberry) Knowlton. (N. J. Formation doubtful.)
Viburnum hollickii Berry. (N. Y. Formation doubtful.)
Viburnum integrifolium Newberry. (N. J.)
Widdringtonites reichii (Ettingshausen) Heer. (N. Y., N. J.)
Widdringtonites subtilis Heer. (N. J.)
Williamsonia riesii Hollick. (N. Y.)
Williamsonia smockii Newberry. (N. J.)

MAGOTHY FORMATION.

Acer paucidentatum Hollick. (N. J.)
Acer sp. Hollick. (Mass.)
Algites americana Berry. (Md.)
Amelanchier whitei Hollick. (Mass.)
Andromeda angustifolia Berry. (Md.)
Andromeda cookii Berry. (Md., N. Y.)
Andromeda grandifolia Berry. (Md.)
Andromeda novaecaesareae Hollick. (N. J., Md.)
Andromeda parlatorii Heer. (N. Y., N. J., Del., Md., Mass.)
Aralia brittoniana Berry. (N. J.)
Aralia coriacea Velenovsky. (N. Y., Mass.)
Aralia groenlandica Heer. (N. J., Mass., Md.)
Aralia mattewanensis Berry. (N. J.)
Aralia nassauensis Hollick. (N. Y.)
Aralia newberryi Berry. (N. J.)
Aralia patens Newberry. (N. Y.)
Aralia ravniana Heer. (N. J., Mass., Md.)
Aralia towneri Lesquereux. (N. J.)
Araliopsoides cretacea (Newberry) Berry. (Mass., Md., D. C.)
Araucaria bladenensis Berry. (Md.)
Araucaria marylandica Berry. (Md.)
Araucarioxylon sp. Holden. (N. J.)
Araucarites ovatus Hollick. (N. J.)
Araucarites zeilleri Berry. (N. J.)
Arisaema cretacea Lesquereux. (N. J.)
Arisaema? mattewanense Hollick. (N. J.)
Asplenium cecilensis Berry. (Md.)
Baiera grandis? Heer. (Mass.)
Banksia pusilla Velenovsky. (N. J.)
Banksites saportanus Velenovsky. (Mass.)

Bauhinia marylandica Berry. (Md.)
Betulites populifolius Lesquereux. (N. J.)
Brachyphyllum macrocarpum Newberry. (N. J., Md., Del.)
Brachyphyllum macrocarpum formosum Berry. (Md.)
Brachyoxylon sp. Holden. (N. J.)
Bumelia praenuntia Berry. (Md.)
Calycites alatus Hollick. (N. Y.)
Calycites obovatus Hollick. (Mass.)
Carex clarkii Berry. (N. J., Del., Md.)
Carpites liriophylli Lesquereux. (Del.)
?Carpites minutulus Lesquereux.
Carpolithus cliffwoodensis Berry. (N. J.)
Carpolithus drupaeformis Hollick. (N. J.)
Carpolithus floribundus Newberry. Mass.)
Carpolithus hirsutus Newberry. (Mass.)
Carpolithus juglandiformis Berry. (N. J.)
Carpolithus mattewanensis Berry. (N. J.)
Carpolithus ostryaeformis Berry. (N. J.)
Carpolithus septoloculus Berry. (Del.)
Carpolithus sp. Hollick. (Mass.)
Cassia insularis Hollick. (N. Y.?)
Cassia sp. Hollick. (Mass.)
Ceanothus constrictus Hollick. (N. Y., Mass.)
Celastrophyllum crassipes Lesquereux. (N. Y.)
Celastrophyllum elegans Berry. (N. J.)
Celastrophyllum grandifolium Newberry?
Celastrophyllum newberryanum Hollick. (N. J.)
Celastrophyllum undulatum Newberry. (Md.)
Celastrus arctica Heer. (N. Y., Md.)
Chondrites flexuosus Newberry. (N. J.)
Cinnamomum crassipetiolatum Hollick. (N. Y.)
Cinnamomum heerii (Lesquereux) Lesquereux. (N. J., Mass.)
Cinnamomum membranaceum (Lesquereux) Hollick. (Mass.)
Cinnamomum newberryi Berry. (N. J., Del., Md., N. Y.)
Cinnamomum sp. Hollick. (Mass.)
Cissites formosus magothiensis Berry. (Md.)
Cissites newberryi Berry. (N. J., Del.)
Coccolobites cretaceus Berry (homonym). (Md.)
Cocculites imperfectus Hollick. (Mass.)
Cocculites inquirendus Hollick. (Mass.)
Cocculus cinnamomeus Velenovsky. (Mass.)
Cocculus minutus Hollick. (N. Y.)
Colutea obovata Berry. (Md.)
Colutea primordialis Heer. (Md., Mass.)
Confervites dubius Berry. (N. J.)
Cordia apiculata (Hollick) Berry. (N. Y., Del.))
Cornus cecilensis Berry. (Md.)
Cornus forchhammeri Heer. (Md.)
Crataegus monmouthensis Berry. (N. J.)
Credneria macrophylla Heer. (N. Y.)
Crotonophyllum cretaceum Velenovsky. (Md.)
Cunninghamites elegans (Corda) Endlicher. (N. J., Mass.)
Cunninghamites squamosus Heer. (N. J.)
Cupressinoxylon? bibbinsi Knowlton. (Md.)

Cyperacites sp. Hollick. (N. Y.)
Czekanowskia dichotoma? (Heer) Heer. (Mass.)
Dalbergia irregularis Hollick. (Mass.)
Dalbergia minor Hollick. (Mass.)
Dalbergia severnensis Berry. (Md.)
Dammara borealis Heer. (Mass., N. Y.)
Dammara? cliffwoodensis Hollick. (N. J., Md.)
Dammara northportensis Hollick. (N. Y.)
Dewalquea grönlandica Heer. (N. J.)
Diospyros apiculata Lesquereux. (Mass., N. Y.)
Diospyros lesquereuxii Knowlton and Cockerell. (N. J., N. Y., Md.)
Diospyros primaeva Heer. (N. J., Md.)
Diospyros prodromus Heer. (N. Y.)
Diospyros provecta Velenovsky. (Mass.)
Diospyros pseudoanceps Lesquereux. (N. Y.)
Doryanthes cretacea Berry. (N. J., Md.)
Dryandroides quercinea Velenovsky. (Mass.)
Elaeodendron marylandicum Berry. (Md.)
Elaeodendron strictum Hollick. (Mass.)
Elaeodendron sp. Hollick. (N. Y.)
Embothriopsis presagita Hollick. (N. Y.)
Eucalyptus? attenuata Newberry. (N. J., Md.)
Eucalyptus geinitzi (Heer) Heer. (N. J., Mass., Md., Del.)
Eucalyptus geinitzi propinqua Hollick. (N. Y.)
Eucalyptus latifolia Hollick. (N. Y., Mass., Md.)
Eucalyptus linearifolia Berry. (N. Y., R. I.)
Eucalyptus schübleri (Heer) Hollick. (Mass.)
Eucalyptus wardiana Berry. (N. J., Del., Md.)
Ficus atavina Heer. (Mass., N. Y.)
Ficus cecilensis Berry. (Md.)
Ficus crassipes (Heer) Heer. (Del., Md.)
Ficus daphnogenoides (Heer) Berry. (N. J., N. Y., Del., Md.)
Ficus fracta Velenovsky. (N. Y.)
Ficus krausiana Heer. (Md., Mass., R. I.)
Ficus krausiana subsimilis Hollick. (N. Y.)
Ficus laurophylla Lesquereux. (N. J.)
Ficus lesquereuxii Knowlton. (Mass.)
Ficus myricoides Hollick. (Mass., N. Y.)
Ficus sapindifolia Hollick. (Mass., N. Y.)
Ficus willisiana Hollick. (N. Y.)
Ficus woolsoni Newberry. (N. J.)
Frenelopsis hoheneggeri (Ettingshausen) Schenk. (N. J., N. Y.)
Geinitzia formosa Heer. (N. J., Del.)
Gleichenia delawarensis Berry. (Del., Md.)
Gleichenia gracilis Heer? (R. I.)
Gleichenia protogaea Debey and Ettingshausen. (Mass.)
Gleichenia saundersii Berry. (N. J., Md.)
Gleichenia zippei (Corda) Heer. (N. J., Del.)
Grewiopsis flabellata (Lesquereux) Knowlton. (Md.)
Guatteria cretacea Hollick. (Mass., N. Y.)
Gyminda primordialis Hollick. (Mass.)
Hedera cecilensis Berry. (Md.)
Hedera cretacea Lesquereux. (Del.)
Hedera simplex Hollick. (Mass.)

Heterofilicites anceps Berry. (N. J.)
Hymenaea dakotana Lesquereux. (N. Y., N. J., Mass.)
Hymenaea primigenia Saporta. (Mass.)
Ilex papillosa Lesquereux. (Mass.)
Ilex severnensis Berry. (Md.)
Ilex strangulata Lesquereux. (N. J.)
Illicium deltoides Berry. (Md.)
Juglans arctica Heer. (N. J., Mass., R. I.)
Juglans crassipes Heer. (Mass., N. Y.)
Juglans elongata Hollick. (Mass., N. Y.)
Juniperus hypnoides Heer. (N. Y., N. J.)
Laurophyllum angustifolium Newberry. (N. J., Md.)
Laurophyllum elegans Hollick. (Md.)
Laurophyllum lanceolatum Newberry. (N. Y.)
Laurophyllum ocotaeoides Hollick. (N. Y.)
Laurus antecedens Lesquereux. (N. Y.)
Laurus atanensis Berry. (N. Y.)
Laurus hollae Heer. (N. J.)
Laurus hollickii Berry. (N. J., Del., Md.)
Laurus nebrascensis (Lesquereux) Lesquereux. (Mass.)
Laurus newberryana Hollick. (N. Y.)
Laurus omalii Saporta and Marion. (N. Y.)
Laurus plutonia Heer. (N. J., N. Y., R. I., Md.)
Laurus proteaefolia Lesquereux. (N. J., Md.)
Laurus teliformis Lesquereux. (Mass.)
Leguminosites canavalioides Berry. (Md.)
Leguminosites convolutus Lesquereux. (Mass., N. Y.)
Leguminosites coronilloides Heer. (Mass., Md.)
Leguminosites omphaloboides Lesquereux. (Md.)
Ligustrum sublite Hollick. (N. Y.)
Liriodendron attenuatum Hollick. (N. Y., Mass.)
Liriodendron morganensis Berry. (N. J.)
Liriodendron oblongifolium Newberry. (N. J.)
Liriodendropsis angustifolia (Newberry) Newberry. (Mass.; N. Y.)
Liriodendropsis constricta Ward. (N. Y., Mass., Md.)
Liriodendropsis retusa (Heer) Hollick. (N. Y.)
Liriodendropsis simplex (Newberry) Newberry. (Mass., N. Y.)
Liriodendropsis spectabilis Hollick. (Mass., N. Y.)
Lycopodium cretaceum Berry. (Md.)
Magnolia alternans Heer. (N. Y., Mass.)
Magnolia amplifolia Heer. (Mass.)
Magnolia boulayana Lesquereux. (Md., N. Y.)
Magnolia capellinii Heer. (N. J., N. Y., Md.)
Magnolia hollicki Berry. (N. Y., Mass., Md.)
Magnolia isbergiana Heer. (N. Y.)
Magnolia lacoeana Lesquereux. (Mass., Md.)
Magnolia longipes Newberry. (N. Y.)
Magnolia obtusata Heer. (N. J., Md.)
Magnolia pseudoacuminata Lesquereux. (Mass.)
Magnolia speciosa Heer. (N. J., N. Y., Mass.)
Magnolia tenuifolia Lesquereux. (N. J., Del., N. Y., Mass.)
Magnolia vaningeni Hollick. (N. Y.)
Magnolia woodbridgensis Hollick. (N. J., R. I.)
Malapoenna falcifolia (Lesquereux) Knowlton. (N. J.)

Malapoenna sp. Hollick. (Mass.)
Marsilea andersoni Hollick. (N. Y., Mass.)
Menispermites acutilobus Lesquereux. (Mass.)
Menispermites brysoniana Hollick. (N. Y.)
Microzamia? dubia Berry. (N. J.)
Moriconia americana Berry. (Del., Md.)
Moriconia cyclotoxon Debey and Ettingshausen. (N. J., Del., Md., R. I.)
Myrica brittoniana Berry. (N. J.)
Myrica cliffwoodensis Berry. (N. J.)
Myrica longa (Heer) Heer. (Md.)
Myrica zenkeri (Ettingshausen) Heer. (N. Y.)
Myrica sp. Hollick. (Mass.)
Myrsine borealis Heer. (N. Y., Mass., Md.)
Myrsine crassula Knowlton and Cockerell. (N. J.)
Myrsine gaudini (Lesquereux) Berry. (Md.)
Myrsine sp. Hollick. (N. Y.)
Nectandra imperfecta Hollick. (Mass.)
Nelumbites primaeva (Berry) Berry. (N. J., Md.)
Nelumbo kempii (Hollick) Hollick. (Mass., N. Y., N. J.)
Ocotea nassauensis Hollick. (N. Y.)
Onoclea inquirenda (Hollick) Hollick. (N. Y., Mass., Md.)
Osmunda delawarensis Berry. (Del.)
Osmunda novaecaesarae Berry. (N. J.)
Paliurus integrifolius Hollick. (N. Y., N. J.)
Paliurus ovalis Dawson. (Mass.)
Paliurus populiferus Berry. (N. J.)
Paliurus sp. Hollick. (N. Y.)
Palmoxylon cliffwoodensis Berry. (N. J.)
Panax cretacea Heer. (Mass.)
Paracupressinoxylon cupressoides Holden. (N. J.)
Paracupressinoxylon sp. Holden. (N. J.)
Periploca cretacea Hollick. (Mass.)
Persea leconteana (Lesquereux) Lesquereux. (Mass.)
Persea valida Hollick. (N. Y.)
Persoonia lesquereuxii Knowlton. (N. J.)
Phaseolites manhassettensis Hollick. (N. Y.)
Phragmites? cliffwoodensis Berry. (N. J.)
Phyllites cliffwoodensis Berry. (N. J.)
Picea cliffwoodensis Berry. (N. J.)
Pinus andraei Coemans. (N. J.?)
Pinus delicatulus Berry. (N. J.)
Pinus mattawanensis Berry. (N. J.)
Pinus protoscleropitys Holden. (N. J.)
Pinus sp. Hollick. (Mass.)
Pistia nordenskiöldi (Heer) Berry. (Md.)
Pityoxylon anomalum Holden. (N. J.)
Pityoxylon foliosum Holden. (N. J.)
Pityoxylon hollicki Knowlton. (N. J.)
Planera betuloides Hollick. (Mass.)
Platanus kümmellii Berry. (N. J.)
?Poacites sp. Hollick. (N. Y., Mass.)
Podozamites knowltoni Berry. (Md.)
Podozamites lanceolatus (Lindley and Hutton) F. Braun. (N. Y.)
Podozamites marginatus Heer. (N. J., Md.)

Podozamites proximans Conrad. (N. Y.)
?Podozamites sp. Hollick. (Mass., N. Y.)
Populites tenuifolius Berry. (N. J.)
Populus stygia Heer? (Mass., Md.)
Populus sp. (aments of) Hollick. (N. Y., Mass.)
Premnophyllum trigonum Velenovsky. (N. Y.)
Prepinus viticetensis Jeffrey. (Mass.)
Protophyllocladus lobatus Berry. (Md.)
Protophyllocladus subintegrifolius (Lesquereux) Berry. (N. J., Mass., Md., R. I.?)
Quercus coprinoides Berry. (N. J.)
Quercus hollickii Berry. (N. J.)
Quercus holmesii (Lesquereux) Lesquereux. (N. J.)
Quercus morrisoniana Lesquereux. (N. Y., N. J., Md.)
Quercus? novaecaesareae Hollick. (N. J.)
Quercus severnensis Berry. (Md.)
Quercus sp. Berry. (N. J.)
?Quercus sp. Hollick. (N. J., Mass.)
Raritania gracilis (Newberry) Hollick and Jeffrey. (Md.)
Rhamnites apiculatus Lesquereux. (Md.)
Rhamnus inaequilateralis Lesquereux. (N. J.)
Rhamnus novae-caesareae Berry. (N. J.)
Rhus cretacea Heer? (N. Y.)
Sabalites magothiensis (Berry) Berry. (N. J., Del., Md.)
Sagenopteris variabilis (Velenovsky) Velenovsky. (Mass.)
Salix flexuosa Newberry. (Mass., R. I., N. J., N. Y., Md., Del.)
Salix inaequalis Newberry. (N. Y.)
Salix lesquereuxii Berry. (D. C., Del.)
Salix mattawanensis Berry. (N. J.)
Salix meekii Newberry. (N. J., Mass.)
Salix purpuroides Hollick. (N. Y.)
Salix raritanensis Berry. (Mass.)
Salix siouxiana Knowlton and Cockerell. (N. Y.)
Salix sp. Hollick. (N. Y.)
Santalum novae-caesareae Berry. (N. J.)
Sapindus apiculatus Velenovsky. (N. Y., N. J.)
Sapindus imperfectus Hollick. (N. Y.)
Sapindus morrisoni Lesquereux. (N. Y., N. J.)
Sapotacites knowltoni Berry. (Del.)
Sassafras acutilobum Lesquereux. (Mass., N. Y., N. J., Md.)
Sassafras angustilobum Hollick. (Mass.)
Sassafras hastatum Newberry. (Mass., N. Y.)
Sassafras progenitor Newberry. (N. J., N. Y.?)
Sequoia ambigua Heer. (Mass., N. J.)
Sequoia concinna Heer. (Mass.)
Sequoia fastigiata (Sternberg) Heer. (Mass.)
Sequoia gracillima (Lesquereux) Newberry. (N. J., Del.)
Sequoia gracillis Heer. (Mass.)
Sequoia heterophylla Velenovsky. (N. J., Del., Md.)
Sequoia reichenbachi (Geinitz) Heer. (N. J., Del., Md.)
Sequoia sp. Hollick. (Mass.)
Sequoia sp. Lesquereux. (N. J. Formation doubtful.)
Smilax raritanensis Berry. (Locality and formation doubtful.)
Sphaerites raritanensis Berry. (Md.)
Sterculia cliffwoodensis Berry. (N. J., Del.)

Sterculia minima Berry. (N. J., Del., Md.)
Sterculia prelabrusca Hollick. (Mass.)
?Sterculia snowii Lesquereux. (Mass.)
Sterculia snowii bilobata Berry. (N. J.)
?Sterculia sp. Hollick. (N. J., Mass.)
Strobilites inquirendus Hollick. (N. J.)
Strobilites perplexus Hollick. (Mass.)
Thuja cretacea (Heer) Newberry. (Md., Del.)
Thyrsopteris grevillioides (Heer) Hollick. (Mass.)
Tricalycites major Hollick. (Mass., R. I., N. Y.)
Tricalycites papyraceus Newberry. (Mass., R. I.?, N. Y.?)
Tricarpellites strictus Newberry. (Mass.)
Typha sp. Hollick. (Mass., N. Y.?)
Viburnum hollickii Berry. (N. J.)
Viburnum integrifolium Newberry. (N. Y.)
Viburnum mattewanense Berry. (N. J.)
Widdringtonites fasciculatus Hollick. (Mass.)
Widdringtonites reichii (Ettingshausen) Heer. (N. J., Del., Md., D. C., Mass.)
Widdringtonites subtilis Heer. (Mass., R. I. Formation doubtful.)
Williamsonia delawarensis Berry. (Del., Md.)
Williamsonia marylandica Berry. (Md.)
Williamsonia riesii Hollick. (Mass. Formation doubtful.)
Williamsonia sp. Hollick. (N. Y.)
Zizyphus cliffwoodensis Berry. (N. J.)
Zizyphus elegans Hollick. (N. Y.)
Zizyphus grönlandicus Heer. (Mass.)
Zizyphus lewisiana Hollick. (N. Y.)
Zizyphus oblongus Hollick. (N. Y.)

MATAWAN FORMATION.

Dammara cliffwoodensis Hollick. (Md.)
Ficus matawanensis Berry. (N. J.)
Paracedroxylon scituatense Sinnott. (Scituate, Mass. Formation doubtful.)
Pinus scituatensiformis Bailey. (Scituate, Mass. Formation doubtful.)
Pityoxylon scituatense Jeffrey and Chrysler. (Scituate, Mass. Formation doubtful.)

MONMOUTH FORMATION.

Palmoxylon anchorus Stephens. (N. J.)

SOUTH ATLANTIC AND GULF REGIONS.

TUSCALOOSA FORMATION.

Abietites foliosus (Fontaine) Berry. (Ala.)
Acerates amboyensis Berry. (Ala.)
Andromeda grandifolia Berry. (Ala.)
Andromeda novaecaesareae Hollick. (Ala.)
Andromeda parlatorii Heer. (Ala.)
Andromeda wardiana Lesquereux. (Ala.)
Androvettia carolinensis Berry. (Miss.)
Androvettia sp. Berry. (Ga.)
Aralia cottondalensis Berry. (Ala.)
Aralia sp. Berry. (Ga.)
Asplenium dicksonianum Heer. (Ala.)
Bauhinia cretacea Newberry. (Ala.)
Bauhinia marylandica Berry. (Ala.)

Calycites sexpartitus Berry. (Ala.)
Capparites cynophylloides Berry. (Ala.)
Capparites orbicularis Berry. (Ala.)
Carpolithus floribundus Newberry. (Ala.)
Carpolithus tuscaloosensis Berry. (Ala.)
Cassia vaughani Berry. (Ala.)
Celastrophyllum alabamensis Berry. (Ala.)
Celastrophyllum brittonianum Hollick. (Ala.)
Celastrophyllum carolinensis Berry. (Ala.)
Celastrophyllum crenatum Heer. (Ala.)
Celastrophyllum decurrens Lesquereux. (Ala.)
Celastrophyllum grandifolium Newberry. (Ala.)
Celastrophyllum gymindaefolium Berry. (Ala.)
Celastrophyllum newberryanum Hollick. (Ala.)
Celastrophyllum praecrassipes Berry. (Ala.)
Celastrophyllum shirleyensis Berry. (Ala.)
Celastrophyllum undulatum Newberry. (Ala.)
Cinnamomum newberryi Berry. (Ala.)
Cissites formosus Heer. (Ala.)
Citrophyllum aligerum (Lesquereux) Berry. (Ala.)
Cladophlebis alabamensis Berry. (Ala.)
Cocculus cinnamomeus Velenovsky. (Ala.)
Cocculus polycarpaefolius Berry. (Ala.)
Cocculus problematicus Berry. (Ala.)
Colutea obovata Berry. (Ala.)
Conocarpites formosus Berry. (Ala.)
Cordia apiculata (Hollick) Berry. (Ala.)
Cornophyllum obtusatum Berry. (Ala.)
Cornophyllum vetustum Newberry. (Ala.)
Crotonophyllum panduraeformis Berry. (Ala.)
Cycadinocarpus circularis Newberry. (Ala.)
Cyperacites sp. Hollick. (Ala.)
Czekanowskia dichotoma? (Heer) Heer. (Ala.)
Dammara borealis Heer. (Ala.)
Dermatophyllites acutus Heer. (Ala.)
Dewalquea smithii Berry. (Ala.)
Dicksonia groenlandica Heer. (Ala.)
Diospyros amboyensis Berry. (Ala.)
Diospyros lesquereuxi Knowlton and Cockerell. (Ala.)
Diospyros primaeva Heer. (Ala.)
Dryopteris stephensoni Berry. (Ala.)
Eorhamnidium cretaceum Berry. (Ala.)
Eorhamnidium platyphylloides (Lesquereux) Berry. (Ala.)
Equisetum? sp. Berry. (Ala.)
Eucalyptus latifolia Hollick. (Ala.)
Eugenia tuscaloosensis Berry. (Ala.)
Ficus alabamensis Berry. (Ala.)
Ficus crassipes (Heer) Heer. (Ala.)
Ficus daphnogenoides (Heer) Berry. (Ala.)
Ficus fontainii Berry. (Ala.)
Ficus inaequalis Lesquereux. (Ala.)
Ficus krausiana Heer. (Ala.)
Ficus shirleyensis Berry. (Ala.)
Ficus woolsoni Hollick. (Ala.)

Geinitzia formosa Heer. (Ala.)
Grewiopsis formosa Berry. (Ala.)
Grewiopsis tuscaloosensis Berry. (Ala.)
Hymenaea fayettensis Berry. (Ala.)
Ilex masoni Lesquereux. (Ala.)
Inga cretacea Lesquereux. (Ala.)
Juglans arctica Heer. (Ala.)
Jungermannites cretaceus Berry. (Ala.)
Kalmia brittoniana Hollick. (Ala.)
Laurophyllum angustifolium Newberry. (Ala.)
Laurophyllum nervillosum Hollick. (Ala.)
Laurus plutonia Heer. (Ala.)
Leguminosites ingaefolia Berry. (Ala.)
Leguminosites omphalobeoides Lesquereux. (Ala.)
Leguminosites shirleyensis Berry. (Ala.)
Leguminosites tuscaloosensis Berry. (Ala.)
Liriodendron meekii Heer. (Ala.)
Liriodendropsis angustifolia Newberry. (Ala.)
Liriodendropsis constricta (Ward) Hollick. (Ala.)
Liriodendropsis simplex (Newberry) Newberry. (Ala.)
Lycopodites tuscaloosensis Berry. (Ala.)
?Lycopodium cretaceum Berry. (Ala.)
Magnolia alternans Heer. (Ala.)
Magnolia boulayana Lesquereux. (Ala.)
Magnolia capellinii Heer. (Ala.)
Magnolia hollicki Berry. (Ala.)
Magnolia lacoana Lesquereux. (Ala.)
Magnolia longipes Hollick. (Ala.)
Magnolia newberryi Berry. (Ala.)
Magnolia obtusata Heer. (Ala.)
Magnolia speciosa Heer. (Ala.)
Malapoenna cottondalensis Berry. (Ala.)
Malapoenna cretacea (Lesquereux) Knowlton. (Ala.)
Malapoenna falcifolia (Lesquereux) Knowlton. (Ala.)
?Marattia cretacea Velenovsky. (Ala.)
Menispermites integrifolius Berry. (Ala.)
Menispermites trilobatus Berry. (Ala.)
Menispermites sp. Berry. (Ga.)
Myrica dakotensis minima Berry. (Ala.)
Myrica emarginata Heer. (Ala.)
Myrica longa (Heer) Heer. (Ala.)
Myrsine borealis Heer. (Ala.)
Myrsine gaudini (Lesquereux) Berry. (Ala.)
Nyssa snowiana Lesquereux. (Ala.)
Oreodaphne alabamensis Berry. (Ala.)
Oreodaphne shirleyensis Berry. (Ala.)
Paleocassia laurinea Lesquereux. (Ala.)
Paliurus sp. Berry. (Ga.)
Panax cretacea Heer. (Ala.)
Persea valida Hollick. (Ala.)
Persoonia lesquereuxii Knowlton. (Ala.)
Persoonia lesquereuxii minor Berry. (Ala.)
Phaseolites formosus Lesquereux. (Ala.)
Phyllites pistiaeformis Berry. (Ala., Miss.)

Phyllites shirleyensis Berry. (Ala.)
Pinus raritanensis Berry. (Ala.)
Piperites tuscaloosensis Berry. (Ala.)
Platanus asperaeformis Berry. (Ala.)
Platanus latior (Lesquereux) Knowlton. (Ala.)
Platanus shirleyensis Berry. (Ala.)
Podozamites marginatus Heer. (Ala.)
Populites tuscaloosensis Berry. (Ala.)
Populus hyperborea Heer. (Ala.)
Proteoides conospermaefolia Berry. (Ala.)
Protodammara speciosa Hollick and Jeffrey. (Ala.)
Protophyllocladus subintegrifolius (Lesquereux) Berry. (Ala.)
Pterospermites carolinensis Berry. (Ala.)
Rhamnus tenax Lesquereux. (Ala.)
Salix flexuosa Newberry. (Ala.)
Salix lesquereuxii Berry. (Ala.)
Salix meekii Newberry. (Ala.)
Sapindus morrisoni Lesquereux. (Ala.)
Sapindus variabilis Berry. (Ala.)
Sapotacites ettingshauseni Berry. (Ala.)
Sapotacites formosus Berry. (Ala.)
Sapotacites shirleyensis Berry. (Ala.)
Sassafras acutilobum Lesquereux. (Ala.)
Sequoia ambigua Heer. (Ala.)
Sequoia fastigiata (Sternberg) Heer. (Ala.)
Sequoia heterophylla Velenovsky. (Ala.)
Sequoia reichenbachi (Geinitz) Heer. (Ala., Miss.)
Sphaerites alabamensis Berry. (Ala.)
Tricalycites papyraceus Hollick. (Ala.)
Widdringtonites reichii (Ettingshausen) Heer. (Ala.)
Widdringtonites subtilis Heer. (Ala.)
Zizyphus lamarensis Berry. (Ala.)

EUTAW FORMATION.

Andromeda cretacea Lesquereux? (Ga.)
Andromeda novaecaesareae Hollick.
Andromeda parlatorii Heer. (Ala.)
Andromeda wardiana Lesquereux. (Ga., Tenn.)
Androvettia elegans Berry. (Ga.)
Aralia eutawensis Berry. (Ga.)
Araucaria bladensis Berry. (Ga.)
Araucaria jeffreyi Berry. (Ga.)
Bauhinia alabamensis Berry. (Ala.)
Bauhinia cretacea Newberry. (Ga.)
Brachyphyllum macrocarpum formosum Berry. (Ga.)
Cephalotaxospermum carolinianum Berry. (Ala.)
Cinnamomum heerii Lesquereux? (Ga., Tenn.)
Cinnamomum newberryi Berry. (Ga.)
Cupressinoxylon sp. Berry. (Tenn.)
Dewalquea smithii Berry. (Tenn.)
Diospyros primaeva Heer. (Tenn.)
Doryanthites cretacea Berry. (Ala.)
Eucalyptus angusta Velenovsky. (Ga.)
Ficus crassipes (Heer) Heer. (Ga., Tenn.)

Ficus krausiana Heer. (Ga., Tenn.)
Ficus ovatifolia Berry. (Ga., Tenn.)
Halymenites major Lesquereux. (Tenn.)
Juglans arctica Heer. (Ga.)
Laurophyllum elegans Hollick. (Tenn.)
Laurus plutonia Heer. (Ala.)
Magnolia boulayana Lesquereux. (Ala., Ga.)
Magnolia capellinii Heer. (Ga.)
Malapoenna horrellensis Berry. (Ga., Tenn.)
Manihotites georgiana Berry. (Ga., Tenn.)
Menispermites variabilis Berry. (Ga.)
Myrcia havanensis Berry. (Ala., Tenn.)
Paliurus upatoiensis Berry. (Ga.)
Phragmites prattii Berry. (Ga., Tenn.)
Phyllites asplenioides Berry.
Pterospermites carolinensis Berry. (Tenn.)
Salix eutawensis Berry. (Ga., Tenn.)
Salix flexuosa Newberry. (Ga.)
Salix lesquereuxii Berry. (Ga.)
Sequoia ambigua Heer. (Ala.)
Sequoia reichenbachi (Geinitz) Heer. (Ala., Ga.)
Tumion carolinianum Berry? (Ga.)
Zizyphus laurifolius Berry. (Ga.)

BLACK CREEK FORMATION.[1]

Acerates amboyensis Berry. (N. C.)
Algites americana Berry. (N. C., S. C.)
Andromeda grandifolia Berry. (N. C., S. C.)
Andromeda novaecaesareae Hollick. (N. C.)
Andromeda parlatorii Heer. (N. C., S. C.)
Androvettia carolinensis Berry. (N. C.)
Aralia newberryi Berry. (N. C.)
Araucaria bladenensis Berry. (N. C., S. C.)
Araucaria clarkii Berry. (N. C.)
Araucaria darlingtonensis Berry. (S. C.)
Araucaria jeffreyi Berry. (N. C.)
Brachyphyllum macrocarpum Newberry. (N. C.)
Celastrophyllum crenatum Heer. (N. C.)
Celastrophyllum undulatum Newberry. (N. C.)
Cephalotaxospermum carolinianum Berry. (N. C., S. C.)
Cinnamomum heerii Lesquereux. (N. C.)
Cunninghamites elegans (Corda) Endlicher. (N. C., S. C.)
Cycadinocarpus circularis Newberry. (N. C.)
Dammara borealis Heer. (N. C.)
Dewalquea grönlandica Heer. (N. C.)
Diospyros primaeva Heer. (N. C.)
Eucalyptus angusta Velenovsky. (S. C.)
Eucalyptus? attenuata Newberry. (N. C.)
Eucalyptus geinitzi (Heer) Heer. (N. C.)
Eucalyptus linearifolia Berry. (N. C.)
Ficus daphnogenoides (Heer) Berry. (N. C.)
Ficus inaequalis Lesquereux. (N. C.)

[1] This list comprises the species listed by Berry (Maryland Geol. Survey, Upper Cretaceous, pp. 210-215, 1916) from the Black Creek formation of North Carolina and South Carolina.

Ficus ovatifolia Berry. (N. C.)
Ficus stephensoni Berry. (N. C.)
Ficus woolsoni Newberry. (N. C.)
Gleditsiaphyllum triacanthoides Berry. (N. C.)
Hedera ovalis Lesquereux. (N. C.)
Hedera primordialis Saporta. (N. C., S. C.)
Juglans arctica Heer. (N. C.)
Kalmia brittoniana Hollick. (N. C.)
Laurophyllum elegans Hollick. (N. C.)
Leguminosites robiniifolia Berry. (N. C., S. C.)
Liriodendron dubium Berry. (N. C.)
Liriodendron cf. L. primaevum Newberry. (N. C.)
Magnolia capellinii Heer. (N. C.)
Magnolia newberryi Berry. (N. C., S. C.?)
Malapoenna horrellensis Berry. (N. C.)
Moriconia americana Berry. (N. C.)
Myrica brittoniana Berry. (S. C.)
Myrica cliffwoodensis Berry. (N. C.)
Myrica elegans Berry. (N. C., S. C.)
Myrsine borealis Heer. (N. C.)
Myrsine gaudini (Lesquereux) Berry. (N. C., S. C.)
Phaseolites formus Berry. (N. C.)
Phragmites pratti Berry. (N. C.)
Phragmites sp. Berry. (N. C.)
Pinus raritanensis Berry. (N. C.)
Pisonia cretacea Berry. (N. C.)
Pistia nordenskiöldi (Heer) Berry. (N. C.)
Planera cretacea Berry. (N. C.)
Podozamites knowltoni Berry. (N. C.)
Pterospermites carolinensis Berry. (N. C.)
Pterospermites credneriifolius Berry. (N. C.)
Quercus pratti Berry. (N. C.)
Rhus darlingtonensis Berry. (S. C.)
Salix eutawensis Berry. (N. C.)
Salix flexuosa Newberry. (N. C.)
Salix lesquereuxii Berry. (N. C.)
Salix newberryana Hollick. (N. C.)
Sassafras acutilobum Lesquereux. (N. C.)
Sequoia heterophylla Velenovsky. (N. C.)
Sequoia minor Velenovsky. (N. C.)
Sequoia reichenbachi (Geinitz) Heer. (N. C.)
Strobilites anceps Berry. (S. C.)
Tumion carolinianum Berry. (N. C.)

Middendorf arkose member of Black Creek formation, South Carolina.

Acaciaphyllites grevilleoides Berry.
Andromeda euphorbiophylloides Berry.
Andromeda novaecaesareae Hollick.
Araucaria jeffreyi Berry.
Arundo grönlandica Heer?
Brachyphyllum macrocarpum Newberry.
Caesalpinia middendorfensis Berry.
Calycites middendorfensis Berry.
Carex clarkii Berry.

Celastrophyllum carolinensis Berry.
Celastrophyllum crenatum Heer.
Celastrophyllum elegans Berry.
Cinnamomum middendorfensis Berry.
Cinnamomum newberryi Berry.
Citrophyllum aligerum (Lesquereux) Berry.
Crotonophyllum panduraeformis Berry.
Dewalquea smithii Berry.
Diospyros primaeva Heer.
Eucalyptus geinitzi (Heer) Heer.
Eucalyptus wardiana Berry.
Ficus atavina Heer.
Ficus celtifolius Berry.
Ficus crassipes (Heer) Heer.
Ficus krausiana Heer.
Ficus stephensoni Berry.
Hamamelites? cordatus Lesquereux.
Heterolepis cretaceus Berry.
Illicium waterensis Berry.
Juglans arctica Heer.
Laurophyllum elegans Hollick.
Laurophyllum nervillosum Hollick.
Laurus atanensis Berry.
Laurus plutonia Heer.
Leguminosites middendorfensis Berry.
Lycopodium cretaceum Berry.
Magnolia capellinii Heer.
Magnolia obtusata Heer.
Magnolia tenuifolia Lesquereux.
Mimosa carolinensis Berry.
Moriconia americana Berry.
Onoclea inquirenda (Hollick) Hollick.
Pachystima? cretacea Berry.
Phragmites pratti Berry.
Pinus raritanensis Berry.
Potamogeton middendorfensis Berry.
Proteoides lancifolius Heer.
Proteoides parvula Berry.
Protophyllocladus lobatus Berry.
Quercus pseudo-westfalica Berry.
Quercus sumpterensis Berry.
Sabalites carolinensis Berry.
Salix flexuosa Newberry.
Salix pseudo-hayei Berry.
Salix lesquereuxii Berry.
Salix sloani Berry.
Sapindus morrisoni Lesquereux.
Sequoia reichenbachi (Geinitz) Heer.
Widdringtonites subtilis Heer.

RIPLEY FORMATION.

Andromeda novaecaesareae Hollick. (Tenn.)
Bauhinia ripleyensis Berry. (Ala., Tenn.)
Cissites newberryi Berry. (Tenn.)
Dryophyllum gracile Debey. (Tenn.)

Dryopteris stephensoni Berry. (Ga.)
Eugenia? anceps Berry. (Tenn.)
Halymenites major Lesquereux. (Tenn.)
Magnolia capellinii Heer. (Tenn.)
Malapoenna horrellensis Berry. (Tenn.)
Manihotites georgiana Berry. (Tenn.)
Myrcia havanensis Berry. (Tenn.)
Myrica ripleyensis Berry. (Tenn.)
Platanus ripleyensis Berry. (Ala.)
Platanus sp. Berry. (Ala.)
Sabalites sp. Berry. (Tenn.)
Sterculia snowii tennesseensis Berry. (Tenn.)

Cusseta sand member of Ripley formation, Georgia.

Andromeda novaecaesareae Hollick.
Araucaria bladenensis Berry.
Araucaria jeffreyi Berry.
Cunninghamites elegans (Corda) Endlicher.
Doryanthes cretacea Berry.
Dryopterites stephensoni Berry.
Dryopteris sp. Berry.
Eucalyptus angusta Velenovsky.
Ficus georgiana Berry.
Ficus sp. Berry.

TRINITY GROUP.

Abietites foliosus (Fontaine) Berry. (Glen Rose, Tex.)
Abietites linkii (Roemer) Dunker. (Glen Rose, Tex.)
Abietites sp. Fontaine. (Glen Rose, Tex.)
Brachyphyllum parceramosum Fontaine. (Glen Rose, Tex.)
Carpolithus harveyi Fontaine. (Glen Rose, Tex.)
Carpolithus obovatus Fontaine. (Glen Rose, Tex.)
Cycadeospermum rotundatum Fontaine. (Glen Rose, Tex.)
Dioonites buchianus (Ettingshausen) Bornemann. (Glen Rose, Tex.)
Dioonites buchianus rarinervis Fontaine. (Glen Rose, Tex.)
Dioonites dunkerianus (Göppert) Miquel. (Glen Rose, Tex.)
Equisetum texense Fontaine. (Glen Rose, Tex.)
Frenelopsis hoheneggeri (Ettingshausen) Schenk. (Glen Rose, Tex.)
Frenelopsis varians Fontaine. (Ark.; Glen Rose, Tex.)
Laricopsis longifolia Fontaine. (Glen Rose, Tex.)
Pagiophyllum dubium Fontaine. (Glen Rose, Tex.)
Paleohillia arkansana Knowlton. (Ark.)
Pinus sp. Fontaine. (Glen Rose, Tex.)
Podozamites acutifolius Fontaine. (Glen Rose, Tex.)
Podozamites sp. Fontaine. (Glen Rose, Tex.)
Ruffordia göpperti (Dunker) Seward? (Glen Rose, Tex.)
Sequoia pagiophylloides Fontaine. (Glen Rose, Tex.)
Sphenolepidium sternbergianum (Dunker) Heer. (Glen Rose, Tex.)
Thuyoxylon americanum Unger. (Age doubtful.)
Widdringtonites ramosus (Fontaine) Berry. (Tex.)
Williamsonia texana Fontaine. (Glen Rose, Tex.)
Zamites tenuinervis Fontaine. (Glen Rose, Tex.)

UPSON CLAY (UPPER CRETACEOUS).

Cycadeoidea uddeni Wieland. (Paloma, Tex.)

BINGEN SAND, ARKANSAS.

Andromeda novaecaesareae Hollick.
Andromeda parlatorii Heer.
Araucaria darlingtonensis Berry.
Cinnamomum newberryi Berry?
Colutea primordialis Heer.
Cordia apiculata (Hollick) Berry.
Dewalquea insigniformis Berry.
Dicksonia groenlandica Heer.
Ficus crassipes (Heer) Heer.
Ficus daphnogenoides (Heer) Berry.
Ficus ovatifolia Berry?
Leguminosites omphalobioides Lesquereux.
Liriodendron meekii Heer.
Manihotites georgiana Berry.
Menispermites integrifolia Berry.
Myrica cinnamomifolia Newberry.
Myrica cliffwoodensis Berry.
Myrica longa (Heer) Heer.
Podozamites marginatus Heer.
Podozamites sp. Berry.
Salix flexuosa Newberry.
Salix lesquereuxii Berry.
Sapindus morrisoni Lesquereux.
Sequoia concinna Heer.
Sequoia heterophylla Velenovsky.
Widdringtonites subtilis Heer.

ROCKY MOUNTAIN REGION.

LOWER CRETACEOUS FLORAS.

KOOTENAI FORMATION.

Abietites foliosus (Fontaine) Berry. (Mont.)
Acrostichopteris fimbriata Knowlton. (Mont.)
Adiantum montanensis Knowlton. (Mont.)
Angiopteridium canmorense Dawson. (B. C.)
Anomozamites acutilobus? Heer. (B. C.)
Anomozamites sp. Dawson. (B. C.)
Antholithes horridus (Dawson) Dawson. (B. C.)
Asplenium dicksonianum Heer. (B. C.)
Asplenium martinianum Dawson. (B. C.)
Baiera longifolia (Pomel) Heer. (B. C.)
Baieropsis sp. Dawson. (B. C.)
Carpolithes sp. Dawson. (B. C.)
Carpolithus virginiensis Fontaine. (Mont.)
Cephalotaxopsis sp. Dawson. (B. C.)
Cheiropteris williamsii Newberry. (Mont.)
Cladophlebis albertsii montanense (Fontaine) Knowlton. (Mont.)
Cladophlebis angustifolia Newberry. (Mont.)
Cladophlebis browniana (Dunker) Seward. (Mont., B. C.)
Cladophlebis constricta Fontaine. (Mont.)
Cladophlebis distans Fontaine. (Mont.)
Cladophlebis fisheri Knowlton. (Mont.)
Cladophlebis heterophylla Fontaine. (Mont.)
Cladophlebis virginiensis Fontaine. (B. C.)

Cladophlebis sp. Dawson. (B. C.)
Cycadeospermum montanense Fontaine. (Mont.)
Cyperacites sp. (Dawson) Knowlton. (B. C.)
Dicksonia montanensis Fontaine. (Mont.)
Dicksonia pachyphylla Fontaine. (Mont.)
Dicksonia sp. Dawson. (B. C.)
Dioonites borealis Dawson. (B. C.)
Dryopteris? kootaniensis Knowlton. (Mont.)
Dryopteris montanensis (Fontaine) Knowlton. (Mont.)
Equisetum lyelli Mantell. (Mont.)
Equisetum phillipsii (Dunker) Brongniart. (Mont.)
Ginkgo lepida (Heer) Heer. (B. C.)
Ginkgo nana (Dawson) Knowlton. (B. C.)
Ginkgo siberica (Heer) Heer. (B. C., Mont.)
Glyptostrobus grönlandicus Heer. (B. C.)
Laricopsis longifolia latifolia Fontaine. (Mont.)
Lycopodites? montanensis Fontaine. (Mont.)
?Nageiopsis longifolia Fontaine. (Mont.)
Nilsonia schaumbergensis (Dunker) Nathorst. (Mont.)
Oleandra arctica Heer. (Mont.)
Oleandra graminaefolia Knowlton. (Mont.)
Onychiopsis göpperti (Schenk) Berry. (Mont.)
Onychiopsis psilotoides (Stokes and Webb) Ward. (Mont.)
Osmunda dicksonioides Fontaine. (Mont.)
Pagiophyllum sp. Dawson. (B. C.)
Pecopteris montanensis Fontaine. (Mont.)
Pinus anthraciticus Dawson. (B. C.)
Pinus susquaensis Dawson. (B. C.)
Podozamites inaequilateralis (Fontaine) Berry. (Mont.)
Podozamites lanceolatus (Lindley and Hutton) F. Braun. (Mont.)
Podozamites latipennis Heer. (Mont.)
Podozamites newberryi Knowlton. (Mont.)
Protorhipis fisheri Knowlton (Mont.)
Pterophyllum montanense (Fontaine) Knowlton. (Mont.)
Sagenopteris elliptica Fontaine. (Mont.)
Sequoia acutifolia Newberry. (Mont.)
Sequoia ambigua Heer. (Mont.)
Sequoia cuneata (Newberry) Newberry. (B. C.)
Sequoia gracillima (Lesquereux) Newberry. (N. W. T.)
Sequoia gracilis Heer. (Mont.)
Sequoia reichenbachi (Geinitz) Heer. (Mont.)
Sequoia regida Heer. (Mont.)
Sequoia smittiana Heer. (Mont., B. C.)
Sphenolepidium pachyphyllum Fontaine. (B. C.)
Sphenolepidium sp. Dawson. (B. C.)
Sphenolepis kurriana (Dunker) Schenk. (Mont.)
Sphenozamites sp. Dawson. (B. C.)
Widdringtonites ramosus (Fontaine) Berry. (Mont.)
Williamsonia? sp. Dawson. (B. C.)
Zamites acutipennis Heer. (Mont., B. C.)
Zamites apertus Newberry. (Mont.)
Zamites arcticus Göppert. (Mont.)
Zamites borealis Heer. (Mont.)
Zamites montana Dawson. (B. C.)

MORRISON FORMATION, FREEZEOUT HILLS, WYO.

Araucarioxylon? obscurum Knowlton.
Cycadella beecheriana Ward.
Cycadella carbonensis Ward.
Cycadella cirrata Ward.
Cycadella compressa Ward.
Cycadella concinna Ward.
Cycadella contracta Ward.
Cycadella crepidaria Ward.
Cycadella exogena Ward.
Cycadella ferruginea Ward.
Cycadella gelida Ward.
Cycadella gravis Ward.
Cycadella jejuna Ward.
Cycadella jurassica Ward.
Cycadella knightii Ward.
Cycadella knowltoniana Ward.
Cycadella nodosa Ward.
Cycadella ramentosa Ward.
Cycadella reedii Ward.
Cycadella utopiensis (Ward) Wieland.
Cycadella verrucosa Ward.
Cycadella wyomingensis Ward.
Nilsonia nigracollensis Wieland. (Big Horn Basin, Wyo.)
Zamites arcticus Göppert. (Big Horn Basin, Wyo.)

CLOVERLY FORMATION, WYOMING.

Cladophlebis heterophylla Fontaine.
Dicksonia pachyphylla Fontaine.
Equisetum lyelli Mantell.

LAKOTA SANDSTONE.

Acrostichopteris adiantifolia (Fontaine) Berry.
Acrostichopteris pluripartita (Fontaine) Berry.
Araucarioxylon hoppertonae Knowlton.
Araucarites? cuneatus Ward.
Araucarites wyomingensis Fontaine.
Asplenium dicksonianum Heer.
Carpolithus barrensis Ward.
Carpolithus fasciculatus Fontaine.
Carpolithus foenarius Ward.
Carpolithus montanum nigrorum Ward.
Carpolithus virginiensis Fontaine.
Cephalotaxopsis magnifolia Fontaine.
Cladophlebis parva Fontaine.
Cycadeoidea aspera Ward.
Cycadeoidea cicatricula Ward.
Cycadeoidea colei Ward.
Cycadeoidea colossalis Ward.
Cycadeoidea dacotensis (McBride) Ward.
Cycadeoidea dartoni Wieland.
Cycadeoidea excelsa Ward.
Cycadeoidea formosa Ward.
Cycadeoidea furcata Ward.
Cycadeoidea heliochorea Ward.

Cycadeoidea ingens Ward.
Cycadeoidea insoleta Ward.
Cycadeoidea jenneyana Ward.
Cycadeoidea marshiana Ward.
Cycadeoidea mcbridei Ward.
Cycadeoidea minima Ward.
Cycadeoidea minnekahtensis Ward
Cycadeoidea nana Ward.
Cycadeoidea occidentalis Ward.
Cycadeoidea paynei Ward.
Cycadeoidea protea Ward.
Cycadeoidea pulcherrima Ward.
Cycadeoidea reticulata Ward.
Cycadeoidea rhombica Ward.
Cycadeoidea stillwelli Ward.
Cycadeoidea superba Ward.
Cycadeoidea turrita Ward.
Cycadeoidea wellsii Ward.
Cycadeoidea wielandi Ward.
Czekanowskia nervosa Heer.
Equisetum burchardti (Dunker) Brongniart?
Gleichenia zippei (Corda) Heer.
Glossozamites fontaineanus Ward.
?Nageiopsis angustifolia Fontaine.
?Nageiopsis longifolia Fontaine.
Nilsonia nigracollensis Wieland.
Onychiopsis brevifolia (Fontaine) Berry.
Onychiopsis göpperti (Schenk) Berry.
Onychiopsis latiloba (Fontaine) Berry.
Onychiopsis nervosa (Fontaine) Berry.
Onychiopsis psilotoides (Stokes and Webb) Ward.
Pinus susquaensis Dawson.
Pityoxylon dakotense (Knowlton) Penhallow.
Scleropteris distantifolia Fontaine.
Scleropteris rotundifolia Fontaine.
Sequoia sp. Fontaine.
Sphenopteris plurinervia Heer?
Williamsonia? phoenicopsoides Ward
Zamites borealis Heer.
Zamites brevipennis Heer.

FUSON FORMATION.

Abietites longifolius (Fontaine) Berry.
Abietites macrocarpus Fontaine.
Acrostichopteris pluripartita (Fontaine) Berry.
Athrotaxopsis grandis Fontaine.
Cephalotaxopsis magnifolia Fontaine.
Cladophlebis wyomingensis Fontaine.
Cycadeospermum rotundatum Fontaine.
Equisetum burchardti (Dunker) Brongniart.
Feistmantelia oblonga Ward.
Ficophyllum? fontainei Knowlton.
Geinitzia jenneyi Fontaine.
Leptostrobus? alatus Ward.
Matonidium althausii (Dunker) Ward.

Pecopteris borealis Brongniart.
Pecopteris geyleriana Nathorst.
Pinus susquaensis Dawson.
Quercophyllum wyomingense Fontaine.
Sapindopsis magnifolia Fontaine.
Sapindopsis variabilis Fontaine.
Sequoia gracilis Heer.
Sequoia reichenbachi (Geinitz) Heer.
Sphenolepis kurriana (Dunker) Schenk.
Ulmophyllum densinerve Fontaine.
Weichselia reticulata (Stokes and Webb) Ward.
Zamites? sp. Fontaine.

CHEYENNE SANDSTONE.

Cupressinoxylon cheyennense Penhallow.
Cycadeoidea munita Cragin.
Ficus daphnogenoides (Heer) Berry.
Laurus plutonia Heer.
Rhus uddeni Lesquereux.
Sequoia gracillima (Lesquereux) Newberry.
Sequoia sp. Knowlton.
Sterculia snowii Lesquereux.

LOWER CRETACEOUS (LAKOTA AGE?), HOVENWEEP CANYON, SOUTHWESTERN COLORADO.

Cycadospadix (?) sp. Cockerell.
Equisetum burchardti (Dunker) Brongniart.
Matonidium althausii (Dunker) Ward.
Sapindopsis variabilis Fontaine.

COMANCHE SERIES, FORMATION NOT IDENTIFIED.

Cupressinoxylon comanchense Penhallow. (Kans.)

LOWER CRETACEOUS ROCKS, FORMATIONS NOT IDENTIFIED.

Cladophlebis skagitensis Penhallow. (B. C.)
Cycadites unjiga Dawson. (B. C.)
Dioonites columbianus (Dawson) Dawson. (B. C.)
Dorstenia? sp. Penhallow. (B. C.)
Gleichenia gilbert-thompsoni Fontaine. (B. C.)
Gleichenia sp. Penhallow. (B. C.)
Myrica serrata Penhallow. (B. C.)
Nilsonia pasaytensis Penhallow. (B. C.)
Osmundites skidegatensis Penhallow. (B. C.)
Pinus sp. Penhallow. (B. C.)
Populites cyclophylla (Heer) Lesquereux. (B. C. Age doubtful.)
Quercus flexuosa Newberry. (B. C. Age doubtful.)
Sagenopteris elliptica Fontaine. (B. C.)
Sagenopteris nilsoniana (Brongniart) Ward. (B. C.)
Sagenopteris oblongifolia Penhallow. (B. C.)
Taxoxylon sp. Dawson. (B. C.)
Zamites crassinervis Fontaine. (B. C.)
Zamites tenuinervis Fontaine. (B. C.)

UPPER CRETACEOUS FLORAS.

DAKOTA SANDSTONE.

Abietites cretacea Newberry. (N. Mex.)
Abietites ernestinae Lesquereux. (Nebr.)
Acerites multiformis Lesquereux. (Kans.)

Alismacites dakotensis Lesquereux. (Kans.)
Alnites crassus Lesquereux. (Minn.)
Alnites grandifolia (Newberry) Newberry. (Nebr.)
Ampelophyllum attenuatum Lesquereux. (Kans.)
Ampelophyllum firmum Lesquereux. (Kans.)
Ampelophyllum ovatum Lesquereux. (Nebr.)
Anacardites antiquus Lesquereux. (Kans.)
Andromeda cretacea Lesquereux. (Kans.)
Andromeda dakotana Knowlton and Cockerell. (Kans.)
Andromeda linifolia Lesquereux. (Kans.)
Andromeda parlatorii Heer. (Kans., Nebr., Minn.)
Andromeda parlatorii longifolia Lesquereux. (Kans.)
Andromeda pfaffiana Heer. (Wyo., Kans.)
Andromeda snowii Lesquereux. (Kans.)
Andromeda tenuinervis Lesquereux. (Kans.)
Andromeda wardiana Lesquereux. (Kans.)
Anisophyllum semialatum (Lesquereux) Lesquereux. (Nebr.)
Anona cretacea Lesquereux. (Kans.)
Apeibopsis cyclophylla Lesquereux. (Kans.)
Apocynophyllum sordidum Lesquereux. (Kans.)
Aralia berberidifolia Lesquereux. (Kans.)
Aralia concreta Lesquereux. (Kans.)
Aralia formosa Heer. (Colo.)
Aralia groenlandica Heer. (Kans.)
Aralia masoni Lesquereux. (Kans.)
Aralia quinquepartita Lesquereux. (Kans.)
Aralia radiata Lesquereux. (Kans.)
Aralia saportana Lesquereux. (Wyo.)
Aralia saportanea deformata Lesquereux. (Kans.)
Aralia subemarginata Lesquereux. (Kans.)
Aralia tenuinervis Lesquereux. (Kans.)
Aralia towneri Lesquereux. (Kans.)
Aralia wellingtoniana Lesquereux. (Kans.)
Araliopsoides cretacea (Newberry) Berry. (Nebr., Kans.)
Araliopsoides cretacea dentata (Lesquereux) Berry. (Kans.)
Araliopsoides cretacea salisburiaefolia (Lesquereux) Berry. (Kans., Nebr., S. Dak.)
Araucaria spatulata Newberry. (Nebr.)
Arisaema cretacea Lesquereux. (Kans.)
Aristolochites dentata Heer. (Nebr.)
Artocarpidium cretaceum Ettingshausen. (Kans.)
Aspidiophyllum dentatum Lesquereux. (Kans.)
Aspidiophyllum platanifolium Lesquereux. (Kans.)
Aspidiophyllum trilobatum Lesquereux. (Kans.)
Asplenium dicksonianum Heer. (Kans., S. Dak.)
Benzoin masoni (Lesquereux) Knowlton. (Kans.)
Benzoin venustum (Lesquereux) Knowlton. (Kans.)
Betula beatriciana Lesquereux. (Kans.)
Betulites denticulata Heer. (Nebr.)
Betulites populifolius Lesquereux. (Kans.)
Betulites rugosus Lesquereux. (Kans.)
Betulites snowii Lesquereux. (Kans.)
Betulites, stipules of Lesquereux. (Kans.)
Betulites westii Lesquereux. (Kans.)
Betulites westii crassus Lesquereux. (Kans.)

Betulites westii cuneatus Lesquereux. (Kans.)
Betulites westii grewiopsideus Lesquereux. (Kans.)
Betulites westii inaequilateralis Lesquereux. (Kans.)
Betulites westii lanceolatus Lesquereux. (Kans.)
Betulites westii latifolius Lesquereux. (Kans.)
Betulites westii multinervis Lesquereux. (Kans.)
Betulites westii oblongus Lesquereux. (Kans.)
Betulites westii obtusus Lesquereux. (Kans.)
Betulites westii populoides Lesquereux. (Kans.)
Betulites westii quadratifolius Lesquereux. (Kans.)
Betulites westii reniformis Lesquereux. (Kans.)
Betulites westii rhomboidalis Lesquereux. (Kans.)
Betulites westii rotundatus Lesquereux. (Kans.)
Betulites westii subintegrifolius Lesquereux. (Kans.)
Brachyphyllum macrocarpum Newberry. (Kans.)
Bromelia? tenuifolia Lesquereux. (Kans.)
Bumelia? rhomboidea Lesquereux. (Kans.)
Callistemophyllum heerii Ettingshausen. (Kans.)
Calycites sp. Lesquereux. (Kans.)
Carpites coniger Lesquereux. (Kans.)
Carpites cordiformis Lesquereux. (Kans.)
Carpites dakotensis Knowlton. (Kans.)
Carpites liriophylli Lesquereux. (Colo.)
Carpites obovatus Lesquereux. (Kans.)
Carpites tiliaceus (Heer) Lesquereux. (Kans.)
Carpites (?) sp. Lesquereux. (Kans.(?))
Cassia polita Lesquereux. (Kans.)
Cassia problematica Lesquereux. (Kans.)
Caudex spinosus (Lesquereux) Lesquereux. (Kans.)
Celastrophyllum acutidens Fontaine. (S. Dak.)
Celastrophyllum crassipes Lesquereux. (Kans.)
Celastrophyllum cretaceum Lesquereux. (Kans.)
Celastrophyllum decurrens Lesquereux. (Kans.)
Celastrophyllum ensifolium (Lesquereux) Lesquereux. (Wyo.)
Celastrophyllum myrsinoides Lesquereux. (Kans.)
Celastrophyllum obliquum Knowlton. (Kans.)
Cinnamomum ellipsoideum Saporta and Marion. (Kans.)
Cinnamomum heerii (Lesquereux) Lesquereux. (Kans.)
Cinnamomum marioni Lesquereux. (Kans.)
Cinnamomum membranaceum (Lesquereux) Hollick. (Kans.)
Cinnamomum newberryi Berry. (Kans.)
Cinnamomum scheuchzeri Heer. (Kans., Minn.)
Cissites acerifolius Lesquereux. (Kans.)
Cissites acuminatus Lesquereux. (Kans.)
Cissites acutiloba Hollick. (Kans.)
Cissites affinis (Lesquereux) Lesquereux. (Kans., Wyo.)
Cissites alatus Lesquereux. (Kans.)
Cissites brownii Lesquereux. (Kans., Minn.)
Cissites dentato-lobatus Lesquereux. (Kans.)
Cissites formosus Heer. (Kans.)
Cissites harkerianus (Lesquereux) Lesquereux. (Kans.)
Cissites heerii Lesquereux. (Kans.)
Cissites ingens Lesquereux. (Kans., S. Dak.)
Cissites ingens parvifolia Lesquereux. (Kans.)

Cissites insignis Heer. (Nebr.)
Cissites obtusilobus Lesquereux. (Kans.)
Cissites platanoidea Hollick. (Kans.)
Cissites populoides Lesquereux. (Kans.)
Cissus browniana Lesquereux. (Minn.)
Citrophyllum aligerum (Lesquereux) Berry. (Kans.)
Colutea primordialis Heer. (Kans.)
Cornus platyphylloides Lesquereux. (Kans.)
Cornus praecox Lesquereux. (Kans.)
Crataegus aceroides Lesquereux. (Kans.)
Crataegus atavina Heer. (Minn.)
Crataegus lacoei Lesquereux. (Kans.)
Crataegus laurenciana Lesquereux. (Kans.)
Crataegus tenuinervis Lesquereux. (Kans.)
Credneria? microphylla Lesquereux. (Kans.)
Cyatheites? nebraskana (Heer) Knowlton. (Kans.)
Cycadeospermum columnare Lesquereux. (Kans.)
Cycadeospermum lineatum Lesquereux. (Kans.)
Cycadites pungens Lesquereux. (Kans.)
Dammarites caudatus Lesquereux. (Kans.)
Dammarites emarginatus (Lesquereux) Lesquereux. (Kans.)
Daphnophyllum angustifolium Lesquereux. (Kans.)
Daphnophyllum dakotense Lesquereux. (Kans.)
Dewalquea dakotensis Lesquereux. (Kans.)
Dewalquea primordialis Lesquereux. (Minn.)
Dioscorea? cretacea Lesquereux. (Kans.)
Diospyros apiculata Lesquereux. (Kans.)
Diospyros? celastroides Lesquereux. (Kans.)
Diospyros lesquereuxi Knowlton and Cockerell. (Kans.)
Diospyros nebraskana Knowlton and Cockerell. (Nebr.)
Diospyros primaeva Heer. (Kans., Nebr.)
Diospyros pseudoanceps Lesquereux. (Kans., Minn.)
Diospyros steenstrupi? Heer. (Kans.)
Elaeodendron speciosum Lesquereux. (Kans.)
Embothrites? daphneoides (Lesquereux) Lesquereux. (Kans.)
Encephalartos cretaceus Lesquereux. (Kans.)
Eremophyllum fimbriatum (Lesquereux) Lesquereux. (Nebr.)
Eucalyptus dakotensis Lesquereux. (Kans.)
Eucalyptus geinitzi (Heer) Heer. (Kans.)
Eucalyptus gouldii Ward. (Kans.)
Eugenia primaeva Lesquereux. (Kans.)
Fagus cretacea Newberry. (Kans.)
Fagus orbiculata Lesquereux. (Kans.)
Fagus polycladus Lesquereux. (Nebr., Kans.)
Ficus austiniana Lesquereux. (Minn.)
Ficus crassipes (Heer) Heer. (Kans.)
Ficus daphnogenoides (Heer) Berry. (Nebr.)
Ficus deflexa Lesquereux. (Kans.)
Ficus distorta Lesquereux. (Kans.)
Ficus glascoeana Lesquereux. (Kans.)
Ficus? halliana Lesquereux. (Minn.)
Ficus harkeriana Knowlton. (Kans.)
Ficus inaequalis Lesquereux. (Kans.)
Ficus? kansasensis Knowlton and Cockerell. (Kans.)

Ficus krausiana Heer. (Kans.)
Ficus lanceolato-acuminata Ettingshausen. (Kans.)
Ficus laurophylla Lesquereux. (Nebr., Kans.)
Ficus leonis Knowlton and Cockerell. (Kans.)
Ficus lesquereuxii Knowlton. (Kans.)
Ficus magnoliaefolia Lesquereux. (Colo., Kans.)
Ficus melanophylla Lesquereux. (Kans.)
Ficus mudgei Lesquereux. (Kans.)
Ficus neurocarpa Hollick. (Kans.)
Ficus praecursor Lesquereux. (Kans.)
Ficus primordialis Heer. (Nebr.)
Ficus undulatiformis Knowlton and Cockerell. (Kans.)
Ficus, fruits of, Lesquereux. (Kans.)
Flabellaria? minima Lesquereux. (Kans.)
Galla quercina Lesquereux. (Kans.)
Geinitzia sp. Heer. (Iowa.)
Gleichenia kurriana Heer. (Kans.; and formation doubtful, Mont.)
Gleichenia nordenskiöldi Heer. (Kans.)
Grewiopsis aequidentata Lesquereux. (Kans.)
Grewiopsis flabellaria (Lesquereux) Knowlton. (Kans.)
Grewiopsis mudgei Lesquereux. (Kans.)
Hamamelites? cordatus Lesquereux. (Kans.)
Hamamelites quadrangularis (Lesquereux) Lesquereux. (Kans.?)
Hamamelites quercifolius Lesquereux. (Kans.)
Hamamelites tenuinervis Lesquereux. (Kans.)
Hedera cretacea Lesquereux. (Kans.)
Hedera decurrens Lesquereux. (Kans.)
Hedera microphylla Lesquereux. (Kans.)
Hedera orbiculata (Heer) Lesquereux. (Kans.)
Hedera ovalis Lesquereux. (Kans.)
Hedera platanoidea Lesquereux. (Kans.)
Hedera schimperi Lesquereux. (Kans.)
Hymenaea dakotana Lesquereux. (Kans.)
Hymenophyllum cretaceum Lesquereux. (Wyo.)
Ilex armata Lesquereux. (Kans.)
Ilex borealis Heer. (Kans.)
Ilex dakotensis Lesquereux. (Kans.)
Ilex masoni Lesquereux. (Kans.)
Ilex papillosa Lesquereux. (Kans.)
Ilex scudderi Lesquereux. (Kans.)
Ilex strangulata Lesquereux. (Colo.)
Inga cretacea Lesquereux. (Kans.)
Inolepis? sp. Lesquereux. (Kans.)
Juglandites ellsworthianus Lesquereux. (Kans.)
Juglandites lacoei Lesquereux. (Kans.)
Juglandites primordialis Lesquereux. (Kans.)
Juglandites sinuatus Lesquereux. (Kans.)
Juglans arctica Heer. (Kans.)
Juglans crassipes Heer. (Kans.)
Juglans debeyana (Heer) Lesquereux. (Nebr., Minn.)
Laurus antecedens Lesquereux. (Kans.)
Laurus atanensis Berry. (Kans.)
Laurus hollae Heer. (Kans.)
Laurus knowltoni Lesquereux. (Kans.)

Laurus macrocarpa Lesquereux. (Nebr.)
Laurus microcarpa Lesquereux. (Kans.)
Laurus modesta Lesquereux. (Colo.)
Laurus nebrascensis (Lesquereux) Lesquereux. (Nebr., Kans., Minn.)
Laurus plutonia Heer. (Kans., Minn.)
Laurus proteaefolia Lesquereux. (Colo., Kans., Wyo.)
Laurus teliformis Lesquereux. (Kans.)
Leguminosites constrictus Lesquereux. (Kans.)
Leguminosites convolutus Lesquereux. (Kans.)
Leguminosites coronilloides Heer. (Kans.)
Leguminosites cultriformis Lesquereux. (Locality?)
Leguminosites dakotensis Lesquereux. (Kans.)
Leguminosites hymenophyllus Lesquereux. (Kans.)
Leguminosites insularis Heer. (Kans.)
Leguminosites omphalobioides Lesquereux. (Kans.)
Leguminosites phaseolites? Heer. (Kans.)
Leguminosites podogonialis Lesquereux. (Kans.)
Leguminosites truncatus Knowlton. (Kans.)
Liquidambar obtusilobatus (Heer) Hollick. (Nebr., Kans.)
Liriodendron acuminatum Lesquereux. (Kans.)
Liriodendron acuminatum bilobatum Lesquereux. (Kans.)
Liriodendron giganteum Lesquereux. (Kans.)
Liriodendron giganteum cruciforme (Lesquereux) Lesquereux. (Kans.)
Liriodendron intermedium Lesquereux. (Kans.)
Liriodendron marcouanum (Heer) Knowlton. (Nebr., Kans., Minn.)
Liriodendron meekii Heer. (Nebr.)
Liriodendron pinnatifidum Lesquereux. (Kans.)
Liriodendron semialatum Lesquereux. (Kans.)
Liriodendron snowii Lesquereux. (Kans.)
Liriodendron wellingtonii Lesquereux. (Kans.)
Liriophyllum beckwithii Lesquereux. (Colo.)
Liriophyllum obcordatum Lesquereux. (Colo., Kans.)
Liriophyllum populoides Lesquereux. (Colo., Kans.)
Lomatia? saportanea (Lesquereux) Lesquereux. (Kans.)
Lomatia? saportanea longifolia Lesquereux. (Colo.)
Lygodium trichomanoides Lesquereux. (Kans.)
Macclintockia cretacea Heer. (Kans.)
Magnolia alternans Heer. (Kans., Nebr., Minn.)
Magnolia amplifolia Heer. (Kans.)
Magnolia boulayana Lesquereux. (Kans.)
Magnolia capellinii Heer. (Kans., Nebr.)
Magnolia lacoeana Lesquereux. (Kans.)
Magnolia obtusata Heer. (Kans.)
Magnolia palaeopetala Hollick. (Kans.)
Magnolia pseudoacuminata Lesquereux. (Kans., Wyo.)
Magnolia speciosa Heer. (Colo., Kans.)
Magnolia tenuifolia Lesquereux. (Nebr., Kans.)
Magnolia sp. Lesquereux. (Colo.)
Malapoenna cretacea (Lesquereux) Knowlton. (Kans.)
Malapoenna falcifolia (Lesquereux) Knowlton. (Kans.)
Menispermites acutilobus Lesquereux. (Kans.)
Menispermites cyclophyllus Lesquereux. (Kans.)
Menispermites grandis Lesquereux. (Kans.)
Menispermites menispermifolius (Lesquereux) Knowlton. (Nebr.)

Menispermites obtusiloba Lesquereux. (Kans.)
Menispermites ovalis Lesquereux. (Kans.)
Menispermites populifolius Lesquereux. (Kans.)
Menispermites rugosus Lesquereux. (Kans.)
Menispermites salinae (Lesquereux) Knowlton. (Kans.)
Myrica aspera Lesquereux. (Kans., Wyo.)
Myrica dakotensis Lesquereux. (Kans.)
Myrica emarginata Heer. (Kans.)
Myrica longa (Heer) Heer. (Wyo., Kans.)
Myrica obliqua Knowlton. (Kans.)
Myrica obtusa Lesquereux. (Kans.)
Myrica schimperi Lesquereux. (Kans.)
Myrica? semina Lesquereux. (Nebr.)
Myrica sternbergii Lesquereux. (Kans.)
Myrica? trifoliata Newberry. (N. Mex.)
Myrsine crassula Knowlton and Cockerell. (Kans.)
Negundo acutifolia (Lesquereux) Pax. (Nebr.)
Nyssa snowiana Lesquereux. (Kans.)
Nyssa vetusta Newberry. (Nebr.)
Oreodaphne cretacea Lesquereux. (Kans.)
Palaeocassia laurinea Lesquereux. (Kans.)
Paliurus anceps Lesquereux. (Kans.)
Paliurus cretaceus Lesquereux. (Kans.)
Paliurus obovatus Lesquereux. (Kans.)
Paliurus ovalis Dawson. (Kans.)
Parrotia canfieldi Lesquereux. (Kans.)
Parrotia grandidentata Lesquereux. (Kans.)
Parrotia? winchelli Lesquereux. (Kans.)
Persea hayana Lesquereux. (Kans.)
Persea leconteana (Lesquereux) Lesquereux. (Kans.)
Persea schimperi Lesquereux. (Kans.)
Persea sternbergii (Lesquereux) Lesquereux. (Kans.)
Persoonia lesquereuxii Knowlton. (Kans.)
Phaseolites formus Lesquereux. (Kans.)
Phragmites cretaceus (Lesquereux) Lesquereux. (Kans.)
Phyllites amissus Lesquereux. (Kans.)
Phyllites amphorus Lesquereux. (Nebr.)
Phyllites aristolochiaeformis Lesquereux. (Kans.)
Phyllites betulaefolius Lesquereux. (Kans., Nebr.)
Phyllites celatus Lesquereux. (Kans.)
Phyllites durescens Lesquereux. (Kans.)
Phyllites erosus Lesquereux. (Kans.)
Phyllites ilicifolius Lesquereux. (Kans.)
Phyllites innectens Lesquereux. (Kans.)
Phyllites lacoei Lesquereux. (Kans.)
Phyllites laurencianus Lesquereux. (Kans.)
Phyllites perplexus Lesquereux. (Kans.)
Phyllites rhoifolius Lesquereux. (Nebr.)
Phyllites rhomboideus (Lesquereux) Lesquereux. (Nebr.)
Phyllites snowii Lesquereux. (Kans.)
Phyllites stipulaeformis Lesquereux. (Kans.)
Phyllites umbonatus Lesquereux. (Nebr.)
Phyllites vanonae Heer. (Nebr., Kans.)
Phyllites zamiaeformis Lesquereux. (Kans.)

Phyllites sp. Lesquereux. (Kans.)
Pinus quenstedti Heer. (Kans.)
Pinus sp. Lesquereux. (Minn.)
Platanus cissoides Lesquereux. (Kans., S. Dak.)
Platanus diminutiva Lesquereux. (Nebr.?)
Platanus heerii Lesquereux. (Kans.)
Platanus latior (Lesquereux) Knowlton. (Nebr., Minn.)
Platanus latior grandidentata (Lesquereux) Knowlton. (Kans.)
Platanus latior integrifolia (Lesquereux) Knowlton. (Kans.)
Platanus latior subintegrifolia (Lesquereux) Knowlton. (Kans.)
Platanus? newberryana Heer. (Nebr.)
Platanus obtusiloba Lesquereux. (Nebr., Kans.)
Platanus sp. Cragin. (Kans. Formation doubtful.)
Podozamites haydenii (Lesquereux) Lesquereux. (Nebr.)
Podozamites knowltoni Berry. (Kans.)
Podozamites lanceolatus (Lindley and Hutton) F. Braun. (Kans.)
Podozamites oblongus Lesquereux. (Kans.)
Podozamites stenopus Lesquereux. (Kans.)
Populites cyclophylla (Heer) Lesquereux. (Nebr., Kans., Minn.)
Populites elegans Lesquereux. (Nebr., Kans., Minn.)
Populites lancastriensis (Lesquereux) Lesquereux. (Nebr., Minn.)
Populites litigiosus (Heer) Lesquereux. (Nebr., Kans., Minn.)
Populites microphyllus Lesquereux. (Kans.)
Populites sternbergii Lesquereux. (Kans.)
Populites winchelli Lesquereux. (Minn.)
Populus aristolochioides Lesquereux. (Kans.
Populus berggreni Heer. (Kans., Minn.)
Populus dakotana Cockerell. (Nebr., Wyo.)
Populus elliptica Newberry. (Nebr.)
Populus harkeriana Lesquereux. (Kans.)
Populus hyperborea Heer. (Kans.)
Populus kansaseana Lesquereux. (Kans.)
Populus leuce (Rossmässler) Unger. (Kans.?)
Populus litigiosus denticulata (Lesquereux) Knowlton. (Kans.?)
Populus microphylla Newberry. (Nebr.)
Populus stygia Heer. (Kans.)
Proteoides acuta Heer. (Nebr., Kans.)
Proteoides grevilleaeformis Heer. (Nebr., Kans.)
Proteoides lancifolius Heer. (Kans.)
Protophyllocladus subintegrifolius (Lesquereux) Berry. (Locality?)
Protophyllum crassum Lesquereux. (Kans.)
Protophyllum crednerioides Lesquereux. (Kans., Minn.)
Protophyllum crenatum Knowlton. (Kans.)
Protophyllum denticulatum Lesquereux. (Kans.)
Protophyllum dimorphum Lesquereux. (Kans.)
Protophyllum haydenii (Lesquereux) Lesquereux. (Kans.)
Protophyllum integerrimum Lesquereux. (Minn.)
Protophyllum leconteanum (Lesquereux) Lesquereux. (Kans.)
Protophyllum minus Lesquereux. (Kans.)
Protophyllum? mudgii (Lesquereux) Knowlton. (Kans.)
Protophyllum multinerve (Lesquereux) Lesquereux. (Kans.)
Protophyllum nebrascensis Lesquereux. (Nebr.)
Protophyllum praestans Lesquereux. (Kans.)
Protophyllum pseudospermoides Lesquereux. (Kans.)

Protophyllum pterospermifolium Lesquereux. (Kans.)
Protophyllum quadratum (Lesquereux) Lesquereux. (Kans.)
Protophyllum querciforme Hollick. (Kans.)
Protophyllum rugosum (Lesquereux) Lesquereux. (Kans.)
Protophyllum sternbergii (Lesquereux) Lesquereux. (Kans.)
Protophyllum? trilobatum Lesquereux. (Kans.)
Protophyllum undulatum Lesquereux. (Kans.)
Prunus antecedens Lesquereux. (Kans.)
Prunus cretacea Lesquereux. (Nebr.)
Ptenostrobus nebrascensis Lesquereux. (N. Dak.? Formation doubtful.)
Pteris dakotensis Lesquereux. (Kans.)
Pterospermites longeacuminatus Lesquereux. (Kans.)
Pterospermites modestus Lesquereux. (Kans.)
Pyrus cretacea Newberry. (Kans.)
Quercus alnoides Lesquereux. (Kans.)
Quercus antiqua Newberry. (Utah. Formation doubtful.)
Quercus dakotensis Lesquereux. (Kans.)
Quercus dryophyllopsis Trelease. (Kans.)
Quercus eamesi Trelease. (Nebr.)
Quercus? ellsworthianus Lesquereux. (Kans., Nebr.)
Quercus glascoena Lesquereux. (Kans.)
Quercus hexagona Lesquereux. (Nebr., Kans.)
Quercus hieracifolia (Debey) Hosius and von d. Marck. (Kans.)
Quercus holmesii (Lesquereux) Lesquereux. (Colo.)
Quercus hosiana Lesquereux. (Kans.)
Quercus kanseana (Lesquereux) Knowlton. (Kans.)
Quercus morrisoniana Lesquereux. (Colo.)
Quercus newberryi Trelease. (Nebr.)
Quercus poranoides Lesquereux. (Kans.)
Quercus prae-undulata Trelease. (Utah. Formation doubtful.)
Quercus primordialis Lesquereux. (Kans.)
Quercus rhamnoides Lesquereux. (Kans.)
Quercus spurio-ilex Knowlton. (Kans.)
Quercus suspecta Lesquereux. (Kans.)
Quercus wardiana Lesquereux. (Kans., S. Dak.)
Rhamnites apiculatus Lesquereux. (Kans.)
Rhamnus inaequilateralis Lesquereux. (Kans.)
Rhamnus kansensis Knowlton and Cockerell. (Kans.)
Rhamnus mudgei Lesquereux. (Kans.)
Rhamnus revoluta Lesquereux. (Kans.)
Rhamnus similis Lesquereux. (Kans.)
Rhamnus tenax Lesquereux. (Kans.)
Rhus powelliana Lesquereux. (Kans., Wyo.)
Rhus uddeni Lesquereux. (Kans.)
Rhus? westii Knowlton. (Kans.)
Salix deleta Lesquereux. (Kans.)
Salix flexuosa Newberry. (Kans., Nebr.)
Salix foliosa Newberry. (N. Mex.)
Salix hayei Lesquereux. (Kans.)
Salix meekii Newberry. (Nebr., Kans.)
Salix nervillosa Heer. (Nebr.)
Salix siouxiana Knowlton and Cockerell. (Nebr.)
Salix sp. Lesquereux. (Kans.)
Sapindus diversifolius Lesquereux. (Kans.)
Sapindus morrisoni Lesquereux. (Kans., Colo., Minn.)

Sapotacites haydenii Heer. (Nebr.)
Sapotacites knowltoni Berry. (Kans.)
Sassafras acutilobum Lesquereux. (Kans., Wyo.)
Sassafras cretaceum grossedentatum Lesquereux. . (Kans.)
Sassafras dissectum Lesquereux. (Kans.)
Sassafras dissectum symmetricum Hollick. (Kans.)
Sassafras giganteum Gould. (Kans.?)
Sassafras latilobum (Newberry) Knowlton. (Nebr., Kans., Wyo.)
Sassafras mudgii Lesquereux. (Kans., S. Dak.)
Sassafras papillosum Lesquereux. (Kans.)
Sassafras platanoides Lesquereux. (Kans.)
Sassafras primordiale Lesquereux. (Kans.)
Sassafras recurvatum Lesquereux. (Kans.)
Sassafras subintegrifolium Lesquereux. (Kans.)
Sassafras sp. Knowlton. (Kans.)
Sclerotites sp. (Lesquereux) Knowlton. (Kans.)
Sequoia condita Lesquereux. (Kans.)
Sequoia fastigiata (Sternberg) Heer. (Kans.)
Sequoia formosa Lesquereux. (Nebr.)
Sequoia gracillima (Lesquereux) Newberry. (Iowa, N. Mex.)
Sequoia reichenbachi (Geinitz) Heer. (Kans.)
Sequoia winchellii Lesquereux. (Minn.)
Smilax grandifolia-cretacea Lesquereux. (Kans.)
Smilax kansana Cockerell. (Kans.)
Sphaerites problematicus (Knowlton) Meschinelli. (Kans.)
Sphenopteris corrugata Newberry. (Nebr., Kans.)
Sterculia aperta Lesquereux. (Colo., Kans.)
Sterculiaineariloba Lesquereux. (Kans.)
Sterculia lugubris Lesquereux. (Colo.)
Sterculia mucronata Lesquereux. (Kans.)
Sterculia reticulata Lesquereux. (Kans.)
Sterculia snowii Lesquereux. (Kans., N. Mex.)
Sterculia snowii disjuncta Lesquereux. (Kans.)
Sterculia tripartita (Lesquereux) Knowlton. (Kans.)
Tumion oblanceolatum (Lesquereux) Knowlton. (Colo.)
Viburnites crassus Lesquereux. (Kans.)
Viburnites evansanus Ward. (S. Dak.)
Viburnites masoni Lesquereux. (Kans.)
Viburnum ellsworthianum Lesquereux. (Kans.)
Viburnum grewiopsideum Lesquereux. (Kans.)
Viburnum inaequilaterale Lesquereux. (Kans.)
Viburnum lesquereuxii Ward. (Kans.)
Viburnum lesquereuxii commune Lesquereux. (Kans.)
Viburnum lesquereuxii cordifolium Lesquereux. (Kans.)
Viburnum lesquereuxii lanceolatum Lesquereux. (Kans.)
Viburnum lesquereuxii latius Lesquereux. (Kans.)
Viburnum lesquereuxii longifolium Lesquereux. (Kans.)
Viburnum lesquereuxii rotundifolium Lesquereux. (Kans.)
Viburnum lesquereuxii tenuifolium Lesquereux. (Kans.)
Viburnum robustum Lesquereux. (Kans.)
Viburnum sphenophyllum Knowlton. (Kans.)
Williamsonia elocata Lesquereux. (Kans.)
Zamites sp. Lesquereux. (Kans.)
Zizyphus dakotensis Lesquereux. (Kans.)
Zonarites digitatus (Brongniart) Sternberg. (Kans.)

WOODBINE SAND, ARTHURS BLUFF, LAMAR COUNTY, TEX.

Andromeda novaecaesareae Hollick.
Andromeda snowii Lesquereux.
Aralia wellingtoniana Lesquereux.
Aralia wellingtoniana vaughanii Knowlton.
Benzoin venustum (Lesquereux) Knowlton.
Brachyphyllum macrocarpum formosum Berry.
Cinnamomum membranaceum (Lesquereux) Hollick.
Colutea primordialis Heer.
Cornophyllum vetustum Newberry.
Diospyros primaeva Heer.
Eucalyptus geinitzi (Heer) Heer.
Ficus daphnogenoides (Heer) Berry.
Laurophyllum minus Newberry.
Laurus plutonia Heer.
Liriodendron quercifolium Newberry.
Magnolia boulayana Lesquereux. (Arthurs Bluff?)
Magnolia speciosa Heer.
Malapoenna falcifolia (Lesquereux) Knowlton.
Myrica emarginata Heer.
Oreodaphne alabamensis Berry.
Palaeocassia laurinea Lesquereux.
Podozamites lanceolatus (Lindley and Hutton) F. Braun.
Populus harkeriana Lesquereux.
Rhamnus tenax Lesquereux.
Rhus redditiformis Berry.
Sapindus morrisoni Lesquereux.
Sterculia lugubris Lesquereux. (Arthurs Bluff?)
Viburnum robustum Lesquereux.
Zizyphus lamarensis Berry.

BEAR RIVER FORMATION, WESTERN WYOMING.

Chara stantoni Knowlton.

COLORADO GROUP.

FRONTIER FORMATION, WYOMING.

Anemia fremonti Knowlton.
Aralia veatchii Knowlton.
Asplenium occidentale Knowlton.
Cinnamomum hesperium Knowlton.
Cinnamomum? sp. Knowlton.
Dennstaedtia? fremonti (Hall) Knowlton.
Dewalquea pulchella Knowlton.
Dryandroides lanceolata Knowlton.
Dryopteris coloradensis Knowlton.
Equisetum? sp. Knowlton.
Ficus fremonti Knowlton.
Ficus? sp. Knowlton.
Microtaenia paucifolia (Hall) Knowlton.
Microtaenia variabilis Knowlton.
Myrica nervosa Knowlton.
Phyllites dentata Knowlton.
Phyllites ficifolia Knowlton.
Phyllites sp. Knowlton.

Quercus stantoni Knowlton.
Salix cumberlandensis Knowlton.
Salix frontierensis Knowlton.
Smilax? coloradensis Knowlton.
Staphylea? fremonti Knowlton.
Tapeinidium undulatum (Hall) Knowlton.

GRANEROS FORMATION, NEBRASKA.

Ovularites barbouri Whitford.
Puccinites whitfordi Knowlton.

MILL CREEK FORMATION, MILL CREEK, BRITISH COLUMBIA.

Alnites insignis Dawson.
Aralia rotundata Dawson.
Aralia westoni Dawson.
Aralia sp. Dawson.
Asplenium albertum Dawson.
Cinnamomum canadense Dawson.
Cissites affinis (Lesquereux) Lesquereux.
Cissites affinis ampla (Dawson) Knowlton.
Dicksonia munda Dawson.
Ficus daphnogenoides (Heer) Berry.
Gleichenia gracilis Heer.
Gleichenia kurriana Heer.
Juglandites cretacea Dawson.
Laurus crassinervis Dawson.
Macclintockia cretacea Heer.
Magnolia magnifica Dawson.
Paliurus ovalis Dawson.
Williamsonia recentior Dawson.

MONTANA GROUP.

EAGLE SANDSTONE, NEAR VIRGELLE, MONT.

Ficus missouriensis Knowlton.
Juglans? missouriensis Knowlton.
Laurus? sp. Knowlton.
Liriodendron alatum Hollick.
Platanus? wardii Knowlton.
Protophyllocladus polymorphus (Lesquereux) Berry.
Quercus? montanensis Knowlton.

PIERRE SHALE, CANADA.

Abietites tyrrellii Dawson.
Betula sp. Dawson.
Hicoria (Dawson sp.) Knowlton. (Formation doubtful.)
Populus sp. Dawson.
Sequoia sp. Dawson.
Ulmus? sp. Dawson.

MESAVERDE FORMATION.

Andromeda sp. Lesquereux. (Wyo.)
Anemia elongata (Newberry) Knowlton. (Wyo.)
Asimina eocenica Lesquereux. (Wyo.)
Asplenium neomexicanum Knowlton. (Wyo.)
Asplenium tenellum Knowlton. (Wyo.)

Asplenium wyomingense Knowlton. (Wyo.)
Brachyphyllum macrocarpum Newberry. (Wyo.)
Castalia? duttoniana Knowlton. (Wyo.)
Celastrus sp. Knowlton. (Wyo.)
Cornus emmonsii Ward. (Wyo.)
Cornus studeri? Heer. (Wyo. Formation doubtful.)
Cunninghamites? sp. Knowlton. (Wyo.)
Cyperacites sp. Knowlton. (Wyo.)
Diospyros brachysepala Al. Braun. (Wyo. Formation doubtful.)
Dryophyllum bruneri Ward. (Wyo.)
Dryophyllum crenatum Lesquereux. (Wyo.)
Dryophyllum falcatum Ward. (Wyo.)
Dryophyllum subfalcatum Lesquereux. (Wyo.)
Ficus asarifolia Lesquereux. (Wyo.)
Ficus asarifolia minor Lesquereux. (Wyo.)
Ficus coloradensis Cockerell. (Wyo.?)
Ficus dalmatica Ettingshausen. (Wyo.)
Ficus firma Knowlton. (Wyo.)
Ficus hesperia Knowlton. (Wyo.)
Ficus incompleta Knowlton. (Wyo.)
Ficus planicostata Lesquereux. (Wyo.)
Ficus populoides Knowlton. (Wyo.)
Ficus problematica Knowlton. (Wyo.)
Ficus regularis Knowlton. (Wyo.)
Ficus rhamnoides Knowlton.)Wyo.)
Ficus speciosissima Ward. (Wyo.)
Ficus squarrosa Knowlton. (Wyo.)
Ficus tiliaefolia (Al. Braun) Heer. (Wyo.)
Ficus trinervis Knowlton. (Wyo. Formation doubtful.)
Ficus wardii Knowlton. (Wyo.)
Fucus lignitum Lesquereux. (Wyo.)
Geinitzia biforma (Lesquereux) Knowlton. (Wyo.)
Geinitzia formosa Heer. (Wyo., Colo.)
Ginkgo laramiensis Ward. (Wyo.)
Gleichenia pulchella Knowlton. (Wyo.)
Glyptostrobus? sp. Knowlton. (Wyo.)
Grewiopsis cleburni Lesquereux. (Wyo.)
Halymenites major Lesquereux. (Common.)
Laurus praestans Lesquereux. (Wyo.)
Lemna? bullata Lesquereux. (Wyo.)
Liriodendron laramiense Ward. (Wyo.)
Lycopodium lesquereuxiana Knowlton. (Wyo.)
Magnolia pulchra Ward. (Wyo.)
Malapoenna sp. Knowlton. (Wyo.)
Marsilea attenuata (Lesquereux) Hollick. (Wyo.)
Myrica torreyi Lesquereux. (Colo., Wyo.)
Myrica? sp. Knowlton. (Wyo.)
Nelumbo intermedia Knowlton. (Wyo.)
Nelumbo? sp. Knowlton. (Wyo.)
Ottelia americana Lesquereux. (Wyo.)
?Pinus quenstedti Heer. (Wyo.)
Pistia corrugata Lesquereux. (Wyo.)
Populus knightii Knowlton. (Wyo.)
Populus melanarioides Lesquereux. (Wyo.)

Populus obovata Knowlton. (Wyo.)
Populus wardii Knowlton. (Wyo.)
Populus sp. Knowlton. (Wyo.)
Pterospermites undulatus Knowlton. (Wyo.)
Pterospermites wardii Knowlton. (Wyo.)
Quercus competens Lesquereux. (Wyo.)
Quercus dentonoides Knowlton. (Wyo.)
Quercus lesquereuxiana Knowlton. (Wyo.)
Quercus valdensis Heer. (Wyo.)
Rhamnus salicifolius Lesquereux. (Wyo.)
Rhus membranacea Lesquereux. (Wyo.)
Sabal? montana Knowlton. (Colo., Wyo.)
Salix angusta Al. Braun. (Wyo.)
Salix sp. Knowlton. (Wyo.)
Selaginella falcata Lesquereux. (Wyo.)
Selaginella laciniata Lesquereux. (Wyo.)
Sequoia heterophylla Velenovsky. (Wyo.)
Sequoia longifolia Lesquereux. (Wyo.)
Sequoia obovata Knowlton. (Colo., Wyo.)
Sequoia reichenbachi (Geinitz) Heer. (Wyo., Colo.)
Spathites sp. Knowlton. (Wyo.)
Trapa? cuneata Knowlton. (Wyo.)
Trapa? microphylla Lesquereux. (Wyo.)
Viburnum anomalinervum Knowlton. (Wyo.)
Viburnum hollickii Berry. (Wyo.)
Viburnum montanum Knowlton. (Wyo.)
Vitis islandica? Heer. (Wyo.)
Widdringtonia? complanata Lesquereux. (Wyo.)
Woodwardia crenata Knowlton. (Wyo.)
Woodwardia sp. Knowlton. (Wyo.)
Zizyphus sp. Knowlton. (Wyo.)

BELLY RIVER FORMATION, CANADA.

Acer saskatchewense Dawson. (Formation doubtful.)
Betula sp. Dawson.
Ginkgo sp. Dawson.
Nelumbo dawsoni Hollick.
Pityoxylon sp. Dawson.
Podocarpites tyrellei Dawson.
Populus latidentata Dawson.
Sequoia albertensis Penhallow.
Sequoia reichenbachi (Geinitz) Heer.
Sequoia sp. Dawson.
Sequoia sp.? Knowlton.
Taxites sp. Dawson.
Thuja sp. Dawson.
Trapa? microphylla Lesquereux.

JUDITH RIVER FORMATION, MONTANA.

Betulites? hatcheri Knowlton.
Carpites alatus Knowlton.
Carpites judithae Knowlton.
Carpites pruni Knowlton.
Castalia stantoni Knowlton.

Cunninghamites elegans (Corda) Endlicher.
Cunninghamites recurvatus? Hosius and Von der Marck.
Cyperacites sp. Knowlton.
Dammara acicularis Knowlton.
Diospyros judithae Knowlton.
Dryopteris lloydii Knowlton.
Osmunda montanensis Knowlton.
Phyllites denticulatus Knowlton.
Phyllites intricata Knowlton.
Pistia corrugata Lesquereux.
Populites amplus Knowlton.
Populus cretacea Knowlton.
Populus sp. Knowlton.
Quercus hatcheri Knowlton.
Quercus judithae Knowlton.
Sapindus inexpectans Knowlton.
Sequoia heterophylla Velenovsky.
Sequoia reichenbachi (Geinitz) Heer.
Thuja cretacea? (Heer) Newberry.
Trapa? cuneata Knowlton.
Trapa? microphylla Lesquereux.

TRINIDAD SANDSTONE.

Caulerpites incrassatus (Lesquereux) Lesquereux.
Caulerpites lingulatus (Lesquereux) Knowlton.
Chondrites bulbosus Lesquereux.
Chondrites subsimplex Lesquereux.
Dryophyllum bruneri Ward.
Halymenites major Lesquereux.
Halymenites striatus Lesquereux
Phaseolites minutus Knowlton.
Phyllites ratonensis Knowlton.
Populus? neomexicana Knowlton.
Rosellinites lapideus (Lesquereux) Knowlton.
Viburnum crassum Knowlton.
Vitus fragmenta Knowlton.

VERMEJO FORMATION.

Abietites dubius Lesquereux.
Acrostichum haddeni Hollick.
Allantodiopsis erosa (Lesquereux) Knowlton and Maxon.
Amelanchier obovata Knowlton.
Anemia robusta Hollick.
Anemia supercretacea Hollick.
Artocarpus dissecta Knowlton.
Asplenium? coloradense Knowlton.
Asplenium sp. Knowlton.
Brachyphyllum cf. B. macrocarpum Newberry.
Canna (?) magnifolia Knowlton.
Canna (?) sp. Knowlton.
Celastrus haddeni Knowlton.
Celastrus? hesperius Knowlton.
Celastrus sp. Knowlton.
Cissites panduratus Knowlton.
Colutea speciosa Knowlton.

Credneria protophylloides Knowlton.
Cupressinoxylon coloradense Knowlton.
Cupressinoxylon? vermejoense Knowlton.
Diospyros? leei Knowlton.
Ficus curta Knowlton.
Ficus dalmatica Lesquereux.
Ficus eucalyptifolia Knowlton.
Ficus gigantea Knowlton.
Ficus haddeni Knowlton.
Ficus leei Knowlton.
Ficus minima Knowlton.
Ficus newberryana Knowlton.
Ficus praetrinervis Knowlton.
Ficus regularis Knowlton.
Ficus rhamnoides Knowlton.
Ficus rockvalensis Knowlton.
Ficus speciosissima Ward.
Ficus? starkvillensis Knowlton.
Ficus tessellata Knowlton.
Ficus wardii Knowlton.
Ficus sp. Knowlton.
Fraxinus? sp. Knowlton.
Geinitzia formosa Heer.
Gleichenia delicatula? Heer.
Gleichenia rhombifolia Hollick.
Halymenites major Lesquereux.
Hedera rotundifolia Knowlton.
Juglans coloradensis Knowlton.
Juglans similis Knowlton.
Laurus coloradensis Knowlton.
Liriodendron alatum Newberry.
Myrica coriacea Knowlton.
Myrica torreyi Lesquereux.
Nelumbo laramiensis Hollick.
Osmunda hollicki Knowlton.
Palaeoaster inquirenda Knowlton.
Phaseolites crassus Knowlton.
Phaseolites leei Knowlton.
Phyllites aurantiacus Knowlton.
Phyllites castalioides Knowlton.
Phyllites nanus Knowlton.
Phyllites populoides Knowlton.
Phyllites protophylloides Knowlton.
Phyllites rosaefolius Knowlton.
Phyllites sapindus Knowlton.
Phyllites vermejoensis Knowlton.
Phyllites walsenbergensis Knowlton.
Platanus sp. Knowlton.
Polystichum hillsianum Hollick.
Pteris russellii Newberry.
Pterospermites nervosus Knowlton.
Pterospermites undulatus Knowlton.
Pterospermites wardii Knowlton.
Quercus gardneri Knowlton.

Quercus rockvalensis Knowlton.
Rhamnus salicifolius Lesquereux.
Sabal? montana Knowlton.
Sabal? ungeri (Lesquereux) Knowlton.
Salix gardneri Knowlton.
Salix plicata Knowlton.
Salix sp. a Knowlton.
Salix sp. b Knowlton.
Salix sp. c Knowlton.
Sequoia obovata Knowlton.
Sequoia reichenbachi (Geinitz) Heer.
Sparganium? sp. Knowlton.
Stenopteris? cretacea Hollick.
Sterculia coriacea Knowlton.
Taxodium? sp. Knowlton.
Viburnum anomalinervum Knowlton.
Viburnum? hesperium Knowlton.
Viburnum montanum Knowlton.
Viburnum? problematicum Knowlton.
Viburnum rhamnifolium Knowlton.
Viburnum simile Knowlton.
Viburnum sp. Knowlton.
Widdringtonia? complanata Lesquereux.
Woodwardia crenata Knowlton.
Zizyphus paliurifolius Knowlton.

FRUITLAND FORMATION, SAN JUAN BASIN, N. MEX.

Anemia hesperia Knowlton.
Anemia sp. Knowlton.
Carpites baueri Knowlton.
Ficus baueri Knowlton.
Ficus curta Knowlton?
Ficus eucalyptifolia Knowlton?
Ficus leei Knowlton.
Ficus praelatifolia Knowlton.
Ficus praetrinervis Knowlton.
Ficus rhamnoides Knowlton.
Ficus squarrosa Knowlton?
Ficus sp. Knowlton.
Geinitzia formosa Heer.
Heteranthera cretacea Knowlton.
Laurus baueri Knowlton.
Laurus coloradensis Knowlton.
Myrica torreyi Lesquereux.
Nelumbo sp. Knowlton.
Phyllites neomexicanus Knowlton.
Phyllites petiolatus Knowlton.
Pterospermites neomexicanus Knowlton.
Pterospermites undulatus Knowlton.
Quercus baueri Knowlton.
Ribes neomexicanus Knowlton.
Sabal? montana Knowlton.
Sabal sp. Knowlton.
Salix baueri Knowlton.

Salix sp. a Knowlton.
Sequoia obovata Knowlton.
Sequoia reichenbachi (Geinitz) Heer.
Unassigned plant (a).
Unassigned plant (b).

KIRTLAND FORMATION, SAN JUAN BASIN, N. MEX.

Asplenium neomexicanum Knowlton.
Ficus leei Knowlton.
Leguminosites? neomexicana Knowlton.
Myrica? neomexicana Knowlton.
Onoclea neomexicana Knowlton.
Pistia corrugata Lesquereux.
Pterospermites sp. Knowlton.

ADAVILLE FORMATION, HODGES PASS, WYO.

Diospyros? ficoidea Lesquereux.
Dryophyllum bruneri Ward.
Dryophyllum falcatum Ward.
Grewiopsis cleburni Lesquereux.
Nyssa buddiana Ward.
Quercus dentonoides Knowlton.
Quercus doljensis Pilar.

FOX HILLS FORMATION, NORTHEASTERN COLORADO.

Anemia sp. Knowlton.
Cephalotaxus? coloradensis Knowlton.
Equisetum sp. Knowlton.
Ficus speciosissima Ward.
Halymenites major Lesquereux.
Myrica torreyi Lesquereux.
Phyllites cockerelli Knowlton.
Podocarpus? stantoni Knowlton.
Rhamnus? williardi Knowlton.
Sequoia magnifolia Knowlton.
Sequoia reichenbachi (Geinitz) Heer.
Viburnum vulpinum Knowlton.

MONTANA GROUP, FORMATIONS NOT IDENTIFIED.

Cinnamomum affine Lesquereux. (Formation doubtful.)
Cinnamomum? stantoni Knowlton. (Utah.)
Ficus multinervis Heer. (Utah.)
Ficus planicostata Lesquereux. (Colo. Formation doubtful.)
Ficus speciosissima Ward. (Colo. Formation doubtful.)
Geinitzia formosa Heer. (Mont.)
Magnolia? coalvillensis Knowlton. (Utah.)
Pistia corrugata Lesquereux. (Wyo.)
Salix angusta Al. Braun. (Utah.)
Salix stantoni Knowlton. (Utah.)
Salix sp. Knowlton. (Utah.)
Sequoia longifolia Lesquereux. (Utah.)
Sequoia sp. Knowlton. (Utah.)
Viburnum? problematicum Knowlton. (Utah.)

LARAMIE FORMATION, DENVER BASIN, COLO.

Anemia elongata (Newberry) Knowlton.
Anemia supercretacea Hollick.
Anemia sp. Knowlton.
Anona coloradensis Knowlton.
Anona robusta Lesquereux.
Apeibopsis? laramiensis Knowlton.
Apocynophyllum? taenifolium Knowlton.
Aristolochia brittoni Knowlton.
Artocarpus lessigiana (Lesquereux) Knowlton.
Artocarpus liriodendroides Knowlton.
Asplenium martini Knowlton.
Carpites lakesii Knowlton.
Carpites lesquereuxiana Knowlton.
Carpites rhomboidalis Lesquereux.
Cassia? laramiensis Knowlton.
Ceanothus eriensis Knowlton.
Ceanothus ovatifolius Knowlton.
Celastrinites alatus Knowlton.
Celastrinites ámbiguus (Lesquereux) Knowlton.
Celastrinites cowanensis Knowlton.
Celastrinites eriensis Knowlton.
Cercis eocenica Lesquereux.
Cinnamomum affine Lesquereux.
Cinnamomum laramiense Knowlton.
Cornus praeimpressa Knowlton.
Cornus suborbifera Lesquereux.
Cornus sp. Knowlton.
Cycadeoidea mirabilis (Lesquereux) Ward.
Cyperacites? hillsii Knowlton.
Cyperacites? tesselatus Knowlton.
Cyperacites? sp. Knowlton.
Dammara sp. Knowlton.
Delesseria fulva Lesquereux.
Diospyros berryana Knowlton.
Dombeyopsis obtusa Lesquereux.
Dombeyopsis ovata Knowlton.
Dombeyopsis? sinuata Knowlton.
Dombeyopsis trivialis Lesquereux.
Dryopteris? carbonensis Knowlton.
Dryopteris georgei Knowlton.
Dryopteris laramiensis Knowlton.
Dryopteris lesquereuxii Knowlton.
Equisetum perlaevigatum Cockerell.
Ficus? apiculatus Knowlton.
Ficus arenacea Lesquereux.
Ficus berryana Knowlton.
Ficus cannoni Knowlton.
Ficus cockerelli Knowlton.
Ficus coloradensis Cockerell.
Ficus cowanensis Knowlton.
Ficus crossii Ward.
Ficus dalmatica Ettingshausen.
Ficus denveriana? Cockerell.

Ficus impressa Knowlton.
Ficus? leyden Knowlton.
Ficus multinervis? Heer.
Ficus navicularis Cockerell.
Ficus neodalmatica Knowlton.
Ficus pealei Knowlton.
Ficus planicostata Lesquereux.
Ficus planicostata magnifolia Knowlton.
Ficus post-trinervis Knowlton.
Ficus praeplanicostata Knowlton.
Ficus? smithsoniana? (Lesquereux) Lesquereux.
Fraxinus libbeyi Knowlton. (Formation doubtful.)
Hedera lucens Knowlton.
Hicoria angulata Knowlton.
Hicoria minutula Knowlton.
Ilex laramiensis Knowlton.
Juglans laramiensis Knowlton.
Juglans leconteana Lesquereux. (Formation doubtful.)
Juglans leydenianus Knowlton.
Juglans newberryi Knowlton.
Juglans praerugosa Knowlton.
Laurus lakesii Knowlton.
Laurus lanceolata Knowlton.
Laurus wardiana Knowlton.
Leguminosites? coloradensis Knowlton.
Leguminosites columbianus Knowlton.
Leguminosites? laramiensis Knowlton.
Lygodium? compactum Lesquereux.
Magnolia lakesii Knowlton.
Magnolia marshalli Knowlton.
Malapoenna louisvillensis Knowlton.
Mimosites marshallanus Knowlton.
Myrica dubia Knowlton.
Myrica oblongifolia Knowlton.
Myrica torreyi Lesquereux.
Negundo brittoni Knowlton.
Nelumbo tenuifolia (Lesquereux) Knowlton.
Onoclea? fecunda (Lesquereux) Knowlton.
Palaeoaster? similis Knowlton.
Paliurus zizyphoides Lesquereux?
Phanerophlebites pealei Knowlton.
Phragmites laramianus Cockerell.
Phyllites dombeyopsoides Knowlton.
Phyllites leydenianus Knowlton.
Phyllites marshallensis Knowlton.
Phyllites trinervis Knowlton.
Phyllites sp. Knowlton.
Pistacia eriensis Knowlton.
Pistacia hollicki Knowlton.
Platanus platanoides (Lesquereux) Knowlton.
Populus? distorta Knowlton.
Pteris goldmani Knowlton.
Pteris? sp. Knowlton.
Quercus eximia Knowlton.

Quercus praeangustiloba Knowlton.
Quercus stramineus Lesquereux.
Quercus viburnifolia? Lesquereux.
Rhamnus belmontensis Knowlton and Cockerell.
Rhamnus brittoni Knowlton.
?Rhamnus goldianus Lesquereux.
Rhamnus marshallensis Knowlton.
Rhamnus minutus Knowlton.
Rhamnus? pealei Knowlton.
Rhamnus salicifolius Lesquereux.
Rhamnus sp. Knowlton.
Robinia mesozoica Cockerell. (Formation doubtful.)
Sabal? montana Knowlton.
Salix brittoniana Knowlton.
Salix myricoides Knowlton.
Salix wyomingensis Knowlton and Cockerell.
Sequoia acuminata? Lesquereux.
Sequoia longifolia Lesquereux.
Sequoia reichenbachi (Geinitz) Heer.
Smilax? inquirenda Knowlton.
Zizyphus coloradensis Knowlton.
Zizyphus corrugatus Knowlton.
Zizyphus hendersoni Knowlton.
Zizyphus minutus Knowlton.

LARAMIE (?) FORMATION, HODGES PASS, WYO.[1]

Alnus grewiopsis Ward.
Cinnamomum wardii Knowlton.

"LARAMIE" FORMATION, YELLOWSTONE NATIONAL PARK.

Asplenium haguei Knowlton.
Diospyros stenosepala Heer.
Paliurus zizyphoides Lesquereux.

MEDICINE BOW FORMATION ("LOWER LARAMIE"), CARBON COUNTY, WYO.[2]

Apeibopsis discolor Lesquereux.
Aristolochia sp.
Artocarpus lessigii (Lesquereux) Knowlton.
Carpites sp.
Cassia marshallensis Knowlton.
Ceanothus? sp.
Cinnamomum affine Lesquereux.
Cyperacites sp.?
Daphnogene elegans Watelet.
Diospyros? ficoidea Lesquereux.
Dombeyopsis obtusa Lesquereux.
Dombeyopsis trivialis Lesquereux.
Dryophyllum cf. D. aquamarum Ward.
Dryophyllum bruneri Ward.
Dryopteris carbonensis Knowlton.
Equisetum, tubers of.

[1] These species, collected in a railroad tunnel at Hodges Pass, are different from those subsequently obtained from the Adaville formation in the same neighborhood. For this reason the lists are kept separate.
[2] This tentative list is compiled from an unpublished manuscript list and has not been recorded in the cards.

Ficus arenacea Lesquereux.
Ficus impressa Knowlton.
Ficus cockerelli Knowlton.
Ficus navicularis Cockerell.
Ficus planicostata? Lesquereux.
Ficus praetrinervis Knowlton.
Ficus sp.
Ficus? sp.
Geonomites cf. G. ungeri Lesquereux.
Ilex? sp.
Juglans praerugosa Knowlton.
Mimosa sp.?
Mimosites? sp.
Myrica torreyi Lesquereux.
Paliurus zizyphoides Lesquereux.
Palmoxylon sp.
Pecopteris sepulta Newberry of Hollick.
Phyllites sp.
Pistia corrugata Lesquereux?
Platanus marginata (Lesquereux) Heer.
Platanus platanoides (Lesquereux) Knowlton.
Rhamnus belmontensis Knowlton and Cockerell.
Rhamnus salicifolius Lesquèreux.
Sabal montana Knowlton.
Salix elongata Al. Braun?
Salix sp.
Sequoia longifolia Lesquereux.
Sequoia reichenbachi (Geinitz) Heer.
Sequoia sp.
Woodwardia new, nearest to W. crenata Knowlton.
Zizyphus minimus.

PACIFIC COAST REGION.

KNOXVILLE FORMATION.

Abietites macrocarpus Fontaine. (Calif.)
Abietites (?) sp. (Calif.)
Acaciaephyllum pachyphyllum Fontaine. (Calif.)
Angiopteridium canmorense Dawson. (Calif.)
Angiopteridium strictinerve Fontaine.
Angiopteridium strictinerve latifolium Fontaine.
Cephalotaxopsis ramosa? Fontaine. (Calif.)
Cephalotaxopsis rhytidodes Ward.
Cladophlebis browniana (Dunker) Seward.
Cladophlebis heterophylla Fontaine. (Calif.)
Cladophlebis parva Fontaine. (Calif.)
Cladophlebis ungeri (Dunker) Ward. (Calif.)
Cupressinoxylon sp. Knowlton. (Calif.)
Cycadeospermum californicum Fontaine. (Calif.)
Dicksonia pachyphylla Fontaine. (Calif.)
Dioönites buchianus (Ettingshausen) Bornemann. (Calif.)
Dioönites buchianus abietinus Ward. (Calif.)
Dioönites buchianus rarinervis? Fontaine. (Calif.)
Equisetum texense Fontaine. (Calif.)
Gleichenia? gilbert-thompsoni Fontaine. (Calif.)

Gleichenia nordenskiöldi Heer. (Calif.)
Hausmannia? californica Fontaine. (Calif.)
Matonidium althausii (Dunker) Ward. (Calif.)
Nageiopsis angustifolia Fontaine. (Calif.)
Nageiopsis longifolia Fontaine. (Calif.)
Nilsonia californica Fontaine. (Calif.)
Nilsonia oregonensis (Fontaine) Berry. (Calif.)
Nilsonia? sambucensis Ward. (Calif.)
Nilsonia schaumbergensis (Dunker) Nathorst. (Calif.)
Nilsonia stantoni Ward. (Calif.)
Onychiopsis psilotoides (Stokes and Webb) Ward. (Calif.)
Osmunda dicksonioides Fontaine. (Calif.)
Phyllites latifolius (Fontaine) Ward.
Proteaephyllum californicum Fontaine. (Calif.)
Pterophyllum? lowryanum Ward. (Calif.)
Sagenopteris elliptica Fontaine. (Calif.)
Sagenopteris mantelli (Dunker) Seward. (Calif.)
Sagenopteris nervosa Fontaine. (Calif.)
Sagenopteris? sp. Fontaine. (Calif.)
Taeniopteris shastensis Knowlton. (Calif.)
Zamites arcticus Göppert. (Calif.)
Zamites tenuinervis Fontaine. (Calif.)

HORSETOWN FORMATION.

Abietites macrocarpus Fontaine. (Calif.)
Acaciaephyllum ellipticum Fontaine. (Calif.)
Angiopteridium canmorense Dawson. (Calif.)
Cephalotaxopsis ramosa Fontaine.
Cladophlebis browniana (Dunker) Seward.
Ctenopsis latifolia (Fontaine) Berry. (Calif. Formation doubtful.)
Dioonites buchianus (Ettingshausen) Bornemann. (Calif.)
Dioonites buchianus abietinus (Göppert) Ward. (Calif.)
Dioonites buchianus rarinervis Fontaine. (Calif.)
Dioonites dunkerianus (Göppert) Miquel. (Calif.)
Menispermites californicus Fontaine. (Calif.)
Nageiopsis angustifolia Fontaine. (Calif.)
Nageiopsis longifolia Fontaine. (Calif.)
Nageiopsis zamoides Fontaine. (Calif.)
Nilsonia oregonensis (Fontaine) Berry. (Calif.)
Nilsonia schaumbergensis (Dunker) Nathorst. (Calif.)
Phyllites latifolius (Fontaine) Berry. (Oreg.)
Pinus shastensis Fontaine. (Calif.)
Populus? ricei Fontaine. (Oreg.)
Sagenopteris elliptica Fontaine. (Calif.)
Sagenopteris mantelli (Dunker) Schenk. (Calif.)
Sagenopteris nervosa Fontaine. (Oreg., Calif.)
Sagenopteris oregonensis (Fontaine) Fontaine. (Oreg., Calif.)
Saliciphyllum californicum Fontaine. (Calif.)
Saliciphyllum pachyphyllum Fontaine. (Calif.)
?Scleropteris elliptica Fontaine. (Calif.)
Sequoia ambigua Heer. (Calif.)
Sequoia reichenbachi (Geinitz) Heer. (Calif.)
Sphenolepidium sternbergianum (Dunker) Heer.
Taeniopteris shastensis Knowlton. (Calif.)
Zamites tenuinervis Fontaine. (Calif.)

CHICO FORMATION.

Cycadeoidea stantoni Ward. (Calif. Formation doubtful.)
Matonidium althausii (Dunker) Ward. (Calif.)
Taeniopteris shastensis Knowlton. (Calif.)

CHIGNIK FORMATION (UPPER CRETACEOUS), ALASKA.

Anomozamites schmidtii Heer.
Osmunda arctica Heer.
Pterophyllum lepidum? Heer.
Quercus johnstrupi Heer.
Sequoia rigida Heer.
Trapa? microphylla Lesquereux.

UPPER CRETACEOUS ROCKS, VANCOUVER ISLAND, BRITISH COLUMBIA.

Adiantites praelongus Dawson.
Alnites insignis Dawson.
Anisophyllum sp. Dawson.
Artocarpophyllum occidentale Dawson.
Betula praeantiqua Dawson.
Betula sp. Dawson.
Carpolithes meridionalis Dawson.
Carpolithes sp. Dawson.
Ceanothus cretaceus Dawson.
Cinnamomum newberryi Berry.
Cladophlebis columbiana Dawson.
Cornus obesus Dawson.
Ctenopteris columbiensis Penhallow.
Dammarites dubius Dawson.
Davallites richardsoni Dawson.
Diospyros eminens Dawson.
Diospyros vancouverensis Dawson.
Diospyros, calyx of, Dawson.
Dryopteris kennerlyi (Newberry) Knowlton.
Fagophyllum nervosum Dawson.
Fagophyllum retosum Dawson.
Ficus contorta Dawson.
Ficus magnoliaefolia Lesquereux.
Ficus rotundata Dawson.
Ficus wellingtoniae Dawson.
Ginkgo baynesiana (Dawson) Knowlton.
Ginkgo dawsoni Knowlton.
Glyptostrobus sp. Dawson.
Juglandites fallax Dawson.
Juglandites? sp. Dawson.
Juglans harwoodensis Dawson.
Laurophyllum insigne Dawson.
Liriodendron succedens Dawson.
Macclintockia trinervis Dawson.
Macrotaeniopteris vancouverensis Dawson.
Magnolia capellinii Heer.
Magnolia occidentalis Dawson.
Menispermites sp. Dawson.
"Neuropteris castor" Dawson.
Nilsonia lata Dawson.

Noeggerathiopsis robinsi Dawson.
Paliurus neillii Dawson.
Pecopteris sp. Dawson.
Persea leconteana Lesquereux.
Phragmites cordaiformis Dawson.
Phyllites sp. Dawson.
Phyllites sp. Dawson.
Populites probalsamifera Dawson.
Populus longior Dawson.
Populus protozaddachi Dawson.
Populus rectinervata Dawson.
Populus rhomboidea Lesquereux.
Populus trinervis Dawson.
Populus sp. Dawson.
Proteoides major Dawson.
Proteoides neillii Dawson.
Proteoides sp. Dawson.
Protophyllum nanaimo Dawson.
Protophyllum sp. Dawson.
Pteris glossopteroides Dawson.
Quercus holmesii (Lesquereux) Lesquereux.
Quercus multinervis Lesquereux.
Quercus platinervis Lesquereux.
Quercus vancouveriana Trelease.
Quercus victoriae Dawson.
Quercus sp. Dawson.
Sabal imperialis Dawson.
Salix pacifica Dawson.
Salix sp. Dawson.
Sassafras sp. Dawson.
Sequoia cuneata (Newberry) Newberry.
Taeniopteris plumosa Dawson.
Taonurus incertus Dawson.
Taxodium sp. Dawson.
Tumion densifolium (Dawson) Knowlton.
Ulmophyllum priscum Dawson.
Ulmus dubia Dawson.

CRETACEOUS ROCKS, FORMATIONS NOT IDENTIFIED.

Antholithes horridus (Dawson). (N. W. T.)
Cupress[in]oxylon dawsoni Penhallow. (Alberta.)
Cupress[in]oxylon sp. Dawson. (Alberta.)
Dewalquea haldemiana (Debey) Saporta and Marion. (Utah.)
Diospyros nitida Dawson. (N. W. T.)
Ficus? sp. Lesquereux. (B. C.)
Laurus crassinervis Dawson. (B. C.)
Liriodendron praetulipiferum Dawson. (B. C.)
Nilsonia gibbsii (Newberry) Hollick. (Wash.)

EOCENE FLORAS.

SOUTH ATLANTIC AND GULF REGIONS.

MIDWAY (?) FORMATION, EARLE, BEXAR COUNTY, TEX.

Asimina eocenica Lesquereux.
Cinnamomum affine Lesquereux.
Dolichites deusseni Berry.
Ficus denveriana Cockerell.
Ficus occidentalis (Lesquereux) Lesquereux.
Ficus sp. Berry.
Laurus wardiana Knowlton.
Magnolia hilgardiana Lesquereux.
Platanus aceroides latifolia Knowlton.
Pourouma texana Berry.

WILCOX GROUP.[1]

Acacia wilcoxensis Berry. (Miss.)
Acacia sp. Knowlton. (Tenn.)
Anacardites falcatus Berry. (Tenn.)
Anacardites grevillaefolia Berry. (Tenn., Miss.)
Anacardites marshallensis Berry. (Miss.)
Anacardites metopifolia Berry. (Miss., Tenn.)
Anacardites minor Berry. (Tenn.)
Anacardites puryearensis Berry. (Tenn.)
Anacardites serratus Berry. (Tenn.)
Anemia eocenica Berry. (Tenn., Ark., La.)
Anona ampla Berry. (Tenn., Ark., La.)
Anona eolignitica Berry. (Tenn., La.)
Anona wilcoxiana Berry. (Tenn., La.)
Antholithus arundites Berry. (Tenn.)
Antholithus marshallensis Berry. (Miss.)
Apocynophyllum constrictum Berry. (Ark.)
Apocynophyllum mississippiensis Berry. (Miss., La.)
Apocynophyllum sapindifolium Hollick. (Tenn., Miss., La.)
Apocynophyllum tabellarum (Lesquereux) Berry. (Miss., Tenn., La., Tex.)
Apocynophyllum wilcoxensis Berry. (Miss., Tenn., La.)
Araceaetes friteli Berry. (La.)
Aralia dakotana Knowlton and Cockerell. (Miss.)
Aralia jorgenseni Heer? (Miss.)
Aralia notata Lesquereux. (Ark., La.)
Aristolochia wilcoxiana Berry. (Tenn.)
Artocarpoides wilcoxensis Berry. (Tenn.)
Artocarpus dubia Hollick. (La.)
Artocarpus pungens (Lesquereux) Hollick. (Miss., Ark., La.)

---

[1] Includes species from beds of Wilcox age in the Lagrange formation of western Tennessee and Kentucky and southern Illinois.

Asimina leiocarpa Lesquereux. (Miss.)
Asplenium eolignitica Berry. (Miss., Ark., Tex.)
Asplenium hurleyensis Berry. (Miss.)
Athrotaxis? eolignitica Berry. (Tenn.)
Avicennia eocenica Berry. (Tenn.)
Avicennia nitidaformis Berry. (Miss.)
Banisteria fructuosa Berry. (Tenn.)
Banisteria pseudolaurifolia Berry. (Ky., Tenn.)
Banisteria repandifolia Berry. (Tenn.)
Banisteria wilcoxiana Berry. (Tenn., Ky.)
Banksia puryearensis Berry. (Tenn.)
Banksia saffordi (Lesquereux) Berry. (Miss., Tenn., Ky.)
Banksia tenuifolia Berry. (Ark., Tenn., Ky.)
Bombacites formosus Berry. (Tenn.)
Bombacites wilcoxianus Berry. (Tenn., La.)
Bumelia americana (Lesquereux) Berry. (Tenn.)
Bumelia grenadensis Berry. (Miss.)
Bumelia hurleyensis Berry. (Miss.)
Bumelia pseudohorrida Berry. (Tenn.)
Bumelia pseudotenax Berry. (Miss., Tex.)
Bumelia wilcoxiana Berry. (Miss., Tenn.)
Caenomyces annulata Berry. (Tenn.)
Caenomyces cassiae Berry. (Tenn.)
Caenomyces laurinea Berry. (Miss.)
Caenomyces pestalozzites Berry. (Miss.)
Caenomyces saportae Berry. (Miss.)
Caesalpinea wilcoxiana Berry. (Miss., Tenn.)
Caesalpinites aculeatafolia Berry. (Tenn.)
Caesalpinites bentonensis Berry. (Ark.)
Caesalpinites mississippiensis Berry. (Miss.)
Caesalpinites pinsonensis Berry. (Tenn.)
Calycites davillaformis Berry. (Tenn.)
Calycites ostryaformis Berry. (Tex.)
Calyptranthes eocenica Berry. (Tenn.)
Canavalia acuminata Berry. (Miss., Tenn.)
Canavalia eocenica Berry. (Miss., Tenn.)
Canna eocenica Berry. (Miss., Tex.)
Carapa eolignitica Berry. (Tenn., Ky.)
Carpolithus dictyolomoides Berry. (Tenn.)
Carpolithus grenadensis Berry. (Miss.)
Carpolithus henryensis Berry. (Tenn.)
Carpolithus hyoseritiformis Berry. (Tenn.)
Carpolithus pilocarpoides Berry. (Miss.)
Carpolithus prangosoides Berry. (Tenn.)
Carpolithus proteoides Berry. (Tenn.)
Carpolithus puryearensis Berry. (Tenn.)
Carpolithus sophorites Berry. (Miss.)
Carpolithus tennesseensis Berry. (Tenn.)
Cassia bentonensis Berry. (Ark., Tex.)
Cassia emarginata Berry. (Miss., Tenn.)
Cassia eolignitica Berry. (Miss., Tenn., La.)
Cassia fayettensis Berry. (Miss., Tenn., Ky.)
Cassia glenni Berry. (Miss., Tenn., Ky.)
Cassia glenni major Berry. (Tenn.)

Cassia lowii Berry. (Miss.)
Cassia marshallensis Berry. (Miss., Tenn., Ky.)
Cassia mississippiensis Berry. (Miss., Tenn.)
Cassia puryearensis Berry. (Tenn.)
Cassia tennesseensis Berry. (Miss., Tenn.)
Cassia wilcoxiana Berry. (Miss.)
Cedrela mississippiensis Berry. (Miss.)
Cedrela odoratifolia Berry. (Tenn.)
Cedrela puryearensis Berry. (Tenn.)
Cedrela wilcoxiana Berry. (Miss., Tenn.)
Celastrus bruckmannifolia Berry. (Miss.)
Celastrus eolignitica Berry. (Miss., Tenn.)
Celastrus minor Berry. (Miss.)
Celastrus taurinensis Ward. (La.)
Celastrus veatchii Hollick. (Tenn., La., Miss.)
Cercis wilcoxiana Berry. (Tenn., Miss.)
Chamaedorea danai (Lesquereux) Berry. (Miss., Ark.)
Chrysobalanus eocenica Berry. (Miss., Tenn.)
Chrysobalanus inaequalis (Lesquereux) Berry. (Miss., Tenn.)
Chrysophyllum ficifolia Berry. (Miss., Tenn.)
Cinnamomum buchii Heer. (La.)
Cinnamomum mississippiensis Berry. (Miss.)
Cinnamomum oblongatum Berry. (La., Tenn.)
Cinnamomum obovatus Berry. (Miss.)
Cinnamomum postnewberryi Berry. (La., Ark.)
Cinnamomum vera Berry. (Miss., Tenn.)
Citharexylon eolignitícum Berry. (Miss.)
Citrophyllum wilcoxianum Berry. (Tenn.)
Coccolobis eolignitica Berry. (Tenn.)
Coccolobis uviferafolia Berry. (Tenn.)
Combretanthites eocenica Berry. (Tenn.)
Combretum obovalis Berry. (Tenn., La.)
Combretum ovalis (Lesquereux) Berry. (Tenn., Miss., Tex.)
Combretum wilcoxensis Berry. (Tenn.)
Conocarpus eolignitica Berry. (La., Tenn.)
Cordia eocenica Berry. (Tenn.)
Cordia? lowii Berry. (Miss.)
Cornus studeri? Heer. (La.)
Crotonophyllum appendiculatum Berry. (Tenn.)
Crotonophyllum eocenicum Berry. (Tenn.)
Cryptocarya eolignitica Hollick. (La.)
Cupanites eolignitícus Berry. (Tenn., La.)
Cupanites loughridgii Berry. (Ky.)
Cupressinoxylon arkansanum Knowlton. (Ark. Formation doubtful.)
Cupressinoxylon calli Knowlton. (Ark.)
Cyperacites sp. Hollick. (La.)
Dalbergia eocenica Berry. (Tenn.)
Dalbergia monospermoides Berry. (Tenn.)
Dalbergia tennesseensis Berry. (Tenn.)
Dalbergia wilcoxiana Berry. (Tenn.)
Dalbergites ellipticifolius Berry. (Miss.)
Dalbergites ovatus Berry. (Miss.)
Dillenites microdentatus (Hollick) Berry. (La.)
Dillenites ovatus Berry. (La., Miss.)

Dillenites serratus Berry. (Miss.)
Dillenites tetraceræfolia Berry. (Miss., Tenn.)
Dillenites texensis Berry. (Tex., Miss.)
Diospyros brachysepala Al. Braun. (Tenn., Tex.)
Diospyros wilcoxiana Berry. (Tenn.)
Dodonaea knowltoni Berry. (Tenn.)
Dodonaea wilcoxiana Berry. (Miss.)
Dryophyllum amplum Berry. (La.)
Dryophyllum anomalum Berry. (Tenn.)
Dryophyllum moorii (Lesquereux) Berry. (Miss., Ky., Tex.)
Dryophyllum puryearensis Berry. (Miss., Ky., Tenn.)
Dryophyllum tennesseensis Berry. (Miss., La., Tenn., Ky.)
Drypetes praelatifolia Berry. (Miss.)
Drypetes prekeyensis Berry. (Tenn.)
Echitonium lanceolatum Ettingshausen. (Tenn.)
Engelhardtia ettingshauseni Berry. (Miss., Tenn.)
Engelhardtia mississippiensis Berry. (Miss.)
Engelhardtia puryearensis Berry. (Tenn.)
Eucalyptus sp. Knowlton. (Ky.)
Eugenia densinervia (Lesquereux) Berry. (Tenn.)
Eugenia grenadensis Berry. (Miss.)
Eugenia hilgardiana Berry. (Miss.)
Eugenia puryearensis Berry. (Tenn.)
Euonymus splendens Berry. (Miss., Tenn., La., Tex.)
Euphorbiophyllum fayettensis Berry. (Tenn.)
Exostema pseudocaribaeum Berry. (Miss., Ky.)
Fagara eocenica Berry. (Tenn.)
Fagara hurleyensis Berry. (Miss.)
Fagara puryearensis Berry. (Tenn.)
Ficus artocarpoides Lesquereux. (La.)
Ficus cinnamomoides Lesquereux. (Miss.)
Ficus cockerelli Knowlton. (La.)
Ficus denveriana Cockerell. (Ky., Ark., La.)
Ficus eolignitica Berry. (Ark.)
Ficus harrisiana Hollick. (La.)
Ficus monodon (Lesquereux) Berry. (Miss., Tenn.)
Ficus myrtifolius Berry. (Miss., Ky.)
Ficus neoplanicostata Knowlton. (La.)
Ficus occidentalis (Lesquereux) Lesquereux. (La., Miss.)
Ficus planicostata Lesquereux. (La.)
Ficus planicostata goldiana (Lesquereux) Knowlton. (La.)
Ficus planicostata maxima Berry. (La., Tex., Tenn.)
Ficus pseudocuspidata Berry. (Tenn.)
Ficus pseudo-populus Lesquereux. (La., Tenn.)
Ficus puryearensis Berry. (Miss., Tenn.)
Ficus schimperi Lesquereux. (Miss., Tenn., La., Tex.)
Ficus vaughani Berry. (Miss., Tenn., Ark., Tex.)
Ficus wilcoxensis Berry. (Tenn., Ky.)
Ficus sp. Berry. (Miss.)
Ficus sp. Knowlton. (Tex.)
Fraxinus johnstrupi Heer. (La., Miss., Tenn.)
Fraxinus wilcoxiana Berry. (Tenn.)
Gleditsiophyllum constrictum Berry. (Tenn.)
Gleditsiophyllum ellipticum Berry. (Tenn.)

Gleditsiophyllum entadaformis Berry. (Miss.)
Gleditsiophyllum eocenicum Berry. (Miss., Tenn., La., Tex.)
Gleditsiophyllum fruticosum Berry. (Miss.)
Gleditsiophyllum hilgardianum Berry. (Miss.)
Gleditsiophyllum minor Berry. (Tenn.)
Gleditsiophyllum ovatum Berry. (Tenn.)
Glyptostrobus europaeus (Brongniart) Heer. (Miss.)
Grewiopsis tennesseensis Berry. (Tenn.)
Guettarda ellipticifolia Berry. (Miss., Tenn.)
Heterocalyx saportana Berry. (Miss.)
Hicoria antiquora (Newberry) Knowlton. (La.)
Hiraea wilcoxiana Berry. (Tenn.).
Icacorea prepaniculata Berry. (Tenn.)
Ilex affinis Lesquereux. (La.)
Ilex eolignitica Berry. (Miss.)
Ilex vomitoriafolia Berry. (Miss.)
Ilex sp. Hollick. (La.)
Inga laurinafolia Berry. (La.)
Inga mississippiensis Berry. (Miss., Tenn.)
Inga puryearensis Berry. (Tenn.)
Inga wickliffensis Berry. (Ky.)
Juglans appressa Lesquereux. (Miss.)
Juglans berryi Knowlton. (Ky., La.)
Juglans rugosa Lesquereux. (La., Ky.)
Juglans saffordiana Lesquereux. (Tenn.)
Juglans schimperi Lesquereux. (Ky., La., Ark., Tex., Miss.)
Juglans sp. Knowlton. (Tex.)
Knightiophyllum wilcoxianum Berry (homonym). (Tenn.)
Laguncularia preracemosa Berry. (Miss., Tenn.)
Laurinoxylon brauneri Knowlton. (Formation doubtful.)
Laurophyllum florum Berry. (Tenn.)
Laurophyllum juvenalis Berry. (Tenn.)
Laurophyllum preflorum Berry. (Tenn.)
Laurus primigenia Unger. (La.)
Laurus ratonensis Knowlton. (Ark., Tenn.)
Laurus sp. Knowlton. (Tex.)
Leguminosites? arachioides (Lesquereux) Lesquereux. (Tex., La.)
Leguminosites perfoliatus Berry. (Tenn.)
Leguminosites reniformis Bowerbank. (Tenn.)
Leguminosites subovatus Bowerbank. (Tenn.)
Leguminosites wickliffensis Berry. (Ky.)
Lycopodites? eoligniticus Berry. (Miss.)
Lygodium binervatum (Lesquereux) Berry. (Miss., La.)
Magnolia angustifolia Newberry. (La., Tenn.)
Magnolia cordifolia Lesquereux. (Miss.)
Magnolia hilgardiana Lesquereux. (Miss., Tenn., La., Tex.)
Magnolia laurifolia Lesquereux. (La.)
Magnolia leei Knowlton. (Miss., Tenn.)
Magnolia lesleyana Lesquereux. (Miss.)
Magnolia ovalis Lesquereux. (Miss.)
Maytenus puryearensis Berry. (Tenn.)
Melastomites americanus Berry. (Miss., La., Tenn.)
Meniphylloides ettingshauseni Berry. (Miss., La., Tex.)
Menispermites wilcoxensis Berry. (La.)

Mespilodaphne coushatta Berry. (Tenn., Ark., La.)
Mespilodaphne eolignitica (Hollick) Berry. (Miss., Tenn., La., Tex.)
Mespilodaphne pseudoglauca Berry. (La., Miss., Tenn., Ky.)
Mespilodaphne puryearensis Berry. (Tenn.)
Metopium wilcoxianum Berry. (Miss., Tenn., Tex.)
Mimosites acaciaefolius Berry. (Tenn.)
Mimosites inaequilateralis Berry. (Tenn.)
Mimosites lanceolatus Berry. (Tenn.)
Mimosites variabilis Berry. (Miss., Tenn., Ky.)
Mimusops mississippiensis Berry. (Miss.)
Mimusops siberifolia Berry. (Tenn.)
Mimusops truncata (Lesquereux) Knowlton. (Miss., Tenn.)
Myrcia bentonensis Berry. (Miss., Tenn., Ark.)
Myrcia granadensis Berry. (Miss.)
Myrcia parvifolia Berry. (Tenn.)
Myrcia puryearensis Berry. (Tenn.)
Myrcia vera Berry. (Miss., Tenn.)
Myrcia worthenii (Lesquereux) Berry. (Tenn., Ill.)
Myrica elaeanoides Lesquereux. (Miss., Ky., Tenn.)
Myrica wilcoxensis Berry. (Miss.)
Nectandra glenni Berry. (Tenn.)
Nectandra lancifolia (Lesquereux) Berry. (Miss., Tenn., Tex.)
Nectandra lowii Berry. (Miss., Ark., La., Tenn.)
Nectandra pseudocoriacea Berry. (Miss., La., Ark., Tenn.)
Nectandra puryearensis Berry. (Tenn., Miss.)
Nectandra sp. Berry. (Tenn., La., Tex.)
Nelumbo protolutea Berry. (Miss.)
Nipadites burtini umbonatus Bowerbank. (Miss.)
Nyssa eolignitica Berry. (Tenn.)
Nyssa wilcoxiana Berry. (La., Tenn.)
Oreodaphne coushatta Berry. (La.)
Oreodaphne mississippiensis Berry. (Miss., Ark.)
Oreodaphne obtusifolia Berry. (Miss., Tenn., La., Tex.)
Oreodaphne pseudoguianensis Berry. (La., Tenn.)
Oreodaphne puryearensis Berry. (Miss., Tenn.)
Oreodaphne salinensis Berry. (Ark.)
Oreodaphne wilcoxensis Berry. (Tenn.)
Oreopanax minor Berry. (Tenn.)
Oreopanax oxfordensis Berry. (Miss., Ark.)
Osmanthus pedatus (Lesquereux) Berry. (Miss., Tenn.)
Palaeodendron americanum Berry. (Miss., Ark.)
Paliurus angustus Berry. (Miss.)
Paliurus mississippiensis Berry. (Miss.)
Paliurus pinsonensis Berry. (Tenn.)
Parrotia cuneata (Newberry) Berry. (Tenn.)
Persea longipetiolatum (Hollick) Berry. (Tenn., La., Tex.)
Persea wilcoxiana Berry. (La.)
Phyllites sp. Berry. (Tenn.)
Pisonia chlorophylloides Berry. (Miss.)
Pisonia eolignitica Berry. (Tenn.)
Pisonia puryearensis Berry. (Tenn.)
Pistia wilcoxensis Berry. (La.)
Pithecolobium eocenicum Berry. (Tenn.)
Pithecolobium oxfordensis Berry. (Miss.)

Planera lingualis Knowlton and Cockerell. (Miss.)
Poacites sp. Hollick. (La.)
Porana sp. Lesquereux. (Ky.)
Praeengelhardtia eocenica Berry. (Tenn.)
Proteoides wilcoxensis Berry. (Miss., La., Tenn.)
Prunus nabortensis Berry. (La.)
Pseudolmedia eocenica Berry. (Tenn.)
Psychotria grandifolia Engelhardt. (Tenn.)
Pteris pseudopinnaeformis Lesquereux. (La., Miss.)
Quercus? ratonensis Knowlton. (La.)
Quercus cf. Q. cuspidata (Rossmässler) Unger. (Ky.)
Quercus sp. Knowlton. (Ky.)
Reynosia praenuntia Berry. (Miss., Tenn.)
Reynosia wilcoxiana Berry. (Tenn.)
Rhamnites berchemiaformis Berry. (Tex.)
Rhamnus cleburni Lesquereux. (Miss., La.)
Rhamnus coushatta Berry. (Miss., La.)
Rhamnus eolignitieus Berry. (Tenn., La.)
Rhamnus marginatus Lesquereux. (Miss., Tenn.)
Rhamnus marginatus apiculatus Berry. (Miss.)
Rhamnus puryearensis Berry. (Tenn.)
Sabalites grayanus (Lesquereux) Lesquereux. (Miss., Ark., Tex., Tenn.)
Salix media Heer. (Ky.)
Sapindus bentonensis Berry. (Ark., Tex.)
Sapindus caudatus Lesquereux. (La.)
Sapindus coushatta Berry. (La.)
Sapindus eolignitieus Berry. (Ky., Tenn.)
Sapindus formosus Berry. (Miss., La., Ky., Tenn.)
Sapindus knowltoni Berry. (Ark., Tenn.)
Sapindus linearifolius Berry. (Miss., Ark., Tenn., Ky., Tex.)
Sapindus mississippiensis Berry. (Miss., La., Ky., Tenn.)
Sapindus oxfordensis Berry. (Miss.)
Sapindus pseudaffinis Berry. (Tenn.)
Sideroxylon ellipticus Berry. (Tenn.)
Sideroxylon premastichodendron Berry. (Tenn.)
Simaruba eocenica Berry. (Tenn.)
Solanites saportana Berry. (Miss.)
Sophora henreyensis Berry. (Tenn.)
Sophora lesquereuxi Berry. (Tenn.)
Sophora mucronata Berry. (Tenn.)
Sophora palaeolobifolia Berry. (Tenn.)
Sophora puryearensis Berry. (Tenn.)
Sophora repandifolia Berry. (Tenn.)
Sophora wilcoxiana Berry. (Tenn., Miss., Ark.)
Sterculia puryearensis Berry. (Miss., Tenn.)
Sterculiocarpus eocenicus Berry. (La.)
Sterculiocarpus sezannelloides Berry. (Tenn.)
Taxodium dubium (Sternberg) Heer. (Tenn.)
Taxodium sp. Berry. (Tenn.)
Terminalia wilcoxiana Berry. (Miss.)
Ternstroemites eolignitieus Berry. (Tenn.)
Ternstroemites lanceolatus Berry. (Miss., Tenn.)
Ternstroemites ovatus Berry. (Miss., La., Tenn.)
Ternstroemites preclaibornensis Berry. (La., Tenn.)

Trapa wilcoxensis Berry. (Tenn.)
Vantanea wilcoxiana Berry. (Tenn., La.)
Zamia mississippiensis Berry. (Miss.)
Zamia? wilcoxensis Berry. (La.)
Zizyphus falcatus Berry. (Tenn.)
Zizyphus meigsii (Lesquereux) Berry. (Miss., Tenn.)

BARNWELL FORMATION (TWIGGS CLAY MEMBER), GEORGIA.

Acrostichum georgianum Berry.
Arundo pseudogöpperti Berry.
Carpolithus najasoides Berry.
Cassia georgiana Berry.
Castanea clairbornensis Berry.
Conocarpus eocenica Berry.
Cupanites nigricans Berry.
Cinnamomum angustum Berry.
Dodonaea viscosoides Berry.
Ficus clairbornensis Berry.
Ficus newtonensis Berry.
Malapoenna sp. Berry.
Mimosa americana Berry.
Mimosites georgianus Berry.
Pestalozzites minor Berry.
Pisonia claibornensis Berry.
Pistia claibornensis Berry.
Potamogeton megaphyllus Berry.
Pteris inquirenda Berry.
Rhizophora eocenica Berry.
Sapindus georgiana Berry.
Sophora claiborneana Berry.
Sophora wilcoxiana Berry.
Sphaerites claibornensis Berry.
Terminalia phaeocarpoides Berry.
Thrinax eocenica Berry.

CLAIBORNE GROUP.

YEGUA FORMATION.

Cupress[in]oxylon dawsoni Penhallow. (Tex.)
Goniopteris claiborniana Berry. (La.)
Laurinoxylon branneri Knowlton. (Tex.)
Reynosia texana Penhallow. (Tex.)
Rhamnacinium texanum Penhallow. (Tex.)
Sequoia langsdorfii (Brongniart) Heer.

LISBON FORMATION.

Goniopteris claiborniana Berry. (Miss.)

FORMATION NOT IDENTIFIED.

Carapa xylocarpoides Berry. (Ga.)
Eocharas eocenica Berry. (Miss.)

ROCKY MOUNTAIN REGION.

LANCE FORMATION.

Acacia sp. Knowlton.
Ampelopsis xantholithensis (Ward) Cockerell. (Mont.)
Aralia notata Lesquereux. (Mont.)

Aralia sp. Knowlton. (Wyo.)
Aralia sp. Knowlton. (Wyo.)
Aralia sp. Knowlton. (Wyo.)
Araucaria hatcheri Weiland. (Wyo.)
Asplenium sp. Knowlton. (Wyo.)
Bauhinia sp. Knowlton. (Wyo.)
Berchemia multinervis (Braun) Heer. (Wyo.)
Betula sp. Knowlton. (Wyo.)
Carpites sp. Knowlton. (Wyo.)
Carpolithes lineatus Newberry. (Mont.)
Castalia sp. Knowlton. (Wyo.)
Celastrus curvinervis Ward. (Mont.)
Celastrus ferrugineus Ward. (Wyo.)
Celastrus montanensis Knowlton and Cockerell. (Mont., N. Dak.)
Celastrus pterospermoides Ward. (Mont.)
Celastrus taurinensis Ward. (Wyo.)
Celastrus wardii Knowlton and Cockerell. (Mont.)
Cissus parrottiaefolia Lesquereux. (Wyo. Formation doubtful.)
Cocculus haydenianus Ward. (Mont.)
Cornus newberryi Hollick. (Mont.)
Corylus americana Walter. (Mont., Wyo.)
Corylus rostrata Ailton. (Mont.)
Credneria? daturaefolia Ward. (Mont.)
Cyperacites sp. Knowlton. (Wyo.)
Cyperus sp. Knowlton. (Wyo.)
Dammara? sp. Knowlton. (Wyo.)
Diospyros haguei Knowlton. (Mont.)
Dombeyopsis sp. Knowlton. (S. Dak.)
Dryopteris sp. Knowlton. (Mont.)
Elaeodendron polymorphum Ward. (Mont.)
Equisetum sp. Knowlton. (Wyo., Mont.)
Ficus artocarpoides Lesquereux. (Mont.)
Ficus ceratops Knowlton. (Wyo., Mont.)
Ficus denveriana Cockerell. (S. Dak.)
Ficus russelli Knowlton. (Mont.)
Ficus sp. Knowlton. (Wyo., Mont.)
Ginkgo adiantoides (Unger) Heer. (S. Dak., Mont., Wyo.)
Glyptostrobus europaeus (Brongniart) Heer. (S. Dak., Wyo., Mont.)
Grewiopsis eocenica (Lesquereux) Knowlton. (Wyo.)
Grewiopsis platanifolia Ward. (Mont.)
Grewiopsis populifolia Ward. (Mont.)
Hicoria antiquora (Newberry) Knowlton. (Mont., N. Dak., Wyo.)
Hicoria sp. Knowlton. (Wyo.)
Juglans rugosa Lesquereux. (N. Dak.)
Juglans sp. Knowlton. (Wyo.)
Laurus sp. Knowlton. (Mont.)
Leguminosites? arachioides (Lesquereux) Lesquereux. (S. Dak., Wyo.)
Leguminosites sp. Knowlton. (S. Dak.)
Lysimachia sp. Knowlton. (Wyo.)
Marchantia pealei Knowlton. (Mont.)
Menispermites sp. Knowlton. (Wyo.)
Musophyllum sp. Knowlton. (Wyo.)
?Myrica torreyi Lesquereux. (Wyo.)
Nelumbo sp. Knowlton. (Mont., Wyo.)

Onoclea sensibilis fossilis Newberry. (Mont.)
Ottelia? sp. Knowlton. (Wyo.)
Paliurus colombi Heer. (S. Dak.)
Paliurus pulcherrimus Ward. (Wyo.)
Palmocarpon palmarum (Lesquereux) Knowlton. (Wyo.)
Phyllites n. sp. Knowlton. (Mont., Wyo.)
Platanus guillelmae Göppert. (Mont., Wyo.)
Platanus haydeni Newberry. (Mont., Wyo.)
Platanus marginata (Lesquereux) Heer. (S. Dak.)
Platanus nobilis Newberry. (Mont., Wyo.)
?Platanus platanoides (Lesquereux) Knowlton. (S. Dak.)
Platanus raynoldsii Newberry. (S. Dak., Mont., Wyo.)
Platanus raynoldsii integrifolia Lesquereux. (Wyo.)
Platanus rhomboidea Lesquereux. (Wyo.)
Platanus sp. Knowlton. (Mont. Colgate sandstone member.)
Polygonum sp. Knowlton. (Mont., S. Dak.)
Populus amblyrhyncha Ward. (S. Dak., Mont., Wyo., and also Colgate sandstone member.)
Populus arctica Heer. (Mont., Wyo.)
Populus cuneata Newberry. (S. Dak., Mont., Wyo., and Colgate sandstone member.)
Populus cyclomorpha Knowlton and Cockerell. (Mont.)
Populus daphnogenoides Ward. (Mont., Wyo.)
Populus genetrix Newberry. (Mont., Wyo.)
Populus inaequalis Ward. (Mont.)
Populus nebrascensis Newberry. (Mont.)
Populus nervosa elongata Newberry. (Mont.)
Populus newberryi Cockerell. (Mont.)
Populus speciosa Ward. (S. Dak., Mont.)
Populus subrotunda Lesquereux. (Wyo.)
Populus sp. Knowlton. (Wyo.)
Protophyllocladus? sp. Knowlton. (Wyo.)
Pterospermites cupanioides (Newberry) Knowlton. (Mont.)
Pterospermites minor Ward. (Mont.)
Pterospermites whitei Ward. (Mont.)
Quercus breweri Lesquereux. (Mont.)
Quercus cinereoides Lesquereux. (Wyo.)
Quercus viburnifolia Lesquereux. (Wyo., Mont.)
Quercus sp. Knowlton. (Mont., Wyo.)
Rhus? sp. Knowlton. (Wyo.)
Sabal? eocenica (Lesquereux) Knowlton. (Wyo.)
Sabal grandifolia Newberry. (Wyo.)
Sabal rigida Hatcher. (Wyo.)
Sabalites grayanus (Lesquereux) Lesquereux. (Wyo.)
Salix angusta Al. Braun. (Mont., Wyo.)
Salix sp. Knowlton. (Wyo.)
Salvinia sp. Knowlton. (Wyo.)
Sapindus affinis Newberry. (Mont., N. Dak., Wyo.)
Sapindus grandifoliolus Ward. (N. Dak., Mont., Wyo.)
Sassafras sp. Knowlton. (N. Dak., Mont.)
Sequoia acuminata Lesquereux. (S. Dak.)
Sequoia heerii Lesquereux. (Mont., Wyo.)
Sequoia langsdorfii (Brongniart) Heer. (S. Dak.)
Sequoia longifolia Lesquereux or new species. (Mont.)
Sequoia nordenskiöldi Heer. (N. Dak., Mont., Wyo.)

Sparganium stygium Heer. (Wyo.)
Taxodium dubium (Sternberg) Heer. (Wyo.)
Taxodium occidentale Newberry. (S. Dak., N. Dak., Mont., Wyo.)
Thuja interrupta Newberry. (N. Dak., Mont.)
Thuja sp. Knowlton. (Wyo.)
Trapa? microphylla Lesquereux. (Mont., N. Dak., Wyo.)
Ulmus sp. Knowlton. (Wyo.)
Viburnum antiquum (Newberry) Hollick. (Mont., Wyo.)
Viburnum elongatum Ward. (Mont., S. Dak.)
Viburnum newberryanum Ward. (Wyo., Mont., Colgate sandstone member.)
Viburnum perplexum Ward. (Wyo.)
Viburnum whymperi Heer. (Mont., N. Dak., Wyo.)
Viburnum sp. Knowlton. (Wyo., Mont., S. Dak.)

POST-LARAMIE FORMATION, BLACK BUTTES, WYO.

Apeibopsis? discolor (Lesquereux) Lesquereux.
Carpites bursaeformis Lesquereux.
Carpites glumaeformis Lesquereux.
Carpites mitratus Lesquereux.
Carpites verrucosus Lesquereux.
Carpites viburni Lesquereux.
Caulinites sparganioides Lesquereux.
Cinnamomum sp. Knowlton.
Cissus lesquereuxii Knowlton.
Cissus lobato-crenata Lesquereux.
Cyperacites sp. Knowlton.
Daphnogene elegans Watelet.
Diospyros brachysepala Al. Braun.
Dryandroides cleburni (Lesquereux) Ettingshausen.
Dryophyllum aquamarum Ward.
Dryopteris nigricans (Lesquereux) Knowlton.
Eucalyptus haeringiana? Ettingshausen.
Ficus asarifolia Lesquereux.
Ficus cockerelli Knowlton.
Ficus coloradensis Cockerell.
Ficus? corylifolia Lesquereux.
Ficus haydenii Lesquereux.
Ficus planicostata Lesquereux.
Ficus praesinuosa Knowlton and Cockerell.
Ficus trinervis Knowlton.
Ginkgo sp. Knowlton.
Grewiopsis eocenica (Lesquereux) Knowlton.
Grewiopsis paliurifolia Ward.
Grewiopsis tenuifolia Lesquereux.
Hedera bruneri Ward.
Juglans baltica? Heer.
Juglans rhamnoides Lesquereux.
Macclintockia lyallii Heer.
Magnolia magnifolia Knowlton.
Myrica torreyi Lesquereux.
Nyssa? racemosa Knowlton.
Opegrapha antiqua Lesquereux.
Paliurus zizyphoides Lesquereux.
Phyllites improbatus Lesquereux.

Pisonia racemosa Lesquereux.
Platanus aceroides Göppert.
Platanus marginata (Lesquereux) Heer.
Platanus platanoides (Lesquereux) Knowlton.
Platanus raynoldsii integrifolia Lesquereux.
Podogonium americanum Lesquereux.
Populus leucophylla Unger.
Quercus aquamara (Ward) Knowlton.
Quercus doljensis Pilar.
Quercus viburnifolia Lesquereux.
Quercus wyomingiana Lesquereux.
Rhamnus dechenii Weber.
Rhamnus? discolor Lesquereux.
Sabal? eocenica (Lesquereux) Knowlton.
Salix wyomingensis Knowlton and Cockerell.
Sequoia acuminata Lesquereux.
Sequoia sp. Lesquereux.
Sphaerites lesquereuxi Meschinelli.
Sphaerites myricae (Lesquereux) Meschinelli.
Viburnum contortum Lesquereux.
Viburnum melaenum Knowlton and Cockerell.
Viburnum rotundifolium Lesquereux.
Vitis sparsa Lesquereux.
Zizyphus meekii Lesquereux.

POST-LARAMIE (?) FORMATION, MEDICINE BOW, WYO.

Populus latior cordifolia Heer.

FORT UNION FORMATION.

Acer arcticum Heer. (N. Dak.)
Acer graciliscens Lesquereux. (N. Dak.)
Acer trilobatum tricuspidatum (Braun) Heer. (Mont., N. Dak.)
Acer sp. Knowlton. (Yel. Nat. Pk.)
Allantodiopsis erosa (Lesquereux) Knowlton and Maxon. (Yel. Nat. Pk.)
Alnus serrata Newberry. (Yel. Nat. Pk.)
Amelanchier similis Newberry. (Mont.)
Ampelopsis montanensis Cockerell. (Mont.)
Ampelopsis xantholithensis (Ward) Cockerell. (Mont.)
Andromeda delicatula Lesquereux. (Mont.)
Anomalofilicites monstrosus Hollick. (Mont.)
Aralia dakotana Knowlton and Cockerell. (N. Dak.)
Aralia looziana Saporta and Marion. (Mont.)
Aralia notata Lesquereux. (N. Dak., Mont., Yel. Nat. Pk., Wyo.)
Aralia serrulata Knowlton. (Yel. Nat. Pk.)
Aralia triloba Newberry. (N. Dak.)
Aralia wardiana Knowlton and Cockerell. (Mont.)
Aralia whitneyii Lesquereux. (Wyo. Formation doubtful.)
Arctostaphylos elliptica Knowlton. (Yel. Nat. Pk.)
Aristolochia cordifolia Newberry. (S. Dak., Wyo.)
Artocarpus? quercoides Knowlton. (Yel. Nat. Pk.)
Asimina eocenica Lesquereux. (Wyo.)
Asplenium iddingsi Knowlton. (Yel. Nat. Pk.)
Asplenium magnum Knowlton. (Yel. Nat. Pk.)
Asplenium remotidens Knowlton. (Yel. Nat. Pk.)
Berchemia multinervis (Al. Braun) Heer. (Mont.)

Betula basiserrata Ward. (Mont.)
Betula coryloides Ward. (Mont.)
Betula iddingsi Knowlton. (Yel. Nat. Pk.)
Betula prisca Ettingshausen. (Mont.)
Betula sp. Knowlton. (Mont.)
Cabomba? gracilis Newberry. (N. Dak.)
Cabomba ineremis (Newberry) Hollick. (N. Dak.)
Calycites polysepala Newberry. (Mont., N. Dak.)
Carpites pedunculatus Knowlton. (Yel. Nat. Pk.)
Carpolithes lineatus Newberry. (N. Dak.)
Carpolithes osseus Lesquereux. (Yel. Nat. Pk.)
Castanea pulchella Knowlton. (Yel. Nat. Pk.)
Celastrus culveri Knowlton. (Yel. Nat. Pk.)
Celastrus curvinervis Ward. (Mont., S. Dak.)
Celastrus ellipticus Knowlton. (Yel. Nat. Pk.)
Celastrus ferrugineus Ward. (Mont., Wyo.)
Celastrus grewiopsis Ward. (Mont.)
Celastrus inaequalis Knowlton. (Yel. Nat. Pk.)
Celastrus montanensis Knowlton and Cockerell. (Mont., N. Dak.)
Celastrus pterospermoides Ward. (Mont.)
Celastrus taurinensis Ward. (Mont., Wyo.)
Celastrus wardii Knowlton and Cockerell. (Mont.)
Celtis ligualis Knowlton and Cockerell. (Wyo.)
Celtis newberryi Knowlton and Cockerell. (Wyo.)
Cercis borealis Newberry. (Mont.)
Cercis truncata Lesquereux. (N. Dak.)
Cinchonidium ovale Lesquereux. (N. Dak.)
Cocculus haydenianus Ward. (N. Dak., Mont.)
Cornus fosteri Ward. (Mont.)
Cornus newberryi Hollick. (Yel. Nat. Pk.)
Corylus americana Walter. (N. Dak., Mont.)
Corylus fosteri Ward. (Mont.)
Corylus macquarrii (Forbes) Heer. (N. Dak., Mont.)
Corylus orbiculata Newberry. (N. Dak.)
Corylus rostrata Aiton. (N. Dak., Mont.)
Credneria? daturaefolia Ward. (Mont.)
Credneria? pachyphylla Knowlton. (Yel. Nat. Pk.)
Cupressinoxylon elongatum Knowlton. (Mont.)
Cyperacites angustior (Braun) Schimper. (Yel. Nat. Pk.)
Cyperacites giganteus Knowlton. (Yel. Nat. Pk.)
Cyperacites sp. Knowlton. (Yel. Nat. Pk.)
Dennstaedtia americana Knowlton. (N. Dak., Wyo.)
Diospyros brachysepala Al. Braun. (Mont. Formation doubtful.)
Diospyros ficoidea Lesquereux. (Mont.)
Diospyros haguei Knowlton. (Yel. Nat. Pk.)
Diospyros? obtusata Ward. (Mont.)
Dryopteris weedii Knowlton. (Yel. Nat. Pk.)
Elaeodendron polymorphum Ward. (Mont., Yel. Nat. Pk.)
Elaeodendron serrulatum Ward. (Mont.)
Equisetum canaliculatum Knowlton. (Yel. Nat. Pk.)
Equisetum deciduum Knowlton. (Yel. Nat. Pk.)
Equisetum globulosum Lesquereux. (N. Dak.)
Euonymus xantholithensis Ward. (Mont.)
Fagus antipofii Abich. (Yel. Nat. Pk.)

Fagus undulata Knowlton. (Yel. Nat. Pk.)
Ficus artocarpoides Lesquereux. (N. Dak.)
Ficus deformata Knowlton. (Yel. Nat. Pk.)
Ficus densifolia Knowlton. (Yel. Nat. Pk.)
Ficus limpida Ward. (Mont.)
Ficus tiliaefolia (Braun) Heer. (Mont., Yel. Nat. Pk.)
Ficus ungeri Lesquereux. (Yel. Nat. Pk. Formation doubtful.)
Ficus viburnifolia Ward. (Mont.)
Ficus sp. Knowlton. (Yel. Nat. Pk.)
Fraxinus wrightii Knowlton. (Yel. Nat. Pk.)
Ficus lignitum Lesquereux. (Mont. Formation doubtful.)
Ginkgo adiantoides (Unger) Heer. (Mont.)
Glyptostrobus europaeus (Brongniart) Heer. (N. Dak., Mont.; widespread.)
Grewia celastroides Ward. (Mont.)
Grewia crenata (Unger) Heer. (Mont.)
Grewia obovata Heer. (Mont.)
Grewia pealei Ward. (Mont.)
Grewiopsis platanifolia Ward. (Mont.)
Grewiopsis populifolia Ward. (Mont.)
Grewiopsis viburnifolia Ward. (Mont.)
Hamamelites fothergilloides Saporta. (Mont.)
Hedera minima Ward. (Mont.)
Hedera parvula Ward. (Mont.)
Hicoria antiquora (Newberry) Knowlton. (Mont., Wyo., Dakotas.)
Hicoria crescentia Knowlton. (Yel. Nat. Pk.)
Hicoria culveri Knowlton. (Yel. Nat. Pk.)
Hicoria? sp. Knowlton. (Mont.)
Ilex microphyllina Knowlton and Cockerell. (N. Dak.)
Juglans crescentia Knowlton. (Yel. Nat. Pk.)
Juglans nigella Heer. (N. Dak., Mont.)
Juglans rugosa Lesquereux. (Yel. Nat. Pk.)
?Juglans ungeri Heer. (Mont.)
Juglans woodiana Heer. (N. Dak.)
Juglans sp. Knowlton. (Wyo.)
Laurus montana Knowlton. (Yel. Nat. Pk.)
Laurus palaeophila Knowlton and Cockerell. (Mont.)
Laurus primigenia Unger. (Yel. Nat. Pk.)
Laurus princeps Heer. (Yel. Nat. Pk. Formation doubtful.)
Laurus resurgens Saporta. (Mont.)
Leguminosites? arachioides (Lesquereux) Lesquereux. (Mont.)
Leguminosites lesquereuxiana Knowlton. (Yel. Nat. Pk.)
Lemna scutata Dawson. (Mont.)
Lygodium kaulfusii Heer. (Yel. Nat. Pk.)
Magnolia elliptica Newberry. (Mont.)
Magnolia? pollardi Knowlton. (Yel. Nat. Pk.)
Magnolia sp. Knowlton. (Wyo.)
Malapoenna cuneata (Knowlton) Knowlton. (Yel. Nat. Pk.)
Malapoenna praecursoria (Lesquereux) Knowlton. (N. Dak.)
Monimiopsis amboraefolia Saporta. (Mont.)
Monimiopsis fraterna Saporta. (Mont.)
Negundo triloba Newberry. (N. Dak.)
Nelumbo sp. Knowlton. (Mont.)
Onoclea sensibilis fossilis Newberry. (N. Dak., Wyo., Mont.)
Paliurus colombi Heer. (Yel. Nat. Pk.)

Paliurus pealei Ward. (N. Dak.)
Parrotia cuneata (Newberry) Berry. (Mont.)
Phragmites alaskana Heer. (Mont.)
Phragmites? latissima Knowlton. (Yel. Nat. Pk.)
Phragmites sp.? Newberry. (N. Dak.)
Phyllites carneosus Newberry. (N. Dak.)
Phyllites unitus Knowlton and Cockerell. (N. Dak.)
Pinus iddingsi Knowlton. (Yel. Nat. Pk.)
Pityoxylon krausei Felix. (N. Dak.)
Pityoxylon sp. Dawson. (49th Par. Formation doubtful.)
Planera lingualis Knowlton and Cockerell. (Wyo.)
Planera microphylla Newberry. (N. Dak.)
Platanus aceroides Göppert. (N. Dak. Formation doubtful.)
Platanus basilobata Ward. (Mont.)
Platanus guillelmae Göppert. (Mont., Yel. Nat. Pk.)
Platanus haydenii Newberry. (Mont. and elsewhere.)
Platanus nobilis Newberry. (N. Dak. and widespread.)
Platanus raynoldsii Newberry. (Mont., Wyo.)
Populus amblyrhyncha Ward. (Mont. and widespread.)
Populus anomala Ward. (Mont.)
Populus balsamoides eximia (Göppert) Lesquereux. (N. Dak.)
Populus cordata Newberry. (Mont.)
Populus craspedodroma Ward. (Mont.)
Populus cuneata Newberry. (Mont.)
Populus cyclomorpha Knowlton and Cockerell. (N. Dak., Mont.)
Populus daphnogenoides Ward. (Mont.)
Populus genetrix Newberry. (Mont.)
Populus glandulifera Heer. (Mont., N. Dak., Yel. Nat. Pk.)
Populus grewiopsis Ward. (Mont.)
Populus hederoides Ward. (Mont.)
Populus inaequalis Ward. (Mont.)
Populus latior truncata Al. Braun. (N. Dak.)
Populus meedsii Knowlton. (Mont.)
Populus nebrascensis Newberry. (Mont.)
Populus nervosa Newberry. (Mont.)
Populus nervosa elongata Newberry. (Mont.)
Populus newberryi Cockerell. (N. Dak.)
Populus oxyrhyncha Ward. (Mont.)
Populus richardsoni Heer. (Mont.)
Populus smilacifolia Newberry. (N. Dak.)
Populus speciosa Ward. (Mont.)
Populus whitei Ward. (Mont.)
Populus xantholithensis Knowlton. (Yel. Nat. Pk.)
Preissites wardii Knowlton. (Mont.)
Protoficus inaequalis Newberry. (Mont.)
Prunus dakotensis Lesquereux. (N. Dak.)
Pterospermites cordatus Ward. (Mont.)
Pterospermites cupanioides (Newberry) Knowlton. (Mont.)
Pterospermites minor Ward. (Mont.)
Pterospermites whitei Ward. (Mont.)
Quercus aemulans Lesquereux. (Wyo. Formation doubtful.)
Quercus bicornis Ward. (Mont.)
Quercus castanopsis Newberry. (Mont.)
Quercus consimilis Newberry? (Yel. Nat. Pk.)

Quercus culveri Knowlton. (Yel. Nat. Pk.)
Quercus dentoni Lesquereux. (N. Dak.)
Quercus dubia Newberry. (Mont.)
Quercus hesperia Knowlton. (Yel. Nat. Pk.)
Quercus? magnifolia Knowlton. (Yel. Nat. Pk.)
Quercus olafseni Heer. (Yel. Nat. Pk., N. Dak.)
Quercus penhallowi Trelease. (N. Dak.)
Quercus sullyi Newberry. (N. Dak.)
Quercus yanceyi Knowlton. (Yel. Nat. Pk.)
Quercus sp. Knowlton. (Yel. Nat. Pk.)
Rhamnites concinnus Newberry. (N. Dak.)
Rhamnus newberryi Knowlton and Cockerell. (Mont.)
Rhus unitus Knowlton and Cockerell. (N. Dak.)
Rhus winchellii Lesquereux. (Mont.)
Rhus sp. Knowlton. (Wyo.)
Sabal grandifolia Newberry. (Mont.)
Salix lavateri Heer. (Yel. Nat. Pk.)
Salpichlaena anceps (Lesquereux) Knowlton. (Yel. Nat. Pk.)
Sapindus affinis Newberry. (Mont., Yel. Nat. Pk.; common.)
?"Sapindus angustifolius" Lesquereux. (Mont.)
Sapindus glendivensis Knowlton. (Mont., Yel. Nat. Pk.)
Sapindus grandifoliolus Ward. (Mont. and elsewhere; common.)
Sapindus? membranaceus Newberry. (N. Dak.)
Sapindus obtusifolius Lesquereux. (S. Dak.)
Sapindus wardii Knowlton. (Yel. Nat. Pk.)
Selaginella collieri Knowlton. (Mont.)
Sequoia couttsiae Heer. (Yel. Nat. Pk.)
Sequoia langsdorfii (Brongniart) Heer. (Mont., Yel. Nat. Pk.)
Sequoia nordenskiöldi Heer. (Mont. and common.)
Sequoia sp.? Knowlton. (Wyo.)
Sparganium stygium Heer. (Mont., Yel. Nat. Pk.)
Spiraxis bivalvis Ward. (Mont.)
Taxites olriki Heer. (Yel. Nat. Pk.)
Taxodium occidentale Newberry. (Mont. and common.)
Thuja interrupta Newberry. (N. Dak. and common.)
Tilia populifolia Lesquereux. (Yel. Nat. Pk.)
Tilia weedii Knowlton. (Mont.)
Trapa? microphylla Lesquereux. (Mont. and common.)
Ulmus grandifolia Newberry. (Wyo.)
Ulmus minima Ward. (Mont., Yel. Nat. Pk.)
Ulmus orbicularis Ward. (Mont.)
Ulmus rhamnifolia Ward. (Mont., Yel. Nat. Pk.)
Ulmus wardii Knowlton and Cockerell. (Mont.)
Ulmus, fruits of, Knowlton. (Yel. Nat. Pk.)
Viburnum antiquum (Newberry) Hollick. (Mont., N. Dak.)
Viburnum asperum Newberry. (Mont., N. Dak.)
Viburnum betulaefolium Ward. (Mont.)
Viburnum castrae Knowlton and Cockerell. (N. Dak.)
Viburnum dakotense Lesquereux. (N. Dak.)
Viburnum dentoni Lesquereux. (N. Dak.)
Viburnum elongatum Ward. (Mont.)
Viburnum erectum Ward. (Mont.)
Viburnum finale Ward. (Mont.)
Viburnum grandedentatum Newberry. (Wyo.)

Viburnum limpidum Ward. (Mont.)
Viburnum macrodontum Ward. (Mont.)
Viburnum newberryanum Ward. (Mont)
Viburnum nordenskiöldi Heer. (Mont., N. Dak.)
Viburnum oppositinerve Ward. (Mont.)
Viburnum paucidentatum Newberry. (Wyo.)
Viburnum perfectum Ward. (Mont.)
Viburnum perplexum Ward. (Mont.)
Viburnum whymperi Heer. (Mont.)
Viburnum sp. Knowlton. (Wyo.)
Winchellia triphylla Lesquereux. (Mont.)
Woodwardia maxoni Knowlton. (Wyo.)
Woodwardia preareolata Knowlton. (Yel. Nat. Pk.)
Zizyphus cinnamomoides (Lesquereux) Lesquereux. (Mont.)
Zizyphus serrulatus Ward. (Mont., Yel. Nat. Pk.)
Zizyphus sp. Knowlton (Wyo.)

PASKAPOO FORMATION, CANADA.

Aesculus antiquus Dawson.
Aesculus sp. Dawson.
Ailanthophyllum incertum Dawson.
Alnites grandiflora (Newberry) Newberry.
Aralia dakotana Knowlton and Cockerell.
Aralia notata Lesquereux.
Aristolochia crassifolia (Newberry) Cockerell.
Betula sp. Dawson.
Carpolithes sp. Dawson.
Cercis parvifolia Lesquereux.
Clintonia oblongifolia Penhallow.
Cornus newberryi Hollick.
Cornus rhamnifolia O. Weber.
Corylus americana Walter.
Corylus macquarrii (Forbes) Heer.
Corylus rostrata Aiton.
Cupress[in]oxylon dawsoni Penhallow.
Cupressoxylon sp. (d) Dawson. (Formation doubtful.)
Dennstaedtia americana Knowlton.
Diospyros sp. Dawson.
Equisetum arcticum Heer.
Equisetum parlatorii (Heer) Schimper.
Equisetum 2 sp. Dawson.
Ginkgo 2 sp. (Dawson) Knowlton.
Ginkgo sp. Dawson.
Glyptostrobus europaeus (Brongniart) Heer.
Hicoria antiqua (Newberry) Knowlton.
Juglans acuminata Al. Braun.
Juglans laurifolia Knowlton.
Juglans leconteana Lesquereux.
Juglans nigella Heer.
Juglans occidentalis Newberry.
Lastrea fischeri Heer.
Majanthemophyllum grandifolium Penhallow.
Nelumbo saskatchuense (Dawson) Knowlton. (Formation doubtful.)
Onoclea sensibilis fossilis Newberry.

?Phegopteris guyottii (Lesquereux) Cockerell.
Phyllites carneosus Newberry.
Phyllites unitus Knowlton and Cockerell.
Phyllites sp. Dawson.
Planera lingualis Knowlton and Cockerell.
Populus cordata Newberry.
Populus daphnogenoides Ward.
Populus nebrascensis Newberry.
Populus richardsoni Heer.
Populus speciosa Ward.
Populus ungeri Lesquereux.
Quercus castanopsis Newberry. (Formation doubtful.)
Quercus consimilis Newberry. (Formation doubtful.)
Quercus dawsoni Knowlton. (Formation doubtful.)
Quercus ellisiana Lesquereux.
Quercus sp. Dawson.
Rhamnites concinnus Newberry.
Rhamnus sp. Dawson. (Formation doubtful.)
Salix laramina Dawson. (Formation doubtful.)
Sassafras burpeana Dawson.
Sassafras selwynii Dawson.
Scirpus 2 sp. Dawson.
Sequoia burgessii Penhallow.
Sequoia couttsiae Heer.
Sequoia langsdorfii (Brongniart) Heer.
Sequoia nordenskiöldi Heer.
Sequoia 2 sp. Dawson.
Sphenozamites? oblanceolatus Penhallow.
Symphorocarpophyllum albertum Dawson.
Symphorocarpophyllum linnaeiforme Dawson.
Symphorocarpophyllum sp. Dawson.
Taxites sp. Dawson.
Taxodium dubium (Sternberg) Heer.
Taxodium laramianum Penhallow.
Taxoxylon sp. Dawson.
Thalictrum sp. Dawson. (Formation doubtful.)
Thuja interrupta Newberry.
Thuja sp. Dawson.
Thuja sp. Penhallow. (Formation doubtful.)
Typha sp. Penhallow.
Ulnus praecursor Dawson.
Viburnum asperum Newberry.
Viburnum calgarianum Dawson.
Viburnum castrae Knowlton and Cockerell.
Viburnum dentoni Lesquereux.
Viburnum ovatum Penhallow.
Viburnum oxycoccoides Dawson. (Formation doubtful.)
Viburnum saskatchuense Dawson.

BATON FORMATION.

Acer fragilis Knowlton.
Alismaphyllum crassifolium Knowlton.
Allantodiopsis erosa (Lesquereux) Knowlton and Maxon.
Andromeda (?) lanceolata Knowlton.
Andromeda scripta Knowlton.

Anemia occidentalis Knowlton.
Apeibopsis? neomexicanus Knowlton.
Apocynophyllum lesquereuxii Ettingshausen.
Apocynophyllum linifolium Knowlton.
Apocynophyllum wilcoxensis Berry.
Aralia coloradense Knowlton.
Aralia? serrata Knowlton.
Aristolochia? elongata Knowlton.
Artocarpus similis Knowlton.
Asplenium? primero Knowlton.
Berchemia multinervis (Braun) Heer.
Carapa eolignitica Berry?
Carpites coffcaeformis Lesquereux.
Carpites ligatus Lesquereux. (Formation doubtful.)
Carpites spiralis (Lesquereux) Lesquereux. (Formation doubtful.)
Carpolithes spinosus Newberry. (Formation doubtful.)
Cassia fisheriana Knowlton.
Cassia richardsoni Knowlton.
Cassia sapindoides Knowlton.
Castalca leei Knowlton.
Castanea intermedia Lesquereux.
Celastrus serratus Knowlton.
Celastrus? sp. Knowlton.
Cercocarpus orestesi Knowlton.
Chionanthus membranaceus Knowlton.
Cinnamomum? ficifolium Knowlton.
Cinnamomum linifolium Knowlton.
Cinnamomum mississippiense? Lesquereux.
Cissus coloradensis Knowlton and Cockerell.
Cissus grossedentata Knowlton.
Cornus neomexicana Knowlton.
Cornus studeri? Heer.
Dombeyopsis magnifolia Knowlton.
Dryophyllum aquamarum? Ward.
Dryophyllum moorii (Lesquereux) Berry.
Dryophyllum tennesseensis Berry.
Dryopteris? cladophleboides Knowlton.
Dryopteris? sp. Knowlton.
Euonymus splendens Berry.
Euphorbocarpum richardsoni Knowlton.
Fagus papyracea Knowlton.
Ficus aguilar Knowlton.
Ficus artocarpoides Lesquereux.
Ficus cockerelli Knowlton. (Formation doubtful.)
Ficus denveriana Cockerell.
Ficus duplicata Knowlton.
Ficus harrisiana Hollick.
Ficus minutidens Knowlton.
Ficus neoplanicostata Knowlton.
Ficus occidentalis (Lesquereux) Lesquereux.
Ficus planicostata goldiana Lesquereux.
Ficus praetrinervis Knowlton.
Ficus pseudopopulus Lesquereux.
Ficus ratonensis Knowlton.

Ficus richardsoni Knowlton.
Ficus schimperi Lesquereux.
Ficus? smithsoniana Lesquereux.
Ficus uncata Lesquereux.
"Geonoma" gigantea Knowlton.
Geonomites tenuirachis Lesquereux.
Inga heterophylla Knowlton.
Juglans acuminata Braun.
Juglans berryi Knowlton.
Juglans leconteana Lesquereux.
Juglans minutidens Knowlton.
Juglans nigella Heer.
Juglans rhamnoides Lesquereux.
Juglans rugosa Lesquereux.
Juglans sapindiformis Knowlton.
Juglans sapindoides Knowlton.
Juglans schimperi Lesquereux.
Laurus? caudata Knowlton.
Laurus ratonensis Knowlton.
Laurus socialis Lesquereux.
Laurus utahensis Lesquereux.
Leguminosites arachioides Lesquereux.
Liquidambar? cucharas Knowlton.
Magnolia angustifolia Newberry.
Magnolia cordifolia Lesquereux.
Magnolia hilgardiana Lesquereux.
Magnolia laurifolia Lesquereux.
Magnolia leei Knowlton.
Magnolia lesleyana Lesquereux.
Magnolia magnifolia Knowlton.
Magnolia regalis? Heer.
Magnolia rotundifolia Newberry.
Nectandra lancifolia (Lesquereux) Berry.
Nelumbo lakesiana (Lesquereux) Knowlton.
Nyssa lanceolata Lesquereux.
Nyssa? racemosa Knowlton.
Oreodaphne? ratonensis Knowlton.
Oreodoxites plicatus Lesquereux.
Osmanthus pedatus (Lesquereux) Berry? (Colo.)
Palaeoaster inquirenda Knowlton.
Paliurus zizyphoides Lesquereux.
Palmocarpon compositum (Lesquereux) Lesquereux. (Formation doubtful.)
Palmocarpon mexicanum (Lesquereux) Knowlton. (Formation doubtful.)
Palmocarpon palmarum (Lesquereux) Knowlton.
Phyllites retusoides Knowlton.
Pisonia chlorophylloides Berry.
Platanus aceroides Göppert.
Platanus aceroides cuneata Knowlton.
Platanus aceroides latifolia Knowlton.
Platanus guillelmae Göppert.
Platanus guillelmae heerii Knowlton.
Platanus platanoides (Lesquereux) Knowlton.
Platanus raynoldsii Newberry.
Platanus? regularis Knowlton. (Formation doubtful.)

Platanus rhomboidea Lesquereux.
Populus neotremuloides Knowlton.
Populus, female ament of, Knowlton.
Prunus coloradensis Knowlton.
Pteris linearis Knowlton.
Pteris russellii Newberry. (Formation doubtful.)
Quercus fisheriana Knowlton.
Quercus? neomexicanus Knowlton.
Quercus? ratonensis Knowlton.
Quercus simplex Newberry.
Rhamnus cleburni Lesquereux.
Rhamnus deletus? Heer. (Formation doubtful.)
Rhamnus fisheri Lesquereux. (Formation doubtful.)
Rhamnus goldianus? Lesquereux.
Rhamnus obovatus Lesquereux. (Formation doubtful.)
Rhamnus? woottonensis Knowlton.
Rhus viburnoides Knowlton.
Sabal? eocenica (Lesquereux) Knowlton.
Sabal? leei Knowlton.
Sabal? rugosa Knowlton.
Sabal? ungeri (Lesquereux) Knowlton.
Sabalites grayanus Lesquereux.
Sapindus affinis Newberry.
Sapindus caudatus Lesquereux.
Sapindus rocklandensis Knowlton.
Sophora nervosa Knowlton.
Sterculia berryana Knowlton.
Tilia speciosissima Knowlton.
Ulmus sp. Knowlton.
Viburnum contortum Lesquereux.
Viburnum lakesii Lesquereux.
Viburnum magnum Knowlton.
Viburnum speciosum Knowlton.
Viburnum woottonianum Knowlton.
Vitis inominata Knowlton.
Vitis leei Knowlton.
Vitis? platanifolia Knowlton.
Zanthoxylum dubium Lesquereux. (Formation doubtful.)
Zizyphus fibrillosus (Lesquereux) Lesquereux.
Zizyphus meiggsii (Lesquereux) Berry.

DENVER FORMATION.

Acorus brachystachys Heer.
Allantodiopsis erosa (Lesquereux) Knowlton and Maxon.
Alnus auraria Knowlton and Cockerell.
Alnus carpinifolia Lesquereux.
Amelanchier typica var. Lesquereux.
Andromeda linearifolia Lesquereux.
Aralia notata Lesquereux?
Artocarpus pungens (Lesquereux) Hollick.
Arundo göpperti? Münster.
Arundo? obtusa Lesquereux.
Asimina eocenica Lesquereux.
Asplenium crossii Knowlton.

Berchemia multinervis (Al. Braun) Heer.
Betula fallax Lesquereux.
Betula gracilis? Ludwig.
Betula schimperi Lesquereux.
Carex berthoudi Lesquereux.
Carpites coffeaeformis Lesquereux.
Carpites costatus Lesquereux.
Carpites laurineus Lesquereux.
Carpites minutulus Lesquereux.
Carpites oviformis Lesquereux.
Carpites triangulosus Lesquereux.
Carpolithes corrugatus (Lesquereux) Cockerell.
Castanea intermedia Lesquereux.
Celastrinites artocarpidioides Lesquereux.
Celastrus gaudini Lesquereux.
Cissus coloradensis Knowlton and Cockerell.
Cissus corylifolia Lesquereux.
Cissus duplicato-serrata Lesquereux.
Cissus lobato-crenata Lesquereux.
Cornus holmesii Lesquereux.
Cornus impressa Lesquereux.
Cornus studeri? Heer.
Crataegus antiqua Heer.
Crataegus betulaefolia Lesquereux.
Crataegus engelhardti Lesquereux.
Crataegus holmesii Lesquereux.
Crataegus myricoides Lesquereux.
Daphnogene anglica? Heer.
Diospyros brachysepala Al. Braun.
Dombeyopsis grandifolia? Unger.
Dombeyopsis obtusa Lesquereux.
Dryopteris arguta (Lesquereux) Knowlton.
Dryopteris nigricans (Lesquereux) Knowlton.
Dryopteris polypodioides (Ettingshausen) Knowlton.
Equisetum sp. (Lesquereux).
Eriocaulon? porosum Lesquereux.
Ficus andraei Lesquereux.
Ficus berthoudi Lesquereux.
Ficus coloradensis Cockerell.
Ficus denveriana Cockerell.
Ficus occidentalis (Lesquereux) Lesquereux.
Ficus planicostata clintoni (Lesquereux) Knowlton.
Ficus subtruncata Lesquereux.
Ficus tiliaefolia (Braun) Heer.
Ficus zizyphoides Lesquereux. (Formation doubtful.)
Fraxinus eocenica Lesquereux.
Fraxinus praedicta Heer.
Geonomites goldianus (Lesquereux) Lesquereux.
Geonomites graminifolius Lesquereux.
Geonomites tenuirachis Lesquereux.
Hymenophyllum confusum . .quereux.
Juglans rugosa Lesquereux.
Juglans schimperi Lesquereux.
Juglans thermalis Lesquereux.

Laurus primigenia Unger.
Laurus schmidtiana Heer.
Laurus socialis Lesquereux.
Leguminosites? arachioides (Lesquereux) Lesquereux.
Magnolia magnifolia Knowlton.
Negundo decurrens Lesquereux.
Nelumbo lakesiana (Lesquereux) Knowlton.
Nelumbo tenuifolia (Lesquereux) Knowlton.
Nyssa denveriana Knowlton.
Nyssa europaea Unger.
Nyssa lanceolata Lesquereux.
Nyssa? racemosa Knowlton.
Oreodoxites plicatus Lesquereux.
Paliurus coloradensis Lesquereux.
Paliurus zizyphoides Lesquereux.
Palmocarpon lineatum Lesquereux.
Palmocarpon palmarum (Lesquéreux) Knowlton.
Palmocarpon subcylindricum Lesquereux.
Palmocarpon truncatum Lesquereux. (Formation doubtful.)
Persea brossiana (Lesquereux) Lesquereux.
Phyllites cyclophyllus (Lesquereux) Hollick.
Piper heerii Lesquereux.
Pisonia chlorophylloides Berry.
Platanus aceroides Göppert.
Platanus guillelmae Göppert.
Platanus haydenii Newberry.
Platanus raynoldsii integrifolia Lesquereux.
Platanus rhomboidea Lesquereux.
Populus nebrascensis acute-dentata Lesquereux.
Populus nebrascensis grandidentata Lesquereux.
Populus nebrascensis longifolia Lesquereux.
Populus nebrascensis rotunda Lesquereux.
Populus subrotunda Lesquereux.
Populus tenuinervata Lesquereux.
Populus ungeri Lesquereux.
Protoficus zeilleri Lesquereux.
Pteris pseudopinnaeformis Lesquereux.
Pterocarya americana Lesquereux. (Formation doubtful.)
Pterocarya retusa Lesquereux.
Pterospermites grandidentatus Lesquereux.
Pterospermites sp. Lesquereux.
Quercus celastrifolia Lesquereux.
Quercus coloradensis Lesquereux.
Quercus crossii Lesquereux.
Quercus eucalyptifolia Ettingshausen.
Quercus haidingeri Ettingshausen. (Formation doubtful.)
Quercus viburnifolia Lesquereux.
Quercus whitei Lesquereux.
Rhamnus alaternoides Heer.
Rhamnus cleburni Lesquereux.
Rhamnus crenulatus Knowlton and Cockerell.
Rhamnus deformatus Lesquereux.
Rhamnus goldianus Lesquereux.
Rhamnus obovatus Lesquereux. (Formation doubtful.)

Rhamnus rectinervis Heer.
Sabal? eocenica (Lesquereux) Knowlton.
Sabalites fructifer Lesquereux.
Saccaloma gardneri (Lesquereux) Knowlton.
Salpichlaena anceps (Lesquereux) Knowlton.
Sapindus caudatus Lesquereux.
Sclerotites rubellus (Lesquereux) Meschinelli.
Selaginella berthoudi Lesquereux.
Sequoia? sp. Lesquereux.
Sterculia saportanea Knowlton.
Styrax ambra Unger.
Styrax laramiense Lesquereux.
Ulmus antecedens Lesquereux.
Ulmus quercifolia Unger.
Vaccinium coloradense Lesquereux.
Viburnum goldianum Lesquereux.
Viburnum lakesii Lesquereux.
Viburnum solitarium Lesquereux.
Woodwardia latiloba Lesquereux.
Zingiberites dubius Lesquereux.
Zizyphus beckwithii Lesquereux.
Zizyphus distortus Lesquereux.
Zizyphus fibrillosus (Lesquereux) Lesquereux.
Zizyphus lesquereuxii Knowlton.
Zizyphus meekii Lesquereux.

DAWSON ARKOSE, COLORADO.

Acer trilobatum productum? (Al. Braun) Heer.
Berchemia multinervis (Al. Braun) Heer.
Cissus coloradensis Knowlton and Cockerell.
Dombeyopsis obtusa Lesquereux.
?Laurus wardiana Knowlton.
Myrica torreyi minor Lesquereux.
Nelumbo lakesiana (Lesquereux) Knowlton.
Nyssa lanceolata Lesquereux.
Platanus marginata (Lesquereux) Heer.
Rhamnus salicifolius Lesquereux.
Sabal? eocenica (Lesquereux) Knowlton.
Saccaloma gardneri (Lesquereux) Knowlton.
Sequoia acuminata Lesquereux.
Sequoia obovata Knowlton.

EVANSTON FORMATION.

Acer trilobatum? (Sternberg) Al. Braun.
Acer secreta Lesquereux.
Alnus americana Ettingshausen.
Betula stevensoni Lesquereux.
Carpites lineatus (Newberry) Lesquereux.
Carpites utahensis Lesquereux.
Cassia evanstonensis Knowlton and Cockerell.
Cyperus chavannesi Heer.
Diospyros hexaphylla (Lesquereux) Knowlton.
Ficus evanstonensis Knowlton and Cockerell.
Ficus pseudo-populus Lesquereux.
Juglans leconteana Lesquereux.

Juglans rugosa Lesquereux.
Laurus primigenia Unger.
Laurus socialis Lesquereux.
Leguminosites? arachioides (Lesquereux) Lesquereux.
Malapoenna sessiliflora (Lesquereux) Knowlton.
Morus affinis Lesquereux.
Populus subrotunda Lesquereux.
Quercus drymeja Unger var.? Lesquereux.
Quercus negundoides Lesquereux.
Rhamnus rectinervis Heer.
Rhus evansii Lesquereux.
Vitis olriki Heer.

LIVINGSTON FORMATION.

Abietites dubius Lesquereux.
Abietites setigera (Lesquereux) Lesquereux.
Andromeda affinis Lesquereux.
Andromeda grayana Heer.
Andromeda reticulata? Ettingshausen.
Anemia elongata (Newberry) Knowlton.
Caulinites sparganioides Lesquereux.
Celastrinites ambiguus (Lesquereux) Knowlton.
Cinnamomum ellipticum Knowlton.
Cornus rhamnifolia Weber.
Dombeyopsis platanoides Lesquereux.
Fraxinus denticulata Heer.
Juglans rhamnoides Lesquereux.
Juglans rugosa Lesquereux.
Laurus socialis Lesquereux.
Leguminosites lesquereuxiana Knowlton.
Malapoenna weediana (Knowlton) Knowlton.
Nyssa lanceolata Lesquereux. (Formation doubtful.)
Protophyllocladus lanceolatus (Knowlton) Berry.
Protophyllocladus polymorphus (Lesquereux) Berry.
Quercus ellisiana Lesquereux.
Quercus? fraxinifolia Lesquereux.
Quercus godeti? Heer.
Sequoia langsdorfii (Brongniart) Heer. (Formation doubtful.)
Sparganium sp. Lesquereux.
Zizyphus meekii Lesquereux.

HANNA FORMATION, CARBON COUNTY, WYO.

Acorus brachystachys Heer.
Ampelopsis bruneri (Ward) Cockerell.
Ampelopsis bruneri carbonensis (Ward) Cockerell.
Apeibopsis? discolor (Lesquereux) Lesquereux.
Asimina eocenica Lesquereux.
Calycites polysepala Newberry.
Carpites cocculoides (Heer) Lesquereux.
Carpites cocculoides major Lesquereux.
Coccoloba laevigata Lesquereux.
Crataegus aequidentata Lesquereux.
Equisetum haydenii Lesquereux.
Ficus coloradensis Cockerell. (Formation doubtful.)
Juglans crossii Knowlton.

Laurus primigenia Unger.
Malapoenna carbonensis (Ward) Knowlton.
Paliurus colombi Heer.
Paliurus pulcherrimus Ward.
Platanus aceroides Göppert.
Platanus cordata Knowlton.
Platanus guillelmae Goppert.
Populus aequalis Lesquereux.
Populus cyclomorpha Knowlton and Cockerell.
Populus decipiens Lesquereux.
Populus mutabilis crenata Heer.
Populus subrotunda Lesquereux.
Quercus carbonensis Ward.
Quercus haydenii Lesquereux.
Sapindus sp. Knowlton.
Smilax carbonensis Cockerell.
Trapa sp. Knowlton.
Zizyphus lesquereuxii Knowlton.

GREEN RIVER FORMATION.

Acer lesquereuxii Knowlton.
Acer sp. Lesquereux. (Wyo. Formation doubtful.)
Acrostichum hesperium Newberry. (Wyo.)
Ailanthus longe-petiolata Lesquereux. (Wyo.)
Alnus inaequilateralis Lesquereux. (Wyo.)
Ampelopsis tertiaria Lesquereux. (Wyo.)
Amygdalus gracilis Lesquereux. (Wyo.)
Andromeda delicatula Lesquereux. (Wyo.)
Antholithes improbus Lesquereux. (Wyo.)
Apocynophyllum scudderi Lesquereux. (Wyo.)
Aralia angustiloba Lesquereux. (Wyo. Formation doubtful.)
Aralia? gracilis (Lesquereux) Lesquereux. (Wyo. Formation doubtful.)
Aralia wyomingensis Knowlton and Cockerell. (Wyo.)
Arundo göpperti? Münster.
Arundo reperta Lesquereux.
Blechnum göpperti Ettingshausen. (Utah. Formation doubtful.)
Brasenia antiqua Newberry. (Wyo.)
Carya heerii (Ettingshausen) Heer. (Wyo.)
Cissus parrottiaefolia Lesquereux. (Wyo.)
Cyperacites haydenii (Lesquereux) Knowlton. (Wyo.)
Cyperites deucalionis Heer. (Wyo. Formation doubtful.)
Cyperus braunianus? Heer. (Wyo. Formation doubtful.)
Cyperus chavannesii Heer. (Wyo.)
Equisetum haydenii Lesquereux. (Wyo. Formation doubtful.)
Equisetum wyomingense Lesquereux. (Wyo.)
Eucalyptus? americana Lesquereux. (Wyo.)
Euonymus flexifolius Lesquereux. (Wyo.)
Ficus alkalina Lesquereux. (Wyo.)
Ficus populina Heer. (Wyo.)
Ficus tenuinervis Lesquereux. (Wyo.)
Ficus ungeri Lesquereux. (Wyo.)
Ficus wyomingiana Lesquereux. (Wyo.)
Flabellaria florissanti Lesquereux. (Wyo.)
Geonomites goldianus (Lesquereux) Lesquereux.
Ilex affinis Lesquereux. (Wyo.)

Ilex dissimilis Lesquereux. (Mont. Formation doubtful.)
Ilex maculata Lesquereux. (Wyo.)
Ilex wyomingiana Lesquereux. (Wyo.)
Juglans alkalina Lesquereux. (Wyo.)
Juglans crossii Knowlton. (Wyo.)
Juglans dentata Newberry. (Wyo.)
Juglans occidentalis Newberry. (Wyo.)
Juglans schimperi Lesquereux. (Wyo.)
Juncus sp. Lesquereux. (Wyo.)
Leguminosites alternans Lesquereux. (Wyo.)
Leguminosites lesquereuxiana Knowlton. (Wyo.)
Lomatia microphylla Lesquereux. (Wyo. Formation doubtful.)
Lygodium dentoni Lesquereux. (Wyo. Formation doubtful.)
Lygodium kaulfusii Heer. (Wyo.)
Magnolia sp. Lesquereux. (Wyo.)
Manicaria haydenii Newberry. (Wyo.)
Musophyllum complicatum Lesquereux. (Wyo.
Myrica ludwigii Schimper. (Wyo.)
Myrica nigricans Lesquereux. (Wyo.)
Myrica salicina Unger. (Wyo.)
Myrica studeri? Heer. (Wyo.? Formation doubtful.)
Nordenskiöldia borealis Heer. (Wyo.)
Pecopteris (Cheilanthes) sepulta Newberry. (Wyo.)
Phaseolites juglandinus? Heer. (Wyo. Formation doubtful.)
Phyllites fremonti Unger. (Wyo. Formation doubtful.)
Phyllites sapindiformis Lesquereux. (Wyo. Formation doubtful.)
Planera nervosa Newberry. (Wyo.)
Planera variabilis Newberry. (Wyo.)
Poacites laevis Al. Braun. (Wyo. Formation doubtful.)
Quercus castaneopsis Lesquereux. (Wyo.)
Quercus castanoides Newberry. (Wyo.)
Quercus lonchitis Unger. (Wyo.)
Rhamnus washakiensis Cockerell. (Wyo.)
Rhus lesquereuxii Knowlton and Cockerell. (Wyo.)
Sabal powellii Newberry. (Wyo.)
Salix angusta Al. Braun. (Wyo.)
Salix media Heer. (Wyo)
Sapindus dentoni Lesquereux. (Utah. Formation doubtful.)
Sequoia heerii Lesquereux. (Mont. Formation doubtful.)
Sphaerites myricae (Lesquereux) Meschinelli. (Wyo.)
Zanthoxylon juglandinum? Al. Braun. (Wyo. Formation doubtful.)
Zizyphus cinnamomoides (Lesquereux) Lesquereux. (Wyo.)
Zizyphus longifolia Newberry. (Wyo.)

WASATCH (?) FORMATION, CLARK FORK BASIN, WYO.[1]

Tithymalus phenacodorum Cockerell.

PACIFIC COAST REGION.

PUGET GROUP.

Acer trilobatum? (Sternberg) Al. Braun.
Equisetum newberryi Knowlton and Cockerell.
Persoonia oviformis Lesquereux.

[1] No other fossil plants have been described as from the Wasatch formation, but at some localities, especially in southern Wyoming, rocks mapped as Wasatch have yielded plants that have been identified as belonging to the Fort Union flora.

Populus flabellum Newberry.
Quercus banksiaefolia Newberry.
Quercus coriacea Newberry.
Quercus evansii Lesquereux.
Quercus flexuosa Newberry.
Quercus gaudini Lesquereux.
Quercus washingtonensis Trelease.
Rhamnus gaudini Heer. (Age doubtful.)
Rhynchostegium knowltoni Britton.
Sabalites campbelli (Newberry) Lesquereux.
Salix islandicus Lesquereux.
Salvinia elliptica Newberry.
Taxodium dubium (Sternberg) Heer.
Thuja interrupta Newberry.

SWAUK FORMATION.[1]

Aspleninum magnum intermedium Duror.
Ficus ungeri Lesquereux.
Glyptostrobus ungeri Heer.
Hicoria antiquora (Newberry) Knowlton.
Juglans acuminata Heer.
Laurus cascadia Duror.
Magnolia nordenskiöldi Heer.
Populus amblyrhyncha Ward.
Populus arctica Heer.
Populus cuneata Newberry.
Populus zaddachi Heer.
Protoficus fossi Duror.
Pteris pinnaeformis Heer.
Pterospermites whitei Ward.
Sabal powelli Newberry.
Sapindus obtusifolius Lesquereux.
Sequoia nordenskiöldi Heer.
Taxodium distichum miocenum.

PAYETTE FORMATION.

Cassia idahoensis Knowlton.
Celastrus lindgreni Knowlton.
Dryopteris idahoensis Knowlton.
Equisetum sp. Knowlton.
Juglans oregoniana Lesquereux.
Myrica? idahoensis Knowlton.
Myrica lanceolata Knowlton.
Phyllites flexuosus Knowlton.
Phyllites payettensis Knowlton.
Pinus sp. Knowlton.
Populus eotremuloides Knowlton.
Populus lindgreni Knowlton.
Populus occidentalis Knowlton.
Quercus idahoensis Knowlton.
Quercus payettensis Knowlton.
Quercus simulata Knowlton.
Quercus? sp. Knowlton.

[1] List from report by Caroline A. Duror, Jour. Geology, vol. 24, pp. 570-580, 1916.

Rhus payettensis Knowlton.
Sequoia sp. Knowlton.
Trapa americana Knowlton.
Trapa? occidentalis Knowlton.

CLARNO FORMATION (LOWER PART).

Allantodiopsis erosa (Lesquereux) Knowlton and Maxon.
Aralia wardiana Knowlton and Cockerell.
Aralia sp. Knowlton.
Aralia? sp. Knowlton.
Cinnamomum dilleri Knowlton.
Cornus ferox? Unger.
Diospyros alaskana Schimper.
Equisetum oregonense Newberry.
Ficus tenuinervis Lesquereux.
Hicoria? oregoniana Knowlton.
Juglans? bendirei Knowlton.
Juglans vetusta Heer.
Lastrea fischeri Heer.
Lygodium kaulfusii Heer.
Magnolia culveri Knowlton.
Phyllites personatus Knowlton.
Phyllites wascoensis Lesquereux.
Phyllites sp. Knowlton.
Pteris elegans Newberry.
Pteris pseudopinneaeformis Lesquereux.
Quercus furcinervis Rossmässler.
Quercus? sp. Knowlton.
?Rhamnus cleburni Lesquereux.
Salix schimperi Lesquereux.
Taxodium dubium (Sternberg) Heer.

CLARNO FORMATION (UPPER PART).

Acer osmonti Knowlton.
Acer sp. Knowlton.
Acer sp.? Lesquereux.
Ailanthus ovata Lesquereux.
Alnus carpinoides Lesquereux.
Alnus macrodonta Knowlton.
Alnus serrulata fossilis Newberry.
Alnus sp.? Knowlton.
Betula angustifolia Newberry.
Betula bendirei Knowlton.
Betula heterodonta Newberry.
Betula sp.(?) Newberry.
Carpinus betuloides Unger.
Cassia sp.? Newberry.
Cinnamomum bendirei Knowlton.
Corylus macquarrii (Forbes) Heer.
Crataegus newberryi Cockerell.
Ficus planicostata Lesquereux.
Fraxinus denticulata Heer?
Fraxinus oregoniana Knowlton and Cockerell.
Grewia auriculata Lesquereux.

Grewia crenata (Unger) Heer.
Hicoria sp.? Knowlton.
Juglans acuminata Al. Braun.
Juglans crassifolia Knowlton.
Juglans cryptata Knowlton.
Juglans schimperi Lesquereux.
Juglans sp. Knowlton.
Liquidambar europaeum Al. Braun.
Magnolia dayana Cockerell.
Myrica? personata Knowlton.
Odostemon simplex (Newberry) Cockerell.
Phyllites oregonianus Knowlton.
Platanus aspera Newberry.
Platanus condoni (Newberry) Knowlton.
Populus polymorpha Newberry.
Quercus berryi Trelease.
Quercus clarnensis Trelease.
Quercus consimilis Newberry.
Quercus drymeja Unger.
Quercus oregoniana Knowlton.
Quercus paucidentata Newberry.
Quercus pseudo-alnus Ettingshausen.
Quercus simplex Newberry.
Rhamnus eridani Unger.
Salix sp. Knowlton.
Sapindus merriami Knowlton.
Sequoia heerii Lesquereux.
Sequoia langsdorfii (Brongniart) Heer.
Ulmus newberryi Knowlton.
Ulmus speciosa Newberry.

IONE FORMATION, CALIFORNIA.

Ficus tiliaefolia (Braun) Heer.
Magnolia californica Lesquereux.
Persea dilleri Lesquereux. (Formation doubtful.)
Platanus appendiculata Lesquereux. (Formation doubtful.)
Populus zaddachi Heer. (Formation doubtful.)
Sabalites californicus Lesquereux.
Sequoia sp. Knowlton.

KENAI FORMATION, ALASKA.

Abies sp. Grewingk. (Kodiak Island.)
Acer grahamensis Knowlton and Cockerell. (Port Graham.)
Acer trilobatum productum? (Al. Braun) Heer. (Herendeen Bay.)
Acer trilobatum var. Knowlton. (Kukak Bay.)
Aesculus arctica Knowlton. (Kukak Bay.)
Alnites grandifolia (Newberry) Newberry. (Kenai (?), Cook Inlet.)
Alnus alaskana Newberry. (Kenai (?), Kootznahov Archipelago.)
Alnus corylina Knowlton and Cockerell. (Kachemak Bay, Cook Inlet, Kukak Bay, Alaska Peninsula.)
Alnus grandiflora Newberry. (Cook Inlet.)
Alnus kefersteinii (Göppert) Heer. (No locality.)
Alnus kefersteinii var. Heer. (Port Graham.)
Alnus sp. Grewingk. (Kenai Peninsula.)
Alnus sp. Knowlton. (Kukak Bay.)

Andromeda grayana Heer. (Port Graham, Kukak Bay.)
Betula alaskana Lesquereux. (Chignik Bay.)
Betula grandifolia Ettingshausen. (Port Graham and Hess Creek.)
Betula prisca Ettingshausen. (Port Graham.)
Betula sp. Knowlton. (Kukak Bay.)
Carex servata Heer. (Port Graham, Herendeen Bay.)
Carex sp. Lesquereux. (Sitka. Formation doubtful.)
Carpinus grandis Unger. (Cook Inlet and Port Graham.)
Carpites sp. Knowlton. (Miller's mine, Yukon River; Ninilchik.)
"Caulinia laevis (Göppert) Göppert." (Cook Inlet.)
Celastrus borealis Heer. (Port Graham.)
Comptonia cuspidata Lesquereux. (Unga Island.)
Comptonia praemissa Lesquereux. (Unga Island.)
Cornus orbifera Heer. (Cook Inlet.)
Corylus harrimani Knowlton. (Kukak Bay.)
Corylus macquarrii (Forbes) Heer. (Yukon River, Cook Inlet, Unga Island, Port Graham, Herendeen Bay, Kukak Bay.)
Corylus macquarrii macrophylla Heer. (Port Graham.)
Corylus? palachei Knowlton. (Kukak Bay.)
Diospyros alaskana Schimper. (Neniltschik.)
Diospyros anceps Heer. (Cook Inlet.)
Diospyros stenosepala Heer. (Neniltschik.)
Ebenoxylon boreale Platen. (Alaska. Formation doubtful.)
Elaeodendron helveticum Heer. (Unga Island.)
Equisetum globulosum Lesquereux. (Kukak Bay and Chignik Bay?)
Fagus antipofii Abich. (Port Graham.)
Fagus deucalionis Unger. (Kachemak Bay.)
Fagus feroniae Unger. (Port Graham.)
Fagus macrophylla Unger. (Port Graham.)
Ficus? alaskana Newberry. (Cook Inlet, Admiralty Inlet, Millers Creek, Yukon River, 25 miles below Mission Creek, above Mynook Creek.)
Ficus dalli Cockerell. (Cook Inlet.)
Fraxinus herendeensis Knowlton. (Herendeen Bay.)
Ginkgo adiantoides (Unger) Heer. (Sitka and Herendeen Bay.)
Glyptostrobus europaeus (Brongniart) Heer. (Kuiu Island near Sitka, Ninilchik, Herendeen Bay, Admiralty Island.)
Hedera auriculata Heer. (Port Graham.)
Hedera macclurii Heer. (Chignik Bay.)
Hicoria magnifica Knowlton. (Kukuk Bay.)
Ilex insignis Heer. (Port Graham.)
Juglans acuminata Al. Braun. (Port Graham and Kukak Bay.)
Juglans nigella Heer. (Port Graham and Admiralty Inlet.)
Juglans picroides Heer. (Port Graham.)
Juglans townsendi Knowlton. (Herendeen Bay.)
Juglans woodiana Heer. (Chignik Bay.)
Liquidambar europaeum Al. Braun. (Port Graham.)
Magnolia nordenskiöldi Heer. (Chignik Bay.)
Myrica banksiaefolia Unger. (Port Graham.)
Myrica vindobonensis (Ettingshausen) Heer. (Ninilchik. Formation doubtful.)
"Neuropteris acutifolia" Grewingk. (Unga Island.)
Nyssa arctica Heer. (Unga Island.)
Osmunda doroschkiana Göppert. (Unga Island.)
Paliurus colombi Heer. (Locality?)
Phragmites alaskana Heer. (Port Graham, Yukon River.)

Phyllites arctica Knowlton. (Herendeen Bay.)
Phyllites saundersi Knowlton. (Kukak Bay)
Picea harrimani Knowlton. (Kukak Bay.)
Picea 2 sp. Knowlton. (Kukak Bay.)
Pinites pannonicus (Unger) Göppert. (Unga Island. Formation doubtful.)
Pinus? 2 sp. Knowlton. (Kukak Bay.)
Pinus sp. Heer. (Port Graham.)
Pityoxylon inaequale Felix. (Danaaha, Alaska. Formation doubtful.)
Planera ungeri Ettingshausen? (Port Graham.)
Platanus sp.? Knowlton. (Miller's mine, Yukon River.)
Poacites tenue-striatus Heer. (Port Graham.)
Populus arctica Heer. (Chignik Bay and elsewhere.)
Populus balsamoides Göppert. (Port Graham, near Sitka, and Yukon River.)
Populus balsamoides eximia (Göppert) Lesquereux. (Kootznahoo.)
Populus glandulifera Heer. (Port Graham.)
Populus latior Al. Braun. (Port Graham.)
Populus leucophylla Unger. (Locality?)
Populus richardsoni Heer. (Chignik Bay.)
?Populus zaddachi Heer. (Port Graham.)
Prunus variabilis Newberry. (Cook Inlet.)
Pteris sitkensis Heer. (Near Sitka. Formation doubtful.)
Pterospermites alaskana Knowlton. (Kukak Bay.)
Pterospermites dentatus Heer. (Below Mission Creek and Yukon River.)
Pterospermites magnifolia Knowlton. (Kukak Bay.)
Quercus alaskana Trelease. (Port Graham.)
Quercus chamissonis Heer. (Port Graham.)
Quercus dallii Lesquereux. (Cook Inlet.)
Quercus furuhjelmi Heer. (Port Graham.)
Quercus grönlandica Heer. (Cook Inlet.)
Quercus pseudo-castanea Göppert. (Port Graham.)
Quercus sp. Knowlton. (Miller's mine, Yukon River.)
Rhus frigida Knowlton. (Herendeen Bay.)
Sagittaria pulchella Heer. (Locality?)
Sagittaria n. sp.? Lesquereux. (Sitka. Formation doubtful.)
Salix lavateri Heer. (Port Graham.)
Salix macrophylla Heer. (Port Graham.)
Salix minuta Knowlton. (Herendeen Bay.)
Salix pilosula Göppert. (Formation and locality doubtful.)
Salix raeana Heer. (Cook Inlet.)
Salix varians Göppert. (English Bay and Ninilchik.)
?Salix wimmeriana Göppert. (Ninilchik.)
Salix wyomingensis Knowlton and Cockerell. (Ninilchik. Formation doubtful.)
Sequoia heerii Lesquereux. (Kukak Bay.)
Sequoia langsdorfii (Brongniart) Heer. (Unga Island, Herendeen Bay, and 25 miles below Mission Creek.)
Sequoia langsdorfii var. Heer. (Sitka. Formation doubtful.)
Sequoia spinosa. Newberry. (Cook Inlet. Formation doubtful.)
Sequoia sp. Knowlton. (Kukak Bay.)
Spiraea andersoni Heer. (Port Graham.)
Taxites microphyllus Heer. (Port Graham.)
Taxites olriki Heer. (Port Graham.)
Taxodium dubium (Sternberg) Heer. (Port Graham, Sitka, Herendeen Bay, Kukak Bay.)
Taxodium tinajorum Heer var. (Port Graham, Chignik Bay, Ninilchik.)

Thuites alaskensis Lesquereux. (Unga Island.)
Tilia alaskana Heer. (Port Graham.)
Trapa borealis Heer. (Port Graham.)
Ulmus braunii Heer. (Kukak Bay.)
Ulmus plurinervia Unger. (Port Graham.)
Ulmus sorbifolia Göppert. (Cook Inlet.)
Vaccinium alaskanum Knowlton. (Kukak Bay.)
Vaccinium friesii Heer. (Port Graham.)
Vaccinium hollicki Knowlton. (Cook Inlet.)
Viburnum nordenskiöldi Heer. (Ninilchik. Formation doubtful.)
Vitis alaskana Cockerell. (Admiralty Inlet. Formation doubtful.)
Vitis heeriana Knowlton and Cockerell. (Port Graham and Miller's mine, Yukon River.)
Zizyphus townsendi Knowlton. (Herendeen Bay.)

EDMONTON FORMATION, CANADA.

Ginkgo sp. (Dawson).
Picea albertensis Penhallow.

EOCENE ROCKS, FORMATION NOT IDENTIFIED.

Acer grossedentatum Heer. (B. C.)
Aesculophyllum hastingsense Dawson. (B. C.)
Alnites curta Dawson. (B. C.)
Alnus alaskana Newberry. (B. C.)
Alnus americana Ettingshausen. (Wyo.)
Amygdalus gracilis Lesquereux. (B. C.)
Andromeda delicatula Lesquereux. (B. C.)
Anemia sp. Knowlton. (Wash.)
Antholithes amissus Heer. (N. W. T.)
Antholithes sp. Dawson. (B. C.)
Aralia whitneyii Lesquereux. (Oreg.)
Arundo sp. Dawson. (B. C.)
Asplenites sp. Dawson. (B. C.)
Betula angustifolia Newberry. (B. C.)
Betula prisca Ettingshausen. (B. C.)
Betula sp. Penhallow. (B. C.)
Callistemophyllum latum Dawson. (N. W. T. Age doubtful.)
Carex burrardiana Penhallow. (B. C.)
Carex vancouverensis Penhallow. (B. C.)
Carex sp. Penhallow. (B. C. Age doubtful.)
Carpinus grandis Unger. (Wash., Mont. Age doubtful.)
Carpolithes dentatus Penhallow. (B. C.)
Carpolithes semierectum Heer. (N. W. T.)
Carpolithes sp. Penhallow. (B. C.)
Castanea intermedia Lesquereux. (B. C.)
Castanea sp. Dawson. (B. C.)
Celastrus sp. Knowlton. (Wash.)
Chara compressa Knowlton. (Utah.)
Cinnamomum dilleri Knowlton. (Oreg.)
Cinnamomum 3 sp. Knowlton. (Wash., Oreg. Age doubtful.)
Cladosporites fasciculatus Berry. (Tex.)
Comptonia dryandroides Unger. (B. C. Age doubtful.)
Comptonia partita (Lesquereux) Berry. (B. C. Age doubtful.)
Comptonia quilchenensis Penhallow. (B. C. Age doubtful.)

Cyperacites paucinervis (Heer) Schimper. (B. C.)
Diospyros alaskana Schimper. (B. C.)
Dryophyllum stanleyanum Dawson. (B. C.)
Dryopteris? civica (Dawson) Knowlton. (B. C.)
Equisetum 2 sp. Dawson.
Ficus occidentalis (Lesquereux) Lesquereux. (B. C. Age doubtful.)
Ficus shastensis Lesquereux. (Burrand Inlet. Age doubtful.)
Ficus tiliaefolia (Braun) Heer.
Ficus 4 sp. Dawson. (Alberta.)
Ficus 3 sp. Knowlton. (Oreg., age doubtful; also Wash.)
Ginkgo adiantoides (Unger) Heer. (Canada.)
Glyptostrobus ungeri Heer. (N. W. T. Age doubtful.)
Glyptostrobus sp. Knowlton. (Wash.)
Grewia crenata (Unger) Heer. (B. C.)
Hedera macclurii Heer. (N. W. T.)
Hemitelites torelli Heer. (Wyo. Age doubtful.)
Hicoria sp. Knowlton. (B. C.)
Juglans cinerea Linné? Dawson. (B. C.)
?Juglans crossii Knowlton. (B. C.)
Juglans leonis Cockerell. (Oreg. Age doubtful.)
Juglans rhamnoides Lesquereux. (B. C. Age doubtful.)
Juglans schimperi Lesquereux. (B. C.)
Juglans woodiana Heer. (B. C. Age doubtful.)
Juglans sp. Knowlton. (Oreg. Age doubtful.)
Lastrea fischeri Heer. (Oreg. Age doubtful; B. C.)
Laurus californica Lesquereux. (Oreg. Age doubtful.)
Laurus sp. Knowlton. (Oreg. Age doubtful.)
Leguminosites? arachioides (Lesquereux) Lesquereux. (B. C.)
Leguminosites? borealis Dawson. (N. W. T.)
Lemna scutata Dawson. (Alberta.)
Lygodium kaulfusii Heer. (B. C.)
Magnolia californica Lesquereux. (Oreg. Formation doubtful.)
Magnolia dayana Cockerell. (Oreg.)
Magnolia inglefieldi Heer. (Calif.)
Nordenskiöldia borealis Heer. (B. C., N. W. T.)
Nyssidium? sp. Dawson. (B. C.)
Oreodaphne litsaeformis Lesquereux. (Calif.)
Osmunda macrophylla Penhallow. (Alberta.)
Pecopteris arctica? Heer. (Alaska. Age doubtful.)
Persea speciosa Heer. (Tex. Age doubtful.)
Phyllites aceroides Heer. (N. W. T.)
Phyllites sp. Knowlton. (Wash.)
Pinus? sp. Dawson. (N. W. T.)
Pinus sp. Heer. (N. W. T.)
Piper sp. Knowlton. (Wash.)
Platanus aceroides Göppert. (N. W. T. Age doubtful.)
Platanus haydenii Newberry. (B. C.)
Platanus nobilis Newberry. (Saskatchewan, Alberta.)
Platanus raynoldsii Newberry. (Canada.)
Platanus sp. Dawson. (B. C.)
Populus arctica Heer. (Canada.)
"Populus arctica latior Heer." (N. W. T.)
Populus balsamoides Göppert. (B. C. Age doubtful.)
Populus cuneata Newberry. (Alberta.)

Populus geneatrix Newberry. (B. C.)
Populus hookeri Heer. (N. W. T.)
Populus nervosa Newberry. (Saskatchewan.)
Populus newberryi Cockerell. (Alberta, Saskatchewan.)
Populus obtrita Dawson. (B. C.)
Populus speciosa Ward. (B. C.)
Populus subrotunda Lesquereux. (B. C.)
Populus zaddachi Heer. (Oreg. Age doubtful.)
Populus sp. Dawson. (B. C.)
Populus sp. Knowlton. (Wash.)
Pteris sitkensis Heer. (N. W. T.)
Pterospermites dentatus Heer. (N. W. T. Age doubtful.)
Pterospermites spectabilis Heer. (Canada.)
Pterospermites sp. Dawson. (B. C.)
Quercus pseudo-castanea Göppert. (B. C.)
Quercus simplex Newberry. (Oreg. Formation doubtful.)
Rhamnacinium porcupinianum Penhallow. (Saskatchewan.)
Rhamnacinium triseriatum Penhallow. (Saskatchewan.)
Rhamnus belmontensis Knowlton and Cockerell. (B. C. Age doubtful.)
Rhamnus eridani Unger. (B. C.)
Rhamnus gaudini Heer. (B. C.)
Rhus sp. Knowlton. (Oreg. Age doubtful.)
?Sabalites californicus Lesquereux. (Oreg.)
Salix kamloopsiana Dawson. (B. C. Age doubtful.)
Salix raeana Heer. (Saskatchewan, N. W. T.)
Salix wyomingensis Knowlton and Cockerell. (B. C. Age doubtful.)
Sapindus affinis Newberry. (Saskatchewan.)
Sterculia sp. Knowlton. (Oreg. Age doubtful.)
Taxites olriki Heer. (N. W. T., Saskatchewan.)
Taxodium occidentale Newberry. (Canada.)
Taxus sp. Dawson. (Alberta.)
Trapa borealis Heer. (Canada.)
Ulmus californica Lesquereux. (Oreg.)
Woodwardia sp. Knowlton. (Oreg. Age doubtful.)
Xylomites borealis Heer. (Canada.)

Later Tertiary Floras.

"LOUP FORK," LONG ISLAND, KANS.

Tithymalus willistoni Cockerell.

"MONUMENT CREEK FORMATION," COLORADO.

Perseoxylon eberi Platen.

CATAHOULA SANDSTONE (OLIGOCENE).

Phoenicites occidentalis Berry. (Tex.)

OLIGOCENE[1] ROCKS, FORMATIONS NOT IDENTIFIED.

Acer dubium Penhallow. (B. C.)
Acer sp. Penhallow. (B. C.)
Acerites negundifolium Dawson. (B. C.)
Aesculus sp. Dawson. (B. C.)
Alnus sp. Penhallow. (B. C.)
Alnus serrulata fossilis Newberry. (B. C.)

[1] This list is incomplete and more or less unreliable. It can be perfected only by a thorough revision.

Azollophyllum primaevum Penhallow. (B. C.)
Betula heterodonta Newberry. (B. C.)
Ceanothus? sp. Dawson. (B. C.)
Cladosporites oligocenicum Berry. (Miss.)
Comptonia diforme (Sternberg) Berry. (B. C.)
Crataegus tranquillensis Penhallow. (B. C.)
Crataegus tulameensis Penhallow. (B. C.)
Cupress[in]oxylon macrocarpoides Penhallow. (B. C.)
Cyperacites haydenii (Lesquereux) Knowlton. (B. C.)
Cyperacites? sp. (Dawson) Knowlton. (B. C.)
Cyperacites 2 sp. Penhallow. (B. C.)
Dombeyopsis islandica Heer. (B. C.)
Equisetum similkamense Dawson. (B. C.)
Fagara spireaefolia (Lesquereux) Cockerell. (B. C.)
Fagus antipofii Abich. (B. C.)
Fagus feroniae Unger. (B. C.)
Ficus asiminaefolia Lesquereux. (B. C.)
Ficus decandolleana Heer. (B. C.)
Ficus populina Heer. (B. C.)
Ficus ungeri Lesquereux. (B. C.)
Ficus 2 sp. Dawson. (B. C.)
Ginkgo adiantoides (Unger) Heer. (B. C.)
Grewia crenata (Unger) Heer. (B. C.)
Juglans occidentalis Newberry. (B. C.)
Magnolia? sp. Dawson. (B. C.)
Myrica? personata Knowlton. (B. C. Age doubtful.)
Myrica sp. Dawson. (B. C. Age doubtful.)
Osmunda heerii Gaudin. (B. C.)
Paliurus? sp. Dawson. (B. C.)
Palmoxylon cellulosum Knowlton. (Miss.)
Palmoxylon quenstedti Felix. (La. Age doubtful.)
Peronosporoides palmi Berry. (Miss.)
Phragmites sp. Penhallow. (B. C.)
Picea columbiensis Penhallow. (B. C.)
Pinus columbiana Penhallow. (B. C.)
Pinus hardyana Heer. (B. C. Age doubtful.)
Pinus steenstrupiana Heer. (B. C. Age doubtful.)
Pinus trunculus Dawson. (B. C.)
Pinus tulameensis Penhallow. (B. C.)
Pinus sp. Penhallow. (B. C.)
Populus arctica Heer. (B. C.)
Populus cuneata Newberry. (B. C.)
Populus latior Al. Braun. (B. C.)
Populus latior cordifolia Heer. (B. C.)
Populus mutabilis oblonga (Braun) Heer. (B. C.)
Populus newberryi Cockerell. (B. C.)
Populus zaddachi Heer. (B. C.)
Potamogeton? verticillatus Lesquereux. (B. C.)
Potamogeton sp. Penhallow, (B. C.)
Pyrus sp. Dawson. (B. C.)
Rhamnus quilchensis Penhallow. (B. C.)
Salix orbicularis Penhallow. (B. C.)
Salix perplexa Knowlton. (B. C.)
Salix tulameenensis Penhallow. (B. C.)

Sapindus sp. Dawson. (B. C.)
Sassafras sp. Penhallow. (B. C.)
Sequoia angustifolia Lesquereux. (B. C.)
Sequoia brevifolia Heer. (B. C. Age doubtful.)
Sequoia heerii Lesquereux. (B. C.)
Taxodium distichum (Linné) Richard. (B. C. Age doubtful.)
Taxodium occidentale Newberry. (B. C.)
Ulmus columbiana Penhallow. (B. C.)
Ulmus protoamericana Penhallow. (B. C.)
Ulmus protoracemosa Penhallow. (B. C.)
Ulmus sp. Penhallow. (B. C.)
Vitis olriki Heer. (B. C.)

MIOCENE FLORAS.

BRANDON, VT.

Apeibopsis gaudini Lesquereux.
Apeibopsis heerii Lesquereux.
Apeibopsis parva Perkins.
Aristolochia obscura Lesquereux.
Aristolochia oeningensis Heer.
Aristolochites acutus Perkins.
Aristolochites apicalis Perkins.
Aristolochites brandonianus Perkins.
Aristolochites coniodeus Perkins.
Aristolochites crassicostatus Perkins.
Aristolochites cuneatus Perkins.
Aristolochites curvatus (Lesquereux) Perkins.
Aristolochites dubius Perkins.
Aristolochites elegans Perkins.
Aristolochites excavatus Perkins.
Aristolochites globosus Perkins.
Aristolochites irregularis Perkins.
Aristolochites latisulcatus Perkins.
Aristolochites majus Perkins.
Aristolochites ovoides Perkins.
Aristolochites rugosus Perkins.
Aristolochites sulcatus Perkins.
Aristolochites sp. Knowlton.
Betuloxylon? sp. Jeffrey and Chrysler.
Bicarpellites abbreviatus Perkins.
Bicarpellites attenuatus Perkins.
Bicarpellites bicarinatus Perkins.
Bicarpellites brevis Perkins.
Bicarpellites carinatus Perkins.
Bicarpellites crassus Perkins.
Bicarpellites crateriformis Perkins.
Bicarpellites grayana (Lesquereux) Perkins.
Bicarpellites inaequalis Perkins.
Bicarpellites knowltoni Perkins.
Bicarpellites lanceolatus Perkins.
Bicarpellites latus Perkins.
Bicarpellites major Perkins.
Bicarpellites medius Perkins.
Bicarpellites minimus Perkins.

Bicarpellites obesus Perkins.
Bicarpellites ovatus Perkins.
Bicarpellites papillosus Perkins.
Bicarpellites parvus Perkins.
Bicarpellites quadrangulatus Perkins.
Bicarpellites rotundus Perkins.
Bicarpellites rugosus Perkins.
Bicarpellites solidus Perkins.
Bicarpellites sulcatus Perkins.
Bicarpellites vermontanus (Lesquereux) Perkins.
Brandonia globulus Perkins.
Carpinus grandis Unger?
Carpites inaequalis Perkins.
Carpites ovalis Perkins.
Carpites trigonus Perkins.
Carpolithes brandoniana Lesquereux.
Carpolithes emarginatus Perkins.
Carpolithes hitchcockii Perkins.
Carpolithes irregularis Lesquereux.
Carpolithes mucronatus Perkins.
Carpolithes ovatus Perkins.
Carpolithes simplex Perkins.
Carpolithes solidus Perkins.
Carpolithes vermontanus Perkins.
Cinnamomum corrugatum Perkins.
Cinnamomum lignitum Perkins.
Cinnamomum novae-angliae Lesquereux.
Cinnamomum ovoides Perkins.
Cucumites lesquereuxii Knowlton.
Drupa rhabdosperma Lesquereux.
Fagus hitchcockii Lesquereux.
Glossocarpellites elongatus (Lesquereux) Perkins
Glossocarpellites grandis (Perkins) Perkins.
Glossocarpellites obtusa (Lesquereux) Perkins.
Glossocarpellites parvus (Perkins) Perkins.
Glossocarpellites? sp. Perkins.
Hicoria biacuminata Perkins.
Hicoria verrucosa (Lesquereux) Knowlton.
Hicoroides angulata Perkins.
Hicoroides ellipsoidea Perkins.
Hicoroides globulus Perkins.
Hicoroides levis Perkins.
Hicoroides parva Perkins.
Hicoroides triangularis Perkins.
Illicium lignitum Lesquereux.
Juglans brandonianus Perkins.
Laurinoxylon brandonianum Jeffrey and Chrysler.
Leguminosites pisiformis? Heer.
Lescuria attenuata Perkins.
Monocarpellites amygdaloidus Perkins.
Monocarpellites elegans Perkins.
Monocarpellites gibbosus Perkins.
Monocarpellites hitchcockii Perkins.
Monocarpellites irregularis Perkins.

Monocarpellites medius Perkins.
Monocarpellites multicostatus Perkins.
Monocarpellites orbicularis Perkins.
Monocarpellites ovalis Perkins.
Monocarpellites pruniformis Perkins.
Monocarpellites pyramidalis Perkins.
Monocarpellites sulcatus Perkins.
Monocarpellites vermontanus Perkins.
Monocarpellites whitfieldii Perkins.
Nyssa acuticostata Perkins.
Nyssa ascoidea Perkins.
Nyssa clarkii Perkins.
Nyssa complanata Lesquereux.
Nyssa crassicostata Perkins.
Nyssa curta Perkins.
Nyssa cylindrica Perkins.
Nyssa elongata Perkins.
Nyssa equicostata Perkins.
Nyssa excavata Perkins.
Nyssa jonesii Perkins.
Nyssa laevigata Lesquereux.
Nyssa lamellosa Perkins.
Nyssa lescurii (Hitchcock) Perkins.
Nyssa microcarpa Lesquereux.
Nyssa multicostata Perkins.
Nyssa ovalis Perkins.
Nyssa ovata Perkins.
Nyssa solea Perkins.
Pinus conoides Perkins.
Pinus cuneatus Perkins.
Pityoxylon microporosum brandonianum Knowlton.
Pityoxylon sp. Jeffrey and Chrysler.
Prunoides bursaeformis (Lesquereux) Perkins.
Prunoides inaequalis Perkins.
Prunoides seeleyi Perkins.
Rubioides lignita Perkins.
Sapindoides americanus (Lesquereux) Perkins.
Sapindoides cylindricus Perkins.
Sapindoides medius Perkins.
Sapindoides minimus Perkins.
Sapindoides parva Perkins.
Sapindoides urceolatus Perkins.
Sapindoides varius Perkins.
Sapindoides vermontanus Perkins.
Sclerotites brandonianus Jeffrey and Chrysler.
Staphidoides ovalis Perkins.
Staphidoides perkinsi Knowlton.
Tricarpellites acuminatus Perkins.
Tricarpellites alatus Perkins.
Tricarpellites amygdaloideus Perkins.
Tricarpellites angularis Perkins.
Tricarpellites brandonianus Perkins.
Tricarpellites carinatus Perkins.
Tricarpellites castanoides Perkins.

Tricarpellites contractus Perkins.
Tricarpellites curtus Perkins.
Tricarpellites daleii Perkins.
Tricarpellites elongatus Perkins.
Tricarpellites fagoides Perkins.
Tricarpellites fissilis (Lesquereux) Perkins.
Tricarpellites hemiovalis Perkins.
Tricarpellites inaequalis Perkins.
Tricarpellites lignitus Perkins.
Tricarpellites major Perkins.
Tricarpellites obesus Perkins.
Tricarpellites ovalis Perkins.
Tricarpellites pringlei Perkins.
Tricarpellites rostratus Perkins.
Tricarpwllites rugosus Perkins.
Tricarpellites seelyi Perkins.
Tricarpellites triangularis Perkins.

CALVERT FORMATION, ATLANTIC COAST.

Berchemia priscaformis Berry. (D. C.)
Carpinus grandis Unger.
Cassia toraformis Berry. (D. C.)
Celastrus bruckmani Braun. (Va.)
Dalbergia calvertensis Berry. (Va.)
Ficus richmondensis Berry. (Va.)
Fraxinus richmondensis Berry. (Va.)
Ilex calvertensis Berry. (D. C.)
Nyssa gracilis Berry. (Va.)
Phyllites cercocarpifolia Berry. (D. C.)
Phyllites sp. Hollick. (D. C.)
Pieris scorbiculata Hollick. (D. C.)
Pinus sp. Berry. (D. C.)
Platanus aceroides Göppert.
Podogonium? virginianum Berry. (Va.)
Quercus chapmanifolia Berry. (D. C.)
Quercus lehmani Hollick. (D. C.)
Quercus milleri Berry. (Va.)
Rhus milleri Hollick. (Va., Md.)
Salix raeana Heer. (Va.)
Salvinia formosa? Heer.
Taxodium dubium (Sternberg) Heer. (Va.)
Ulmus basicordata Hollick. (D. C.)
Vaccinium cf. V. textum Heer. Berry. (D. C.)

ELKO, NEV.

Comptonia partita (Lesquereux) Berry.
Carpinus elkoana Lesquereux.
Fagopsis longifolia (Lesquereux) Hollick.
Fagus feroniae Unger.
Ficus jynx Heer.
Lycopodium prominens Lesquereux.
Myrica brongniarti? Ettingshausen.
Myrica drymeja (Lesquereux) Knowlton.
Myrica undulata (Heer) Schimper.

Planera ungeri Ettingshausen:
Populus richardsoni? Heer.
Salix elongata O. Weber.
Salix media Heer.
Sapotacites coriaceus (Lesquereux) Knowlton.
Sequoia affinis Lesquereux.
Sequoia angustifolia Lesquereux.
Thuja garmani Lesquereux.

FLORISSANT LAKE BEDS, COLORADO.

Acacia septentrionalis Lesquereux.
Acalypha myricina Cockerell.
Acer florigerum Cockerell.
Acer florissanti Kirchner.
Acer kirchnerianum Knowlton.
Acer mysticum Kirchner.
Acer perditum Cockerell.
Acer sp. Lesquereux.
Acerates fructifer Cockerell.
Acorus affinis Lesquereux.
Ailanthus americana Cockerell.
Alnus praecordata Cockerell.
Alnus sp. Knowlton.
Amelanchier peritula Cockerell.
Amelanchier scudderi Cockerell.
Amelanchier typica Lesquereux.
Amygdalus gracilis Lesquereux.
Anemia? gracillima (Lesquereux) Cockerell.
Anona spoliata Cockerell.
Antholithes amoenus Lesquereux.
Antholithes obtusilobus Lesquereux.
Antholithes pediloides Cockerell.
Apocynophyllum pealii Ettingshausen.
Aralia dissecta Lesquereux.
Aristolochia mortua Cockerell.
Aristolochia williardiana Knowlton.
Aster florissantia Cockerell.
Banksites lineatus Lesquereux.
Bauhinia pseudocotyledon Cockerell.
Betula deltoides Knowlton.
Betula florissanti Lesquereux.
Betula truncata Lesquereux.
Buettneria? perplexans Cockerell.
Bumelia florissanti Lesquereux.
Carpinus attenuata Lesquereux.
Carpinus fraterna Lesquereux.
Carpinus grandis Unger.
Carpites gemmaceus Lesquereux.
Carpites milioides Lesquereux.
Carpites pealei Lesquereux.
Carpolithes macrophyllus Cockerell.
Carpolithes sp. Lesquereux.
Castanea dolicophylla Cockerell.
Celastrinites elegans Lesquereux.

Celastrus fraxinifolius Lesquereux.
Celastrus lacoei Lesquereux.
Celtis mccoshii Lesquereux.
Cercis parvifolia Lesquereux.
Chara? glomerata Lesquereux.
Chara peritula Cockerell.
Comptonia acutiloba (Lesquereux) Cockerell.
Comptonia insignis (Lesquereux) Cockerell.
Cotinus fraterna (Lesquereux) Cockerell.
Crataegus sp. Knowlton.
Croton? furcatulum Cockerell.
Cytisus florissantinus Lesquereux.
Dalbergia? coloradensis Knowlton.
Dalbergia cuneifolia Heer.
Dalbergia lesquereuxii Ettingshausen.
Dalbergia? minuta Knowlton.
Didymosphaeria betheli Cockerell.
Diospyros brachysepala Al. Braun?
Dodonaea sp. Lesquereux.
Dryopteris scansa Cockerell.
Engelhardtia oxyptera Saporta.
Equisetum florissantense Cockerell.
Fagara? delicatula Cockerell.
Fagara spireaefolia (Lesquereux) Cockerell.
Fagopsis longifolia (Lesquereux) Hollick.
Ficus arinaceaeformis Cockerell.
Ficus bruesi Cockerell.
Ficus florissantella Cockerell.
Ficus florissantia Knowlton.
Florissantia physalis Knowlton.
Fontinalis pristina Lesquereux. [Not a plant.]
Fraxinus abbreviata Lesquereux.
Fraxinus heerii Lesquereux.
Fraxinus libbeyi Lesquereux.
Fraxinus mespilifolia Lesquereux.
Fraxinus? myricaefolia Lesquereux.
Fraxinus palaeophila Cockerell.
Fraxinus praedicta Heer?
Fraxinus ungeri Lesquereux.
Geaster florissantensis Cockerell.
Glyphomitrium cockerelleae E. G. Britton and Hollick.
Hedera marginata Lesquereux.
Hicoria juglandiformis (Sternberg) Knowlton.
Hicoria princetonia Cockerell.
Hicoria rostrata (Göppert) Knowlton.
Hydrangea florissantia Cockerell.
Hydrangea? subincerta Cockerell.
Hypnum brownii Kirchner.
Hypolepis coloradensis Cockerell.
Ilex florissantensis Knowlton and Cockerell.
Ilex grandifolia Lesquereux.
Ilex knightiaefolia Lesquereux.
Ilex leonis Cockerell.
Ilex pseudo-stenophylla Lesquereux.

Ilex sphenophylla? Heer.
Ilex subdenticulata Lesquereux.
Isoetes brevifolius Lesquereux.
Juglans affinis Kirchner.
Juglans florissanti Lesquereux.
Juglans magnifica Knowlton.
Juglans? sepultus Cockerell.
Juncus crassulus Cockerell.
Leucaena coloradensis Cockerell.
Libocedrus coloradensis (Cockerell) Cockerell.
Limnobium obliteratum Cockerell.
Liquidambar convexum Cockerell.
Lomatia abbreviata Lesquereux.
Lomatia acutiloba Lesquereux.
Lomatia interrupta Lesquereux.
Lomatia terminalis Lesquereux.
Lomatia tripartita Lesquereux.
Lomatites hakeaefolia (Lesquereux) Cockerell.
Lomatites spinosa (Lesquereux) Cockerell.
Magnolia florissanticola Cockerell.
Malvastrum? exhumatum Cockerell.
Melia? expulsa Cockerell.
Melothria? coloradensis Cockerell.
Menyanthes coloradensis Cockerell.
Mimosites linearis (Lesquereux) Knowlton.
Mimulus saxorum Cockerell.
Morus symmetrica Cockerell.
Muhlenbergia florissanti Knowlton.
Myrica coloradensis Knowlton.
Myrica copeana Lesquereux.
Myrica drymeja (Lesquereux) Knowlton.
Myrica hendersoni Cockerell.
Myrica obscura Lesquereux.
Myrica scottii Lesquereux.
Myrsine laminarum Cockerell.
Najadopsis rugulosa Lesquereux.
Odostemon florissantensis Cockerell.
Osmanthus praemissa (Lesquereux) Cockerell.
Oxypolis destructus Cockerell.
Pachistima? integra Cockerell.
Palaeopotamogeton florissanti Knowlton.
Paliurus florissanti Lesquereux.
Paliurus haydeni Cockerell.
Palmocarpon? globosum Lesquereux.
Panax andrewsii Cockerell.
Parthenocissus osborni Cockerell.
Pellaea antiquella Cockerell.
Persea coloradica Cockerell.
Persicaria tertiaria (Small) Cockerell.
Phaca wilamattae Cockerell.
Phalaris? geometrorum Cockerell.
Phegopteris guyottii (Lesquereux) Cockerell.
Phenanthera petalifera Hollick.
Philadelphus palaeophilus Cockerell.

Pimelia delicatula Lesquereux.
Pinus florissanti Lesquereux.
Pinus hambachi Knowlton.
Pinus sturgisi Cockerell.
Pinus wheeleri Cockerell.
Plagiopodopsis scudderi Britton and Hollick.
Planera myricaefolia (Lesquereux) Cockerell.
Podogonium acuminatum Lesquereux.
Polytrichum? florissanti Knowlton.
Populus arctica decipiens Cockerell.
Populus crassa (Lesquereux) Cockerell.
Populus lesquereuxii Cockerell.
Populus microtremuloides Knowlton.
Populus? pseudocredneria Cockerell.
Populus pyrifolia Kirchner.
Populus scudderi Cockerell.
Populus sp. Knowlton.
Porana cockerelli Knowlton.
Porana similis Knowlton.
Porana speirii Lesquereux.
Porana tenuis Lesquereux.
Potamogeton geniculatus Al. Braun.
Potamogeton? verticillatus Lesquereux.
Ptelia modesta (Lesquereux) Cockerell.
Quercus balaninorum Cockerell.
Quercus drymeja Unger.
Quercus elaena Unger.
Quercus florissantensis Cockerell.
Quercus knowltoniana Cockerell.
Quercus lyratiformis Cockerell.
Quercus osbornii Lesquereux.
Quercus peritula Cockerell.
Quercus ramaleyi Cockerell.
Quercus scudderi Knowlton.
Quercus serra Unger.
Rhamnus florissantensis Cockerell.
Rhamnus kirchneri Cockerell.
Rhamnus lesquereuxi Berry.
Rhus cassioides Lesquereux.
Rhus coriarioides Lesquereux.
Rhus hilliae Lesquereux.
Rhus lesquereuxii Knowlton and Cockerell.
Rhus subrhomboidalis Lesquereux.
Rhus? trifolioides Lesquereux.
Ribes? florissanti Knowlton.
Ribes protomelaenum Cockerell.
Robinia brittoni Cockerell.
Rosa hilliae Lesquereux.
Rosa? inquirenda Knowlton.
Rosa ruskiniana Cockerell.
Rosa scudderi Knowlton.
Rosa wilmattae Cockerell.
Sabina linguaefolia (Lesquereux) Cockerell.
Sabina linguaefolia gracilis (Lesquereux).

Salix florissanti Knowlton and Cockerell.
Salix libbeyi Lesquereux.
Salix ramaleyi Cockerell.
Salix ramaleyi rohweri Cockerell.
Salix sp. Knowlton.
Sambucus amabilis Cockerell.
Sambucus ellisiae Cockerell.
Sambucus newtoni Cockerell.
Santalum americanum Lesquereux.
"Sapindus angustifolius" Lesquereux.
Sapindus coloradensis Cockerell.
Sapindus inflexus Lesquereux.
Sapindus lancifolius Lesquereux.
Sapindus leonis Cockerell.
Sapindus stellariaefolius Lesquereux.
Saxifraga? peritula Cockerell.
Schmaltzia vexans (Lesquereux) Cockerell.
Sequoia affinis Lesquereux.
Sequoia langsdorfii (Brongniart) Heer.
Sicyos? florissanti Cockerell.
Smilax labidurommae Cockerell.
Sorbus diversifolia (Lesquereux) Cockerell.
Sorbus megaphylla Cockerell.
Sorbus nupta Cockerell.
Spirodella penicillata (Lesquereux) Cockerell.
Staphylea acuminata Lesquereux.
Sterculia engleri Kirchner.
Sterculia rigida Lesquereux.
Stipa laminarium Cockerell.
Tilia populifolia Lesquereux.
Tmesipteris alleni (Lesquereux) Hollick.
Typha lesquereuxii Cockerell.
Ulmus braunii Heer.
Ulmus brownellii Lesquereux.
Ulmus hilliae Lesquereux.
Ulmus tenuinervis Lesquereux.
Viborquia nigrostipellata (Cockerell) Cockerell.
Vicia sp. Knowlton.
Vitis florissantella Cockerell.
Vitis hesperia Knowlton.
Weinmannia? dubiosa Cockerell.
Weinmannia haydenii (Lesquereux) Lesquereux.
Weinmannia integrifolia Lesquereux.
Weinmannia lesquereuxi Cockerell.
Weinmannia obtusifolia Lesquereux.
Woodwardia florissantia Cockerell.
Zizyphus obtusa Kirchner.

INTERMEDIATE FLORA, YELLOWSTONE NATIONAL PARK.

Cinnamomum spectabile Heer.
Corylus macquarrii (Forbes) Heer.
Cupressinoxylon eutreron Felix.
Diospyros brachysepala Al. Braun.
Elaeodendroxylon polymorphum Platen.

Equisetum canaliculatum Knowlton.
Equisetum haguei Knowlton.
Equisetum lesquereuxii Knowlton. (Formation doubtful.)
Ficus asiminaefolia Lesquereux.
Ficus densifolia Knowlton.
Ficus haguei Knowlton.
Ficus sordida Lesquereux.
Grewiopsis? aldersoni Knowlton.
Juglans laurifolia Knowlton.
Juglans leonis Cockerell.
Laurinoxylon pulchrum Knowlton.
Laurus californica Lesquereux.
Laurus grandis Lesquereux.
Laurus perdita Knowlton.
Magnolia? pollardi Knowlton.
Perseoxylon aromaticum (Felix) Felix.
Pinus sp. Knowlton.
Pityoxylon amethystinum Knowlton.
Pityoxylon fallax Felix.
Plataninium haydeni Felix.
Plataninium knowltoni Platen.
Platanus montana Knowlton.
Pruinium gummosum Platen.
Quercinium knowltoni Felix.
Quercinium lamarense Knowlton.
Rhamnicinium radiatum Felix.
Salix elongata O. Weber.
Spegazzinites cruciformis Felix.

FLORA OF LAMAR RIVER, YELLOWSTONE NATIONAL PARK.

Acacia lamarensis Knowlton.
Acacia macrosperma Knowlton.
Acacia wardii Knowlton.
Acer vivarium Knowlton.
Aralia whitneyi Lesquereux.
Aralia wrightii Knowlton.
Aralia sp. Knowlton.
Castanea pulchella Knowlton.
Cissus haguei Knowlton.
Cornus wrightii Knowlton.
Davallia? montana Knowlton.
Diospyros lamarensis Knowlton.
Dryophyllum longipetiolatum Knowlton.
Dryopteris xantholithensis Knowlton.
Fagopsis longifolia (Lesquereux) Hollick.
Ficus shastensis Lesquereux?
Juglans schimperi Lesquereux.
Leguminosites lamarensis Knowlton.
Magnolia californica Lesquereux.
Magnolia culveri Knowlton.
Magnolia microphylla Knowlton.
Magnolia spectabilis Knowlton.
Malapoenna lamarensis Knowlton.
Musophyllum complicatum Lesquereux.

Myrica lamarensis Knowlton.
Myrica wardii Knowlton.
Persea pseudo-carolinensis Lesquereux.
Phyllites crassifolia Knowlton.
Pinus gracilistrobus Knowlton.
Pinus macrolepis Knowlton.
Pinus premurrayana Knowlton.
Pinus wardii Knowlton.
Pityoxylon aldersoni Knowlton.
Populus balsamoides Göppert.
Populus? vivaria Knowlton.
Populus zaddachi Heer.
Pterospermites haguei Knowlton.
Quercus furcinervis Americana Knowlton.
Quercus grossedentata Knowlton.
Quercus weedii Knowlton.
Rhamnus eorectinervis Knowlton.
Rhus mixta Lesquereux?
Salix angusta Al. Braun.
Sapindus grandifolioloides Knowlton.
Sequoia magnifica Knowlton.
Sequoia sp. Knowlton.
Smilax lamarensis Knowlton.
Ulmus pseudo-fulva Lesquereux.

ESMERALDA FORMATION, NEVADA.

Cercis? nevadensis Knowlton.
Chrysobalanus polliardiana Knowlton.
Cinchonidium? turneri Knowlton.
Dryopteris? gleichenoides Knowlton.
Ficus lacustris Knowlton.
Gleichenia? obscura Knowlton.
Quercus argentum Knowlton.
Quercus prae-dumosa Trelease.
Rhus? nevadensis Knowlton.
Salix angusta Al. Braun.
Salix vaccinifolia Knowlton.
Salix 2 sp. Knowlton.
Spathyema? nevadensis Knowlton.

MASCALL FORMATION, JOHN DAY BASIN, OREG.

Acacia oregoniana Lesquereux.
Acer bendirei Lesquereux.
Acer dimorphum Lesquereux.
Acer gigas Knowlton.
Acer medianum Knowlton.
Acer merriami Knowlton.
Acer minor Knowlton.
Acer oregonianum Knowlton.
Acer sp. Lesquereux.
Aesculus? simulata Knowlton.
Alnus kefersteinii (Göppert) Heer.
Andromeda crassa Lesquereux.
Aralia whitneyi Lesquereux.

Artocarpus californica Knowlton.
Berberis? gigantea Knowlton.
Betula? dayana Knowlton.
Carpinus grandis Unger.
Carpites fragariaeformis Lesquereux.
Castanea castaneaefolia (Unger) Knowlton.
Celastrus confluens Knowlton.
Celastrus dignatus Knowlton.
Crataegus imparilis Knowlton.
Cyperacites sp. Knowlton.
Diospyros elliptica Knowlton.
Equisetum sp. Knowlton.
Fagus? sp. Knowlton.
Ficus? oregoniana Lesquereux.
Ginkgo sp. Knowlton.
Glyptostrobus ungeri Heer.
Grewia crenata (Unger) Heer.
Hicoria elaenoides (Unger) Knowlton.
Hydrangea bendirei (Ward) Knowlton.
Juglans oregoniana Lesquereux.
Laurus oregoniana Knowlton.
Liquidambar europaeum patulum Knowlton.
Liquidambar pachyphyllum Knowlton.
Liquidambar protensum Unger?
Liquidambar sp.? Knowlton.
Magnolia inglefieldi Heer.
Myrica oregoniana Knowlton.
Phyllites bifurcies Knowlton.
Phyllites inexpectans Knowlton.
Planera ungeri Ettingshausen.
Platanus aceroides Göppert.
?Platanus nobilis Newberry.
Planatus sp. Knowlton.
Populus lindgreni Knowlton.
Prunus? merriami Knowlton.
Prunus? tuffacea Knowlton.
Quercus dayana Knowlton.
Quercus duriuscula Knowlton.
Quercus horniana Lesquereux.
Quercus merriami Knowlton.
Quercus pseudo-lyrata Lesquereux.
Quercus ursina Knowlton.
Quercus? sp. Knowlton.
Rhus benderei Lesquereux.
Rhus? sp. Lesquereux.
Rulac crataegifolium Knowlton.
Salix angusta Al. Braun.
Salix dayana Knowlton.
Salix engelhardti Lesquereux.
Salix florissanti Knowlton and Cockerell.
Salix mixta Knowlton.
Salix perplexa Knowlton.
Salix pseudo-argentea Knowlton.
Salix raeana Heer.

Saix varians Göppert.
"Sapindus angustifolius" Lesquereux.
Sapindus obtusifolius Lesquereux.
Sapindus oregonianus Knowlton.
Sequoia angustifolia Lesquereux.
Sequoia langsdorfii (Brongniart) Heer.
Sequoia sp. Knowlton.
Smilax wardii Lesquereux.
Taxodium dubium (Sternberg) Heer.
Taxodium sp. Knowlton.
Thuites sp. Knowlton.
Ulmus californica Lesquereux.
Ulmus plurinervia Unger.

ELLENSBURG FORMATION, WASHINGTON.

Alnus sp. Knowlton.
Diospyros elliptica Knowlton.
Ficus? oregoniana Lesquereux.
Magnolia dayana Cockerell.
Platanus aceroides Göppert.
Platanus dissecta Lesquereux.
Populus glandulifera Heer.
Populus russellii Knowlton.
Quercus dayana Knowlton.
Quercus pseudo-lyrata Lesquereux.
Salix engelhardti Lesquereux.
Salix pseudo-argentea Knowlton.
Salix varians Göppert.
Ulmus californica Lesquereux.

AURIFEROUS GRAVELS, CALIFORNIA.

Acer aequidentatum Lesquereux.
Acer arcticum Heer.
Acer bolanderi Lesquereux.
Acer pseudo-chrysophylla Lesquereux.
Acer sp. Knowlton.
Acer sp. Lesquereux.
Aesculus sp. Knowlton.
Anacardioxylon magniporosum Platen.
Aralia angustiloba Lesquereux.
Aralia whitneyi Lesquereux.
Aralia zaddachi? Heer.
Aralinium excellens Platen.
Aralinium lindgreni Platen. (Formation doubtful.)
Aralinium multiradiatum Platen.
Aralinium pachychymoticum Platen.
Arisaemites sp. Knowlton.
Artocarpus californica Knowlton.
Betula aequalis Lesquereux.
Castanea castaneaefolia (Unger) Knowlton. (Formation doubtful.)
Castanea sp. Knowlton.
Castanopsis chrysophylloides Lesquereux.
Cercocarpus antiquus Lesquereux.
Colutea boweniana Lesquereux.

Cornus kelloggii Lesquereux.
Cornus ovalis Lesquereux.
Cornus sp. Knowlton.
Corylus sp. Knowlton.
Ebenoxylon speciosum Platen.
Fagara diversifolia (Lesquereux) Cockerell.
Fagus pseudo-ferruginea Lesquereux.
Ficus asiminaefolia Lesquereux.
Ficus mensae Cockerell.
Ficus shastensis Lesquereux?
Ficus sordida Lesquereux.
Ficus tiliaefolia (Braun)Heer.
Ficus 2 sp. Knowlton.
Ilex prunifolia Lesquereux.
Juglans egregia Lesquereux.
Juglans laurinea Lesquereux.
Juglans leonis Cockerell.
Juglans sp. Knowlton.
Laurus californica Lesquereux.
Laurus förstenbergii Al. Braun.
Laurus grandis Lesquereux.
Laurus princeps Heer.
Laurus resurgens Saporta.
Laurus saliciformis Knowlton and Cockerell.
Liquidambar californicum Lesquereux.
Liquidambar sp. Knowlton.
Magnolia californica Lesquereux.
Magnolia dayana Cockerell.
Myrica ungeri Heer.
Myrtus oregonensis Lesquereux.
Persea pseudo-carolinensis Lesquereux.
Persea punctulata Lesquereux.
Perseoxylon californicum Platen.
Plataninium pacificum Platen.
Platanus appendiculata Lesquereux.
Platanus dissecta Lesquereux.
Populus zaddachi Heer.
Quercinium anomalum Platen.
Quercus boweniana Lesquereux
Quercus castanopsis Newberry.
Quercus chrysolepis montana Lesquereux.
Quercus convexa Lesquereux.
Quercus distincta Lesquereux.
Quercus elaenoides Lesquereux.
Quercus göpperti Lesquereux.
Quercus nevadensis Lesquereux.
Quercus pseudo-chrysophylla Lesquereux.
Quercus steenstrupiana? Heer.
Quercus transgressus Lesquereux.
Quercus sp. Eames.
Quercus sp. Knowlton.
Rhus boweniana Lesquereux.
Rhus dispersa Lesquereux.
Rhus mensae Cockerell.

Rhus mixta Lesquereux.
Rhus myricaefolia Lesquereux.
Rhus typhinoides Lesquereux.
Rhus sp. Knowlton.
Sabalites californicus Lesquereux.
Salix californica Lesquereux.
Salix merriami Cockerell.
Simarubinium crystallophorum Platen.
Simarubinium engelhardti Platen.
Tmesipteris alleni (Lesquereux) Hollick.
Ulmus affinis Lesquereux.
Ulmus californica Lesquereux.
Ulmus pseudo-fulva Lesquereux.
Ulmus sp. Knowlton.
?Viburnum elongatum Ward.
?Viburnum whymperi Heer.
Viburnum sp. Knowlton.
Zizyphus californicus Knowlton and Cockerell.
Zizyphus piperoides Lesquereux.
Zizyphus sp. Knowlton.

SAN PABLO FORMATION, CALIFORNIA.

Alnus sp. Knowlton.

CALISTOGA, CALIF.

Quercinium abromeiti Platen.
Quercinium lesquereuxi Platen.
Ulmoxylon simrothi Platen. (Age doubtful.)

MIOCENE ROCKS, FORMATIONS NOT IDENTIFIED.

Acer benderei Lesquereux. (B. C.)
Acer sp. Knowlton. (Wash.)
Alnus carrollina Lesquereux. (Calif.)
Betula elliptica Saporta. (Calif.)
Betula macrophylla Göppert. (B. C.)
Betula parcedentata Lesquereux. (Calif.)
Bumelia oklahomensis Berry. (Okla.)
Caulinites sp. Berry. (Okla.)
Cinchonidium copeanum (Lesquereux) Ettingshausen. (Nev. Age doubtful.)
Cinnamomum? sp. Dawson. (B. C. Age doubtful.)
Cinnamomum sp. Knowlton. (Wash.)
Comptonia insignis (Lesquereux) Cockerell. (Wyo.)
Cupressinoxylon dubium Cramer. (Baring Island.)
Cupressinoxylon polyommatum Cramer. (Banks Land.)
Cupressinoxylon pulchrum Cramer. (Banks Land.)
Cyperacites sp. Berry. (Okla.)
Diospyros brachysepala Al. Braun. (Okla.)
Diospyros virginiana turneri Lesquereux. (Contra Costa, Calif.)
Fagopsis longifolia (Lesquereux) Hollick. (B. C.)
Ficoxylon helictoxyloides Platen. (Calif.)
Ficus appendiculata Heer. (Calif.)
Ficus? applegatei Knowlton. (Oreg.)
Ficus sp. Knowlton. (Wash.)
Fraxinus brownellii Lesquereux. (Colo.)
Glyptostrobus sp. Dawson. (B. C.)

Grewia crenata (Unger) Heer. (B. C.)
Gymnocladus casei Berry. (Okla.)
Hypnum columbianum Penhallow. (B. C. Age doubtful.)
Juglans rhamnoides Lesquereux. (B. C. Age doubtful.)
Juniperus? haydenii (Lesquereux) Knowlton.
Lebephyllum reineckei Wilson. (B. C.)
Leguminosites? arachioides (Lesquereux) Lesquereux. (B. C.)
Magnolia hilgardiana Lesquereux. (Calif.)
Nelumbo pygmaea (Dawson) Knowlton. (B. C.)
Nelumbo sp. (Dawson). (B. C.)
Physematopitys goepperti Platen. (Tex.)
Picea quilchensis Penhallow. (B. C.)
Picea tranquillensis Penhallow. (B. C.)
Pinus trunculus Dawson. (B. C.)
Pityoxylon pealei Knowlton. (Mont.)
Platanus aceroides Göppert. (Okla.)
Populus cuneata? Newberry. (B. C.)
Populus mutabilis oblonga (Braun) Heer. (B. C.)
Populus newberryi Cockerell. (B. C.)
Populus zaddachi Heer. (B. C.)
Prunus? merriami Knowlton. (B. C.)
Pseudotsuga miocena Penhallow. (B. C. and Alberta?)
Quercinium abromeiti Platen. (Calif.)
Quercinium lesquereuxi Platen. (Calif.)
Quercus? sp. Dawson. (B. C.)
Rhamnus lesquereuxi Berry.
Rhus heüfleri Heer. (Calif.)
Salix varians Göppert. (B. C. Age doubtful.)
Salix sp. Berry. (Okla.)
Sapindus oklahomensis Berry. (Okla.)
Sequoia angustifolia Lesquereux.
Sequoia penhallowi Jeffrey. (Calif.)
Smilax franklini Heer. (N. W. T. Age doubtful.)
Taxodioxylon credneri Platen. (Nev. Age doubtful.)
Taxodium dubium (Sternberg) Heer. (B. C.)
?Taxodium occidentale Newberry. (B. C.)
Thuja gracilis Newberry. (Locality?)
Typha lesquereuxi Cockerell. (B. C.)
Ulmites pusillus Dawson. (B. C.)
Ulmoxylon simrothi Platen. (Calif. Age doubtful.)
Ulmus tenuinervis Lesquereux. (B. C.)
Vaccinophyllum quaestum Dawson. (B. C.)
Viburnum pubescens Pursh. (Saskatchewan. Age doubtful.)
Vitis sp.? Lesquereux. (Calif. Age doubtful.)

PLIOCENE FLORAS.

CITRONELLE FORMATION, ALABAMA.

Betula prenigra Berry.
Bumelia preangustifolia Berry.
Caesalpinia citronellensis Berry.
Fagus lambertensis Berry.
Fraxinus sp. Berry.
Hicoria pretexana Berry.
Nyssa aquaticaformis Berry.

Pinus sp. Berry.
Planera aquatica (Walter) Gmelin.
Prunus sp. Berry.
Quercus catesbaeifolia Berry.
Quercus lambertensis Berry.
Quercus nigra Linné.
Quercus previrginiana Berry.
Taxodium distichum (Linné) Richard.
Trapa alabamensis Berry.
Vitis sp. Berry.
Yucca sp. Berry.

MERCED FORMATION, CALIFORNIA.

Pinus radiata Don.

PLIOCENE ROCKS, FORMATIONS NOT IDENTIFIED.

Alnus rhamnifolia Nuttall. (Calif.)
Amelanchier alnifolia Nuttall. (Calif.)
Arbutus menziesii Pursh. (Calif.)
Arctostaphylos manzanita Parry. (Calif.)
Cephalanthus occidentalis Linné. (Calif.)
Cercidoxylon zirkeli Platen. (Nebr.)
Cercocarpus betulaefolius Nuttall. (Calif.)
Cladosporites ligni-perditor Whitford. (Nebr.)
Cornus glabrata Bentham. (Calif.)
Cyperacites borealis (Heer) Heer. (State? Age doubtful.)
Padus demissa (Nuttall) Roemer. (Calif.)
Pasania densiflora (Hooker and Arnott) Orst. (Calif.)
Pinus lindgrenii Knowlton. (Idaho.)
Pinus macclurii Heer. (Banks Land, 35 miles below Tanana, Yukon River. Age doubtful.)
Poacites mengeanus Heer. (Calif. Age doubtful.)
Populus trichocarpa Torrey and Gray. (Calif.)
Pseudotsuga taxifolia (Poiret) Britton. (Calif.)
Psoralea physodes Douglas. (Calif.)
Quercus agrifolia Née. (Calif.)
Quercus chrysolepis Liebman. (Calif.)
Rhamnus californica Eschscholtz. (Calif.)
Rhamnus purshiana De Candolle. (Calif.)
Salix fluviatilis Nuttall. (Calif.)
Salix laevigata Bebb. (Calif.)
Sequoia sempervirens (Lamb) Endlicher. (Calif.)
Sequoia sp. Knowlton. (Calif. Age doubtful.)
Trapa alabamensis Berry. (Ala.)
Woodwardia columbiana Knowlton. (Oreg. Age doubtful.)

TERTIARY ROCKS, FORMATIONS NOT IDENTIFIED.

Agaricites conwentzi Platen. (Calif.)
Betula macclintockia Cramer. (Banks Land.)
Cupressinoxylon glasgowi Knowlton. (Iowa.)
Cupressinoxylon taxodioides Conwentz. (Calif.)
Laurinoxylon lesquereuxiana Knowlton. (Ark.)
Nyssa denveriana Knowlton. (N. Mex.)
Palmoxylon cellulosum Knowlton. (La.)
Pityoxylon columbina Penhallow. (B. C.)
Salix angusta Al. Braun. (Mont., Wyo., Calif., Ky., etc.)

PLEISTOCENE FLORAS.

TALBOT FORMATION, MIDDLE ATLANTIC COAST.

Alnus rugosa (Du Roi) K. Koch. Hollick. (Md.)
Fagus americana Sweet. (Md., Va.)
Hicoria sp.? Hollick. (Md.)
Juglans nigra Linné. (Md.)
Nyssa sp.? Hollick. (Md.)
Osmunda sp.? Hollick. (Md.)
Pinus echinata Miller. (Md.)
Pinus strobus Linné. (Md.)
Pinus sp. Hollick. (Md.)
Polygonum sp.? Hollick. (Md.)
Quercus michauxii Nuttall. (Md.)
Robinia pseudacacia Linné. (Md.)
Ulmus americana Linné. Hollick. (Md.)
Ulmus sp.? Hollick. (Md.)
Vaccinium corymbosum Linné. Hollick. (Md.)
Vitis sp. Berry. (Va.)
Vitis sp.? Berry. (Md.)

SUNDERLAND FORMATION, MIDDLE ATLANTIC COAST.

Acer sp.? Hollick.
Bumelia pseudo-lanuginosa Hollick.
Carpinus pseudo-caroliniana Hollick.
Cassia sp.? Hollick.
Celtis pseudo-crassifolia Hollick.
Fagus sp.? Hollick. (Md.)
Hicoria pseudo-glabra Hollick. (Md.)
Hicoria sp.? Hollick. (Md.)
Juglans acuminata Al. Braun. (Md. Formation doubtful.)
Planera ungeri Ettingshausen. (Md.)
Platanus sp.? Hollick. (Md.)
Populus clarkiana Hollick. (Md.)
Populus latior? Al. Braun. (Md.)
Populus pseudo-tremuloides Hollick. (Md.)
Prunus? merriami Knowlton. (Md.)
Pterocarya denticulata (Weber) Heer.
Quercus glennii Hollick. (Md.)
Quercus pseudo-alba Hollick. (Md.)
Sapindus marylandicus Hollick. (Md.)
Sequoia angustifolia Lesquereux. (Md.)
Ulmus betuloides Hollick. (Md.)
Ulmus pseudo-racemosa Hollick. (Md.)

CHOWAN FORMATION, NORTH CAROLINA.

Quercus michauxii Nuttall.

PLEISTOCENE DEPOSITS,[1] FORMATION NOT IDENTIFIED.

Abies balsamea (Linné) Miller. (Ont.)
Acer pleistocenicum Penhallow. (Toronto.)
Acer rubrum Linné. (Ala., Fla.)

[1] There is still so much uncertainty regarding the identification of horizons within the Pleistocene that it has been thought best to give the species in a single list without attempt at differentiation.

Acer saccharinum Linné. (Ala.)
Acer saccharum Marshall. (Canada.)
Acer spicatum Lamarck. (Toronto.)
Acer torontonensis Penhallow. (Toronto.)
Acer vitiphyllum Knowlton and Cockerell. (Va.)
Acer sp. Berry. (Va.)
Acorus calamus Linné. (Ky.)
Alga gen. et sp.? (Canada.)
Alnus rubra Bongard. (Alaska.)
Alnus sp. Penhallow. (Ont.)
Amelanchier rotundifolia (Michaux) Roemer. (N. J.)
Ampelopsis quinquefolia Michaux. (Pa.)
Anomalophyllites bridgetonensis Hollick. (N. J.)
Anona glabra Linné. (Fla.)
Arundinaria macrosperma Michaux. (Ala.)
Arundinaria sp. Berry. (Ala.)
Asimina triloba (Linné) Dunal. (N. J., Toronto.)
Benzoin cf. B. melissaefolium (Walter) Nees. (Fla.)
Betula lutea Michaux f. (Canada.)
Betula nigra Linné. (W. Va., N. C., Ala., Va.)
Betula pseudo-fontinalis Berry. (N. C.)
Betula sp. Berry. (Ky.)
Brasenia purpurea (Michaux) Caspary. (Canada, Fla.)
Bromus ciliatus Linné. (Canada.)
Bryum sp. Hinde. (Canada.)
Carex aquatilis Wahlenberg. (Ont.)
Carex magellanica Lamarck. (Canada.)
Carex utriculata Boott. (Ont.)
Carex sp. Berry. (Fla.)
Carpinus caroliniana Walter.
Castanea pumila (Linné) Miller. (Ky., W. Va.)
Castanea sp. Hollick. (N. J.)
Caulinites beckeri Lesquereux.
Ceanothus americanus? Linné. (Ky.)
Cebatha carolina (Linné) Britton. (Ky.)
Cercis canadensis Linné. (Toronto, N. C.)
Chaetochloa sp. Berry. (N. C.)
Chamaecyparis nutkaensis Spach. (Alaska.)
Chamaecyparis thyoides (Linné) Britton, Sterns, and Poggenberg. (Toronto.)
Chara springerae Knowlton. (N. Mex.)
Chara sp. Hinde. (Ont.)
Cinnamomum sp. Hollick. (N. J.)
Cissites microphyllus Lesquereux. (Calif.)
Clethra alnifolia Linné. (N. J., Toronto.)
Cornus sp. Penhallow. (N. J.)
Corylus americana Walter. (Pa.)
Crataegus coccineaefolia Berry. (N. C.)
Crataegus crus-galli? Linné. (Pa.)
Crataegus punctata Jacques. (Toronto.)
Crataegus spathulatoides Berry. (N. C.)
Crataegus sp. Berry. (Va.)
Cupressus macrocarpa Hartwig. (Calif.)
Cyperus sp. Knowlton. (Va.)
Cyperus sp. Penhallow. (Toronto.)

Dendrium pleistocenicum Berry. (N. C., Va.)
Distichium capillaceum Bruchman and Schimper. (Canada.
Drosera rotundifolia Linné. (Canada.)
Equisetum arvense Linné. (W. Va.)
Equisetum fluviatile Linné. (Toronto.)
Equisetum scirpoides Michaux. (Canada.)
Equisetum sylvaticum Linné. (Canada.)
Equisetum sp. Paisley. (N. B.)
Equisetum sp. Penhallow. (Ont.)
Eriocaulon sp. Penhallow. (Toronto.)
Fagus americana Sweet. (Tenn., W. Va., N. C., Ala., Va., Md., Ky.)
Festuca ovina Linné. (Toronto.)
Ficus interglacialis Hollick. (B. C.)
Fontinalis sp. Dawson. (Canada.)
Fontinalis sp. Hinde. (Ont.)
Fontinalis sp. Penhallow. (Ont.)
Fraxinus americana Linné. (Toronto, Ky., Md.)
Fraxinus nigra Marshall. (Toronto.)
Fraxinus quadrangulata Michaux. (Toronto.)
Fucus digitatus Penhallow. (Canada.)
Gaylussacia dumosa (Andre) Torrey and Gray. (Maine.)
Gaylussacia resinosa (Aiton) Torrey and Gray. (Canada.)
Gleditschia donensis Penhallow. (Toronto.)
Gleditschia triacanthos Linné. (Ky.)
Grossularia menziesii (Pursh) Coville. (Calif.)
Hicoria alba (Linné) Britton. (Toronto, Pa.)
Hicoria aquatica (Michaux f.) Britton. (N. C., Ala.)
Hicoria glabra (Miller) Britton. (N. J., Md., Va., N. C., Ky., Pa.)
Hicoria ovata (Miller) Britton. (N. C.)
Hicoria pecan (Marshall) Britton. (N. J., Ky.)
Hicoria villosa (Sargent) Ashe. (Ala.)
Hicoria sp. cf. H. microcarpa (Nuttall) Britton. (N. C.)
Hippuris vulgaris Linné. (Toronto.)
Hypnum commutatum Hedwig. (Ont.)
Hypnum fluitans Linné. (Canada.)
Hypnum fluitans brachydictyon Holzinger. (Iowa.)
Hypnum fluitans glaciale Holzinger. (Iowa.)
Hypnum recurvans Schwegler. (Ont.)
Hypnum revalvens Swartz. (Ont., Iowa.)
Hypnum richardsoni. (Iowa.)
Hypnum sp. Penhallow. (Toronto.)
Ilex cassine Linné. (N. J., Va.)
Ilex glabra (Linné) A. Gray. (Fla.)
Ilex integrifolia (Elliott) Chapman. (Ky.)
Ilex opaca Aiton. (N. J., Ala., N. C.)
Ilex verticillata (Linné) A. Gray. (Maine.)
Ilex? sp. Berry. (Ky.)
Juglans nigra Linné. (Ala.)
Juniperus californica Carrier. (Calif.)
Juniperus virginiana Linné. (Toronto and other places in Canada, Manitoba, N. J.)
Juniperus sp. Penhallow. (Toronto.)
Larix churchbridgensis Penhallow. (Manitoba.)
Larix larcinia (De Roy) Koch. (Toronto, Manitoba, N. Y., Ga.)
Laurus sp. Hollick. (N. J.)

Leitneria floridana Chapman? (Fla.)
Leucothoe racemosa (Linné) Gray. (N. J.)
Liquidambar styraciflua Linné. (Va., N. C., Ala., N. J.)
Liriodendron tulipifera Linné. (Ala., N. C., Md.)
Lonchocarpus novae-caesarae Hollick. (N. J.)
Lycopodium sp. Hinde. (Ont.)
Lycopodium sp.? Penhallow. (Canada.)
Magnolia acuminata Linné. (N. J.)
Magnolia virginiana Linné. (Fla., N. J.)
Malus coronarifolia Berry. (N. C.)
Malus pseudo-angustifolia Berry. (N. C.)
Menyanthes trifoliata Linné. (Canada.)
Mezoneurum bridgetonense Hollick. (N. J.)
Morus sp. Hollick. (N. J.)
Myrica cerifera Linné. (Fla.)
Nyssa aquatica Marshall. (N. J.)
Nyssa biflora Walter. (Va., N. C., Md., N. J., Ala.)
Nyssa sylvatica Marshall. (Ky.)
Nyssa uniflora Walter. (N. J.)
Oryzopsis asperifolia Michaux. (Canada.)
Osmunda spectabilis Willdenow. (Ala.)
Ostrya walkeri? Heer. (Va. Age doubtful.)
Ostrya virginiana (Miller) Willdenow. (N. J., Ala.)
Oxycoccus palustris Persoon. (Ont.)
Persea borbonia (Linné) Spreng. (N. J., Tenn.)
Persea pubescens (Pursh) Sargent. (N. C., Ala.)
Philotria canadensis (Michaux) Britton. (Ont., Manitoba.)
Phoradendron flavescens (Pursh) Nuttall. (Ala.)
Phyllites fraxineus Lesquereux. (N. J.)
Phyllites mimusopsoides Lesquereux. (N. J.)
Picea canadensis (Miller) Britton, Sterns, and Poggenberg. (Ill., Ont., Iowa.)
Picea mariana (Miller) Britton, Sterns, and Poggenberg. (Toronto, Ont., Montreal, Manitoba, Iowa.)
Picea sp.? Penhallow. (Toronto.)
Pinus attenuata Lemmon. (Calif.)
Pinus caribaea Morelet. (Fla.)
Pinus echinata Miller. (Ala.)
Pinus rigida Miller. (N. Y., Pa., N. C.)
Pinus strobus Linné. (Toronto.)
Pinus taeda Linné. (Ala., N. J., Fla.)
Pinus 3 sp. Berry. (Va., Fla.)
Pistia spathulata Michaux. (Fla.)
Planera aquatica (Walter) Gmelin. (N. J., N. C., Ky.)
Platanus occidentalis Linné. (Toronto, Md., W. Va., N. C., Ala.)
Polygonum 2 sp. Berry. (N. C., Fla.)
Populus balsamifera Linné. (Toronto and elsewhere in Canada; Maine.)
Populus deltoides Marshall. (Ala.)
Populus cf. P. deltoides Marshall. (N. C.)
Populus grandidentata Michaux. (Toronto, Montreal.)
Populus sp. Berry. (Ky.)
Potamogeton natans Linné. (Toronto.)
Potamogeton pectinatus Linné. (Canada.)
Potamogeton perfoliatus Linné. (Canada.)
Potamogeton pusillus Linné. (Canada.)

Potamogeton robbinsii Oakes. (Carmichaels clay, W. Va.)
Potamogeton rutilus Wolfg. (Canada.)
Potentilla anserina Linné. (Canada.)
Prunus sp. Penhallow. (Toronto.)
Pseudotsuga macrocarpa Mayr. (Calif.)
Pseudotsuga taxifolia (Poiret) Britton. (Mont.)
Quercus abnormalis Berry. (N. C.)
Quercus acuminata (Michaux) Sargent. (Toronto.)
Quercus alba Linné. (Toronto, N. C., Va.)
Quercus brevifolia (Lamarck) Sargent. (Fla.)
Quercus chapmani Sargent? (Fla.)
Quercus digitata (Marshall) Sudworth. (W. Va.)
Quercus imbricaria fossilis Lesquereux. (N. J.)
Quercus laurifolia Michaux. (Fla.)
Quercus lyrata Walter. (N. C.)
Quercus macrocarpa Michaux. (Toronto, Pa.)
Quercus marylandica Muench. (N. C.)
Quercus minor (Marshall) Sargent. (Toronto.)
Quercus nigra Linné. (N. C., Ala.)
Quercus patustris Du Roi. (Pa., Md., N. C.)
Quercus phellos Linné. (N. C., Ala., Miss., N. J.?)
Quercus platanoides (Lamarck) Sudworth. (N. C.)
Quercus predigitata Berry. (Va., N. C.)
Quercus prinoides Willdenow. (N. C.)
Quercus prinus Linné. (N. C., Ala.)
Quercus rubra Linné. (Toronto.)
Quercus velutina Lamarck. (Toronto.)
Quercus virginiana Miller. (Ala., Ky., Fla.)
Quercus 2 sp. Berry. (Va., Ky.)
Quercus sp. Hollick. (N. J.)
Robinia pseudacacia Linné. (Toronto.)
Rubus sp. Berry. (N. C.)
Sabal palmetto (Walter) Roemer and Schultes. (Fla.)
Salix viminalifolia Berry. (Ky.)
Salix sp. Penhallow. (Toronto.)
Serenoa serrulata (Michaux) Hooker. (Fla.)
Sparganium sp. Berry. (N. C.)
Taxodium distichum (Linné) Richard. (Md., N. J., Va., N. C., Fla., Ky., Miss.)
Taxus minor (Michaux) Britton. (Toronto, Ill., Iowa, Manitoba, Cape Breton.)
Tacoma preradicans Berry. (Ky.)
Thuja occidentalis Linné. (Toronto, Montreal, Manitoba, Ohio, Va.)
Tilia americana Linné. (Toronto.)
Tilia dubia (Newberry) Berry. (N. J.)
Toxylon pomiferum Rafinesque. (Toronto, N. J.)
Typha latifolia? Linné. (Canada.)
Ulmus alata Michaux. (Ky., N. C., Ala.)
Ulmus americana Linné. (Toronto, N. J.)
Ulmus protoracemosa Penhallow. (N. Y.)
Ulmus racemosa Thomas. (Toronto.)
Vaccinium arboreum Marshall. (N. C., Va., Ala.)
Vaccinium corymbosum Linné. (N. C., Ala., Maine.)
Vaccinium spathulatum Berry. (N. C.)
Vaccinium uliginosum Linné. (Ont.)

Vallisneria spiralis Linné. (Canada.)
Viburnum bridgetonense Britton. (N. J.)
Viburnum cf. V. dentatum Linné. (Fla.)
Viburnum cf. V. molle Michaux. (N. C.)
Viburnum nudum Linné. (Fla.)
Vitis pseudo-rotundifolia Berry. (N. J.)
Vitis cf. V. rotundifolia Michaux. (Fla.)
Vitis sp. Berry. (N. C., Fla.)
Vitis cf. V. aestivalis Michaux. (N. J.)
Vitis cf. V. rotundifolia Michaux. (Fla.)
Xanthium sp. Berry. (Fla.)
Xolisma ligustrina (Linné) Britton. (Md., N. C., Ala., Ky.)
Zizyphus 2 sp. Berry. (N. J., Fla.)
Zostera marina Linné. (Canada.)

ADDITIONAL COPIES
OF THIS PUBLICATION MAY BE PROCURED FROM
THE SUPERINTENDENT OF DOCUMENTS
GOVERNMENT PRINTING OFFICE
WASHINGTON, D. C.
AT
60 CENTS PER COPY

Lightning Source UK Ltd.
Milton Keynes UK
UKHW010938061118
331792UK00011B/2254/P

9 780331 521665